PREFACE

The objective of this book is to make the concepts and methods of physical chemistry clear and interesting to students who have had a year of calculus and a year of physics. The underlying theory of chemical phenomena is complicated, and so it is a challenge to make the most important concepts and methods understandable to undergraduate students. However, these basic ideas are accessible to students, and they will find them useful whether they are chemistry majors, biologists, engineers, or earth scientists. The basic theory of chemistry is presented from the viewpoint of academic physical chemists, but many applications of physical chemistry to practical problems are described.

One of the important objectives of a course in physical chemistry is to learn how to solve numerical problems. The problems in physical chemistry help emphasize features in the underlying theory, and they illustrate practical applications.

There are two types of problems: problems that can be solved with a hand-held calculator and **COMPUTER PROBLEMS** that require a personal computer with a mathematical application installed. There are two sets of problems of the first type. The answers to problems in the first set are given in the back of the textbook, and worked-out solutions to these problems are given in the *Solutions Manual for Physical Chemistry*. The answers for the second set of problems are given in the *Solutions Manual*. In the two sets of problems that can be solved using hand-held calculators, some problems are marked with an icon to indicate that they may be more conveniently solved on a personal computer with a mathematical program. There are 170 COMPUTER PROBLEMS that require a personal computer with a mathematical application such as Mathematica™, MathCad™, MATLAB™, or MAPLE™ installed. The recent development of these mathematical applications makes it possible to undertake problems that were previously too difficult or too time consuming. This is particularly true for two- and three-dimensional plots, integration and differentiation of complicated functions, and solving differential equations. The *Solutions Manual for Physical Chemistry* provides Mathematica™ programs and printouts for the COMPUTER PROBLEMS.

The Mathematica™ solutions of the 170 COMPUTER PROBLEMS in digital form are available on the web at http://www.wiley.com/college/silbey. They can be downloaded into a personal computer with Mathematica™ installed. Students can obtain Mathematica at a reduced price from Wolfram Research, 100 Trade Center Drive, Champaign, Illinois, 61820-7237. A password is required and will be available in the *Solutions Manual,* along with further information about how to access the Mathematica solutions in digital form. Emphasis in the COMPUTER PROBLEMS has been put on problems that do not require complicated programming, but do make it possible for students to explore important topics more deeply. Suggestions are made as to how to vary parameters and how to apply these programs to other substances and systems. As an aid to showing how commands are used, there is an index in the *Solutions Manual* of the major commands used.

Mathematica™ plots are used in some 60 figures in the textbook. The legends for these figures indicate the COMPUTER PROBLEM where the program is given. These programs make it possible for students to explore changes in the ranges of variables in plots and to make calculations on other substances and systems.

One of the significant changes in the fourth edition is increased emphasis on the thermodynamics and kinetics of biochemical reactions, including the denaturation of proteins and nucleic acids. In this edition there is more discussion of the uses of statistical mechanics, nuclear magnetic relaxation, nano science, and oscillating chemical reactions.

This edition has 32 new problems that can be solved with a hand-held calculator and 35 new problems that require a computer with a mathematical application. There are 34 new figures and eight new tables.

Because the number of credits in physical chemistry courses, and therefore the need for more advanced material, varies at different universities and colleges, more topics have been included in this edition than can be covered in most courses.

The Appendix provides an alphabetical list of symbols for physical quantities and their units. The use of nomenclature and units is uniform throughout the book. SI (Système International d'Unités) units are used because of their advantage as a coherent system of units. That means that when SI units are used with all of the physical quantities in a calculation, the result comes out in SI units without having to introduce numerical factors. The underlying unity of science is emphasized by the use of seven base units to represent all physical quantities.

HISTORY

Outlines of Theoretical Chemistry, as it was then entitled, was written in 1913 by Frederick Getman, who carried it through 1927 in four editions. The next four editions were written by Farrington Daniels. In 1955, Robert Alberty joined Farrington Daniels. At that time, the name of the book was changed to *Physical Chemistry,* and the numbering of the editions was started over. The collaboration ended in 1972 when Farrington Daniels died. It is remarkable that this textbook traces its origins back 91 years.

Over the years this book has profited tremendously from the advice of physical chemists all over the world. Many physical chemists who care how their subject is presented have written to us with their comments, and we hope that will continue. We are especially indebted to colleagues at MIT who have reviewed various sections and given us the benefit of advice. These include Sylvia T. Ceyer, Robert W. Field, Carl W. Garland, Mario Molina, Keith Nelson, and Irwin Oppenheim.

The following individuals made very useful suggestions as to how to improve this fourth edition: Kenneth G. Brown (Old Dominion University), Thandi Buthelezi (Western Kentucky University), Susan Collins (California State University Northridge), Jon Gold (East Straudsburg University), Keith J. Stine (University of Missouri–St. Louis), Ronald J. Terry (Western Illinois University), and Worth E. Vaughan (University of Wisconsin, Madison). We are also indebted to reviewers of earlier editions and to people who wrote us about the third edition.

The following individuals made very useful suggestions as to how to improve the Mathematica™ solutions to COMPUTER PROBLEMS: Ian Brooks (Wolfram Research), Carl W. David (U. Connecticut), Robert N. Goldberg (NIST), Mark R. Hoffmann (University of North Dakota), Andre Kuzniarek (Wolfram Research), W. Martin McClain (Wayne State University), Kathryn Tomasson (University of North Dakota), and Worth E. Vaughan (University of Wisconsin, Madison).

We are indebted to our editor Deborah Brennan and to Catherine Donovan and Jennifer Yee at Wiley for their help in the production of the book and the solutions manual. We are also indebted to Martin Batey for making available the web site, and to many others at Wiley who were involved in the production of this fourth edition.

Cambridge, Massachusetts
January 2004

Robert J. Silbey
Robert A. Alberty
Moungi G. Bawendi

CONTENTS

PART ONE
THERMODYNAMICS

PART TWO
QUANTUM CHEMISTRY

PART THREE
KINETICS

PART FOUR
MACROSCOPIC AND MICROSCOPIC STRUCTURES

APPENDIX

PART

ONE

Thermodynamics

Thermodynamics deals with the interconversion of various kinds of energy and the changes in physical properties that are involved. Thermodynamics is concerned with equilibrium states of matter and has nothing to do with time. Even so, it is one of the most powerful tools of physical chemistry; because of its importance, the first part of this book is devoted to it. The first law of thermodynamics deals with the amount of work that can be done by a chemical or physical process and the amount of heat that is absorbed or evolved. On the basis of the first law it is possible to build up tables of enthalpies of formation that may be used to calculate enthalpy changes for reactions that have not yet been studied. With information on heat capacities of reactants and products also available, it is possible to calculate the heat of a reaction at a temperature where it has not previously been studied.

The second law of thermodynamics deals with the natural direction of processes and the question of whether a given chemical reaction can occur by itself. The second law was formulated initially in terms of the efficiencies of heat engines, but it also leads to the definition of entropy, which is important in determining the direction of chemical change. The second law provides the basis for the definition of the equilibrium constant for a chemical reaction. It provides an answer to the question, "To what extent will this particular reaction go before equilibrium is reached?" It also provides the basis for reliable predictions of the effects of temperature, pressure, and concentration on chemical and physical equilibrium. The third law provides the basis for calculating equilibrium constants from calorimetric measurements only. This is an illustration of the way in which thermodynamics interrelates apparently unrelated measurements on systems at equilibrium.

After discussing the laws of thermodynamics and the various physical quantities involved, our first applications will be to the quantitative treatment of chemical equilibria. These methods are then applied to equilibria between different phases. This provides the basis for the quantitative treatment of distillation and for the interpretation of phase changes in mixtures of solids. Then thermodynamics is applied to electrochemical cells and biochemical reactions.

1

Zeroth Law of Thermodynamics and Equations of State

Physical chemistry is concerned with understanding the quantitative aspects of chemical phenomena. To introduce physical chemistry we will start with the most accessible properties of matter—those that can readily be measured in the laboratory. The simplest of these are the properties of matter at equilibrium. Thermodynamics deals with the properties of systems at equilibrium, such as temperature, pressure, volume, and amounts of species; but it also deals with work done on a system and heat absorbed by a system, which are not properties of the system but measures of changes. The amazing thing is that the thermodynamic properties of systems at equilibrium obey all the rules of calculus and are therefore interrelated. The principle involved in defining temperature was not recognized until the establishment of the first and second laws of thermodynamics, and so it is referred to as the zeroth law. This leads to a discussion of the thermodynamic properties of gases and liquids. After discussing the ideal gas, we consider the behavior of real gases. The thermodynamic properties of a gas or liquid are represented by an equation of state, such as the virial equation or the van der Waals equation. The latter has the advantage that it provides a description of the critical region, but much more complicated equations are required to provide an accurate quantitative description.

1.1 STATE OF A SYSTEM

A thermodynamic system is that part of the physical universe that is under consideration. A system is separated from the rest of the universe by a real or idealized **boundary.** The part of the universe outside the boundary of the system is referred to as the **surroundings,** as illustrated in Fig. 1.1. The boundary between the system and its surroundings may have certain real or idealized characteristics. For example, the boundary may conduct heat or be a perfect insulator. The boundary may be rigid or it may be movable so that it can be used to apply a specified pressure. The boundary may be impermeable to the transfer of matter between the system and its surroundings, or it may be permeable to a specified species. In other words, matter and heat may be transferred between system and surroundings, and the surroundings may do work on the system, or vice versa. If the boundary around a system prevents interaction of the system with its surroundings, the system is called an **isolated** system.

If matter can be transferred from the surroundings to the system, or vice versa, the system is referred to as an **open** system; otherwise, it is a **closed** system.

When a system is under discussion it must be described precisely. A system is **homogeneous** if its properties are uniform throughout; such a system consists of a single phase. If a system contains more than one phase, it is **heterogeneous.** A simple example of a two-phase system is liquid water in equilibrium with ice. Water can also exist as a three-phase system: liquid, ice, and vapor, all in equilibrium.

Experience has shown that the macroscopic state of a system at equilibrium can be specified by the values of a small number of macroscopic variables. These variables, which include, for example, temperature T, pressure P, and volume V, are referred to as **state variables** or **thermodynamic variables.** They are called state variables because they specify the state of a system. Two samples of a substance that have the same state variables are said to be in the same state. It is remarkable that the state of a homogeneous system at equilibrium can be specified by so few variables. When a sufficient number of state variables are specified, all of the other properties of the system are fixed. It is even more remarkable that these state variables follow all of the rules of calculus; that is, they can be treated as mathematical functions that can be differentiated and integrated. Thermodynamics leads to the definition of additional properties, such as internal energy and entropy, that can also be used to describe the state of a system, and are themselves state variables.

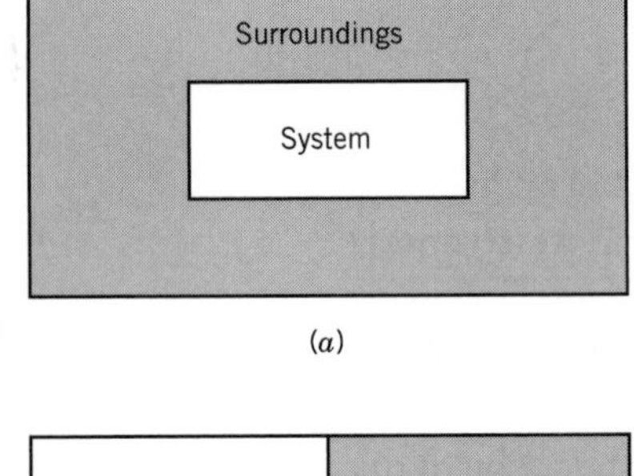

Figure 1.1 (*a*) A system is separated from its surroundings by a boundary, real or idealized. (*b*) As a simplification we can imagine the system to be separated from the surroundings by a single wall that may be an insulator or a heat conductor. Later, in Section 6.7 and Section 8.3 (see Fig. 8.6), we will consider semipermeable boundaries so that the system is open to the transfer of matter.

The thermodynamic state of a specified amount of a pure substance in the fluid state can be described by specifying properties such as temperature T, pressure P, and volume V. But experience has shown that only two of these three properties have to be specified when the amount of pure substance is fixed. If T and P, or P and V, or T and V are specified, all the other thermodynamic properties (including those that will be introduced later) are fixed and the system is at equilibrium. More properties have to be specified to describe the thermodynamic state of a homogeneous mixture of different species.

Note that the description of the microscopic state of a system containing many molecules requires the specification of a very large number of variables. For example, to describe the microscopic state of a system using classical mechanics, we would have to give the three coordinates and three components of the momentum of each molecule, plus information about its vibrational and rotational motion. For one mole of gas molecules, this would mean more than 6×10^{23} numbers. An important thing to notice is that we can use a small number of state variables to describe the equilibrium thermodynamic state of a system that is too complicated to describe in a microscopic way.

Thermodynamic variables are either intensive or extensive. **Intensive variables** are independent of the size of the system; examples are pressure, density, and temperature. **Extensive variables** do depend on the size of the system and double if the system is duplicated and added to itself; examples are volume, mass, internal energy, and entropy. Note that the ratio of two extensive variables is an intensive variable; density is an example. Thus we can talk about the **intensive state of the system,** which is described by intensive variables, or the **extensive state of a system,** which is described by intensive variables plus at least one extensive variable. The intensive state of the gas helium is described by specifying its pressure and density. The extensive state of a certain amount of helium is described by specifying the amount, the pressure, and the density; the extensive state of one mole of helium might be represented by 1 mol He(P, ρ), where P and ρ represent the pressure and density, respectively. We can generalize this by saying that the intensive state of a pure substance in the fluid state is specified by $N_s + 1$ variables, where N_s is the number of different kinds of species in the system. The extensive state is specified by $N_s + 2$ variables, one of which has to be extensive.

In chemistry it is generally more useful to express the size of a system in terms of the amount of substance it contains, rather than its mass. **The amount of substance *n* is the number of entities (atoms, molecules, ions, electrons, or specified groups of such particles) expressed in terms of moles.** If a system contains N molecules, the amount of substance $n = N/N_A$, where N_A is the Avogadro constant (6.022×10^{23} mol^{-1}). The ratio of the volume V to the amount of substance is referred to as the molar volume: $\overline{V} = V/n$. The volume V is expressed in SI units of m^3, and the molar volume $\overline{V}$ is expressed in SI units of m^3 mol^{-1}. We will use the overbar regularly to indicate molar thermodynamic quantities.

Comment:

Since this is our first use of physical quantities, we should note that the value of a physical quantity is equal to the product of a numerical factor and a unit:

$$\textit{physical quantity} = \textit{numerical value} \times \textit{unit}$$

The values of all physical quantities can be expressed in terms of SI base units (see Appendix A). However, some physical quantities are dimensionless, and so the symbol for the SI unit is taken as 1 because this is what you get when units cancel. Note that, in print, physical quantities are represented by italic type and units are represented by roman type.

When a system is in a certain state with its properties independent of time and having no fluxes (e.g., no heat flowing through the system), then the system is said to be at **equilibrium.** When a thermodynamic system is at equilibrium its state is defined entirely by the state variables, and **not by the history of the system.** By history of the system, we mean the previous conditions under which it has existed.

Since the state of a system at equilibrium can be specified by a small number of state variables, it should be possible to express the value of a variable that has not been specified as a function of the values of other variables that have been specified. The simplest example of this is the ideal gas law.

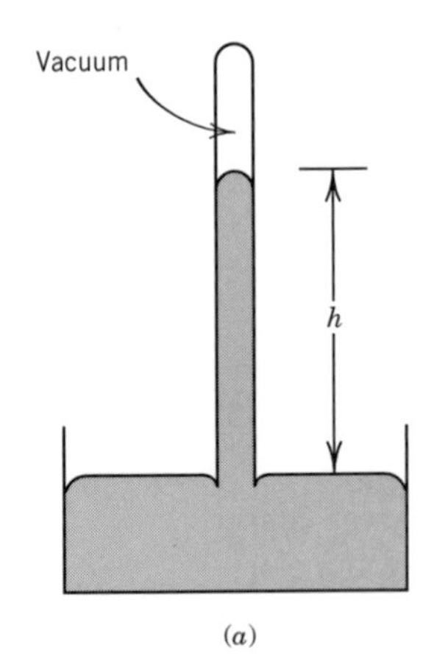

(a)

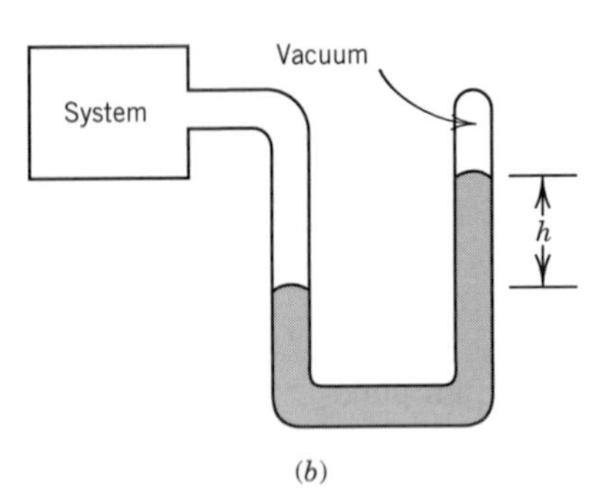

(b)

Figure 1.2 (*a*) The pressure exerted by the atmosphere on the surface of mercury in a cup is given by $P = h\rho g$ (see Example 1.1). (*b*) The pressure of a system is given by the same equation when a closed-end manometer is used.

For some systems, more than two intensive variables must be stated to specify the state of the system. If there is more than one species, the composition has to be given. If a liquid system is in the form of small droplets, the surface area has to be given. If the system is in an electric or magnetic field, this may have an effect on its properties, and then the electric field strength and magnetic field strength become state variables. We will generally ignore the effect of the earth's gravitational field on a system, although this can be important, as we will see in the special topic at the end of this chapter. Note that the properties used to describe the state of a system must be independent; otherwise they are redundant. Independent properties are separately controllable by the investigator.

The pressure of the atmosphere is measured with a barometer, as shown in Fig. 1.2*a*, and the pressure of a gaseous system is measured with a closed-end manometer, as shown in Fig. 1.2*b*.

1.2 THE ZEROTH LAW OF THERMODYNAMICS

Although we all have a commonsense notion of what temperature is, we must define it very carefully so that it is a useful concept in thermodynamics. If two closed systems with fixed volumes are brought together so that they are in thermal contact, changes may take place in the properties of both. Eventually a state is reached in which there is no further change, and this is the state of **thermal equilibrium.** In this state, the two systems have the same temperature. Thus, we can readily determine whether two systems are at the same temperature by bringing them into thermal contact and seeing whether observable changes take place in the properties of either system. If no change occurs, the systems are at the same temperature.

Now let us consider three systems, A, B, and C, as shown in Fig. 1.3. It is an experimental fact that if system A is in thermal equilibrium with system C, and system B is also in thermal equilibrium with system C, then A and B are in thermal equilibrium with each other. It is not obvious that this should be true, and so this empirical fact is referred to as the **zeroth law of thermodynamics.**

To see how the zeroth law leads to the definition of a temperature scale, we need to consider thermal equilibrium between systems A, B, and C in more detail. Assume that A, B, and C each consist of a certain mass of a different fluid. We use the word **fluid** to mean either a gas or a compressible liquid. Our experience is that if the volume of one of these systems is held constant, its pressure may vary over a range of values, and if the pressure is held constant, its volume may vary over a range of values. Thus, the pressure and the volume are independent thermodynamic variables. Furthermore, suppose that the experience with these systems is that their intensive states are specified completely when the pressure and volume are specified. That is, when one of the systems reaches equilibrium at a certain pressure and volume, all of its macroscopic properties have certain characteristic values. It is quite remarkable and fortunate that the macroscopic state of a given mass of fluid of a given composition can be fixed by specifying only the pressure and the volume.*

If there are further constraints on the system, there will be a smaller number of independent variables. An example of an additional constraint is thermal

*This is not true for water in the neighborhood of 4 °C, but the state is specified by giving the temperature and the volume or the temperature and the pressure. See Section 6.1.

equilibrium with another system. Experience shows that if a fluid is in thermal equilibrium with another system, it has only one independent variable. In other words, if we set the pressure of system A at a particular value P_A, we find that there is thermal equilibrium with system C, in a specified state, only at a particular value of V_A. Thus, system A in thermal equilibrium with system C is characterized by a **single** independent variable, pressure or volume; one or the other can be set arbitrarily, but not both. The plot of all the values of P_A and V_A for which there is equilibrium with system C is called an **isotherm.** Figure 1.4 gives this isotherm, which we label Θ_1. Since system A is in thermal equilibrium with system C at any P_A, V_A on the isotherm, we can say that each of the pairs P_A, V_A on this isotherm corresponds with the same temperature Θ_1.

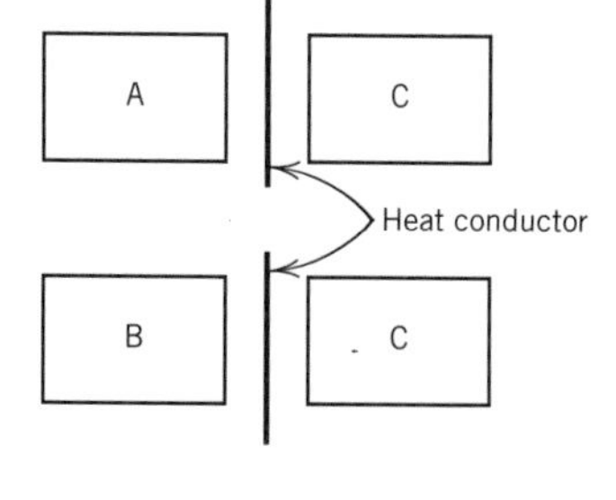

If A and C are in thermal equilibrium, and B and C are in thermal equilibrium, then

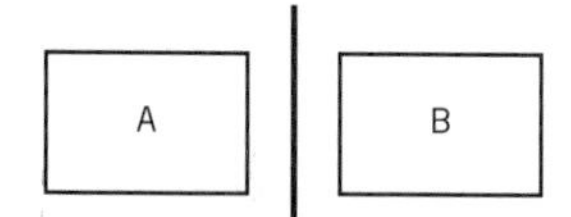

A and B will be found to be in thermal equilibrium when connected by a heat conductor.

Figure 1.3 The zeroth law of thermodynamics is concerned with thermal equilibrium between three bodies.

When heat is added to system C and the experiment is repeated, a different isotherm is obtained for system A. In Fig. 1.4, the isotherm for the second experiment is labeled Θ_2. If still more heat is added to system C and the experiment is repeated again, the isotherm labeled Θ_3 is obtained.

Figure 1.4 illustrates Boyle's law, which states that PV = constant for a specified amount of gas at a specified temperature. Experimentally, this is strictly true only in the limit of zero pressure. Charles and Gay-Lussac found that the volume of a gas varies linearly with the temperature at specified pressure when the temperature is measured with a mercury in glass thermometer, for example. Since it would be preferable to have a temperature scale that is independent of the properties of particular materials like mercury and glass, it is better to say that the ratio of the P_2V_2 product at temperature Θ_2 to P_1V_1 at temperature Θ_1 depends only on the two temperatures:

$$\frac{P_2V_2}{P_1V_1} = \phi(\Theta_1, \Theta_2) \tag{1.1}$$

where ϕ is an unspecified function. The simplest thing to do is to take the ratio of the PV products to be equal to the ratio of the temperatures, thus defining

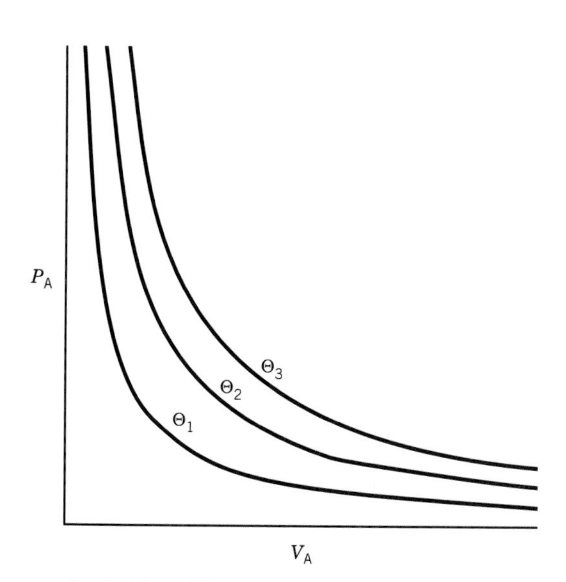

Figure 1.4 Isotherms for fluid A. This plot, which is for a hypothetical fluid, might look quite different for some other fluid.

a temperature scale:

$$\frac{P_2V_2}{P_1V_1} = \frac{T_2}{T_1} \quad \text{or} \quad \frac{P_2V_2}{T_2} = \frac{P_1V_1}{T_1} \tag{1.2}$$

Here we have introduced a new symbol T for the temperature because we have made a specific assumption about the function ϕ. Equations 1.1 and 1.2 are exact only in the limit of zero pressure, and so T is referred to as the ideal gas temperature.

Since, according to equation 1.2, PV/T is a constant for a fixed mass of gas and since V is an extensive property,

$$PV/T = nR \tag{1.3}$$

where n is the amount of gas and R is referred to as the **gas constant.** Equation 1.3 is called the ideal gas **equation of state.** An equation of state is a relation between the thermodynamic properties of a substance at equilibrium.

1.3 THE IDEAL GAS TEMPERATURE SCALE

The ideal gas temperature scale can be defined more carefully by taking the temperature T to be proportional to $P\overline{V} = PV/n$ in the limit of zero pressure. Since different gases give slightly different scales when the pressure is about one bar (1 bar $= 10^5$ pascal $= 10^5$ Pa $= 10^5$ N m^{-2}), it is necessary to use the limit of the $P\overline{V}$ product as the pressure approaches zero. When this is done, all gases yield the same temperature scale. We speak of gases under this limiting condition as **ideal.** Thus, the **ideal gas temperature** T is defined by

$$T = \lim_{P \to 0} (P\overline{V}/R) \tag{1.4}$$

The proportionality constant is called the gas constant R. The unit of thermodynamic temperature, 1 kelvin or 1 K, is defined as the fraction 1/273.16 of the temperature of the triple point of water.* Thus, the temperature of an equilibrium system consisting of liquid water, ice, and water vapor is 273.16 K. The temperature 0 K is called absolute zero. According to the current best measurements, the freezing point of water at 1 atmosphere (101 325 Pa; see below) is 273.15 K, and the boiling point at 1 atmosphere is 373.12 K; however, these are experimental values and may be determined more accurately in the future. The Celsius scale t is formally defined by

$$t/°\text{C} = T/\text{K} - 273.15 \tag{1.5}$$

The reason for writing the equation in this way is that temperature T on the Kelvin scale has the unit K, and temperature t on the Celsius scale has the unit °C, which need to be divided out before temperatures on the two scales are compared. In Fig. 1.5, the molar volume of an ideal gas is plotted versus the Celsius temperature t at two pressures.

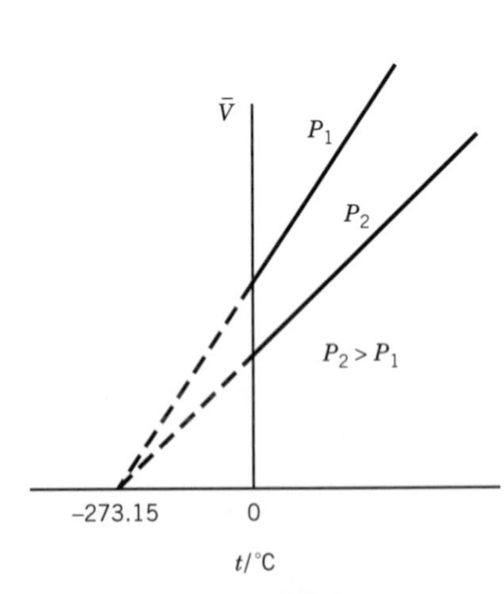

Figure 1.5 Plots of $\overline{V}$ versus temperature for a given amount of a real gas at two low pressures P_1 and P_2, as given by Gay-Lussac's law.

*The triple point of water is the temperature and pressure at which ice, liquid, and vapor are in equilibrium with each other in the absence of air. The pressure at the triple point is 611 Pa. The freezing point in the presence of air at 1 atm is 0.0100 °C lower because (1) the solubility of air in liquid water at 1 atm (101 325 Pa) is sufficient to lower the freezing point 0.0024 °C (Section 6.7), and (2) the increase of pressure from 611 to 101 325 Pa lowers the freezing point 0.0075 °C, as shown in Example 6.2. Thus, the ice point is at 273.15 K.

We will find later that the ideal gas temperature scale is identical with one based on the second law of thermodynamics, which is independent of the properties of any particular substance (see Section 3.9). In Chapter 16 the ideal gas temperature scale will be identified with that which arises in statistical mechanics.

The gas constant R can be expressed in various units, but we will emphasize the use of SI units. The SI unit of **pressure** (P) is the pascal, Pa, which is the pressure produced by a force of 1 N on an area of 1 m^2. In addition to using the prefixes listed in the back cover of the book to express larger and smaller pressures, it is convenient to have a unit that is approximately equal to the atmospheric pressure. This unit is the bar, which is 10^5 Pa. Earlier the atmosphere, which is defined as 101 325 Pa, had been used as a unit of pressure.

Example 1.1 ***Express one atmosphere pressure in SI units***

Calculate the pressure of the earth's atmosphere at a point where the barometer reads 76 cm of mercury at 0 °C and the acceleration of gravity g is 9.806 65 m s^{-2}. The density of mercury at 0 °C is 13.5951 g cm^{-3}, or 13.5951×10^3 kg m^{-3}.

Pressure P is force f divided by area A:

$$P = f/A$$

The force exerted by a column of air over an area A is equal to the mass m of mercury in a vertical column with a cross section A times the acceleration of gravity g:

$$f = mg$$

The mass of mercury raised above the flat surface in Fig. 1.2*a* is ρAh so that

$$f = \rho Ahg$$

Thus, the pressure of the atmosphere is

$$P = h\rho g$$

If h, ρ, and g are expressed in SI units, the pressure P is expressed in pascals. Thus, the pressure of a standard atmosphere may be expressed in SI units as follows:

$$\begin{aligned} 1 \text{ atm} &= (0.76 \text{ m})(13.5951 \times 10^3 \text{ kg m}^{-3})(9.806\,65 \text{ m s}^{-2}) \\ &= 101\,325 \text{ N m}^{-2} = 101\,325 \text{ Pa} = 1.013\,25 \text{ bar} \end{aligned}$$

This equality is expressed by the conversion factor 1.013 25 bar atm^{-1}.

To determine the value of the gas constant we also need the definition of a mole. A **mole** is the amount of substance that has as many atoms or molecules as 0.012 kg (exactly) of ^{12}C. The **molar mass** M of a substance is the mass divided by the amount of substance n, and so its SI unit is kg mol^{-1}. Molar masses can also be expressed in g mol^{-1}, but it is important to remember that in making calculations in which all other quantities are expressed in SI units, the molar mass must be expressed in kg mol^{-1}. The molar mass M is related to the molecular mass m by $M = N_A m$, where N_A is the **Avogadro constant** and m is the mass of a single molecule.

Until 1986 the recommended value of the gas constant was based on measurements of the molar volumes of oxygen and nitrogen at low pressures. The accuracy of such measurements is limited by problems of sorption of gas on the walls of the glass vessels used. In 1986 the recommended value* of the gas constant

$$R = 8.314\,51 \text{ J K}^{-1} \text{ mol}^{-1} \tag{1.6}$$

was based on measurements of the speed of sound in argon. The equation used is discussed in Section 17.4. Since pressure is force per unit area, the product of pressure and volume has the dimensions of force times distance, which is work or energy. Thus, the gas constant is obtained in joules if pressure and volume are expressed in pascals and cubic meters; note that 1 J = 1 Pa m^3.

Example 1.2 ***Express the gas constant in various units***

Calculate the value of R in cal K^{-1} mol^{-1}, L bar K^{-1} mol^{-1}, and L atm K^{-1} mol^{-1}.

Since the calorie is defined as 4.184 J,

$$\begin{aligned} R &= 8.314\,51 \text{ J K}^{-1} \text{ mol}^{-1}/4.184 \text{ J cal}^{-1} \\ &= 1.987\,22 \text{ cal K}^{-1} \text{ mol}^{-1} \end{aligned}$$

Since the liter is 10^{-3} m^3 and the bar is 10^5 Pa,

$$\begin{aligned} R &= (8.314\,51 \text{ Pa m}^3 \text{ K}^{-1} \text{ mol}^{-1})(10^3 \text{ L m}^{-3})(10^{-5} \text{ bar Pa}^{-1}) \\ &= 0.083\,145\,1 \text{ L bar K}^{-1} \text{ mol}^{-1} \end{aligned}$$

Since 1 atm is 1.013 25 bar,

$$\begin{aligned} R &= (0.083\,145\,1 \text{ L bar K}^{-1} \text{ mol}^{-1})/(1.013\,25 \text{ bar atm}^{-1}) \\ &= 0.082\,057\,8 \text{ L atm K}^{-1} \text{ mol}^{-1} \end{aligned}$$

1.4 IDEAL GAS MIXTURES AND DALTON'S LAW

Equation 1.3 applies to a mixture of ideal gases as well as a pure gas, when n is the total amount of gas. Since $n = n_1 + n_2 + \cdots$, then

$$\begin{aligned} P &= (n_1 + n_2 + \cdots)RT/V \\ &= n_1RT/V + n_2RT/V + \cdots \\ &= P_1 + P_2 + \cdots = \sum_i P_i \end{aligned} \tag{1.7}$$

where P_1 is the partial pressure of species 1. Thus, the total pressure of an ideal gas mixture is equal to the sum of the partial pressures of the individual gases; this is **Dalton's law.** The partial pressure of a gas in an ideal gas mixture is the pressure that it would exert alone in the total volume at the temperature of the mixture:

$$P_i = n_iRT/V \tag{1.8}$$

A useful form of this equation is obtained by replacing RT/V by P/n:

$$P_i = n_iP/n = y_iP \tag{1.9}$$

*E. R. Cohen and B. N. Taylor, The 1986 Adjustment of the Fundamental Physical Constants, *CODATA Bull.* **63**:1 (1986); *J. Phys. Chem. Ref. Data* **17**:1795 (1988).

The dimensionless quantity y_i is the mole fraction of species i in the mixture, and it is defined by n_i/n. Substituting equation 1.9 in 1.7 yields

$$1 = y_1 + y_2 + \cdots = \sum_i y_i \tag{1.10}$$

so that the sum of the mole fractions in a mixture is unity.

Figure 1.6 shows the partial pressures P_1 and P_2 of two components of a binary mixture of ideal gases at various mole fractions and at constant total pressure. The various mixtures are considered at the same total pressure P.

The behavior of real gases is more complicated than the behavior of an ideal gas, as we will see in the next section.

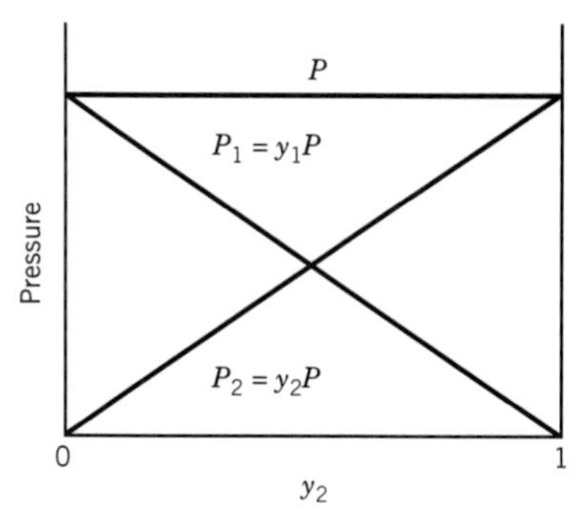

Figure 1.6 Total pressure P and partial pressures P_1 and P_2 of components of binary mixtures of gases as a function of the mole fraction y_2 of the second component at constant total pressure. Note that $y_1 = 1 - y_2$.

Example 1.3 ***Calculation of partial pressures***

A mixture of 1 mol of methane and 3 mol of ethane is held at a pressure of 10 bar. What are the mole fractions and partial pressures of the two gases?

$$y_m = 1 \text{ mol}/4 \text{ mol} = 0.25$$
$$P_m = y_m P = (0.25)(10 \text{ bar}) = 2.5 \text{ bar}$$
$$y_e = 3 \text{ mol}/4 \text{ mol} = 0.75$$
$$P_e = y_e P = (0.75)(10 \text{ bar}) = 7.5 \text{ bar}$$

Example 1.4 ***Express relative humidity as mole fraction of water***

The maximum partial pressure of water vapor in air at equilibrium at a given temperature is the vapor pressure of water at that temperature. The **actual** partial pressure of water vapor in air is a percentage of the maximum, and that percentage is called the relative humidity. Suppose the relative humidity of air is 50% at a temperature of 20 °C. If the atmospheric pressure is 1 bar, what is the mole fraction of water in the air? The vapor pressure of water at 20 °C is 2330 Pa. Assuming the gas mixture behaves as an ideal gas, the mole fraction of H_2O in the air is given by

$$y_{H_2O} = P_i/P = (0.5)(2330 \text{ Pa})/10^5 \text{ Pa} = 0.0117$$

1.5 REAL GASES AND THE VIRIAL EQUATION

Real gases behave like ideal gases in the limits of low pressures and high temperatures, but they deviate significantly at high pressures and low temperatures. The **compressibility factor** $Z = P\overline{V}/RT$ is a convenient measure of the deviation from ideal gas behavior. Figure 1.7 shows the compressibility factors for N_2 and O_2 as a function of pressure at 298 K. Ideal gas behavior, indicated by the dashed line, is included for comparison. As the pressure is reduced to zero, the compressibility factor approaches unity, as expected for an ideal gas. At very high pressures the compressibility factor is always greater than unity. This can be understood in terms of the finite size of molecules. At very high pressures the molecules of the gas are pushed closer together, and the volume of the gas is larger than expected

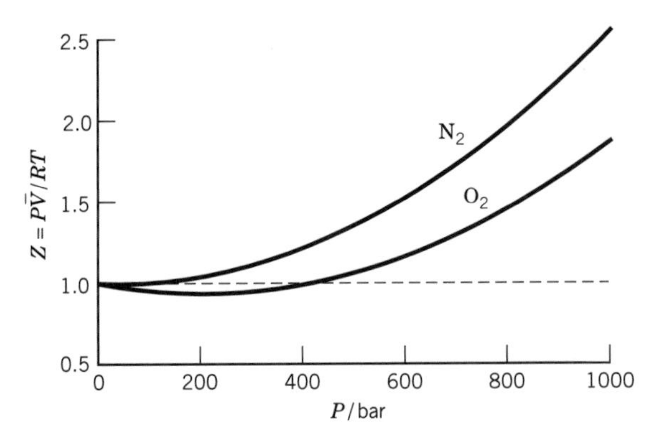

Figure 1.7 Influence of high pressure on the compressibility factor, $P\overline{V}/RT$, for N_2 and O_2 at 298 K. (See Computer Problem 1.D.)

for an ideal gas because a significant fraction of the volume is occupied by the molecules themselves. At low pressure a gas may have a smaller compressibility factor than an ideal gas. This is due to intermolecular attractions. The effect of intermolecular attractions disappears in the limit of zero pressure because the distance between molecules approaches infinity.

Figure 1.8 shows how the compressibility factor of nitrogen depends on temperature, as well as pressure. As the temperature is reduced, the effect of intermolecular attraction at pressures of the magnitude of 100 bar increases because the molar volume is smaller at lower temperatures and the molecules are closer together. All gases show a minimum in the plot of compressibility factor versus pressure if temperature is low enough. Hydrogen and helium, which have very low boiling points, exhibit this minimum only at temperatures much below 0 °C.

A number of equations have been developed to represent P–V–T data for real gases. Such an equation is called an **equation of state** because it relates state properties for a substance at equilibrium. Equation 1.3 is the equation of state for an ideal gas. The first equation of state for real gases that we will discuss is closely related to the plots in Figs. 1.7 and 1.8, and is called the virial equation.

In 1901 H. Kamerlingh-Onnes proposed an equation of state for real gases, which expresses the compressibility factor Z as a power series in $1/\overline{V}$ for a pure gas:

$$Z = \frac{P\overline{V}}{RT} = 1 + \frac{B}{\overline{V}} + \frac{C}{\overline{V}^2} + \cdots \tag{1.11}$$

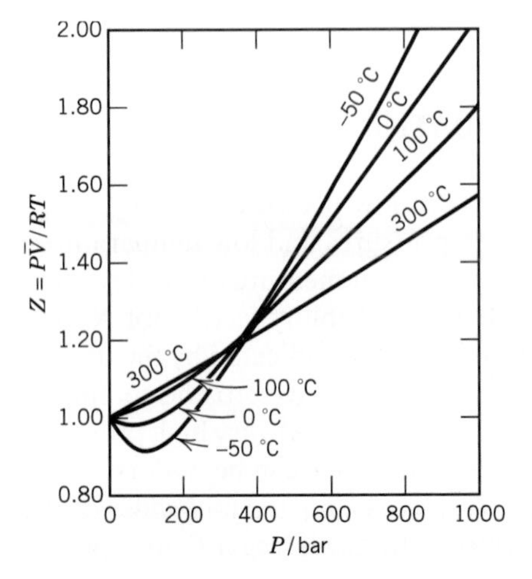

Figure 1.8 Influence of pressure on the compressibility factor, $P\overline{V}/RT$, for nitrogen at different temperatures (given in °C).

Table 1.1 Second and Third Virial Coefficients at 298.15 K

Gas	$B/10^{-6}$ m^3 mol^{-1}	$C/10^{-12}$ m^6 mol^{-2}
H_2	14.1	350
He	11.8	121
N_2	−4.5	1100
O_2	−16.1	1200
Ar	−15.8	1160
CO	−8.6	1550

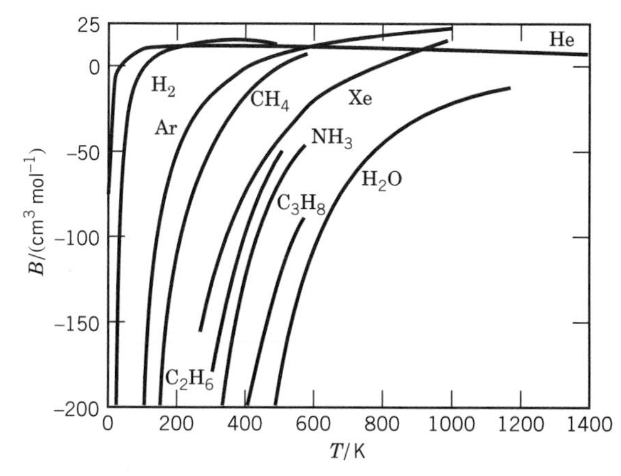

Figure 1.9 Second virial coefficient *B*. (From K. E. Bett, J. S. Rowlinson, and G. Saville, *Thermodynamics for Chemical Engineers*. Cambridge, MA: MIT Press, 1975. Reproduced by permission of The Athlone Press.) (See Computer Problem 1.E.)

The coefficients B and C are referred to as the second and third virial coefficients, respectively.* For a particular gas these coefficients depend only on the temperature and not on the pressure. The word *virial* is derived from the Latin word for force.

The second and third virial coefficients at 298.15 K are given in Table 1.1 for several gases. The variation of the second virial coefficient with temperature is illustrated in Fig. 1.9.

For many purposes, it is more convenient to use P as an independent variable and write the virial equation as

$$Z = \frac{P\overline{V}}{RT} = 1 + B'P + C'P^2 + \cdots \tag{1.12}$$

Example 1.5 ***Derive the relationships between two types of virial coefficients***

Derive the relationships between the virial coefficients in equation 1.11 and the virial coefficients in equation 1.12.

*Statistical mechanics shows that the term $B/\overline{V}$ arises from interactions involving two molecules, the $C/\overline{V}^2$ term arises from interactions involving three molecules, etc. (Section 16.11).

The pressures can be eliminated from equation 1.12 by use of equation 1.11 in the following forms:

$$P = \frac{RT}{\overline{V}} + \frac{BRT}{\overline{V}^2} + \frac{CRT}{\overline{V}^3} + \cdots \tag{1.13}$$

$$P^2 = \left(\frac{RT}{\overline{V}}\right)^2 + \frac{2B(RT)^2}{\overline{V}^3} + \cdots \tag{1.14}$$

Substituting these expressions into equation 1.12 yields

$$Z = 1 + B'\left(\frac{RT}{\overline{V}}\right) + \frac{B'BRT + C'(RT)^2}{\overline{V}^2} + \cdots \tag{1.15}$$

When we compare this equation with equation 1.11 we see that

$$B = B'RT \tag{1.16}$$

$$C = BB'RT + C'(RT)^2 \tag{1.17}$$

Thus

$$B' = B/RT \tag{1.18}$$

$$C' = \frac{C - B^2}{(RT)^2} \tag{1.19}$$

The second virial coefficient B for nitrogen is zero at 54 °C, which is consistent with Fig. 1.8. A real gas may behave like an ideal gas over an extended range in pressure when the second virial coefficient is zero, as shown in Fig. 1.10. The temperature at which this occurs is called the **Boyle temperature** T_B. The Boyle temperatures of a number of gases are given in Table 1.2.

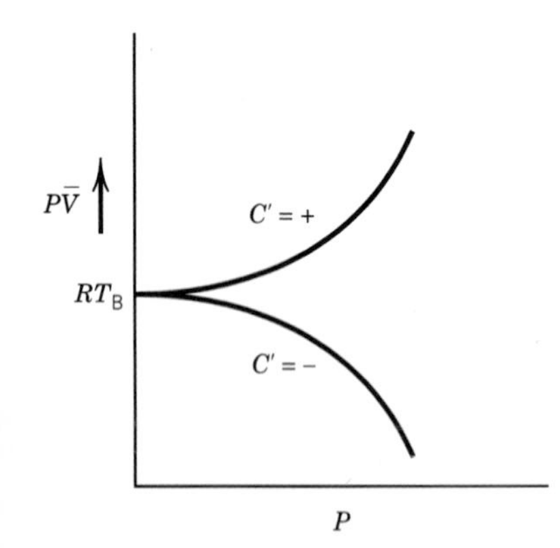

Figure 1.10 At the Boyle temperature ($B = 0$), a gas behaves nearly ideally over a range of pressures. The curvature at higher pressures depends on the sign of the third virial coefficient.

Table 1.2 Critical Constants and Boyle Temperatures

Gas	T_c/K	P_c/bar	$\overline{V}_c$/L mol^{-1}	Z_c	T_B/K
Helium-4	5.2	2.27	0.0573	0.301	22.64
Hydrogen	33.2	13.0	0.0650	0.306	110.04
Nitrogen	126.2	34.0	0.0895	0.290	327.22
Oxygen	154.6	50.5	0.0734	0.288	405.88
Chlorine	417	77.0	0.124	0.275	
Bromine	584	103.0	0.127	0.269	
Carbon dioxide	304.2	73.8	0.094	0.274	714.81
Water	647.1	220.5	0.056	0.230	
Ammonia	405.6	113.0	0.0725	0.252	995
Methane	190.6	46.0	0.099	0.287	509.66
Ethane	305.4	48.9	0.148	0.285	
Propane	369.8	42.5	0.203	0.281	
n-Butane	425.2	38.0	0.255	0.274	
Isobutane	408.1	36.5	0.263	0.283	
Ethylene	282.4	50.4	0.129	0.277	624
Propylene	365.0	46.3	0.181	0.276	
Benzene	562.1	49.0	0.259	0.272	
Cyclohexane	553.4	40.7	0.308	0.272	

1.6 P–$\overline{V}$–T SURFACE FOR A ONE-COMPONENT SYSTEM

To discuss more general equations of state, we will now look at the possible values of P, $\overline{V}$, and T for a pure substance. The state of a pure substance is represented by a point in a Cartesian coordinate system with P, $\overline{V}$, and T plotted along the three axes. Each point on the surface of the three-dimensional model in Fig. 1.11 describes the state of a one-component system that contracts on freezing. We will not be concerned here with the solid state, but will consider that part of the surface later (Section 6.2). Projections of this surface on the P–$\overline{V}$ and P–T planes are shown. There are three two-phase regions on the surface: S + G, L + G, and S + L (S is solid, G gas, and L liquid). These three surfaces intersect at the **triple point** t where vapor, liquid, and solid are in equilibrium.

The projection of the three-dimensional surface on the P–T plane is shown to the right of the main diagram in Fig. 1.11. The vapor pressure curve goes from the triple point t to the **critical point** c (see Section 1.7). The sublimation pressure curve goes from the triple point t to absolute zero. The melting curve rises from the triple point. Most substances contract on freezing, and for them the slope $\mathrm{d}P/\mathrm{d}T$ for the melting line is positive.

At high temperatures the substance is in the gas state, and as the temperature is raised and the pressure is lowered the surface is more and more closely represented by the ideal gas equation of state $P\overline{V} = RT$. However, much more complicated equations are required to describe the rest of the surface that represents gas and liquid. Before discussing equations that can represent this part of the surface, we will consider the unusual phenomena that occur near the critical point. Any realistic equation of state must be able to reproduce this behavior at least qualitatively.

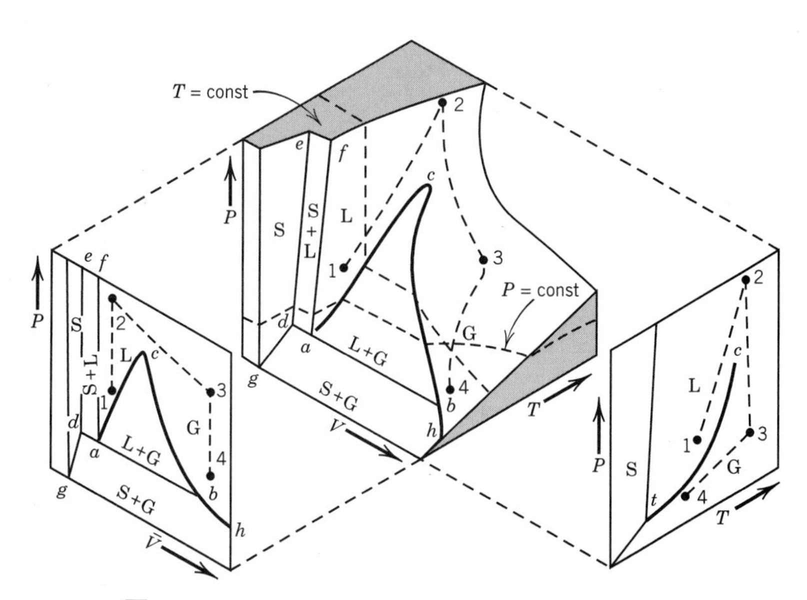

Figure 1.11 P–$\overline{V}$–T surface for a one-component system that contracts on freezing. (From K. E. Bett, J. S. Rowlinson, and G. Saville, *Thermodynamics for Chemical Engineers*. Cambridge, MA: MIT Press, 1975. Reproduced by permission of The Athlone Press.)

1.7 CRITICAL PHENOMENA

For a pure substance there is a critical point (P_c, T_c) at the end of the liquid–gas coexistence curve where the properties of the gas and liquid phases become so nearly alike that they can no longer be distinguished as separate phases. Thus, T_c is the highest temperature at which condensation of a gas is possible, and P_c is the highest pressure at which a liquid will boil when heated.

The critical pressures P_c, volumes $\overline{V}_c$, and temperatures T_c of a number of substances are given in Table 1.2, along with the compressibility factor at the critical point $Z_c = P_c\overline{V}_c/RT_c$, and the Boyle temperature T_B.

Critical phenomena are most easily discussed using the projection of the three-dimensional surface in Fig. 1.11 on the P–$\overline{V}$ plane. Figure 1.12 shows only the parts of the P–$\overline{V}$ plot labeled L, G, and L + G. When the state of the system is represented by a point in the L + G region of this plot, the system contains two phases, one liquid and one gas, in equilibrium with each other. The molar volumes of the liquid and gas can be obtained by drawing a horizontal line parallel to the $\overline{V}$ axis through the point representing the state of the system and noting the intersections with the boundary line for the L + G region. Such a line, which connects the state of one phase with the state of another phase with which it is in equilibrium, is called a **tie line.** Two tie lines are shown in Fig. 1.12. The pressure in this case is the equilibrium vapor pressure of the liquid. As the temperature is

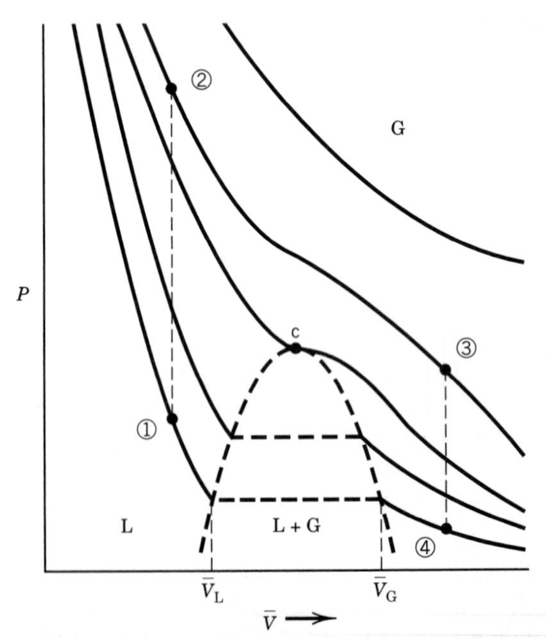

Figure 1.12 Pressure–molar volume relations (e.g., isotherms) in the region of the critical point. The dashed horizontal lines in the two-phase region are called tie lines. The path 1–2–3–4 shows how a liquid can be converted to a gas without the appearance of a meniscus. If liquid at point 4 is compressed isothermally, the volume decreases until the two-phase region is reached. At this point there is a large decrease in volume at constant pressure (the **vapor pressure** of the liquid) until all of the gas has condensed to liquid. As the liquid is compressed, the pressure rises rapidly.

raised, the tie line becomes shorter, and the molar volumes of the liquid and gas approach each other. At the critical point c the tie line vanishes and the distinction between liquid and gas is lost. At temperatures above the critical temperature, there is a single fluid phase. Above the critical point a gas may have a very high density, and it may be characterized as a supercritical fluid.

The isotherm that goes through the critical point has the following two properties: It is horizontal at the critical point,

$$\left(\frac{\partial P}{\partial \overline{V}}\right)_{T=T_c} = 0 \tag{1.20}$$

and it has a point of inflection at the critical point,

$$\left(\frac{\partial^2 P}{\partial^2 \overline{V}}\right)_{T=T_c} = 0 \tag{1.21}$$

Figures 1.11 and 1.12 also show how a liquid at point 1 can be converted to a gas at point 4 without the appearance of an interface between two phases. To do this, liquid at point 1 is heated at constant volume to point 2, then expanded at constant temperature to point 3, and finally cooled at constant volume to point 4, where it is a gas. Thus, liquid and vapor phases are really the same in terms of molecular organization, and so when the densities of these two phases for a substance become equal, they cannot be distinguished and there is a critical point. On the other hand, a solid and a liquid have different molecular organizations, and the two phases do not become identical even if their densities are equal. Therefore, solid–liquid, solid–gas, and solid–solid equilibrium lines do not have critical points as do gas–liquid lines.

At the critical point the **isothermal compressibility** [$\kappa = -\overline{V}^{-1}(\partial \overline{V}/\partial P)_T$; see Problem 1.17] becomes infinite because $(\partial P/\partial \overline{V})_{T_c} = 0$. If the isothermal compressibility is very large, as it is in the neighborhood of the critical point, very little work is required to compress the fluid. Therefore, gravity sets up large differences in density between the top and bottom of the container, as large as 10% in a column of fluid only a few centimeters high. This makes it difficult to determine $P\overline{V}$ isotherms near the critical point. These large differences, or **spontaneous fluctuations,** in the density can extend over macroscopic distances. The distance may be as large as the wavelength of visible light or larger. Since fluctuations in density are accompanied by fluctuations in refractive index, light is strongly scattered, and this is called **critical opalescence.**

1.8 THE VAN DER WAALS EQUATION

Although the virial equation is very useful, it is important to have approximate equations of state with only a few parameters. We turn now to the equation that was introduced by van der Waals in 1877, which is based on plausible reasons that real gases do not follow the ideal gas law. The ideal gas law can be derived for point particles that do not interact except in elastic collisions (see Chapter 17, Kinetic Theory of Gases). The first reason that van der Waals modified the ideal gas law is that molecules are not point particles. Therefore $\overline{V}$ is replaced by $\overline{V} - b$, where b is the volume per mole that is occupied by the molecules. This leads to

$$P(\overline{V} - b) = RT \tag{1.22}$$

which corresponds to equation 1.12 with $B' = b/RT$ and C' and higher constants equal to zero. This equation can represent compressibility factors greater than unity, but it cannot yield compressibility factors less than unity.

The second reason for modifying the ideal gas law is that gas molecules attract each other and that real gases are therefore more compressible than ideal gases. The forces that lead to condensation are still referred to as van der Waals forces, and their origin is discussed in Section 11.10. Van der Waals provided for intermolecular attraction by adding to the observed pressure P in the equation of state a term $a/\overline{V}^2$, where a is a constant whose value depends on the gas.

The **van der Waals equation** is*

$$(P + a/\overline{V}^2)(\overline{V} - b) = RT \tag{1.23}$$

When the molar volume $\overline{V}$ is large, b becomes negligible in comparison with $\overline{V}$, $a/\overline{V}^2$ becomes negligible with respect to P, and the van der Waals equation reduces to the ideal gas law, $P\overline{V} = RT$.

The van der Waals constants for a few gases are listed in Table 1.3. They can be calculated from experimental measurements of P, $\overline{V}$, and T or from the critical constants, as shown later in equations 1.32 and 1.33. The van der Waals equation is very useful because it exhibits phase separation between gas and liquid phases.

Figure 1.13 shows three isotherms calculated using the van der Waals equation. At the critical temperature the isotherm has an inflection point at the critical point. At temperatures below the critical temperature each isotherm passes through a minimum and a maximum. The locus of these points shown by the dotted line has been obtained from $(\partial P/\partial \overline{V})_T = 0$. The states within the dotted line have $(\partial P/\partial \overline{V})_T > 0$, that is, the volume increases when the pressure increases. These states are therefore mechanically unstable and do not exist. Maxwell showed that states corresponding to the points between A and B and

Table 1.3 Van der Waals Constants

Gas	a/L^2 bar mol^{-2}	b/L mol^{-1}	Gas	a/L^2 bar mol^{-2}	b/L mol^{-1}
H_2	0.247 6	0.026 61	CH_4	2.283	0.042 78
He	0.034 57	0.023 70	C_2H_6	5.562	0.063 80
N_2	1.408	0.039 13	C_3H_8	8.779	0.084 45
O_2	1.378	0.031 83	$C_4H_{10}(n)$	14.66	0.122 6
Cl_2	6.579	0.056 22	C_4H_{10}(iso)	13.04	0.114 2
NO	1.358	0.027 89	$C_5H_{12}(n)$	19.26	0.146 0
NO_2	5.354	0.044 24	CO	1.505	0.039 85
H_2O	5.536	0.030 49	CO_2	3.640	0.042 67

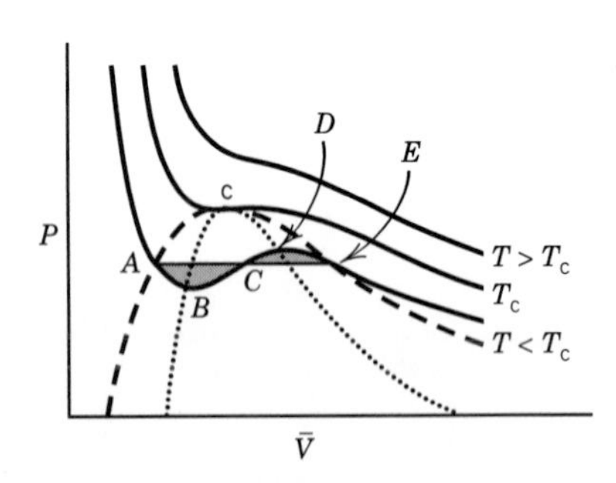

Figure 1.13 Isotherms calculated from the van der Waals equation. The dashed line is the boundary of the L + G region.

*The van der Waals equation can also be written in the form

$$(P + an^2/V^2)(V - nb) = nRT$$

those between D and E are metastable, that is, not true equilibrium states. The dashed line is the boundary of the two-phase region; the part of the isotherm to the left of A represents the liquid and that to the right of E, gas. The horizontal line ACE that produces two equal areas (ABC and CDE) is referred to as the Maxwell construction. It connects the thermodynamic properties of the liquid phase (A) with the properties of the gas phase (E) that is in equilibrium with it.

The compressibility factor for a van der Waals gas is given by

$$\begin{aligned} Z &= \frac{P\overline{V}}{RT} = \frac{\overline{V}}{\overline{V} - b} - \frac{a}{RT\overline{V}} \\ &= \frac{1}{1 - b/\overline{V}} - \frac{a}{RT\overline{V}} \end{aligned} \tag{1.24}$$

At low pressures, $b/\overline{V} \ll 1$ so that we can expand the first term using $(1-x)^{-1} = 1 + x + x^2 + \cdots$.

Example 1.6 ***Expansion of*** $(1 - x)^{-1}$ ***using the Maclaurin series***

Since we will use series like

$$1/(1 - x) = 1 + x + x^2 + \cdots$$

a number of times, it is important to realize that functions can often be expressed as series by use of the Maclaurin series

$$f(x) = f(0) + \left(\frac{\mathrm{d}f}{\mathrm{d}x}\right)_{x=0} x + \frac{1}{2!}\left(\frac{\mathrm{d}^2 f}{\mathrm{d}x^2}\right)_{x=0} x^2 + \cdots$$

In this case,

$$f(0) = 1$$

$$\left(\frac{\mathrm{d}f}{\mathrm{d}x}\right) = \frac{1}{(1 - x)^{-2}} \quad \text{and} \quad \left(\frac{\mathrm{d}f}{\mathrm{d}x}\right)_{x=0} = 1$$

$$\left(\frac{\mathrm{d}^2 f}{\mathrm{d}x^2}\right) = 2(1 - x)^{-3} \quad \text{and} \quad \left(\frac{\mathrm{d}^2 f}{\mathrm{d}x^2}\right)_{x=0} = 2$$

Equation 1.24 then yields the virial equation in terms of volume:

$$Z = 1 + \left(b - \frac{a}{RT}\right)\frac{1}{\overline{V}} + \left(\frac{b}{\overline{V}}\right)^2 + \cdots \tag{1.25}$$

From this equation we can see that the value of a is relatively more important at low temperatures, and the value of b is relatively more important at high temperatures. To obtain the virial equation in terms of pressure, we can replace $\overline{V}$ in the second term by the ideal gas value to obtain, to first order in P,

$$Z = 1 + \frac{1}{RT}\left(b - \frac{a}{RT}\right)P + \cdots \tag{1.26}$$

but this approximation is not good enough to give the correct coefficient for the P^2 term. At the Boyle temperature the second virial coefficient is zero, and so for a van der Waals gas

$$T_B = \frac{a}{bR} \tag{1.27}$$

The values of the van der Waals constants may be calculated from the critical constants for a gas, as shown in the following example.

Example 1.7 ***Van der Waals constants expressed in terms of critical constants***

Derive the expressions for the van der Waals constants in terms of the critical constants for a gas.

The van der Waals equation may be written

$$P = \frac{RT}{\overline{V} - b} - \frac{a}{\overline{V}^2} \tag{1.28}$$

Differentiating with respect to molar volume and evaluating these equations at the critical point yields

$$\left(\frac{\partial P}{\partial \overline{V}}\right)_{T_c} = \frac{-RT_c}{(\overline{V}_c - b)^2} + \frac{2a}{\overline{V}_c^3} = 0 \tag{1.29}$$

$$\left(\frac{\partial^2 P}{\partial \overline{V}^2}\right)_{T_c} = \frac{2RT_c}{(\overline{V}_c - b)^3} - \frac{6a}{\overline{V}_c^4} = 0 \tag{1.30}$$

A third simultaneous equation is obtained by writing equation 1.28 for the critical point:

$$P_c = \frac{RT_c}{\overline{V}_c - b} - \frac{a}{\overline{V}_c^2} \tag{1.31}$$

These three simultaneous equations may be combined to obtain expressions for a and b in terms of T_c and P_c or T_c and $\overline{V}_c$:

$$a = \frac{27R^2T_c^2}{64P_c} = \frac{9}{8}RT_c\overline{V}_c \tag{1.32}$$

$$b = \frac{RT_c}{8P_c} = \frac{\overline{V}_c}{3} \tag{1.33}$$

Example 1.8 ***Critical constants expressed in terms of van der Waals constants***

Derive the expressions for the molar volume, temperature, and pressure at the critical point in terms of the van der Waals constants.

Equation 1.33 shows that

$$\overline{V}_c = 3b$$

Equation 1.32 shows that

$$T_c = \frac{8a}{9R\overline{V}_c} = \frac{8a}{27Rb}$$

Equation 1.33 shows that

$$P_c = \frac{RT_c}{8b} = \frac{a}{27b^2}$$

Example 1.9 ***Calculation of the molar volume using the van der Waals equation***

What is the molar volume of ethane at 350 K and 70 bar according to (*a*) the ideal gas law and (*b*) the van der Waals equation?

$$(a)\ \overline{V} = RT/P = (0.083\,145\ \text{L bar K}^{-1}\ \text{mol}^{-1})(350\ \text{K})/(70\ \text{bar})$$
$$= 0.416\ \text{L mol}^{-1}$$

(*b*) The van der Waals constants are given in Table 1.3.

$$P = \frac{RT}{\overline{V} - b} - \frac{a}{\overline{V}^2}$$

$$70 = \frac{(0.083\,15)(350)}{\overline{V} - 0.063\,80} - \frac{5.562}{\overline{V}^2}$$

This is a cubic equation, but we know it has a single real, positive solution because the temperature is above the critical temperature. This cubic equation can be solved using a personal computer with a mathematical application. This yields two complex roots and one real root, namely 0.2297 L mol^{-1} (see Computer Problem 1.G).

We will see later that equations of state are very important in the calculation of various thermodynamic properties of gases. Therefore, a variety of them have been developed. To represent the P–V–T properties of a one-component system over a wide range of conditions it is necessary to use an equation with many more parameters. As more parameters are used they lose any simple physical interpretation. The van der Waals equation does not fit the properties of any gas exactly, but it is very useful because it does have a simple interpretation and the qualitatively correct behavior.

The van der Waals equation fails in the immediate neighborhood of the critical point. The coexistence curve (see Fig. 1.12) is not parabolic in the neighborhood of the critical point. The van der Waals equation indicates that near T_c, $\overline{V}_c - \overline{V} = k(T_c - T)^{1/2}$, but experiments show that the exponent is actually 0.32. Other properties in the neighborhood of the critical point vary with $(T_c - T)$ with exponents that differ from what would be expected from the van der Waals equation. These exponents are the same for all substances, which shows that the properties in the neighborhood of the critical point are universal.

1.9 DESCRIPTION OF THE STATE OF A SYSTEM WITHOUT CHEMICAL REACTIONS

In Section 1.1 we observed that the intensive state of a one-phase system can be described by specifying $N_s + 1$ intensive variables, where N_s is the number of species. The intensive state of a solution containing species A and species B is completely described by specifying T, P, and n_A/n_B, and so three intensive variables are required. Now that we have discussed several systems, it is time to think about the numbers of intensive variables required to define the thermodynamic states of these more complicated systems. The number of independent variables required is represented by F, which is referred to as the **number of degrees of freedom.** Therefore, for a one-phase system without chemical reactions, $F = N_s + 1$. As we have seen, if $N_s = 1$, the independent intensive properties can be chosen to be T and P. If $N_s = 1$, but the system has two phases at equilibrium, Fig. 1.12 shows that it is sufficient to specify either T or P, but not both, so that $F = 1$. Thus the intensive state of this system is described completely by saying that two phases are at equilibrium and specifying T or P. In defining the ideal gas temperature scale, we saw that water vapor, liquid water, and ice are in equilibrium at a particular T and P. Thus the intensive state of this three-phase system is completely described by saying that three phases are at equilibrium. There are no independent intensive variables, and so $F = 0$.

Earlier we contrasted the thermodynamic description of a system with the classical description of a system in terms of molecules, and now we can see that the description of the thermodynamic state of a system is really quite different. Another interesting aspect of specifying degrees of freedom is that the choice of variables is not unique, although the number is. For example, the intensive state of a binary solution can be described by T, P, and the mole fraction of one of the species.

The preceding paragraph has discussed the intensive state of a system, but it is often necessary to describe the extensive state of a system. The number of variables required to describe the extensive state of a system is given by $D = F + p$, where p is the number of different phases, because the amount of each phase must be specified. For a one-phase system with one species and no reactions, $D = 2 + 1 = 3$, and so a complete description requires T, P, and the amount of the species (n). For a two-phase system with one species, $D = 1 + 2 = 3$, and so it is necessary to specify T or P and the amounts of the two phases. For a three-phase system with one species, $D = 0 + 3 = 3$, and so it is necessary to specify the amounts of the three phases. For a one-phase binary solution, $D = 3 + 1 = 4$, and so it is necessary to specify T, P, n_A/n_B, and the amount of the solution. Phase equilibria and chemical equilibria introduce constraints, and we will see in the next several chapters how these constraints arise and how they are treated quantitatively in thermodynamics.

Comment:

It is a good thing that this issue of the number of variables required to describe the state of a system has come up before we discuss the laws of thermodynamics because the conclusions in this section cannot be derived from the laws of thermodynamics. The fact that $N_s + 2$ variables are required to describe the extensive state of a homogeneous one-phase system at equilibrium is a generalization of experimental observations, and we will consider it to be a postulate. It is a postulate that has stood the test of time, and we will use it often in discussing thermodynamic systems.

1.10 PARTIAL MOLAR PROPERTIES

This chapter has mostly been about pure gases, but we need to be prepared to consider mixtures of gases and mixtures of liquids. There is an important mathematical difference between extensive properties and intensive properties of mixtures. These properties can be treated as mathematical functions. A function $f(x_1, x_2, \ldots, x_N)$ is said to be homogeneous of degree k if

$$f(\lambda x_1, \lambda x_2, \ldots, \lambda x_N) = \lambda^k f(x_1, x_2, \ldots, x_N) \tag{1.34}$$

All extensive properties are homogeneous of degree 1. This is illustrated by the volume for which

$$V(\lambda n_1, \lambda n_2, \ldots, \lambda n_N) = \lambda^1 V(n_1, n_2, \ldots, n_N) = \lambda V(n_1, n_2, \ldots, n_N) \tag{1.35}$$

where $n_1, n_2, \ldots$ are amounts of substances. That is, if we increase the amounts of every substance λ-fold, the total volume increases λ-fold. All intensive properties are homogeneous of degree zero. This is illustrated by the temperature for which

$$T(\lambda n_1, \lambda n_2, \ldots, \lambda n_N) = \lambda^0 T(n_1, n_2, \ldots, n_N) = T(n_1, n_2, \ldots, n_N) \tag{1.36}$$

According to **Euler's theorem,** when equation 1.34 applies,

$$kf(x_1, x_2, \ldots, x_N) = \sum_{i=1}^{N} x_i \left(\frac{\partial f}{\partial x_i}\right)_{x_j \neq x_i} \tag{1.37}$$

Thus for the volume of a mixture ($k = 1$),

$$\begin{aligned} V &= \left(\frac{\partial V}{\partial n_1}\right)_{T,P,n_j} n_1 + \left(\frac{\partial V}{\partial n_2}\right)_{T,P,n_j} n_2 + \cdots + \left(\frac{\partial V}{\partial n_N}\right)_{T,P,n_j} n_N \\ &= \overline{V}_1 n_1 + \overline{V}_2 n_2 + \cdots + \overline{V}_N n_N \end{aligned} \tag{1.38}$$

where the subscript n_j indicates that the amounts of all other substances are held constant when the amount of one of the substances is changed. These derivatives are referred to as **partial molar volumes.** Since we will use such equations a lot, partial molar properties are indicated by the use of an overbar:

$$\overline{V}_i = \left(\frac{\partial V}{\partial n_i}\right)_{T,P,\{n_{j\neq i}\}} \tag{1.39}$$

This definition for the partial molar volume can be stated in words by saying that $\overline{V}_i \mathrm{d}n_i$ is the change in V when an infinitesimal amount ($\mathrm{d}n_i$) of this substance is added to the solution at constant T, P, and all other n_j. Alternatively, it can be said that $\overline{V}_i$ is the change in V when 1 mol of i is added to an infinitely large amount of the solution at constant T and P.

Note that the partial molar volume depends on the composition of the solution. When the amount of substance 1 is changed by $\mathrm{d}n_1$, the amount of substance 2 is changed by $\mathrm{d}n_2$, etc., and the volume of the solution is changed by

$$dV = \overline{V}_1 \, dn_1 + \overline{V}_2 \, dn_2 + \cdots + \overline{V}_N \, dn_N \tag{1.40}$$

Dividing equation 1.38 by the total number of moles in the solution yields

$$\overline{V} = \overline{V}_1 x_1 + \overline{V}_2 x_2 + \cdots + \overline{V}_N x_N \tag{1.41}$$

where $\overline{V}$ is the molar volume of the solution and x_i is the mole fraction of substance i in the solution. In Chapter 6 we will discuss the determination of the partial molar volume of a species in a solution, and we will also see that in ideal solutions the partial molar volume of a substance is equal to its molar volume in the pure liquid.

Example 1.10 ***The partial molar volume of a gas in an ideal gas mixture***

Calculate the partial molar volume of a gas in an ideal gas mixture.

The volume of an ideal gas mixture is

$$V = \frac{RT}{P}(n_1 + n_2 + \cdots)$$

Using equation 1.39 to find the partial molar volume of gas i yields

$$\overline{V}_i = \left(\frac{\partial V}{\partial n_i}\right)_{T,P,\{n_{j \neq i}\}} = \frac{RT}{P}$$

Thus all of the gases in a mixture of ideal gases have the same partial molar volume. This is not true for nonideal gases or for liquids.

Comment:

Calculus is used so much in physical chemistry that we have included a section on calculus in Appendix D for quick reference. Since the properties of a system depend on a number of variables, it is important to be clear about which properties are held constant for a measurement or a process and to use subscripts on partial derivatives.

1.11 SPECIAL TOPIC: BAROMETRIC FORMULA

In applying thermodynamics we generally ignore the effect of the gravitational field, but it is important to realize that if there is a difference in height there is a difference in gravitational potential. For example, consider a vertical column of a gas with a unit cross section and a uniform temperature T, as shown in Fig. 1.14. The pressure at any height h is simply equal to the mass of gas above that height per unit area times the gravitational acceleration g. The standard acceleration due to gravity is defined as 9.806 65 m s^{-2}. The difference in pressure dP between h and $h + dh$ is equal to the mass of the gas between these two levels times g and divided by the area. Thus,

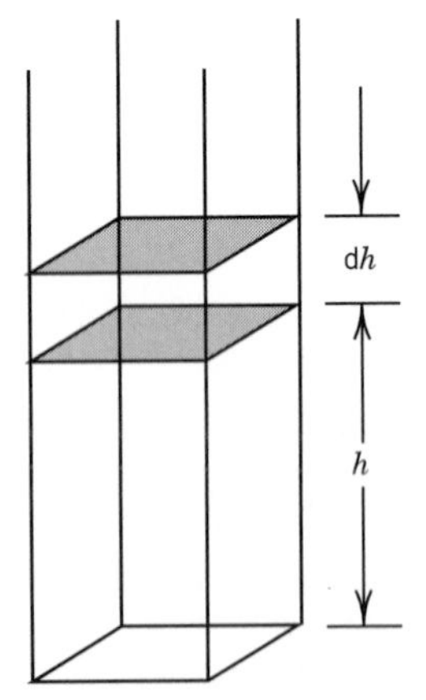

Figure 1.14 Column of an ideal gas of uniform temperature and unit cross section.

$$dP = -\rho g \, dh \tag{1.42}$$

where ρ is the density of the gas. If the gas is an ideal gas, then $\rho = PM/RT$, where M is the molar mass, so that

$$\mathrm{d}P = -\frac{PMg}{RT}\,\mathrm{d}h \tag{1.43}$$

Separating variables and integrating from $h = 0$, where the pressure is P_0, to h, where the pressure is P, yields

$$\int_{P_0}^{P} \frac{\mathrm{d}P}{P} = -\int_0^h \frac{gM}{RT}\,\mathrm{d}h \tag{1.44}$$

$$\ln\frac{P}{P_0} = -\frac{gMh}{RT} \tag{1.45}$$

$$P = P_0\,\mathrm{e}^{-gMh/RT} \tag{1.46}$$

This relation is known as the **barometric formula.**

Example 1.11 ***Pressure and composition of air at 10 km***

Assuming that air is 20% O_2 and 80% N_2 at sea level and that the pressure is 1 bar, what are the composition and pressure at a height of 10 km, if the atmosphere has a temperature of 0 °C independent of altitude?

$$P = P_0 \exp\left(-\frac{gMh}{RT}\right)$$

For O_2,

$$P_{O_2} = (0.20\text{ bar})\exp\left[-\frac{(9.8\text{ m s}^{-2})(32\times10^{-3}\text{ kg mol}^{-1})(10^4\text{ m})}{(8.3145\text{ J K}^{-1}\text{ mol}^{-1})(273\text{ K})}\right]$$

$$= 0.0503\text{ bar}$$

For N_2,

$$P_{N_2} = (0.80)\exp\left(-\frac{9.8\times28\times10^{-3}\times10^4}{8.3145\times273}\right)$$

$$= 0.239\text{ bar}$$

The total pressure is 0.289 bar, and $y_{O_2} = 0.173$ and $y_{N_2} = 0.827$.

Figure 1.15 gives the partial pressures of oxygen, nitrogen, and the total pressure as a function of height in feet, assuming the temperature is 273.15 K independent of height.

Comment:

This is our first contact with exponential functions, but there will be many more. The barometric formula can also be regarded as an example of a Boltzmann distribution, which will be derived in Chapter 16 (Statistical Mechanics). The temperature determines the way particles distribute themselves over various energy levels in a system.

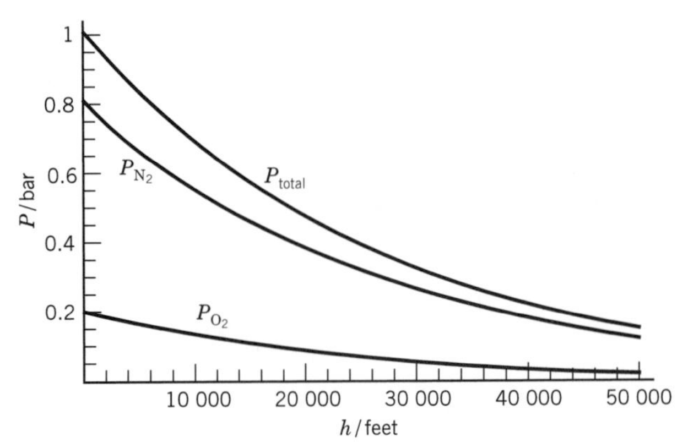

Figure 1.15 Partial pressures of oxygen, nitrogen, and the total pressure of the atmosphere as a function of height in feet, assuming the temperature is 273.15 K independent of height (see Computer Problem 1.H).

Nine Key Ideas in Chapter 1

1. The state of a macroscopic system at equilibrium can be specified by the values of a small number of macroscopic variables. For a system in which there are no chemical reactions, the intensive state of a one-phase system can be specified by $N_s + 1$ intensive variables, where N_s is the number of different species.
2. According to the zeroth law of thermodynamics, if systems A and B are individually in thermal equilibrium with system C, then A and B are in thermal equilibrium with each other.
3. The ideal gas temperature scale is based on the behavior of gases in the limit of low pressures. The unit of thermodynamic temperature, the kelvin, represented by K, is defined as the fraction 1/273.16 of the temperature of the triple point of water.
4. The total pressure of a mixture of ideal gases is equal to the sum of the partial pressures of the gases in the mixture.
5. The virial equation of state, which expresses the compressibility factor Z for a gas in terms of powers of the reciprocal molar volume or of the pressure, is useful for expressing experimental data on a gas provided the pressure is not too high or the gas too close to its critical point.
6. The van der Waals equation is useful because it exhibits phase separation between gas and liquid phases, but it does not represent experimental data exactly.
7. For a one-phase system without chemical reactions, we have seen that the number of degrees of freedom F is equal to $N_s + 1$. But if the system contains two phases at equilibrium, $F = N_s$, and if the system contains three phases at equilibrium, $F = N_s - 1$. The number of variables D required to describe the extensive state of a multiphase macroscopic system at equilibrium is $F + p$, where p is the number of phases.
8. The volume of a mixture is equal to the sum of the partial molar volumes of the species it contains each multiplied by the amount of that species.
9. For an isothermal atmosphere, the pressure decreases exponentially with the height above the surface of the earth.

REFERENCES

M. Bailyn, *A Survey of Thermodynamics.* New York: American Institute of Physics, 1944.

K. E. Bett, J. S. Rowlinson, and G. Saville, *Thermodynamics for Chemical Engineers.* Cambridge, MA: MIT Press, 1975.

J. H. Dymond and E. B. Smith, *The Virial Coefficients of Pure Gases and Mixtures.* Oxford, UK: Oxford University Press, 1980.

K. S. Pitzer, *Thermodynamics,* 3rd ed. New York: McGraw-Hill, 1995.

J. M. Smith, H. C. Van Ness, and M. M. Abbott, *Introduction to Chemical Engineering Thermodynamics.* New York: McGraw-Hill, 1996.

J. W. Tester and M. Modell, *Thermodynamics and Its Applications.* Upper Saddle River, NJ: Prentice PTR, 1997.

PROBLEMS

(M) Problems marked with an icon may be more conveniently solved on a personal computer with a mathematical program.

1.1 The intensive state of an ideal gas can be completely defined by specifying (1) T, P, (2) $T, \overline{V}$, or (3) $P, \overline{V}$. The extensive state of an ideal gas can be specified in four ways. What are the combinations of properties that can be used to specify the extensive state of an ideal gas? Although these choices are deduced for an ideal gas, they also apply to real gases.

1.2 The ideal gas law also represents the behavior of mixtures of gases at low pressures. The molar volume of the mixture is the volume divided by the amount of the mixture. The partial pressure of gas i in a mixture is defined as $y_i P$ for an ideal gas mixture, where y_i is its mole fraction and P is the total pressure. Ten grams of N_2 is mixed with 5 g of O_2 and held at 25 °C at 0.750 bar. (*a*) What are the mole fractions of N_2 and O_2? (*b*) What are the partial pressures of N_2 and O_2? (*c*) What is the volume of the ideal mixture?

1.3 A mixture of methane and ethane is contained in a glass bulb of 500 cm^3 capacity at 25 °C. The pressure is 1.25 bar, and the mass of gas in the bulb is 0.530 g. What is the average molar mass, and what is the mole fraction of methane?

1.4 Nitrogen tetroxide is partially dissociated in the gas phase according to the reaction

$$N_2O_4(g) = 2NO_2(g)$$

A mass of 1.588 g of N_2O_4 is placed in a 500-cm^3 glass vessel at 298 K and dissociates to an equilibrium mixture at 1.0133 bar. (*a*) What are the mole fractions of N_2O_4 and NO_2? (*b*) What percentage of the N_2O_4 has dissociated? Assume that the gases are ideal.

1.5 Although a real gas obeys the ideal gas law in the limit as $P \to 0$, not all of the properties of a real gas approach the values for an ideal gas as $P \to 0$. The second virial coefficient of an ideal gas is zero, and so $\mathrm{d}Z/\mathrm{d}P = 0$ at all pressures. But calculate $\mathrm{d}Z/\mathrm{d}P$ for a real gas as $P \to 0$.

(M) **1.6** Show how the second virial coefficient of a gas and its molar mass can be obtained by plotting P/ρ versus P, where ρ is the density of the gas. Apply this method to the following data on ethane at 300 K.

P/bar	1	10	20
$\rho/10^{-3}$ g cm^{-3}	1.2145	13.006	28.235

(M) **1.7** Calculate the second and third virial coefficients for hydrogen at 0 °C from the fact that the molar volumes at 50.7, 101.3, 202.6, and 303.9 bar are 0.4634, 0.2386, 0.1271, and 0.090 04 L mol^{-1}, respectively.

(M) **1.8** The critical temperature of carbon tetrachloride is 283.1 °C. The densities in g/cm^3 of the liquid ρ_l and vapor ρ_v at different temperatures are as follows:

t/°C	100	150	200	250	270	280
ρ_l	1.4343	1.3215	1.1888	0.9980	0.8666	0.7634
ρ_v	0.0103	0.0304	0.0742	0.1754	0.2710	0.3597

What is the critical molar volume of CCl_4? It is found that the mean of the densities of the liquid and vapor does not vary rapidly with temperature and can be represented by

$$\frac{\rho_l + \rho_v}{2} = A + Bt$$

where A and B are constants. The extrapolated value of the average density at the critical temperature is the critical density. The molar volume $\overline{V}_c$ at the critical point is equal to the molar mass divided by the critical density.

1.9 Show that for a gas of rigid spherical molecules, b in the van der Waals equation is four times the molecular volume times Avogadro's constant. If the molecular diameter of Ne is 0.258 nm (Table 17.4), approximately what value of b is expected?

1.10 What is the molar volume of n-hexane at 660 K and 91 bar according to (*a*) the ideal gas law and (*b*) the van der Waals equation? For n-hexane, $T_c = 507.7$ K and $P_c = 30.3$ bar.

1.11 Derive the expressions for van der Waals constants a and b in terms of the critical temperature and pressure; that is, derive equations 1.32 and 1.33 from 1.29–1.31.

1.12 Calculate the second virial coefficient of methane at 300 K and 400 K from its van der Waals constants, and compare these results with Fig. 1.9.

1.13 You want to calculate the molar volume of O_2 at 298.15 K and 50 bar using the van der Waals equation, but you don't want to solve a cubic equation. Use the first two terms of equation 1.26. The van der Waals constants of O_2 are a = 0.138 Pa m^6 mol^{-1} and b = 31.8 × 10^{-6} m^3 mol^{-1}. What is the molar volume in L mol^{-1}?

1.14 The isothermal compressibility κ of a gas is defined in Problem 1.17, and its value for an ideal gas is shown to be $1/P$. Use implicit differentiation of V with respect to P at constant T to obtain the expression for the isothermal compressibility of a van der Waals gas. Show that in the limit of infinite volume, the value for an ideal gas is obtained.

1.15 Calculate the second and third virial coefficients of O_2 from its van der Waals constants in Table 1.3.

1.16 Calculate the critical constants for ethane using the van der Waals constants in Table 1.3.

1.17 The cubic expansion coefficient α is defined by

$$\alpha = \frac{1}{V}\left(\frac{\partial V}{\partial T}\right)_P$$

and the isothermal compressibility κ is defined by

$$\kappa = -\frac{1}{V}\left(\frac{\partial V}{\partial P}\right)_T$$

Calculate these quantities for an ideal gas.

1.18 What is the equation of state for a liquid for which the coefficient of cubic expansion α and the isothermal compressibility κ are constant?

1.19 For a liquid the cubic expansion coefficient α is nearly constant over a narrow range of temperature. Derive the expression for the volume as a function of temperature and the limiting form for temperatures close to T_0.

1.20 (*a*) Calculate $(\partial P/\partial V)_T$ and $(\partial P/\partial T)_V$ for a gas that has the following equation of state:

$$P = \frac{nRT}{V - nb}$$

(*b*) Show that $(\partial^2 P/\partial V \partial T) = (\partial^2 P/\partial T \partial V)$. These are referred to as mixed partial derivatives.

1.21 Assuming that the atmosphere is isothermal at 0 °C and that the average molar mass of air is 29 g mol^{-1}, calculate the atmospheric pressure at 20 000 ft above sea level.

1.22 Calculate the pressure and composition of air on the top of Mt. Everest, assuming that the atmosphere has a temperature of 0 °C independent of altitude (h = 29 141 ft). Assume that air at sea level is 20% O_2 and 80% N_2.

1.23 Calculate the pressure due to a mass of 100 kg in the earth's gravitational field resting on an area of (*a*) 100 cm^2 and (*b*) 0.01 cm^2. (*c*) What area is required to give a pressure of 1 bar?

1.24 A mole of air (80% nitrogen and 20% oxygen by volume) at 298.15 K is brought into contact with liquid water, which has a vapor pressure of 3168 Pa at this temperature. (*a*) What is the volume of the dry air if the pressure is 1 bar? (*b*) What is the final volume of the air saturated with water vapor if the total pressure is maintained at 1 bar? (*c*) What are the mole fractions of N_2, O_2, and H_2O in the moist air? Assume the gases are ideal.

1.25 Using Fig. 1.9, calculate the compressibility factor Z for NH_3(g) at 400 K and 50 bar.

1.26 In this chapter we have considered only pure gases, but it is important to make calculations on mixtures as well. This requires information in addition to that for pure gases. Statistical mechanics shows that the second virial coefficient for an N-component gaseous mixture is given by

$$B = \sum_{i=1}^{N}\sum_{j=1}^{N} y_i y_j B_{ij}$$

where y is mole fraction and i and j identify components. Both indices run over all components of the mixture. The bimolecular interactions between i and j are characterized by B_{ij}, and so $B_{ij} = B_{ji}$. Use this expression to derive the expression for B for a binary mixture in terms of y_1, y_2, B_{11}, B_{12}, and B_{22}.

(M) **1.27** The densities of liquid and vapor methyl ether in g cm^{-3} at various temperatures are as follows:

t/°C	30	50	70	100	120
ρ_l	0.6455	0.6116	0.5735	0.4950	0.4040
ρ_v	0.0142	0.0241	0.0385	0.0810	0.1465

The critical temperature of methyl ether is 299 °C. What is the critical molar volume? (See Problem 1.8.)

1.28 Use the van der Waals constants for CH_4 in Table 1.3 to calculate the initial slopes of the plots of the compressibility factor Z versus P at 300 and 600 K.

1.29 A gas follows the van der Waals equation. Derive the relation between the third and fourth virial coefficients and the van der Waals constants.

1.30 Using the van der Waals equation, calculate the pressure exerted by 1 mol of carbon dioxide at 0 °C in a volume of (*a*) 1.00 L and (*b*) 0.05 L. (*c*) Repeat the calculations at 100 °C and 0.05 L.

1.31 A mole of n-hexane is confined in a volume of 0.500 L at 600 K. What will be the pressure according to (*a*) the ideal gas law and (*b*) the van der Waals equation? (See Problem 1.10.)

1.32 A mole of ethane is contained in a 200-mL cylinder at 373 K. What is the pressure according to (*a*) the ideal gas law and (*b*) the van der Waals equation? The van der Waals constants are given in Table 1.3.

1.33 When pressure is applied to a liquid, its volume decreases. Assuming that the isothermal compressibility

$$\kappa = -\frac{1}{V}\left(\frac{\partial V}{\partial P}\right)_T$$

is independent of pressure, derive an expression for the volume as a function of pressure.

1.34 Calculate α and κ for a gas for which

$$P(\overline{V} - b) = RT$$

1.35 What is the molar volume of $N_2(g)$ at 500 K and 600 bar according to (*a*) the ideal gas law and (*b*) the virial equation? The virial coefficient B of $N_2(g)$ at 500 K is 0.0169 L mol^{-1}.

1.36 What is the mean atmospheric pressure in Denver, Colorado, which is a mile high, assuming an isothermal atmosphere at 25 °C? Air may be taken to be 20% O_2 and 80% N_2.

1.37 Calculate the pressure and composition of air 100 miles above the surface of the earth assuming that the atmosphere has a temperature of 0 °C independent of altitude.

1.38 The density $\rho = m/V$ of a mixture of ideal gases A and B is determined and is used to calculate the average molar mass M of the mixture; $M = \rho RT/P$. How is the average molar mass determined in this way related to the molar masses of A and B?

(M) **1.39** Figure 1.13 shows the Maxwell construction for calculating the vapor pressure of a liquid from its equation of state. Since this requires an iterative process, a computer is needed, and J. H. Noggle and R. H. Wood have shown how to write a computer program in Mathematica (Wolfram Research, Inc., Champaign, IL 61820-7237) to do this. Use this method with the van der Waals equation to calculate the vapor pressure of nitrogen at 120 K.

Computer Problems

1.A Problem 1.7 yields $B = 0.135$ L mol^{-1} and $C = 4.3 \times 10^{-4}$ L^2 mol^{-2} for $H_2(g)$ at 0 °C. Calculate the molar volumes of molecular hydrogen at 75 and 150 bar and compare these molar volumes with the molar volume of an ideal gas.

1.B (*a*) Plot the pressure of ethane versus its molar volume in the range $0 < P < 200$ bar and molar volumes up to 0.5 mol L^{-1} using the van der Waals equation at 265, 280, 310.671, 350, and 400 K, where 310.671 K is the critical temperature calculated with the van der Waals constants. (*b*) Discuss the significance of the plots and the extent to which they represent reality. (*c*) Calculate the molar volumes at 400 K and $P = 150$ bar and at 265 K and 20 bar.

1.C This is a follow-up to Computer Problem 1.B on the van der Waals equation. (*a*) Plot the derivative of the pressure with respect to the molar volume for ethane at 265 K. (*b*) Plot the derivative at the critical temperature. (*c*) Plot the second derivative of the pressure with respect to the molar volume at the critical temperature. In each case, what is the significance of the maxima and minima?

1.D (*a*) Express the compressibility factors for N_2 and O_2 at 298.15 K as a function of pressure using the virial coefficients in Table 1.1. (*b*) Plot these compressibility factors versus P from 0 to 1000 bar.

1.E The second virial coefficients of N_2 at a series of temperatures are given by

T/K	75	100	125	150	200	250	300	400	500	600	700
B'/cm^3 mol^{-1}	−274	−160	−104	−71.5	−35.2	−16.2	−4.2	9	16.9	21.3	24

(*a*) Fit these data to the function

$$B' = \alpha + \beta T + \gamma T^2$$

(*b*) Plot this function versus temperature. (*c*) Calculate the Boyle temperature of molecular nitrogen.

1.F Nitrogen tetroxide (N_2O_4) gas is placed in a 500-cm^3 glass vessel, and the reaction $N_2O_4 = 2NO_2$ goes to equilibrium at 25 °C. The density of the gas at equilibrium at 1.0133 bar is 3.176 g L^{-1}. Assuming that the gas mixture is ideal, what are the partial pressures of the two gases at equilibrium?

1.G Calculate the molar volume of ethane at 350 K and 70 bar using the van der Waals constants in Table 1.3.

1.H Plot the partial pressures of oxygen, nitrogen, and the total pressure in bars versus height above the surface of the earth from zero to 50 000 feet assuming that the temperature is constant at 273 K.

2 First Law of Thermodynamics

In this chapter we begin to emphasize processes that take a chemical system from one state to another. The first law of thermodynamics, which is often referred to as the law of conservation of energy, leads to the definition of a new thermodynamic state function, the internal energy U. An additional state function, the enthalpy H, is defined in terms of U, P, and V for reasons of convenience.

Thermochemistry, which deals with the heat produced by chemical reactions and solution processes, is based on the first law. If heat capacities of reactants and products are known, the heat of a reaction may be calculated at other temperatures once it is known at one temperature.

2.1 WORK AND HEAT

Force is a vector quantity; that is, it has direction as well as magnitude. Other examples of vector quantities are displacement, velocity, acceleration, and electric field strength. In this book vector quantities are represented by boldface italic type. The magnitude of the vector is represented with lightface italic type. **Force** is defined by

$$\boldsymbol{f} = m\boldsymbol{a} \tag{2.1}$$

where $\boldsymbol{f}$ is the force that will give a mass m an acceleration $\boldsymbol{a}$.

Work (w) is a scalar quantity defined by

$$w = \boldsymbol{f} \cdot \boldsymbol{L} \tag{2.2}$$

where $\boldsymbol{f}$ is the vector force, $\boldsymbol{L}$ is the vector length of path, and the dot indicates a scalar product (i.e., the product is taken of the magnitude of one vector by the projection of the second vector along the direction of the first). If the force vector of magnitude f and the vector length of magnitude L are separated by the angle θ, the work is given by $fL\cos\theta$.*

The SI unit of force is the newton N, which is equal to kg m s^{-2}. The SI unit of work is the joule J, which is N m or kg m^2 s^{-2}. The differential quantity of work dw done by a force f operating over a distance dL in the direction of the force is $f\,dL$.

Since pressure P is force per unit area, the force on a piston is PA, where A is the surface area perpendicular to the direction of the motion of the piston. Thus, the differential quantity of work done by an expanding gas that causes the piston to move distance dL is $PA\,dL$. But $A\,dL = dV$, the increase in gas volume, and so the differential quantity of pressure–volume work is $P\,dV$.

Work w can be positive or negative since work may be done on a system or a system may do work on its surroundings, as shown in Fig. 2.1. **The convention on w is that it is positive when work is done on the system of interest and negative when the system does work on the surroundings.** (As we will see later, a similar convention is applied to heat q; q is positive when heat is transferred from the surroundings to a system, and q is negative when heat is transferred from the system to the surroundings.) Thus, the differential of the PV work done on a system is given by

$$đw = -P_{ext}\,dV \tag{2.3}^{\dagger}$$

where P_{ext} is the external or applied pressure.

Work is often conveniently measured by the lifting or falling of masses. The work required to lift a mass m in the earth's gravitational field, which has an acceleration g, is mgh, where h is the height through which the mass is lifted.

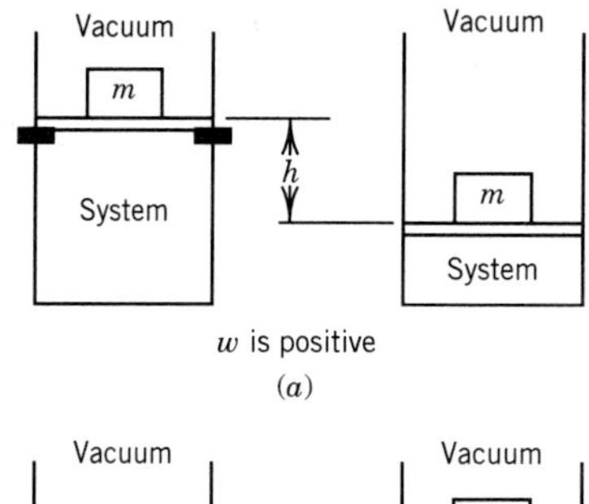

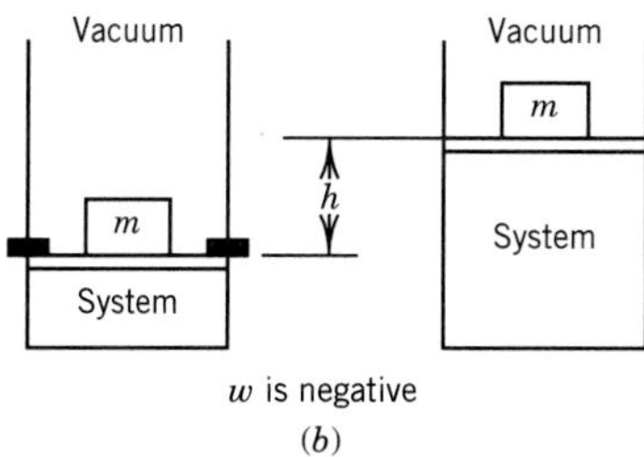

Figure 2.1 (*a*) Work is done on a system by the surroundings. In this case the stops are pulled out, and the system is compressed to a new equilibrium state. (*b*) Work is done on the surroundings by the system. When the stops are pulled out, the system expands to a new equilibrium state.

*Since this is our first contact with vectors and matrices, we want to note that they are represented by boldface italic type. They may have units like other physical quantities. Sections D.7 and D.8 of Appendix D give information about the mathematical properties of vectors and matrices.

†We use $đw$ rather than dw as a reminder that work is not an exact differential (Section 2.3), and so the value of its integral depends on the path.

The work done by a system in lifting a kilogram 0.1 m is

$$w = -mgh = -(1\ \text{kg})(9.807\ \text{m s}^{-2})(0.1\ \text{m}) = -0.9807\ \text{J}$$

where g is the acceleration of gravity. The negative sign indicates that work has been done by the system.

The total work w on a system when there is a finite change in volume is obtained by summing the infinitesimal amounts of work given by equation 2.3:

$$w = -\int_1^2 P_{\text{ext}}\,\mathrm{d}V \tag{2.4}$$

To make this calculation of the work for a finite change in state, P_{ext} must have a definite value at each volume.

If the expansion or compression of a gas is carried out very slowly, the pressure throughout the gas will be uniform and equal to P_{ext} (within an infinitesimal amount) and the maximum work of expansion (negative) or of compression (positive) will be obtained, as we will see in Section 2.4. When a process is carried out in this way, the pressure given by the equation of state can be used in equation 2.4. Such a process is said to be **quasistatic.** When the gas is allowed to expand rapidly or is compressed rapidly, the pressure is not uniform and so such a substitution cannot be made.

The integral in equation 2.4 is called a **line integral** because its value depends on the path. Line integrals are discussed in greater detail in Section 2.3. In the quasistatic case $P_{\text{ext}} = P$ and the pressure is a function of temperature and volume, and so equation 2.4 should be written

$$w = -\int_1^2 P(T, V)\,\mathrm{d}V \tag{2.5}$$

In an ordinary definite integral, the integrand is a function of one variable. Later, in Section 2.4, we will replace P with nRT/V for an ideal gas and integrate equation 2.5 at constant temperature, but now we want to take a more general point of view and consider the two processes in Fig. 2.2. The state of a mole of gas can be changed from $(2P_0, V_0)$ to $(P_0, 2V_0)$ by an infinite number of quasistatic paths, but we will consider only the two paths shown. In the upper path, the pressure is held constant at $2P_0$ and the gas is heated until it reaches $2V_0$. Then the volume is held constant while the gas is cooled until the pressure reaches P_0. For this path, $w = -2P_0V_0$. In the lower path, the volume is held constant at V_0 and the gas is cooled until it reaches P_0. Then the gas is heated at constant pressure until it reaches $2V_0$. For this path, $w = -P_0V_0$. Note that in both cases the work is the negative of the area under the path. This example shows that w depends on the path.

When work is done on a system that is thermally insulated so that there is no exchange of heat with the surroundings, the thermodynamic state of the system is changed. This type of process is referred to as an **adiabatic process.** Joule performed experiments in 1840–1849 showing that the change in state of water in an adiabatic process is independent of the path, that is, whether the work is used to turn a paddle wheel (Fig. 2.3) or is dissipated by an electrical current flowing through a resistance or by the friction of rubbing two objects together. Since a

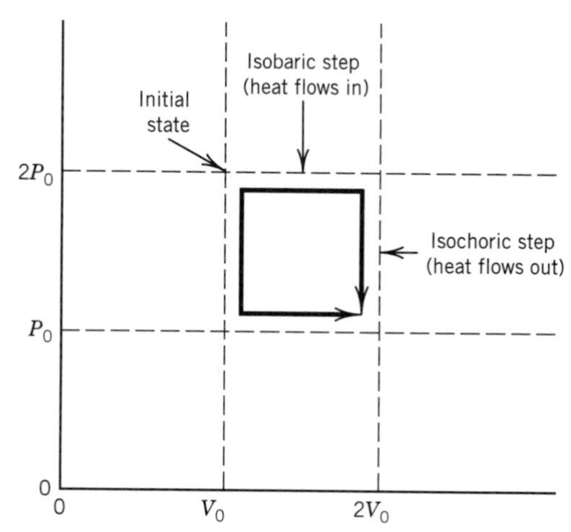

Figure 2.2 For the change in state of a mole of gas from $(2P_0, V_0)$ to $(P_0, 2V_0)$, the work done on the gas depends on the path. By the upper path, $w = -2P_0V_0$. By the lower path, $w = -P_0V_0$. For a clockwise cyclic process, $w = -2P_0V_0 + P_0V_0 = -P_0V_0$. In the cyclic process, the gas is returned to its initial state, and so $\Delta U = 0 = q - P_0V_0$. Thus, $q = P_0V_0$ and heat is absorbed by the system in the cyclic process.

given change in state of the water in the calorimeter can be accomplished in different ways involving the same amount of work, or by different sequences of steps, the change in state is independent of the path and is dependent only on the total amount of work. This makes it possible to express the change in state of a system in an adiabatic process in terms of the work required, without stating the type of work or the sequence of steps used. The property of the system whose change is calculated in this way is called the **internal energy** U. Since the internal energy U of a system may be increased by doing work on it, we may calculate the increase in internal energy from the work w done on a system to change it from one state to another in an adiabatic process:

$$\Delta U = w \qquad \text{(in an adiabatic process)} \tag{2.6}$$

In words, **the work done on a closed system in an adiabatic process is equal to the increase in internal energy of the system.** The symbol Δ indicates the value of the quantity in the final state minus the value of the quantity in the initial state; $\Delta U = U_2 - U_1$, where U_1 is the internal energy in the initial state and U_2 is the internal energy in the final state. **If the system does work on its surroundings, w is negative and, furthermore, ΔU is negative (i.e., the internal energy of the system decreases) if the process is adiabatic.**

Although equation 2.6 provides a way to determine the change in internal energy of a system, it does not provide a way to determine the absolute magnitude of the internal energy of the system. However, the internal energy can be fixed arbitrarily for some given equilibrium state of the system, and equation 2.6 can be used to determine the internal energy with respect to that reference state.

When equation 2.6 is applied to a system of arbitrary size, the internal energy is an extensive quantity, but in working problems we will often deal with molar quantities and express the change in molar internal energy $\Delta\overline{U}$ in J mol^{-1}.

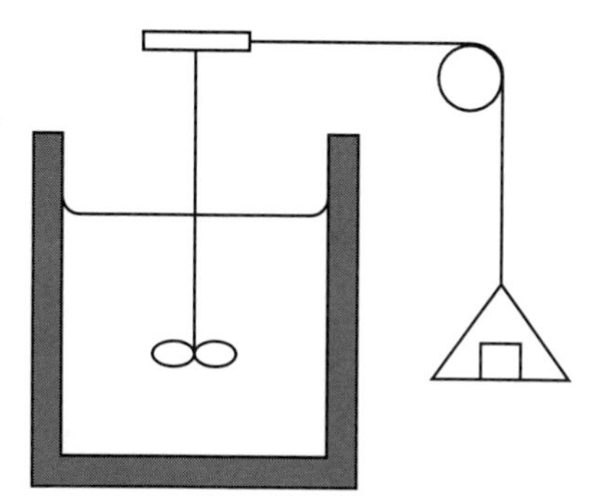

Figure 2.3 Joule heated water by performing work on it, in this case by rotating a paddle wheel, and found that the temperature rise depends only on the amount of work done on the system.

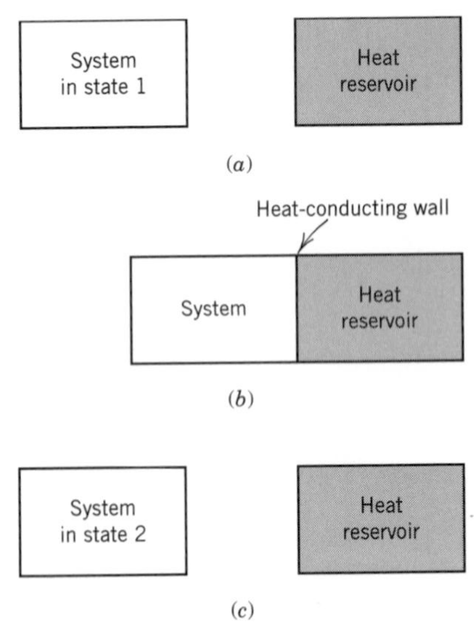

Figure 2.4 (*a*) A system in state 1 is insulated from the heat reservoir. (*b*) The system is brought into contact with the heat reservoir through a heat-conducting wall. (*c*) The system is then insulated from the heat reservoir and is found to be in state 2.

A given change in state of a system can be accomplished in ways other than by the performance of work under adiabatic conditions. A change equivalent to that in the Joule experiment may be obtained by immersing a hot object in the water. We should not say, however, that the water now has more heat any more than we would say it has more work after it has been heated with moving paddle wheels. In other words, heat and work are forms of energy crossing a boundary. After the experiment, the temperature of the water is higher, and it has a greater internal energy U. Heat is transferred when there is a gradient in temperature, as shown in Fig. 2.4.

Since the same change in state (as determined by measuring properties such as temperature, pressure, and volume) may be produced by doing work on the system or by allowing heat to flow in, the amount of heat q may be expressed in mechanical units. When Joule was doing his experiments the unit of heat was the calorie, which is the heat required to raise the temperature of a gram of water 1 °C, from 14.5 to 15.5 °C. Joule was able to determine the mechanical equivalent of heat, which is now known to be 1 calorie $= 4.184 \text{ kg m}^2 \text{ s}^{-2} = 4.184$ J. Now we find it more convenient to express heat in joules and to **define** the calorie as 4.184 J. A joule of heat is the amount of heat that produces the same change in a system as a joule of work. The dietary calorie is actually a kilocalorie.

Since heat is an algebraic quantity, it is important to adopt a sign convention. **The convention is that a positive value of q indicates that heat is absorbed by the system from its surroundings. A negative value of q means that the system gives up heat to its surroundings.** The change in internal energy U produced by the transfer of heat q to a system when no work is done is given by

$$\Delta U = q \qquad \text{(no work done)} \tag{2.7}$$

In words, **the heat absorbed by a closed system in a process in which no work is done is equal to the increase in internal energy of the system.** Or, put another way, if no work is done, the heat evolved is equal to the decrease in the internal energy of the system.

It is important to understand that it is the measurement of work in the surroundings that makes it possible to determine the quantity of heat q transferred to a system. In Section 2.5 we will find that there are a number of different kinds of work, and each of them can be readily measured by measuring the raising or lowering of weights in the gravitational field of the earth.

2.2 FIRST LAW OF THERMODYNAMICS AND INTERNAL ENERGY

Since the internal energy of a system can be changed a given amount by either heat or work, these quantities are in this sense equivalent. They are both usually expressed in joules. If both heat and work are added to a system,

$$\Delta U = q + w \tag{2.8}$$

For an infinitesimal change in state,

$$\mathrm{d}U = đq + đw \tag{2.9}$$

The đ indicates that q and w are not exact differentials, as discussed in the next section.

Equations 2.8 and 2.9 are statements of the **first law of thermodynamics.** This law is the postulate that there exists a property U, referred to as the internal energy, (1) that is a function of the state variables for the system and (2) for which the change ΔU for a process in a closed system may be calculated using equation 2.8. The first law is not restricted to reversible processes.

This mathematical form of the first law seems obvious to us now, but prior to 1850 it was not obvious at all. Before 1850 the principle of conservation of energy in mechanical systems was understood, but the role of heat in this principle was not clear until Joule's experiments led to equation 2.8.

If ΔU is negative, we may say that the system loses energy in heat that is evolved and work that is done by the system. The first law has nothing to say about how much heat is evolved and how much work is done except that equation 2.8 is obeyed. In other words, the entire decrease in internal energy could show up as work ($q = 0$). Another possibility is that even more than this amount of work would be done and heat would be absorbed ($q > 0$), so that equation 2.8 is obeyed. Although the first law has nothing to say about the relative amounts of heat and work, the second law does, as we will see in Chapter 3. Since the internal energy is a function of the state of a system, there is no change in internal energy when a system is taken through a series of changes that return it to its **initial** state. This is expressed by setting the **cyclic integral** equal to zero:

$$\oint \mathrm{d}U = 0 \tag{2.10}$$

The circle indicates integration around a cycle, that is, where the initial and final states are the same. The cyclic integrals of q and w are not generally equal to zero, and their values depend on the path followed.

The first law is frequently stated in the form that energy may be transferred in one form or another, but it cannot be created or destroyed. Thus, the total energy of an isolated system is constant.

The internal energy U of a system is an extensive property (Section 1.1); thus, if we double a system, the internal energy is doubled. However, the molar internal energy is an intensive property. We will use U for the extensive property and $\overline{U}$ for the intensive property.

The quantity of heat transferred to an object can be calculated using $q = \Delta U - w$, where w is the measured quantity of work done on the system. The change in internal energy ΔU in the process can be calculated from the quantity w of work required in an adiabatic process (see Section 2.1).

Comment:

The statement of the first law of thermodynamics in mathematical form was a great achievement, and actually did not occur until after the statement of the second law. A key idea is that the quantity of work required to produce the same temperature rise in the system as an unknown quantity of heat can be used as a measure of the quantity of heat; thus heat is measured in terms of joules, just like work.

Now that we have established that the internal energy is a state function, we want to be sure that we know how many variables have to be specified to describe the state of the system. If the system involves only PV work, the internal energy of a mass of a pure substance can be described by a mathematical function of T, V, and n or T, P, and n; these functions are represented by $U(T, V, n)$ and $U(T, P, n)$. We will give the precise form of these functions only for ideal gases because the functions for real substances are very complicated. The internal energy of a homogeneous binary mixture can be specified by a function $U(T, V, n_1, n_2)$, $U(T, P, n_1, n_2)$, or $U(T, P, x_1, n_t)$, where x_1 is the mole fraction of substance 1 and n_t is the total amount of material in the system. Thus the description of the *extensive* state of a homogeneous mixture of N species requires $N + 2$ variables, one of which must be extensive. The *intensive* state of a pure substance is determined by two intensive variables (T and P), and the intensive state of a homogeneous binary mixture is determined by three intensive variables (T, P, and x_1). Thus the intensive state of a homogeneous mixture of N species is specified by $N + 1$ independent intensive variables. In Section 5.4 we will discuss the change in this rule when chemical reactions are involved and are at equilibrium.

2.3 EXACT AND INEXACT DIFFERENTIALS

The internal energy U is a state function, like V, because it depends only on the state of the system. The integral of the differential of a state function along any arbitrary path is simply the difference between values of the function at two limits. For example, if a system goes from state a to state b, we can write

$$\int_a^b dU = U_b - U_a = \Delta U \tag{2.11}$$

Since the integral is path independent, the differential of a state function is called an **exact differential.** The quantities q and w are not state functions. The integrals of their differentials in going from state a to state b **depend on the path chosen.** Therefore, their differentials are called **inexact differentials.** We will use đ instead of d to indicate inexact differentials. In going from state a to state b the work w done is represented by

$$\int_a^b đw = w \tag{2.12}$$

Note that the result of the integration is not written $w_b - w_a$, because the amount of work done depends on the particular path that is followed between state a and state b. For example, when a gas is allowed to expand, the amount of work obtained may vary from zero (if the gas is allowed to expand into a vacuum) to a maximum value that is obtained if the expansion is carried out reversibly, as described in Section 2.4.

If an infinitesimal quantity of heat $đq$ is absorbed by a system, and an infinitesimal amount of work $đw$ is done on the system, the infinitesimal change in the internal energy is given by

$$dU = đq + đw \tag{2.13}$$

where the d is used with U since dU is an **exact** differential and đ is used with q and w because they are **inexact** differentials. In other words, U is a function of the state of the system, and q and w for a process depend on the path.

It is interesting to note that the sum of two inexact differentials can be an exact differential. To illustrate this point further, we consider the path from a to b in Fig. 2.5. We may define the path by a curve $y = y(x)$ connecting a and b.

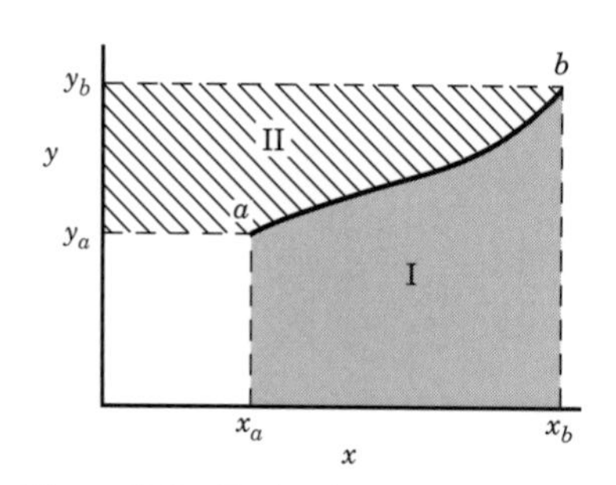

Figure 2.5 Path of a system in going from state a to state b.

The differential $\text{đ}z = y\,dx$ is not an exact differential,

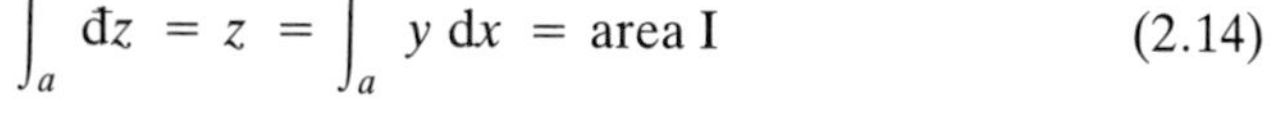

$$\int_a^b \text{đ}z = z = \int_a^b y\,dx = \text{area I} \tag{2.14}$$

because this area depends on the path between a and b, as may be seen from Fig. 2.5.

The differential $dz = y\,dx + x\,dy$ is an exact differential. Since $dz = d(xy)$,

$$\int_a^b dz = \Delta z$$

$$= \int_a^b d(xy) = x_b y_b - x_a y_a \tag{2.15}$$

The reason $dz = y\,dx + x\,dy$ is an exact differential may be seen from Fig. 2.5. The integral of dz from state a to state b may be written

$$\int_a^b dz = \Delta z = \int_a^b y\,dx + \int_a^b x\,dy = \text{area I} + \text{area II} \tag{2.16}$$

The sum of these areas is independent of the shape of the curve (path) between a and b. If $\int dz$ does not depend on the path taken between the points, then dz is said to be an exact differential. Thermodynamic quantities like U, H, S, and G (all of which will soon be introduced) form exact differentials, since their values are dependent on the state variables and not on the path by which the system got there. There is a simple test to see whether a differential is exact.

For a system with just two independent degrees of freedom, the total differential dz of a quantity z may be determined by the differentials dx and dy in two other quantities x and y. In general,

$$dz = M(x, y)\,dx + N(x, y)\,dy \tag{2.17}$$

where M and N are functions of the independent variables x and y.

To show the test for exactness we now consider a function z that has an exact differential. If z has a definite value at each point in the xy plane, then it must be a function of x and y. If $z = f(x, y)$, then

$$dz = \left(\frac{\partial z}{\partial x}\right)_y dx + \left(\frac{\partial z}{\partial y}\right)_x dy \tag{2.18}$$

Comparing equations 2.17 and 2.18, we find

$$M(x, y) = \left(\frac{\partial z}{\partial x}\right)_y \tag{2.19}$$

$$N(x, y) = \left(\frac{\partial z}{\partial y}\right)_x \tag{2.20}$$

Since the mixed partial derivatives are equal,

$$\left[\frac{\partial}{\partial y}\left(\frac{\partial z}{\partial x}\right)_y\right]_x = \left[\frac{\partial}{\partial x}\left(\frac{\partial z}{\partial y}\right)_x\right]_y \tag{2.21}$$

then

$$\left(\frac{\partial M}{\partial y}\right)_x = \left(\frac{\partial N}{\partial x}\right)_y \tag{2.22}$$

This equation must be satisfied if $\mathrm{d}z$ is an exact differential. **It is Euler's criterion for exactness.** This relation is also very useful for obtaining relations between the derivatives of thermodynamic functions.

To illustrate the use of equation 2.22 let us reconsider the differential $\mathrm{d}z = y\,\mathrm{d}x$. Since $M = y$ and $N = 0$, $(\partial M/\partial y)_x = 1$ and $(\partial N/\partial x)_y = 0$, so that equation 2.22 is not satisfied. Therefore, $\mathrm{d}z = y\,\mathrm{d}x$ is not an exact differential. On the other hand, $\mathrm{d}z = y\,\mathrm{d}x + x\,\mathrm{d}y$ is an exact differential: $M = y$, $N = x$ so that $(\partial M/\partial y)_x = 1$, $(\partial N/\partial x)_y = 1$, satisfying equation 2.22.

Example 2.1 ***An inexact differential and an exact differential***

Suppose

$$\mathrm{d}z = xy^3\,\mathrm{d}x + 3x^2y^2\,\mathrm{d}y$$

Can z be a function of the state of a system (i.e., a function of x and y)? The partial derivatives of z are

$$\left(\frac{\partial z}{\partial x}\right)_y = xy^3, \qquad \left(\frac{\partial z}{\partial y}\right)_x = 3x^2y^2$$

The mixed partial derivatives of z are

$$\frac{\partial^2 z}{\partial x\,\partial y} = 3xy^2, \qquad \frac{\partial^2 z}{\partial y\,\partial x} = 6xy^2$$

Thus $\mathrm{d}z$ is not an exact differential, and so z cannot be a function of the state of the system.

However, consider the total differential of z':

$$\mathrm{d}z' = 2xy^3\,\mathrm{d}x + 3x^2y^2\,\mathrm{d}y$$

Can z' be a function of the state of a system? The partial derivatives of z' are

$$\left(\frac{\partial z'}{\partial x}\right)_y = 2xy^3, \qquad \left(\frac{\partial z'}{\partial y}\right)_x = 3x^2y^2$$

The mixed partial derivatives of z' are

$$\frac{\partial^2 z'}{\partial x\, \partial y} = 6xy^2, \qquad \frac{\partial^2 z'}{\partial y\, \partial x} = 6xy^2$$

Since the mixed partial derivatives are equal, dz' is an exact differential, and so z' can be a function of the state of the system. In fact, $z' = x^2y^3 + \text{const.}$

When the differential of a physical quantity is inexact, it may be possible to use it to define another physical quantity that has an exact differential by multiplying by an **integrating factor.** For example, if the differential of the physical quantity $f(x, y)$ is given by

$$df(x, y) = y(xy + 1)\,dx - x\,dy \tag{2.23}$$

$df(x, y)$ is an inexact differential because

$$\left[\frac{\partial[y(xy+1)]}{\partial y}\right]_x = 2xy + 1 \tag{2.24}$$

$$\left[\frac{\partial(-x)}{\partial x}\right]_y = -1 \tag{2.25}$$

However, multiplying equation 2.23 by the integrating factor $1/y^2$ yields

$$df' \equiv \frac{df}{y^2} = \left(x + \frac{1}{y}\right)dx - \frac{x}{y^2}\,dy \tag{2.26}$$

The mixed partial derivatives are

$$\left[\frac{\partial(x + 1/y)}{\partial y}\right]_x = -\frac{1}{y^2} \tag{2.27}$$

$$\left[\frac{\partial(-x/y^2)}{\partial x}\right]_y = -\frac{1}{y^2} \tag{2.28}$$

The function f' is given by

$$f'(x, y) = \frac{x^2}{2} + \frac{x}{y} + \text{const.} \tag{2.29}$$

Integrating factors are useful for obtaining exact differentials from inexact differentials and in solving first-order differential equations.

2.4 WORK OF COMPRESSION AND EXPANSION OF A GAS AT CONSTANT TEMPERATURE

Since work done in compressing a gas is positive, we start by considering the compression of a gas **at constant temperature** using the idealized apparatus shown in Fig. 2.6. The gas is contained in a rigid cylinder by a frictionless and weightless

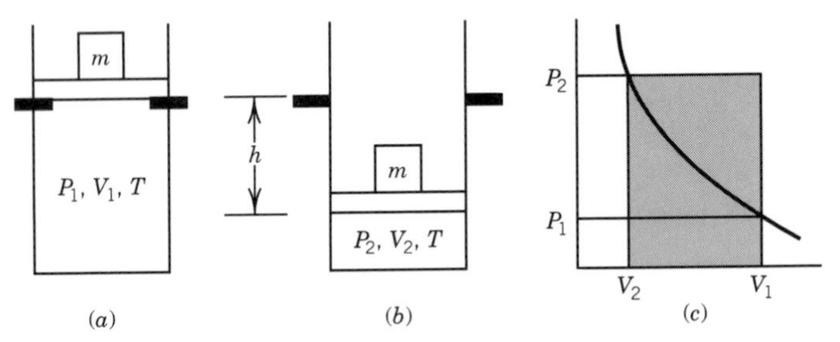

Figure 2.6 Compression of a gas from P_1, V_1, T to P_2, V_2, T in a single step.

piston. The cylinder is immersed in a thermostat at temperature T, and the space above the cylinder is evacuated so that the final pressure P_2 is due only to the mass m. The gas is initially confined to volume V_1 because the piston is held up by stops. When the stops are pulled out, the piston falls to the equilibrium position, and the gas is compressed to volume V_2. The pressure of the gas at the end of the process is given by

$$P_2 = \frac{mg}{A} \tag{2.30}$$

where g is the acceleration due to gravity, and A is the area of the piston. The amount of work lost in the surroundings is mgh (Section 2.1), where h is the difference in height, and so the work done on the gas is

$$w = mgh = -P_2(V_2 - V_1) \tag{2.31}$$

Since $V_2 < V_1$, the work done on the gas is positive. This is the smallest amount of work that can be used to compress the gas from V_1 to V_2 in a single step at constant temperature. The work done is given by the shaded area in the P–V plot of Fig. 2.6c. Notice that the pressure used in calculating the work is not the pressure of the gas but the external pressure determined by the mass m, cross-sectional area A, and acceleration of gravity g.

However, we can carry out the compression with less work if we do it in two or more steps, as shown in Fig. 2.7. We can compress the gas in two steps by first using a mass m' just large enough to compress the gas to volume $(V_1 + V_2)/2$ in

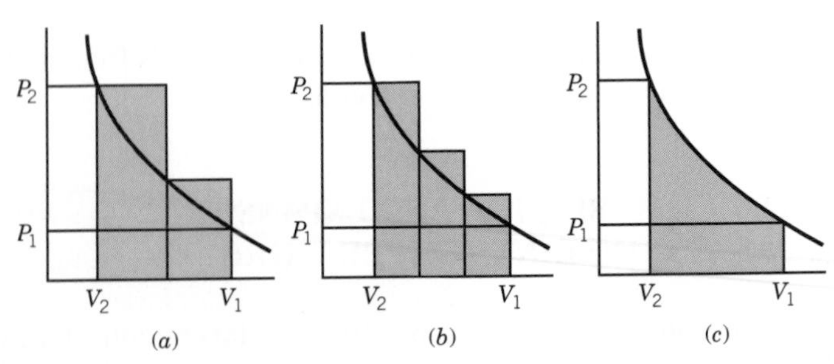

Figure 2.7 Compression of a gas from P_1, V_1, T to P_2, V_2, T in two, three, and an infinite number of steps.

the first step, and then using the larger mass m for the second step. The work lost to the surroundings is given by the shaded area in Fig. 2.7*a*. It is clear that by using more and more steps we arrive eventually at the diagram in Fig. 2.7*c*, which shows that the minimum amount of work is required in the limit of an infinite number of steps. In this case the pressure is changed by an infinitesimal amount for each infinitesimal step, and the work is given by the integral of equation 2.3 at constant temperature.

$$w = \int đw = -\int_{V_1}^{V_2} P\,dV \tag{2.32}$$

In case c it is not necessary to distinguish between the external pressure and the gas pressure because they differ at most by an infinitesimal amount.

The work w done on a gas in an **expansion** can be determined using the idealized frictionless piston arrangement shown in Fig. 2.8. The gas is initially confined to volume V_1 because the piston is held by stops. When the stops are pulled out, the gas expands to volume V_2. The mass is chosen so that $P_2 = mg/A$; in other words, this is the maximum mass that the gas will raise to this height. The work gained in the surroundings is mgh, and so the work done on the gas is

$$w = -mgh = -P_2(V_2 - V_1) \tag{2.33}$$

This is the negative of the largest amount of work that can be obtained in the surroundings by the expansion of the gas from V_1 to V_2 at constant temperature in a single step. The work done on the gas is given by the negative of the shaded area in Fig. 2.8*c*.

More work can be obtained in the surroundings by using two, three, or an infinite number of steps, as shown in Fig. 2.9. The largest amount of work in the surroundings and the largest negative work on the gas are obtained in the limit of an infinite number of steps. The work done on the gas in the limiting case is given by equation 2.32.

The work obtained in the surroundings in the single-step expansion is clearly not great enough to compress the gas back to its initial state in a single-step compression; this is evident from the shaded areas in Figs. 2.8*c* and 2.6*c*. (The subscripts are different in the two figures because we have followed the usual convention of labeling the initial state with a 1 and the final state with a 2.) However, the work obtained in the surroundings in the infinite-step expansion is exactly the

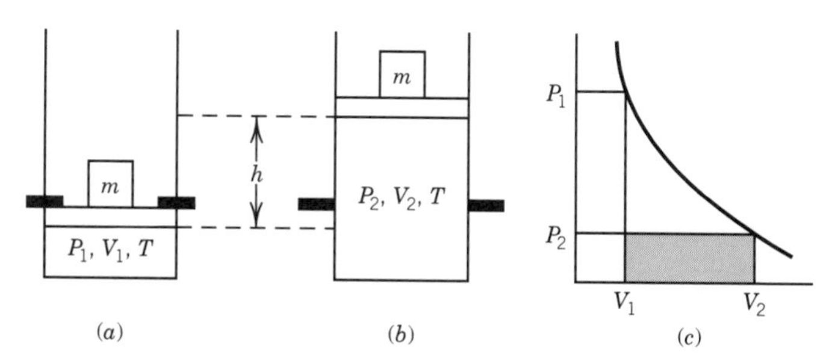

Figure 2.8 Expansion of a gas from P_1, V_1, T to P_2, V_2, T in a single step.

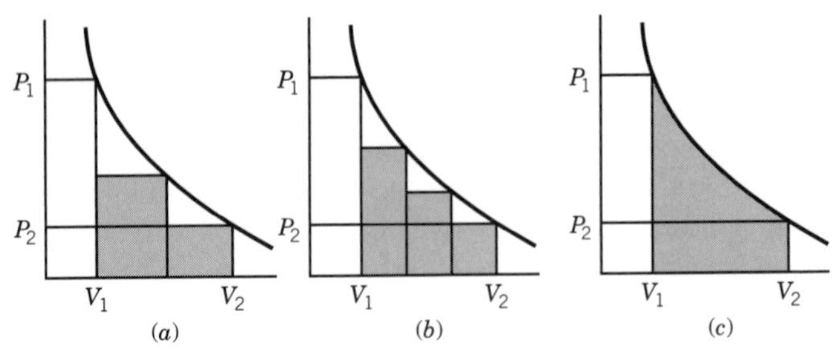

Figure 2.9 Expansion of a gas from P_1, V_1, T to P_2, V_2, T in two, three, and an infinite number of steps.

amount required to compress the gas back to its initial state by an infinite-step compression; this is evident from the shaded areas in Figs. 2.9*c* and 2.7*c*.

The infinite-step compression described by Fig. 2.7*c* and the infinite-step expansion described by Fig. 2.9*c* are referred to as **reversible processes.** These idealized processes at constant temperature are reversible because the energy accumulated in the surroundings in the expansion is exactly the amount required to compress the gas back to the initial state. This can also be seen by applying equation 2.32 to the gas for a complete cycle from P_1, V_1, T to P_2, V_2, T and back again. (Note that here we are using the same subscripts for the expansion and the compression.)

$$\begin{aligned} w_{\text{cycle}} &= -\int_{V_1}^{V_2} P\,\mathrm{d}V - \int_{V_2}^{V_1} P\,\mathrm{d}V \\ &= -\int_{V_1}^{V_2} P\,\mathrm{d}V + \int_{V_1}^{V_2} P\,\mathrm{d}V = 0 \qquad \text{(reversible)} \end{aligned} \tag{2.34}$$

Another important point about reversible processes is that they can be reversed at any point in the process by making an infinitesimal change, in this case in the pressure. Thus, a reversible expansion or compression requires an absence of friction, a balancing of internal and external pressures, and time to reestablish equilibrium after each infinitesimal step. When these conditions are not met the process is **irreversible,** and the system and its surroundings cannot both be restored to their initial conditions. Remember that these are isothermal processes, and there is heat flow that we have not talked about.

All real processes are irreversible, yet it is possible to approach reversibility closely in some real processes. Heat may be transferred nearly reversibly if the temperature gradient across which it is transferred is made very small. Electrical charge may be transferred nearly reversibly from a battery if a potentiometer is used so that the difference in electrical potential is very small. A liquid may be vaporized nearly reversibly if the external pressure is made only very slightly less than the equilibrium vapor pressure.

The concept of a reversible process is important because certain thermodynamic calculations can be made only for reversible processes. For processes in the chemical industry, the greater the irreversibility, the greater is the loss in capacity to do work; thus, literally every irreversibility has its cost.

Example 2.2 ***Work of compressing an ideal gas at constant pressure***

Two moles of gas at 1 bar and 298 K are compressed at constant temperature by use of a constant pressure of 5 bar. How much work is done on the gas? If the compression is driven by a 100-kg mass, how far will the mass fall in the earth's gravitational field?

$$\begin{aligned} w &= -P_2(V_2 - V_1) \\ &= -P_2\left(\frac{nRT}{P_2} - \frac{nRT}{P_1}\right) \\ &= -nRT\left(1 - \frac{P_2}{P_1}\right) \\ &= -(2\ \text{mol})(8.3145\ \text{J K}^{-1}\ \text{mol}^{-1})(298\ \text{K})(1 - 5) \\ &= 19\,820\ \text{J} \end{aligned}$$

$$\begin{aligned} h &= -\frac{w}{mg} \\ &= -(19\,820\ \text{J})/(100\ \text{kg})(9.8\ \text{m s}^{-2}) \\ &= -20.22\ \text{m} \end{aligned}$$

Now we consider the work required for a reversible isothermal compression of a gas and the work that can be obtained from a reversible isothermal expansion. In a reversible process, the change is accomplished in infinitesimal steps. Such a reversible process is often spoken of as one that consists of a series of successive equilibria. Since the gas is at its equilibrium pressure (within an infinitesimal amount) at each step in the expansion, we may substitute the pressure given by an equation of state into equation 2.32 and integrate it. If the gas were allowed to expand rapidly, the pressure and temperature would not be uniform throughout the volume of the gas, and so such a substitution could not be made. If the expansion is carried out reversibly at constant temperature for an ideal gas, the external pressure is always given by $P = nRT/V$. Substituting in equation 2.32, we obtain

$$\begin{aligned} w_{\text{rev}} &= -\int_{V_1}^{V_2} P\,\text{d}V \\ &= -\int_{V_1}^{V_2} \frac{nRT}{V}\,\text{d}V = -nRT\ln\frac{V_2}{V_1} \end{aligned} \qquad (2.35)^*$$

since the temperature is constant.

In integration the lower limit always refers to the initial state and the upper limit to the final state. If the gas is compressed, the final volume is smaller and w_{rev} is positive. The positive value means that work is done on the gas. The isothermal expansion of one mole of an ideal gas by a factor of 10 yields $w_{\text{rev}} = -(1\ \text{mol})RT\ln 10 = -5229$ J at 273.15 K.

*ln represents the natural logarithm, and log represents the base 10 logarithm; $\ln x = \ln(10)\log x = 2.303\log x$.

Since for an ideal gas at constant temperature $P_1V_1 = P_2V_2$, the reversible work is also given by

$$w_{\text{rev}} = nRT \ln \frac{P_2}{P_1} \tag{2.36}$$

The equation for the maximum work of isothermal expansion of a van der Waals gas is obtained by using equation 1.28:

$$\begin{aligned} w_{\text{rev}} &= -\int_{V_1}^{V_2} \left(\frac{nRT}{V - nb} - \frac{an^2}{V^2} \right) \mathrm{d}V \\ &= -nRT \ln \frac{V_2 - nb}{V_1 - nb} + an^2 \left(\frac{1}{V_1} - \frac{1}{V_2} \right) \end{aligned} \tag{2.37}$$

Example 2.3 ***Work of reversible expansion of an ideal gas***

One mole of an ideal gas expands from 5 to 1 bar at 298 K. Calculate w (a) for a reversible expansion and (b) for an expansion against a constant external pressure of 1 bar.

$$\begin{aligned} (a) \quad w_{\text{rev}} &= nRT \ln \frac{P_2}{P_1} \\ &= (1\ \text{mol})(8.3145\ \text{J K}^{-1}\ \text{mol}^{-1})(298\ \text{K}) \ln \frac{1\ \text{bar}}{5\ \text{bar}} \\ &= -3988\ \text{J} \end{aligned}$$

$$\begin{aligned} (b) \quad w_{\text{irrev}} &= -P_2(V_2 - V_1) = -P_2\left(\frac{nRT}{P_2} - \frac{nRT}{P_1}\right) \\ &= -(1\ \text{bar})(1\ \text{mol})(8.3145\ \text{J K}^{-1}\ \text{mol}^{-1})(298\ \text{K})\left(\frac{1}{1\ \text{bar}} - \frac{1}{5\ \text{bar}}\right) \\ &= -1982\ \text{J} \end{aligned}$$

More work is done on the surroundings when the expansion is carried out reversibly.

2.5 VARIOUS KINDS OF WORK

There are a number of ways that work can be done on a system, or a system can do work on its surroundings, other than PV work. If a system has a surface, surface work may be involved. If the system is a solid, there may be work of elongation. If the system involves electric charges, there may be work of transport of electric charge from a phase at one electric potential to a phase at a different electric potential. If the system is in a gravitational field, there may be work of transport of mass from one height in the field to another height in the field. If the system involves electric or magnetic dipoles, there may be work of an electric or magnetic field in orienting these dipoles. Here we consider only surface work, work of elongation, and work of transport of electric charge.

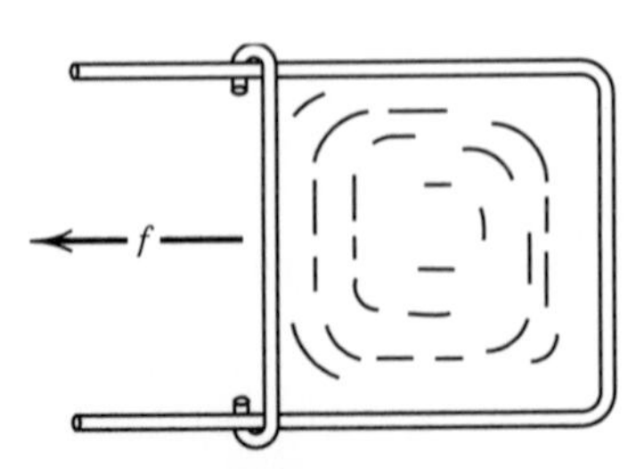

Figure 2.10 Idealized experiment for the determination of the surface tension of a liquid.

Let us consider the work required to increase the area of a surface. The force f required to increase the area of a liquid film, as illustrated in Fig. 2.10, is

$$f = 2L\gamma \tag{2.38}$$

where γ is the **surface tension** of the liquid and L is the length of the movable bar. The factor 2 is involved because there are two liquid–gas interfaces in this experiment. The surface tension is force per unit length and is usually expressed in $N\,m^{-1}$. It is a temperature-dependent quantity in general. The surface tension of water at 25 °C is 71.97×10^{-3} $N\,m^{-1}$ or 71.97 $mN\,m^{-1}$.* The surface tensions of liquid metals and molten salts are large in comparison with those of other liquids, as shown in Table 6.6. The work required to move the bar in Fig. 2.10 to the left by a distance Δx is

$$w = f\Delta x = 2L\,\Delta x\gamma = \gamma\,\Delta A_s \qquad (2.39)$$

where ΔA_s is the change in surface area ($2L\,\Delta x$). This is the amount of work done on the liquid system. According to this equation, surface tension is equal to the ratio of work to change in area, so it can also be expressed in $J\,m^{-2}$. The differential of surface work is given by

$$đw = \gamma\,dA_s \qquad (2.40)$$

Surface tension arises from the fact that the molecules in the surface of a liquid are attracted into the body of the liquid by the molecules in the body. This inward attraction causes the surface to contract if it can and gives rise to a force in the plane of the surface. Surface tension is responsible for the formation of spherical droplets, the rise of water in a capillary, and the movement of a liquid through a porous solid. Solids also have surface tensions, but it is harder to measure them.

Two other forms of work are more familiar, so we do not need to discuss them in detail. When a piece of rubber is stretched, the differential work done on the rubber is given by

$$đw = f\,dL \qquad (2.41)$$

where f is the force and dL is the differential increase in length. When a small charge dQ is moved through an electric potential difference ϕ, the work done on the charge is given by

$$đw = \phi\,dQ \qquad (2.42)$$

These differential work terms become a part of the first law if surface, elongational, and electrical work are involved:

$$dU = đq - P_{ext}\,dV + \gamma\,dA_s + f\,dL + \phi\,dQ \qquad (2.43)$$

This is an important equation because it shows how surface tension, surface area, force, elongation, electric potential, and electric charge come into thermodynamics.

The variables involved in work form **conjugate pairs** of intensive and extensive variables, as shown in Table 2.1. If both the intensive variable and the extensive variable are expressed in SI units (see symbols in Appendix G), the work is expressed in joules.

In this section and the preceding one we have considered processes in thermostats without saying anything about the quantity of heat flowing into or out of the gas. Now it is time to talk about the flow of heat that accompanies such processes.

*In older literature surface tensions are usually expressed in dynes cm^{-1}. In these units the surface tension of water at 25 °C is $(71.97 \times 10^{-3}\ N\,m^{-1})(10^5\ dynes\ N^{-1})(10^{-2}\ m\,cm^{-1}) = 71.97$ dynes cm^{-1}.

Table 2.1 Some Conjugate Pairs of Thermodynamic Variables

Type of Work	*Intensive Variable*	*Extensive Variable*	*Differential Work*
Hydrostatic	Pressure, P	Volume, V	$-P\,\mathrm{d}V$
Surface	Surface tension, γ	Area, A_s	$\gamma\,\mathrm{d}A_s$
Elongation	Force, f	Length, L	$f\,\mathrm{d}L$
Electrical	Potential difference, ϕ	Electric charge, Q	$\phi\,\mathrm{d}Q$

Example 2.4 ***Calculating other kinds of work***

(*a*) A piece of stretched rubber exerts a force of 1 N. How much work has to be done on the rubber to stretch it one centimeter? (*b*) The surface tension of water is 0.072 N/m at 25 °C. How much work has to be done to increase the water surface by one square meter? (*c*) A mole of electrons is transported across a potential difference of 1 V from the positive electrode to the negative electrode. How much work is required?

$$(a)\,w = f\,\Delta L = (1\text{ N})(0.01\text{ m}) = 0.01\text{ J}$$

$$(b)\,w = \gamma\,\Delta A_s = (0.072\text{ N m}^{-1})(1\text{ m}^2) = 0.072\text{ J}$$

$$(c)\,w = \phi\,\Delta Q = (1\text{ V})(96\,500\text{ coulombs}) = 96\,500\text{ J}$$

2.6 CHANGE IN STATE AT CONSTANT VOLUME

The quantity of heat q can be measured by determining the change in temperature of a mass of material that absorbs the heat. The heat capacity C is defined by the derivative $C = đq/\mathrm{d}T$, but $đq$ is an inexact differential because heat is not a state function. Therefore, the path has to be specified; for example, we may consider a constant-volume path or a constant-pressure path. First we consider changes in state at constant volume.

When a system changes from one state to another at constant volume, the change in internal energy U may be calculated from the heat q evolved and the work w done on the system by the surroundings. For a chemically inert system of fixed mass the internal energy U may be taken to be a function of any two of T, V, and P. It is most convenient to take it as a function of T and V. Since U is a state function, the differential $\mathrm{d}U$ is given by

$$\mathrm{d}U = \left(\frac{\partial U}{\partial T}\right)_V \mathrm{d}T + \left(\frac{\partial U}{\partial V}\right)_T \mathrm{d}V \tag{2.44}$$

The first term is the change in internal energy due to the temperature change alone, and the second term is the change in internal energy due to the volume change alone. Since the differential of the internal energy is given by $đq - P_{\text{ext}}\,\mathrm{d}V$, if only pressure–volume work is involved, then

$$đq = \left(\frac{\partial U}{\partial T}\right)_V \mathrm{d}T + \left[P_{\text{ext}} + \left(\frac{\partial U}{\partial V}\right)_T\right] \mathrm{d}V \tag{2.45}$$

If the change in state of system X takes place at constant volume, it may be represented by

$$X(V_1, T_1) \rightarrow X(V_1, T_2)$$

In this case equation 2.45 reduces to

$$đq_V = \left(\frac{\partial U}{\partial T}\right)_V dT \tag{2.46}$$

Since the change in temperature and the heat transferred are readily measured, it is convenient to define the **heat capacity C_V at constant volume** as

$$C_V \equiv \frac{đq_V}{dT} = \left(\frac{\partial U}{\partial T}\right)_V \tag{2.47}$$

This equation may be applied to a system of any size, but frequently we will be concerned with the intensive thermodynamic quantity $\overline{C_V}$, which has the SI units J K^{-1} mol^{-1}. Since the heat capacity at constant volume is readily measured, equation 2.47 may be integrated to obtain the change in internal energy for a finite change in temperature at constant volume:

$$\Delta U_V = \int_{T_1}^{T_2} C_V \, dT = q_V \tag{2.48}$$

This is illustrated in Fig. 2.11. Over a small temperature range C_V may be nearly constant so that

$$\Delta U_V = C_V(T_2 - T_1) = C_V \Delta T \tag{2.49}$$

In principle, the quantity $(\partial U/\partial V)_T$ may be measured in an experiment devised by Joule. Imagine two gas bottles connected with a valve and enclosed in a thermally isolated container, as shown in Fig. 2.12. The two bottles constitute the system under consideration. The first bottle is filled with a gas under pressure, and the second is evacuated. When the valve is opened, gas rushes from the first bottle into the second. Joule found that there was no discernible change in the temperature once thermal equilibrium had been established, and so $đq = 0$. No work is done in this expansion since $P_{ext} = 0$, and so $đw = 0$ and $dU = đq + đw = 0$. Since the temperature is constant, equation 2.44 becomes

$$dU = \left(\frac{\partial U}{\partial V}\right)_T dV = 0 \tag{2.50}$$

Since $dV \neq 0$,

$$\left(\frac{\partial U}{\partial V}\right)_T = 0 \tag{2.51}$$

Thus, Joule concluded (incorrectly) that the internal energy of the gas is independent of the volume. However, this method is not very sensitive because of the large heat capacity of the gas bottles relative to the gas. Equation 2.51 actually

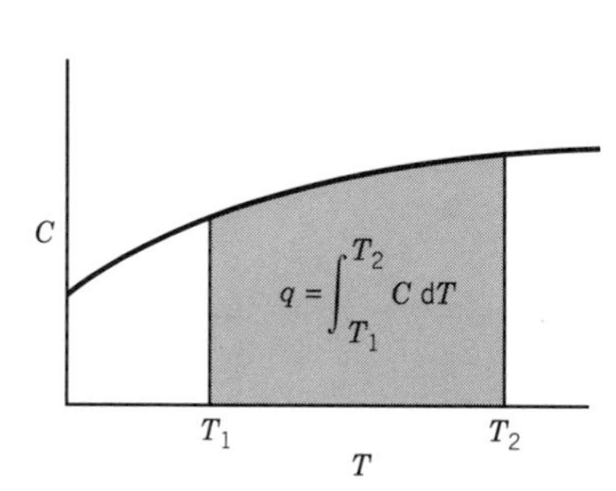

Figure 2.11 The heat q absorbed by a substance when it is heated is equal to the integral of $C\,dT$ from the initial temperature T_1 to the final temperature T_2. If an amount n is heated at constant volume, $C = n\overline{C_V}$, and if an amount n is heated at constant pressure, $C = n\overline{C_P}$.

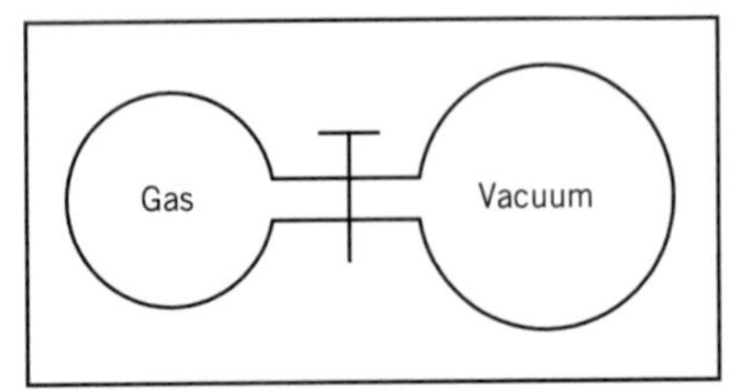

Figure 2.12 Joule's experiment in which a gas expands into a vacuum. Joule found there was no discernible change in temperature and concluded that $(\partial U/\partial V)_T = 0$. We now know that this applies to ideal gases but not to real gases.

applies only to an ideal gas. The molecular interpretation of this relation is that there is no interaction between the molecules of an ideal gas, and so the internal energy does not change with the distance between molecules. On the other hand, the internal energy of a real gas depends on the volume at constant temperature, but the second law of thermodynamics is needed to derive an equation that can be used to obtain $(\partial U/\partial V)_T$ experimentally.*

Example 2.5 ***Heat absorbed in expansion of an ideal gas***

Calculate the heat absorbed and the changes in internal energy for the two expansions of an ideal gas described in Example 2.3.

(*a*) According to Joule's experiment, $\Delta U = 0$ for the reversible isothermal expansion of an ideal gas. Therefore,

$$\begin{aligned} q_{\text{rev}} &= \Delta U - w_{\text{rev}} \\ &= 0 - (-3988\ \text{J}) \\ &= 3988\ \text{J} \end{aligned}$$

(*b*)

$$\begin{aligned} q_{\text{irrev}} &= \Delta U - w_{\text{irrev}} \\ &= 0 - (-1982\ \text{J}) \\ &= 1982\ \text{J} \end{aligned}$$

Thus, more heat is absorbed by the gas in the reversible isothermal expansion.

2.7 ENTHALPY AND CHANGE OF STATE AT CONSTANT PRESSURE

Constant-pressure processes are more common in chemistry than constant-volume processes because many operations are carried out in open vessels. If only pressure–volume work is done and the pressure is constant and equal to the applied pressure, the work w done on the system equals $-P\Delta V$, so that equation 2.7 may be written

$$\Delta U = q_P - P\Delta V \tag{2.52}$$

where q_P is the heat for the isobaric (constant pressure) process. If the initial state is designated by 1 and the final state by 2, then

$$U_2 - U_1 = q_P - P(V_2 - V_1) \tag{2.53}$$

*In equation 4.112 we will find that for a van der Waals gas

$$\left(\frac{\partial U}{\partial V}\right)_T = \frac{a}{\overline{V}^2}$$

This should not be surprising because van der Waals added $a/\overline{V}^2$ to the pressure to provide for intermolecular attractions. The quantity $(\partial U/\partial V)_T$ is often called the internal pressure. It has the dimensions of pressure and is due to intermolecular attractions and repulsions. The internal pressure changes with the volume because as the volume is increased, the average intermolecular distances increase and the average intermolecular potential energy changes.

so that the heat absorbed is given by

$$q_P = (U_2 + PV_2) - (U_1 + PV_1) \tag{2.54}$$

Since the heat absorbed is given by the difference of two quantities that are functions of the state of the system, it is convenient to introduce a new state function, the **enthalpy** H, defined by

$$H = U + PV \tag{2.55}$$

Equation 2.54 may be written

$$q_P = H_2 - H_1 \tag{2.56}$$

In words, **the heat absorbed in a process at constant pressure is equal to the change in enthalpy.** For an infinitesimal change at constant pressure

$$đq_P = \mathrm{d}H \tag{2.57}$$

where $\mathrm{d}H$ is an exact differential since the enthalpy is a function of the state of the system.

When pressure–volume work is the only kind of work (electrical and other kinds being excluded), it is easy to visualize ΔU and ΔH; in a constant-volume calorimeter (Section 2.13) the evolution of heat is a measure of the decrease in internal energy U, and in a constant-pressure calorimeter the evolution of heat is a measure of the decrease in enthalpy H.

The enthalpy H is an extensive property. Therefore, for a homogeneous mixture of N_s species involving only PV work, the enthalpy can be specified by $N_s + 2$ variables, one of which must be extensive. The intensive state of a homogeneous mixture of N_s species involving only PV work can be specified by $N_s + 1$ intensive variables.

Changes in state at constant pressure are of special interest in the laboratory, where processes generally take place at constant pressure. For a chemically inert system of fixed mass, it is most convenient to take enthalpy H as a function of temperature and pressure. Since H is a state function, the differential $\mathrm{d}H$ is given by

$$\mathrm{d}H = \left(\frac{\partial H}{\partial T}\right)_P \mathrm{d}T + \left(\frac{\partial H}{\partial P}\right)_T \mathrm{d}P \tag{2.58}$$

If the change in state of a mole of X takes place reversibly at constant pressure, it may be represented by

$$\mathrm{X}(P_1, T_1) \rightarrow \mathrm{X}(P_1, T_2) \tag{2.59}$$

For such a change equations 2.57 and 2.58 may be combined to obtain

$$đq_P = \left(\frac{\partial H}{\partial T}\right)_P \mathrm{d}T \tag{2.60}$$

Since the change in temperature and the heat transferred are readily measured, it is convenient to define the **heat capacity at constant pressure** C_P as

$$C_P \equiv \frac{đq_P}{\mathrm{d}T} = \left(\frac{\partial H}{\partial T}\right)_P \tag{2.61}$$

Since the heat capacity at constant pressure is readily measured, this equation may be integrated to obtain the change in enthalpy for a finite change in temperature at constant pressure:

$$\Delta H_P = \int_{T_1}^{T_2} C_P \, \mathrm{d}T \tag{2.62}$$

Comment:

The enthalpy H is not entirely new because it is defined in terms of U, P, and V. From one point of view, the enthalpy is redundant, and so why is it introduced? The answer is convenience. Use of H is more convenient in considering measurements and processes at constant pressure, and U is more convenient in considering measurements and processes at constant volume. Later, in Section 4.2, we will formalize this process of changing independent variables by use of Legendre transforms.

2.8 HEAT CAPACITIES

Values of $\overline{C}_P$ at 25 °C for about 200 substances are given in Table C.2, and values for 298 to 3000 K are given for a smaller number of substances in Table C.3 in Appendix C. The dependence of $\overline{C}_P$ on temperature is shown for a number of gases in Fig. 2.13. In general, the more complex the molecule, the greater is its molar heat capacity, and the greater the increase with increasing temperature.

Power series in temperature may be used to represent $\overline{C}_P$ as a function of temperature:

$$\overline{C}_P = \alpha + \beta T + \gamma T^2 \tag{2.63}$$

Parameters for several gases are given in Table 2.2. The change in enthalpy with temperature at constant pressure is then given by

$$\overline{H}_2 - \overline{H}_1 = \int_{\overline{H}_1}^{\overline{H}_2} \mathrm{d}\overline{H} = \int_{T_1}^{T_2} \overline{C}_P \, \mathrm{d}T \tag{2.64}$$

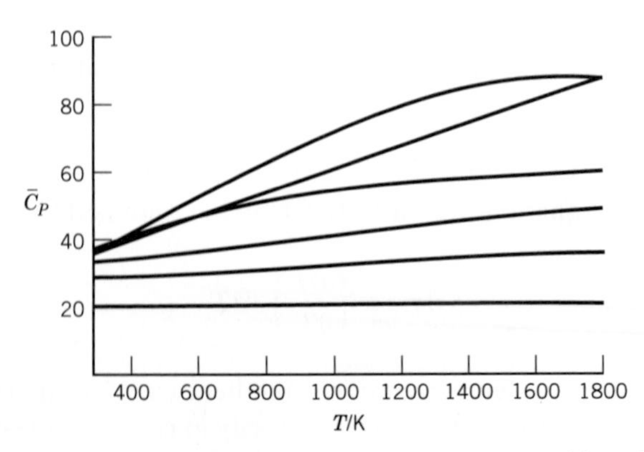

Figure 2.13 The effect of temperature on the molar heat capacities of several gases at constant pressure. Starting with the highest curve at 1200 K, the gases are CH_4, NH_3, CO_2, H_2O, N_2, and He. (See Computer Problem 2.B.)

Table 2.2 Molar Heat Capacity at Constant Pressure as a Function of Temperature from 300 to 1800 K: $\overline{C}_P = \alpha + \beta T + \gamma T^2 + \delta T^3$

	α	β	γ	δ
	$\text{J K}^{-1}\text{ mol}^{-1}$	$10^{-2}\text{ J K}^{-2}\text{ mol}^{-1}$	$10^{-5}\text{ J K}^{-3}\text{ mol}^{-1}$	$10^{-9}\text{ J K}^{-4}\text{ mol}^{-1}$
$N_2(g)$	28.883	−0.157	0.808	−2.871
$O_2(g)$	25.460	1.519	−0.715	1.311
$H_2(g)$	29.088	−0.192	0.400	−0.870
$CO(g)$	28.142	0.167	0.537	−2.221
$CO_2(g)$	22.243	5.977	−3.499	7.464
$H_2O(g)$	32.218	0.192	1.055	−3.593
$NH_3(g)$	24.619	3.75	−0.138	—
$CH_4(g)$	19.875	5.021	1.268	−11.004

Source S. I. Sandler, *Chemical and Engineering Thermodynamics,* 3rd ed. Copyright ©1999 Wiley, Hoboken, NJ. This material is used by permission of John Wiley & Sons, Inc.

$$\overline{H}_2 - \overline{H}_1 = \alpha(T_2 - T_1) + \frac{\beta}{2}(T_2^2 - T_1^2) + \frac{\gamma}{3}(T_2^3 - T_1^3) \tag{2.65}$$

JANAF Thermochemical Tables and Stull, Westrum, and Sinke, in The Chemical Thermodynamics of Organic Compounds, give values of $\overline{H}_T^\circ - \overline{H}_{298}^\circ$ for various temperatures so that $\overline{H}_2^\circ - \overline{H}_1^\circ$ is readily calculated for the substances listed. The superscript indicates that the substance is in its standard state; standard states are discussed in Section 2.11.

Example 2.6 ***The change in molar enthalpy on heating***

Using data in Table C.3, calculate the change in the molar enthalpy of methane in going from 500 to 1000 K.

$$\begin{aligned}\overline{H}_{1000}^\circ - \overline{H}_{500}^\circ &= (\overline{H}_{1000}^\circ - \overline{H}_{298}^\circ) - (\overline{H}_{500}^\circ - \overline{H}_{298}^\circ)\\ &= (38.179 - 8.200)\text{ kJ mol}^{-1}\\ &= 29.979\text{ kJ mol}^{-1}\end{aligned}$$

(Note that this is not the change in the enthalpy of formation of methane; see Problem 2.29.)

The relation between heat capacities at constant pressure and constant volume can be derived using equation 2.45 at constant pressure ($P = P_{\text{ext}}$), where it can be written as

$$đq_P = C_V\,\mathrm{d}T + \left[P + \left(\frac{\partial U}{\partial V}\right)_T\right]\mathrm{d}V \tag{2.66}$$

Dividing by $\mathrm{d}T$ and setting $đq_P/\mathrm{d}T = C_P$, we obtain

$$C_P - C_V = \left[P + \left(\frac{\partial U}{\partial V}\right)_T\right]\left(\frac{\partial V}{\partial T}\right)_P \tag{2.67}$$

The quantity on the right-hand side is positive so that $C_P > C_V$. The two terms on the right-hand side may be interpreted as follows: $P(\partial V/\partial T)_P$ is the work produced per unit increase in temperature at constant pressure, and

$$\left(\frac{\partial U}{\partial V}\right)_T \left(\frac{\partial V}{\partial T}\right)_P$$

is the energy per unit temperature required to separate the molecules against intermolecular attraction.

Equation 2.67 takes on a particularly simple form for an ideal gas because $(\partial U/\partial V)_T = 0$ and $(\partial V/\partial T)_P = nR/P$. Thus,

$$C_P - C_V = nR \qquad \text{or} \qquad \overline{C}_P - \overline{C}_V = R \tag{2.68}$$

This relationship may be visualized as follows. When an ideal gas is heated at constant pressure, the work done in pushing back the piston is $P\Delta V = nR\Delta T$. For a 1 K change in temperature, the work done is $nR(1\text{ K})$, and this is just the extra energy required to heat an ideal gas 1 K at constant pressure over that required at constant volume.

We will see later, in equation 4.120, that by use of the second law the difference between $\overline{C}_P$ and $\overline{C}_V$ for any material may be expressed in terms of the cubic expansion coefficient α and the isothermal compressibility κ (see Problems 1.17 and 1.18). The values of $\overline{C}_P$ and $\overline{C}_V$ for liquids and solids are nearly the same.

Thermodynamics does not deal with molecular models, and it is unnecessary even to discuss molecules in connection with thermodynamics. This is one of the strengths of thermodynamics, but it is also a weakness because thermodynamics, by itself, does not provide the means for predicting the numerical values of thermodynamic properties of particular substances. We will see later that kinetic theory and statistical mechanics do lead to quantitative predictions of thermodynamic properties.

Kinetic theory (Chapter 17) shows that the molar translational energy of a monatomic ideal gas is $\frac{3}{2}RT$. The translational energy $\overline{U}_t$ is independent of pressure or molar mass for a monatomic ideal gas so that

$$\overline{U}_t = \tfrac{3}{2}RT \tag{2.69}$$

According to equation 2.55, the molar enthalpy of a monatomic ideal gas is larger than the internal energy by $P\overline{V}$ (or RT), so that

$$\overline{H}_t = \tfrac{3}{2}RT + RT = \tfrac{5}{2}RT \tag{2.70}$$

Thus, the translational contributions to the molar heat capacities of monatomic ideal gases are expected to be

$$\overline{C}_V = \left(\frac{\partial \overline{U}}{\partial T}\right)_V = \tfrac{3}{2}R = 12.472\text{ J K}^{-1}\text{ mol}^{-1} \tag{2.71}$$

$$\overline{C}_P = \left(\frac{\partial \overline{H}}{\partial T}\right)_P = \tfrac{5}{2}R = 20.786\text{ J K}^{-1}\text{ mol}^{-1} \tag{2.72}$$

Tables C.2 and C.3 in Appendix C show that values of $\overline{C}_P$ for monatomic gases are constant at 20.786 J K^{-1} mol^{-1} independent of temperature, except for cases where electrons in the atom can be excited to low-lying levels [see especially O(g)].

2.9 JOULE–THOMSON EXPANSION

A gas flowing along an insulated pipe through a porous plate that separates two regions of different constant pressures may be heated up or cooled down. This **Joule–Thomson expansion** is shown in Fig. 2.14, where $P_1 > P_2$. To push one mole of gas through the porous plate, work amounting to $P_1\overline{V}_1$ has to be done on one mole of the gas by the piston on the left. Work amounting to $P_2\overline{V}_2$ is done on the surroundings by one mole of gas pushing the piston on the right, and so the net work on the gas is

$$w = P_1\overline{V}_1 - P_2\overline{V}_2 \tag{2.73}$$

Since the pipe is insulated, $q = 0$, and

$$\overline{U}_2 - \overline{U}_1 = P_1\overline{V}_1 - P_2\overline{V}_2 \tag{2.74}$$

or

$$\overline{U}_2 + P_2\overline{V}_2 = \overline{U}_1 + P_1\overline{V}_1 \tag{2.75}$$

$$\overline{H}_2 = \overline{H}_1 \tag{2.76}$$

Thus we see that there is no change in the enthalpy of the gas in a Joule–Thomson expansion.

The **Joule–Thomson coefficient** μ_{JT} is defined as the derivative of the temperature with respect to pressure in this process:

$$\mu_{JT} = \lim_{\Delta P \to 0} \frac{T_2 - T_1}{P_2 - P_1} = \left(\frac{\partial T}{\partial P}\right)_H \tag{2.77}$$

The Joule–Thomson coefficient is zero for an ideal gas, but for real gases $(\partial T/\partial P)_H$ is positive at low temperatures and negative at high temperatures; that means that a cooling effect is obtained below the inversion temperature and a heating effect is obtained above the inversion temperature. Below the inversion temperature, this effect can be used for refrigeration, but we will not discuss it

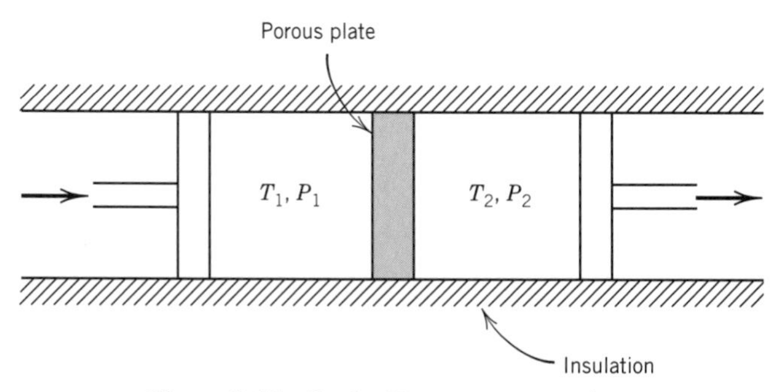

Figure 2.14 Joule–Thomson expansion.

further. The inversion temperature for nitrogen is 607 K, for hydrogen 204 K, and for helium 43 K.

2.10 ADIABATIC PROCESSES WITH GASES

In Section 2.4 we discussed the work of compression and expansion of gases in contact with a heat reservoir. Now we consider the compression and expansion of gases in isolated systems. No heat is gained or lost by the gas, so the process is adiabatic and the first law becomes simply $dU = đw$. If only pressure–volume work is involved, $dU = -P_{ext}\,dV$. If the system expands adiabatically, dV is positive and dU is negative; thus, if the expansion is opposed by an external pressure P_{ext}, work is done on the surroundings at the expense of the internal energy. The relation $dU = đw = -P_{ext}\,dV$ applies to any adiabatic process, reversible or irreversible, if PV work is the only kind of work involved. If the external pressure is zero (adiabatic expansion into a vacuum), no work is done and there is no change in the internal energy for all gases. If the expansion is opposed by an external pressure, work is done on the surroundings and the temperature drops as internal energy is converted to work. Integrating yields

$$\int_{U_1}^{U_2} dU = -\int_{V_1}^{V_2} P_{ext}\,dV$$

$$\Delta U = U_2 - U_1 = w \tag{2.78}$$

where w is the work done on the gas.

For an ideal gas, the internal energy is a function only of temperature and so $dU = C_V\,dT$ (equation 2.44). Thus, when an ideal gas expands adiabatically against an external pressure, the temperature drop is simply related to the change in internal energy. If C_V is independent of temperature for an ideal gas in the temperature range of interest, then

$$\int_{U_1}^{U_2} dU = C_V \int_{T_1}^{T_2} dT$$

$$\Delta U = U_2 - U_1 = C_V(T_2 - T_1) \tag{2.79}$$

Since $q = 0$, then $\Delta U = w$ and

$$w = \int_{T_1}^{T_2} C_V\,dT = C_V(T_2 - T_1) \tag{2.80}$$

where the second form applies when C_V is independent of temperature. This relation applies to the adiabatic expansion of an ideal gas with C_V independent of temperature whether the process is reversible or irreversible. If the gas expands, the final temperature T_2 will be lower than the initial temperature T_1, and the work done on the gas is negative. If the gas is compressed adiabatically, it will heat up.

When an adiabatic expansion is carried out reversibly, the equilibrium pressure is substituted for the external pressure, and so for an ideal gas

$$\overline{C}_V \, \mathrm{d}T = -P \, \mathrm{d}\overline{V} = -\frac{RT}{\overline{V}} \mathrm{d}\overline{V}$$

$$\overline{C}_V \frac{\mathrm{d}T}{T} = -R \frac{\mathrm{d}\overline{V}}{\overline{V}} \tag{2.81}$$

If the heat capacity is independent of temperature, then

$$\overline{C}_V \int_{T_1}^{T_2} \frac{\mathrm{d}T}{T} = -R \int_{\overline{V}_1}^{\overline{V}_2} \frac{\mathrm{d}\overline{V}}{\overline{V}}$$

$$\overline{C}_V \ln \frac{T_2}{T_1} = R \ln \frac{\overline{V}_1}{\overline{V}_2} \tag{2.82}$$

This equation is a good approximation only if the temperature range is small enough so that $\overline{C}_V$ does not change very much.

Since $\overline{C}_P - \overline{C}_V = R$, equation 2.82 may be written

$$\frac{T_2}{T_1} = \left(\frac{\overline{V}_1}{\overline{V}_2}\right)^{\gamma-1} \tag{2.83}$$

where $\gamma = \overline{C}_P/\overline{C}_V$. By use of the ideal gas law we can obtain the following alternative forms of this equation:

$$\frac{T_2}{T_1} = \left(\frac{P_2}{P_1}\right)^{(\gamma-1)/\gamma} \tag{2.84}$$

$$P_1\overline{V}_1^{\gamma} = P_2\overline{V}_2^{\gamma} \tag{2.85}$$

Thus, when a gas expands adiabatically to a larger volume and a lower pressure, the volume is smaller than it would be after an isothermal expansion to the same final pressure. Plots of pressure versus volume for adiabatic and isothermal expansions are shown in Fig. 2.15.

Example 2.7 ***Reversible adiabatic expansion of a monatomic ideal gas***

Figure 2.15 shows that when one mole of an ideal monatomic gas is allowed to expand adiabatically and reversibly from 22.7 L mol^{-1} at 1 bar and 0 °C (at point A on the graph) to a volume of 45.4 L mol^{-1} (at point C), the pressure drops to 0.315 bar. Confirm this pressure and calculate the temperature at C. How much work is done in the adiabatic expansion?

$$\gamma = \frac{\frac{5}{2}R}{\frac{3}{2}R} = \frac{5}{3}$$

$$P_2 = P_1\left(\frac{\overline{V}_1}{\overline{V}_2}\right)^{\gamma} = (1\ \text{bar})\left(\frac{22.7\ \text{L mol}^{-1}}{45.4\ \text{L mol}^{-1}}\right)^{5/3} = 0.315\ \text{bar}$$

$$T_2 = T_1\left(\frac{\overline{V}_1}{\overline{V}_2}\right)^{\gamma-1} = (273.15\ \text{K})\left(\frac{22.7\ \text{L mol}^{-1}}{45.4\ \text{L mol}^{-1}}\right)^{2/3} = 172.07\ \text{K or } -101.08\ ^\circ\text{C}$$

$$w = \int_{T_1}^{T_2} \overline{C}_V \, \mathrm{d}T = \tfrac{3}{2}R(172.07\ \text{K} - 273.15\ \text{K}) = -1261\ \text{J mol}^{-1}$$

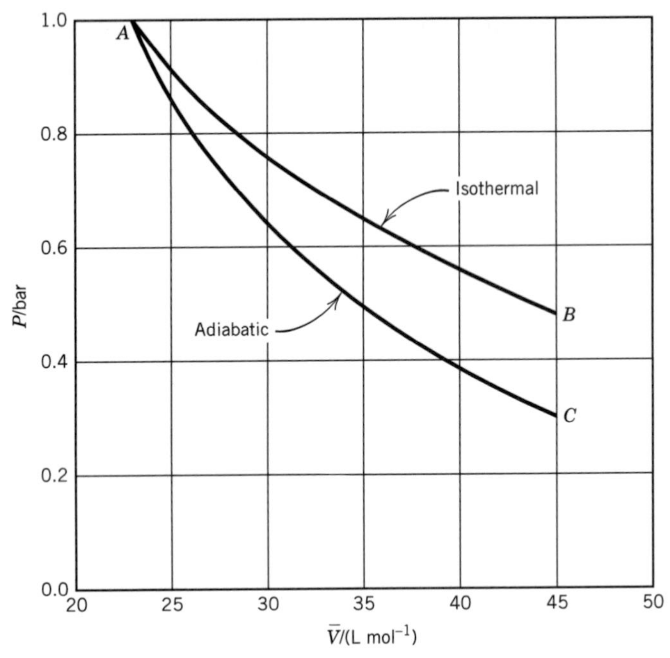

Figure 2.15 Isothermal and reversible adiabatic expansions of one mole of an ideal monatomic gas.

2.11 THERMOCHEMISTRY

The quantity of heat evolved or absorbed in a chemical reaction or a phase change can be determined by measuring the temperature change in an adiabatic process. Since very small temperature changes can be measured, this provides a sensitive method for studying the thermodynamics of chemical reactions and phase changes. If the temperature rises when a reaction occurs in an isolated system, then in order to restore that system to its initial temperature, heat must be allowed to flow to the surroundings. Such a reaction is said to be **exothermic,** and the heat q is negative. If the temperature falls when a reaction occurs in an isolated system, heat must flow from the surroundings to the system to restore the system to its initial temperature. Such a reaction is said to be **endothermic,** and the heat q is positive.

Since the enthalpy is an extensive property that is a function of the state of the system, its differential (at constant T and P) can be written in terms of the partial molar enthalpies of the species in the system (see equation 1.37 in Section 1.10):

$$\mathrm{d}H = \sum_{i=1}^{N_s} \overline{H}_i \,\mathrm{d}n_i \tag{2.86}$$

where N_s is the number of species and $\overline{H}_i$ is the molar enthalpy of species i. When the temperature and the pressure are constant, equation 2.57 yields

$$\mathrm{d}H = đq_P = \sum_{i=1}^{N} \overline{H}_i \,\mathrm{d}n_i \tag{2.87}$$

Now let us apply this equation to a system in which a single chemical reaction occurs.

To connect heat absorbed or evolved with a chemical reaction, it is of course necessary to know what the chemical change is and to have a measure of its amount. To discuss the thermodynamics of chemical reactions, we will find it convenient to represent chemical reactions in general by

$$0 = \sum_{i=1}^{N_s} \nu_i \mathrm{B}_i \tag{2.88}$$

where the ν_i are the **stoichiometric numbers** and the B_i are the molecular formulas for the N_s species involved in the reaction. The stoichiometric numbers, which are dimensionless, are positive for products and negative for reactants. Thus, according to this way of writing a reaction equation, the reaction $H_2 + \frac{1}{2}O_2 = H_2O$ would be written

$$0 = -1\mathrm{H_2} - \tfrac{1}{2}\mathrm{O_2} + 1\mathrm{H_2O} \tag{2.89}$$

The reason for using this convention is that it makes it easier to write thermodynamic equations for chemical reactions.

The amount of reaction that has occurred up to some time is expressed by the **extent of reaction** ξ, which is defined by

$$n_i = n_{i0} + \nu_i \xi \tag{2.90}$$

Here n_{i0} is the amount of substance i present initially, and n_i is the amount at some later time. Since n is expressed in moles and ν_i is dimensionless, we see that the extent of reaction is expressed in moles. The concept of extent of reaction is important because it provides a connection between the amount of reaction and a particular balanced chemical equation. We will also use the extent of reaction later in calculating equilibrium compositions.

Equation 2.90 shows that $\mathrm{d}n_i = \nu_i \,\mathrm{d}\xi$, and when we substitute this relation into equation 2.87, we obtain

$$\mathrm{d}H = đq_P = \sum_{i=1}^{N} \nu_i \overline{H}_i \,\mathrm{d}\xi \tag{2.91}$$

Dividing by $\mathrm{d}\xi$ gives

$$\Delta_r H = \left(\frac{\partial H}{\partial \xi}\right)_{T,P} = \frac{đq_P}{\mathrm{d}\xi} = \sum \nu_i \overline{H}_i \tag{2.92}$$

The quantity $\Delta_r H$ is the **reaction enthalpy.**

The reaction enthalpy is the derivative of the enthalpy of the system with respect to the extent of reaction. This is perhaps easiest to visualize for a very large system for which we can write $\Delta H/\Delta\xi = \Delta_r H$. If one mole of reaction occurs, $\Delta\xi = 1$ mol and $\Delta H = (1\ \mathrm{mol})\Delta_r H$. To know what a mole of reaction is, we must have a balanced chemical equation, since the way an equation is written is arbitrary with respect to direction and with respect to multiplying or dividing by an integer. Thus, the enthalpy of reaction for $2H_2 + O_2 = 2H_2O$ is twice that of the reaction $H_2 + \frac{1}{2}O_2 = H_2O$. To distinguish between the extensive property

ΔH and the change in enthalpy for a specified chemical reaction, we will write $\Delta_r H$ for a reaction. It is evident from equation 2.92 that the reaction enthalpy has the SI units J mol^{-1}. Here the mol^{-1} refers to a mole of reaction for the **reaction as written.** An overbar is not used on $\Delta_r H$ because the subscript r indicates that the mol^{-1} unit is involved.

As a further specification of the states of the reactants and products, we will usually consider reactions in which the reactants in their standard states are converted to the products in their standard states. When substances are in their standard states, thermodynamic quantities are labeled with superscript zeros (actually degree signs). Thus, if reactants and products are in their standard states, equation 2.95 becomes

$$\Delta_r H^\circ = \sum_{i=1}^{N} \nu_i \overline{H}_i^\circ \tag{2.93}$$

THERMODYNAMIC STANDARD STATES

The standard states that are used in chemical thermodynamics are defined as follows:

1. The standard state of a pure gaseous substance, denoted by g, at a given temperature is the (hypothetical) ideal gas at 1 bar pressure.
2. The standard state of a pure liquid substance, denoted by l, at a given temperature is the pure liquid at 1 bar pressure.
3. The standard state of a pure crystalline substance at a given temperature is the pure crystalline substance, denoted by s, at 1 bar pressure.
4. The standard state of a substance in solution is the hypothetical 1 of the substance in ideal solution of standard state molality (1 mol kg^{-1}) at 1 bar pressure, at each temperature. To indicate the standard state of an electrolyte, the NBS Tables of Chemical Thermodynamic Properties (1982) use two symbols. The thermodynamic properties of completely dissociated electrolytes in water are designated by ai. The thermodynamic properties of undissociated molecules in water are designated by ao. The thermodynamic properties of ions in water are also designated ao to indicate that no further ionization occurs.

Lavoisier and Laplace recognized in 1780 that the heat absorbed in decomposing a compound must be equal to the heat evolved in its formation under the same conditions. Thus, if the reverse of a chemical reaction is written, the sign of ΔH is changed. Hess pointed out in 1840 that the overall heat of a chemical reaction at constant pressure is the same, regardless of the intermediate steps involved. These principles are both corollaries of the first law of thermodynamics and are a consequence of the fact that enthalpy is a state function. This makes it possible to calculate the enthalpy changes for reactions that cannot be studied directly. For example, it is not practical to measure the heat evolved when carbon burns to carbon monoxide in a limited amount of oxygen, because the product will be an uncertain mixture of carbon monoxide and carbon dioxide. However, carbon may be burned completely to carbon dioxide in an excess of oxygen and the heat of reaction measured. Thus, for graphite at 25 °C,

$$C(\text{graphite}) + O_2(g) = CO_2(g) \qquad \Delta_r H^\circ = -393.509 \text{ kJ mol}^{-1}$$

The heat evolved when carbon monoxide burns to carbon dioxide can be readily measured also:

$$CO(g) + \tfrac{1}{2}O_2(g) = CO_2(g) \qquad \Delta_r H^\circ = -282.984 \text{ kJ mol}^{-1}$$

Writing these equations in such a way as to obtain the desired reaction, adding, and canceling, we have

$$\begin{array}{rlr} C(\text{graphite}) + O_2(g) = CO_2(g) & & \Delta_r H^\circ = -393.509 \text{ kJ mol}^{-1} \\ CO_2(g) = CO(g) + \tfrac{1}{2}O_2(g) & & \Delta_r H^\circ = 282.984 \text{ kJ mol}^{-1} \\ \hline C(\text{graphite}) + \dfrac{1}{2}O_2(g) = CO(g) & & \Delta_r H^\circ = -110.525 \text{ kJ mol}^{-1} \end{array}$$

In this way an accurate value can be obtained for the heat evolved when graphite burns to CO.

These data may be represented in the form of an enthalpy level diagram, as shown in Fig. 2.16. In addition, this diagram shows the enthalpy changes that are involved in vaporizing graphite to atoms and dissociating oxygen into atoms at 25 °C:

$$C(\text{graphite}) = C(g) \qquad \Delta_r H^\circ = 716.682 \text{ kJ mol}^{-1}$$
$$O_2(g) = 2O(g) \qquad \Delta_r H^\circ = 498.340 \text{ kJ mol}^{-1}$$

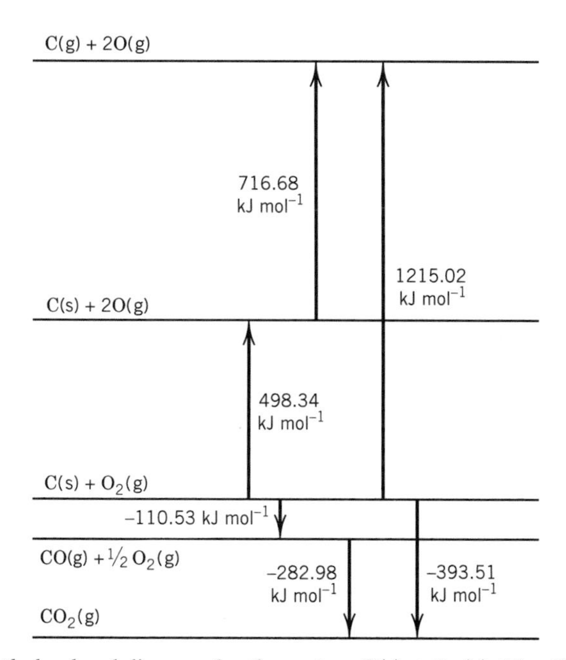

Figure 2.16 Enthalpy level diagram for the system $C(s) + O_2(g)$. The differences in level are standard enthalpy changes at 25 °C and 1 bar.

2.12 ENTHALPY OF FORMATION

Since absolute enthalpies are not known, enthalpies relative to a defined **reference state** are used instead. The defined reference state for each substance is made up of the stoichiometric amounts of the elements in the substance, each in its standard state and at the temperature under consideration. These "relative" enthalpies of substances are called **enthalpies of formation** and are represented by $\Delta_f H°$. Since the same reference state is used for reactants and products, the same $\Delta_r H°$ is obtained as would be obtained with equation 2.93 and absolute enthalpies. Thus, standard enthalpy changes for reactions may be calculated using enthalpies of formation as follows:

$$\Delta_r H° = \sum_{i=1}^{N} \nu_i \Delta_f H_i° \tag{2.94}$$

Note that the enthalpy of formation does not have an overbar because the subscript f, for formation, indicates that the mol^{-1} unit is involved.

The enthalpy of formation of a substance at a given temperature is the change in enthalpy for the reaction in which one mole of the substance in its standard state at the given temperature is formed from its elements, each in its standard state at that temperature. If there is more than one solid form of an element, one must be selected as a reference. For thermodynamic tables at 25 °C the reference form is usually the most stable form of the element at 25 °C, 1 bar pressure. Thus, the reference form of hydrogen is $H_2(g)$ instead of H(g), the reference form of carbon is graphite, and the reference form of sulfur is rhombic sulfur. For thermodynamic tables that cover a wide range of temperatures, different reference states may be used in various temperature ranges. In any case, **the enthalpy of formation of an element in its standard state is zero at every temperature.**

From reactions given previously we can see how the following enthalpies of formation at 25 °C are obtained:

	$\Delta_f H°/kJ\ mol^{-1}$
$CO_2(g)$	−393.509
CO(g)	−110.525
C(g)	716.682
O(g)	249.170

These enthalpies of formation should be identified in Fig. 2.16. Enthalpies of formation $\Delta_f H°$ at 25 °C for some 200 substances are given in Table C.2. These values are from the NBS Tables of Chemical Thermodynamic Properties (1982). Table C.3 gives enthalpies of formation from 0 to 3000 K for a smaller group of substances from the JANAF Thermochemical Tables (1985).

The values of enthalpies of formation given in these tables come from four sources: (1) calorimetrically measured enthalpies of reaction, fusion, vaporization, sublimation, transition, solution, and dilution; (2) temperature variation of equilibrium constants (see Section 5.5); (3) spectroscopically determined dissociation energies (see Section 14.3); (4) calculation from Gibbs energies and entropies (see the third-law method in Section 3.8).

You may calculate $\Delta_r H^\circ$ for any reaction for which the reactants and products are listed in tables, but the reaction will not necessarily occur spontaneously in the direction written. The question as to whether or not the reaction can occur is answered by calculations based on the second law of thermodynamics.

Example 2.8 ***Calculating the standard enthalpy of reaction at constant temperature***

What are the standard enthalpy changes at 298.15 K and 2000 K for the following reaction?

$$CO_2(g) + C(graphite) = 2CO(g)$$

Using Table C.3, at 298.15 K,

$$\begin{aligned}\Delta_r H^\circ &= 2\Delta_f H^\circ(CO) - \Delta_f H^\circ(CO_2)\\ &= 2(-110.527\ \text{kJ mol}^{-1}) - (-393.522\ \text{kJ mol}^{-1})\\ &= 172.468\ \text{kJ mol}^{-1}\end{aligned}$$

At 2000 K,

$$\begin{aligned}\Delta_r H^\circ &= 2(-118.896\ \text{kJ mol}^{-1}) - (-396.784\ \text{kJ mol}^{-1})\\ &= 158.992\ \text{kJ mol}^{-1}\end{aligned}$$

So far we have talked mainly about the reaction enthalpy at 298.15 K. To calculate the standard enthalpy change at some other temperature, given the value at 298.15 K, it is necessary to have heat capacity data on the reactants and the products. Since enthalpy is a state function, we can use the paths indicated below to calculate the standard enthalpy change at any desired temperature.

$$\begin{array}{ccc} \text{reactants} & \xrightarrow{\Delta_r H_T^\circ} & \text{products} \\ \int_T^{298} C_{P,\text{react}}\,dT \downarrow & & \uparrow \int_{298}^{T} C_{P,\text{prod}}\,dT \\ \text{reactants} & \xrightarrow{\Delta_r H_{298}^\circ} & \text{products} \end{array} \tag{2.95}$$

$$\Delta_r H_T^\circ = \int_T^{298} C_{P,\text{react}}\,dT + \Delta_r H_{298}^\circ + \int_{298}^{T} C_{P,\text{prod}}\,dT \tag{2.96}$$

$$\begin{aligned}\Delta_r H_T^\circ &= \Delta_r H_{298}^\circ + \int_{298}^{T} (C_{P,\text{prod}}^\circ - C_{P,\text{react}}^\circ)\,dT\\ &= \Delta_r H_{298}^\circ + \int_{298}^{T} \Delta_r C_P^\circ\,dT\end{aligned} \tag{2.97}$$

where

$$\Delta_r C_P^\circ = \sum_i \nu_i \overline{C}_{P,i}^\circ \tag{2.98}$$

The **reaction heat capacity** $\Delta_r C_P^\circ$ does not have an overbar because the subscript r indicates that the unit mol^{-1} is involved.

If data are available on the heat capacities of reactants and products in the form of power series in T (see Table 2.2), $\Delta_r H_T^\circ$ may be expressed as a function of T as follows:

$$\Delta_r C_P^\circ = \Delta_r \alpha + (\Delta_r \beta)T + (\Delta_r \gamma)T^2 \tag{2.99}$$

where $\Delta_r \alpha = \sum \nu_i \alpha_i$, and so on. Substituting in equation 2.97 gives

$$\begin{aligned}\Delta_r H_T^\circ &= \Delta_r H_{298}^\circ + \int_{298}^{T} [\Delta_r \alpha + (\Delta_r \beta)T + (\Delta_r \gamma)T^2]\,\mathrm{d}T \\ &= \Delta_r H_{298}^\circ + \Delta_r \alpha (T - 298) + \frac{\Delta_r \beta}{2}(T^2 - 298^2) + \frac{\Delta_r \gamma}{3}(T^3 - 298^3) \\ &= \Delta_r H_0^\circ + (\Delta_r \alpha)T + (\Delta_r \beta/2)T^2 + (\Delta_r \gamma/3)T^3 \end{aligned} \tag{2.100}$$

In the last form of this equation the constant terms have been added together to obtain a hypothetical enthalpy of reaction at 0 K, hypothetical because the power-series representations of C_P are for a limited temperature range. Within this temperature range equation 2.100 does represent the standard enthalpy of reaction as a function of temperature.

The JANAF tables give standard enthalpies of formation at a series of temperatures, and so these values may be used directly to calculate enthalpies of reaction. Some values from the JANAF tables are given in Table C.3.

Some thermodynamic tables give values of $\overline{H}_T^\circ - \overline{H}_{298}^\circ$ to assist in the calculation of $\Delta_r H_T^\circ$ for a chemical reaction or phase transition:

$$\overline{H}_T^\circ - \overline{H}_{298}^\circ = \int_{298\ \text{K}}^{T} \overline{C}_P^\circ\,\mathrm{d}T \tag{2.101}$$

Depending on the table, $\Delta_r H^\circ$ for phase transitions in the intervening temperature range may be added to the right-hand side of the equation.

Example 2.9 ***Calculation of the bond energy of molecular hydrogen from the enthalpy of formation of hydrogen atoms at 298 K***

What is the value of $\Delta_r H^\circ$ at 0 K for the following reaction?

$$H_2(g) = 2H(g)$$

The calculation using $\Delta_r H^\circ(298\ \text{K})$ illustrates the use of $\overline{H}_0^\circ - \overline{H}_{298}^\circ$ from Table C.3:

$$\Delta_r H^\circ(298\ \text{K}) = 2(217.999\ \text{kJ mol}^{-1}) = 435.998\ \text{kJ mol}^{-1}$$

$$\begin{array}{ccccc} & H_2(g) & \xrightarrow{298\ \text{K}} & 2H(g) & \Delta H_{298}^\circ = 435.998\ \text{kJ mol}^{-1} \\ \overline{H}_{298}^\circ - \overline{H}_0^\circ = 8.467\ \text{kJ mol}^{-1} & \uparrow & & \downarrow & \overline{H}_0^\circ - \overline{H}_{298}^\circ = -(2)(6.197\ \text{kJ mol}^{-1}) \\ & H_2(g) & \xrightarrow[0\ \text{K}]{} & 2H(g) & \end{array}$$

$$\begin{aligned}\Delta_r H^\circ(0\ \text{K}) &= (8.467 + 435.998 - 12.394)\ \text{kJ mol}^{-1} \\ &= 432.071\ \text{kJ mol}^{-1}\end{aligned}$$

Alternatively, this value may be calculated from the enthalpy of formation of H(g) at 0 K in Table C.3:

$$\Delta_r H^\circ(0\text{ K}) = 2(216.035\text{ kJ mol}^{-1}) = 432.070\text{ kJ mol}^{-1}$$

This value is often referred to as the H—H bond energy. In Chapter 11 we will see how this value can be calculated theoretically; there this energy is referred to as the dissociation energy D_0.

Comment:

The concept of a thermodynamic cycle will be used in many ways. The important idea is that it may be easier to measure changes in a thermodynamic property, such as enthalpy, along three sides of a cycle than along the fourth. For example, it is easier to make calorimetric measurements of enthalpy changes at room temperature and use heat capacity measurements to calculate the enthalpy change at a higher temperature than it would be to make the calorimetric measurement at the higher temperature.

2.13 CALORIMETRY

Heats of reaction are determined using **adiabatic calorimeters;** that is, the reaction or solution process occurs in a container, which is immersed in a weighed quantity of water and is surrounded by insulation or an adiabatic shield that is kept at the same temperature as the calorimeter so that no heat is gained or lost. A simple adiabatic calorimeter operated at constant-pressure is illustrated in Fig. 2.17. Thus, ΔH_A for this adiabatic process is zero. When a certain amount of reactants R are converted completely to products P in a constant-pressure calorimeter, the changes in state involved may be represented as follows:

$$\begin{array}{ccc} R(T_2) + \text{Cal}(T_2) & \xrightarrow{\Delta H(T_2)} & P(T_2) + \text{Cal}(T_2) \\ \uparrow \Delta H_R & \nearrow \Delta H_A = 0 & \uparrow \Delta H_P \\ R(T_1) + \text{Cal}(T_1) & \xrightarrow[\Delta H(T_1)]{} & P(T_1) + \text{Cal}(T_1) \end{array} \tag{2.102}$$

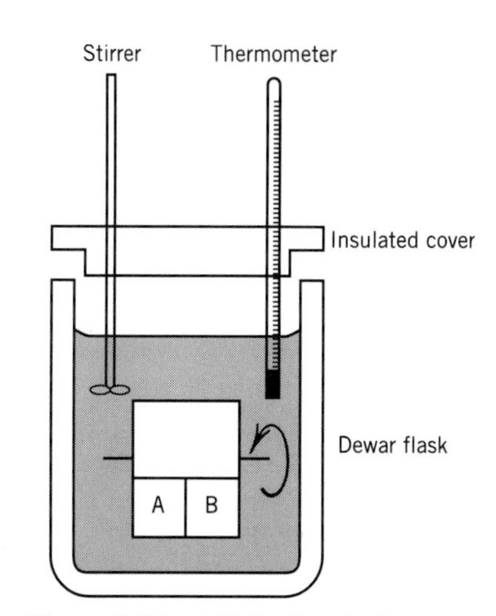

Figure 2.17 Adiabatic calorimeter operated at constant pressure. A reaction between solutions A and B is initiated by rotating the reaction vessel around the axis indicated.

The adiabatic container, thermometer, stirrer, and weighed quantity of water are represented by Cal. Since the enthalpy is a state function, the enthalpy change for the actual process may be written two ways:

$$\Delta H_A = \Delta H(T_1) + \Delta H_P = 0 \tag{2.103}$$

$$\Delta H_A = \Delta H_R + \Delta H(T_2) = 0 \tag{2.104}$$

Since the heat capacities of the reactants, products, and calorimeter may be assumed constant over the range T_1 to T_2, these equations become

$$\Delta H(T_1) = -\Delta H_{\mathrm{P}} = -[C_P(\mathrm{P}) + C_P(\mathrm{Cal})](T_2 - T_1) \quad (2.105)$$

$$\Delta H(T_2) = -\Delta H_{\mathrm{R}} = -[C_P(\mathrm{R}) + C_P(\mathrm{Cal})](T_2 - T_1) \quad (2.106)$$

where these C_P's are extensive properties. Thus, the results of the calorimetric experiments can be interpreted to obtain ΔH for the conversion of a certain amount of R to P either at T_1 or T_2. The heat capacity term in the first equation can be determined by using a calibrated electric heater coil and measuring I^2Rt with only products present, and the heat capacity term in the second equation can be determined with only reactants present. In this expression I is a constant current that flows through resistor R for time t.

Once ΔH has been determined in a calorimetric experiment, $\Delta_r H$ for a balanced chemical reaction can be calculated using $\Delta_r H = \Delta H/\Delta\xi$.

When a reaction is carried out in a sealed bomb (see Fig. 2.18), no PV work is done, and the first law may be written $\Delta U = q_V$. Thus, the change in internal energy for the reaction is obtained. When the reaction is carried out at constant pressure, the first law may be written $\Delta H = q_P$. Chemists are usually more interested in ΔH than ΔU because chemical reactions are generally carried out at constant pressure. If ΔU is determined in a bomb calorimeter, the value of ΔH may be calculated using equation 2.55:

$$\Delta_r H = \Delta_r U + RT\sum \nu_g \quad (2.107)$$

where $\sum \nu_g$ is the sum of stoichiometric numbers of gaseous products and gaseous reactants. Remember that stoichiometric numbers are positive for products and

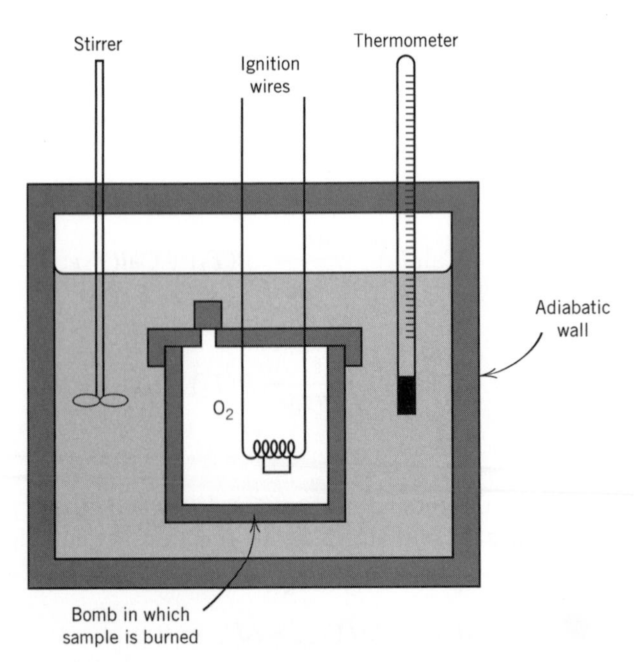

Figure 2.18 Adiabatic bomb calorimeter for carrying out combustions at constant volume.

negative for reactants. In writing equation 2.107 in this way we are ignoring the volume change due to solid and liquid reactants, because this is negligible in comparison with the change in gas volume. We are also assuming that the gases are ideal.

Example 2.10 ***Calculating the enthalpy of reaction from the heat of combustion***

The combustion of ethanol in a constant-volume calorimeter produces 1364.34 kJ mol^{-1} at 25 °C. What is the value of $\Delta_r H^\circ$ for the following combustion reaction?

$$C_2H_5OH(l) + 3O_2(g) = 2CO_2(g) + 3H_2O(l)$$

$$\begin{aligned}\Delta_r H^\circ &= \Delta_r U^\circ + RT \sum \nu_g \\ &= -1364.34 \text{ kJ mol}^{-1} + (8.3145 \times 10^{-3} \text{ kJ K}^{-1} \text{ mol}^{-1})(298.15 \text{ K})(-1) \\ &= -1366.82 \text{ kJ mol}^{-1}\end{aligned}$$

This is the quantity of heat that would be evolved at 25 °C and a constant pressure of 1 bar.

Example 2.11 ***Calculating the molar internal energy of combustion***

In an adiabatic bomb calorimeter, the combustion of 0.5173 g of ethanol causes the temperature to rise from 25.0 to 29.289 °C. The heat capacity of the bomb, the reactants, and the other contents of the calorimeter is 3576 J K^{-1}. What is the molar internal energy of combustion of ethanol at 25.0 °C?

The change in the state can be written

$$\begin{Bmatrix} C_2H_5OH + 3O_2 \\ + \text{ other contents} \end{Bmatrix} [T = 25\,^\circ\text{C}, V] = \begin{Bmatrix} 2CO_2 + 3H_2O \\ + \text{ other contents} \end{Bmatrix} [T = 29.289\,^\circ\text{C}, V]$$

for which $q = 0$ (adiabatic) and $w = 0$ (constant volume V) so that $\Delta U = 0$. This change in state can be written as the sum of

$$\begin{Bmatrix} C_2H_5OH + 3O_2 \\ + \text{ other contents} \end{Bmatrix} [T = 25\,^\circ\text{C}, V] = \begin{Bmatrix} 2CO_2 + 3H_2O \\ + \text{ other contents} \end{Bmatrix} [T = 25\,^\circ\text{C}, V] \qquad \Delta U_1$$

$$\begin{Bmatrix} 2CO_2 + 3H_2O \\ + \text{ other contents} \end{Bmatrix} [T = 25\,^\circ\text{C}, V] = \begin{Bmatrix} 2CO_2 + 3H_2O \\ + \text{ other contents} \end{Bmatrix} [T = 29.289\,^\circ\text{C}, V] \qquad \Delta U_2$$

where $\Delta U_1 + \Delta U_2 = \Delta U = 0$, so that

$$\Delta U_1 = -\Delta U_2$$

or

$$\begin{aligned}\Delta_r U &= -\frac{(3.576 \text{ kJ K}^{-1})(4.289 \text{ K})(46.0 \text{ g mol}^{-1})}{0.5173 \text{ g}} \\ &= -1364 \text{ kJ mol}^{-1}\end{aligned}$$

When a solute is dissolved in a solvent, heat may be absorbed or evolved; in general, the heat of solution depends on the concentration of the final solution. The **integral heat of solution** is the enthalpy change for the solution of 1 mol of solute in n mol of solvent. The solution process may be represented by a chemical equation such as

$$HCl(g) + 5H_2O(l) = HCl \text{ in } 5H_2O \qquad \Delta_{sol}H°(298 \text{ K}) = -63.467 \text{ kJ mol}^{-1} \tag{2.108}$$

at 1 bar pressure, where "HCl in $5H_2O$" represents a solution of 1 mol of HCl in 5 mol of H_2O. As the amount of water is increased, the integral heats of solution approach asymptotic values.

The solution of liquid acetic acid in water to form an aqueous solution in which undissociated acetic acid is in its standard state is represented by

$$CH_3CO_2H(l) = CH_3CO_2H(ao) \qquad \Delta_{sol}H°(298 \text{ K}) = -1.3 \text{ kJ mol}^{-1} \tag{2.109}$$

where the ao indicates that the ions do not dissociate further.

When a solute is dissolved in a solvent that is chemically quite similar to it and there are no complications of ionization or solvation, the heat of solution may be nearly equal to the heat of fusion of the solute. It might be expected that heat would always be absorbed in overcoming the attraction between the molecules or ions of the solid solute when the solute is dissolved. Another process that commonly occurs, however, is a strong interaction with the solvent, referred to as solvation, which evolves heat. In the case of water the solvation is called hydration.

The importance of this attraction of the solvent for the solute in the process of solution is illustrated by the dissolving of sodium chloride in water. In the crystal lattice of sodium chloride, positive sodium ions and negative chloride ions attract each other strongly. The energy required to separate them is so great that nonpolar solvents like benzene and carbon tetrachloride do not dissolve sodium chloride; but a solvent like water, which has a high dielectric constant and a large dipole moment, has a strong attraction for the sodium and chloride ions and solvates them with a large decrease in the energy of the system. When the energy required to separate the ions from the crystal is about the same as the solvation energy, as it is for dissolving NaCl in water, ΔH for the net process is close to zero. When NaCl is dissolved in water at 25 °C, there is only a small cooling effect; q is positive. When Na_2SO_4 is dissolved in water at 25 °C, there is an evolution of heat because the energy of hydration of the ions is greater than the energy required to separate the ions from the crystal.

Example 2.12 ***Calculating the enthalpy of neutralization in aqueous solution***

Calculate $\Delta_r H°(298 \text{ K})$ for the following reaction using Table C.2.

$$HCl \text{ in } 100H_2O + NaOH \text{ in } 100H_2O = NaCl \text{ in } 200H_2O + H_2O(l)$$

$$\Delta_r H°(298 \text{ K}) = -406.923 - 285.830 + 165.925 + 469.646$$

$$= -57.182 \text{ kJ mol}^{-1}$$

For dilute solutions it is found that the heat of reaction of strong bases, such as NaOH and KOH, with strong acids, such as HCl and HNO_3, is independent of the nature of the acid or base. This constancy of the heat of neutralization is a result of the complete ionization of strong acids and bases and the salts formed by neutralization. Thus, when a dilute solution of a strong acid is added to a dilute solution of a strong base, the only chemical reaction is

$$OH^-(ao) + H^+(ao) = H_2O(l) \qquad \Delta_r H°(298\ K) = -55.835\ kJ\ mol^{-1}$$

When a dilute solution of a weak acid or base is neutralized, the heat of neutralization is somewhat less because of the absorption of heat in the dissociation of the weak acid or base.

Since for strong electrolytes in dilute solution the thermal properties of the ions are essentially independent of the accompanying ions, it is convenient to use enthalpies of formation of individual ions. The sum of the enthalpies of formation of H^+ and OH^- ions may be calculated from

$$\begin{array}{rcll} H_2O(l) &=& H^+(ao) + OH^-(ao) & \Delta_r H° = 55.835\ kJ\ mol^{-1} \\ H_2(g) + \frac{1}{2}O_2(g) &=& H_2O(l) & \Delta_r H° = -285.830\ kJ\ mol^{-1} \\ \hline H_2(g) + \frac{1}{2}O_2(g) &=& H^+(ao) + OH^-(ao) & \Delta_r H° = -229.995\ kJ\ mol^{-1} \end{array}$$

The separate enthalpies of formation of H^+ and OH^- cannot be calculated, and so, to construct a table of enthalpies of formation of individual ions, it is necessary to adopt an arbitrary convention. Enthalpies of formation of aqueous ions in Table C.2 are based on the convention that $\Delta_f H° = 0$ for $H^+(ao)$. In other words, by convention,

$$\tfrac{1}{2}H_2(g) = H^+(ao) + e^- \qquad \Delta_f H° = 0$$

where e^- is the electron, which is assigned $\Delta_f H°(e^-) = 0$. This electron is not dissolved in water, but is a formal electron required to balance the equation. Therefore, the enthalpy of formation of OH^- is given by

$$\tfrac{1}{2}H_2(g) + \tfrac{1}{2}O_2(g) + e^- = OH^-(ao) \qquad \Delta_f H° = -229.995\ kJ\ mol^{-1}$$

On the basis of these values for the enthalpies of formation of H^+ and OH^-, the enthalpies of formation of other ions of strong electrolytes may be calculated.

From the enthalpy of formation of HCl(ai) it is possible to calculate the enthalpy of formation of $Cl^-(ao)$.

$$\tfrac{1}{2}H_2(g) + \tfrac{1}{2}Cl_2(g) = H^+(ao) + Cl^-(ao) \qquad \Delta_r H° = -167.159\ kJ\ mol^{-1}$$

$$\tfrac{1}{2}Cl_2(g) + e^- = Cl^-(ao) \qquad \Delta_f H° = -167.159\ kJ\ mol^{-1}$$

Eleven Key Ideas in Chapter 2

1. In thermodynamics work w is a signed quantity, and it is positive when work is done on the system of interest and is negative when the system does work on the surroundings. The work in a process depends on the path, even when the process is reversible. The differential of the work is represented by $đw$ as a reminder that work is not an exact differential.
2. The change in the internal energy U of a system in an adiabatic process is equal to the work done on the system. This provides a way to determine the difference ΔU in internal energy of two states of a system.

3. A system can also undergo a specified change in internal energy by absorbing heat or evolving heat q. Since the change in internal energy can be expressed in joules, the quantity of heat can also be expressed in joules.
4. According to the first law of thermodynamics, (1) there exists a property U that is a function of the state variables of a system and (2) the change in internal energy for a closed system can be calculated from $\Delta U = q + w$. However, the first law by itself does not provide any information as to whether a given process will proceed in the forward direction or the reverse direction.
5. The mathematical test for whether a variable is exact or inexact is whether the mixed partial derivatives are equal or unequal.
6. When a gas is allowed to expand, the maximum work is obtained when the process is carried out reversibly, that is, when the process is carried out with an infinite number of infinitesimal steps.
7. In addition to pressure–volume work, there is surface work, elongation work, electric charge displacement, and other kinds of work such as work of electric and magnetic polarization, which are not discussed here.
8. The heat capacity at constant volume is defined as the partial derivative of the internal energy with respect to temperature when the volume is held constant. The enthalpy is defined by $H = U + PV$, and the heat capacity at constant pressure is defined as the partial derivative of the enthalpy with respect to temperature when the pressure is held constant.
9. A cooling effect is obtained in the adiabatic expansion of an ideal gas, and the maximum cooling is obtained when the expansion is carried out reversibly.
10. The change in standard enthalpy in a chemical reaction is equal to the summation of the standard enthalpies of formation of the reacting species, each multiplied by its stoichiometric number in a specified chemical equation. If the standard enthalpy of reaction is known at one temperature, its value at any other temperature can be calculated if the molar heat capacities of the species involved are known throughout the temperature range involved.
11. Calorimeters are useful for determining standard enthalpies of formation of species, but we will see later that these values can also be determined in other ways.

REFERENCES

M. Bailyn, *A Survey of Thermodynamics*. New York: American Institute of Physics, 1994.

K. E. Bett, J. S. Rowlinson, and G. Saville, *Thermodynamics for Chemical Engineers.* Cambridge, MA: MIT Press, 1975.

H. B. Callen, *Thermodynamics and an Introduction to Thermostatistics,* 2nd ed. Hoboken, NJ: Wiley, 1985.

K. S. Pitzer, *Thermodynamics,* 3rd ed. New York: McGraw-Hill, 1995.

J. M. Smith, H. C. Van Ness, and M. M. Abbott, *Introduction to Chemical Engineering Thermodynamics*, 5th ed. New York: McGraw-Hill, 1996.

D. R. Stull, E. F. Westrum, and G. C. Sinke, *The Chemical Thermodynamics of Organic Compounds.* Hoboken, NJ: Wiley-Interscience, 1969.

J. W. Tester and M. Modell, *Thermodynamics and Its Applications*. Upper Saddle River, NJ: Prentice-Hall, 1997.

PROBLEMS

(M) Problems marked with an icon may be more conveniently solved on a personal computer with a mathematical program.

2.1 How high can a person (assume a weight of 70 kg) climb on one ounce of chocolate, if the heat of combustion (628 kJ) can be converted completely into work of vertical displacement?

2.2 A mole of sodium metal is added to water. How much work is done on the atmosphere by the subsequent reaction if the temperature is 25 °C?

2.3 You want to heat 1 kg of water 10 °C, and you have the following four methods under consideration. The heat capacity of water is 4.184 J K^{-1} g^{-1}.

(a) You can heat it with a mechanical eggbeater that is powered by a 1-kg mass on a rope over a pulley. How far does the mass have to descend in the earth's gravitational field to supply enough work?

(b) You can send 1 A through a 100-Ω resistor. How long will it take?

(c) You can send the water through a solar collector that has an area of 1 m^2. How long will it take if the sun's intensity on the collector is 4 J cm^{-2} min^{-1}?

(d) You can make a charcoal fire. The heat of combustion of graphite is −393 kJ mol^{-1}. That is, 12 g of graphite will produce 393 kJ of heat when it is burned to $CO_2(g)$ at constant pressure. How much charcoal will have to burn?

2.4 Show that the differential df is inexact.

$$df = dx - \frac{x}{y}\,dy$$

Thus, the integral $\int df$ depends on the path. However, we can define a new function g by

$$dg = \frac{1}{y}\,df$$

which has the property that dg is exact. Show that dg is exact, so that

$$\oint dg = 0$$

2.5 Show that the function $f(x, y)$ defined by

$$df(x, y) = (x + 2y)\,dx - x\,dy$$

is inexact. Test to see whether the integrating factor $1/x^3$ makes it an exact differential.

2.6 Show that the function defined by

$$df(x, y) = (y^2 - xy)\,dx - x^2\,dy$$

is inexact. Test the integrating factor $1/xy^2$ to see whether it produces an exact differential.

2.7 What are the partial derivatives $(\partial z/\partial x)_y$ and $(\partial z/\partial y)_x$ of the following functions? (a) $z = xy$, (b) $z = x/y$, (c) $z = \ln(xy)$, (d) $z = \ln(x/y)$, and (e) $z = \exp(xy)$.

2.8 (a) The surface tension of water at 25°C is 0.072 N m^{-1}. How much work is required to form a surface 100 m by 100 m? (b) The force on a wire is due to a 75-kg person in the earth's gravitational field. If the wire stretches 1 m, how much work is done on the wire? (c) A gas expands 1 L against a constant external pressure of 1 bar. How much work is done on the gas?

2.9 Over narrow ranges of temperature and pressure, the differential expression for the volume of a fluid as a function of temperature and pressure can be integrated to obtain

$$V = K e^{\alpha T} e^{-\kappa P}$$

(α and κ are defined in Section 4.10). Show that V is a state function.

2.10 One mole of nitrogen at 25 °C and 1 bar is expanded reversibly and isothermally to a pressure of 0.132 bar. (a) What is the value of w? (b) What is the value of w if the nitrogen is expanded against a constant pressure of 0.132 bar?

2.11 (a) Derive the equation for the work of reversible isothermal expansion of a van der Waals gas from V_1 to V_2. (b) A mole of CH_4 expands reversibly from 1 to 50 L at 25 °C. Calculate the work in joules assuming (1) the gas is ideal and (2) the gas obeys the van der Waals equation. For $CH_4(g)$, a = 2.283 L^2 bar mol^{-2} and b = 0.042 78 L mol^{-1}.

2.12 Liquid water is vaporized at 100 °C and 1.013 bar. The heat of vaporization is 40.69 kJ mol^{-1}. What are the values of (a) w_{rev} per mole, (b) q per mole, (c) $\Delta\overline{U}$, and (d) $\Delta\overline{H}$?

2.13 An ideal gas expands reversibly and isothermally from 10 bar to 1 bar at 298.15 K. What are the values of (a) w per mole, (b) q per mole, (c) $\Delta\overline{U}$, and (d) $\Delta\overline{H}$? (e) The ideal gas expands isothermally against a constant pressure of 1 bar. How much work is done on the gas?

2.14 Calculate $\overline{H}°(2000\ K) - \overline{H}°(0\ K)$ for H(g).

2.15 The heat capacities of a gas may be represented by

$$\overline{C_P} = \alpha + \beta T + \gamma T^2 + \delta T^3$$

For N_2, α = 28.883 J $K^{-1}mol^{-1}$, β = -1.57×10^{-3} J K^{-2} mol^{-1}, $\gamma = 0.808 \times 10^{-5}$ J K^{-3} mol^{-1}, and $\delta = -2.871 \times 10^{-9}$ J K^{-4} mol^{-1}. How much heat is required to heat a mole of N_2 from 300 to 1000 K?

2.16 **(a)** In a reversible adiabatic expansion of an ideal gas with $\gamma = C_P/C_V$ independent of temperature, the pressure and volume are related by

$$PV^{\gamma} = \text{constant}$$

Show that the work of adiabatic expansion from P_1, V_1 to P_2, V_2 is

$$w = (P_2V_2 - P_1V_1)/(\gamma - 1)$$

(b) Check this equation to be sure it gives the same amount of work as Example 2.7.

2.17 Calculate the temperature increase and final pressure of helium if a mole is compressed adiabatically and reversibly from 44.8 L at 0 °C to 22.4 L.

2.18 A mole of argon is allowed to expand adiabatically and reversibly from a pressure of 10 bar and 298.15 K to 1 bar. What is the final temperature, and how much work is done on the argon?

2.19 A tank contains 20 L of compressed nitrogen at 10 bar and 25 °C. Calculate w when the gas is allowed to expand reversibly to 1 bar pressure (a) isothermally and (b) adiabatically.

2.20 An ideal monatomic gas at 298.15 K and 1 bar is expanded in a reversible adiabatic process to a final pressure of $\frac{1}{2}$ bar. Calculate q per mole, w per mole, and $\Delta\overline{U}$.

2.21 An ideal monatomic gas at 1 bar and 300 K is expanded adiabatically against a constant pressure of $\frac{1}{2}$ bar until the final pressure is $\frac{1}{2}$ bar. What are the values of q per mole, w per mole, $\Delta\overline{U}$, and $\Delta\overline{H}$? Given: $\overline{C}_V = \frac{3}{2}R$.

2.22 Derive the equation for calculating the work involved in a reversible, adiabatic pressure change of one mole of an ideal gas so that the work can be calculated from the initial temperature T_1, initial pressure P_1, and final pressure P_2.

2.23 Calculate $\Delta_r H^\circ_{298}$ for

$$H_2(g) + F_2(g) = 2HF(g)$$
$$H_2(g) + Cl_2(g) = 2HCl(g)$$
$$H_2(g) + Br_2(g) = 2HBr(g)$$
$$H_2(g) + I_2(g) = 2HI(g)$$

2.24 The following reactions might be used to power rockets:

$$(1)\quad H_2(g) + \tfrac{1}{2}O_2(g) = H_2O(g)$$
$$(2)\quad CH_3OH(l) + 1\tfrac{1}{2}O_2(g) = CO_2(g) + 2H_2O(g)$$
$$(3)\quad H_2(g) + F_2(g) = 2HF(g)$$

(a) Calculate the enthalpy changes at 25 °C for each of these reactions per kilogram of reactants. (b) Since the thrust is greater when the molar mass of the exhaust gas is lower, divide the heat per kilogram by the molar mass of the product (or the average molar mass in the case of reaction 2) and arrange the above reactions in order of effectiveness on the basis of thrust.

2.25 Calculate $\Delta_r H$ for the dissociation

$$O_2(g) = 2O(g)$$

at 0, 298, and 3000 K. In Section 14.3 the enthalpy change for dissociation at 0 K will be found to be equal to the spectroscopic dissociation energy D_0.

2.26 Methane may be produced from coal in a process represented by the following steps, where coal is approximated by graphite:

$$2C(s) + 2H_2O(g) = 2CO(g) + 2H_2(g)$$
$$CO(g) + H_2O(g) = CO_2(g) + H_2(g)$$
$$CO(g) + 3H_2(g) = CH_4(g) + H_2O(g)$$

The sum of the three reactions is

$$2C(s) + 2H_2O(g) = CH_4(g) + CO_2(g)$$

What is $\Delta_r H^\circ$ at 500 K for each of these reactions? Check that the sum of the $\Delta_r H^\circ$ values of the first three reactions is equal to $\Delta_r H^\circ$ for the fourth reaction. From the standpoint of heat balance, would it be better to develop a process to carry out the overall reactions in three separate reactors or in a single reactor?

2.27 What is the heat evolved in freezing water at −10 °C given that

$$H_2O(l) = H_2O(s) \qquad \Delta H^\circ(273\text{ K}) = -6004\text{ J mol}^{-1}$$
$$\overline{C}_P(H_2O, l) = 75.3\text{ J K}^{-1}\text{ mol}^{-1}$$
$$\overline{C}_P(H_2O, s) = 36.8\text{ J K}^{-1}\text{ mol}^{-1}$$

2.28 What is the enthalpy change for the vaporization of water at 0 °C? This value may be estimated from Table C.2 by assuming that the heat capacities of $H_2O(l)$ and $H_2O(g)$ are independent of temperature from 0 to 25 °C.

2.29 Calculate the standard enthalpy of formation of methane at 1000 K from the value at 298.15 K using the $\overline{H}^\circ_T - \overline{H}^\circ_{298}$ data in Table C.3.

2.30 For a diatomic molecule the bond energy is equal to the change in internal energy for the reaction

$$X_2(g) = 2X(g)$$

at 0 K. Of course, the change in internal energy and the change in enthalpy are the same at 0 K. Calculate the enthalpy of dissociation of $O_2(g)$ at 0 K. The enthalpy of formation of O(g) at 298.15 K is 249.173 kJ mol^{-1}. In the range of 0–298 K the average value of the heat of capacity of $O_2(g)$ is 29.1 J K^{-1} mol^{-1} and the average value of the heat capacity of O(g) is 22.7 J K^{-1} mol^{-1}. What is the value of the bond energy in electron volts? (When the changes in heat capacities in the range of 0–298 K are taken into account, the enthalpy of dissociation at 0 K is 493.58 kJ mol^{-1}.)

2.31 One gram of liquid benzene is burned in a bomb calorimeter. The temperature before ignition was 20.826 °C, and the temperature after the combustion was 25.000 °C. This was an adiabatic calorimeter. The heat capacity of the bomb, the water around it, and the contents of the bomb before the combustion was 10 000 J K^{-1}. Calculate $\Delta_f H^\circ$ for $C_6H_6(l)$ at 298.15 K from these data. Assume that the water produced in the combustion is in the liquid state and the carbon dioxide produced in the combustion is in the gas state.

2.32 An aqueous solution of unoxygenated hemoglobin containing 5 g of protein (M = 64 000 g mol^{-1}) in 100 cm^3 of solution is placed in an insulated vessel. When enough molecular oxygen is added to the solution to completely saturate the hemoglobin, the temperature rises 0.031 °C. Each mole of hemoglobin binds 4 mol of oxygen. What is the enthalpy of reaction per mole of oxygen bound? The heat capacity of the solution may be assumed to be 4.18 J K^{-1} cm^{-3}.

2.33 Calculate the heat of hydration of $Na_2SO_4(s)$ from the integral heats of solution of $Na_2SO_4(s)$ and $Na_2SO_4 \cdot 10H_2O(s)$ in infinite amounts of H_2O, which are -2.34 and 78.87 kJ mol^{-1}, respectively. Enthalpies of hydration cannot be measured directly because of the slowness of the phase transition.

2.34 We want to determine the enthalpy of hydration of $CaCl_2$ to form $CaCl_2 \cdot 6H_2O$:

$$CaCl_2(s) + 6H_2O(l) = CaCl_2 \cdot 6H_2O(s)$$

We cannot do this directly for a couple of reasons: (1) reactions in the solid state are slow, and (2) there is a series of hydrates and so a mixture of different hydrates would probably be obtained. We can, however, determine the heats of solution of $CaCl_2(s)$ and $CaCl_2 \cdot 6H_2O(s)$ in water at 298 K and take the difference. The experimental heats of solution are as follows:

$$CaCl_2(s) + Aq = CaCl_2(ai) \qquad \Delta_r H = -81.33 \text{ kJ mol}^{-1}$$

$$CaCl_2 \cdot 6H_2O(s) + Aq = CaCl_2(ai) + 6H_2O(l)$$

$$\Delta_r H = 15.79 \text{ kJ mol}^{-1}$$

where Aq represents a large amount of water. What is the enthalpy of hydration?

2.35 The change in internal energy $\Delta_c U°$ in the combustion of $C_{60}(s)$ is $-25\ 968$ kJ mol^{-1} at 298.15 K [Kolesov et al., *J. Chem. Thermodyn.* **28**:1121 (1996)]. (*a*) What is the enthalpy of combustion $\Delta_c H°$? (*b*) What is the enthalpy of formation $\Delta_f H°[C_{60}(s)]$? (*c*) What is the enthalpy of vaporization of $C_{60}(s)$ to C(g) per mole of C(g)? (*d*) How does this compare with the enthalpy of vaporization of graphite and diamond to C(g)?

2.36 How much work is done when a person weighing 75 kg (165 lb) climbs the Washington Monument, 555 ft high? How many kilojoules must be supplied to do this muscular work, assuming that 25% of the energy produced by the oxidation of food in the body can be converted to muscular mechanical work?

2.37 The average person generates about 2500 kcal of heat a day. How many kilowatt-hours of energy is this? If walking briskly dissipates energy at 500 W, what fraction of the day's energy does walking one hour represent? How many kilograms of water would have to be evaporated if this were the only means of heat loss? (The heat of vaporization of water at 35 °C is 2400 J g^{-1}.)

2.38 The surface tension of water is 71.97×10^{-3} N m^{-1} or 71.97×10^{-3} J m^{-2} at 25 °C. Calculate the surface energy in joules of 1 mol of water dispersed as a mist containing droplets of 1 μm (10^{-4} cm) in radius. The density of water may be taken as 1.00 g cm^{-3}.

2.39 Are the following expressions exact differentials?

(*a*) $xy^2\, dx - x^2 y\, dy$ (*b*) $\dfrac{dx}{y} - \dfrac{x}{y^2}\, dy$

2.40 Show that

$$dq = dU + P\, dV = C_V\, dT + RT\, d\ln V$$

is not an exact differential, but

$$\frac{dq}{T} = C_V\, d\ln T + R\, d\ln V$$

is an exact differential.

2.41 Show that the differential $d\overline{V}$ of the molar volume of an ideal gas is an exact differential.

2.42 Calculate w for a reversible isothermal (298.15 K) expansion of a mole of N_2 from 1 to 10 L assuming it is (*a*) an ideal gas and (*b*) a van der Waals gas (see Table 1.3).

2.43 An ideal gas at 25 °C and 100 bar is allowed to expand reversibly and isothermally to 5 bar. Calculate (*a*) w per mole, (*b*) the heat absorbed per mole, (*c*) $\Delta\overline{U}$, and (*d*) $\Delta\overline{H}$.

2.44 Ammonia gas is condensed at its boiling point at 1.013 25 bar at −33.4 °C by the application of a pressure infinitesimally greater than 1 bar. To evaporate ammonia at its boiling point requires the absorption of 23.30 kJ mol^{-1}. Calculate (*a*) w_{rev} per mole, (*b*) q per mole, (*c*) $\Delta\overline{U}$, and (*d*) $\Delta\overline{H}$.

2.45 What is w per mole for a reversible isothermal expansion of ethane from 5 to 10 L mol^{-1} at 298 K assuming (*a*) ethane is an ideal gas and (*b*) it follows the van der Waals equation? (Van der Waals constants are in Table 1.3.)

2.46 According to Table C.3, how much heat is required to raise the temperature of a mole of oxygen from 298 to 3000 K at constant pressure?

Ⓜ **2.47** From the following data calculate the value of $(\overline{H}^\circ_{298} - \overline{H}^\circ_0)$ for $Al_2O_3(s)$.

T/K	$\overline{C}^\circ_P$/J K^{-1} mol^{-1}	T	$\overline{C}^\circ_P$	T	$\overline{C}^\circ_P$	T	$\overline{C}^\circ_P$
10	0.009	90	9.69	170	39.94	250	67.01
20	0.076	100	12.84	180	43.79	260	69.76
30	0.263	110	16.32	190	47.53	270	72.37
40	0.691	120	20.06	200	51.14	280	74.84
50	1.492	130	23.96	210	54.60	290	77.19
60	2.779	140	27.96	220	57.92	298.16	79.01
70	4.582	150	31.98	230	61.10	273.16	73.16
80	6.895	160	35.99	240	64.13		

2.48 One mole of hydrogen at 25 °C and 1 bar is compressed adiabatically and reversibly into a volume of 5 L. Assuming ideal gas behavior, calculate (*a*) the final temperature, (*b*) the final pressure, and (*c*) the work done on the gas.

2.49 One mole of argon at 25 °C and 1 bar pressure is allowed to expand reversibly to a volume of 50 L (*a*) isothermally and (*b*) adiabatically. Assuming ideal gas behavior, calculate the final pressure in each case and the work done on the gas.

2.50 A mole of monatomic ideal gas at 1 bar and 273.15 K is allowed to expand adiabatically against a constant pressure of 0.395 bar until equilibrium is reached. (*a*) What is the final temperature? (*b*) What is the final volume? (*c*) How much work is done on the gas in this process? (*d*) What is the change in the internal energy of the gas in this process?

2.51 A tank contains 20 L of compressed nitrogen at 10 bar and 25 °C. Calculate the maximum work that can be obtained when the gas is allowed to expand reversibly to 1 bar pressure (*a*) isothermally and (*b*) adiabatically. The heat capacity of nitrogen at constant volume can be taken to be 20.8 J K^{-1} mol^{-1} independent of temperature.

2.52 Compare the enthalpies of combustion of $CH_4(g)$ to $CO_2(g)$ and $H_2O(g)$ at 298 and 2000 K.

$$CH_4(g) + 2O_2(g) = CO_2(g) + 2H_2O(g)$$

2.53 Compare the enthalpy of combustion of $CH_4(g)$ to $CO_2(g)$ and $H_2O(l)$ at 298 K with the sum of the enthalpies of combustion of graphite and $2H_2(g)$, from which $CH_4(g)$ can, in principle, be produced.

2.54 The enthalpy change for the combustion of toluene to $H_2O(l)$ and $CO_2(g)$ is −3910.0 kJ mol^{-1} at 25 °C. Calculate the enthalpy of formation of toluene.

2.55 Calculate ΔH (298 K) per gram of fuel (exclude oxygen) for

(a) $H_2(g) + \frac{1}{2}O_2(g) = H_2O(g)$

(b) $CH_4(g) + 2O_2(g) = CO_2(g) + 2H_2O(g)$

(c) $CH_3OH(l) + \frac{3}{2}O_2(g) = CO_2(g) + 2H_2O(g)$

(d) $C_6H_{14}(g) + 9\frac{1}{2}O_2(g) = 6CO_2(g) + 7H_2O(g)$

2.56 A 1:3 mixture of CO and H_2 is passed through a catalyst to produce methane at 500 K.

$$CO(g) + 3H_2(g) = CH_4(g) + H_2O(g)$$

How much heat is liberated in producing a mole of methane? How does this compare with the heat obtained from burning a mole of methane at this temperature? How does the heat of combustion of CH_4 compare with the heat of combustion of $CO + 3H_2$?

2.57 Calculate $\Delta_r H°$ for the dissociation

$$H_2(g) = 2H(g)$$

at 0, 298, and 3000 K. The value at 0 K is equal to the spectroscopic dissociation energy D_0.

2.58 In principle, methanol can be produced from methane in two steps or one:

I. $CH_4(g) + H_2O(g) = CO(g) + 3H_2(g)$

$CO(g) + 2H_2(g) = CH_3OH(g)$

II. $CH_4(g) + H_2O(g) = CH_3OH(g) + H_2(g)$

What is $\Delta_r H°$ at 500 K for each of these reactions? From the standpoint of heat balance, would it be better to develop a process to carry out the overall reaction in two separate reactors or in a single reactor?

2.59 Calculate the heat of vaporization of water at 25 °C. The specific heat of water may be taken as 4.18 J K^{-1} g^{-1}. The heat capacity of water vapor at constant pressure in this temperature range is 33.5 J K^{-1} mol^{-1}, and the heat of vaporization of water at 100 °C is 2258 J g^{-1}.

2.60 Calculate the enthalpy of dissociation of $H_2(g)$ at 3000 K using $\Delta_f H_T^\circ[H(g), 298.15 \text{ K}] = 217.999$ kJ mol^{-1} and $\overline{H}_T^\circ - \overline{H}_{298}^\circ$ values in Appendix C.3.

2.61 Calculate the dissociation energy of CH_4 into atoms at 298.15 K using $\Delta H°$ for the dissociation reaction at 0 K and $\overline{H}_T^\circ - \overline{H}_{298}^\circ$ values from Appendix C.3.

2.62 The reaction of heated coal (approximated here by graphite) with superheated steam absorbs heat. This heat is usually provided by burning some of the coal. Calculate $\Delta_r H°$ (500 K) for both reactions.

2.63 Ammonia is to be oxidized to $NO_2(g)$ to make nitric acid. What temperature will be reached if the only reaction is

$$NH_3(g) + \tfrac{7}{4}O_2(g) = NO_2(g) + \tfrac{3}{2}H_2O(g)$$

and a stoichiometric amount of oxygen is used?

2.64 In an adiabatic bomb calorimeter, oxidation of 0.4362 g of naphthalene ($C_{10}H_8$) caused a temperature rise of 1.707 °C. The final temperature was 298 K. The heat capacity of the calorimeter and water was 10 290 J K^{-1}, and the heat capacity of the products can be neglected. If corrections for the oxidation of the wire and residual nitrogen are neglected, what is the molar internal energy of combustion of naphthalene? What is its enthalpy of formation?

2.65 The combustion of oxalic acid in a bomb calorimeter yields 2816 J g^{-1} at 25 °C. Calculate (*a*) $\Delta_c U°$ and (*b*) $\Delta_c H°$ for the combustion of 1 mol of oxalic acid (M = 90.0 g mol^{-1}).

2.66 A mole of liquid sulfuric acid (98 g) is added to a certain quantity of water at 25 °C, and it is found that the temperature is 100 °C! The NBS Tables of Chemical Thermodynamic Properties yield the following information:

	$\Delta_f H°$/kJ mol^{-1}
$H_2SO_4(l)$	−813.99
$H_2SO_4(ai)$	−909.27

The ai indicates this value is for sulfuric acid that is completely ionized in water. What is the enthalpy of solution of sulfuric acid in enough water to completely ionize it? Let's make some approximations and estimate the mass of the "certain quantity of water" in the first line. Assume that the solution has the same heat capacity as water ($\overline{C}_P$ = 75.3 J K^{-1} mol^{-1}), independent of temperature, and ignore the mass of the sulfuric acid added. How much water was used?

2.67 When nitroglycerin explodes, the chemical reaction that occurs can be assumed to be

$$C_3H_5N_3O_9(l) = 3CO_2(g) + \tfrac{5}{2}H_2O(g) + \tfrac{3}{2}N_2(g) + \tfrac{1}{4}O_2(g)$$

(*a*) Calculate $\Delta_r H°$ and $\Delta_r U°$ for this reaction at 298 K, given that the enthalpy of formation of liquid nitroglycerin is -372.4 kJ mol^{-1}. (*b*) Consider 0.20 mol of nitroglycerin at 25 °C completely filling a constant-volume cell of 0.030 L. Calculate the *maximum* temperature and pressure that would be generated by the explosion of the nitroglycerin if the constant-volume cell did not burst (or vaporize). You may assume that (1) the explosion occurs so rapidly that the conditions are adiabatic, (2) the pressure cell comes to immediate thermal equilibrium with the products, (3) the products are ideal gases, and (4) the total constant-volume heat capacity of products plus cell has the temperature-independent value of 100 J K^{-1}.

Computer Problems

2.A In Section 2.8 we have seen that the knowledge of the temperature dependence of $C_P°$, which can be represented by a polynomial in T (see Table 2.2), makes it possible to calculate the change in molar enthalpy $\overline{H}°$ with temperature. In Section 2.13 we have seen that the standard enthalpy of reaction at other temperatures can be calculated from $\Delta_r H°(298.15 \text{ K})$ by integration of $\Delta_r C_P° \, dT$. It is difficult to make these calculations with a hand-held calculator, but they can be conveniently made using a mathematical application that can integrate. By putting in empirical equations for $C_P°$, we can solve the following problems:

(a) Print out the values of $C_P°$ at 298.15 K and 1000 K for the following gases: N_2, O_2, H_2, CO, CO_2, H_2O, NH_3, and CH_4. Compare these values with those in Table C.3 of Appendix C.

(b) Plot $C_P°$ for CO_2, H_2O, NH_3, and CH_4 versus temperature from 298.15 K to 1500 K.

(c) Calculate $\overline{H}°(1000 \text{ K}) - \overline{H}°(298.15 \text{ K})$ for N_2 and compare it with the values in Table C.3.

(d) Calculate $\Delta_r H°(1000 \text{ K})$ for the following gas reactions:

$$H_2 + \tfrac{1}{2}O_2 = H_2O$$
$$\tfrac{1}{2}N_2 + \tfrac{3}{2}H_2 = NH_3$$
$$CO_2 + 4H_2 = CH_4 + 2H_2O$$
$$CO + 3H_2 = CH_4 + H_2O$$

2.B Plot the molar heat capacities in J K^{-1} mol^{-1} at constant pressure for gaseous He, N_2, H_2O, CO_2, NH_3, and CH_4 from 300 to 1800 K using the parameters in Table 2.2.

2.C Calculate the standard enthalpy of reaction at 298 K and 1000 K for the gas reaction $CO_2 + 4H_2 = CH_4 + 2H_2O$.

2.D Calculate the standard enthalpy of formation of CO_2 at 1000 K from the standard enthalpy of formation at 298.15 K and empirical equations for the heat capacities. The standard molar heat capacities of graphite are given in Table C.3 as a function of temperature. It is convenient to fit these data to a function of temperature.

2.E **(a)** Calculate the work of reversible isothermal expansion of a mole of carbon dioxide at 298.15 K from an initial volume of 5 L to a final volume of 20 L on the assumption that carbon dioxide is an ideal gas.

(b) Calculate the reversible work using the van der Waals constants for carbon dioxide.

(c) Explain the difference.

3

Second and Third Laws of Thermodynamics

The first law of thermodynamics states that when one form of energy is converted to another, the total energy is conserved. It does not indicate any other restriction on this process. However, we know that many processes have a natural direction, and it is with the question of direction that the second law is concerned. For example, a gas expands into a vacuum, but the reverse never occurs, although it would not violate the first law. For a metal bar at uniform temperature to become hot at one end and cold at the other would not be a violation of the first law, yet we know this never occurs spontaneously.

The second law of thermodynamics is one of the most important generalizations in science. It is important in chemistry because it can tell us whether a process or reaction will occur in the forward or backward direction. The quantity that tells us whether a chemical reaction or a physical change can occur spontaneously in an isolated system is the entropy S. Entropy is a function of the state of the system, as is the internal energy U. Thermodynamics does not deal with the rate of approach to equilibrium, only with the equilibrium state. Some time is required even for a gas to expand into another container, and for some chemical reactions the rate of approach to equilibrium is very slow.

The third law of thermodynamics allows us to obtain the absolute value of the entropy of a substance.

3.1 ENTROPY AS A STATE FUNCTION

Historically, the second law of thermodynamics and the concept of entropy arose from considering the efficiencies of heat engines, but heat engines are quite different from chemical systems. Therefore, we are going to introduce the state function entropy from the more mathematical viewpoint of the preceding two chapters.

To understand the need for a second law of thermodynamics, consider what the first law of thermodynamics has not provided. According to the first law, energy is conserved in a process that takes a system from one state to another, but this provides no information as to whether the process of chemical reaction can proceed spontaneously or not. Yet we know that processes and chemical reactions proceed spontaneously in one direction and not in the opposite direction. For example, gases always expand into a vacuum. Heat is always transferred from a hot body to a cold body. In the presence of a catalyst, molecular hydrogen and molecular oxygen react to form water. These are referred to as **spontaneous processes.** The reverse processes, **nonspontaneous processes,** can be accomplished only by performing work on the system. A gas can be compressed to a smaller volume by use of mechanical work. Heat can be transferred to a hot body from a cold body by use of a refrigeration device. Water can be electrolyzed to molecular hydrogen and molecular oxygen by use of electrical work from a battery.

When we see a movie run backward, we often laugh because we know the event could not happen that way. The first law of thermodynamics does not tell us the direction in which a process can occur spontaneously. The laws of classical mechanics and quantum mechanics also do not tell us the direction in which time increases. When a movie of a collision between two particles is run backward we don't laugh, because it looks as reasonable as the movie run in the forward direction. Indeed, the mechanical equations of motion are invariant under time reversal. It is useful to be able to predict whether a physical change or a chemical reaction will go spontaneously in the forward or backward direction, and so the second law is very important.

To see how a state function can be introduced to identify a spontaneous process, let us consider the transfer of heat, which obviously involves the issue of the temperature. In the preceding chapter we saw that heat is not a state property of a system, even though it can be expressed in terms of state properties for a specified process, i.e., along a particular path. For example, the first law shows that when an ideal gas is heated reversibly,

$$dU = C_V\,dT = đq_{\text{rev}} + dw = đq_{\text{rev}} - P\,dV = đq_{\text{rev}} - \frac{nRT}{V}\,dV \tag{3.1}$$

The reversible heat is not a state function because $đq_{\text{rev}}$ is an inexact differential (Section 2.3). This is indicated by the symbol used here, but it can be confirmed by applying the test of exactness, $(\partial M/\partial y)_x = (\partial N/\partial x)_y$, to equation 3.1 written in the form

$$đq_{\text{rev}} = C_V\,dT + \frac{nRT}{V}\,dV \tag{3.2}$$

Taking the derivative of the coefficient of dT with respect to V at constant T yields

$$\left(\frac{\partial C_V}{\partial V}\right)_T = 0 \tag{3.3}$$

since for an ideal gas C_V is independent of the volume. Taking the derivative of the coefficient of $\mathrm{d}V$ with respect to T at constant V yields

$$\left[\frac{\partial(nRT/V)}{\partial T}\right]_V = \frac{nR}{V} \tag{3.4}$$

Since the mixed partial derivatives are not equal, q_{rev} is not a state function.

However, we saw in Section 2.3 that there may be an integrating factor that will convert an inexact differential to an exact differential. Since the transfer of heat depends on the temperature, let us try $1/T$ as an integrating factor. This is accomplished by multiplying both sides of equation 3.2 by $1/T$.

$$\frac{đq_{\text{rev}}}{T} = \frac{C_V}{T}\,\mathrm{d}T + \frac{nR}{V}\,\mathrm{d}V \tag{3.5}$$

Applying the test for exactness to $đq_{\text{rev}}/T$ yields

$$\left[\frac{\partial(C_V/T)}{\partial V}\right]_T = 0 \tag{3.6}$$

$$\left[\frac{\partial(nR/V)}{\partial T}\right]_V = 0 \tag{3.7}$$

Since both of these derivatives are equal to zero, $đq_{\text{rev}}/T$ is the exact differential of a state function. This proof applies to an ideal gas, but the fact that this is a general conclusion is shown in Section 3.9 on heat engines.

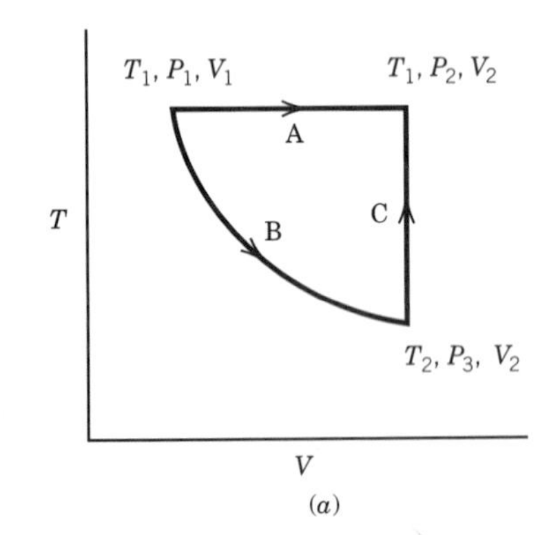

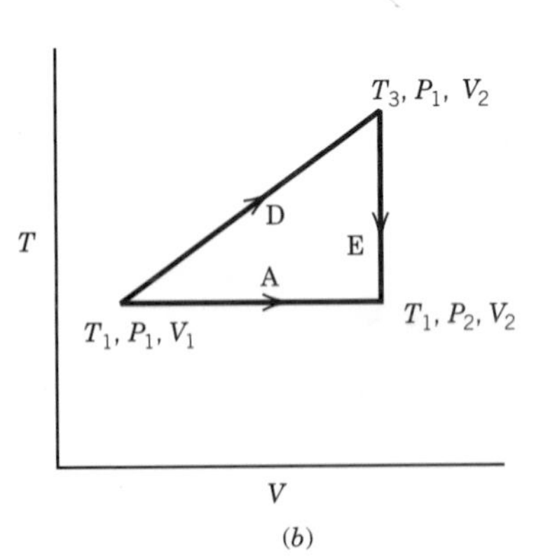

Figure 3.1 Reversible changes in the state of an ideal gas.

Example 3.1 ***Demonstration that the integral of the differential of the reversible heat divided by T is a state function***

Show that $\int đq_{\text{rev}}/T$ is independent of path, by considering the following **reversible** changes in the state of an ideal gas:

(a)	ideal gas(T_1, P_1, V_1) = ideal gas(T_1, P_2, V_2)	reversible, isothermal
(b)	ideal gas(T_1, P_1, V_1) = ideal gas(T_2, P_3, V_2)	reversible, adiabatic
(c)	ideal gas(T_2, P_3, V_2) = ideal gas(T_1, P_2, V_2)	reversible, constant volume
(d)	ideal gas(T_1, P_1, V_1) = ideal gas(T_3, P_1, V_2)	reversible, constant pressure
(e)	ideal gas(T_3, P_1, V_2) = ideal gas(T_1, P_2, V_2)	reversible, constant volume

Our first law knowledge is sufficient to calculate $\int đq_{\text{rev}}/T$ for each of these changes, and plots for an ideal gas are given in Fig. 3.1. The value of $\int đq_{\text{rev}}/T$ is calculated for each of these changes in state as follows:

(a) For an ideal gas, the internal energy depends only on the temperature, and so $\mathrm{d}U = 0$. Therefore, $đq_A = -đw = P\,\mathrm{d}V = RT\,\mathrm{d}V/V$. Thus $\int đq_A/T = R\ln(V_2/V_1)$.

(b) Since the change is adiabatic, $đq_B = 0$ and $\int đq_B/T = 0$, and from the expression derived in Chapter 2,

$$\frac{T_1}{T_2} = \left(\frac{V_2}{V_1}\right)^{R/C_V}$$

(c) Since the volume is constant, $w = 0$, and from the first law $đq_C = \mathrm{d}U = C_V\,\mathrm{d}T$. Thus $\int đq_C/T = \int C_V\,\mathrm{d}T/T = C_V\ln(T_1/T_2)$.

As indicated in Fig. 3.1*a*, the sum of B and C should equal A for state functions; that is, $\int đq_A/T = \int đq_B/T + \int đq_C/T$ or $R\ln(V_2/V_1) = C_V\ln(T_1/T_2)$. This is in agreement with the expression for a reversible adiabatic expansion. This confirms that $\int dq_{rev}/T$ is a state function.

(*d*) At constant pressure, $\int dq_D/T = \int dH_D/T$, which is equal to $\int C_P\, dT/T = C_P \ln(T_3/T_1)$. Note that since the pressure is the same in the initial and final states, $T_3/T_1 = V_2/V_1$.

(*e*) At constant V, $dw_E = 0$, and so from the first law, $\int dq_E/T = \int dU_E/T = \int C_V\, dT/T = C_V \ln(T_1/T_3)$.

As indicated in Fig. 3.1*b*, the sum of D and E should equal A for state functions; that is, $\int dq_D/T + \int dq_E/T = \int dq_A/T$ or $C_P \ln(T_3/T_1) + C_V \ln(T_1/T_3) = R \ln(T_3/T_1)$ should be equal to $R \ln(V_2/V_1)$. This is true, as indicated in considering *d*.

3.2 THE SECOND LAW OF THERMODYNAMICS

We have seen that $đq_{rev}/T$ is the differential of a state function. Clausius named this state function the **entropy** S, and so for an infinitesimal change in state

$$dS = \frac{đq_{rev}}{T} \tag{3.8}$$

and for a finite change in state,

$$\Delta S = \int \frac{đq_{rev}}{T} \tag{3.9}$$

Summing around a closed cycle yields

$$\sum \frac{q_{rev}}{T} = 0 \qquad \text{(finite isothermal steps)} \tag{3.10}$$

$$\oint \frac{đq_{rev}}{T} = 0 \qquad \text{(infinitesimal steps)} \tag{3.11}$$

$$\sum \Delta S = 0 \qquad \text{(finite isothermal steps)} \tag{3.12}$$

$$\oint dS = 0 \qquad \text{(infinitesimal steps)} \tag{3.13}$$

We are indebted to Clausius for much more than a name, because in 1854 he extended these equations to cycles containing irreversible steps by showing that

$$\oint \frac{đq}{T} \leq 0 \tag{3.14}$$

This Clausius theorem is the mathematical statement of the second law of thermodynamics. The cyclic integral is to be understood in the following way: (a) If any part of the cyclic process is irreversible (spontaneous), the inequality applies and the cyclic integral is negative. (b) If the cyclic process is reversible, the equality applies. (c) It is impossible for the cyclic integral to be greater than zero. Note that the temperature that appears in the cyclic integral in equation 3.14 is that of the heat reservoir or surroundings. When the process is reversible, the temperature of the system is equal to the temperature of the surroundings. Equation 3.14 leads to the following inequality for a noncyclic process:

$$dS \geq \frac{đq}{T} \tag{3.15}$$

If the process is reversible, $dS = đq_{rev}/T$, and if the process is irreversible, $dS > đq_{irrev}/T$.

Thus there are two parts of the **second law of thermodynamics:**

1. There is a state function called the entropy S that can be calculated from $dS = đq_{rev}/T$.
2. The change in entropy in any process is given by $dS \geq dq/T$, where the $>$ applies to a spontaneous process (irreversible process) and the equality applies to a reversible process. This means that in order to calculate ΔS for a change in state, one *must* use a reversible process.

Example 3.2 ***Demonstration that dS is greater than the differential heat for an irreversible process divided by T***

Apply the Clausius theorem to the following isothermal irreversible cycle to obtain $dS > đq_{irrev}/T$.

$$\text{state 1} \xrightarrow[\text{irreversible}]{} \text{state 2} \xrightarrow[\text{reversible}]{} \text{state 1}$$

Since this cycle is irreversible, equation 3.14 yields

$$\int_1^2 \frac{đq_{irrev}}{T} + \int_2^1 \frac{đq_{rev}}{T} < 0$$

Since the second step is reversible, $đq_{rev}/T$ can be replaced by dS, and the limits can be interchanged, with a change in sign.

$$\int_1^2 \frac{đq_{irrev}}{T} - \int_1^2 dS < 0$$

Thus,

$$\Delta S = S_2 - S_1 = \int_1^2 dS > \int_1^2 \frac{đq_{irrev}}{T}$$

We can also write

$$dS > \frac{đq_{irrev}}{T}$$

Thus, for an infinitesimal irreversible process the change in entropy is greater than the differential of the heat divided by the temperature.

We can restate the second law by saying that **the entropy increases in a spontaneous process in an isolated system** because for an isolated system, $đq = 0$, and therefore $\Delta S > 0$ for a spontaneous process. The entropy of an isolated system (in which the internal energy is constant) can continue to increase as long as spontaneous processes occur. When there are no more possible spontaneous processes, the entropy is at a maximum; for any further infinitesimal process in the system, $dS = 0$. Thus the entropy change tells us whether a process or chemical reaction can occur spontaneously in an isolated system.

This reasoning can be applied to a system that is not isolated by treating the system plus its surroundings as an isolated system. When a spontaneous change occurs in the system of interest, the entropy change dS_{total} in the system plus the surroundings is given by

$$dS_{total} = dS_{syst} + dS_{surr} \tag{3.16}$$

First, consider a reversible process in which the system gains heat $\mathrm{d}q_{\mathrm{rev}}$ from the surroundings. Since the surroundings gain heat equal to $-q_{\mathrm{rev}}$, equation 3.16 becomes

$$0 = \mathrm{d}S_{\mathrm{syst}} - \frac{\mathrm{d}q_{\mathrm{rev}}}{T_{\mathrm{surr}}} \tag{3.17}$$

since $\mathrm{d}S_{\mathrm{total}} = 0$ for a reversible process in an isolated system. Thus, for a reversible process in the system plus surroundings,

$$\mathrm{d}S_{\mathrm{syst}} = \frac{\mathrm{d}q_{\mathrm{rev}}}{T_{\mathrm{surr}}} = \frac{\mathrm{d}q_{\mathrm{rev}}}{T} \tag{3.18}$$

since for a reversible process, the temperature of the system is equal to the temperature of the surroundings.

Second, consider an irreversible process in which the system gains heat $\mathrm{d}q_{\mathrm{irrev}}$ from its surroundings. In this case, it is convenient to write equation 3.16 as

$$\mathrm{d}S_{\mathrm{syst}} = \mathrm{d}S_{\mathrm{total}} - \mathrm{d}S_{\mathrm{surr}} \tag{3.19}$$

If the transfer of heat occurs reversibly in the surroundings, the entropy change in the surroundings is $-\mathrm{d}q_{\mathrm{irrev}}/T_{\mathrm{surr}}$ and equation 3.19 becomes

$$\mathrm{d}S_{\mathrm{syst}} > \frac{\mathrm{d}q_{\mathrm{irrev}}}{T_{\mathrm{surr}}} \tag{3.20}$$

since $\mathrm{d}S_{\mathrm{total}} > 0$ for a spontaneous change in the total (isolated) system. Equations 3.18 and 3.20 can be combined to obtain

$$\mathrm{d}S_{\mathrm{syst}} \geq \frac{\mathrm{d}q}{T_{\mathrm{surr}}} \tag{3.21}$$

where the inequality applies when the process is irreversible and the equality applies when the process is reversible. For a finite process,

$$\Delta S_{\mathrm{syst}} \geq \int \frac{\mathrm{d}q}{T_{\mathrm{surr}}} \tag{3.22}$$

As a spontaneous change occurs in an isolated system, the entropy increases with time as shown in Fig. 3.2 and eventually levels off. When a process occurs reversibly in an isolated system, the entropy does not change. When a spontaneous process occurs in an isolated system, the entropy increases. The direction of spontaneous change in any system is in the direction of increasing the entropy of the universe, and thus **the increase in entropy indicates the time sequence of a spontaneous process.** The entropy is sometimes referred to as the "arrow of time." The second law of thermodynamics contrasts with the equations of classical mechanics and quantum mechanics, which are reversible in time.

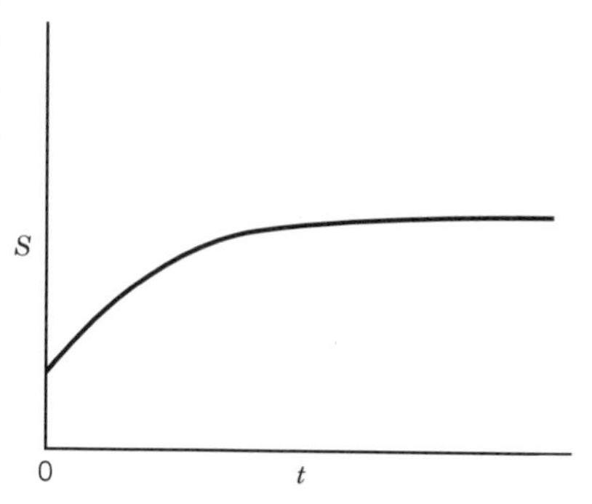

Figure 3.2 Change in entropy of an isolated system with time. The entropy of the system increases spontaneously until equilibrium is reached. Thermodynamics does not deal with the question of how long it will take to reach equilibrium.

Equation 3.21 describes three types of processes:

$$\begin{aligned} &\mathrm{d}S > đq/T &&\text{spontaneous and irreversible process}\\ &\mathrm{d}S = đq/T &&\text{reversible process}\\ &\mathrm{d}S < đq/T &&\text{impossible process} \end{aligned} \tag{3.23}$$

The simplest place to apply equation 3.21 is to an isolated system, because $đq = 0$. For a finite change the three possibilities are

$$\begin{aligned} &\Delta S > 0 &&\text{spontaneous and irreversible process}\\ &\Delta S = 0 &&\text{reversible process}\\ &\Delta S < 0 &&\text{impossible process} \end{aligned} \tag{3.24}$$

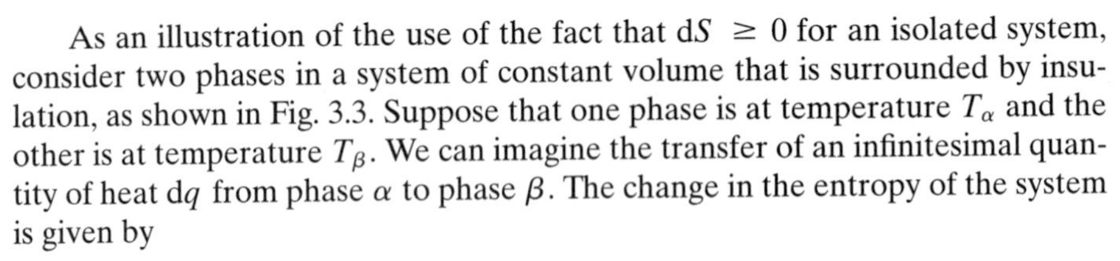

As an illustration of the use of the fact that $dS \geq 0$ for an isolated system, consider two phases in a system of constant volume that is surrounded by insulation, as shown in Fig. 3.3. Suppose that one phase is at temperature T_α and the other is at temperature T_β. We can imagine the transfer of an infinitesimal quantity of heat dq from phase α to phase β. The change in the entropy of the system is given by

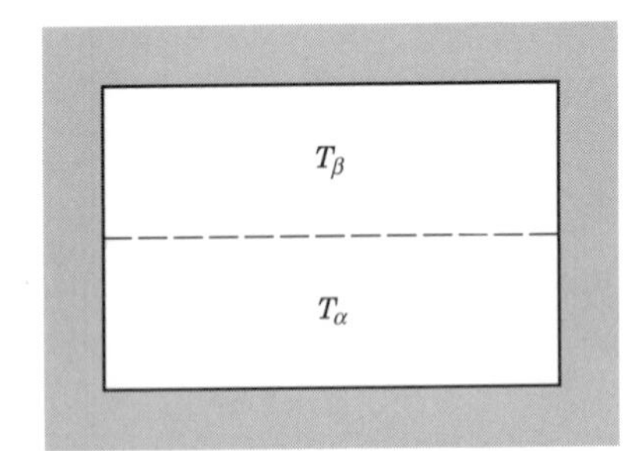

Figure 3.3 Two phases in an isolated system of constant volume. The container is surrounded by insulation.

$$dS = \frac{dq}{T_\beta} - \frac{dq}{T_\alpha} = dq\left(\frac{1}{T_\beta} - \frac{1}{T_\alpha}\right) \tag{3.25}$$

If this process occurs spontaneously, $dS > 0$, and so this equation shows that $T_\alpha > T_\beta$. In other words, heat flows spontaneously to the phase with the lower temperature. If the two phases are at thermal equilibrium, $dS = 0$ and therefore $T_\alpha = T_\beta$. Thus the second law has led us to the conclusion that for two phases to be in equilibrium, they must be at the same temperature.

Since S is a state function, we can integrate dS between two states of a system. For the change in state

$$\text{A(state 1)} = \text{A(state 2)} \tag{3.26}$$

$$\int_{S_1}^{S_2} dS = \int_1^2 \frac{đq_{rev}}{T} = S_2 - S_1 \tag{3.27}$$

Thus to determine the change in entropy in a process, we have to integrate along the path of a *reversible* process connecting states 1 and 2. The ΔS for an irreversible process can be calculated if we can devise a reversible path and use it for the integration in equation 3.27. There is no change in entropy in a reversible adiabatic process.

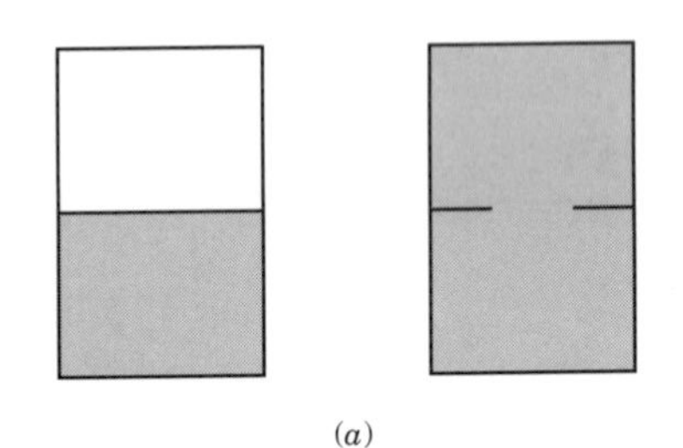

(*a*)

(*b*)

Figure 3.4 (*a*) Irreversible expansion of an ideal gas at 298 K into twice the initial volume with no heat or work. (*b*) Reversible isothermal expansion of an ideal gas to twice the initial volume at 298 K. In (*b*) the piston must be surrounded by a heat reservoir.

Example 3.3 ***The entropy of expansion of an ideal gas in an isolated system***

Is the expansion of an ideal gas into a larger volume thermodynamically spontaneous in an isolated system? More specifically, consider the expansion of an isolated ideal gas initially at 298 K into a volume that is twice as large as its initial volume as shown in Fig. 3.4*a*.

Remember that Joule found that there is no change in temperature when a dilute gas is allowed to expand in an isolated system. To determine the change in entropy between the initial state and the final state, the reversible isothermal expansion described in Fig. 3.4*b* can be used. As we have seen earlier, the work per mole done on the gas is $-RT \ln 2$, so that the heat absorbed by the gas is

$$q_{rev} = RT \ln 2$$

since $\Delta U = 0$. Thus, the change in entropy of the gas is

$$\Delta \overline{S}_b = R \ln 2 = 5.76 \text{ J K}^{-1} \text{ mol}^{-1}$$

Since the change in state of the gas for the process described in Fig. 3.4*a* is the same as that for Fig. 3.4*b*, $\Delta \overline{S}_a > 0$. Thus, we can conclude that the expansion is spontaneous, as we knew all along. The reverse of the process described in Fig. 3.4*a*, that is, the gas flowing back spontaneously into the initial volume, is impossible since $\Delta \overline{S} < 0$. It is important to note that this problem has nothing to do with minimizing the energy, which is constant.

3.3 ENTROPY CHANGES IN REVERSIBLE PROCESSES

We will now consider some simple processes for which entropy changes are readily calculated. It is especially easy to calculate entropy changes for reversible isothermal processes.

The transfer of heat from one body to another at an infinitesimally lower temperature is a reversible change, since the direction of heat flow can be reversed by an infinitesimal change in the temperature of one of the bodies. For example, consider the vaporization of a pure liquid into its vapor at the equilibrium vapor pressure P:

$$\text{liquid}(T, P) \rightarrow \text{vapor}(T, P) \tag{3.28}$$

Since T is constant, the integration of equation 3.18 yields

$$S_2 - S_1 = \Delta S = \frac{q_{\text{rev}}}{T} \tag{3.29}$$

where q_{rev} represents the heat absorbed in the reversible change. Since the pressure is constant, the reversible heat is equal to the change in enthalpy ΔH, so that

$$\Delta S = \frac{\Delta H}{T} \tag{3.30}$$

This equation may also be used to calculate the entropy of sublimation, the entropy of melting, and the entropy change for a transition between two forms of a solid.

Example 3.4 ***The molar entropy of vaporization of n-hexane***

What is the change in the molar entropy of n-hexane when it is vaporized at its boiling point?

n-Hexane boils at 68.7 °C at 1.013 25 bar, and the molar enthalpy of vaporization is 28 850 J mol^{-1} at this temperature. If n-hexane is vaporized into the saturated vapor at this temperature, the process is reversible and the molar entropy change is given by

$$\Delta\overline{S} = \frac{\Delta\overline{H}}{T} = \frac{28\,850 \text{ J mol}^{-1}}{341.8 \text{ K}} = 84.41 \text{ J K}^{-1} \text{ mol}^{-1}$$

The overbars indicate that we are dealing with molar changes.

The molar entropy of a vapor is always greater than that of the liquid with which it is in equilibrium, and the molar entropy of the liquid is always greater than that of the solid at the melting point. According to the statistical interpretation of entropy to be discussed in Section 3.6, in which the entropy is a measure of the disorder of the system, the molecules of the gas are more disordered than those of the liquid, and the molecules of the liquid are more disordered than those of the solid.

Now let us apply the second law to the vaporization of a liquid into its saturated vapor. To apply equations 3.24, we must consider an isolated system. In this case the liquid and vapor (the system) and the heat reservoir at T (the surroundings) form an isolated overall system. The total entropy change is given by

$$\Delta S = \Delta S_{\text{sys}} + \Delta S_{\text{surr}} \tag{3.31}$$

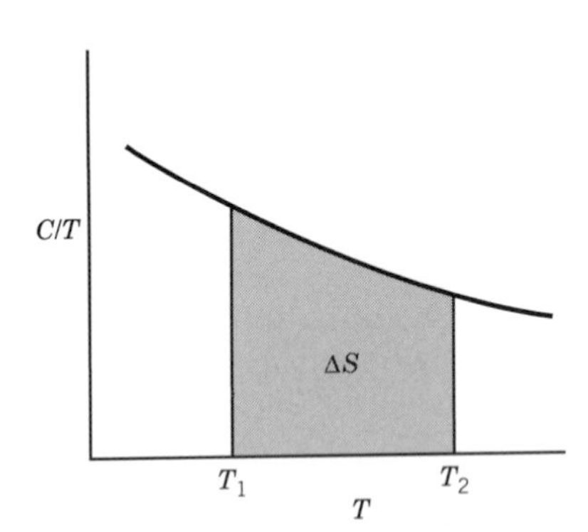

Figure 3.5 When a substance is heated from T_1 to T_2, without a phase change, the increase in entropy is given by the indicated area. If the volume is constant, $n\overline{C}_V$ is used for C; and if the pressure is constant, $n\overline{C}_P$ is used.

Since the heat gained by the system is equal to that lost by the surroundings, the entropy change for the surroundings is the negative of the entropy change for the system if the vaporization is carried out reversibly; for both the system and surroundings taken together, the total change of entropy ΔS is zero if the transfer of heat is carried out reversibly. This is in agreement with the second form in equations 3.24.

Heating and cooling a substance are other examples of processes that can be carried out reversibly. The change in entropy when a substance is heated or cooled can be calculated using

$$dS = \frac{đq_{\text{rev}}}{T} = \frac{C\,dT}{T} \tag{3.32}$$

where C is C_P for a process at constant pressure and C_V for a process at constant volume. If the heat capacity is independent of temperature, and the temperature is changed from T_1 to T_2, then

$$\Delta S = \int_{T_1}^{T_2} \frac{C}{T} dT = C \ln \frac{T_2}{T_1} \tag{3.33}$$

If the heat capacity is a function of temperature, this function can be substituted into the integral form of equation 3.33, or the entropy change can be obtained from a numerical integration, as shown in Fig. 3.5.

Example 3.5 ***The change in molar entropy on heating an ideal gas***

Oxygen is heated from 300 to 500 K at a constant pressure of 1 bar. What is the increase in molar entropy?

The coefficients in an empirical equation for the heat capacity at constant pressure as a function of temperature are given in Table 2.2. Using equation 3.33, we find

$$\Delta S = \int_{T_1}^{T_2} \frac{\overline{C}_P}{T} dT = \int_{T_1}^{T_2} \left(\frac{\alpha}{T} + \beta + \gamma T \right) dT$$

$$= \alpha \ln \frac{T_2}{T_1} + \beta(T_2 - T_1) + \frac{\gamma}{2}(T_2^2 - T_1^2)$$

$$= 25.503 \ln \frac{500}{300} + (13.612 \times 10^{-3})(200) - \frac{1}{2}(42.553 \times 10^{-7})(500^2 - 300^2)$$

$$= 15.41 \text{ J K}^{-1} \text{ mol}^{-1}$$

The units have been omitted in this calculation, but you should check that they cancel properly.

The entropy change is readily calculated for a reversible isothermal expansion of an ideal gas. Since the internal energy of an ideal gas is independent of volume at constant temperature (Section 2.6), $đq = -đw = P\,dV$. For a reversible isothermal expansion from V_1 to V_2,

$$\text{ideal gas}(T, V_1, P_1) \rightarrow \text{ideal gas}(T, V_2, P_2) \tag{3.34}$$

$$\begin{aligned}
\Delta S &= \int_{V_1}^{V_2} \frac{P}{T} \mathrm{d}V = nR \int_{V_1}^{V_2} \frac{1}{V} \mathrm{d}V \\
&= nR \ln \frac{V_2}{V_1} \\
&= -nR \ln \frac{P_2}{P_1}
\end{aligned} \tag{3.35}$$

where the final form simply comes from the fact that the volume is inversely proportional to pressure at constant temperature.

Equation 3.35 can be applied to the isothermal expansion of a mole of an ideal gas from its standard pressure P° to some other pressure P:

$$\overline{S} = \overline{S}^\circ - R \ln\left(\frac{P}{P^\circ}\right) \tag{3.36}$$

where $\overline{S}^\circ$ is the molar entropy in the standard state.

Example 3.6 ***The change in entropy for a system and its surroundings in a reversible process***

Half a mole of an ideal gas expands isothermally and **reversibly** at 298.15 K from a volume of 10 L to a volume of 20 L. (*a*) What is the change in the entropy of the gas? (*b*) How much work is done on the gas? (*c*) What is q_{surr}? (*d*) What is the change in the entropy of the surroundings? (*e*) What is the change in the entropy of the system plus the surroundings?

(*a*) $\Delta S = nR \ln(V_2/V_1) = (0.5 \text{ mol})(8.3145 \text{ J K}^{-1} \text{ mol}^{-1}) \ln 2 = 2.88 \text{ J K}^{-1}$

(*b*) $w_{rev} = -nRT \ln(V_2/V_1)$
$w_{rev} = -(0.5 \text{ mol})(8.3145 \text{ J K}^{-1} \text{ mol}^{-1})(298.15 \text{ K}) \ln 2 = -859 \text{ J}$

(*c*) Since the gas is ideal there is no change in its internal energy; $\Delta U = q + w = 0$. Thus $q_{sys} = 859$ J, and $q_{surr} = -859$ J.

(*d*) Since heat flows out of the surroundings, it has a decrease in entropy.

$$\Delta S_{surr} = -859 \text{ J}/298.15 \text{ K} = -2.88 \text{ J K}^{-1}$$

(*e*) Since the entropy of the gas increases and the entropy of the surroundings decreases by the same amount, there is no change in entropy for the system and its surroundings. The gas and the surroundings can be considered to be an isolated system. The process is reversible, and so we expect $\Delta S = 0$.

Example 3.7 ***The change in entropy for a system and its surroundings in an irreversible process***

Now consider that the expansion in the preceding example occurs **irreversibly** by simply opening a stopcock and allowing the gas to rush into an evacuated bulb of 10-L volume. (*a*) What is the change in the entropy of the gas? (*b*) How much work is done on the gas? (*c*) What is q_{surr}? (*d*) What is the change in the entropy of the surroundings? (*e*) What is the change in the entropy of the system plus the surroundings?

(*a*) The change in entropy is the same as above because entropy is a state function.
(*b*) No work is done in the expansion.
(*c*) No heat is exchanged with the surroundings.
(*d*) The entropy of the surroundings does not change.
(*e*) The entropy of the system plus surroundings increases by 2.88 J K^{-1}. Since this is an irreversible process we expect the entropy to increase.

CALCULATION OF ΔS FOR VARIOUS CHANGES IN STATE

General

$$\Delta S = \int \frac{dq_{rev}}{T}$$

One must find a **reversible** path to go from the initial state to the final state in order to calculate the change in entropy; however, since entropy is a state function, the change in entropy for the **irreversible** path between the same initial and final states is the same.

Specific Cases

(a) Constant V heating of one mole of substance

$$\text{substance}(T_1, V) = \text{substance}(T_2, V)$$

Then, $\Delta S = \int_{T_1}^{T_2} C_V \, dT/T$, where C_V is the constant-volume heat capacity of the substance. If C_V is not a function of T, then $\Delta S = C_V \ln(T_2/T_1)$.

(b) Constant P heating of one mole of substance

$$\text{substance}(T_1, P) = \text{substance}(T_2, P)$$

Then, $\Delta S = \int_{T_1}^{T_2} C_P dT/T$, where C_P is the constant-pressure heat capacity of the substance.

(c) Phase change at constant T and P

$$H_2O(l, 373\,K, 1\,atm) = H_2O(g, 373\,K, 1\,atm)$$

Then $\Delta S = \Delta H/T$, where ΔH is the heat of vaporization in this case.

(d) Ideal gas changes in state at constant T

$$\text{ideal gas}(P, V, T) = \text{ideal gas}(P', V', T)$$

Then $\Delta S = R \ln(V'/V) = R \ln(P/P')$. Note the position of the P and P', and the position of the V and V'. Remember that ΔS is positive if the volume increases at constant T.

(e) Mixing of ideal systems at constant T and P

$$n_A\,A(T, P) + n_B\,B(T, P) = n\,\text{mixture}(T, P)$$

Here, $n = n_A + n_B$ is the total number of moles in the mixture. Then,

$$\Delta S = -nR(y_A \ln y_A + y_B \ln y_B)$$

where the y terms are the mole fractions: $y_A = n_A/(n_A + n_B)$ and $y_B = n_B/(n_A + n_B)$. Note that since the mole fractions are less than 1, the change in entropy is positive, as expected for a spontaneous process.

Example 3.8 ***The molar entropy of an ideal gas as a function of T and P***

Calculate the change in entropy of an ideal monatomic gas B in changing from P_1, T_1 to P_2, T_2. What does this indicate about the form of the expression for the molar entropy of the gas?

A reversible path is

$$\mathrm{B}(T_1, P_1) \rightarrow \mathrm{B}(T_2, P_1) \rightarrow \mathrm{B}(T_2, P_2)$$

For the first step,

$$\Delta S = C_P \int_{T_1}^{T_2} \frac{\mathrm{d}T}{T} = C_P \ln \frac{T_2}{T_1} = n\overline{C}_P \ln \frac{T_2}{T_1}$$

For the second step,

$$\Delta S = -nR \ln \frac{P_2}{P_1}$$

For the sum of the two steps,

$$\Delta S = nR \left\{ \ln \left[\left(\frac{T_2}{T_1} \right)^{5/2} \right] - \ln \left(\frac{P_2}{P_1} \right) \right\}$$

since $\overline{C}_P = 5/2\,R$.

This indicates that the expression for the molar entropy of an ideal monatomic gas is of the form

$$\overline{S} = R \left\{ \ln \left[\left(\frac{T}{T^\circ} \right)^{5/2} \right] - \ln \left(\frac{P}{P^\circ} \right) \right\} + \text{const}$$

where T° is the reference temperature, P° is the reference pressure, and the constant is a characteristic of the particular gas. In Chapter 16 on statistical mechanics we will see that for an ideal monatomic gas the constant term is equal to $-1.151\,693R + \frac{3}{2}R \ln A_\mathrm{r}$ when $P^\circ = 1$ bar, where A_r is the relative atomic mass of the gas. This equation, which will be derived from statistical mechanics in Section 16.3, is referred to as the **Sackur–Tetrode equation.**

Example 3.9 ***The change in entropy of a monatomic gas when both T and P are changed***

What is the change in molar entropy of helium in the following process?

$$1\ \mathrm{He}(298\ \mathrm{K}, 1\ \mathrm{bar}) \rightarrow 1\ \mathrm{He}(100\ \mathrm{K}, 10\ \mathrm{bar})$$

For an ideal monatomic gas, $\overline{C}_P = \frac{5}{2}R$. Using an equation from the preceding example,

$$\Delta\overline{S} = \tfrac{5}{2}R \ln \frac{100}{298} - R \ln \frac{10}{1} = -41.84\ \mathrm{J\ K^{-1}\ mol^{-1}}$$

3.4 ENTROPY CHANGES IN IRREVERSIBLE PROCESSES

To obtain the change in entropy in an irreversible process we have to calculate ΔS along a reversible path between the initial state and the final state. We have already illustrated this in Section 3.2 by the expansion of an isolated ideal gas into a vacuum.

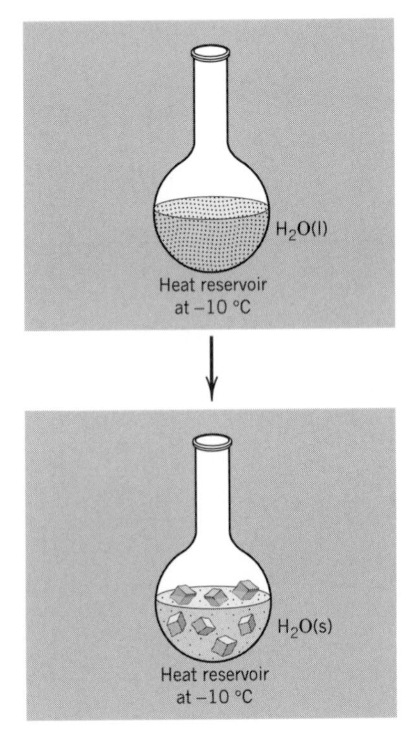

Figure 3.6 When a mole of liquid water freezes at $-10\,^{\circ}\mathrm{C}$, its entropy changes by $-20.54\ \mathrm{J\ K^{-1}\ mol^{-1}}$, corresponding to the increase in structural order. The entropy of the reservoir increases by $21.37\ \mathrm{J\ K^{-1}\ mol^{-1}}$ because of the heat transferred to it. The system as a whole increases in entropy, as expected for an irreversible process in an isolated system.

As a second illustration we will consider the spontaneous (irreversible) freezing of water below its freezing point (see Fig. 3.6). The freezing of a mole of supercooled water at $-10\,^{\circ}\mathrm{C}$ is an irreversible change, but it can be carried out reversibly by means of the three steps for which the entropy changes are indicated:

$$
\begin{array}{ccc}
H_2O(\mathrm{l},\ 0\,^{\circ}\mathrm{C}) & \xrightarrow{\Delta\overline{S}=\Delta\overline{H}/T} & H_2O(\mathrm{s},\ 0\,^{\circ}\mathrm{C}) \\
\Delta\overline{S} = \int_{263}^{273} \frac{\overline{C}_{\mathrm{liq}}}{T}\,\mathrm{d}T \ \Big\uparrow & & \Big\downarrow \ \Delta\overline{S} = \int_{273}^{263} \frac{\overline{C}_{\mathrm{ice}}}{T}\,\mathrm{d}T \\
H_2O(\mathrm{l},\ -10\,^{\circ}\mathrm{C}) & \longrightarrow & H_2O(\mathrm{s},\ -10\,^{\circ}\mathrm{C})
\end{array}
\tag{3.37}
$$

For the crystallization of liquid water at $0\,^{\circ}\mathrm{C}$, $\Delta\overline{H} = -6004\ \mathrm{J\ mol^{-1}}$. The heat capacity of water may be taken to be $75.3\ \mathrm{J\ K^{-1}\ mol^{-1}}$, and that of ice may be taken to be $36.8\ \mathrm{J\ K^{-1}\ mol^{-1}}$ over this range. Then the total entropy change of the water when 1 mol of liquid water at $-10\,^{\circ}\mathrm{C}$ changes to ice at $-10\,^{\circ}\mathrm{C}$ is simply the sum of the three entropy changes:

$$
\begin{aligned}
\Delta\overline{S} &= (75.3\ \mathrm{J\ K^{-1}\ mol^{-1}})\ln\frac{273}{263} + \frac{(-6004\ \mathrm{J\ mol^{-1}})}{273\ \mathrm{K}} \\
&\quad + (36.8\ \mathrm{J\ K^{-1}\ mol^{-1}})\ln\frac{263}{273} \\
&= -20.54\ \mathrm{J\ K^{-1}\ mol^{-1}}
\end{aligned}
\tag{3.38}
$$

The decrease in entropy corresponds to the increase in structural order when water freezes.

The statement that the entropy of an isolated system increases in a spontaneous process may be illustrated by considering supercooled water at $-10\,^{\circ}\mathrm{C}$ in contact with a large heat reservoir at this temperature. The entropy change for the isolated system upon freezing includes the entropy change of the reservoir as well as the entropy change of the water. Since the heat reservoir is large, the heat evolved by the water on freezing is absorbed by the reservoir with only an infinitesimal change in temperature. The transfer of heat to a reservoir at the same temperature is a reversible process. Since the heat of fusion of water at $-10\,^{\circ}\mathrm{C}$ is $\Delta H(263\ \mathrm{K}) = (75.3\ \mathrm{J\ K^{-1}\ mol^{-1}})(10\ \mathrm{K}) - 6004\ \mathrm{J\ mol^{-1}} - (36.8\ \mathrm{J\ K^{-1}\ mol^{-1}})(10\ \mathrm{K}) = -5619\ \mathrm{J\ mol^{-1}}$, the entropy change of the reservoir in the transfer of heat is

$$
\Delta\overline{S} = \frac{(5619\ \mathrm{J\ mol^{-1}})}{263\ \mathrm{K}} = 21.37\ \mathrm{J\ K^{-1}\ mol^{-1}}
\tag{3.39}
$$

The entropy change of the water is $-20.54\ \mathrm{J\ K^{-1}\ mol^{-1}}$, and the total entropy change of the system water plus reservoir is

$$
\Delta\overline{S} = (21.37 - 20.54)\ \mathrm{J\ K^{-1}\ mol^{-1}} = 0.83\ \mathrm{J\ K^{-1}\ mol^{-1}}
\tag{3.40}
$$

Thus, the total entropy of the isolated system, including water and reservoir, increases, as required by inequality 3.24.

3.5 ENTROPY OF MIXING IDEAL GASES

Figure 3.7 shows that amount n_1 of gas 1 at P and T is separated from amount n_2 of gas 2 at the same pressure and temperature. When the partition is withdrawn the gases will diffuse into each other at constant temperature and pressure if they are ideal gases. To calculate the change in entropy in this irreversible process we need to find a way to carry it out reversibly. This can be done in two steps. First, we expand each gas isothermally and reversibly to the final volume $V = V_1 + V_2$. These volumes are not molar volumes, but are actual volumes. When extensive volumes are used, the ideal gas law is written $PV = nRT$.

Using equation 3.35, the entropy changes for the two gases are

$$\Delta S_1 = -n_1 R \ln \frac{V_1}{V} = -n_1 R \ln \frac{n_1}{n_1 + n_2} = -n_1 R \ln y_1 \tag{3.41}$$

$$\Delta S_2 = -n_2 R \ln \frac{V_2}{V} = -n_2 R \ln \frac{n_2}{n_1 + n_2} = -n_2 R \ln y_2 \tag{3.42}$$

where y_i is the mole fraction. The **entropy of mixing** $\Delta_{\text{mix}} S$ is the sum of the entropy changes for the two gases:

$$\Delta_{\text{mix}} S = -n_1 R \ln y_1 - n_2 R \ln y_2 \tag{3.43}$$

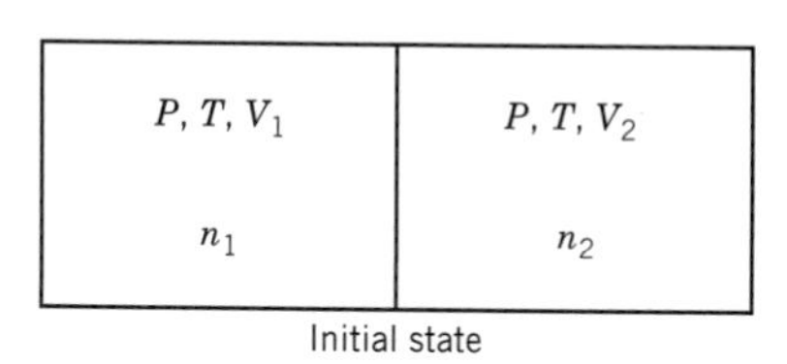

Figure 3.7 Mixing of ideal gases. The partition between n_1 mol of gas 1 at P and T and n_2 mol of gas 2 at P and T is withdrawn so that the gases can mix.

In the second step the expanded gases are mixed reversibly at constant volume. To see how this might be done we have to imagine two semipermeable membranes, arranged as shown in Fig. 3.8, one of which (represented by dashes) is permeable only to gas 1, and the other (represented by dots) is permeable only to gas 2. The membrane that is permeable to gas 1 and an impermeable membrane, which is separated from it by volume V, are moved at the same infinitesimally slow rate to the left. As shown by the diagram for the intermediate stage in the reversible mixing process, the ⋮ | membrane combination has to be moved to the left against a pressure of P_1 (to the left of ⋮ |) plus P_2 (to the left of |). But the pressure on the right-hand side of ⋮ is also $P_1 + P_2$. Therefore, no work is required to move the membrane in this frictionless device. The internal energies of the two ideal gases are functions only of T (Section 2.6), and so according to the first law no heat is absorbed by the gas in this step. Consequently, there is no entropy change associated with the second step, and so equation 3.43 gives the total entropy change for the isothermal mixing of the two ideal gases.

Equation 3.43 can be generalized to

$$\Delta_{\text{mix}} S = -R \sum n_i \ln y_i = -n_t R \sum y_i \ln y_i \tag{3.44}$$

Since $y_i < 1$, $\ln y_i < 0$, and $\Delta_{\text{mix}} S$ is always positive. A plot of the entropy of mixing two ideal gases is shown in Fig. 3.9.

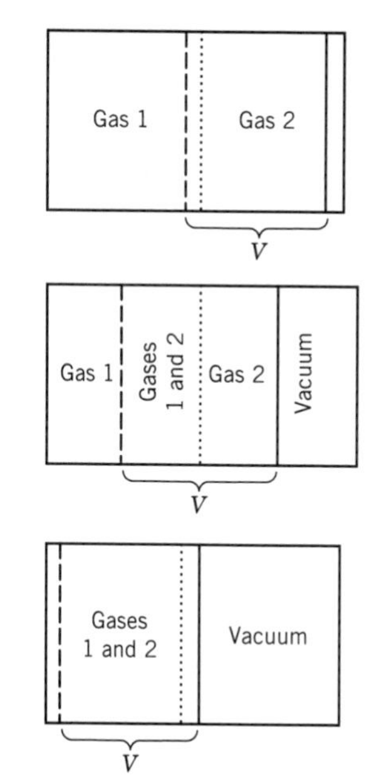

Figure 3.8 Reversible isothermal mixing of ideal gases 1 and 2 using a membrane permeable only to gas 1 (shown by dashes), a membrane permeable to gas 2 (shown by dots), and an impermeable membrane (shown by the solid line).

Example 3.10 ***Entropy of mixing ideal gases***

What is the entropy of mixing of 1 mol of oxygen with 1 mol of nitrogen at 25 °C, assuming that they are ideal gases?

Equation 3.44 becomes

$$\begin{aligned}\Delta_{\text{mix}} S &= -(2\ \text{mol})(8.3145\ \text{J K}^{-1}\ \text{mol}^{-1})\left(\tfrac{1}{2}\ln\tfrac{1}{2} + \tfrac{1}{2}\ln\tfrac{1}{2}\right)\\ &= 11.526\ \text{J K}^{-1}\end{aligned}$$

You might try other proportions and see what happens.

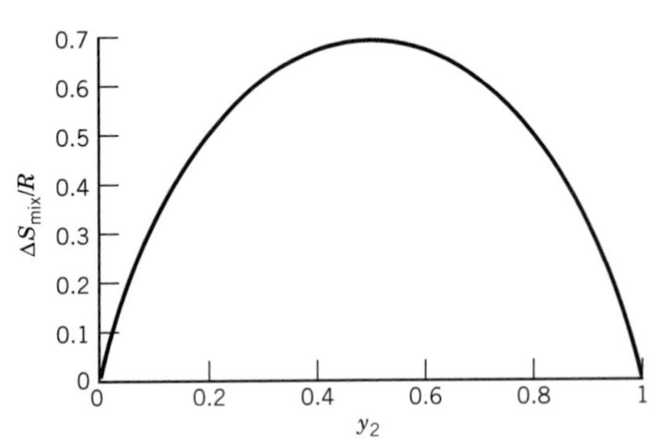

Figure 3.9 Plot of the entropy of mixing of two ideal gases to produce one mole of an ideal mixture. (See Computer Problem 3.C.)

Example 3.11 ***The Gibbs paradox***

For the mixing of two gases that have the same temperature and pressure, equation 3.43 can be written in the form

$$\Delta_{\text{mix}} S = n_1 R \ln \frac{V_{\text{f}}}{V_1} + n_2 R \ln \frac{V_{\text{f}}}{V_2}$$

where V_1 and V_2 are the volumes of the two gases and $V_{\text{f}} = V_1 + V_2$. Suppose that two equal volumes of the same gas are mixed. What is the change in entropy?

If we apply this equation to the mixing of two equal volumes of the same gas, we obtain

$$\Delta_{\text{mix}} S = (n_1 + n_2) R \ln 2$$

This answer is wrong because there is no change of state in this process, and so $\Delta S = 0$. This is known as the **Gibbs paradox.** The answer to this paradox was not properly understood until the development of quantum mechanics. According to quantum mechanics, the molecules of a single species are indistinguishable; therefore, equation 3.44 does not apply. If two different species have properties that are very nearly the same, they are still different from a quantum mechanical point of view, and equation 3.44 does apply.

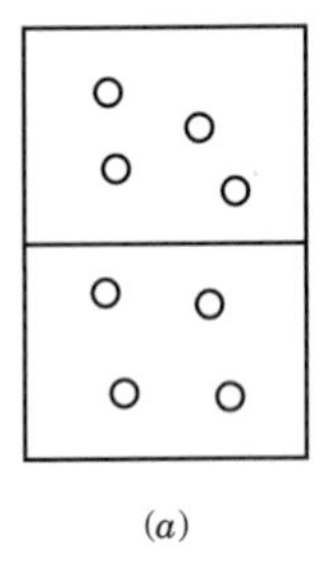

(*a*)

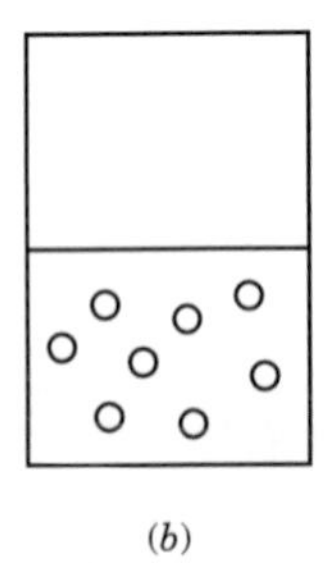

(*b*)

Figure 3.10 (*a*) Equilibrium gaseous system after a hole is punched in the diaphragm. (*b*) Highly improbable state of the gaseous system after the hole has been punched.

3.6 ENTROPY AND STATISTICAL PROBABILITY

Using the macroscopic approach of thermodynamics, we have found that the equilibrium state of an isolated system is that state in which the entropy has its maximum value. From a microscopic point of view, we might expect that the equilibrium state of an isolated system would be the state with the maximum statistical probability. For simple systems we can use the molecular point of view to calculate the statistical probabilities of different final states. For example, assume that one mole of an ideal gas is in a container that is connected to a container of equal volume through a stopcock; this expansion has already been discussed in Example 3.3. Actually, it is a little easier to think about the reverse process, and so we will do that. As shown in Fig. 3.10, we will start with an opening between the two chambers and ask, what is the statistical probability that all of the molecules will be in the original chamber? The probability that a particular molecule will be in the original chamber is 1/2. The probability that two particular molecules will be in the original chamber is $(1/2)^2$, and the probability that all the molecules will be in the original chamber is $(1/2)^N$, where N is the number of molecules in the system. If the system contains a mole of gas molecules, the statistical probability that all the molecules will be in the original chamber is

$$(1/2)^{6.022\times10^{23}} = e^{-4.174\times10^{23}} \tag{3.45}$$

Boltzmann postulated that

$$S = k \ln \Omega \tag{3.46}$$

where k is **Boltzmann's constant** (R/N_A) and Ω is the number of equally probable microscopic arrangements for the system. This relation can be used to calculate ΔS for the transformation from an initial state with entropy S and Ω equally probable arrangements to a final state with entropy S' and Ω' equally probable arrangements:

$$\Delta S = S' - S = k \ln(\Omega'/\Omega) \tag{3.47}$$

It is often difficult to count the number of equally probable arrangements in the final state and in the initial state, but the ratio Ω'/Ω in this case is equal to the ratio of the probability that all the molecules are in one chamber to the probability that they are all in one chamber or the other. The probability that all the molecules are in the original chamber or the other chamber is, of course, unity. Thus, for the system in the preceding paragraph, we have already calculated the ratio of the probabilities that is equal to Ω'/Ω. The change in entropy in going from the state with the gas distributed between the two chambers to the state with all the molecules in the original chamber is

$$\begin{aligned}\Delta S &= k \ln e^{-4.174\times10^{23}} \\ &= (1.381 \times 10^{-23}\ \mathrm{J\ K^{-1}})(-4.174 \times 10^{23}) \\ &= -5.76\ \mathrm{J\ K^{-1}}\end{aligned} \tag{3.48}$$

Since we are considering one mole of an ideal gas, the change in entropy for the expansion process in Example 3.3 is $\Delta \overline{S} = 5.76\ \mathrm{J\ K^{-1}\ mol^{-1}}$, in agreement with the result using Boltzmann's hypothesis. This confirms that the Boltzmann constant is indeed given by $k = R/N_A$.

If the gas molecules were all to be found in one chamber after having been distributed between the two chambers, we would say that the second law had been violated. We have just seen that the probability that such a thing might happen is not zero. It is, however, so small that we could never expect to be able to observe all the molecules in one chamber, even for systems containing much, much less than one mole of gas. If, however, we considered a system of only two molecules, then we could find both molecules in one chamber with reasonable probability. This shows that the laws of thermodynamics are based on the fact that macroscopic systems contain very large numbers of molecules.

The equation $S = k \ln \Omega$ embodies an important concept, but it is not used very often because it is difficult to calculate Ω. In Chapter 16 on statistical mechanics we will use other equations to calculate the entropy.

The collision of two gas molecules is reversible in the sense that the reverse process can also happen. If, after the molecules are moving away from each other, we could simply reverse the direction of the velocity vectors, the molecules would move along the same trajectories in the reverse direction. In short, the movie of the reverse process is just as reasonable as the movie of the forward process. This is true for both classical mechanics and quantum mechanics. If molecular collisions are reversible, then why is the expansion of a gas into a vacuum, or the mixing of two gases, irreversible? If we could take a movie of the expansion of a gas into a vacuum that would show the locations of all the molecules, we could tell whether the movie was being run forward or backward. However, if we were to look at each molecular collision we would find that each followed the laws of mechanics, and was reversible. If we were to look at the movie being run backward, we would feel that it was depicting something that could not happen. But why couldn't it happen? As a matter of fact, it could, but only if we could give all the molecules the positions they have at the end of the movie but then reverse their velocity vectors. If we then looked at the individual collisions, we would find that they would take all the molecules to the region from which they had expanded. The reason this does not happen in real life is that it takes an extraordinarily special set of molecular coordinates and velocities. This set of coordinates and velocity vectors is so unlikely that thermodynamics says that the reverse process can never happen.

Since $\mathrm{d}S \geq \mathrm{d}q/T$, the entropy is a measure of the flow of heat between a system and its environment. When heat is absorbed by the system from its surroundings, q is positive and the entropy of the system increases. The energy flowing into the sysem is "dispersed" in the sense that it goes into increasing the energy of various molecular motions in the system. This concept of the dispersal of energy also applies to the expansion of an ideal gas into a vacuum. In this case q is zero, but the total energy of the gas is dispersed over a larger volume. Thus entropy is a measure of the dispersal of energy among the possible microstates of molecules in a system.

Sometimes entropy is referred to as "disorder," and a messy desk is referred to as a state of high entropy. Or shuffling a deck of playing cards is said to result in an increase in entropy of the cards. But this is misleading from a scientific viewpoint because moving macroscopic objects around does not involve an increase in entropy.* Another source of confusion about entropy comes from the use of this term in information theory, which was introduced by Shannon in 1948. The quantity entropy in information theory is not the entropy of thermodynamics because it does not deal with the transfer of heat and the dispersal of energy among the microstates of a system.

The concept of temperature is necessarily involved in understanding thermodynamic entropy because it indicates the thermal environment of the particles in a system. These particles are involved in the ever-present thermal motion that makes spontaneous change possible because it is the mechanism by which molecules can occupy new microstates when the external conditions are altered.

Comment:

The entropy of mixing of ideal gases brings up a very important idea, namely, that some processes happen spontaneously even though they do not reduce the energy of a system. Our experience with mechanics leads us to expect that if something happens spontaneously, there is necessarily a decrease in energy. Now we know that is not true, and we can expect to find chemical reactions that occur because of the contribution of a positive ΔS.

*F. L. Lambert, *J. Chem. Educ.* **79:**187–192 (2002).

3.7 CALORIMETRIC DETERMINATION OF ENTROPIES

The entropy of a substance at any desired temperature relative to its entropy at absolute zero may be obtained by integrating $đq_{rev}/T$ from absolute zero to the desired temperature. This requires heat capacity measurements down to the neighborhood of 0 K as well as enthalpy of transition measurements for all transitions in this temperature range. Since measurements of C_P cannot be carried to 0 K, the Debye function (equation 16.104) is used to represent C_P below the temperature of the lowest measurements.

If data on the enthalpy of fusion at the melting point T_m and the enthalpy of vaporization at the boiling point T_b are available, the entropy at a temperature T above the boiling point T_b relative to that at 0 K may be calculated from

$$\overline{S}_T^\circ - \overline{S}_0^\circ = \int_0^{T_m} \frac{\overline{C}_P^\circ(\mathrm{s})}{T}\mathrm{d}T + \frac{\Delta_{fus}H^\circ}{T_m} + \int_{T_m}^{T_b} \frac{\overline{C}_P^\circ(\mathrm{l})}{T}\mathrm{d}T + \frac{\Delta_{vap}H^\circ}{T_b} + \int_{T_b}^{T} \frac{\overline{C}_P^\circ(\mathrm{g})}{T}\mathrm{d}T \tag{3.49}$$

If there are various solid forms with enthalpies of transition between the forms, the corresponding entropies of transition would have to be included in this sum.

Heat capacity measurements down to these very low temperatures are made with special calorimeters in which the substance is heated electrically in a carefully insulated system and the input of electrical energy and the temperature are measured accurately.

The attainment of very low temperatures in the laboratory involves successive application of different methods. Vaporization of liquid helium (b.p. 4.2 K at 1 bar) at low pressures produces temperatures down to about 0.3 K. Lower temperatures may be reached by use of adiabatic demagnetization. A paramagnetic (Section 22.6) salt such as gadolinium sulfate is cooled with liquid helium in the presence of a strong magnetic field. The salt is thermally isolated from its surroundings, and the magnetic field is removed. The salt undergoes a reversible adiabatic process in which the atomic spins become disordered. Since the energy must come from the crystal lattice, the salt is cooled. Temperatures of about 0.001 K may be reached in this way. Adiabatic demagnetization of nuclear spins can then be used to obtain temperatures of the order of 10^{-6} K.

As an illustration of the determination of the entropy of a substance relative to its entropy at 0 K, the measured heat capacities for SO_2 are shown as a function of T and of $\log T$ in Fig. 3.11. Solid SO_2 melts at 197.64 K, and the heat of fusion is 7402 J mol^{-1}. Liquid SO_2 vaporizes at 263.08 K at 1.013 25 bar, and the heat of vaporization is 24 937 J mol^{-1}. The calculation is summarized in Table 3.1.

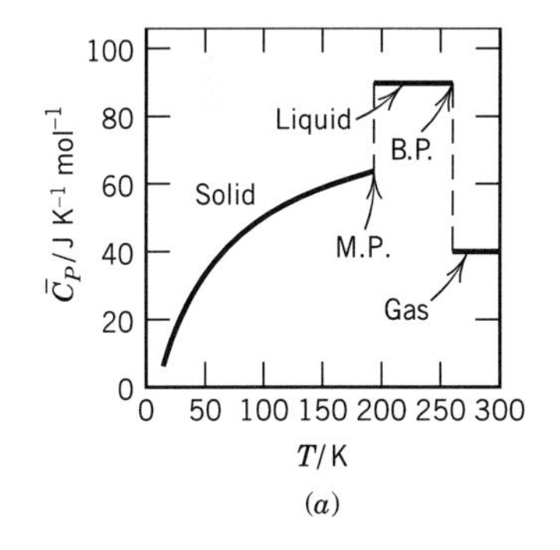

(a)

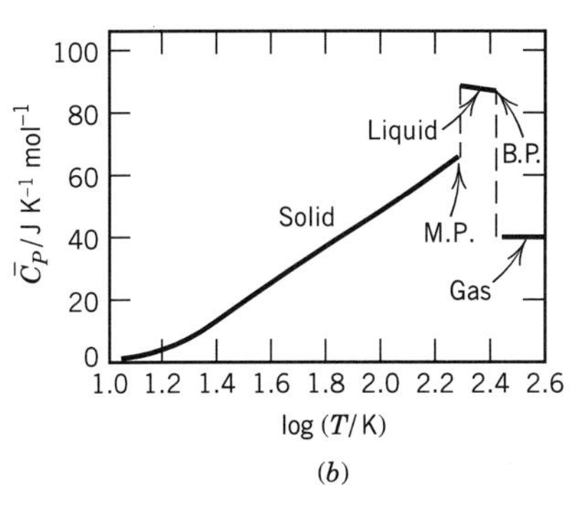

(b)

Figure 3.11 Heat capacity of sulfur dioxide at a constant pressure of 1 bar at different temperatures. [Graph *a* redrawn from W. F. Giauque and C. C. Stephenson, *J. Am. Chem. Soc.* **60:**1389 (1938).]

Table 3.1 Entropy of Sulfur Dioxide

T/K	*Method of Calculation*	$\Delta\overline{S}^\circ$/J K^{-1} mol^{-1}
0–15	Debye function ($\overline{C}_P$ = constant T^3)	1.26
15–197.64	Graphical, solid	84.18
197.64	Fusion, 7402/197.64	37.45
197.64–263.08	Graphical, liquid	24.94
263.08	Vaporization, 24 937/263.08	94.79
263.08–298.15	From $\overline{C}_P$ of gas	5.23
	$\overline{S}^\circ$(298.15 K) − $\overline{S}^\circ$(0 K) =	247.85

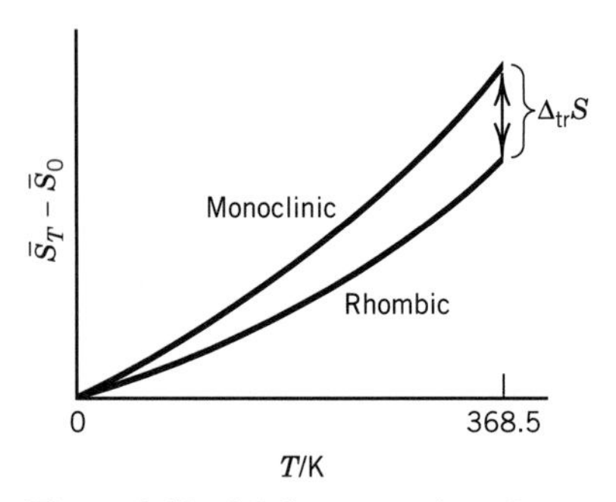

Figure 3.12 Molar entropies of monoclinic and rhombic sulfur from absolute zero to the transition point (368.5 K).

3.8 THE THIRD LAW OF THERMODYNAMICS

In the early years of the twentieth century T. W. Richards and W. Nernst independently studied the entropy changes of certain isothermal chemical reactions and found that the change in entropy approached zero as the temperature was reduced. The entropy change for a chemical reaction cannot be determined calorimetrically using $\Delta S = q_{rev}/T$ because reactions do not occur reversibly. We will see in Chapter 5 how the entropy change for a chemical reaction may be determined. In the meantime we can discuss the measurement of the change in entropy for a phase change for a single substance. As an example we will consider the phase change

$$\mathrm{S(rhombic)} = \mathrm{S(monoclinic)} \tag{3.50}$$

As for chemical reactions, the change in entropy for a phase change approaches zero as the temperature is reduced to absolute zero.

Figure 3.12 shows the molar entropies of monoclinic and rhombic sulfur down to absolute zero as determined from

$$\bar{S}_T - \bar{S}_0 = \int_0^T \frac{\bar{C}_P}{T} \mathrm{d}T \tag{3.51}$$

Rhombic sulfur is the stable form below the transition temperature of 368.5 K. However, monoclinic sulfur may be supercooled below this temperature, and its heat capacity may be measured down to the neighborhood of absolute zero. The entropy change for the phase change (equation 3.50) at the transition temperature calculated on the assumption that $\bar{S}_0$ is zero for both forms is 1.09 J K^{-1} mol^{-1}. This is in agreement with the entropy change at 368.5 K calculated from the difference in enthalpy of 401 J mol^{-1} between the two forms at 368.5 K:

$$\Delta_{tr}S = \frac{401 \text{ J mol}^{-1}}{368.5 \text{ K}} = 1.09 \text{ J K}^{-1} \text{ mol}^{-1}$$

where $\Delta_{tr}S$ is the entropy change for the transition per mole. The difference in enthalpy between the two forms is simply equal to the difference between their heats of combustion at 368.5 K.

It can be seen from Fig. 3.12 that as $T \to 0$, $\Delta_{tr}S \to 0$. It has been found that $\Delta_{tr}S \to 0$ as $T \to 0$ for many other isothermal phase transitions and chemical reactions. In 1905 these observations led Nernst to the conclusion that as temperature approaches 0 K, $\Delta_r S$ for all reactions approaches zero.

$$\lim_{T \to 0} \Delta_r S = 0 \tag{3.52}$$

In 1913 Max Planck took this idea one step further, and we will take his statement as the third law of thermodynamics: **The entropy of each pure element or substance in a perfect crystalline form is zero at absolute zero.**

We will see later (Chapter 16) that statistical mechanics gives a reason for picking this value. As the derivation in Section 3.6 suggests, this corresponds to a single quantum state ($\Omega = 1$) for a perfect crystal at absolute zero. Thus, according to the third law, $\bar{S}_0°$ of the preceding section can be taken as zero if the substance has a perfect crystalline form in the neighborhood of absolute zero. Heat capacity measurements down to temperatures of nearly 0 K are therefore often said to yield "third-law entropies." Thus, the calculation of the preceding section yielded the third-law entropy of $SO_2(g)$ at 298.15 K.

Third-law entropies can be tested against what we would expect from two other types of measurements, namely measurements of equilibrium constants and of spectroscopic data. As we will see in Chapter 5, if the equilibrium constant for a reaction is measured over a range of temperatures, then both ΔH° and ΔS° can be calculated. This ΔS° can be compared with the value expected from the third law if the heat capacities of all reactants and products have been measured down to the neighborhood of absolute zero.

In Chapter 16, on statistical mechanics, we will see that the entropies of relatively simple gases at any desired temperature may be calculated from molar masses and certain spectroscopic information. The molar entropy of a gas at a certain temperature calculated using statistical mechanics may be compared with the molar entropy obtained from calorimetric measurements, assuming that the entropy of the pure crystalline substance is zero at absolute zero.

In general the tests of the third law described in the preceding two paragraphs confirm the third law, but there are some apparent violations. For example, the entropy of $N_2O(g)$ at 298.15 K determined from heat capacity measurements is 5.8 J K^{-1} mol^{-1} smaller than that calculated from spectroscopic data. This indicates an entropy of 5.8 J K^{-1} mol^{-1} at absolute zero for crystals of N_2O. This is an asymmetric linear molecule, NNO. The residual entropy of the crystal is due to disorder in the arrangement of N_2O molecules. In solid N_2O, the molecules are arranged with random head–tail alignments (such as NNO, ONN, NNO, NNO, ONN) instead of being perfectly ordered (NNO, NNO, NNO, NNO, NNO). If the orientation were perfectly random, the crystal might be regarded as a mixed crystal with equal mole fractions of NNO and ONN. The entropy of the mixed crystal would then be the entropy change of mixing. Using equation 3.44, which applies to ideal crystals as well as ideal gases, we see that

$$\Delta_{\text{mix}}S = -R\left(\tfrac{1}{2}\ln\tfrac{1}{2} + \tfrac{1}{2}\ln\tfrac{1}{2}\right) = 5.76 \text{ J K}^{-1}\text{ mol}^{-1} \tag{3.53}$$

Therefore, the statistical mechanical value is taken as the correct entropy, and this is the value that will be found in tables. Thus the apparent violation of the third law is understood and can be calculated for ideal crystals.

Another example of an imperfect crystal from the standpoint of the third law is H_2O. Crystals of H_2O have a residual entropy of 3.35 J K^{-1} mol^{-1} at 0 K. In ice, the hydrogen atoms are arranged around each oxygen atom in a tetrahedral manner. Two of the four atoms are covalently bonded to that oxygen atom, and two are hydrogen bonded (Section 11.10) to that oxygen atom. Since the arrangements of the two types of hydrogen atoms around the oxygen atoms in the crystal are random, it may be shown that the entropy of the crystal should approach $R \ln \frac{3}{2} = 3.37$ J K^{-1} mol^{-1} at absolute zero.

There are two types of randomness in crystals at absolute zero that are not considered in calculating entropies at absolute zero if these entropies are to be used only for chemical purposes:

1. Most crystals are made up of a mixture of isotopic species, but the entropy of mixing isotopes is ignored because the reactants and products in a reaction or phase change contain the same mixtures of isotopes.
2. There is a nuclear spin degeneracy at absolute zero that is ignored because it exists in both reactants and products.

There is a corollary to the third law that is very much in the spirit of the Clausius statement of the second law, since it states an impossibility. According to this corollary, it is impossible to reduce the temperature of a system to 0 K in a finite number of steps. This conclusion that absolute zero is unattainable may be derived from the third law.

The third law is important because it makes possible the calculation of the equilibrium constant for a chemical reaction purely from calorimetric measurements on the reactants and products. The entropies of a number of substances at 298.15 K are given in Table C.2. The entropies of a smaller number of substances are given in Table C.3 for 0, 298, 500, 1000, 2000, and 3000 K. These values and others in the much larger tables from which they have been taken come from four sources:

1. Heat capacities and enthalpies as a function of temperature
2. Statistical mechanical calculations using molecular structure and energy levels
3. Temperature variation of equilibrium constants (Section 5.5)
4. Calculations from enthalpies and Gibbs energies from other sources such as electromotive force measurements (Section 7.6)

Where the entropies have been determined calorimetrically, corrections have been made for imperfections of crystals encountered in the neighborhood of absolute zero in the few cases where this occurs and for gas nonideality at 1 bar. The standard states are, of course, the same as discussed in Section 2.11. The entropy of $H^+(ao)$ is arbitrarily assigned the value of zero, and this makes it possible to calculate entropies of other aqueous ions. Some ions have negative entropies because of this arbitrary convention.

In comparing standard entropy values from various sources it is important to be aware of the standard pressure used. The adjustment of standard entropies from a standard state pressure of 1 atm to 1 bar is discussed in Problem 3.35.

3.9 SPECIAL TOPIC: HEAT ENGINES

A heat engine is an engine that uses heat to generate mechanical work by carrying a "working substance" through a cyclic process. The arrangements for a Carnot heat engine are shown in Fig. 3.13. The arrows indicate that in one cycle this engine receives heat $|q_1|$ from the high-temperature reservoir, rejects heat $|q_2|$ to the low-temperature reservoir, and does work $|w|$ on its surroundings. The absolute value signs are used because the signs of these algebraic quantities are set by conventions that require in this case that q_1 is positive, q_2 is negative, and w is negative. In the operation of the Carnot heat engine a working fluid, which we will refer to as a gas, is taken through a sequence of four steps that return it to its initial state. The engine itself consists of an idealized cylinder with a piston that can slide without friction and can do work on the surroundings or have the surroundings do work on the gas. The temperatures of the two reservoirs are represented by T_1 and T_2.

The cycle for the Carnot heat engine consists of the following four steps, which are represented in Fig. 3.14:

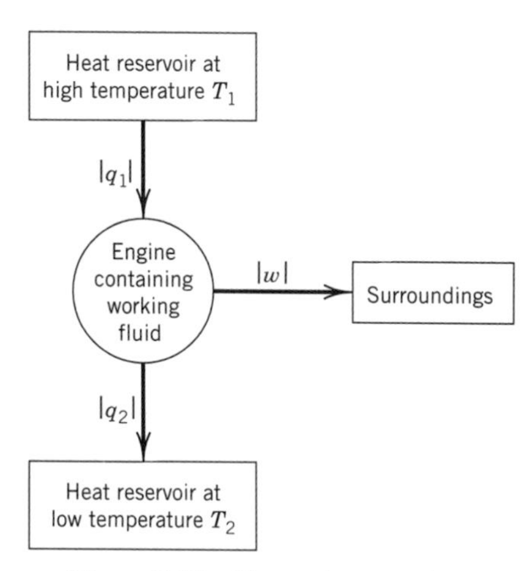

Figure 3.13 Carnot heat engine.

1. **Reversible isothermal expansion of the gas from state A to state B.** During this expansion step, the piston does work $|w_1|$ on the surroundings, and heat $|q_1|$ is absorbed by the gas from the high-temperature reservoir. Note that according to the convention that w is work done on the gas, w_1 is negative. To ensure that the heat transfer is reversible, the temperature of the high-temperature reservoir is only infinitesimally higher than the temperature T_1 of the gas.
2. **Reversible adiabatic expansion of the gas from state B to state C.** For this step we assume that the piston and cylinder are thermally insulated so that no heat is gained or lost. The expansion continues until the temperature of the gas has dropped to T_2. During this expansion step, the piston does work $|w_2|$ on the surroundings.
3. **Reversible isothermal compression of the gas from state C to state D.** During this compression step the surroundings do work $|w_3|$ on the gas, and heat $|q_2|$ flows out of the gas to the low-temperature heat reservoir. Note that according to the convention that q is heat absorbed by the gas, q_2 is negative. The temperature of the low-temperature reservoir is only infinitesimally lower than the temperature T_2 of the gas so that the heat transfer is reversible.
4. **Reversible adiabatic compression of the gas from state D to state A.** This step completes the cycle by bringing the gas back to its initial state at temperature T_1. During this compression step, the surroundings do work $|w_4|$ on the gas, but no heat is gained or lost.

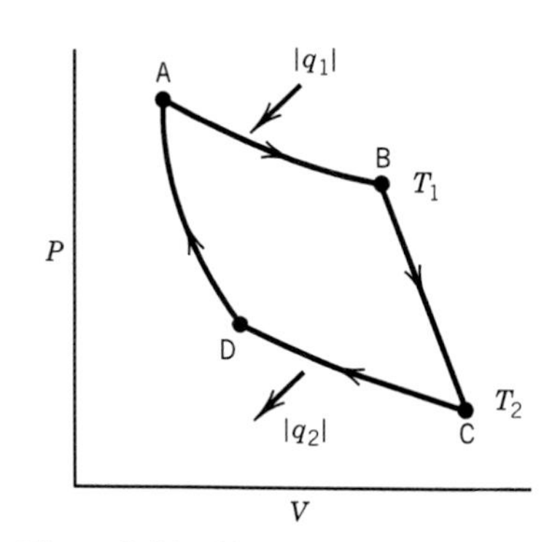

Figure 3.14 Plot of P versus V for the working fluid in a Carnot engine. Heat $|q_1|$ is absorbed in the isothermal expansion compression at T_1, and heat $|q_2|$ is evolved in the isothermal compression at T_2. The other two steps are adiabatic.

A first-law analysis of a Carnot cycle can be summarized as follows:

Step 1. Isothermal expansion, $\Delta U_1 = q_1 + w_1$
Step 2. Adiabatic expansion, $\Delta U_2 = w_2$
Step 3. Isothermal compression, $\Delta U_3 = q_2 + w_3$
Step 4. Adiabatic compression, $\Delta U_4 = w_4$

The change in the internal energy of the gas for the cycle ΔU_{cy} is the sum of the changes in the four steps, and it is equal to zero because the gas is returned to its initial state. Thus,

$$\begin{aligned} \Delta U_{cy} &= 0 = (q_1 + q_2) + (w_1 + w_2 + w_3 + w_4) \\ &= q_{cy} + w_{cy} \end{aligned} \tag{3.54}$$

Thus, the work $|w_{cy}|$ done by the engine in a cycle is equal to the difference between the "heat in" $|q_1|$ and the "heat out" $|q_2|$:

$$-w_{cy} = q_{cy} = q_1 + q_2 \tag{3.55}$$

$$|w_{cy}| = |q_1| - |q_2| \tag{3.56}$$

The engine absorbs heat q_1 from the high-temperature reservoir and does work on the surroundings so that w is negative, but the first law does not tell us the relative amounts of work and of heat rejected at T_2.

In practice, the low-temperature reservoir of a heat engine is often the atmosphere, so that the economic cost of producing work in the surroundings is mainly that of supplying q_1. The **efficiency** ϵ of a heat engine is defined as the ratio of the work done on the surroundings to the heat input at the higher temperature. Thus, the efficiency is defined by

$$\epsilon = \frac{-w_{cy}}{q_1} = \frac{q_1 + q_2}{q_1} \tag{3.57}$$

where the negative sign is required by the fact that w_{cy} is negative and the efficiency is, of course, positive. It is convenient to express the efficiency in terms of the magnitudes of the heats,

$$\epsilon = 1 - \frac{|q_2|}{|q_1|} \tag{3.58}$$

because this reminds us that $0 < \epsilon < 1$.

From equation 3.58 it is clear that to improve the efficiency of a Carnot heat engine one would like to reduce the magnitude of the heat rejected to the cold reservoir, $|q_2|$. But it has been found impossible to reduce $|q_2|$ to zero.

Kelvin showed that a temperature scale could be set up by taking the temperature of a heat reservoir in a reversible Carnot engine to be proportional to the heat transferred in a cycle. Thus

$$\frac{q_{\mathrm{h}}}{q_{\mathrm{c}}} = \frac{T_{\mathrm{h}}}{T_{\mathrm{c}}} \tag{3.59}$$

where h indicates the hot reservoir and c indicates the cold reservoir, and the heats are absolute values. This equation can be derived by writing the equation for the cyclic integral of the entropy for the Carnot cycle. Figure 3.14 shows that this cyclic integral is given by

$$\frac{q_{\mathrm{h}}}{T_{\mathrm{h}}} - \frac{q_{\mathrm{c}}}{T_{\mathrm{c}}} = 0 \tag{3.60}$$

since there are no changes in entropy in the adiabatic steps. The temperatures in Fig. 3.14 are based on the ideal gas scale, but since equations 3.59 and 3.60 are the same, these two scales agree for an ideal gas. The advantage of the Kelvin definition of the temperature scale, however, is that a reversible Carnot cycle can be carried out with any fluid and so the definition is not dependent on the special properties of an ideal gas. Because of his contributions in establishing the absolute temperature scale, the unit of temperature is named the kelvin. By taking $T = 273.16$ K at the triple point of water, the Kelvin scale becomes identical with the ideal gas scale. However, the ideal gas scale is more practical for laboratory use and is used in establishing secondary standards for temperature.

Note that when equation 3.59 is used in equation 3.58, we find that the efficiency ϵ of a heat engine is given by

$$\epsilon = 1 - \frac{T_2}{T_1} \tag{3.61}$$

Thus, high efficiencies are obtained if the ratio T_2/T_1 is small; in practice T_2 is generally close to room temperature and so T_1 is made as high as possible.

Eight Key Ideas in Chapter 3

1. Since the first law provides no information as to whether a process or chemical reaction can proceed spontaneously or not, thermodynamics needs a state function that can be used for this purpose. We know that when there is a temperature difference in a system, heat is transferred, but heat is not a state function. However, the differential heat divided by temperature is an exact differential, and so $đq/T$ is the differential of a state function. This state function is the entropy S, and so $\sum \Delta S = 0$ around a closed cycle, where ΔS is the entropy change in a step.
2. According to the second part of the second law, the change in entropy in any process is given by $\mathrm{d}S \geq đq/T$, where the inequality applies to a spontaneous process and the equality applies to a reversible process. This means that the entropy increases in a spontaneous process in an isolated system.
3. Entropy changes can be determined using reversible processes, like the vaporization of a liquid, the expansion of a gas, or the heating of a gas, liquid, or solid. To obtain the change in entropy in an irreversible process, we have to calculate ΔS along a reversible path.
4. When ideal gases are mixed, there is an entropy of mixing that is always positive.
5. The entropy of a system is a measure of the dispersal of energy among the molecules in microstates. The relation between the entropy of a system and its microstates will be discussed in Chapter 16, on statistical mechanics.
6. The entropy of a substance at any desired temperature relative to its entropy at absolute zero can be obtained by integrating $đq_{\text{rev}}/T$ from absolute zero to the desired temperature.
7. According to the third law of thermodynamics, the entropy of each pure element or substance in a perfect crystalline form is zero at absolute zero.
8. Historically the second law was first discovered in considerations of the efficiencies of heat engines. The maximum efficiency of a heat engine in converting heat to work is given by $1 - T_2/T_1$, where T_1 is the temperature of the high-temperature reservoir and T_2 is the temperature of the low-temperature reservoir. Kelvin showed that a temperature scale can be set up by taking the temperature of a heat reservoir in a reversible Carnot engine to be proportional to the heat transferred in a cycle. This has the advantage over the ideal gas temperature scale introduced in Chapter 1 that it is independent of the working substance.

REFERENCES

M. Bailyn, *A Survey of Thermodynamics.* New York: American Institute of Physics, 1994.

K. E. Bett, J. S. Rowlinson, and G. Saville, *Thermodynamics for Chemical Engineers.* Cambridge, MA: MIT Press, 1975.

H. B. Callen, *Thermodynamics and an Introduction to Thermostatistics,* 2nd ed. Hoboken, NJ: Wiley, 1985.

J. W. Gibbs, *The Collected Works of J. Willard Gibbs.* New Haven, CT: Yale University Press, 1948.

K. S. Pitzer, *Thermodynamics,* 3rd ed. New York: McGraw-Hill, 1995.

J. M. Smith, H. C. Van Ness, and M. M. Abbott, *Introduction to Chemical Engineering Thermodynamics,* 5th ed. New York: McGraw-Hill, 1996.

J. W. Tester and M. Modell, *Thermodynamics and Its Applications.* Upper Saddle River, NJ: Prentice-Hall, 1997.

PROBLEMS

Ⓜ Problems marked with an icon may be more conveniently solved on a personal computer with a mathematical program.

3.1 Show that $(\partial C_V/\partial V)_T = 0$ for an ideal gas, a gas following $P = nRT/(V - nb)$, and a van der Waals gas.

3.2 Show that q_{rev} is not a state function for a gas obeying the equation of state $P(V - nb) = nRT$, but that q_{rev}/T is.

3.3 Show that q_{rev} is not an exact differential for a gas obeying the van der Waals equation, but that q_{rev}/T is.

3.4 An ideal gas initially at P_1, V_1, T_1 undergoes a reversible isothermal expansion to P_2, V_2, T_1. The same change in state of the gas can be accomplished by allowing it to expand adiabatically to P_3, V_2, T_2 and then heating it at constant volume to P_2, V_2, T_1. Show that the entropy change for the reversible isothermal expansion is the same as the sum of the entropy changes in the reversible adiabatic expansion and the reversible heating to P_2, V_2, T_1. This shows that ΔS is independent of path and is therefore a state function.

3.5 Water is vaporized reversibly at 100 °C and 1.013 25 bar. The heat of vaporization is 40.69 kJ mol^{-1}. (*a*) What is the value of ΔS for the water? (*b*) What is the value of ΔS for the water plus the heat reservoir at 100 °C?

3.6 Assuming that CO_2 is an ideal gas, calculate ΔH° and ΔS° for the following process:

$$1\,CO_2(g,\ 298.15\ K,\ 1\ bar) \rightarrow 1\,CO_2(g,\ 1000\ K,\ 1\ bar)$$

Given: $\overline{C}_P^\circ = 26.648 + 42.262 \times 10^{-3}T - 142.4 \times 10^{-7}T^2$ in J K^{-1} mol^{-1}.

3.7 The temperature of an ideal monatomic gas is increased from 300 to 500 K. What is the change in molar entropy of the gas (*a*) if the volume is held constant and (*b*) if the pressure is held constant?

3.8 Ammonia (considered to be an ideal gas) initially at 25 °C and 1 bar pressure is heated at constant pressure until the volume has trebled. Calculate (*a*) q per mole, (*b*) w per mole, (*c*) $\Delta\overline{H}$, (*d*) $\Delta\overline{U}$, and (*e*) $\Delta\overline{S}$. Given: $\overline{C}_P = 25.895 + 32.999 \times 10^{-3}T - 30.46 \times 10^{-7}T^2$ in J K^{-1} mol^{-1}.

3.9 Two blocks of the same metal are the same size but are at different temperatures, T_1 and T_2. These blocks of metal are brought together and allowed to come to the same temperature. Show that the entropy change is given by

$$\Delta S = C_P \ln \frac{(T_1 + T_2)^2}{4T_1T_2}$$

if C_P is constant. How does this equation show that the change is spontaneous?

3.10 In the reversible isothermal expansion of an ideal gas at 300 K from 1 to 10 L, where the gas has an initial pressure of 20.27 bar, calculate (*a*) ΔS for the gas and (*b*) ΔS for all systems involved in the expansion.

3.11 A mole of oxygen is expanded reversibly from 1 to 0.1 bar at 298 K. What is the change in entropy of the gas, and what is the change in entropy for the gas plus the heat reservoir with which it is in contact?

3.12 Three moles of an ideal gas expand isothermally and reversibly from 90 to 300 L at 300 K. (*a*) Calculate ΔU, ΔS, w, and q for this system. (*b*) Calculate $\Delta\overline{U}$, $\Delta\overline{S}$, w per mole, and q per mole. (*c*) If the expansion is carried out irreversibly by allowing the gas to expand into an evacuated container, what are the values of $\Delta\overline{U}$, $\Delta\overline{S}$, w per mole, and q per mole?

3.13 (*a*) A system consists of a mole of ideal gas that undergoes the following change in state:

$$1X(g,\ 298\ K,\ 10\ bar) \rightarrow 1X(g,\ 298\ K,\ 1\ bar)$$

What is the value of $\Delta \overline{S}$ if the expansion is reversible? What is the value of $\Delta \overline{S}$ if the gas expands into a larger evacuated container so that the final pressure is 1 bar? (*b*) The same change in state takes place, but we now consider the gas plus the heat reservoir at 298 K to be our system. What is the value of $\Delta \overline{S}$ if the expansion is reversible? What is the value of $\Delta \overline{S}$ if the gas expands into a larger container so that the final pressure is 1 bar?

3.14 An ideal gas at 298 K expands isothermally from a pressure of 10 bar to 1 bar. What are the values of w per mole, q per mole, $\Delta \overline{U}$, $\Delta \overline{H}$, and $\Delta \overline{S}$ in the following cases? (*a*) The expansion is reversible. (*b*) The expansion is free. (*c*) The gas and its surroundings form an isolated system, and the expansion is reversible. (*d*) The gas and its surroundings form an isolated system, and the expansion is free.

3.15 An ideal monatomic gas is heated from 300 to 1000 K and the pressure is allowed to rise from 1 to 2 bar. What is the change in molar entropy?

3.16 The purest acetic acid is often called glacial acetic acid because it is purified by fractional freezing at its melting point of 16.6 °C. A flask containing several moles of acetic acid at 16.6 °C is lowered into an ice–water bath briefly. When it is removed, it is found that exactly 1 mol of acetic acid has frozen. Given: $\Delta_{fus}H(CH_3CO_2H) = 11.45$ kJ mol^{-1} and $\Delta_{fus}H(H_2O) = 5.98$ kJ mol^{-1}. (*a*) What is the change in entropy of the acetic acid? (*b*) What is the change in entropy of the water bath? (*c*) Now consider that the water bath and acetic acid are in the same system. What is the entropy change for the combined system? Is the process reversible or irreversible? Why?

3.17 In Problem 2.21 one mole of an ideal monatomic gas at 1 bar and 300 K was expanded adiabatically against a constant pressure of $\frac{1}{2}$ bar until the final pressure was $\frac{1}{2}$ bar; a temperature of 240 K was reached. What is the value of $\Delta \overline{S}$ for this process?

3.18 Ten moles of H_2 and two moles of D_2 are mixed at 25 °C and 1 bar. What is the value of $\Delta S°$? Assume ideal gases.

3.19 (*a*) Write the expression for the entropy of a mixture of ideal gases A, B, and C at T and P using y_i for the mole fraction of gas i. (*b*) Now let us carry out the combination of terms contributing to the entropy of the system in two steps. First, imagine that gases A and C are mixed to form a mixture with mole fractions r_A and r_C within the A–C mixture, but that B remains unmixed. Write the equation for the entropy of the system with two terms, one for the $n_I = n_A + n_C$ moles of the A plus C mixture, which contributes pressure $P_I = P_A + P_C$, and the other $n_B \overline{S}_B$. (*c*) Second, imagine that B is mixed with mixture I, considered as one species, and show that this equation is the same as that obtained in (*a*).

3.20 One mole of A at 1 bar and one mole of B at 2 bar are separated by a partition and surrounded by a heat reservoir. When the partition is withdrawn, how much does the entropy change?

3.21 Use the microscopic point of view of Section 3.6 to show that for the expansion of amount n of an ideal gas by a factor of 2, $\Delta S = nR \ln 2$. In this expression S is an extensive property.

3.22 Calculate the change in molar entropy of aluminum that is heated from 600 to 700 °C. The melting point of aluminum is 660 °C, the heat of fusion is 393 J g^{-1}, and the heat capacities of the solid and the liquid may be taken as 31.8 and 34.4 J K^{-1} mol^{-1}, respectively.

3.23 Steam is condensed at 100 °C, and the water is cooled to 0 °C and frozen to ice. What is the molar entropy change of the water? Consider that the average specific heat of liquid water is 4.2 J K^{-1} g^{-1}. The enthalpy of vaporization at the boiling point and the enthalpy of fusion at the freezing point are 2258.1 and 333.5 J g^{-1}, respectively.

Ⓜ **3.24** Calculate the molar entropy of carbon disulfide at 25 °C from the following heat capacity data and the heat of fusion, 4389 J mol^{-1}, at the melting point (161.11 K):

T/K	15.05	20.15	29.76	42.22	
$\overline{C}_P$/J K^{-1} mol^{-1}	6.90	12.01	20.75	29.16	
T/K	57.52	75.54	89.37	99.00	
$\overline{C}_P$/J K^{-1} mol^{-1}	35.56	40.04	43.14	45.94	
T/K	108.93	119.91	131.54	156.83	161–298
$\overline{C}_P$/J K^{-1} mol^{-1}	48.49	50.50	52.63	56.62	75.48

3.25 Ten grams of molecular hydrogen at 1 bar expand to triple the volume (*a*) isothermally and reversibly and (*b*) adiabatically and reversibly. In each case what are $\Delta S(H_2)$, ΔS(surr), and $\Delta S(H_2 + \text{surr})$?

3.26 Theoretically, how high could a gallon of gasoline lift an automobile weighing 2800 lb against the force of gravity, if it is assumed that the cylinder temperature is 2200 K and the exit temperature 1200 K? (Density of gasoline = 0.80 g cm^{-3}; 1 lb = 453.6 g; 1 ft = 30.48 cm; 1 L = 0.2642 gal; heat of combustion of gasoline = 46.9 kJ g^{-1}.)

3.27 (*a*) What is the maximum work that can be obtained from 1000 J of heat supplied to a steam engine with a high-temperature reservoir at 100 °C if the condenser is at 20 °C? (*b*) If the boiler temperature is raised to 150 °C by the use of superheated steam under pressure, how much more work can be obtained?

3.28 The term *adiabatic lapse rate* used by meteorologists is the decrease in temperature with height that results from the adiabatic expansion of an air mass as it is pushed up a mountain by the wind. Similarly, the wind coming down the mountain slope warms up. This adiabatic expansion is represented by

$$P_0, V_0, T_0 \rightarrow P, V, T \qquad \Delta S_3 = 0$$

where P_0, V_0, T_0 represents the sea level conditions. The calculation of the entropy change can be carried out in two steps (the first at constant pressure, the second at constant temperature):

$$P_0, V_0, T_0 \rightarrow P_0, V^*, T \qquad \Delta S_1 = n\overline{C}_P \ln \frac{T}{T_0}$$

$$P_0, V^*, T \rightarrow P, V, T \qquad \Delta S_2 = nR \ln \frac{P_0}{P}$$

We have seen in Section 1.11 that the pressure of the atmosphere drops off exponentially if temperature is independent of height h. (*a*) Since $\Delta S_1 + \Delta S_2 = \Delta S_3$, what is the expression for $\Delta T/\Delta h$? (*b*) If the temperature at the foot of a 14,000-foot mountain is 25 °C, what temperature would you expect at the summit from this adiabatic lapse rate? For this calculation the molar mass of air can be taken to be $M = 29 \text{ g mol}^{-1}$ and its heat capacity can be taken as $29.1 \text{ J K}^{-1} \text{ mol}^{-1}$.

3.29 When an ideal gas is allowed to expand isothermally in a piston, $\Delta U = q + w = 0$. Thus, the work done by the system on the surroundings is equal to the heat transferred from the reservoir to the gas, and the efficiency of turning heat into work is 100%. Explain why this is not a violation of the second law.

3.30 In Problem 2.50 we found that when an ideal monatomic gas at 1 bar and 273.15 K is allowed to expand adiabatically against a constant pressure of 0.395 bar until it reaches equilibrium, the final temperature is 207.04 K. What is the value of $\Delta \overline{S}$ for this process? Given: $\overline{C}_V = \frac{3}{2}R$ and therefore $\overline{C}_P = \frac{5}{2}R$.

3.31 Compare the entropy difference between 1 mol of liquid water at 25 °C and 1 mol of vapor at 100 °C and 1.013 25 bar. The average specific heat of liquid water may be taken as $4.2 \text{ J K}^{-1} \text{ g}^{-1}$, and the heat of vaporization is 2259 J g^{-1}.

3.32 Calculate the increase in the molar entropy of nitrogen when it is heated from 25 to 1000 °C (*a*) at constant pressure and (*b*) at constant volume. Given: $\overline{C}_P = 26.9835 + 5.9622 \times 10^{-3}T - 3.377 \times 10^{-7}T^2$ in $\text{J K}^{-1} \text{ mol}^{-1}$.

3.33 Assuming that the heat capacity of water is independent of temperature, calculate the net change in entropy when 1 mol of water at 0 °C is mixed with 1 mol of water at 100 °C. Assume that the heat capacity of water is $(4.184 \text{ J K}^{-1} \text{ g}^{-1})(18 \text{ g mol}^{-1}) = 75.3 \text{ J K}^{-1} \text{ mol}^{-1}$ and that the heat capacity of the calorimeter is negligible.

3.34 In Section 3.3 we derived the expression

$$S_2 - S_1 = nR \ln \frac{V_2}{V_1}$$

for the reversible isothermal expansion of an ideal gas. We ought to derive this same expression by another reversible path. Do this by imagining that the gas is first expanded adiabatically and reversibly from V_1 to V_2. Since the temperature falls from T to T', heat must be added reversibly and at constant volume in a second step to restore the temperature to T.

3.35 Show that the standard molar entropy $S°(T)$ for an ideal gas at 1 bar can be calculated from the standard molar entropy $S^*(T)$ at 1 atm using

$$\overline{S}°(T) - \overline{S}^*(T) = R \ln \frac{P^*}{P°} = 0.109 \text{ J K}^{-1} \text{ mol}^{-1}$$

The conversion for liquids and solids is negligible. Why?

3.36 Argon undergoes the following change in state:

$$P_1 = 20 \text{ bar}, T_1 = 300 \text{ K} \rightarrow P_2 = 1 \text{ bar}, T_2 = 200 \text{ K}$$

What is the change in molar entropy, assuming that argon is an ideal gas?

3.37 A flask containing several moles of liquid benzene at its freezing temperature (5.5 °C) is placed in thermal contact with an ice–water bath. When the flask is removed from the ice–water bath, it is found that 1 mol of benzene has frozen. Given: $\Delta_{\text{fus}}H(C_6H_6) = 9.87 \text{ kJ mol}^{-1}$, $\Delta_{\text{fus}}H(H_2O) = 5.98 \text{ kJ mol}^{-1}$. (*a*) What is the change in entropy of benzene? (*b*) What is the change in entropy of the water bath? (*c*) Considering the flask of benzene and the water bath as an isolated system, what is the change in entropy of the isolated system? Is the process reversible or irreversible?

3.38 One mole each of H_2(g), N_2(g), and O_2(g) are mixed at 25 °C. What is $\Delta_{\text{mix}}S$?

3.39 According to the Debye equation (Section 16.12) the heat capacity of a solid is proportional to the temperature cubed at low temperatures:

$$C_P(T) = (\text{const})T^3$$

Show that the entropy at a temperature T' is given by

$$S(T') = \frac{C_P(T')}{3}$$

when the Debye equation holds.

3.40 Calculate the molar entropy of liquid chlorine at its melting point, 172.12 K, from the following data obtained by W. F. Giauque and T. M. Powell:

T/K	15	20	25	30	35
$\overline{C}_P$/J K^{-1} mol^{-1}	3.72	7.74	12.09	16.69	20.79

T/K	40	50	60	70	90
$\overline{C}_P$/J K^{-1} mol^{-1}	23.97	29.25	33.47	36.32	40.63

T/K	110	130	150	170	172.12
$\overline{C}_P$/J K^{-1} mol^{-1}	43.81	47.24	51.04	55.10	m.p.

The heat of fusion is 6406 J mol^{-1}. Below 15 K it may be assumed that $\overline{C}_P$ is proportional to T^3.

3.41 In running a Carnot cycle backward to produce refrigeration, the objective is to remove as much heat $|q_2|$ from the cold reservoir as possible for a given amount of work, and so the coefficient of performance β is defined as

$$\beta = \frac{|q_2|}{|w|} = \frac{|q_2|}{|q_1| - |q_2|} = \frac{T_2}{T_1 - T_2}$$

A household refrigerator operates between 35 and $-10\,°C$. How many joules of heat can in principle be removed per kilowatt-hour of work?

3.42 A heat pump is used to heat a home in the winter when the temperature in the ground is 0 °C and the temperature of the radiator in the house is 35 °C. When a Carnot cycle is run backward for this purpose, the objective is to obtain as much heat in the radiator as possible for a given amount of electrical work. The coefficient of performance β' for a heat pump is defined by

$$\beta' = \frac{|q_1|}{|w|} = \frac{|q_1|}{|q_1| - |q_2|} = \frac{T_1}{T_1 - T_2}$$

What is the minimum amount of electrical work needed to produce a kilowatt-hour of heat?

Computer Problems

3.A Calculate the standard molar entropies at 1000 K for the eight gases in Computer Problem 2.A using the empirical equations given there for the molar heat capacities as a function of temperature and the values of the standard molar entropies at 298.15 K in Table C.2. Compare these calculated values with values in Table C.3 in Appendix C.

3.B Plot the standard molar entropy of water vapor versus temperature from 300 to 1000 K by use of the empirical equation for the molar heat capacity of water vapor.

3.C Plot the entropy of mixing of two ideal gases to form a mole of mixture versus the mole fraction of one of the species. Investigate the slope of this plot as $y_2 \to 1$ and as $y_2 \to 0$.

3.D Plot the molar entropy in J K^{-1} mol^{-1} of a monatomic ideal gas versus pressure P from 1 bar to 100 bar and T from 298.15 K to 500 K, assuming that its molar entropy is zero at 298.15 K and 1 bar.

4

Fundamental Equations of Thermodynamics

With the definitions of T, U, and S we have completed the set of necessary thermodynamic properties. Although the entropy provides a criterion of whether a change in an isolated system is spontaneous, it does not provide a convenient criterion at constant T and V or constant T and P, the usual conditions in the laboratory. We need two additional thermodynamic properties (functions) to make calculations at constant T and V or constant T and P more convenient than they are with entropy. Fortunately, there is a general way to do this using Legendre transforms. These two new thermodynamic properties are the Helmholtz energy A and the Gibbs energy G. At constant T and V, spontaneous processes occur with a decrease in A, and at constant T and P, spontaneous processes occur with a decrease in G.

In order to discuss equilibria in systems with more than two phases and chemical equibrium, we need to introduce the chemical potential μ_i of a species. The chemical potential of a species is the same in all of the phases of a system at equilibrium. The chemical potential is also the property that determines whether a species will undergo chemical reaction. There are many relations between the thermodynamic properties of a system, and the fundamental equations for the various thermodynamic potentials (U, H, A, and G) are the source of these very useful relations. In the next four chapters, these fundamental equations will lead us to the quantitative treatment of chemical equilibrium, phase equilibrium, electrochemical equilibrium, and biochemical equilibrium.

4.1 FUNDAMENTAL EQUATION FOR THE INTERNAL ENERGY

The first law is given in equation 2.9 as

$$dU = đq + đw \tag{4.1}$$

and the second law is given by equation 3.15, which is

$$dS \geq đq/T \tag{4.2}$$

where the inequality applies when $đq$ is for an irreversible process and the equal sign applies when $đq$ is for a reversible process. If we restrict consideration to closed systems in which only reversible PV work is involved,

$$đw = -P\,dV \tag{4.3}$$

and equation 4.2 becomes

$$dS = đq/T \tag{4.4}$$

so that the **combined first and second law** is

$$dU = T\,dS - P\,dV \tag{4.5}$$

Since equation 4.5 involves only state functions, it applies to both reversible and irreversible processes. We saw in Table 2.1 that $-P$ and V are conjugate variables. Now we can see that T and S are also conjugate variables.

In 1876 Gibbs made a very important addition to this equation in order to discuss phase equilibrium and reaction equilibrium. He introduced the concept of the **chemical potential** μ_i of a species and wrote the fundamental equation for U as

$$dU = T\,dS - P\,dV + \mu_1\,dn_1 + \mu_2\,dn_2 + \cdots \tag{4.6}$$

where the additional terms are "chemical work" terms for each species and n_i is the amount of species i. The chemical potential is a measure of the potential a species has to move from one phase to another or undergo a chemical reaction. Equation 4.6 shows that if more of species i is added to a system at constant S and V, there is a contribution $\mu_i\,dn_i$ to the internal energy. Note that for a system in which there are no chemical reactions, μ_i and n_i are conjugate variables. If a system contains N_s different species, equation 4.6 can be written

$$dU = T\,dS - P\,dV + \sum_{i=1}^{N_s} \mu_i\,dn_i \tag{4.7}$$

This equation shows that U is a function of S, V, and $\{n_i\}$, where $\{n_i\}$ is the set of amounts of species i; this can be represented by $U(S, V, \{n_i\})$. In view of the importance of the variables S, V, and $\{n_i\}$ in equation 4.7, they are called the **natural variables** of U. Note that the natural variables for the interal energy are all extensive. The total differential of U can be written in terms of the differentials of these natural variables.

$$\mathrm{d}U = \left(\frac{\partial U}{\partial S}\right)_{V,\{n_i\}} \mathrm{d}S + \left(\frac{\partial U}{\partial V}\right)_{S,\{n_i\}} \mathrm{d}V + \sum_{i=1}^{N_s} \left(\frac{\partial U}{\partial n_i}\right)_{S,V,n_j \neq n_i} \mathrm{d}n_i \tag{4.8}$$

Comparison of equation 4.7 with equation 4.8 shows that

$$T = \left(\frac{\partial U}{\partial S}\right)_{V,\{n_i\}} \tag{4.9}$$

$$P = -\left(\frac{\partial U}{\partial V}\right)_{S,\{n_i\}} \tag{4.10}$$

$$\mu_i = \left(\frac{\partial U}{\partial n_i}\right)_{S,V,n_j \neq n_i} \tag{4.11}$$

where $n_j \neq n_i$ means that species other than i are held constant. These three equations are often referred to as **equations of state** because they give relations between state properties. Note that the derivatives of extensive properties with respect to extensive properties are intensive properties. Equations 4.9 to 4.11 are very important because they show that if U of a system can be determined as a function of its natural variables (S, V, and $\{n_i\}$), all the other thermodynamic properties of the system—including the chemical potentials of all the species—can be calculated. In Chapters 1 and 2 we saw that the extensive state of a system can be described by specifying T, P, and $\{n_i\}$ or T, V, and $\{n_i\}$, but now we have to understand that determination of U as a function of these variables is not enough to allow us to calculate all the other thermodynamic properties of the system.

A fundamental equation such as 4.7 leads to relations of another type between thermodynamic properties, which are referred to as **Maxwell relations.** They are obtained by equating second cross-partial derivatives. We have already used equations like this in the test for exactness in Section 2.3. The Maxwell equations for equation 4.7 are

$$\left(\frac{\partial T}{\partial V}\right)_{S,\{n_i\}} = -\left(\frac{\partial P}{\partial S}\right)_{V,\{n_i\}} \tag{4.12}$$

$$\left(\frac{\partial T}{\partial n_i}\right)_{S,V,n_j \neq n_i} = \left(\frac{\partial \mu_i}{\partial S}\right)_{V,\{n_i\}} \tag{4.13}$$

$$-\left(\frac{\partial P}{\partial n_i}\right)_{S,V,n_j \neq n_i} = \left(\frac{\partial \mu_i}{\partial V}\right)_{S,\{n_i\}} \tag{4.14}$$

$$\left(\frac{\partial \mu_i}{\partial n_j}\right)_{S,V,n_i \neq n_j} = \left(\frac{\partial \mu_j}{\partial n_i}\right)_{S,V,n_j \neq n_i} \tag{4.15}$$

This shows that these properties are related in interesting ways.

The internal energy provides a criterion for whether a process can occur spontaneously at constant S, V, and $\{n_i\}$. If we substitute the second law in the form $\mathrm{d}S \geq đq/T$ and $đw = -P_{\text{ext}}\,\mathrm{d}V + \sum \mu_i\,\mathrm{d}n_i$ in equation 4.1, we obtain

$$\mathrm{d}U \leq T\,\mathrm{d}S - P_{\text{ext}}\,\mathrm{d}V + \sum_{i=1}^{N_s} \mu_i\,\mathrm{d}n_i \tag{4.16}$$

Thus if an infinitesimal change takes place in a system of constant entropy, volume, and $\{n_i\}$,

$$(\mathrm{d}U)_{S,V,\{n_i\}} \leq 0 \tag{4.17}$$

This is the criterion for spontaneous change and equilibrium in the system involving PV work and specified amounts of N_s species. At equilibrium, U at constant S, V, and $\{n_i\}$ must be at a minimum.

At constant values of the intensive properties T, P, and μ_i, equation 4.7 can be integrated to obtain

$$U = TS - PV + \sum_{i=1}^{N_s} \mu_i n_i \tag{4.18}$$

This equation can also be considered as a consequence of Euler's theorem (see Section 1.10).

The equations in this section are not very useful in the laboratory because the entropy and volume are not easily controlled. Fortunately, however, more useful thermodynamic properties can be defined, based on the internal energy.

4.2 DEFINITIONS OF ADDITIONAL THERMODYNAMIC POTENTIALS USING LEGENDRE TRANSFORMS

The internal energy and other thermodynamic properties defined starting with the internal energy are referred to as **thermodynamic potentials.** To define new thermodynamic potentials, we use the method of Legendre transforms. We have already seen an example of this with the enthalpy H, defined by $H = U + PV$ (see Section 2.7). We did not emphasize that this is a Legendre transform, but it is, and now we are going to use two more Legendre transforms to define two thermodynamic potentials. A Legendre transform is a linear change in variables that starts with a mathematical function and defines a new function by subtracting one or more products of conjugate variables. This is different from the usual change in variables in that a partial derivative of a thermodynamic potential becomes an independent variable in the new thermodynamic potential. As explained in Appendix D.5, no information is lost in this process. Thus a new thermodynamic potential defined in terms of the internal energy contains all of the information that is in $U(S, V, \{n_i\})$.*

Now we can make a more complete treatment of the enthalpy H than in Section 2.7. The total differential of the enthalpy is

$$\mathrm{d}H = \mathrm{d}U + P\,\mathrm{d}V + V\,\mathrm{d}P \tag{4.19}$$

Substituting equation 4.7 for $\mathrm{d}U$ yields the fundamental equation for the enthalpy of a system involving pressure–volume work and N_s different species:

$$\mathrm{d}H = T\,\mathrm{d}S + V\,\mathrm{d}P + \sum_{i=1}^{N_s} \mu_i\,\mathrm{d}n_i \tag{4.20}$$

*H. B. Callen, *Thermodynamics*, 2nd ed. Hoboken, NJ: Wiley, 1985; R. A. Alberty, J. M. G. Barthel, E. R. Cohen, R. N. Goldberg, and E. Wilhelm, Use of Legendre Transforms in Chemical Thermodynamics, (an IUPAC Technical Report), *Pure Appl. Chem.* **73,** 8 (2001).

Thus we see that

$$T = \left(\frac{\partial H}{\partial S}\right)_{P,\{n_i\}} \tag{4.21}$$

$$V = \left(\frac{\partial H}{\partial P}\right)_{S,\{n_i\}} \tag{4.22}$$

$$\mu_i = \left(\frac{\partial H}{\partial n_i}\right)_{S,P,\{n_{j\neq i}\}} \tag{4.23}$$

If H can be determined as a function of S, P, and the amounts of all the species, then T, V, and μ_i can be determined by taking partial derivatives of H. Thus all the thermodynamic properties of a system can in principle be obtained using H, just as all the thermodynamic properties of the system were determined using U in the preceding section. This shows that no information is lost in making a Legendre transform. Equation 4.20 provides a number of Maxwell equations.

The enthalpy provides a criterion for whether a process can occur spontaneously for a system at constant S, P, and n_i. Following the same reasoning we used to obtain equation 4.17 we obtain

$$(\mathrm{d}H)_{S,P,\{n_i\}} \leq 0 \qquad (P = P_{\mathrm{ext}}) \tag{4.24}$$

This shows that a change can take place spontaneously at constant entropy, pressure, and amounts of species if the enthalpy decreases.

At constant values of T, P, and μ_i, equation 4.20 can be integrated to obtain

$$H = TS + \sum_{i=1}^{N} \mu_i n_i \tag{4.25}$$

Note that using $H = U + PV$, we can obtain equation 4.18.

The internal energy and the enthalpy do not provide very useful criteria for spontaneous change because the entropy has to be held constant. This problem is avoided by the use of Legendre transforms in which the product TS of conjugate variables is subtracted from the internal energy and the enthalpy. This introduces the intensive variable T as a natural variable in place of the extensive variable S.

The two Legendre transforms that define the **Helmholtz energy** A and the **Gibbs energy** G are*

$$A = U - TS \tag{4.26}$$

$$G = U + PV - TS = H - TS \tag{4.27}$$

The total differential of the Helmholtz energy is given by

$$\mathrm{d}A = \mathrm{d}U - T\mathrm{d}S - S\mathrm{d}T \tag{4.28}$$

*The Helmholtz energy A is named in honor of Hermann L. F. von Helmholtz. The Gibbs energy G is named in honor of J. Willard Gibbs of Yale University, whose many important generalizations in thermodynamics have given him the position of one of the great geniuses of science. The Gibbs energy is sometimes referred to as the Gibbs free energy or the free energy. See J. W. Gibbs, *Trans. Conn. Acad. Arts Sci.*, 1876–1878; *The Collected Works of J. Willard Gibbs*, Vol. 1. New Haven, CT: Yale University Press, reprinted 1948.

Substituting equation 4.7 for $\mathrm{d}U$ yields the fundamental equation for the Helmholtz energy:

$$\mathrm{d}A = -S\,\mathrm{d}T - P\,\mathrm{d}V + \sum_{i=1}^{N} \mu_i\,\mathrm{d}n_i \tag{4.29}$$

Thus the natural variables for A are T, V, and n_i. Thus we see that

$$S = -\left(\frac{\partial A}{\partial T}\right)_{V,\{n_i\}} \tag{4.30}$$

$$P = -\left(\frac{\partial A}{\partial V}\right)_{T,\{n_i\}} \tag{4.31}$$

$$\mu_i = \left(\frac{\partial A}{\partial n_i}\right)_{T,V,\{n_{j\neq i}\}} \tag{4.32}$$

Equation 4.29 also provides a number of Maxwell relations.

The Helmholtz energy A provides the criterion for spontaneous change at specified T, V, and $\{n_i\}$:

$$(\mathrm{d}A)_{T,V,\{n_i\}} \leq 0 \tag{4.33}$$

Thus a change can take place spontaneously at constant temperature, volume, and amounts of species if the Helmholtz energy decreases (see Table 4.1). Integration of the fundamental equation for A at constant values of the properties T, P, and $\{\mu_i\}$ yields

$$A = -PV + \sum_{i=1}^{N_s} \mu_i n_i \tag{4.34}$$

The Helmholtz energy is less useful in chemistry than the Gibbs energy because processes and reactions are more often carried out at constant pressure rather than constant volume.

The total differential of the Gibbs energy is

$$\mathrm{d}G = \mathrm{d}U + P\,\mathrm{d}V + V\,\mathrm{d}P - T\,\mathrm{d}S - S\,\mathrm{d}T \tag{4.35}$$

Substituting equation 4.7 for $\mathrm{d}U$ yields the fundamental equation for the Gibbs energy:

$$\mathrm{d}G = -S\,\mathrm{d}T + V\,\mathrm{d}P + \sum_{i=1}^{N_s} \mu_i\,\mathrm{d}n_i \tag{4.36}$$

Thus, we see that

$$S = -\left(\frac{\partial G}{\partial T}\right)_{P,\{n_i\}} \tag{4.37}$$

$$V = \left(\frac{\partial G}{\partial P}\right)_{T,\{n_i\}} \tag{4.38}$$

$$\mu_i = \left(\frac{\partial G}{\partial n_i}\right)_{T,P,\{n_{j\neq i}\}} \tag{4.39}$$

Table 4.1 Criteria for Irreversibility and Reversibility for Processes Involving Only Pressure–Volume Work

For Irreversible Processes	*For Reversible Processes*
$(\mathrm{d}S)_{V,U,\{n_i\}} > 0$	$(\mathrm{d}S)_{V,U,\{n_i\}} = 0$
$(\mathrm{d}U)_{V,S,\{n_i\}} < 0$	$(\mathrm{d}U)_{V,S,\{n_i\}} = 0$
$(\mathrm{d}H)_{P,S,\{n_i\}} < 0$	$(\mathrm{d}H)_{P,S,\{n_i\}} = 0$
$(\mathrm{d}A)_{T,V,\{n_i\}} < 0$	$(\mathrm{d}A)_{T,V,\{n_i\}} = 0$
$(\mathrm{d}G)_{T,P,\{n_i\}} < 0$	$(\mathrm{d}G)_{T,P,\{n_i\}} = 0$

If G can be determined as a function of its natural variables T, P, and $\{n_i\}$, then S, V, and μ_i can be calculated by taking partial derivatives of G. Note that the chemical potential of species i is equal to the partial molar Gibbs energy. Thus equation 4.36 can be written with $\mu_i = \overline{G}_i$, where $\overline{G}_i$ is the **partial molar Gibbs energy** of species i. The partial molar volume $\overline{V}_i$ of a species was introduced in Section 1.10.

Equations 4.37 and 4.38 tell us something interesting about the dependence of G on T and P. Since the entropy of a system is always positive, G decreases with increasing temperature at constant pressure. Since S is greater for a gas than for the corresponding solid, the temperature coefficient of the Gibbs energy is more negative for a gas than for the corresponding solid. Since the volume of a system is always positive, G increases with increasing P at constant T. Since V is greater for a gas than for the corresponding solid, the pressure coefficient of G is larger for a gas than for the corresponding solid.

Now we are in a position to illustrate the fact that if a thermodynamic potential is known as a function of its natural variables, we can calculate all of the thermodynamic properties of the system. Suppose that G for a system containing a single species has been determined as a function of temperature and pressure. The entropy and volume of the system can be calculated from

$$S = -\left(\frac{\partial G}{\partial T}\right)_P \quad \text{and} \quad V = \left(\frac{\partial G}{\partial P}\right)_T \tag{4.40}$$

Then U, H, and A can be calculated using the equations

$$U = G - PV + TS = G - P\left(\frac{\partial G}{\partial P}\right)_T - T\left(\frac{\partial G}{\partial T}\right)_P \tag{4.41}$$

$$H = G + TS = G - T\left(\frac{\partial G}{\partial T}\right)_P \tag{4.42}$$

$$A = G - PV = G - P\left(\frac{\partial G}{\partial P}\right)_T \tag{4.43}$$

This is not possible when G is known as a function of V and T or P and V.

The Gibbs energy provides a criterion for whether a process can occur spontaneously at constant T, P, and $\{n_i\}$. Following the same reasoning we used to obtain equation 4.17, we obtain

$$(\mathrm{d}G)_{T,P,\{n_i\}} \leq 0 \qquad (T = T_{\mathrm{surr}},\ P = P_{\mathrm{ext}}) \tag{4.44}$$

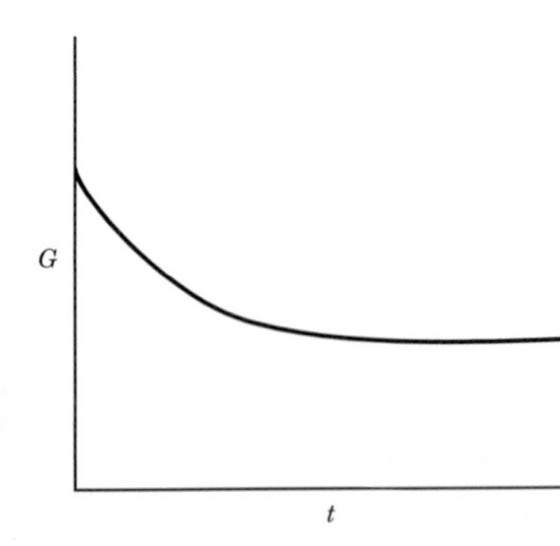

Figure 4.1 When a system undergoes spontaneous change at constant T and P, the Gibbs energy decreases until equilibrium is reached.

where the subscript $\{n_i\}$ indicates that the amounts of all species are held constant. **A change can take place spontaneously at constant temperature, pressure, and amounts of species if the Gibbs energy decreases.** This equation provides the means for discussing phase equilibrium and chemical equilibrium. Figure 4.1 illustrates the way G changes when a system spontaneously goes from an initial state to the equilibrium state.

Since the Gibbs energy decreases in an irreversible process at constant T and P, it becomes a minimum in the final equilibrium state. We can, however, *imagine* a reversible process occurring at equilibrium; for example, we may imagine the evaporation of an infinitesimal amount of water from the liquid into the vapor phase that is saturated with water vapor at constant temperature and pressure. For such a process, $dG = 0$.

These same relations may be applied to finite changes as well as infinitesimal changes, replacing the d's by Δ's. It must be remembered, however, that spontaneous changes always go to the minimum (as in the case of the Gibbs energy at constant T and P) or to the maximum (as in the case of the entropy of an isolated system).

Integration of the fundamental equation for G at constant values of the intensive properties T, P, and μ_i yields

$$G = \sum_{i=1}^{N_s} \mu_i n_i \tag{4.45}$$

This is a very interesting and important equation because it shows that the Gibbs energy is made up of a sum of terms with one for each of the N species when the intensive properties for G are constant. The corresponding equations for U, H, and A when their intensive properties are held constant are given in equations 4.18, 4.25, and 4.34. The importance of equation 4.45 will be discussed in Section 4.7.

Equation 4.36 yields the following Maxwell relations:

$$-\left(\frac{\partial S}{\partial P}\right)_{T,\{n_j\}} = \left(\frac{\partial V}{\partial T}\right)_{P,\{n_i\}} \tag{4.46}$$

$$-\left(\frac{\partial S}{\partial n_i}\right)_{T,P,\{n_j\}} = \left(\frac{\partial \mu_i}{\partial T}\right)_{P,\{n_i\}} = -\overline{S}_i \tag{4.47}$$

$$\left(\frac{\partial V}{\partial n_i}\right)_{T,P,\{n_{j\neq i}\}} = \left(\frac{\partial \mu_i}{\partial P}\right)_{T,\{n_i\}} = \overline{V}_i \tag{4.48}$$

where $\overline{S}_i$ is the **partial molar entropy** of species i, and $\overline{V}_i$ is the **partial molar volume** of species i. There is also a Maxwell equation for each pair of species, which will be discussed later.

The Gibbs energy has been defined so that we will have a criterion for spontaneous change at constant T and P that will not require us to specifically consider what is happening in the surroundings, as we did in the derivation in Section 3.2. For an isolated system consisting of the system of interest and the surroundings at constant T and P, the entropy criterion is that $\Delta S_{sys} + \Delta S_{surr}$ must increase for a spontaneous process. Since the temperature is constant, $\Delta S_{surr} = -\Delta H_{sys}/T$, so that $\Delta S_{sys} - \Delta H_{sys}/T = -\Delta G_{sys}/T$ must increase in a spontaneous process at constant T and P, or ΔG_{sys} must decrease. Thus, the Gibbs energy G simply provides a more convenient thermodynamic property than the entropy for the application of the second law at constant T and P.

Although these criteria show whether a certain change is irreversible, it does not necessarily follow that the change will take place with an appreciable speed. Thus, a mixture of 1 mol of carbon and 1 mol of oxygen at 1 bar pressure and 25 °C has a Gibbs energy greater than that of 1 mol of carbon dioxide at 1 bar and 25 °C, and so it is possible for the carbon and the oxygen to combine to form carbon dioxide at this constant temperature and pressure. Although carbon may exist for a very long time in contact with oxygen, the reaction is theoretically possible. The reverse of a thermodynamically spontaneous change is, of course, a nonspontaneous change. Thus, the decomposition of carbon dioxide to carbon and oxygen at room temperature, which involves an increase in Gibbs energy, is nonspontaneous. It can occur only with the aid of an outside agency.

When work other than PV work occurs in a system, it contributes a term to the fundamental equation for the internal energy, as indicated in equation 2.43. These terms carry forward into the fundamental equation for the Gibbs energy, so that if, for example, extension work and surface work are involved,

$$dG = -S\,dT + V\,dP + \sum_{i=1}^{N_s} \mu_i\,dn_i + f\,dL + \gamma\,dA_s \tag{4.49}$$

where f is the force of extension, L is the length, γ is the surface tension, and A_s is the surface area. Thus we see that

$$f = -\left(\frac{\partial G}{\partial L}\right)_{T,P,\{n_i\},A_s} \tag{4.50}$$

$$\gamma = \left(\frac{\partial G}{\partial A_s}\right)_{T,P,\{n_i\},L} \tag{4.51}$$

This opens up the possibility of many more Legendre transforms and Maxwell equations.

Comment:

Legendre transforms look pretty mathematical, so we should stop and think about why they are used. The first law introduces the internal energy U, which is related to heat and work, and the second law introduces the entropy S, which increases when a spontaneous change occurs in an isolated system. That is really all we know about thermodynamic systems. Any thermodynamic question can be answered by using U and S, but these explanations become very complicated when we want to use P and T as independent variables. The natural variables of U are S and V, and S and V may be impossible or very difficult to control in the laboratory. And so we use Legendre transforms to introduce intensive variables that are more convenient. The Gibbs energy G is especially important for chemistry because its natural variables are T and P.

If various kinds of work are involved, the first law is

$$dU = đq + đw \tag{4.52}$$

Using equation 3.21 yields

$$-dU + T\,dS \geq -đw \tag{4.53}$$

At constant temperature, this becomes

$$-d(U - TS) \geq -đw \tag{4.54}$$

or

$$(dA)_T \leq đw \tag{4.55}$$

where A is the Helmholtz energy. The symbol A actually comes from *arbeit*, the German word for work. Thus, in a reversible process at constant temperature, the work done on the system is equal to the increase in the Helmholtz energy. In general, equation 4.54 shows that *the decrease in A is an upper bound on the total work done in the surroundings*. When the system does work on the surroundings in a real process, the work done on the surroundings (remember that it is negative) is less than the decrease in the Helmholtz energy of the system.

The Gibbs energy is especially useful when non-PV work is involved. In this case the first law can be written

$$dU = đq - P_{ext}\, dV + đw_{nonpv} \tag{4.56}$$

so that the inequality $T\, dS \geq đq$ can be written

$$-dU - P_{ext}\, dV + T\, dS \geq -đw_{nonpv} \tag{4.57}$$

The external pressure is represented by P_{ext}. At constant temperature and $P = P_{ext} =$ constant, this can be written

$$-d(U + PV - TS) \geq -đw_{nonpv} \tag{4.58}$$

$$(dG)_{T,P} \leq đw_{nonpv} \tag{4.59}$$

For a reversible process at constant T and P the change in Gibbs energy is equal to the non-PV work done on the system by the surroundings. Thus, when work is done on the system, the Gibbs energy increases, and when the system does work on the surroundings, the Gibbs energy decreases. In general, equation 4.58 shows that the decrease in G is an *upper bound* on the non-PV work done on the surroundings. When the system does work on the surroundings, the work done (remember that it is negative) is less than the decrease in Gibbs energy.

Let us consider inequality 4.59 in more detail by applying it to the charging and discharging of an electrochemical cell at constant temperature and pressure. The electrochemical cell is charged by an electrical generator, and we will imagine a perfect direct-current generator that consumes a known amount of mechanical work. When the electrochemical cell is charged, the increase in the Gibbs energy of the cell is less than the electrical work done by the generator on the system in a real process and is equal to the electrical work in the theoretical limit of a reversible process.

When the electrochemical cell is discharged by operating an idealized electrical motor that does mechanical work, the Gibbs energy of the cell decreases and the work done is negative. According to inequality 4.59, the work done by the electrical motor is more positive than the decrease in G in a real process. Thus, the amount of work done in the surroundings is less than the decrease in the Gibbs energy of the electrochemical cell, except in the theoretical limit of a reversible process. This provides a simple interpretation for the change in Gibbs energy for a system. For a reversible process at constant temperature and pressure, the change in the Gibbs energy for the system is equal to the non-PV work done on the system by the surroundings.

4.3 EFFECT OF TEMPERATURE ON THE GIBBS ENERGY

For an open system, equation 4.37 shows that $(\partial G/\partial T)_{P,\{n_i\}} = -S$. Since S is a positive quantity, the Gibbs energy G necessarily decreases as the temperature increases at constant pressure and constant amounts of species. An important equation is obtained by using this equation to eliminate S from $G = H - TS$:

$$G = H + T\left(\frac{\partial G}{\partial T}\right)_{P,\{n_i\}} \tag{4.60}$$

Since this equation involves both the Gibbs energy and its temperature derivative, it is more convenient to rearrange it so that only a temperature derivative appears. This can be accomplished by first differentiating G/T with respect to temperature at constant pressure and constant amounts of species:

$$\left[\frac{\partial G/T}{\partial T}\right]_{P,\{n_i\}} = -\frac{G}{T^2} + \frac{1}{T}\left(\frac{\partial G}{\partial T}\right)_{P,\{n_i\}} \tag{4.61}$$

Eliminating G from the right-hand side by use of equation 4.60 yields

$$H = -T^2\left[\frac{\partial G/T}{\partial T}\right]_{P,\{n_i\}} \tag{4.62}$$

This is referred to as the **Gibbs–Helmholtz equation.** For a change between state 1 and state 2, this equation can be written

$$\Delta H = -T^2\left[\frac{\partial(\Delta G/T)}{\partial T}\right]_{P,\{n_i\}} \tag{4.63}$$

This equation is very useful because if we can determine ΔG for a process or a reaction as a function of temperature, the enthalpy change for the process or reaction can be calculated without using a calorimeter. It is also useful because if ΔH and ΔG are known at one temperature, equation 4.63 can be integrated to calculate ΔG at another temperature assuming that ΔH is independent of temperature (see Section 5.5).

4.4 EFFECT OF PRESSURE ON THE GIBBS ENERGY

The equation $(\partial G/\partial P)_{T,\{n_i\}} = V$ (equation 4.38) may be integrated to obtain the value of G at another pressure, provided that its value is known at one pressure and V is known as a function of pressure at constant temperature:

$$\int_{G_1}^{G_2} \mathrm{d}G = \int_{P_1}^{P_2} V\,\mathrm{d}P \tag{4.64}$$

$$G_2 = G_1 + \int_{P_1}^{P_2} V\,\mathrm{d}P \tag{4.65}$$

This equation always applies, and, as we have noted, the Gibbs energy of a single substance always increases with the pressure. There are two special cases where equation 4.65 leads to simple relationships. If the volume is nearly independent of pressure, as it is for a liquid or solid, then

$$G_2 = G_1 + V(P_2 - P_1) \tag{4.66}$$

The Gibbs energy of a gas is more dependent on pressure than that of a liquid. For an ideal gas the dependence of Gibbs energy on pressure is obtained by substituting $V = nRT/P$ in equation 4.64:

$$\int_{G^\circ}^{G} dG = nRT \int_{P^\circ}^{P} d\ln P \tag{4.67}$$

$$G = G^\circ + nRT \ln \frac{P}{P^\circ} \tag{4.68}$$

where P° is the standard state pressure. The Gibbs energy is like the internal energy and the enthalpy in that its absolute value is not known, but the Gibbs energy of a species can be determined with respect to the elements it contains. Thus equation 4.68 can be written $\Delta_f G_i = \Delta_f G_i^\circ + RT \ln(P_i/P^\circ)$. The standard Gibbs energy G° has different values at different temperatures. The logarithmic dependence of the molar Gibbs energy of an ideal gas on the pressure of the gas is illustrated in Fig. 4.2.

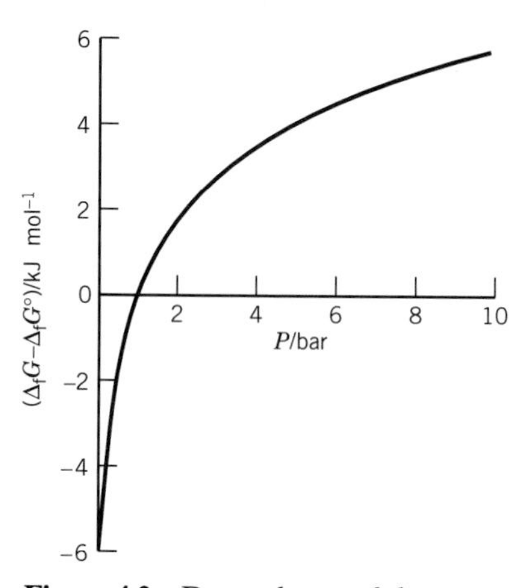

Figure 4.2 Dependence of the Gibbs energy of formation of an ideal gas on the pressure of the gas. (See Computer Problem 4.D.)

Alternatively, equation 4.65 may be integrated between any two pressures to obtain

$$\Delta G = G_2 - G_1 = nRT \ln \frac{P_2}{P_1} \tag{4.69}$$

Example 4.1 ***Calculation of molar thermodynamic properties for an ideal gas***

Since the molar Gibbs energy of an ideal gas is given by $\overline{G} = \overline{G}^\circ + RT \ln(P/P^\circ)$, derive the corresponding expressions for $\overline{V}$, $\overline{U}$, $\overline{H}$, $\overline{S}$, and $\overline{A}$.

Using equations 4.38, 4.41, 4.42, 4.37, and 4.43, respectively,

$$\overline{V} = \frac{RT}{P}$$

$$\overline{U} = \overline{U}^\circ = \overline{H}^\circ - RT$$

$$\overline{H} = \overline{H}^\circ = \overline{G}^\circ + T\overline{S}^\circ$$

$$\overline{S} = \overline{S}^\circ - R \ln \frac{P}{P^\circ}$$

$$\overline{A} = \overline{A}^\circ + RT \ln \frac{P}{P^\circ}$$

where $\overline{S}^\circ = -(\partial \overline{G}^\circ/\partial T)_P$ and $\overline{U}^\circ = \overline{G}^\circ + T\overline{S}^\circ - RT$. Note that the internal energy U and enthalpy H of an ideal gas are independent of pressure and volume.

Example 4.2 ***Calculation of changes in thermodynamic properties in the reversible isothermal expansion of an ideal gas***

An ideal gas at 27 °C expands isothermally and reversibly from 10 to 1 bar against a pressure that is gradually reduced. Calculate q per mole and w per mole and each of the thermodynamic quantities $\Delta\overline{U}$, $\Delta\overline{H}$, $\Delta\overline{G}$, $\Delta\overline{A}$, and $\Delta\overline{S}$.

Since the process is carried out isothermally and reversibly,

$$w = -RT \ln \frac{V_2}{V_1} = -RT \ln \frac{P_1}{P_2} = -(8.3145 \text{ J K}^{-1} \text{ mol}^{-1})(300.15 \text{ K}) \ln \frac{10}{1}$$

$$= -5746 \text{ J mol}^{-1}$$

$$\Delta\overline{A} = w = -5746 \text{ J mol}^{-1}$$

Since the internal energy of an ideal gas is not affected by a change in volume,

$$\Delta\overline{U} = 0$$

$$q = \Delta\overline{U} - w = 0 + 5746 = 5746 \text{ J mol}^{-1}$$

$$\Delta\overline{H} = \Delta\overline{U} + \Delta(P\overline{V}) = 0 + 0 = 0$$

and since $P\overline{V}$ is constant for an ideal gas at constant temperature,

$$\Delta\overline{G} = \int_{10}^{1} \overline{V}\, \mathrm{d}P$$

$$= RT \ln\frac{1}{10} = (8.3145 \text{ J K}^{-1} \text{ mol}^{-1})(300.15 \text{ K})(-2.3026)$$

$$= -5746 \text{ J mol}^{-1}$$

$$\Delta\overline{S} = \frac{q_{\text{rev}}}{T} = \frac{5746 \text{ J mol}^{-1}}{300.15 \text{ K}} = 19.14 \text{ J K}^{-1} \text{ mol}^{-1}$$

Also,

$$\Delta\overline{S} = \frac{\Delta\overline{H} - \Delta\overline{G}}{T} = \frac{0 + (5746 \text{ J mol}^{-1})}{300.15 \text{ K}} = 19.14 \text{ J K}^{-1} \text{ mol}^{-1}$$

Example 4.3 ***Calculation of changes in thermodynamic properties in the irreversible isothermal expansion of an ideal gas***

An ideal gas expands isothermally at 27 °C into an evacuated vessel so that the pressure drops from 10 to 1 bar; that is, it expands from a vessel of 2.463 L into a connecting vessel such that the total volume is 24.63 L. Calculate the change in thermodynamic quantities.

This process is isothermal, but is not reversible.

$w = 0$ because the system as a whole is closed and no external work can be done.

$\Delta\overline{U} = 0$ because the gas is an ideal gas.

$$q = \Delta\overline{U} - w = 0 + 0 = 0$$

$\Delta\overline{U}, \Delta\overline{H}, \Delta\overline{G}, \Delta\overline{A}$, and $\Delta\overline{S}$ are the same as in Example 4.2 because the initial and final states are the same.

Example 4.4 ***Calculation of the Gibbs energy of formation of gaseous and liquid methanol as a function of pressure***

The effect on the Gibbs energy of a gas due to changing the pressure is much greater than the effect on the Gibbs energy of the corresponding liquid because the molar volume of the gas is much larger. To see how this works for a specific substance, consider the liquid and gas forms of methanol, for which the standard Gibbs energy of formation and standard enthalpy of formation at 298.15 K are given in Table C.2. The standard Gibbs energy of formation for liquid CH_3OH at 298.15 K is -166.27 kJ mol^{-1}, and that for gaseous CH_3OH is -161.96 kJ mol^{-1}. The density of liquid methanol at 298.15 K is 0.7914 g/cm^3. (*a*) Calculate $\Delta_f G(CH_3OH, g)$ at 10 bar at 298.15 K assuming methanol vapor is an ideal gas. (Note that the superscript ° has been deleted because the pressure is not the standard pressure of 1 bar.) (*b*) Calculate $\Delta_f G(CH_3OH, l)$ at 10 bar at 298.15 K.

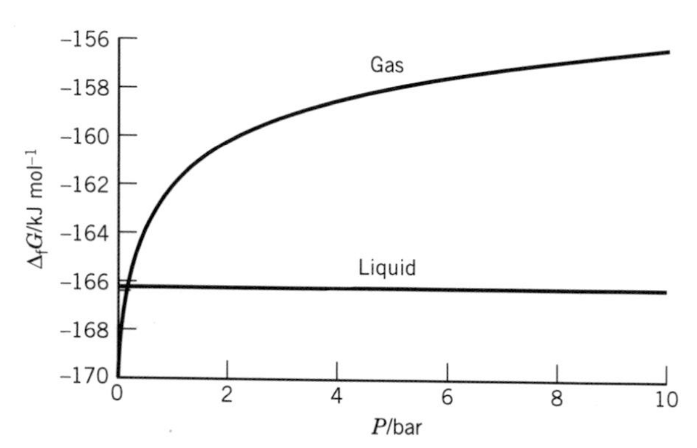

Figure 4.3 Plots of the Gibbs energies of formation of gaseous and liquid methanol at various pressures at 298.15 K. Note that the Gibbs energy of formation of the gas increases with pressure much more rapidly than the Gibbs energy of formation of the liquid. The curves cross at the vapor pressure of liquid methanol at this temperature. (See Computer Problem 4.D.)

(*a*) The effect of pressure on $\Delta G(\text{g})$ is given by equation 4.68:

$$\begin{aligned}\Delta_f G &= \Delta_f G^\circ + RT\ln(P/P^\circ)\\ &= -161.96\ \text{kJ mol}^{-1} + (8.3145\times 10^{-3}\ \text{kJ K}^{-1}\ \text{mol}^{-1})(298.15\ \text{K})\ln 10\\ &= -156.25\ \text{kJ mol}^{-1}\end{aligned}$$

This point is plotted in Fig. 4.3 along with points at lower pressures.

(*b*) The effect of pressure on $\Delta G(\text{l})$ is given by equation 4.66:

$$\Delta_f G = \Delta_f G^\circ + \overline{V}(P - P^\circ)$$

The molar volume of methanol is

$$\begin{aligned}\overline{V} &= (32.04\ \text{g mol}^{-1})/(0.7914\ \text{g cm}^{-3}) = 40.49\ \text{cm}^3\ \text{mol}^{-1}\\ &= (40.49\ \text{cm}^3\ \text{mol}^{-1})(10^{-2}\ \text{m/cm})^3 = 40.49\times 10^{-6}\ \text{m}^3\ \text{mol}^{-1}\end{aligned}$$

and

$$\begin{aligned}\Delta_f G &= -166.27\ \text{kJ mol}^{-1} + (40.49\times 10^{-6}\ \text{m}^3\ \text{mol}^{-1})(9\times 10^5\ \text{Pa})/(10^3\ \text{J kJ}^{-1})\\ &= -166.23\ \text{kJ mol}^{-1}\end{aligned}$$

This point is plotted in Fig. 4.3 along with points at lower pressures. Note that the liquid and gas curves cross at a pressure less than 1 bar. The intersection is at the vapor pressure of methanol at 298.15 K; at this pressure the liquid and gas have the same molar Gibbs energy of formation $\Delta_f G$. Note that $\Delta_f G(\text{g})$ of an ideal gas becomes $-\infty$ at zero pressure.

4.5 FUGACITY AND ACTIVITY

The Gibbs energy of a real gas is not given by equation 4.68, which we derived for an ideal gas. However, G. N. Lewis recognized that it would be convenient to keep the same form of equation for a real gas. He accomplished this by defining the **fugacity** f, which is a function of T and P, using

$$\overline{G} = \overline{G}^\circ + RT \ln \frac{f}{P^\circ} \tag{4.70}$$

$$\lim_{P \to 0} \frac{f}{P} = 1 \tag{4.71}$$

The fugacity has the units of pressure. As the pressure approaches zero, the gas approaches ideal behavior and the fugacity approaches the pressure. The fugacity is simply a measure of the molar Gibbs energy of a real gas, but it has the following advantage over the molar Gibbs energy: f goes from 0 to ∞, while $\overline{G}$ goes from $-\infty$ to $+\infty$ (see Fig. 4.2).

The fugacity of a real gas at a particular temperature and pressure can be calculated if the equation of state (Section 1.5) of the gas is known. As we will see in the following derivation, it is convenient to use the virial equation of state written in terms of pressure (equation 1.12). Since $(\partial \overline{G}/\partial P)_T = \overline{V}$, the differential Gibbs energy at constant temperature is $\mathrm{d}\overline{G} = \overline{V}\,\mathrm{d}P$ for a real gas and $\mathrm{d}\overline{G}^{\mathrm{id}} = \overline{V}^{\mathrm{id}}\,\mathrm{d}P$ for an ideal gas. The difference in Gibbs energy between a real gas and an ideal gas can be integrated from some low pressure P^* to the pressure at which we would like to know the fugacity:

$$\int_{P^*}^{P} \mathrm{d}(\overline{G} - \overline{G}^{\mathrm{id}}) = \int_{P^*}^{P} (\overline{V} - \overline{V}^{\mathrm{id}})\,\mathrm{d}P \tag{4.72}$$

$$(\overline{G} - \overline{G}^{\mathrm{id}})_P - (\overline{G}^* - \overline{G}^{*\mathrm{id}})_{P^*} = \int_{P^*}^{P} (\overline{V} - \overline{V}^{\mathrm{id}})\,\mathrm{d}P \tag{4.73}$$

Now, if we let $P^* \to 0$, then $\overline{G}^* \to \overline{G}^{*\mathrm{id}}$, and

$$(\overline{G} - \overline{G}^{\mathrm{id}})_P = \int_{0}^{P} (\overline{V} - \overline{V}^{\mathrm{id}})\,\mathrm{d}P \tag{4.74}$$

Introducing equation 4.70 for $\overline{G}$ and equation 4.68 for $\overline{G}^{\mathrm{id}}$ yields

$$\ln\left(\frac{f}{P}\right) = \frac{1}{RT}\int_{0}^{P} (\overline{V} - \overline{V}^{\mathrm{id}})\,\mathrm{d}P \tag{4.75}$$

or

$$\frac{f}{P} = \exp\left[\frac{1}{RT}\int_{0}^{P} (\overline{V} - \overline{V}^{\mathrm{id}})\,\mathrm{d}P\right] \tag{4.76}$$

The ratio of the fugacity to the pressure is called the **fugacity coefficient** ϕ; $\phi = f/P$. The fugacity coefficient is frequently used as a measure of the nonideality of a gas in connection with its phase equilibrium or chemical equilibrium properties. When PVT data are available on a gas, a plot may be prepared of the difference between its molar volume and the molar volume of an ideal gas versus pressure at the temperature of interest. The integral of this plot up to the pressure of interest is then used in equation 4.76 to calculate $\phi = f/P$. Equation 4.76 may be written in terms of the compressibility factor Z. Since $\overline{V} = RTZ/P$,

$$\frac{f}{P} = \exp\left[\frac{1}{RT}\int_{0}^{P}\left(\frac{RTZ}{P} - \frac{RT}{P}\right)\mathrm{d}P\right] = \exp\left[\int_{0}^{P}\frac{Z-1}{P}\,\mathrm{d}P\right] \tag{4.77}$$

Thus, the fugacity of a gas is readily calculated at some pressure P if Z is known as a function of pressure up to that particular pressure.

Example 4.5 ***Expression of the fugacity in terms of virial coefficients***

Given the expression for the compressibility factor Z as a power series in P (equation 1.12), what is the expression for the fugacity in terms of the virial coefficients?

Using equation 4.77, we find

$$\ln \frac{f}{P} = \int_0^P (B' + C'P + \cdots)\,dP = B'P + \frac{C'P^2}{2} + \cdots$$

Example 4.6 ***The fugacity of a van der Waals gas***

Using the expression for the compressibility factor Z of a van der Waals gas given in equation 1.26, what is the expression for fugacity of a van der Waals gas?

As an approximation, terms in P^2 and higher in the series expansion are omitted.

$$Z = 1 + \left[b - \left(\frac{a}{RT}\right)\right]\frac{P}{RT}$$

$$\ln \frac{f}{P} = \int_0^P \left(\frac{Z-1}{P}\right) dP$$

$$= \int_0^P \left[b - \left(\frac{a}{RT}\right)\right]\frac{1}{RT}\,dP$$

$$= \left[b - \left(\frac{a}{RT}\right)\right]\frac{P}{RT}$$

$$f = P \exp\left[\left(b - \frac{a}{RT}\right)\frac{P}{RT}\right]$$

Example 4.7 ***Estimating the fugacity of nitrogen gas at 50 bar and 298 K***

Given that the van der Waals constants of nitrogen are $a = 1.408$ L^2 bar mol^{-2} and $b = 0.03913$ L mol^{-1}, estimate the fugacity of nitrogen gas at 50 bar and 298 K.

$$f = P \exp\left[\left(b - \frac{a}{RT}\right)\frac{P}{RT}\right]$$

$$= (50\text{ bar}) \exp\left\{\left[0.03913 - \frac{1.408}{(0.083\,145)(298)}\right]\frac{50}{(0.083\,145)(298)}\right\}$$

$$= 48.2\text{ bar}$$

Now we are in a better position to understand the standard state of a gas that is used for thermodynamic tables, such as Tables C.2 and C.3. The standard state is

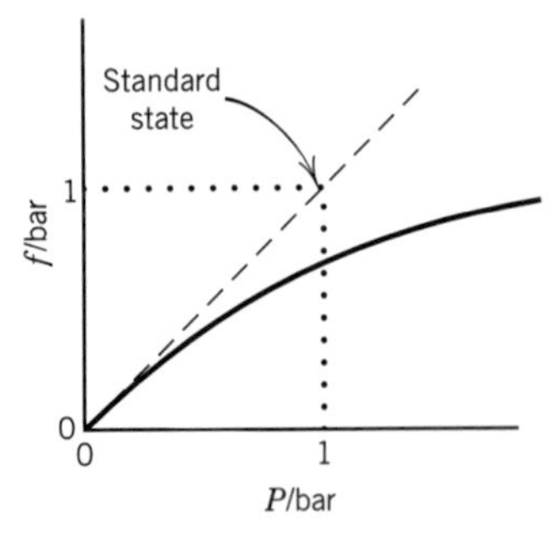

Figure 4.4 Plot of fugacity versus pressure for a real gas. The dashed line is for an ideal gas. The standard state is the pure substance at a pressure of 1 bar in a hypothetical state in which it exhibits ideal gas behavior.

the pure substance at a pressure of 1 bar in a hypothetical state in which it exhibits ideal gas behavior as shown in Fig. 4.4. The solid line gives the behavior of a real gas. As the pressure is reduced, the real gas approaches ideal behavior. This ideal gas is then compressed to 1 bar along the dashed line as a hypothetical ideal gas.

G. N. Lewis introduced the **activity** a as a means of dealing with real substances in the gas, liquid, and solid state. In analogy to equation 4.70, the activity of a pure substance, or its activity in a mixture, is defined by

$$\mu_i = \mu_i^\circ + RT \ln a_i \tag{4.78}$$

Thus, the activity a_i is simply a means for expressing the chemical potential of a species in a mixture. The activity is dimensionless, and $a_i = 1$ in the reference state for which $\mu_i = \mu_i^\circ$. For a real gas $a_i = f_i/P^\circ$, where f_i is the fugacity. For an ideal gas, $a_i = P_i/P^\circ$. We will see later in dealing with solutions that it is convenient to write the activity a_i as the product of an activity coefficient γ_i and a concentration.

The activity of a pure solid or liquid can be taken as unity if the pressure is close enough to the standard state pressure so that the effect of pressure on the chemical potential is negligible. If the effect of pressure is not negligible, the activity of a solid or liquid can be readily calculated because the molar volume $\overline{V}$ can be assumed to be constant at all reasonable pressures. For a pure solid or liquid, equation 4.66 can be written

$$\mu(T, P) = \mu^\circ(T) + \overline{V}(P - P^\circ) \tag{4.79}$$

Comparison with equation 4.78 shows that $RT \ln a = \overline{V}(P - P^\circ)$ or

$$a = e^{\overline{V}(P - P^\circ)/RT} \tag{4.80}$$

Small changes in pressure do not have a significant effect on the activity of a solid or liquid because of the smallness of the exponent.

Example 4.8 ***Calculating the activity of liquid water at 10 and 100 bar***

What is the activity of liquid water at 1, 10, and 100 bar at 25 °C, assuming that $\overline{V}$ is constant?

At $P = 1$ bar,

$$a = 1$$

At $P = 10$ bar,

$$a = \exp \frac{\overline{V}(P - P^\circ)}{RT}$$

$$= \exp \frac{(0.018 \text{ kg mol}^{-1})(9 \text{ bar})}{(0.083\,145 \text{ L bar K}^{-1} \text{ mol}^{-1})(298 \text{ K})}$$

$$= 1.007$$

At $P = 100$ bar, $a = 1.075$.

4.6 THE SIGNIFICANCE OF THE CHEMICAL POTENTIAL

The chemical potential μ_i of a species was introduced in equation 4.6, and we saw in equation 4.11 that it is equal to the partial derivative of the internal energy of a homogeneous mixture with respect to the amount of species i in the system at constant S, V, and $\{n_j\}$. Later we found in equations 4.23, 4.32, and 4.39 that this property of a species can be obtained in three other ways:

$$\mu_i = \left(\frac{\partial U}{\partial n_i}\right)_{S,V,\{n_{j\neq i}\}} = \left(\frac{\partial H}{\partial n_i}\right)_{S,P,\{n_{j\neq i}\}} = \left(\frac{\partial A}{\partial n_i}\right)_{T,V,\{n_{j\neq i}\}} = \left(\frac{\partial G}{\partial n_i}\right)_{T,P,\{n_{j\neq i}\}} \quad (4.81)$$

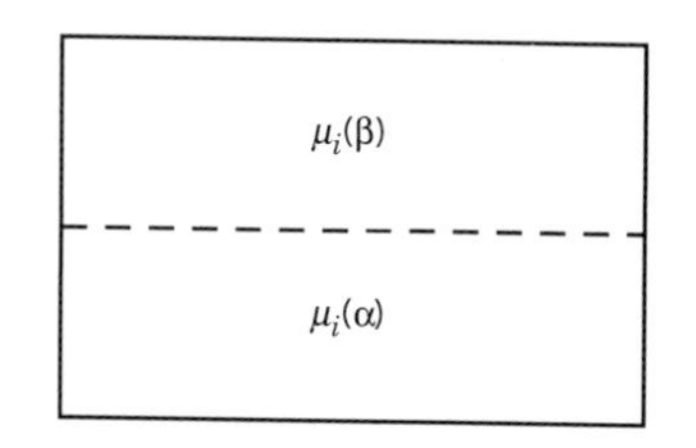

Figure 4.5 Two phases at the same temperature and pressure. Many species may be present, but we will focus on species i.

Since it is impossible to hold S constant (except in reversible adiabatic processes) and since experiments are not often carried out at constant volume, it is this last definition of the chemical potential that is most often used, and at constant T and P the chemical potential can be referred to as the partial molar Gibbs energy; $\mu_i = \overline{G}_i$.

The concept of the chemical potential μ_i of a species is extremely important, and we can see that in considering two phases that are at constant T and P, as shown in Fig. 4.5. We have already seen that at equilibrium these phases have the same temperature (Section 3.2) and the same pressure. The two phases may contain many species, and we will consider transferring an infinitesimal amount dn_i of species i from phase α to phase β at constant T and P. The total differential of the Gibbs energy for this two-phase system is given by equation 4.36, which becomes

$$(dG)_{T,P} = -\mu_i(\alpha)\,dn_i + \mu_i(\beta)\,dn_i = dn_i[\mu_i(\beta) - \mu_i(\alpha)] \quad (4.82)$$

For this transfer to occur spontaneously, $dG < 0$ and therefore $\mu_i(\alpha) > \mu_i(\beta)$; in other words, a species diffuses spontaneously from the phase where its chemical potential is higher to the phase where its chemical potential is lower. In this way the chemical potential is analogous to the electric potential and mechanical potential. If the phases are in equilibrium (i.e., no spontaneous change can take place), then there is no change in Gibbs energy in the transfer, $dG = 0$, and

$$\mu_i(\alpha) = \mu_i(\beta) \quad (4.83)$$

Thus, at equilibrium, the chemical potential of a species is the same in all of the phases of a system.

We will later see that at equilibrium the chemical potential of a species is the same in all of the phases in a system, even if the phases are at different pressures, as in a small liquid droplet in equilibrium with its vapor (Section 6.9) or in an osmotic pressure experiment (Section 6.7). In the next chapter, we will see that it is the chemical potentials of the species involved that determine whether they will undergo a chemical reaction. In Chapter 7 we will see that the chemical potential of an ion in a multiphase system with phases at different electric potentials is the same in each phase at equilibrium.

The partial molar entropy and partial molar volume of a species can be calculated from measurements of the chemical potential of the species as a function of temperature and pressure by use of two Maxwell relations for the fundamental equation for G (see equations 4.47 and 4.48):

$$-\overline{S}_i = \left(\frac{\partial \mu_i}{\partial T}\right)_{P,\{n_i\}} \quad (4.84)$$

$$\overline{V}_i = \left(\frac{\partial \mu_i}{\partial P}\right)_{T,\{n_i\}} \quad (4.85)$$

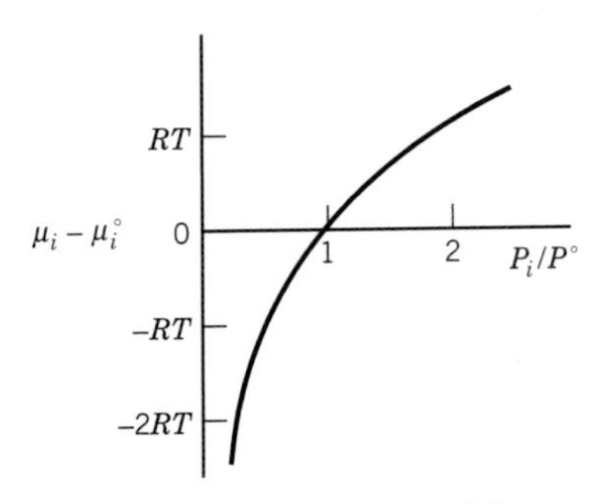

Figure 4.6 Chemical potential μ_i of an ideal gas as a function of pressure relative to the chemical potential μ_i° of the gas at the standard pressure of 1 bar. This plot is closely related to Figure 4.2, but it is given because of the importance of the chemical potential.

As a simple example, consider a mixture of ideal gases. Since the partial molar volume for a species in an ideal gas is equal to the molar volume of the mixture (i.e., $\overline{V_i} = \overline{V}$),

$$\overline{V} = \frac{RT}{P} = \left(\frac{\partial \mu_i}{\partial P}\right)_{T,\{n_i\}} = \left[\left(\frac{\partial \mu_i}{\partial P_i}\right)\left(\frac{\partial P_i}{\partial P}\right)\right]_{T,\{n_i\}} = x_i \left(\frac{\partial \mu_i}{\partial P_i}\right)_{T,\{n_i\}} \tag{4.86}$$

Dividing both sides by x_i yields

$$\left(\frac{\partial \mu_i}{\partial P_i}\right)_{T,\{n_i\}} = \frac{RT}{P_i} \tag{4.87}$$

Thus,

$$\int_{\mu_i^\circ}^{\mu_i} \mathrm{d}\mu_i = RT \int_{P^\circ}^{P_i} \frac{\mathrm{d}P_i}{P_i} \tag{4.88}$$

Integration yields

$$\mu_i = \mu_i^\circ + RT \ln \frac{P_i}{P^\circ} \tag{4.89}$$

where P° is the standard state pressure. The standard state pressure is taken as 10^5 Pa $= 1$ bar. A similar equation was given earlier (cf. equation 4.68) for a pure gas. See Fig. 4.6.

We can derive the expression for the partial molar entropy of an ideal gas in a mixture by using equations 4.84 and 4.89. Substituting equation 4.89 in equation 4.84 yields

$$\overline{S_i} = \overline{S_i^\circ} - R \ln \frac{P_i}{P^\circ} \tag{4.90}$$

The equation for the molar entropy of a pure ideal gas was derived in Example 4.1.

Comment:

The chemical potential is one of the most important concepts in chemical thermodynamics. In both chemical reactions and phase changes, the chemical potential of a species times its differential amount (reacted or transferred) determines the change in U, H, A, or G, depending on the variables that are held constant during the change. Although the chemical potential is a very general concept, we will use it most frequently in discussing systems at specified T and P. The chemical potential of a species is determined by its partial pressure in an ideal gas mixture and by its concentration in an ideal solution. For real mixtures, whether gaseous or liquid, the chemical potential of a species is a much more complicated function of the composition, but we can always think of it as a simple function of the activity a, which was introduced in Section 4.5.

4.7 ADDITIVITY OF PARTIAL MOLAR PROPERTIES WITH APPLICATIONS TO IDEAL GASES

Since $G = \sum n_i \mu_i = \sum n_i \overline{G_i}$ (equation 4.45), all the other extensive properties of a one-phase system are also additive. One way of looking at this is that if the

Gibbs energy of a system is known as a function of T, P, and amounts of species, the entropy of the system can be calculated by taking the negative temperature derivative of the Gibbs energy (see equation 4.37):

$$S = -\sum_{i=1}^{N_s} \left(\frac{\partial \mu_i}{\partial T}\right)_{P,\{n_i\}} n_i = \sum_{i=1}^{N_s} \overline{S}_i n_i \tag{4.91}$$

where $\overline{S}_i$ is the **partial molar entropy** of species i.

Equation 4.38 shows that the volume of a system is equal to the derivative of the Gibbs energy with respect to pressure. Substituting equation 4.48 in equation 4.38 yields

$$V = -\sum_{i=1}^{N_s} \left(\frac{\partial \mu_i}{\partial P}\right)_{P,\{n_i\}} n_i = \sum_{i=1}^{N_s} \overline{V}_i n_i \tag{4.92}$$

where $\overline{V}_i$, the **partial molar volume** of species i, is defined in equation 4.38. The additivity of the partial molar volumes of a mixture of ideal gases was discussed in Chapter 1.

Equation 4.61 shows the relation between the Gibbs energy and the enthalpy for a system. Substituting equation 4.45 in equation 4.62 yields

$$H = -T^2 \sum_{i=1}^{N_s} \left[\frac{\partial(\mu_i/T)}{\partial T}\right]_{P,\{n_i\}} n_i = \sum_{i=1}^{N_s} \overline{H}_i n_i \tag{4.93}$$

where the **partial molar enthalpy** of species i is defined by

$$\overline{H}_i = \left(\frac{\partial H}{\partial n_i}\right)_{T,P,\{n_{j\neq i}\}} \tag{4.94}$$

Example 4.9 ***Derivations of relations between partial molar properties***

Take the derivatives of $G = H - TS$ and $-S = (\partial G/\partial T)_{P,\{n_i\}}$ with respect to n_i to obtain the corresponding equations for the partial molar properties.

$$\left(\frac{\partial G}{\partial n_i}\right)_{T,P,\{n_{j\neq i}\}} = \left(\frac{\partial H}{\partial n_i}\right)_{T,P,\{n_{j\neq i}\}} - T\left(\frac{\partial S}{\partial n_i}\right)_{T,P,\{n_{j\neq i}\}}$$

$$\mu_i \equiv \overline{G}_i = \overline{H}_i - T\overline{S}_i$$

$$-\left(\frac{\partial S}{\partial n_i}\right)_{T,P,\{n_{j\neq i}\}} = \left[\frac{\partial}{\partial n_i}\left(\frac{\partial G}{\partial T}\right)_{P,n}\right]_{T,P,\{n_{j\neq i}\}} = \left[\frac{\partial}{\partial T}\left(\frac{\partial G}{\partial n_i}\right)_{T,P,\{n_{j\neq i}\}}\right]_{P,\{n_{j\neq i}\}}$$

$$-\overline{S}_i = \left(\frac{\partial \overline{G}_i}{\partial T}\right)_{P,\{n_{j\neq i}\}} = \left(\frac{\partial \mu_i}{\partial T}\right)_{P,\{n_{j\neq i}\}}$$

This last equation is actually a Maxwell relation from equation 4.36. Note that whereas V and S for a system are always positive, $\overline{V}_i$ and $\overline{S}_i$ may be negative.

There are similar equations for U and A, which can be obtained from Legendre transforms. These additivity equations are general, but they are most easily applied to mixtures of ideal gases.

Thermodynamics alone cannot lead to the conclusion that a mixture of ideal gases will behave as an ideal gas. Nevertheless, it is found that at low pressures mixtures of real gases do behave as ideal gases; that is, they behave as if there are no interactions between the molecules of the gas. Such mixtures are referred to as **ideal** mixtures to indicate the additional assumption involved.

The expression for the chemical potential of a species in an ideal gas mixture is given by equation 4.89, but we will find it convenient to write it as

$$\mu_i = \mu_i^\circ + RT \ln \frac{y_i P}{P^\circ} \tag{4.95}$$

since the partial pressure P_i of any species i in an ideal gas mixture is defined by

$$P_i \equiv y_i P \tag{4.96}$$

where y_i is the mole fraction and P is the total pressure. We will use y to represent the mole fraction in the gas phase, and x to represent the mole fraction in the liquid phase.

When we substitute equation 4.95 in $G = \sum n_i \mu_i$, we obtain the expression for the Gibbs energy of an ideal mixture of ideal gases:

$$\begin{aligned} G &= \sum n_i \mu_i^\circ + RT \sum n_i \ln y_i + \sum n_i RT \ln(P/P^\circ) \\ &= n_t \left[\sum y_i \mu_i^\circ + RT \sum y_i \ln y_i + RT \ln(P/P^\circ) \right] \\ &= n_t \overline{G} \end{aligned} \tag{4.97}$$

where $n_t = \sum n_i$ is the total amount of gas in the system and $\overline{G} = G/n_t$ is the molar Gibbs energy of the mixture. This expression for the Gibbs energy of an ideal gas mixture is written in terms of the natural variables T and P. Therefore, we can use equations 4.40–4.43 to derive the expressions for S, V, U, H, and A for an ideal gas mixture.

$$\begin{aligned} S &= -\left(\frac{\partial G}{\partial T}\right)_{P,\{n_i\}} = \sum n_i \overline{S}_i^\circ - R \sum n_i \ln y_i - \sum n_i R \ln(P/P^\circ) \\ &= n_t \left[\sum y_i \overline{S}_i^\circ - R \sum y_i \ln y_i - R \ln(P/P^\circ) \right] \\ &= n_t \overline{S} \end{aligned} \tag{4.98}$$

where $\overline{S} = S/n_t$ is the molar entropy of the mixture.

The enthalpy of the ideal gas mixture can be calculated from $H = G + TS$ since we have expressions for both G and S:

$$\begin{aligned} H &= \sum n_i (\mu_i^\circ + T\overline{S}_i^\circ) = \sum n_i \overline{H}_i^\circ \\ &= n_t \left[\sum y_i \overline{H}_i^\circ \right] = n_t \overline{H}^\circ \end{aligned} \tag{4.99}$$

since $\overline{H}_i^\circ = \mu_i^\circ + T\overline{S}_i^\circ$. Note that the enthalpy of a mixture of ideal gases is independent of the pressure. This is a result of the fact that the molecules in a mixture of ideal gases do not interact with each other.

The volume of an ideal gas mixture is given by

$$V = \left(\frac{\partial G}{\partial P}\right)_{T,\{n_i\}} = \sum n_i RT/P = \sum n_i \overline{V}_i \tag{4.100}$$

where the last form simply comes from the definition of the partial molar volume. The partial molar volume of each species in the mixture is the same: $\overline{V_i} = RT/P$, as we saw in Section 1.10. Gibbs commented that "every gas is as a vacuum to every other gas." These statements apply only to ideal gas mixtures.

Example 4.10 ***Calculation of changes in thermodynamic properties on mixing ideal gases***

In Section 3.5 we considered the mixing of two ideal gases that are initially at the same temperature and pressure but are separated from each other by a partition. Use equations 4.97–4.100 to calculate the changes in Gibbs energy, entropy, enthalpy, and volume when the partition is withdrawn. Note that the final total pressure is the same as the initial pressure of each gas.

The initial values of these quantities are

$$\begin{aligned} G &= n_1\mu_1^\circ + n_1 RT\ln(P/P^\circ) + n_2\mu_2^\circ + n_2 RT\ln(P/P^\circ) \\ &= n_1\mu_1^\circ + n_2\mu_2^\circ + (n_1+n_2)RT\ln(P/P^\circ) \end{aligned}$$

$$\begin{aligned} S &= n_1\overline{S}_1^\circ - n_1 R\ln(P/P^\circ) + n_2\overline{S}_2^\circ - n_2 R\ln(P/P^\circ) \\ &= n_1\overline{S}_1^\circ + n_2\overline{S}_2^\circ - (n_1+n_2)R\ln(P/P^\circ) \end{aligned}$$

$$H = n_1\overline{H}_1^\circ + n_2\overline{H}_2^\circ$$

$$V = n_1 RT/P + n_2 RT/P = (n_1+n_2)RT/P$$

The values after mixing are given by equations 4.97–4.100. Therefore the changes upon pulling out the partition are

$$\begin{aligned} \Delta_{\text{mix}} G &= RT(n_1\ln y_1 + n_2\ln y_2) \\ \Delta_{\text{mix}} S &= -R(n_1\ln y_1 + n_2\ln y_2) \\ \Delta_{\text{mix}} H &= 0 \\ \Delta_{\text{mix}} V &= 0 \end{aligned}$$

Since the mole fractions in a mixture are less than unity, the logarithmic terms are negative, and $\Delta_{\text{mix}} G < 0$. This corresponds to the fact that the mixing of gases is a spontaneous process at constant temperature and pressure. In other words, if two gases at the same pressure and temperature are brought into contact, they will spontaneously diffuse into each other until the gas phase is macroscopically homogeneous.

The Gibbs energy change for mixing two ideal gases is plotted versus the mole fraction of one of the gases in Fig. 4.7*a*. The greatest Gibbs energy change on mixing is obtained for $y_1 = y_2 = \frac{1}{2}$. The dependence of the entropy of mixing on the mole fraction of one of the two gases is shown in Fig. 4.7*b*.

When ideal gases are mixed at constant temperature and pressure, no heat is produced or consumed. This corresponds to the fact that molecules of ideal gases do not attract or repel each other. Thus, from an energy standpoint, it makes no difference whether the gases are separated or mixed. The driving force for mixing arises exclusively from the change in entropy. From the viewpoint of statistical

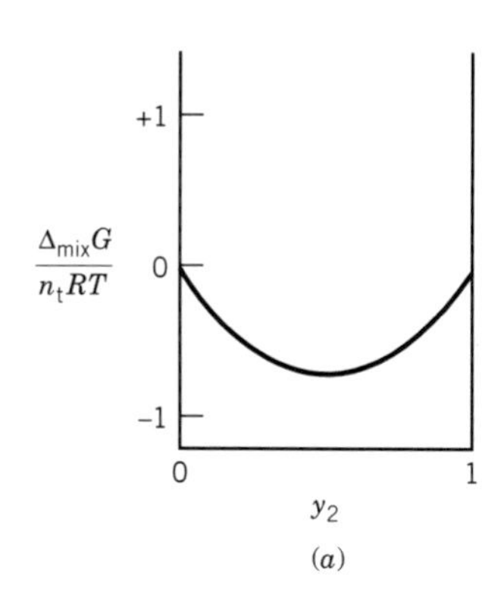

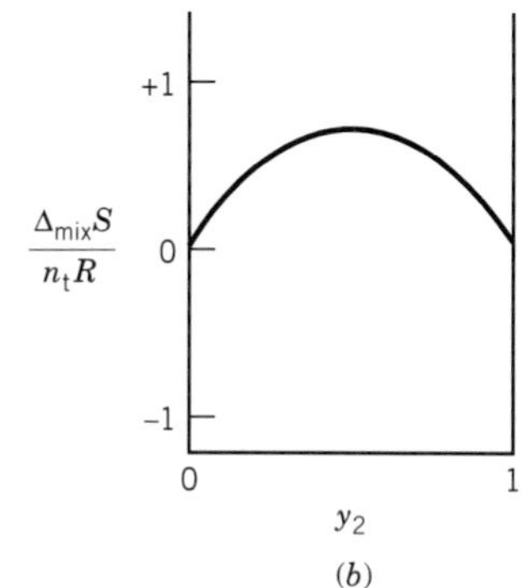

Figure 4.7 Thermodynamic quantities for the mixing of two ideal gases to form an ideal mixture. The total amount of gas is represented by n_t.

mechanics (Chapter 16), the mixed state is found at equilibrium because it is more probable, as discussed in Section 3.6 for the expansion of a gas.

4.8 GIBBS–DUHEM EQUATION

In making Legendre transforms of the internal energy, we introduced the intensive variables P and T as natural variables. This process can be continued to introduce the intensive variables $\mu_1, \mu_2, \ldots, \mu_N$ by making the **complete Legendre transform**

$$U' = U + PV - TS - \sum_{i=1}^{N_s} n_i \mu_i = 0 \tag{4.101}$$

It is evident from equation 4.18 that this transformed internal energy U' is equal to zero. The differential of U' is

$$\mathrm{d}U + P\,\mathrm{d}V + V\,\mathrm{d}P - T\,\mathrm{d}S - S\,\mathrm{d}T - \sum_{i=1}^{N_s} n_i\,\mathrm{d}\mu_i - \sum_{i=1}^{N_s} \mu_i\,\mathrm{d}n_i = 0 \tag{4.102}$$

Subtracting this from the fundamental equation (4.17) for U yields

$$V\,\mathrm{d}P - S\,\mathrm{d}T - \sum_{i=1}^{N_s} n_i\,\mathrm{d}\mu_i = 0 \tag{4.103}$$

which is known as the **Gibbs–Duhem equation.** Note that it deals with changes only of the intensive variables for the system. Because of this relation, the intensive variables for a system are not independent. This is in agreement with the conclusion discussed in Section 1.9 that there are $N_s + 1$ independent intensive variables for a one-phase system. In a multiphase system there is a separate Gibbs–Duhem equation for each phase.

For a system with two species at constant temperature and pressure that contains 1 mol of material,

$$x_1\,\mathrm{d}\mu_1 + x_2\,\mathrm{d}\mu_2 = 0 \tag{4.104}$$

$$x_1\,\mathrm{d}\mu_1 + (1 - x_1)\,\mathrm{d}\mu_2 = 0 \tag{4.105}$$

where x_1 is the mole fraction of component 1 and $(1 - x_1)$ is the mole fraction of component 2. Thus, when the composition is changed at constant T and P, the change in the chemical potential of species 2 is not independent of the change in the chemical potential of species 1. Later, in Section 6.6, we will use this form of the Gibbs–Duhem equation to show that if Henry's law holds for the solute (species 2), Raoult's law holds for the solvent (species 1).

Comment:

We can generalize on what we have here by pointing out that the Gibbs–Duhem equation is the complete Legendre transform for a system with a certain set of intensive variables. You may have thought of the intensive variables of a system as being independent, as the extensive variables are, but they are not. Suppose that the intensive variables for a system are T, P, μ_1, μ_2, and μ_3. According to the Gibbs–Duhem equation, if you specify the values of T, P, μ_1, and μ_2, then μ_3 can have only a particular value. This will be even more evident when we discuss equilibrium constants in the next chapter.

4.9 SPECIAL TOPIC: ADDITIONAL APPLICATIONS OF MAXWELL RELATIONS

A number of applications have already been made of Maxwell relations, but some others are of special interest. For the purpose of making calculations about a mole of a substance, the Maxwell relations from the fundamental equations for U, H, A, and G can be written

$$\left(\frac{\partial T}{\partial \overline{V}}\right)_{\overline{S}} = -\left(\frac{\partial P}{\partial \overline{S}}\right)_{\overline{V}} \tag{4.106}$$

$$\left(\frac{\partial T}{\partial P}\right)_{\overline{S}} = \left(\frac{\partial \overline{V}}{\partial \overline{S}}\right)_P \tag{4.107}$$

$$\left(\frac{\partial \overline{S}}{\partial \overline{V}}\right)_T = \left(\frac{\partial P}{\partial T}\right)_{\overline{V}} \tag{4.108}$$

$$-\left(\frac{\partial \overline{S}}{\partial P}\right)_T = \left(\frac{\partial \overline{V}}{\partial T}\right)_P \tag{4.109}$$

In Section 2.6, we found that $(\partial \overline{U}/\partial \overline{V})_T = 0$ for an ideal gas. Now we can calculate this partial derivative for any gas. Dividing the combined first and second law for a mole of a substance by $d\overline{V}$ and imposing constant temperature yields

$$\left(\frac{\partial \overline{U}}{\partial \overline{V}}\right)_T = T\left(\frac{\partial \overline{S}}{\partial \overline{V}}\right)_T - P \tag{4.110}$$

Using one of the Maxwell equations (4.108) yields

$$\left(\frac{\partial \overline{U}}{\partial \overline{V}}\right)_T = T\left(\frac{\partial P}{\partial T}\right)_{\overline{V}} - P \tag{4.111}$$

As an example, this equation can be applied to a van der Waals gas. Differentiation of $P = RT/(\overline{V} - b) - a/\overline{V}^2$ with respect to T at constant $\overline{V}$ yields $(\partial P/\partial T)_{\overline{V}} = R/(\overline{V} - b)$. Substituting this relation in equation 4.111 yields

$$\begin{aligned}\left(\frac{\partial \overline{U}}{\partial \overline{V}}\right)_T &= \frac{RT}{\overline{V} - b} - P = \frac{RT}{\overline{V} - b} - \left(\frac{RT}{\overline{V} - b} - \frac{a}{\overline{V}^2}\right) \\ &= \frac{a}{\overline{V}^2}\end{aligned} \tag{4.112}$$

Thus, for a van der Waals gas the internal pressure is inversely proportional to the square of the molar volume.

Example 4.11 ***Calculation of the change in molar internal energy in the expansion of propane gas, assuming it is a van der Waals gas***

Propane gas is allowed to expand isothermally from 10 to 30 L. What is the change in molar internal energy?

The change in internal energy for a given change in volume at constant temperature is given by

$$\int_{\overline{U}_1}^{\overline{U}_2} \mathrm{d}\overline{U} = \int_{\overline{V}_1}^{\overline{V}_2} \frac{a}{\overline{V}^2}\,\mathrm{d}\overline{V} = a\left(-\frac{1}{\overline{V}}\right)_{\overline{V}_1}^{\overline{V}_2}$$

$$\Delta\overline{U} = a\left(\frac{1}{\overline{V}_1} - \frac{1}{\overline{V}_2}\right)$$

According to Table 1.3, $a = 8.779\ \mathrm{L^2\ bar\ mol^{-2}}$, but we need to convert this to SI base units to calculate $\Delta\overline{U}$ in $\mathrm{J\ mol^{-1}}$:

$$\begin{aligned} a &= (8.779\ \mathrm{L^2\ bar\ mol^{-2}})(10^5\ \mathrm{Pa\ bar^{-1}})(10^{-3}\ \mathrm{m^3\ L^{-1}})^2 \\ &= 0.8779\ \mathrm{Pa\ m^6\ mol^{-2}} \end{aligned}$$

$$\begin{aligned} \Delta\overline{U} &= a\left(\frac{1}{\overline{V}_1} - \frac{1}{\overline{V}_2}\right) \\ &= (0.8779\ \mathrm{Pa\ m^6\ mol^{-2}})\left(\frac{1}{10\times 10^{-3}\ \mathrm{m^3\ mol^{-1}}} - \frac{1}{30\times 10^{-3}\ \mathrm{m^3\ mol^{-1}}}\right) \\ &= 58.5\ \mathrm{J\ mol^{-1}} \end{aligned}$$

Example 4.12 ***Calculation of the molar entropy of the isothermal expansion of a van der Waals gas***

Derive the equation for the molar entropy of isothermal expansion of a van der Waals gas.

$$P = \frac{RT}{\overline{V} - b} - \frac{a}{\overline{V}^2}$$

$$\left(\frac{\partial\overline{S}}{\partial\overline{V}}\right)_T = \left(\frac{\partial P}{\partial T}\right)_{\overline{V}} = \frac{R}{\overline{V} - b}$$

$$\int_{\overline{S}_1}^{\overline{S}_2} \mathrm{d}\overline{S} = R\int_{\overline{V}_1}^{\overline{V}_2} \frac{1}{\overline{V} - b}\,\mathrm{d}\overline{V}$$

$$\Delta\overline{S} = R\ln\frac{\overline{V}_2 - b}{\overline{V}_1 - b}$$

In dealing with derivatives of the volume of a fluid with respect to T and P, it is convenient to use α for the **cubic expansion coefficient,**

$$\alpha = \frac{1}{V}\left(\frac{\partial V}{\partial T}\right)_P = \frac{1}{\overline{V}}\left(\frac{\partial\overline{V}}{\partial T}\right)_P \tag{4.113}$$

and κ for the **isothermal compressibility,**

$$\kappa = -\frac{1}{V}\left(\frac{\partial V}{\partial P}\right)_T = -\frac{1}{\overline{V}}\left(\frac{\partial\overline{V}}{\partial P}\right)_T \tag{4.114}$$

For an ideal gas these quantities are $\alpha = 1/T$ and $\kappa = 1/P$. The derivative of $\overline{U}$ with respect to $\overline{V}$ at constant T can be written in terms of α and κ. According to the cyclic rule (Appendix D.3),

$$\left(\frac{\partial P}{\partial T}\right)_{\overline{V}} = -\frac{(\partial\overline{V}/\partial T)_P}{(\partial\overline{V}/\partial P)_T} = \frac{\alpha}{\kappa} \tag{4.115}$$

Substituting this in equation 4.111 yields

$$\left(\frac{\partial \overline{U}}{\partial \overline{V}}\right)_T = \frac{\alpha T - \kappa P}{\kappa} \tag{4.116}$$

The dependence of the enthalpy on the pressure can be obtained by dividing $\mathrm{d}H = T\mathrm{d}S + V\mathrm{d}P$ by the differential of the pressure to obtain

$$\left(\frac{\partial \overline{H}}{\partial P}\right)_T = T\left(\frac{\partial \overline{S}}{\partial P}\right)_T + \overline{V} \tag{4.117}$$

Substituting equation 4.109 yields

$$\left(\frac{\partial \overline{H}}{\partial P}\right)_T = -T\left(\frac{\partial \overline{V}}{\partial T}\right)_P + \overline{V} \tag{4.118}$$

We found in Section 2.8 that

$$\overline{C}_P - \overline{C}_V = \left[P + \left(\frac{\partial \overline{U}}{\partial \overline{V}}\right)_T\right]\left(\frac{\partial \overline{V}}{\partial T}\right)_P \tag{4.119}$$

Inserting $(\partial \overline{V}/\partial T)_P = \overline{V}\alpha$ and using equation 4.116 we obtain

$$\overline{C}_P - \overline{C}_V = \frac{T\overline{V}\alpha^2}{\kappa} \tag{4.120}$$

Since it is difficult to measure $\overline{C}_V$, its value is generally calculated from measurements of $\overline{C}_P$, the molar volume $\overline{V}$, the cubic expansion coefficient α, and the isothermal compressibility κ.

Nine Key Ideas in Chapter 4

1. The combined first and second law plus the introduction of chemical work terms by Gibbs yields the fundamental equation for the internal energy. This equation provides equations of state for T, P, and $\{\mu_i\}$, Maxwell equations, and an integrated equation for U.
2. The internal energy provides a criterion for spontaneous change and equilibrium when its natural variables S, V, and $\{n_i\}$ are held constant. However, S and V are usually not convenient independent variables in the laboratory. Legendre transforms can be used to define more useful thermodynamic potentials for this purpose. A Legendre transform is the definition of a new thermodynamic potential by subtracting the product of conjugate variables from another thermodynamic potential.
3. The Helmholtz energy is defined by $A = U - TS$ and the Gibbs energy is defined by $G = H - TS$. The Helmholtz energy provides the criterion for spontaneous change at specified T, V, and amounts of species. The Gibbs energy provides the criterion for spontaneous change at specified T, P, and amounts of species. Since T and P are convenient independent variables, the Gibbs energy is widely used as a criterion for phase equilibrium and chemical equilibrium.
4. The Gibbs energy decreases in a spontaneous process at constant T and P, and for a reversible process, the change in Gibbs energy is equal to the non-PV work that can be done on the surroundings by the system.

5. The determination of ΔG for a process or chemical reaction at a series of temperatures makes it possible to calculate ΔH by use of the Gibbs–Helmholtz equation.
6. The molar Gibbs energy for an ideal gas depends on its partial pressure, and the molar Gibbs energy for a nonideal gas depends on its fugacity. The activity of a real gas is equal to the ratio of its fugacity to the standard pressure.
7. At equilibrium the chemical potential of a species is the same in all of the phases of the system.
8. Since the Gibbs energy of a phase is equal to the sum of the products of amounts times molar Gibbs energies of species, all the other extensive thermodynamic properties of a one-phase system are also additive.
9. The complete Legendre transform for a system provides a relation between the differentials of all of the intensive variables for a system, and so the intensive variables for a system are not independent. This remarkable equation is known as the Gibbs–Duhem equation.

REFERENCES

M. Bailyn, *A Survey of Thermodynamics*. New York: American Institute of Physics, 1994.
J. A. Beattie and I. Oppenheim, *Principles of Thermodynamics.* New York: Elsevier Scientific, 1979.
K. E. Bett, J. S. Rowlinson, and G. Saville, *Thermodynamics for Chemical Engineers.* Cambridge, MA: MIT Press, 1975.
H. B. Callen, *Thermodynamics and an Introduction to Thermostatistics.* Hoboken, NJ: Wiley, 1985.
J. W. Gibbs, *The Collected Works of J. Willard Gibbs.* New Haven, CT: Yale University Press, 1948.
W. Greiner, L. Neise, and H. Stöcker, *Thermodynamics and Statistical Mechanics*. New York: Springer-Verlag, 1995.
K. S. Pitzer, *Thermodynamics,* 3rd ed. New York: McGraw-Hill, 1995.
S. I. Sandler, *Chemical and Engineering Thermodynamics.* Hoboken, NJ: Wiley, 1999.
J. M. Smith, H. C. Van Ness, and M. M. Abbott, *Introduction to Chemical Engineering Thermodynamics*, 5th ed. New York: McGraw-Hill, 1996.
J. W. Tester and M. Modell, *Thermodynamics and Its Applications*. Upper Saddle River, NJ: Prentice-Hall, 1997.

PROBLEMS

(M) Problems marked with an icon may be more conveniently solved on a personal computer with a mathematical program.

4.1 One mole of nitrogen gas is allowed to expand from 0.5 to 10 L. Calculate the change in molar entropy using (*a*) the ideal gas law and (*b*) the van der Waals equation.

4.2 Derive the relation for $\overline{C_P} - \overline{C_V}$ for a gas that follows the van der Waals equation.

4.3 Earlier we derived the expression for the entropy of an ideal gas as a function of T and P. Now that we have the Maxwell relations, derive the expression for dS for any fluid.

4.4 What is the change in molar entropy of liquid benzene at 25 °C when the pressure is raised to 1000 bar? The coefficient of thermal expansion α is 1.237×10^{-3} K^{-1}, the density is 0.879 g cm^{-3}, and the molar mass is 78.11 g mol^{-1}.

4.5 Derive the expression for $\overline{C_P} - \overline{C_V}$ for a gas with the following equation of state:

$$(P + a/\overline{V}^2)\overline{V} = RT$$

4.6 What is the difference between the molar heat capacity of iron at constant pressure and constant volume at 25 °C? Given:

$\alpha = 35.1 \times 10^{-6}\ \text{K}^{-1}$, $\kappa = 0.52 \times 10^{-6}\ \text{bar}^{-1}$, and the density is 7.86 g cm^{-3}.

4.7 In equation 1.26 we saw that the compressibility factor of a van der Waals gas can be written as

$$Z = 1 + \frac{1}{RT}\left(b - \frac{a}{RT}\right)P + \cdots$$

(*a*) To this degree of approximation, derive the expression for $(\partial\overline{H}/\partial P)_T$ for a van der Waals gas. (*b*) Calculate $(\partial\overline{H}/\partial P)_T$ for $CO_2(g)$ in J bar^{-1} mol^{-1} at 298 K. Given: $a = 3.640\ \text{L}^2\ \text{bar mol}^{-2}$ and $b = 0.042\,67\ \text{L mol}^{-1}$.

4.8 Derive the expression for $(\partial U/\partial V)_T$ (the internal pressure) for a gas following the virial equation with $Z = 1 + B/\overline{V}$.

4.9 In Section 3.4 we calculated that the enthalpy of freezing water at -10 °C is -5619 J mol^{-1}, and we calculated that the entropy of freezing water is -20.54 J K^{-1} mol^{-1} at -10 °C. What is the Gibbs energy of freezing water at -10 °C?

4.10 (*a*) Integrate the Gibbs–Helmholtz equation to obtain an expression for ΔG_2 at temperature T_2 in terms of ΔG_1 at T_1, assuming that ΔH is independent of temperature. (*b*) Obtain an expression for ΔG_2 using the more accurate approximation that $\Delta H = \Delta H_1 + (T - T_1)\Delta C_P$, where T_1 is an arbitrary reference temperature.

4.11 When a liquid is compressed its Gibbs energy is increased. To a first approximation the increase in molar Gibbs energy can be calculated using $(\partial\overline{G}/\partial P)_T = \overline{V}$, assuming a constant molar volume. What is the change in molar Gibbs energy for liquid water when it is compressed to 1000 bar?

4.12 An ideal gas is allowed to expand reversibly and isothermally (25 °C) from a pressure of 1 bar to a pressure of 0.1 bar. (*a*) What is the change in molar Gibbs energy? (*b*) What would be the change in molar Gibbs energy if the process occurred irreversibly?

4.13 The standard entropy of $O_2(g)$ at 298.15 K and 1 bar is listed in Table C.2 as 205.138 J K^{-1} mol^{-1}, and the standard Gibbs energy of formation is listed as 0 kJ mol^{-1}. Assuming that O_2 is an ideal gas, what will be the molar entropy and molar Gibbs energy of formation at 100 bar?

4.14 Helium is compressed isothermally and reversibly at 100 °C from a pressure of 2 to 10 bar. Calculate (*a*) q per mole, (*b*) w per mole, (*c*) $\Delta\overline{G}$, (*d*) $\Delta\overline{A}$, (*e*) $\Delta\overline{H}$, (*f*) $\Delta\overline{U}$, and (*g*) $\Delta\overline{S}$, assuming that helium is an ideal gas.

4.15 Toluene is vaporized at its boiling point, 111 °C. The heat of vaporization at this temperature is 361.9 J g^{-1}. For the vaporization of toluene, calculate (*a*) w per mole, (*b*) q per mole, (*c*) $\Delta\overline{H}$, (*d*) $\Delta\overline{U}$, (*e*) $\Delta\overline{G}$, and (*f*) $\Delta\overline{S}$.

4.16 If the Gibbs energy varies with temperature according to

$$G/T = a + b/T + c/T^2$$

how will the enthalpy and entropy vary with temperature? Check that these three equations are consistent.

4.17 Calculate the change in molar Gibbs energy $\overline{G}$ when supercooled water at -3 °C freezes at constant T and P. The density of ice at -3 °C is 0.917×10^3 kg m^{-3}, and its vapor pressure is 475 Pa. The density of supercooled water at -3 °C is 0.9996×10^3 kg m^{-3}, and its vapor pressure is 489 Pa.

4.18 Calculate the molar Gibbs energy G of fusion when supercooled water at -3 °C freezes at constant T and P. The enthalpy of fusion of ice is 6000 J mol^{-1} at 0 °C. The heat capacities of water and ice in the vicinity of the freezing point are 75.3 and 38 J K^{-1} mol^{-1}, respectively.

4.19 At 298.15 K and a particular pressure, a real gas has a fugacity coefficient ϕ of 2.00. At this pressure, what is the difference in the chemical potential of this real gas and an ideal gas?

4.20 As shown in Example 4.6, the fugacity of a van der Waals gas is given by a fairly simple expression if only the second virial coefficient is used. To this degree of approximation, derive the expressions $\overline{G}$, $\overline{S}$, $\overline{A}$, $\overline{U}$, $\overline{H}$, and $\overline{V}$.

4.21 A mole of a van der Waals gas is expanded isothermally from V_1 to V_2. Derive the expressions for the changes in Helmoltz energy and internal energy.

4.22 A one-component system has three natural variables, as is evident from

$$\text{d}U = T\,\text{d}S - P\,\text{d}V + \mu\,\text{d}n$$

We have seen how three additional potentials can be defined by making Legendre transforms and how a complete Legendre transform yields the Gibbs–Duhem equation, which in a certain sense is like a thermodynamic potential but has the value zero. The total number of thermodynamic potentials for a system with D natural variables is 2^D. This is the number of Legendre transforms that can be made taking all possible pairs of conjugate variables, two pairs, three pairs, ... Since $2^3 = 8$, there are three more thermodynamic potentials for this system that can be defined by Legendre transforms. Write their fundamental equations; let's call them X, Y, and Z.

4.23 Using the relation derived in Example 4.6, calculate the fugacity of $H_2(g)$ at 100 bar at 298 K.

4.24 Show that if the compressibility factor is given by $Z = 1 + BP/RT$ the fugacity is given by $f = P\,\text{e}^{Z-1}$. If Z is not very different from unity, $\text{e}^{Z-1} = 1 + (Z-1) + \cdots \cong Z$ so that $f = PZ$. Using this approximation, what is the fugacity of $H_2(g)$ at 50 bar and 298 K using its van der Waals constants?

Ⓜ **4.25** Calculate the partial molar volume of zinc chloride in 1 molal $ZnCl_2$ solution using the following data:

% by weight of $ZnCl_2$	2	6	10	14	18
Density/g cm^{-3}	1.0167	1.0532	1.0891	1.1275	1.1665

4.26 Calculate $\Delta_{\text{mix}}G$ and $\Delta_{\text{mix}}S$ for the formation of a quantity of air containing 1 mol of gas by mixing nitrogen and oxygen at 298.15 K. Air may be taken to be 80% nitrogen and 20% oxygen.

4.27 A mole of gas A is mixed with a mole of gas B at 1 bar and 298 K. How much work is required to separate these gases to produce a container of each at 1 bar and 298 K?

4.28 The fundamental equation for the enthalpy is given by equation 4.20. Show that the fundamental equation for the internal energy can be obtained by using the inverse Legendre transform $U = H - PV$. This is an example of what is meant by saying that there is no loss of information in making a Legendre transform.

4.29 Derive

$$C_P = -T\left(\frac{\partial^2 G}{\partial T^2}\right)$$

4.30 In studying statistical mechanics we will find (see Table 16.1) that for a monatomic ideal gas, the molar Gibbs energy is given by

$$\overline{G} = -T \ln \frac{T^{5/2}}{P}$$

where numerical constants have been omitted so that only the functional dependence on the natural variables of $\overline{G}$, that is, T and P, is shown. Derive the corresponding equations for $\overline{S}$, $\overline{H}$, $\overline{V}$, $\overline{U}$, and $\overline{A}$.

4.31 Statistical mechanics shows that for a monatomic ideal gas, the molar Gibbs energy is given by

$$\overline{G} = -\tfrac{5}{2}T \ln T + T \ln P$$

where the numerical factors have been omitted so that only the functional dependence on the natural variables, T and P, is shown. If we want to treat the thermodynamics of an ideal monatomic gas at specified T and $\overline{V}$ without losing any information, we cannot simply replace P with $T/\overline{V}$ and use

$$\overline{G} = -\frac{5}{2}T \ln T + T \ln \frac{T}{\overline{V}}$$

even though this relation is correct. If we want to treat the thermodynamics of an ideal monatomic gas at specified T and $\overline{V}$ without losing any information, we have to use the following Legendre transform to define the molar Helmholtz energy $\overline{A}$:

$$\overline{A} = \overline{G} - P\overline{V}$$

Use the expression for $\overline{A}$ obtained in this way to calculate $\overline{S}$, $\overline{V}$, $\overline{H}$, and $\overline{U}$ for an ideal monatomic gas as a function of T and $\overline{V}$. Show that these expressions agree with the expressions obtained in the preceding problem.

4.32 We already know enough about the thermodynamics of a monatomic ideal gas to express V, U, and S in terms of the natural variables of G, namely T, P, and n.

$$V = nRT/P$$

$$U = \tfrac{3}{2}nRT$$

$$S = nR\left\{\frac{\overline{S}^\circ}{R} + \ln\left[\left(\frac{T}{T^\circ}\right)^{5/2}\left(\frac{P^\circ}{P}\right)\right]\right\}$$

The last equation is the Sackur–Tetrode equation, where $\overline{S}^\circ$ is the molar entropy at the standard temperature T° (298.15 K) and standard pressure P° (1 bar). The Gibbs energy $G(T, P, n)$ of the ideal monatomic gas can be calculated by using the Legendre transform

$$G = U + PV - TS$$

The fundamental equation for the Gibbs energy is

$$dG = -S\, dT + V\, dP + \mu\, dn$$

Show that the correct expressions for S, V, and μ are obtained by using the partial derivatives of G indicated by this fundamental equation.

4.33 Show that

$$\left(\frac{\partial U}{\partial S}\right)_V = \left(\frac{\partial H}{\partial S}\right)_P \qquad \left(\frac{\partial H}{\partial P}\right)_S = \left(\frac{\partial G}{\partial P}\right)_T$$

4.34 Earlier we derived the expression for the entropy of an ideal gas as a function of T and V. Now that we have the Maxwell relations, derive the expression for dS for any fluid.

4.35 The coefficient of thermal expansion α of Fe(s) at 25 °C is $355 \times 10^{-7}\ K^{-1}$. What is the change in molar entropy of iron when the pressure is raised to 1000 bar? (The density of iron at 25 °C is 7.86 g cm^{-3}.)

4.36 Show that C_P and C_V for an ideal gas are independent of volume and pressure.

4.37 Derive the thermodynamic equation of state

$$\left(\frac{\partial \overline{U}}{\partial P}\right)_T = \overline{V}(\kappa P - \alpha T)$$

4.38 Derive the expression for $(\partial \overline{H}/\partial P)_T$ for a gas following the virial equation

$$P\overline{V} = RT + B(T)P$$

4.39 Assuming that the density of water is independent of pressure in the range 1 to 50 bar, what is the change in molar Gibbs energy of water when the pressure is raised this amount?

4.40 An ideal gas is compressed isothermally from 1 to 5 bar at 100 °C. (*a*) What is the molar Gibbs energy change? (*b*) What would have been the change in molar Gibbs energy if the compression had been carried out at 0 °C?

4.41 At 298 K, for H_2(g), $S^\circ = 130.684\ J\ K^{-1}$, $\Delta_f H^\circ = 0\ kJ\ mol^{-1}$, and $\Delta_f G^\circ = 0\ kJ\ mol^{-1}$. What are the values of the molar entropy, enthalpy of formation, and Gibbs energy of formation at 10^{-2} bar, assuming that H_2 is an ideal gas?

4.42 The heat of vaporization of liquid oxygen at 1.013 25 bar is 6820 J mol^{-1} at its boiling point, −183 °C, at that pressure. For the reversible vaporization of liquid oxygen, calculate (*a*) q per mole, (*b*) $\Delta\overline{U}$, (*c*) $\Delta\overline{S}$, and (*d*) $\Delta\overline{G}$.

4.43 An ideal gas is expanded isothermally and reversibly at 0 °C from 1 to $\frac{1}{10}$ bar. Calculate (*a*) w per mole, (*b*) q per mole, (*c*) $\Delta\overline{H}$, (*d*) $\Delta\overline{G}$, and (*e*) $\Delta\overline{S}$ for the gas. One mole of an ideal gas in 22.71 L is allowed to expand irreversibly into an evacuated vessel such that the final total volume is 227.1 L. Calculate (*f*) w per mole, (*g*) q per mole, (*h*) $\Delta \overline{H}$, (*i*) $\Delta \overline{G}$, and (*j*) $\Delta \overline{S}$ for the gas. Calculate (*k*) $\Delta \overline{S}$ for the system and its surroundings involved in the reversible isothermal expansion and calculate (*l*) $\Delta \overline{S}$ for the system and its surroundings involved in an irreversible isothermal expansion in which the gas expands into an evacuated vessel such that the final total volume is 227.1 L.

4.44 Steam is compressed reversibly to liquid water at the boiling point 100 °C. The heat of vaporization of water at 100 °C and 1.013 25 bar is 2258 J g^{-1}. Calculate w per mole and q per mole and each of the thermodynamic quantities $\Delta \overline{H}$, $\Delta \overline{U}$, $\Delta \overline{G}$, $\Delta \overline{A}$, and $\Delta \overline{S}$.

4.45 An ideal gas at 300 K has an initial pressure of 15 bar and is allowed to expand isothermally to a pressure of 1 bar. Calculate (*a*) the maximum work that can be obtained from the expansion, (*b*) $\Delta \overline{U}$, (*c*) $\Delta \overline{H}$, (*d*) $\Delta \overline{G}$, and (*e*) $\Delta \overline{S}$.

4.46 Calculate the molar entropy changes for the gas plus reservoir in Examples 4.2 and 4.3.

4.47 An ideal gas is compressed reversibly at 100 °C from 2 to 10 bar. Calculate $\Delta \overline{H}$, $\Delta \overline{S}$, and $\Delta \overline{G}$. Show that you can get the same value of $\Delta \overline{G}$ in two ways.

4.48 Toluene is vaporized at its boiling point (111 °C). The heat of vaporization at this temperature is 361.9 J g^{-1}. Calculate $\Delta \overline{H}$, $\Delta \overline{S}$, and $\Delta \overline{G}$.

4.49 What is the change in molar Gibbs energy for the freezing of water at -10 °C? The vapor pressure of H_2O(l, -10 °C) is 286.5 Pa, and the vapor pressure of H_2O(s, -10 °C) is 260 Pa. At -10 °C the molar volume of supercooled water is 1.80×10^{-5} m^3 mol^{-1} and the molar volume of ice is 2.00×10^{-5} m^3 mol^{-1}.

4.50 At low pressures the compressibility factor for a van der Waals gas is given by

$$Z = \frac{P\overline{V}}{RT} = 1 + \left(b - \frac{a}{RT}\right)\frac{P}{RT}$$

Derive the expression for $\Delta \overline{G}$ for a change in pressure from P_1 to P_2.

4.51 For a gas that follows the equation of state $P(\overline{V} - b) = RT$, show that the fugacity is given by

$$f = P\,e^{bP/RT}$$

4.52 The apparent specific volume v of a solute (i.e., volume contributed to the solution by one kilogram of solute) is equal to the volume V of the solution minus the volume of pure solvent it contains divided by the mass of solute:

$$v = \frac{V - (V\rho_s - m)/\rho_0}{m}$$

where m is the mass of solute, ρ_s is the density of the solution, and ρ_0 is the density of the solvent. The density of a solution of serum albumin containing 15.4 g of protein per liter is 1.0004×10^3 kg m^{-3} at 25 °C ($\rho_0 = 0.977\,07 \times 10^3$ kg m^{-3}). Calculate the apparent specific volume.

4.53 A solution of magnesium chloride, $MgCl_2$, in water containing 41.24 g L^{-1} has a density of 1.0311×10^3 kg m^{-3} at 20 °C. The density of water at this temperature is 0.998×10^3 kg m^{-3}. Calculate (*a*) the apparent specific volume (see the equation in Problem 4.52) and (*b*) the apparent molar volume of $MgCl_2$ in this solution.

4.54 Calculate ΔG and ΔS for mixing 2 mol of H_2 with 1 mol of O_2 at 25 °C under conditions where no chemical reaction occurs.

4.55 Liquid water can be superheated to 120 °C at 1.01325 bar. Calculate the changes in entropy, enthalpy, and Gibbs energy for the process of superheated water at 120 °C and 1.01325 bar changing to steam at the same temperature and pressure. The enthalpy of vaporization is 40.58 kJ mol^{-1} at 100 °C and 1.01325 bar. Given: $\overline{C}_P(H_2O, l) = 75.3$ J K^{-1} mol^{-1} and $\overline{C}_P(H_2O, g) = (36 + 0.013T)$ J K^{-1} mol^{-1}, where T is in kelvins.

Computer Problems

4.A We know a good deal about the thermodynamics of monatomic ideal gases, as we have already seen. Plot (*a*) μ, (*b*) $\overline{S}$, and (*c*) $\overline{V}$ versus temperature from 100 to 500 K and pressure from 0.1 to 20 bar and discuss the slopes of each of the plots of μ in the two directions. The necessary equations are

(a) $\mu = -T\ln(T^{5/2}/P)$
(b) $\overline{S} = \ln(T^{5/2}/P)$
(c) $\overline{V} = T/P$

4.B Given the virial equation for N_2 in terms of pressure at 298.15 K, plot the compressibility factor and the fugacity from 0 to 1000 bar versus the pressure.

4.C Given the virial coefficients for O_2 at 298.15 K in Table 1.1, plot the compressibility factor and the fugacity of the gas versus its pressure up to 1000 bar.

4.D Plot the Gibbs energy of formation of an ideal gas relative to its standard Gibbs energy at 298.15 K from 0 to 10 bar.

4.E The 3D plots of molar entropy and molar volume for a monatomic ideal gas made in Computer Problem 4.A can be made in another way by using

$$S_m = -(\partial\mu/\partial T)_P$$

$$V_m = (\partial\mu/\partial P)_T$$

(*a*) Plot S_m, (*b*) plot V_m, (*c*) plot $-(\partial\mu/\partial T)_P$, and (*d*) plot $(\partial\mu/\partial P)_T$.

4.F (*a*) Calculate the fugacity of molecular hydrogen at 100 bar and 25 °C using the virial coefficients in Table 1.1. (*b*) Plot the fugacity of molecular hydrogen from $P = 0$ to 1000 bar at 25 °C.

5

Chemical Equilibrium

In 1864 Guldberg and Waage showed experimentally that in chemical reactions an equilibrium is reached that can be approached from either direction. They were apparently the first to realize that there is a mathematical relation between the concentrations of reactants and products at equilibrium. In 1877 van't Hoff suggested that in the equilibrium expression for the hydrolysis of ethyl acetate, the concentration of each reactant should appear to the first power, corresponding to the stoichiometric numbers in the balanced chemical equation.

The fundamental equation provides the basis for understanding chemical equilibrium. The basic equations in terms of the chemical potential are completely general, but we will emphasize ideal gas reactions because of the simple relation between the chemical potential of a species and its partial pressure. There are brief discussions of multireaction equilibria and of gas–solid reactions. Chemical equilibrium in the liquid phase is discussed in later chapters. In discussing systems at chemical equilibrium, it is important to know how many independent variables there are; this question is answered by the phase rule, which is derived in this chapter. The choice of independent variables is somewhat arbitrary, but the number is specified by the phase rule. The measurement of an equilibrium constant yields the standard reaction Gibbs energy, and the measurement of the temperature coefficient yields the standard reaction enthalpy.

5.1 DERIVATION OF THE GENERAL EQUILIBRIUM EXPRESSION

When discussing chemical equilibrium at a specified temperature and pressure, we must add terms to the fundamental equation for $\mathrm{d}U$ (equation 4.5) similar to those we added in Section 4.1 to get equation 4.16. These terms allow for changes in the number of moles of all the species in the chemical reaction. This leads to the fundamental equation for the Gibbs energy for a closed system when chemical reactions are considered:

$$\mathrm{d}G = -S\,\mathrm{d}T + V\,\mathrm{d}P + \sum_{i=1}^{N_s} \mu_i\,\mathrm{d}n_i \tag{5.1}$$

where N_s is the number of species. When chemical reactions are involved, we have already seen in Section 2.11 that the various $\mathrm{d}n_i$ in equation 5.1 are not independent variables. If there is a single reaction, the amounts n_i of the various species at any time are given by

$$n_i = n_{i0} + \nu_i \xi \tag{5.2}$$

where n_{i0} is the initial amount of species i, ν_i is the stoichiometric number of the species in the reaction, and ξ is the extent of reaction. Note that the extent of reaction for a system is expressed in moles. Since $\mathrm{d}n_i = \nu_i\,\mathrm{d}\xi$, substituting this relation into equation 5.1 yields

$$\mathrm{d}G = -S\,\mathrm{d}T + V\,\mathrm{d}P + \left(\sum_{i=1}^{N_s} \nu_i \mu_i\right)\mathrm{d}\xi \tag{5.3}$$

Thus at specified T and P,

$$\left(\frac{\partial G}{\partial \xi}\right)_{T,P} = \sum_{i=1}^{N_s} \nu_i \mu_i \tag{5.4}$$

At chemical equilibrium and constant T and P G has its minimum value; thus, the derivative in equation 5.4 is zero:

$$\sum_{i=1}^{N_s} \nu_i \mu_{i,\mathrm{eq}} = 0 \tag{5.5}$$

This equilibrium condition applies to all chemical equilibria, whether they involve gases, liquids, solids, or solutions.

It is convenient to refer to the derivative in equation 5.4 as the **reaction Gibbs energy** and represent it with $\Delta_r G$:

$$\Delta_r G = \left(\frac{\partial G}{\partial \xi}\right)_{T,P} \tag{5.6}$$

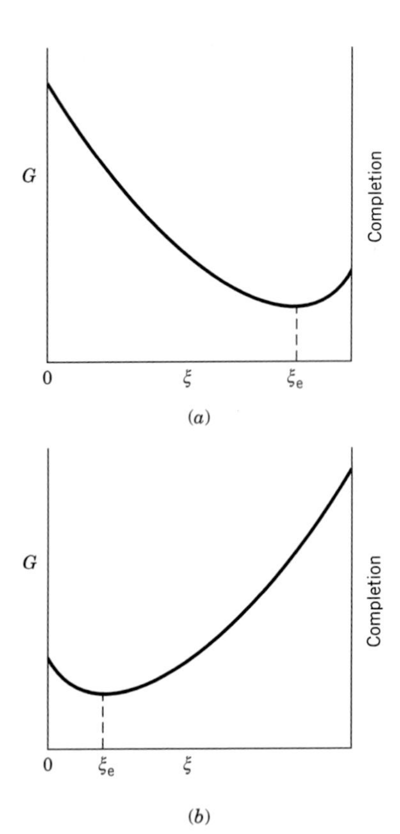

Figure 5.1 (*a*) Gibbs energy as a function of extent of reaction ξ at constant T and P for a reaction in a single phase that goes nearly to completion. (*b*) Gibbs energy as a function of ξ for a reaction that does not go very far toward completion.

Therefore, we will use equation 5.4 in the form

$$\Delta_r G = \sum_{i=1}^{N_s} \nu_i \mu_i \tag{5.7}$$

The reaction Gibbs energy is the change in Gibbs energy when the extent of reaction ξ changes by 1 mol, as specified by a balanced chemical equation, at specified partial pressures or concentrations of the species involved.

When a reaction takes place at constant temperature and pressure, the Gibbs energy decreases, and the reaction continues until the Gibbs energy has reached a minimum value, as shown by Fig. 5.1. It is of interest to note that the thermodynamic condition for chemical equilibrium (equation 5.5) has the same form as the reaction to which it applies (equation 2.88), except that molecular formulas are replaced by the corresponding chemical potentials of the reactant and product species. Substituting the expression for the chemical potential of a species at equilibrium,

$$\mu_{i,\text{eq}} = \mu_i^\circ + RT \ln a_{i,\text{eq}} \tag{5.8}$$

into equation 5.5 yields

$$\sum_{i=1}^{N_s} \nu_i \mu_i^\circ = -RT \sum_{i=1}^{N_s} \nu_i \ln a_{i,\text{eq}} \tag{5.9}$$

The stoichiometric numbers can be put in the exponential position to obtain

$$\sum_{i=1}^{N_s} \nu_i \mu_i^\circ = -RT \sum_{i=1}^{N_s} \ln(a_{i,\text{eq}})^{\nu_i} \tag{5.10}$$

and the sum of logarithms can be replaced by the logarithm of a product:

$$\sum_{i=1}^{N_s} \nu_i \mu_i^\circ = -RT \ln \prod_{i=1}^{N_s} a_{i,\text{eq}}^{\nu_i} \tag{5.11}$$

Example 5.1 ***Changing a summation of logarithmic terms to a product of exponential terms***

Show how equation 5.10 is converted to equation 5.11. For a simple example, the sum is given by

$$\text{Sum} = \ln(a_1^{\nu_1}) + \ln(a_2^{\nu_2})$$

The product is given by

$$\ln(a_1^{\nu_1} a_2^{\nu_2}) = \ln(a_1^{\nu_1}) + \ln(a_2^{\nu_2})$$

This shows that the sum of logarithmic terms is equal to the logarithm of the product given in equation 5.11.

Since the product of activities at equilibrium in equation 5.11 is so useful, it is defined as the **equilibrium constant** K of the reaction:

$$K = \prod_{i=1}^{N_s} a_{i,\text{eq}}^{\nu_i} \tag{5.12}$$

The equilibrium constant is a dimensionless quantity, but its magnitude depends on the way the chemical equation is written because of the stoichiometric numbers. To interpret an equilibrium constant it is necessary to know the balanced chemical equation to which it applies and the standard states of the species on which the activities are based (see Section 4.5).

The quantity $\sum \nu_i \mu_i^\circ$ in equation 5.11 is equal to $\Delta_r G^\circ$, the **standard reaction Gibbs energy,** and so equation 5.11 can be written

$$\Delta_r G^\circ = -RT \ln K \tag{5.13}$$

This is a very important equation because the equilibrium constant, which can be determined experimentally, tells us the change in the standard Gibbs energy for the reaction. Conversely, since $\Delta_r G^\circ$ can be evaluated by other methods, K can be calculated. Note that since $\Delta_r G^\circ$ is a function only of T, K is also a function only of T.

Now we go back to equation 5.7 to see how to calculate the reaction Gibbs energy $\Delta_r G$ under a particular set of conditions when we know the equilibrium constant. If we substitute $\mu_i = \mu_i^\circ + RT \ln a_i$ for the chemical potential of a species in equation 5.7, we obtain

$$\Delta_r G = \sum_{i=1}^{N_s} \nu_i \mu_i^\circ + RT \sum_{i=1}^{N_s} \nu_i \ln a_i = \Delta_r G^\circ + RT \ln \prod_{i=1}^{N_s} a_i^{\nu_i} \tag{5.14}$$

As we can see from this equation, $\Delta_r G^\circ$ is the change in Gibbs energy for the reaction when the activities of reactants and products are all unity. In other words, $\Delta_r G^\circ$ is the change in Gibbs energy when separate reactants in *their standard states* are converted to the separate products in *their standard states.* $\Delta_r G$ is the change in the Gibbs energy in a specified reaction when separated reactants at specified activities are converted to separated products at specified activities. The product in the last term is much like an equilibrium constant, except that the activities of reactants and products can have any values we want. This product of activities is called the **reaction quotient** and is represented by Q:

$$Q = \prod_{i=1}^{N_s} a_i^{\nu_i} \tag{5.15}$$

so that

$$\Delta_r G = \Delta_r G^\circ + RT \ln Q \tag{5.16}$$

This equation gives the change in Gibbs energy for a specified chemical reaction when the reactants and products have activities a_i, so it can be used to test for spontaneity in the forward direction ($\Delta_r G < 0$) or backward direction ($\Delta_r G > 0$).

We can substitute the definition of the Gibbs energy ($G = H - TS$) in equation 5.6 to obtain

$$\begin{aligned}\Delta_r G &= \left(\frac{\partial H}{\partial \xi}\right)_{T,P} - T\left(\frac{\partial S}{\partial \xi}\right)_{T,P} \\ &= \Delta_r H - T\Delta_r S \end{aligned} \tag{5.17}$$

which provides a logical introduction of the **reaction enthalpy** $\Delta_r H$, in agreement with equation 2.93, and the **reaction entropy** $\Delta_r S$ that is given later, in equation 5.42. This equation also applies when the reactants and products are each in their standard states, so that $\Delta_r G^\circ = \Delta_r H^\circ - T\Delta_r S^\circ$.

Example 5.2 ***The general expression for an equilibrium constant***

What is the most general equilibrium constant expression for the following reaction?

$$3\text{C(graphite)} + 2\text{H}_2\text{O(g)} = \text{CH}_4\text{(g)} + 2\text{CO(g)}$$

$$K = \left(\frac{a_{\text{CH}_4} a_{\text{CO}}^2}{a_{\text{C}}^3 a_{\text{H}_2\text{O}}^2} \right)_{\text{eq}}$$

If the pressure is not too high, the graphite can be considered to be in its standard state so that $a_{\text{C}} = 1$. The activities of the gases can be replaced by f_i/P° or, if the pressure is low enough, by P_i/P°.

5.2 EQUILIBRIUM CONSTANT EXPRESSIONS FOR GAS REACTIONS

For real gases the activity is given by $a_i = f_i/P^\circ$, where f_i is the fugacity of the ith species and P° is the standard state pressure.

In Section 4.5 we saw that the partial molar Gibbs energy, which we can now refer to as the chemical potential, of a gas is given by

$$\mu_i = \mu_i^\circ + RT \ln \frac{f_i}{P^\circ} \tag{5.18}$$

This relation can be substituted into equation 5.5 and the same operations carried out to obtain

$$K = \prod_{i=1}^{N_s} \left(\frac{f_{i,\text{eq}}}{P^\circ} \right)^{\nu_i} \tag{5.19}$$

This equation is not used very often because of the difficulty in evaluating f_i in a mixture of gases, but it is the most general expression for the equilibrium constant of a reaction involving real gases.

For ideal gases, we have seen earlier that

$$\mu_i = \mu_i^\circ + RT \ln \frac{P_i}{P^\circ} \tag{5.20}$$

Substituting this relation in equation 5.5 and carrying out the operations in Section 5.1 yields

$$K = \prod_{i=1}^{N_s} \left(\frac{P_{i,\text{eq}}}{P^\circ} \right)^{\nu_i} \tag{5.21}$$

This equilibrium constant is also a function only of temperature. For real gases the right-hand side of equation 5.21 will depend on pressure since in general for real gases $f_i \neq P_i$. The term **thermodynamic equilibrium constant** is often used for the equilibrium constant obtained by use of equation 5.19 or by using equation 5.21 at low pressure. Calculations of K using tables of Gibbs energies of formation yield thermodynamic equilibrium constants.

The value of an equilibrium constant cannot be interpreted unless it is accompanied by a balanced chemical equation and a specification of the standard state of each reactant and product. The values of the stoichiometric numbers are arbitrary to the extent that a chemical equation may be multiplied or divided by a positive or negative integer. In this section we are considering gas reactions, and so the standard state of each reactant and product is the pure gas at 1 bar in the ideal gas state. Later, in Section 7.6, we will discuss the standard states of substances in liquid solution in more detail. In using equations 5.12, 5.19, 5.21, and similar equations, we will omit the subscript eq.

The equilibrium extent of a chemical reaction in ideal gas mixtures depends on just three independent variables: (1) pressure, (2) initial composition, and (3) temperature. We will examine the effects of each of these variables and also the effect of adding inert gases to the reaction mixture.

Example 5.3 ***How partial pressures determine whether a reaction goes forward or backward***

(*a*) A mixture of $CO(g)$, $H_2(g)$, and $CH_3OH(g)$ at 500 K with $P_{CO} = 10$ bar, $P_{H_2} = 1$ bar, and $P_{CH_3OH} = 0.1$ bar is passed over a catalyst. Can more methanol be formed? Given: $\Delta_r G^\circ = 21.21$ kJ mol^{-1}.

$$CO(g, 10\text{ bar}) + 2H_2(g, 1\text{ bar}) = CH_3OH(g, 0.1\text{ bar})$$

$$\begin{aligned}\Delta_r G &= \Delta_r G^\circ + RT \ln Q \\ &= 21.21\text{ kJ mol}^{-1} + (0.008\,314\,5\text{ kJ K}^{-1}\text{ mol}^{-1})(500\text{ K}) \ln \frac{(0.1)}{(10)(1)^2} \\ &= 2.07\text{ kJ mol}^{-1}\end{aligned}$$

Thus, the reaction as written is not spontaneous. (*b*) Can the following conversion occur at 500 K?

$$CO(g, 1\text{ bar}) + 2H_2(g,\ 10\text{ bar}) = CH_3OH(g, 0.1\text{ bar})$$

$$\begin{aligned}\Delta_r G &= 21.21\text{ kJ mol}^{-1} + (0.008\,314\,5\text{ kJ K}^{-1}\text{ mol}^{-1})(500\text{ K}) \ln \frac{(0.1)}{(1)(10)^2} \\ &= -7.51\text{ kJ mol}^{-1}\end{aligned}$$

Under these conditions the reaction is thermodynamically spontaneous.

To illustrate why gas reactions never go to completion, let us consider a simple isomerization of ideal gas A to ideal gas B at constant pressure:

$$A(g) = B(g) \tag{5.22}$$

According to equation 4.47, the Gibbs energy of the reaction mixture at any extent of reaction is

$$G = n_A \mu_A + n_B \mu_B \tag{5.23}$$

where n_A is the amount of A and n_B is the amount of B. If the reaction is started with 1 mol of A, the amounts of A and B at a later time are given in terms of extent of reaction ξ by

$$n_A = 1 - \xi \tag{5.24}$$

$$n_B = \xi \tag{5.25}$$

Thus,

$$G = (1 - \xi)\mu_A + \xi\mu_B \tag{5.26}$$

From equation 5.20, the chemical potentials of A and B in the mixture of ideal gases are given by

$$\begin{aligned}
\mu_A &= \mu_A^\circ + RT \ln(P_A/P^\circ) \\
&= \mu_A^\circ + RT \ln y_A + RT \ln(P/P^\circ) \\
&= \mu_A^\circ + RT \ln(1 - \xi) + RT \ln(P/P^\circ) \\
\mu_B &= \mu_B^\circ + RT \ln y_B + RT \ln(P/P^\circ) \\
&= \mu_B^\circ + RT \ln \xi + RT \ln(P/P^\circ)
\end{aligned} \tag{5.27}$$

where P is the total pressure at equilibrium. Substituting these equations into equation 5.26,

$$\begin{aligned}
G &= (1 - \xi)\mu_A^\circ + \xi\mu_B^\circ + RT \ln(P/P^\circ) + RT[(1 - \xi)\ln(1 - \xi) + \xi \ln \xi] \\
&= \mu_A^\circ - (\mu_A^\circ - \mu_B^\circ)\xi + RT \ln(P/P^\circ) + \Delta_{mix}G^\circ
\end{aligned} \tag{5.28}$$

where $\Delta_{mix}G^\circ$ is the Gibbs energy of mixing $(1 - \xi)$ mol of A with ξ mol of B. Figure 5.2 gives a plot of G versus ξ. The first three terms give the linear function. It is the mixing term that causes the minimum in the plot of G versus extent of reaction ξ. At constant temperature and pressure, the criterion of equilibrium is that the Gibbs energy is a minimum (see Section 4.2). Thus, starting with A, the Gibbs energy can decrease along the curve until ξ_{eq} mol of B have been formed. Starting with B, the Gibbs energy can decrease until $(1 - \xi_{eq})$ mol of A have formed.

Even though B has the lower value of the standard molar Gibbs energy $(\mu_A^\circ > \mu_B^\circ)$, the system can achieve a lower Gibbs energy by having some A present at equilibrium with the resulting Gibbs energy of mixing. Generalizing from this example, we can say that no chemical reaction of gases goes to completion; nevertheless, it may be very difficult to detect reactants at equilibrium if the products have a very much lower Gibbs energy.

A surprising feature of Fig. 5.2 is that the slope $dG/d\xi$ approaches $-\infty$ as $\xi \to 0$ and $+\infty$ as $\xi \to 1$. This can be shown by differentiating equation 5.28 with respect to ξ and looking at the limits.

5.3 DETERMINATION OF EQUILIBRIUM CONSTANTS

If the initial concentrations of the reactants are known and only one reaction occurs, it is necessary to determine the concentration of only one reactant or product at equilibrium to be able to calculate the concentrations or pressures of the others by means of the balanced chemical equation. Chemical methods based on chemical reaction with one of the reactants or products can be used for such analyses only when the reaction being studied can be stopped at equilibrium, as by a

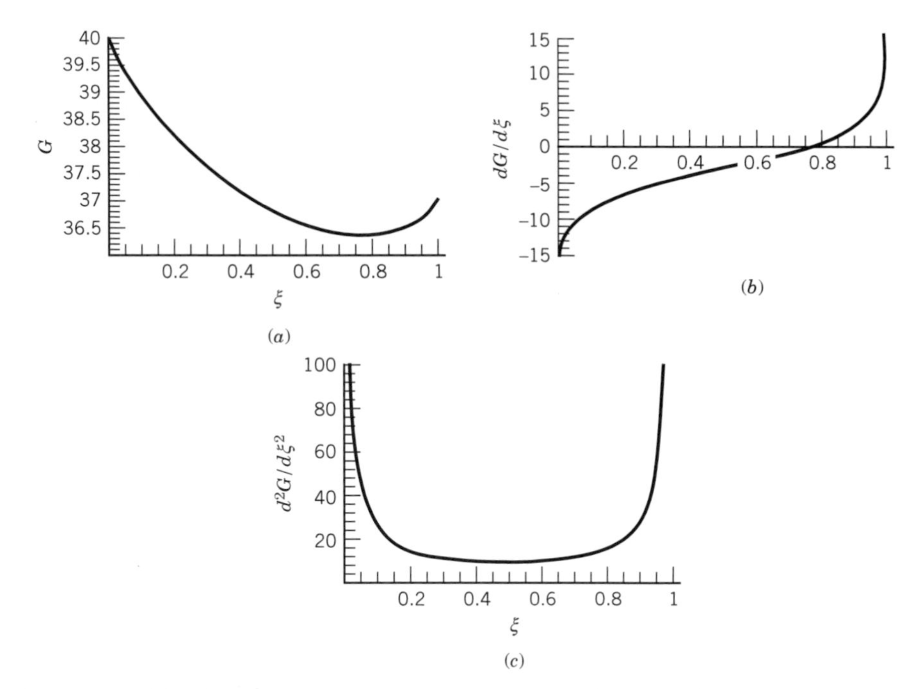

Figure 5.2 (a) Gibbs energy of the reaction system A(g) = B(g) versus the extent of reaction ξ at constant temperature. (b) Derivative of the Gibbs energy with respect to extent of reaction, which is zero at the minimum Gibbs energy. (c) Second derivative of the Gibbs energy with respect to the extent of reaction, which is positive over the whole range. Note that the second derivative has to be positive for the equilibrium to be stable. (See Computer Problem 5.N.)

very sudden chilling to a temperature where the rate of further chemical change is negligible, or by destruction of a catalyst. Otherwise, the concentrations will shift during the chemical analysis.

Measurements of physical properties, such as density, pressure, light absorption, refractive index, electromotive force, and electrical conductivity, are especially useful for determining the concentrations of reactants at equilibrium since, for these methods, it is unnecessary to "stop" the reaction.

It is essential to know that equilibrium has been reached before the analysis of the mixture can be used for calculating the equilibrium constant. The following criteria for the attainment of equilibrium at constant temperature are useful:

1. The same value of the equilibrium constant should be obtained when the equilibrium is approached from either side.
2. The same value of the equilibrium constant should be obtained when the initial concentrations of reacting material are varied over a wide range.

The determination of the density of a partially dissociated gas provides one of the simplest methods for measuring the extent to which the gas is dissociated. When a gas dissociates, more molecules are produced, and at constant temperature and pressure the volume increases.

If the equilibrium extent of reaction ξ has been determined, we need the expression for the equilibrium constant in terms of the equilibrium extent of reaction and the pressure. We will assume that the gases are ideal. The first step is to express the mole fractions of reactants and products, in terms of the extent of reaction ξ.

For example, consider the dissociation of an initial amount $n_0(N_2O_4)$ of N_2O_4 to NO_2 at a given temperature and pressure. If the equilibrium extent of reaction is ξ, the equilibrium amount of N_2O_4 is $n_0(N_2O_4) - \xi$ and the equilibrium amount of NO_2 is 2ξ. However, since the equilibrium composition at a given temperature and pressure is independent of the size of the system, we might as well divide these expressions by $n_0(N_2O_4)$ and represent the equilibrium amounts by the dimensionless quantities $1 - \xi'$ for N_2O_4 and $2\xi'$ for NO_2. The quantity ξ' is a **dimensionless extent of reaction** defined by $\xi' = \xi/n_0(N_2O_4)$. In working equilibrium problems we will leave off the prime to simplify the notation and write $1 - \xi$ rather than 1 mol $-\xi$. The following format is recommended:

$$\begin{array}{rlll} & N_2O_4(g) & = 2NO_2(g) & \quad (5.29) \\ \text{Initial amount} & 1 & 0 & \\ \text{Equilibrium amount} & 1-\xi & 2\xi & \text{Total amount} = 1+\xi \\ \text{Equilibrium mole fraction} & \dfrac{1-\xi}{1+\xi} & \dfrac{2\xi}{1+\xi} & \end{array}$$

$$K = \frac{(P_{NO_2}/P^\circ)^2}{P_{N_2O_4}/P^\circ} = \frac{\{[2\xi/(1+\xi)](P/P^\circ)\}^2}{[(1-\xi)/(1+\xi)](P/P^\circ)} = \frac{4\xi^2 P/P^\circ}{1-\xi^2} \tag{5.30}$$

At equilibrium the amount of $N_2O_4(g)$ is $1 - \xi$ and the amount of $NO_2(g)$ is 2ξ, so that the total amount is $1 + \xi$. The partial pressures of the reactants at equilibrium are obtained by multiplying their equilibrium mole fractions by the total pressure P. Equation 5.30 gives the relation between the equilibrium constant K, the equilibrium extent of reaction ξ, and the total pressure P. This may be solved for the equilibrium extent of reaction to obtain

$$\xi = \frac{1}{[1 + (4/K)(P/P^\circ)]^{1/2}} \tag{5.31}$$

This is the dimensionless extent of reaction obtained by dividing the extent of reaction ξ by $n_0(N_2O_4)$. Equations 5.30 and 5.31 apply to any dissociation reaction of the type $A(g) = 2B(g)$ (see Fig. 5.3), but not to dissociation reactions of the type $A(g) = B(g) + C(g)$. For reactions of the latter type, the factor 4 in equation 5.31 is replaced by 1.

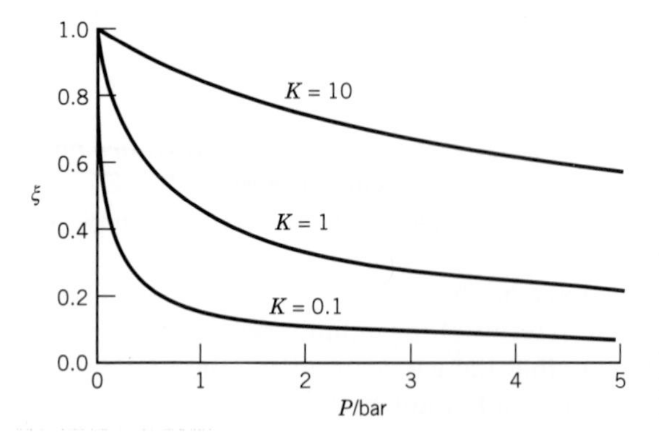

Figure 5.3 Extent of reaction ξ at equilibrium for the reaction $A(g) = 2B(g)$ as a function of P/P°, for various values of K, using equation 5.31. (See Computer Problem 5.H.)

The equilibrium extent of reaction 5.29 is readily determined by measuring the density of the partially dissociated gas. For example, assume that we start with a mass m of N_2O_4(g). The initial volume is $V_1 = mRT/M_1P$, where M_1 is the molar mass of N_2O_4 (92.01 g mol^{-1}). If the gas is held at a constant pressure and temperature, the equilibrium volume is given by $V_2 = mRT/M_2P$, where M_2 is the average molar mass of the partially dissociated gas, which is defined by $M_2 = M_{N_2O_4}y_{N_2O_4} + M_{NO_2}y_{NO_2}$. Thus, $V_1/V_2 = M_2/M_1$. This ratio is equal to $1/(1+\xi)$, and so

$$\xi = \frac{M_1 - M_2}{M_2} \tag{5.32}$$

Example 5.4 ***Calculation of the equilibrium constant for a gas reaction from the equilibrium density***

When nitrogen tetroxide is held in a container at constant T and P near room temperature or higher, it rather quickly reaches an equilibrium degree of dissociation. If 1.588 g of nitrogen tetroxide gives a total pressure of 1.0133 bar when partially dissociated in a 500$-$cm^3 glass vessel at 25 °C, what is the extent of reaction? What is the value of K? What is the extent of reaction at a total pressure of 0.5 bar?

The average molar mass of the gas at equilibrium is given by

$$M_2 = \frac{RT}{P}\frac{m}{V} = \frac{(0.083\,145\text{ L bar K}^{-1}\text{ mol}^{-1})(298.15\text{ K})(1.588\times10^{-3}\text{ kg})}{(1.0133\text{ bar})(0.5\text{ L})}$$

$$= 77.70\text{ g mol}^{-1}$$

$$\xi = \frac{92.01 - 77.70}{77.70} = 0.1842$$

$$K = \frac{4\xi^2(P/P^\circ)}{1-\xi^2} = \frac{(4)(0.1842)^2(1.0133)}{1-(0.1842)^2} = 0.143$$

The extent of reaction at 0.5 bar is calculated using equation 5.31:

$$\xi = \left[\frac{0.143}{0.143 + 4(0.5)}\right]^{1/2} = 0.258$$

As an example of a more complicated gas reaction, consider that ammonia is produced by holding an initial mixture containing equal amounts of nitrogen and hydrogen at constant temperature and high pressure in contact with a catalyst. Note that reactants are not always added in stoichiometric proportions. In the actual production of ammonia, hydrogen is, of course, the more expensive reactant. Again we use the dimensionless extent of reaction:

	N_2(g) +	$3H_2$(g) =	$2NH_3$(g)	(5.33)
Initial amount	1	1	0	
Equilibrium amount	$1-\xi$	$1-3\xi$	2ξ	Total amount $= 2-2\xi$
Equilibrium mole fraction	$\dfrac{1-\xi}{2-2\xi}$	$\dfrac{1-3\xi}{2-2\xi}$	$\dfrac{2\xi}{2-2\xi}$	

Note that when ξ moles of N_2 have been used up, 3ξ moles of H_2 will have reacted, and 2ξ moles of NH_3 will have been formed. Thus, the total amount of reactants and products at equilibrium is $2 - 2\xi$. The equilibrium constant is given by

$$K = \frac{16\xi^2(1-\xi)}{(1-3\xi)^3(P/P^\circ)^2} \tag{5.34}$$

If the balanced chemical equation is divided by 2, the equilibrium constant for it will be the square root of K as expressed by equation 5.34. If the balanced chemical equation is reversed, the equilibrium constant for it will be the reciprocal of K in equation 5.34.

Example 5.5 ***The total pressure to obtain 10% conversion of nitrogen to ammonia***

What total pressure must be used to obtain a 10% conversion of nitrogen to ammonia at 400 °C, assuming an initially equimolar mixture of nitrogen and hydrogen and ideal gas behavior? The equilibrium constant for the formation of $NH_3(g)$ according to reaction 5.33 with a standard state pressure of 1 bar is 1.60×10^{-4} at 400 °C.

We use equation 5.34 to obtain

$$1.60 \times 10^{-4} = \frac{16(0.10)^2(0.90)}{(0.70)^3(P/P^\circ)^2}$$

$$\frac{P}{P^\circ} = 51.2$$

$$P = 51.2 \text{ bar}$$

Example 5.6 ***Calculation of the equilibrium constant for the forward reaction and the reverse reaction***

(*a*) What is the value of the standard Gibbs energy for reaction 5.33 at 400 °C? (*b*) What is the value of the equilibrium constant and the standard reaction Gibbs energy when reaction 5.33 is divided by 2? (*c*) What is the value of the equilibrium constant and the standard reaction Gibbs energy when reaction 5.33 is reversed?

(*a*)
$$\begin{aligned}\Delta_r G^\circ &= -RT\ln K \\ &= -(8.315\text{ J K}^{-1}\text{ mol}^{-1})(673\text{ K})\ln(1.60\times 10^{-4}) \\ &= 48.91\text{ kJ mol}^{-1}\end{aligned}$$

(*b*)
$$\begin{aligned}K &= (1.60\times 10^{-4})^{1/2} = 0.012\,65 \\ \Delta_r G^\circ &= -RT\ln 0.012\,65 \\ &= 24.46\text{ kJ mol}^{-1}\end{aligned}$$

(*c*)
$$\begin{aligned}K &= \frac{1}{(1.60\times 10^{-4})} = 6250 \\ \Delta_r G^\circ &= -RT\ln 6250 \\ &= -48.91\text{ kJ mol}^{-1}\end{aligned}$$

The calculation of the equilibrium composition of a reaction system involving two or more reactions is more complicated, because the equilibrium composition is the one that satisfies a set of independent equilibrium constant expressions for the system and a set of independent conservation equations for components (see Sections 5.9 and 5.11). This solution of simultaneous nonlinear equations cannot be obtained analytically, but requires an iterative process in which some kind of initial guess is improved using the Newton–Raphson process. Computer programs such as equcalc (see Computer Problem 5.K) have been written to do this, and Mathematica also provides Solve, which can do this.

5.4 USE OF STANDARD GIBBS ENERGIES OF FORMATION TO CALCULATE EQUILIBRIUM CONSTANTS

There are three ways in which $\Delta_r G^\circ$ for a reaction may be obtained: (1) $\Delta_r G^\circ$ may be calculated from a measured equilibrium constant using equation 5.13, (2) $\Delta_r G^\circ$ may be calculated from

$$\Delta_r G^\circ = \Delta_r H^\circ - T\Delta_r S^\circ \tag{5.35}$$

using $\Delta_r H^\circ$ obtained calorimetrically and $\Delta_r S^\circ$ obtained from the third law entropies, and (3) for gas reactions $\Delta_r G^\circ$ may be calculated using statistical mechanics (Chapter 16) and certain information about molecules obtained from spectroscopic data. Methods 2 and 3 make it possible to calculate equilibrium constants of reactions that have never been studied in the laboratory. In method 2 the necessary data are obtained solely from thermal measurements, including heat capacity measurements down to the neighborhood of absolute zero. The calculation of equilibrium constants for gas reactions using statistical mechanics is even more remarkable in that only properties of the individual molecules are used to calculate equilibrium constants for reactions of ideal gases. For the simplest reactions $\Delta_r H^\circ$ may be calculated from spectroscopic data; however, for more complicated reactions, the calculation of $\Delta_r H^\circ$ requires calorimetric data.

Rather than tabulating values of equilibrium constants of reactions, or of $\Delta_r G^\circ$ values calculated using equation 5.13, it is more convenient to tabulate values of the **standard Gibbs energy of formation** $\Delta_f G_i^\circ$, which is the standard Gibbs energy for the formation of a mole of i from its elements. The standard Gibbs energy of formation of i is related to the standard enthalpy of formation of i and the standard entropy of i by

$$\Delta_f G_i^\circ = \Delta_f H_i^\circ - T\left(\overline{S}_i^\circ - \sum \nu_e \overline{S}_e^\circ\right) \tag{5.36}$$

where $\sum \nu_e \overline{S}_e^\circ$ is the sum of the standard entropies of the elements in the formation reaction for species i. Since values of $\Delta_f G_i^\circ$ are tabulated, the equation for $\Delta_r G^\circ$ used in equation 5.14 ($\Delta_r G^\circ = \sum \nu_i \mu_i^\circ$) can be written as

$$\Delta_r G^\circ = \sum_{i=1}^{N_s} \nu_i \Delta_f G_i^\circ \tag{5.37}$$

where the ν_i's are the stoichiometric numbers from the balanced chemical equation. The standard Gibbs energies of formation of elements in their reference states are zero at all temperatures.

Standard Gibbs energies of formation at 298.15 K and 1 bar are provided for about 15 000 species in the NBS Tables of Chemical Thermodynamic Properties.* Some values for those tables are given in Table C.2. Standard Gibbs energies of formation at temperatures up to 6000 K are given in the JANAF tables.† Some values from those tables are given in Table C.3. Gibbs energies of formation of several hundred organic compounds up to 1000 K are given in Stull et al.‡ The values in the latter table need to be converted to joules, and to a standard state pressure of 1 bar, before being used with values in the other tables.§

Most chemical reactions that occur are exothermic; that is, $\Delta_r H^\circ$ is negative. However, some endothermic reactions do occur. Endothermic reactions have equilibrium constants greater than unity only if $T\Delta_r S^\circ$ is sufficiently positive to give a negative $\Delta_r G^\circ$. In the case of gaseous reactions, this may happen if there are more molecules on the right-hand side of the chemical equation than the left. Note that in the limit of high temperature, reactions with positive $\Delta_r S^\circ$ have equilibrium constants greater than unity, independent of $\Delta_r H^\circ$.

Example 5.7 *Calculation of the standard Gibbs energy of formation for H_2O (g)*

Given the following calorimetric information, calculate the standard Gibbs energy of formation of H_2O(g) at 298.15 K.

	$\Delta_f H^\circ$/kJ mol^{-1}	$\overline{S}^\circ$/J K^{-1} mol^{-1}
H_2O(g)	−241.818	188.825
H_2(g)	0	130.684
O_2(g)	0	205.138

The standard Gibbs energy of formation is the standard reaction Gibbs energy for the following reaction.

$$H_2(g) + \tfrac{1}{2}O_2(g) = H_2O(g)$$

$$\begin{aligned}\Delta_f G^\circ(H_2O,\ g) &= \Delta_f H^\circ(H_2O,\ g) - T\Delta_f S^\circ(H_2O,\ g)\\ &= -241.818 - (298.15)[188.825 - 130.684 - (0.5)(205.138)](10^{-3})\\ &= -228.572 \text{ kJ mol}^{-1}\end{aligned}$$

Example 5.8 *Use of tables to calculate equilibrium constants*

Calculate the equilibrium constants for the following reactions at the indicated temperature.

(*a*) $3O_2(g) = 2O_3(g)$ at 25 °C

(*b*) $CO(g) + 2H_2(g) = CH_3OH(g)$ at 500 K

*D. D. Wagman et al., The NBS Tables of Chemical Thermodynamic Properties, *J. Phys. Chem. Ref. Data* **11** (suppl. 2) (1982).

†M W. Chase et al., JANAF Thermochemical Tables, *J. Phys. Chem. Ref. Data* **14** (suppl. 1) (1985).

‡D. R. Stull, E. F. Westrum, and G. C. Sinke, *The Chemical Thermodynamics of Organic Compounds.* New York: Wiley, 1969.

§R. D. Freeman, *J. Chem. Educ.* **62**:681 (1985).

(*a*) Use Table C.2 to obtain

$$\begin{aligned}\Delta_r G^\circ &= 2\Delta_f G^\circ(O_3,\ g) - 3\Delta_f G^\circ(O_2,\ g)\\ &= 2(163.2) = 326.4\ \text{kJ mol}^{-1}\\ K &= \exp\left(-\frac{\Delta_r G^\circ}{RT}\right)\\ &= \exp\left[-\frac{326\,400}{(8.3145)(298.15)}\right]\\ &= 6.62 \times 10^{-58}\end{aligned}$$

(*b*) Use Table C.3 to obtain

$$\begin{aligned}\Delta_r G^\circ &= -134.27 - (-155.41) = 21.14\ \text{kJ mol}^{-1}\\ K &= \exp\left[-\frac{21\,140}{(8.3145)(500)}\right]\\ &= 6.19 \times 10^{-3}\end{aligned}$$

Comment:

There are two ways to store information on equilibrium constants and use it to calculate equilibrium constants of reactions that have not been studied. One way is to tabulate reactions and equilibrium constants and then calculate new equilibrium constants by adding and subtracting reactions from the tabulation. When two reactions are added, the equilibrium constant of the new reaction is equal to the product of the two equilibrium constants. When one reaction is subtracted from the other, the equilibrium constant of the new reaction is equal to the ratio of the two equilibrium constants. The other way, which is used in thermodynamic tables, makes use of equation 5.37 to calculate standard Gibbs energies of formation for species and tabulate them. This has the advantage that although a species can be involved in a large number of reactions, it requires only one entry in a table. Note that equilibria between different phases can be handled in the same way.

5.5 EFFECT OF TEMPERATURE ON THE EQUILIBRIUM CONSTANT

The effect of temperature on chemical equilibrium is determined by $\Delta_r H^\circ$, as shown by the Gibbs–Helmholtz equation 4.63. If $\Delta_r G^\circ = -RT \ln K$ is substituted into this equation, we obtain

$$\Delta_r H^\circ = -T^2\left[\frac{\mathrm{d}(\Delta_r G^\circ/T)}{\mathrm{d}T}\right] = RT^2\left(\frac{\mathrm{d}\ln K}{\mathrm{d}T}\right) \tag{5.38}$$

or

$$\left(\frac{\mathrm{d}\ln K}{\mathrm{d}T}\right) = \frac{\Delta_r H^\circ}{RT^2} \tag{5.39}$$

This equation is often called the **van't Hoff equation.** Note that since K is independent of P for an ideal gas, the left-hand side need not be written as a partial derivative.

Thus, for an endothermic reaction the equilibrium constant increases as the temperature is increased, and for an exothermic reaction the equilibrium constant decreases as the temperature is increased.

According to **Le Châtelier's principle,** when an equilibrium system is perturbed, the equilibrium will always be displaced in such a way as to oppose the applied change. When the temperature of an equilibrium system is raised, this change cannot be prevented by the system, but what happens is that the equilibrium shifts in such a way that more heat is required to heat the reaction mixture to the higher temperature than would have been required if the mixture were inert. In other words, when the temperature is raised, the equilibrium shifts in the direction that causes an absorption of heat.

If $\Delta_r H^\circ$ is independent of temperature, the integral of equation 5.39 from T_1 to T_2 yields

$$\ln \frac{K_2}{K_1} = \frac{\Delta_r H^\circ (T_2 - T_1)}{R T_1 T_2} \tag{5.40}$$

If $\Delta_r H^\circ$ is independent of temperature, then $\Delta_r C_P^\circ$ is zero. If $\Delta_r C_P^\circ$ is zero, then $\Delta_r S^\circ$ is also independent of temperature, as shown later in equation 5.46. Thus, when $\Delta_r C_P^\circ = 0$, the temperature dependence of the equilibrium constant is given by $\Delta_r G^\circ = -RT \ln K = \Delta_r H^\circ - T\Delta_r S^\circ$, or

$$\ln K = -\frac{\Delta_r H^\circ}{RT} + \frac{\Delta_r S^\circ}{R} \tag{5.41}$$

According to this equation a plot of $\ln K$ versus $1/T$ is linear over a temperature range in which $\Delta_r H^\circ$ and $\Delta_r S^\circ$ for the reaction are constant.

Example 5.9 ***Calculation of the standard enthalpy of reaction and standard entropy of reaction from the equilibrium as a function of temperature***

Calculate $\Delta_r H^\circ$ and $\Delta_r S^\circ$ for the reaction

$$N_2(g) + O_2(g) = 2NO(g)$$

from the following values of K:

T/K	1900	2000	2100	2200	2300	2400	2500	2600
$K/10^{-4}$	2.31	4.08	6.86	11.0	16.9	25.1	36.0	50.3

These data are plotted in Fig. 5.4.

Since the plot is linear, we can use equation 5.41. The slope of the plot is -2.19×10^4 K and so

$$\Delta_r H^\circ = -\text{slope} \times R = -(-2.19 \times 10^4\ \text{K})(8.3145\ \text{J K}^{-1}\ \text{mol}^{-1})$$
$$= 182\ \text{kJ mol}^{-1}$$

The intercept of Fig. 5.4 at $1/T = 0$ can be calculated from the experimental value of K at some temperature and the slope. The intercept may be used to calculate the standard entropy change $\Delta_r S^\circ$ according to equation 5.41.

$$\frac{\Delta_r S^\circ}{R} = 3.13$$

$$\Delta_r S^\circ = (3.13)(8.3145\ \text{J K}^{-1}\ \text{mol}^{-1}) = 26.0\ \text{J K}^{-1}\ \text{mol}^{-1}$$

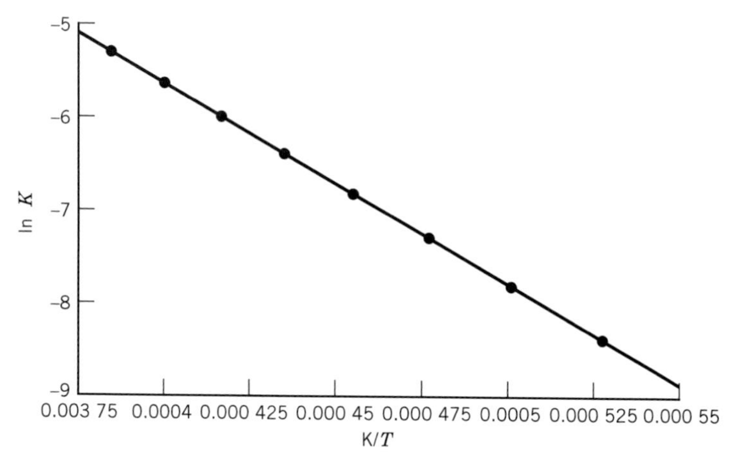

Figure 5.4 Plot of ln K against reciprocal absolute temperature for the reaction $N_2(g) + O_2(g) = 2NO(g)$. The standard enthalpy of reaction is calculated from the slope of the straight line. (See Computer Problem 5.B.)

The **standard reaction entropy** $\Delta_r S^\circ$ for a reaction $\sum \nu_i B_i = 0$ is equal to the change in entropy when the separated reactants, each in its standard state, are completely converted to separated products, each in its standard state, at the specified temperature. $\Delta_r S^\circ$ can be calculated by taking the partial derivative of $\Delta_r G^\circ$ with respect to temperature T, as indicated by equation 4.39. The standard reaction entropy $\Delta_r S^\circ$ at a particular temperature is given by

$$\Delta_r S^\circ = \sum_{i=1}^{N_s} \nu_i \overline{S}_i^\circ \tag{5.42}$$

where N_s is the number of species involved in the reaction.

Example 5.10 ***Calculation of standard reaction entropies using tables***

Calculate the standard reaction entropies for the following reactions at 298 K:

(*a*) $H_2(g) + \frac{1}{2}O_2(g) = H_2O(l)$

(*b*) $N_2(g) + 3H_2(g) = 2NH_3(g)$

(*c*) $CaCO_3(\text{s, calcite}) = CaO(s) + CO_2(g)$

(*d*) $N_2O_4(g) = 2NO_2(g)$

The following reaction entropies are calculated using Table C.2:

(*a*) $\Delta_r S^\circ = 69.91 - 130.68 - \frac{1}{2}(205.14)$

$= -163.34 \text{ J K}^{-1} \text{ mol}^{-1}$

(*b*) $\Delta_r S^\circ = 2(192.45) - 191.61 - 3(130.68)$

$= -198.75 \text{ J K}^{-1} \text{ mol}^{-1}$

(*c*) $\Delta_r S^\circ = 39.75 + 213.74 - 92.9$

$= 160.59 \text{ J K}^{-1} \text{ mol}^{-1}$

(*d*) $\Delta_r S^\circ = 2(240.06) - 304.29$

$= 175.83 \text{ J K}^{-1} \text{ mol}^{-1}$

Note that since the molar entropies of gases are greater than those of liquids and solids, the entropy always increases when the reaction produces more moles of gaseous products than reactants.

In general, $\Delta_r H^\circ$ and $\Delta_r S^\circ$ depend on the temperature because the heat capacities of the reactants depend on temperature, and we have seen earlier that

$$\Delta_f H_i^\circ(T) = \Delta_f H_i^\circ(298.15\ \text{K}) + \int_{298.15}^{T} \overline{C}_{Pi}^\circ\, \mathrm{d}T' \tag{5.43}$$

$$\overline{S}_i^\circ(T) = \overline{S}_i^\circ(298.15\ \text{K}) + \int_{298.15}^{T} \frac{\overline{C}_{Pi}^\circ}{T'}\, \mathrm{d}T' \tag{5.44}$$

where $\overline{C}_{Pi}^\circ$ can be represented by a power series in T (see Table 2.2). Since $\Delta_r H^\circ = \sum_{\nu_i} \Delta_f H_i^\circ$ and $\Delta_r S^\circ = \sum \nu_i \overline{S}_i^\circ$,

$$\Delta_r H^\circ(T) = \Delta_r H^\circ(298.15\ \text{K}) + \int_{298.15}^{T} \Delta_r C_P^\circ\, \mathrm{d}T' \tag{5.45}$$

$$\Delta_r S^\circ(T) = \Delta_r S^\circ(298.15\ \text{K}) + \int_{298.15}^{T} \frac{\Delta_r C_P^\circ}{T'}\, \mathrm{d}T' \tag{5.46}$$

Substituting these relations in $\Delta_r G^\circ(T) = \Delta_r H^\circ(T) - T\Delta_r S^\circ(T)$ yields

$$\Delta_r G^\circ(T) = \Delta_r G^\circ(298.15) + \int_{298.15}^{T} \Delta_r C_P^\circ\, \mathrm{d}T' - T\int_{298.15}^{T} \frac{\Delta_r C_P^\circ}{T'}\, \mathrm{d}T' \tag{5.47}$$

Since $\ln K = -\Delta_r G^\circ/RT$,

$$\ln K(T) = \frac{(298.15)}{T} \ln K(298.15\ \text{K}) - \frac{1}{RT}\int_{298.15}^{T} \Delta_r C_P^\circ\, \mathrm{d}T' + \frac{1}{R}\int_{298.15}^{T} \frac{\Delta_r C_P^\circ}{T'}\, \mathrm{d}T' \tag{5.48}$$

The calculation of $\ln K(T)$ in this way is pretty tedious without a computer, but it can easily be made using a mathematical program that can integrate.

Another way to do this calculation is to use the Gibbs–Helmholtz equation (see equation 4.62):

$$\left[\frac{\mathrm{d}(\Delta_r G^\circ/T)}{\mathrm{d}T}\right] = -\frac{\Delta_r H^\circ}{T^2} \tag{5.49}$$

The effect of temperature on an endothermic reaction $A(g) = 2B(g)$ is shown in Fig. 5.5 for three total pressures. As the temperature is increased, the equilibrium extent of reaction increases. At a given temperature the extent of reaction is greater at a lower pressure.

Example 5.11 ***The dependence of the equilibrium constant on temperature when $\Delta_r C_P^\circ$ is constant***

When $\Delta_r H^\circ$ and $\Delta_r S^\circ$ depend only slightly on T, it may be sufficient to assume that $\Delta_r C_P^\circ$ is constant. Derive the expression for $\ln K(T)$ when $\Delta_r C_P^\circ$ is constant.

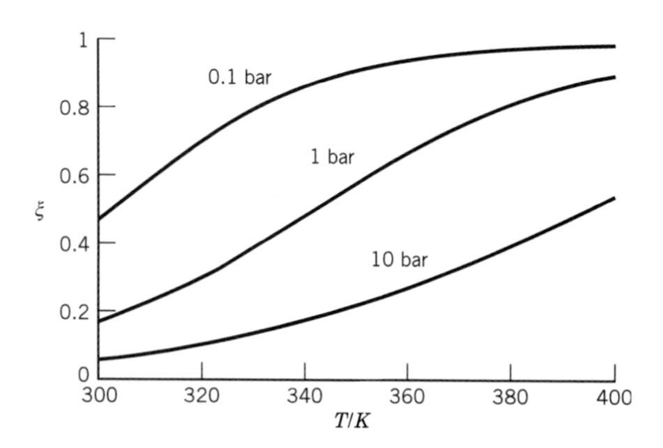

Figure 5.5 Extent of reaction ξ versus temperature for an endothermic reaction A(g) = $2B(g)$, $\Delta_r H^\circ = 50$ kJ mol^{-1}, at three total pressures. (See Computer Problem 5.I.)

The temperature dependencies of the standard enthalpy of reaction and standard entropy of reaction are given by

$$\Delta_r H_T^\circ = \Delta_r H_{298}^\circ + \Delta_r C_P^\circ (T - 298.15 \text{ K})$$

$$\Delta_r S_T^\circ = \Delta_r S_{298}^\circ + \Delta_r C_P^\circ \ln \frac{T}{298.15 \text{ K}}$$

Substituting these relations in $-RT \ln K = \Delta_r H_T^\circ - T \Delta_r S_T^\circ$ yields

$$\ln K = -\frac{\Delta_r H_{298}^\circ}{RT} + \frac{\Delta_r S_{298}^\circ}{R} - \frac{\Delta_r C_P^\circ}{R}\left(1 - \frac{298.15 \text{ K}}{T} - \ln \frac{T}{298.15 \text{ K}}\right)$$

Example 5.12 ***Dependence of the standard Gibbs energy of reaction on temperature when the standard enthalpy of reaction is independent of temperature***

At any temperature,

$$\Delta_r G^\circ(T) = \Delta_r H^\circ(T) - T \Delta \text{r} S^\circ(T)$$

When the standard enthalpy of reaction is independent of temperature because $\Delta_r C_P^\circ$ is independent of temperature, the standard entropy of reaction is also independent of temperature (see equations 5.43 and 5.44). Thus the standard Gibbs energy of a reaction at a temperature other than 298.15 K can be calculated using

$$\Delta_r G^\circ(T) = \Delta_r H^\circ(298.15 \text{ K}) - T \Delta_r S^\circ(298.15 \text{ K})$$

When $\Delta_r G^\circ(298.15)$ K and $\Delta_r H^\circ(298.15)$ K are available, the standard entropy of reaction can be calculated using

$$\Delta_r S^\circ(298.15 \text{ K}) = \frac{\Delta_r H^\circ(298.15 \text{ K}) - \Delta_r G^\circ(298.15 \text{ K})}{298.15 \text{ K}}$$

Substituting this into the expression for $\Delta G^\circ(T)$ yields

$$\Delta_r G^\circ(T) = (T/298.15 \text{ K}) \Delta_r G^\circ(298.15 \text{ K}) + (1 - T/298.15 \text{ K}) \Delta_r H^\circ(298.15 \text{ K})$$

5.6 EFFECT OF PRESSURE, INITIAL COMPOSITION, AND INERT GASES ON THE EQUILIBRIUM COMPOSITION

For an ideal mixture of ideal gases the equilibrium partial pressures of the reactants and products can be expressed in terms of their equilibrium mole fractions y_i and the total pressure P of reactants and products:

$$K = \prod_i \left(\frac{y_i P}{P^\circ}\right)^{\nu_i} = \prod_i y_i^{\nu_i} \prod_i \left(\frac{P}{P^\circ}\right)^{\nu_i} = \left(\frac{P}{P^\circ}\right)^{\nu} K_y \tag{5.50}$$

In this equation $\nu = \sum_i \nu_i$, and K_y is the equilibrium constant written in terms of mole fractions at a particular total pressure.

The value of K_y is a function only of temperature at a constant total pressure P:

$$K_y = \prod_{i=1}^{N_s} y_i^{\nu_i} = \left(\frac{P}{P^\circ}\right)^{-\nu} K \tag{5.51}$$

The equilibrium constant written in terms of mole fractions depends on the pressure as well as the temperature, but, as we will see, it is very useful in calculating the equilibrium extent of reaction because it is written in terms of amounts. If the amount of gaseous products is equal to the amount of gaseous reactants, then $\nu = \sum_i \nu_i = 0$, $K_y = K$, and changing the total pressure of reactants does not affect the equilibrium mole fractions of the reactants and products. If a reaction causes an increase in the number of molecules, then $\nu > 0$, and K_y decreases as the pressure is increased at constant temperature. Thus, raising the pressure decreases the equilibrium mole fractions of the products and increases the equilibrium mole fractions of the reactants; in short, raising the pressure pushes the reaction backward.

Le Châtelier's principle provides a quick way to check conclusions like this about the effects of changes in independent variables on chemical equilibrium. According to Le Châtelier, when an independent variable of a system at equilibrium is changed, the equilibrium shifts in the direction that tends to reduce the effect of the change. When pressure is increased, the equilibrium shifts in the direction to reduce the number of molecules.

If a reaction involves only solids and liquids, the effect of pressure on the equilibrium is small.

Example 5.13 ***The effect of pressure on the ammonia synthesis reaction***

In Example 5.5, we saw that when an initially equimolar mixture of nitrogen and hydrogen is placed in contact with an ammonia catalyst at 400 °C and 51.2 bar, there is a 10% conversion to ammonia at equilibrium. What pressure is required to obtain a 15% conversion?

Using the equation for the equilibrium constant, we obtain

$$1.60 \times 10^{-4} = \frac{16(0.15)^2(0.85)}{(0.65)(P/P^\circ)^2}$$

$$\frac{P}{P^\circ} = 83.45$$

$$P = 83.45 \text{ bar}$$

According to Le Châtelier's principle, raising the total pressure will cause the reaction to shift in the direction of the product NH_3 because there is a decrease in moles of gas in the forward reaction.

To discuss the effect of initial composition on the equilibrium composition for a reaction, we will use the equilibrium constant expressed in terms of mole fractions. At any time during a reaction the amounts of each reactant and product may be expressed in terms of the initial amount n_{i0} and the extent of reaction ξ:

$$K_y = \prod_i \left(\frac{n_{i0} + \nu_i \xi}{n_0 + \nu\xi}\right)^{\nu_i} = \left(\frac{1}{n_0 + \nu\xi}\right)^{\nu} \prod_i (n_{i0} + \nu_i \xi)^{\nu_i} \tag{5.52}$$

In this equilibrium expression the amount of gas is represented by

$$\sum n_i = \sum (n_{i0} + \nu_i \xi) = \sum n_{i0} + \xi \sum_i \nu_i = n_0 + \xi\nu \tag{5.53}$$

where the initial amount of gaseous reactants and products is represented by n_0 and $\nu = \sum_i \nu_i$. Thus, the calculation of the amounts of reactants and products at equilibrium from the initial composition and the value of K_y simply comes down to the solution of a polynomial in ξ. The polynomials arising here always have one positive real root. Quadratic equations are readily solved, and higher-order polynomials can be solved by iterative methods or by using a mathematical application in a computer.

If an inert gas is added to an equilibrium mixture of gases at constant temperature and volume, there is no effect on the equilibrium. But adding an inert gas at constant temperature and pressure has the same effect as lowering the pressure. When inert gases are present, equation 5.52 has to be modified to include the number of moles of inert gas in the denominator of each mole fraction; thus $n_0 + \nu\xi$ becomes $n_0 + \nu\xi + n_{\text{inerts}}$. Substituting the modified form of equation 5.52 in equation 5.50 yields

$$K = \left(\frac{P/P^\circ}{n_0 + \nu\xi + n_{\text{inerts}}}\right)^{\nu} \prod_{i=1}^{N_s} (n_{i0} + \nu_i \xi)^{\nu_i} \tag{5.54}$$

If $\nu < 0$, the addition of an inert gas at constant pressure reduces the sum of the partial pressures of the reactants and products, and the reaction shifts to the left to compensate for this. Equation 5.54 applies generally to ideal mixtures of ideal gases, but if the partial pressure of the inert gases is known, this partial pressure may be subtracted from the total pressure, and equation 5.50 may be used with P equal to the sum of the partial pressures of the reactants and products.

Example 5.14 ***The effect of initial composition, pressure, and inert gases on equilibrium***

As an illustration of the effect of initial composition, pressure, and the addition of an inert gas, consider the equilibrium for the production of methanol from CO and H_2:

$$CO(g) + 2H_2(g) = CH_3OH(g)$$

The value of K at 500 K is 6.23×10^{-3}. (*a*) A gas stream containing equimolar amounts of CO and H_2 is passed over a catalyst at 1 bar. What is the extent of reaction at equilibrium?

	CO	H_2	CH_3OH	
Initial amount	1	1	0	
Equilibrium amount	$1-\xi$	$1-2\xi$	ξ	Total amount $= 2(1-\xi)$
Mole fraction	$\frac{1}{2}$	$\frac{1-2\xi}{2(1-\xi)}$	$\frac{\xi}{2(1-\xi)}$	

$$K = \frac{4\xi(1-\xi)}{(1-2\xi)^2 1^2} = 6.23 \times 10^{-3}$$

Solving this quadratic equation indicates that $\xi = 0.001\ 55$, so that $y_{CO} = 0.5000$, $y_{H_2} = 0.4992$, and $y_{CH_3OH} = 0.0008$.

(*b*) To attain a more complete reaction the pressure is raised to 100 bar and 2 mol of hydrogen is used per mole of CO. What is the equilibrium extent of reaction?

	CO	H_2	CH_3OH	
Initial amount	1	2	0	
Equilibrium amount	$1-\xi$	$2-2\xi$	ξ	Total amount $= 3-2\xi$
Mole fraction	$\frac{1-\xi}{3-2\xi}$	$\frac{2-2\xi}{3-2\xi}$	$\frac{\xi}{3-2\xi}$	

$$K = \frac{\xi(3-2\xi)^2}{(1-\xi)(2-2\xi)^2(100)^2} = 6.23 \times 10^{-3}$$

Solution of this equation for the extent of reaction by successive approximation yields $\xi = 0.817$, so that, $y_{CO} = 0.134$, $y_{H_2} = 0.268$, and $y_{CH_3OH} = 0.598$. (You might check that this gives the right value for the equilibrium constant.)

(*c*) If the reactant gases contain a mole of nitrogen in addition to 1 mol of CO and 2 mol of hydrogen, what is the equilibrium extent of reaction at 100 bar?

The first two lines of the table in (*b*) are unchanged. In the third line, the total number of moles is now $4 - 2\xi$.

$$K = \frac{\xi(4-2\xi)^2}{(1-\xi)(2-2\xi)^2(100)^2} = 6.23 \times 10^{-3}$$

Solution of this equation yields $\xi = 0.735$, so that $y_{CO} = 0.105$, $y_{H_2} = 0.210$, $y_{CH_3OH} = 0.291$, and $y_{N_2} = 0.395$. Here the presence of an inert gas reduces the equilibrium conversion to product, but for a reaction for which $\nu = \sum \nu_i$ is positive, addition of an inert gas will cause the reaction to go further to the right.

5.7 EQUILIBRIUM CONSTANTS FOR GAS REACTIONS WRITTEN IN TERMS OF CONCENTRATIONS

Since thermodynamic tables for gases are based on a standard state pressure of 1 bar for the ideal gas, the $\Delta_f G_i^\circ$ values lead directly to equilibrium constants in terms of pressure (or fugacities). However, in connection with chemical kinetics it is useful to express equilibrium constants of gas reactions in terms of concentrations, since rate equations are written in terms of concentrations (Section 18.2). These two types of equilibrium constants will be represented by K_P and K_c. To obtain a general expression for the equilibrium constant K_c

in terms of concentrations for ideal gases, we replace P_i in equation 5.21 with $P_i = n_i RT/V = c_i RT$:

$$K_P = \prod_{i=1}^{N_s} \left(\frac{P_i}{P^\circ}\right)^{\nu_i} = \prod_i \left(\frac{c_i RT}{P^\circ}\right)^{\nu_i} \tag{5.55}$$

To define a dimensionless equilibrium constant in terms of concentration, we introduce the standard concentration c°, which represents one mole per liter. Introducing this standard concentration into each term of equation 5.55 yields

$$K_P = \prod_{i=1}^{N_s} \left[\left(\frac{c_i}{c^\circ}\right)\left(\frac{c^\circ RT}{P^\circ}\right)\right]^{\nu_i} = \left(\frac{c^\circ RT}{P^\circ}\right)^{\sum \nu_i} \prod_i \left(\frac{c_i}{c^\circ}\right)^{\nu_i}$$

$$= \left(\frac{c^\circ RT}{P^\circ}\right)^{\sum \nu_i} K_c \tag{5.56}$$

where the **equilibrium constant expressed in terms of concentration,**

$$K_c = \prod_i \left(\frac{c_i}{c^\circ}\right)^{\nu_i} \tag{5.57}$$

is a function only of temperature for a mixture of ideal gases. If $c^\circ = 1$ mol L^{-1} and $P^\circ = 1$ bar, then $c^\circ RT/P^\circ = 24.79$ at 298.15 K.

Example 5.15 ***A gas equilibrium constant expressed in concentrations***

What is the value of the equilibrium constant K_c for the dissociation of ethane into methyl radicals at 1000 K?

$$C_2H_6(g) = 2CH_3(g)$$

$$\Delta_r G^\circ = 2\Delta_f G^\circ(CH_3) - \Delta_f G^\circ(C_2H_6)$$

$$= 2(159.82) - 109.55 = 210.09 \text{ kJ mol}^{-1}$$

$$K_P = \exp\left(-\frac{\Delta_r G^\circ}{RT}\right)$$

$$= \exp \frac{(-210.09)}{(8.3145 \times 10^{-3})(1000)}$$

$$= 1.062 \times 10^{-11}$$

$$K_c = \frac{([CH_3]/c^\circ)^2}{[C_2H_6]/c^\circ} = K_P \frac{P^\circ}{c^\circ RT}$$

$$= (1.062 \times 10^{-11}) \frac{(1 \text{ bar})}{(1 \text{ mol L}^{-1})(0.083145)(1000 \text{ K})}$$

$$= 1.278 \times 10^{-13}$$

Thus, at equilibrium $[CH_3]^2/[C_2H_6] = 1.278 \times 10^{-13}$ mol L^{-1}, where the brackets indicate concentrations in moles per liter.

5.8 HETEROGENEOUS REACTIONS

A reaction involving more than one phase that does not involve equilibria of a species between phases is referred to as a **heterogeneous reaction.** Examples are

$$CaCO_3(s) = CaO(s) + CO_2(g) \tag{5.58}$$

$$CH_4(g) = C(s) + 2H_2(g) \tag{5.59}$$

Depending on the initial conditions, reactions of this type can go to completion, whereas reactions in a single phase do not go to completion because of the entropy of mixing (Section 3.5). The equilibrium constant for reaction 5.58 is equal to the partial pressure of CO_2 gas that is measured at equilibrium when all three phases are present. If the pressure applied to the system is less than K, reaction 5.58 will go to completion to the right. If the pressure applied to the system is greater than K, reaction 5.58 will go to completion to the left.

Equilibrium constants for reactions like 5.58 and 5.59 **can be written without terms for the pure solids, provided they are present.** The reason for doing this is that the activities of the pure solid phases are very nearly equal to unity for moderate pressures. If the gases are ideal, then the equilibrium constant expressions for reactions 5.58 and 5.59 are

$$K = \frac{P_{CO_2}}{P^\circ} \tag{5.60}$$

$$K = \frac{P_{H_2}^2}{P_{CH_4} P^\circ} \tag{5.61}$$

These equilibrium constants are independent of the amount of pure solid phase present. As long as graphite is present, the second reaction behaves like a homogeneous reaction in that the reaction does not go to completion because of the entropy of mixing in the gas phase.

Table 5.1 gives the pressure of $CO_2(g)$ in equilibrium with $CaCO_3(s)$ and $CaO(s)$ at a series of temperatures. The natural logarithm of the equilibrium pressure is plotted versus $1/T$ in Fig. 5.6. The three phases are at equilibrium only along the line.

Example 5.16 ***Calculation of reaction properties at high temperatures assuming $\Delta_r C_P^\circ$ is constant***

Calculate $\Delta_r G^\circ$, $\Delta_r H^\circ$, and $\Delta_r S^\circ$ for reaction 5.58 at 1000 K using data in (*a*) Table 5.1 and (*b*) Table C.2 with the assumption that $\Delta_r C_P^\circ$ is independent of temperature.

(*a*) $\Delta_r G^\circ = -RT \ln K = -(8.3145 \text{ J K}^{-1} \text{ mol}^{-1})(1000 \text{ K})(-3.00) = 24.9 \text{ kJ mol}^{-1}$

$\Delta_r H^\circ = -R(\text{slope}) = -(8.3145 \text{ J K}^{-1} \text{ mol}^{-1})(-2.055 \times 10^4 \text{ K})$

$= 171 \text{ kJ mol}^{-1}$

Table 5.1 Pressures of $CO_2(g)$ in Equilibrium with $CaCO_3(s)$ and $CaO(s)$

$t/°C$	500	600	700	800
P_{CO_2}/P°	9.2×10^{-5}	2.39×10^{-3}	2.88×10^{-2}	0.2217
$t/°C$	897	1000	1100	1200
P_{CO_2}/P°	0.987	3.820	11.35	28.31

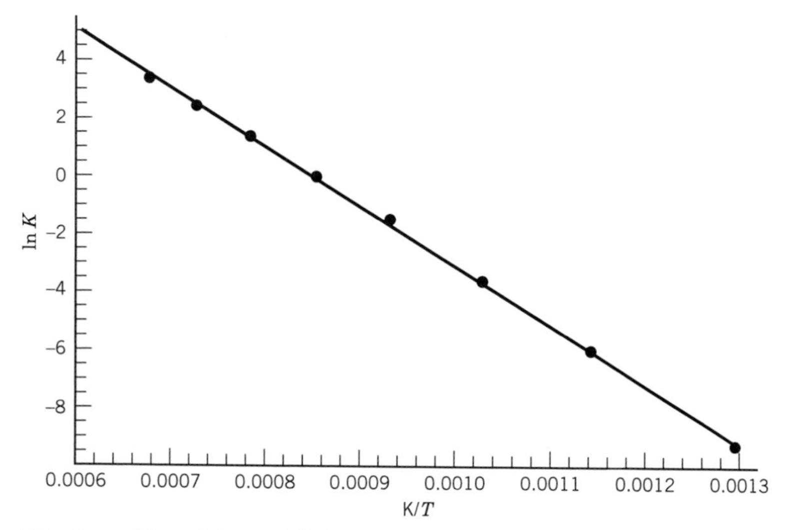

Figure 5.6 Logarithm of the equilibrium constant for reaction 5.58 as a function of reciprocal temperature. The equilibrium partial pressuer of $CO_2(g)$ is essentially atmospheric pressure when $\ln K = 0$. Above the line $CaCO_3(s)+CO_2(g)$ are stable, and below the line $CaO(s)+CO_2(g)$ are stable. (See Computer Problem 5.L.)

$$\begin{aligned}\Delta_r S^\circ &= \frac{\Delta_r H^\circ - \Delta_r G^\circ}{T} \\ &= \frac{(171 - 24.9) \times 10^3 \text{ J mol}^{-1}}{1000 \text{ K}} \\ &= 146 \text{ J K}^{-1} \text{ mol}^{-1}\end{aligned}$$

(*b*) According to Table C.2 the values at 298 K are $\Delta_r G^\circ = 130.40$ kJ mol^{-1}, $\Delta_r H^\circ = 78.32$ kJ mol^{-1}, and $\Delta_r S^\circ = 160.59$ J K^{-1} mol^{-1} when calcite is the reactant.

$$\begin{aligned}\Delta_r C_P &= \overline{C}_P^\circ(\text{CaO}) + \overline{C}_P^\circ(\text{CO}_2) - \overline{C}_P^\circ(\text{CaCO}_3) \\ &= 42.80 + 37.11 - 81.88 = -1.97 \text{ J K}^{-1} \text{ mol}^{-1}\end{aligned}$$

Using the equations in Example 5.11,

$$\begin{aligned}\Delta_r H_{1000}^\circ &= 178.32 \text{ kJ mol}^{-1} - (1.97 \times 10^{-3} \text{ kJ K}^{-1} \text{ mol}^{-1})(701.85 \text{ K}) \\ &= 176.94 \text{ kJ mol}^{-1} \\ \Delta_r S_{1000}^\circ &= 160.59 \text{ J K}^{-1} \text{ mol}^{-1} - (1.97 \text{ J K}^{-1} \text{ mol}^{-1}) \ln(1000/298.15) \\ &= 158.21 \text{ J K}^{-1} \text{ mol}^{-1} \\ \Delta_r G_{1000}^\circ &= \Delta H_{1000}^\circ - T\Delta S_{1000}^\circ \\ &= 176.94 \text{ kJ mol}^{-1} - (1000 \text{ K})(158.21 \times 10^{-3} \text{ kJ K}^{-1} \text{ mol}^{-1}) \\ &= 18.73 \text{ kJ mol}^{-1}\end{aligned}$$

5.9 DEGREES OF FREEDOM AND THE PHASE RULE

In Section 1.9 we discussed the fact that for a one-phase system without chemical reactions, $F = N_s + 1$ variables have to be specified in order to describe the

intensive state of the system at equilibrium, where F is the **number of degrees of freedom** and N_s is the number of different kinds of species. For a one-phase system involving a single species, $F = 2$ so that it is sufficient to specify T and P, but other properties could be used. For a two-phase system containing a single species, $F = 1$, so only T or P has to be specified. For a three-phase system containing a single species, $F = 0$, so that no variables have to be specified. We also discussed the fact that for these systems containing a single species, $D = F + p = 2 + p$ variables are required to describe the **extensive state** of the system, where p is the **number of different phases.** Thus the descriptions for the one-phase, two-phase, and three-phase systems containing a single species each require the specification of $D = 3$ variables since amounts of phases have to be specified. In Section 2.2 these ideas were confirmed in our discussion of the internal energy U of a system in a certain intensive state or a certain extensive state. In Sections 4.1 and 4.2, we noted that the $D = N_s + 2$ variables in the fundamental equations are also involved in the criteria, such as $(dG)_{T,P,\{n_i\}} \leq 0$, for spontaneous change and equilibrium. In Section 4.8 we saw that the Gibbs–Duhem equation for a nonreaction system, which gives a relation between the intensive variables for a phase in a system, is in agreement with $F = N_s + 1$ of these variables being independent.

However, when a chemical reaction occurs and is at equilibrium, this provides a relationship between the chemical potentials of the species involved, and so the intensive state of the system is described by specifying one fewer intensive variable. For example, to describe the intensive state of a system containing N_2, H_2, and NH_3 at chemical equilibrium, it is necessary to specify only T, P, and $y(H_2)/y(N_2)$ or T, P, and $n(NH_3)/n(N_2)$. The choice of independent intensive variables is optional, but the number is not. We can generalize this discussion by stating that the number of intensive variables that have to be specified to describe the intensive state of a one-phase system is $N_s - R + 1$, where R is the number of **independent chemical reactions** that are at equilibrium. The reactions in a set are independent if no reaction in the set can be obtained by adding and subtracting other reactions in the set. (We will discuss this in more detail in Section 5.11.)

In using $N_s - R + 1$ to count the number of intensive variables for a reaction system, it is convenient to introduce a new concept, which is component. **Components are the things that are conserved in a chemical reaction system.** An obvious choice of components are the amounts of atoms of elements, but we will see later, in Section 5.11, that this is not the only choice; molecules or groups of atoms can be used as components. We will use C to represent the number of components in a reaction system, and for each of these C components there is a conservation equation. But only an independent set of conservation equations can be used in an equilibrium calculation, just as only an independent set of R chemical reactions can be used. Thus C is the number of independent components that is given by $C = N_s - R$. The number of intensive variables that have to be specified to describe the intensive state of a one-phase reaction system is $C + 1$. We can look at it this way: When there are reactions at equilibrium, we do not have to specify the concentrations of all species because they can be calculated using the equilibrium constant expressions. The number of intensive variables that have to be specified to describe the intensive state of a one-phase system is $N_s + 1$ if there are no reactions and $C + 1$ if there are chemical reactions at equilibrium. The number of variables required to describe the extensive state of a one-phase reaction system is $D = C + 2$.

Example 5.17 ***The choices of independent intensive properties***

(*a*) Oxygen is in equilibrium with ozone: $3O_2(g) = 2O_3(g)$. How many intensive variables have to be specified to describe the intensive state of the system? (*b*) The water gas shift reaction, $H_2O(g) + CO(g) = H_2(g) + CO_2(g)$, is at equilibrium. How many intensive variables have to be specified to describe the state of the system? Give three possible choices of intensive variables. (*c*) The equilibrium behavior of a system involving the following two reactions is to be investigated:

$$H_2O(g) + CO(g) = H_2(g) + CO_2(g)$$
$$3H_2(g) + CO(g) = CH_4(g) + H_2O(g)$$

How many intensive variables have to be specified to describe the state of the system? Give a possible choice. Someone suggests that the equilibrium calculations be made with the following two reactions:

$$H_2O(g) + CO(g) = H_2(g) + CO_2(g)$$
$$2H_2(g) + 2CO(g) = CH_4(g) + CO_2(g)$$

Is that all right?

(*a*) There is one component; $C = N_s - R = 2 - 1 = 1$. Therefore, two intensive variables have to be specified.

(*b*) There are three components; $C = N_s - R = 4 - 1 = 3$. Therefore, four intensive variables have to be specified. Three possible choices of intensive variables are T, P, $n(H_2O)/n(CO)$, $n(H_2)/n(CO)$; T, P, $n(CO)/n(CO_2)$, $n(H_2)/n(CO_2)$; T, P, $n(H_2O)/n(CO)$, $P(H_2)$.

(*c*) There are three components; $C = N_s - R = 5 - 2 = 3$. Thus four intensive variables have to be specified. Note that this is the same as in the preceding example, and so any of the choices in (*b*) can be used. Since the suggested reactions are independent, they are just as good as the original set for equilibrium calculations.

In 1876, Gibbs derived a general expression for the number of intensive variables (the number of degrees of freedom) that have to be specified for a multiphase system at equilibrium. It is important to emphasize that we are only interested in describing the intensive state of the system; thus we are not concerned with the relative amounts of the various phases. The **number of degrees of freedom** F of a system is the number of intensive variables that must be specified to describe the state of the system completely.

To derive an expression for F, we will consider a system that consists of p phases. If a phase contains C components, its composition may be specified by stating $(C - 1)$ mole fractions—one less than the number of components because the mole fraction of one component can be obtained from $\sum x_i = 1$, where x_i represents the mole fraction of component i. Thus, the total number of concentrations to be specified for the whole system is $(C - 1)$ for each of the p phases or $p(C - 1)$ concentrations. In general, there are two more intensive variables that have to be considered, temperature and pressure. Thus,

$$\text{Number of intensive variables} = p(C - 1) + 2 \tag{5.62}$$

Next we consider the number of relationships that must be satisfied for phase equilibrium. The chemical potential μ for each component is the same in each phase α, β, γ, and so on, and so $\mu_{i,\alpha} = \mu_{i,\beta} = \mu_{i,\gamma} = \cdots$ for component i. There are p phases but only $(p - 1)$ equilibrium relationships of the type $\mu_{i,\alpha} = \mu_{i,\beta}$ for each component. For example, if there are two phases, there is only one equilibrium relationship for each component that gives its distribution between the two phases. Altogether there are C components, each one of which can be involved in an equilibrium between phases. Thus,

$$\text{Number of independent equations} = C(p-1) \tag{5.63}$$

The difference between the number of variables and the number of independent equations is the **number of independent variables,** which is referred to as the number of degrees of freedom:

$$F = [p(C-1)+2] - C(p-1)$$

or

$$F = C - p + 2 \tag{5.64}$$

This is the important **phase rule** of Gibbs. The number of degrees of freedom F is equal to the number of intensive variables that can be set arbitrarily. For example, for a one-component system $F = 3 - p$. Under conditions where a single phase is present, $F = 2$ and the pressure and temperature can both be set arbitrarily. Under conditions where two phases are in equilibrium, $F = 1$ and either the temperature or pressure may be set arbitrarily. Under conditions where three phases are in equilibrium, $F = 0$ and the pressure and temperature are fixed by the equilibrium. Such a system is said to be **invariant** and is represented by a point in a plot of pressure versus temperature.

It can be seen from the phase rule that the greater the number of components in a system, the greater the number of degrees of freedom. On the other hand, the greater the number of phases, the smaller is the number of variables such as temperature, pressure, and mole fraction that must be specified to describe the system completely.

Equations 5.62 and 5.63 are based on the assumption that pressure and temperature are both variables. If the pressure, for example, is fixed, the phase rule becomes $F = C - p + 1$. On the other hand, if the system is affected by both temperature and pressure and another variable, such as magnetic field strength, the phase rule becomes $F = C - p + 3$.

When there are special conditions involved in the specification of the system, the number of these special constraints s must be included in the phase rule to give $F = C - p + 2 - s$. An example of a special constraint would be taking the initial amounts of two reactants in the ratio of their stoichiometric numbers in the reaction.

The number D of independent properties required to describe the extensive state of a system can be readily counted in the fundamental equation for a thermodynamic potential: D is the number of differential terms on the right-hand side. The number F of properties required to describe the intensive state of a system can be readily counted in the Gibbs–Duhem equation for the system: F is the number of differential terms in the Gibbs–Duhem equation. This is illustrated first by a one-phase system containing one species at a specified T and P. The appropriate thermodynamic potential is the Gibbs energy, and the fundamental equation is

$$\mathrm{d}G = -S\,\mathrm{d}T + V\,\mathrm{d}P + \mu\,\mathrm{d}n \tag{5.65}$$

The differential terms on the right-hand side indicate that the natural variables are T, P, and n, and so $D = 3$. The corresponding Gibbs–Duhem equation is

$$0 = -S\,\mathrm{d}T + V\,\mathrm{d}P - n\,\mathrm{d}\mu \tag{5.66}$$

The intensive variables T, P, and μ for the system are indicated by the differential terms, but only two of them are independent according to equation 5.66. Any two of the intensive variables can be chosen as degrees of freedom, and $F = 2$. This is in agreement with the phase rule, which gives $F = 1 - 1 + 2 = 2$. The relation $D = F + p = 2 + 1 = 3$ is satisfied. The criterion for equilibrium is $(\mathrm{d}G)_{T,P,n} \leq 0$.

Now consider a one-phase system at specified T and P involving the chemical reaction A + B = C. The appropriate thermodynamic potential is the Gibbs energy, and the fundamental equation for the system before equilibrium is reached is

$$\mathrm{d}G = -S\,\mathrm{d}T + V\,\mathrm{d}P + \mu_{\mathrm{A}}\,\mathrm{d}n_{\mathrm{A}} + \mu_{\mathrm{B}}\mathrm{d}n_{\mathrm{B}} + \mu_{\mathrm{C}}\,\mathrm{d}n_{\mathrm{C}} \tag{5.67}$$

We now substitute the equilibrium relation $\mu_{\mathrm{A}} + \mu_{\mathrm{B}} = \mu_{\mathrm{C}}$ into equation 5.67 to eliminate μ_{C}, yielding

$$\mathrm{d}G = -S\,\mathrm{d}T + V\,\mathrm{d}P + \mu_{\mathrm{A}}\,\mathrm{d}n_{\mathrm{cA}} + \mu_{\mathrm{B}}\,\mathrm{d}n_{\mathrm{cB}} \tag{5.68}$$

where $n_{\mathrm{cA}} = n_{\mathrm{A}} + n_{\mathrm{C}}$ and $n_{\mathrm{cB}} = n_{\mathrm{B}} + n_{\mathrm{C}}$ are the amounts of the A and B **components.** The differential terms on the right-hand side of equation 5.68 indicate that when the reaction is at equilibrium the natural variables are T, P, n_{cA}, and n_{cB}, so that $D = 4$ for this one-phase, three-species system involving one reaction. Note that the amounts of components A and B are independent variables for the system at equilibrium, but the amounts of species n_{A}, n_{B}, and n_{C} are not. The corresponding Gibbs–Duhem equation is

$$0 = -S\,\mathrm{d}T + V\,\mathrm{d}P - n_{\mathrm{cA}}\,\mathrm{d}\mu_{\mathrm{A}} - n_{\mathrm{cB}}\,\mathrm{d}\mu_{\mathrm{B}} \tag{5.69}$$

The intensive variables T, P, μ_{A}, and μ_{B} for the system are indicated by the differential terms, but because they are related by equation 5.69 only three of them are independent, so that $F = 3$. This is in agreement with the phase rule, which gives $F = N - R - p + 2 = 3 - 1 - 1 + 2 = 3$. The relation $D = F + p = 3 + 1 = 4$ is satisfied. The criterion for equilibrium is $(\mathrm{d}G)_{T,P,n_{\mathrm{cA}},n_{\mathrm{cB}}} \leq 0$.

The relationship between D and the fundamental equation and the relationship between F and the Gibbs–Duhem equations are illustrated by a two-phase system containing one species at specified T and P. The appropriate thermodynamic potential is the Gibbs energy, and the fundamental equation is

$$\mathrm{d}G = -S\,\mathrm{d}T + V\,\mathrm{d}P + \mu_{\alpha}\,\mathrm{d}n_{\alpha} + \mu_{\beta}\,\mathrm{d}n_{\beta} \tag{5.70}$$

where the two phases are labeled α and β. The equilibrium relation $\mu_{\alpha} = \mu_{\beta}$ (see Section 4.6) can be substituted in equation 5.70 to obtain

$$\mathrm{d}G = -S\,\mathrm{d}T + V\,\mathrm{d}P + \mu\,\mathrm{d}n \tag{5.71}$$

where $\mu = \mu_{\alpha} = \mu_{\beta}$ and $n = n_{\alpha} + n_{\beta}$. The differential terms on the right-hand side indicate that the natural variables are T, P, and n, and so $D = 3$. There is a separate Gibbs–Duhem equation for each phase:

$$0 = -S_{\alpha}\,\mathrm{d}T + V_{\alpha}\,\mathrm{d}P - n_{\alpha}\,\mathrm{d}\mu_{\alpha} \tag{5.72}$$

$$0 = -S_{\beta}\,\mathrm{d}T + V_{\beta}\,\mathrm{d}P - n_{\beta}\,\mathrm{d}\mu_{\beta} \tag{5.73}$$

Since the equilibrium relation indicates that $\mathrm{d}\mu_{\alpha} = \mathrm{d}\mu_{\beta}$, there are only three intensive variables. Equations 5.72 and 5.73 provide two relationships between these three intensive variables, and so only one is independent. The independent variable can be taken to be T, P, or μ; thus $F = 1$, as expected from the phase rule. The relation $D = F + p = 1 + 2 = 3$ is also satisfied. The criterion for equilibrium is $(\mathrm{d}G)_{T,n_{\alpha},n_{\beta}} \leq 0$.

Example 5.18 ***Identifying properties to describe the intensive state of a system and the extensive state of a system***

Identify sets of properties to specify the extensive and intensive states of the following three systems by examining their fundamental equations for the Gibbs energy. In each case, show that the number of properties required to specify the intensive state is given by $F = N_s - R - p + 2 = C - p + 2$ and the number of properties required to specify the extensive state is given by $D = F + p$. Also state the criterion for equilibrium and spontaneous change for each system.

(*a*) One-phase system containing two species

(*b*) One-phase system containing A and B, which are in equilibrium with each other

(*c*) Two-phase system containing two species

(*a*) The fundamental equation for G is

$$dG = -S\,dT + V\,dP + \mu_1\,dn_1 + \mu_2\,dn_2$$

The differential terms on the right-hand side indicate that the natural variables are T, P, n_1, and n_2, and so $D = 4$. The corresponding Gibbs–Duhem equation is

$$0 = -S\,dT + V\,dP - n_1\,d\mu_1 - n_2\,d\mu_2$$

The intensive variables T, P, μ_1, and μ_2 for the system are indicated by the differential terms, but only three of them are independent. Any three of them can be chosen, and $F = 3$. This agrees with the phase rule, which gives $F = 2 - 1 + 2 = 3$. The relation $D = F + p = 3 + 1 = 4$ is satisfied. The criterion for equilibrium is $(dG)_{T,P,n_1,n_2} \leq 0$.

(*b*) The fundamental equation for G is

$$dG = -S\,dT + V\,dP + \mu_A\,dn_A + \mu_B\,dn_B$$

This form of the fundamental equation can be used to derive the equilibrium relation $\mu_A = \mu_B$. Substituting this relation into the fundamental equation for G yields

$$dG = -S\,dT + V\,dP + \mu_c\,dn_c$$

where $\mu_c = \mu_A = \mu_B$ and $n_c = n_A + n_B$. The differential terms on the right-hand side indicate that the natural variables are T, P, and n_c, and so $D = 3$. Note that the amount of the component $(n_A + n_B)$ is an independent variable for the system, but n_A and n_B are not. The corresponding Gibbs–Duhem equation is

$$0 = -S\,dT + V\,dP - n\,d\mu$$

The intensive variables T, P, and μ for the system are indicated by the differential terms, but only two of them are independent so that $F = 2$. This is in agreement with the phase rule, which gives $F = 2 - 1 - 1 + 2 = 2$. The relation $D = F + p = 2 + 1 = 3$ is satisfied. The criterion for equilibrium is $(dG)_{T,P,n_c} \leq 0$, where n_c is the amount of the component.

(*c*) The fundamental equation for G is

$$dG = -S\,dT + V\,dP + \mu_{1\alpha}\,dn_{1\alpha} + \mu_{2\alpha}\,dn_{2\alpha} + \mu_{1\beta}\,dn_{1\beta} + \mu_{2\beta}\,dn_{2\beta}$$

This fundamental equation can be used to derive the equilibrium relations $\mu_{1\alpha} = \mu_{1\beta}$ and $\mu_{2\alpha} = \mu_{2\beta}$. Substituting these relations into the fundamental equation yields

$$dG = -S\,dT + V\,dP + \mu_{1\alpha}\,dn_1 + \mu_{2\alpha}\,dn_2$$

where $n_1 = n_{1\alpha} + n_{1\beta}$ and $n_2 = n_{2\alpha} + n_{2\beta}$. The differential terms on the right-hand side indicate that the natural variables can be taken as T, P, n_1, and n_2, and so $D = 4$. There is a Gibbs–Duhem equation for each phase:

$$0 = -S_\alpha\,dT + V_\alpha\,dP - n_{1\alpha}\,d\mu_{1\alpha} - n_{2\alpha}\,d\mu_{2\alpha}$$

$$0 = -S_\beta\,dT + V_\beta\,dP - n_{1\beta}\,d\mu_{1\beta} - n_{2\beta}\,d\mu_{2\beta}$$

But the equilibrium relations indicate that $d\mu_{1\alpha} = d\mu_{1\beta}$ and $d\mu_{2\alpha} = d\mu_{2\beta}$, so that there are only four intensive variables remaining. The two Gibbs–Duhem equations provide two relationships between these four intensive variables, and so only two are independent. They can be taken to be T and P. This is in agreement with the phase rule $F = 2 - 2 + 2 = 2$. The relation $D = F + p = 2 + 2 = 4$ is satisfied. The criterion for equilibrium is $(dG)_{T,P,n_1,n_2} \leq 0$ or $(dG)_{T,P,n_\alpha,n_\beta} \leq 0$, where $n_\alpha = n_{1\alpha} + n_{2\alpha}$ and $n_\beta = n_{1\beta} + n_{2\beta}$.

Example 5.19 ***Calculating degrees of freedom***

The reaction $CaCO_3(s) = CaO(s) + CO_2(g)$ is at equilibrium. (*a*) How many degrees of freedom are there when all three phases are present at equilibrium? (*b*) How many degrees of freedom are there when only $CaCO_3(s)$ and $CO_2(g)$ are present? (See the introduction to components early in this section.)

(*a*) $C = N_s - R = 3 - 1 = 2$

$F = C - p + 2 = 2 - 3 + 2 = 1$

Therefore, only the temperature or pressure may be varied independently.

(*b*) $F = 2 - 2 + 2 = 2$

Therefore, both the temperature and pressure may be varied without destroying a phase.

5.10 SPECIAL TOPIC: ISOMER GROUP THERMODYNAMICS

There is a special case of chemical equilibrium that is of sufficient importance to be discussed separately, and that is the chemical equilibrium between isomers. We have already had an example of isomerism A = B in Section 5.2. In Section 5.9, we saw that the number of components is given by $C = N_s - R$, where N_s is the number of species in the system and R is the number of independent reactions. If N_s isomers are in equilibrium, there are $N_s - 1$ independent reactions, and $C = N_s - (N_s - 1) = 1$. Thus when a group of isomers is in equilibrium, they form a single component, and we can calculate the standard thermodynamic properties of the isomer group as a whole if we know the standard thermodynamic properties of the individual isomers. When a system involving isomers is in chemical equilibrium, the isomer group can be treated as a pseudospecies in a larger system because the distribution of isomers in an isomer group is a function of temperature only and is not affected by the presence of other reactants. Thus an isomer group has a standard Gibbs energy of formation, a standard enthalpy of formation, a standard entropy, and a standard heat capacity at constant pressure. This greatly simplifies equilibrium calculations on organic mixtures, where the number of isomers in a homologous series increases approximately exponentially with carbon number.

As an example, consider the three isomers of pentane (*n*-pentane, isopentane, and neopentane) in equilibrium with each other. For ideal gases, the equilibrium constants of the three formation reactions are of the form $K_i = (P_i/P^\circ)/(P_{H_2}/P^\circ)^6 = \exp(-\Delta_f G_i^\circ/RT)$. The equilibrium mole fractions r_i of the three isomers within the isomer group are given by

$$r_i = \frac{P_i}{\sum P_i} = \frac{\exp(-\Delta_f G_i^\circ/RT)}{\sum \exp(-\Delta_f G_i^\circ/RT)} \tag{5.74}$$

since the partial pressure of molecular hydrogen cancels. We can represent the denominator by $\exp[-\Delta_f G^\circ(\text{iso})/RT]$, where $\Delta_f G^\circ(\text{iso})$ is the **standard Gibbs energy of formation of the isomer group,** and write

$$\begin{aligned} r_i &= \frac{\exp(-\Delta_f G_i^\circ/RT)}{\exp[-\Delta_f G^\circ(\text{iso})/RT]} \\ &= \exp\{[\Delta_f G^\circ(\text{iso}) - \Delta_f G_i^\circ]/RT\} \end{aligned} \tag{5.75}$$

where

$$\Delta_f G^\circ(\text{iso}) = -RT \ln \sum_{i=1}^{N_I} \exp(-\Delta_f G_i^\circ/RT) \tag{5.76}$$

The number of isomers in the isomer group is represented by N_I.

Note that the standard Gibbs energy of formation of the isomer group is not the mole fraction average of $\Delta_f G_i^\circ$, but is actually more negative than the standard Gibbs energy of formation of the most stable isomer. This is to be expected because in any equilibrium calculation on a larger system, the equilibrium mole fraction of the isomer group has to be larger than that of the most stable isomer. The value of $\Delta_f G^\circ(\text{iso})$ can also be calculated from the mole fraction average Gibbs energy of formation plus the entropy of mixing.

The standard enthalpy of formation $\Delta_f H^\circ(\text{iso})$ of the isomer group is the mole fraction weighted average of the $\Delta_f H_i^\circ$, and it can be calculated using

$$\Delta_f H^\circ(\text{iso}) = \sum_{i=1}^{N_I} r_i \Delta_f H_i^\circ \tag{5.77}$$

The expressions for $\Delta_f S^\circ(\text{iso})$ and $\Delta_f C_P^\circ(\text{iso})$ are readily derived.*

The fact that species in equilibrium can be treated as a single species in an equilibrium calculation is illustrated by the fact that $H_2O(l)$ and $H^+(aq)$ are shown as single species in thermodynamic tables. In these cases we do not know the thermodynamic properties of the various species, which are rapidly interconverted, but we can treat the sum of species as a single species. Another example is glucose(aq), which is made up of α and β forms. Later, in treating biochemical reactions, we will see that thermodynamic properties can be calculated for adenosine triphosphate at a specified pH, although it is made up of several species.

Example 5.20 ***The standard Gibbs energy of an isomer group at equilibrium***

The standard Gibbs energies of formation of the gaseous pentanes at 298.15 K are as follows: *n*-pentane, -8.33 kJ mol^{-1}; isopentane, -13.27 kJ mol^{-1}; and neopentane, -17.37 kJ mol^{-1}. Calculate the standard Gibbs energy of formation of the isomer group at chemical equilibrium. What are the equilibrium mole fractions of the isomers at this temperature?

$$\begin{aligned} \Delta_f G^\circ(\text{iso}) &= -RT \ln[\exp(8.33/RT) + \exp(13.27/RT) + \exp(17.37/RT)] \\ &= -17.86 \text{ kJ mol}^{-1} \end{aligned}$$

where RT is in kJ mol^{-1}.

*R. A. Alberty, *Ind. and Eng. Chem. Fund.* **22:**318 (1983).

The equilibrium mole fractions are given by

$$r_n = \exp[(-17.86 + 8.33)/RT] = 0.021$$
$$r_{\text{iso}} = \exp[(-17.86 + 13.27)/RT] = 0.157$$
$$r_{\text{neo}} = \exp[(-17.86 + 17.37)/RT] = 0.821$$

Comment:

When isomers are in equilibrium with one another, the mole fraction of an isomer within the isomer group is dependent only on the temperature, for ideal gases. Thus we can calculate a standard Gibbs energy of formation of the isomer group if we know the standard Gibbs energies of formation of the individual isomers. For ideal gases, the standard enthalpy of formation of the isomer group is simply a weighted average of the various isomers. If isomers are interconverted sufficiently rapidly, we may not be aware of the existence of isomers, so that the substance is treated as a single species with standard formation properties determined in the usual way.

5.11 SPECIAL TOPIC: CHEMICAL EQUATIONS AS MATRIX EQUATIONS

So far, we have considered systems in which there is a single chemical reaction. However, in many practical applications of chemical thermodynamics a number of chemical reactions occur simultaneously. The equilibrium composition can be calculated by using a set of independent chemical reactions that will represent all possible chemical changes in the system. A set of reactions is independent if no member of the set may be obtained by adding and subtracting other members of the set. Different sets of reactions may be chosen to describe a given system, but the *number* R of independent reactions is always the same.

For a simple system, a set of independent reactions can be found by inspection, but for a more complex system, matrix operations are required. As we saw in Section 5.9, the number R of independent reactions is needed to calculate the number C of components in a system at chemical equilibrium because $C = N_s - R$, where N_s is the number of species. The determination of the number of independent reactions and the number of components depends on the fact that atoms of each element have to be conserved in a chemical reaction.

Chemical equations are really matrix equations, and the recognition of their mathematical character is especially useful in considering multireaction systems.* As a first step, consider the following chemical equation:

$$\begin{array}{cccc} & CO + \tfrac{1}{2}O_2 & = & CO_2 \\ \begin{matrix} C \\ O \end{matrix} & \begin{bmatrix} 1 \\ 1 \end{bmatrix} \quad \begin{bmatrix} 0 \\ 2 \end{bmatrix} & & \begin{bmatrix} 1 \\ 2 \end{bmatrix} \end{array} \tag{5.78}$$

Molecular formulas can be interpreted as column matrices, as indicated below the chemical equation. In this case, the top integer of the column matrix gives the number of carbon atoms, and the bottom integer gives the number of oxygen

*R. A. Alberty, *J. Chem. Educ.* **68:**984 (1991); R. A. Alberty, *J. Chem. Educ.* **69:**493 (1992).

atoms. We have been representing a general chemical reaction as $0 = \sum \nu_i B_i$, and so equation 5.78 can be written as

$$0 = -1\text{CO} - \tfrac{1}{2}\text{O}_2 + 1\text{CO}_2 \tag{5.79}$$

When the molecular formulas are replaced by the corresponding column matrices, we obtain

$$-1\begin{bmatrix}1\\1\end{bmatrix} - \tfrac{1}{2}\begin{bmatrix}0\\2\end{bmatrix} + 1\begin{bmatrix}1\\2\end{bmatrix} = \begin{bmatrix}0\\0\end{bmatrix} \tag{5.80}$$

This is one way to write the conservation equations for carbon and oxygen. Textbooks on linear algebra* show that this equation can be written in the form of a matrix product.

$$\begin{bmatrix}1 & 0 & 1\\1 & 2 & 2\end{bmatrix}\begin{bmatrix}-1\\-\frac{1}{2}\\+1\end{bmatrix} = \begin{bmatrix}0\\0\end{bmatrix} \tag{5.81}$$

Notice that the stoichiometric numbers in equation 5.79 form a column matrix. When you multiply a 2×3 matrix times a 3×1 matrix, you obtain a 2×1 matrix (see Appendix D.8). This equation is a form of the conservation equations for the two elements involved in this reaction.

We can generalize equation 5.81 by writing it as

$$\boldsymbol{A\nu} = \mathbf{0} \tag{5.82}$$

where $\boldsymbol{A}$ is the **conservation matrix,** $\boldsymbol{\nu}$ is the **stoichiometric number matrix,** and $\mathbf{0}$ is a **zero matrix.** This equation can be used to calculate the $\boldsymbol{\nu}$ matrix from the $\boldsymbol{A}$ matrix. The $\boldsymbol{\nu}$ matrix is called the **null space** of the $\boldsymbol{A}$ matrix. For small matrices the null space can be calculated by hand, and for larger matrices a computer with the operations of linear algebra can be used (e.g., Mathematica†). The first step in calculating the null space by hand is to make a Gaussian reduction of the $\boldsymbol{A}$ matrix to get it into row echelon form, that is, a matrix with an **identity matrix** (see Appendix D.8) on the left. Rows in the $\boldsymbol{A}$ matrix can be multiplied by integers and added or subtracted from other rows to obtain this row echelon form. If we subtract the first row of the $\boldsymbol{A}$ matrix in equation 5.81 from the second row, we obtain

$$\boldsymbol{A} = \begin{bmatrix}1 & 0 & 1\\0 & 2 & 1\end{bmatrix} \tag{5.83}$$

If we then divide the second row by 2, we obtain

$$\boldsymbol{A} = \begin{bmatrix}1 & 0 & 1\\0 & 1 & \frac{1}{2}\end{bmatrix} \tag{5.84}$$

The rule for calculating the null space is to change the sign of the entries in the

*G. Strang, *Linear Algebra and Its Applications.* San Diego: Harcourt Brace Jovanovich, 1988.

†Wolfram Research, Inc., 100 Trade Center Drive, Champaign, IL 61820-7237.

column(s) to the right of the identity matrix and append an appropriate-size unit matrix below it. This yields

$$\boldsymbol{\nu} = \begin{bmatrix} -1 \\ -\frac{1}{2} \\ 1 \end{bmatrix} \tag{5.85}$$

The rows here are in the order of the columns in the $\mathbf{A}$ matrix, and so this yields chemical equation 5.79.

The use of row reduction is important for another reason, namely, that the rank of the $\mathbf{A}$ matrix is equal to the number C of components (see Section 5.9).

$$C = \text{rank } \mathbf{A} \tag{5.86}$$

Equation 5.84 shows that the rank of the $\mathbf{A}$ matrix is 2. The number of components is often equal to the number of elements, as it is in this case, but there are important exceptions. The rank of the $\boldsymbol{\nu}$ matrix is equal to the number of independent columns, and so this gives us the number R of independent reactions for a system:

$$R = \text{rank } \boldsymbol{\nu} \tag{5.87}$$

The rank of the $\boldsymbol{\nu}$ matrix in equation 5.85 is unity, which corresponds to the fact that there is a single reaction. We have seen earlier (Section 5.9) that

$$N_s = C + R \tag{5.88}$$

and so

$$N_s = \text{rank } \mathbf{A} + \text{rank } \boldsymbol{\nu} \tag{5.89}$$

We can see from our simple example that the Gaussian reduction divides the species in a system into C components and $N_s - C$ noncomponents, where N_s is the number of species, that is, the number of columns in the $\mathbf{A}$ matrix. Thus the conservation matrix $\mathbf{A}$ is $C \times N_s$, and the stoichiometric number matrix $\boldsymbol{\nu}$ is $N_s \times R$.

To see how these operations work out for a larger system, let us consider a reaction system containing CO, H_2, CO_2, H_2O, and CH_4. The conservation matrix is

$$\mathbf{A} = \begin{array}{c} \\ \text{C} \\ \text{H} \\ \text{O} \end{array} \begin{array}{c} \begin{array}{ccccc} \text{CO} & \text{H}_2 & \text{CO}_2 & \text{H}_2\text{O} & \text{CH}_4 \end{array} \\ \begin{bmatrix} 1 & 0 & 1 & 0 & 1 \\ 0 & 2 & 0 & 2 & 4 \\ 1 & 0 & 2 & 1 & 0 \end{bmatrix} \end{array} \tag{5.90}$$

Row reduction yields the following row echelon form:

$$\mathbf{A} = \begin{array}{c} \\ \text{CO} \\ \text{H}_2 \\ \text{CO}_2 \end{array} \begin{array}{c} \begin{array}{ccccc} \text{CO} & \text{H}_2 & \text{CO}_2 & \text{H}_2\text{O} & \text{CH}_4 \end{array} \\ \begin{bmatrix} 1 & 0 & 0 & -1 & 2 \\ 0 & 1 & 0 & 1 & 2 \\ 0 & 0 & 1 & 1 & -1 \end{bmatrix} \end{array} \tag{5.91}$$

Note that the components are now taken to be CO, H_2, and CO_2. Thus the $\boldsymbol{\nu}$ matrix is

$$\boldsymbol{\nu} = \begin{array}{c} \\ CO \\ H_2 \\ CO_2 \\ H_2O \\ CH_4 \end{array} \begin{array}{c} \begin{array}{cc} R_1 & R_2 \end{array} \\ \begin{bmatrix} 1 & -2 \\ -1 & -2 \\ -1 & 1 \\ 1 & 0 \\ 0 & 1 \end{bmatrix} \end{array} \tag{5.92}$$

This corresponds to the chemical equations

$$H_2 + CO_2 = H_2O + CO \tag{5.93}$$

$$2CO + 2H_2 = CH_4 + CO_2 \tag{5.94}$$

Equation 5.91 shows that CO, H_2, and CO_2 become components, and H_2O and CH_4 become noncomponents. The fact that the $\boldsymbol{A}$ matrix came out this way indicates that the composition of the system can be expressed in terms of CO, H_2, and CO_2. If we had put other species first, they might have been chosen as components, but of course the components must contain all of the elements. The choice of components is somewhat arbitrary, but the number is not.

Equation 5.82 shows that the conservation matrix and the stoichiometric number matrix are equivalent because the stoichiometric number matrix can be calculated from the conservation matrix, as we have seen. Equation 5.82 can also be written as

$$\boldsymbol{\nu}^{T} \boldsymbol{A}^{T} = \mathbf{0} \tag{5.95}$$

where the superscript T indicates the transpose (see Appendix D.8). Thus $\boldsymbol{A}^T$ is the null space of $\boldsymbol{\nu}^T$ so that we can start with a chemical equation, or a system of independent chemical equations, and calculate the conservation matrix for the system. Because of the nature of conservation equations and chemical equations, neither the $\boldsymbol{A}$ matrix nor the $\boldsymbol{\nu}$ matrix is unique. Conservation equations can be written in various ways that are all equivalent, and a set of independent reactions for a system can be written in different ways that are all equivalent, even though we may prefer to see chemical reactions written in a particular way. Any set of independent reactions for a system can be used to calculate the equilibrium composition. However, if we have two $\boldsymbol{A}$ matrices, or two $\boldsymbol{\nu}$ matrices, and want to know whether they are equivalent, we can make a Gaussian reduction to see if they give the same row echelon form.

Usually the number C of components is equal to the number of elements in the system. However, there are two types of situations where this is not true. If two elements are always in the same ratio, they can be replaced by a pseudoelement, and the number of components is smaller than the number of elements. Consider the isomerization reaction

$$C_4H_{10}(n\text{-butane}) = C_4H_{10}(\text{isobutane}) \tag{5.96}$$

The $\boldsymbol{A}$ matrix is

$$\boldsymbol{A} = \begin{bmatrix} 4 & 4 \\ 10 & 10 \end{bmatrix} \tag{5.97}$$

which reduces to

$$A = \begin{bmatrix} 1 & 1 \\ 0 & 0 \end{bmatrix} \tag{5.98}$$

which has a rank of 1. Thus there is a single component. The intensive state of the system at equilibrium is specified by giving T and P. In systems involving ions the charge balance may be redundant.

The rank of $\boldsymbol{A}$ is larger than the number of elements when there are additional constraints in a reaction system. This does not happen very often in chemistry, but it is illustrated by the following example.

Example 5.21 ***The conservation matrix and the stoichiometric number matrix for a reaction system***

Consider a system in which only the following reaction occurs because of the catalyst used:

$$C_6H_6(g) + CH_4(g) = C_6H_5CH_3(g) + H_2(g)$$

Write out the $\boldsymbol{A}$ matrix and the $\boldsymbol{\nu}$ matrix, and show that they are compatible. How many degrees of freedom does this system have, and what is a possible choice of degrees of freedom?

Note that aromatic rings are conserved as well as C and H. Thus the $\boldsymbol{A}$ matrix has three rows, where the third row is for the aromatic component.

$$\boldsymbol{A} = \begin{array}{c} \\ \text{C} \\ \text{H} \\ \text{ar} \end{array} \begin{array}{cccc} C_6H_6 & CH_4 & C_6H_5CH_3 & H_2 \\ \hline 6 & 1 & 7 & 0 \\ 6 & 4 & 8 & 2 \\ 1 & 0 & 1 & 0 \end{array}$$

This matrix has a rank of 3, and so $C = 3$. Thus the number of components is one more than the number of elements. This matrix can be used to calculate the stoichiometric numbers in the chemical equation that occurs. Row reduction yields

$$\boldsymbol{A} = \begin{array}{c} \\ C_6H_6 \\ CH_4 \\ C_6H_5CH_3 \end{array} \begin{array}{cccc} C_6H_6 & CH_4 & C_6H_5CH_3 & H_2 \\ \hline 1 & 0 & 0 & 1 \\ 0 & 1 & 0 & 1 \\ 0 & 0 & 1 & -1 \end{array}$$

The last column indicates the stoichiometric numbers of the four reactants in the order listed, $\boldsymbol{\nu}^T = [-1, -1, 1, 1]$, in agreement with the balanced chemical equation. The number of intensive variables that have to be specified to describe the intensive state of this system at equilibrium is $F = C - p + 2 = 3 - 1 + 2 = 4$, which can be taken to be T, P, $n(\text{H})/n(\text{C})$, and $n(\text{ar})/n(\text{C})$.

Example 5.22 ***Calculation of the number of independent reactions***

A system contains $CaO(s)$, $CO_2(g)$, and $CaCO_3(s)$. How many independent reactions are there?

The system formula matrix is

$$\boldsymbol{A} = \begin{bmatrix} 1 & 0 & 1 \\ 2 & 1 & 3 \\ 0 & 1 & 1 \end{bmatrix}$$

if the elements (rows) are in the order C, O, Ca and the species (columns) are in the order CO_2, CaO, $CaCO_3$. Gaussian elimination yields

$$\mathbf{A} = \begin{bmatrix} 1 & 0 & 1 \\ 0 & 1 & 1 \\ 0 & 0 & 0 \end{bmatrix}$$

This corresponds to the stoichiometric number matrix

$$\boldsymbol{\nu} = \begin{bmatrix} -1 \\ -1 \\ 1 \end{bmatrix}$$

which corresponds to the equation

$$\mathrm{CaO(s)} + \mathrm{CO_2(g)} = \mathrm{CaCO_3(s)}$$

The number of independent reactions is given by

$$R = N_s - \mathrm{rank}\,\mathbf{A} = 3 - 2 = 1$$

Ten Key Ideas in Chapter 5

1. The fundamental equation for the Gibbs energy can be used to derive the expression for the equilibrium constant for a chemical reaction in terms of the activities of the reactants and shows that $-RT \ln K$ is equal to the standard Gibbs energy of reaction $\Delta_r G^\circ$. The value of an equilibrium constant cannot be interpreted unless it is accompanied by a balanced chemical equation and a specification of the standard state of each reactant and product.
2. For gas reactions the activity can be expressed in terms of the fugacity ($a_i = f_i/P^\circ$), but most of our calculations have been made for reactions in mixtures of ideal gases. Because of the entropy of mixing, a plot of the Gibbs energy versus extent of reaction always has a minimum at the equilibrium extent of reaction.
3. For a mixture of ideal gases in which there is a single reaction, the expression for the equilibrium constant can always be written in terms of the extent of reaction and the total pressure. Thus if the equilibrium constant is known, the equilibrium composition can always be calculated by solving a polynomial equation.
4. The standard Gibbs energy of reaction is made up of contributions from the standard enthalpy of reaction $\Delta_r H^\circ$ and the standard entropy of reaction: $\Delta_r G^\circ = \Delta_r H^\circ - T\,\Delta_r S^\circ$. Therefore, the equilibrium constant for a reaction can be calculated from calorimetric measurements that can be used to obtain $\Delta_r H^\circ$ and third law calorimetric measurements that can be used to obtain $\Delta_r S^\circ$.
5. Equilibrium constants can be calculated for many reactions for which $\Delta_f H_i^\circ$ and $\overline{S}_i^\circ$ have been tabulated for all the species since $\Delta_r H^\circ = \sum \nu_i \Delta_f H_i^\circ$ and $\Delta_r S^\circ = \sum \nu_i \overline{S}_i^\circ$, where the ν_i are stoichiometric numbers.
6. For an endothermic reaction, the equilibrium constant increases as the temperature is increased, and for an exothermic reaction the equilibrium constant decreases as the temperature is increased.
7. According to Le Châtelier's principle, when an independent variable of a system at equilibrium is changed, the equilibrium shifts in the direction that tends to reduce the effect of the change.

8. According to the phase rule, the number F of degrees of freedom for a system at equilibrium is given by $N_s - R - p + 2$, where N_s is the number of species, R is the number of independent reactions, and p is the number of different phases. This is in agreement with the Gibbs–Duhem equation.
9. When a group of isomers is in equilibrium, they form a single component, and we can calculate the standard thermodynamic properties of the isomer group as a whole if we know the standard thermodynamic properties of the individual isomers.
10. The conservation of atoms in a system of reactions can be represented in two ways, by the conservation equations for each element and by an independent set of chemical equations. A Gaussian reduction of the conservation equation yields a set of independent chemical equations.

REFERENCES

K. E. Bett, J. S. Rowlinson, and G. Saville, *Thermodynamics for Chemical Engineers.* Cambridge, MA: MIT Press, 1975.

K. S. Pitzer, *Thermodynamics,* 3rd ed. New York: McGraw-Hill, 1995.

S. I. Sandler, *Chemical and Engineering Thermodynamics.* Hoboken, NJ: Wiley, 1999.

J. M. Smith, H. C. Van Ness, and M. M. Abbott, *Introduction to Chemical Engineering Thermodynamics*, 5th ed. New York: McGraw-Hill, 1996.

W. R. Smith and R. W. Missen, *Chemical Reaction Equilibrium Analysis.* Hoboken, NJ: Wiley, 1982.

J. W. Tester and M. Modell, *Thermodynamics and Its Applications*. Upper Saddle River, NJ: Prentice-Hall, 1997.

PROBLEMS

(M) Problems marked with an icon may be more conveniently solved on a personal computer with a mathematical program.

5.1 For the reaction $N_2(g)+3H_2(g) = 2NH_3(g)$, $K = 1.60\times 10^{-4}$ at 400 °C. Calculate (*a*) $\Delta_r G°$ and (*b*) $\Delta_r G$ when the pressures of N_2 and H_2 are maintained at 10 and 30 bar, respectively, and NH_3 is removed at a partial pressure of 3 bar. (*c*) Is the reaction spontaneous under the latter conditions?

5.2 A 1:3 mixture of nitrogen and hydrogen was passed over a catalyst at 450 °C. It was found that 2.04% by volume of ammonia gas was formed when the total pressure was maintained at 10.13 bar. Calculate the value of K for $\frac{3}{2}H_2(g) + \frac{1}{2}N_2(g) = NH_3(g)$ at this temperature.

5.3 At 55 °C and 1 bar the average molar mass of partially dissociated N_2O_4 is 61.2 g mol^{-1}. Calculate (*a*) ξ and (*b*) K for the reaction $N_2O_4(g) = 2NO_2(g)$. (*c*) Calculate ξ at 55 °C if the total pressure is reduced to 0.1 bar.

5.4 A 1-liter reaction vessel containing 0.233 mol of N_2 and 0.341 mol of PCl_5 is heated to 250 °C. The total pressure at equilibrium is 29.33 bar. Assuming that all gases are ideal, calculate K for the only reaction that occurs:

$$PCl_5(g) = PCl_3(g) + Cl_2(g)$$

5.5 An evacuated tube containing 5.96×10^{-3} mol L^{-1} of solid iodine is heated to 973 K. The experimentally determined pressure is 0.496 bar. Assuming ideal gas behavior, calculate K for $I_2(g) = 2I(g)$.

5.6 Nitrogen trioxide dissociates according to the reaction

(M)
$$N_2O_3(g) = NO_2(g) + NO(g)$$

When one mole of $N_2O_3(g)$ is held at 25 °C and 1 bar total pressure until equilibrium is reached, the extent of reaction is 0.30. What is $\Delta_r G°$ for this reaction at 25 °C?

(M)
$$2HI(g) = H_2(g) + I_2(g)$$

at 698.6 K, $K = 1.83\times 10^{-2}$. (*a*) How many grams of hydrogen iodide will be formed when 10 g of iodine and 0.2 g of hydrogen are heated to this temperature in a 3-L vessel? (*b*) What will be the partial pressures of H_2, I_2, and HI?

5.8 Express K for the reaction

$$CO(g) + 3H_2(g) = CH_4(g) + H_2O(g)$$

in terms of the equilibrium extent of reaction ξ when one mole of CO is mixed with one mole of hydrogen.

5.9 What are the percentage dissociations of $H_2(g)$, $O_2(g)$, and $I_2(g)$ at 2000 K and a total pressure of 1 bar?

5.10 To produce more hydrogen from "synthesis gas" ($CO + H_2$) the water gas shift reaction is used:

$$CO(g) + H_2O(g) = CO_2(g) + H_2(g)$$

Calculate K at 1000 K and the equilibrium extent of reaction starting with an equimolar mixture of CO and H_2O.

5.11 Calculate the extent of reaction ξ of 1 mol of $H_2O(g)$ to form $H_2(g)$ and $O_2(g)$ at 2000 K and 1 bar. (Since the extent of reaction is small, the calculation may be simplified by assuming that $P_{H_2O} = 1$ bar.)

5.12 At 500 K CH_3OH, CH_4, and other hydrocarbons can be formed from CO and H_2. Until recently the main source of the CO mixture for the synthesis of CH_3OH was methane:

$$CH_4(g) + H_2O(g) = CO(g) + 3H_2(g)$$

When coal is used as the source, the "synthesis gas" has a different composition:

$$C(\text{graphite}) + H_2O(g) = CO(g) + H_2(g)$$

Suppose we have a catalyst that catalyzes only the formation of CH_3OH. (*a*) What pressure is required to convert 25% of the CO to CH_3OH at 500 K if the synthesis gas comes from CH_4? (*b*) If the synthesis gas comes from coal?

5.13 Many equilibrium constants in the literature were calculated with a standard state pressure of 1 atm (1.013 25 bar). Show that the corresponding equilibrium constant with a standard pressure of 1 bar can be calculated using

$$K(\text{bar}) = K(\text{atm})(1.01325)^{\sum \nu_i}$$

where the ν_i are the stoichiometric numbers of the gaseous reactants.

5.14 Older tables of chemical thermodynamic properties are based on a standard state pressure of 1 atm. Show that the corresponding $\Delta_f G_j^\circ$ with a standard state pressure of 1 bar can be calculated using

$$\Delta_f G_j^\circ(\text{bar}) = \Delta_f G_j^\circ(\text{atm}) - (0.1094 \times 10^{-3}\ \text{kJ K}^{-1}\ \text{mol}^{-1})T \sum \nu_i$$

where the ν_i are the stoichiometric numbers of the gaseous reactants and products in the formation reaction.

5.15 Show that the equilibrium mole fractions of n-butane and isobutane are given by

$$y_n = \frac{e^{-\Delta_f G_n^\circ/RT}}{(e^{-\Delta_f G_n^\circ/RT} + e^{-\Delta_f G_{iso}^\circ/RT})}$$

$$y_{iso} = \frac{e^{-\Delta_f G_{iso}^\circ/RT}}{(e^{-\Delta_f G_n^\circ/RT} + e^{-\Delta_f G_{iso}^\circ/RT})}$$

(M) **5.16** Calculate the molar Gibbs energy of butane isomers for extents of reaction of 0.2, 0.4, 0.6, and 0.8 for the reaction

$$n\text{-butane} = \text{isobutane}$$

at 1000 K and 1 bar. At 1000 K

$$\Delta_f G^\circ(n\text{-butane}) = 270\ \text{kJ mol}^{-1}$$
$$\Delta_f G^\circ(\text{isobutane}) = 276.6\ \text{kJ mol}^{-1}$$

Make a plot and show that the minimum corresponds to the equilibrium extent of reaction.

5.17 In the synthesis of methanol by $CO(g) + 2H_2(g) = CH_3OH(g)$ at 500 K, calculate the total pressure required for a 90% conversion to methanol if CO and H_2 are initially in a 1:2 ratio. Given: $K = 6.09 \times 10^{-3}$.

5.18 At 1273 K and at a total pressure of 30.4 bar the equilibrium in the reaction $CO_2(g) + C(s) = 2CO(g)$ is such that 17 mol % of the gas is CO_2. (*a*) What percentage would be CO_2 if the total pressure were 20.3 bar? (*b*) What would be the effect on the equilibrium of adding N_2 to the reaction mixture in a closed vessel until the partial pressure of N_2 is 10 bar? (*c*) At what pressure of the reactants will 25% of the gas be CO_2?

5.19 When alkanes are heated up, they lose hydrogen and alkenes are produced. For example,

$$C_2H_6(g) = C_2H_4(g) + H_2(g)$$
$$K = 0.36 \text{ at } 1000\ \text{K}$$

If this is the only reaction that occurs when ethane is heated to 1000 K, at what total pressure will ethane be (*a*) 10% dissociated and (*b*) 90% dissociated to ethylene and hydrogen?

5.20 At 2000 °C water is 2% dissociated into oxygen and hydrogen at a total pressure of 1 bar. (*a*) Calculate K for $H_2O(g) = H_2(g) + \frac{1}{2}O_2(g)$. (*b*) Will the extent of reaction increase or decrease if the pressure is reduced? (*c*) Will the extent of reaction increase or decrease if argon gas is added, when the total pressure is held equal to 1 bar? (*d*) Will the extent of reaction change if the pressure is raised by the addition of argon at constant volume to the closed system containing partially dissociated water vapor? (*e*) Will the extent of reaction increase or decrease if oxygen gas is added while holding the total pressure constant at 1 bar?

5.21 At 250 °C, PCl_5 is 80% dissociated at a pressure of 1.013 bar, and so $K = 1.80$. What is the extent of reaction at equilibrium after sufficient nitrogen has been added at constant pressure to produce a nitrogen partial pressure of 0.9 bar? The total pressure is maintained at 1 bar.

5.22 The following exothermic reaction is at equilibrium at 500 K and 10 bar:

$$CO(g) + 2H_2(g) = CH_3OH(g)$$

Assuming that the gases are ideal, what will happen to the amount of methanol at equilibrium when (*a*) the temperature is raised, (*b*) the pressure is increased, (*c*) an inert gas is pumped in at constant volume, (*d*) an inert gas is pumped in at constant pressure, and (*e*) hydrogen gas is added at constant pressure?

5.23 The following reaction is nonspontaneous at room temperature and endothermic:

$$3C(graphite) + 2H_2O(g) = CH_4(g) + 2CO(g)$$

As the temperature is raised, the equilibrium constant will become equal to unity at some point. Estimate this temperature using data from Table C.3.

5.24 The measured density of an equilibrium mixture of N_2O_4 and NO_2 at 15 °C and 1.103 bar is 3.62 g L^{-1}, and the density at 75 °C and 1.013 bar is 1.84 g L^{-1}. What is the enthalpy change of the reaction $N_2O_4(g) = 2NO_2(g)$?

5.25 Calculate K_c for the reaction in Problem 5.19 at 1000 K and describe what it is equal to.

 **5.26** The equilibrium constant for the reaction

$$N_2(g) + 3H_2(g) = 2NH_3(g)$$

is 35.0 at 400 K when partial pressures are expressed in bars. Assume that the gases are ideal. (*a*) What is the equilibrium composition and equilibrium volume when 0.25 mol N_2 is mixed with 0.75 mol H_2 at a temperature of 400 K and a pressure of 1 bar? (*b*) What is the equilibrium composition and equilibrium pressure if this mixture is held at a constant volume of 33.26 L at 400 K?

5.27 Show that to a first approximation the equation of state of a gas that dimerizes to a small extent is given by

$$\frac{P\overline{V}}{RT} = 1 - \frac{K_c}{\overline{V}}$$

5.28 Water vapor is passed over coal (assumed to be pure graphite in this problem) at 1000 K. Assuming that the only reaction occurring is the water gas reaction

$$C(graphite) + H_2O(g) = CO(g) + H_2(g) \qquad K = 2.52$$

calculate the equilibrium pressures of H_2O, CO, and H_2 at a total pressure of 1 bar. [Actually, the water gas shift reaction

$$CO(g) + H_2O(g) = CO_2(g) + H_2(g)$$

also occurs, but it is considerably more complicated to take this additional reaction into account.]

5.29 What is the equilibrium partial pressure of NO in air at 1000 K at a pressure of 1 bar?

$$\tfrac{1}{2}N_2(g) + \tfrac{1}{2}O_2(g) = NO(g)$$

5.30 Starting with the fundamental equation for G in the form

$$dG = -S\,dT + V\,dP + \Delta_r G\,d\xi$$

derive equations for calculating $\Delta_r S$, $\Delta_r V$, and $\Delta_r H$ from experimental data on $\Delta_r G$ for a chemical reaction as a function of T and P.

5.31 What is $\Delta_r S°(298\text{ K})$ for the following reaction?

$$H_2O(l) = H^+(ao) + OH^-(ao)$$

Why is this change negative and not positive?

5.32 Mercuric oxide dissociates according to the reaction $2HgO(s) = 2Hg(g)+O_2(g)$. At 420 °C the dissociation pressure is 5.16×10^4 Pa, and at 450 °C it is 10.8×10^4 Pa. Calculate (*a*) the equilibrium constants and (*b*) the enthalpy of dissociation per mole of HgO.

5.33 The decomposition of silver oxide is represented by

$$2Ag_2O(s) = 4Ag(s) + O_2(g)$$

Using data from Table C.2 and assuming $\Delta_r C_P° = 0$, calculate the temperature at which the equilibrium pressure of O_2 is 0.2 bar. This temperature is of interest because Ag_2O will decompose to yield Ag at temperatures above this value if it is in contact with air.

5.34 The dissociation of ammonium carbamate takes place according to the reaction

$$(NH_2)CO(ONH_4)(s) = 2NH_3(g) + CO_2(g)$$

When an excess of ammonium carbamate is placed in a previously evacuated vessel, the partial pressure generated by NH_3 is twice the partial pressure of the CO_2, and the partial pressure of $(NH_2)CO(ONH_4)$ is negligible in comparison. Show that

$$K = \left(\frac{P_{NH_3}}{P°}\right)^2 \frac{P_{CO_2}}{P°} = \frac{4}{27}\left(\frac{P}{P°}\right)^3$$

where P is the total pressure.

5.35 At 1000 K methane at 1 bar is in the presence of hydrogen. In the presence of a sufficiently high partial pressure of hydrogen, methane does not decompose to form graphite and hydrogen. What is this partial pressure?

5.36 For the reaction

$$Fe_2O_3(s) + 3CO(g) = 2Fe(s) + 3CO_2(g)$$

the following values of K are known.

t/°C	250	1000
K	100	0.0721

At 1120 °C for the reaction $2CO_2(g) = 2CO(g) + O_2(g)$, $K = 1.4 \times 10^{-12}$ bar. What equilibrium partial pressure of O_2 would have to be supplied to a vessel at 1120 °C containing solid Fe_2O_3 just to prevent the formation of Fe?

5.37 When a reaction is carried out at constant pressure, the entropy change can be used as a criterion of equilibrium by including a heat reservoir as part of an isolated system containing the reaction chamber. Show that $-\Delta_r G/T$ is the global increase in entropy for the reaction system plus heat reservoir.

5.38 The effect of temperature on K_P is given by equation 5.51, and the effect of temperature on K_c is given by

$$\Delta U^\circ = RT^2 \left(\frac{\partial \ln K_c}{\partial T}\right)_V$$

Is it possible for a gas reaction to have K_P increase with increasing temperature, but K_c decrease with increasing T? If so, what has to be true?

5.39 Calculate the partial pressure of $CO_2(g)$ over $CaCO_3$(calcite)—$CaO(s)$ at 500 °C using the equation in Example 5.11 and data in Table C.2.

5.40 The NBS tables contain the following data at 298 K:

	$\Delta_f H^\circ$/kJ mol^{-1}	$\Delta_f G^\circ$/kJ mol^{-1}
$CuSO_4(s)$	−771.36	−661.8
$CuSO_4 \cdot H_2O(s)$	−1085.83	−918.11
$CuSO_4 \cdot 3H_2O(s)$	−1684.31	−1399.96
$H_2O(g)$	−241.818	−228.572

(*a*) What is the equilibrium partial pressure of H_2O over a mixture of $CuSO_4(s)$ and $CuSO_4 \cdot H_2O(s)$ at 25 °C? (*b*) What is the equilibrium partial pressure of H_2O over a mixture of $CuSO_4 \cdot H_2O(s)$ and $CuSO_4 \cdot 3H_2O(s)$ at 25 °C? (*c*) What are the answers to (*a*) and (*b*) if the temperature is 100 °C and ΔC_P° is assumed to be zero?

5.41 One micromole of $CuO(s)$ and 0.1 μmol of $Cu(s)$ are placed in a 1-L container at 1000 K. Determine the identity and quantity of each phase present at equilibrium if $\Delta_f G^\circ$ of $CuO(s)$ is −66.66 kJ mol^{-1} and that of $Cu_2O(s)$ is −77.94 kJ mol^{-1} at 1000 K. [From H. F. Franzen, *J. Chem. Educ.* **65**:146 (1988).]

5.42 For the heterogeneous reaction

$$CH_4(g) = C(s) + 2H_2(g)$$

derive the expression for the extent of reaction in terms of the equilibrium constant and the applied pressure, when graphite is in equilibrium with the gas mixture. Is this the same expression (equation 5.31) that was obtained for the reaction $N_2O_4(g) = 2NO_2(g)$?

5.43 Calculate the equilibrium extent of the reaction $N_2O_4(g) = 2NO_2(g)$ at 298.15 K and a total pressure of 1 bar if the $N_2O_4(g)$ is mixed with an equal volume of $N_2(g)$ before the reaction occurs. As shown by Example 5.4, $K = 0.143$. Do you expect the same equilibrium extent of reaction as in Example 5.4 If not, do you expect a larger or smaller equilibrium extent of reaction?

5.44 (*a*) A system contains $CO(g)$, $CO_2(g)$, $H_2(g)$, and $H_2O(g)$. How many chemical reactions are required to describe chemical changes in this system? Give an example. (*b*) If solid carbon is present in the system in addition, how many independent chemical reactions are there? Give a suitable set.

(M) **5.45** For a closed system containing C_2H_2, H_2, C_6H_6, and $C_{10}H_8$, use Gaussian elimination to obtain a set of independent chemical reactions. Perform the matrix multiplication to verify $\boldsymbol{A\nu} = \boldsymbol{0}$.

5.46 The reaction A + B = C is at equilibrium at a specified T and P. Derive the fundamental equation for G in terms of components by eliminating μ_C.

(M) **5.47** The article by C. A. L. Figueiras in *J. Chem. Educ.* **69**:276 (1992) illustrates an interesting problem you can get into in trying to balance a chemical equation. Consider the following reaction without stoichiometric numbers:

$$ClO_3^- + Cl^- + H^+ = ClO_2 + Cl_2 + H_2O$$

There are actually an infinite number of ways to balance this equation. The following steps in unraveling this puzzle can be carried out using a personal computer with a program such as *Mathematica*, which can do matrix operations. Write the conservation matrix **A** and determine the number of components. How many independent reactions are there for this system of six species? What are the stoichiometric numbers for a set of independent reactions? These steps show that chemical change in this system is represented by two chemical reactions, not one.

5.48 A chemical reaction system contains three species: C_2H_4 (ethylene), C_3H_6 (propene), and C_4H_8 (butene). (*a*) Write the **A** matrix. (*b*) Row-reduce the **A** matrix. (*c*) How many components are there? (*d*) Derive a set of independent reactions from the **A** matrix.

5.49 How many degrees of freedom are there for the following systems, and how might they be chosen?

(a) $CuSO_4 \cdot 5H_2O(cr)$ in equilibrium with $CuSO_4(cr)$ and $H_2O(g)$

(b) N_2O_4 in equilibrium with NO_2 in the gas phase

(c) CO_2, CO, H_2O, and H_2 in chemical equilibrium in the gas phase

(d) The system described in (*c*) made up with stoichiometric amounts of CO and H_2

5.50 Graphite is in equilibrium with gaseous H_2O, CO, CO_2, H_2, and CH_4. How many degrees of freedom are there? What degrees of freedom might be chosen for an equilibrium calculation?

5.51 A gaseous system contains CO, CO_2, H_2, H_2O, and C_6H_6 in chemical equilibrium. (*a*) How many components are there? (*b*) How many independent reactions? (*c*) How many degrees of freedom are there?

5.52 At 500 °C, $K = 5.5$ for the reaction

$$CO(g) + H_2O(g) = CO_2(g) + H_2(g)$$

If a mixture of 1 mol of CO and 5 mol of H_2O is passed over a catalyst at this temperature, what will be the equilibrium mole fraction of H_2O?

5.53 At 400 °C, $K = 79.1$ for the reaction

$$NH_3(g) = \tfrac{1}{2}N_2(g) + \tfrac{3}{2}H_2(g)$$

Show that the fraction α of NH_3 dissociated at a total pressure P is given by

$$\alpha = \frac{1}{\sqrt{1 + kP}}$$

and calculate the value of k in this equation.

5.54 For the reaction $N_2O_4(g) = 2NO_2(g)$, K at 25 °C is 0.143. What pressure would be expected if 1 g of liquid N_2O_4 were allowed to evaporate into a 1-liter vessel at this temperature? Assume that N_2O_4 and NO_2 are ideal gases.

5.55 The dissociation of N_2O_4 is represented by $N_2O_4(g) = 2NO_2(g)$. If the density of the equilibrium gas mixture is 3.174 g L^{-1} at a total pressure of 1.013 bar at 24 °C, what minimum pressure would be required to keep the degree of dissociation of N_2O_4 below 0.1 at this temperature?

5.56 At 250 °C, 1 L of partially dissociated phosphorus pentachloride gas at 1.013 bar weighs 2.690 g. Calculate the extent of reaction ξ and the equilibrium constant.

5.57 Derive the analogue of equation 5.34 for the reaction $N_2(g)+3H_2(g) = 2NH_3(g)$ when stoichiometric amounts of nitrogen and hydrogen are used.

5.58 Hydrogen is produced on a large scale from methane. Calculate the equilibrium constant K for the production of H_2 from CH_4 at 1000 K using the reaction $CH_4(g) + H_2O(g) = CO(g) + 3H_2(g)$.

5.59 (*a*) What is the equilibrium constant for the formation of CH_4 from CO and H_2 at 500 and 1000 K?

$$CO(g) + 3H_2(g) = CH_4(g) + H_2O(g)$$

(*b*) What is the equilibrium constant for the formation of CH_4 from graphite and H_2O at 500 and 1000 K?

$$2C(\text{graphite}) + 2H_2O(g) = CH_4(g) + CO_2(g)$$

5.60 (*a*) Calculate the extent of dissociation of $H_2(g)$ at 3000 K and 1 bar. A value of 0.072 was obtained experimentally by Langmuir. (*b*) Calculate the extent of dissociation of $O_2(g)$ at 3000 K at 1 bar.

5.61 What is K for

$$I_2(g) = 2I(g)$$

at 1000 K, and what is the degree of dissociation at 1 bar? At 0.1 bar?

5.62 Propene and cyclopropane are isomers. Their standard thermodynamic properties in the gas phase are given in Table C.2. If they were in equilibrium, what would be the standard Gibbs energy of formation and the standard enthalpy of the isomer group at 25 °C? Show that the same value of the standard Gibbs energy of formation is obtained by calculating the mole fraction average Gibbs energy and adding the entropy of mixing.

5.63 Show that in going from a standard state pressure of 1 atm to 1 bar, thermodynamic quantities for reactions are corrected as follows:

$$\Delta C_P^\circ(\text{bar}) = \Delta C_P^\circ(\text{atm})$$

$$\Delta H^\circ(\text{bar}) = \Delta H^\circ(\text{atm})$$

$$\Delta S^\circ(\text{bar}) = \Delta S\,(\text{atm}) + (0.109\ \text{J K}^{-1}\ \text{mol}^{-1})\sum \nu_i$$

$$\Delta G^\circ(\text{bar}) = \Delta G^\circ(\text{atm}) - (0.109 \times 10^{-3}\ \text{kJ K}^{-1}\ \text{mol}^{-1})T \sum \nu_i$$

$$K\,(\text{bar}) = K\,(\text{atm})(1.013\,25)^{\sum \nu_i}$$

where $\sum \nu_i$ is the difference between the stoichiometric numbers of gaseous products and gaseous reactants in the balanced chemical equation.

5.64 The reaction

$$2NOCl(g) = 2NO(g) + Cl_2(g)$$

comes to equilibrium at 1 bar total pressure and 227 °C when the partial pressure of the nitrosyl chloride, NOCl, is 0.64 bar. Only NOCl was present initially. (*a*) Calculate $\Delta_r G^\circ$ for this reaction. (*b*) At what total pressure will the partial pressure of Cl_2 be 0.1 bar?

5.65 Acetic acid is produced on a large scale by the carbonylation of methanol at about 500 K and 25 bar using a rhodium catalyst. What is K_y under these conditions? ($\Delta_f G^\circ$ for acetic acid gas at 500 K is −335.28 kJ mol^{-1}.)

5.66 Calculate the total pressure that must be applied to a mixture of three parts of hydrogen and one part nitrogen to give a mixture containing 10% ammonia at equilibrium at 400 °C. At 400 °C, $K = 1.60 \times 10^{-4}$ for the reaction $N_2(g) + 3H_2(g) = 2NH_3(g)$.

5.67 A mixture of one mole of nitrogen and one mole of hydrogen is equilibrated over a catalyst for the ammonia reaction at 500 K and 1 bar. (*a*) What are the equilibrium mole fractions of N_2, H_2, and NH_3? (*b*) The experiment is repeated with 1.2 mol of nitrogen and 1 mol of hydrogen. What is the equilibrium partial pressure of ammonia? (*c*) How do you explain this result in terms of Le Châtelier's principle?

5.68 Assume that the following reaction is in equilibrium at 1000 K:

$$3C(\text{graphite}) + 2H_2O(g) = CH_4(g) + 2CO(g)$$

$\Delta_r H^\circ(1000\ \text{K}) = 182$ kJ mol^{-1}. (*a*) What will be the effect on the equilibrium composition of raising the temperature at a total pressure of 1 bar? (*b*) What will be the effect of raising the pressure to 5 bar? (*c*) What will be the effect of adding nitrogen at a constant pressure of 1 bar?

5.69 When N_2O_4 is allowed to dissociate into NO_2 at 25 °C at a total pressure of 1 bar, it is 18.5% dissociated at equilibrium, and so $K = 0.141$. (*a*) If N_2 is added to the system at constant volume, will the equilibrium shift? (*b*) If the system is allowed to expand as N_2 is added at a constant total pressure of 1 bar, what will be the equilibrium degree of dissociation when the N_2 partial pressure is 0.6 bar?

5.70 For the formation of nitric oxide

$$N_2(g) + O_2(g) = 2NO(g)$$

K at 2126.9 °C is 2.5×10^{-3}. (*a*) In an equilibrium mixture containing 0.1 bar partial pressure of N_2 and 0.1 bar partial pressure of O_2, what is the partial pressure of NO? (*b*) In an equilibrium mixture of N_2, O_2, NO, CO_2, and other inert gases at 2126.9 °C and 1 bar total pressure, 80% by volume of the gas is N_2 and 16% O_2. What is the percentage by volume of NO? (*c*) What is the total partial pressure of inert gases?

5.71 Prove that for $2C_2 = C_4$ in the presence of other gases in the gas phase, the equilibrium constant expression

$$K = \frac{y_4}{y_2^2(P/P°)}$$

may be interpreted in two ways:

(i) The mole fractions are expressed only in terms of C_2 and C_4 and $P = P_2 + P_4$.

(ii) The mole fractions are expressed in terms of all gases present and P is the total pressure.

(M) **5.72** The following data apply to the reaction $Br_2(g) = 2Br(g)$:

T/K	1123	1172	1223	1273
$K/10^{-3}$	0.408	1.42	3.32	7.2

Determine by graphical means the reaction enthalpy at 1200 K.

(M) **5.73** The average molar mass M of equilibrium mixtures of NO_2 and N_2O_4 at 1.013 bar total pressure is given in the following table at three temperatures:

t/°C	25	45	65
M/g mol^{-1}	77.64	66.80	56.51

(*a*) Calculate the degree of dissociation of N_2O_4 and the equilibrium constant at each of these temperatures. (*b*) Plot log K against $1/T$ and calculate $\Delta_r H°$ for the dissociation of N_2O_4. (*c*) Calculate the equilibrium constant at 35 °C. (*d*) Calculate the degree of dissociation for NO_2 at 35 °C when the total pressure is 0.5 bar.

5.74 For a chemical reaction, $\ln K = a + b/T + c/T^2$. Derive the corresponding expressions to calculate $\Delta_r G°$, $\Delta_r H°$, $\Delta_r S°$, and $\Delta_r C_P°$.

5.75 One mole of carbon monoxide is mixed with one mole of hydrogen and passed over a catalyst for the following reaction:

$$CO(g) + 2H_2(g) = CH_3OH(g)$$

At 500 K and a total pressure of 100 bar, 0.40 mol of CH_3OH is found at equilibrium. What is the value of the equilibrium constant expressed in terms of $P/P°$, where $P°$ is the reference pressure of 1 bar? What is the equilibrium constant expressed in terms of $c/c°$, where $c°$ is the reference concentration of 1 mol L^{-1}? Assume ideal gases.

5.76 Calculate $\Delta_r S°$ for the reaction

$$CH_4(g) + 2O_2(g) = CO_2(g) + 2H_2O(g)$$

at 298 and 1000 K.

5.77 What is $\Delta_r S°$ for $H_2(g) = 2H(g)$ at 298, 1000, and 3000 K?

5.78 Calculate the entropy changes for the following reactions at 25 °C:

(a) $Ag^+(ao) + Cl^-(ao) = AgCl(s)$

(b) $HS^-(ao) = H^+(ao) + S^{2-}(ao)$

5.79 From electromotive force measurements it has been found that $\Delta_r S°$ for the reaction

$$\tfrac{1}{2}H_2(g) + AgCl(s) = HCl(aq) + Ag(s)$$

is -62.4 J K^{-1} mol^{-1} at 298.15 K. What is the value of $S°[Cl^-(aq)]$?

5.80 The solubility of hydrogen in a molten iron alloy is found to be proportional to the square root of the partial pressure of hydrogen. How can this be explained?

5.81 Is magnetite (Fe_3O_4) or hematite (Fe_2O_3) the more stable ore thermodynamically at 25 °C in contact with air?

5.82 An equimolar mixture of CO(g), H_2(g), and H_2O(g) at 1000 K is compressed. At what total pressure will solid carbon start to precipitate out if there is chemical equilibrium? (See Problem 5.28.)

5.83 At 50 °C the partial pressure of H_2O(g) over a mixture of $CuSO_4 \cdot 3H_2O(s)$ and $CuSO_4 \cdot H_2O(s)$ is 4.0×10^3 Pa and that over a mixture of $CuSO_4 \cdot 3H_2O(s)$ and $CuSO_4 \cdot 5H_2O(s)$ is 6.3×10^3 Pa. Calculate the change in Gibbs energy for the reaction

$$CuSO_4 \cdot 5H_2O(s) = CuSO_4 \cdot H_2O(s) + 4H_2O(g)$$

5.84 Calculate (*a*) K and (*b*) $\Delta_r G°$ for the following reaction at 20 °C:

$$CuSO_4 \cdot 4NH_3(s) = CuSO_4 \cdot 2NH_3(s) + 2NH_3(g)$$

The equilibrium pressure of NH_3 is 8.26 kPa.

5.85 The vapor pressure of water above mixtures of $CuCl_2 \cdot H_2O(s)$ and $CuCl_2 \cdot 2H_2O(s)$ is given as a function of temperature in the following table:

t/°C	17.9	39.8	60.0	80.0
P/bar	0.0049	0.0250	0.122	0.327

(*a*) Calculate $\Delta_r H°$ for the reaction

$$CuCl_2 \cdot 2H_2O(s) = CuCl_2 \cdot H_2O(s) + H_2O(g)$$

(*b*) Calculate $\Delta_r G°$ for the reaction at 60 °C. (*c*) Calculate $\Delta_r S°$ for the reaction at 60 °C.

5.86 The equilibrium constant for the association of benzoic acid to a dimer in dilute benzene solutions is as follows at 43.9 °C:

$$2C_6H_5COOH = (C_6H_5COOH)_2 \qquad K_c = 2.7 \times 10^2$$

Molar concentrations are used in expressing the equilibrium constant. Calculate $\Delta_r G°$, and state its meaning.

5.87 Superheated steam is passed over coal, represented in these calculations by graphite, at 1000 K to produce CO, CO_2, H_2, and CH_4. Since there are six species and three element balance equations, there are three independent chemical reactions, which may be written as follows:

$$\begin{aligned} \text{C(graphite)} + H_2O(g) &= CO(g) + H_2(g) \\ 3\text{C(graphite)} + 2H_2O(g) &= CH_4(g) + 2CO(g) \\ 2\text{C(graphite)} + 2H_2O(g) &= CH_4(g) + CO_2(g) \end{aligned}$$

What are the equilibrium constants for these reactions? If the total pressure is raised to 100 bar, one of these reactions will predominate. Neglecting the other two reactions, calculate the equilibrium mole fractions of the gases present.

5.88 The following molecular species are in equilibrium in the gas phase: CH_4, C_2H_6, C_2H_4, and C_3H_6. How many independent chemical reactions are required to represent all possible changes in this system? Derive a set of independent reactions.

5.89 Derive equation 5.31 for calculating the extent of reaction for a gas dissociation reaction of the type

$$A(g) = 2B(g)$$

at constant temperature and pressure.

5.90 (*a*) The three pentane isomers are in equilibrium with one another. How many degrees of freedom does this system have, and what would you choose as the independent variables? (*b*) An equilibrium system involves the following two gas reactions:

$$\begin{aligned} 2CH_2CH_2 &= CH_3CH{=}CHCH_3 \\ 3CH_2CH_2 &= 2CH_3CH{=}CH_2 \end{aligned}$$

How many degrees of freedom does this system have, and what would you choose as independent variables?

Computer Problems

5.A For the reaction A(g) = B(g), assume that the standard Gibbs energy of formation of A is 50 kJ mol^{-1} and that of B is 45 kJ mol^{-1} at 298.15 K. (*a*) For a reaction starting with a mole of A at a pressure of 1 bar, plot the Gibbs energy of the mixture versus extent of reaction from zero to unity and identify the approximate extent of reaction at equilibrium. (*b*) Identify the equilibrium extent of reaction more precisely by plotting the derivative of the Gibbs energy of the mixture with respect to extent of reaction. Note that the slope of this plot goes to minus infinity at $\xi = 0$ and positive infinity at $\xi = 1$. (*c*) Calculate the equilibrium constant to verify the equilibrium extent of reaction.

5.B Calculate $\Delta_r H°$ and $\Delta_r S°$ for the reaction

$$N_2(g) + O_2(g) = 2NO(g)$$

from the following values of the equilibrium constant:

T/K	1900	2000	2100	2200	2300	2400	2500	2600
$K/10^{-4}$	2.31	4.08	6.86	11.0	16.9	25.1	36.0	50.3

5.C Calculate the equilibrium composition for the reaction

$$N_2(g) + 3H_2(g) = 2NH_3(g)$$

at 600 K and 15 bar, starting with two moles of N_2 and one mole of H_2. The equilibrium constant at 600 K is 0.001 715.

5.D Calculate the equilibrium constants for the following four reactions at 500, 1000, and 2000 K using Table C.3 and make a table.

$$\begin{aligned} H_2(g) &= 2H(g) \\ 2HI(g) &= H_2(g) + I_2(g) \\ I_2(g) &= 2I(g) \\ HI(g) &= H(g) + I(g) \end{aligned}$$

5.E Sufficient data to calculate $\Delta_r G°$ and $\Delta_r H°$ for the reaction 2HI(g) + I_2(g) are given in Table C.3. Calculate the standard reaction Gibbs energy at 500, 1000, 2000, and 3000 K. Use these data to calculate the standard reaction enthalpy at these temperatures by use of the Gibbs–Helmholtz equation, and compare the values calculated in this way with the values calculated from Table C.3.

5.F Consider the four following gas reactions:

$$\begin{aligned} H_2 + \tfrac{1}{2}O_2 &= H_2O \\ \tfrac{1}{2}N_2 + \tfrac{3}{2}H_2 &= NH_3 \\ CO_2 + 4H_2 &= CH_4 + 2H_2O \\ CO + 3H_2 &= CH_4 + H_2O \end{aligned}$$

This problem continues the use of the empirical dependencies of $C_P°$ on temperature used in problems in Chapters 2 and 3. (*a*) Calculate $\Delta_r H°$(1000 K) for these reactions. (*b*) Calculate $\Delta_r S°$(1000 K) for these reactions. (*c*) Use the values of $\Delta_r H°$(1000 K) and $\Delta_r S°$(1000 K) for these reactions to calculate K (1000 K) for each reaction. (*d*) Make a table of the values calculated in (*a*)–(*c*) and discuss the relative importance of $\Delta_r H°$ and $\Delta_r S°$ in determining whether a reaction goes forward or backward as written.

5.G Plot the equilibrium extent of the reaction A(g) = 2B(g) as a function of pressure for K = 0.1, 1, and 10. Equation 5.31 can be used to calculate the equilibrium extent of reaction.

5.H Plot the equilibrium extent of reaction for the reaction A(g) = 2B(g) versus pressure for reactions with equilibrium constants of 0.1, 1, and 10 by solving for the extent of reaction, rather than using the analytic expression for the extent of reaction. This method can be used for more complicated reactions.

5.I Plot the equilibrium extent of the reaction A(g) = 2B(g) versus temperature at pressures of 0.1, 1, and 10 bar for a reaction having an equilibrium constant of 0.1 at 298 K and a standard enthalpy of reaction of 50 kJ mol^{-1} that is independent of temperature.

5.J Calculate the equilibrium composition of a reaction mixture containing CO, CO_2, H_2, H_2O, and CH_4 at 1000 K and a total pressure of 1 bar. In doing that, also calculate the volume. The mixture initially consists of 2 mol CH_4 and 3 mol H_2O. There are two independent reactions, which can be written as

$$CH_4 + H_2O = CO + 3H_2$$
$$CO + H_2O = CO_2 + H_2$$

5.K There is another way in which the equilibrium composition of a multireaction system can be calculated, and that is by use of the Newton–Raphson method. This method starts with an estimate of the equilibrium composition and improves it in an iterative process. A very efficient program for carrying this out, called equcal in Mathematica, has been written by F. Krambeck of Mobil Research and Development [A. M. Sapre and F. J. Krambeck (eds.), *Chemical Reactions in Complex Systems*, New York: Van Nostrand Reinhold, 1991]. The input consists of a conservation matrix, a vector of Gibbs energies of formation at the desired pressure multiplied by $(-1/RT)$, and a vector of initial amounts of species. Calculate the equilibrium amounts for Computer Problem 5.J using equcalc.

5.L (*a*) Calculate the standard enthalpy of dissociation of calcite from the data in Table 5.1 by fitting the data to $\ln P = A + B/T$. Compare $\Delta_r H°$ with NBS value at 298.15 K. (*b*) Calculate the standard Gibbs energy of the reaction at 298.15 K using the parameters obtained in (*a*).

5.M The standard Gibbs energies of the isomers of the pentanes at 700 K are as follows:

n-pentane	193.26 kJ mol^{-1}
2-methylbutane	190.75
2,2-dimethylpropane	202.05

What are the equilibrium mole fractions of the three isomers? If the gases are ideal, will changing the pressure change this distribution?

5.N The reaction A = B, where A and B are ideal gases, is studied at 298.15 K. The standard chemical potential of A at this temperature is 40 kJ mol$_{-1}$ and that for B is 37 kJ mol$_{-1}$. (*a*) Plot $G_{avg} = (1-\xi)\mu_A^\circ + \xi\mu_B^\circ$ versus ξ from 0 to 1. (*b*) Plot the Gibbs energy of mixing versus ξ. (*c*) Plot the sum of (*a*) and (*b*) versus ξ. (*d*) Plot $dG/d\xi$ versus ξ. (*e*) Plot $d^2G/d\xi^2$ versus ξ.

6 Phase Equilibrium

A **phase** is a part of a system, uniform in chemical composition and physical properties, that is separated from other homogeneous parts of the system by boundary surfaces. The phase behavior exhibited by pure substances is quite varied and complicated, but there are powerful generalizations from thermodynamics that help us to understand these phenomena. The Clapeyron equation expresses $\mathrm{d}P/\mathrm{d}T$ for a two-phase system containing one component at equilibrium in terms of other thermodynamic quantities.

We have seen earlier that when two phases are at equilibrium they have the same temperature and pressure, and the chemical potential of each species, which is in both phases, is the same in each phase. Surface tension is involved in equilibrium between phases when there is curvature because the curved interface exerts a pressure so that the pressure is higher in the phase on the concave side of the interface.

Two-component, liquid–liquid systems may show very complicated behavior. The concept of ideal solutions provides a standard against which real solutions may be compared. Ideal solutions follow Raoult's law, and so it is convenient to express the deviations of real solutions from ideality by use of activity coefficients calculated from deviations from Raoult's law. Real solutions of nonelectrolytes follow Henry's law at low concentrations, and activity coefficients can also be calculated from deviations from Henry's law. Ideal solubility, freezing-point depression, and osmotic pressure are also discussed in this chapter.

6.1 PHASE DIAGRAMS OF ONE-COMPONENT SYSTEMS

The P–$\overline{V}$–T surface for a one-component system was introduced in Section 1.6, and we saw that it is convenient to use projections onto the P–$\overline{V}$ plane and the P–T plane. Projections onto the P–T plane are shown for water, carbon dioxide, and carbon in Fig. 6.1.

The phase diagram for water at low pressures is given in Fig. 6.1*a*, and the phase diagram at high pressures is given in Fig. 6.1*b*. In Fig. 6.1*a*, the vapor pres-

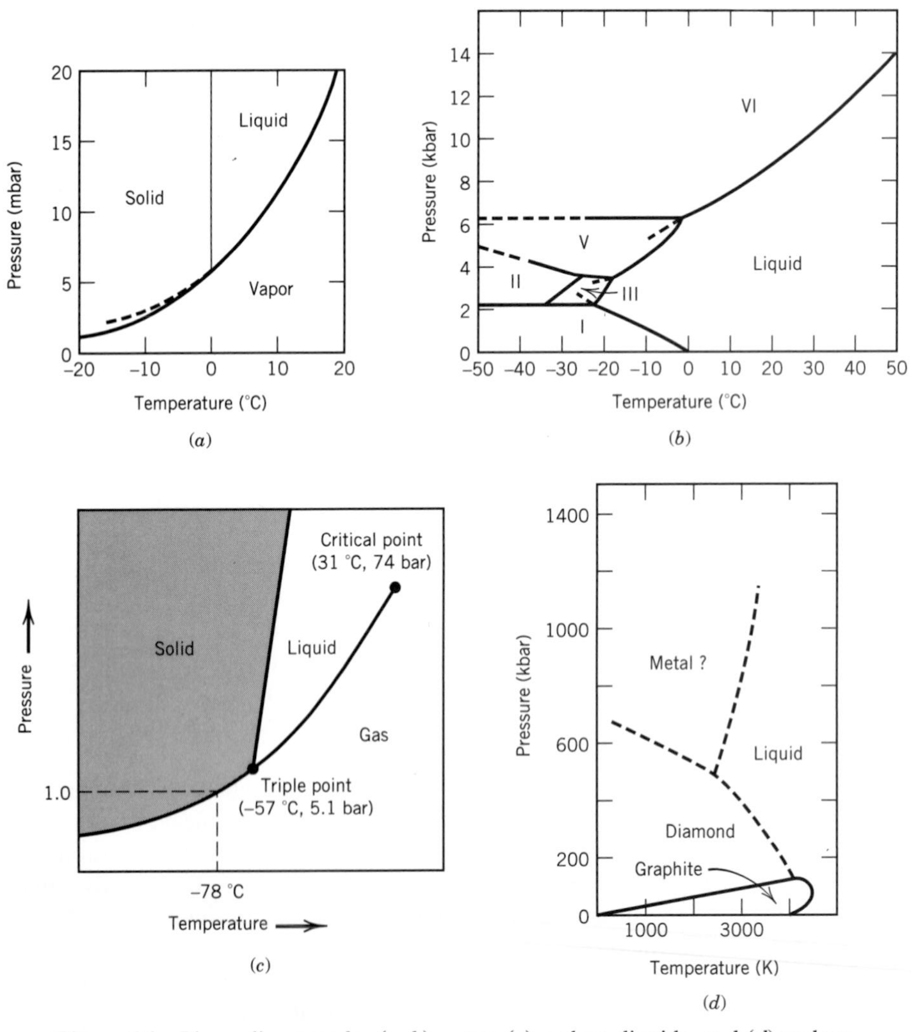

Figure 6.1 Phase diagrams for (*a*, *b*) water, (*c*) carbon dioxide, and (*d*) carbon.

sure of water is given as a function of temperature by the line between the liquid and vapor areas. The dashed extension of this line gives the vapor pressure of supercooled water. The curve for the sublimation pressure of ice goes down to zero at 0 K. At higher pressures, four other crystal forms of ice are formed. Starting at the triple point, the freezing point of ice I is lowered to −22 °C when the pressure is raised to 2000 bar, but higher pressures lead to other crystal forms of ice for which dP/dT is positive, as shown in Fig. 6.1*b*.

The phase diagram for carbon dioxide in Fig. 6.1*c* shows the equilibrium between solid and gas at 1 bar at −78 °C. Liquid carbon dioxide is produced only above 5.1 bar.

The phase diagram for carbon in Fig. 6.1*d* shows that graphite and diamond are in equilibrium at room temperature only at pressures above 10 000 bar. Diamonds for industrial use are produced at high pressures and temperatures using catalysts. The details of this phase diagram are not well known because of the difficulty in obtaining equilibrium.

In Section 5.9, we saw that the number of degrees of freedom for the description of the intensive state of the system is $F = C - p + 2$ if only pressure–volume work is involved. For a one-component system, $F = 3 - p$, so more than three phases cannot be in equilibrium. In the areas of Fig. 6.1, $p = 1$, so $F = 2$ and the intensive state of the system is completely described by specifying T and P.* Along a line, there are two phases at equilibrium, so the system can be completely described by specifying either T or P. At a triple point, three phases are in equilibrium, so $F = 0$. If the temperature or pressure is changed, two phases will disappear because the point representing the system will then lie in the solid, gas, or liquid area of the phase diagram.

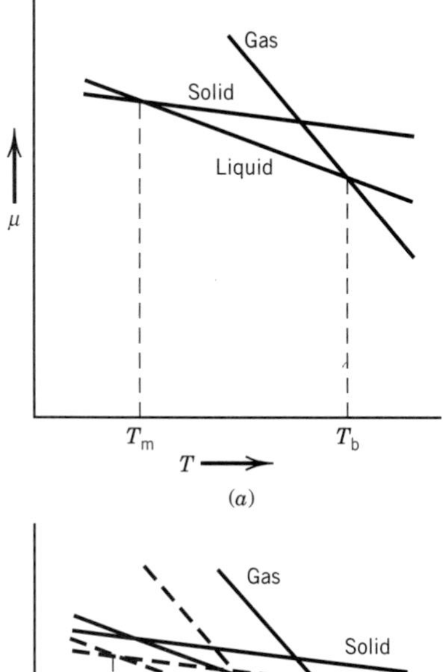

Comment:

We represent a critical point with a dot in a phase diagram, but in the neighborhood of that dot there are some phenomena, such as the turbidity mentioned in Section 1.7, that are very interesting in their own right. We do not have the space to pursue these remarkable phenomena, but we should remember that they also occur in connection with critical points of mixtures of liquids, which we will encounter later in this chapter.

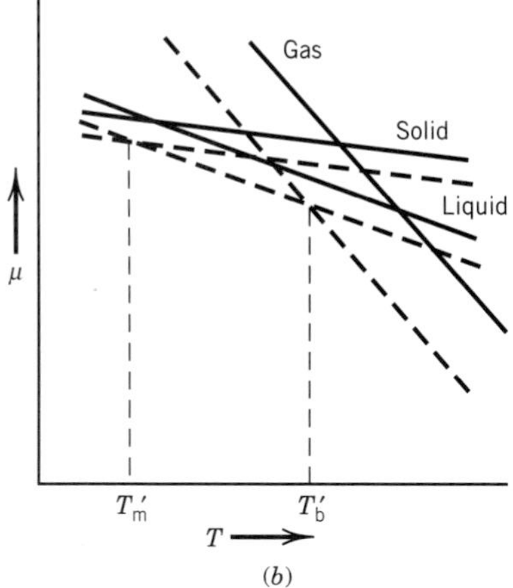

Figure 6.2 (*a*) Dependence of the chemical potentials of solid, liquid, and gas phases on temperature at constant pressure. The dashed lines in (*b*) are for a lower pressure. The plots should be slightly concave downward, since entropy increases with increasing temperature, but they have been drawn as straight lines here for simplicity.

To understand the change from solid to liquid to gas phase when a solid is heated at constant pressure, we may consider a plot of chemical potential versus temperature at constant pressure for the various phases, as shown in Fig. 6.2*a*. The stable phase is that with the lowest value of the chemical potential.

If two or three phases of a single component have the same chemical potential at a certain temperature and pressure, they will coexist at equilibrium as at the melting point T_m, boiling point T_b, or triple point. Below the melting point T_m the solid has the lowest chemical potential and is therefore the stable phase. Between T_m and T_b the liquid is the stable phase. It may be seen from Fig. 6.2 that the

*According to the phase rule, the intensive state of liquid water is characterized by two intensive variables. It is all right if we pick temperature and pressure. However, there is a problem if we pick pressure and molar volume since the temperature is not uniquely determined by these variables. In the case of water, the molar volume has a minimum in the neighborhood of 4 °C. The lesson from this is that we should pick one conjugate variable from each pair, and not two conjugate variables from the same pair.

phase transitions are sudden, but there are no indications of drastic change in the properties of the system as the temperature approaches the transition point.

For a single phase of a pure substance, the chemical potential is a function of temperature and pressure. Thus, the chemical potential can be represented as a surface in μ–P–T space. There are chemical potential surfaces of this type for each phase of a substance: gas, liquid, and one or more solid phases. The surfaces representing any two phases will intersect along a line, and three surfaces will intersect at a point, called the triple point. The phase diagram for a one-component system is the projection of these intersections onto the P–T plane.

Rather than looking at surfaces in three dimensions, it is more convenient to consider the chemical potential as a function of temperature at specified pressures, as in Fig. 6.2*b*.

The slopes of the lines giving the chemical potentials of solid, liquid, and gas in Fig. 6.2*a* are given by (see equation 4.84)

$$\left(\frac{\partial \mu}{\partial T}\right)_P = -\overline{S} \tag{6.1}$$

Since the entropy is positive, the slopes are negative, and since $\overline{S}_g > \overline{S}_l > \overline{S}_s$, the slope is more negative for the gas than for the liquid and more negative for the liquid than for the solid.

At a lower pressure the plots of μ versus T are displaced, as shown in Fig. 6.2*b*. The effect of pressure on the chemical potential of a pure substance at constant temperature is given by (see equation 4.85)

$$\left(\frac{\partial \mu}{\partial P}\right)_T = \overline{V} \tag{6.2}$$

Since the molar volume is always positive, the chemical potential μ decreases as the pressure is decreased at constant temperature. Since $\overline{V}_g \gg \overline{V}_l, \overline{V}_s$, this effect is much greater for a gas than for a liquid or solid. As shown in Fig. 6.2*b*, reducing the pressure lowers the boiling point and normally lowers the melting point. The effect on the boiling point is much greater because of the large difference in the molar volumes of gas and liquid. As a result, the range of temperature over which the liquid is the stable phase has been reduced. It is evident that at a sufficiently low pressure the curve for the chemical potential of the gas will intercept the solid curve below the temperature where the solid and liquid have the same chemical potential. At this low pressure the solid will sublime instead of melt; that is, it passes directly into the vapor state without going through the liquid state, as illustrated by dry ice in Fig. 6.1*c*.

At some particular pressure the solid, liquid, and vapor curves will intersect at a point; the temperature and pressure at which these three phases coexist is referred to as the triple point. If a substance can exist in more than one solid phase, the phase diagram will have more than one triple point, as illustrated in Fig. 6.1*b*,*d*.

The transitions we discussed in the preceding section are referred to as **first-order phase transitions** because there is a discontinuity in the first derivatives of the chemical potential. Since $(\partial\mu/\partial T)_P$ is different on the two sides of the transition temperature, the two phases have different entropies, and thus different enthalpies. Since $(\partial\mu/\partial P)_T$ is different on the two sides, the two phases have different volumes. The behavior of $\overline{C}_P$ is more complicated in that $\mathrm{d}q_P/\mathrm{d}T$ is infinite when two phases coexist at the transition temperature.

In a **second-order phase transition** there is no discontinuity in the first derivatives of μ, but discontinuities occur in the second derivatives (curvature) of the chemical potential. Since $(\partial\mu/\partial T)_P$ is the same on two sides of the transition, there is no discontinuity in the entropy at the transition temperature. Therefore, there is no heat of transition. There is a discontinuity in $\overline{C}_P$ because $\overline{C}_P = -T(\partial\mu^2/\partial T^2)_P$, as may be derived readily from equation 4.42. However, $\overline{C}_P$ does not become infinite at the critical point as in the case of first-order transitions. The sudden appearance of superconductivity in certain metals when they are cooled to low temperatures is an example of such a second-order transition.

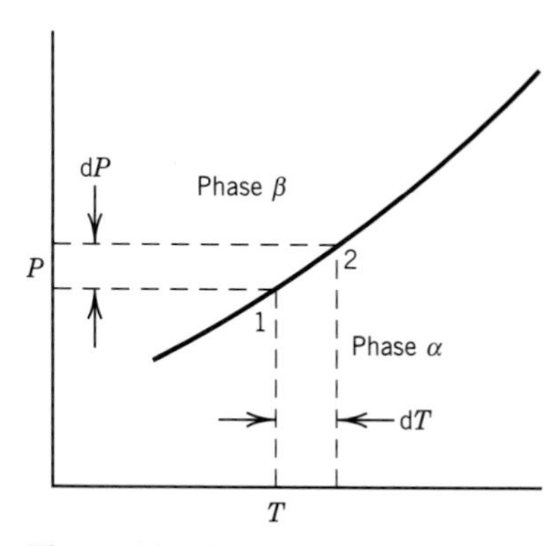

Figure 6.3 Coexistence curve for a one-component system. Along this line $\mu_\alpha = \mu_\beta$. If the temperature is changed by $\mathrm{d}T$, the pressure has to be changed by $\mathrm{d}P$ as indicated to maintain equilibrium. At point 1, $\mu_{1\alpha} = \mu_{1\beta}$. At point 2, $\mu_{1\alpha} + \mathrm{d}\mu_\alpha = \mu_{1\beta} + \mathrm{d}\mu_\beta$, so that $\mathrm{d}\mu_\alpha = \mathrm{d}\mu_\beta$.

6.2 THE CLAPEYRON EQUATION

Consider a one-component system with two phases (α and β). At equilibrium, the pressure, temperature, and chemical potential must be the same in the two phases. For the chemical potentials,

$$\mu_\alpha = \mu_\beta \tag{6.3}$$

When the temperature is changed at constant pressure, or the pressure is changed at constant temperature, one of the phases will disappear. However, if the temperature and pressure are both changed in such a way as to keep the two chemical potentials equal to each other, the two phases will continue to coexist. The plot of pressure versus temperature along which the two phases coexist is referred to as the **coexistence curve** (see Fig. 6.3). The necessary relation for $\mathrm{d}P/\mathrm{d}T$ was derived by Clapeyron.

For a change of pressure and temperature along the coexistence curve,

$$\mathrm{d}\mu_\alpha = \mathrm{d}\mu_\beta \tag{6.4}$$

since the chemical potential is equal to the molar Gibbs energy for a one-component system, $\mathrm{d}\mu = \mathrm{d}\overline{G} = \overline{V}\,\mathrm{d}P - \overline{S}\,\mathrm{d}T$ according to equation 4.36. Thus,

$$\overline{V}_\alpha\,\mathrm{d}P - \overline{S}_\alpha\,\mathrm{d}T = \overline{V}_\beta\,\mathrm{d}P - \overline{S}_\beta\,\mathrm{d}T \tag{6.5}$$

or

$$\frac{\mathrm{d}P}{\mathrm{d}T} = \frac{\overline{S}_\beta - \overline{S}_\alpha}{\overline{V}_\beta - \overline{V}_\alpha} = \frac{\Delta\overline{S}}{\Delta\overline{V}} = \frac{\Delta\overline{H}}{T\Delta\overline{V}} \tag{6.6}$$

Although we have considered molar quantities, the quantity unit cancels in this equation. Thus, ΔS and ΔV per unit mass can also be used. This equation is referred to as the **Clapeyron equation,** and it may be applied to vaporization, sublimation, fusion, or the transition between two solid phases of a pure substance. Note that $\mathrm{d}P/\mathrm{d}T$ is a total derivative, not a partial derivative; however, there is a constraint: The two phases remain in equilibrium so that $\Delta G = 0$. The molar enthalpies of sublimation, fusion, and vaporization at the triple point are related by

$$\Delta_{\text{sub}}H = \Delta_{\text{fus}}H + \Delta_{\text{vap}}H \tag{6.7}$$

since the heat required to vaporize a given amount of the solid is the same whether this process is carried out directly or by first melting the solid and then vaporizing the liquid.

Water is unusual in that it expands on freezing, so that ΔV and thus $\mathrm{d}P/\mathrm{d}T$ for melting is negative.

Example 6.1 ***Change in the boiling point of water with pressure***

What is the change in the boiling point of water at a 100 °C per Pa change in atmospheric pressure?

The molar enthalpy of vaporization is 40.69 kJ mol^{-1}, the molar volume of liquid water is 0.019×10^{-3} m^3 mol^{-1}, and the molar volume of steam is 30.199×10^{-3} m^3 mol^{-1}, all at 100 °C and 1.01325 bar:

$$\frac{\mathrm{d}P}{\mathrm{d}T} = \frac{\Delta_{\mathrm{vap}}H}{T(\overline{V}_{\mathrm{g}} - \overline{V}_{\mathrm{l}})} = \frac{(40\,690\ \mathrm{J\ mol^{-1}})}{(373.15\ \mathrm{K})(30.180 \times 10^{-3}\ \mathrm{m^3\ mol^{-1}})}$$

$$= 3613\ \mathrm{Pa\ K^{-1}}$$

Thus, $\mathrm{d}T/\mathrm{d}P = 2.768 \times 10^{-4}\ \mathrm{K\ Pa^{-1}}$.

Example 6.2 ***Change in the freezing point of water with pressure***

Calculate the change in pressure required to change the freezing point of water 1 °C. At 0 °C the heat of fusion of ice is 333.5 J g^{-1}, the density of water is 0.9998 g cm^{-3}, and the density of ice is 0.9168 g cm^{-3}.

The reciprocals of the densities, 1.0002 and 1.0908, are the volumes in cubic centimeters of 1 g. The volume change on freezing $(\overline{V}_{\mathrm{l}} - \overline{V}_{\mathrm{s}})$ is therefore -9.06×10^{-8} m^3 g^{-1}. For small changes ΔH_{fus}, T, and $(\overline{V}_{\mathrm{l}} - \overline{V}_{\mathrm{s}})$ are virtually constant, so that

$$\frac{\Delta P}{\Delta T} = \frac{\Delta_{\mathrm{fus}}H}{T(\overline{V}_{\mathrm{l}} - \overline{V}_{\mathrm{s}})} = \frac{333.5\ \mathrm{J\ g^{-1}}}{(273.15\ \mathrm{K})(-9.06 \times 10^{-8}\ \mathrm{m^3\ g^{-1}})}$$

$$= -1.348 \times 10^7\ \mathrm{Pa\ K^{-1}}$$

The change in the freezing point of water per bar pressure is

$$\frac{\Delta T}{\Delta P} = \frac{10^5\ \mathrm{Pa\ bar^{-1}}}{-1.348 \times 10^7\ \mathrm{Pa\ K^{-1}}} = -0.0075\ \mathrm{K\ bar^{-1}}$$

This shows that an increase in pressure of 1 bar lowers the freezing point 0.0075 K. The negative sign indicates that an increase in pressure causes a decrease in temperature. The change in pressure required to change the freezing point of water 1 °C is

$$\frac{\Delta P}{\Delta T} = -\frac{1}{0.0075\ \mathrm{K\ bar^{-1}}} = -133\ \mathrm{bar\ K^{-1}}$$

Example 6.3 ***Calculation of vapor pressures from thermodynamic tables***

Using the data in Table C.2, determine the vapor pressures of $H_2O(l)$ and $Br_2(l)$ at 298.15 K. These vaporization processes may be handled as if they were chemical reactions.

$$\begin{aligned}
H_2O(l) &= H_2O(g) \\
\Delta_{vap}G^\circ_{298} &= -RT \ln \frac{P}{P^\circ} \\
&= -228.572 - (-237.129) = 8.557 \text{ kJ mol}^{-1} \\
P &= 0.031\,69 \text{ bar} \\
Br_2(l) &= Br_2(g) \\
\Delta_{vap}G^\circ_{298} &= -RT \ln \frac{P}{P^\circ} \\
&= 3.110 \text{ kJ mol}^{-1} \\
P &= 0.2852 \text{ bar}
\end{aligned}$$

Example 6.4 ***Calculation of the pressure to make diamonds from graphite***

Calculate the equilibrium pressure for the conversion of graphite to diamond at 25 °C. The densities of graphite and diamond may be taken to be 2.25 and 3.51 g cm^{-3}, respectively, independent of pressure, in calculating the change of ΔG with pressure.

$$\text{C(graphite)} = \text{C(diamond)}$$

From Table C.2,

$$\begin{aligned}
\Delta G^\circ &= 2900 \text{ J mol}^{-1} \\
\Delta V &= 12\left(\frac{1}{3.51} - \frac{1}{2.25}\right) \times 10^{-6} \text{ m}^3 \text{ mol}^{-1} = -1.91 \times 10^{-6} \text{ m}^3 \text{ mol}^{-1}
\end{aligned}$$

Since $(\partial \Delta G / \partial P)_T = \Delta V$,

$$\begin{aligned}
\int_1^2 \mathrm{d}\Delta G &= \int_{P_1}^{P_2} \Delta V \,\mathrm{d}P = \Delta G_2 - \Delta G_1 = \Delta V(P_2 - P_1) \\
P_2 &= \frac{\Delta G_2 - \Delta G_1}{\Delta V} + P_1 \\
&= \frac{0 - 2900 \text{ J mol}^{-1}}{-1.91 \times 10^{-6} \text{ m}^3 \text{ mol}^{-1}} + 10^5 \text{ Pa} \\
&= 1.52 \times 10^9 \text{ Pa or } 1.52 \times 10^4 \text{ bar}
\end{aligned}$$

6.3 THE CLAUSIUS–CLAPEYRON EQUATION

For vaporization and sublimation Clausius showed how the Clapeyron equation may be simplified by assuming that the vapor obeys the ideal gas law and by neglecting the molar volume of the liquid V_l in comparison with the molar volume of the gas V_g. Substituting RT/P for V_g, we have

$$\frac{\mathrm{d}P}{\mathrm{d}T} = \frac{\Delta_{vap}H}{TV_g} = \frac{P\,\Delta_{vap}H}{RT^2} \tag{6.8}$$

On rearrangement equation 6.8 becomes

$$\frac{\mathrm{d}P}{P} = \mathrm{d}\ln\frac{P}{P^\circ} = \frac{\Delta_{vap}H}{RT^2}\,\mathrm{d}T \tag{6.9}$$

Table 6.1 Vapor Pressures of Ice and Water

t/°C	P/kPa
−40	0.013
−30	0.038
−20	0.103
−10	0.260
0	0.611
10	1.228
20	2.338
30	4.245
40	7.381
60	19.933
100	101.325
140	361.21
180	1001.9

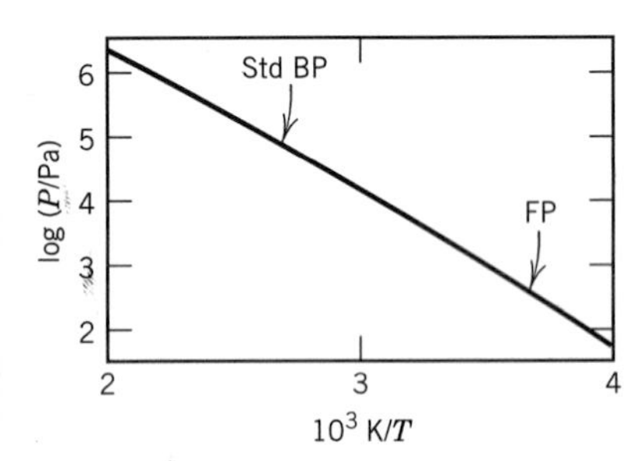

Figure 6.4 Vapor pressure of water (FP = freezing point, BP = boiling point).

where $P°$ is the standard pressure used. Integrating on the assumption that ΔH_{vap} is independent of temperature and pressure yields

$$\int \mathrm{d}\ln\frac{P}{P°} = \frac{\Delta_{vap}H}{R}\int T^{-2}\,\mathrm{d}T \tag{6.10}$$

$$\ln\frac{P}{P°} = -\frac{\Delta_{vap}H}{RT} + C \tag{6.11}$$

where C is the integration constant. This suggests that a plot of $\ln(P/P°)$ versus $1/T$ should be linear, and this is borne out by data on both vaporization and sublimation, as shown in Fig. 6.4. Vapor pressure data for ice and water are given in Table 6.1.

Frequently, it is more convenient to use the equation obtained by integrating between limits, P_2 at T_2 and P_1 at T_1, as follows:

$$\int_{P_1/P°}^{P_2/P°} \mathrm{d}\ln\frac{P}{P°} = \frac{\Delta_{vap}H}{R}\int_{T_1}^{T_2} T^{-2}\,\mathrm{d}T \tag{6.12}$$

$$\ln\frac{P_2}{P_1} = \frac{\Delta_{vap}H}{R}\left[-\frac{1}{T_2} - \left(-\frac{1}{T_1}\right)\right] \tag{6.13}$$

$$\ln\frac{P_2}{P_1} = \frac{\Delta_{vap}H\,(T_2 - T_1)}{RT_1T_2} \tag{6.14}$$

To represent the vapor pressure as a function of temperature over a wide range of temperature, it is necessary to take the temperature dependence of $\Delta_{vap}H$ into account. Another deficiency in this simple equation is that the vapor has been assumed to be an ideal gas.

Over narrow ranges of temperatures, the enthalpy of vaporization can be taken to be a linear function of temperature (see equation 5.43). However, in calculating vapor pressures over a wider range of temperature, we have to recognize that the enthalpy of vaporization approaches zero as the temperature approaches the critical temperature, as shown in Fig. 6.5.

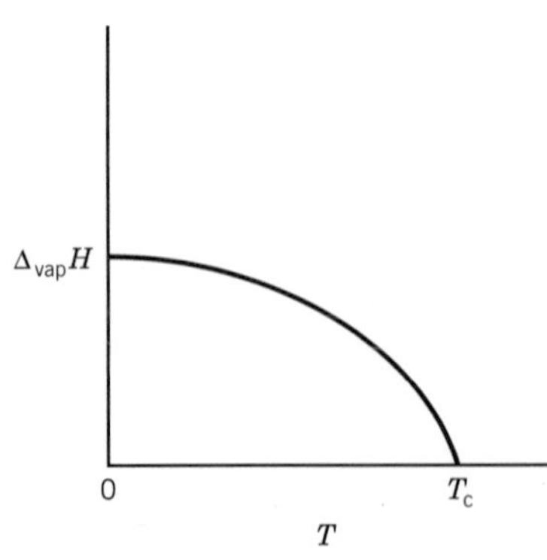

Figure 6.5 The enthalpy of vaporization approaches zero as the temperature approaches the critical temperature.

The dependence of the heat of vaporization on temperature can be approximately represented by

$$\Delta_{vap}H = A + BT + CT^2 \tag{6.15}$$

Thus equation 6.9 can be written

$$\frac{\mathrm{d}P_{vap}}{P_{vap}} = \frac{\Delta_{vap}H}{RT^2}\mathrm{d}T = \frac{1}{R}\left(\frac{A}{T^2} + \frac{B}{T} + C\right)\mathrm{d}T \tag{6.16}$$

Integration yields

$$\ln P_{vap} = \frac{1}{R}\left(-\frac{A}{T} + B\ln T + CT + D\right) \tag{6.17}$$

where D is the integration constant. Thus vapor pressures determined experimentally over a range of temperatures can be represented by values of A, B, C, and D determined by curve fitting to minimize the sum of the squares of the deviations from the experimental values. Since the enthalpy of vaporization is given by

$$\Delta_{vap}H = RT^2\left(\frac{\mathrm{d}\ln P_{vap}}{\mathrm{d}T}\right)_P = -R\left[\frac{\mathrm{d}\ln P_{vap}}{\mathrm{d}(1/T)}\right]_P \tag{6.18}$$

Table 6.2 Enthalpies and Entropies of Fusion at the Melting Point and Vaporization at the Boiling Point[a]

Substance	$\frac{T_{\text{fus}}}{\text{K}}$	$\frac{\Delta_{\text{fus}}H}{\text{kJ mol}^{-1}}$	$\frac{\Delta_{\text{fus}}S}{\text{J K}^{-1}\text{ mol}^{-1}}$	$\frac{T_{\text{b}}}{\text{K}}$	$\frac{\Delta_{\text{vap}}H}{\text{kJ mol}^{-1}}$	$\frac{\Delta_{\text{vap}}S}{\text{J K}^{-1}\text{ mol}^{-1}}$
N_2	63.3	0.720	11.37	77.4	5.577	72.1
C_2H_6	89.9	2.86	31.81	184.52	14.71	79.7
NH_3	195.4	5.653	28.93	239.72	23.33	97.3
CCl_4	250.3	2.5	10.00	349.9	30.0	85.7
H_2O	273.2	6.01	22.00	373.15	40.66	109.0

[a]The boiling points are at 1 atm.

where P is the total pressure on the surface of the liquid, the heat of vaporization as a function of temperature can be obtained by differentiating $\ln P_{\text{vap}}$ with respect to T or $1/T$. The use of these equations is illustrated by Computer Problems 6.C and 6.D. If the vapor pressures are measured over a range of temperatures up to near to the critical point, a plot of $\Delta_{\text{vap}}H$ like Fig. 6.5 can be obtained. However, to do this successfully requires a more complicated empirical equation than equation 6.15.

A number of enthalpies and entropies of fusion and vaporization are given in Table 6.2.

6.4 VAPOR–LIQUID EQUILIBRIUM OF BINARY LIQUID MIXTURES

In this section, we will discuss only ideal liquid mixtures because their vapor pressures follow simple equations and the basic principles can be easily understood. First, we will consider partial pressures of the vapors in equilibrium with binary liquid mixtures. Then we will show that the thermodynamic properties of ideal solutions are readily calculated from the properties of the pure liquids. For ideal solutions, boiling point diagrams can also be calculated, and distilling columns can readily be designed to achieve any desired degree of separation of the two liquids. Since these systems do not involve chemical reactions, the number of components is equal to the number of species (see Section 5.9). In the following discussions, we will use the term *components,* rather than *species,* since that is customary in discussing phase equilibria.

The calculation of properties of both ideal and nonideal solutions is based on the fact that the chemical potential of a species is the same in different phases when the phases are at equilibrium. When a binary liquid mixture is in equilibrium with its vapor at a constant temperature, the chemical potential of each component is the same in the gas and liquid phases:

$$\mu_i(\text{g}) = \mu_i(\text{l}) \tag{6.19}$$

Rather than using fugacities for the components in the vapor phase, we will assume the vapor is an ideal gas. Thus, the chemical potential of a component in the gas phase is given by

$$\mu_i(\text{g}) = \mu_i^\circ(\text{g}) + RT \ln \frac{P_i}{P^\circ} \tag{6.20}$$

where P° is the standard state pressure of 1 bar. The most general way to express the chemical potential of a component in a liquid mixture is to use the activity, which was introduced in Section 4.5:

$$\mu_i(\mathrm{l}) = \mu_i^\circ(\mathrm{l}) + RT \ln a_i \tag{6.21}$$

Thus, equation 6.20 can be written

$$\mu_i^\circ(\mathrm{g}) + RT \ln \frac{P_i}{P^\circ} = \mu_i^\circ(\mathrm{l}) + RT \ln a_i \tag{6.22}$$

This equation can also be applied to pure liquid, which, of course, has an activity of unity in the liquid phase:

$$\mu_i^\circ(\mathrm{g}) + RT \ln \frac{P_i^*}{P^\circ} = \mu_i^\circ(\mathrm{l}) \tag{6.23}$$

The equilibrium vapor pressure of pure i at temperature T is P_i^*. Now we subtract this equation from equation 6.22 to eliminate the standard chemical potentials of gas and liquid:

$$RT \ln \frac{P_i}{P_i^*} = RT \ln a_i \tag{6.24}$$

or

$$a_i = \frac{P_i}{P_i^*} \tag{6.25}$$

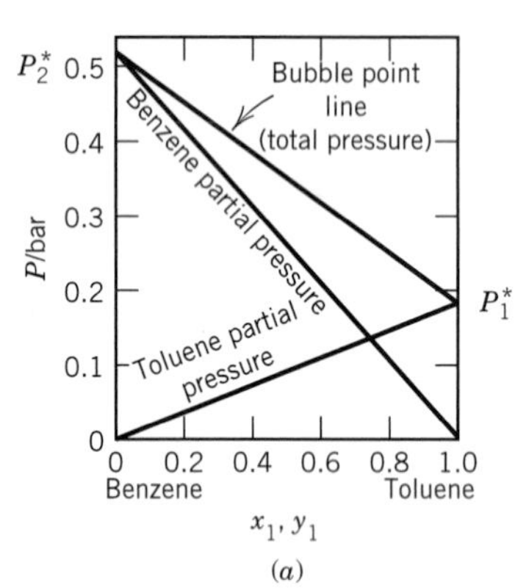

(*a*)

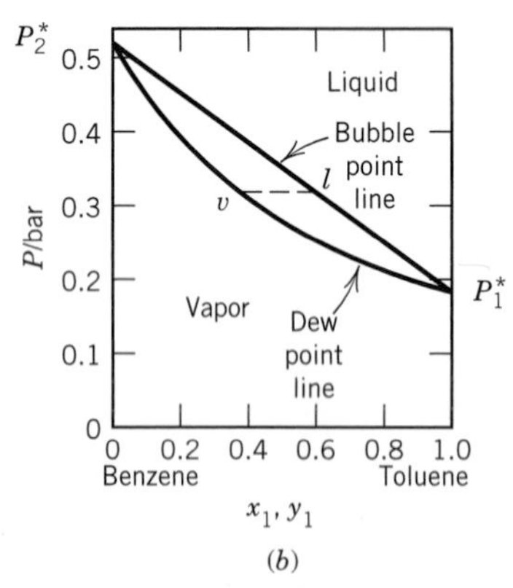

(*b*)

Figure 6.6 Benzene (2)–toluene (1) system at 60 °C. (*a*) Partial and total pressures. (*b*) Liquid and vapor compositions.

Thus, if the vapor is an ideal gas, **the activity of a component of a solution is equal to the ratio of its partial pressure above the solution to the vapor pressure of the pure liquid.**

This discussion has been quite general, but so far it has not shown us how to predict the partial pressure P_i of a component of an actual solution. In 1884 Raoult found that for certain solutions the partial pressure of a component is equal to the mole fraction of that component times the vapor pressure of the pure component:

$$P_i = x_i P_i^* \tag{6.26}$$

This equation is not exact, but it is so useful that it is referred to as **Raoult's law.** It is obeyed most closely when the components are quite similar. Mixtures of benzene and toluene obey Raoult's law, as illustrated by Fig. 6.6. In molecular terms Raoult's law is obeyed by pairs of liquids A and B where A–A, A–B, and B–B interactions are all the same. Since the gas phase has been assumed to be ideal, Raoult's law can also be written

$$y_i P = x_i P_i^* \tag{6.27}$$

where y_i is the mole fraction of i in the gas phase.

When equation 6.26 is substituted in equation 6.25, we see that for solutions with very similar components, the activity of a component is equal to its mole fraction; thus, $a_i = x_i$ for an ideal solution. When this relation is substituted in equation 6.21, we obtain

$$\mu_i(\mathrm{l}) = \mu_i^\circ(\mathrm{l}) + RT \ln x_i \tag{6.28}$$

We take this equation as the definition of an **ideal solution.**

By use of Raoult's law we can calculate the phase diagram for an ideal liquid mixture. The total vapor pressure of an ideal binary liquid mixture is given by

$$\begin{aligned} P &= P_1 + P_2 = x_1 P_1^* + x_2 P_2^* \\ &= P_2^* + (P_1^* - P_2^*) x_1 \end{aligned} \tag{6.29}$$

This equation for the **bubble point line** is plotted in Fig. 6.6 and in Fig. 6.7*a*. At points in the phase diagram above the bubble point line, the system is in the liquid state. Suppose the pressure on a specific binary solution of benzene and toluene is reduced from some high value. When the pressure reaches the bubble point line, it is equal to the total vapor pressure of the solution given by equation 6.29, and a further lowering of the pressure will cause bubbles of vapor to form.

The composition of the vapor in equilibrium with a binary solution can readily be calculated using Raoult's law. The mole fraction of component 1 in the vapor is given by

$$y_1 = \frac{P_1}{P_1 + P_2} = \frac{x_1 P_1^*}{x_1 P_1^* + x_2 P_2^*} = \frac{x_1 P_1^*}{P_2^* + (P_1^* - P_2^*) x_1} \tag{6.30}$$

This equation can be solved for x_1 to obtain the following expression for the mole fraction of component 1 in solution that corresponds to a certain mole fraction y_1 in the equilibrium vapor:

$$x_1 = \frac{y_1 P_2^*}{P_1^* + (P_2^* - P_1^*) y_1} \tag{6.31}$$

Now we can use equation 6.23 to calculate the total pressure P that corresponds to a certain mole fraction of component 1 in the vapor phase. Substituting equation 6.31 in equation 6.27 yields

$$P = \frac{P_1^* P_2^*}{P_1^* + (P_2^* - P_1^*) y_1} \tag{6.32}$$

This equation for the **dew point line** is plotted in Figs. 6.6*b* and 6.7*b*. At points in the phase diagram below the dew point, the system is in the vapor state. Suppose the pressure on a specific binary vapor of benzene and toluene is raised from some low value. When the pressure reaches the dew point line, it is equal to that given by equation 6.32, and raising the pressure further causes condensation of the first droplets of liquid. In the region of the phase diagram between the dew point line and the bubble point line, two phases are present at equilibrium.

Points on the dew point line and bubble point line at the same pressure represent the compositions of vapor and liquid phases that are in equilibrium. These points are connected by a horizontal line referred to as a **tie line,** which is shown in Fig 6.6*b*. The overall composition of the two-phase system can range from v to l. At v the system is all vapor, and at l the system is all liquid. If the mole fraction of toluene in the system is halfway between v and l, the amount of liquid is equal to the amount of vapor. If the mole fraction of toluene in the system is x, the ratio of the number of moles of liquid to the amount of vapor is equal to $(x - v)/(l - x)$. This rule, which is readily derived from the conservation of moles, is referred to as the **lever rule.**

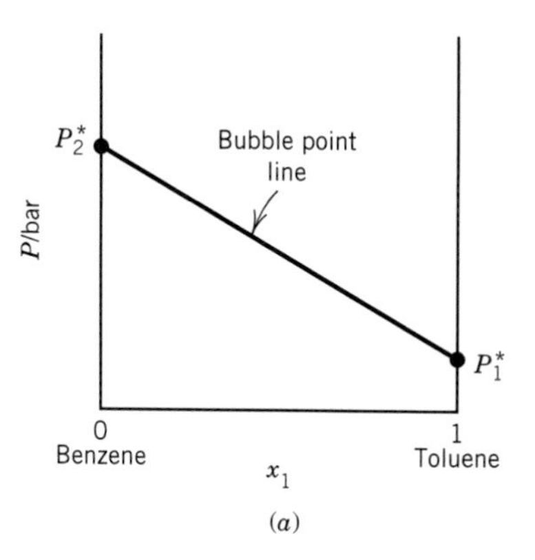

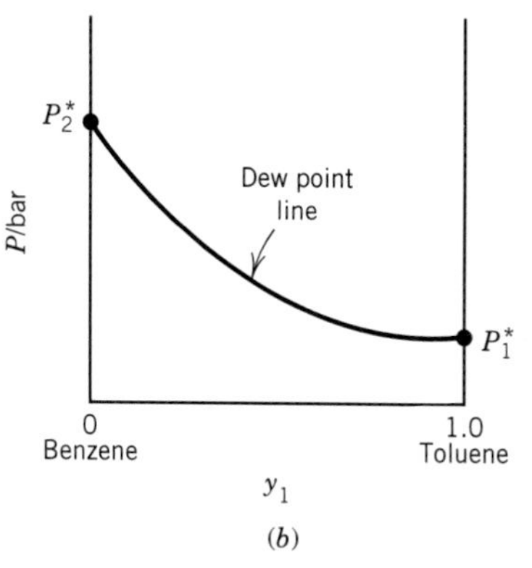

Figure 6.7 (*a*) Plot of the total pressure for the benzene (2)–toluene (1) system versus x_1, as given by equation 6.29. (*b*) Plot of the total pressure for the benzene (2)–toluene (1) system versus y_1, as given by equation 6.32. These plots are superimposed in Fig. 6.6.

Example 6.5 ***Calculation of partial pressures for a solution of benzene and toluene***

At 60 °C the vapor pressures of pure benzene and toluene are 0.513 and 0.185 bar, respectively. What are the equations of the bubble point line and dew point line? For a solution with 0.60 mole fraction toluene, what are the partial pressures of toluene and benzene, and what is the mole fraction of toluene in the vapor?

We will consider toluene to be component 1.

Bubble point line:

$$\begin{aligned} P &= P_2^* + (P_1^* - P_2^*)x_1 \\ &= 0.513 \text{ bar} - (0.328 \text{ bar})x_1 \end{aligned}$$

Dew point line:

$$\begin{aligned} P &= \frac{P_1^* P_2^*}{P_1^* + (P_2^* - P_1^*)y_1} \\ &= \frac{0.0949 \text{ bar}^2}{0.185 \text{ bar} - (0.328 \text{ bar})y_1} \end{aligned}$$

$$\begin{aligned} P_1 &= x_1 P_1^* = (0.60)(0.185 \text{ bar}) = 0.111 \text{ bar} \\ P_2 &= x_2 P_2^* = (0.40)(0.513 \text{ bar}) = 0.205 \text{ bar} \\ P &= 0.513 \text{ bar} - (0.328 \text{ bar})(0.60) = 0.316 \text{ bar} \end{aligned}$$

$$\begin{aligned} y_1 &= \frac{x_1 P_1^*}{P_2^* + (P_1^* - P_2^*)x_1} \\ &= \frac{(0.60)(0.185 \text{ bar})}{0.513 \text{ bar} - (0.328 \text{ bar})(0.60)} \\ &= 0.351 \end{aligned}$$

Example 6.6 ***Calculation of activities of benzene and toluene in a solution***

According to the data in the previous example, what are the activities of toluene (component 1) and benzene (component 2) in a solution containing 0.600 mole fraction toluene according to equation 6.25?

$$a_1 = \frac{P_1}{P_1^*} = \frac{0.111 \text{ bar}}{0.185 \text{ bar}} = 0.600$$

$$a_2 = \frac{P_2}{P_2^*} = \frac{0.205 \text{ bar}}{0.513 \text{ bar}} = 0.400$$

Since these are ideal solutions, the activities are equal to the mole fractions.

Before turning to boiling point diagrams, we consider the thermodynamic consequences of equation 6.28, which defines an ideal solution. The molar Gibbs energy of an ideal solution is given by

$$\overline{G} = \sum x_i \mu_i = \sum x_i \mu_i^\circ + RT \sum x_i \ln x_i \tag{6.33}$$

Since this equation gives $\overline{G}$ as a function of temperature and pressure, it contains all of the thermodynamic information about an ideal solution. The molar entropy of an ideal solution is obtained from $-(\partial\overline{G}/\partial T)_P$:

$$\overline{S} = \sum x_i \overline{S}_i^\circ - R \sum x_i \ln x_i \tag{6.34}$$

The molar enthalpy of an ideal solution is obtained from $-T^2[\partial(\overline{G}/T)/\partial T]_P$:

$$\overline{H} = \sum x_i \overline{H}_i^\circ \tag{6.35}$$

and the molar volume is obtained from $(\partial\overline{G}/\partial P)_T$:

$$\overline{V} = \sum x_i \overline{V}_i^\circ \tag{6.36}$$

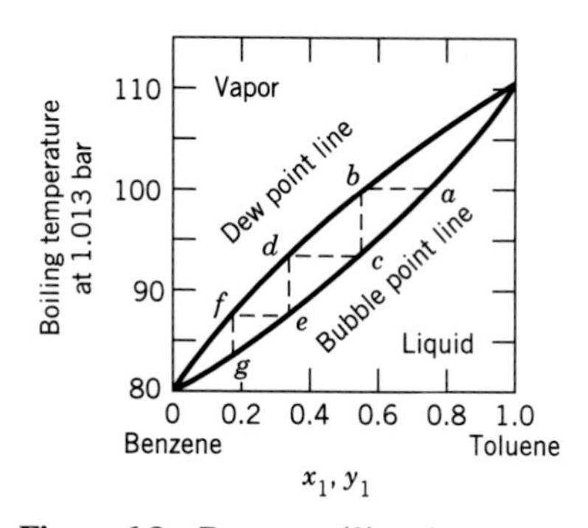

Figure 6.8 Benzene (2)–toluene (1) boiling points: liquid and vapor compositions. The liquid boils at the temperature given by the lower curve.

The first terms in these equations give the thermodynamic properties of the amounts of pure liquids required to form the mixture. Therefore, the changes in these thermodynamic properties per mole of solution are given by

$$\Delta_{\text{mix}} G = RT \sum_{i=1}^{N} x_i \ln x_i \tag{6.37}$$

$$\Delta_{\text{mix}} S = -R \sum_{i=1}^{N} x_i \ln x_i \tag{6.38}$$

$$\Delta_{\text{mix}} H = 0 \tag{6.39}$$

$$\Delta_{\text{mix}} V = 0 \tag{6.40}$$

These are the same equations that were obtained for ideal gas mixtures (Example 4.10), except that these equations are written for 1 mol of mixture. **Thus, there is no volume change or heat evolution when liquids are mixed to form ideal solutions at constant temperature and pressure.**

We have been considering ideal binary liquid mixtures at constant temperature, and now we will consider them at constant pressure because this is of special interest in connection with distillation. Phase diagrams at constant pressure are referred to as **boiling point diagrams.** The boiling point diagram for benzene–toluene solutions at 1.013 bar (1 atm) is shown in Fig. 6.8. At points in the phase diagram above the dew point line, the system is in the vapor state. At points below the bubble point line, the system is in the liquid state. Between the two lines, two phases are present, and the relative amounts of the two phases are given by the lever rule. Since benzene and toluene form ideal solutions, this diagram may be calculated from the information given in Table 6.3 on the vapor pressures of benzene and toluene at temperatures between their boiling points of 80.1 and 110.6 °C, respectively, at 1.013 bar.

Table 6.3 Vapor Pressures of Toluene (1) and Benzene (2)

	t/°C							
	80.1	*88*	*90*	*94*	*98*	*100*	*104*	*110.6*
P_1^*/bar	—	0.508	0.543	0.616	0.698	0.742	0.836	1.013
P_2^*/bar	1.013	1.285	1.361	1.526	1.705	1.800	2.004	—

Example 6.7 ***Calculation of mole fractions of benzene and toluene in a solution that boils at 100 °C***

What is the mole fraction x_1 of toluene in the toluene–benzene solution that boils at 100 °C, and what is the mole fraction y_1 of toluene in the vapor?

Equation 6.29 may be solved for x_1:

$$x_1 = \frac{P - P_2^*}{P_1^* - P_2^*} = \frac{1.013\ \text{bar} - 1.800\ \text{bar}}{0.742\ \text{bar} - 1.800\ \text{bar}} = 0.744$$

Now that the mole fraction of toluene in the liquid phase is known, its mole fraction y_1 in the vapor phase is readily calculated using equation 6.27:

$$y_1 = \frac{x_1 P_1^*}{P} = \frac{(0.744)(0.742\ \text{bar})}{1.013\ \text{bar}}$$
$$= 0.545$$

These points are labeled a and b in Fig. 6.8. For nonideal solutions the points have to be obtained experimentally.

The relationship between the vapor pressure diagram of Fig. 6.6 and the boiling point diagram of Fig. 6.8 is shown in Fig. 6.9. It is possible to plot vapor–liquid data for a two-component system in a three-dimensional diagram because the maximum number of degrees of freedom is $F = C - p + 2 = 2 - 1 + 2 = 3$.

In the P–T–composition diagram the states of pairs of phases in equilibrium with each other define surfaces. These two surfaces come together in the planes that represent the vapor pressures of the two components as a function of temperature. Note that the vapor pressure curve for component 1 ends at the critical point c_1, and the vapor pressure curve for component 2 ends at the critical point c_2. The upper and lower surfaces also come together along a critical locus between c_1 and c_2 made up of the points at which the vapor and liquid phases in equilibrium become identical. In Fig. 6.9 the equilibrium pressures and equilibrium mole

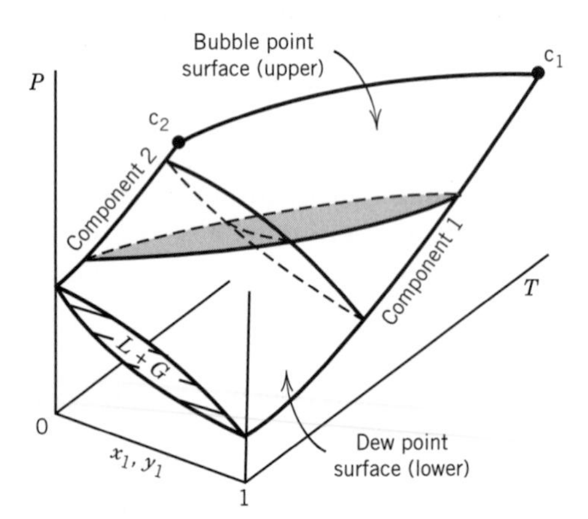

Figure 6.9 Three-dimensional plot for a two-component liquid and vapor system. The plot consists of two surfaces (bubble point surface and dew point surface). (From K. E. Bett, J. S. Rowlinson, and G. Saville, *Thermodynamics for Chemical Engineers.* Cambridge, MA: MIT Press, 1975. Reproduced by permission of The Athlone Press.)

fractions of component 1 in the liquid phase x_1 form the upper surface. The lower surface in Fig. 6.9 is made up of points representing the equilibrium pressures and equilibrium mole fractions of component 1 in the vapor phase y_1.

Between the two surfaces two phases are present at equilibrium, saturated vapor and saturated liquid. Since there are two phases, the number of degrees of freedom between the surfaces is $F = 2 - 2 + 2 = 2$. If the temperature and pressure are specified, the compositions of the two phases are fixed, and they are given by the abscissas of the two ends of a horizontal tie line. The relative amounts of the two phases are not fixed by specifying only temperature and pressure, but the phase rule is not concerned with the relative amounts of phases. If the temperature and the composition of the liquid phase are given, the pressure is given by the ordinate and the composition of the vapor phase is given by the other end of the horizontal tie line.

When a binary solution is partially vaporized, the component that has the higher vapor pressure is concentrated in the vapor phase, thus producing a difference in composition between the liquid and the equilibrium vapor. This vapor may be condensed, and the vapor obtained by partially vaporizing this condensate is still further enriched in the more volatile component. In **fractional distillation** this process of successive vaporization and condensation is carried out in a fractionating column. Figure 6.8 shows that a solution of 0.75 mole fraction toluene and 0.25 mole fraction benzene boils at 100 °C under 1 atm pressure, as indicated by point a. The equilibrium vapor is richer in the more volatile compound, benzene, and has the composition b. This vapor may be condensed by lowering the temperature along the line bc. If a small fraction of this condensed liquid is vaporized, the first vapor formed will have the composition corresponding to d. This process of vaporization and condensation may be repeated many times, with the result that a vapor fraction rich in benzene is obtained.

Each vaporization and condensation represented by the line *abcde* corresponds to an idealized process in that only a small fraction of the vapor is condensed and only a small fraction of the condensate is revaporized. It is more practical to effect the separation by means of a distillation column, such as a bubble-cap column illustrated in Fig. 6.10.

Each layer of liquid on the plates of the column is equivalent to the boiling liquid in a distilling flask, and the liquid on the plate above it is equivalent to the condenser. The vapor passes upward through the bubble caps, where it is partially condensed in the liquid and mixed with it. Part of the resulting solution is vaporized in this process and is condensed in the next higher layer, while part of the liquid overflows and runs down the tube to the next lower plate. In this way there is a continuous flow of redistilled vapor coming out the top and a continuous flow of recondensed liquid returning to the boiler at the bottom. To make up for this loss of material from the distilling column, fresh solution is fed into the column, usually at the middle. The column is either well insulated or surrounded by a controlled heating jacket so that there will not be too much condensation on the walls. The whole system reaches a steady state in which the composition of the solution on each plate remains unchanged as long as the composition of the liquid in the distilling pot remains unchanged.

A distillation column may alternatively be packed with material that provides efficient contact between liquid and vapor and occupies only a small volume, so that there is free space to permit a large throughput of vapor. Helices of glass, spirals of screen, and different types of packing are used with varying degrees of efficiency.

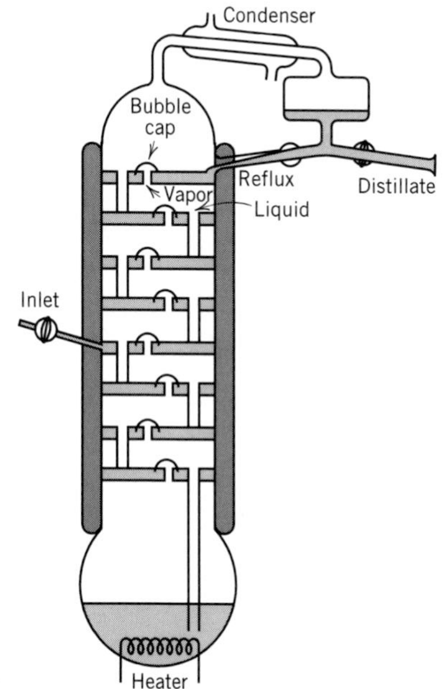

Figure 6.10 Bubble-cap fractionating column.

The efficiency of a column is expressed in terms of the equivalent number of theoretical plates. The number of **theoretical plates** in a column is equal to the number of successive infinitesimal vaporizations at equilibrium required to give the separation that is actually achieved. The number of theoretical plates depends somewhat on the reflux ratio, the ratio of the rate of return of liquid to the top of the column to the rate of distilling liquid off. The number of theoretical plates in a distillation column under actual operating conditions may be obtained by counting the number of equilibrium vaporizations required to achieve the separation actually obtained with the column.

Suppose that in distilling a solution of benzene and toluene with a certain distillation column it is found that distillate of composition g is obtained when the composition of the liquid in the boiler is given by a, in Fig. 6.8. Such a distillation is equivalent to three simple vaporizations and condensations, as indicated by steps abc, cde, and efg. Since the distilling pot itself corresponds to one theoretical plate, the column has two theoretical plates.

6.5 VAPOR PRESSURE OF NONIDEAL MIXTURES AND HENRY'S LAW

Both negative and positive deviations from Raoult's law are found. Figure 6.11*a* shows a system with pronounced negative deviations, and Fig. 6.11*b* shows a system with pronounced positive deviations. Note that in both cases the bubble point line and dew point line are horizontally tangent to each other at the maximum or minimum. Systems with a maximum or minimum are referred to as **azeotropes.** Note that at the azeotropic composition, the vapor has the same composition as the liquid. When a system has a minimum in the vapor pressure plot, it will have a maximum in the boiling point plot, as shown in Fig. 6.12*a*. When a system has a maximum in the vapor pressure plot, it has a minimum in the boiling point plot, as shown in Fig. 6.12*b*. When a system forms an azeotrope, its components cannot be separated by simple fractional distillation. For example, in Fig. 6.12 to the left of the maximum, solutions can be separated into component 2 and the azeotrope by fractional distillation, but pure component 1 cannot be obtained. Azeotropes can be "broken" by distilling at another pressure where the system does not form an azeotrope, or by adding a third component.

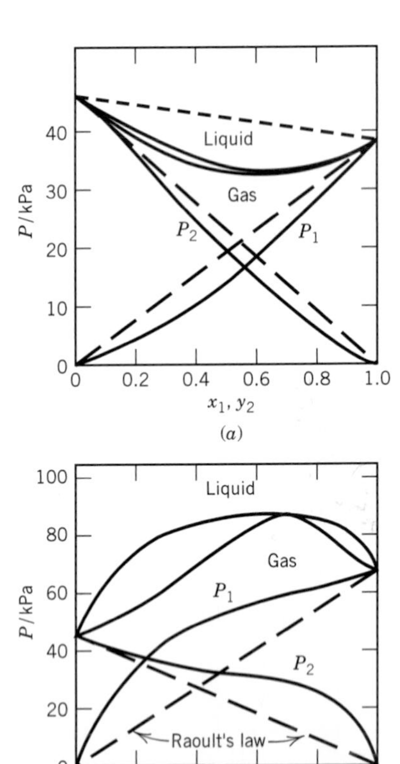

Figure 6.11 (*a*) Liquid mixture with a negative azeotrope: chloroform (1)–acetone (2) at 35.17 °C. (*b*) Liquid mixture with a positive azeotrope: carbon disulfide (1)–acetone (2) at 35.17 °C.

Consider a two-phase region in Fig. 6.12*a*; the number of degrees of freedom is given by $F = 2-2+2 = 2$. We have assumed that *any* two intensive properties may be used to specify the intensive state of the system. However, specifying T and P does not uniquely determine the composition in this case, as can be seen from Fig. 6.12*a*, which is for fixed pressure. In this case, T and x_1 or P and x_1 should be used rather than T and P.

Now how can we understand systems with a minimum in the vapor pressure plot in molecular terms?

The system acetone–chloroform has a minimum in its vapor pressure curve because of the formation of a weak hydrogen bond between the oxygen of the acetone and the hydrogen of the chloroform:

```
    Cl           CH3
    |            |
Cl—C—H···O=C
    |            |
    Cl           CH3
```

A hydrogen bond is a bond between two molecules, or two parts of one molecule, that results from the sharing of a proton between two atoms, one of which is usually fluorine, oxygen, or nitrogen (Section 11.10). Because of hydrogen bonding, the vapor pressures of both components are less than would be expected if there were no interaction and the mixture obeyed Raoult's law.

Positive deviations from Raoult's law result when A–A and B–B interactions are stronger than A–B interactions. If these deviations are large enough, immiscibility results. When the positive deviations from Raoult's law are larger than in Fig 6.11*b*, phase separation occurs. Phase separation occurs when the Gibbs energy of the two-phase system is lower than that of the homogeneous system.

In all of these phase diagrams Raoult's law is approached for a component as its mole fraction approaches unity. As the mole fraction of a component approaches zero, its partial pressure is given by

$$P_i = y_i P = K_i x_i \tag{6.41}$$

which is known as **Henry's law.** These two statements are illustrated in Fig. 6.13. Henry's law results from the circumstance that in sufficiently dilute solutions the environment of the minor component is constant, and its partial pressure is proportional to its mole fraction. The value of the Henry's law constant K_i (see equation 6.41) is obtained by plotting the ratio P_i/x_i versus x_i and extrapolating to $x_i = 0$. Such a plot is shown later in Fig. 6.15.

It is convenient to express the solubilities of gases in liquids by use of Henry's law constants. A few gas solubilities at 25 °C are summarized in this way in Table 6.4. Up to a pressure of 1 bar Henry's law holds within 1 to 3% for many slightly soluble gases.

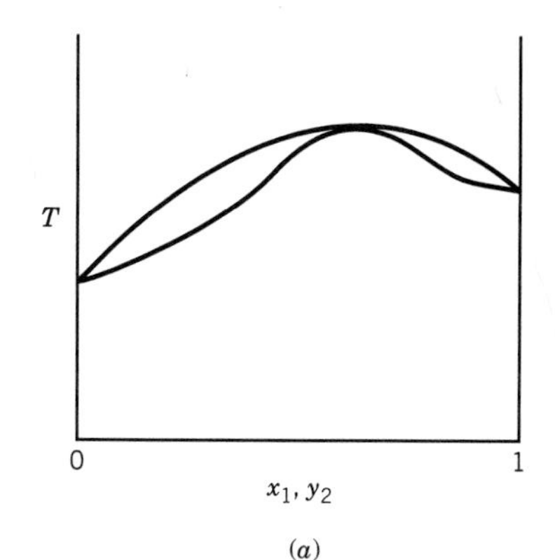

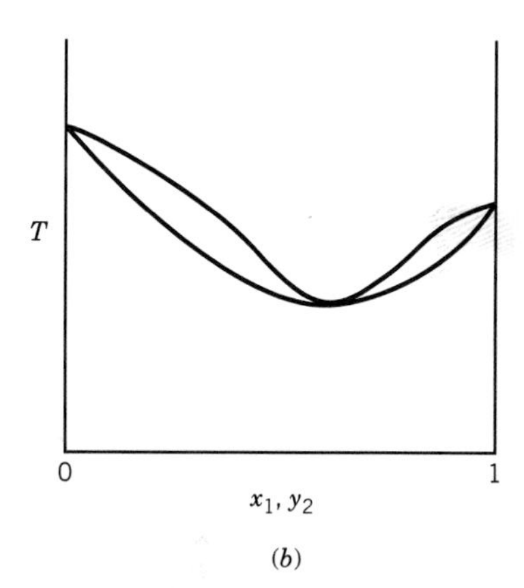

Figure 6.12 (*a*) Boiling point curve at constant pressure for a maximum boiling point azeotrope. (*b*) Boiling point curve at constant pressure for a minimum boiling point azeotrope.

Example 6.8 *The solubility of carbon dioxide in water*

Using the Henry's law constant, calculate the solubility of carbon dioxide in water at 25 °C in moles per liter (represented by using square brackets) at a partial pressure of CO_2 over the solution of 1 bar. Assume that 1 liter of solution contains practically 1000 g of water.

$$K_i = \frac{P_i}{x_i} = 0.167 \times 10^9 \text{ Pa} = \frac{10^5 \text{ Pa}}{[CO_2]}\left([CO_2] + \frac{1000}{18.02}\right)$$

Since $[CO_2]$ may be considered negligible in comparison with the number of moles of water, 1000/18.02,

$$[CO_2] = \frac{(10^5 \text{ Pa})(55.5 \text{ mol L}^{-1})}{0.167 \times 10^9 \text{ Pa}} = 3.32 \times 10^{-2} \text{ mol L}^{-1}$$

The solubility of a gas in liquids usually decreases with increasing temperature, since heat is generally evolved in the solution process. There are numerous exceptions, however, especially with the solvents liquid ammonia, molten silver, and many organic liquids. It is a common observation that a glass of cold water, when warmed to room temperature, shows the presence of many small air bubbles.

The solubility of an unreactive gas is due to intermolecular attractive forces between gas molecules and solvent molecules. There is a good correlation between the boiling points of gases and their solubilities in solvents at room temperature. Substances with low boiling points (He, H_2, N_2, Ne, etc.) have weak intermolecular attractions and are therefore not very soluble in liquids.

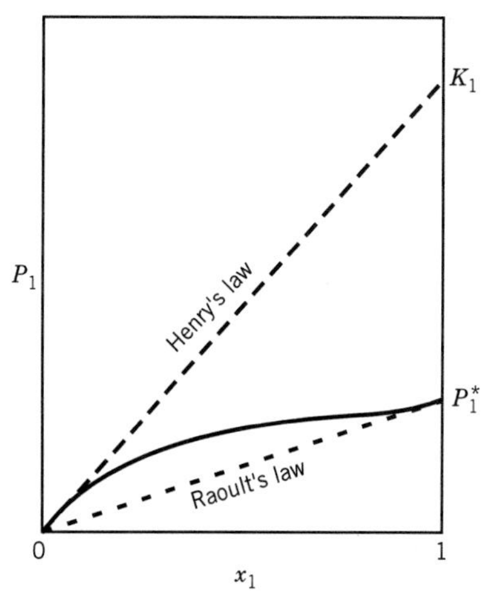

Figure 6.13 Vapor pressure curve for one component of a binary liquid mixture at constant temperature.

Table 6.4 Henry's Law Constants ($K_i/10^9$ Pa) for Gases at 25 °C[a]

	Solvent	
Gas	*Water*	*Benzene*
H_2	7.12	0.367
N_2	8.68	0.239
O_2	4.40	
CO	5.79	0.163
CO_2	0.167	0.0114
CH_4	4.19	0.0569
C_2H_2	0.135	
C_2H_4	1.16	
C_2H_6	3.07	

[a] $K_i = P_i/x_i$. The partial pressure of the gas is expressed in pascals.

The solubility of gases in water is usually decreased by the addition of other solutes, particularly electrolytes. The extent of this "salting out" varies considerably with different salts, but with a given salt the relative decrease in solubility is nearly the same for different gases. The solubility of liquids in water also shows this salting out phenomenon.

Henry's law gives rise to the concept of **dilute real solutions.** If the vapor pressure of a solute follows Henry's law, then in dilute solutions its chemical potential follows an equation very much like that for an ideal solution (equation 6.28). We can show that as follows. Substituting Henry's law into equation 6.20 yields

$$\begin{aligned} \mu_i(\mathrm{l}) &= \mu_i^\circ(\mathrm{g}) + RT \ln \frac{K_i x_i}{P^\circ} \\ &= \mu_i^\circ(\mathrm{g}) + RT \ln \frac{K_i}{P^\circ} + RT \ln x_i \\ &= \mu_i^*(\mathrm{l}) + RT \ln x_i \end{aligned} \tag{6.42}$$

where

$$\mu_i^*(\mathrm{l}) = \mu_i^\circ(\mathrm{g}) + RT \ln \frac{K_i}{P^\circ} \tag{6.43}$$

is the chemical potential of solute i in its standard state in the liquid. In this standard state, the solute at $x_i = 1$ has the same properties as in very dilute solutions, where each molecule is surrounded only by molecules of solvent. This standard state is useful, even if it is hypothetical.

It is important to understand that dilute real solutions behave in a simple way, but they are not ideal solutions. For the binary mixture illustrated in Fig. 6.13, the vapor pressure of component 1 in its standard state for a dilute real solution is equal to K_1.

Example 6.9 ***Proof that if Henry's law holds for the solute, Raoult's law holds for the solvent***

Show that if Henry's law holds for the solute (component 2), Raoult's law holds for the solvent (component 1).

The Gibbs–Duhem equation (4.103) provides a relationship between the differentials of the chemical potentials of components 1 and 2 at constant temperature and pressure:

$$\begin{aligned} \text{If } \mu_2 &= \mu_2^\circ + RT \ln a_2 \\ &= \mu_2^\circ + RT \ln \frac{K x_2}{P_2^*} \end{aligned}$$

$$\mathrm{d}\mu_2 = \frac{RT}{x_2}\,\mathrm{d}x_2$$

Using equation 4.104,

$$\mathrm{d}\mu_1 = -\frac{x_2}{x_1}\,\mathrm{d}\mu_2 = -\frac{RT}{x_1}\,\mathrm{d}x_2$$

Since $x_1 + x_2 = 1$, $\mathrm{d}x_2 = -\mathrm{d}x_1$, then

$$\mathrm{d}\mu_1 = RT\,\frac{\mathrm{d}x_1}{x_1} = RT\,\mathrm{d}\ln x_1$$

$$\mu_1 = RT \ln x_1 + \text{constant}$$

If $x_1 = 1$, then $\mu_1 = \mu_1^\circ$. Therefore,

$$\mu_1 = \mu_1^\circ + RT \ln x_1$$

which can be used to derive Raoult's law. Thus, the range of applicability of Henry's law for the solute is identical to the range of applicability of Raoult's law for the solvent.

Comment:

Phase separation in liquid mixtures (i.e., the formation of gas–liquid interfaces and liquid–liquid interfaces) is of tremendous practical importance, but the equations required to describe and predict the behavior of nonideal systems become rather complicated and require solution by computer. Some of the basic theory of phase equilibria in liquids is given in the References at the end of the chapter.

6.6 ACTIVITY COEFFICIENTS

To make quantitative calculations on nonideal solutions, it is convenient to introduce the **activity coefficient** γ_i. Several different activity coefficients can be defined, but we will start with the one based on deviations from Raoult's law. This is the one that is normally used when both components are liquids. One way to look at this is to simply insert an activity coefficient into equation 6.28 for the chemical potential of an ideal solution:

$$\mu_i(\mathrm{l}) = \mu_i^\circ(\mathrm{l}) + RT \ln \gamma_i x_i \tag{6.44}$$

This equation gives the correct chemical potential for a component of a real solution. Another way of looking at this is to say that we have set the activity of a component equal to $\gamma_i x_i$:

$$a_i = \gamma_i x_i \tag{6.45}$$

Since the activity of a component always approaches its mole fraction as x_i approaches unity, we can see that

$$\gamma_i \to 1 \quad \text{as} \quad x_i \to 1 \tag{6.46}$$

If there are positive deviations from Raoult's law, γ_i is greater than unity; and if there are negative deviations from Raoult's law, γ_i is less than unity.

According to equation 6.25, $a_i = P_i/P_i^*$, where P_i^* is the vapor pressure of component i, and according to equation 6.45, $a_i = \gamma_i x_i$, so that we can set these two expressions equal to each other:

$$a_i = \gamma_i x_i = \frac{P_i}{P_i^*} \tag{6.47}$$

Now we have a way to calculate the activity coefficient of i from experimental data.

$$\gamma_i = \frac{P_i}{x_i P_i^*} \tag{6.48}$$

Thus, the activity coefficient of component i is equal to the ratio of the partial pressure of i above the solution to the partial pressure of i expected from Raoult's law. Since, for ideal gases, $P_i = y_i P$, this equation can also be written as

$$\gamma_i = \frac{y_i P}{x_i P_i^*} \tag{6.49}$$

Example 6.10 ***Calculation of activity coefficients in an ether–acetone solution***

Calculate the activity coefficients for ether (1) and acetone (2) in ether–acetone solutions at 30 °C. The experimental data are given in Table 6.5 and are plotted in Fig. 6.14.

At 0.5 mole fraction acetone, the activity coefficients of the two components are given by

$$\gamma_1 = \frac{P_1}{x_1 P_1^*} = \frac{52.1 \text{ kPa}}{(0.5)(86.1 \text{ kPa})} = 1.21$$

$$\gamma_2 = \frac{P_2}{x_2 P_2^*} = \frac{22.4 \text{ kPa}}{(0.5)(37.7 \text{ kPa})} = 1.19$$

The activity coefficients of both components, calculated in this way at other mole fractions, are summarized in Table 6.5. It will be noted that as the mole fraction of either component approaches unity, its activity coefficient approaches unity, since the vapor pressure asymptotically approaches that given by Raoult's law.

Activity coefficients can also be calculated from deviations from Henry's law. In fact, this is necessary when the solute is a gas above its critical temperature. Activity coefficients of the other component, usually referred to as the solvent, can continue to be based on its deviations from Raoult's law.

When Henry's law is used, the activity coefficient γ_i' is introduced into equation 6.42 for the chemical potential of solute i in a dilute real solution:

$$\mu_i(1) = \mu_i^*(1) + RT \ln \gamma_i' x_i \tag{6.50}$$

Table 6.5 Activity Coefficients for Acetone–Ether Solutions at 30 °C

Mole Fraction Acetone	*Raoult's Law*						*Henry's Law*	
	Ether[a]			*Acetone*			*Acetone*	
x_2	P_1/kPa	$x_1 P_1^*$/kPa	γ_1	P_2/kPa	$x_2 P_2^*$/kPa	γ_2	$K_2 x_2$/kPa	γ_2'
0	86.1	86.1	1.0	0	0	...	0	(1.000)
0.2	71.3	68.9	1.04	12.0	7.5	1.60	15.7	0.77
0.4	58.7	51.7	1.14	19.7	15.1	1.31	31.4	0.63
0.5	52.1	43.1	1.21	22.4	18.9	1.19	39.2	0.57
0.6	44.3	34.4	1.28	25.3	22.7	1.12	47.0	0.54
0.8	26.9	17.3	1.56	31.3	30.1	1.04	62.7	0.50
1.0	0	0	...	37.7	37.7	1.00	78.4	(0.48)

[a]The activity coefficients for ether are those calculated from Raoult's law.

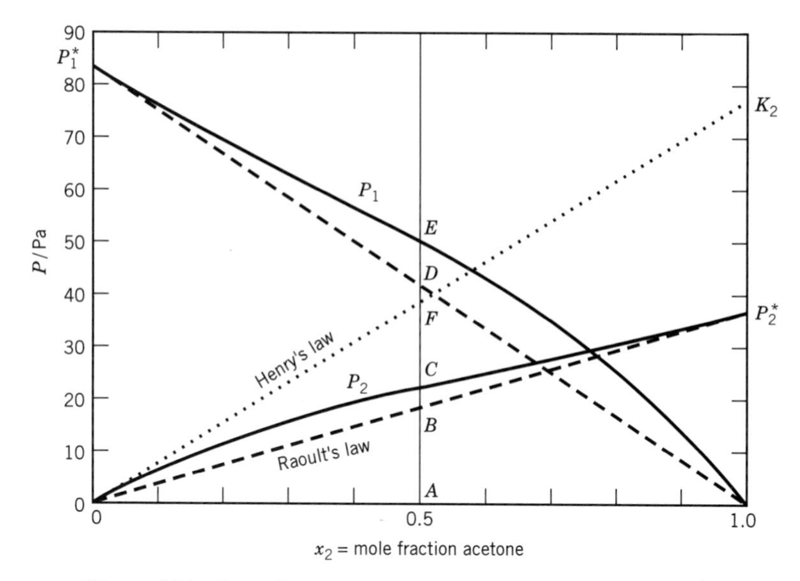

Figure 6.14 Partial pressure of ether–acetone solutions at 30 °C.

If we replace $\mu_i(\mathrm{l})$ with $\mu_i^\circ(\mathrm{g}) + RT\ln(P_i/P^\circ)$ (equation 6.20), and $\mu_i^*(\mathrm{l})$ with $\mu_i^\circ(\mathrm{g}) + RT\ln(K_i/P^\circ)$ (equation 6.43), we obtain

$$P_i = \gamma_i' K_i x_i \tag{6.51}$$

which is the modified form of Henry's law. Thus, the activity coefficient based on Henry's law is calculated from

$$\gamma_i' = \frac{P_i}{x_i K_i} \tag{6.52}$$

Since $P_i = y_i P$, this can also be written

$$\gamma_i' = \frac{y_i P}{x_i K_i} \tag{6.53}$$

This activity coefficient is greater than 1 when there are positive deviations from Henry's law, and it is less than 1 when there are negative deviations from Henry's law. Note that

$$\gamma_i' \to 1 \quad \text{as} \quad x_i \to 0 \tag{6.54}$$

because Henry's law is approached as the concentration of the solute goes to zero. In molecular terms this means that the standard state for the solute, in which x_i for the solute is unity, is one in which each molecule of solute has the same interactions that it experiences in very dilute solutions.

To calculate the activity coefficients for acetone based on deviations from Henry's law, we have to calculate the Henry's law constant K_2 from the data in Table 6.5. This is done by calculating the apparent Henry's law constant $K_2' = P_2/x_2$ versus x_2 and extrapolating to $x_2 = 0$:

$$K_2' = \frac{P_2}{x_2} \tag{6.55}$$

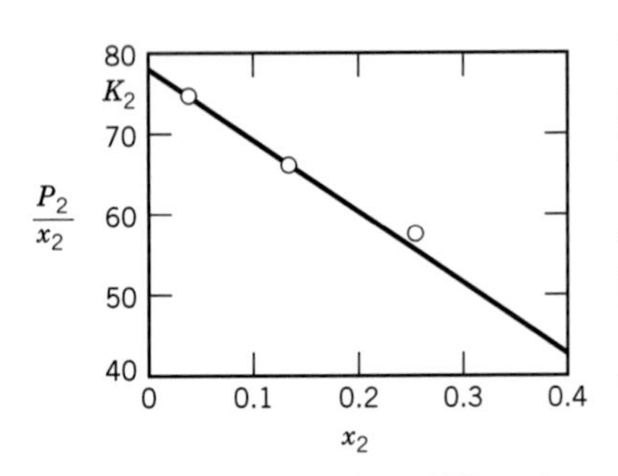

Figure 6.15 Evaluation of Henry's law constant K_2 for acetone in ether–acetone solutions at 30 °C.

The extrapolation of this ratio is illustrated in Fig. 6.15, where values of P_2/x_2 are plotted versus x_2. It is found that the Henry's law constant at infinite dilution (K_2) has a value of 78.4 kPa at 30 °C.

If acetone obeyed Henry's law with this value of the constant over the entire concentration range, its vapor pressure for a solution with 0.5 mole fraction would be given by point F in Fig. 6.14. The actual partial pressure is given by point C.

At 0.5 mole fraction acetone, the activity coefficient of acetone based on deviations from Henry's law is

$$\gamma_2' = \frac{P_2}{K_2 x_2} = \frac{22.4\ \text{kPa}}{(0.5)(78.4\ \text{kPa})} = 0.572 \tag{6.56}$$

The activity coefficients of acetone calculated in this way are summarized in Table 6.5. The activity coefficients for the "solvent" ether remain the same as calculated before on the basis of deviations from Raoult's law.

Since the activity of a component of a real solution is given by P_i/P_i^* (equation 6.25), substituting equation 6.52 yields

$$a_i = \frac{\gamma_i' K_i x_i}{P_i^*} \tag{6.57}$$

The relation between the two types of activity coefficients can be obtained by substituting $a_i = \gamma_i x_i$,

$$\gamma_i = \frac{\gamma_i' K_i}{P_i^*} \quad \text{or} \quad \gamma_i' = \frac{\gamma_i P_i^*}{K_i} \tag{6.58}$$

For dilute real solutions, the solvent is usually treated on the basis of deviations from Raoult's law and the solute is usually treated on the basis of deviations from Henry's law.

For mixtures of two liquids, mole fractions provide a natural way of expressing concentrations, but for other types of solutions, other concentration scales are used. Henry's law may also be written $P_i = K_i m_i$, where m_i is the molal concentration of i (moles per kilogram of solvent), or $P_i = K_i c_i$, where c_i is the molar concentration of i (moles per liter of solution), but we will use only $P_i = K_i x_i$ here.

Although the numerical values of the activity coefficients of acetone depend on which method is employed, the same result is obtained in any thermodynamic calculation using these activity coefficients, independent of method or concentration scale. These thermodynamic calculations involve the comparison of initial and final states, and the standard reference state cancels out.

6.7 COLLIGATIVE PROPERTIES

Now we return to a consideration of ideal solutions and a group of properties that are referred to as **colligative properties.** These properties are freezing point depression, boiling point elevation, osmotic pressure, and the lowering of the vapor pressure by a nonvolatile solute. The Latin root of the word *colligative* means to bind together. The thing that binds these four properties together is that, for ideal solutions, they all depend on the number of particles. Thus, these properties are useful for determining molar masses of solutes. We will consider only two of these properties, freezing point depression and osmotic pressure.

Suppose we want to determine the molar mass of a solute B in a solvent A by the depression of the freezing point of A. Assuming that solution is ideal and that pure crystalline A freezes out of solution, the equation for equilibrium is

$$\begin{aligned}\mu_A^\circ(\text{s}, T) &= \mu_A(\text{l}, T, x_A) \\ &= \mu_A^\circ(\text{l}, T) + RT \ln x_A \end{aligned} \tag{6.59}$$

Thus, at the T at which the two phases are in equilibrium,

$$\begin{aligned}\ln x_A &= \frac{\mu_A^\circ(\text{s}, T) - \mu_A^\circ(\text{l}, T)}{RT} \\ &= -\frac{\Delta_{\text{fus}} G_A^\circ(T)}{RT} \end{aligned} \tag{6.60}$$

where $\Delta_{\text{fus}} G_A^\circ(T)$ is the Gibbs energy of fusion of the solvent at temperature T. Now we make a further assumption that $\Delta_{\text{fus}} C_{P,A}^\circ = 0$; in other words, we assume that $\Delta_{\text{fus}} H_A^\circ$ and $\Delta_{\text{fus}} S_A^\circ$ are independent of temperature in the range near the freezing point. In that case, the Gibbs energy of fusion at temperature T is given by

$$\begin{aligned}\Delta_{\text{fus}} G_A^\circ(T) &= \Delta_{\text{fus}} H_A^\circ - T \Delta_{\text{fus}} S_A^\circ \\ &= \Delta_{\text{fus}} H_A^\circ - T \frac{\Delta_{\text{fus}} H_A^\circ}{T_{\text{fus,A}}} \\ &= \Delta_{\text{fus}} H_A^\circ \left(1 - \frac{T}{T_{\text{fus,A}}}\right) \end{aligned} \tag{6.61}$$

Substituting this equation into equation 6.60 yields

$$\begin{aligned}\ln x_A &= -\left(\frac{\Delta_{\text{fus}} H_A^\circ}{R}\right)\left(\frac{1}{T} - \frac{1}{T_{\text{fus,A}}}\right) \\ &= -\left(\frac{\Delta_{\text{fus}} H_A^\circ}{R}\right)\left(\frac{T_{\text{fus,A}} - T}{T_{\text{fus,A}} T}\right) \end{aligned} \tag{6.62}$$

If the freezing point depression is small, equation 6.81 can be written as follows:

$$\ln x_A = \ln(1 - x_B) = -\frac{\Delta_{\text{fus}} H_A^\circ \Delta T_f}{R T_{\text{fus,A}}^2} \tag{6.63}$$

The logarithmic term may be expanded according to

$$\ln(1 - x) = -x - \frac{1}{2}x^2 - \frac{1}{3}x^3 - \cdots \qquad (-1 < x < 1) \tag{6.64}$$

For dilute solutions the first term is an adequate approximation so that equation 6.63 may be written

$$\Delta T_f = \left(\frac{R T_{\text{fus,A}}^2}{\Delta_{\text{fus}} H_A^\circ}\right) x_B \tag{6.65}$$

In discussing the depression of the freezing point, the concentration of the solute is generally given in terms of molal concentration m (i.e., moles of solute per kilogram of solvent) rather than the mole fraction. The relation between these concentrations is

$$x_B = \frac{m_B}{1/M_A + m_B} \approx m_B M_A \quad (6.66)$$

where M_A is the molar mass of the solvent, and the approximation applies to dilute solutions. The molal concentration m_B has the units mol kg^{-1}, and $1/M_A$ also has the units mol kg^{-1} when SI units are used.

Substituting equation 6.66 into equation 6.65 yields

$$\Delta T_f = \frac{RT^2_{\text{fus,A}} M_A m_B}{\Delta_{\text{fus}} H^\circ_A} = K_f m_B \quad (6.67)$$

where the **freezing point constant** is given by

$$K_f = \frac{RT^2_{\text{fus,A}} M_A}{\Delta_{\text{fus}} H^\circ_A} \quad (6.68)$$

The foregoing relations apply only to dilute solutions. Information about activity coefficients may be obtained by studying the freezing of more concentrated solutions.

Example 6.11 ***Calculation of the freezing point constant for water***

What is the value of the freezing point constant for water? The enthalpy of fusion at 273.15 K is 6.00 kJ mol^{-1}.

$$K_f = \frac{RT^2_{\text{fus,A}} M_A}{\Delta_{\text{fus}} H^\circ_A} = \frac{(8.3145\ \text{J K}^{-1}\ \text{mol}^{-1})(273.15\ \text{K})^2(18.02 \times 10^{-3}\ \text{kg mol}^{-1})}{6000\ \text{J mol}^{-1}}$$
$$= 1.86\ \text{K (mol kg}^{-1})^{-1}$$

According to this value of K_f, 0.1 mol of solute added to 1 kg of water will lower the freezing point 0.186 K, but the relation holds only for dilute solutions. Even a 1-molal solution is too concentrated, and the depression is something less than 1.86 K.

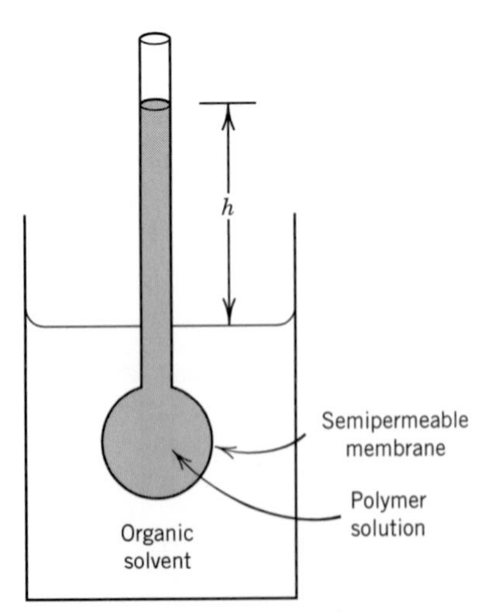

Figure 6.16 Determination of the molar mass of a high polymer by use of osmotic pressure.

When a solution is separated from the solvent by **a semipermeable membrane** that is permeable to solvent but not to solute, the solvent flows through the membrane into the solution, where the chemical potential of the solvent is lower. This process is known as osmosis. This flow of solvent through the membrane can be prevented by applying a sufficiently high pressure to the solution. The **osmotic pressure** Π is the pressure difference across the membrane required to prevent the spontaneous flow of solvent in either direction across the membrane. This is illustrated in Fig. 6.16.

At equilibrium the chemical potential $\mu_1^\circ(P, T)$ of pure solvent at pressure P is equal to the chemical potential of the solvent in the solution at pressure $P + \Pi$.

$$\mu_1^\circ(P, T) = \mu_1(P + \Pi, T, x_1) \quad (6.69)$$

The osmotic pressure Π that is applied to the solution exactly compensates for the lowering of the chemical potential of the solvent that is caused by the solute. For an ideal solution, equation 6.69 may be written

$$\mu_1^\circ(P, T) = \mu_1^\circ(P + \Pi, T) + RT \ln x_1 \quad (6.70)$$

where $\mu_1^\circ(P + \Pi, T)$ is the chemical potential of the pure solvent at temperature T and pressure $P + \Pi$. According to equation 4.38,

$$d\mu_1 = \overline{V}_1 \, dP \qquad \text{(constant } T \text{ and composition)} \tag{6.71}$$

where $\overline{V}_1$ is the partial molar volume of the solvent. Thus, the effect on the chemical potential of the solvent of raising the pressure is given by

$$\mu_1^\circ(P + \Pi, T) = \mu_1^\circ(P, T) + \int_P^{P+\Pi} \overline{V}_1 \, dP \tag{6.72}$$

Assuming that $\overline{V}_1$ is constant, we obtain

$$\mu_1^\circ(P + \Pi, T) = \mu_1^\circ(P, T) + \overline{V}_1^* \Pi \tag{6.73}$$

Substituting equation 6.73 into equation 6.70 yields

$$\overline{V}_1^* \Pi = -RT \ln x_1 = -RT \ln(1 - x_2) \tag{6.74}$$

This equation is, of course, applicable only to ideal solutions, since equation 6.70 has been used in the derivation.

At a sufficiently low mole fraction of the solute, the logarithmic term may be expanded according to equation 6.64. When only the first term in the series is retained, equation 6.74 becomes

$$\overline{V}_1^* \Pi = RT x_2 \tag{6.75}$$

Since the solution is dilute, $x_2 = n_2/(n_1 + n_2) \cong n_2/n_1$ and $\overline{V}_1^* = V/n_1$, where V is the volume of the solution, and n_2 is the amount of solute, $n_2 = m_2/M_2$, where m_2 represents mass of solute and M_2 is molar mass. Thus, equation 6.75 may be written

$$\Pi V = n_2 RT \tag{6.76}$$

or

$$\Pi = \frac{m_2}{V} \frac{RT}{M_2} \tag{6.77}$$

where m_2/V is the concentration in mass per unit volume, and M_2 is the molar mass of the solute. This is the approximate equation that van't Hoff found empirically. It is evident from the approximations introduced why this equation cannot hold for concentrated solutions.

To represent osmotic pressure data on high polymers over a wider range of concentration it is necessary to add terms in higher powers of the concentration, as in the virial equation for gases (Section 1.5):

$$\frac{\Pi}{C} = \frac{RT}{M} + BC + \cdots \tag{6.78}$$

where C is the concentration (mass per volume). Synthetic polymers have a distribution of molar masses, and in the chapter on macromolecules (Chapter 21) we will see that M in equation 6.78 is the number average molar mass.

Example 6.12 ***Determination of the number average molar mass of a sample of polystyrene***

A solution of polystyrene in benzene contains 10 g/L. The equilibrium height of the column of solution (density 0.88 g cm^{-3}) in the osmometer (Fig. 6.16) corrected for capillary rise is 11.6 cm at 25 °C. What is the number average molar mass of the polystyrene, assuming the solution is ideal?

$$P = h\rho g = (0.116 \text{ m})(0.88 \times 10^3 \text{ kg m}^{-3})(9.8 \text{ m s}^{-2})$$
$$= 1000 \text{ Pa}$$
$$\Pi = \frac{mRT}{MV}$$
$$M = \frac{mRT}{\Pi V}$$
$$M = \frac{(10 \text{ g})(8.3145 \text{ J K}^{-1} \text{ mol}^{-1})(298 \text{ K})}{(10^3 \text{ Pa})(10^{-3} \text{ m}^3)}$$
$$= 24.8 \times 10^3 \text{ g mol}^{-1}$$

6.8 TWO-COMPONENT SYSTEMS CONSISTING OF SOLID AND LIQUID PHASES

First we will consider systems in which the components are completely miscible in the liquid state and completely immiscible in the solid state, so that only the pure solid phases separate out on cooling solutions. Such a phase diagram is illustrated in Fig. 6.17. When molten bismuth or molten cadmium is cooled, the plot of temperature versus time has a nearly constant slope. At the temperature at which the solid crystallizes out, however, the cooling curve becomes horizontal if the cooling is slow enough. The halt in the cooling curve results from the heat evolved when the liquid solidifies. This is shown by the cooling curves for bismuth (labeled 0% Cd) and cadmium in Fig. 6.17 at 273 and 323 °C, respectively.

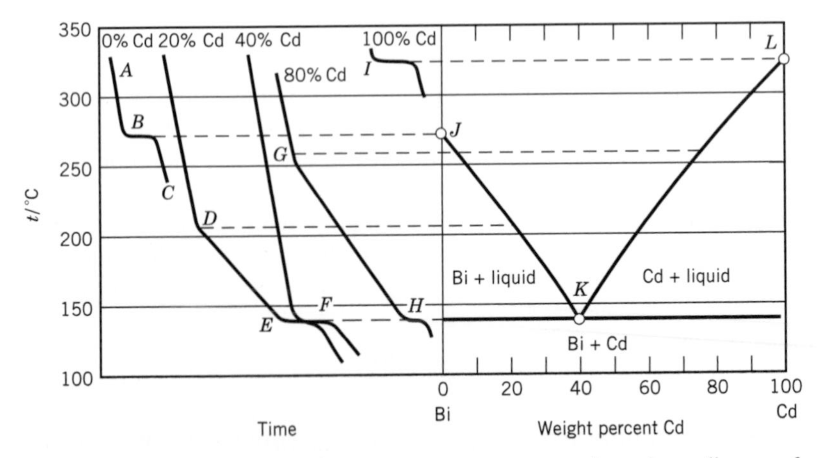

Figure 6.17 Cooling curves and the temperature–concentration phase diagram for the system bismuth–cadmium at constant pressure. The eutectic point is at K.

When a *solution* is cooled, there is a change in slope of the cooling curve at the temperature at which one of the components begins to crystallize out. The change in slope is due to the evolution of heat by the progressive crystallization of the solid as the solution is cooled and to the change in heat capacity. Such changes in slope are evident in the cooling curves for 20% cadmium and 80% cadmium. These curves also show horizontal sections, both at 140 °C. At this temperature both solid cadmium and solid bismuth crystallize out at the same time. The temperature at which this occurs is called a **eutectic temperature.** A solution of cadmium and bismuth containing 40% cadmium shows a single plateau F at 140 °C, and so this is the eutectic composition. The eutectic is not a phase; it is a mixture of two solid phases and has a fine grain structure.

The temperatures at which new phases appear, as indicated by the cooling curves, are then transferred to the temperature–composition diagram, as shown at the right in Fig. 6.17. In the area above JKL there is one liquid phase. For a two-component system without chemical reaction at constant pressure, the phase rule is $F = 2 - p + 1 = 3 - p$. Therefore, if there is a single phase, there are two degrees of freedom, which can be taken as temperature and one mole fraction. Thus, in the solution region above JKL, temperature and composition may be varied without changing the number of phases.

Along JK, bismuth freezes out; along KL, cadmium freezes out. Thus, in the area under JK and down to the eutectic temperature K there are two phases: solid bismuth and a solution having a composition that is determined by the temperature. Similarly, under KL and down to the eutectic temperature K, two phases are in equilibrium: solid cadmium and a solution having a composition given by KL. In these two regions there is one degree of freedom. Thus, in these regions only the temperature or the composition of the liquid need be specified. If the temperature is specified, the composition of the liquid can be read off of the curved line. If the composition of the liquid is given, the equilibrium temperature can be read off the curved line. Within the two-phase regions there are horizontal tie lines (not shown), and for any given point in either region the lever rule (Section 6.4) can be used to calculate the relative amounts of the two phases.

At the eutectic point K there are three phases: solid bismuth, solid cadmium, and liquid solution containing 40% cadmium. Then $F = 3 - 3 = 0$, so this is an invariant point. There is only one temperature and one composition of solution at which these three phases can exist together at equilibrium at a given constant pressure.

The area below the eutectic temperature K is a two-phase area in which solid bismuth and solid cadmium are present, and $F = 3-2 = 1$. Only the temperature need be specified to describe the system completely at a given constant pressure. The ratio of bismuth to cadmium may change, but there is only a mixture of pure solid bismuth and pure solid cadmium; therefore, there is no need to specify any concentration.

It is evident from Fig. 6.17 that the addition of cadmium lowers the freezing point of bismuth along line JK, and that the addition of bismuth lowers the freezing point of cadmium along line LK. Alternatively, we may consider that JK is the solubility curve for bismuth in liquid cadmium, and LK is the solubility curve for cadmium in liquid bismuth. If the solutions are ideal and if the phases that separate are pure solids, solubilities may be calculated.

Earlier, in Section 6.7, we discussed the determination of the molar mass of solute B in solvent A in a liquid solution in equilibrium with pure solid A, and derived equation 6.62, which gives the mole fraction of the solvent A as a function of temperature. We can change our point of view and think of A as the solute and B as the solvent. Thus, the solubility x_A of A in solvent B is given by

$$x_A = \exp\left[-\frac{\Delta_{fus}H_A^\circ}{R}\left(\frac{1}{T} - \frac{1}{T_{fus,A}}\right)\right] \tag{6.79}$$

The remarkable thing about this equation is that it does not contain any parameters for the solvent B. **Thus, the solubility of A is the same in all solvents that form ideal solutions.** It is evident that a high melting point and a large enthalpy of fusion lead to a low solubility. Since the enthalpy of fusion is positive, the ideal solubility increases as the temperature increases.

The components of a binary system may react to form a solid compound that exists in equilibrium with liquid over a range of composition. If the formation of a compound leads to a maximum in the temperature–composition diagram, as illustrated by Fig. 6.18 for the zinc–magnesium system, we say there is a **congruently melting compound.** The composition that corresponds to the maximum temperature is the composition of the compound. On the mole percent scale such maxima may be achieved at 50%, 33%, 25%, and so on, corresponding to integer ratios of the components of 1:1, 1:2, 1:3, and so on. Figure 6.18 looks very much like two phase diagrams of the type we have discussed placed side by side, but there is a difference. The liquid curve has a horizontal tangent (zero slope) at the melting point of the congruently melting compound $MgZn_2$, while the slope is not zero at the melting points of the pure components. Thus, additions of small amounts of zinc and magnesium to the compound will not lower the melting or freezing point.

Example 6.13 ***Phase transitions in a magnesium–zinc system***

Six-tenths mol of Mg and 0.40 mol of Zn are heated to 650 °C, represented by point J in Fig. 6.18. Describe what happens when this solution is cooled to 200 °C, as indicated by

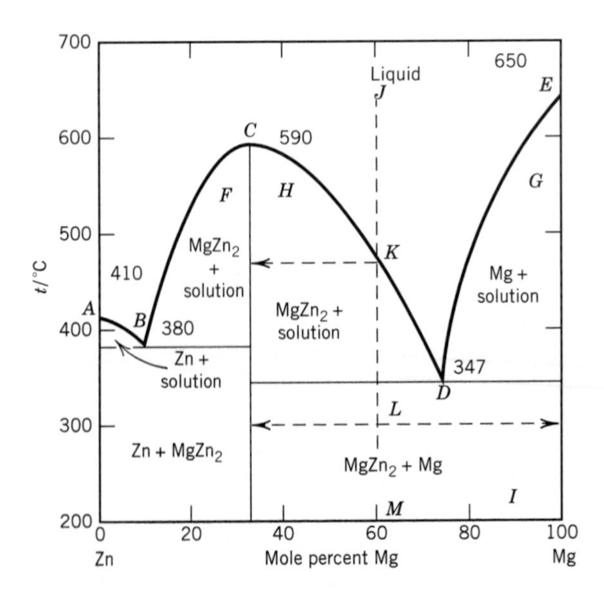

Figure 6.18 Temperature–composition diagram, showing a maximum for the system zinc–magnesium.

the vertical dashed line. (The experiment would have to be done in an inert atmosphere to prevent oxidation by air.)

At 470 °C point K is reached and solid $MgZn_2$ separates from solution. The freezing point is gradually lowered as the solution becomes richer in Mg. Finally, at 347 °C, when the liquid is 74 mol% in Mg and 26 mol% in Zn, the whole solution freezes, and solid $MgZn_2$ and solid Mg come out together. From this temperature down there is no further change in the phases. At all temperatures below 347 °C there are pure solids Mg and $MgZn_2$.

Often pure solid freezes out of a solution, but for some systems a solid solution freezes out. A continuous series of solid solutions may be formed, as illustrated in Fig. 6.19 for platinum and gold. The two lines in this diagram give the compositions of the liquid solutions (upper line) and solid solutions (lower line) that are in equilibrium with each other. When these diagrams are studied, it is convenient to remember that the liquid phase is richer in that component of the mixture that has the lower melting point.

Above the upper line of Fig. 6.19 the two metals exist in liquid solution; below the lower line the two metals exist in solid solutions. The upper curve is the freezing point curve for the liquid, and the lower one is the melting point curve for the solid. The space between the two curves represents mixtures of the two: one liquid solution and one solid solution in equilibrium. For example, a mixture containing 50 mol% gold and 50 mol% platinum, when brought to equilibrium at 1400 °C, will consist of two phases, a solid solution containing 70 mol% platinum and a liquid solution containing 28 mol% platinum. If the original mixture contained 60 mol% platinum, there would still be the same two liquid and solid solutions at 1400 °C of the same compositions, 70 and 28 mol%, but there would be a relatively greater amount of the solid solution that contains 70 mol% platinum.

The fractional crystallization of solid solutions is seriously complicated by the fact that the attainment of equilibrium is much slower in solid solutions than in liquid solutions. It takes a considerable length of time, particularly at low temperatures, for a change in concentration at the surface to affect the concentration at a point in the interior of the solid solution.

In view of the use of the freezing point as a criterion of purity, it is important to note that when solid solutions are formed the freezing point may be *raised* by the presence of the other component.

Figure 6.19 is analogous to the phase diagram for two miscible liquids and vapor, as shown in Fig. 6.8. For substances forming ideal solid solutions, the phase diagram may be calculated theoretically. Systems exhibiting nonideal solid solution behavior may show maxima or minima in their melting curves (analogous to Fig. 6.12) that have nothing to do with the formation of compounds.

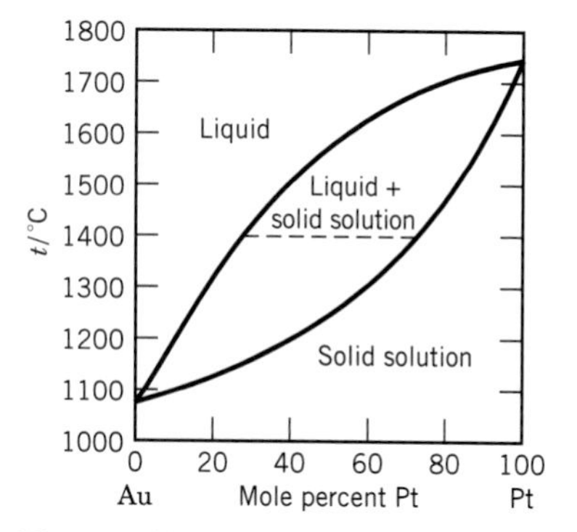

Figure 6.19 Phase diagram for gold–platinum showing solid solutions.

6.9 SPECIAL TOPIC: EFFECT OF SURFACE TENSION ON THE VAPOR PRESSURE

The surface tension γ was introduced in Section 2.5 in connection with surface work. The surface tension of a liquid may be measured by a variety of methods. For example, since the equilibrium shape of liquid surfaces is determined by a balance of surface tension and gravitational forces, analysis of drop or bubble shape may be used to determine surface tension. The rise of liquid in a capillary or the pull on a thin vertical plate partially immersed in the liquid may be determined and used to calculate the surface tension quite accurately. Less accurate values of the surface tension may be obtained from measurements on moving liquid surfaces. These methods include studies of liquid jets, ripples, drop weight, and the force required to rupture a surface. The surface tensions of some liquids are given in Table 6.6.

A curved surface of a liquid, or a curved interface between phases, exerts a pressure so that the pressure is higher in the phase on the concave side of the interface. This is analogous to the fact that the pressure inside of a rubber balloon is higher than the atmospheric pressure because of the pressure exerted by the tension of the rubber. However, the difference between a sheet of rubber and the surface of a pure substance is that the tension of the sheet of rubber is roughly proportional to the distance stretched, but the surface tension of a pure substance is independent of area.

To derive the relation between the radius of curvature of a surface, the pressure difference, and the surface tension, we consider a spherical droplet of a pure liquid in contact with its vapor in a closed container at temperature T, as illustrated in Fig. 6.20.

The solid line in Fig. 6.20 shows the equilibrium position of the surface, and the dashed line shows the effect of an infinitesimal expansion of the droplet. Since the volume of the system and the temperature are specified, the Helmholtz energy A is the thermodynamic potential to use. Since there are two phases, we first write the fundamental equations at constant temperature for the two phases separately:

$$(\mathrm{d}A_\mathrm{l})_T = -P_\mathrm{l}\,\mathrm{d}V_\mathrm{l} + \mu_\mathrm{l}\,\mathrm{d}n_\mathrm{l} + \gamma\,\mathrm{d}A_\mathrm{s} \tag{6.80}$$

$$(\mathrm{d}A_\mathrm{v})_T = -P_\mathrm{v}\,\mathrm{d}V_\mathrm{v} + \mu_\mathrm{v}\,\mathrm{d}n_\mathrm{v} \tag{6.81}$$

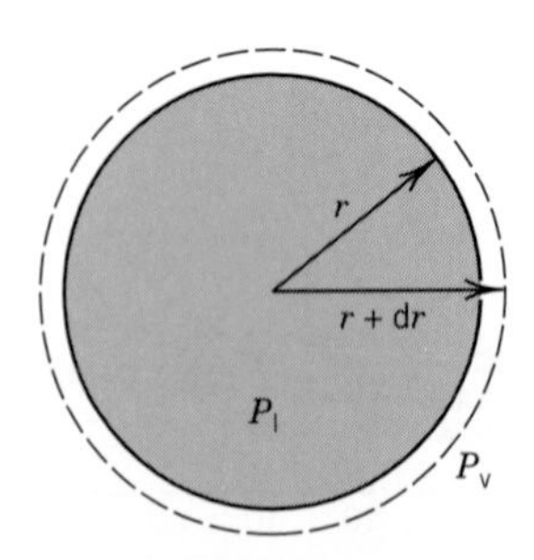

Figure 6.20 Spherical droplet of a pure liquid in contact with its vapor in a closed container at temperature T.

Table 6.6 Surface Tensions of Some Liquids

Substance	$t/°\mathrm{C}$	$\gamma/\mathrm{mN\ m^{-1}}$
Acetone	20	23.7
Benzene	20	28.88
Ethanol	20	22.8
Diethyl ether	20	16.96
Glycerol	20	63.4
Water	0	75.7
	15	73.5
	25	72.0
	50	67.9
	100	58.8

Source: G. W. C. Kaye and T. H. Laby, *Tables of Physical and Chemical Constants,* 1973. © Reprinted by permission of Pearson Education, Inc., Upper Saddle River, NJ.

where A_s is the area of the surface of the droplet. In this system, the surface area is related to the volume V_l of the liquid and to the amount n_l of the liquid, and so surface area is really not an independent variable. We can eliminate A_s from equation 6.80 by expressing it in terms of V_l or n_l. In this section we will eliminate A_s by expressing it in terms of V_l, and later we will learn something different by expressing it in terms of n_l. The volume of liquid and the area of the spherical surface are given by

$$V_l = \frac{4}{3}\pi r^3 \qquad \text{and} \qquad A_s = 4\pi r^2 \tag{6.82}$$

so that

$$\mathrm{d}V_l = 4\pi r^2\,\mathrm{d}r \qquad \text{and} \qquad \mathrm{d}A_s = 8\pi r\,\mathrm{d}r \tag{6.83}$$

Thus

$$\mathrm{d}A_s = \frac{2}{r}\,\mathrm{d}V_l \tag{6.84}$$

If we insert this in equation 6.80 and add equations 6.80 and 6.81, we obtain the following fundamental equation for the two-phase system:

$$(\mathrm{d}A)_T = (P_v - P_l)\,\mathrm{d}V_l + \frac{2\gamma}{r}\,\mathrm{d}V_l \tag{6.85}$$

where $A = A_l + A_v$. We have used the relation $\mathrm{d}V_v = -\mathrm{d}V_l$, and the fact that the chemical potential terms cancel since $\mu_l = \mu_v$ at equilibrium and $\mathrm{d}n_l = -\mathrm{d}n_v$. Since the system is at equilibrium, $(\mathrm{d}A)_{T,V} = 0$, and

$$P_l - P_v = \frac{2\gamma}{r} \tag{6.86}$$

Thus the pressure inside the droplet is higher than the pressure of the vapor. When the radius of curvature is increased to infinity, this pressure difference disappears, as it must for a planar surface.

Equation 6.86 can be applied to a bubble in a liquid, but to do that we have to change the sign of the radius of curvature. Thus, for a bubble in a liquid,

$$P_l - P_v = -\frac{2\gamma}{r} \qquad \text{or} \qquad P_v - P_l = \frac{2\gamma}{r} \tag{6.87}$$

where r is the radius of the bubble. In this case, the pressure in the bubble is higher than the pressure in the liquid; the way to remember this is that the pressure is always greater on the concave side of the surface.

Since the liquid inside of a droplet is under pressure, we want to calculate what this does to the equilibrium vapor pressure. There we expressed A_s in terms of the volume of the liquid, but now we will express it in terms of the amount of liquid n_l. The volume of liquid in the droplet is equal to the molar volume of the liquid $\overline{V}_l$ times the amount n_l, and so equation 6.84 can be written

$$\mathrm{d}A_s = (2/r)\,\mathrm{d}V_l = (2/r)\overline{V}_l\,\mathrm{d}n_l \tag{6.88}$$

In the preceding discussion we used the fundamental equation for A because we wanted to consider the system at constant volume. Now we will use the fundamental equation for G because we want to consider the system in Fig. 6.20 at equilibrium at a specified pressure P of the vapor phase. First we write the fundamental equations separately for the two phases. The fundamental equation for G for the liquid phase is

$$\begin{aligned}(\mathrm{d}G_{\mathrm{l}})_{T,P} &= \mu_{\mathrm{l}}' \,\mathrm{d}n_{\mathrm{l}} + (2\gamma/r)\overline{V}_{\mathrm{l}}\,\mathrm{d}n_{\mathrm{l}} \\ &= (\mu_{\mathrm{l}}' + 2\gamma\overline{V}_{\mathrm{l}}/r)\,\mathrm{d}n_{\mathrm{l}} \\ &= \mu_{\mathrm{l}}\,\mathrm{d}n_{\mathrm{l}}\end{aligned} \tag{6.89}$$

where the chemical potential μ_{l} of the liquid is given by

$$\mu_{\mathrm{l}} = \mu_{\mathrm{l}}' + 2\gamma\overline{V}_{\mathrm{l}}/r \tag{6.90}$$

and the chemical potential is defined by

$$\mu_i = \left(\frac{\partial G}{\partial n_i}\right)_{T,P,\{n_{j\neq i}\}} \tag{6.91}$$

as in equation 4.39. A prime has been put on μ_{l}' in equation 6.89 so that μ_{l} can have its usual meaning.

The fundamental equation for G for the vapor phase is simply

$$(\mathrm{d}G_{\mathrm{v}})_{T,P} = \mu_{\mathrm{v}}\,\mathrm{d}n_{\mathrm{v}} \tag{6.92}$$

Adding equations 6.89 and 6.92 yields the fundamental equation for the two-phase system:

$$(\mathrm{d}G)_{T,P} = (\mu_{\mathrm{l}}' + 2\gamma\overline{V}_{\mathrm{l}}/r)\,\mathrm{d}n_{\mathrm{l}} + \mu_{\mathrm{v}}\,\mathrm{d}n_{\mathrm{v}} \tag{6.93}$$

Since the system is at equilibrium, this differential is equal to zero, and

$$\mu_{\mathrm{v}} = \mu_{\mathrm{l}}' + 2\gamma\overline{V}_{\mathrm{l}}/r = \mu_{\mathrm{l}} \tag{6.94}$$

since $\mathrm{d}n_{\mathrm{l}} = -\mathrm{d}n_{\mathrm{v}}$. The chemical potentials are the same in the two phases, even though they are at different pressures. Assuming that the vapor is an ideal gas,

$$\mu_{\mathrm{v}}^{\circ} + RT\,\ln(P/P^{\circ}) = \mu_{\mathrm{l}}' + 2\gamma\overline{V}_{\mathrm{l}}/r \tag{6.95}$$

This equation is general, and P corresponds to the radius r of the droplet. As the radius increases, the vapor pressure P of the droplet approaches the vapor pressure P^* determined for a flat surface. For a flat surface,

$$\mu_{\mathrm{v}}^{\circ} + RT\,\ln(P^*/P^{\circ}) = \mu_{\mathrm{l}}' \tag{6.96}$$

Subtracting equation 6.96 from equation 6.95 yields

$$RT\,\ln(P/P^*) = 2\gamma\overline{V}_{\mathrm{l}}/r \tag{6.97}$$

which is referred to as the **Kelvin equation.** The Kelvin equation is based on the assumption that the surface tension is independent of the radius of curvature; therefore, it is not accurate when the radius of curvature becomes quite small. Nevertheless, the Kelvin equation is helpful in understanding the nucleation of condensation and boiling. Figure 6.21 shows the magnitude of the effect on the vapor pressure of water at 25 °C.

We can think of small droplets as having higher vapor pressures than liquids with planar interfaces because surface molecules are not drawn into the interior by so many near neighbors. For a vapor to condense in the absence of foreign surfaces it is necessary for small clusters of molecules to form and to grow and finally coalesce to form the bulk phase. This does not happen if the pressure of the vapor is only slightly higher than the equilibrium vapor pressure, because the very small droplets that are formed first have a higher vapor pressure. However, when the pressure has been increased sufficiently over the equilibrium value, general condensation of droplets occurs.*

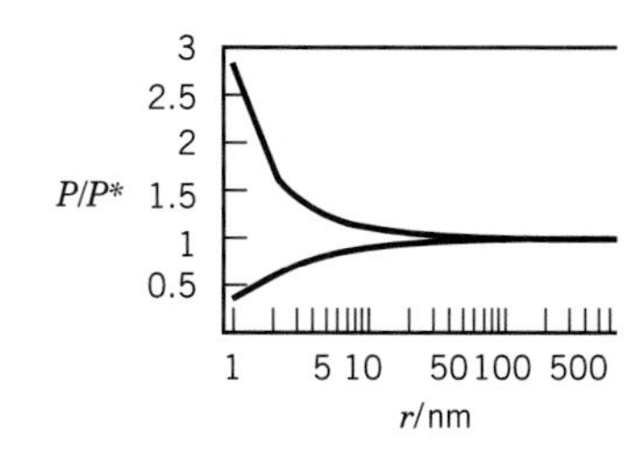

Figure 6.21 Effect of radius of curvature of a surface on the vapor pressure of water at 25 °C. (See Computer Problem 6.H.)

Example 6.14 ***The size of a water droplet when condensation starts***

Water vapor is rapidly cooled to 25 °C to find the degree of supersaturation required to nucleate water droplets spontaneously. It is found that the vapor pressure of water must be four times its equilibrium vapor pressure at 25 °C. (*a*) Calculate the radius of a stable water droplet formed at this degree of supersaturation. (*b*) How many water molecules are there in the droplet?

$$(a) \quad r = \frac{2\overline{V}_1\gamma}{RT\ln(P/P_{\text{sat}})} = \frac{2(18\times10^{-6}\text{ m}^3\text{ mol}^{-1})(0.071\,97\text{ N m}^{-1})}{(8.314\text{ J K}^{-1}\text{ mol}^{-1})(298\text{ K})\ln 4} = 0.75\text{ nm}$$

$$(b) \quad N = \frac{\frac{4}{3}\pi r^3\rho}{M/N_A} = \frac{\frac{4}{3}\pi(0.75\times10^{-9}\text{ m})^3(1\times10^3\text{ kg m}^{-3})}{(18\times10^{-3}\text{ kg mol}^{-1})/(6.022\times10^{23}\text{ mol}^{-1})} = 59$$

In view of the assumption that the surface tension is independent of the radius of curvature, these values must be considered approximations.

Since the radius of curvature r is a signed quantity, the Kelvin equation can be used to calculate the vapor pressure of a concave surface as in a bubble in a liquid by simply changing the sign of one side of the equation. We can think of the surface of the liquid in a capillary that it wets or in a small bubble as having a lower vapor pressure than the bulk liquid because molecules in the surface are drawn into the interior by more near neighbors than in a flat surface. The Kelvin equation helps us to understand why liquids have a tendency to superheat at their boiling points. If a small bubble starts to form at the boiling point, equation 6.97 (with a sign change) is not satisfied, and the bubble will be squeezed out of existence by the force of surface tension. At a temperature above the boiling point, the vapor pressure will be higher enough that a bubble of a certain radius will be thermodynamically stable.

*Similar phenomena are involved in the freezing of liquids. Water without impurities or foreign surfaces may be cooled to −40 °C before nucleation begins spontaneously.

Six Key Ideas in Chapter 6

1. For a single phase of a pure substance, the chemical potential can be represented as a surface in μ–P–T space. The slope of the surface in the temperature direction is equal to the negative of the molar entropy. The slope in the pressure direction is equal to the molar volume. The projection onto the P–T plane of the intersections of these surfaces is the phase diagram.
2. If the temperature and pressure are both changed in such a way as to keep the chemical potentials of two phases equal to each other, the rate of change of pressure with temperature is given by the Clapeyron equation. When the vapor obeys the ideal gas law, the rate of change of pressure gives the enthalpy of vaporization.
3. A curved surface of a liquid, or a curved interface between phases, exerts a pressure so that the pressure is higher in the phase on the concave side of the interface. As a result, small droplets have higher vapor pressures than flat surfaces.
4. For ideal liquid–liquid solutions, the partial pressure of a component is equal to the mole fraction of that component times the vapor pressure of the pure component (Raoult's law). For nonideal solutions, Raoult's law is approached for a component as its mole fraction approaches unity. As the mole fraction of a component in a nonideal solution approaches zero, its partial pressure becomes proportional to its mole fraction (Henry's law).
5. Activity coefficients represent the deviation of the chemical potential of a component from Raoult's law or Henry's law.
6. The colligative properties freezing point depression, boiling point elevation, osmotic pressure, and vapor pressure lowering are used to determine molar masses of solutes. They are useful for this purpose because they all depend on the concentrations of particles.

REFERENCES

J. Prausnitz, T. Anderson, E. Grens, C. Eckert, R. Hsieh, and J. O'Connell, *Computer Calculations for Multicomponent Vapor–Liquid and Liquid–Liquid Equilibria.* Englewood Cliffs, NJ: Prentice Hall, 1980.

J. S. Rowlinson, *Liquids and Liquid Mixtures.* London: Butterworths, 1969.

S. I. Sandler, *Chemical and Engineering Thermodynamics.* Hoboken, NJ: Wiley, 1999.

J. M. Smith, H. C. Van Ness, and M. M. Abbott, *Introduction to Chemical Engineering Thermodynamics,* 5th ed. New York: McGraw-Hill, 1996.

J. W. Tester and M. Modell, *Thermodynamics and Its Applications*. Upper Saddle River, NJ: Prentice-Hall, 1997.

H. C. Van Ness and M. M. Abbott, *Classical Thermodynamics of Nonelectrolyte Solutions.* New York: McGraw-Hill, 1982.

PROBLEMS

Problems marked with an icon may be more conveniently solved on a personal computer with a mathematical program.

6.1 The boiling point of hexane at 1 atm is 68.7 °C. What is the boiling point at 1 bar? Given: The vapor pressure of hexane at 49.6 °C is 53.32 kPa.

6.2 What is the boiling point of water 2 miles above sea level? Assume that the atmosphere follows the barometric formula (equation 1.46) with $M = 0.0289$ kg mol^{-1} and $T = 300$ K. Assume the enthalpy of vaporization of water is 44.0 kJ mol^{-1} independent of temperature.

6.3 The barometric equation 1.46 and the Clausius–Clapeyron equation 6.8 can be used to estimate the boiling point of a liquid at a higher altitude. Use these equations to derive a single equation to make this calculation. Use this equation to solve Problem 6.2.

6.4 Liquid mercury has a density of 13.690 g cm^{-3}, and solid mercury has a density of 14.193 g cm^{-3}, both being measured at the melting point, −38.87 °C, at 1 bar pressure. The heat of fusion is 9.75 J g^{-1}. Calculate the melting points of mercury under a pressure of (*a*) 10 bar and (*b*) 3540 bar. The observed melting point under 3540 bar is −19.9 °C.

6.5 From the $\Delta_f G°$ of $Br_2(g)$ at 25 °C, calculate the vapor pressure of $Br_2(l)$. The pure liquid at 1 bar and 25 °C is taken as the standard state.

6.6 Calculate $\Delta G°$ for the vaporization of water at 0 °C using data in Table C.2 and assuming that $\Delta H°$ for the vaporization is independent of temperature. Use $\Delta G°$ to calculate the vapor pressure of water at 0 °C.

6.7 The change in Gibbs energy for the conversion of aragonite to calcite at 25 °C is −1046 J mol^{-1}. The density of aragonite is 2.93 g cm^{-3} at 25 °C, and the density of calcite is 2.71 g cm^{-3}. At what pressure at 25 °C would these two forms of $CaCO_3$ be in equilibrium?

6.8 *n*-Propyl alcohol has the following vapor pressures:

t/°C	40	60	80	100
P/kPa	6.69	19.6	50.1	112.3

Plot these data so as to obtain a nearly straight line, and calculate (*a*) the enthalpy of vaporization, (*b*) the boiling point at 1 bar, and (*c*) the boiling point at 1 atm.

6.9 For uranium hexafluoride the vapor pressure (in Pa) for the solid and liquid are given by

$$\ln P_s = 29.411 - 5893.5/T$$
$$\ln P_l = 22.254 - 3479.9/T$$

Calculate the temperature and pressure of the triple point.

6.10 The heats of vaporization and of fusion of water are 2490 J g^{-1} and 333.5 J g^{-1} at 0 °C. The vapor pressure of water at 0 °C is 611 Pa. Calculate the sublimation pressure of ice at −15 °C, assuming that the enthalpy changes are independent of temperature.

6.11 The sublimation pressures of solid Cl_2 are 352 Pa at −112 °C and 35 Pa at −126.5 °C. The vapor pressures of liquid Cl_2 are 1590 Pa at −100 °C and 7830 Pa at −80 °C. Calculate (*a*) $\Delta_{sub}H$, (*b*) $\Delta_{vap}H$, (*c*) $\Delta_{fus}H$, and (*d*) the triple point.

6.12 The vapor pressure of solid benzene, C_6H_6, is 299 Pa at −30 °C and 3270 Pa at 0 °C, and the vapor pressure of liquid C_6H_6 is 6170 Pa at 10 °C and 15 800 Pa at 30 °C. From these data, calculate (*a*) the triple point of C_6H_6 and (*b*) the enthalpy of fusion of C_6H_6.

6.13 The surface tension of toluene at 20 °C is 0.0284 N m^{-1}, and its density at this temperature is 0.866 g cm^{-3}. What is the radius of the largest capillary that will permit the liquid to rise 2 cm?

6.14 Mercury does not wet a glass surface. Calculate the capillary depression if the diameter of the capillary is (*a*) 0.1 mm and (*b*) 2 mm. The density of mercury is 13.5 g cm^{-3}. The surface tension of mercury at 25 °C is 0.520 N m^{-1}.

6.15 If the surface tension of a soap solution is 0.05 N m^{-1}, what is the difference in pressure across the film for (*a*) a soap bubble of 2 mm in diameter and (*b*) a bubble 2 cm in diameter?

6.16 From tables giving $\Delta_f G°$, $\Delta_f H°$, and $\overline{C}_P°$ for $H_2O(l)$ and $H_2O(g)$ at 298 K, calculate (*a*) the vapor pressure of $H_2O(l)$ at 25 °C and (*b*) the boiling point at 1 atm.

6.17 What is the maximum number of phases that can be in equilibrium in one-, two-, and three-component systems?

6.18 The vapor pressure of water at 25 °C is 23.756 mm Hg. What is the vapor pressure of water when it is in a container with an air pressure of 100 bar, assuming the dissolved gases do not affect the vapor pressure? The density of water is 0.997 07 g cm^{-3}.

6.19 A binary liquid mixture of A and B is in equilibrium with its vapor at constant temperature and pressure. Prove that $\mu_A(g) = \mu_A(l)$ and $\mu_B(g) = \mu_B(l)$ by starting with

$$G = G(g) + G(l)$$

and the fact that $dG = 0$ when infinitesimal amounts of A and B are simultaneously transferred from the liquid to the vapor.

6.20 Ethanol and methanol form very nearly ideal solutions. At 20 °C, the vapor pressure of ethanol is 5.93 kPa, and that of methanol is 11.83 kPa. (*a*) Calculate the mole fractions of methanol and ethanol in a solution obtained by mixing 100 g of each. (*b*) Calculate the partial pressures and the total vapor pressure of the solution. (*c*) Calculate the mole fraction of methanol in the vapor.

6.21 One mole of benzene (component 1) is mixed with two moles of toluene (component 2). At 20°, the vapor pressures of benzene and toluene are 51.3 and 18.5 kPa, respectively. (*a*) As the pressure is reduced, at what pressure will boiling begin? (*b*) What will be the composition of the first bubble of vapor?

6.22 The vapor pressures of benzene and toluene have the following values in the temperature range between their boiling points at 1 bar:

t/°C	79.4	88	94	100	110.0
$P^*_{C_6H_6}$/bar	1.000	1.285	1.526	1.801	
$P^*_{C_7H_8}$/bar		0.508	0.616	0.742	1.000

(*a*) Calculate the compositions of the vapor and liquid phases at each temperature and plot the boiling point diagram. (*b*) If a solution containing 0.5 mole fraction benzene and 0.5 mole fraction toluene is heated, at what temperature will the first bubble of vapor appear, and what will be its composition?

6.23 At 1.013 bar pressure propane boils at −42.1 °C and *n*-butane boils at −0.5 °C; the following vapor–pressure data are available:

t/°C	−31.2	−16.3
P/kPa (propane)	160.0	298.6
P/kPa (*n*-butane)	26.7	53.3

Assuming that these substances form ideal binary solutions with each other, (*a*) calculate the mole fractions of propane at which the solution will boil at 1.013 bar pressure at −31.2 and −16.3 °C. (*b*) Calculate the mole fractions of propane in the equilibrium vapor at these temperatures. (*c*) Plot the temperature–mole fraction diagram at 1.013 bar, using these data, and label the regions.

6.24 The following table gives mole percent acetic acid in aqueous solutions and in the equilibrium vapor at the boiling point of the solution at 1.013 bar:

b.p., °C	118.1	113.8	107.5	104.4	102.1	100.0
Mol% of acetic acid						
In liquid	100	90.0	70.0	50.0	30.0	0
In vapor	100	83.3	57.5	37.4	18.5	0

Calculate the minimum number of theoretical plates for the column required to produce an initial distillate of 28 mol% acetic acid from a solution of 80 mol% acetic acid.

6.25 If two liquids (1 and 2) are completely immiscible, the mixture will boil when the sum of the two partial pressures exceeds the applied pressure: $P = P_1^* + P_2^*$. In the vapor phase the ratio of the mole fractions of the two components is equal to the ratio of their vapor pressures:

$$\frac{P_1^*}{P_2^*} = \frac{x_1}{x_2} = \frac{m_1 M_2}{m_2 M_1}$$

where m_1 and m_2 are the masses of components 1 and 2 in the vapor phase, and M_1 and M_2 are their molar masses. The boiling point of the immiscible liquid system naphthalene–water is 98 °C under a pressure of 97.7 kPa. The vapor pressure of water at 98 °C is 94.3 kPa. Calculate the weight percent of naphthalene in the distillate.

6.26 A regular binary solution is defined as one for which

$$\mu_1 = \mu_1^\circ + RT \ln x_1 + w x_2^2$$

$$\mu_2 = \mu_2^\circ + RT \ln x_2 + w x_1^2$$

Derive $\Delta_{\text{mix}}G$, $\Delta_{\text{mix}}S$, $\Delta_{\text{mix}}H$, and $\Delta_{\text{mix}}V$ for the mixing of x_1 moles of component 1 with x_2 moles of component 2. Assume that the coefficient w is independent of temperature.

6.27 From the data given in the following table, construct a complete temperature–composition diagram for the system ethanol–ethyl acetate for 1.013 bar. A solution containing 0.8 mole fraction of ethanol, EtOH, is distilled completely at 1.013 bar. (*a*) What is the composition of the first vapor to come off, and (*b*) what is that of the last drop of liquid to evaporate? (*c*) What would be the values of these quantities if the distillation were carried out in a cylinder provided with a piston so that none of the vapor could escape?

x_{EtOH}	y_{EtOH}	b.p., °C	x_{EtOH}	y_{EtOH}	b.p., °C
0	0	77.15	0.563	0.507	72.0
0.025	0.070	76.7	0.710	0.600	72.8
0.100	0.164	75.0	0.833	0.735	74.2
0.240	0.295	72.6	0.942	0.880	76.4
0.360	0.398	71.8	0.982	0.965	77.7
0.462	0.462	71.6	1.000	1.000	78.3

6.28 The Henry's law constants for oxygen and nitrogen in water at 0 °C are 2.54×10^4 bar and 5.45×10^4 bar, respectively. Calculate the lowering of the freezing point of water by dissolved air with 80% N_2 and 20% O_2 by volume at 1 bar pressure.

6.29 Use the Gibbs–Duhem equation to show that if one component of a binary liquid solution follows Raoult's law, the other component will, too.

6.30 The following data on ethanol–chloroform solutions at 35 °C were obtained by G. Scatchard and C. L. Raymond [*J. Am. Chem. Soc.* **60**:1278 (1938)]:

$x_{\text{EtOH,liq}}$	0	0.2	0.4
$y_{\text{EtOH,vap}}$	0.0000	0.1382	0.1864
Total pressure, kPa	39.345	40.559	38.690

$x_{\text{EtOH,liq}}$	0.6	0.8	1.0
$y_{\text{EtOH,vap}}$	0.2554	0.4246	1.0000
Total pressure, kPa	34.387	25.357	13.703

Calculate the activity coefficients of ethanol and chloroform based on the deviations from Raoult's law.

6.31 Show that the equations for the bubble point line and dew point line for nonideal solutions are given by

$$x_1 = \frac{P - \gamma_2 P_2^*}{\gamma_1 P_1^* - \gamma_2 P_2^*}$$

$$y_1 = \frac{P\gamma_1 P_1^* - \gamma_1 \gamma_2 P_1^* P_2^*}{P\gamma_1 P_1^* - P\gamma_2 P_2^*}$$

6.32 A regular binary solution is defined as one for which

$$\mu_1 = \mu_1^\circ + RT \ln x_1 + w x_2^2$$

$$\mu_2 = \mu_2^\circ + RT \ln x_2 + w x_1^2$$

Derive the expressions for the activity coefficients γ_1 and γ_2 in terms of w.

6.33 The expressions for the activity coefficients of the components of a regular binary solution were derived in the preceding problem. Derive the expressions for γ_1 in terms of the experimentally measured total pressure P, the vapor pressures of the two components, and the composition of the solution for the case in which the deviations from ideality are small.

6.34 Using the data in Problem 6.75, calculate the activity coefficients of water (1) and n-propanol (2) at 0.20, 0.40, 0.60, and 0.80 mole fraction of n-propanol, based on deviations from Henry's law and considering water to be the solvent.

6.35 If 68.4 g of sucrose ($M = 342$ g mol^{-1}) is dissolved in 1000 g of water, (*a*) what is the vapor pressure at 20 °C? (*b*) What is the freezing point? The vapor pressure of water at 20 °C is 2.3149 kPa.

6.36 The protein human plasma albumin has a molar mass of 69 000 g mol^{-1}. Calculate the osmotic pressure of a solution of this protein containing 2 g per 100 cm^3 at 25 °C in (*a*) pascals and (*b*) millimeters of water. The experiment is carried out using a salt solution for solvent and a membrane permeable to salt as well as water.

(M) **6.37** The following osmotic pressures were measured for solutions of a sample of polyisobutylene in benzene at 25 °C.

C/kg m^{-3}	5	10	15	20
Π/Pa	49.5	101	155	211

Calculate the number average molar mass from the value of Π/C extrapolated to zero concentration of the polymer.

6.38 Calculate the osmotic pressure of a 1 mol L^{-1} sucrose solution in water from the fact that at 30° C the vapor pressure of the solution is 4.1606 kPa. The vapor pressure of water at 30 °C is 4.2429 kPa. The density of pure water at this temperature (0.995 64 g cm^{-3}) may be used to estimate V_1 for a dilute solution. To do this problem, Raoult's law is introduced into equation 6.70.

6.39 Calculate the solubility of p-dibromobenzene in benzene at 20 and 40 °C assuming that ideal solutions are formed. The enthalpy of fusion of p-dibromobenzene is 13.22 kJ mol^{-1} at its melting point (86.9 °C).

6.40 Calculate the solubility of naphthalene at 25 °C in any solvent in which it forms an ideal solution. The melting point of naphthalene is 80 °C, and the enthalpy of fusion is 19.29 kJ mol^{-1}. The measured solubility of naphthalene in benzene is $x_1 = 0.296$.

6.41 The addition of a nonvolatile solute to a solvent increases the boiling point above that of the pure solvent. The elevation of the boiling point is given by

$$\Delta T_b = \frac{RT_{b,A}^2 M_A m_B}{\Delta_{vap}H_A^\circ} = K_b m_B$$

where $T_{b,A}$ is the boiling point of the pure solvent and M_A is its molar mass. The derivation of this equation parallels that of equation 6.67 very closely, and so it is not given. What is the elevation of the boiling point when 0.1 mol of nonvolatile solute is added to 1 kg of water? The enthalpy of vaporization of water at the boiling point is 40.6 kJ mol^{-1}.

6.42 The NBS Tables of Chemical Thermodynamic Properties list $\Delta_f G^\circ$ for I_2 in C_6H_6:x as 7.1 kJ mol^{-1}. The x indicates that the standard state for I_2 in C_6H_6 is on the mole fraction scale. What is the solubility of I_2 in C_6H_6 at 298 K on the mole fraction scale? A chemical handbook lists the solubility as 16.46 g I_2 in 100 cm^3 of C_6H_6. Are these solubilities consistent?

6.43 The following cooling curves have been found for the system antimony–cadmium:

Cd, wt%	0	20	37.5	47.5	50	58	70	93	100
First break in curve, °C	—	550	461	—	419	—	400	—	—
Continuing constant temperature, °C	630	410	410	410	410	439	295	295	321

Construct a phase diagram, assuming that no breaks other than these actually occur in any cooling curve. Label the diagram completely and give the formula of any compound formed. How many degrees of freedom are there for each area and at each eutectic point?

6.44 The phase diagram for magnesium–copper at constant pressure shows that two compounds are formed: $MgCu_2$, which melts at 800 °C, and Mg_2Cu, which melts at 580 °C. Copper melts at 1085 °C, and Mg at 648 °C. The three eutectics are at 9.4% by weight Mg (680 °C), 34% by weight Mg (560 °C), and 65% by weight Mg (380 °C). Construct the phase diagram. How many degrees of freedom are there for each area and at each eutectic point?

6.45 The Gibbs–Duhem equation in the form

$$\left(\frac{\partial M}{\partial T}\right)_{P,x} dT + \left(\frac{\partial M}{\partial P}\right)_{T,x} dP - \sum(x_i \, dM_i) = 0$$

applies to any molar thermodynamic property M in a homogeneous phase. If this applied to G_E, it may be shown that if the vapor is an ideal gas,

$$x_1 \frac{d\ln(y_1 P)}{dx_1} + x_2 \frac{d\ln(y_2 P)}{dx_1} = 0 \qquad \text{(constant } T\text{)}$$

Show that this can be rearranged to the coexistence equation

$$\frac{dP}{dy_1} = \frac{P(y_1 - x_1)}{y_1(1 - y_1)}$$

Thus, if P versus y_1 is measured, there is no need for measurements of x_1.

6.46 For a solution of ethanol and water at 20 °C that has 0.2 mole fraction ethanol, the partial molar volume of water is 17.9 cm^3 mol^{-1} and the partial molar volume of ethanol is 55.0 cm^3 mol^{-1}. What volumes of pure ethanol and water are required to make 1 liter of this solution? At 20 °C the density of ethanol is 0.789 g cm^{-3} and the density of water is 0.998 g cm^{-3}.

6.47 Since the average entropy of vaporization at the standard boiling point (at 1 atm) is 88 J K^{-1} mol^{-1} (see Table 6.2), the vapor pressure of a liquid can be estimated using

$$\Delta_{vap}S = \Delta_{vap}H/T_b = 88 \text{ J K}^{-1} \text{ mol}^{-1}$$

where T_b is the temperature at the boiling point. This equation is often referred to as Trouton's rule. Estimate the vapor pressure of benzene at 25 °C from the fact that its boiling point is 80.1 °C.

6.48 Calculate the solubility of bismuth in an ideal solution at 150 °C and 200 °C and compare the results with Fig. 6.17. The ethalpy of fusion of bismuth at its melting point (273 °C) is is 10.5 kJ mol^{-1}.

6.49 The molar volume of a binary solution is given by

$$\overline{V} = x_1\overline{V}_1 + x_2\overline{V}_2$$

This kind of additive equation applies to other thermodynamic properties at constant T and P as well. A convenient way to treat the data on the molar volume or other thermodynamic property of a solution is to fit it to a function (for example, a function of x_2) and then calculate the molar volumes of the substances involved by differentiation of the polynomial. Show that

$$\overline{V}_1 = \overline{V} - x_2\left(\frac{\partial \overline{V}}{\partial x_2}\right)$$

$$\overline{V}_2 = \overline{V} - (1 - x_2)\left(\frac{\partial \overline{V}}{\partial x_2}\right)$$

6.50 Derive the Gibbs–Duhem equation for the volume of a binary solution, and show that if the partial molar volume for substance 1 can be determined as a function of x_2, the partial molar volume of substance 2 can be calculated by integrating the relation obtained from the Gibbs–Duhem equation.

6.51 Calculate the partial molar volumes of water and glycerol in solutions at 20 °C. The molar volumes are given as a function of the molar volume of glycerol in the following table:

x_2	$\overline{V}/(cm^3\ mol^{-1})$
0	18.05
0.0212	19.18
0.0466	20.53
0.1153	24.18
0.2269	30.21
0.4390	41.82
0.6923	55.87
1.000	73.02

(*a*) Fit these data to $\overline{V} = A + Bx_2 + Cx_2^2$, where x_2 is the mole fraction of glycerol. (*b*) Calculate the two partial molar volumes as a function of x_2. (*c*) Show that the molar volumes in the table can be calculated using the partial molar volumes. (*d*) Show that the partial molar volume of water can be calculated by using the function for the partial molar volume of glycerol and the Gibbs–Duhem equation. (See Problems 6.49 and 6.50.)

6.52 Ice has the unusual property of a melting point that is lowered by increasing pressure. If this is the reason we can skate on ice, would a 75-kg skater whose skates contact the ice with an area of 0.1 cm^2 be able to skate at −3 °C?

6.53 Calculate the vapor pressure of liquid mercury at 25 °C using data in Table C.2.

6.54 Calculate an approximate value for the transition temperature for

$$CaCO_3(\text{calcite}) = CaCO_3(\text{aragonite})$$

using data in Table C.2 and assuming $\Delta C_P^\circ = 0$.

6.55 The vapor pressure of Hg(l) is 0.133 bar at 260 °C and 0.533 bar at 330 °C. Assume that $\Delta C_P^\circ = 0$ and that mercury vapor is an ideal gas. What are the values of $\Delta_{vap}H^\circ$, $\Delta_{vap}G^\circ$, and S°(g) at 25 °C? The entropy of Hg(l) is 76.0 J K^{-1} mol^{-1} at 25 °C.

6.56 The enthalpy of vaporization of toluene is 38.1 kJ mol^{-1}. Given that the boiling point of toluene at 1 atm is 110.6 °C, what is the boiling point at 1 bar, and what is the change in boiling point with this change in pressure?

6.57 What is the boiling point of water on a mountain where the barometer reading is 88 kPa? The heat of vaporization of water may be taken to be 40.67 kJ mol^{-1}.

6.58 The sublimation pressure of solid CO_2 is 133 Pa at −134.3 °C and 2660 Pa at −114.4 °C. Calculate the enthalpy of sublimation.

6.59 Estimate the vapor pressure of ice at the temperature of solid carbon dioxide (−78 °C at 1 bar pressure of CO_2), assuming that the heat of sublimation is constant. The heat of sublimation of ice is 2.83 kJ g^{-1}, and the vapor pressure of ice is 611 Pa at 0 °C.

6.60 Given the thermodynamic information on H_2O(l) and H_2O(g) in Table C.2, calculate the vapor pressure of water at 500 °C using the equation in Example 5.11 with and without the $\Delta_r C_P^\circ$ term.

6.61 The vapor pressure of toluene is 8.00 kPa at 40.3 °C and 2.67 kPa at 18.4 °C. Calculate (*a*) the heat of vaporization and (*b*) the vapor pressure at 25 °C.

6.62 At 0 °C ice absorbs 333.5 J g^{-1} in melting; water absorbs 2490 J g^{-1} in vaporizing. (*a*) What is the enthalpy of sublimation of ice at this temperature? (*b*) At 0 °C the vapor pressure of both ice and water is 611 Pa. What is the rate of change of vapor pressure with temperature dP/dT for ice and liquid water at this temperature? (*c*) Estimate the vapor pressure of ice and of liquid water at −5 °C.

6.63 According to Trouton's rule, the entropy of vaporization of a liquid at its boiling point is 88 J K^{-1} mol^{-1}. What is the change in boiling point expected for a liquid with a boiling point of (*a*) 100 °C and (*b*) 200 °C at 101 325 Pa in going to a reference state of 1 bar?

6.64 Calculate ΔG° for

$$H_2O(g, 25\,°C) = H_2O(l, 25\,°C)$$

The vapor pressure of water at 25 °C is 3168 Pa.

6.65 An aqueous solution contains NaCl, NaBr, KCl, and KBr. How many components are there in the solution?

6.66 How many degrees of freedom do the following systems have? (*a*) $NH_4Cl(s)$ is allowed to dissociate to $NH_3(g)$ and $HCl(g)$ until equilibrium is reached. (*b*) A solution of alcohol exists in equilibrium with its vapor.

6.67 The vapor pressure of mercury is 133.3 Pa at 126.2 °C. When mercury is enclosed in a container with an air pressure of 500 bar at 126.2 °C, what is the vapor pressure of the mercury? The density of mercury can be taken as 13.6 g cm^{-3}.

6.68 Ethylene dibromide and propylene dibromide form very nearly ideal solutions. Plot the partial vapor pressure of ethylene dibromide ($P^* = 22.9$ kPa), the partial vapor pressure of propylene dibromide ($P^* = 16.9$ kPa), and the total vapor pressure of the solution versus the mole fraction of ethylene dibromide at 80 °C. (*a*) What will be the composition of the vapor in equilibrium with a solution containing 0.75 mole fraction of ethylene dibromide? (*b*) What will be the composition of the liquid phase in equilibrium with ethylene dibromide–propylene dibromide vapor containing 0.50 mole fraction of each?

6.69 At 25 °C the vapor pressures of chloroform and carbon tetrachloride are 26.54 and 15.27 kPa, respectively. If the liquids form ideal solutions, (*a*) what is the composition of the vapor in equilibrium with a solution containing 1 mol of each and (*b*) what is the total vapor pressure of the mixture?

6.70 Benzene and toluene form very nearly ideal solutions. At 80 °C, the vapor pressures of benzene and toluene are as follows: benzene, $P^* = 100.4$ kPa; toluene, $P^* = 38.7$ kPa. (*a*) For a solution containing 0.5 mole fraction of benzene and 0.5 mole fraction of toluene, what is the composition of the vapor and the total vapor pressure at 80 °C? (*b*) What is the composition of the liquid phases in equilibrium at 80 °C with benzene–toluene vapor having 0.75 mole fraction benzene?

6.71 At 140 °C the vapor pressure of pure C_6H_5Cl is 1.237 bar and that of pure C_6H_5Br is 0.658 bar. These two liquids form ideal solutions to a very high degree of approximation. (*a*) What is the mole fraction of C_6H_5Cl in a C_6H_5Cl–C_6H_5Br solution that just boils at 140 °C at 1 bar? (*b*) What is the mole fraction of C_6H_5Cl in the vapor produced in (*a*)? (*c*) Suppose this vapor is condensed. What is its total vapor pressure at 140 °C?

6.72 At 100 °C benzene has a vapor pressure of 180.9 kPa, and toluene has a vapor pressure of 74.4 kPa. Assuming that these substances form ideal binary solutions with each other, calculate the composition of the solution that will boil at 1 bar at 100 °C and the vapor composition.

6.73 The vapor pressure of the immiscible liquid system diethylaniline–water is 1.013 bar at 99.4 °C. The vapor pressure of water at that temperature is 99.2 kPa. How many grams of steam are necessary to distill 100 g of diethylaniline? (See Problem 6.25.)

6.74 What are the entropy change and Gibbs energy change on mixing to produce a benzene–toluene solution with $\frac{1}{3}$ mole fraction benzene at 25 °C?

6.75 For a solution of *n*-propanol and water, the following partial pressures in kPa are measured at 25 °C. Draw a complete pressure–composition diagram, including the total pressure. What is the composition of the vapor in equilibrium with a solution containing 0.5 mole fraction of *n*-propanol?

$x_{n\text{-propanol}}$	P_{H_2O}	$P_{n\text{-propanol}}$
0	3.168	0.00
0.020	3.13	0.67
0.050	3.09	1.44
0.100	3.03	1.76
0.200	2.91	1.81
0.400	2.89	1.89
0.600	2.65	2.07
0.800	1.79	2.37
0.900	1.08	2.59
0.950	0.56	2.77
1.000	0.00	2.901

6.76 Plot the following boiling point data for benzene–ethanol solutions at 1.013 bar and estimate the azeotropic composition.

b.p., °C	78	75	70	70	75	80
Mole fraction of benzene						
In liquid	0	0.04	0.21	0.86	0.96	1.00
In vapor	0	0.18	0.42	0.66	0.83	1.00

State the range of mole fractions of benzene for which pure benzene could be obtained by fractional distillation at 1.013 bar.

6.77 The following table gives the mole percent of *n*-propanol ($M = 60.1$ g mol^{-1}) in aqueous solutions and in the vapor at the boiling point of the solution at 1.013 bar pressure:

b.p., °C	100.0	92.0	89.3	88.1	87.8	88.3	90.5	97.3
Mol% of *n*-propanol								
In liquid	0	2.0	6.0	20.0	43.2	60.0	80.0	100.0
In vapor	0	21.6	35.1	39.2	43.2	49.2	64.1	100.0

With the aid of a graph of these data, calculate the mole fraction of *n*-propanol in the first drop of distillate when the following solutions are distilled with a simple distilling flask that gives one theoretical plate: (*a*) 87 g of *n*-propanol and 211 g of water; (*b*) 50 g of *n*-propanol and 5.02 g of water.

6.78 Using the Henry's law constants in Table 6.4, calculate the percentage (by volume) of oxygen and nitrogen in air dissolved in water at 25 °C. The air in equilibrium with the water at 1 bar pressure may be considered to be 20% oxygen and 80% nitrogen by volume.

6.79 By use of the data of the following table, which gives pressures in kPa at 35.2 °C for carbon disulfide–acetone solutions, calculate the activity coefficients based on deviations from Raoult's law for acetone and carbon disulfide at 32.5 °C for a solution containing 0.6 mole fraction of carbon disulfide.

x_{CS_2}	0	0.2	0.4	0.6	0.8	1.0
P_{CS_2}/kPa	0	37.3	50.4	56.7	61.3	68.3
$P_{acetone}$/kPa	45.9	38.7	34.0	30.7	25.3	0

6.80 Using the data of the following table, which gives vapor pressures at 35.2 °C, calculate the activity coefficients of acetone (2) and chloroform (1) at 35.2 °C, based on the deviations from Raoult's law for mole fractions of chloroform of 0.2, 0.4, 0.6, and 0.8.

x_1	0	0.2	0.4	0.6	0.8	1.00
P_1/kPa	0	4.5	10.9	19.7	30.0	39.1
P_2/kPa	45.9	36.0	24.4	13.6	5.6	0

6.81 What is the change in Gibbs energy for the transfer of 1 mol of acetone from pure liquid acetone to an infinite amount of a solution of an equimolar mixture of acetone and ether at 303 K? Make the calculation with activity coefficients of acetone in the final solution based on deviations from Raoult's law ($\gamma_2 = 1.19$ and $P_2^* = 37.7$ kPa) and based on deviations from Henry's law ($\gamma_2' = 0.572$ and $K_2 = 78.3$ kPa).

6.82 The solubility of I_2(s) in CCl_4 is listed in a chemical handbook as 29.1 g in 100 cm^3 of CCl_4 at 25 °C. What is $\Delta_f G°$ for I_2 in CCl_4 on the mole fraction scale?

6.83 (*a*) Use the following data to calculate the Henry's law constant for the solute chloroform in the solvent acetone at 35.2 °C:

x_{CHCl_3}	0	0.0603	0.1853	0.2910
P_{CHCl_3}/kPa	0	1.26	4.25	7.39

(*b*) Using this value of the Henry's law constant and the data in Problem 6.80, calculate the activity coefficients of $CHCl_3$ from deviations from Henry's law. The activity coefficients for acetone, considered as the solvent, are the same as in Problem 6.80.

6.84 Using the data in Problem 6.75, calculate the activity coefficients of water and *n*-propanol at 0.20, 0.40, 0.60, and 0.80 mole fraction *n*-propanol, based on deviations from Henry's law and considering *n*-propanol to be the solvent.

6.85 The logarithms of the activity coefficients for a binary solution may be expressed as power series:

$$\ln \gamma_1 = a_1 x_2 + b_1 x_2^2 + c_1 x_2^3 + \cdots$$

$$\ln \gamma_2 = a_2 x_1 + b_2 x_1^2 + c_2 x_1^3 + \cdots$$

Using the Gibbs–Duhem equation,

$$x_1 \, d\ln \gamma_1 + x_2 \, d\ln \gamma_2 = 0 \qquad \text{(fixed } P, T\text{)}$$

show that

$$a_2 = a_1 = 0 \qquad b_2 = b_1 + \tfrac{3}{2} c_1 \qquad c_2 = -c_1$$

6.86 The vapor pressure of a solution containing 13 g of a nonvolatile solute in 100 g of water at 28 °C is 3.6492 kPa. Calculate the molar mass of the solute, assuming that the solution is ideal. The vapor pressure of water at this temperature is 3.7417 kPa.

6.87 In acidic aqueous solutions acetic acid exists in the monomer form, but in nonpolar solvents, such as benzene, it exists in the form of a dimer. Derive the following expression for the distribution coefficient:

$$K = \frac{x^2_{CH_3CO_2H \text{ in } H_2O}}{x_{CH_3CO_2H \text{ in } C_6H_6}}$$

(M) **6.88** The following osmotic pressures of polyvinyl acetate in dioxane were measured by G. V. Browning and J. D. Ferry at 25 °C:

$C/10^{-2}$g cm^{-3}	0.292	0.579	0.810	1.140
Π/cm of solvent	0.73	1.76	2.73	4.68

Calculate the number average molar mass. The density of dioxane is 1.035 g cm^{-3}.

6.89 Calculate the solubility of anthracene (M = 178.2 g mol^{-1}) in toluene (M = 92.1 g mol^{-1}) at 100 °C. The enthalpy of fusion of anthracene is 28.9 kJ mol^{-1}, and the melting point of anthracene is 217 °C. The actual solubility is 0.0592 on the mole fraction scale. How do you explain the difference?

6.90 Calculate the solubility of cadmium in bismuth at 250 °C. The melting point of cadmium is 323 °C, and the enthalpy of fusion at the melting point is 6.07 kJ mol^{-1}.

6.91 The following data are obtained by cooling solutions of magnesium and nickel:

Ni, wt%	0	10	28	38	60	83	88	100
Inflection in cooling curve, °C	—	608	—	770	1050	—	—	—
Plateau in cooling curve, °C	651	510	510	510	770	1180	1080	1450

It is found that, in addition, cooling solutions between 28% and 38% Ni deposit Mg_2Ni, whereas solutions containing between 38% and 82% Ni deposit $MgNi_2$. Plot the phase diagram.

Computer Problems

6.A The vapor pressure of ice from –40°C to 0°C is given in Table 6.1. Fit these data to the equation

$$\ln P = -a/T + b \ln T + cT + d$$

The form of this equation is suggested by the equation in Example 5.11. Plot this equation as $\ln P$ versus $1/T$. Calculate $\Delta_{sub}H°$ as a function of temperature, and calculate its value at each of the temperatures in Table 6.1.

6.B The vapor pressure of water from 0 to 100 °C is given in Table 6.1. Fit these data to the equation

$$\ln P = -a/T + b \ln T + cT + d$$

Plot the data as $\ln P$ versus $1/T$. Determine the parameters a, b, c, and d, and test this equation to see how well it represents the vapor pressure data. Calculate $\Delta_{vap}H°$ as a function of temperature, and calculate its value at each of the temperatures in Table 6.1. Calculate the standard enthalpy of vaporization at 298.15 K using Table C.2 and compare it with the value calculated with this empirical equation.

6.C The vapor pressure of methanol in bars is given in the following table as a function of temperature. Fit these vapor pressures to equation 6.17 using a computer and calculate the heat of vaporization at each of these temperatures using equation 6.18.

Calculate the standard enthalpy of vaporization of methanol using Table C.2 and compare it with the value calculated with the empirical equation.

t/°C	−44.0	−16.2	5.0	21.2	49.9	64.7
P_{vap}/bar	0.001333	0.01333	0.0533	0.1333	0.533	1.0132

6.D The vapor pressure of ethane in bars is given in the following table as a function of temperature. Fit these vapor pressures to equation 6.17 using a computer and calculate the heat of vaporization at each of these temperatures using equation 6.18.

t/°C	−159.5	−142.9	−129.8	−119.3	−99.7	−88.6
P_{vap}/bar	0.001333	0.01333	0.0533	0.1333	0.533	1.0132

6.E The molar volumes of solutions of water and methanol at 0 °C and 1 bar are as follows:

x(methanol)	0	0.114	0.197	0.249	0.495	0.692	0.785	0.892	1
Molar volume	0.0181	0.0203	0.0219	0.0230	0.0283	0.0329	0.0352	0.0379	0.0407

where the molar volumes are in m^3 $kmol^{-1}$. (a) Plot the molar volumes versus the mole fraction of methanol and fit these data to a quadratic polynomial. (b) Calculate the polynomials that give the partial molar volumes of water and methanol and calculate the partial molar volumes in each of the nine solutions. (c) Show that the molar volumes of the solutions are given by the sum of the contributions of the two liquids. (d) Show that the equation for the partial molar volume of water can be calculated using the equation for the partial molar volume of methanol by use of the Gibbs–Duhem equation for the molar volume.

6.F Calculate the standard and normal boiling points of water using the expression calculated in Computer Problem 6.B for $\ln P_{vap}$ for water. The standard boiling point is at the standard pressure of 1 bar, and the normal boiling point is at a pressure of 1 atm (1.013 25 bar).

6.G When pressure is applied to the surface of a liquid, the vapor pressure of the liquid is increased because molecules are squeezed out of the liquid into the vapor phase. This pressure can be applied by a gas. The following equation for the ratio of vapor pressure P at the higher pressure to the vapor pressure P^* in the absence of the pressurizing gas can be calculated using

$$\ln(P/P^*) = \overline{V}\Delta P/RT$$

This equation is based on the assumption that the pressurizing gas is insoluble. Calculate P/P^* at 298.15 K and $\Delta P = 100$ bar for n-octane, which has a density of 0.7036 g mL at this temperature.

6.H The ratio of the vapor pressure P of a liquid droplet is greater than the vapor pressure P^* of the liquid over a flat surface. Conversely, the ratio of the vapor pressure P inside a bubble is smaller than the vapor pressure P^* of the liquid over a flat surface. (a) Plot P/P^* of water at 25 °C versus $\log r$ of the droplet, where r is the radius in nm. (b) Plot P/P^* of water at 25 °C versus $\log r$ of the bubble, where r is the radius in nm. The surface tension of water at 25 °C is 0.071 97 N m^{-1}.

6.I (a) Fit the data in Table 6.1 on the vapor pressure of liquid water to equation 6.17, and plot $\ln P_{vap}$ calculated with this equation between −40 and 180 °C. (b) Fit the data in Table 6.1 on the vapor pressure of ice to equation 6.17 without the C term, and plot $\ln P_{ice}$ calculated with this euqation between −40 and 180 °C. (c) Superimpose these two plots to see the relation between the vapor pressure of supercooled water and ice below 0°C.

7 Electrochemical Equilibrium

Electrochemical reactions involve electrons that are transferred from a metal to a species in a solution. The equilibria of electrochemical reactions are important in galvanic cells, which produce an electric current, and electrolytic cells, which consume an electric current.

The measurement of the electromotive force of an electrochemical cell over a range of temperature makes it possible to obtain the thermodynamic quantities for the reaction that occurs in the cell. The activity coefficients of electrolytes may also be calculated from these measurements; this is illustrated in this chapter by the determination of the activity coefficient of hydrochloric acid.

Electrochemical cells are of practical interest in that they offer the means to convert the Gibbs energy change of a chemical reaction to work without the second-law losses of heat engines.

An understanding of the conversion of chemical energy to electrical energy is important for work with batteries, fuel cells, electroplating, corrosion, electrorefining (e.g., the production of aluminum), and electroanalytical techniques. This chapter discusses the thermodynamics of such processes.

7.1 COULOMB'S LAW, ELECTRIC FIELD, AND ELECTRIC POTENTIAL

Of the four kinds of interactions recognized in physics (strong nuclear interactions, weak interactions, electromagnetic interactions, and gravitation), only electromagnetic interactions are of importance in chemistry. The most elementary of these is the attractive or repulsive interaction between two charges, Q_1 and Q_2. Since the direction of the force $\boldsymbol{f}$ is along the line connecting the charges, it is convenient to write **Coulomb's law** in vector notation:

$$\boldsymbol{f} = \frac{1}{4\pi\epsilon_0\epsilon_r}\frac{Q_1Q_2}{r^2}\hat{\boldsymbol{r}} \tag{7.1}$$

where r is the distance between charges, $\hat{\boldsymbol{r}}$ is the unit vector in the direction of the force, ϵ_0 is the **permittivity of vacuum** ($8.854\ 187\ 817 \times 10^{-12}\ \mathrm{C^2\ N^{-1}\ m^{-2}}$), and ϵ_r is the **relative permittivity** (dielectric constant). The relative permittivities of a number of gases and liquids are given later in Table 22.1. When the direction of the force is not being considered, Coulomb's law may be written in the form

$$f = \frac{Q_1Q_2}{4\pi\epsilon_0\epsilon_r r^2} \tag{7.2}$$

The electric field strength $\boldsymbol{E}$ at a certain point is defined as the electrical force per unit charge. The electric field strength is a vector because it has direction as well as magnitude. If a small test charge Q_1 is used, the **electric field strength** is equal to the ratio of the force to the charge:

$$\boldsymbol{E} = \frac{\boldsymbol{f}}{Q_1} \tag{7.3}$$

From equation 7.2 the magnitude of the electric field strength in a vacuum due to charge Q_2 is given by

$$E = \frac{Q_2}{4\pi\epsilon_0 r^2} \tag{7.4}$$

The electric field strength has the SI units of $\mathrm{V\ m^{-1}}$.

The electric field $\boldsymbol{E}$ is the negative gradient of the electric potential ϕ:

$$\boldsymbol{E} = -\nabla\phi \tag{7.5}$$

As we can see from Appendix D.7, if the electric potential is a function only of x,

$$E = -\frac{\partial\phi}{\partial x} \tag{7.6}$$

The difference between the electric potential at two points is equal to the work per unit charge required to move a charge from one point to the other. Thus, the unit of potential difference is joules per coulomb; this unit is referred to as a volt ($1\ \mathrm{V} = 1\ \mathrm{J\ C^{-1}}$). The choice of zero potential is arbitrary, but it is customary to define the potential as zero when the particles are at infinite distance. Thus, the **electric potential** at a point is the work required to bring a unit positive charge from infinity to the point in question. The electric potential is given by integrating equation 7.5 using equation 7.4:

$$\phi = -\int_\infty^r \frac{Q_2\,\mathrm{d}r}{4\pi\epsilon_0\epsilon_r r^2} = \frac{Q_2}{4\pi\epsilon_0\epsilon_r r} \tag{7.7}$$

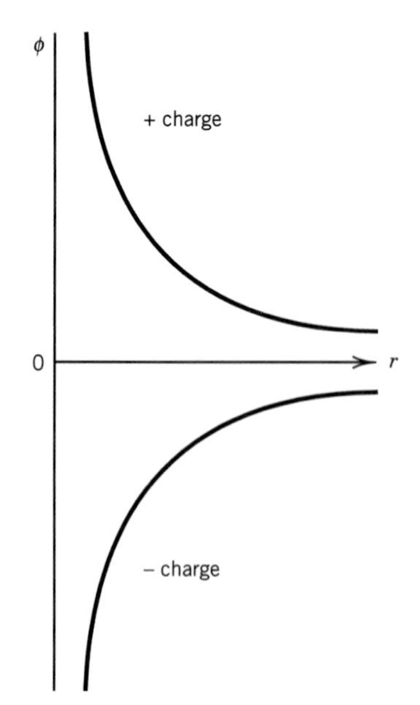

Figure 7.1 Electric potential ϕ as a function of distance from an electric charge.

Figure 7.1 shows the electric potential as a function of distance from an electric charge. The electromotive force is the difference in electric potential between two points and is expressed in volts. In the remainder of this chapter we will use E to represent electromotive force, rather than electric field strength.

As we begin to treat electrolyte solutions we need to remember that there is an additional constraint on the composition of a phase, and that is the **electroneutrality condition,**

$$\sum_i n_i z_i = 0 \tag{7.8}$$

where n_i is the amount of ions of charge $z_i e$ in the phase. The charge number z_i is positive for cations and negative for anions; e is the charge on a proton, 1.6022×10^{-19} C. For a phase to have a nonzero electric potential ϕ, there must be a small deviation from the electroneutrality condition, but these deviations are so small that we can neglect them in using equation 7.8.

Example 7.1 ***The electric potential of the surface of a sphere containing a charge***

What is the electric potential on the surface of a sphere 10 cm in diameter if it contains an excess of 10^{-10} mol of a monovalent cation?

The charge on the sphere is 10^{-10} F, where F is the Faraday constant ($F = N_A e = 96\,485$ C mol^{-1}). The potential on the surface can be calculated by assuming that this charge is located at the center of the sphere. Assuming a relative permittivity of unity,

$$\phi = \frac{(10^{-10}\ \text{mol})(96\,485\ \text{C mol}^{-1})}{4\pi(8.854\,187 \times 10^{-12}\ \text{C}^2\ \text{N}^{-1}\ \text{m}^{-2})(0.05\ \text{m})}$$
$$= 1.734 \times 10^6\ \text{V}$$

Since potential differences between phases are much smaller than this, the deviations from equation 7.8 are much smaller than 10^{-10} mol in a volume of $\frac{4}{3}\pi(0.05\ \text{m})^3$. Thus, it is a good approximation to say that equation 7.8 is obeyed, and two phases can have the same chemical composition but different electric potential.

7.2 EQUILIBRIA INVOLVING POTENTIAL DIFFERENCES

In discussing the differential form of the first law of thermodynamics, we saw that when a small charge dQ is moved through an electric potential difference ϕ, the work done on the charge is given by $đw = \phi\, \text{d}Q$ (equation 2.42). This term carries over to the equation for the differential of the Gibbs energy (equation 4.36). If ions with charge number z_i are transferred, the differential charge dQ can be expressed in terms of the differential amount dn_i of ion i by

$$\text{d}Q = z_i F\, \text{d}n_i \tag{7.9}$$

where F is the Faraday constant. **The charge number of an ion z_i is the charge in terms of the proton charge with the sign of the ion.** The **Faraday constant** is equal to the product of the Avogadro constant and the proton charge:

$$F = N_A e = (6.022\,136\,7 \times 10^{23}\ \text{mol}^{-1})(1.602\,177\,33 \times 10^{-19}\ \text{C})$$
$$= 96\,485.309\ \text{C mol}^{-1} \tag{7.10}$$

The charge number z_i of an ion is dimensionless, so the differential charge dQ is expressed in coulombs.

When we consider a multiphase system with phases at different electric potentials, the electrical work term is of the form $\sum z_i F\phi_i n_i$, where ϕ_i is the **electric potential of the phase containing species** i. In using the fundamental equation, a species in another phase counts as another species. The electric potentials ϕ_i of the various phases in a system are not natural variables of the Gibbs energy G, so if we want to make them natural variables (see Section 4.2), it is necessary to define a **transformed Gibbs energy** G' by making the Legendre transform

$$G' = G - \sum z_i F\phi_i n_i \tag{7.11}$$

Substituting $G = \sum \mu_i n_i$ (equation 4.45) yields

$$G' = \sum(\mu_i - z_i F\phi_i)n_i = \sum \mu_i' n_i \tag{7.12}$$

where the **transformed chemical potential** μ_i' of species i is given by

$$\mu_i' = \mu_i - z_i F\phi_i \tag{7.13}$$

Note that the transformed Gibbs energy is additive in the transformed chemical potentials μ_i', just as the Gibbs energy is additive in the chemical potentials μ_i.

For a multiphase system involving electric potential differences between the phases, both the electric potentials of the phases ϕ_i and the amounts of species n_i can be taken as independent variables, so the differential of the transformed Gibbs energy G' is given by

$$\mathrm{d}G' = \mathrm{d}G - \sum z_i F\phi_i \,\mathrm{d}n_i - \sum z_i F n_i \,\mathrm{d}\phi_i \tag{7.14}$$

Substituting $\mathrm{d}G = -S\,\mathrm{d}T + V\,\mathrm{d}P + \sum \mu_i\,\mathrm{d}n_i$ (equation 4.36) yields the fundamental equation for the transformed Gibbs energy:

$$\mathrm{d}G' = -S\,\mathrm{d}T + V\,\mathrm{d}P + \sum \mu_i'\,\mathrm{d}n_i - \sum z_i F n_i\,\mathrm{d}\phi_i \tag{7.15}$$

Thus, we can see that the transformed chemical potential μ_i' of species i is defined by

$$\left(\frac{\partial G'}{\partial n_i}\right)_{T,P,n_j,\phi_i} = \mu_i' \tag{7.16}$$

where $j \neq i$. As we can see from the subscript on the partial derivative of G', the transformed chemical potential of an ion depends on the electric potential ϕ_i of the phase that it is in. At equilibrium, that is not true for the chemical potential μ_i, which is independent of the type of phase (solid, liquid, or gas) and the pressure and electric potential of the phase.

To contrast the fundamental equations for G and G' for this multiphase system with phases at different electric potentials, we can derive the fundamental equation for G by moving the last two terms of equation 7.14 to the other side and substituting equation 7.15. This yields

$$\begin{aligned}\mathrm{d}G &= -S\,\mathrm{d}T + V\,\mathrm{d}P + \sum(\mu_i' + z_i F\phi_i)\,\mathrm{d}n_i \\ &= -S\,\mathrm{d}T + V\,\mathrm{d}P + \sum \mu_i\,\mathrm{d}n_i \end{aligned} \tag{7.17}$$

where the chemical potential of i is defined by the usual

$$\left(\frac{\partial G}{\partial n_i}\right)_{T,P,n_j} = \mu_i \tag{7.18}$$

where $j \neq i$. The chemical potential of i is given by

$$\mu_i = \mu_i' + z_i F\phi_i \tag{7.19}*$$

which is the same as equation 7.13.

We have seen earlier, in Section 6.1, that at equilibrium the **chemical potential** μ_i of a species is equal in all of the phases where it is present. Following the same arguments used in Section 6.1, equation 7.17 can be used to show that when phases α and β have different electric potentials, we still have

$$\mu_i(\alpha) = \mu_i(\beta) \tag{7.20}$$

as in equation 6.3. This equation will be used in Section 7.10 to derive a relation for the membrane potential that exists between two solutions at equilibrium if the membrane separating them is permeable to some ions and impermeable to others. For a chemical reaction involving species in different phases, equation 7.17 can also be used to derive $\sum \nu_i \mu_i = 0$, where $\mu_i = \mu_i' + z_i F\phi_i$. We will use this equation to derive the equilibrium relation for an electrochemical cell.

7.3 EQUATION FOR AN ELECTROCHEMICAL CELL

Electrochemical cells can be classified as **galvanic cells,** in which chemical reactions occur spontaneously, and **electrolytic cells,** in which chemical reactions are caused by an externally applied potential difference. Galvanic cells of commercial importance include the Leclanché Zn/MnO_2 cell and the Zn/Ag_2O_3 cell used, for example, in watches. Fuel cells, such as the H_2–O_2 cell used in spacecraft, are galvanic cells in which the oxidizable and reducible fuels are supplied continuously. Electrolytic cells are used in the commercial production of chlorine and aluminum and in the electrorefining of copper. The Pb–PbO_2–H_2SO_4 storage cell used in automobile batteries is an electrolytic cell when it is being charged and a galvanic cell when it is being used as a battery. To connect what happens in the laboratory to convention, **we must always write the electrode reaction for the electrode on the right in a schematic representation of a galvanic cell as a reduction.**† The mnemonic for this is "reduction at the electrode on the right":

$$\text{Ox} + n\text{e}^- = \text{Red} \tag{7.21}$$

The **cathode** is the electrode at which a **reduction reaction** occurs. This electrode is positive in a galvanic cell because electrons are flowing into it and reacting with an

*This equation is often found in the literature of electrochemistry as $\tilde{\mu}_i = \mu_i + z_i F\phi_i$, where $\tilde{\mu}_i$ is referred to as the electrochemical potential. This has two disadvantages. The first is that $\tilde{\mu}_i$ is defined by equation 7.18, which we previously used for μ_i. The second is that the μ_i used in electrochemistry is not the chemical potential that has the same value for a species throughout a system at equilibrium.

† Electrochemical Nomenclature, *Pure Appl. Chem.* **37:**501 (1974).

oxidized species to form a reduced species, as shown in equation 7.21. The **anode** is the electrode at which an oxidation reaction occurs:

$$\text{Red}' = \text{Ox}' + ne^- \tag{7.22}$$

The primes indicate that the reactions at the two electrodes are generally different. Ox and Ox′ can be referred to as oxidants or oxidizing agents. Red and Red′ can be referred to as reductants or reducing agents. Note that the reductant is the electron donor and the oxidant is the electron acceptor.

The difference in potential between the electrodes of a cell can be measured using a **potentiometer,** as shown in Fig. 7.2. In a potentiometer a steady current from a battery flows through a resistor. A sliding contact is used to apply some fraction of the potential difference across the slide wire to an electrochemical cell. If the applied potential difference is less than the electromotive force of the galvanic cell, the cell discharges spontaneously as shown in Fig. 7.2*a*.

As the potential difference applied to the cell is increased, a point will be reached at which the current through the cell is zero, as shown in Fig. 7.2*b*. This is the equilibrium potential difference that we will be primarily concerned with in this chapter. This is a thermodynamic measurement because the direction of the current through the cell can be changed by an infinitesimal change in the applied potential, if the electrode reactions are fast under the conditions in the electrochemical cell.

When the applied potential is further increased, as shown in Fig. 7.2*c*, it drives the cell reaction in the reverse direction, and the cell is referred to as an electrolytic cell. Now oxidation occurs at the right electrode, and so it is called the anode. (Note that our rule "reduction at the electrode on the right" applies to a galvanic cell, not to an electrolytic cell.) The right electrode supplies electrons to the external circuit according to reaction 7.22. The definitions of anode and cathode in terms of oxidation and reduction are independent of whether a cell is a galvanic cell or an electrolytic cell. The terms *anode* and *cathode* should not be used

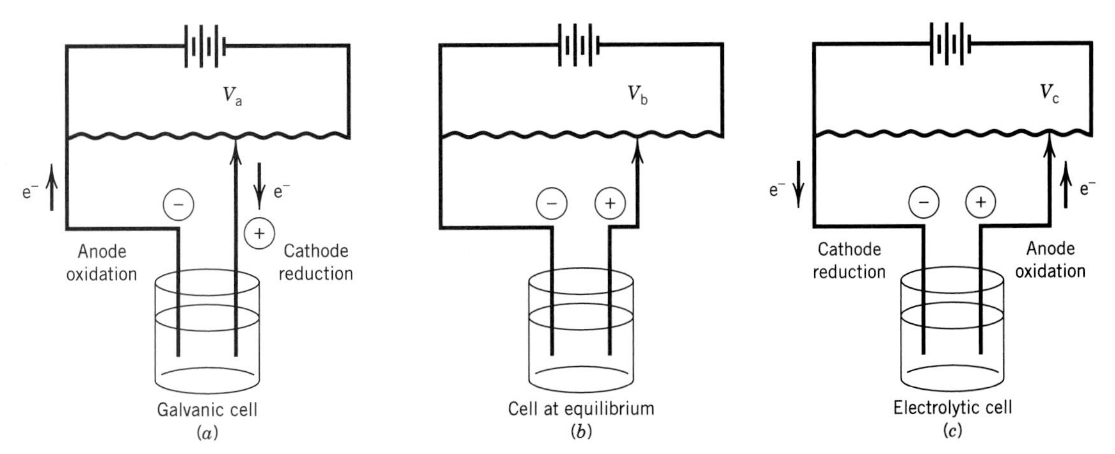

Figure 7.2 Use of a potentiometer to measure the potential difference E for an electrochemical cell with no current flowing. (*a*) The cell operates spontaneously with reduction on the right since $V_a < E$. (*b*) No current passes through the cell since $V_b = E$. (*c*) A nonspontaneous reaction in the cell is driven by the battery in the potentiometer since $V_c > E$.

for the electrodes in an electrochemical cell that is at equilibrium, as in Fig. 7.2*b*.

A very wide variety of different types of galvanic cells can be constructed. A metal electrode can be immersed in a solution containing its ions. An amalgam (solution of a metal in liquid mercury) electrode can be in contact with a solution containing ions of the metal. A nonmetal can be used by bubbling gas over a platinum electrode. Or a platinum electrode can simply be in contact with a solution containing oxidized and reduced forms of other species (for example, Fe^{3+} and Fe^{2+}).

In considering electrochemical cells it is important to make a clear distinction between cells without liquid junctions (see Fig. 7.3*a*) and cells with liquid junctions (see Fig. 7.3*b*).

1. **Cells without liquid junctions.** Examples are represented by

$$\mathrm{Pt(s)} \mid \mathrm{H_2(g)} \mid \mathrm{HCl}(m) \mid \mathrm{AgCl(s)} \mid \mathrm{Ag(s)} \tag{7.23}$$

$$\mathrm{Pt(s)} \mid \mathrm{H_2(g)} \mid \mathrm{HCl}(m) \mid \mathrm{Cl_2(g)} \mid \mathrm{Pt(s)} \tag{7.24}$$

$$\mathrm{Hg\text{-}Na}(x_{\mathrm{Na}}) \mid \mathrm{NaOH}(m) \mid \mathrm{Hg\text{-}Na}(x_{\mathrm{Na}}) \tag{7.25}$$

The vertical lines represent phase boundaries. Hg-Na(x_{Na}) is a sodium amalgam with x_{Na} mole fraction sodium. Cells of this type can be held at equilibrium indefinitely, and therefore they can be given exact thermodynamic treatments. The electromotive force of the cell depends on the electrodes and the activity of the electrolyte solution.

2. **Cells with liquid junctions.** Examples are represented by

$$\mathrm{Zn(s)} \mid \mathrm{Zn^{2+}}(m_1) \vdots \mathrm{Cu^{2+}}(m_2) \mid \mathrm{Cu(s)} \tag{7.26}$$

$$\mathrm{Zn(s)} \mid \mathrm{Zn^{2+}}(m_1) \vdots \mathrm{Zn^{2+}}(m_2) \mid \mathrm{Zn(s)} \tag{7.27}$$

$$\mathrm{Ag(s)} \mid \mathrm{AgCl(s)} \mid \mathrm{Cl^-} \vdots\vdots \mathrm{Ag^+}(m) \mid \mathrm{Ag(s)} \tag{7.28}$$

The symbol $\vdots$ represents a junction between two liquids, and $\vdots\vdots$ represents a salt bridge made up of a concentrated solution of potassium chloride or ammonium nitrate, in which the anion and cation have nearly equal mobilities (Section 20.3). It is necessary to use a salt bridge when the solutions in contact can react with each other, as in cell 7.28. Cells with liquid junctions are never completely at equilibrium because diffusion always occurs at the liquid junction and contributes an unknown potential or electromotive force; however, these contributions are often small compared with experimental errors. Cells with liquid junctions depend on the activities of ions, and so ionic species are used in representing the cells.

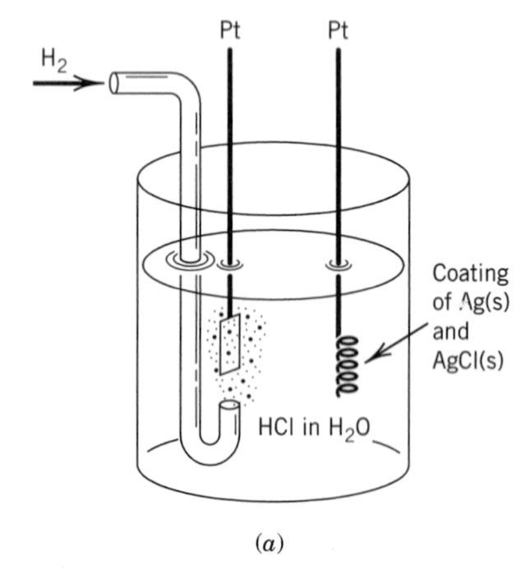

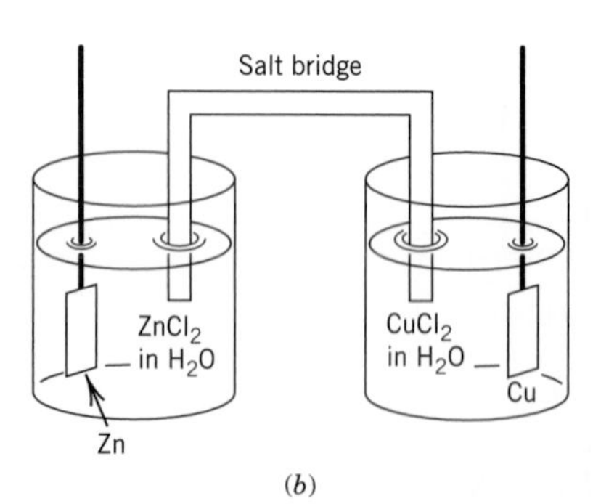

Figure 7.3 (*a*) Galvanic cell without liquid junction. (*b*) Galvanic cell with liquid junction.

To obtain the relationship between the electromotive force for a cell and the chemical potentials or activities of the reactants and products, we will consider the following cell without a liquid junction:

$$\mathrm{Pt_L} \mid \mathrm{H_2(g)} \mid \mathrm{HCl}(m) \mid \mathrm{AgCl(s)} \mid \mathrm{Ag(s)} \mid \mathrm{Pt_R} \tag{7.29}$$

where m is the molality. According to the international convention we assume that a reduction reaction occurs in the right electrode and an oxidation occurs in the left electrode. These two electrode reactions are

$$2\mathrm{AgCl(s)} + 2\mathrm{e^-(Pt_R)} = 2\mathrm{Ag(s)} + 2\mathrm{Cl^-}(m) \tag{7.30}$$

$$\mathrm{H_2(g)} = 2\mathrm{H^+}(m) + 2\mathrm{e^-(Pt_L)} \tag{7.31}$$

The sum of these electrode reactions is the cell reaction.

$$H_2(g) + 2AgCl(s) + 2e^-(Pt_R) = 2H^+(m) + 2Cl^-(m) + 2Ag(s) + 2e^-(Pt_L) \tag{7.32}$$

This reaction can also be written

$$H_2(g) + 2AgCl(s) + 2e^-(Pt_R) = 2HCl(m) + 2Ag(s) + 2e^-(Pt_L) \tag{7.33}$$

It is convenient to refer to such a reaction as an **electrochemical reaction** to differentiate it from the **chemical reaction**

$$H_2(g) + 2AgCl(s) = 2HCl(aq) + 2Ag(s) \tag{7.34}$$

Consider the cell in Fig. 7.3*a*, in which the electrodes are not connected electrically (open circuit). Only minute quantities of electric charge have to be transported into the electrodes to bring the electric potential ϕ_L of the platinum electrode on the left and the electric potential ϕ_R of the electrode on the right to their equilibrium values. Although several phases are involved, the equilibrium relation is $\sum \nu_i \mu_i = 0$, which is

$$2\mu(HCl, aq) + 2\mu(Ag, s) + 2\mu(e^-, Pt_L) - \mu(H_2, g) - 2\mu(AgCl, s) - 2\mu(e^-, Pt_R) = 0 \tag{7.35}$$

The chemical potentials in this equation are each given by $\mu_i = \mu_i' + z_i F\phi_i$ (equation 7.19), where μ_i' is the transformed chemical potential of species i. Since HCl, Ag, H_2, and AgCl are electrically neutral, $\mu_i = \mu_i'$ for each of them, and we can use the usual symbol μ_i. Hydrochloric acid is, of course, dissociated, but the $z_i F\phi_i$ terms for the hydrogen and chloride ions cancel each other when they are included. The chemical potential of the electrons in the right electrode is given by $\mu(e^-, Pt_R) = \mu'(e^-, Pt_R) - F\phi_R$, and the chemical potential of the electrons in the left electrode is given by $\mu(e^-, Pt_L) = \mu'(e^-, Pt_L) - F\phi_L$. The terms $\mu'(e^-, Pt_R)$ and $\mu'(e^-, Pt_L)$ cancel. This can be understood in the following way. If two pieces of platinum are in contact and at equilibrium, $\mu(e^-, Pt_R) = \mu(e^-, Pt_L)$ or $\mu'(e^-, Pt_R) - F\phi_R = \mu'(e^-, Pt_L) - F\phi_L$ and $\mu'(e^-, Pt_R) = \mu'(e^-, Pt_L)$ since the two pieces of platinum are at the same electric potential (i.e., in contact). Therefore, when two pieces of platinum are in contact, the electrons in the two pieces have the same chemical potential, but in an electrochemical cell they have different chemical potentials.

Thus equilibrium relation 7.35 for the galvanic cell on an open circuit becomes

$$2\mu(HCl, aq) + 2\mu(Ag, s) - 2F\phi_L - \mu(H_2, g) - 2\mu(AgCl, s) + 2F\phi_R = 0 \tag{7.36}$$

or

$$\Delta_r G = -2F(\phi_R - \phi_L) = -2FE \tag{7.37}$$

where

$$E = \phi_R - \phi_L \tag{7.38}$$

is the **electromotive force**, which can be measured by use of a potentiometer. **In this chapter, we will always consider E to be the potential difference when the current is zero.** The potential difference of the electrochemical cell depends on T, P, and the molality of the HCl.

The change in Gibbs energy in chemical reaction 7.34 is represented by

$$\Delta_r G = 2\mu(\text{HCl, aq}) + 2\mu(\text{Ag, s}) - \mu(\text{H}_2\text{, g}) - 2\mu(\text{AgCl, s}) \tag{7.39}$$

Although equation 7.39 applies to a special case, it should be clear that this equation can be generalized to any galvanic cell by writing it as

$$\Delta_r G = -|\nu_e|FE \tag{7.40}$$

where $|\nu_e|$ is the absolute value of the **stoichiometric number of the electron in the electrochemical reaction;** electrons appear with both positive and negative signs in the electrochemical reaction. $|\nu_e|$ is sometimes referred to as the charge number for a cell reaction and is represented by n.

Note that when the right-hand electrode has a more positive potential than the left-hand electrode, the electromotive force E for the cell is positive. If E is positive, $\Delta_r G$ for the cell reaction is negative (i.e., the cell reaction is spontaneous at constant temperature and pressure). If the right-hand electrode is more positive, its electrode reaction is a reduction reaction when the cell operates spontaneously.

If the schematic representation for a cell is reversed, the electrode reactions and cell reaction are written in the opposite directions, and the signs of E and $\Delta_r G$ are reversed. The cell reaction can be multiplied or divided by a number, and that is the reason equation 7.40 contains the factor $|\nu_e|$. The electromotive force may be expressed in terms of the activities a_i of the reactants and products by use of $\mu_i = \mu_i^\circ + RT \ln a_i$ in equation 7.40, to obtain

$$\begin{aligned} -|\nu_e|FE &= \sum_i \nu_i \mu_i^\circ + RT \sum_i \nu_i \ln a_i \\ &= -|\nu_e|FE^\circ + RT \ln \prod_i a_i^{\nu_i} \end{aligned} \tag{7.41}$$

where E° is the **standard electromotive force of the cell,** the electromotive force when the activities of all reactants and products are equal to unity. Equation 7.41 is the **Nernst equation** and is usually written

$$E = E^\circ - \frac{RT}{|\nu_e|F} \ln \prod_i a_i^{\nu_i} \tag{7.42}$$

At 25 °C

$$\begin{aligned} E &= E^\circ - \frac{(8.3145 \text{ J K}^{-1} \text{ mol}^{-1})(298.15 \text{ K})}{|\nu_e|(96\,485 \text{ C mol}^{-1})} \ln \prod_i a_i^{\nu_i} \\ &= E^\circ - \frac{(0.025\,69 \text{ V})}{|\nu_e|} \ln \prod_i a_i^{\nu_i} \end{aligned} \tag{7.43}$$

Later we will see in detail how E° is determined, but for now we can note that it is the electromotive force of a cell in which all reactants and products are at unit activities.

If the activities of the reactants and products correspond to those of an equilibrium mixture, $E = 0$, and equation 7.42 becomes

$$E^\circ = \frac{RT}{|\nu_e|F} \ln K \qquad \text{or} \qquad K = e^{|\nu_e|FE^\circ/RT} \tag{7.44}$$

where K is the equilibrium constant for the cell reaction.

Example 7.2 ***Relation between the standard electromotive force and the equilibrium constant of the cell reaction***

Three different galvanic cells have standard electromotive forces E° of 0.01, 0.1, and 1.0 V, respectively, at 25 °C. Calculate the equilibrium constants of the reactions that occur in these cells assuming the charge number $|\nu_e|$ for each reaction is unity.

$$K = e^{|\nu_e|FE^\circ/RT}$$

For $E^\circ = 0.01$ V,

$$K = \exp\frac{(96\,485\ \mathrm{C\ mol^{-1}})(0.01\ \mathrm{V})}{(8.3145\ \mathrm{J\ K^{-1}\ mol^{-1}})(298.15\ \mathrm{K})}$$
$$= 1.476$$

For $E^\circ = 0.1$ V, $K = 49.0$, and for $E^\circ = 1$ V, $K = 8.02 \times 10^{16}$.

In deriving the Nernst equation, a_i refers to the activity of an electrically neutral reactant such as HCl(aq), and that meaning of a_i will be used in working with cells without liquid junctions. However, in working with cells with liquid junctions, we will consider that the reactants are ionic species and use a_i to represent the activity of an ionic species such as H^+ or Cl^-.

7.4 ACTIVITY OF ELECTROLYTES

Electrolytes have to be treated in a different way from nonelectrolytes because they dissociate, but the ions cannot be studied separately because the condition of electric neutrality applies. In work with electrolyte solutions it is customary to use the molal scale. The molality m_i is equal to the amount of electrolyte per kilogram of solvent, which is given by n_i/M_1n_1, where n_i is the amount of electrolyte in n_1 mol of solvent and M_1 is the molar mass of the solvent. Thus, the molality has the units mol kg^{-1}.

The relation between molality and mole fraction is obtained by dividing numerator and denominator by the total amount in the system to obtain $m_i = x_i/x_1M_1$. In dilute solutions the molality of species i is approximately x_i/M_1. The molality has the interesting property, as compared with the mole fraction, that the addition of a second solute does not change the molality of the first. The molality also has the advantage that it is not a function of temperature.

Activity coefficients may be given on the molality scale for nonelectrolytes as well. On the m scale the **activity** of a solute substance i is defined by*

$$a_i = \frac{\gamma_i m_i}{m^\circ} \tag{7.45}$$

*Activity coefficients may also be given on the molar concentration scale (moles per liter). On the c scale the activity of solute substance i is defined by

$$a_i = \frac{\gamma_i c_i}{c^\circ}$$

where c° is the standard value of the molar concentration (1 mol L^{-1}), and

$$\lim_{c_i \to 0} \gamma_i = 1$$

The various activity coefficients have different numerical values, but we will use γ_i for all of them to avoid confusion with other symbols. When there is danger of confusing activity coefficients on various scales, they may be given subscripts as in $\gamma_{x,i}$, $\gamma_{m,i}$, and $\gamma_{c,i}$.

where m° is the standard value of the molality (1 mol/kg solvent), and

$$\lim_{m_i \to 0} \gamma_i = 1 \tag{7.46}$$

When an infinitesimal quantity of an electrolyte is added to a kilogram of solvent, the differential change in the Gibbs energy is given by

$$dG = \mu_+ \, dm_+ + \mu_- \, dm_- \tag{7.47}$$

However, we cannot add cations and anions separately because the solution must be electrically neutral. If the strong electrolyte is $A_{\nu_+}B_{\nu_-}$, where ν_+ is the number of cations and ν_- is the number of anions, electroneutrality requires that

$$m = \frac{m_+}{\nu_+} = \frac{m_-}{\nu_-} \tag{7.48}$$

Equation 7.47 can be written

$$\begin{aligned} dG &= (\nu_+\mu_+ + \nu_-\mu_-)\, dm \\ &= \mu \, dm \end{aligned} \tag{7.49}$$

where

$$\mu = \nu_+\mu_+ + \nu_-\mu_- \tag{7.50}$$

is the chemical potential for the electrolyte, which can be determined experimentally. The chemical potentials of the cation and anion are given by

$$\mu_+ = \mu_+^\circ + RT \ln \gamma_+ m_+ \tag{7.51}$$

$$\mu_- = \mu_-^\circ + RT \ln \gamma_- m_- \tag{7.52}$$

where μ_+° and μ_-° are the standard state chemical potentials and γ_+ and γ_- are the activity coefficients of the cation and anion. We will omit the standard values m° of the molality in the remainder of this chapter to simplify the notation. When equations 7.51 and 7.52 are substituted in equation 7.50,

$$\mu = (\nu_+\mu_+^\circ + \nu_-\mu_-^\circ) + RT \ln \gamma_+^{\nu_+} \gamma_-^{\nu_-} m_+^{\nu_+} m_-^{\nu_-} \tag{7.53}$$

To make the argument of the logarithm proportional to the molality m of the electrolyte in the logarithmic term, a **mean ionic molality** $m_\pm$ and a **mean ionic activity coefficient** $\gamma_\pm$ are defined as

$$m_\pm = (m_+^{\nu_+} m_-^{\nu_-})^{1/\nu_\pm} = m(\nu_+^{\nu_+} \nu_-^{\nu_-})^{1/\nu_\pm} \tag{7.54}$$

$$\gamma_\pm = (\gamma_+^{\nu_+} \gamma_-^{\nu_-})^{1/\nu_\pm} \tag{7.55}$$

where

$$\nu_\pm = \nu_+ + \nu_- \tag{7.56}$$

Then equation 7.53 becomes

$$\mu = \mu^\circ + \nu_\pm RT \ln \gamma_\pm m_\pm \tag{7.57}$$

and the **activity of the electrolyte** is given by

$$\begin{aligned} a_{A_{\nu+}B_{\nu-}} &= (\gamma_\pm m_\pm)^{\nu_\pm} \\ &= \gamma_\pm^{\nu_\pm} m^{\nu_\pm} (\nu_+^{\nu_+} \nu_-^{\nu_-}) \end{aligned} \tag{7.58}$$

The standard chemical potential μ° of the electrolyte is the chemical potential in a solution of unit activity on the molality scale. It can be determined using deviations from Henry's law (Section 6.5) and extrapolating to $m = 0$ where $\gamma_\pm = 1$. This extrapolation is guided by the Debye–Hückel limiting law (Section 7.5).

The mean ionic molality $m_\pm$ of a 1–1 electrolyte such as NaCl is equal to m, that of a 2–1 electrolyte such as $CaCl_2$ is equal to $4^{1/3}m$, that of a 2–2 electrolyte such as $CuSO_4$ is equal to m, and that of a 3–1 electrolyte such as $LaCl_3$ is equal to $27^{1/4}m$, as may be deduced from equation 7.58. The numbers 1, 2, and 3 refer to the number of charges on the cation and anion. Equation 7.58 shows that $a(\mathrm{NaCl}) = m^2\gamma_\pm^2$, $a(\mathrm{CaCl_2}) = 4m^3\gamma_\pm^3$, $a(\mathrm{CuSO_4}) = m^2\gamma_\pm^2$, and $a(\mathrm{LaCl_3}) = 27m^4\gamma_\pm^4$.

7.5 DEBYE–HÜCKEL THEORY*

Electrolytes containing ions with multiple charges have greater effects on the activity coefficients of ions than electrolytes containing only singly charged ions. To express electrolyte concentrations in a way that takes this into account, G. N. Lewis introduced the ionic strength I, defined by

$$I = \frac{1}{2}\sum_i m_i z_i^2 = \frac{1}{2}(m_1 z_1^2 + m_2 z_2^2 + \cdots) \tag{7.59}$$

where z_i is the charge (signed) of the ion in units of the charge on a proton. The summation is continued over all the different ionic species in the solution, and m is the molal concentration. The greater effectiveness of ions of higher charge in reducing the activity coefficient is provided for by multiplying their concentrations by the square of their charges. According to equation 7.59, the ionic strength of a 1–1 electrolyte is equal to its molality. The ionic strength for a 1–2 electrolyte is $3m$, and that for a 2–2 electrolyte is $4m$.

The Coulomb forces between ions are of much longer range than van der Waals forces (Section 11.9), and therefore it is almost impossible to make measurements on electrolyte solutions at sufficiently low concentrations to obtain dilute solution behavior, in the sense of Henry's law. At infinite dilution the distribution of ions in an electrolytic solution can be considered to be completely random because the ions are too far apart to exert any attraction on each other, and the activity coefficient of the electrolyte is unity. At higher concentrations, where the ions are closer together, however, the Coulomb attractive and repulsive forces become important. Because of this interaction of ions the concentration of positive ions is higher in the neighborhood of a negative ion, and the concentration of negative ions is slightly higher in the neighborhood of a positive ion, than in the bulk solution. Because of the attractive forces between an ion and its surrounding ionic atmosphere, the activity coefficient of the electrolyte is reduced. This effect is greater for ions of high charge and is greater in solvents of lower dielectric constant, where the electrostatic interactions are stronger.

Debye and Hückel were able to show that in dilute solutions the activity coefficient γ_i of an ion species i with a charge number of z_i is given by

$$\log \gamma_i = -Az_i^2 I^{1/2} \tag{7.60}$$

*P. Debye and E. Hückel, *Phys. Z.* **24**:185, 305 (1923).

where I is the ionic strength and

$$A = \frac{1}{2.303}\left(\frac{2\pi N_A m_{solv}}{V}\right)^{1/2}\left(\frac{e^2}{4\pi\epsilon_0\epsilon_r kT}\right)^{3/2} \tag{7.61}$$

where m_{solv} is the mass of solvent in volume V and ϵ_r is the relative permittivity.

Example 7.3 ***Calculation of the Debye–Hückel constant at 298.15 K***

Calculate the value of the coefficient A in the Debye–Hückel equation for aqueous solutions at 298.15 K. The relative permittivity ϵ_r of water at this temperature is 78.54. The value of $1/4\pi\epsilon_0$ is 0.8988×10^{10} N m^2 C^{-2}.

$$A = \frac{1}{2.303}\left[\frac{2\pi(6.022\times10^{23}\ \text{mol}^{-1})(997\ \text{kg})}{1.000\ \text{m}^3}\right]^{1/2}$$

$$\times\left[\frac{(1.602\times10^{-19}\ \text{C})^2(0.8988\times10^{10}\ \text{N m}^2\ \text{C}^{-2})}{(78.54)(1.3807\times10^{-23}\ \text{J K}^{-1})(298.15\ \text{K})}\right]^{3/2}$$

$$= 0.509\ \text{kg}^{1/2}\ \text{mol}^{-1/2}$$

Since the ionic strength I has the units mol kg^{-1}, the units of $I^{1/2}$ and A cancel, as they must, to give a logarithm.

Equation 7.60 gives the activity coefficient of a single ion, but the quantity that is accessible to experimental determination is the mean ionic activity coefficient, which for the electrolyte $A_{\nu_+}B_{\nu_-}$ is given by equation 7.55.

Taking the logarithm of equation 7.55, we have

$$\log\gamma_\pm = \frac{1}{\nu_+ + \nu_-}(\nu_+\log\gamma_+ + \nu_-\log\gamma_-) \tag{7.62}$$

Substituting equation 7.60 for each activity coefficient, we have

$$\log\gamma_\pm = -A\left(\frac{\nu_+z_+^2 + \nu_-z_-^2}{\nu_+ + \nu_-}\right)I^{1/2} \tag{7.63}$$

Introducing $\nu_+z_+ = -\nu_-z_-$, we see that

$$\log\gamma_\pm = Az_+z_-I^{1/2} \tag{7.64}$$

The charge number z has the sign of the ion, and so we see that the effect of the ion atmosphere is to lower the activity coefficient of the electrolyte.

Example 7.4 ***Calculation of activity coefficients using the Debye–Hückel equation***

Use the Debye–Hückel theory to calculate γ_+, γ_-, and $\gamma_\pm$, and a_{NaCl} for 0.001 molal sodium chloride in water at 25 °C.

$$\log\gamma_i = -Az_i^2I^{1/2}$$

$$= -(0.509)(0.001)^{1/2}$$

$$\gamma_+ = \gamma_- = 0.964$$

$$\log \gamma_\pm = Az_+z_-I^{1/2}$$
$$= (0.509)(1)(-1)(0.001)^{1/2}$$
$$\gamma_\pm = (\gamma_+\gamma_-)^{1/2} = 0.964$$
$$a_{\text{NaCl}} = m^2\gamma_\pm^2 = (10^{-3})^2(0.964)^2$$
$$= 9.29 \times 10^{-7}$$

The Debye–Hückel theory has been of great value in interpreting the properties of electrolyte solutions. It is a limiting law at low ionic strengths. At high values of the ionic strength the activity coefficient of an electrolyte usually increases with increasing ionic strength. Equation 7.64 is in excellent agreement with experiment up to an ionic strength of about 0.01, but large deviations are encountered even at this ionic strength if the product of the charge of the highest charged ion of the salt and the charge of the oppositely charged ion of the electrolyte medium is greater than about 4.

In working with electrolyte solutions at higher ionic strength, the following empirical extension is often useful:

$$\log \gamma_\pm = Az_+z_-I^{1/2}/(1 + BI^{1/2}) \tag{7.65}$$

where $B = 1.6\ (\text{kg/mol})^{1/2}$ at 25 °C. The dependencies of the mean ionic activity coefficients of two salts on ionic strength are shown in Fig. 7.4 for the extended Debye–Hückel equation (i.e., equation 7.65). The Debye–Hückel theory is indispensable in interpreting measurements of electromotive force in electrochemical cells.

The cell discussed in Section 7.3 may be used to determine the activity coefficient of hydrochloric acid. For this purpose, the cell reaction will be written

$$\tfrac{1}{2}H_2(g) + AgCl(s) = HCl(m) + Ag(s) \tag{7.66}$$

Since the charge number $|\nu_e|$ is equal to unity, the electromotive force, given by equation 7.42, is

$$E = E^\circ - \frac{RT}{F}\ln\frac{a_{\text{HCl}}}{(P_{H_2}/P^\circ)^{1/2}} \tag{7.67}$$

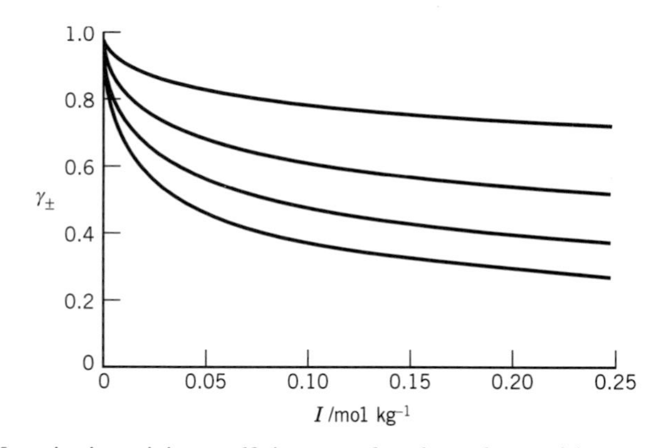

Figure 7.4 Mean ionic activity coefficients $\gamma_\pm$ for electrolytes with $z_+z_- = -1, -2, -3,$ and -4 as a function of ionic strength I according to the extended Debye–Hückel equation. (See Computer Problem 7.E.)

assuming that H_2 is an ideal gas. If the pressure of hydrogen is 1 bar and equation 7.58 is introduced, then

$$E = E^\circ - \frac{2.303RT}{F}\log(\gamma_\pm^2 m^2)$$

$$= E^\circ - 0.059\,16\log(\gamma_\pm^2 m^2) \tag{7.68}$$

The mean ionic activity coefficient of hydrochloric acid is represented by $\gamma_\pm$, and m is the molality.

As it stands, equation 7.68 contains two unknown quantities, E° and $\gamma_\pm$. These may be obtained by determining the electromotive force of this cell over a range of hydrochloric acid concentrations, including dilute solutions. Rearranging equation 7.68 and substituting numerical values for 25 °C gives

$$E + 0.1183\log m = E^\circ - 0.1183\log\gamma_\pm \tag{7.69}$$

The exponents in equation 7.68 have been placed in front of the logarithmic term, giving (2)(0.059 16) = 0.1183. Since at infinite dilution $m = 0$, $\gamma_\pm = 1$, and $\log\gamma_\pm = 0$, it can be seen that when $E + 0.1183\log m$ is plotted against m, the extrapolation of $E + 0.1183\log m$ to $m = 0$ will give E°.

To make a satisfactory extrapolation, use is made of the extended Debye–Hückel equation to furnish a function that will give a nearly straight line. The following expression is a useful empirical extension of equation 7.64 for the mean ionic activity coefficient of a 1–1 electrolyte in dilute aqueous solutions at 25 °C:

$$\log\gamma_\pm = -0.509\sqrt{m} + bm$$

where b is an empirical constant.

Substituting into equation 7.69 and rearranging terms, we have

$$E + 0.1183\log m - 0.0602m^{1/2} = E' = E^\circ - (0.1183b)m \tag{7.70}$$

According to this equation, the left-hand side, which we will designate as E', will give a straight line when it is plotted against m, and the intercept at $m = 0$ is E°.

In Fig. 7.5, E' is plotted against m. The extrapolated value is $E^\circ = 0.2224$ V when the straight line is drawn through the points at the lower molalities. This is the electromotive force that the cell would deliver with the hydrochloric acid at unit activity.

The value of E° having been determined, the activity coefficient of hydrochloric acid at any other concentration may be calculated from the electromotive force of the cell containing hydrochloric acid at that concentration.

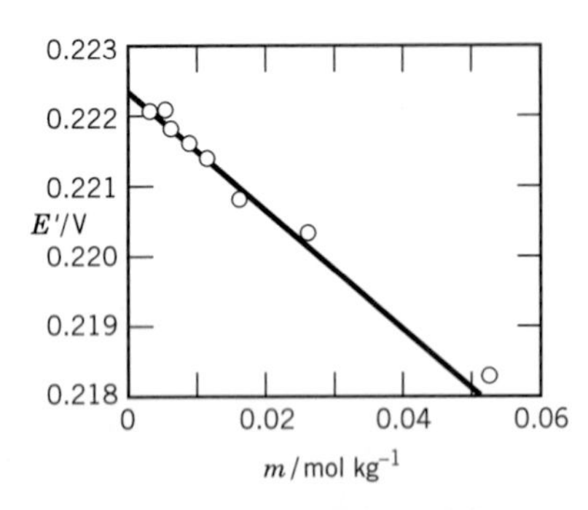

Figure 7.5 Determination of the silver–silver chloride electrode potential by extrapolation of a function of the potential of the cell Pt | H_2(1 bar) | HCl(m) | AgCl | Ag to infinite dilution.

Example 7.5 ***Determination of activity coefficients using electromotive force cells***

Calculate the mean ionic activity coefficient of 0.1 mol kg^{-1} hydrochloric acid at 25 °C from the fact that the electromotive force of the cell described in this section is 0.3524 V at 25 °C.

Table 7.1 Mean Ionic Activity Coefficients $\gamma_{\pm}$ in Water at 25 °C

$m/(\text{mol kg}^{-1})$	HCl	LiCl	NaCl	CsCl
0.01	0.905	0.904	0.902	0.899
0.02	0.875	0.873	0.870	0.865
0.05	0.830	0.825	0.820	0.807
0.10	0.796	0.790	0.778	0.756
0.20	0.767	0.757	0.735	0.718
0.40	0.755	0.740	0.693	0.628
1.0	0.809	0.774	0.657	0.544
2.0	1.009	0.921	0.668	0.495
3.0	1.316	1.156	0.714	0.478
4.0	1.762	1.510	0.783	0.473
5.0	2.38	2.02	0.874	0.474

Source: G. W. C. Kaye and T. H. Laby, *Tables of Physical and Chemical Constants,* 1973. © Reprinted by permission of Pearson Education, Inc., Upper Saddle River, NJ.

Substituting into equation 7.69, we have

$$0.3524 = 0.2224 - 0.1183 \log \gamma_{\pm} - 0.1183 \log 0.1$$

$$\log \gamma_{\pm} = \frac{-0.3524 + 0.2224 + 0.1183}{0.1183} = -0.0989$$

$$\gamma_{\pm} = 0.796$$

In this general manner the activity coefficients of the electrolytes shown in Table 7.1 and Fig. 7.6 have been determined. It should be noted that at high

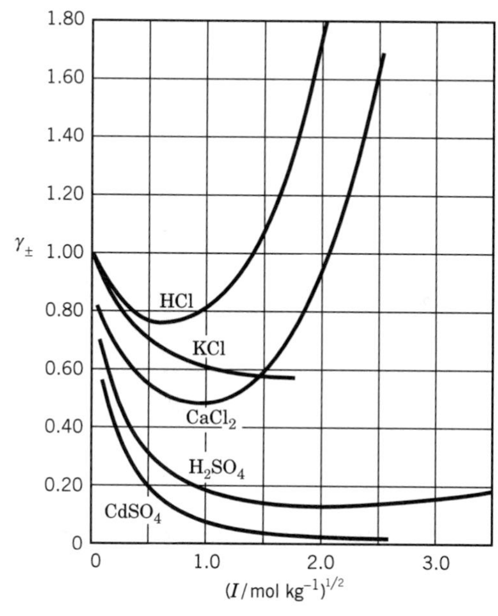

Figure 7.6 Dependence of the mean ionic activity coefficient $\gamma_{\pm}$ on $I^{1/2}$ for electrolytes at 25 °C.

concentrations of electrolytes, activity coefficients may be considerably greater than unity.

The mean ionic activity coefficient for an electrolyte can also be determined by measuring the vapor pressure of the solvent and using the Gibbs–Duhem relation. However, electrochemical cells generally provide much more accurate values of $\gamma_\pm$ for electrolytes. The same value of the activity coefficient of an electrolyte is, of course, obtained whether the equilibrium data come from measurements of vapor pressure, freezing point lowering, boiling point elevation, osmotic pressure, distribution coefficients, equilibrium constants, solubility, or electromotive force.

7.6 DETERMINATION OF STANDARD THERMODYNAMIC PROPERTIES OF IONS

The thermodynamic treatment of data from the Pt(s) | H_2(g) | HCl(m) | AgCl(s) | Ag(s) cell in the preceding section shows how the standard electromotive force for the cell can be determined. As shown by equation 7.44, the standard electromotive force of a cell yields the equilibrium constant for the cell reaction since

$$\Delta_r G^\circ = -|\nu_e| F E^\circ = -RT \ln K \tag{7.71}$$

Thus the equilibrium constant for reaction 7.66 is given by

$$\begin{aligned} K &= \exp \frac{(96\,485 \text{ C mol}^{-1})(0.2224 \text{ V})}{(8.3145 \text{ J K}^{-1} \text{ mol}^{-1})(298.15 \text{ K})} \\ &= 5745 \\ &= \frac{a_{\text{HCl}}}{(P_{\text{H}_2}/P^\circ)^{1/2}} \end{aligned} \tag{7.72}$$

where hydrogen is assumed to be an ideal gas.

If the standard electromotive force of a cell is measured as a function of temperature, then $\Delta_r S^\circ$, $\Delta_r H^\circ$, and $\Delta_r C_P^\circ$ can be calculated using

$$\Delta_r S^\circ = |\nu_e| F \left(\frac{\partial E^\circ}{\partial T} \right)_P \tag{7.73}$$

$$\Delta_r H^\circ = -|\nu_e| F E^\circ + |\nu_e| F T \left(\frac{\partial E^\circ}{\partial T} \right)_P \tag{7.74}$$

$$\Delta_r C_P^\circ = |\nu_e| F T \left(\frac{\partial^2 E^\circ}{\partial T^2} \right)_P \tag{7.75}$$

Example 7.6 ***Determination of the changes in standard thermodynamic properties of a reaction using an electromotive force cell***

The standard electromotive force of the cell Pt | H_2(g) | HCl(ai) | AgCl(s) | Ag has been determined from 0 to 90 °C by R. G. Bates and V . E. Bower, *J. Res. Nat. Bur. Stand.* **53:**283

(1954). Their data may be represented by

$$\frac{E^\circ}{\text{V}} = 0.236\,59 - 4.8564 \times 10^{-4}\left(\frac{t}{^\circ\text{C}}\right) - 3.4205 \times 10^{-6}\left(\frac{t}{^\circ\text{C}}\right)^2 + 5.869 \times 10^{-9}\left(\frac{t}{^\circ\text{C}}\right)^3$$

Determine $\Delta_r G^\circ$, $\Delta_r S^\circ$, $\Delta_r H^\circ$, and $\Delta_r C_P^\circ$ at 25 °C for the reaction

$$\tfrac{1}{2}H_2(g) + AgCl(s) = HCl(ai) + Ag(s)$$

Substituting $t = 25$ °C yields $E^\circ = 0.222\,40$ V. Thus,

$$\begin{aligned}
\Delta_r G^\circ &= -|\nu_e|FE^\circ = -(96\,485\ \text{C mol}^{-1})(0.222\,40\ \text{V}) \\
&= -21.458\ \text{kJ mol}^{-1} \\
\Delta_r S^\circ &= |\nu_e|F\left(\frac{\partial E^\circ}{\partial T}\right)_P \\
&= (96\,485\ \text{C mol}^{-1})\left[-4.8564 \times 10^{-4} - 2(3.4205 \times 10^{-6})\left(\frac{t}{^\circ\text{C}}\right)\right. \\
&\quad \left. + 3(5.869 \times 10^{-9})\left(\frac{t}{^\circ\text{C}}\right)^2\right] \\
&= -62.297\ \text{J K mol}^{-1} \\
\Delta_r H^\circ &= \Delta G^\circ + T\,\Delta S^\circ = -21.458 - (298.15)(62.297 \times 10^{-3}) \\
&= -40.032\ \text{kJ mol}^{-1} \\
\Delta_r C_P^\circ &= |\nu_e|FT\left(\frac{\partial^2 E^\circ}{\partial T^2}\right)_P \\
&= (96\,485\ \text{C mol}^{-1})(298.15\ \text{K}) \\
&\quad \times\left[-2(3.4205 \times 10^{-6}) + 6(5.869 \times 10^{-9})\left(\frac{t}{^\circ\text{C}}\right)\right] \\
&= -171.4\ \text{J K}^{-1}\ \text{mol}^{-1}
\end{aligned}$$

Since the standard thermodynamic properties of $H_2(g)$, AgCl(s), and Ag(s) are known from other sources, the standard electromotive forces of this cell over a range of temperatures yields the standard thermodynamic properties of aqueous HCl.

Example 7.7 ***Determination of the standard thermodynamic properties of HCl(ai) at 25 °C***

What are the standard thermodynamic properties of HCl($a = 1$) at 25 °C?

The properties of $H_2(g)$, AgCl(s), and Ag(s) at 25 °C are given in Table C.2. According to Example 7.6:

$$\begin{aligned}
\Delta_r G^\circ &= \Delta_f G^\circ[\text{HCl(ai)}] - \Delta_f G^\circ[\text{AgCl(s)}] \\
\Delta_f G^\circ[\text{HCl(ai)}] &= -21.458 - 109.789 \\
&= -131.247\ \text{kJ mol}^{-1} \\
\Delta_r H^\circ &= \Delta_f H^\circ[\text{HCl(ai)}] - \Delta_f H^\circ[\text{AgCl(s)}]
\end{aligned}$$

$$\Delta_f H^\circ[\mathrm{HCl(ai)}] = -40.032 - 127.068$$
$$= -167.100 \text{ kJ mol}^{-1}$$
$$\Delta_r S^\circ = \overline{S}^\circ[\mathrm{HCl(ai)}] + \overline{S}^\circ[\mathrm{Ag(s)}] - \overline{S}^\circ[\mathrm{AgCl(s)}] - \tfrac{1}{2}\overline{S}^\circ[\mathrm{H_2(g)}]$$
$$\overline{S}^\circ[\mathrm{HCl(ai)}] = -62.297 - 42.55 + 96.2 + \tfrac{1}{2}(130.684)$$
$$= 56.7 \text{ J K}^{-1} \text{ mol}^{-1}$$

$$\Delta_r C_P^\circ = \overline{C}_P^\circ[\mathrm{HCl(ai)}] + \overline{C}_P^\circ[\mathrm{Ag(s)}] - \overline{C}_P^\circ[\mathrm{AgCl(s)}] - \tfrac{1}{2}\overline{C}_P^\circ[\mathrm{H_2(g)}]$$
$$\overline{C}_P^\circ[\mathrm{HCl(ai)}] = -171.4 - 25.4 + 50.8 + \tfrac{1}{2}(28.8)$$
$$= -131.6 \text{ J K}^{-1} \text{ mol}^{-1}$$

It is important to understand that these thermodynamic properties apply to the hypothetical standard state of HCl in water illustrated in Fig. 7.7. They are the properties that HCl would have in aqueous solution at $m = 1$ mol kg^{-1} if the interactions of the ions with each other and water were the same as at infinite dilution. The activity of a 1 molal solution of HCl is less than 1, as shown by the solid line in Fig. 7.7. Notice that Fig. 7.7 is similar to Fig. 4.4, which was used to explain the concept of the standard state of a gas.

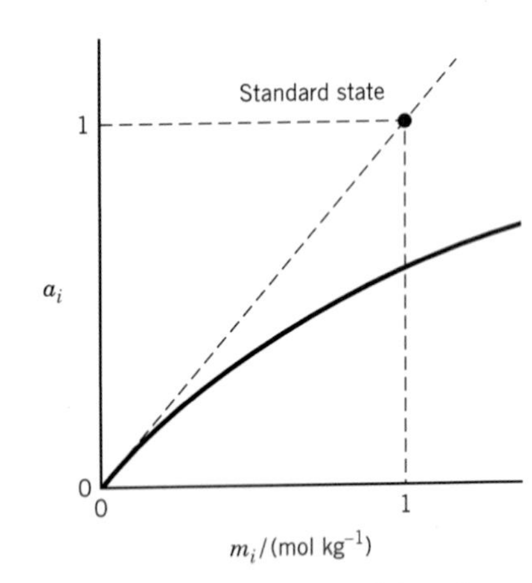

Figure 7.7 Plot of the activity of a solute versus its molality. The dashed line is for an ideal solution. The standard state is a hypothetical $m = 1$ mol kg^{-1} solution in which the solute would have unit activity if it were ideal.

The NBS Tables of Chemical Thermodynamic Properties (see Table C.2) give the standard thermodynamic properties of many electrolytes. The standard state of an electrolyte or ion is indicated by (ai) if the electrolyte is assumed to be completely ionized (such as NaCl or HCl). For a weak electrolyte (such as acetic acid) properties are given for both the completely ionized standard state (ai) and for the un-ionized (not dissociated) standard state. The designation ao is also used for an ion when no further ionization is considered. For an un-ionized solute in aqueous solution, the standard state is the ideal solution at unit molality.

Since the thermodynamic properties of completely dissociated electrolytes in water are made up of sums of contributions of ions, it would be convenient to have tables of standard thermodynamic properties of ions. However, there is no way that ions can be studied separately because of the electroneutrality condition. Nevertheless, if the properties of one ion are set by convention, the properties of other ions can be calculated. The following convention is used in preparing thermodynamic tables: $\Delta_f G^\circ(\mathrm{H^+}) = \Delta_f H^\circ(\mathrm{H^+}) = 0$ at each temperature.

For a strong electrolyte the standard Gibbs energy of formation is equal to the sum of the standard Gibbs energies of formation of the ions, and the standard enthalpy of formation is equal to the sum of the standard enthalpies of formation of the ions. Thus for HCl(ai), $\Delta_f G^\circ(\mathrm{HCl}) = \Delta_f G^\circ(\mathrm{H^+}) + \Delta_f G^\circ(\mathrm{Cl^-})$ and $\Delta_f H^\circ(\mathrm{HCl}) = \Delta_f H^\circ(\mathrm{H^+}) + \Delta_f H^\circ(\mathrm{Cl^-})$. Since $\Delta_f G^\circ(\mathrm{H^+}) = \Delta_f H^\circ(\mathrm{H^+}) = 0$, $\Delta_f G^\circ(\mathrm{HCl}) = \Delta_f G^\circ(\mathrm{Cl^-})$ and $\Delta_f H^\circ(\mathrm{HCl}) = \Delta_f H^\circ(\mathrm{Cl^-})$, as seen in Table C.2.

The situation is a little more complicated when we come to the entropies of aqueous ions. Since $\Delta_f G^\circ(\mathrm{H^+}) = \Delta_f H^\circ(\mathrm{H^+}) = 0$, it is expected that $\Delta_f S^\circ(\mathrm{H^+}) = 0$, but the convention established a long time ago, and used in current tables, is that $\overline{S}^\circ[\mathrm{H^+(ao)}] = 0$. If we consider the formation reaction for the hydrogen ion,

$$\tfrac{1}{2}\mathrm{H_2(g)} = \mathrm{H^+(ao)} + \mathrm{e^-} \tag{7.76}$$

we can see that both $\Delta_f S^\circ[\mathrm{H^+(ao)}] = 0$ and $\overline{S}^\circ[\mathrm{H^+(ao)}] = 0$ will be satisfied if the formal electron is treated as a reactant and assigned $\overline{S}^\circ(\mathrm{e^-}) = \tfrac{1}{2}\overline{S}^\circ[\mathrm{H_2(g)}]$ at each temperature. This formal electron, which is used in balancing half-cell reactions, is not an aqueous electron. Treating equation 7.76 as a formation reaction yields

$$\begin{aligned}\Delta_f S^\circ[\mathrm{H}^+(\mathrm{ao})] &= \overline{S}^\circ(\mathrm{H}^+) + \overline{S}^\circ(\mathrm{e}^-) - \tfrac{1}{2}\overline{S}^\circ[\mathrm{H}_2(\mathrm{g})] \\ &= 0 + \tfrac{1}{2}\overline{S}^\circ[\mathrm{H}_2(\mathrm{g})] - \tfrac{1}{2}\overline{S}^\circ[\mathrm{H}_2(\mathrm{g})] = 0 \end{aligned} \tag{7.77}$$

Formation reactions like equation 7.76 can be used to calculate standard Gibbs energies of formation and standard enthalpies of formation by use of the convention that $\Delta_f G^\circ(\mathrm{e}^-) = \Delta_f H^\circ(\mathrm{e}^-) = 0$ at each temperature. The reason that this problem arises with the entropy is that whereas the standard Gibbs energies and enthalpies of formation of the elements are taken as zero, the entropies of the elements have positive values. (Note that the standard molar entropies of ions can be negative.)

Example 7.8 ***Calculation of the standard molar entropy of chloride ion in aqueous solution***

Calculate the standard molar entropy of chloride ion in aqueous solution at 298.15 K starting with the Gibbs energy of formation and the enthalpy of formation from Table C.2.

First we calculate the entropy of formation of $\mathrm{Cl}^-(\mathrm{ao})$ using

$$\begin{aligned}\Delta_f S^\circ(\mathrm{Cl}^-) &= \frac{\Delta_f H^\circ(\mathrm{Cl}^-) - \Delta_f G^\circ(\mathrm{Cl}^-)}{T} = \frac{(-167\,159 + 131\,228)\ \mathrm{J\ mol^{-1}}}{298.15\ \mathrm{K}} \\ &= -120.51\ \mathrm{J\ K^{-1}\ mol^{-1}}\end{aligned}$$

Then we use the formation reaction

$$\tfrac{1}{2}\mathrm{Cl}_2(\mathrm{g}) + \mathrm{e}^- = \mathrm{Cl}^-(\mathrm{ao})$$

to calculate the entropy of the Cl^- ion. The conventional value of $\frac{1}{2}\overline{S}^\circ[\mathrm{H}_2(\mathrm{g})]$ is used for $\overline{S}^\circ(\mathrm{e}^-)$.

$$\begin{aligned}\Delta_f \overline{S}^\circ(\mathrm{Cl}^-) &= \overline{S}^\circ(\mathrm{Cl}^-) - \tfrac{1}{2}\overline{S}^\circ(\mathrm{Cl}_2) - \overline{S}^\circ(\mathrm{e}^-) \\ -120.51 &= \overline{S}^\circ(\mathrm{Cl}^-) - \frac{223.066}{2} - \frac{130.684}{2} \\ \overline{S}^\circ(\mathrm{Cl}^-) &= 56.36\ \mathrm{J\ K^{-1}\ mol^{-1}}\end{aligned}$$

Table C.2 is very useful for calculating changes in standard thermodynamic properties and equilibrium constants for chemical reactions involving aqueous ions at 298.15 K, but it is important to remember that these properties apply to solutions of zero ionic strength. In order to make calculations at higher ionic strengths, the extended Debye–Hückel equation can be used to calculate activity coefficients at low ionic strengths. It is more complicated to calculate the effects of ionic strength on $\Delta_r H^\circ$ and $\Delta_r S^\circ$. A convenient way to make quantitative calculations of the effects of ionic strength is to consider that $\Delta_f G^\circ$, $\Delta_f H^\circ$, and $\Delta_f S^\circ$ are functions of the ionic strength, in other words, to calculate the values of these properties that apply at the desired ionic strength.

Equation 7.60, for the activity coefficient of an ion, can be written as

$$\ln \gamma_i = -\alpha z_i^2 I^{1/2}/(1 + BI^{1/2}) \tag{7.78}$$

Note that ln is used rather than log. The effect of ionic strength on the standard Gibbs energy of formation of ion i in kJ mol^{-1} is given by

$$\begin{aligned} \Delta_f G_i^\circ(I) &= \Delta_f G_i^\circ(I = 0) + RT \ln \gamma_I \\ &= \Delta_f G_i^\circ(I = 0) - RT\alpha z_i^2 I^{1/2}/(1 + BI^{1/2}) \\ &= \Delta_f G_i^\circ(I = 0) - 2.91482 z_i^2 I^{1/2}/(1 + BI^{1/2}) \end{aligned} \tag{7.79}$$

where the last form applies at 298.15 K when $\Delta_f G_i^\circ$ is expressed in kJ mol^{-1}. The corresponding equation for the effect of ionic strength on the standard enthalpy of formation of an ion can be obtained by use of the Gibbs–Helmholtz equation. This yields

$$\begin{aligned} \Delta_f H_i^\circ &= \Delta_f H_i^\circ(I = 0) + RT^2(\mathrm{d}\alpha/\mathrm{d}T)_P z_i^2 I^{1/2}/(1 + BI^{1/2}) \\ &= \Delta_f H_i^\circ(I = 0) + 1.4775 z_i I^{1/2}/(1 + BI^{1/2}) \end{aligned} \tag{7.80}$$

where the last form applies at 298.15 K when the $\Delta_f H_i^\circ$ is expressed in kJ mol^{-1}. The empirical parameter B is usually taken to be 1.6 kg$^{1/2}$ mol$^{-1/2}$.

The effects of ionic strength on the Gibbs energy of reaction and the enthalpy of reaction are given by

$$\Delta_r G^\circ(I) = \Delta_r G^\circ(I = 0) - 2.914\,82\left(\sum \nu_i z_i^2\right) I^{1/2}/(1 + BI^{1/2}) \tag{7.81}$$

$$\Delta_r H^\circ(I) = \Delta_r H^\circ(I = 0) + 1.4775\left(\sum \nu_i z_i^2\right) I^{1/2}/(1 + BI^{1/2}) \tag{7.82}$$

The effect of ionic strength on the equilibrium constant of a reaction is given by

$$K(I) = K(I = 0) \times 10^{0.5107 I^{1/2} \sum \nu_i z_i^2/(1+1.6 I^{1/2})} \tag{7.83}$$

Example 7.9 ***Calculation of formation properties of H^+ (ao) and Cl^- (ao) at 298.15 K***

(*a*) Calculate $\Delta_f G^\circ$ and $\Delta_f H^\circ$ of H^+(ao) and Cl^-(ao) at 298.15 K at an ionic strength of 0.10 molal using the extended Debye–Hückel equation. (*b*) Calculate $\Delta_r G^\circ$ and $\Delta_r H^\circ$ for the following reaction:

$$\tfrac{1}{2}H_2(g) + AgCl(s) = H^+ + Cl^- + Ag(s)$$

(*c*) Calculate $E^\circ(I = 0.10\ m)$ for the corresponding galvanic cell. (*d*) Calculate $K(I = 0.10\ m)$ for the reaction.

(*a*) For the hydrogen ion, equation 7.79 yields

$$\Delta_f G^\circ(H^+, I = 0.10) = 0 - 2.914\,82(0.209\,98) = -0.612 \text{ kJ mol}^{-1}$$

where

$$\frac{I^{1/2}}{1 + 1.6 I^{1/2}} = 0.209\,98 \qquad \text{for} \qquad I = 0.10\ m$$

Equation 7.78 yields

$$\Delta_f H^\circ(H^+, I = 0.10) = 0 + 1.4775(0.209\,98) = 0.310 \text{ kJ mol}^{-1}$$

For the chloride ion,

$$\Delta_f G°(\mathrm{Cl}^-, I = 0.10) = -131.228 - 2.914\,82(0.209\,98) = -131.840 \text{ kJ mol}^{-1}$$

$$\Delta_f H°(\mathrm{Cl}^-, I = 0.10) = -167.159 + 1.4775(0.209\,98) = -166.849 \text{ kJ mol}^{-1}$$

(*b*)

$$\Delta_r G° = -0.612 - 131.84 + 0 + 109.789 = -22.663 \text{ kJ mol}^{-1}$$

$$\Delta_r H° = 0.310 - 166.849 + 0 + 127.068 = -39.471 \text{ kJ mol}^{-1}$$

(*c*)

$$E° = -\frac{\Delta_r G°}{|\nu_e| F} = \frac{226\,63 \text{ J mol}^{-1}}{964\,85 \text{ C mol}^{-1}} = 0.234\,89 \text{ V}$$

(*d*)

$$K = \exp\left[\frac{226\,63 \text{ J mol}^{-1}}{(8.314\,51 \text{ J K}^{-1} \text{ mol}^{-1})(298.15 \text{ K})}\right] = 9340$$

7.7 STANDARD ELECTRODE POTENTIALS

Rather than simply tabulating the standard electromotive forces of many electrochemical cells, it is more useful to define standard electrode potentials $E°$ and tabulate them. **The standard electrode potential is the potential of a cell in which the hydrogen electrode is on the left and all components of the cell are at unit activity.** The standard electrode potential is a signed quantity, and its value for the cell

$$\tfrac{1}{2}\mathrm{H_2(g)} + \mathrm{AgCl(s)} = \mathrm{HCl}(m) + \mathrm{Ag(s)} \tag{7.84}$$

is 0.2224 V.

Now we adopt the convention that this potential difference is to be attributed entirely to the electrode reaction in the right-hand electrode; that is, the standard electrode potential of the standard hydrogen electrode $\mathrm{H^+}(a = 1) \mid \mathrm{H_2}(1 \text{ bar}) \mid$ Pt is arbitrarily assigned the value zero so that its electrode reaction and standard electrode potential are given by

$$\mathrm{H^+(aq)} + \mathrm{e^-} = \tfrac{1}{2}\mathrm{H_2(g)} \qquad E° = 0.0000 \text{ V} \tag{7.85}$$

Thus, the electrode reaction and the standard electrode potential for the $\mathrm{Cl^-} \mid \mathrm{AgCl(s)} \mid \mathrm{Ag}$ electrode are

$$\mathrm{AgCl(s)} + \mathrm{e^-} = \mathrm{Ag(s)} + \mathrm{Cl^-(aq)} \qquad E° = 0.2224 \text{ V} \tag{7.86}$$

Notice that these electrode reactions are both written as reduction reactions, and that standard electrode potentials may be called **reduction potentials.** Thus, electrode potentials are a measure of the tendency of an electrode reaction to occur in the direction of reduction. Since the formal electron can be treated like a species in thermodynamic calculations, thermodynamic tables such as Table C.2 can be used to calculated electrode potentials using $E° = -\Delta_r G°/|v_e|F$. A brief listing of electrode potentials is given in Table 7.2. Notice that the fluorine electrode has the most positive electrode potential; that is, the reaction

$$\tfrac{1}{2}\mathrm{F_2(g)} + \mathrm{e^-} = \mathrm{F^-}(a = 1) \qquad E° = 2.87 \text{ V} \tag{7.87}$$

has the greatest tendency to go to the right of all of the reactions listed. Thus $F_2(g)$ has the greatest tendency to pull electrons out of a platinum electrode and leave it positively charged. The lithium electrode has the most negative standard electrode potential; that is, the reaction

$$Li^+(a = 1) + e^- = Li(s) \qquad E° = -3.045\ V \tag{7.88}$$

has the least tendency to go to the right of all of the reduction reactions listed. Thus $Li^+(a = 1)$ has the least tendency to pull electrons out of a Li(s) electrode and leave it positively charged.

Table 7.2 Standard Electrode Potentials at 25 °C[a,e]

Electrode	*E°/V*	*Electrode Reaction*
$F^- \mid F_2(g) \mid Pt$	2.87	$\frac{1}{2}F_2(g) + e^- = F^-$
$Au^{3+} \mid Au$	1.50	$\frac{1}{3}Au^{3+} + e^- = \frac{1}{3}Au$
$Pb^{2+} \mid PbO_2 \mid Pb$	1.455	$\frac{1}{2}PbO_2 + 2H^+ + e^- = \frac{1}{2}Pb^{2+} + H_2O$
$Cl^- \mid Cl_2(g) \mid Pt$	1.3604	$\frac{1}{2}Cl_2(g) + e^- = Cl^-$
$H^+ \mid O_2(g) \mid Pt$	1.2288	$H^+ + \frac{1}{4}O_2(g) + e^- = \frac{1}{2}H_2O$
$Ag^+ \mid Ag$	0.7992	$Ag^+ + e^- = Ag$
$Fe^{3+}, Fe^{2+} \mid Pt$	0.771	$Fe^{3+} + e^- = Fe^{2+}$
$I^- \mid I_2(s) \mid Pt$	0.5355	$\frac{1}{2}I_2 + e^- = I^-$
$Cu^+ \mid Cu$	0.521	$Cu^+ + e^- = Cu$
$OH^- \mid O_2(g) \mid Pt$[b]	0.4009	$\frac{1}{4}O_2(g) + \frac{1}{2}H_2O + e^- = OH^-$
$Cu^{2+} \mid Cu$	0.3394	$\frac{1}{2}Cu^{2+} + e^- = \frac{1}{2}Cu$
$Cl^- \mid Hg_2Cl_2(s) \mid Hg$[c]	0.268	$\frac{1}{2}Hg_2Cl_2 + e^- = Hg + Cl^-$
$Cl^- \mid AgCl(s) \mid Ag$	0.2224	$AgCl + e^- = Ag + Cl^-$
$Cu^{2+}, Cu^+ \mid Pt$[d]	0.153	$Cu^{2+} + e^- = Cu^+$
$Br^- \mid AgBr(s) \mid Ag$	0.0732	$AgBr + e^- = Ag + Br^-$
$H^+ \mid H_2(g) \mid Pt$	0.0000	$H^+ + e^- = \frac{1}{2}H_2(g)$
$D^+ \mid D_2(g) \mid Pt$	−0.0034	$D^+ + e^- = \frac{1}{2}D_2(g)$
$Pb^{2+} \mid Pb$	−0.126	$\frac{1}{2}Pb^{2+} + e^- = \frac{1}{2}Pb$
$Sn^{2+} \mid Sn$	−0.140	$\frac{1}{2}Sn^{2+} + e^- = \frac{1}{2}Sn$
$Ni^{2+} \mid Ni$	−0.250	$\frac{1}{2}Ni^{2+} + e^- = \frac{1}{2}Ni$
$Cd^{2+} \mid Cd$	−0.4022	$\frac{1}{2}Cd^{2+} + e^- = \frac{1}{2}Cd$
$Fe^{2+} \mid Fe$	−0.440	$\frac{1}{2}Fe^{2+} + e^- = \frac{1}{2}Fe$
$Zn^{2+} \mid Zn$	−0.763	$\frac{1}{2}Zn^{2+} + e^- = \frac{1}{2}Zn$
$OH^- \mid H_2(g) \mid Pt$	−0.8279	$H_2O + e^- = \frac{1}{2}H_2(g) + OH^-$
$Mg^{2+} \mid Mg$	−2.37	$\frac{1}{2}Mg^{2+} + e^- = \frac{1}{2}Mg$
$Na^+ \mid Na$	−2.714	$Na^+ + e^- = Na$
$Li^+ \mid Li$	−3.045	$Li^+ + e^- = Li$

[a] All ions are at unit activity (on the molal scale) in water, and all gases are at 1 bar.

[b] See Problem 7.61.

[c] The electrode potential of the normal calomel electrode is 0.2802 V and that of the calomel electrode containing saturated KCl is 0.2415 V.

[d] The order of writing the ions in the electrolyte solution is immaterial.

[e] Many more standard electrode potentials at 298.15 K are given in S. G. Bratsch, *J. Phys. Chem. Ref. Data* **18**:1 (1989).

We saw earlier that the electromotive force of a cell is given by $E = \phi_R - \phi_L$ (equation 7.38). We can express the electromotive force of a cell in terms of the electrode potentials of the right and left electrodes, $E = E_R - E_L$, and the standard electromotive force E° of a cell in terms of the standard electrode potentials of the right and left electrodes:

$$E^\circ = E_R^\circ - E_L^\circ \tag{7.89}$$

The advantage of having a list of standard electrode potentials like Table 7.2 is that we can calculate the standard electrode potentials of any cell involving the electrodes listed. By use of the Nernst equation 7.42 we can calculate the electromotive force E of a cell for specified concentrations or partial pressures of reactants. It is also possible to identify which electrode will be positive and which will be negative when the cell delivers a current. This table is also a source of information about equilibrium constants since $K = \exp(|\nu_e| FE^\circ/RT)$ (equation 7.44).

To use Table 7.2 without making mistakes, it is a good idea to follow some simple rules. The first rule is that the half-cell reactions that yield the desired reaction are written down as reductions with their standard electrode potentials (reduction potentials). For example, consider the cell Pt | $P(H_2)$ | HCl(m) | $P(Cl_2)$ | Pt. The half-cell reactions are

Right: $\quad Cl_2(g) + 2e^- = 2Cl^- \qquad E_R^\circ = 1.3604$ V $\quad$ (7.90)

Left: $\quad 2H^+ + 2e^- = H_2(g) \qquad E_L^\circ = 0$ V $\quad$ (7.91)

The two half-cell reactions should be written with the same number of electrons, but E° is not affected by $|\nu_e|$. The standard electromotive force for the cell is obtained by subtracting E_L° from E_R° (equation 7.89), and the cell reaction is obtained by subtracting the half-cell reaction for the left electrode from the half-cell reaction for the right electrode.

$$H_2(g) + Cl_2(g) = 2H^+ + 2Cl^- = 2HCl(ai) \qquad E^\circ = 1.3604 \text{ V} \tag{7.92}$$

If the standard electromotive force is positive, the reaction will spontaneously go to the right if the reactants are in their standard states, and the right electrode will be the positive electrode when the cell is operated as a galvanic cell. If the standard electrode potential is negative, the cell reaction will go spontaneously to the left, and the right electrode will be the negative electrode when the cell is operated as a galvanic cell.

The equilibrium constant for reaction 7.92 is given by

$$K = \exp\frac{|\nu_e|FE^\circ}{RT} = \exp\left[\frac{2(96\,485 \text{ C})(1.3604 \text{ V})}{(8.3145 \text{ J K}^{-1} \text{ mol}^{-1})(298.15 \text{ K})}\right]$$

$$= 9.78 \times 10^{45} = \frac{a(\text{HCl})^2(P^\circ)^2}{P(H_2)P(Cl_2)} \tag{7.93}$$

The electromotive force for the cell with different partial pressures of the gases and HCl not in its standard state can be calculated using the Nernst equation:

$$E = E^\circ - \frac{RT}{|\nu_e|F}\ln\frac{a(\text{HCl})^2(P^\circ)^2}{P(H_2)P(Cl_2)} \tag{7.94}$$

If the cell has a liquid junction, we write the Nernst equation in terms of ion species, as shown in the following example.

Example 7.10 ***Calculation of an equilibrium constant from the standard electromotive force of a galvanic cell***

Consider the following galvanic cell:

$$\mathrm{Zn(s)} \mid \mathrm{Zn^{2+}} \,\vdots\vdots\, \mathrm{Cu^{2+}} \mid \mathrm{Cu(s)}$$

What are (*a*) the cell reaction, (*b*) the standard electromotive force, (*c*) the equilibrium constant, and (*d*) the equilibrium constant expression?

$$\begin{array}{lll} (a,b)\ \mathrm{R} & \mathrm{Cu^{2+}} + 2\mathrm{e^-} = \mathrm{Cu(s)} & E^\circ = 0.339\ \mathrm{V} \\ \quad\ \ \mathrm{L} & \mathrm{Zn^{2+}} + 2\mathrm{e^-} = \mathrm{Zn(s)} & E^\circ = -0.763\ \mathrm{V} \\ & \mathrm{Zn(s)} + \mathrm{Cu^{2+}} = \mathrm{Cu(s)} + \mathrm{Zn^{2+}} & E^\circ = 0.337 - (-0.763) = 1.102\ \mathrm{V} \end{array}$$

$$(c)\ K = \exp\left[\frac{2(96\,485)(1.102)}{(8.3145)(298.15)}\right] = 1.80 \times 10^{37}$$

$$(d)\ K = \frac{a(\mathrm{Zn^{2+}})}{a(\mathrm{Cu^{2+}})} = \frac{[\mathrm{Zn^{2+}}]\gamma(\mathrm{Zn^{2+}})}{[\mathrm{Cu^{2+}}]\gamma(\mathrm{Cu^{2+}})} = \frac{[\mathrm{Zn^{2+}}]}{[\mathrm{Cu^{2+}}]}$$

where the last form would be obtained if the solutions on the opposite sides of the salt bridge have the same ionic strength.

Example 7.11 ***Calculation of standard electrode potentials using a thermodynamic table***

Use Table C.2 to calculate the standard electrode potentials for the following three electrodes: $\mathrm{Cd^{2+}} \mid \mathrm{Cd}$, $\mathrm{Cl^-} \mid \mathrm{Cl_2(g)} \mid \mathrm{Pt}$, $\mathrm{Cl^-} \mid \mathrm{AgCl(s)} \mid \mathrm{Ag}$.

$$\mathrm{Cd^{2+}(ao)} + 2\mathrm{e^-} = \mathrm{Cd(s)}$$

$$\Delta_r G^\circ = -(-77.612) = 77.612\ \mathrm{kJ\ mol^{-1}}$$

$$E^\circ = -\frac{\Delta G^\circ}{|\nu_e|F} = -\frac{77\,612\ \mathrm{J\ mol^{-1}}}{2(96\,485\ \mathrm{C\ mol^{-1}})}$$

$$= -0.4022\ \mathrm{V}$$

$$\mathrm{Cl_2(g)} + 2\mathrm{e^-} = 2\mathrm{Cl^-(ao)}$$

$$\Delta_r G^\circ = 2(-131.228\ \mathrm{kJ\ mol^{-1}})$$

$$E^\circ = \frac{2(131\,228\ \mathrm{J\ mol^{-1}})}{2(96\,485\ \mathrm{C\ mol^{-1}})}$$

$$= 1.360\,1\ \mathrm{V}$$

$$\mathrm{AgCl(s)} + \mathrm{e^-} = \mathrm{Ag(s)} + \mathrm{Cl^-(ao)}$$

$$\Delta_r G^\circ = -131.228 - (-109.789) = -21.439\ \mathrm{kJ\ mol^{-1}}$$

$$E^\circ = -\frac{-21\,439}{96\,485} = 0.2222\ \mathrm{V}$$

In concluding our discussion of equilibrium constants of solution reactions, there are two more points to mention. The first concerns the convention on reactions in dilute solution that involve the solvent as a reactant. In this case the solvent is usually treated on the mole fraction scale, rather than the molal or molar concentration scale used for the other reactants. In this case the equilibrium constant expression is written

$$K = (\gamma_{x,A} x_A)^{\nu_A} \prod_{i \neq A} \left(\frac{\gamma_{m,i} m_i}{m^\circ} \right)^{\nu_i} \tag{7.95}$$

where A is the solvent, $\gamma_{x,A}$ is its activity coefficient on the mole fraction scale, $\gamma_{m,i}$ is the activity coefficient of reactant i on the molal scale, and m° is the standard value of the molality (1 mol kg^{-1}). For dilute solutions $\gamma_{x,A} x_A$ is close to unity, and so the equilibrium constant expression is written

$$K = \prod_{i \neq A} \left(\frac{\gamma_{m,i} m_i}{m^\circ} \right)^{\nu_i} \tag{7.96}$$

which simplifies further if the $\gamma_{m,i}$ are close to unity. Although the solvent is left out of equation 7.96, the Gibbs energy of formation of the solvent must be included in calculating ΔG° for a reaction in liquid solution.

The second point is that some species in aqueous solution may be listed in thermodynamic tables in more than one way—versus for example, NH_3(ao) or NH_4OH(ao), CO_2(ao) versus H_2CO_3(ao). In many of these cases we do not know the extent of hydration because of the difficulty in distinguishing between the species in solution. The convention in the NBS tables is that $\Delta_f G^\circ = \Delta_f H^\circ = \Delta_f S^\circ = 0$ for the hydration reactions $B(ao) + nH_2O = B \cdot (H_2O)_n(ao)$ of these species. In dilute solutions, the concentrations of the two forms are proportional to each other, so equilibrium constant expressions may be written in terms of either one of the pair.

Comment:

The table of electrode potentials is a convenience rather than a necessity, in the sense that this information is in the table of standard thermodynamic properties. In fact, the table of standard thermodynamic properties can be used to calculate electrode potentials of electrodes that do not actually yield potentials in the laboratory because the reactions involved are not fast enough. Note the tremendous range of equilibrium constants that is covered by the table of electrode potentials. The strongest oxidizing agent is F_2(g) *because it takes up electrons so strongly. The strongest reducing agent is* Li(s) *because it gives up electrons the most readily. The equilibrium constant at room temperature for the reaction* $\frac{1}{2}F_2(g) + Li(s) = Li^+(aq) + F^-(aq)$ *is about* 10^{100}.

7.8 DETERMINATION OF pH

The concentrations of hydrogen ions in aqueous solutions range from about 1 mol L^{-1} in 1 mol L^{-1} HCl to about 10^{-14} mol L^{-1} in 1 mol L^{-1} NaOH. Because of this wide range of concentrations, Sorenson adopted an exponential notation in 1909. He defined pH as the negative exponent of 10 that gives the hydrogen ion concentration. Now the pH is defined to be as close as possible to the negative base 10 logarithm of the hydrogen ion activity:

$$\mathrm{pH} = -\log a_{H^+} \tag{7.97}$$

Strictly speaking, the activity of a single ion cannot be determined, but pH meters are calibrated with buffers for which the pH has been calculated using the extended Debye–Hückel equation.

The pH may be measured with a hydrogen electrode connected with a calomel electrode through a salt bridge:

$$\mathrm{Pt} \mid \mathrm{H_2}(P_{\mathrm{H_2}}) \mid \mathrm{H^+}(a_{\mathrm{H^+}}) \,\vdots\vdots\, \mathrm{Cl^-} \mid \mathrm{Hg_2Cl_2} \mid \mathrm{Hg}$$

The electromotive force of this cell may be considered to be made up of three contributions:

$$E = 0.2802 - 0.0591 \log\left[\frac{a_{\mathrm{H^+}}}{(P_{\mathrm{H_2}}/P^\circ)^{1/2}}\right] + E_{\text{liquid junction}} \tag{7.98}$$

where the contribution by the normal calomel electrode is 0.2802 V at 25 °C. Although the activities of single ions cannot be determined, equation 7.98 is often used with the assumption that $E_{\text{liquid junction}} = 0$. If $P_{\mathrm{H_2}} = 1$ bar, then

$$E - 0.2802 = -0.0591 \log a_{\mathrm{H^+}} \tag{7.99}$$

Hydrogen ion activities obtained in this way are of great practical use, even though they are based on an approximation.

When $\mathrm{pH} = -\log a_{\mathrm{H^+}}$, equation 7.99 becomes

$$E - 0.2802 = 0.0591\ \mathrm{pH} \tag{7.100}$$

or

$$\mathrm{pH} = \frac{E - 0.2802}{0.0591} \tag{7.101}$$

Usually the pH is measured with a glass electrode because this avoids the use of hydrogen and the possibility of poisoning the platinized platinum surface. A glass electrode consists of a reversible electrode, such as a calomel or Ag–AgCl electrode, in a solution of constant pH inside a thin membrane of a special glass. The thin glass bulb of this electrode is immersed in the solution to be studied along with a reference calomel electrode to form the cell indicated by

$$\mathrm{Ag} \mid \mathrm{AgCl} \mid \mathrm{Cl^-}, \mathrm{H^+} \mid \text{glass membrane} \mid \text{solution} \,\vdots\vdots\, \text{calomel electrode}$$

and by Fig. 7.8. It is found experimentally that the potential of such a glass electrode varies with the activity of hydrogen ions in the same way as the hydrogen electrode, that is, 0.0591 V per pH unit at 25 °C. An ordinary potentiometer cannot be used to measure the voltage of such a cell because of the high resistance of the glass membrane, so an electronic voltmeter must be employed. Electronic devices using the glass electrode have been developed that make it possible to measure pH values to ±0.01 pH unit with an easily portable apparatus. The pH meter, as it is often called, is calibrated by means of a buffer of known pH before it is used to measure the pH of an unknown solution.*

The glass electrode has become the most useful electrode for determining the pH of a solution. It is not affected by oxidizing or reducing agents and is not easily poisoned. It is especially useful in biochemical investigations.

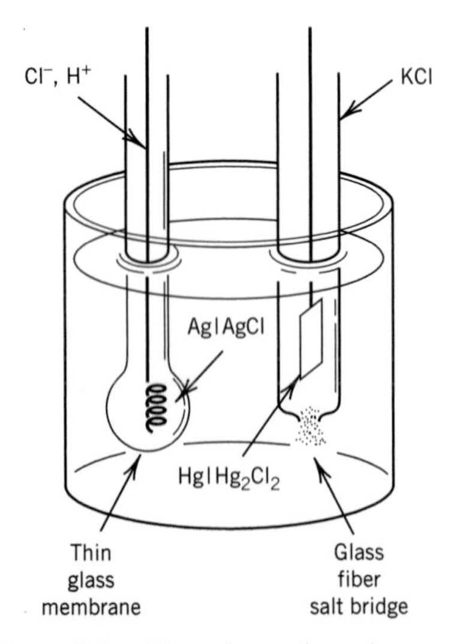

Figure 7.8 Glass electrode and calomel electrode for pH meters.

*Definition of pH Scales, Standard Reference Values, Measurement of pH and Related Terminology. *Pure Appl. Chem.* **57**:531 (1985).

The carbon dioxide electrode is another useful clinical tool. This electrode consists of a glass electrode in contact with a solution of fixed bicarbonate concentration that is separated from the sample solution by a polymer membrane permeable to CO_2. When CO_2 diffuses into the bicarbonate solution, it is hydrated to H_2CO_3 and then rapidly ionizes to $HCO_3^- + H^+$ with a consequent change in pH.

There are a growing number of ion-selective electrodes that use semipermeable membranes to obtain a cell potential that depends on a particular ion.

7.9 SPECIAL TOPIC: FUEL CELLS

A fuel cell is a cell that is continuously supplied with an oxidant and a reductant so that it can deliver a current indefinitely. Fuel cells offer the possibility of achieving high thermodynamic efficiency in the conversion of Gibbs energy to mechanical work. Internal combustion engines at best convert only a small part of the fraction $(T_2 - T_1)/T_2$ of the heat of combustion to mechanical work. In this relation, which comes from the second law of thermodynamics, T_2 is the temperature of the gas during expansion and T_1 is the temperature of the exhaust (see Section 3.9).

Fuel cells may be classified according to the temperature range in which they operate: low temperature (25 to 100 °C), medium temperature (100 to 500 °C), high temperature (500 to 1000 °C), and very high temperature (above 1000 °C). The advantage of using high temperatures is that catalysts for the various steps in the process are not so necessary. Polarization of a fuel cell reduces the current. Polarization is the result of slow reactions or processes such as diffusion in the cell.

Figure 7.9 indicates the construction of a hydrogen–oxygen fuel cell with a solid electrolyte, which is an ion-exchange membrane. The membrane is impermeable to the reactant gases but is permeable to hydrogen ions, which carry the current between the electrodes. To facilitate the operation of the cell at 40 to 60 °C, the electrodes are covered with finely divided platinum that functions as a catalyst. Water is drained out of the cell during operation. Fuel cells of this general type have been used successfully in the space program and are quite efficient. Their disadvantages for large-scale commercial application are that hydrogen presents storage problems, and platinum is an expensive catalyst. Less expensive catalysts have been found for higher-temperature operation of hydrogen–oxygen fuel cells.

Fuel cells that use hydrocarbons and air have been developed, but their power per unit weight is too low to make them practical in ordinary automobiles. Better catalysts are needed.

A hydrogen–oxygen fuel cell may have an acidic or alkaline electrolyte. The half-cell reactions are

$$\begin{array}{rcll} \frac{1}{2}O_2(g) + 2H^+ + 2e^- &=& H_2O(l) & E^\circ = 1.2288\ V \\ 2H^+ + 2e^- &=& H_2(g) & E^\circ = 0 \\ \hline H_2(g) + \frac{1}{2}O_2(g) &=& H_2O(l) & E^\circ = 1.2288\ V \end{array}$$

or

$$\begin{array}{rcll} \frac{1}{2}O_2(g) + H_2O(l) + 2e^- &=& 2OH^- & E^\circ = 0.4009\ V \\ 2H_2O(l) + 2e^- &=& H_2(g) + 2OH^- & E^\circ = -0.8279\ V \\ \hline H_2(g) + \frac{1}{2}O_2(g) &=& H_2O(l) & E^\circ = 1.2288\ V \end{array}$$

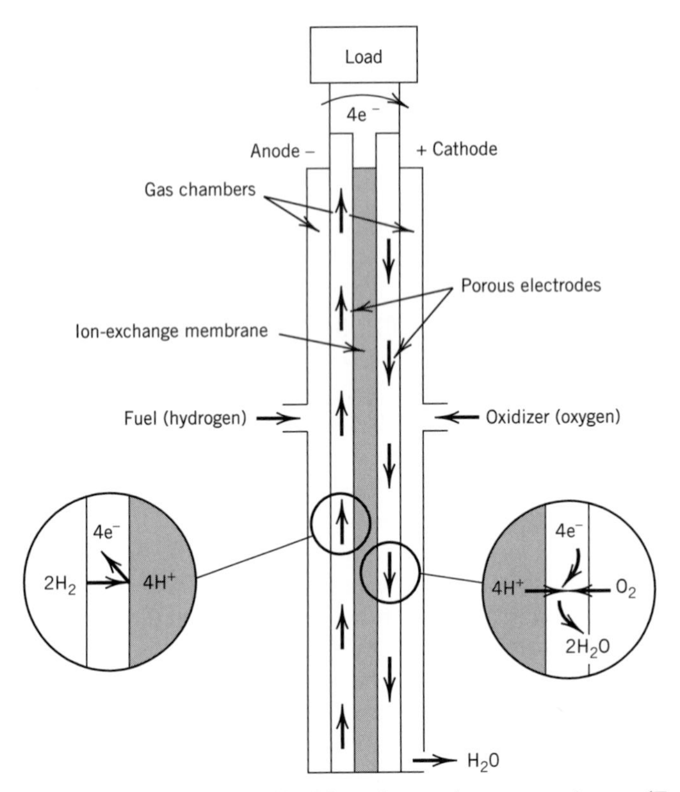

Figure 7.9 Hydrogen–oxygen fuel cell with an ion-exchange membrane. (From J. O'M. Bockris and S. Srinivasan, *Fuel Cells, Their Electrochemistry.* New York: McGraw-Hill. Copyright ©1969 by McGraw-Hill, Inc. Used with permission of McGraw-Hill Book Co.)

To maximize the power per unit mass of an electrochemical cell, the electronic and electrolytic resistances of the cell must be minimized. Since fused salts have lower electrolytic resistances than aqueous solutions, high-temperature electrochemical cells are of special interest for practical applications. High temperatures also allow the use of liquid metal electrodes, which enable higher current densities than solid electrodes.

7.10 SPECIAL TOPIC: MEMBRANE POTENTIAL

If two different electrolyte solutions are separated by a membrane, a potential difference will be set up between the two solutions if the membrane is permeable to some ions and impermeable to others. For example, consider two KCl solutions separated by a membrane permeable to K^+ but impermeable to Cl^-. If solution α is more concentrated than solution β, K^+ ions will diffuse through the membrane from α to β. This will cause solution β to become positively charged relative to solution α and to have a higher electric potential. Actually, the amount of K^+ that has to diffuse through the membrane to produce the potential difference is chemically insignificant, as we have seen in Section 7.1. As K^+ diffuses through the membrane, the electrical potential difference that is set up across the membrane retards the diffusion of more K^+, and eventually an equilibrium is reached.

At equilibrium, equation 7.20 applies to each ion that can pass through the membrane, and so insertion of equation 7.19 yields

$$\mu_i'(\alpha) + z_i F\phi(\alpha) = \mu_i'(\beta) + z_i F\phi(\beta) \tag{7.102}$$

The transformed chemical potential depends on the activity a_i as the chemical potential does, so we have

$$\mu_i' = \mu_i'^{\circ} + RT \ln a_i \tag{7.103}$$

Substituting this relation in equation 7.102, we obtain

$$\mu_i'^{\circ}(\alpha) + RT \ln a_i(\alpha) + z_i F\phi(\alpha) = \mu_i'^{\circ}(\beta) + RT \ln a_i(\beta) + z_i F\phi(\beta) \tag{7.104}$$

Since the solvent is the same on both sides of the membrane and the electric potentials have to be the same in the definition of the standard states for the transformed chemical potential of i, $\mu_i'^{\circ}(\alpha) = \mu_i'^{\circ}(\beta)$, equation 7.102 becomes

$$RT \ln a_i(\alpha) + z_i F\phi(\alpha) = RT \ln a_i(\beta) + z_i F\phi(\beta) \tag{7.105}$$

Thus

$$\Delta\phi = \phi(\beta) - \phi(\alpha) = -\frac{RT}{z_i F} \ln \frac{a_i(\beta)}{a_i(\alpha)} \tag{7.106}$$

where $\Delta\phi$ is referred to as the **membrane potential.** This equation may be written in terms of molal concentrations if the activity coefficients of ion i are nearly the same in solutions α and β. The membrane potential $\Delta\phi$ may be measured by placing identical reversible electrodes in solutions α and β.

Example 7.12 ***Calculation of a membrane potential***

The membrane potential for a resting nerve cell is given by $\Delta\phi = \phi_{\text{int}} - \phi_{\text{ext}} = -70$ mV, where ϕ_{int} is the potential internal to the cell and ϕ_{ext} is the potential external to the cell. Given the fact that the concentration of K^+ inside a resting nerve cell is about 35 times that outside the cell, what membrane potential is expected?

$$\Delta\phi = \phi_{\text{int}} - \phi_{\text{ext}} = -\frac{(8.3145 \text{ J K}^{-1} \text{ mol}^{-1})(298 \text{ K})}{96\,485 \text{ C mol}^{-1}} \ln 35$$

$$= -91 \text{ mV}$$

This is perhaps as close as equation 7.106 should be expected to come to the observed -70 mV because the resting nerve cell is not actually at equilibrium.

While K^+ is concentrated inside a nerve cell, Na^+ is at about a 10-fold higher concentration outside. The concentration differences for these ions across the cell membrane are maintained by "pumps" utilizing energy from ATP (Section 8.6) and under the control of enzymes.

When a nerve impulse starts at one end of a nerve cell, the membrane potential becomes momentarily positive. When this happens, the membrane's permeability to Na^+ momentarily increases and $\Delta\phi$ moves toward the equilibrium value for Na^+ of about $+60$ mV. This pulse propagates along the nerve cell with a speed of 10–100 m s^{-1}. After the peak, the permeability to Na^+ decreases and the permeability to K^+ temporarily increases so that $\Delta\phi$ returns to its resting value of -70 mV.

Ten Key Ideas in Chapter 7

1. The electric work of transferring a mole of electric charge through a potential difference E is $z_i FE$, where z_i is the charge number of the ion and F is the Faraday constant.
2. In a galvanic cell a chemical reaction occurs spontaneously and produces an electric potential difference between electrical conductors. In an electrolytic cell, the application of an electric potential difference between the electrodes produces chemical reactions in the cell.
3. Cells without liquid junctions can be given exact thermodynamic treatments. Cells with liquid junctions involve irreversible processes, and so they cannot be given exact treatments. However, cells with salt bridges are widely used, and the error due to the liquid junction potential is believed to be small.
4. The Nernst equation gives the electromotive force of a cell in terms of the standard electromotive force for the cell reaction and the activities of the species in the cell reaction. The standard electromotive force is directly related to the equilibrium constant for the cell reaction.
5. The activities of single ions cannot be determined, but the mean ionic activities of a neutral electrolyte can. The Debye–Hückel equation predicts that the logarithm of the activity coefficient is proportional to the square root of the ionic strength in dilute solutions.
6. Measurements of electromotive forces of cells without transference yield standard Gibbs energies of formation, and measurements at a series of temperatures yield in addition standard enthalpies of formation, standard molar entropies, and standard molar heat capacities.
7. Standard electrode potentials (reduction potentials) can be used to calculate electromotive forces of galvanic cells and the equilibrium constants of reactions occurring in galvanic cells.
8. A useful application of electromotive force measurements is the determination of the pH with a glass electrode. Because of the resistance of the glass membrane an electronic voltmeter has to be used to measure the electromotive force.
9. Fuel cells offer the opportunity to achieve high thermodynamic efficiency in the conversion of Gibbs energy to mechanical work. Better catalysts may make fuel cells that use hydrocarbons and air practical.
10. When different electrolyte solutions are separated by a membrane that is permeable by some ions and not others, an electric potential is set up between the two phases. Membrane potentials are involved in the propagation of nerve impulses.

REFERENCES

A. Bard and L. Faulkner, *Electrochemical Methods.* Hoboken, NJ: Wiley, 1980.
A. J. Bard, R. Parsons, and J. Jordan, *Standard Potentials in Aqueous Solutions.* New York: Dekker, 1985.
J. Goodisman, *Electrochemistry: Theoretical Foundations.* Hoboken, NJ: Wiley, 1987.
A. McDougall, *Fuel Cells.* Hoboken, NJ: Wiley, 1976.
P. Rieger, *Electrochemistry.* New York: Chapman & Hall, 1994.

PROBLEMS

(M) Problems marked with an icon may be more conveniently solved on a personal computer with a mathematical program.

7.1 How much work is required to bring two protons from an infinite distance of separation to 0.1 nm? Calculate the answer in joules using the protonic charge 1.602×10^{-19} C. What is the work in kJ mol^{-1} for a mole of proton pairs?

7.2 How much work in kJ mol^{-1} can in principle be obtained when an electron is brought to 0.5 nm from a proton?

7.3 A small dry battery of zinc and ammonium chloride weighing 85 g will operate continuously through a 4-Ω resistance for 450 min before its voltage falls below 0.75 V. The initial voltage is 1.60, and the effective voltage over the whole life of the battery is taken to be 1.00. Theoretically, how many kilometers above the earth could this battery be raised by the energy delivered under these conditions?

7.4 (*a*) The mean ionic activity coefficient of 0.1 molar HCl(aq) at 25 °C is 0.796. What is the activity of HCl in this solution? (*b*) The mean ionic activity coefficient of 0.1 molar H_2SO_4 is 0.265. What is the activity of H_2SO_4 in this solution?

7.5 The solubility of Ag_2CrO_4 in water is 8.00×10^{-5} mol kg^{-1} at 25 °C, and its solubility in 0.04 mol kg^{-1} $NaNO_3$ is 8.84×10^{-5} mol kg^{-1}. What is the mean ionic activity coefficient of Ag_2CrO_4 in 0.04 mol kg^{-1} $NaNO_3$?

7.6 A solution of NaCl has an ionic strength of 0.24 mol kg^{-1}. (*a*) What is its molality? (*b*) What molality of Na_2SO_4 would have the same ionic strength? (*c*) What molality of $MgSO_4$?

7.7 Using the limiting law, calculate the mean ionic activity coefficients at 25 °C in water of the following electrolytes at 10^{-3} m: (*a*) NaCl, (*b*) $CaCl_2$, (*c*) $LaCl_3$.

7.8 Estimate the electromotive force of the cell Zn(s) | $ZnCl_2$(aq, 0.02 mol kg^{-1}) | AgCl(s) | Ag(s) at 25 °C using the Debye–Hückel equation.

7.9 The cell Pt | H_2(1 bar) | HBr(m) | AgBr | Ag has been studied by H. S. Harned, A. S. Keston, and J. G. Donelson [*J. Am. Chem. Soc.* **58**:989 (1936)]. The following table gives the electromotive forces obtained at 25 °C:

$m/\text{mol kg}^{-1}$	0.01	0.02	0.05	0.10
E/V	0.3127	0.2786	0.2340	0.2005

Calculate (*a*) $E°$ and (*b*) the activity coefficient for a 0.10 mol kg^{-1} solution of hydrogen bromide.

7.10 Design cells without a liquid junction that could be used to determine the activity coefficients of aqueous solutions of (*a*) NaOH and (*b*) H_2SO_4. Give the equations relating electromotive force to the mean ionic activity coefficient at 25 °C.

7.11 The electromotive force of the cell

$$Pb(s) \mid PbSO_4(s) \mid Na_2SO_4 \cdot 10H_2O(sat) \mid Hg_2SO_4(s) \mid Hg(l)$$

is 0.9647 V at 25 °C. The temperature coefficient is 1.74×10^{-4} V K^{-1}. (*a*) What is the cell reaction? (*b*) What are the values of $\Delta_r G$, $\Delta_r S$, and $\Delta_r H$?

7.12 For the galvanic cell

$$H_2(1\ \text{bar}) \mid HCl(ai) \mid Cl_2(1\ \text{bar})$$

the standard electromotive force at 298.15 K is 1.3604 V, and $(\partial E°/\partial T)_P = -1.247 \times 10^{-3}$ V K^{-1}. (*a*) For the cell reaction, what are the values of $\Delta_r G°$, $\Delta_r H°$, and $\Delta_r S°$? (*b*) For Cl^-(ao), what are the values of $\Delta_f G°$, $\Delta_f H°$, and $\Delta \overline{S}°$?

7.13 In Problem 4.10 two equations were derived for calculating $\Delta_r G$ at another temperature if it is known at one. Compare the values of $\Delta_r G°(323\ \text{K})$ and K_w calculated with these equations for

$$H_2O(l) = H^+(ao) + OH^-(ao)$$

7.14 Calculate $E°$ for the half-cell $OH^- \mid H_2 \mid Pt$ at 25 °C using the value of the ion product for water, which is 1.006×10^{-14} (Section 8.1).

7.15 What are the values of $\Delta_r G°$ and K for the following reactions at 298 K from Table C.2?

(*a*) $Cu(s) + Zn^{2+}(ao) = Cu^{2+}(ao) + Zn(s)$
(*b*) $H_2(g) + Cl_2(g) = 2HCl(ai)$
(*c*) $Ca^{2+}(ao) + CO_3{}^{2-}(ao) = CaCO_3(s, \text{calcite})$
(*d*) $\frac{1}{2}Cl_2(g) + Br^-(ao) = \frac{1}{2}Br_2(g) + Cl^-(ao)$
(*e*) $Ag^+(ao) + Fe^{2+}(ao) = Fe^{3+}(ao) + Ag(s)$

7.16 Derive the equation giving the effect of ionic strength at 298.15 K on (*a*) $\Delta_f S_i°$ of an ionic species and (*b*) $E°$ for a cell reaction according to the extended Debye–Hückel equation.

7.17 From the standard electrode potentials in Table 7.2, what are the standard Gibbs energies of formation at 25 °C for Cl^-(ao), OH^-(ao), and Na^+(ao)?

7.18 According to Table 7.2, what are the equilibrium constants for the following reactions at 25 °C?

(*a*) $H^+(ao) + Li(s) = Li^+(ao) + \frac{1}{2}H_2(g)$
(*b*) $2H^+(ao) + Pb(s) = Pb^{2+}(ao) + H_2(g)$
(*c*) $3H^+(ao) + Au(s) = Au^{3+}(ao) + \frac{3}{2}H_2(g)$

7.19 Use Table C.2 to calculate the standard electrode potential for $Cl^- \mid AgCl(s) \mid Ag$ at 90 °C if $\Delta_r C_P° = 0$.

7.20 The phase rule for an electrochemical cell is $F = C - p + 3$. (*a*) Why is this so? (*b*) Calculate the number of degrees of freedom of the following reaction considered as a chemical reaction.

$$H_2(g) + 2AgCl(s) = 2HCl(aq) + 2Ag(s)$$

(*c*) Calculate the number of degrees of freedom for the following electrochemical reaction.

$$H_2(g) + 2AgCl(s) + 2e^-(Pt_R) = 2HCl(ai) + 2Ag(s) + 2e^-(Pt_L)$$

7.21 The NBS tables have entries for $H_2CO_3(ao)$, $HCO_3^-(ao)$, and CO_3^{2-} (ao), where ao means "not dissociated" and ai means "completely ionized," with a note explaining that the table is based on the convention that for

$$A(ao) + nH_2O(l) = A \cdot nH_2O(ao) \quad (a)$$

$\Delta_r G° = \Delta_r H° = \Delta_r S°$. This means that for the reaction

$$CO_2(ao) + H_2O(l) = H_2CO_3(ao) \quad (b)$$

the equilibrium constant is taken equal to unity. The reason for this convention is that in dilute solutions in water it is impossible to determine the equilibrium constants for these reactions by varying the concentration of water. Thus the properties for $H_2CO_3(ao)$ apply to the sum $H_2CO_3(ao) + CO_2(ao)$, where these are interpreted as species. (*a*) To see how this works, calculate K_1 and K_2 for carbonic acid at 298.15 K and zero ionic strength. (*b*) Since the hydration of $CO_2(ao)$ in the neutral pH range is slow (half-life about 1 second), it has been possible to determine the equilibrium constant K_h for equation b.

$$K_h = \frac{[H_2CO_3(ao)]}{[CO_2(ao)]} = 2.6 \times 10^{-3} \quad (c)$$

Given this information, calculate K_a for $H_2CO_3(ao)$:

$$H_2CO_3(ao) = H^+ + HCO_3^-(ao) \quad (d)$$

$$K_a = \frac{[H^+][HCO_3^-(ao)]}{[H_2CO_3(ao)]} \quad (e)$$

7.22 At 25 °C the standard electrode potential for the $Ag^+ \mid Ag$ electrode is 0.7991 V, and the solubility product for AgI is 8.2×10^{-17}. What is the standard electrode potential for $I^- \mid AgI \mid Ag$?

7.23 Using data from Table C.2, calculate the solubility of AgCl(s) in water at 298.15 K. The salt is completely dissociated in the aqueous phase.

7.24 Calculate the standard electrode potentials at 25 °C for the following electrodes using Table C.2: (*a*) $Li^+(ao) \mid Li(s)$, (*b*) $F^-(ao) \mid F_2(g)$, and (*c*) $Pb^{2+}(ao) \mid PbO_2(s) \mid Pb$.

7.25 Using Table C.2 calculate the values of $\Delta_r G°$, $\Delta_r H°$, $\Delta_r S°$, and $\Delta_r C_P°$ at 25 °C for the electrode reaction for the $Na^+ \mid Na$ electrode.

7.26 The standard electrode potentials $E°$ in the earlier literature are based on a standard state pressure of 1 atm. Show that when the bar is used as the standard state pressure, standard electrode potentials $E°(atm)$ need to be corrected to $E°(bar)$ using

$$E°(bar) = E°(atm) + (0.000\,169\ V)\,\Delta\nu$$

where $\Delta\nu$ is the increase in the number of gaseous molecules as the cell reaction (including hydrogen) proceeds as written.

7.27 Calculate $\Delta_f S°$ for $Na^+(ao)$ at 298.15 K from (*a*) $\Delta_f G°(Na^+)$ and $\Delta_f H°(Na^+)$ and (*b*) $\overline{S}°(Na^+)$.

7.28 When a hydrogen electrode and a normal calomel electrode are immersed in a solution at 25 °C, a potential of 0.664 V is obtained. Calculate (*a*) the pH and (*b*) the hydrogen ion activity.

7.29 Calculate the equilibrium constant at 25 °C for the reaction

$$2H^+ + D_2(g) = H_2(g) + 2D^+$$

from the electrode potential for $D^+ \mid D_2 \mid Pt$, which is −3.4 mV at 25 °C.

7.30 A water electrolysis cell operated at 25 °C consumes 25 kWh/lb of hydrogen produced. Calculate the cell efficiency using $\Delta_r G°$ for the decomposition of water.

7.31 Calculate $E°$ at 25 °C for fuel cells utilizing the reactions

(*a*) $C_2H_6(g) + 3\frac{1}{2}O_2(g) = 2CO_2(g) + 3H_2O(l)$
(*b*) $C_2H_4(g) + 3O_2(g) = 2CO_2(g) + 2H_2O(l)$

Catalysts have not yet been developed to make these fuel cells possible.

7.32 (*a*) When methane is oxidized completely to $CO_2(g)$ and $H_2O(l)$ at 25 °C, how much electrical energy can be produced using a fuel cell, assuming that there are no electrical losses? What is the electromotive force of the fuel cell? (*b*) When one mole of methane is oxidized completely in a Carnot engine that operates between 500 and 300 K, how much electrical energy can be produced, assuming that the mechanical energy can be converted completely to electrical energy?

7.33 Calculate the electromotive force of

$$Li(l) \mid LiCl(l) \mid Cl_2(g)$$

at 900 K for $P_{Cl_2} = 1$ bar. This high-temperature battery is attractive because of its high electromotive force and low atomic masses. Lithium chloride melts at 883 K and lithium at 453.69 K. [The $\Delta_f G°$ for LiCl(l) at 900 K in JANAF Thermochemical Tables is −336.140 kJ mol^{-1}.]

7.34 A membrane permeable only by Na^+ is used to separate the following two solutions:

α 0.10 mol kg^{-1} NaCl, 0.05 mol kg^{-1} KCl
β 0.05 mol kg^{-1} NaCl, 0.10 mol kg^{-1} KCl

What is the membrane potential at 25 °C, and which solution has the highest positive potential?

7.35 Since Table 7.2 does not give $E°[Fe^{3+}, Fe(s)]$, calculate it from other data in the table.

7.36 In an electrolysis experiment, 0.1575 g of copper is placed on the cathode from a solution of copper sulfate when a current of 0.400 amperes is passed for 1200 s. (*a*) Calculate the value of the Faraday constant. (*b*) Given that the charge on an electron is 1.602×10^{-19} C, calculate the Avogadro constant. [This experiment is described by C. A. Seiglie, *J. Chem. Educ.* **80:**668 (2003).]

7.37 What is the electric field strength 0.5 nm from a proton?

7.38 Calculate the energy in kJ mol^{-1} required to separate a positive and negative charge from 0.3 nm to infinity in (*a*) a vacuum, (*b*) a solvent of dielectric constant 10, and (*c*) water at 25 °C, which has a dielectric constant of approximately 80.

7.39 What is the expression for the activity of Na_2SO_4 in terms of the mean ionic activity coefficient and the molality?

7.40 Give the expressions for the mean ionic activity coefficients of LiCl, $AlCl_3$, and $MgSO_4$ in terms of the activity of the electrolyte and its molality.

7.41 Determine the ionic strength of each of the following solutions: (*a*) 0.1 mol kg^{-1} NaCl, (*b*) 0.1 mol kg^{-1} $Na_2C_2O_4$, (*c*) 0.1 mol kg^{-1} $CuSO_4$, (*d*) a solution containing 0.1 mol kg^{-1} Na_2HPO_4 and 0.1 mol kg^{-1} NaH_2PO_4.

7.42 For 0.002 mol kg^{-1} $CaCl_2$ at 25 °C use the Debye–Hückel limiting law to calculate the activity coefficients of Ca^{2+} and Cl^-. What is the mean ionic activity coefficient for the electrolyte?

7.43 According to Table C.2, what is the value of the equilibrium constant for the reaction

$$\tfrac{1}{2}H_2(g) + AgCl(s) = Ag(s) + H^+(ao) + Cl^-(ao)$$

at 25 °C and how is it defined?

7.44 Derive the expression for the electromotive force of the cell

$$\text{Pt} \mid H_2(g, 1\ \text{bar}) \mid KH_2PO_4(m_1),$$
$$Na_2HPO_4(m_2) \,\vdots\vdots\, NaX(m_3) \mid AgX(s) \mid Ag$$

Substitute the equilibrium expression for the second dissociation of phosphoric acid and describe how the thermodynamic dissociation constant K for that dissociation could be obtained from electromotive force measurements at constant temperature.

7.45 (*a*) Write the reaction that occurs when the cell

$$\text{Zn} \mid ZnCl_2(0.555\ \text{mol kg}^{-1}) \mid AgCl \mid Ag$$

delivers current and calculate (*b*) $\Delta_r G$, (*c*) $\Delta_r S$, and (*d*) $\Delta_r H$ at 25 °C for this reaction. At 25 °C, E = 1.015 V and $(\partial E/\partial T)_P = -4.02 \times 10^{-4}$ V K^{-1}.

7.46 The electromotive force of the cell

$$\text{Cd} \mid CdCl_2 \cdot 2\tfrac{1}{2}H_2O, \text{sat. solution} \mid AgCl \mid Ag$$

at 25 °C is 0.675 33 V, and the temperature coefficient is -6.5×10^{-4} V K^{-1}. Calculate the values of $\Delta_r G$, $\Delta_r S$, and $\Delta_r H$ at 25 °C for the reaction

$$Cd(s) + 2AgCl(s) + 2\tfrac{1}{2}H_2O(l) = 2Ag(s) + CdCl_2 \cdot 2\tfrac{1}{2}H_2O(s)$$

7.47 A thallium amalgam of 4.93% Tl in mercury and another amalgam of 10.02% Tl are placed in separate legs of a glass cell and covered with a solution of thallous sulfate to form a concentration cell. The voltage of the cell is 0.029 480 V at 20 °C and 0.029 071 V at 30 °C. (*a*) Which is the negative electrode? (*b*) What is the heat of dilution per mole of Tl when Hg is added at 30 °C to change the concentration from 10.02% to 4.93%? (*c*) What is the voltage of the cell at 40 °C?

7.48 What are the values of $\Delta_r G°$, $\Delta_r H°$, $\Delta_r S°$, and K at 298.15 K for

$$H_2O(l) = H^+(ao) + OH^-(ao)$$

calculated from Table C.2? Compare $\Delta_r G°$ with the value obtained in Problem 7.57.

7.49 We found in Section 7.5 that E = 0.2224 V for

$$\text{Pt} \mid H_2 \mid HCl \mid AgCl \mid Ag$$

at 25 °C. Using the value of $\Delta_f G°[Cl^-(ao)]$ given in Table C.2, what is the value of $\Delta_f G°[AgCl(s)]$?

7.50 Consider the following cell.

$$\text{Pt} \mid Cu^{2+}, Cu^+ \,\vdots\vdots\, Fe^{2+}, Fe^{3+} \mid \text{Pt}$$

(*a*) What is the cell reaction? (*b*) What is the standard electromotive force of the cell at 298.15 K? (*c*) Calculate $\Delta_r G°$ for the cell reaction from the standard electromotive force. (*d*) Calculate $\Delta_r G°$ for the cell reaction using the $\Delta_f G°$ values for the ions in Table C.2. (*e*) Calculate $\Delta_r G°$ for the cell reaction using the $\Delta_f H°$ values and $\overline{S}°$ values in Table C.2.

7.51 The electrode potential for the electrode $Cl^- \mid Cl_2(g) \mid$ Pt is given by

$$E°/\text{V} = 1.484\,867 + (3.958\,492 \times 10^{-4})(T/\text{K}) - (2.750\,639 \times 10^{-6})(T/\text{K})^2$$

in the range 273–373 K. Calculate $E°$, $\Delta_r G°$, $\Delta_r H°$, $\Delta_r S°$, and $\Delta_r C_P°$ at 298.15 K.

7.52 The standard electrode potential for $Cl^- \mid Cl_2 \mid$ Pt is 1.2604 V. Calculate the standard Gibbs energy of formation of $Cl^-(ao)$.

7.53 Given the following electrode potentials at 25 °C,

$$Fe^{3+} + e^- = Fe^{2+} \qquad E_1° = 0.771\ \text{V}$$
$$\tfrac{1}{2}Fe^{2+} + e^- = \tfrac{1}{2}Fe(s) \qquad E_2° = -0.440\ \text{V}$$

calculate the electrode potential for

$$\tfrac{1}{3}Fe^{3+} + e^- = \tfrac{1}{3}Fe \qquad E_3° = ?$$

7.54 Calculate the solubility of AgCl(s) in water at 298 K from data in Table C.2.

7.55 Calculate the thermodynamic properties of the following strong electrolytes from those of the constituent ions at 25 °C and check that the same values are tabulated in Table C.2 under the following entries: HCl(ai), NaCl(ai), and NaOH(ai).

7.56 Calculate the standard electromotive force of the cell

$$\text{Li} \mid LiCl(ai) \mid Cl_2(g) \mid \text{Pt}$$

at 25 °C using (*a*) electrode potentials and (*b*) standard Gibbs energies of formation.

7.57 Devise a cell for which the cell reaction is

$$H_2O(l) = H^+(ao) + OH^-(ao)$$

Calculate $\Delta_r G°$ at 25 °C from electrode potentials. What is the value of the equilibrium constant at this temperature?

7.58 Devise an electromotive force cell for which the cell reaction is

$$AgBr(s) = Ag^+ + Br^-$$

Calculate the equilibrium constant (usually called the solubility product) for this reaction at 25 °C.

7.59 Using data in Table 7.2 and the Gibbs energies of formation of Ag(ao) + Cl^-(ao), calculate the solubility of AgCl(s) in water at 25 °C.

7.60 What are the differences between the standard electrode potentials for a standard state pressure of 1 bar and 1 atm for the following electrodes at 25 °C?

	$E°$(1 atm)
Cl^- | AgCl(s) | Ag	0.2224 V
Cl^- | Cl_2(g) | Pt	1.3604 V

See Problem 7.26.

7.61 The value of $E°$ for the electrode Pt | O_2(g) | OH^- cannot be measured directly because the electrode is not reversible. Calculate $\Delta_r G°$ and $E°$ at 25 °C for the electrode reaction

$$\tfrac{1}{4}O_2(g) + \tfrac{1}{2}H_2O + e^- = OH^-(ao)$$

from $\Delta_f G°$ values from Table C.2.

7.62 Calculate the standard Gibbs energy of formation of NO_3^-(ao) from its $\Delta_f H°$ and $\overline{S}°$ values at 25 °C using data from Table C.2.

7.63 Calculate the standard Gibbs energy of formation of SO_4^{2-}(ao) from $\Delta_f H°$ and $\overline{S}°$ values in Table C.2.

7.64 A hydrogen–oxygen fuel cell is operated at 25 °C and a total pressure of 5 bar. What is the electromotive force, assuming the gases are ideal?

7.65 A hydrogen electrode and a normal calomel electrode give an electromotive force of 0.435 V when placed in a certain solution at 25 °C. (*a*) What is the pH of the solution? (*b*) What is the value of a_{H^+}?

7.66 A mole of H_2O(l) is electrolyzed at 298 K and 1 bar. (*a*) How much electrical energy is required if there are no losses due to electrical resistance and overvoltage? (*b*) The hydrogen and oxygen are then burned at constant pressure to produce one mole of H_2O(l) at 298 K and 1 bar. How much heat is produced? (*c*) Can this heat be used in a heat engine to produce the amount of electrical energy that was used to electrolyze the water initially? If so, what condition has to be met?

7.67 Ammonia may be used as the anodic reactant in a fuel cell. The reactions occurring at the electrodes are

$$NH_3(g) + 3OH^-(ao) = \tfrac{1}{2}N_2(g) + 3H_2O(l) + 3e^-$$
$$O_2(g) + 2H_2O(l) + 4e^- = 4OH^-(ao)$$

What is the electromotive force of this fuel cell at 25 °C?

7.68 For a membrane potential of −70 mV for a resting nerve cell, to what ratio of the concentrations of K^+ inside and outside would this correspond if there was equilibrium at 25 °C?

7.69 (*a*) What are the half-cell reactions for the cell Pt(s) | Cu^+, Cu^{2+} ⁞⁞ Cu^+ | Cu(s)? (*b*) What is the cell reaction? (*c*) What is the standard electromotive force of the cell at 25 °C? (*d*) What is the value of the equilibrium constant and the equilibrium constant expression in general?

7.70 In connection with the preceding problem, another investigator, who is also interested in the copper disproportionation reaction, studies the cell Cu(s) | Cu^{2+} ⁞⁞ Cu^+ | Cu(s). (*a*) What are the half-cell reactions? (*b*) What is the cell reaction and the standard electromotive force of the cell at 298.15 K? (*c*) What is the equilibrium constant for the cell reaction? How does this equilibrium constant compare with that in the preceding problem?

Computer Problems

7.A Calculate the values of B in equation 7.65 that fit each data point for the mean ionic activity coefficient of (*a*) HCl, (*b*) NaCl, and (*c*) CsCl up to 0.40 molal in Table 7.1. Also plot the data as the base 10 logarithms of the mean ionic activity coefficients versus the square root of the ionic strength. (*d*) Calculate the percent error in the mean ionic activity coefficients calculated for HCl, NaCl, and CsCl using equation 7.65 using 1.6 $kg^{1/2}$ $mol^{-1/2}$ at each ionic strength.

7.B Calculate the values of B in equation 7.65 that give the best fit of the data on the mean ionic activity coefficients of (*a*) $CaCl_2$ and (*b*) $LaCl_3$ up to 0.05 molal. The values of the mean ionic activity coefficients are given by

$m/m°$	0.001	0.005	0.01	0.05
$\gamma_\pm(CaCl_2)$	0.888	0.789	0.732	0.584
$\gamma_\pm(LaCl_3)$	0.790	0.636	0.560	0.388

In addition, do two things for these two electrolytes: (i) Plot the experimental data as base 10 logarithms of the mean ionic activity coefficients versus the square root of the ionic strength and compare the data with the extended Debye–Hückel equation with B = 1.6. (ii) Calculate the percent errors in the mean ionic activity coefficients calculated for $CaCl_2$ and $LaCl_3$ using equation 7.65 with B = 1.6 $kg^{1/2}$ $mol^{-1/2}$ at each ionic strength.

7.C Calculate the standard electrode potential of (*a*) Cd^{2+} | Cd, (*b*) Cl^- | Cl_2(g) | Pt, and (*c*) Cl^- | AgCl(s) | Ag(s) at

0.25 m ionic strength using the extended Debye–Hückel equation. See the values at $I = 0$ in Example 7.11.

7.D Plot $E°(I)$, $\Delta_r G°(I)$, and $K(I)$ for the reaction

$$\frac{1}{2}H_2(g) + AgCl(s) = H^+ + Cl^- + Ag(s)$$

versus the ionic strength from 0 to 0.25 m using the extended Debye–Hückel equation.

7.E (*a*) Plot the log (base 10) of the mean ionic activity coefficient versus the square root of the ionic strength, up to 0.5, at 25 °C for $z_+z_- = -1, -2, -3$, and -4 using the extended Debye–Hückel equation. (*b*) Plot the mean ionic activity coefficients versus the ionic strength up to 0.25 mol kg^{-1} for these cases.

7.F In the range 0–100 °C the Debye–Hückel coefficient varies with temperature according to
$\alpha = 1.10708 - 1.54508 \times 10^{-3}T + 5.95584 \times 10^{-6}T^2$
(*a*) Calculate the temperature dependencies of the ionic strength coefficients in the equations for $\Delta_f G°$, $\Delta_f H°$, and $\Delta_f S°$. (*b*) Calculate the values of these coefficients at 0 °C, 25 °C, and 40 °C and make a table.

7.G Calculate and plot versus ionic strength the ionic strength contributions to $\Delta_f G°$, $\Delta_f H°$, and $\Delta_f S°$ for ions with charges of 1, 2, 3, and 4 up to $I = 0.25$ M at 298.15 K.

8

Thermodynamics of Biochemical Reactions

The dissociations of weak acids and complex ions are interesting and important examples of equilibria in aqueous solutions and are frequently involved in biochemical reactions. Two types of equilibrium equations are useful in biochemistry, chemical equations and biochemical equations. Biochemical equations are written in terms of sums of species at a specified pH. The corresponding equilibrium constant is referred to as an apparent equilibrium constant and is represented by K' because it is a function of pH. The thermodynamic properties calculated from K' and its temperature coefficient depend on the pH and are referred to as transformed thermodynamic properties because they are defined by Legendre transforms of G, H, and S. Since apparent equilibrium constants are frequently needed in biochemistry at pH 7, it is convenient to make tables of standard transformed Gibbs energies of formation and standard transformed enthalpies of formation of reactants at pH 7. In contrast with the binding properties of smaller molecules, including myoglobin, the affinity of hemoglobin for oxygen increases as more oxygen is bound; this is a consequence of structural changes in the hemoglobin molecule. This chapter closes with discussions of denaturation equilibria of proteins and DNA.

8.1 EXACT TREATMENT OF THE DISSOCIATION OF WEAK ACIDS

Some acids, such as hydrochloric acid, are believed to be completely dissociated in water, but others, such as acetic acid, are only partially dissociated and are referred to as **weak acids.** The dissociation of a monoprotic weak acid in water is represented by

$$HA = H^+ + A^- \tag{8.1}$$

In water, all ions are hydrated to a greater or lesser extent, and this has a significant effect on their thermodynamic properties. In water the proton may form a hydronium ion, H_3O^+, or other complex species. A hydronium ion H_3O^+ may be hydrated by three water molecules so that $H_9O_4^+$ is formed. Mass spectroscopic studies show that this is a stable species in the gas phase. Since the state of the proton in aqueous solution is not exactly known, the symbol H^+ will be used to represent the hydrated hydrogen ion in aqueous solution.

The **acid dissociation constant** K_a is defined by

$$K_a = \frac{a_{H^+}a_{A^-}}{a_{HA}} = \frac{m_{H^+}m_{A^-}\gamma_\pm^2}{m^\circ m_{HA}\gamma_{HA}} \tag{8.2}$$

where m is molal concentration (mol kg^{-1} of solvent) and m° is the standard molality (1 mol/kg of solvent). The mean ionic activity coefficient for the dissociated acid is represented by $\gamma_\pm$, and γ_{HA} is the activity coefficient for the undissociated acid. Since the undissociated acid is a nonelectrolyte, its activity coefficient is close to unity in dilute solutions and may be taken as unity to a good approximation. Since acid dissociation constants range over many powers of 10, they are often expressed as pK values, where p$K = -\log K$.

Table 8.1 gives the pK values and other thermodynamic quantities at 25 °C for equilibrium constants expressed in terms of activities. Equilibrium constants for these acids at zero ionic strength have been determined by use of extrapolations similar to those described in Section 7.5. From the temperature dependence of K it is possible to calculate $\Delta_r H^\circ$, $\Delta_r S^\circ$, and $\Delta_r C_P^\circ$ for the dissociation.

The weakest acid listed in Table 8.1 is water itself. Its acid dissociation is represented by

$$H_2O = H^+ + OH^- \tag{8.3}$$

Thus, the acid dissociation constant of water, usually referred to as the **ion product,** is defined by

$$K_w = a_{H^+}a_{OH^-} = \frac{m_{H^+}m_{OH^-}\gamma_\pm^2}{(m^\circ)^2} \tag{8.4}$$

The activity of water, which can be written in the denominator of this expression, is taken as unity since it is the solvent. The value of K_w for water at 25 °C is 1.007×10^{-14}. This is listed in Table 8.1 as a pK value. The ion product of water is given at a series of temperatures in Table C.4. The enthalpy of dissociation of water is the negative of the enthalpy of neutralization of a strong acid with a strong base (Section 2.13). Weaker acids than H_2O are known. For example, the pK for

Table 8.1 Standard Thermodynamic Quantities for Acid Dissociation at 25 °C[a]

Weak Acid	pK	$\Delta_r G°$ / kJ mol^{-1}	$\Delta_r H°$ / kJ mol^{-1}	$\Delta_r S°$ / J K^{-1} mol^{-1}	$\Delta_r C_P°$ / J K^{-1} mol^{-1}
Water (K_w)	13.997	79.868	56.563	−78.2	−197
Acetic acid	4.756	27.137	−0.385	−92.5	−155
Chloroacetic acid	2.861	16.322	−4.845	−71.1	−167
Butyric acid	4.82	27.506	−2.900	−102.1	0
Succinic acid, pK_1	4.207	24.016	3.188	−69.9	−134
Succinic acid, pK_2	5.636	31.188	−0.452	−109.2	−218
Carbonic acid, pK_1	6.352	36.259	9.372	−90.4	−377
Carbonic acid, pK_2	10.329	58.961	15.075	−147.3	−272
Phosphoric acid, pK_1	2.148	12.259	−7.648	−66.9	−155
Phosphoric acid, pK_2	7.198	41.099	4.130	−123.8	−226
Glycerol-2-phosphoric acid, pK_1	1.335	7.615	12.103	−66.1	−326
Glycerol-2-phosphoric acid, pK_2	6.650	37.945	−1.724	−133.1	−226
Ammonium ion	9.245	52.777	52.216	−1.7	0
Methylammonium ion	10.615	60.601	54.760	−19.7	33
Dimethylammonium ion	10.765	49.618	49.618	−39.7	96
Trimethylammonium ion	9.791	55.890	36.882	−63.6	184
Tris(hydroxymethyl)aminomethane	8.076	46.099	45.606	−1.3	0
Glycine, pK_1	2.350	13.410	4.837	−28.9	−134
Glycine, pK_2	9.780	55.815	44.141	−39.3	−50
Glycylglycine, pK_1	3.148	17.322	3.607	−54.0	−167
Glycylglycine, pK_2	8.252	47.112	44.350	−8.4	−42

[a]These values apply at zero ionic strength and are obtained by extrapolation of experimental data at higher ionic strengths.
Source: J. Edsall and J. Wyman, *Biophysical Chemistry*. New York: Academic, 1958.

methanol is 15.53. Special methods, not described here, are required to study such weak acids.

The fact that $\Delta_r S°$ for dissociation for most weak acids is negative is surprising at first. Gas dissociation reactions have positive entropy changes. However, the experimental results for acid dissociations in aqueous solution show that there is a decrease in entropy and, correspondingly, an increase in "order" in the dissociation. This results from the participation of water molecules in the reaction, which is not indicated by the balanced equation 8.1. Water molecules, being dipoles, tend to be oriented in the neighborhood of ions. Thus, the dissociation of an uncharged acid results in the orientation of a number of water molecules about the ions formed, and the consequent decrease in $\Delta_r S°$ overshadows the increase in $\Delta_r S°$ resulting from the formation of two particles from one. The effects that lead to a negative value of $\Delta_r S°$ tend to oppose dissociation. For a number of weak acids in Table 8.1 the value of $\Delta_r H°$ is very small, and so the standard entropy change largely determines the value of pK according to $2.303RT\,\mathrm{p}K = \Delta_r H° - T\Delta_r S°$.

There is almost no entropy change in the dissociation of ammonium ion

$$NH_4^+ = H^+ + NH_3 \tag{8.5}$$

because there is no change in the number or charge of ions in the reaction, so that the dissociation causes little change in water structure. Water molecules are not so organized around methylammonium ion as around NH_4^+, and therefore, there is a decrease in entropy on dissociation of the methylammonium ion. The entropy decrease on dissociation is even greater for the dimethylammonium and trimethylammonium ions, suggesting decreasing water organization as methyl substitution increases.

The data in Table 8.1 can be used to calculate acid dissociation constants at other temperatures using the equation in Example 5.11, which is based on the assumption that $\Delta_r C_P^\circ$ is independent of temperature.

8.2 PRACTICAL CALCULATIONS WITH WEAK ACIDS

In previous chapters and in the preceding section, activity coefficients have been emphasized, but in the remainder of this chaper we will find that it is more convenient to write expressions for equilibrium constants in terms of concentrations and to use thermodynamic properties that are functions of ionic strength, as well as temperature. In other words, we will use activity coefficients implicitly, rather than explicitly. The equation for the chemical potential of a species j can be written as

$$\mu_j = \mu_j^\circ + RT \ln \gamma_j c_j = \mu_j^\circ + RT \ln \gamma_j + RT \ln c_j \tag{8.6}$$

where c_j is the molar concentration. In this equation μ_j° is the standard chemical potential at the standard state, where the ionic strength is zero. The contribution of the activity coefficient γ_j to the chemical potential of the species is $RT \ln \gamma_j$, where γ_j is a function of the ionic strength. However, in studying biochemical equilibria, the ionic strength is generally under the control of the investigator and is in the low concentration range, where the extended Debye–Hückel equation should be a good approximation. When the ionic strength is specified, the terms μ_j° and $RT \ln \gamma_j$ can be combined, and equation 8.6 can be written as

$$\mu_j = \mu_j^\circ + RT \ln c_j \tag{8.7}$$

where μ_j° is now a function of ionic strength as well as temperature. Note that the symbol μ_j° in equations 8.6 and 8.7 has different meanings. Equation 8.7 has the advantage that expressions for equilibrium constants can be written in terms of concentrations of species, and so we will use it in the rest of the book. This makes it possible to treat dilute solutions as ideal solutions even when there are significant ionic strength effects. For example, the dissociation constant for a weak acid can be written in terms of concentrations as

$$K = [H^+][A^-]/[HA] \tag{8.8}$$

It is also useful to write this equation in the form

$$K = 10^{-pH}[A^-]/[HA] = 10^{-pK} \tag{8.9}$$

where $pH = -\log[H^+]$ and $pK = -\log K$. We could use the symbol pH_c to distinguish this pH from $pH_a = -\log a(H^+)$ determined using a glass electrode. In this chapter we will use pH to refer to pH_c, but in making comparisons between calculations and experimental data, we need to know the difference $pH_a - pH_c$. Since $pH_a = -\log\{\gamma(H^+)[H^+]\} = -\log \gamma(H^+) + pH_c$, substituting the extended Debye–Hückel equation yields

$$pH_a - pH_c = \frac{AI^{1/2}}{1 + 1.6I^{1/2}} \tag{8.10}$$

Table 8.2 $pH_a - pH_c$ as a Function of Ionic Strength and Temperature

I/M	10 °C	25 °C	40 °C
0	0	0	0
0.05	0.082	0.084	0.086
0.1	0.105	0.107	0.110
0.15	0.119	0.122	0.125
0.2	0.130	0.133	0.137
0.25	0.138	0.142	0.146

Source: R. A. Alberty, *J. Phys. Chem. B* **105**:7870 (2001). Reprinted with permission of the American Chemical Society.

This difference depends on the temperature, but since the temperature dependence of A is known, $pH_a - pH_c$ is given in Table 8.2. These are the adjustments to be subtracted from pH_a, obtained with a pH meter, to obtain pH_c, which is used in the equations in this chapter. pH_c is lower than pH_a because the ion atmosphere of H^+ reduces its activity. It is also convenient to use $pMg = -\log[Mg^{2+}]$.

Table 8.3 gives the pK's for acid dissociation for some weak acids of interest in biochemistry at 289.15 K at three ionic strengths. In this table certain low and high

Table 8.3 pK's of Weak Acids at 298.15 K in Dilute Aqueous Solution at Three Ionic Strengths

	$I = 0\ M$	$I = 0.10\ M$	$I = 0.25\ M$
Acetate	4.75	4.54	4.47
Ammonia	9.25	9.25	9.25
ATP K_1	7.60	6.74	6.47
ATP K_2	4.68	4.04	3.83
ADP K_1	7.18	6.54	6.33
ADP K_2	4.36	3.93	3.79
AMP K_1	6.73	6.30	6.16
AMP K_2	3.99	3.78	3.71
Adenosine	3.50	3.50	3.50
HCO_3^- K_1	10.30	9.90	9.76
H_2CO_3 K_2	6.37	6.15	6.08
Glucose 6-phosphate K_1	6.42	5.99	5.85
Phosphate K_2	7.22	6.79	6.65
Pyrophosphate K_1	9.38	8.52	8.25
Pyrophosphate K_2	6.74	6.10	5.89
Succinate K_1	5.64	5.21	5.07
Succinate K_2	4.21	3.99	3.92
Fumarate K_1	4.60	4.17	4.03
Fumarate K_2	3.09	2.88	2.81

Source: R. A. Alberty, *Thermodynamics of Biochemical Reactions*, © Wiley, Hoboken, NJ, 2003. This material is used by permission of John Wiley & Sons, Inc.

pK's have been omitted because they are unimportant in the usually considered range of pH 5 to 9.

In using acid dissociation constants to discuss buffers, it is convenient to write equation 8.9 as

$$\mathrm{pH} = \mathrm{p}K + \log \frac{[\mathrm{A}^-]}{[\mathrm{HA}]} \tag{8.11}$$

where pK = $-\log K$. This useful equation is often referred to as the **Henderson–Hasselbalch equation.** It shows that the apparent pK of a weak acid in a particular electrolyte solution can be calculated from the pH of a solution containing known concentrations of weak acid [HA] and weak base [A^-]. For an aqueous solution containing NH_4Cl and NH_3, the ratio in equation 8.11 would be [NH_3]/[NH_4Cl]. The Henderson–Hasselbalch equation can be used to calculate the ratio of base and acid forms required to make a buffer of a particular pH. A buffer is most effective in a range of pH values between pK -1 and pK $+1$ because the [base]/[acid] ratio goes from 0.1 to 10 in this range.

We are now in a position to discuss the titration curve of a weak acid with a strong base. As shown in Fig. 8.1, the pH of 0.1 molar acetic acid is about 2.9. As a concentrated solution of sodium hydroxide is added, the pH rises rapidly at first and then moves slowly through the buffering region around pH 4.7 (reaching pH 4.7 when one-half of the acid has been neutralized); it then rises very rapidly at the equivalence point at pH 8.8. When sodium hydroxide is added beyond the equivalence point, the pH corresponds to that of a mixture of sodium hydroxide and sodium acetate. The titration curve can be calculated most easily using the approximation that the electrolyte concentration

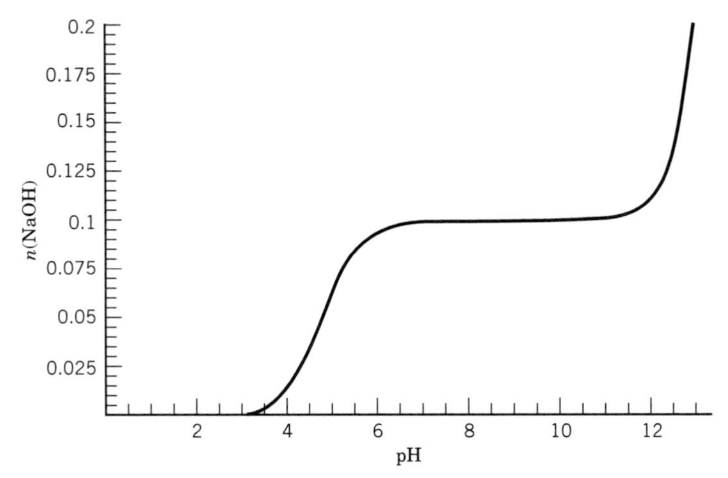

Figure 8.1 Titration of acetic acid with a concentrated solution of sodium hydroxide. The number of moles of NaOH added to a liter of 0.10 M acetic acid is represented by n. (See Computer Problem 8.I.)

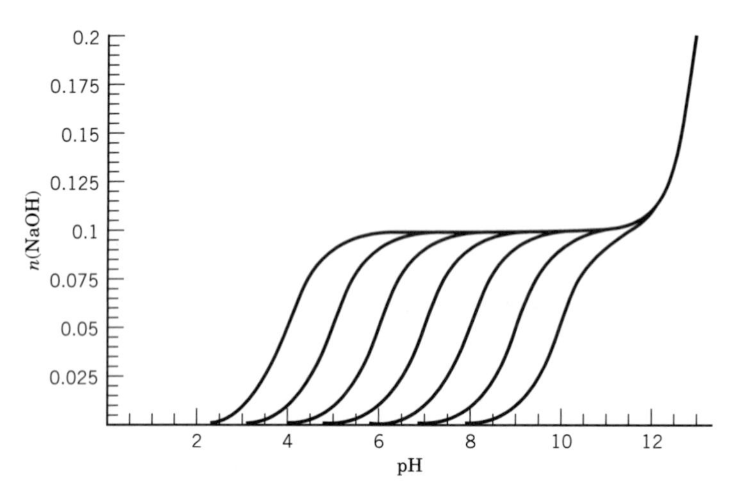

Figure 8.2 Titration curves for monoprotic acids with pK's of 4, 5, 6, 7, 8, 9, and 10 at 298.15 K. The number of molds of NaOH added to a liter 0.10 M weak acid is represented by n. (See Computer Problem 8.I.)

is constant so that an apparent acid dissociation constant for that electrolyte concentration can be used.

The calculation of the titration curve for acetic acid involves the four unknowns $[H^+]$, $[A^-]$, $[HA]$, and $[OH^-]$. To calculate these unknown concentrations we have two equilibrium expressions,

$$K = \frac{[H^+][A^-]}{[HA]} \tag{8.12}$$

$$K_w = [H^+][OH^-] \tag{8.13}$$

We also have two conservation equations (conservation of mass and charge),

$$[\text{acetate}] = [HA] + [A^-] \tag{8.14}$$

$$[Na^+] + [H^+] = [A^-] + [OH^-] \tag{8.15}$$

where [acetate] is the total concentration of acetate and $[Na^+]$ is the concentration of sodium ions, which is equal to the concentration of sodium hydroxide added up to that point in the titration. These simultaneous equations can be solved for the amount of sodium hydroxide that has to be added to a liter (see Problem 8.2). Figure 8.2 shows the titration curves for 0.10 M monoprotic weak acids with pK's of 4, 5, 6, 7, 8, 9, and 10 at 298.15 K. Note that each titration curve starts at a different pH and that the endpoints are each at a different pH. When weaker monoprotic acids are titrated, their buffering regions are at higher pH values, as shown in Fig. 8.2.

Example 8.1 ***Preparation of an acetate buffer***

You want to prepare 1 liter of an acetate buffer of 0.1 ionic strength and pH 5.0 at 25 °C. How many moles of sodium acetate and acetic acid should you add if the pK of acetic acid at this ionic strength is 4.54?

$$\text{pH} = \text{p}K + \log \frac{[\text{NaA}]}{[\text{HA}]}$$

$$5.0 = 4.54 + \log \frac{0.1}{[\text{HA}]}$$

$$[\text{HA}] = 0.0347 \text{ mol L}^{-1}$$

Thus, you would add 0.1 mol of sodium acetate and 0.0347 mol of acetic acid to prepare a liter of buffer.

The values of $\text{p}K_1$ are given for a large number of acids in Table C.5. When the acid is diprotic, $\text{p}K_2$ is also given. The dissociation constant for the second proton is smaller, and so $\text{p}K_2 > \text{p}K_1$.

In discussing H_3PO_4 we are going to number the pK expressions in the opposite direction to make a point about the average number $\overline{N}_H$ of protons bound at a particular pH. For H_3PO_4, the acid dissociation constants at a specified ionic strength are defined by

$$HPO_4^{2-} = H^+ + PO_4^{3-} \qquad K_1 = \frac{[H^+][PO_4^{3-}]}{[HPO_4^{2-}]} \tag{8.16}$$

$$H_2PO_4^{-} = H^+ + HPO_4^{2-} \qquad K_2 = \frac{[H^+][HPO_4^{2-}]}{[H_2PO_4^{-}]} \tag{8.17}$$

$$H_3PO_4 = H^+ + H_2PO_4^{-} \qquad K_3 = \frac{[H^+][H_2PO_4^{-}]}{[H_3PO_4]} \tag{8.18}$$

As acid is added to a solution of Na_3PO_4, the phosphate ions pick up protons as indicated by equations 8.16–8.18. At any specified pH, there is an equilibrium between the various species, and we can talk about the average number $\overline{N}_H$ of protons per phosphorus atom in the collection of species. At very high pH, this number approaches zero, and at very low pH, it approaches 3. The average number of protons bound can be calculated using $\overline{N}_H = \sum i f_i$, where f_i is the fraction of the phosphate ions with i hydrogen atoms; for example, $f_2 = [H_2PO_4^-]/([PO_4^{3-}] + [HPO_4^{2-}] + [H_2PO_4^-] + [H_3PO_4])$. The average number of protons bound by phosphate is given by

$$\overline{N}_H = \frac{[HPO_4^{2-}] + 2[H_2PO_4^{-}] + 3[H_3PO_4]}{[PO_4^{3-}] + [HPO_4^{2-}] + [H_2PO_4^{-}] + [H_3PO_4]} \tag{8.19}$$

Substituting the equilibrium relations in equations 8.16 to 8.18 yields

$$\overline{N}_H = \frac{[H^+]/K_1 + 2[H^+]^2/K_1K_2 + 3[H^+]^3/K_1K_2K_3}{1 + [H^+]/K_1 + [H^+]^2/K_1K_2 + [H^+]^3/K_1K_2K_3} \tag{8.20}$$

The number of protons bound by inorganic phosphate is shown as a function of pH in Fig. 8.3. This is really a titration curve, but we are focusing on what it tells us, rather than on the volume of base, or acid, added. At 25 °C the successive dissociation constants are $K_1 \cong 10^{-12}$, $K_2 = 6.34 \times 10^{-8}$, and $K_3 = 7.11 \times 10^{-3}$.

The way to look at this binding curve is to start at the far right. There phosphate is in the form of PO_4^{3-}, and the hydrogen ion concentration is, let us say,

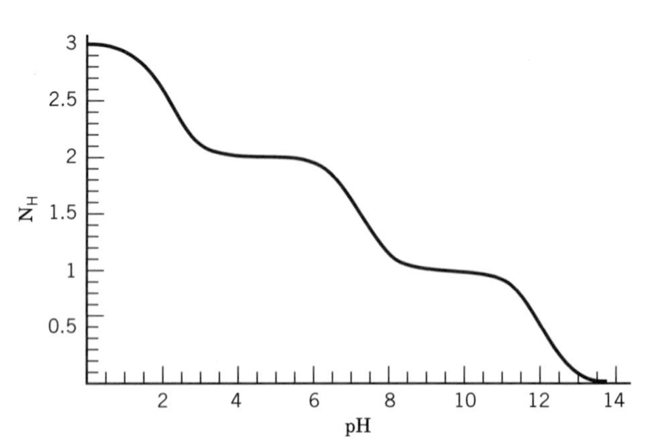

Figure 8.3 Average number of hydrogen ions bound by phosphate at 298.15 K as a function of pH at zero ionic strength. (See Computer Problem 8.J.)

10^{-14} mol/L. As we increase the concentration of hydrogen ion (decrease the pH), the PO_4^{3-} ion begins to bind protons, and the acid dissociation constant of the HPO_4^{2-} that is formed is about 10^{-12} ($pK_1 = 12$). As the concentration of hydrogen ions is further increased, nearly all of the PO_4^{3-} ions become protonated and we reach the first equivalence point at $\overline{N}_H = 1$. But as the pH approaches the pK for $H_2PO_4^-$ (about pH 7), the HPO_4^{2-} ion begins to bind a second proton. As the concentration of hydrogen ions is further increased, we reach the second equivalence point at $\overline{N}_H = 2$. In more strongly acidic solutions, $H_2PO_4^-$ begins to bind a third proton to produce H_3PO_4, which has an acid dissociation constant of 7.11×10^{-3} ($pK_3 = 2.1$). The inorganic phosphate is in the form of H_3PO_4 only in strongly acidic solutions. Thus, as PO_4^{3-} binds protons, its proton affinity decreases. For this example, the successive dissociation constants differ by large factors, and so this binding curve is very nearly the sum of three titration curves of monoprotic acids.

The polynomial in the denominator of equation 8.20 is referred to as the **binding polynomial** P. It is actually a kind of partition function because it gives the partition of a reactant between the various species that make it up. Note that the numerator of equation 8.20 is the derivative of the denominator multiplied by $[H^+]$. The average binding of hydrogen ions $\overline{N}_H$ is given by

$$\overline{N}_H = \frac{[H^+]}{P}\frac{dP}{d[H^+]} = \frac{d\ln P}{d\ln[H^+]} = \frac{d\log P}{d\log[H^+]} = -\frac{d\log P}{d\,pH} \tag{8.21}$$

This relation has been used to generate the plot in Fig. 8.3.

Because of their negative charges, PO_4^{3-} and HPO_4^{2-} also bind cations such as Mg^{2+}. For example, $MgHPO_4$ is formed, and its dissociation is represented by

$$MgHPO_4 = Mg^{2+} + HPO_4^{2-} \tag{8.22}$$

The **dissociation constant** for this complex ion is defined by

$$K_{MgP} = \frac{[Mg^{2+}][HPO_4^{2-}]}{[MgHPO_4]} \tag{8.23}$$

Protons and Mg^{2+} compete in their binding to HPO_4^{2-}. Since highly charged phosphate ions also tend to bind Na^+ and K^+, acid titrations and binding experiments are often carried out in the presence of cations such as $(CH_3CH_2)_4N^+$, where the bulky substituent groups prevent the close approach of the positive charge to a negatively charged ion. The study of the competition of metal ions and hydrogen ions is often used to determine dissociation constants for complex ions since we usually do not have metal ion electrodes that can be used to determine the concentration of unbound metal ions. In the neutral pH region we can neglect the H_3PO_4 concentration and consider a system consisting of Mg^{2+}, H^+, HPO_4^{2-}, $H_2PO_4^-$, and $MgHPO_4$ only. The number of additional protons bound by HPO_4^{2-} per mole of phosphate is given by

$$\begin{aligned}\overline{N}_H &= \frac{[H_2PO_4^-]}{[HPO_4^{2-}] + [H_2PO_4^-] + [MgHPO_4]} \\ &= \frac{[H^+]/K_2}{1 + [H^+]/K_2 + [Mg^{2+}]/K_{MgP}}\end{aligned} \tag{8.24*}$$

The number of $\overline{N}_{Mg}$ of magnesium ions bound by HPO_4^{2-} per mole of phosphate is given by

$$\begin{aligned}\overline{N}_{Mg} &= \frac{[MgHPO_4]}{[HPO_4^{2-}] + [H_2PO_4^-] + [MgHPO_4]} \\ &= \frac{[Mg^{2+}]/K_{MgP}}{1 + [H^+]/K_2} + [Mg^{2+}]/K_{MgP}\end{aligned} \tag{8.25}$$

Figure 8.4 shows the binding curve for the second proton of phosphate in the presence and absence of Mg^{2+}. The shape of the curve is independent of the concentration of Mg^{2+}, but the curve is displaced to lower pH values as the concentration of Mg^{2+} is increased. This suggests that equation 8.24 might be written in the form

$$\overline{N}_H = \frac{[H^+]/K_2'}{1 + [H^+]/K_2'} \tag{8.26}$$

where the apparent second acid dissociation constant for phosphate is given by

$$K_2' = K_2\left(1 + \frac{[Mg^{2+}]}{K_{MgP}}\right) \tag{8.27}$$

This provides a means for obtaining the value of the dissociation constant K_{MgP} from acid titration curves.

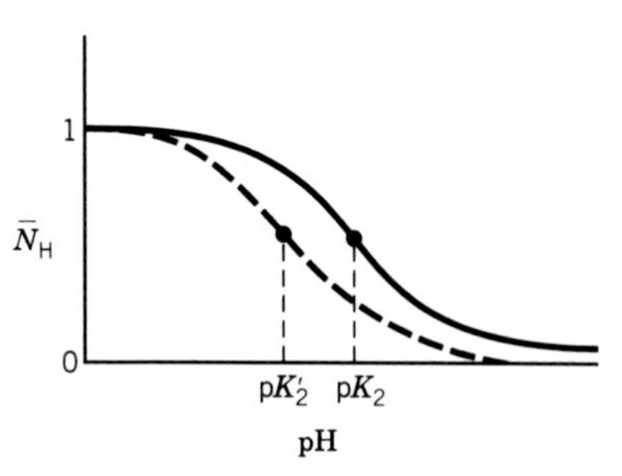

Figure 8.4 Proton binding curve for HPO_4^{2-} in the absence (—) and presence (---) of Mg^{2+}.

Example 8.2 ***Determination of the dissociation constant of $MgHPO_4$***

A solution is prepared that is 0.01 M in NaH_2PO_4 and 0.2 M in $(CH_3CH_2)_4NCl$ and is titrated with $(CH_3CH_2)_4NOH$ at 25 °C. The midpoint of the titration curve is at pH 6.80. An identical solution is made that is 0.05 mol L^{-1} in $MgCl_2$ and the titration is repeated. This time the midpoint is pH 6.37. What is the dissociation constant of $MgHPO_4$?

*Note that this equation gives the number of protons bound by HPO_4^{2-}, rather than the number of protons bound by PO_4^{3-}, which is given by equation 8.26.

Using equation 8.27,

$$10^{-6.37} = 10^{-6.80}\left(1 + \frac{0.05}{K_{\mathrm{MgP}}}\right)$$

$$K_{\mathrm{MgP}} = 2.96 \times 10^{-2}$$

8.3 THERMODYNAMICS OF ENZYME-CATALYZED REACTIONS

About 4000 enzyme-catalyzed reactions have names recommended by the International Union of Biochemistry and Molecular Biology,* but some of the names apply to classes of reactions, and so the number of reactions is much larger. The equilibrium constants and other thermodynamic properties of these reactions are independent of the enzyme and the mechanisms by which the reaction is catalyzed. The thermodynamics of biochemical reactions can be discussed in terms of species as in the preceding three chapters, but there is a more convenient way that is used in discussing metabolism. The problem involved in using chemical reactions is that many of the reactants in enzyme-catalyzed reactions are made up of multiple species. For example, adenosine triphosphate (ATP), which is the energy storage reactant shown in Fig. 8.5, is made up of two species, ATP^{4-} and $HATP^{3-}$, in the range pH 6 to 8. When it is hydrolyzed to adenosine diphosphate (ADP) and inorganic phosphate (P_i) in the neighborhood of pH 7, the following chemical reactions are involved:

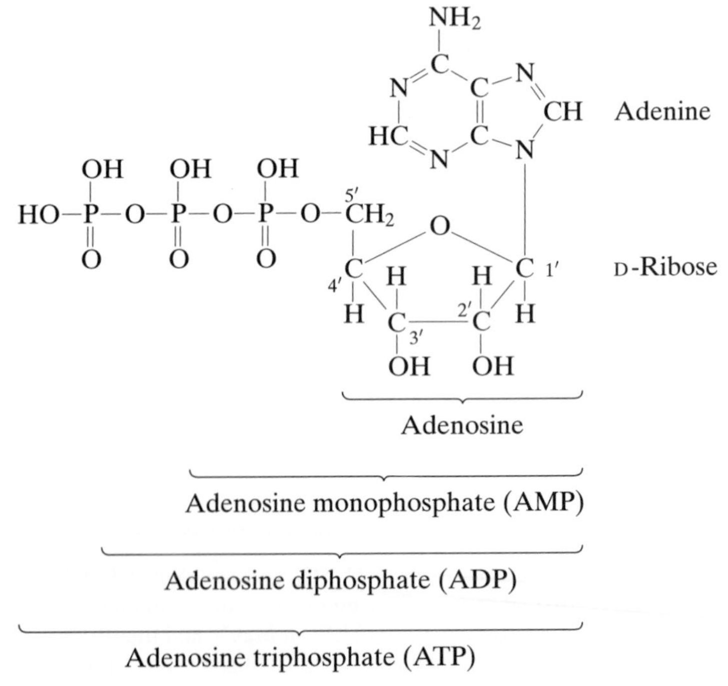

Figure 8.5 Structure of adenosine triphosphate (ATP).

*E. C. Webb, *Enzyme Nomenclature,* San Diego: Academic Press, 1992.

$$ATP^{4-} + H_2O = ADP^{3-} + HPO^{3-} + H^+ \qquad K_{ref} = [ADP^{3-}][HPO_4^{2-}][H^+]/[ATP^{4-}] \tag{8.28}$$

$$HATP^{3-} = H^+ + ATP^{4-} \qquad K_{ATP} = [H^+][ATP^{4-}]/[HATP^{3-}] \tag{8.29}$$

$$HADP^{2-} = H^+ + ADP^{3-} \qquad K_{ADP} = [H^+][ATP^{3-}]/[HADP^{2-}] \tag{8.30}$$

$$H_2PO_4^- = H^+ + HPO_4^{2-} \qquad K_{Pi} = [H^+][H_2PO_4^-]/[HPO_4^{3-}] \tag{8.31}$$

At lower pH, the acid dissociation of the adenine group has to be taken into account, and in the presence of Mg^{2+} and Ca^{2+}, the dissociation reactions of various complex ions have to be included in the calculation of the equilibrium composition. The equilibrium constants for reactions involving ions are taken to be functions of the ionic strength, as well as the temperature. Note that the activity of water in reaction 8.28 has been taken as unity in dilute aqueous solutions.

Calculations can be made on the thermodynamics of reactions in terms of species if the standard Gibbs energies of formation and standard enthalpies of formation are known. Table 8.4 gives the basic thermodynamic data at 298.15 K and zero ionic strength on species in biochemical reactions that will be discussed later in this chapter. Some of these data come from the NBS and CODATA tables, and some come from measurements on enzyme-catalyzed reactions. For NAD_{ox}^- and $ferredoxin_{ox}^+$, $\Delta_r G^\circ$ and $\Delta_r H^\circ$ are not known all of the way to the elements involved, so these properties have been assigned zero values by convention, as we have already seen for H^+. Some species that are not important in the pH range 5 to 9 have been excluded from Table 8.4. The charge numbers z_j and number $N_H(j)$ of hydrogen atoms in a species will be used later. The data in Table 8.4 can be used to calculate the equilibrium constants for reactions 8.28 to 8.31, so the equilibrium composition of this system of reactions can be calculated in terms of species.

However, in discussing metabolism it is more convenient to specify the pH and to write the hydrolysis of ATP to ADP and P_i (inorganic phosphate) as

$$ATP + H_2O = ADP + P_i \qquad K' = \frac{[ADP][P_i]}{[ATP]} \tag{8.32}$$

where the abbreviations represent sums of species; for example $[ATP] = [ATP^{4-}] + [HATP^{3-}]$. The equilibrium constant K' is referred to as an **apparent equilibrium constant** because it is a function of pH, in addition to temperature and ionic strength. The expression for the apparent equilibrium constant for reaction 8.32 can be written in terms of the concentrations of species as follows:

$$K' = \frac{([ADP^{3-}] + [HADP^{2-}])(HPO_4^{2-}] + [H_2PO_4^-])}{([ATP^{4-}] + [HATP^{3-}])} \tag{8.33}$$

Multiplying the right-hand side by $[ADP^{3-}][HPO_4^{2-}]/[ATP^{4-}]$, dividing by the same quantity, and introducing the acid dissociation constants K_{ATP}, K_{ADP}, and K_{Pi} yields

$$K' = \frac{[ADP^{3-}][HPO_4^{2-}]}{[ATP^{4-}]} \frac{(1 + [H^+]/K_{ADP})(1 + [H^+]/K_{Pi})}{(1 + [H^+]/K_{ATP})} \tag{8.34}$$

The first factor is close to being the equilibrium constant for a chemical reaction between species; in fact, it is equal to $K_{ref}/[H^+]$, where K_{ref} is the equilibrium

Table 8.4 Standard Gibbs Energies of Formation, Standard Enthalpies of Formation, Charge Numbers, and Numbers of Hydrogen Atoms in Species at 298.15 K and Zero Ionic Strength

Species	$\Delta_f G°$kJ/mol^{-1}	$\Delta_f H°$kJ/mol^{-1}	z_j	$N_H(j)$
H^+	0	0	1	1
ATP^{4-}	−2768.10	−3619.21	−4	12
$HADP^{3-}$	−2811.48	−3612.91	−3	23
H_2ATP^{2-}	2838.18	−3627.91	−2	14
ADP^{3-}	−1906.13	−2626.54	−3	12
$HADP^{2-}$	−1947.10	−2620.94	−2	13
H_2ADP^-	−1971.98	−2638.54	−1	14
$HPO_4{}^{2-}$	−1096.10	−1299.00	−2	1
$H_2PO_4{}^-$	−1137.3	−1302.60	−1	2
H_2O	−237.19	−285.83	0	2
Glucose 6-phosphate^{2-}	−1763.94	−2276.44	−2	11
Hglucose 6-phosphate$^-$	−1800.59	−2274.64	−1	12
Glucose	−915.9	−1262.19	0	12
$NAD_{ox}{}^-$	0	0	−1	26
$NAD_{red}{}^{2-}$	22.65	−31.94	−2	27
Ferrodoxin$_{ox}{}^+$	0	—	1	0
Ferrodoxin$_{red}$	38.07	—	0	0
Ethanol	−181.64	−288.30	0	6
Acetaldehyde	−139.00	−212.23	0	4
Formate	−351.00	−425.55	−1	1
$CO_3{}^{2-}$	−527.81	−677.14	−2	0
$HCO_3{}^-$	−568.77	−691.99	−1	1
H_2CO_3	−623.11	−699.63	0	2
$N_2(g)$	0	0	0	0
$H_2(g)$	0	0	0	2
NH_3	−26.5	−80.29	0	3
$NH_4{}^+$	−79.31	−132.51	1	4

Note: $NAD_{ox}{}^-$ (nicotinamide adenine dinucleotide, oxidized form) and ferredoxin$_{ox}{}^+$ (a protein) have been assigned zero values by convention.

The corresponding properties of the species of 131 biochemical reactants are available in R. A. Alberty, *Thermodynamics of Biochemical Reactions,* © Wiley, Hoboken, NJ, 2003. This material is used by permission of John Wiley & Sons, Inc. It is also on the Web at http://www.mathsource.com/cgi-bin/msitem?0211-622 as BasicBiochemData2.

constant for reaction 8.28. K_{ref} is independent of pH, but it depends on T, P, and ionic strength I, so these variables must be specified. Replacing the first factor in equation 8.33 with $K_{ref}/[H^+]$ yields

$$K' = \frac{K_{ref}}{[H^+]} \frac{(1 + [H^+]/K_{ADP})(1 + [H^+]/K_{Pi})}{(1 + [H^+]/K_{ATP})} \tag{8.35}$$

Since K_{ref} is independent of pH, this equation gives the pH dependence of the apparent equilibrium constant for biochemical reaction 8.32. Since $K_{ref} = 0.15$ at 298.15 K and $I = 0.25$ M, this equation can be used to calculate the value of K' at any pH in the range 5–9.

This approach is very useful for discussing the effect of pH on the apparent equilibrium constants of enzyme-catalyzed reactions, but what is lacking in

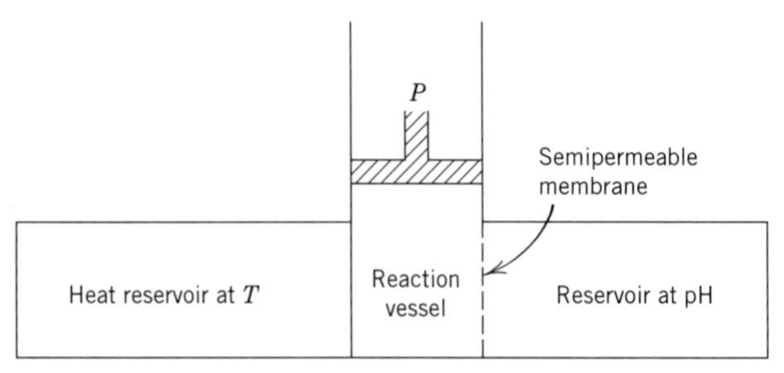

Figure 8.6 Thought experiment in which the reaction vessel is connected to a pH reservoir through a semipermeable membrane that is permeable to H^+.

the discussion so far is the introduction of other thermodynamic properties. It might appear that the discussion of other thermodynamic properties of biochemical reactions like reaction 8.32 can be based on the Gibbs energy G, but that is wrong because the criterion for spontaneous change and equilibrium provided by the Gibbs energy for a reaction system is $(\mathrm{d}G)_{T,P,\{n_c\}} \le 0$, where $\{n_c\}$ is the set of amounts of components. The amounts of components are usually taken to be the amounts of elements involved. The reason G cannot be used in discussing reaction 8.32 is that we specified the pH at equilibrium. The fact that an additional intensive variable is held constant is emphasized by Fig. 8.6. Biochemical equilibrium measurements are not carried out in this way, but the results are interpreted as if they were. According to this thought experiment, when a biochemical reaction produces hydrogen ions, as reaction 8.32 does in the neutral pH range, they diffuse into the pH reservoir. When a biochemical reaction consumes hydrogen ions, they diffuse into the reaction system from the pH reservoir. Thus hydrogen atoms are not conserved in the reaction vessel.

To see how to introduce the pH as an independent variable, let us review previous introductions of intensive variables. The criterion for spontaneous change and equilibrium that was provided by the combined first and second law for a reaction system is $(\mathrm{d}U)_{S,V,\{n_c\}} \le 0$, where $\{n_c\}$ is the set of amounts of components. To introduce the pressure P as an independent value, the Legendre transform $H = U + PV$ was used to define the enthalpy for which $(\mathrm{d}H)_{S,P,\{n_c\}} \le 0$. To introduce the intensive property T as an independent variable, the Legendre transform $A = U - TS$ was used to define the Helmholtz energy for which $(\mathrm{d}A)_{T,V,\{n\}} \le 0$. To introduce the intensive properties T and P as independent variables, the Legendre transform $G = H - TS$ was used to define the Gibbs energy for which $(\mathrm{d}G)_{T,P,\{n_c\}} \le 0$.

Now we need the thermodynamic potential that provides the criterion for spontaneous change and equilibrium at specified T, P, and $\mu(\mathrm{H}^+)$. This is accomplished by defining the transformed Gibbs energy G' with the Legendre transform

$$G' = G - n_c(\mathrm{H})\mu(\mathrm{H}^+) \tag{8.36}$$

where $n_c(\mathrm{H})$ is the amount of the hydrogen component (that is, the total amount of hydrogen atoms in the system). $n_c(\mathrm{H})$ and $\mu(\mathrm{H}^+)$ are conjugate variables (Section 2.5). This leads to the following criterion for spontaneous change and equilibrium: $(\mathrm{d}G')_{T,P,\{n_{c\neq\mathrm{H}}\},\mathrm{pH}} \le 0$, where $\{n_{c\neq\mathrm{H}}\}$ is the set of components excluding hydrogen, because hydrogen atoms are not conserved at constant pH. Derivations based on the fundamental equation for a thermodynamic system are usually carried out using the chemical potential, but specifying the pH is equivalent to specifying $\mu(H^+)$. In making Legendre transform 8.36, the chemical potential of one species (H^+) is changed from a dependent variable to an independent variable.

The amount $n_c(H)$ of the hydrogen component is given by the sum of the amounts of hydrogen atoms in various species in the reaction system:

$$n_c(H) = \sum_{j=1}^{N_s} N_H(j) n_j \tag{8.37}$$

In this equation $N_H(j)$ is the number of hydrogen atoms in species j and N_s is the number of different species in the system. The index number used for species is j, so the index number introduced later for reactants (sums of species) can be i. Substituting equation 8.37 and $G = \sum n_j \mu_j$ (equation 4.45) into the Legendre transform (equation 8.36) yields

$$G' = \sum_{j=1}^{N_s} n_j \mu_j - \sum_{j=1}^{N_s} n_j N_H(j) \mu(H^+) = \sum_{j=1}^{N_s} n_j \{\mu_j - N_H(j) \mu(H^+)\} = \sum_{j=1}^{N_s - 1} n_j \mu_j' \tag{8.38}$$

where the **transformed chemical potential** μ_j' of species j is given by

$$\mu_j' = \mu_j - N_H(j) \mu(H^+) \tag{8.39}$$

This is the effective chemical potential of a species when the chemical potential of hydrogen ions is $\mu(H^+)$. Note that the transformed chemical potential of the hydrogen ion is equal to zero, so there is one less term in the last summation in equation 8.38.

Equation 8.38 shows that the transformed Gibbs energy G' of a system is additive in the transformed chemical potentials μ_j' of $N_s - 1$ species, just as the Gibbs energy G is additive in the chemical potentials μ_j of N_s species (see equation 4.45). The roles of $n_c(H)$ and $\mu(H^+)$ in the fundamental equation are interchanged by making use of Legendre transform 8.36, as shown in the next section.

8.4 FUNDAMENTAL EQUATION OF THERMODYNAMICS FOR THE TRANSFORMED GIBBS ENERGY

The fundamental equation of thermodynamics for the Gibbs energy of a one-phase system is

$$dG = -S\,dT + V\,dP + \sum_{j=1}^{N_s} \mu_j\,dn_j \tag{8.40}$$

as given in equation 4.36. To obtain the fundamental equation for dG', we must first get the contribution for the hydrogen component into a separate term. This can be done by using equation 8.39 to eliminate μ_j from equation 8.40:

$$dG = -S\,dT + V\,dP + \sum_{j=1}^{N_s - 1} \mu_j\,dn_j + \sum_{j=1}^{N_s} N_H(j) \mu(H^+)\,dn_j \tag{8.41}$$

Equation 8.37 shows that $dn_c(\mathrm{H}) = \sum N_\mathrm{H}(j)\, dn_j$, so equation 8.41 can be written

$$dG = -S\, dT + V\, dP + \sum_{j=1}^{N_s-1} \mu_j'\, dn_j + \mu(\mathrm{H}^+)\, dn_c(\mathrm{H}) \tag{8.42}$$

The differential of the transformed Gibbs energy (equation 8.36) is

$$dG' = dG - n_c(\mathrm{H})\, d\mu(\mathrm{H}^+) - \mu(\mathrm{H}^+)\, dn_c(\mathrm{H}) \tag{8.43}$$

Substituting equation 8.42 into this equation yields a form of the fundamental equation for G':

$$dG' = -S\, dT + V\, dP + \sum_{j=1}^{N_s-1} \mu_j'\, dn_j - n_c(\mathrm{H})\, d\mu(\mathrm{H}^+) \tag{8.44}$$

The chemical potential of hydrogen ions is now an independent variable, like T, P, and $\{n_{c\neq\mathrm{H}}\}$, instead of the amount of the hydrogen component (equation 8.42).

Since the chemical potential of hydrogen ions $\mu(\mathrm{H}^+)$ depends on both the temperature and the concentration of hydrogen ions, it is not a very convenient variable when the temperature is changed. The hydrogen ion concentration can be made an independent intensive variable in the fundamental equation for G' by use of the expression for the differential of the chemical potential of H^+:

$$d\mu(\mathrm{H}^+) = \left\{\frac{\partial \mu(\mathrm{H}^+)}{\partial T}\right\}_{[\mathrm{H}^+]} dT + \left\{\frac{\partial \mu(\mathrm{H}^+)}{\partial [\mathrm{H}^+]}\right\}_T d[\mathrm{H}^+] \tag{8.45}$$

The first partial derivative in this equation is equal to $-S_\mathrm{m}(\mathrm{H}^+)$, where $S_\mathrm{m}(\mathrm{H}^+)$ is the **molar entropy** of hydrogen ions. To evaluate the second partial derivative in equation 8.45, we need to recall that the chemical potential of hydrogen ions is given by $\mu(\mathrm{H}^+) = \mu°(\mathrm{H}^+) + RT\ \ln[\mathrm{H}^+]$. Since the thermodynamic properties are taken to be functions of the ionic strength, we do not have to deal with activity coefficients explicitly. Since $d\mu(\mathrm{H}^+)/d[\mathrm{H}^+] = RT/[\mathrm{H}^+]$ and $d\mathrm{pH}/d[\mathrm{H}^+] = -1/\{\ln(10)[\mathrm{H}^+]\}$, equation 8.45 can be written

$$d\mu(\mathrm{H}^+) = -S_\mathrm{m}(\mathrm{H}')\, dT - RT\ \ln(10)\, d\mathrm{pH} \tag{8.46}$$

Substituting this in equation 8.44 yields

$$dG' = -S\, dT + V\, dP + \sum_{j=1}^{N_s-1} \mu_j' dn_j + n_c(\mathrm{H})RT\ \ln(10)\, d\mathrm{pH} \tag{8.47}$$

where the **transformed entropy** S' of the system at a specified pH is given by

$$S' = S - n_c(\mathrm{H})S_\mathrm{m}(\mathrm{H}^+) \tag{8.48}$$

Note that this has the same form as the Legendre transform and that the **transformed enthalpy** of the system is given by

$$H' = H - n_c(\mathrm{H})H_\mathrm{m}(\mathrm{H}^+) \tag{8.49}$$

When the pH is specified, some of the $N_s - 1$ terms in the summation in equation 8.47 can be aggregated—for example, ATP^{4-} and $HATP^{3-}$, which are pseudoisomers at a specified pH. Isomers have the same atomic compositions, and at a specified pH pseudoisomers have the same atomic compositions, except for the number of hydrogen atoms. At a specified pH, these pseudoisomers have the same transformed chemical potential μ_i', and so their terms in equation 8.47 can be replaced with $\mu_i'\,dn_i'$, where n_i' is the amount of pseudoisomer group i. Thus equation 8.47 becomes

$$dG' = -S'\,dT + V\,dP + \sum_{i=1}^{N'} \mu_i'\,dn_i' + n_c(\mathrm{H})RT\,\ln(10)\,\mathrm{dpH} \qquad (8.50)$$

where N' is the number of pseudoisomer groups in the system, which may be considerably less than $N_s - 1$. Note that the subscripts have been changed to i, which apply to pseudoisomer groups rather than the species. This is the fundamental equation for a reaction system at a specified pH. This fundamental equation has a number of equations of state and Maxwell equations, and leads to a Gibbs–Duhem equation.

Example 8.3 ***Derivation of the expression for the apparent equilbrium constant for an enzyme–catalyzed reaction***

Use equation 8.50 to derive the expression for the apparent equilibrium constant of a biochemical reaction $\sum \nu_i'\mathrm{B}_i = 0$, where B_i represents a pseudoisomer group such as ATP. This follows the steps in the derivation of the expression for the equilibrium constant of a chemical reaction.

If a single biochemical reaction is catalyzed, the amounts n_i' of the pseudoisomer groups at each stage of the reaction are each given by

$$n_i' = (n_i')_0 + \nu_i'\xi' \qquad (8.51)$$

where $(n_i')_0$ is the initial amount of reactant i (pseudoisomer group i), ν_i' is the stoichiometric number of reactant i in the biochemical reaction, and ξ' is the apparent extent of the biochemical reaction. It is necessary to put primes on these quantities to differentiate them from the stoichiometric numbers ν_j and extents of reaction ξ of the underlying chemical reactions written in terms of species. Substituting $dn_i' = \nu_i'\,d\xi'$ in equation 8.50 yields

$$dG' = -S\,dT + V\,dP + \left(\sum_{i=1}^{N'} \nu_i'\mu_i'\right)d\xi' + n_c(\mathrm{H})RT\,\ln(10)\,\mathrm{dpH} \qquad (8.52)$$

so that

$$\Delta_r G' = \left(\frac{\partial G'}{\partial \xi'}\right)_{T,P,\mathrm{pH}} = \sum_{i=1}^{N'} \nu_i'\mu_i' \qquad (8.53)$$

where $\Delta_r G'$ is referred to as the transformed reaction Gibbs energy. At chemical equilibrium, $\Delta_r G'$ is equal to zero so that

$$\sum_{i=1}^{N'} \upsilon_i'(\mu_i')_{\mathrm{eq}} = 0 \qquad (8.54)$$

This is the equilibrium condition. In Chapter 5 we saw that the corresponding condition for a chemical reaction is equation 5.5. The subscript eq is frequently omitted in discussions of chemical equilibrium.

The expression for the transformed chemical potential of a reactant (pseudoisomer group) is given by

$$\mu_i' = \mu_i'^{\circ} + RT\,\ln[\mathrm{B}_i] \qquad (8.55)$$

where $\mu_i'^{\circ}$ is the standard transformed chemical potential of reactant i at a specified pH and ionic strength and B_i represents the ith reactant. Substituting equation 8.55 in equation 8.54 yields

$$\begin{aligned} \sum \nu_i' \mu_i'^{\circ} &= -RT \sum \nu_i' \ln[B_i] \\ &= -RT \sum \ln([B_i]^{\nu_i'}) \\ &= -RT \ln \prod [B_i]^{\nu_i'} = -RT \ln K' \end{aligned} \tag{8.56}$$

where $\prod$ is the product sign. The apparent equilibrium constant is given by

$$K' = \prod [B_i]^{\nu_i'} \tag{8.57}$$

This confirms that K' is written in terms of concentrations of pseudoisomer groups and that there is no term for hydrogen ions.

8.5 CALCULATION OF STANDARD TRANSFORMED FORMATION PROPERTIES OF REACTANTS IN BIOCHEMICAL REACTIONS

Equation 8.56 shows how to calculate the apparent equilibrium constant for a biochemical reaction when we know the standard transformed chemical potentials of the reactants at the desired temperature, pressure, pH, and ionic strength. In discussing the calculation of these properties, we will replace standard chemical potentials with standard Gibbs energies of formation because these are the properties that have been tabulated for species, as shown in Table 8.4. Thus equation 8.39 can be written

$$\Delta_f G_j' = \Delta_f G_j - N_H(j) \Delta_f G(H^+) \tag{8.58}$$

There is a corresponding equation for the transformed enthalpy of formation of a species.

$$\Delta_f H_j' = \Delta_f H_j - N_H(j) \Delta_f H(H^+) \tag{8.59}$$

These adjustments for the number of hydrogen atoms in a species also apply to the standard transformed properties, so that

$$\Delta_f G_j'^{\circ} = \Delta_f G_j^{\circ} - N_H(j)(\Delta_f G^{\circ}(H^+) + RT \ln 10^{-pH}) \tag{8.60}$$

$$\Delta_f H_j'^{\circ} = \Delta_f H_j^{\circ} - N_H(j) \Delta_f H^{\circ}(H^+) \tag{8.61}$$

We have seen earlier (equations 7.79 and 7.80) that the standard Gibbs energy of formation and the standard enthalpy of formation of a species at 298.15 K can be adjusted to the desired ionic strength using the extended Debye–Hückel equation:

$$\Delta_f G_j^{\circ}(I) = \Delta_f G_j^{\circ}(I = 0) - 2.91482 z_j^2 I^{1/2}/(1 + BI^{1/2}) \tag{8.62}$$

$$\Delta_f H_j^{\circ}(I) = \Delta_f H_j^{\circ}(I = 0) + 1.4775 z_j^2 I^{1/2}/(1 + BI^{1/2}) \tag{8.63}$$

where z_j is the charge number of ion j and $B = 1.6\ L^{1/2}\ mol^{-1/2}$. Substituting these equations in equations 8.60 and 8.61 yields the following two equations for calculating the standard thermodynamic properties of species in kJ mol^{-1} at 298.15 K:

$$\Delta_f G_j'^\circ = \Delta_f G_j^\circ(I = 0) + N_H(j)RT \ln(10)\text{pH} - 2.914\,82(z_j^2 - N_H(j))I^{1/2}/(1 + 1.6I^{1/2}) \tag{8.64}$$

$$\Delta_f H_j'^\circ = \Delta_f H_j^\circ(I = 0) + 1.4775(z_j^2 - N_H(j))I^{1/2}/(1 + 1.6I^{1/2}) \tag{8.65}$$

Example 8.4 ***Calculation of the standard transformed formation properties of*** HPO_4^{2-} ***and*** $H_2PO_4^-$

Calculate the standard transformed formation properties of HPO_4^{2-} and $H_2PO_4^-$ at 298.15 K, pH 7, and $I = 0.25$ M.

For HPO_4^{2-} equations 8.64 and 8.65 with values from Table 8.4 yield

$$\Delta_f G'^\circ(HPO_4^{2-}) = -1096.10 - 7RT \ln(10) - 2.914\,82(4 - 1)(0.25)^{1/2}/[1 + 1.6(0.25)^{1/2}]$$
$$= -1058.57 \text{ kJ mol}^{-1}$$

$$\Delta_f H'^\circ(HPO_4^{2-}) = -1299.00 + 1.4775(4 - 1)(0.25)^{1/2}/[1 + 1.6(0.25)^{1/2}]$$
$$= -1297.77 \text{ kJ mol}^{-1}$$

For $H_2PO_4^-$ equations 8.64 and 8.65 yield

$$\Delta_f G'^\circ(H_2PO_4^-) = -1137.30 - 14RT \ln(10) - 2.914\,82(1 - 2)(0.25)^{1/2}/[1 + 1.6(0.25)^{1/2}]$$
$$= -1056.58 \text{ kJ mol}^{-1}$$

$$\Delta_f H'^\circ(H_2PO_4^-) = -1302.60 + 1.4775(1 - 2)(0.25)^{1/2}/[1 + 1.6(0.25)^{1/2}]$$
$$= -1303.91 \text{ kJ mol}^{-1}$$

Now that we have the standard transformed Gibbs energies of formation and standard transformed enthalpies of formation of HPO_4^{2-} and $H_2PO_4^-$ at pH 7 and 0.25 M ionic strength, the question is, What are these properties for inorganic phosphate P_i under these conditions? At specified pH these two species are pseudoisomers, and so we can use the equation derived earlier (Section 5.10) for calculating the standard Gibbs energy of formation of an isomer group. An equation of just this form applies to a pseudoisomer group, and it is

$$\Delta_f G'^\circ(\text{pseudoisomer group}) = -RT \ln \sum_{i=1}^{N_I} \exp(-\Delta_f G_i'^\circ/RT) \tag{8.66}$$

The number of pseudoisomers in the pseudoisomer group is represented by N_I. The equilibrium mole fraction of a pseudoisomer in a pseudoisomer group is given by

$$r_i = \exp\{[\Delta_f G'^\circ(\text{pseudoisomer group}) - \Delta_f G_i'^\circ]/RT\} \tag{8.67}$$

The value of $\Delta_f H'^\circ$(pseudoisomer group) can be calculated using

$$\Delta_f H'^\circ(\text{pseudoisomer group}) = \sum_{i=1}^{N_I} r_i \Delta_f H_i'^\circ \tag{8.68}$$

Since we know the standard transformed formation properties of HPO_4^{2-} and $H_2PO_4^-$ at pH 7 (see Example 8.4), we can calculate the standard transformed formation properties of P_i at pH 7.

Example 8.5 ***Calculation of the standard transformed properties of inorganic phosphate at 25 °C, pH 7, and I = 0.25 M***

What are the standard transformed formation properties of inorganic phosphate at 25 °C, pH 7, and $I = 0.25$ M?

The standard transformed Gibbs energy of formation is given by equation 8.66.

$$\Delta_f G'^{\circ}(P_i) = -RT \ln[\exp(1058.57/RT) + \exp(1056.58/RT)]$$

To avoid exponential overflow with a hand-held calculator, we may have to take one of the exponential terms out and write

$$\begin{aligned} \Delta_f G'^{\circ}(P_i) &= -1058.57 - RT \ln\{\exp[1 + \exp(1056.58 - 1058.57)/RT]\} \\ &= -1059.49 \text{ kJ mol}^{-1} \\ r(HPO_4^{2-}) &= \exp[(-1059.49 + 1058.57)/RT] = 0.6906 \\ r(H_2PO_4^{-}) &= \exp[(-1059.49 + 1056.58)/RT] = 0.3094 \\ \Delta_f H'^{\circ}(P_i) &= (0.6906)(-1297.77) + (0.3094)(-1303.01) \\ &= -1299.39 \text{ kJ mol}^{-1} \end{aligned}$$

Values of standard transformed Gibbs energies of formation of reactants for which species are given in Table 8.4 are listed in Table 8.5 at 298.15 K, 0.25 M ionic strength, and pH 5, 6, 7, 8, and 9. As with thermodynamic tables for species, the information in Table 8.5 lies in differences between values, so the number of digits does not indicate the accuracy of the measurements. An error of 0.01 kJ mol^{-1} in a value in Table 8.5 leads to about a 1% error in an equilibrium constant.

Equation 8.56 shows how the apparent equilibrium constant K' for an enzyme-catalyzed reaction can be calculated from $\mu_i'^{\circ}$ values, but now we will use that equation in the form

$$\Delta_r G'^{\circ} = \sum \nu_i' \Delta_f G_i'^{\circ} = -RT \ln K' \tag{8.69}$$

The corresponding equations for $\Delta_r H'^{\circ}$ and $\Delta_r S'^{\circ}$ are

$$\Delta_r H'^{\circ} = \sum \nu_i' \Delta_f H_i'^{\circ} \tag{8.70}$$

and

$$\Delta_r S'^{\circ} = \sum \nu_i' \Delta_f S_i'^{\circ} \tag{8.71}$$

Note that $\Delta_r S'^{\circ} = (\Delta_r H'^{\circ} - \Delta_r G'^{\circ})/T$.

Table 8.6 gives the apparent equilibrium constants of six enzyme-catalyzed reactions at 298.15 K, $I = 0.25$ M, and five pH values that have been calculated using the standard transformed Gibbs energies of formation of the reactants given in Table 8.5. The first two reactions show that ATP has a much higher potential

Table 8.5 Standard Transformed Gibbs Energies of Formation in kJ mol^{-1} of Reactants at 298.15 K, 0.25 M Ionic Strength, and Various pH Values

	pH 5	*pH 6*	*pH 7*	*pH 8*	*pH 9*
ATP	−2437.46	−2363.76	−2292.50	−2223.44	−2154.88
ADP	−1569.05	−1495.55	−1424.70	−1355.78	−1287.24
P_i	−1079.46	−1068.49	−1059.49	−1052.97	−1047.17
H_2O	−178.49	−157.07	−155.66	−144.24	−132.83
Glucose 6-phosphate	−1449.53	−1382.88	−1318.92	−1255.98	−1193.18
Glucose	−563.70	−495.20	−426.71	−358.21	−289.72
NAD_{ox}	762.29	910.70	1059.11	1207.51	1355.92
NAD_{red}	811.86	965.98	1120.09	1274.21	1428.33
$Ferredoxin_{ox}$	−0.81	−0.81	−0.81	−0.81	−0.81
$Ferredoxin_{red}$	38.07	38.07	38.07	38.07	38.07
Ethanol	−5.54	28.71	62.96	97.20	131.45
Acetaldehyde	−21.60	1.23	24.06	46.90	69.73
Formate	−322.46	−316.75	−311.04	−305.34	−299.63
CO_2tot	−564.61	−554.49	−547.10	−541.18	−535.80
$N_2(g)$	0.00	0.00	0.00	0.00	0.00
$H_2(g)$	58.70	70.12	81.53	92.95	104.36
Ammonia	37.28	60.11	82.93	105.64	127.51

Note: NAD_{ox}^- (nicotinamide adenine dinucleotide, oxidized form) and $ferredoxin_{ox}^+$ (a protein) have been assigned zero values by convention. CO_2tot is the sum of the various species of carbon dioxide in aqueous solution.
Source: R. A. Alberty, *Thermodynamics of Biochemical Reactions*, © Wiley, Hoboken, NJ, 2003. This material is used by permission of John Wiley & Sons, Inc.

Table 8.6 Apparent Equilibrium Constants of Six Enzyme-Catalyzed Reactions at 298.15 K, 0.25 M Ionic Strength, and Five pH Values

Reaction	*pH 5*	*pH 6*	*pH 7*	*pH 8*	*pH 9*
ATP + H_2O = ADP + P_i	5.1×10^5	6.6×10^5	2.1×10^6	1.6×10^7	1.5×10^8
Glucose 6-phosphate + H_2O = glucose + P_i	4.5×10^2	2.6×10^2	1.1×10^2	0.8×10^2	0.8×10^2
ATP + glucose = glucose 6-phosphate + ADP	1.1×10^3	2.6×10^3	1.9×10^4	1.9×10^5	1.9×10^6
NAD_{ox} + ethanol = NAD_{red} + acetaldehyde	1.3×10^{-6}	1.3×10^{-5}	1.3×10^{-4}	1.3×10^{-3}	1.3×10^{-2}
NAD_{ox} + formate + H_2O = NAD_{red} + CO_2tot	2.9×10^2	5.0×10^2	2.5×10^3	2.3×10^4	2.6×10^5
$N_2(g)$ + 8 $ferredoxin_{red}$ = 2 ammonia + $H_2(g)$ + 8 $ferredoxin_{ox}$	1.4×10^{31}	1.4×10^{21}	1.4×10^{11}	16	3.4×10^{-9}

to phosphorylate than glucose 6-phosphate. The third reaction is an example of coupling, and the apparent equilibrium constant is equal to the ratio of the apparent equilibrium constants of the first two reactions. NADox is often involved in accepting electrons to produce the oxidized product of a reaction. Since the oxidation of formate produces carbon dioxide, which is hydrated in aqueous solution, H_2O has to appear on the left side of the reaction to balance oxygen atoms. The reduction of molecular hydrogen to ammonia is the only known biochemical pathway for fixing molecular nitrogen. Note that the apparent equilibrium constant for this reaction decreases by a factor of 10^{10} per pH unit, and that the reaction becomes nonspontaneous above pH 8.

In closing this section on biochemical reactions, it is worth noting that the fundamental equation for G' contains a new type of term that is proportional to dpH. This significantly increases the number of Maxwell equations. Since equation 8.52 can be written as

$$dG' = -S'\,dT + V\,dP + \Delta_r G'\,d\xi' + n_c(\mathrm{H})RT\,\ln(10)\,\mathrm{dpH} \qquad (8.72)$$

one of the new Maxwell equations is

$$\left(\frac{\partial \Delta_r G'}{\partial \mathrm{pH}}\right)_{T,P,\xi'} = RT\,\ln(10)\left(\frac{\partial n_c(\mathrm{H})}{\partial \xi'}\right)_{T,P,\mathrm{pH}} \qquad (8.73)$$

The partial derivative on the right-hand side is the change $\Delta_r N_H$ in the amount of the hydrogen component in the system when the biochemical reaction occurs. This change can be calculated using

$$\Delta_r N_H = \frac{1}{RT\,\ln(10)}\left(\frac{\partial \Delta_r G'^\circ}{\partial \mathrm{pH}}\right)_{T,P} \qquad (8.74)$$

where $\Delta_r G'^\circ$ has been used because its derivative has the same value as for $\Delta_f G'$. This property can be measured by use of a pHstat. For example, when ATP is hydrolyzed at pH 9, equation 8.74 shows that $\Delta_r N_H = -1$ and in the thought experiment in Fig 8.6, one mole of H^+ diffuses into the pH reservoir for each mole of ATP hydrolyzed. Table 8.7 gives $\Delta_r N_H$ as a function of pH for the six reactions in Table 8.6.

Other properties of biochemical reations that can be calculated are $\Delta_r G'^\circ$, $\Delta_r H'^\circ$, and $\Delta_r S'^\circ$. Thermodynamic properties at other temperatures can be calculated if $\Delta_f H^\circ$ values are known for species and can be assumed to be independent of temperature.

Comment:

At a specified pH, species have standard transformed thermodynamic properties that depend on the pH. The effect of specifying the pH is to make the various species of a reactant, such as ATP, *behave like isomers in the sense that their distribution within the group is dependent only on temperature, and so they are referred to as pseudoisomers. A pseudoisomer group, such as ATP, has a standard transformed Gibbs energy of formation $\Delta_f G'^\circ$ and a standard transformed enthalpy of formation $\Delta_f H'^\circ$ at a specified pH that can be used like $\Delta_f G^\circ$ and $\Delta_f H^\circ$ to calculate apparent equilibrium constants and standard transformed*

Table 8.7 Changes in the Binding of Hydrogen Ions in Six Enzyme-Catalyzed Reactions at 298.15 K, 0.25 M Ionic Strength, and Five pH Values

Reaction	*pH 5*	*pH 6*	*pH 7*	*pH 8*	*pH 9*
ATP + H_2O = ADP + P_i	−0.04	−0.24	−0.74	−0.96	−1.00
Glucose 6-phosphate + H_2O = glucose + P_i	0.10	0.40	0.24	0.04	0.00
ATP + glucose = glucose 6-phosphate + ADP	−0.14	−0.65	−0.98	−1.00	−1.00
NAD_{ox} + ethanol = NAD_{red} + acetaldehyde	1.00	1.00	1.00	1.00	1.00
NAD_{ox} + formate + H_2O = NAD_{red} + CO_2tot	−0.08	−0.45	−0.89	−1.00	−1.14
N_2(g) + 8 ferredoxin$_{red}$ = 2 ammonia + H_2(g) + 8 ferredoxin$_{ox}$	10.00	10.00	9.99	9.89	9.28

enthalpies of reaction at the specified pH. The new thermodynamic table can be used like the familiar thermodynamic table for species, but we do not have to be concerned with all of the species.

8.6 COUPLING OF BIOCHEMICAL REACTIONS

Two or more biochemical reactions may be combined by an enzyme catalyst. This effect, which is called **coupling,** makes it possible for a spontaneous reaction (such as the hydrolysis of ATP) to drive a nonspontaneous reaction (such as the phosphorylation of glucose). Consider a system that contains ATP, ADP, glucose, glucose 6-phosphate, and inorganic phosphate at 298.15 K, pH 7, pMg 3, and $I = 0.25$ M. If we add the enzymes glucose-6-phosphatase and ATPase, the organic phosphates will both be almost completely hydrolyzed by the reactions

$$\text{Glc6P} + H_2O = \text{glucose} + P_i \tag{8.75}$$

$$\text{ATP} + H_2O = \text{ADP} + P_i \tag{8.76}$$

because the glucose-6-phosphatase catalyzes the first reaction and the ATPase catalyzes the second reaction. The apparent equilibrium constants of these reactions can be calculated by use of Table 8.5. If the initial solution contains 1 mM ATP and 1 mM Glc6P, the equilibrium solution will contain 2.02×10^{-9} mM ATP, 1 mM ADP, 2 mM P_i, 1 mM glucose, and 9.2×10^{-6} mM Glc6P.

However, if the enzyme glucokinase is added to a mixture of these reactants, the concentration of inorganic phosphate does not change because this enzyme catalyzes the difference of reactions 8.75 and 8.76:

$$\text{ATP} + \text{glucose} = \text{Glc6P} + \text{ADP} \tag{8.77}$$

The apparent equilibrium constant of this reaction is the equilibrium constant for reaction 8.76 divided by the apparent equilibrium constant for reaction 8.82. If the initial solution contains 1 mM adenosine triphosphate and 1 mM glucose and no other reactants, the equilibrium solution will contain [ATP] = [glucose] = 0.014 mM and [ADP] = [Glc6P] = 0.98 mM. In other words, organic phosphate is not hydrolyzed, but the phosphate is transferred stoichiometrically from ATP to glucose. This can be accomplished if the mechanism of the reaction is

$$\text{ATP} + \text{E} = \text{EP} + \text{ADP} \tag{8.78}$$

$$\text{EP} + \text{glucose} = \text{Glc6P} + \text{E} \tag{8.79}$$

where E is the enzyme catalyst. According to this mechanism, phosphate is transferred from ATP to the enzyme and then from the enzyme to glucose. This is referred to as **coupling** because equations 8.75 and 8.76 are coupled together by EP. This is not the only type of mechanism that will do this; in another possible mechanism, the reactants ATP and glucose are both bound on the enzyme and react with each other according to reaction 8.77 on the surface of the enzyme. The standard transformed Gibbs energy change for reaction 8.77 is the difference of the standard transformed Gibbs energies of reactions 8.75 and 8.76 or the sum of the standard transformed Gibbs energies of reactions 8.78 and 8.79.

Biochemical reactants can be ranked according to their tendency to transfer phosphate, as shown in Table 8.8. Organic phosphates higher up in the table can

Table 8.8 Standard Transformed Gibbs Energies of Hydrolysis at 25 °C, pH 7, pMg 4, and $I = 0.2$ M

	$\dfrac{\Delta_r G'^{\circ}}{\text{kJ mol}^{-1}}$
Phosphoenolpyruvate + H_2O = enol pyruvate + P_i	−61.9
Creatine phosphate + H_2O = creatine + P_i	−43.5
Acetyl phosphate + H_2O = acetate + P_i	−43.1
CoA-S-phosphate + H_2O = CoA-SH + P_i	−37.7
ATP + H_2O = ADP + P_i	−39.7
ADP + H_2O = AMP + P_i	−36.8
Pyrophosphate (PP) + H_2O = 2P_i	−34.3
Arginine phosphate + H_2O = arginine + P_i	−29.3
Glucose 6-phosphate + H_2O = glucose + P_i	−12.6
Fructose 1-phosphate + H_2O = fructose + P_i	−12.6
AMP + H_2O = adenosine + P_i	−12.6
Glycerol 1-phosphate + H_2O = glycerol + P_i	−9.2

transfer phosphate to nonphosphorylated substances lower in the table if reactants and products are in their standard states. Similar tables may be constructed for the transfer of pyrophosphate (pyrophosphoric acid is $H_4P_2O_7$) or any other group. There is an analogy between the transfer of phosphate in these reactions and the transfer of electrons in reactions studied in electrochemistry or protons in acid–base equilibria.

In discussing equilibria in multicomponent systems in Chapter 5, we noted that any independent set of chemical reactions could be used. In biochemistry, the situation is different because biochemical reactions occur only if the corresponding enzyme is present. Therefore, the number of biochemical reactions in a pathway may be less than the number of independent reactions between the reactants.

Example 8.6 ***Calculation of the apparent equilibrium constant for a biochemical reaction that can be considered to be the sum of the two reactions***

Determine the equilibrium constant K' at pH 7, pMg 4, $I = 0.2$ M, and 25 °C for

$$\text{creatine phosphate} + \text{ADP} = \text{creatine} + \text{ATP}$$

Since

$$\text{creatine phosphate} + H_2O = \text{creatine} + P_i \qquad \Delta_r G^{\circ\prime} = -43.5\ \text{kJ mol}^{-1}$$

$$\text{ADP} + P_i = \text{ATP} + H_2O \qquad \Delta_r G^{\circ\prime} = 39.7\ \text{kJ mol}^{-1}$$

for the given reaction, $\Delta_r G^{\circ\prime} = -43.5 + 39.7 = -3.8\ \text{kJ mol}^{-1}$. Using $\Delta_r G^{\circ\prime} = -RT \ln K'$,

$$K' = \frac{[\text{Cr}][\text{ATP}]}{[\text{CrP}][\text{ADP}]} = 4.6$$

The biochemical equilibria we have been discussing involve small molecules, but many biochemical equilibria involve macromolecules such as proteins and nucleic acids. As an example we will now consider the binding of small molecules by a protein.

8.7 BINDING OF OXYGEN BY MYOGLOBIN AND HEMOGLOBIN

Myoglobin is an oxygen storage protein that is found in muscle tissue in many species. Its molar mass is 16 000 g mol^{-1}, and each molecule contains one heme group and one atom of iron, and binds one molecule of oxygen when it is saturated.

The shape of the binding curve for myoglobin is exactly what is expected for the simple reaction

$$MbO_2 = Mb + O_2 \tag{8.80}$$

where Mb represents a molecule of myoglobin. The apparent dissociation constant at a specified pH is defined by

$$K' = \frac{[Mb]P_{O_2}}{[MbO_2]} \tag{8.81}$$

where P_{O_2} is the partial pressure of oxygen in the gas phase. Conservation of myoglobin requires that

$$[Mb]_t = [Mb] + [MbO_2] \tag{8.82}$$

where $[Mb]_t$ is the total molar concentration of myoglobin. Combining equations 8.81 and 8.82 yields

$$Y = \frac{[MbO_2]}{[Mb]_t} = \frac{P_{O_2}/K'}{1 + P_{O_2}/K'} \tag{8.83}$$

where Y is the fractional saturation. The plot of Y versus P_{O_2} is given in Fig. 8.7.

Hemoglobin is the oxygen transport protein in many species. It has a molar mass of 64 000 g mol^{-1}, and each molecule contains four heme groups and four atoms of iron, and binds four molecules of oxygen when it is saturated. Hemoglobin may be reversibly dissociated into four molecules of molar mass 16 000 g mol^{-1}, each of which contains one heme and one atom of iron. These smaller molecules are of two types, represented by α and β, and hemoglobin has the composition $\alpha_2\beta_2$. Many enzymes have a similar subunit structure, and they and hemoglobin have remarkable binding properties, which are a consequence of this subunit structure.

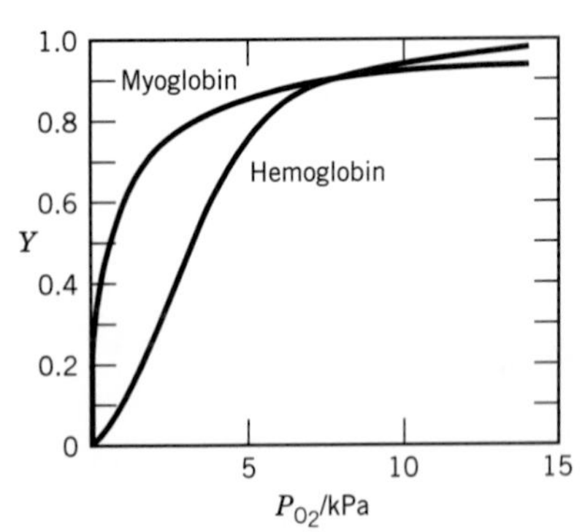

Figure 8.7 Fractional saturation Y of myoglobin and hemoglobin with oxygen at pH 7.4 and 38 °C.

The oxygen binding properties of myoglobin and hemoglobin are distinctively different, as shown by Fig. 8.7. Hemoglobin's S-shaped (sigmoid) binding curve is a great advantage for its physiological function, because the amount of oxygen bound changes rapidly between the partial pressure of oxygen in the lungs (about 13.3 kPa) and in the tissues (about 2.0 kPa).

The equation for the average number $\overline{N}$ of oxygen molecules bound by hemoglobin can be expressed in terms of the partial pressure of oxygen and the successive dissociation constants in the same way that we expressed the number of protons bound by inorganic phosphate in equation 8.24. To compare the binding of hemoglobin with myoglobin it is convenient to use the fractional saturation Y, which in this case is $\overline{N}/4$ since four oxygen molecules may be bound:

$$Y = \frac{P_{O_2}/K_1' + 2P_{O_2}^2/K_1'K_2' + 3P_{O_2}^3/K_1'K_2'K_3' + 4P_{O_2}^4/K_1'K_2'K_3'K_4'}{4[1 + P_{O_2}/K_1' + P_{O_2}^2/K_1'K_2' + P_{O_2}^3/K_1'K_2'K_3' + P_{O_2}^4/K_1'K_2'K_3'K_4']} \tag{8.84}$$

Equation 8.84 can represent the sigmoid binding curve for hemoglobin but, remarkably, only if the successive dissociation constants *decrease* instead of increase. This is contrary to the usual observation that the dissociation constants for successive ligand molecules increase. You might ask, If some of the binding sites have higher affinities for oxygen, shouldn't they be filled first? The answer is that the binding sites with higher affinity do not have this high affinity until some oxygen has been bound. This is called cooperativity.

When binding at the middle of the titration curve increases more rapidly with ligand concentration than can be accounted for by equation 8.83, it is said to be **cooperative.** Cooperativity arises when the binding of the first ligands causes changes in the structure of the protein that increase the affinities of the remaining sites. The origin of this effect in hemoglobin was quite mysterious until the structures of oxygenated and deoxygenated hemoglobin were determined by X-ray diffraction. Myoglobin was the first protein for which the detailed molecular structure was obtained by X-ray diffraction. When the structure of hemoglobin was obtained, it was found that each of its four subunits has a three-dimensional configuration much like that of myoglobin. When oxygen is bound, groups near the heme shift slightly, and these structural changes affect the configurations of the four subunits so that the binding properties of the other heme groups are enhanced.* The cooperativity of the oxygen binding by hemoglobin is treated quantitatively in terms of the Monod–Wyman–Changeux (MWC) concerted mechanism.† According to this mechanism, a small fraction of hemoglobin exists in the quaternary oxy structure that binds oxygen more strongly. When the first oxygen molecule is bound, it is bound preferentially to this structure. As more oxygen molecules are bound, the oxy structure is sufficiently stabilized to be the major structure, and so subsequent binding is strong.

In addition to the positive cooperativity of hemoglobin, there is also negative cooperativity. For example, some enzymes have identical subunits, but the binding of the first substrate molecule causes molecular changes that reduce the affinities of the enzymatic sites in neighboring subunits.

Hemoglobin has another remarkable property that makes it even more effective in the oxygen transport system. When hemoglobin is oxygenated at pH 7.4, it produces 0.6 mol of H^+ for each mole of oxygen molecule bound. A corollary of this so-called Bohr effect is that the affinity of hemoglobin for oxygen depends on pH. In the neighborhood of pH 7.4, the partial pressure of oxygen required to half-saturate hemoglobin decreases as the pH is increased. The explanation of the Bohr effect is that the binding of oxygen affects the acid dissociation constants of certain acid groups in hemoglobin. This effect is of considerable physiological importance because in the lungs H^+, liberated by hemoglobin on oxygenation, reacts with HCO_3^- to make H_2CO_3. The carbonic acid then dehydrates and CO_2 diffuses into the air space of the lungs. If hemoglobin did not dissociate H^+, the blood would become alkaline in the lungs as CO_2 was exhaled. In the capillaries this process is reversed; hemoglobin absorbs H^+ as it loses oxygen, and this converts H_2CO_3, produced metabolically, to HCO_3^-.

*See R. E. Dickerson and I. Geis, *Hemoglobin.* Menlo Park, CA: Benjamin/Cummings, 1983.

†J. Monod, J. Wyman, and J. P. Changeux, *J. Mol. Biol.* **12:**88 (1965); J. Wyman and S. J. Gill, *Binding and Linkage.* Mill Valley, CA: University Science Books, 1990.

Example 8.7 ***On the relationship between macroscopic dissociation constants and microscopic dissociation constants***

Suppose that a particular hemoprotein has two oxygen binding sites that are independent. Express the fractional saturation Y in terms of (a) the macroscopic dissociation constants K_1' and K_2' and (b) the microscopic dissociation constants k_1' and k_2'. (c) How are K_1' and K_2' related to k_1' and k_2'?

(a)
$$Y = \frac{P_{O_2}/K_1' + 2P_{O_2}^2/K_1'K_2'}{2[1 + P_{O_2}/K_1' + P_{O_2}^2/K_1'K_2']}$$

(b)
$$Y = \frac{1}{2}\left[\frac{P_{O_2}/k_1'}{1 + P_{O_2}/k_1'} + \frac{P_{O_2}/k_2'}{1 + P_{O_2}/k_2'}\right]$$

(c)
$$K_1' = \frac{1}{1/k_1' + 1/k_2'}$$

$$K_2' = k_1' + k_2'$$

8.8 PROTEIN DENATURATION

Proteins are polymers of amino acids that are covalently linked by peptide bonds

$$\begin{array}{cc} \text{O} & \text{H} \\ \| & | \\ -\text{C}\!-\!\!\!-\!\!\!- & \text{N}- \end{array}$$

so that a single chain has an amino terminus and a carboxyl terminus. The four atoms in a peptide bond are in a relatively rigid planar configuration. In naturally occurring proteins there are about 20 amino acids, some of which have nonpolar groups and some of which have additional polar groups. The sequence of amino acids in a chain is referred to as the **primary structure.** The **secondary structure** is the local structure that may involve hydrogen bond interactions (Section 11.1) within a chain or with neighboring chains. An example of secondary structure is the alpha helix that is shown in right- and left-handed versions in Fig. 8.8. The helix is held together by hydrogen bonds between a carbonyl oxygen and the N—H of the fourth residue along the chain. This forms an especially stable structure with 3.6 residues per turn. In the direction of the axis, there is a residue every 0.15 nm. Hydrogen bonds can also be formed between chains to make parallel or antiparallel sheets.

The **tertiary structure** of a protein is the overall conformation of the polypeptide chain. In addition to hydrogen bonds and electrostatic interactions, the tertiary structure of a protein is affected by the tendency of hydrophobic groups to aggregate in the interior of the molecule and the charged groups to be on the outside. The charged groups interact strongly with water molecules because

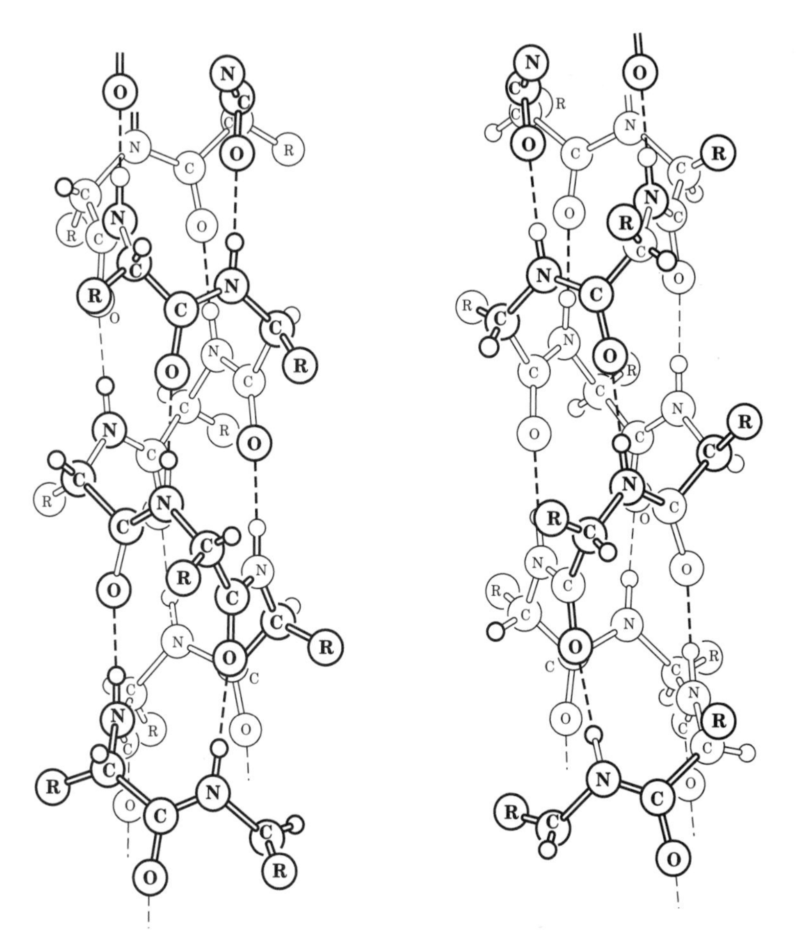

Figure 8.8 Alpha helix in a polypeptide chain. A left-handed helix is shown on the left and a right-handed helix is shown on the right. It is primarily the right-handed helix that exists in proteins.

the dipolar water molecules become oriented in the neighborhood of electric charges. The hydrocarbon groups prefer an organic environment. The pK's of acid groups in proteins may be significantly different from the pK's of these groups in small molecules because of the electrostatic and other interactions in the protein molecule.

Some proteins contain more than one polypeptide chain, and the arrangement of these polypeptide chains is referred to as the **quaternary structure** of the protein. For example, hemoglobin contains four polypeptide chains (Section 8.7).

The three-dimensional structure of a protein may be disrupted by changing the temperature, adding a denaturant such as urea, or making a significant change in pH. Sometimes these changes are irreversible, and sometimes they are reversible. Denaturation may involve a very large number of steps, but the transition from folded to unfolded states may occur over a small change in temperature or denaturant concentration. Such changes are referred to as **cooperative** because once a small change has occurred, the remaining steps take place rapidly. The binding of ligands by a protein may also be cooperative, as in the case of hemoglobin.

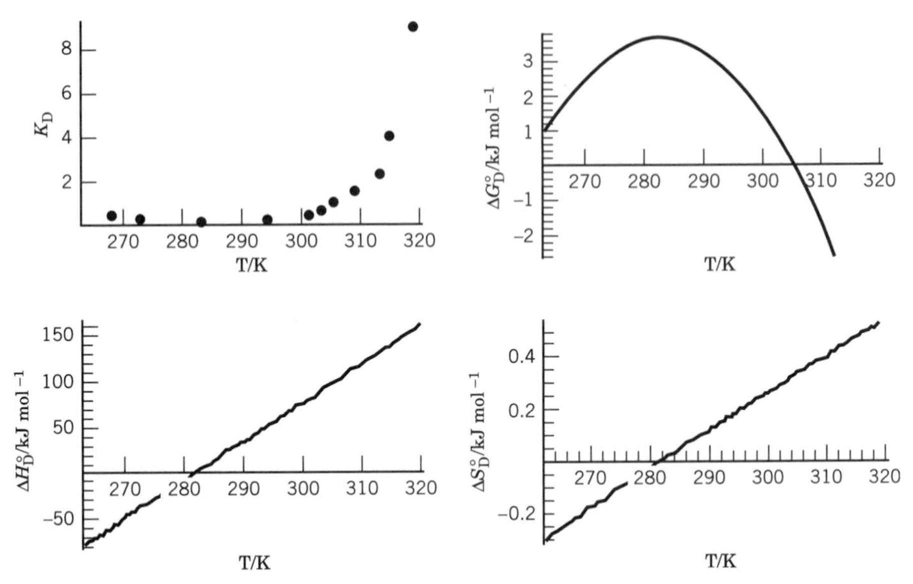

Figure 8.9 Denaturation of a repressor protein by 3 M urea and temperature at pH 8 in 0.1 M NaCl (G. S. Huang and T. G. Oas, *Biochemistry* 35, 6173 (1996)). (a) K_D as a function of temperature. (b) $\Delta_r G^\circ$ as a function of temperature. (c) $\Delta_r H^\circ$ as a function of temperature. Note the change in sign at 10 °C. (d) $\Delta_r S^\circ$ as a function of temperature. $\Delta_r C_P$ is 4.13 kJ K^{-1} mol^{-1} independent of temperature. (See Computer Problem 8.G.)

When the transition between native (N) and denatured (D) forms is a two-state process, it can be represented by the reaction

$$N = D \qquad K_D = [D]/[N] = f_D/(1 - f_D) \tag{8.85}$$

where f_D is the equilibrium fraction in the denatured form. The fraction denatured can be determined using circular dichroism, NMR, or calorimetry. The determination of f_D as a function of temperature makes it possible to calculate $\Delta_r G^\circ$, $\Delta_r H^\circ$, and $\Delta_r S^\circ$ as functions of temperature. Since the denatured form has a much higher heat capacity than the native form, it is useful to assume that $\Delta_r C_P^\circ$ is independent of temperature and use the equation in Example 5.11. The data and analysis in the denaturation of a repressor protein by urea and temperature at pH 8 in 0.1 M NaCl are given in Fig. 8.9. It is especially interesting that the denaturation increases when the temperature is reduced below 10 °C, but this is not the only protein that shows cold denaturation.

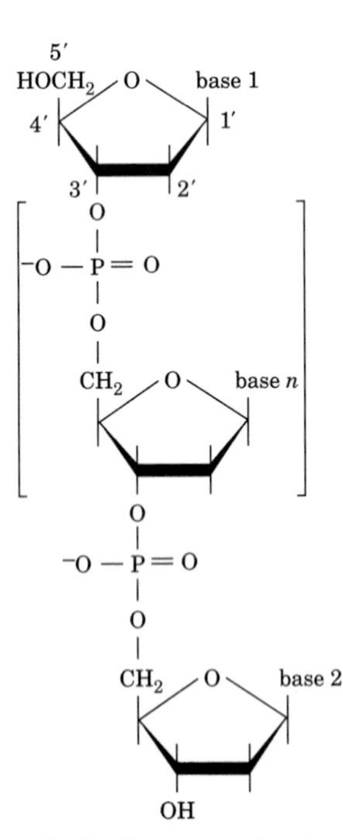

Figure 8.10 Structure of a single strand of DNA.

8.9 DNA DENATURATION

Deoxyribonucleic acid (DNA) is a polynucleotide of the four nucleotides with thymidylic acid (T), deoxyadenylic acid (A), deoxycytidylic acid (C), and deoxyguanylic acid (G). The nucleotides have phosphate groups attached to the 5′ position of the sugar deoxyribose. The structure of a single strand of DNA is shown in Fig. 8.10. In this strand, the phosphodiester linkage is through the 5′ and 3′ positions on the ribose sugars. The convention is that the DNA chain is written so that the 5′ end of the molecule is on the left and the 3′ end on the right. In the double-stranded form of DNA, the two strands form a right-handed helical structure in which the two strands go in opposite directions. In this structure T is always hydrogen-bonded to A and C is hydrogen-bonded to G. The singly charged phosphate groups repel each other, but the two strands are held together by the hydrogen bonding shown in Fig. 8.11 and by the stacking of the nucleotide bases. The planar bases interact through van der Waals forces, but the stacking of the bases in an aqueous environment is favored by the hydrophobic effect. DNA also exists in other forms that are not discussed here.

When DNA is heated or exposed to high pH or certain solvents, the double strand yields two complementary single strands. The dissociation is referred to as **denaturation,** and the association is referred to as **renaturation.** The thermodynamics and kinetics of these reactions are important for both practical and theoretical reasons. The temperature at which the fraction of single strands is 50% is referred to as the **melting temperature.** Since the CG hydrogen bonds are stronger than the AT hydrogen bonds, it might be expected that the melting temperature would be higher for DNA with more CG bonds, but this is not the case. However, studies utilizing small single-stranded DNA with known sequences has shown that

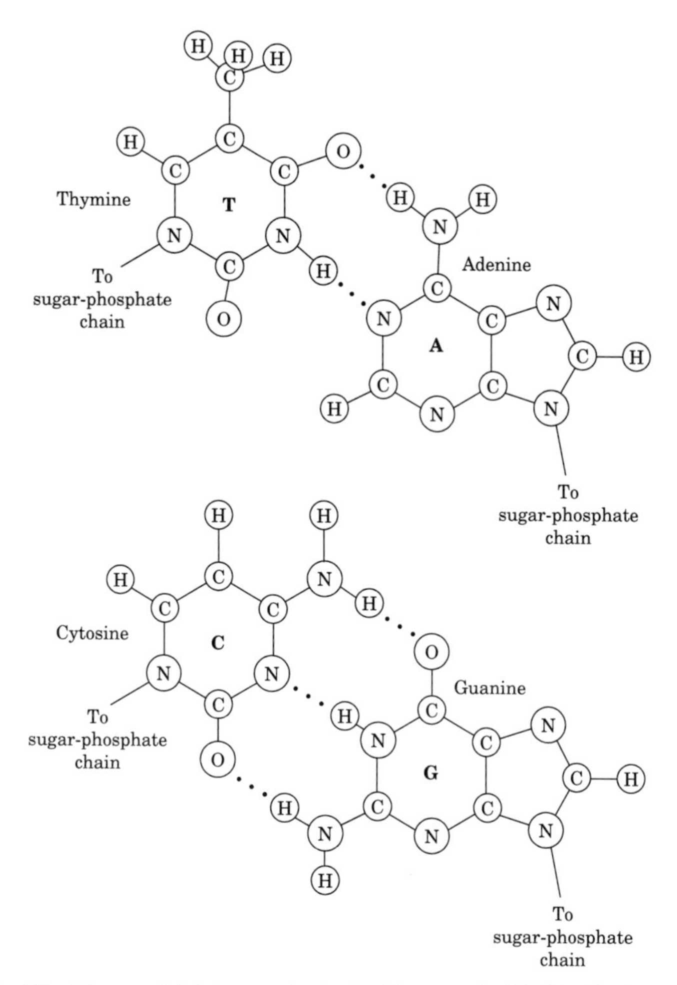

Figure 8.11 The Watson-Crick base pairs in double-stranded DNA. The hydrogen bonds are represented by dotted lines.

nearest neighbor contributions are important in determining the equilibrium constant for

$$A + A' = AA' \tag{8.86}$$

where A and A′ are polynucleotides with the same number of units and AA′ is the double-stranded helical structure. If the initial concentrations of A and A′ are c and the equilibrium concentration of AA′ is represented by x,

$$K = x/(c - x)^2 \tag{8.87}$$

The fraction f_S of the single-strand form is given by $1 - x/c$. Studies of a wide variety of short segments has made it possible to develop parameters for calculating $\Delta_r G°$, $\Delta_r H°$, and $\Delta_r S°$ for 10 DNA pairs.* To calculate these properties for a given reaction, these parameters are added with an additional parameter for double helix initiation. The phenomena involved in very long strands of DNA are more complicated because of the unwinding that has to take place. Related structures are encountered in RNA, but there the hydrogen bonding is intramolecular.

8.10 SPECIAL TOPIC: STATISTICAL EFFECTS IN POLYPROTIC ACIDS

Thermodynamics ordinarily deals with what we call **macroscopic** equilibrium constants. In formulating such equilibrium constant expressions, the existence of isomeric forms that are rapidly interconverted is ignored and the sum of their concentrations is inserted in the equilibrium constant expression. However, in some situations it is important to distinguish between isomeric forms and to define **microscopic** equilibrium constants that involve particular forms. Examples arise in the acid titration of proteins and of synthetic polymers with acidic groups. Here we will discuss the simplest possible example, the titration of a dicarboxylic acid, $HO_2C(CH_2)_nCO_2H$, **where n is so large that the carboxyl groups are independent of each other.** The macroscopic dissociation constants of this dibasic acid can be written as follows:

$$HA^- = H^+ + A^{2-} \qquad K_1 = \frac{[H^+][A^{2-}]}{[HA^-]} \tag{8.88}$$

$$H_2A = H^+ + HA^- \qquad K_2 = \frac{[H^+][HA^-]}{[H_2A]} \tag{8.89}$$

We can also look at this dissociation from a microscopic point of view, and represent the intermediate form by HA^- if the "right" proton dissociates and by AH^- if the "left" proton dissociates. Since the acid groups are identical and independent of one another by assumption,

$$K = \frac{[H^+][HA^-]}{[HAH]} = \frac{[H^+][AH^-]}{[HAH]} = \frac{[H^+][A^{2-}]}{[HA^-]} = \frac{[H^+][A^{2-}]}{[AH^-]} \tag{8.90}$$

*J. SantaLucia, Jr., H. T. Allawi, and P. A. Seneviratne, *Biochemistry,* **35:**3555 (1996).

where K is the microscopic or intrinsic dissociation constant. Thus,

$$K_1 = \frac{[H^+][A^{2-}]}{[HA^-] + [AH^-]} = \frac{1}{1/K + 1/K} = \frac{K}{2} \tag{8.91}$$

$$K_2 = \frac{[H^+]([HA^-] + [AH^-])}{[H_2A]} = 2K \tag{8.92}$$

so that $K_2 = 4K_1$. The second acid dissociation constant is four times larger than the first because of this statistical effect.

One way to look at this is to say that when HAH dissociates, the proton can come off of either end, and so HAH has twice the tendency to dissociate as the corresponding monoprotic acid, $CH_3(CH_2)_nCO_2H$. Conversely, when the base form A^{2-} binds a proton, there are two sites where the proton can be bound.

In contrast with a dibasic acid with different microscopic equilibria, the titration curve for a diprotic acid of the type we are discussing looks exactly like that of a monoprotic acid, except that two moles of base are required to neutralize one mole of acid. The midpoint of this titration curve yields the *microscopic dissociation constant K,* but we would use $K_1 = K/2$ and $K_2 = 2K$ for the *macroscopic constants.* The average number of protons bound by this diprotic acid is

$$\begin{aligned} \overline{N}_H &= \frac{[H^+]/K_1 + 2[H^+]^2/K_1K_2}{1 + [H^+]/K_1 + [H^+]^2/K_1K_2} = \frac{2[H^+]/K + 2[H^+]^2/K^2}{1 + 2[H^+]/K + [H^+]^2/K^2} \\ &= \frac{2[H^+]/K(1 + [H^+]/K)}{(1 + [H^+]/K)^2} = \frac{2[H^+]/K}{1 + [H^+]/K} \end{aligned} \tag{8.93}$$

This result is expected because the acidic groups are identical.

It can be shown* that for a polyprotic acid with N independent and equivalent groups

$$K_i = \frac{iK}{N - i + 1} \tag{8.94}$$

These same considerations apply to the binding of metal ions or to the binding of oxygen to a type of hemoglobin with four equivalent binding sites. Assuming that the four oxygen binding sites of hemoglobin are identical yields $K_1 = K/4$, $K_2 = 2K/3$, $K_3 = 3K/2$, and $K_4 = 4K$.

Nine Key Ideas in Chapter 8

1. The fact that the standard entropy of dissociation of most weak acids is negative shows that there is an increase in "order" in the dissociation; this is a consequence of the creation of electric charges that become hydrated by dipolar water molecules.
2. The calculation of the titration curve for a weak acid involves satisfying the equilibrium relations (acid dissociation and the ion product for water), conservation of mass, and conservation of charge.

*C. Tanford, *Physical Chemistry of Macromolecules.* New York: Wiley, 1961, p. 532.

3. Two types of equilibrium constants are needed in biochemistry. It is necessary to use the usual equilibrium constant expressions in terms of species to study the details of the chemical changes. But it is also useful to use apparent equilibrium constant expressions in terms of reactants (sums of species) at a specified pH to discuss the metabolism of reactants such as adenosine triphoshate (ATP). Apparent equilibrium constants vary with the pH as well as the temperature and ionic strength.
4. When the pH is an independent variable like T and P, the Gibbs energy G is not minimized at equilibrium and it is necessary to use a Legendre transform to define a transformed Gibbs energy, G', that is. This leads to a full set of standard transformed thermodynamic properties of biochemical reactants, such as ATP, and a transformed chemical potential μ_i'.
5. The fundamental equation for the transformed Gibbs energy for a system at a specified pH has the same general form as that for the Gibbs energy, but the amounts deal with reactants (sums of species) such as ATP. Thus biochemical reactions at specified pH are written in terms of sums of species as in ATP + H_2O = ADP + P_i, which balance all atoms except hydrogen.
6. Tables can be made of standard transformed thermodynamic properties for biochemical reactants at a specified pH, and these tables can be used like the standard thermodynamic properties of species studied earlier.
7. The binding of a ligand by protein can also be expressed in terms of an apparent equilibrium constant at a specified pH. The binding of oxygen by hemoglobin is positively cooperative, which means that successive equilibrium constants increase, rather than decrease, as more oxygen molecules are bound.
8. Thermodynamics deals with what we call macroscopic equilibrium constants, but it is sometimes useful to treat systems with microscopic equilibrium constants. This is illustrated by polyprotic acids with N independent and equivalent groups and by the acid dissociations of glycine.
9. When the denaturation of a protein or a short double helix of DNA is a two-state process, the standard thermodynamic properties can be calculated from the temperature dependence of the equilibrium constant measured using circular dichroism, light absorption, NMR, or calorimetry.

REFERENCES

R. A. Alberty, *Thermodynamics of Biochemical Reactions.* Hoboken, NJ: Wiley, 2003.
C. R. Cantor and P. R. Schimmel, *Biophysical Chemistry, Part III: The Behavior of Biological Macromolecules.* San Francisco: Freeman, 1980.
R. E. Dickerson and I. Geis, *Hemoglobin.* Menlo Park, CA: Benjamin/Cummings, 1983.
J. Edsall and J. Wyman, *Biophysical Chemistry.* New York: Academic, 1958.
G. G. Hammes, *Thermodynamics and Kinetics for the Biological Sciences.* Hoboken, NJ: Wiley, 2000.
I. M. Klotz, *Ligand–Receptor Energetics.* Hoboken, NJ: Wiley, 1997.
D. Poland, *Cooperative Equilibria in Physical Biochemistry.* Oxford, UK: Clarendon, 1978.
J. Wyman and S. J. Gill, *Binding and Linkage: Functional Chemistry of Biological Macromolecules.* Mill Valley, CA: University Science Books, 1990.

PROBLEMS

Problems marked with an icon may be more conveniently solved on a personal computer with a mathematical program.

8.1 Show that the slope of the titration curve of a monobasic acid is given by

$$\frac{d\alpha}{dpH} = \frac{2.303K[H^+]}{(K + [H^+])^2}$$

where α is the degree of neutralization.

8.2 A liter of 0.10 M solution of a monoprotic weak acid is titrated with a sufficiently concentrated NaOH solution that there is not a significant change in volume. Derive the expression for the amount n of NaOH that would have to be added to reach a specified pH, assuming that the ionic strength can be taken as zero.

8.3 According to Table C.2, what are the values of $\Delta_r G^\circ$, $\Delta_r H^\circ$, and $\Delta_r S^\circ$ at 298 K for

$$H_2O(l) = H^+(ao) + OH^-(ao)$$

Show that the same value of $\Delta_r S^\circ$ is obtained from $\Delta_r G^\circ$ and $\Delta_r H^\circ$ by using $\Delta_r S^\circ = \sum \nu_i S_i^\circ$. Calculate K_w at 298 K.

8.4 For the acid dissociation of acetic acid, $\Delta_r H^\circ$ is approximately zero at room temperature in H_2O. For the acidic form of aniline, which is approximately as strong an acid as acetic acid, $\Delta_r H^\circ$ is approximately 21 kJ mol^{-1}. Calculate $\Delta_r S^\circ$ for each of the following reactions:

$$CH_3CO_2H = H^+ + CH_3CO_2^- \qquad pK = 4.75$$

$$C_6H_5NH_3^+ = H^+ + C_6H_5NH_2 \qquad pK = 4.63$$

How do you interpret these entropy changes? What compensates for the increase in entropy expected from the increase in number of molecules in the balanced chemical reaction?

8.5 Estimate pK_3 and pK_2 for H_3PO_4 at 25 °C and 0.1 mol L^{-1} ionic strength. The values at zero ionic strength are $pK_3 = 2.148$ and $pK_2 = 7.198$.

8.6 In a strong acid solution, the amino acid histidine binds three protons. The acid dissociation constants numbered from the weakest acid dissociation are 6.92×10^{-10}, 1.00×10^{-6}, and 1.51×10^{-2} at 25 °C. Calculate the concentrations of the four forms of histidine (His^-, HisH, $HisH_2^+$, and $HisH_3^{2+}$) in a 0.1 M solution of histidine at pH 7, assuming that these constants apply at the ionic strength of the solution.

8.7 The pK for the dissociation of $CaATP^{2-}$ at 25 °C in 0.2 mol L^{-1} (n-propyl)$_4$NCl is 3.60. The pK for $HATP^{3-} = H^+ + ATP^{4-}$ is 6.95. Calculate the apparent pK of this ATP ionization when ATP is titrated in a solution containing 0.1 mol L^{-1} $CaCl_2$. Assume that the Ca^{2+} concentration is much larger than total ATP concentrations.

8.8 To illustrate what we mean by a component in a solution at a specified pH, consider a very simple system, namely, a monoprotic weak acid HA and its salt in aqueous solution. Write the fundamental equation for G for this system and use the equilibrium expression in terms of chemical potentials for the acid dissociation to write the fundamental equation in terms of two components, the hydrogen component and the A component. The cation of the salt can be omitted from the fundamental equation because there are always enough cations for charge balance.

8.9 Since we began dealing with dilute solutions, we have assumed that the chemical potential of a species is given by

$$\mu_i = \mu_i^\circ + RT \ln c_i \qquad \text{(a)}$$

and then later, in equation 8.55, we assumed that the transformed chemical potential μ_i' of a reactant made up of two species, for example, $HPO_4^{2-} + H_2PO_4^-$, is given by

$$\mu_i' = \mu_i'^\circ + RT \ln(c_1 + c_2) \qquad \text{(b)}$$

at a specified pH. This looks reasonable, but it is a good idea to write out the mathematical steps. The transformed Gibbs energy of a reactant that is made up of two species is given by

$$G' = n_1\mu_1' + n_2\mu_2' \qquad \text{(c)}$$

The amounts of the two species can be replaced with $n_1 = r_1 n'(P_i)$ and $n_2 = r_2 n'(P_i)$, where $n'(P_i)$ is the amount of inorganic phosphate and r_1 and r_2 are the equilibrium mole fractions of HPO_4^{2-} and $H_2PO_4^-$. Thus equation c can be rewritten as

$$G' = n'(P_i)\{r_1\mu_1'^\circ + r_2\mu_2'^\circ + RT[r_1 \ln c_1 + r_2 \ln c_2]\} \qquad \text{(d)}$$

The last term looks a lot like an entropy of mixing, so we add $RT \ln([P_i]/c^\circ)$ and subtract $(r_1 + r_2)RT \ln[(c_1 + c_2)/c^\circ]$, which are equal. Show that this leads to

$$G' = n'(P_i)\{\mu'^\circ(P_i) + RT \ln([P_i]/c^\circ)\} = n'(P_i)\mu_i'(P_i) \qquad \text{(e)}$$

where

$$\mu'^\circ(P_i) = r_1\mu_1'^\circ + r_2\mu_2'^\circ + RT \ln(r_1 \ln r_1 + r_2 \ln r_2) \qquad \text{(f)}$$

This confirms equation b and shows that the standard transformed chemical potential of a reactant with two species at a specified pH is equal to a mole fraction average transformed chemical potential for the two species plus an entropy of mixing. In making numerical calculations, the standard transformed chemical potentials are replaced by standard transformed Gibbs energies of formation.

8.10 Write out the equations for calculating the standard transformed Gibbs energy of formation and standard transformed enthalpy of formation of a partially neutralized weak acid (HA) at a specified pH.

8.11 Will 0.01 mol L^{-1} creatine phosphate react with 0.01 mol L^{-1} adenosine diphosphate to produce 0.04 mol L^{-1}

creatine and 0.02 mol L^{-1} adenosine triphosphate at 25 °C, pH 7, pMg 4? What concentration of ATP can be formed if the other reactants are maintained at the indicated concentrations?

8.12 The cleavage of fructose 1,6-diphosphate (FDP) to dihydroxyacetone phosphate (DHP) and glyceraldehyde 3-phosphate (GAP) is one of a series of reactions most organisms use to obtain energy. At 37 °C and pH 7, $\Delta_r G^{\circ\prime}$ for the reaction FDP = DHP + GAP is $23.97 \text{ kJ mol}^{-1}$. What is $\Delta_r G^{\circ\prime}$ in an erythrocyte in which [FDP] $= 3 \times 10^{-6} \text{ mol L}^{-1}$, [DHP] $= 138 \times 10^{-6} \text{ mol L}^{-1}$, and [GAP] $= 18.5 \times 10^{-6} \text{ mol L}^{-1}$?

8.13 How many grams of ATP have to be hydrolyzed to ADP to lift 100 lb 100 ft if the available Gibbs energy can be converted to mechanical work with 100% efficiency? It is assumed that [ATP] = [ADP] = [P_i] = 0.01 mol L^{-1} and that $\Delta G^{\circ\prime}$ is $-39.8 \text{ kJ mol}^{-1}$ at 25 °C.

8.14 Biochemistry textbooks give $\Delta_r G^{\circ\prime} = -20.1 \text{ kJ mol}^{-1}$ for the hydrolysis of ethyl acetate at pH 7 and 25 °C. Experiments in acid solution show that

$$\frac{[CH_3CH_2OH][CH_3CO_2H]}{[CH_3CO_2CH_2CH_3]} = 14$$

where the equilibrium concentrations are in moles per liter. What is the value of $\Delta_r G^{\circ\prime}$ obtained from this equilibrium constant? The pK of acetic acid is 4.60 at 25 °C.

8.15 Fumarase catalyzes the reaction

$$\text{fumarate} + H_2O = \text{L-malate}$$

At 25 °C and pH 7

$$K' = 4.4 = \frac{[\text{L-malate}]}{[\text{fumarate}]}$$

What is the value of K' at pH 4? Given: For fumaric acid $K_1 = 10^{-4.18}$. For L-malic acid $K_1 = 10^{-4.73}$.

(M) **8.16** Given $\Delta_r G^\circ = 49.4 \text{ kJ mol}^{-1}$ for

$$ATP^{4-} + H_2O = AMP^{2-} + {P_2O_4}^{7-} + 2H^+$$

calculate $\Delta G^{\circ\prime}$ at pH 7 and 25 °C and 0.2 mol L^{-1} ionic strength. The pKs that are needed are: for ATP, pK_1 = 6.95; for ADP, pK_1 = 6.88; for AMP, pK_1 = 6.45; for pyrophosphoric acid, pK_1 = 8.95 and pK_2 = 6.11.

8.17 Calculate the enthalpy of ionization for ${H_2PO_4}^- = H^+ + {HPO_4}^{2-}$ at $(a)\, I = 0$ and $(b)\, I = 0.25$ M, given that CODATA shows

	$\Delta_f H^\circ/\text{kJ mol}^{-1}$
H^+	0
${HPO_4}^{2-}$	−1299.0
${H_2PO_4}^-$	−1302.6

8.18 If n molecules of a ligand A combine with a molecule of protein to form PA_n without intermediate steps, derive the relation between the fractional saturation Y and the concentration of A.

8.19 A protein M can bind two molecules of a ligand L, which is a gas. The macroscopic equilibrium constants, written in terms of the partial pressures of the ligand, are defined by

$$M + L = ML \qquad K_1 = [ML]/[M]P_L$$
$$ML + L = ML_2 \qquad K_2 = [ML_2]/[ML]P_L$$

Assume that the two binding sites are different and that ML can be distinguished from LM. How are the microscopic dissociation constants

$$M + L = ML \qquad K_1* = [ML]/[M]P_L$$
$$M + L = LM \qquad K_2* = [LM]/[M]P_L$$
$$ML + L = LML \qquad K_3* = [LML]/[ML]P_L$$
$$LM + L = LML \qquad K_4* = [LML]/[LM]P_L$$

related to the macroscopic dissociation constants K_1 and K_2? How many of the microscopic dissociation constants are independent? If there is a relation between them, what is it?

8.20 Since it is difficult to determine the values of the four dissociation constants in equation 8.84, the empirical Hill equation

$$Y = \frac{1}{1 + K_h/P_{O_2}^h}$$

is frequently used to characterize binding of oxygen by hemoglobin. Show that the Hill coefficient h may be obtained by plotting $\log[Y/(1 - Y)]$ versus $\log P_{O_2}$.

8.21 Hemoglobin is made up of two alpha chains and two beta chains, and so it can be represented by $(\alpha\beta)_2$. Hemoglobin dissociates into $\alpha\beta$ subunits. The association constant K' for the reaction $2\alpha\beta = (\alpha\beta)_2$ depends on the partial pressure of molecular oxygen, but at relatively high concentrations of molecular oxygen at pH 7 and 21.5 °C, $K' = 9.47 \times 10^5$, when molar concentrations are used. If a solution is 0.0025 M in hemoglobin, that is, $(\alpha\beta)_2$, what are the concentrations of the dimer and tetramer at equilibrium? What if the hemoglobin solution is 0.000 25M?

(M) **8.22** The percent saturation of a sample of human hemoglobin was measured at a series of oxygen partial pressures at 20 °C, pH 7.1, 0.3 mol L^{-1} phosphate buffer, and $3 \times 10^{-4} \text{ mol L}^{-1}$ heme:

P_{O_2}/Pa	Percent Saturation
393	4.8
787	20
1183	45
2510	78
2990	90

Calculate the values of h and K_h in the Hill equation. (See problem 8.20.)

8.23 Use the fundamental equations and Gibbs–Duhem equations to determine the number F of degrees of freedom and number D of variables required to define the extensive state of a system at constant T and P and with the reaction A + B = C at equilibrium from the following two points of view. (a) Start with the fundamental equation for G in terms of species and derive the form in terms of components to obtain D. Then derive the Gibbs–Duhem equation to obtain F. State the criterion for spontaneous change. (b) Define a transformed Gibbs energy G' to introduce μ_B as an intensive variable. Use the corresponding fundamental equation to obtain D and the Gibbs–Duhem equation to obtain F. State the criterion for spontaneous change.

8.24 The equation for $\ln K$ when $\Delta_r C_P^\circ$ is constant (see Example 5.11) is often used in a different form in treating the equilibrium constant for the thermal denaturation of a protein, which can be represented by N = D and $K_D = [D]/[N]$. The reason for this is that $\Delta_r C_P^\circ$ is generally quite large and nearly independent of temperature. The denaturation reaction is often highly cooperative, so that the reaction can be treated in terms of two states. Show that

$$\Delta_r G_D^\circ(T) = \Delta_r H^\circ(T_g)\left(1 - \frac{T}{T_g}\right) - \Delta_r C_P^\circ\left[T_g - T + T\ln\left(\frac{T}{T_g}\right)\right]$$

where T_g is the temperature at which $K_D = 1$. This equation was first derived by W. J. Bectel and J. A. Schellman, *Biopolymers,* **26**:1858–1877 (1987).

8.25 The transformed Gibbs energy G', transformed enthalpy H', and transformed entropy S' of a system are given by

$$G' = G - n_c(\mathrm{H})G_m(\mathrm{H}^+)$$

$$H' = H - n_c(\mathrm{H})H_m(\mathrm{H}^+)$$

$$S' = S - n_c(\mathrm{H})S_m(\mathrm{H}^+)$$

Show that $G' = H' - TS'$.

8.26 Solutions of two single-stranded DNAs are mixed and a double-stranded DNA is formed:

$$\begin{matrix} 5'\text{-A-G-C-T-G-}3' \\ + \\ 5'\text{-C-A-G-C-T-}3' \end{matrix} = \begin{matrix} 5'\text{-A-G-C-T-G-}3' \\ 3'\text{-T-C-G-A-C-}5' \end{matrix}$$

Use of the parameters of J. SantaLucia, Jr., H. T. Allawi, and P. A. Seneviratne [*Biochemistry,* **35**:3555 (1996)] for 298 K, pH 7, and 1 M NaCl yields $\Delta_r G^\circ = -20.5$ kJ mol^{-1} and $\Delta_r H^\circ = -128.4$ kJ mol^{-1} for this reaction (G. G. Hammes, *Thermodynamics and Kinetics for the Biological Sciences,* Hoboken, NJ: Wiley 2000). (a) Calculate the equilibrium constant for this reaction at 298 K and the equilibrium fraction of single strands when the initial concentrations of the two single strands are 10^{-4} M. (b) Calculate the equilibrium constant for this reaction at 313 K and the equilibrium fraction of single strands when the initial concentrations of the two single strands are 10^{-4} M.

8.27 What is the ion product of water at 0°C, assuming $\Delta_r C_P^\circ$ is independent of temperature? What is the pH of pure water at 0°C?

8.28 What is the ion product of water at 50 °C, assuming that $\Delta_r C_P^\circ$ is independent of temperature? What is the pH of neutrality at this temperature?

8.29 At 298.15 K, for $H_2O(l) = H^+(ao) + OH^-(ao)$, $K = 1.008 \times 10^{-14}$, and $\Delta_r H^\circ = 55.836$ kJ mol^{-1}. Given that $S^\circ[H_2O(l)] = 69.92$ J K^{-1} mol^{-1}, what is the value of $S^\circ[OH^-(ao)]$?

8.30 Using Table C.2, calculate $\Delta_r H^\circ$, $\Delta_r G^\circ$, and $\Delta_r S^\circ$ for

$$CH_3CO_2H(ao) = H^+(ao) + CH_3CO_2^-(ao)$$

and compare the values in Table 8.1 for 298.15 K. Calculate the acid dissociation constant of acetic acid at 298.15 K.

8.31 What is the composition of a 0.1 M solution of acetic acid in water at 25 °C? In a second step take into account the effect of ionic strength.

8.32 Calculate the pK for the acid dissociation

$$HADP^{2-} = H^+ + ADP^{3-}$$

at 298.15 K and $I = 0.25$ M.

8.33 Calculate the chemical equilibrium constant, $\Delta_r H^\circ$, and $\Delta_r S^\circ$ for the hydrolysis of glucose 6-phosphate:

$$GlcP^{2-} + H_2O = \text{glucose} + HPO_4{}^{2-}$$

at 298.15 K and zero ionic strength. These values will apply in the range pH 8 to 10. What will be the effect of increasing the ionic strength?

8.34 An aqueous solution of 0.01 M phosphoric acid is titrated at 25 °C with such a concentrated solution of strong alkali that the solution is not significantly diluted during the titration. Plot $[H_3PO_4]$, $[H_2PO_4{}^-]$, $[HPO_4{}^{2-}]$, and $[PO_4{}^{3-}]$ as a function of pH.

8.35 A protein molecule has two different and independent sites; one binds A and the other binds B. What are the probabilities of P, PA, PB, and PAB?

8.36 At pH 7 and pMg 4 what value of pCa is required to put half the ATP in the form $CaATP^{2-}$? At 0.2 mol L^{-1} ionic strength and 25 °C the following constants are known:

$$HATP^{3-} = H^+ + ATP^{4-} \qquad pK = 6.95$$

$$MgATP^{2-} = Mg^{2+} + ATP^{4-} \qquad pK = 4.00$$

$$CaATP^{2-} = Ca^{2+} + ATP^{4-} \qquad pK = 3.60$$

8.37 For the weak acid $CH[(CH_2)_5CO_2H]_3$, the carboxyl groups are essentially identical and independent. (*a*) If the intrinsic acid dissociation constant is K, what are the values of K_1, K_2, and K_3? (*b*) At $[H^+] = K$, what are the relative proportions of the minus three ion, minus two ion, minus one ion, and uncharged molecules?

8.38 In a series of biochemical reactions the product in one reaction is a reactant in the next. This has the effect that spontaneous reactions drive nonspontaneous reactions. For example, reaction 2 follows reaction 1:

1. L-malate = fumarate + H_2O $\quad \Delta_r G^{\circ\prime} = 2.9\,\text{kJ mol}^{-1}$
2. fumarate + ammonia = aspartate $\quad \Delta_r G^{\circ\prime} = -15.6\,\text{kJ mol}^{-1}$

The $\Delta_r G^{\circ\prime}$ values are for pH 7 and 37 °C, and the state of ionization of the reactants is ignored. In reaction 1, the activity of H_2O is to be taken as 1. If the ammonia concentration is $10^{-2}\,\text{mol L}^{-1}$, calculate [aspartate]/[L-malate] at equilibrium.

8.39 In the living cell two reactions may be coupled by having a common intermediate. This is true for the following two reactions, which are enzyme catalyzed:

$$\text{creatine} + P_i = \text{creatine phosphate} + H_2O \qquad \Delta_r G^{\circ\prime} = 46\,\text{kJ mol}^{-1}$$

$$\text{ATP} + H_2O = \text{ADP} + P_i \qquad \Delta_r G^{\circ\prime} = -33\,\text{kJ mol}^{-1}$$

The $\Delta_r G^{\circ\prime}$ values are for pH 7.5 and 25 °C. If in a steady state in a living cell $[\text{ATP}] = 10^{-3}\,\text{mol L}^{-1}$ and $[\text{ADP}] = 10^{-4}\,\text{mol L}^{-1}$, calculate the steady-state ratio [creatine phosphate]/[creatine].

8.40 From the data of Table 8.8 calculate $\Delta_r G^{\circ\prime}$ for

$$\text{ATP} + H_2O = \text{AMP} + \text{PP}$$

8.41 What is the maximum concentration of ATP that can be formed enzymatically from acetyl phosphate and ADP each at $0.01\,\text{mol L}^{-1}$, pH 7, and pMg 4 at 25 °C, assuming that the ambient concentration of acetate is also $0.01\,\text{mol L}^{-1}$? Given:

$$\text{acetylP} + H_2O = \text{acetate} + P_i \qquad \Delta_r G^{\circ\prime} = -43.1\,\text{kJ mol}^{-1}$$

$$\text{ADP} + P_i = \text{ATP} + H_2O \qquad \Delta_r G^{\circ\prime} = 39.8\,\text{kJ mol}^{-1}$$

8.42 The hydrolysis of adenosine triphosphate (ATP) to adenosine diphosphate (ADP) and inorganic phosphate at pH 8 and 25 °C

$$\text{ATP}^{4-} + H_2O = \text{ADP}^{3-} + HPO_4{}^{2-} + H^+$$

has a standard enthalpy change of $-20.5\,\text{kJ mol}^{-1}$. The standard enthalpy changes of acid dissociation of HATP^{3-}, HADP^{2-}, and $H_2PO_4{}^-$ are -6.3, -5.6, and $3.6\,\text{kJ mol}^{-1}$, respectively. Calculate the standard enthalpy change for the reaction

$$\text{HATP}^{3-} + H_2O = \text{HADP}^{2-} + H_2PO_4{}^-$$

8.43 If $\Delta_r G^{\circ\prime}$ for the hydrolysis of acetyl phosphate ($CH_3CO_2PO_3H_2$) is $-43.1\,\text{kJ mol}^{-1}$ at 25 °C and pH 7, what is the value at pH 4? It may be assumed that the pK of acetyl phosphate in the neighborhood of pH 7 is identical with to pK_2 of orthophosphate.

8.44 Calculate the chemical equilibrium constant K and $\Delta_r H^\circ$ for the alcohol dehydrogenase reaction

$$CH_3CHO + \text{NADH}^{2-} + H^+ = CH_3CH_2OH + \text{NAD}^-$$

at 298.15 K and zero ionic strength. Since none of the species have pK values in the neutral region, it is simple to calculate the apparent equilibrium constant K' at pH 7. What is its value?

$$K' = \frac{[CH_3CH_2OH][\text{NAD}^-]}{[CH_3CHO][\text{NADH}^{2-}]}$$

8.45 Consider the following system of reactions,

$$\begin{array}{ccc} \text{AH} + \text{B} & \overset{K_1}{\rightleftharpoons} & \text{CH} \\ \updownarrow & & \updownarrow \\ \text{A} + \text{B} & \underset{K_2}{\rightleftharpoons} & \text{C} \end{array}$$

where the K's represent equilibrium constants. (*a*) Calculate the dependence on hydrogen ion concentration of the apparent equilibrium constant

$$K' = \frac{([\text{C}] + [\text{CH}])}{([\text{A}] + [\text{AH}])[\text{B}]}$$

(*b*) What is the relationship between the four equilibrium constants?

8.46 Nucleosides associate in aqueous solution to form dimers. For adenosine (A) at 25 °C the equilibrium constant for $2A = A_2$ is 4.5 when concentrations are expressed in mol L^{-1}. What is the concentration of dimers in a $0.5\,\text{mol L}^{-1}$ solution of adenosine? (This association is due to the stacking of bases.)

8.47 The partial pressure of oxygen required to half-saturate hemoglobin at pH 7.4 is 3.7 kPa. If the partial pressure of oxygen in the alveolar spaces of the lungs is 13.3 kPa, and the partial pressure in the capillaries is 5.3 kPa, what percentage of the total oxygen carrying capacity of hemoglobin is being used if h in the Hill equation is 2.7? (See Problem 8.20.)

8.48 When myoglobin is in contact with air, how many parts per million of CO are required to tie up 10% of the myoglobin? The partial pressure of oxygen required to half-saturate myoglobin at 25 °C is 3.7 kPa. The partial pressure of CO required to half-saturate myoglobin in the absence of oxygen is 0.009 kPa.

Computer Problems

8.A Write a program to calculate $\Delta_f G'^{\circ}$ and $\Delta_f H'^{\circ}$ for inorganic phosphate at 298.15 K and pH 7 using data in Example 8.5. The program should be written to handle any number of pseudoisomers so that it can be used for other reactants.

8.B (*a*) Write a program to calculate the apparent equilibrium constant K' for ATP + H_2O = ADP + P_i at 298.15 K, pH 7, and ionic strength 0.25 M. The acid dissociation constants needed are given in Table 8.3. (*b*) Plot K' versus pH. (*c*) Plot $\Delta_r G^\circ$ versus pH.

8.C It can be shown that the change in binding of hydrogen ions in a biochemical reaction at a specified pH is given by the negative of the derivative of log K' with respect to pH at constant T and P. (*a*) Use a mathematical program to take the derivative of log K' for the hydrolysis of ATP to ADP and inorganic phosphate with respect to pH from the previous problem. (*b*) Plot the production of H^+ versus pH in the range pH 5–9. (*c*) Interpret this plot in terms of the predominant chemical reactions at pH 5 and pH 9.

8.D (*a*) Write a program to calculate the apparent equilibrium constant K' for glucose 6-phosphate + H_2O = glucose + P_i at 298.15 K, pH 7, and ionic strength 0.25 M. The standard Gibbs energies of formation of the species involved at 298.15 K and ionic strength 0.25 M are given in the following table:

	$\Delta_f G^\circ$/kJ mol^{-1}
Glucose 6-phosphate^{2-}	−1767.18
Hglucose 6-phosphate^{-}	−1801.4
H_2O	−237.19
Glucose	−915.9
HPO_4^{2-}	−1099.34
$H_2PO_4^-$	−1138.11
H^+	−0.81

(*b*) Plot K' versus pH. (*c*) Plot $\Delta_r G'^\circ$ versus pH.

8.E It can be shown that the change in binding of hydrogen ions in a biochemical reaction at a specified pH is given by the negative of the derivative of log K' with respect to pH at constant T and P. (*a*) Use a mathematical program to take the derivative of log K' for glucose 6-phosphate + H_2O = glucose + P_i with respect to pH from the previous problem. (*b*) Plot the production of H^+ versus pH on the range pH 5–9. (*c*) Interpret this plot in terms of the predominant chemical reactions.

8.F The reaction of two single strands of DNA to form a double helix AA′ is represented by

$$A + A' + AA'$$

The standard reaction Gibbs energy for the reaction in Problem 8.25 is −20.5 kJ mol^{-1}, and the standard reaction enthalpy is −128.4 kJ mol^{-1} at 298 K and pH 7 in 1 M NaCl (G. G. Hammes, *Thermodynamics and Kinetics for the Biological Sciences*, Hoboken, NJ: Wiley 2000). (*a*) Calculate the equilibrium constants at a series of temperatures between 273 K and 313 K on the assumption that the standard reaction enthalpy is independent of temperature. (*b*) Calculate the equilibrium concentration of AA′ at each of these temperatures for initial concentrations of A and A′ of 10^{-4} M. (*c*) Calculate the fraction f_s of the DNA in the single strand form at each temperature and plot f_s versus temperature. (*d*) According to this plot, what is the melting temperature of the double strand? Verify that the equilibrium constant at the melting temperature is given by $K = x/(c-x)^2$.

8.G The fraction of a repressor protein that is denatured at equilibrium in 3 M urea at pH 8 in 0.1 M NaCl is given by the following table [G. S. Huang and T. G. Oas, *Biochemistry* **35:**6173 (1996)].

t/°C	−5	0	10	21	28	30	32	36	40	42	46
f_D	0.30	0.20	0.17	0.20	0.30	0.40	0.50	0.60	0.70	0.80	0.90

These data are especially interesting because they show that when the temperature is reduced below 10 °C, the denaturation increases. (*a*) Calculate K_D at each temperature and plot it versus absolute temperature. (*b*) Calculate ΔG_D° and plot it versus T. (*c*) Fit the data in (*b*) with equation 8.85, and use this function to plot ΔG_D° versus temperature. (*d*) Calculate ΔH_D° using the Gibbs–Duhem equation and plot it versus T. (*e*) Calculate ΔS_D° using $-d\,\Delta G_D^\circ/dT$ and plot it versus T. (*f*) Calculate ΔC_D° using $-d\Delta H_D^\circ/dT$ and plot it versus T.

8.H Since the hydrolysis of a sodium salt of a monoprotic weak acid produces equal concentrations of HA and OH^-, $K_h = [OH^-]^2/c$, where c is the molar concentration of the salt. We know that $K_h = K_w/K_a$ and $[OH^-] = K_h/10^{-pH}$. Therefore it can be shown that

$$pH = \tfrac{1}{2}(14.00 + pK_a + \log c)$$

at 298.15 K. Make a table of the pHs of sodium salts of weak acids with pK's of 5 to 11 and concentrations of 0.10, 0.01, 0.001, and 0.0001 M at 298.15 K.

8.I (*a*) Plot the titration curve for a liter of 0.10 M acetic acid (pK = 4.756) with concentrated NaOH at 298.15 K on the assumption that the ionic strength can be taken as zero and that the NaOH solution used is so concentrated that there is not a significant change in volume (see Problem 8.2). (*b*) Plot the corresponding titration curves for monoprotic weak acids with pK's of 4, 5, 6, 7, 8, 9, and 10.

8.J Calculate the plot of $\overline{N}_H$ versus pH for phosphoric acid at 298.15 K using its binding polynomial.

PART

TWO

Quantum Chemistry

The development of quantum mechanics began in the early twentieth century, when scientists started studying atomic and molecular phenomena and discovered that Newtonian classical mechanics and the wave theory of light did not explain the results of their experiments. Quantum mechanics enables scientists to calculate energy levels and other properties of atoms and molecules. In this section we will consider the electronic orbitals of the hydrogen atom in detail and show how these calculations can be extended to describe atoms with more than one electron. From these data we are able to understand why properties such as ionization potential, electron affinity, and atom size, among others, vary in a periodic manner.

The application of quantum mechanics to molecules made it possible to understand the nature of the chemical bond. We consider bonding in H_2^+ and H_2 in Chapter 11, and then go on to study how to describe bonding in larger molecules using the methods of quantum chemistry. The energy levels, bond lengths, and bond angles can be calculated quite accurately for many molecules. The increasing power of computers has made these calculations more and more accurate.

In Chapter 12 we discuss the symmetry of molecules in their equilibrium configurations. Using symmetry ideas can greatly simplify quantum mechanical calculations and can answer qualitative questions such as whether a molecule has a dipole moment or not.

Quantum mechanics also provides the basis for understanding the results of spectroscopic measurements. Spectroscopy is useful for the identification of molecules and the determination of their concentrations, and is especially important in physical chemistry because it yields information about molecular properties. Microwave and infrared spectroscopy yield information about bond lengths and angles. Infrared and Raman spectroscopy provide information about vibrational frequencies of molecules. Visible and ultraviolet spectroscopy provide information on dissociation energies, bond energies, and electronic excited states.

The development of both laser methods and magnetic resonance methods have revolutionized spectroscopy. The latter has become so important in chemistry and biology that we discuss it in a separate chapter.

Part Two closes with statistical mechanics, the science that connects the properties of individual molecules with the thermodynamic properties of bulk matter, using information obtained from spectroscopic methods. We illustrate the general methods by calculating equilibrium constants of small molecule reactions in the ideal gas state. The calculation of the thermodynamic properties of dense gases, liquids, and solids is more difficult because of the importance of intermolecular forces. Advances in computation have made these calculations possible as well.

9

Quantum Theory

The early years of the twentieth century saw a revolution in physics: the birth of quantum mechanics, which replaces classical mechanics as the description of motion on an atomic scale. In 1900, Planck showed that the description of the distribution of energies of electromagnetic radiation in a cavity requires the quantization of energy. This was quickly followed by the application of quantization to atomic and molecular phenomena. Modern chemistry relies on quantum mechanics for the description of most phenomena. In this chapter we consider the basic concepts of quantum mechanics and apply them to a few simple problems such as the harmonic oscillator and rigid rotator.

9.1 CLASSICAL MECHANICS FAILED TO DESCRIBE EXPERIMENTS ON ATOMIC AND MOLECULAR PHENOMENA

At the end of the nineteenth and beginning of the twentieth century, a number of experimental observations were made that could not be reconciled or explained by the laws of classical physics. For example, Planck measured the emission of radiation from a hot mass (called blackbody radiation) and found that it did not fit the formula derived from classical physics. To derive the correct equation, he had to assume, in contrast to classical ideas, that radiation of frequency $-\nu$ is absorbed and emitted only in multiples of $h\nu$, where h is a universal constant. In another experiment, it was discovered that the energy of an electron ejected from metals by the absorption of radiation (the photoelectric effect) depended only on the frequency of the radiation, and not on intensity, again in contrast with classical ideas. Einstein explained this in 1905 by suggesting that light of frequency ν consists of quanta of energy $h\nu$, called **photons**. When one photon strikes an electron in the metal, the electron is ejected with a kinetic energy that is the difference between the energy of the photon and the minimum energy needed to eject the electron. In 1911 Rutherford showed that an atom has all its positive charge in a tiny nucleus with the electrons surrounding it, but this could not be understood using classical physics, which predicted that the electrons would radiate energy and fall into the nucleus. Bohr in 1913 postulated the existence of stable orbits in atoms and the quantization of angular momentum. This theory marked the beginning of quantum mechanics applied to atoms, but was unable to describe atoms with more than one electron.

The underlying problem that emerged from these and other experiments was that electromagnetic radiation shows properties that are both wavelike and particle-like. Experiments showing the interference of light must be explained with wave theory, whereas phenomena such as the photoelectric effect reveal particle-like properties. In his 1924 doctoral thesis, de Broglie developed an equation for the wavelength of a particle by reasoning in analogy with light. In 1926 Schrödinger published the wave equation for atomic and molecular systems, which is the principal subject of this chapter. In 1927 Heisenberg put forward an uncertainty principle implying that if the momentum of a particle is known precisely, the position of that particle is completely unknown. This new mechanics called **quantum mechanics,** challenged classical mechanics, according to which the position and momentum of a particle can be calculated precisely at all times from knowledge of the forces on the particle. In the next few paragraphs, we consider these early ideas and developments in a little more detail before going on to study the Schrödinger equation.

The distribution of frequencies of electromagnetic radiation from a heated solid depends on the nature of the solid, but the radiation from a container with a small window is independent of the solid because the radiation is in equilibrium with the walls of the container. This radiation is called **blackbody radiation** because any radiation falling on the small window will be completely absorbed. It is most convenient to discuss the distribution of frequencies in terms of the **energy density** ρ_ν, which is the energy per unit volume of the radiation between ν and $\nu + d\nu$. Classical electrodynamic theory indicates that the energy density ρ_ν should be given by

$$\rho_\nu = \frac{8\pi\nu^2}{c^3}kT \tag{9.1}$$

where c is the **speed of light** (299 792 458 m s^{-1}, exactly) and k is the **Boltzmann constant** ($R/N_A = 1.380\ 658 \times 10^{-23}$ J K^{-1}). This equation works at very low frequencies, but it has the fatal flaw that the experimentally determined ρ_ν does not increase to infinity as the frequency increases, but instead approaches zero asymptotically at high frequencies. In 1900, Planck derived

$$\rho_\nu = \frac{8\pi h(\nu/c)^3}{e^{h\nu/kT} - 1} \tag{9.2}$$

by making the radical assumption that the energy of a quantum of radiation (photon) is given by $h\nu$, where $h = 6.626\ 075\ 5 \times 10^{-34}$ J s is now known as **Planck's constant.** Equation 9.2 is in agreement with experiment. Figure 9.1 shows the energy density as a function of frequency for three temperatures. More information about blackbody radiation is given in a Special Topic section at the end of this chapter.

The wave nature of light is demonstrated by interference phenomena, but when light is absorbed by a metal, the total energy of a photon $h\nu$ is given to a single electron within the metal. If this quantity of energy is sufficiently large, the electron may penetrate the potential barrier at the surface of the metal (called the work function) and still retain some energy as kinetic energy. The kinetic energy retained by the electron depends on the energy and, therefore, the frequency of the photon that ejected it. The number of electrons ejected depends on the number of incident photons, and therefore is related to the intensity of the light.

Since these experiments showed that light has wave and particle aspects, they raised the question as to whether small particles have wave aspects. Photons, which have energies given by $E = h\nu$, are unusual particles in that they have zero rest mass and travel with the speed of light. However, Einstein suggested that photons have a relativistic mass given by $E = mc^2$. Equating these two expressions for the energy of a photon yields

$$mc^2 = h\nu = \frac{hc}{\lambda} \quad \text{or} \quad mc = p = \frac{h}{\lambda} \tag{9.3}$$

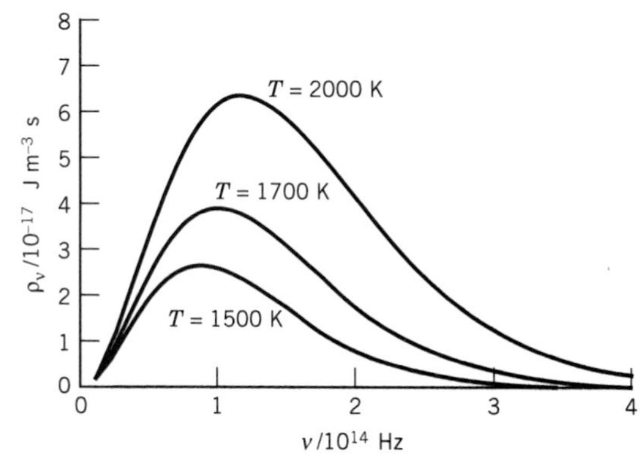

Figure 9.1 Radiant energy density ρ_ν as a function of frequency at 1500, 1700, and 2000 K. Note that the frequency of maximum intensity is directly proportional to the temperature. (See Computer Problem 9.F.)

where p is the **momentum of a photon.** By analogy, de Broglie (1924) suggested that the momentum of a particle with finite rest mass is given by

$$mv = p = \frac{h}{\lambda} \quad \text{or} \quad \lambda = \frac{h}{p} = \frac{h}{mv} \tag{9.4}$$

In this equation m is rest mass plus the relativistic correction, but for particles at low momentum with finite rest mass, the relativistic correction is usually negligible in comparison with the rest mass. Thus the wavelength associated with a particle of finite rest mass was expected to be inversely proportional to its momentum. For macroscopic particles the de Broglie wavelength is so short that it does not lead to observable phenomena. However, for electrons, neutrons, and other microscopic particles the wavelengths may be of the order of interatomic distances in solids.

It was recognized that de Broglie's hypothesis could be tested by scattering a beam of electrons from a crystalline solid, as a way of getting a grating of suitable dimensions to match the wavelength of electrons of accessible energies. In 1928 Davisson and Germer obtained a diffraction pattern from electrons impinging on the face of a nickel crystal that confirmed equation 9.4. This suggested that all particles have a wavelike property with a wavelength that is inversely proportional to the momentum.

Example 9.1 ***de Broglie wavelength of the electron***

What is the de Broglie wavelength λ of an electron that has been accelerated through a potential difference of 100 V?

To use equation 9.4, we need to calculate the momentum after the acceleration process. We do this by first calculating the energy of the electron. The energy of an electron of mass m moving with a velocity v well below the velocity of light is given by

$$E = \frac{1}{2}mv^2 = \frac{p^2}{2m}$$

Thus, the momentum is given by

$$p = \sqrt{2mE}$$

The energy of the electron is $(1.602 \times 10^{-19}\ \text{C})(100\ \text{V}) = 1.602 \times 10^{-17}$ J. Thus, the momentum is

$$p = \sqrt{(2)(9.109 \times 10^{-31}\ \text{kg})(1.602 \times 10^{-17}\ \text{J})}$$
$$= 5.403 \times 10^{-24}\ \text{kg m s}^{-1}$$

and the wavelength is

$$\lambda = \frac{h}{p} = \frac{6.626 \times 10^{-34}\ \text{J s}}{5.403 \times 10^{-24}\ \text{kg m s}^{-1}}$$
$$= 1.226 \times 10^{-10}\ \text{m} = 0.1226\ \text{nm}$$

Note that this λ is of the same order of magnitude as the distance between atoms in a crystal.

The total energy E of a particle is equal to the sum of its kinetic energy $(1/2)mv^2$ and its potential energy V:

$$E = \frac{1}{2}mv^2 + V = \frac{p^2}{2m} + V \tag{9.5}$$

Solving this equation for the momentum and substituting it into equation 9.4 yields

$$\lambda = \frac{h}{[2m(E - V)]^{1/2}} \tag{9.6}$$

This equation shows that the wavelength for a particle with constant energy E will change as it moves into a region with different potential energy. Thus as an electron with constant energy moves into a region where it has a higher potential energy V, its energy decreases and its associated wavelength increases according to equation 9.4.

9.2 THE HEISENBERG UNCERTAINTY PRINCIPLE

Classical mechanics does not involve any limitations in the accuracy with which observables may be measured. For example, the position and momentum of a particle may be simultaneously measured to any desired accuracy. This does require an interaction of the observer with the system that can disturb the system, but the disturbance can be made negligible or can be taken into account by suitable calculations. In 1927 Heisenberg formulated his principle that values of particular pairs of observables cannot be determined simultaneously with arbitrarily high precision in quantum mechanics. Examples of pairs of observables that are restricted in this way are momentum and position, and energy and time; such pairs are referred to as **complementary.** The quantitative expressions of the Heisenberg uncertainty principle can be derived by combining the de Broglie relation $p = h/\lambda$ and the Einstein relation $E = h\nu$ with properties of all waves.

The de Broglie wave for a particle is made up of a superposition of an infinitely large number of waves of the form

$$\begin{aligned} \psi(x, t) &= A \sin 2\pi\left(\frac{x}{\lambda} - \nu t\right) \\ &= A \sin 2\pi(\kappa x - \nu t) \end{aligned} \tag{9.7}$$

where A is amplitude and κ is the reciprocal wavelength. We consider one spatial dimension for simplicity. The waves that are added together have infinitesimally different wavelengths. This superposition of waves produces a **wave packet** as shown in Fig. 9.2. By use of Fourier integral* methods it is possible to show that for wave motion of any type

$$\Delta x\, \Delta\kappa = \Delta x\, \Delta\frac{1}{\lambda} \geq \frac{1}{4\pi} \tag{9.8}$$

$$\Delta t\, \Delta\nu \geq \frac{1}{4\pi} \tag{9.9}$$

where Δx is the extent of the wave packet in space, $\Delta\kappa$ is the range in reciprocal wavelength, $\Delta\nu$ is the range in frequency, and Δt is a measure of the time required for the packet to pass a given point. The Δ's in these equations are actually standard deviations.† We can understand equations 9.8 and 9.9 in the following way. If at a given time the wave packet extends over a short range of x values, there is a

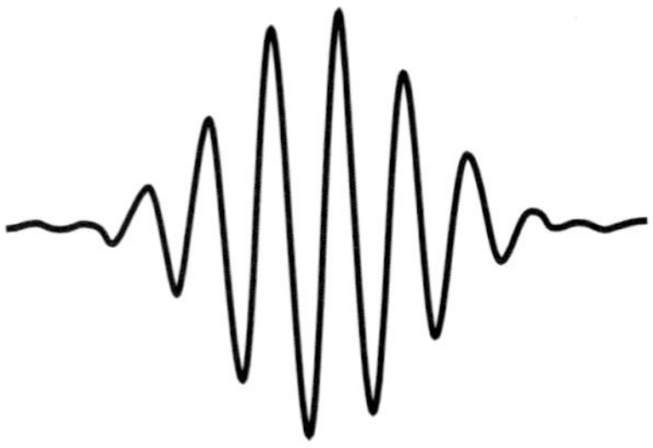

(a)

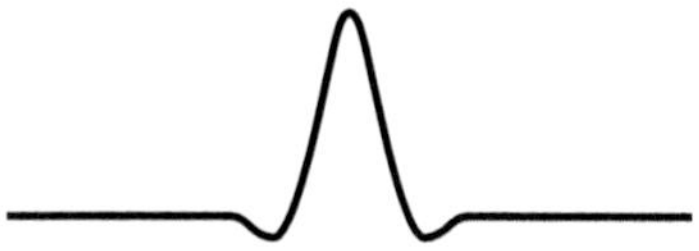

(b)

Figure 9.2 Superposition of waves to give (*a*) a weakly localized and (*b*) a strongly localized wave packet.

*R. Eisberg and R. Resnick, *Quantum Physics,* 2nd ed. New York: Wiley, 1985.

†Calculation of the standard deviation is discussed in Section 9.5.

limit to the accuracy with which we can measure the wavelength. If a wave packet is of short duration, there is a limit to the accuracy with which we can measure the frequency.

One form of the Heisenberg uncertainty principle may be derived by substituting the de Broglie relation in equation 9.8. Since $1/\lambda = p_x/h$ for motion in the x direction, then

$$\Delta x \, \Delta \frac{p_x}{h} \geq \frac{1}{4\pi} \tag{9.10}$$

$$\Delta x \, \Delta p_x \geq \frac{\hbar}{2} \tag{9.11}$$

where $\hbar = h/2\pi$ ($\hbar$ is called "h bar"). We can understand this limitation on our ability to determine the simultaneous position and momentum of an electron in this way. To determine the position of the electron at least one photon would have to strike the electron, and the momentum of the electron would inevitably be altered in the process. This would limit our ability to measure the momentum. If we use a photon of shorter wavelength to determine the position of the electron more accurately, the disturbance of the momentum is greater and Δp_x is greater, according to relation 9.11. Of course, the same uncertainty applies to $\Delta y \, \Delta p_y$ and $\Delta z \, \Delta p_z$.

Another form of the Heisenberg uncertainty principle may be derived by substituting $E = h\nu$ in equation 9.9. This yields

$$\Delta t \, \Delta \frac{E}{h} \geq \frac{1}{4\pi} \tag{9.12}$$

$$\Delta t \, \Delta E \geq \frac{\hbar}{2} \tag{9.13}$$

We can understand this limitation on our ability to measure the energy level of an electron in an atom in the following way. Suppose that excited atoms emit electromagnetic radiation in going to a lower energy state. If these excited atoms live a long time, the radiation will be nearly monochromatic and the spectral line will be sharp. If the excited atoms have a very short half-life, the electromagnetic radiation will have a wider range in frequencies, in accord with equation 9.13. Thus, spectral lines have natural widths determined by the lifetime of the excited state, and they are further broadened by collisions and the Doppler effect. If the frequency is uncertain by $\Delta\nu$, the energy of the excited atom is uncertain by $\Delta E = h \, \Delta\nu$.

It is important to realize that the uncertainties in equations 9.11 and 9.13 are not experimental errors that are dependent on the quality of the measuring apparatus, but are inherent in quantum mechanics. Because of this uncertainty the results of quantum mechanical calculations are expressed in terms of probabilities. The success of a classical mechanical description for macroscopic systems can be understood from the fact that the de Broglie wavelength of macroscopic systems is extremely small (see the problems at the end of the chapter for examples).

Example 9.2 ***The uncertainty of velocity and position for an electron***

What is the minimum uncertainty in the velocity of an electron if the uncertainty in its position is 100 pm?

$$\Delta p_x \geq \frac{h}{4\pi\,\Delta x}$$
$$\geq \frac{6.626 \times 10^{-34}\ \text{J s}}{4\pi(10^{-10}\ \text{m})}$$
$$\geq 5.272 \times 10^{-25}\ \text{kg m s}^{-1}$$

$$\Delta v_x = \frac{\Delta p_x}{m} \geq \frac{5.272 \times 10^{-25}\ \text{kg m s}^{-1}}{9.109 \times 10^{-31}\ \text{kg}}$$
$$\geq 5.79 \times 10^{5}\ \text{m s}^{-1}$$

9.3 THE SCHRÖDINGER EQUATION

After de Broglie's hypothesis that a particle has wavelike properties associated with it, many physicists attempted to derive the correct wave equation for a particle. In 1926, Schrödinger and Heisenberg independently developed theories that looked very different, but were later shown to be equivalent by Dirac, a British scientist. We will consider only the Schrödinger equation. The general form of this equation contains time as a variable, but we will be concerned primarily with the time-independent form because most chemical applications of quantum mechanics need only time-independent states.

Schrödinger's equation is a postulate of quantum mechanics, to be judged by its ability to describe experimental results. However, we can give an argument as to why the form of the equation is plausible by considering the wave equation for the harmonic motion of a one-dimensional string. If the displacement of the string at position x is represented by the **wavefunction** $\psi(x)$, and the wavelength of the displacement is λ, then the classical time-independent wave equation for $\psi(x)$ is

$$\frac{\mathrm{d}^2\psi(x)}{\mathrm{d}x^2} = -\left(\frac{2\pi}{\lambda}\right)^2 \psi(x) \tag{9.14}$$

This wave equation can be generalized to three dimensions for an isotropic and uniform medium as

$$\left(\frac{\partial^2}{\partial x^2} + \frac{\partial^2}{\partial y^2} + \frac{\partial^2}{\partial z^2}\right)\psi(x, y, z) = -\left(\frac{2\pi}{\lambda}\right)^2 \psi(x, y, z) \tag{9.15}$$

This combination of second partial derivatives is called the **Laplacian** and is represented by ∇^2 (del squared) so that equation 9.15 can be written

$$\nabla^2\psi(x, y, z) = -\left(\frac{2\pi}{\lambda}\right)^2 \psi(x, y, z) \tag{9.16}$$

Now, to consider a particle wavefunction $\psi(x)$, we substitute equation 9.6 for the wavelength of a particle with energy E in a region of potential energy V into this equation to obtain

$$-\left(\frac{h^2}{8\pi^2 m}\right)\nabla^2\psi(x, y, z) + V(x, y, z)\psi(x, y, z) = E\psi(x, y, z) \tag{9.17}$$

This equation for a quantum mechanical particle of mass m can be compared with the classical equation for a particle:

$$\frac{1}{2m}(p_x^2 + p_y^2 + p_z^2) + V = E \tag{9.18}$$

Equation 9.17 can be written in the form

$$-\left(\frac{h^2}{8\pi^2 m}\right)\left(\frac{\partial^2}{\partial x^2} + \frac{\partial^2}{\partial y^2} + \frac{\partial^2}{\partial z^2}\right)\psi(x, y, z) + V(x, y, z)\psi(x, y, z) = E\psi(x, y, z) \tag{9.19}$$

This is the time-independent Schrödinger equation.

Comparison of equations 9.18 and 9.19 suggests that the momenta in equation 9.18 are linked to partial derivatives in equation 9.19 as indicated by

$$p_x \leftrightarrow \frac{h}{2\pi i}\frac{\partial}{\partial x} \tag{9.20}$$

We will soon see that this is an important postulate in quantum mechanics. Note that the wavefunction ψ, like the amplitude of an electromagnetic wave, can be complex; that is, it may involve $\mathrm{i} = \sqrt{-1}$ (see Appendix D.9). The interpretation of ψ was provided by Born, who suggested, based on an analogy to electromagnetic waves, that the **probability** of finding the particle between x and $x + \mathrm{d}x$ is given by $\psi^*(x)\psi(x)\,\mathrm{d}x$, where ψ^* is the **complex conjugate** of ψ. (The complex conjugate is found by changing i to $-$i everywhere in ψ.) This means that $\psi^*(x)\psi(x)$ is a **probability density.** For any ψ, $\psi^*\psi$ is both real and nonnegative, as it must be to have the interpretation of a probability density. For example, if ψ is a complex number, it can be written as $a + ib$; then $\psi^* = a - ib$ and $\psi^*\psi = a^2 + b^2$, which is clearly positive and real.

We often write $|\psi|^2$ for $\psi^*\psi$. With this interpretation of ψ, the probability of finding the particle between x_1 and x_2 is

$$\text{Probability}\ (x_1 \leq x \leq x_2) = \int_{x_1}^{x_2} \psi^*(x)\psi(x)\,\mathrm{d}x \tag{9.21}$$

and, since the probability of finding the particle anywhere on the x axis must be 1,

$$\int_{-\infty}^{+\infty} \psi^*(x)\psi(x)\,\mathrm{d}x = 1 \tag{9.22}$$

For this one-dimensional example, the units of ψ are $\mathrm{m}^{-1/2}$ to ensure that the probability is a pure number. If we were considering a three-dimensional system, then the integral of $|\psi|^2$ over three dimensions would be the probability of finding the particle anywhere in the space, which is 1. Then the wavefunction would have units $\mathrm{m}^{-3/2}$.

The wavefunctions that are solutions of the time-independent Schrödinger equation are called **stationary state wavefunctions.** An atom or a molecule can be in any one of the stationary energy states, say, the nth, represented by its own wavefunction ψ_n with energy E_n.

The wavefunction contains all the information we can have about a particle in quantum mechanics; methods to find this information will be presented in Section 9.4. However, for $|\psi|^2$ to be a probability density, all the ψ's must be "well behaved," that is, have certain general properties: (a) They are continuous, (b) they are finite, (c) they are single-valued, and (d) their integral $\int \psi^* \psi \, d\tau$ over the entire range of variables is equal to unity. The differential volume is represented by $d\tau$.

A wavefunction ψ_i is said to be **normalized** if

$$\int \psi_i^* \psi_i \, d\tau = 1 \tag{9.23}$$

Two functions, ψ_i and ψ_j, are said to be **orthogonal** if

$$\int \psi_i^* \psi_j \, d\tau = 0 \tag{9.24}$$

These relations can be combined by writing

$$\int \psi_i^* \psi_j \, d\tau = \delta_{ij} \tag{9.25}$$

where δ_{ij} is called the **Kronecker delta,** which is defined by

$$\delta_{ij} = \begin{cases} 0 & \text{for } i \neq j \\ 1 & \text{for } i = j \end{cases} \tag{9.26}$$

Wavefunctions that satisfy equation 9.26 are said to be **orthonormal.**

Example 9.3 ***Calculation of a normalization factor***

Given that the wavefunction for the hydrogen atom in the ground state ($n = 1$) is of the form $\psi = N\,e^{-r/a_0}$, where r is the distance from the nucleus to the electron and a_0 is the Bohr radius, calculate the normalization factor N.

The element of volume in spherical polar coordinates is $d\tau = r^2 \sin\theta \, dr \, d\theta \, d\phi$. The probability that the electron is in volume $d\tau$ is $\psi^*\psi \, d\tau$, and so

$$1 = \int \psi^* \psi \, d\tau = N^2 \int_0^\infty e^{-2r/a_0} r^2 \, dr \int_0^\pi \sin\theta \, d\theta \int_0^{2\pi} d\phi$$

Integral tables or successive integration by parts shows that

$$\int_0^\infty x^2 e^{-kx} \, dx = \frac{2}{k^3}$$

Therefore,

$$1 = N^2 \frac{a_0^3}{2} 2\pi$$

$$N = \left(\frac{1}{\pi a_0^3} \right)^{1/2}$$

9.4 OPERATORS

Now that we have discussed the wavefunctions that are important in quantum mechanics, we turn to the other basic quantum mechanical concept, the operator. An **operator** is a mathematical operation that is applied to a function. For example, $\partial/\partial x$ is the operator that indicates that the function is to be differentiated with respect to x, and $\hat{x}$ is the operator that indicates that the function is to be multiplied by x. We will designate operators with a caret, as in $\hat{A}$ or $\hat{H}$. The symbol of the operator is placed to the left of the function to which it is applied. The operators of quantum mechanics are linear. A **linear operator** has the following properties:

$$\hat{A}(f_1 + f_2) = \hat{A}f_1 + \hat{A}f_2 \tag{9.27}$$

$$\hat{A}(cf) = c\hat{A}f \tag{9.28}$$

where c is a number. The simplest operator is the identity operator $\hat{E}$, for which $\hat{E}f = f$. There is an algebra of linear operators, and we can write $\hat{A}_3 = \hat{A}_1 + \hat{A}_2$ or $\hat{A}_4 = \hat{A}_1\hat{A}_2$, but operator multiplication is different from the multiplication of numbers, as we will see.

Example 9.4 ***Applying differential operators to functions***

(a) Apply the operator $\hat{A} = \mathrm{d}/\mathrm{d}x$ to the function x^2. (b) Apply the operator $\hat{A} = \mathrm{d}^2/\mathrm{d}x^2$ to the function $4x^2$. (c) Apply the operator $\hat{A} = (\partial/\partial y)_x$ to the function xy^2. (d) Apply the operator $\hat{A} = -\mathrm{i}\hbar\,\mathrm{d}/\mathrm{d}x$ to the function e^{-ikx}. (e) Using the same operator as in (d), apply the operator $\hat{A}\hat{A} = \hat{A}^2 = (-\mathrm{i}\hbar\,\mathrm{d}/\mathrm{d}x)(-\mathrm{i}\hbar\,\mathrm{d}/\mathrm{d}x) = -\hbar^2\,\mathrm{d}^2/\mathrm{d}x^2$ to the function e^{-ikx}.

(a) $$\hat{A}(x^2) = \frac{\mathrm{d}}{\mathrm{d}x}x^2 = 2x$$

(b) $$\hat{A}(4x^2) = \frac{\mathrm{d}^2}{\mathrm{d}x^2}4x^2 = \frac{\mathrm{d}}{\mathrm{d}x}8x = 8$$

(c) $$\hat{A}(xy^2) = \left\{\frac{\partial}{\partial y}(xy^2)\right\}_x = 2xy$$

(d) $$\hat{A}(\mathrm{e}^{-ikx}) = -\mathrm{i}\hbar\frac{\mathrm{d}}{\mathrm{d}x}\mathrm{e}^{-ikx} = \mathrm{i}^2k\hbar\mathrm{e}^{-ikx} = -k\hbar\mathrm{e}^{-ikx}$$

(e) $$\hat{A}^2(\mathrm{e}^{-ikx}) = -\hbar^2\frac{\mathrm{d}^2}{\mathrm{d}x^2}\mathrm{e}^{-ikx} = \mathrm{i}k\hbar^2\frac{\mathrm{d}}{\mathrm{d}x}\mathrm{e}^{-ikx} = -\mathrm{i}^2k^2\hbar^2\,\mathrm{e}^{-ikx} = k^2\hbar^2\mathrm{e}^{-ikx}$$

With this concept of operators, we may now return to the Schrödinger equation, 9.17, and rewrite it in a slightly different form:

$$\left[-\frac{\hbar^2}{2m}\nabla^2 + V(x, y, z)\right]\psi(x, y, z) = E\psi(x, y, z) \tag{9.29}$$

The quantity in square brackets is called the **Hamiltonian operator** and is designated $\hat{H}$. When an operator, e.g., $\hat{A}$, operating on a function, e.g., ϕ_n, yields a constant, a_n, multiplied by that function,

$$\hat{A}\phi_n = a_n\phi_n \tag{9.30}$$

we say that ϕ_n is an eigenfunction of $\hat{A}$ with eigenvalue a_n. Thus for the Schrödinger equation (9.29), $\psi(x, y, z)$ is the eigenfunction of $\hat{H}$ with eigenvalue E.

Example 9.5 ***The eigenfunctions and eigenvalues of a simple operator***

What are the eigenfunctions and eigenvalues of the operator d/dx?

$$\frac{\mathrm{d}}{\mathrm{d}x}f(x) = kf(x) \qquad \frac{\mathrm{d}f(x)}{f(x)} = k\,\mathrm{d}x \qquad \ln f(x) = kx + c \qquad f(x) = \mathrm{e}^c \mathrm{e}^{kx} = c'\mathrm{e}^{kx}$$

where c and c' are constants. For each different value of k there is an eigenfunction $c'\mathrm{e}^{kx}$. Or, to put it another way, the eigenfunction $c'\mathrm{e}^{kx}$ has the eigenvalue k. Note that k can be a complex number.

The name *Hamiltonian* comes from Sir William Hamilton, who developed an alternative form of classical mechanics involving a function H, called the Hamiltonian function. For a system for which the potential energy is a function of the coordinates only—a so-called conservative system because this ensures that the energy is conserved—the Hamiltonian function is equal to the total energy of the system expressed in terms of coordinates and conjugate momenta. For a Cartesian coordinate system the conjugate momenta are the components of the linear momentum p_x, p_y, and p_z in the x, y, and z directions. The Hamiltonian form of classical mechanics is most easily transformed to quantum mechanics.

For a particle of mass m moving in one dimension subject to a potential energy $V(x)$ the classical Hamiltonian function is

$$H = \frac{p_x^2}{2m} + V(x) \tag{9.31}$$

(Note the absence of a caret over H because this is the classical function, not the quantum mechanical operator.) When we compare this equation with equation 9.29 for the quantum mechanical operator for this system, we see that there is a resemblance. The process of converting a function for a classical system to the corresponding operator for the quantum mechanical system is formalized by the following rules:

1. Each Cartesian coordinate in the Hamiltonian function is replaced by the operator multiplication by that coordinate:

$$\hat{q} = q \tag{9.32}$$

2. Each Cartesian component of linear momentum p_q in the Hamiltonian function is replaced by the operator

$$\hat{p}_q = \frac{\hbar}{\mathrm{i}}\frac{\partial}{\partial q} = -\mathrm{i}\hbar\frac{\partial}{\partial q} \tag{9.33}$$

where $\mathrm{i} = \sqrt{-1}$. The quantity 1/i is equal to $-\mathrm{i}$ because $\mathrm{i}(-\mathrm{i}) = 1$.

In converting the classical Hamiltonian to the quantum mechanical Hamiltonian operator, the potential energy function is not changed because of the first of these postulates. In converting the kinetic energy part of the classical Hamiltonian we have to calculate the operator for p_x^2:

$$\hat{p}_x^2 = \left(\frac{\hbar}{\mathrm{i}}\frac{\partial}{\partial x}\right)^2 = \frac{\hbar}{\mathrm{i}}\frac{\partial}{\partial x}\frac{\hbar}{\mathrm{i}}\frac{\partial}{\partial x} = -\hbar^2\frac{\partial^2}{\partial x^2} \tag{9.34}$$

Replacing p_x^2 in the classical Hamiltonian with this operator and replacing the potential energy function by "multiply by $V(x)$" yields the Hamiltonian operator given in equation 9.29.

Each classical observable is associated with a quantum mechanical operator. The rules given in equations 9.32 and 9.33 are used to convert a classical observable to the quantum operator. Another postulate of quantum mechanics is that the only possible measured values of an observable are the eigenvalues of the operator representing that observable.

The correspondence between a number of quantum mechanical observables and quantum mechanical operators is shown in Table 9.1. We are not ready to consider some of these observables yet, but in Section 9.6 we will calculate the energy, the average value of x, the average value of x^2, the momentum p_x, and the average value of p_x^2 for a particle in a one-dimensional box.

It is a postulate of quantum mechanics that the average value $\langle a\rangle$ of an observable corresponding with an operator $\hat{A}$ is given by

$$\langle a\rangle = \int \psi^*\hat{A}\psi\,\mathrm{d}\tau \tag{9.35}$$

where ψ^* is the complex conjugate of ψ. The complex conjugate of a function is obtained by replacing i by $-$i. Taking the energy of a one-dimensional system as an example of an observable, energy eigenvalues are obtained from the Schrödinger equation

$$\hat{H}\psi_n(x) = E_n\psi_n(x) \tag{9.36}$$

where n is an index that labels the different eigenfunctions and eigenvalues. Multiplying from the left by the complex conjugate of the wavefunction and integrating over all values of x yields

$$\int \psi_n^*(x)\hat{H}\psi_n(x)\,\mathrm{d}x = \int \psi_n^*(x)E_n\psi_n(x)\,\mathrm{d}x = E_n\int \psi_n^*(x)\psi_n(x)\,\mathrm{d}x = E_n \tag{9.37}$$

since the wavefunction is normalized.

Note that operators and wavefunctions may be complex; however, eigenvalues of quantum mechanical operators must be real because they are the only possible measured values, and measured values of observables are real. This places a restriction on possible quantum mechanical operators. Operators that have the property that they yield real eigenvalues are called **Hermitian operators.** A Hermitian operator $\hat{A}$ has the following property:

$$\int \psi^*\hat{A}\phi\,\mathrm{d}\tau = \int \phi(\hat{A}\psi)^*\,\mathrm{d}\tau \tag{9.38}$$

for any two well-behaved functions ψ and ϕ. $(\hat{A}\psi)^*$ is the complex conjugate of $\hat{A}\psi$. Suppose that $\psi = \phi$ is an eigenfunction of $\hat{A}$ with eigenvalue a, $\hat{A}\psi = a\psi$.

Table 9.1 Classical Mechanical Observables and Corresponding Quantum Mechanical Operators[a]

Observables		*Operators*	
Name	*Symbol*	*Symbol*	*Operation*
For one-dimensional systems			
Position	x	$\hat{x}$	Multiply by x
Position squared	x^2	$\hat{x}^2$	Multiply by x^2
Momentum	p_x	$\hat{p}_x$	$-i\hbar\dfrac{\partial}{\partial x}$
Momentum squared	p_x^2	$\hat{p}_x^2$	$-\hbar^2\dfrac{\partial^2}{\partial x^2}$
Kinetic energy	$T = \dfrac{p_x^2}{2m}$	$\hat{T}_x$	$-\dfrac{\hbar^2}{2m}\dfrac{\partial^2}{\partial x^2}$
Potential energy	$V(x)$	$\hat{V}(x)$	Multiply by $V(x)$
Total energy	$E = T_x + V(x)$	$\hat{H}$	$-\dfrac{\hbar^2}{2m}\dfrac{\partial^2}{\partial x^2} + V(x)$
For three-dimensional systems			
Position	$\boldsymbol{r}$	$\hat{\boldsymbol{r}}$	Multiply by $\boldsymbol{r}$
Momentum	$\boldsymbol{p}$	$\hat{\boldsymbol{p}}$	$-i\hbar\left(\boldsymbol{i}\dfrac{\partial}{\partial x} + \boldsymbol{j}\dfrac{\partial}{\partial y} + \boldsymbol{k}\dfrac{\partial}{\partial z}\right)$
Kinetic energy	T	$\hat{T}$	$-\dfrac{\hbar}{2m}\nabla^2 = -\dfrac{\hbar^2}{2m}\left(\dfrac{\partial^2}{\partial x^2} + \dfrac{\partial^2}{\partial y^2} + \dfrac{\partial^2}{\partial z^2}\right)$
Potential energy	$V(x, y, z)$	$\hat{V}(x, y, z)$	Multiply by $V(x, y, z)$
Total energy	$E = T + V$	$\hat{H}$	$-\dfrac{\hbar^2}{2m}\nabla^2 + V(x, y, z)$
Angular momentum	$\ell_x = yp_z - zp_y$	$\hat{L}_x$	$-i\hbar\left(y\dfrac{\partial}{\partial z} - z\dfrac{\partial}{\partial y}\right)$
	$\ell_y = zp_x - xp_z$	$\hat{L}_y$	$-i\hbar\left(z\dfrac{\partial}{\partial x} - x\dfrac{\partial}{\partial z}\right)$
	$\ell_z = xp_y - yp_x$	$\hat{L}_z$	$-i\hbar\left(x\dfrac{\partial}{\partial y} - y\dfrac{\partial}{\partial x}\right)$

[a] Actually, the rules given here relating the operators to classical observables are only one of the many possible ways of constructing a set of rules. We call a given set of rules a particular representation (here the coordinate representation) of quantum mechanics. There is an equally valid representation (called the momentum representation) in which the operator for $\hat{p}_x$ is multiply-by the number p_x and the operator for x is $-(\hbar/i)(\partial/\partial p_x)$. Although there are cases for which these other representations are useful, we will use only the most common one here, the coordinate representation.

Source: D. A. McQuarrie, *Quantum Chemistry.* Copyright(1983) University Science Books, Sausalito, CA.

Then

$$\int \psi^* \hat{A} \psi \, d\tau = a \tag{9.39}$$

$$\int \psi (\hat{A}\psi)^* \, d\tau = a^* \tag{9.40}$$

Thus the Hermitian property (equation 9.38) requires $a = a^*$, proving that the eigenvalues are real.

Example 9.6 ***Proof that the momentum is a Hermitian operator***

Show that the momentum operator $\hat{p}_x$ is Hermitian.

Substituting $\hat{p}_x = -i\hbar\,\mathrm{d}/\mathrm{d}x$ in the left-hand side of equation 9.38 yields

$$\int_{-\infty}^{\infty} \psi^*\left(-i\hbar\frac{\mathrm{d}\phi}{\mathrm{d}x}\right)\mathrm{d}x = -i\hbar\int_{-\infty}^{\infty}\psi^*\frac{\mathrm{d}\phi}{\mathrm{d}x}\,\mathrm{d}x = i\hbar\int_{-\infty}^{\infty}\phi\frac{\mathrm{d}\psi^*}{\mathrm{d}x}\,\mathrm{d}x = \int_{-\infty}^{\infty}\phi\left(-i\hbar\frac{\mathrm{d}\psi^*}{\mathrm{d}x}\right)^*\mathrm{d}x$$

where the next to last form is obtained by integrating by parts and using the fact that, for a well-behaved function, $\psi(\pm\infty) = 0$. Thus equation 9.38 is satisfied, and so the momentum operator $\hat{p}_x$ is Hermitian.

We can also use the Hermitian property (9.38) to prove that the eigenfunctions of a Hermitian operator corresponding to different eigenvalues are **orthogonal.** In equation 9.38 choose ψ to be ψ_n, the eigenfunction of $\hat{A}$ with eigenvalue a_n, and ϕ to be ψ_m, the eigenfunction of $\hat{A}$ with eigenvalue a_m. Then the left-hand side of equation 9.38 is equal to

$$\int \psi_n^*\hat{A}\psi_m\,\mathrm{d}\tau = \int \psi_n^* a_m\psi_m\,\mathrm{d}\tau = a_m\int\psi_n^*\psi_m\,\mathrm{d}\tau \tag{9.41}$$

while the right-hand side of equation 9.38 is equal to

$$\int \psi_m(\hat{A}\psi_n)^*\,\mathrm{d}\tau = \int \psi_m(a_n\psi_n)^*\,\mathrm{d}\tau = a_n\int\psi_m\psi_n^*\,\mathrm{d}\tau \tag{9.42}$$

where we have used the fact we just proved that a_n is real, so that $a_n^* = a_n$. Substituting the two results into equation 9.38, we see that if $a_n \neq a_m$, then $\int \psi_m\psi_n^*\,\mathrm{d}\tau$ must be zero:

$$\int \psi_n^*\psi_m\,\mathrm{d}\tau = 0 \qquad \text{if } n \neq m \tag{9.43}$$

This means that the eigenfunctions of a Hermitian operator are orthogonal to one another. They can, in addition, always be normalized, and so form an **orthonormal** set of functions with the property

$$\int \psi_n^*\psi_m\,\mathrm{d}\tau = \delta_{nm} \tag{9.44}$$

(δ_{nm} is the Kronecker delta defined in equation 9.26).

The product of two operators $\hat{A}$ and $\hat{B}$ is formed by first operating on a function with $\hat{B}$ to produce a new function and then operating on that with $\hat{A}$. Thus

$$\hat{A}\hat{B}f = \hat{A}(\hat{B}f) \tag{9.45}$$

The multiplication of operators is associative.

$$\hat{A}\hat{B}\hat{C} = (\hat{A}\hat{B})\hat{C} = \hat{A}(\hat{B}\hat{C}) \tag{9.46}$$

However, the multiplication of operators is in general not **commutative;** in other words, the resulting function may depend on the order in which the operators are applied. For example, if the operators are $\hat{A} \equiv \hat{x}$ and $\hat{B} \equiv \mathrm{d}/\mathrm{d}x$, the application of the operators $\hat{A}\hat{B}$ and $\hat{B}\hat{A}$ yields

$$\hat{A}\hat{B}f(x) = x[(\mathrm{d}/\mathrm{d}x)f(x)] = xf'(x) \tag{9.47}$$

$$\hat{B}\hat{A}f(x) = (\mathrm{d}/\mathrm{d}x)[xf(x)] = xf'(x) + f(x) \tag{9.48}$$

where the prime indicates the derivative of the function. In this case the operators $\hat{A}$ and $\hat{B}$ do not commute. In this respect the algebra of operators is like the algebra of matrices. The **commutator** of two operators is defined by

$$[\hat{A}, \hat{B}] = \hat{A}\hat{B} - \hat{B}\hat{A} \tag{9.49}$$

When $[\hat{A}, \hat{B}] = 0$, the operators are said to commute.

Example 9.7 ***The operators $\hat{x}$ and d/dx do not commute***

What is the commutator for the two operators $\hat{A} = \hat{x}$ and $\hat{B} = \mathrm{d}/\mathrm{d}x$?

An arbitrary function $f(x)$ is operated on by $\hat{A}\hat{B}$ and $\hat{B}\hat{A}$:

$$\hat{A}\hat{B}f(x) = xf'(x)$$

$$\hat{B}\hat{A}f(x) = \frac{\mathrm{d}}{\mathrm{d}x}[xf(x)] = xf'(x) + f(x)$$

The application of the commutator to the function $f(x)$ yields

$$[\hat{A}, \hat{B}]f(x) = \hat{A}\hat{B}f(x) - \hat{B}\hat{A}f(x) = xf'(x) - xf'(x) - f(x) = -f(x)$$

Therefore, since $f(x)$ was an arbitrary function,

$$[\hat{A}, \hat{B}] = -1$$

Thus these operators do not commute.

In Section 9.8 we will see that there is a relation between the commutability of two operators and the maximum precision with which their corresponding observables can be measured.

9.5 EXPECTATION VALUES AND SUPERPOSITION

As indicated above, the only possible measured values of an observable are the eigenvalues of the operator representing that observable. In addition, we noted in equation 9.35 that the average value of an observable A when the system is in the state ψ is $\langle a \rangle = \int \psi^* A \psi \, \mathrm{d}\tau$. In most cases, ψ is not an eigenfunction of A, but it can always be written as a **linear superposition** of the eigenfunctions of A, which we label ϕ_n with eigenvalue a_n $(A\phi_n = a_n \phi_n)$,

$$\psi = \sum_n c_n \phi_n \tag{9.50}$$

where the c_n are constants. Since the ϕ_n are orthonormal (equation 9.44) and ψ has to be normalized, we have

$$\begin{aligned} 1 &= \int \psi^* \psi \, \mathrm{d}\tau = \int \left(\sum_n c_n \phi_n \right)^* \left(\sum_m c_m \phi_m \right) \mathrm{d}\tau \\ &= \sum_n c_n^* c_n \int \phi_n^* \phi_n \, \mathrm{d}\tau = \sum_n |c_n|^2 \end{aligned} \tag{9.51}$$

The average value of A is given by

$$\langle a \rangle = \int \left(\sum c_n \phi_n\right)^* \left(\sum c_m \phi_m\right) \mathrm{d}\tau = \sum c_n^* c_m \int \phi_n^* A \phi_m \, \mathrm{d}\tau \tag{9.52}$$

and using equations 9.41 and 9.43,

$$\langle a \rangle = \sum |\, c_n \,|^2 \, a_n \tag{9.53}$$

Thus the average value $\langle a \rangle$ is the sum of possible measured values (a_n) multiplied by $|\, c_n \,|^2$, a nonnegative number. We interpret $|\, c_n \,|^2$ as the *probability* of measuring the value of a_n.

We can calculate c_n using equations 9.50 and 9.43. Multiplying ψ by ϕ_m^* and integrating, we find

$$\int \phi_m^* \psi \, \mathrm{d}\tau = \int \phi_m^* \sum c_n \phi_n \mathrm{d}\tau = c_m \tag{9.54}$$

Therefore, the probability of measuring the eigenvalue a_m is given by

$$|\, c_m \,|^2 = |\int \phi_m^* \psi \, \mathrm{d}\tau \,|^2 \tag{9.55}$$

Note that the probability amplitude, c_m, may be complex, but the probability $|\, c_m \,|^2$ is always real, as it should be. When ψ is an eigenfunction of A, say ϕ_k, then the probability of measuring a_k is 1 and the probability of measuring any other eigenvalue is 0. Equation 9.55 can be taken as a basic postulate of quantum mechanics.

When the observable A has a continuous set of eigenvalues (such as the observable for the position of a particle, for example), the formula for the average becomes (for a one-dimensional system)

$$\langle x \rangle = \int \psi^*(x) x \psi(x) \, \mathrm{d}x = \int x \,|\, \psi(x) \,|^2 \, \mathrm{d}x \tag{9.56}$$

In this case, we can interpret $|\, \psi(x) \,|^2$ as the **probability density** for the position variable. The probability that x lies between x and $x + \mathrm{d}x$ is $|\, \psi(x) \,|^2 \, \mathrm{d}x$; the probability that the position lies between $x = a$ and $x = b$ is prob$(a \le x \le b)$ $= \int_a^b |\, \psi(x) \,|^2 \, \mathrm{d}x$.

Example 9.8 ***Calculating the average energy and standard deviation of the energy for a particle in a superposition state***

Consider a particle in a quantum state ϕ that is the superposition of two eigenfunctions of energy ψ_1 and ψ_2, with energy eigenvalues E_1 and E_2:

$$\phi = c_1 \psi_1 + c_2 \psi_2$$

What is the probability of measuring E_1 or E_2? What is the average energy and the standard deviation in energy?

Since ϕ is normalized and ψ_1 and ψ_2 are orthogonal, we have $|c_1|^2 + |c_2|^2 = 1$. The probability of measuring E_1 is $|c_1|^2$, and the probability of measuring E_2 is $|c_2|^2$. The average energy is given by

$$\langle E \rangle = \int \phi^* \hat{H} \phi \, \mathrm{d}\tau = |c_1|^2 E_1 + |c_2|^2 E_2$$

and the standard deviation, $\sigma_E = [\langle E^2\rangle - \langle E\rangle^2]^{1/2}$ is found by calculating $\langle E^2\rangle$:

$$\langle E^2\rangle = \int \phi^* \hat{H}^2 \phi \, d\tau = |c_1|^2 E_1^2 + |c_2|^2 E_2^2$$

Therefore,

$$\sigma_E = [(|c_1|^2 E_1^2 + |c_2|^2 E_2^2) - (|c_1|^2 E_1 + |c_2|^2 E_2)^2]^{1/2}$$

Now that we can calculate the mean value of an observable x, there is another question of considerable interest. That is, what is the spread around the mean? The spread around the mean is measured by the **variance** σ_x^2, which is defined by

$$\sigma_x^2 = \langle (x - \langle x\rangle)^2\rangle = \sum (x_i - \langle x\rangle)^2 p_i \tag{9.57}$$

where p_i is the probability of x_i. The average deviation from the mean is equal to zero, but by squaring the deviations from the mean we get an inherently positive quantity that is zero if all the values of x are identical, is small if the distribution of x values is narrow, and is large if the distribution of x values is broad. The variance is represented by the symbol σ_x^2 because the square root of the variance is equal to the **standard deviation** σ_x. The standard deviation is especially useful for representing the breadth of a distribution because it has the units of x.

There is a simple way to calculate the variance that is more readily extended to continuous distributions. Equation 9.57 may be rearranged as follows:

$$\begin{aligned}
\sigma_x^2 &= \sum (x_i^2 - 2\langle x\rangle x_i + \langle x\rangle^2) p_i \\
&= \sum x_i^2 p_i - 2\langle x\rangle \sum x_i p_i + \langle x\rangle^2 \sum p_i \\
&= \langle x^2\rangle - 2\langle x\rangle^2 + \langle x\rangle^2 \\
&= \langle x^2\rangle - \langle x\rangle^2
\end{aligned} \tag{9.58}$$

In the second line $\langle x\rangle$ and $\langle x\rangle^2$ are taken outside the summations because they are simply numbers.

9.6 PARTICLE IN A ONE-DIMENSIONAL BOX

The simplest problem to treat in quantum mechanics is that of a particle of mass m constrained to move in a **one-dimensional box** of length a. The potential energy $V(x)$ is taken to be 0 for $0 < x < a$ and infinite outside this region (Fig. 9.3). We will see that this leads to quantized energy levels.

In the region between $x = 0$ and $x = a$, the Schrödinger equation 9.29 can be written

$$-\frac{\hbar^2}{2m}\frac{d^2\psi}{dx^2} = E\psi \tag{9.59}$$

or

$$\frac{d^2\psi}{dx^2} = -\frac{2mE}{\hbar^2}\psi \equiv -k^2\psi \tag{9.60}$$

where

$$k = \frac{\sqrt{2mE}}{\hbar}$$

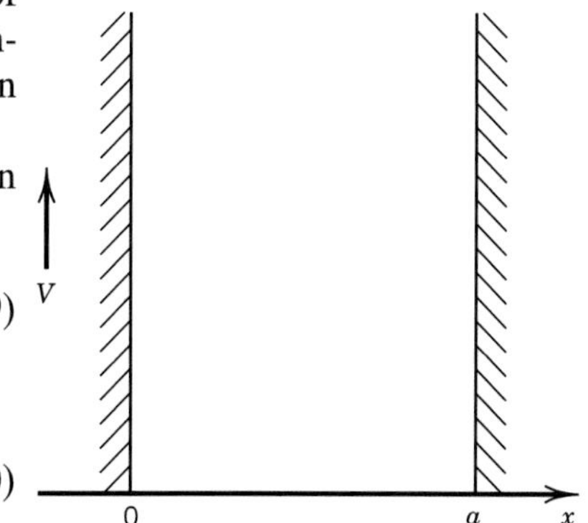

Figure 9.3 Potential for a particle in a one-dimensional box. The potential becomes infinite for $x > a$ and $x < 0$, and is zero for $0 < x < a$.

The general solution to this equation is

$$\psi(x) = A\cos kx + B\sin kx \tag{9.61}$$

In the region outside $0 < x < a$, the only physical solution to the Schrödinger equation is $\psi = 0$, because only then $V\psi = 0$, even for V infinite. This implies that the probability of finding the particle outside the box is zero (i.e., $|\psi|^2 = 0$ outside the box). To avoid a discontinuity in ψ at $x = 0$ and $x = a$, ψ must be zero at those points. To satisfy this condition at $x = 0$, A must be equal to zero, since $\cos 0 = 1$. The condition at $x = a$ can be satisfied only if $\sin ka = 0$, or

$$ka = n\pi \qquad n = 1, 2, \ldots \tag{9.62}$$

This forces the quantization

$$E_n = \frac{h^2n^2}{8ma^2} \qquad n = 1, 2, \ldots \tag{9.63}$$

Therefore, a particle constrained to be between $x = 0$ and $x = a$ has **quantized** energy levels, given by equation 9.63. Notice that as a gets large, the energy levels get closer together. In the limit of a very large box (or a very heavy particle), the energy levels are so close together that the quantization may be unnoticeable. In the limit that a becomes very large, all energies become allowed (i.e., the allowed energies get very close together so that any energy is an eigenvalue), so the perfectly free particle can have any energy.

A particle in a box cannot have zero energy because the lowest energy $h^2/8ma^2$ is given by equation 9.63 for $n = 1$. Although $n = 0$ satisfies the boundary conditions, the corresponding wavefunction is zero everywhere. The **zero-point energy** associated with the state $n = 1$ is found whenever a particle is constrained to a finite region; if this were not so the uncertainty principle would be violated. The next higher energy levels are at four times ($n = 2$) and nine times ($n = 3$) this energy, as shown in Fig. 9.4. The wavefunctions are superimposed on this plot, and we can see that the wavelength is equal to $2a/n$.

Example 9.9 ***Calculating the energy of an electron in a box***

What is the ground-state energy for an electron that is confined to a potential well with a width of 0.2 nm?

Using the formula for the allowed energies for a particle in a box,

$$E_n = \frac{h^2n^2}{8ma^2}$$

we find for the lowest energy ($n = 1$),

$$\begin{aligned} E_1 &= \frac{(6.626 \times 10^{-34}\ \text{J s})^2(1)^2}{8(9.109 \times 10^{-31}\ \text{kg})(0.2 \times 10^{-9}\ \text{m})^2} \\ &= 1.506 \times 10^{-18}\ \text{J} \\ &= \frac{(1.506 \times 10^{-18}\ \text{J})(6.022 \times 10^{23}\ \text{mol}^{-1})}{(10^3\ \text{J kJ}^{-1})} \\ &= 907\ \text{kJ mol}^{-1} \end{aligned}$$

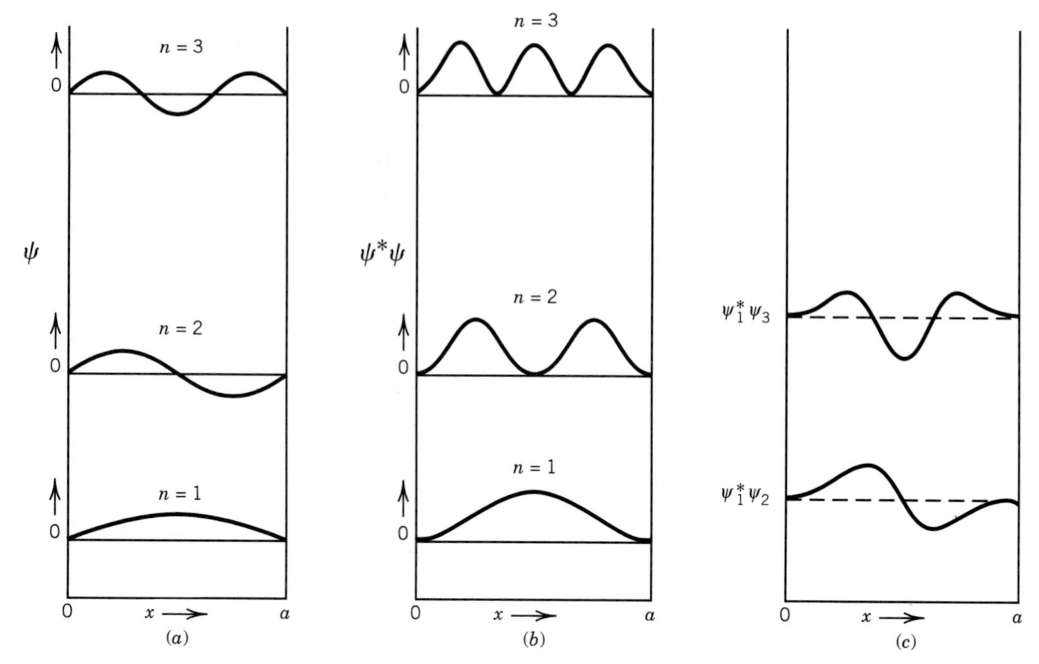

Figure 9.4 (*a*) Wavefunction ψ and (*b*) probability density function $\psi^*\psi$ for the lowest three energy levels for a particle in a box. The plots are placed at vertical heights that correspond to the energies of the levels. As the number of nodes goes up, the energy goes up. (*c*) The product of wavefunctions $\psi_1^*\psi_2$ and $\psi_1^*\psi_3$ plotted against x. (See Computer Problem 9.A.)

Using equation 9.62 for k in the wavefunction yields

$$\psi_n(x) = B \sin \frac{n\pi x}{a} \tag{9.64}$$

In order that the square of the wavefunction can be interpreted as a probability, it is necessary to **normalize** it so that the probability that the particle lies between $x = 0$ and $x = a$ is unity (i.e., the wavefunction is normalized):

$$\int_0^a \psi_n^*(x)\psi_n(x)\,\mathrm{d}x = 1 \tag{9.65}$$

$$|B|^2 \int_0^a \sin^2 \frac{n\pi x}{a}\,\mathrm{d}x = 1 \tag{9.66}$$

Since the value of the integral is $a/2$, $B = (2/a)^{1/2}$, and the normalized wavefunction for a particle in a one-dimensional box is

$$\psi_n = \left(\frac{2}{a}\right)^{1/2} \sin \frac{n\pi x}{a} \tag{9.67}$$

Figure 9.4*b* gives the probability densities $\psi^*\psi$ for a particle in an infinitely deep box. These are the probabilities per unit distance that the particle will be found at a given position. The most probable position for a particle in the zero-point level ($n = 1$) is in the center of the box. Note that the ψ_n are waves with wavelength $\lambda_n = 2a/n$. This means that ψ_n is zero at values of x equal to an integral number of $\lambda_n/2$. These zeros are called nodes of the wavefunction. In one-dimensional problems, the more nodes in an eigenfunction, the higher its eigenvalue of energy. For this problem, the number of nodes is $n - 1$.

As the value of the quantum number n increases, the probability density oscillates more and more. For very high values of n there are so many oscillations we would not expect to observe anything other than a constant value for the probability density. Classically this is just what we would expect for a particle in a box, where the probability density is $1/a$. This is an example of the **correspondence principle** of Bohr, according to which the quantum mechanical predictions approach the predictions of classical mechanics as the quantum number approaches infinity.

The wavefunctions ψ_i have been normalized so that

$$\int_{-\infty}^{\infty} \psi_i^* \psi_j \, dx = 1 \qquad \text{if } i = j \tag{9.68}$$

Particle-in-a-box wavefunctions are orthonormal,

$$\int_{-\infty}^{\infty} \psi_i^* \psi_j \, dx = 0 \qquad \text{if } i \neq j \tag{9.69}$$

which can be seen if we plot $\psi_i^* \psi_j$ for $i \neq j$ as a function of x (see Fig. 9.4c). We see that the negative contribution to the integral just cancels the positive contribution.

We can see from Fig. 9.4 that the most probable position for the particle is in the middle of the box if the system is in the ground state, but it is more likely to be at $a/4$ and $3a/4$ in the first excited state ($n = 2$). Notice that the observable "position" is not an eigenvalue of the wavefunctions for a particle in a one-dimensional box. The operator $\hat{x}$ does not give an eigenvalue when it operates on the particle-in-a-box energy eigenfunctions. This means that if we measured the position of a particle in a box we would get different answers in different trials. If we performed this measurement many times, we would confirm the probability densities $\psi^* \psi$ shown in Fig. 9.4. However, there is another type of question that does have a definite answer. We can ask for the average position of the particle, or its average x coordinate squared, or its average momentum squared.

Example 9.10 ***Calculating the probability of finding a particle in a small region of space***

A particle is in a linear box that is 1 nm in length. What is the probability that a particle is between 0.45×10^{-9} m and 0.55×10^{-9} m in the ground state (i.e., $n = 1$)?

We have seen that the probability density at point x is given by the square of the wavefunction:

$$\psi^2 = \frac{2}{a} \sin^2 \frac{\pi x}{a}$$

Therefore, the total probability of being in the above region is given by

$$P(0.45 \times 10^{-9} \text{ m} \leq x \leq 0.55 \times 10^{-9} \text{ m}) = \int_{0.45\times 10^{-9}\text{ m}}^{0.55\times 10^{-9}\text{ m}} \frac{2}{a} \sin^2 \frac{\pi x}{a} \, dx$$

Since $\sin^2(\pi x/a) \approx 1$ in this region, we find

$$P(0.45 \times 10^{-9} \text{ m} \leq x \leq 0.55 \times 10^{-9} \text{ m}) \approx \frac{2}{a}(0.55 \times 10^{-9} - 0.45 \times 10^{-9}) \text{ m} \approx 0.2$$

Example 9.11 ***Calculation of*** $\langle x\rangle$ ***and*** $\langle x^2\rangle$ ***for a particle in a one-dimensional box***

The value of $\langle x\rangle$ for a particle in any of the eigenstates of a one-dimensional box is

$$\langle x\rangle_n = \frac{2}{a}\int_0^a x \sin^2\frac{n\pi x}{a}\,\mathrm{d}x$$

By use of a table of definite integrals we obtain

$$\langle x\rangle = \frac{a}{2}$$

(i.e., the middle of the box) for all values of the principal quantum number n, which is reasonable. The value of $\langle x^2\rangle$ in the nth eigenstate is

$$\begin{aligned}\langle x^2\rangle_n &= \frac{2}{a}\int_0^a x^2 \sin^2\frac{n\pi x}{a}\,\mathrm{d}x\\ &= \left(\frac{a}{2\pi n}\right)^2\left(4\frac{\pi^2 n^2}{3} - 2\right)\end{aligned}$$

Example 9.12 ***Calculation of*** $\langle p_x\rangle$ ***and*** $\langle p_x^2\rangle$ ***for a particle in a one-dimensional box***

Since the operator for momentum in the x direction is $-i\hbar\,\mathrm{d}/\mathrm{d}x$, the value of $\langle p_x\rangle$ for a particle in any of the eigenstates of a one-dimensional box is given by

$$\begin{aligned}\langle p_x\rangle_n &= \int_0^a\left[\left(\frac{2}{a}\right)^{1/2}\sin\frac{n\pi x}{a}\right]\left(-i\hbar\frac{\mathrm{d}}{\mathrm{d}x}\right)\left[\left(\frac{2}{a}\right)^{1/2}\sin\frac{n\pi x}{a}\right]\mathrm{d}x\\ &= -i\hbar\frac{2\pi n}{a^2}\int_0^a \sin\frac{n\pi x}{a}\cos\frac{n\pi x}{a}\,\mathrm{d}x\\ &= 0\end{aligned}$$

where the last result is obtained by using integral tables. Now that we have derived this result we can observe that the average momentum has to be zero because the particle in a box cannot continue to travel in one direction. The value of $\langle p_x^2\rangle$ is given by

$$\begin{aligned}\langle p_x\rangle_n^2 &= \int_0^a\left[\left(\frac{2}{a}\right)^{1/2}\sin\frac{n\pi x}{a}\right]\left(-i\hbar\frac{\mathrm{d}}{\mathrm{d}x}\right)^2\left[\left(\frac{2}{a}\right)^{1/2}\sin\frac{n\pi x}{a}\right]\mathrm{d}x\\ &= \frac{n^2\pi^2\hbar^2}{a^2}\end{aligned}$$

Example 9.13 ***Standard deviations of*** x ***and*** p_x ***for a particle in a one-dimensional box***

What is the standard deviation in x for the particle in an eigenstate of a one-dimensional box? What is the standard deviation in p_x? How do these results compare with the Heisenberg uncertainty principle? In this example we will use Δx and Δp_x for these standard deviations as we did in Section 9.2.

Using the results of Example 9.11 we find

$$\Delta x = (\langle x^2 \rangle - \langle x \rangle^2)^{1/2}$$
$$= \left(\frac{a}{2\pi n}\right)\left(4\frac{\pi^2 n^2}{3} - 2\right)^{1/2}$$

Using the results of Example 9.12, we find

$$\Delta p_x = (\langle p_x^2 \rangle - \langle p_x \rangle^2)^{1/2}$$
$$= \frac{n\pi\hbar}{a}$$

The product of these standard deviations is

$$\Delta x\ \Delta p_x = \left(\frac{a}{2\pi n}\right)\left(4\frac{\pi^2 n^2}{3} - 2\right)^{1/2}\left(\frac{n\pi\hbar}{a}\right) = \frac{\hbar}{2}\left(4\frac{\pi^2 n^2}{3} - 2\right)^{1/2}$$

The product $\Delta x\ \Delta p_x$ for a particle in a box is in agreement with the Heisenberg uncertainty principle since it is always greater than $\hbar/2$. Now we can see what happens to Δx and Δp_x when we change the dimensions of the box. As the length of the box is increased, the standard deviation in x increases, and we say the position of the particle becomes more uncertain. As the length of the box is increased, the standard deviation in p decreases, and we say the momentum becomes more definite.

Example 9.14 ***Writing the eigenfunctions for a particle in a box in complex form***

The general solution for a particle in a box in the $n = 2$ level

$$\psi(x) = A\cos(2\pi x/a) + B\sin(2\pi x/a) \tag{1}$$

can be written

$$\psi(x) = C\,\mathrm{e}^{2\pi \mathrm{i}x/a} + D\,\mathrm{e}^{-2\pi \mathrm{i}x/a} \tag{2}$$

Since

$$\mathrm{e}^{\pm ikx} = \cos(kx) \pm \mathrm{i}\sin(kx)$$

then

$$\cos(kx) = \frac{1}{2}(\mathrm{e}^{ikx} + \mathrm{e}^{-ikx})$$
$$\sin(kx) = \frac{1}{2\mathrm{i}}(\mathrm{e}^{ikx} - \mathrm{e}^{-ikx})$$

Substituting these two relations in the first equation in this example yields

$$\psi(x) = \left(\frac{A}{2} + \frac{B}{2\mathrm{i}}\right)\mathrm{e}^{ikx} + \left(\frac{A}{2} - \frac{B}{2\mathrm{i}}\right)\mathrm{e}^{-ikx}$$

Taking $k = 2\pi/a$ yields equation 2 with $C = (A - \mathrm{i}B)/2$ and $D = (A + \mathrm{i}B)/2$.

9.7 PARTICLE IN A THREE-DIMENSIONAL BOX

There are several more things we can learn from the particle in a box by extending the box to three-dimensions. The particle is confined to a rectangular parallelepiped with sides of lengths a, b, and c by having an infinite potential outside the box.

The classical Hamiltonian for a particle that can move in three dimensions is

$$H = \frac{p_x^2}{2m} + \frac{p_y^2}{2m} + \frac{p_z^2}{2m} \tag{9.70}$$

The time-independent Schrödinger equation for a single particle of mass m moving in three dimensions is

$$\hat{H}\psi(x, y, z) = E\psi(x, y, z) \tag{9.71}$$

where the Hamiltonian operator is

$$\hat{H} = -\frac{\hbar^2}{2m}\nabla^2 + V(x, y, z) \tag{9.72}$$

and

$$\nabla^2 \equiv \frac{\partial^2}{\partial x^2} + \frac{\partial^2}{\partial y^2} + \frac{\partial^2}{\partial z^2} \tag{9.73}$$

is referred to as the **Laplacian operator** or del squared. The wavefunction is normalized so that

$$\int_{-\infty}^{\infty}\int_{-\infty}^{\infty}\int_{-\infty}^{\infty} \psi^*(x, y, z)\psi(x, y, z)\,\mathrm{d}x\,\mathrm{d}y\,\mathrm{d}z = 1 \tag{9.74}$$

If a particle can move in three dimensions, its probability density $p(x, y, z)$ is given by

$$p(x, y, z) = \psi^*(x, y, z)\psi(x, y, z) \tag{9.75}$$

The probability that the x coordinate is between x and $x + \mathrm{d}x$, the y coordinate is between y and $y + \mathrm{d}y$, and the z coordinate is between z and $z + \mathrm{d}z$ is $p(x, y, z)\,\mathrm{d}x\,\mathrm{d}y\,\mathrm{d}z = \psi^*(x, y, z)\psi(x, y, z)\,\mathrm{d}x\,\mathrm{d}y\,\mathrm{d}z$. This last expression can be shortened to $\psi^*\psi\,\mathrm{d}\tau$, where $\mathrm{d}\tau$ represents the differential element of volume $\mathrm{d}x\,\mathrm{d}y\,\mathrm{d}z$.

Since the potential within the box is zero, we obtain the following partial differential equation for the region inside the box:

$$-\frac{\hbar^2}{2m}\left(\frac{\partial^2}{\partial x^2} + \frac{\partial^2}{\partial y^2} + \frac{\partial^2}{\partial z^2}\right)\psi = E\psi \tag{9.76}$$

Partial differential equations are solved, where possible, by using the technique of separation of variables to obtain a set of ordinary differential equations. In this case we assume that the wavefunction ψ is the product of three functions, each depending on just one coordinate:

$$\psi(x, y, z) = X(x)Y(y)Z(z) \tag{9.77}$$

By substituting this for ψ in equation 9.76 and then dividing by $X(x)Y(y)Z(z)$, we obtain

$$-\frac{\hbar^2}{2m}\left[\frac{1}{X(x)}\frac{d^2X(x)}{dx^2}+\frac{1}{Y(y)}\frac{d^2Y(y)}{dy^2}+\frac{1}{Z(z)}\frac{d^2Z(z)}{dz^2}\right]=E \qquad (9.78)$$

Since the terms on the left-hand side of the equation are each a function of a different independent variable and thus can be varied independently of one another, each must equal a constant in order that the sum of the three terms equals a constant for all values of x, y, and z:

$$E_x+E_y+E_z=E \qquad (9.79)$$

This converts the partial differential equation 9.78 into three ordinary differential equations that can be easily solved:

$$-\frac{\hbar^2}{2m}\left[\frac{1}{X(x)}\frac{d^2X(x)}{dx^2}\right]=E_x \qquad (9.80)$$

$$-\frac{\hbar^2}{2m}\left[\frac{1}{Y(y)}\frac{d^2Y(y)}{dy^2}\right]=E_y \qquad (9.81)$$

$$-\frac{\hbar^2}{2m}\left[\frac{1}{Z(z)}\frac{d^2Z(z)}{dz^2}\right]=E_z \qquad (9.82)$$

These equations are just like equation 9.59 and may be solved in the same way to obtain

$$X(x)=A_x\sin\frac{n_x\pi x}{a}=A_x\sin\left(\frac{2mE_x}{\hbar^2}\right)^{1/2}x \qquad (9.83)$$

$$Y(y)=A_y\sin\frac{n_y\pi y}{b}=A_y\sin\left(\frac{2mE_y}{\hbar^2}\right)^{1/2}y \qquad (9.84)$$

$$Z(z)=A_z\sin\frac{n_z\pi z}{c}=A_z\sin\left(\frac{2mE_z}{\hbar^2}\right)^{1/2}z \qquad (9.85)$$

where a, b, and c are the lengths of the sides in the x, y, and z directions, respectively; n_x, n_y, and n_z are nonzero integers, called quantum numbers; and $E_x = h^2n_x^2/8ma^2$ and so on. Thus, there is a quantum number for each coordinate. When the wavefunction is normalized, we obtain

$$\psi(x,y,z)=\left(\frac{8}{abc}\right)^{1/2}\sin\frac{n_x\pi x}{a}\sin\frac{n_y\pi y}{b}\sin\frac{n_z\pi z}{c} \qquad (9.86)$$

When this eigenfunction is substituted in equation 9.76, we obtain

$$E=\frac{h^2}{8m}\left(\frac{n_x^2}{a^2}+\frac{n_y^2}{b^2}+\frac{n_z^2}{c^2}\right) \qquad (9.87)$$

Later, in statistical mechanics (Section 16.3) we will use this equation for the translational energy of a molecule in a container. The three quantum numbers are independent, and for a given set of three quantum numbers, there is in general a unique value for the energy if $a \neq b \neq c$.

A new feature arises when the sides of the box are equal. If $a = b = c$, the energy levels are given by

$$E = \frac{h^2}{8ma^2}(n_x^2 + n_y^2 + n_z^2) \tag{9.88}$$

In contrast with the three-dimensional box with $a \neq b \neq c$, there may be several combinations (n_x, n_y, n_z) that yield the same energy. For example, (2,1,1), (1,2,1), and (1,1,2) have the same energy. These three states of the system (different wavefunctions) make up a level that we can refer to as the 211 level. Such an energy level is said to be **degenerate,** and the degeneracy is equal to the number of independent wavefunctions associated with a given energy level. As shown in the following table, the 111 level is nondegenerate.

$n_x n_y n_z$	111	211	221	311	222	321	322	411	331
Degeneracy	1	3	3	3	1	6	3	3	3

Degeneracies arise in quantum mechanics when there is some element of symmetry. If the symmetry is "broken" by giving the sides of the box different lengths, the degeneracy is "lifted." We will see later that the spherical symmetry of certain atomic states is broken by the application of magnetic or electric fields, and this lifts some of the degeneracies.

The degeneracy of a translational energy level increases rapidly with energy. If we define $n^2 = n_x^2 + n_y^2 + n_z^2$, then $E = (h^2/8ma^2)n^2$. If we think of the allowed values of n_x as points along the x axis, the allowed values of n_y as points along the y axis, and those of n_z as points along the z axis, then n can be thought of as the length of a vector in this three-dimensional space. All such vectors with the same length have the same energy; that is, they represent degenerate states. For very large n, the number of degenerate states is proportional to the surface area of the sphere of radius n in this space; therefore, the degeneracy is proportional to n^2. Actually, since the allowed values of n_x, n_y, and n_z are *integers,* we have counted states as degenerate that are not *exactly* degenerate. However, for large enough n, these states will be so close in energy that for practical (experimental) purposes we can take them to be degenerate.

Example 9.15 ***The degeneracy of quantum levels at thermal energies***

We shall see (in Chapter 16) that the most probable translational energy for an atom in a gas at temperature T is equal to $\frac{3}{2}kT$, where $k = R/N_A$ is Boltzmann's constant. Calculate the degeneracy of the most probable energy level for an argon atom at 300 K and 1 bar pressure, assuming that the atom can be treated as a particle in a three-dimensional box.

We must find the value of n for the most probable energy level at this temperature. To use equation 9.88 we must find a; we can assume a cubical box. The volume of a gas at these conditions is approximately 0.022 m^3, so that the side of the box a can be taken as $(0.022)^{1/3}$ m. We have

$$E = \frac{h^2}{8ma^2}n^2 = \frac{3}{2}kT$$

or

$$n^2 = \frac{\frac{3}{2}kT8ma^2}{h^2}$$

$$= \frac{\frac{3}{2}(1.38 \times 10^{-23})(300)(8)(40 \times 1.67 \times 10^{-27})(0.022)^{2/3}}{(6.6 \times 10^{-34})^2}$$

$$= 6 \times 10^{20}$$

Thus, the most probable value of n is 2.25×10^{10} and the degeneracy is of the order 6×10^{20} for this level.

The particle in a three-dimensional box illustrates a general point about the separability of the Hamiltonian for a system. The Hamiltonian operator for a particle in a three-dimensional box can be written as the sum of three independent terms (i.e., it depends on different independent variables):

$$\hat{H} = \hat{H}_x + \hat{H}_y + \hat{H}_z \tag{9.89}$$

When the Hamiltonian is separable in this way, we can factor the wavefunction into a product of three wavefunctions, an eigenfunction for each coordinate. This leads to separation of the eigenvalue problem in three variables into three separate equations, each in one variable. The sum of the three eigenvalues obtained is equal to the eigenvalue for the original problem, and the product of the three eigenfunctions is equal to the eigenfunction for the original problem.

Another place where separation of variables works is with systems of independent particles; by this we mean that there are no interactions between the particles. If the n particles are independent, $\hat{H} = \hat{H}_1 + \hat{H}_2 + \cdots + \hat{H}_n$, where $\hat{H}_i$ involves only the coordinates of particle i. The wavefunction for the system is separable,

$$\psi = f_1(x_1,y_1,z_1)f_2(x_2,y_2,z_2)\cdots f_n(x_n,y_n,z_n) \tag{9.90}$$

Their energy is the sum of independent energies,

$$E = E_1 + E_2 + \cdots + E_n \tag{9.91}$$

and each particle obeys its own Schrödinger equation:

$$\hat{H}_1 f_1 = E_1 f_1 \qquad \hat{H}_2 f_2 = E_2 f_2 \qquad \ldots \qquad \hat{H}_n f_n = E_n f_n \tag{9.92}$$

Comment:

The particle in a box provides an introduction to the behavior of atoms or molecules that are constrained to move within a finite distance. Electrons within an atom are constrained to move within a small distance, but their behavior is different because they are not constrained by walls with ***infinitely*** *high potential energies. On the other hand, electrons in a long molecule with conjugated double bonds are confined within this length with pretty abrupt changes in potential energy at the end (Section 14.7). The equation for the possible energies of atoms or molecules in a three-dimensional box will be used later in statistical mechanics to calculate the energy levels of an atom or molecule in a macroscopic container (Section 16.3).*

9.8 RELATION BETWEEN COMMUTABILITY AND PRECISION OF MEASUREMENT

We have seen earlier (equations 9.45–9.49) that operators are different from the usual algebraic quantities in that they do not necessarily commute. As an example of two operators that do not commute, consider the momentum operator $\hat{p}_x$ and the position operator $\hat{x}$. When these two operators act on a wavefunction $\psi(x)$, the result depends on the order (see Example 9.7):

$$\hat{p}_x \hat{x}\, \psi(x) = \left(\frac{\hbar}{\mathrm{i}}\frac{\mathrm{d}}{\mathrm{d}x}\right) x\psi(x) = \frac{\hbar x}{\mathrm{i}}\frac{\mathrm{d}\psi(x)}{\mathrm{d}x} + \frac{\hbar}{\mathrm{i}}\psi(x) \tag{9.93}$$

$$\hat{x}\hat{p}_x\, \psi(x) = x\left(\frac{\hbar}{\mathrm{i}}\frac{\mathrm{d}}{\mathrm{d}x}\right)\psi(x) = \frac{\hbar x}{\mathrm{i}}\frac{\mathrm{d}\psi(x)}{\mathrm{d}x} \tag{9.94}$$

The difference between these two results is

$$(\hat{p}_x\hat{x} - \hat{x}\hat{p}_x)\psi(x) = \frac{\hbar}{\mathrm{i}}\psi(x) = \frac{\hbar}{\mathrm{i}}\hat{I}\,\psi(x) \tag{9.95}$$

where $\hat{I}$ is the identity operator. Thus the commutator (equation 9.49) for these two operators is given by

$$\hat{p}_x\hat{x} - \hat{x}\hat{p}_x = [\hat{p}_x, \hat{x}] = \frac{\hbar}{\mathrm{i}}\hat{I} \tag{9.96}$$

In contrast, the kinetic energy operator $\hat{T}_x$ and the momentum operator $\hat{p}_x$ do commute. When these two operators act on a wavefunction, the same result is obtained for either ordering of the operators:

$$\begin{aligned}\hat{T}_x\hat{p}_x\,\psi(x) &= \left(-\frac{\hbar^2}{2m}\frac{\mathrm{d}^2}{\mathrm{d}x^2}\right)\left(\frac{\hbar}{\mathrm{i}}\frac{\mathrm{d}}{\mathrm{d}x}\right)\psi(x)\\ &= \left(-\frac{\hbar^2}{2m}\frac{\mathrm{d}^2}{\mathrm{d}x^2}\right)\frac{\hbar}{\mathrm{i}}\frac{\mathrm{d}\psi(x)}{\mathrm{d}x}\\ &= -\frac{\hbar^3}{\mathrm{i}2m}\frac{\mathrm{d}^3\psi(x)}{\mathrm{d}x^3}\end{aligned} \tag{9.97}$$

$$\begin{aligned}\hat{p}_x\hat{T}_x\,\psi(x) &= \left(\frac{\hbar}{\mathrm{i}}\frac{\mathrm{d}}{\mathrm{d}x}\right)\left(-\frac{\hbar^2}{2m}\frac{\mathrm{d}^2}{\mathrm{d}x^2}\right)\psi(x)\\ &= \left(\frac{\hbar}{\mathrm{i}}\frac{\mathrm{d}}{\mathrm{d}x}\right)\left(-\frac{\hbar^2}{2m}\frac{\mathrm{d}^2\psi(x)}{\mathrm{d}x^2}\right)\\ &= -\frac{\hbar^3}{\mathrm{i}2m}\frac{\mathrm{d}^3\psi(x)}{\mathrm{d}x^3}\end{aligned} \tag{9.98}$$

The difference between these two results is

$$(\hat{T}_x\hat{p}_x - \hat{p}_x\hat{T}_x)\psi(x) = \hat{0}\psi(x) \tag{9.99}$$

where $\hat{0}$ is the zero operator. Thus the commutator for these two operators is given by

$$\hat{T}_x\hat{p}_x - \hat{p}_x\hat{T}_x = [\hat{T}_x, \hat{p}_x] = \hat{0} \tag{9.100}$$

It can be shown* that if operators $\hat{A}$ and $\hat{B}$ do not commute, then the accuracy with which A and B can be measured **simultaneously** is limited by an uncertainty principle.

The fact that the operators $\hat{p}_x$ and $\hat{x}$ do not commute (and that their commutator is $-i\hbar$) tells us that the accuracy with which p_x and x can be simultaneously determined experimentally is limited by the Heisenberg uncertainty principle. In contrast, there is no limitation on the accuracy of the simultaneous measurement of kinetic energy T_x and momentum p_x of a particle since they commute. We can generalize this observation and say that when two operators commute, the corresponding observables can be measured to any precision, and when they do not commute, the corresponding observables cannot be measured simultaneously to arbitrary precision. We can go further and say that eigenfunctions of one operator will also be eigenfunctions of another operator that commutes with the first operator. We will see how these generalizations apply in our consideration of the harmonic oscillator and the rigid rotor.

Example 9.16 ***Operators with simultaneous eigenfunctions commute***

Show that if all the eigenfunctions of two operators $\hat{A}$ and $\hat{B}$ are the same functions, $\hat{A}$ and $\hat{B}$ commute with each other. The eigenvalues of $\hat{A}$ and $\hat{B}$ are represented by a_i and b_i and the eigenfunctions are ψ_i, so that

$$\hat{A}\psi_i = a_i\psi_i \quad \text{and} \quad \hat{B}\psi_i = b_i\psi_i \tag{a}$$

The eigenfunction of the operator $\hat{A}\hat{B}$ is obtained as follows:

$$\hat{A}\hat{B}\psi_i = \hat{A}(\hat{B}\psi_i) = \hat{A}b_i\psi_i = b_i\hat{A}\psi_i = b_i a_i\psi_i \tag{b}$$

The operator $\hat{B}\hat{A}$ has eigenvalue $a_i b_i$, as may be seen from

$$\hat{B}\hat{A}\psi_i = \hat{B}(\hat{A}\psi_i) = \hat{B}a_i\psi_i = a_i\hat{B}\psi_i = a_i b_i\psi_i \tag{c}$$

Since $a_i b_i = b_i a_i$, $\hat{A}$ and $\hat{B}$ commute with each other.

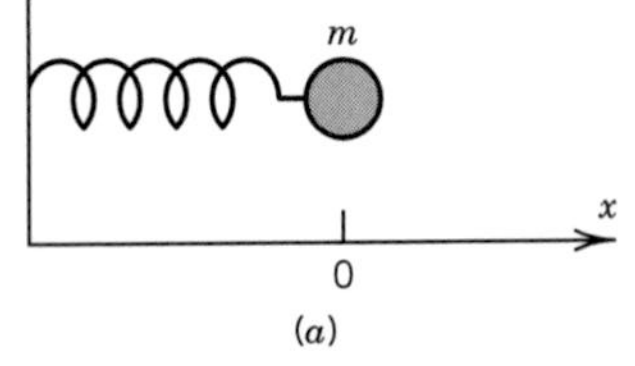

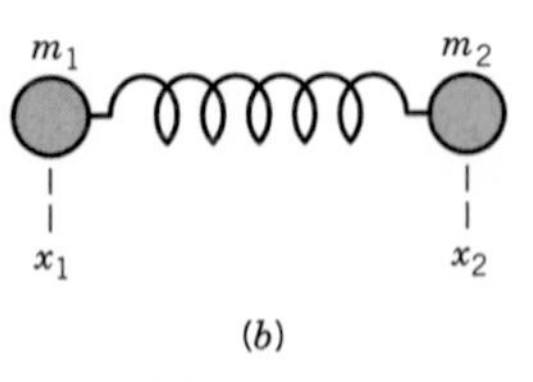

Figure 9.5 (*a*) Mass m connected to a wall by a spring in the absence of gravity. The equilibrium position of the mass is $x = 0$. (*b*) Masses m_1 and m_2 connected by a spring.

9.9 CLASSICAL HARMONIC OSCILLATOR

To understand the vibrations of molecules we need to understand the quantum mechanical treatment of a harmonic oscillator, and as background for that we need to review the classical treatment of a harmonic oscillator. The simplest example of a harmonic oscillator is a mass connected to a wall by means of an idealized spring, in the absence of gravity. As shown in Fig. 9.5*a*, the displacement of the mass is shown by its x coordinate, and the origin of the coordinate system is taken at the equilibrium position. The mass oscillates about its equilibrium position, and the motion is said to be **harmonic** if the force F due to the spring is directly proportional to the displacement x from its equilibrium position x_{eq}, which we can define as the origin of the x axis:

$$F = -kx \tag{9.101}$$

*C. Cohen-Tannoudjii, B. Diu, and F. Laloë, *Quantum Mechanics*. Hoboken, NJ: Wiley, 1977.

The negative sign comes from the fact that F is opposite to the displacement x. The proportionality constant k, referred to as a **force constant,** is small for a weak spring and large for a stiff spring. Since force is expressed by mass times acceleration, the equation for motion in the x direction is

$$m\frac{\mathrm{d}^2x}{\mathrm{d}t^2} + kx = 0 \tag{9.102}$$

The general solution of this differential equation is (compare with equation 9.61)

$$x(t) = A\sin\omega t + B\cos\omega t \tag{9.103}$$

where $\omega = (k/m)^{1/2}$ is the **angular frequency** in radians per second. If we initially stretch the spring so that the mass is at position x_0 and its velocity is zero, and then let go, the time course of the motion is represented by

$$x(t) = x_0\cos\omega t \tag{9.104}$$

The mass oscillates between x_0 and $-x_0$ with a frequency of ω radians per second or $\nu = \omega/2\pi$ cycles per second.

The energy of a harmonic oscillator is equal to the sum of its potential energy and its kinetic energy. When the mass is at $+x_0$ or $-x_0$, the energy is all potential energy, and as the mass goes through $x = 0$, the energy is all kinetic energy. We can calculate the potential energy V from the fact that the force is the negative derivative of the potential energy

$$F = -\frac{\mathrm{d}V}{\mathrm{d}x} \tag{9.105}$$

so that

$$V = -\int F\,\mathrm{d}x + \text{constant} \tag{9.106}$$

Since $F = -kx$, integration of equation 9.106 yields

$$V = \frac{kx^2}{2} \tag{9.107}$$

if we take the potential to be zero when $x = 0$. This is the equation for a parabola. Substituting equation 9.104 yields

$$V = \frac{kx_0^2}{2}\cos^2\omega t \tag{9.108}$$

Thus, the potential energy of the harmonic oscillator varies between zero and $kx_0^2/2$ during each period of oscillation.

The kinetic energy of the moving mass is

$$T = \tfrac{1}{2}m\left(\frac{\mathrm{d}x}{\mathrm{d}t}\right)^2 \tag{9.109}$$

Using equation 9.104 yields

$$\begin{aligned} T &= \tfrac{1}{2}m\omega^2x_0^2\sin^2\omega t \\ &= \tfrac{1}{2}kx_0^2\sin^2\omega t \end{aligned} \tag{9.110}$$

where the second form has been obtained using $\omega = (k/m)^{1/2}$.

The total energy is

$$\begin{aligned} E &= T + V \\ &= \tfrac{1}{2}kx_0^2 \sin^2 \omega t + \tfrac{1}{2}kx_0^2 \cos^2 \omega t \\ &= \tfrac{1}{2}kx_0^2 \end{aligned} \tag{9.111}$$

Thus, the total energy is constant. The potential energy and the kinetic energy each oscillate between zero and a maximum value of $\frac{1}{2}kx_0^2$, but the total energy is conserved. The harmonic oscillator is a conservative system, and it may be shown that any system for which the force can be expressed as the derivative of a potential energy is **conservative.**

Example 9.17 ***Two masses connected by a spring exhibit harmonic motion***

Show that equation 9.102 also applies to mass m_1 connected to mass m_2 by a spring, as shown by Fig. 9.5*b*.

When two masses are connected with a spring, there are two equations of motion:

$$m_1 \frac{d^2 x_1}{dt^2} = k(x_2 - x_1 - l_0) \tag{a}$$

$$m_2 \frac{d^2 x_2}{dt^2} = -k(x_2 - x_1 - l_0) \tag{b}$$

where l_0 is the equilibrium length of the spring. Note that the force on m_1 is equal and opposite to the force on m_2, as required by Newton's law that reaction is equal and opposite to action.

The oscillatory motion of the idealized diatomic molecule in Fig. 9.5*b* depends only on the relative coordinate $x = x_2 - x_1 - l_0$. The equation of motion in terms of that coordinate is obtained by dividing equation a by m_1 and adding the negative of it to equation b divided by m_2. This yields

$$\frac{d^2(x_2 - x_1)}{dt^2} = -k\left(\frac{1}{m_1} + \frac{1}{m_2}\right)(x_2 - x_1 - l_0) \tag{c}$$

$$\frac{d^2 x}{dt^2} = -k\left(\frac{1}{m_1} + \frac{1}{m_2}\right)x \tag{d}$$

where $x = x_2 - x_1 - l_0$. The mass term can be taken equal to the reciprocal of a **reduced mass** μ defined by

$$\frac{1}{\mu} = \frac{1}{m_1} + \frac{1}{m_2} \quad \text{or} \quad \mu = \frac{m_1 m_2}{m_1 + m_2} \tag{e}$$

so that

$$\mu \frac{d^2 x}{dt^2} + kx = 0 \tag{f}$$

This equation is of the same form as equation 9.102 with m replaced by the reduced mass and x being interpreted as the relative coordinate. This change of coordinates reduces the two-body problem to two one-body problems (one for the center of mass and one for the relative motion).

Since equation 9.103 shows that $\omega = (k/m)^{1/2}$, we will take the fundamental vibration frequency for a diatomic molecule to be

$$\omega = \left(\frac{k}{\mu}\right)^{1/2} \tag{9.112}$$

or

$$\nu = \frac{1}{2\pi}\left(\frac{k}{\mu}\right)^{1/2} \tag{9.113}$$

9.10 QUANTUM MECHANICAL HARMONIC OSCILLATOR

To obtain the energy levels for the quantum mechanical harmonic oscillator of mass μ we start with the classical Hamiltonian function found from equations 9.102, 9.109, and 9.112,

$$H = p_x^2/2\mu + \tfrac{1}{2}\omega^2\mu x^2 \tag{9.114}$$

and convert it to the quantum mechanical Hamiltonian operator by replacing x^2 by $\hat{x}^2$ and p_x by $\hat{p}_x = -\mathrm{i}\hbar\,\mathrm{d}/\mathrm{d}x$ to obtain

$$\hat{H} = -\frac{\hbar^2}{2\mu}\frac{\mathrm{d}^2}{\mathrm{d}x^2} + 2\pi^2\nu^2\mu\hat{x}^2 \tag{9.115}$$

where we have put $\omega = 2\pi\nu$. The solution of the Schrödinger equation is too complicated to discuss here in detail, but when it is solved it is found that there are well-behaved solutions only if the harmonic oscillator has energies given by

$$E_v = (v + \tfrac{1}{2})h\nu \qquad v = 0, 1, 2, \ldots \tag{9.116}$$

where v is the **vibrational quantum number** and $\nu = (1/2\pi)(k/\mu)^{1/2}$. The energy levels are equally spaced with a separation of $h\nu$. These levels are shown in Fig. 9.6*b*, *c*. It is especially important to notice that the energy of the ground state is not zero, as it is classically, but $E_0 = h\nu/2$. This **zero-point energy** is in accordance with the Heisenberg uncertainty principle, as shown later in Example 9.18.

The wavefunctions for the first two energy levels are given by

$$\psi_0 = \left(\frac{\alpha}{\pi}\right)^{1/4} \mathrm{e}^{-\alpha x^2/2} \tag{9.117}$$

$$\psi_1 = \left(\frac{4\alpha^3}{\pi}\right)^{1/4} x\,\mathrm{e}^{-\alpha x^2/2} \tag{9.118}$$

where $\alpha = (k\mu/\hbar^2)^{1/2}$. The wavefunction for the ground state has the shape of the Gaussian probability function.

In general, the wavefunctions for the harmonic oscillator are given by

$$\psi_v(x) = N_v H_v(\alpha^{1/2}x)\,\mathrm{e}^{-\alpha x^2/2} \tag{9.119}$$

$$N_v = \frac{1}{(2^v v!)^{1/2}}\left(\frac{\alpha}{\pi}\right)^{1/4} \tag{9.120}$$

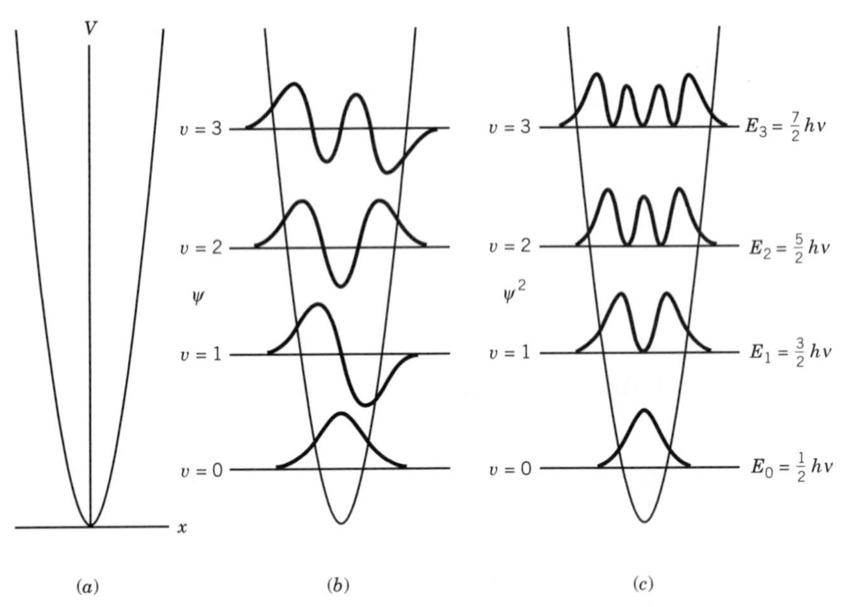

Figure 9.6 (*a*) Potential energy curve for a classical harmonic oscillator. (*b*) Allowed energy levels and wavefunctions for a quantum mechanical harmonic oscillator. (*c*) Probability density functions for a quantum mechanical harmonic oscillator. (See Computer Problems 9.B and 9.C.)

and the $H_v(\alpha^{1/2}x)$ are polynomials called **Hermite polynomials.** The wavefunctions for the first four levels are plotted in Fig. 9.6*b*, and the probability densities $|\psi_v|^2$ are plotted in Fig. 9.6*c*.

The wavefunctions given above have been normalized. The probability that the x coordinate of the harmonic oscillator is between x and $x + \mathrm{d}x$ is given by $\psi^2\,\mathrm{d}x$. If a large number of identically prepared systems are examined, the fraction having coordinates between x and $x + \mathrm{d}x$ is equal to this probability. The **probability densities** ψ^2 are plotted versus x for the first four energy levels in Fig. 9.6*c*. In the ground state ($v = 0$) the most probable distance occurs at the position of the minimum in the potential well. This is in distinct contrast with that for a classical simple harmonic oscillator, which would spend the longest times at the turning points where the velocity goes to zero. As the quantum number increases, however, the quantum mechanical probability density function approaches that for the classical harmonic oscillator. This is an example of Bohr's "correspondence principle," according to which the quantum mechanical result must approach the classical result in the limit of very large quantum numbers.

It is of special interest to note in Fig. 9.6*c* that the quantum mechanical oscillator has a certain probability of being at a greater distance than is allowed for the classical harmonic oscillator at the same energy. For the classical oscillator to be outside the parabolic potential curve, the kinetic energy would have to be negative (since $V > E$ in that region). In quantum mechanics, a particle can have a nonzero probability of being in the classically forbidden region. The probability of being outside the classical turning points given by the parabolic potential energy curve is 0.16 for the ground state. When a particle passes through a region that is classically prohibited, the process is called **tunneling** (Section 9.15). Tunneling is important in determining the rates of chemical reactions involving the transfer of

e^-, H^+, and H (Section 9.15), especially at low temperatures. Electron tunneling occurs in the scanning tunneling microscope (STM), which makes possible atomic resolution of surfaces (see Section 24.6).

The Hermite polynomials, which were known before the development of quantum mechanics, for the first four levels of the harmonic oscillator are as follows:

$$H_0(\xi) = 1 \tag{9.121}$$

$$H_1(\xi) = 2\xi \tag{9.122}$$

$$H_2(\xi) = 4\xi^2 - 2 \tag{9.123}$$

$$H_3(\xi) = 8\xi^3 - 12\xi \tag{9.124}$$

Where ξ is a dummy variable. A property of these polynomials is that $H_v(\xi)$ is an even function of ξ if v is even and odd if v is odd. An even function is a function that satisfies

$$f(x) = f(-x) \qquad \text{(even)} \tag{9.125}$$

and an odd function is a function that satisfies

$$f(x) = -f(-x) \qquad \text{(odd)} \tag{9.126}$$

An even function such as x^2 or $\cos x$ is symmetric when reflected across the x axis, and an odd function such as x or $\sin x$ changes sign. Since $\exp(-\alpha x^2/2)$ in the wavefunctions for the harmonic oscillator is even, the even–odd character of the wavefunctions is determined by the Hermite polynomials. Thus, the harmonic oscillator wavefunctions are even when v is even and odd when v is odd. This even–odd character makes it easy to evaluate integrals, as shown in the following example.

Example 9.18 ***Harmonic oscillator eigenfunctions obey the uncertainty principle***

Show that the probability density for the zero-point level of the harmonic oscillator is in accord with the Heisenberg uncertainty principle.

To calculate the standard deviation σ_x for the x coordinate using equation 9.58, we need to calculate $\langle x \rangle$ and $\langle x^2 \rangle$:

$$\langle x \rangle = \int_{-\infty}^{\infty} \psi_0^* x \psi_0 \, dx = 0$$

Since the wavefunction is real, $\psi_0^* = \psi_0$. Since ψ_0^2 is an even function and x is an odd function, the integrand is odd and so this integral is zero.

$$\begin{aligned}\langle x^2 \rangle &= \int_{-\infty}^{\infty} \psi_0^* x^2 \psi_0 \, dx \\ &= \left(\frac{\alpha}{\pi}\right)^{1/2} \int_{-\infty}^{\infty} x^2 e^{-\alpha x^2} \, dx \\ &= \left(\frac{\alpha}{\pi}\right)^{1/2} \left[\frac{1}{2\alpha}\left(\frac{\pi}{\alpha}\right)^{1/2}\right] \\ &= \frac{1}{2\alpha} = \frac{1}{2}\frac{\hbar}{(\mu k)^{1/2}}\end{aligned}$$

To calculate the standard deviation σ_p for the momentum using equation 9.58, we must calculate $\langle p \rangle$ and $\langle p^2 \rangle$:

$$\langle p \rangle = \int_{-\infty}^{\infty} \psi_0^* \left(-\mathrm{i}\hbar \frac{\mathrm{d}}{\mathrm{d}x}\right)\psi_0 \, \mathrm{d}x = 0$$

Since ψ_0 is an even function, its derivative is odd. Therefore, the integrand is odd, and the above integral is zero.

$$\begin{aligned}
\langle p^2 \rangle &= \int_{-\infty}^{\infty} \psi_0^* \left(-\hbar^2 \frac{\mathrm{d}^2}{\mathrm{d}x^2}\right)\psi_0 \, \mathrm{d}x \\
&= \left(\frac{\alpha}{\pi}\right)^{1/2} \int_{-\infty}^{\infty} \mathrm{e}^{-\alpha x^2/2} \left(-\hbar^2 \frac{\mathrm{d}^2}{\mathrm{d}x^2}\right) \mathrm{e}^{-\alpha x^2/2} \mathrm{d}x \\
&= \hbar^2 \left(\frac{\alpha}{\pi}\right)^{1/2} \int_{-\infty}^{\infty} (\alpha - \alpha^2 x^2) \mathrm{e}^{-\alpha x^2} \mathrm{d}x \\
&= \frac{\hbar^2 \alpha}{2} = \frac{\hbar(\mu k)^{1/2}}{2}
\end{aligned}$$

Using equation 9.58, $\Delta x = \langle x^2 \rangle^{1/2}$, and $\Delta p_x = \langle p^2 \rangle^{1/2}$, we obtain

$$\Delta x \, \Delta p_x = \frac{\hbar}{2}$$

which is in accord with the Heisenberg uncertainty principle and shows that, in its ground state, the harmonic oscillator has the minimum value of $\Delta x \ \Delta p_x$ allowed by the uncertainty principle.

Example 9.19 ***The standard deviations of a bond length can be calculated using harmonic oscillator wavefunctions***

(*a*) Derive the expression for the standard deviation of the bond length of a diatomic molecule when it is in its ground state. (*b*) What percentage of the equilibrium bond length is this standard deviation for carbon monoxide in its ground state? For $^{12}C^{16}O$, $\tilde{\nu} = 2170 \text{ cm}^{-1}$ and $R_e = 113$ pm.

(*a*) Since $\langle x \rangle = 0$ for a harmonic oscillator, the standard deviation or root-mean-square displacement $\sigma_x = \langle x^2 \rangle^{1/2}$ in the $v = 0$ level is given by

$$\sigma_x = \frac{\hbar^{1/2}}{(4\mu k)^{1/4}}$$

In considering spectroscopic data it is more convenient to write this equation in terms of the fundamental vibration frequency in wave numbers $\tilde{\nu}$ rather than k. Make this conversion using

$$\nu = \frac{1}{2\pi}\left(\frac{k}{\mu}\right)^{1/2}$$

where ν is the vibration frequency. Since $\nu = c\,\tilde{\nu}$,

$$k = (2\pi c\,\tilde{\nu})^2 \mu$$

this is used to eliminate the force constant k from the expression for σ_x to obtain

$$\sigma_x = \left(\frac{\hbar}{4\pi c\,\tilde{\nu}\mu}\right)^{1/2}$$

(b) The reduced mass for $^{12}C^{16}O$ is given by

$$\mu = \frac{m_1 m_2}{m_1 + m_2} = \frac{(12 \times 10^{-3}\ \text{kg mol}^{-1})(15.995 \times 10^{-3}\ \text{kg mol}^{-1})}{[(12 + 15.995) \times 10^{-3}\ \text{kg mol}^{-1}](6.022 \times 10^{23}\ \text{mol}^{-1})}$$
$$= 1.139 \times 10^{-26}\ \text{kg}$$

The standard deviation of the bond length for CO is given by

$$\sigma_x = \left[\frac{1.055 \times 10^{-34}\ \text{J s}}{4\pi(2.998 \times 10^{10}\ \text{cm s}^{-1})(2170\ \text{cm}^{-1})(1.139 \times 10^{-26}\ \text{kg})}\right]^{1/2}$$
$$= 3.37\ \text{pm}$$

Since the equilibrium bond length is 113 pm, the standard deviation is 2.98% of this average internuclear separation.

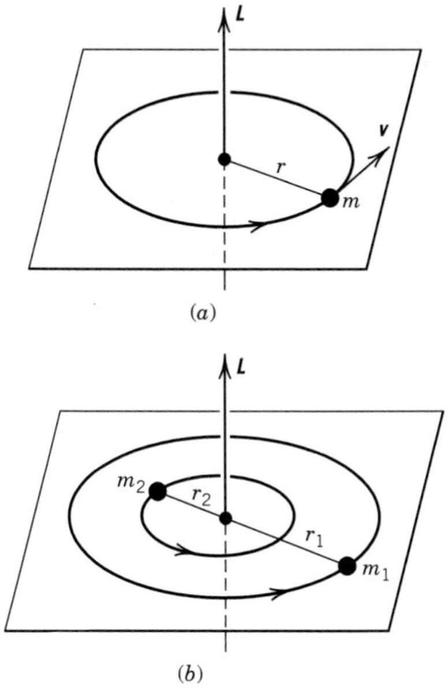

9.11 THE RIGID ROTOR

A particle rotating around a fixed axis as shown in Fig. 9.7a has angular momentum and rotational kinetic energy. The kinetic energy of the revolving particle is given by $T = \frac{1}{2}mv^2 = p^2/2m$, but it is more convenient to express the kinetic energy in terms of the angular velocity ω. If the particle is rotating about a fixed point at a radius r with a frequency of ν, the velocity of the particle is given by

$$v = 2\pi r\nu = r\omega \tag{9.127}$$

where the angular velocity $\omega = \mathrm{d}\theta/\mathrm{d}t$ is equal to $2\pi\nu$. The frequency has units of s^{-1} or Hz. The angular velocity is expressed in radians per second.

The kinetic energy T of a particle in circular motion about a fixed point is usually expressed in terms of the angular velocity or, as we will soon see, in terms of angular momentum:

$$T = \tfrac{1}{2}mv^2 = \tfrac{1}{2}mr^2\omega^2 = \tfrac{1}{2}I\omega^2 \tag{9.128}$$

The **moment of inertia** I, which is introduced in this equation, is defined by

$$I = mr^2 \tag{9.129}$$

for the rotation of a classical particle about an axis. In the expression for the rotational kinetic energy, the moment of inertia plays the role that mass plays in the expression for the kinetic energy of the linear motion, and the angular velocity ω plays the role of the linear velocity v. This suggests that the **angular momentum** L should be defined as

$$L = I\omega = mvr = pr \tag{9.130}$$

so that the kinetic energy of rotational motion can be expressed in terms of the angular momentum,

$$T = \tfrac{1}{2}I\omega^2 = \frac{L^2}{2I} \tag{9.131}$$

Figure 9.7 (a) Rotation of a mass about a fixed point. (b) Rotation of a diatomic molecule about its center of mass.

just as translational kinetic energy can be expressed in terms of mass and linear momentum. If no torque is applied, angular momentum is conserved.

Figure 9.7*b* shows a model of a rigid diatomic molecule that is an example of a rigid rotor. The two masses rotate about their center of mass, satisfying the condition

$$r_1 m_1 = r_2 m_2 \tag{9.132}$$

where r_1 is the distance of m_1 from the center of mass and r_2 is the distance of m_2 from the center of mass. The equilibrium distance R between the nuclei is $R = r_1 + r_2$, so that

$$r_1 = \frac{m_2}{m_1 + m_2} R \qquad \text{and} \qquad r_2 = \frac{m_1}{m_1 + m_2} R \tag{9.133}$$

The rotational kinetic energy is

$$\begin{aligned} T &= \tfrac{1}{2} m_1 r_1^2 \omega^2 + \tfrac{1}{2} m_2 r_2^2 \omega^2 \\ &= \tfrac{1}{2}(m_1 r_1^2 + m_2 r_2^2)\omega^2 \\ &= \tfrac{1}{2} I \omega^2 \end{aligned} \tag{9.134}$$

where I is the moment of inertia of the diatomic molecule:

$$I = m_1 r_1^2 + m_2 r_2^2 \tag{9.135}$$

Using equation 9.133 to eliminate r_1 and r_2, we obtain

$$\begin{aligned} I &= \frac{m_1 m_2}{m_1 + m_2} R^2 \\ &= \mu R^2 \end{aligned} \tag{9.136}$$

where μ is the reduced mass (cf. Example 9.17):

$$\frac{1}{\mu} = \frac{1}{m_1} + \frac{1}{m_2} = \frac{m_1 + m_2}{m_1 m_2} \tag{9.137}$$

The rotational kinetic energy of a diatomic molecule may also be written in terms of the angular momentum L,

$$T = \frac{L^2}{2I} = \frac{L^2}{2\mu R^2} \tag{9.138}$$

by using equation 9.136. Because there is no potential energy, the classical Hamiltonian of a rigid rotor is just the kinetic energy. Applying the correspondence between the classical kinetic energy and the quantum mechanical Hamiltonian operator, we obtain (see Table 9.1)

$$\hat{H} = -\frac{\hbar^2}{2\mu}\nabla^2 \tag{9.139}$$

for rotational energy, where the Laplacian operator ∇^2 was introduced in equation 9.73. In discussing rotation it is more convenient to use the spherical coordinates defined in Fig. 9.8. In spherical coordinates the Laplacian operator is

$$\nabla^2 = \frac{1}{R^2}\frac{\partial}{\partial R}\left(R^2 \frac{\partial}{\partial R}\right) + \frac{1}{R^2 \sin^2\theta}\frac{\partial^2}{\partial \phi^2} + \frac{1}{R^2 \sin\theta}\frac{\partial}{\partial \theta}\left(\sin\theta \frac{\partial}{\partial \theta}\right) \tag{9.140}$$

Table 9.2 First Several Spherical Harmonics

$Y_0^0 = \frac{1}{(4\pi)^{1/2}}$	$Y_2^0 = \left(\frac{5}{16\pi}\right)^{1/2}(3\cos^2\theta - 1)$
$Y_1^0 = \left(\frac{3}{4\pi}\right)^{1/2}\cos\theta$	$Y_2^1 = \left(\frac{15}{8\pi}\right)^{1/2}\sin\theta\cos\theta\, e^{i\phi}$
$Y_1^1 = \left(\frac{3}{8\pi}\right)^{1/2}\sin\theta\, e^{i\phi}$	$Y_2^{-1} = \left(\frac{15}{8\pi}\right)^{1/2}\sin\theta\cos\theta\, e^{-i\phi}$
$Y_1^{-1} = \left(\frac{3}{8\pi}\right)^{1/2}\sin\theta\, e^{-i\phi}$	$Y_2^2 = \left(\frac{15}{32\pi}\right)^{1/2}\sin^2\theta\, e^{2i\phi}$
	$Y_2^{-2} = \left(\frac{15}{32\pi}\right)^{1/2}\sin^2\theta\, e^{-2i\phi}$

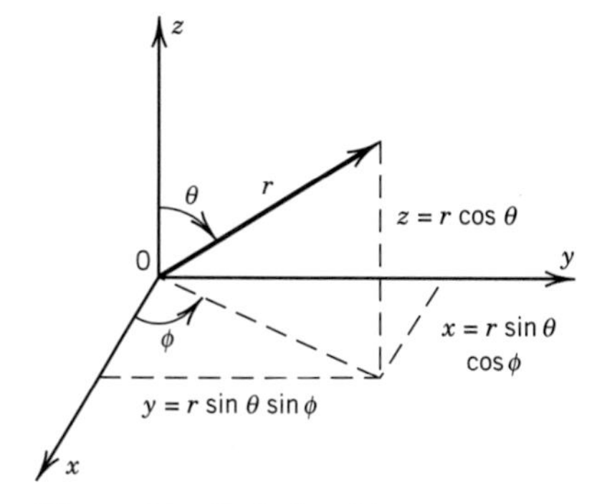

Figure 9.8 Relationship between Cartesian coordinates (x, y, z) and spherical coordinates (r, θ, ϕ). The differential volume in this coordinate system is $d\tau = r^2 \sin\theta\, dr\, d\theta\, d\phi$.

Since the two masses of the rigid rotor are at fixed distances from the origin, R is constant and so we can ignore the derivatives with respect to R in ∇^2. Substitution of equation 9.140 into equation 9.139 yields

$$\hat{H} = -\frac{\hbar^2}{2I}\left[\frac{1}{\sin\theta}\frac{\partial}{\partial\theta}\left(\sin\theta\frac{\partial}{\partial\theta}\right) + \frac{1}{\sin^2\theta}\frac{\partial^2}{\partial\phi^2}\right] \tag{9.141}$$

where $I = \mu R^2$. The rigid rotor wavefunction is a function of the two angles θ and ϕ, and so the eigenvalue problem to be solved is

$$-\frac{\hbar^2}{2I}\left[\frac{1}{\sin\theta}\frac{\partial}{\partial\theta}\left(\sin\theta\frac{\partial}{\partial\theta}\right) + \frac{1}{\sin^2\theta}\frac{\partial^2}{\partial\phi^2}\right]Y(\theta,\phi) = EY(\theta,\phi) \tag{9.142}$$

This equation is a standard differential equation, whose solutions $Y(\theta,\phi)$ are called **spherical harmonics.** The first several spherical harmonics are given in Table 9.2. Two quantum numbers, ℓ and m, arise in the solution of this eigenvalue equation, and so the wavefunctions are represented by $Y_\ell^m(\theta,\phi)$. It is found that

$$\hat{H}\, Y_\ell^m(\theta,\phi) = \frac{\ell(\ell+1)\hbar^2}{2I} Y_\ell^m(\theta,\phi) \tag{9.143}$$

so that the rigid rotor can have only the energies given by

$$E = \frac{\ell(\ell+1)\hbar^2}{2I} \qquad \ell = 0, 1, 2, \ldots \tag{9.144}$$

where ℓ is the **angular momentum quantum number.** Note there is no zero-point energy for rotation. Unlike the harmonic oscillator the uncertainty principle can be satisfied even when the rotational energy is zero, because that wavefunction (Y_0^0) gives equal probability for all θ and ϕ (i.e., maximum uncertainty in those angles). Note that the possible energies are independent of m. The degeneracies of the rotational levels will be discussed in the next section.

9.12 ANGULAR MOMENTUM

So far we have neglected the fact that the angular momentum is a vector and has components in the x, y, and z directions. To develop the quantum mechanical operators for the angular momentum in the x, y, and z directions, we need to review the classical expressions for angular momentum in three dimensions. In the preceding section we saw that the rotational energy of a diatomic molecule may be expressed in terms of its angular momentum. In the next chapter we will find that an electron may have angular momentum. We will also find that electrons and certain nuclei have intrinsic (spin) angular momentum. We will not discuss spin angular momentum here, but will leave that to Section 10.5. Angular momentum is a very important property because it is conserved for many systems we will be interested in.

In classical mechanics the angular momentum of a particle rotating about a fixed point is represented by a vector $\boldsymbol{L}$ in the direction perpendicular to the plane of the circular motion. If a mass m is rotating about a fixed point with linear velocity $\boldsymbol{v}$, the **angular momentum** $\boldsymbol{L}$ is given by the cross product of the radius $\boldsymbol{r}$ and the linear momentum vector $\boldsymbol{p}$:

$$\boldsymbol{L} = \boldsymbol{r} \times m\boldsymbol{v} = \boldsymbol{r} \times \boldsymbol{p} \tag{9.145}$$

The cross product of the two vectors $\boldsymbol{r}$ and $\boldsymbol{p}$ is a vector of magnitude $|\mathbf{r}||\mathbf{p}|\sin\theta$, where θ is the angle between $\boldsymbol{r}$ and $\boldsymbol{p}$, having the direction that a right-hand screw would travel as $\boldsymbol{r}$ is rotated to $\boldsymbol{p}$. Thus, the angular momentum vector for the circular motion shown in Fig. 9.7*a* points up.

The vectors $\boldsymbol{r}$ and $\boldsymbol{p}$ may be expressed in terms of their components and unit vectors $\boldsymbol{i}$, $\boldsymbol{j}$, and $\boldsymbol{k}$ pointing along the positive x, y, and z axes:

$$\boldsymbol{r} = x\boldsymbol{i} + y\boldsymbol{j} + z\boldsymbol{k} \tag{9.146}$$

$$\boldsymbol{p} = p_x\boldsymbol{i} + p_y\boldsymbol{j} + p_z\boldsymbol{k} \tag{9.147}$$

The cross product of $\boldsymbol{r}$ and $\boldsymbol{p}$ may conveniently be calculated from a determinant (see Appendix D.6):

$$\begin{aligned} \boldsymbol{L} = \boldsymbol{r} \times \boldsymbol{p} &= \begin{vmatrix} \boldsymbol{i} & \boldsymbol{j} & \boldsymbol{k} \\ x & y & z \\ p_x & p_y & p_z \end{vmatrix} \\ &= (yp_z - zp_y)\boldsymbol{i} + (zp_x - xp_z)\boldsymbol{j} + (xp_y - yp_x)\boldsymbol{k} \end{aligned} \tag{9.148}$$

Thus, the three components of the classical angular momentum of a particle rotating about a fixed point are

$$L_x = yp_z - zp_y \tag{9.149}$$

$$L_y = zp_x - xp_z \tag{9.150}$$

$$L_z = xp_y - yp_x \tag{9.151}$$

The square of the angular momentum is given by the scalar product of $\boldsymbol{L}$ with itself:

$$\boldsymbol{L} \cdot \boldsymbol{L} = L^2 = L_x^2 + L_y^2 + L_z^2 \tag{9.152}$$

The square of the angular momentum is a scalar (see the vector discussion in Appendix D.7). If no torque acts on a particle, its angular momentum is constant (conserved). In classical mechanics all possible values of L and E are permitted.

The quantum mechanical operators for the angular momentum are obtained by replacing the quantities in equations 9.149–9.151 with their corresponding quantum mechanical operators—specifically, $\hat{p}_x = (-i\hbar)(\partial/\partial x)$, and so on.

$$\hat{L}_x = -i\hbar\left(y\frac{\partial}{\partial z} - z\frac{\partial}{\partial y}\right) \tag{9.153}$$

$$\hat{L}_y = -i\hbar\left(z\frac{\partial}{\partial x} - x\frac{\partial}{\partial z}\right) \tag{9.154}$$

$$\hat{L}_z = -i\hbar\left(x\frac{\partial}{\partial y} - y\frac{\partial}{\partial x}\right) \tag{9.155}$$

The operator for the square of the angular momentum is given by

$$\hat{L}^2 = |\hat{L}|^2 = \hat{L}\cdot\hat{L} = \hat{L}_x^2 + \hat{L}_y^2 + \hat{L}_z^2 \tag{9.156}$$

It is often more convenient to use the angular momentum operators given in equations 9.153–9.155 in spherical coordinates r, θ, ϕ, which are defined in Fig. 9.8. In these new coordinates,

$$\hat{L}_x = i\hbar\left(\sin\phi\frac{\partial}{\partial\theta} + \cot\theta\cos\phi\frac{\partial}{\partial\phi}\right) \tag{9.157}$$

$$\hat{L}_y = i\hbar\left(-\cos\phi\frac{\partial}{\partial\theta} + \cot\theta\sin\phi\frac{\partial}{\partial\phi}\right) \tag{9.158}$$

$$\hat{L}_z = -i\hbar\frac{\partial}{\partial\phi} \tag{9.159}$$

$$\hat{L}^2 = -\hbar^2\left[\frac{1}{\sin\theta}\frac{\partial}{\partial\theta}\left(\sin\theta\frac{\partial}{\partial\theta}\right) + \frac{1}{\sin^2\theta}\frac{\partial^2}{\partial\phi^2}\right] \tag{9.160}$$

Note that the terms in ∇^2 that depend on the angle (equation 9.140) are equal to $-(1/\hbar^2R^2)\hat{L}^2$.

It is readily shown that $\hat{L}_x$ and $\hat{L}_y$, $\hat{L}_y$ and $\hat{L}_z$, and $\hat{L}_x$ and $\hat{L}_z$ do not commute with each other, but $\hat{L}_x$, $\hat{L}_y$, and $\hat{L}_z$ all commute with $\hat{L}^2$. Therefore, we can measure precisely the square of total angular momentum and one, but only one, of its components. Thus, if the magnitude of the total angular momentum $|\boldsymbol{L}| = \sqrt{L^2} = \sqrt{L_x^2 + L_y^2 + L_z^2}$ is measured and L_z is measured, it is not possible to measure L_x or L_y precisely. That is, the eigenfunction of L^2 is also an eigenfunction of L_z but it is not an eigenfunction of L_x or L_y since neither L_x nor L_y commutes with L_z. This is an essential difference between classical and quantum mechanical systems and is in accord with the Heisenberg uncertainty principle.

Since $\hat{L}^2$ and $\hat{L}_z$ commute, it is possible to construct a function that is an eigenfunction of both operators. In fact, we have already seen these wavefunctions, the spherical harmonics, in the eigenvalue equation (see equation 9.143) for the energy of the rigid rotor. Since $L^2 = 2IT$ for the classical rigid rotor, the quantum mechanical operator $\hat{L}^2$ is equal to $2I\hat{H}$. Thus, equation 9.143 can be written

$$\hat{L}^2 Y_\ell^m(\theta,\phi) = \ell(\ell+1)\hbar^2 Y_\ell^m(\theta,\phi) \qquad \ell = 0, 1, 2, \ldots \tag{9.161}$$

According to this eigenvalue equation, the total angular momentum for a rigid rotor can only have the values

$$L^2 = \ell(\ell+1)\hbar^2 \qquad \ell = 0, 1, 2, \ldots \tag{9.162}$$

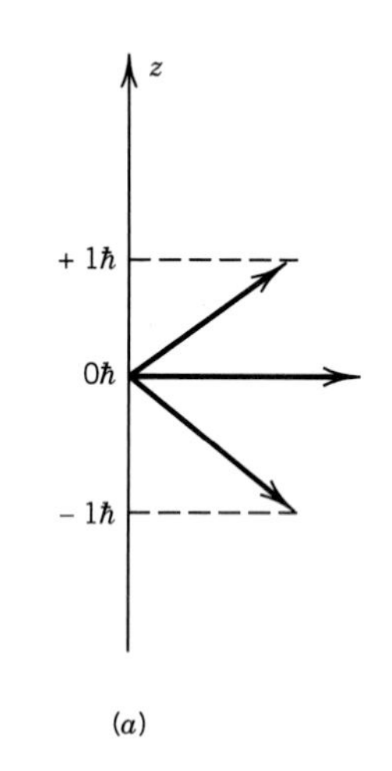

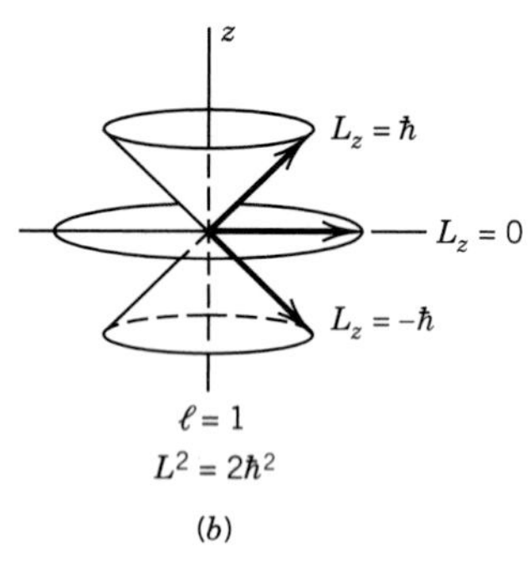

Figure 9.9 (*a*) Possible orientations for angular momentum vector for $\ell = 1$. (*b*) Since the x and y components of angular momentum are indeterminant, the vector can be rotated about the z axis to lie anywhere on the conical surface.

Operating on the spherical harmonics with $\hat{L}_z$,

$$\hat{L}_z Y_\ell^m(\theta,\phi) = m\hbar Y_\ell^m(\theta,\phi) \qquad m = -\ell, -\ell+1, \ldots, \ell-1, \ell \tag{9.163}$$

yields the following eigenvalues for z component L_z of the angular momentum:

$$L_z = m\hbar \qquad m = -\ell, -\ell+1, \ldots, \ell-1, \ell \tag{9.164}$$

where the quantum number ℓ is referred to as the **angular momentum quantum number** and m is referred to as the **magnetic quantum number.** Equations 9.163 and 9.164 will recur a number of times in quantum mechanics. They will appear in connection with the hydrogen atom, and again in several places in spectroscopy. Note that the spherical harmonics are eigenfunctions for two different commuting operators, $\hat{L}^2$ and $\hat{L}_z$.

The possible orientations of the angular momentum vector $\boldsymbol{L}$ with respect to a particular direction are shown in Fig. 9.9 for $\ell = 1$ and in Fig. 9.10 for $\ell = 2$. In the absence of an external electric or magnetic field, the choice of the z axis is entirely arbitrary, but when such a field is applied (either by the experimenter or by placing the particle in a molecule or a crystal) a unique direction is defined that becomes the axis of quantization. Since the L_x and L_y components are unknown, $\boldsymbol{L}$ can be described only as being in the surface of a cone, as illustrated in the figures. The magnitude of the orbital angular momentum $\boldsymbol{L}$ is $[\ell(\ell + 1)]^{1/2}\hbar$, and the *maximum* component L_z in a particular direction is $\ell\hbar$. Thus, the magnitude of the angular momentum $\boldsymbol{L}$ is greater than its z component so that the angular momentum vector cannot point in the direction of an applied magnetic field or along a unique axis.

In the absence of an electric or magnetic field, there is a degeneracy of $2\ell + 1$ since, for an angular momentum ℓ, there are $2\ell + 1$ values of m.

In considering the rotational energy levels of molecules, the rotational quantum number ℓ is denoted by J so that

$$E = \frac{\hbar^2}{2I} J(J+1) \tag{9.165}$$

The square of the total angular momentum is given by

$$L^2 = J(J+1)\hbar^2 \qquad J = 0, 1, 2, \ldots \tag{9.166}$$

The magnetic quantum number m is replaced by M so that $L_z = M\hbar$.

The energy depends only on the quantum number J, but the wavefunction depends on J and M (see equations 9.161 and 9.163). Since the values of M range from $-J$ to J, the rotational levels are $(2J + 1)$-fold degenerate. The degeneracy corresponds to the different possible orientations of the angular momentum vector.

Example 9.20 ***The angular momentum and kinetic energy of a rotor***

Consider a system that has the wavefunction $N \cos\phi$, where N is a normalization constant. What are (*a*) the average angular momentum L_z in the z direction and (*b*) the kinetic energy E of rotation about the z axis?

(*a*)

$$\hat{L}_z = \mathrm{i}\hbar\frac{\partial}{\partial\phi}$$

$$L_z = N^2\int_0^{2\pi}\cos\phi\left(-\mathrm{i}\hbar\frac{\partial}{\partial\phi}\cos\phi\right)\mathrm{d}\phi$$

$$= -N^2\int_0^{2\pi}\cos\phi\sin\phi\,\mathrm{d}\phi = 0$$

(*b*) The operator for the kinetic energy of rotation about the z axis is given by

$$\frac{1}{2I}\hat{L}_z^2 = -\frac{\hbar^2}{2I}\frac{\mathrm{d}^2}{\mathrm{d}\phi^2}$$

Note that $\cos\phi$ is an eigenfunction of $\hat{L}_z^2$ with eigenvalue $\hbar^2$:

$$-\hbar^2\frac{\mathrm{d}^2}{\mathrm{d}\phi^2}N\cos\phi = \hbar^2(N\cos\phi)$$

Therefore,

$$\frac{1}{2I}\hat{L}_z^2\cos\phi = \frac{\hbar^2}{2I}\cos\phi = E_z\cos\phi$$

and

$$E_z = \hbar^2/2I$$

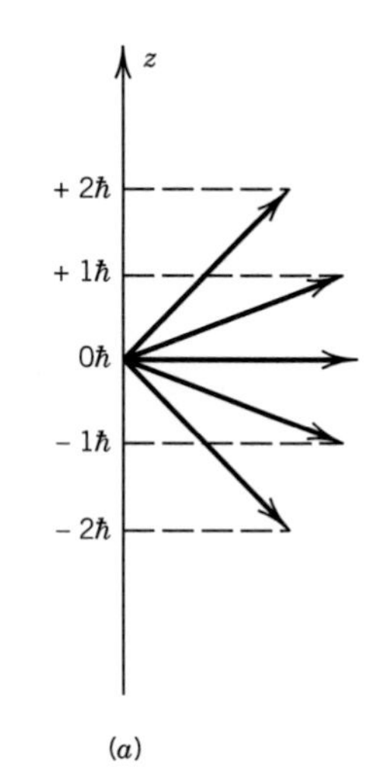

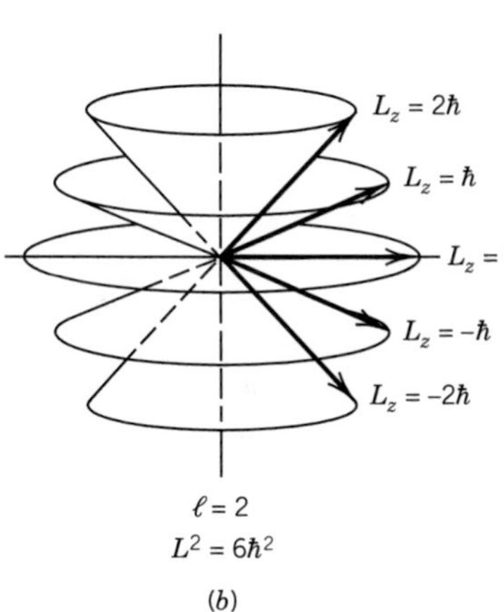

Figure 9.10 (*a*) Possible orientations for angular momentum vector for $\ell = 2$. (*b*) Since the x and y components of angular momentum are indeterminant, the vector can be rotated about the z axis to lie anywhere on the conical surface.

Example 9.21 ***Calculation of the reduced mass and moment of inertia of HCl***

What are the reduced mass and moment of inertia of $H^{35}Cl$? The equilibrium internuclear distance R_e is 127.5 pm. What are the values of L, L_z, and E for the state with $J = 1$? Atomic masses of some isotopes are given inside the back cover.

$$\mu = \frac{(1.007\,825\times10^{-3}\ \mathrm{kg\ mol^{-1}})(34.968\,85\times10^{-3}\ \mathrm{kg\ mol^{-1}})}{[(1.007\,825+34.968\,85)\times10^{-3}\ \mathrm{kg\ mol^{-1}}](6.022\,367\times10^{23}\ \mathrm{mol^{-1}})}$$

$$= 1.626\,65\times10^{-27}\ \mathrm{kg}.$$

$$I = \mu R_e^2$$

$$= (1.626\times10^{-27}\ \mathrm{kg})(127.5\times10^{-12}\ \mathrm{m})^2$$

$$= 2.644\times10^{-47}\ \mathrm{kg\ m^2}$$

$$L = \sqrt{J(J+1)}\,\hbar$$

$$= \frac{\sqrt{2}(6.626\times10^{-34}\ \mathrm{J\ s})}{2\pi}$$

$$= 1.491\times10^{-34}\ \mathrm{J\ s}$$

$$L_z = -\hbar, 0, \hbar$$

$$= -1.054\times10^{-34}\ \mathrm{J\ s}, 0, 1.054\times10^{-34}\ \mathrm{J\ s}$$

$$E = \frac{\hbar^2}{2I}J(J+1)$$

$$= \frac{(6.626\times10^{-34}\ \mathrm{J\ s})^2(2)}{8\pi^2(2.644\times10^{-47}\ \mathrm{kg\ m^2})}$$

$$= 4.206\times10^{-22}\ \mathrm{J}$$

Comment:

We will soon see that angular momentum is one of the most important physical properties of an electron, atom, or molecule. We have seen here that a rotating molecule has quantized angular momentum. In the next chapter, we will see that electrons in orbitals in atoms have quantized angular momentum. Electrons and other subatomic particles have spin, a property with no classical analogy, that follows equations such as 9.164 and 9.166. Spin is important in all kinds of spectroscopy, and it provides the basis for nuclear magnetic resonance and electron spin resonance. Angular momentum is conserved in many circumstances, for example, when a photon is absorbed by an atom or molecule and when two molecules collide.

9.13 POSTULATES OF QUANTUM MECHANICS

A number of postulates of quantum mechanics have been introduced in this chapter as they have been needed. Now is the time to bring them together.

Postulate 1

The state of a quantum mechanical system is completely specified by a wavefunction $\Psi(r, t)$ that is a function of the coordinates of the particles and the time. If time is not a variable, its state is completely specified by a time-independent wavefunction $\psi(\boldsymbol{r})$. These wavefunctions are single-valued, continuous, and square integrable. The wavefunction for a single particle may be interpreted as follows: $\Psi^*(\boldsymbol{r}, t)\Psi(\boldsymbol{r}, t)\, \mathrm{d}x\, \mathrm{d}y\, \mathrm{d}z$ is the probability that the particle is in the volume $\mathrm{d}x\, \mathrm{d}y\, \mathrm{d}z$ located at $\boldsymbol{r}$ at time t.

Postulate 2

For every observable in classical mechanics there is a linear quantum mechanical operator. The operator is obtained from the classical mechanical expression for the observable written in terms of Cartesian coordinates and conjugate momenta by replacing each coordinate q by itself and the conjugate momentum component by $-\mathrm{i}\hbar\, \partial/\partial q$.

Postulate 3

The possible measured values of the physical observable A are the eigenvalues a_i of the equation

$$\hat{A}\Psi_i = a_i \Psi_i \tag{9.167}$$

where $\hat{A}$ is the operator corresponding to the observable.

Postulate 4

If the wavefunction of the system is Φ, the probability of measuring the eigenvalue a_i of observable A is $|\, c_i \,|^2 = |\, \int_{\infty}^{\infty} \Psi_i^* \Phi \, \mathrm{d}\tau \,|^2$.

Postulate 5

The wavefunction of a system changes with time according to

$$\hat{H}\Psi(\boldsymbol{r},t) = \mathrm{i}\hbar\frac{\partial\Psi(\boldsymbol{r},t)}{\partial t} \tag{9.168}$$

where $\hat{H}$ is the Hamiltonian operator for the system.

Postulate 6

The wavefunction of a system of electrons must be antisymmetric to the interchange of any two electrons. We will not discuss this postulate until the next chapter, but it is given here for completeness. This postulate arises in connection with spin and is a more fundamental statement of what is called the Pauli exclusion principle.

It is perhaps worth emphasizing again that we have not derived these postulates; rather, they are to be judged like the laws of classical mechanics or the laws of thermodynamics on the basis of whether they are in accord with experimental results.

9.14 SPECIAL TOPIC: THE TIME-DEPENDENT SCHRÖDINGER EQUATION

To describe the evolution of a one-dimensional quantum mechanical system with time, it is necessary to use the time-dependent Schrödinger equation:

$$\hat{H}\Psi(x,t) = \mathrm{i}\hbar\frac{\partial\Psi}{\partial t} \tag{9.169}$$

In general it is difficult to find solutions to the time-dependent Schrödinger equation. However, it is possible to separate variables for conservative systems, that is, systems in which the potential energy is a function of distance but not of time. For a conservative system the time-dependent Schrödinger equation is

$$\left[-\frac{\hbar^2}{2m}\frac{\partial^2}{\partial x^2} + V(x)\right]\Psi(x,t) = -\frac{\hbar}{\mathrm{i}}\frac{\partial\Psi}{\partial t} \tag{9.170}$$

Special solutions may be found by writing the wavefunction as a product of a function of distance and a function of time:

$$\Psi(x,t) = \psi(x)f(t) \tag{9.171}$$

Substituting this relation into equation 9.170 and dividing by $\psi(x)f(t)$ yields

$$\frac{1}{\psi(x)}\left[-\frac{\hbar^2}{2m}\frac{\mathrm{d}^2}{\mathrm{d}x^2} + V(x)\right]\psi(x) = -\frac{\hbar}{\mathrm{i}}\frac{1}{f(t)}\frac{\mathrm{d}f(t)}{\mathrm{d}t} \tag{9.172}$$

Since the left-hand side of the equation does not depend on t and the right-hand side of the equation does not depend on x, they must each be equal to a constant that is independent of x and t. If we take this constant to be equal to E, we obtain the following two differential equations:

$$\left[-\frac{\hbar^2}{2m}\frac{d^2}{dx_2} + V(x)\right]\psi(x) = E \tag{9.173}$$

$$-\frac{\hbar}{i}\frac{df(t)}{dt} = Ef(t) \tag{9.174}$$

The first is the time-independent Schrödinger equation that we have been using, and we can see that the constant that was introduced is equal to the energy. The second equation is readily integrated to obtain, with the initial condition $\Psi(x, 0) = \psi(x)$,

$$f(t) = e^{-iEt/\hbar} \tag{9.175}$$

which oscillates harmonically in time. Thus, the time-dependent wavefunction for a conservative one-dimensional system in an eigenstate of H is

$$\Psi(x, t) = \psi(x)\,e^{-iEt/\hbar} \tag{9.176}$$

This wavefunction is complex, but the probability $\Psi^*(x, t)\Psi(x, t)$ is real:

$$\begin{aligned}\Psi^*(x, t)\Psi(x, t) &= [\psi(x)\,e^{-iEt/\hbar}]^*[\psi(x)\,e^{-iEt/\hbar}]\\ &= \psi^*(x)\psi(x)\end{aligned} \tag{9.177}$$

Thus, for a conservative system the probability density for an eigenstate of H is independent of time, and we refer to the system as being in a stationary state.

9.15 SPECIAL TOPIC: TUNNELING AND REFLECTION

When discussing the harmonic oscillator, we saw that the wavefunction can be nonzero in classically forbidden regions of space. This is a general property associated with wave phenomena and hence with quantum mechanics. An important consequence of this is that there is a finite probability that a particle can pass through a potential energy barrier in quantum mechanics even though a classical particle would be unable to do so. This is called tunneling and can be illustrated by considering the situation in Fig. 9.11 of a particle of energy E moving in one dimension and encountering a barrier whose height is V such that $V > E$.

In the region to the left of the barrier the Schrödinger equation for the particle is

$$-\frac{\hbar^2}{2m}\frac{d^2}{dx^2}\phi_L = E\phi_L \tag{9.178}$$

with the general solution

$$\phi_L = A\,e^{+ikx} + B\,e^{-ikx} \qquad k^2 = \frac{2mE}{\hbar^2} \tag{9.179}$$

The term A represents the amplitude of the incoming wave, and the term B represents that of the reflected wave. In the barrier ($0 < x < a$), the Schrödinger

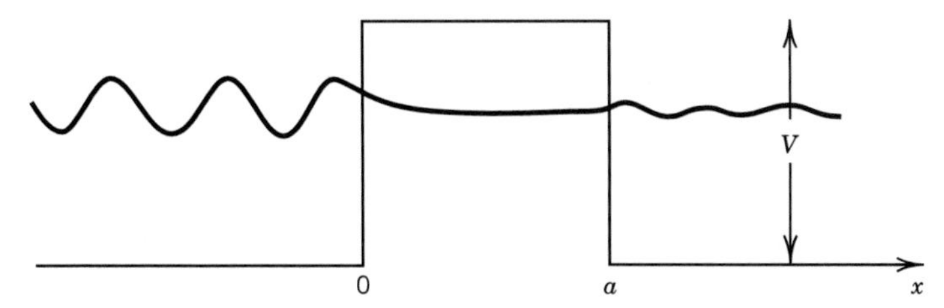

Figure 9.11 Wavefunction for a particle of energy E incident from the left on a potential barrier of height V and width a.

equation is

$$-\frac{\hbar^2}{2m}\frac{d^2}{dx^2}\phi_M + V\phi_M = E\phi_M \tag{9.180}$$

with the general solution

$$\phi_M = A'\,e^{+\kappa x} + B'\,e^{-\kappa x} \qquad \kappa^2 = \frac{2m(V-E)}{\hbar^2} \tag{9.181}$$

To the right of the barrier the wavefunction is

$$\phi_R = F\,e^{+ikx} \tag{9.182}$$

since the particle is incident from the left, so there is no reflected wave here. The amplitude for tunneling through the barrier is then F/A, and the probability of tunneling P_T is then $|F/A|^2$. By making the wavefunction and its derivative continuous at the boundaries of the potential step, we can solve for F/A. For barriers such that $\kappa a \gg 1$, P_T is proportional to $e^{-2\kappa a}$, which is small but nonzero. In Fig. 9.12, the probability of a proton tunneling through a barrier of height 1 eV and width 10 pm is shown.

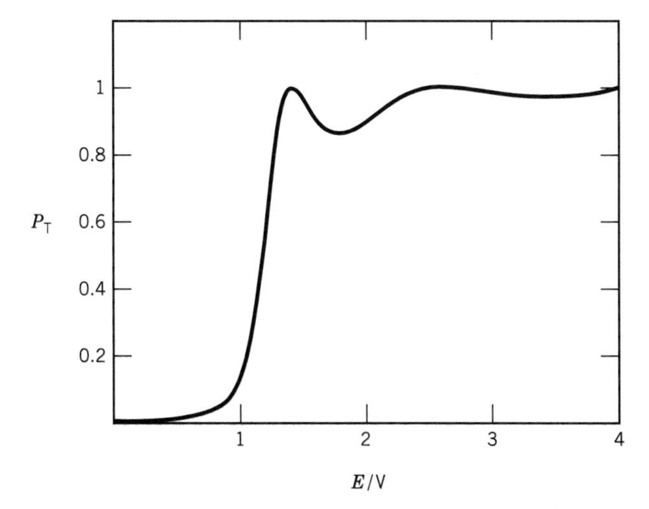

Figure 9.12 Probability of a proton tunneling through a 10-pm-wide potential barrier of height 1 eV.

Tunneling processes are important for light particles such as electrons and protons. They are much less important for heavier particles, since κ is proportional to the square root of mass.

Another quantum mechanical phenomenon is the reflection of particles (or waves) from a barrier even when the energy is higher than the barrier. For particles incident from the left on the barrier of Fig. 9.11, the probability of transmission for $E > V$ is also given in Fig. 9.12.

9.16 SPECIAL TOPIC: BLACKBODY RADIATION

Section 9.1 gave the Planck equation for the energy density $\rho_\nu(\nu, T)$ of blackbody radiation that was based on the assumption that the energy of a photon is given by $h\nu$, but in view of the importance of radiation in chemistry, this Special Topic section is provided to give more information. Note that we have explicitly indicated the dependence of ρ_ν on ν and T. A fundamental quantity is the **total radiant energy density** $\rho(T)$, which is the radiant energy per unit volume (J m^{-3}):

$$\rho(T) = \int_0^\infty \rho_\nu(\nu, T)\,\mathrm{d}\nu = \int_0^\infty \frac{8\pi h(\nu/c)^3\,\mathrm{d}\nu}{e^{h\nu/kT} - 1} \tag{9.183}$$

Thus $\rho_\nu(\nu, T)\,\mathrm{d}\nu$ is the energy density of radiation in the frequency range ν to $\nu + \mathrm{d}\nu$, and it has SI units of J m^{-3} s or J m^{-3} Hz^{-1}. The dependence of the total radiant energy density on temperature was deduced from thermodynamics by Stefan and Boltzmann,* and can easily be derived from equation 9.183 by a change in variable, to yield

$$\rho(T) = \beta T^4 \tag{9.184}$$

where $\beta = 7.56 \times 10^{-16}$ J m^{-3} K^{-4}.

The energy density as a function of wavelength can be calculated from the energy density as a function of frequency by noting that $\nu = c/\lambda$, so $|\mathrm{d}\nu/\mathrm{d}\lambda| = c/\lambda^2$. Therefore,

$$\rho_\lambda(\lambda, T) = \rho_\nu\left(\frac{c}{\lambda}, T\right)\left|\frac{\mathrm{d}\nu}{\mathrm{d}\lambda}\right| = \frac{8\pi hc}{\lambda^5}\frac{1}{e^{hc/\lambda kT} - 1} \tag{9.185}$$

where $\rho_\lambda(\lambda, T)$ has the units J m^{-4}. Thus $\rho_\lambda(\lambda, T)\,\mathrm{d}\lambda$ is the energy density of radiation in the wavelength range λ to $\lambda + \mathrm{d}\lambda$. The radiant energy density as a function of wavelength is plotted in Fig. 9.13. By differentiating $\rho_\lambda(\lambda, T)$ with respect to T, we find

$$\lambda_{\max} T = 2.898 \times 10^{-3}\ \text{K m} \tag{9.186}$$

which is the **Wien displacement law,** known experimentally since the late nineteenth century.

9.17 SPECIAL TOPIC: SUPERPOSITION OF VIBRATIONAL STATES AND WAVE PACKETS

There is a striking example of superposition (Section 9.5) that will be needed in the discussion of femtosecond transition-state spectroscopy in Section 19.10. A fem-

*D. Kondepudi and I. Prigogine, *Modern Thermodynamics*. Hoboken, NJ: Wiley, 1998.

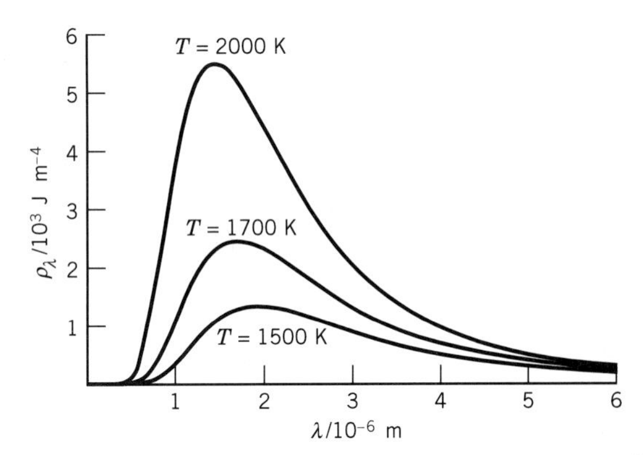

Figure 9.13 Radiant energy density ρ_λ as a function of wavelength at 1500, 1700, and 2000 K. (See Computer Problem 9.F.)

tosecond (fs) is 10^{-15} seconds. Consider the photodissociation of molecular iodine with an ultrashort laser pulse. Pulses with Δt as short as 3 fs can be produced. Short pulses have an energy width ΔE that is given by the Heisenberg uncertainty principle: $\Delta E \geq \hbar/2\Delta t$. This broad energy width of a short pulse might appear to be a disadvantage, but it is actually an advantage in the study of the transition state of a reaction because the irradiated molecule is raised to a superposition of vibrational states coherently (that is, in phase). The fact that a number of levels are excited coherently is extremely important because this leads to interference effects. This superposition of vibrational states propagates back and forth in the parabolic potential well as a **wave packet.** The motion of this wave packet can be observed by use of ultrashort probe pulses, which are slightly delayed from the initial pulse. Figure 9.14 shows the magnitude of the wavefunction and the square of the wavefunction (probability density) for a superposition of vibrational states

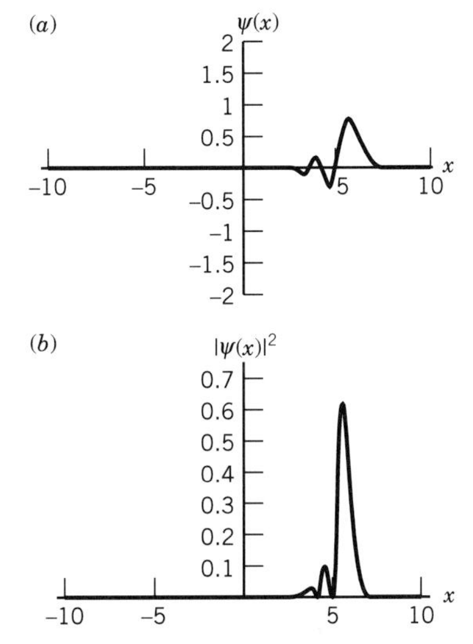

Figure 9.14 (a) Magnitude of the wavefunction and (b) probability density (at a particular instant of time) for the superposition of vibrational states for v = 14 to 22 where the states are weighted according to a Gaussian distribution (see Computer Problem 9.K).

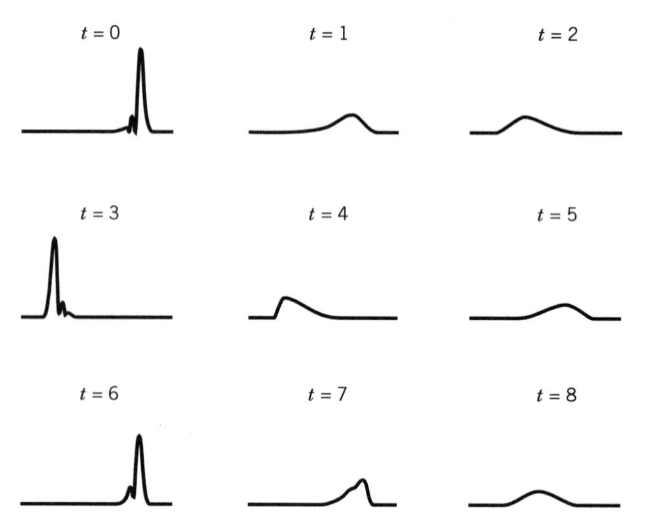

Figure 9.15 Probability densities for a wave packet in a parabolic barrier at a series of times (see Computer Problem 9.K).

for $v = 14$ to 22, where the states are weighted according to a Gaussian distribution (error function) centered at $v = 18$. At this instant in time, the quantum mechanical oscillator is very much like a particle at the right-hand end of its oscillatory motion.

In order to calculate the motion of this particle, we have to use the time-dependent wavefunction for the system:

$$\Psi(x, t) = \sum c_v \psi_v(x) \exp(-iE_v t/\hbar) \tag{9.187}$$

where the summation is over a range of vibrational quantum numbers and the c_v are weighting factors (according to a Gaussian distribution for the case being discussed). The probability density is given by

$$\Psi^*(x, t)\Psi(x, t) = \sum c_v^* \Psi_v^*(x) \exp(iE_v t/\hbar) \sum c_\mu \Psi_\mu(x) \exp(-iE_\mu t/\hbar) \tag{9.188}$$

For the case described in the preceding paragraph, the probability densities at a series of times are given in Fig. 9.15. Thus as time passes, the probability density oscillates from the parabolic barrier on the right to the parabolic barrier on the left and back again like a classical harmonic oscillator. However, when a molecule is excited with an ultrashort pulse to a potential well that allows dissociation, in may oscillate only a few times before it dissociates. Probe pulses can even be used to study what happens on the time scale of the lifetime of a transition state. These studies provide insights into the factors that determine the rate of a chemical reaction, as discussed later in Section 19.10.

Ten Key Ideas in Chapter 9

1. For classical wave motion, $\Delta x\, \Delta(1/\lambda) \geq 1/4\pi$ and $\Delta t\, \Delta\nu \geq 1/4\pi$, but Heisenberg showed that for microscopic systems, $\Delta x\, \Delta p_x \geq \hbar/2$ and $\Delta t\, \Delta E \geq \hbar/2$.
2. The time-independent Schrödinger equation provides a means for calculating the wavefunction ψ for a quantum mechanical particle, and the probability

density is given by the product of the wavefunction with its complex conjugate.

3. An operator is a mathematical operation that is applied to a function, and in quantum mechanics there is a linear operator for each classical mechanical observable. When two operators commute, the corresponding variables can be simultaneously measured to any precision, and when they do not commute, the corresponding observables cannot be measured to arbitrary precision.
4. The average value of a variable (the expectation value) can be calculated by integrating the product of the complex conjugate of the wavefunction times the operator for the variable times the wavefunction.
5. A particle in a box cannot have zero energy, but the correspondence principle indicates that quantum mechanical predictions approach the predictions of classical mechanics as the quantum number approaches infinity. For a particle in a three-dimensional box, many of the energy levels are degenerate.
6. A quantum mechanical harmonic oscillator also has a zero-point energy of $h\nu/2$, but in contrast to a particle in a box, the energy levels are equally spaced.
7. Although the bond distance of a diatomic molecule in its ground vibrational level varies, the standard deviation of the internuclear distance is usually a small percentage of the average internuclear separation.
8. The rigid rotor wavefunctions are spherical harmonics, and the energy levels are proportional to $\ell(\ell + 1)$, where ℓ is the angular momentum quantum number, and inversely proportional to the moment of inertia I.
9. The quantum mechanical angular momentum operators for the x, y, and z directions do not commute with each other, but they each commute with the operator for the square of the angular momentum. Therefore, we can measure precisely the square of the total angular momentum and one, but only one, of its components. The angular momentum in the z direction is equal to the magnetic quantum number times $\hbar$.
10. This chapter introduces five postulates of quantum mechanics, and the next chapter introduces the additional postulate that the wavefunction of a system of electrons must be antisymmetric to the interchange of any two electrons.

REFERENCES

P. F. Bernath, *Spectra of Atoms and Molecules.* Oxford: Oxford University Press, 1995.

R. E. Christofferson, *Basic Principles and Techniques of Molecular Quantum Mechanics.* New York: Springer, 1989.

C. Cohen-Tannoudjii, B. Diu, and F. Laloë, *Quantum Mechanics.* Hoboken, NJ: Wiley, 1977.

R. Eisberg and R. Resnick, *Quantum Physics,* 2nd ed. Hoboken, NJ: Wiley, 1985.

M. D. Fayer, *Elements of Quantum Mechanics*. Oxford: Oxford University Press, 2001.

H. Haken and H. C. Wolf, *Atomic and Quantum Physics.* New York: Springer, 1987.

I. N. Levine, *Quantum Chemistry,* 5th ed. Upper Saddle River, NJ: Prentice-Hall, 1999.

J. P. Lowe, *Quantum Chemistry.* Boston: Academic, 1993.

L. Pauling and E. B. Wilson, *Introduction to Quantum Mechanics.* New York: McGraw-Hill, 1935.

M. A. Ratner and G. C. Schatz, *Introduction to Quantum Mechanics in Chemistry.* Upper Saddle River, NJ: Prentice Hall, 2001.
G. C. Schatz and M. A. Ratner, *Quantum Mechanics in Chemistry*. Englewood Cliffs, NJ: Prentice-Hall, 1993.
J. Simons and J. Nichols, *Quantum Mechanics in Chemistry.* New York: Oxford University Press, 1997.
R. N. Zare, *Angular Momentum.* Hoboken, NJ: Wiley, 1988.

PROBLEMS

Problems marked with an icon may be more conveniently solved on a personal computer with a mathematical program.

9.1 A detector is exposed to a monochromatic source of radiation for 40 ms and indicates that the power level is 10 μW. If 10^9 photons are incident on the detector in this time, what is the frequency of the radiation? What type of electromagnetic radiation is this?

9.2 Calculate the energy per photon and the number of photons emitted per second from (*a*) a 100-W yellow lightbulb (λ = 550 nm) and (*b*) a 1-kW microwave source (λ = 1 cm).

9.3 In the photoelectric effect an electron is emitted from a metal as the result of absorption of a photon of light. Part of the energy of the photon is required to release the electron from the metal; this energy ϕ is called the work function or binding energy. The kinetic energy of the ejected electron is given by

$$\frac{1}{2}mv^2 = h\nu - \phi$$

where m and v are the mass and velocity of the electron. For the 100 face of silver metal (see Chapter 23) the velocity of electrons emitted using 200-nm photons is 7.42×10^5 m s^{-1}. Calculate the work function of this face in eV.

9.4 Photoelectron spectroscopy utilizes the photoelectron effect to measure the binding energy of electrons in molecules and solids, by measuring the kinetic energy of the emitted electrons and using the relation in Problem 9.3 between kinetic energy, wavelength, and binding energy. One variant of photoelectron spectroscopy is X-ray photoelectron spectroscopy (XPS). If the X-ray wavelength is 0.2 nm, calculate the velocity of electrons emitted from molecules in which the binding energies are 10, 100, and 500 eV.

9.5 Electrons are accelerated by a 1000-V potential drop. (*a*) Calculate the de Broglie wavelength. (*b*) Calculate the wavelength of the X-rays that could be produced when these electrons strike a solid.

9.6 An ultraviolet photon (λ = 58.4 nm) from a helium gas discharge tube is absorbed by a hydrogen molecule that is at rest. Since momentum is conserved, what is the velocity of the hydrogen molecule after absorbing the photon? What is the translational energy of the hydrogen molecule in J mol^{-1}?

9.7 What is the de Broglie wavelength of an oxygen molecule at room temperature? Compare this to the average distance between oxygen molecules in a gas at 1 bar at room temperature.

9.8 What is the de Broglie wavelength of a thermal neutron at 300 K?

9.9 Calculate the de Broglie wavelengths of the following:
(a) A 1-g bullet with velocity 300 m s^{-1}
(b) A 10^{-6}-g particle with velocity 10^{-6} m s^{-1}
(c) A 10^{-10}-g particle with velocity 10^{-10} m s^{-1}
(d) An H_2 molecule with energy of $\frac{3}{2}kT$ at T = 20 K

9.10 The lifetime of a molecule in a certain electronic state is 10^{-10} s. What is the uncertainty in energy of this state? Give the answer in J and in J mol^{-1}.

9.11 Show that the function $f = 8\,e^{5x}$ is an eigenfunction of the operator d/dx. What is the eigenvalue?

9.12 What are the results of operating on the following functions with the operator d/dx and d^2/dx^2: (*a*) e^{-ax^2}, (*b*) $\cos bx$, and (*c*) e^{ikx}? Which functions are eigenfunctions of these operators? What are the corresponding eigenvalues?

9.13 Show that the operators for the x coordinate and for the momentum in the x direction p_x do not commute. Calculate the operator representing the commutator of x and p_x.

9.14 For a particle in a one-dimensional box, the ground-state wavefunction is

$$\phi = \left(\frac{2}{a}\right)^{1/2} \sin\frac{\pi x}{a}$$

(*a*) What is the probability that the particle is in the right-hand half of the box? (*b*) What is the probability that the particle is in the middle third of the box?

9.15 (*a*) Calculate the energy levels for n = 1, 2, and 3 for an electron in a potential well of width 0.25 nm with infinite barriers on either side. The energies should be expressed in kJ mol^{-1}. (*b*) If an electron makes a transition from n = 2 to n = 1, what will be the wavelength of the emitted radiation?

9.16 For a helium atom in a one-dimensional box calculate the value of the quantum number of the energy level for which the energy is equal to $\frac{3}{2}kT$ at 25 °C (*a*) for a box 1 nm long, (*b*) for a box 10^{-6} m long, and (*c*) for a box 10^{-2} m long.

9.17 Show that the wavefunctions for a particle in a one dimensional box are orthogonal using

$$\sin\alpha \sin\beta = \frac{1}{2}\cos(\alpha+\beta) - \frac{1}{2}\cos(\alpha-\beta)$$

9.18 Calculate the degeneracies of the first three levels for a particle in a cubical box.

9.19 (*a*) Calculate $\Delta x\,\Delta p_x$ for a particle in a linear box for $n = 1, 2$, and 3 using the equation in Example 9.13. Compare these values with the minimum product of uncertainties from the Heisenberg uncertainty principle. (*b*) What is the uncertainty in x for a particle in a 0.2-nm box when its quantum number is unity? In a 2-nm box?

9.20 Calculate the standard deviation for the x coordinate of a harmonic oscillator at $v = 1$. Since $\langle x \rangle = 0$, it is only necessary to calculate $\langle x^2 \rangle$.

9.21 Calculate the standard deviation for the momentum p of a harmonic oscillator at $v = 1$. Since $\langle p \rangle = 0$, it is only necessary to calculate $\langle p^2 \rangle$.

9.22 Using the results of the two previous problems, calculate $\sigma_x \sigma_p$ and compare it with the Heisenberg uncertainty principle.

9.23 The fundamental vibration frequency of $^{12}C^{16}O$ is 2169.814 cm^{-1}. Calculate the force constant.

9.24 Using data from the previous problem, calculate the fundamental vibration frequency for $^{12}C^{18}O$, assuming the force constant is the same.

9.25 Calculate the root-mean-square displacement of the nuclei of $^{12}C^{16}O$ in the $v = 0$ state and compare it with the equilibrium bond length of 112.832 pm.

9.26 Later, in Table 13.4, we will find that the following molecules have the indicated vibrational frequencies:

$$^{35}Cl_2\ (560\ cm^{-1}) \qquad ^{39}K^{35}Cl\ (281\ cm^{-1})$$
$$^{1}H_2\ (4401\ cm^{-1})$$

(*a*) What are the force constants for these molecules if we treat them as harmonic oscillators? (*b*) Assuming that the force constant for $^{37}Cl_2$ is the same as for $^{35}Cl_2$, predict the fundamental vibrational frequency of $^{37}Cl_2$.

9.27 Check the normalization of ψ_0 and ψ_1 for the harmonic oscillator and show that they are orthogonal.

9.28 Substitute the $v = 1$ eigenfunction for the harmonic oscillator into the Schrödinger equation for the harmonic oscillator, and obtain the expression for the eigenvalue (energy).

9.29 In the vibrational motion of HI, the iodine atom essentially remains stationary because of its large mass. Assuming that the hydrogen atom undergoes harmonic motion and that the force constant k is 317 N m^{-1}, what is the fundamental vibration frequency ν_0? What is ν_0 if H is replaced by D?

9.30 What are the expectation values for $\langle x \rangle$ and $\langle x^2 \rangle$ for a quantum mechanical harmonic oscillator in the $v = 1$ state? What is the standard deviation Δx?

9.31 $^{12}C^{16}O$ is an example of a stiff diatomic molecule, and it has a vibration frequency of 2170 cm^{-1}. (*a*) What is the value of the force constant k? (*b*) What is the value of the standard deviation Δx in the internuclear distance? (*c*) What is the standard deviation Δp_x of the momentum of the vibrational motion? (*d*) Check that the product $\Delta x\,\Delta p_x$ yields $\hbar/2$ in accordance with the Heisenberg uncertainty principle.

9.32 Use information in Example 9.21 to calculate the frequency and wavenumber of the radiation required to take the $H^{35}Cl$ molecule from $J = 1$ to $J = 2$.

(M) **9.33** Use the Schrödinger equation for a rigid rotor in three dimensions to calculate the rotational energy when the wavefunction is given by the spherical harmonic Y_1^0. What is the magnitude of the angular momentum?

9.34 What are the reduced mass and moment of inertia of $^{23}Na^{35}Cl$? The equilibrium internuclear distance R_e is 236 pm. What are the values of E for the states with $J = 1$ and $J = 2$?

9.35 The commutator of two operators, $\hat{A}$ and $\hat{B}$, is defined as the operator $\hat{A}\hat{B} - \hat{B}\hat{A}$. Using the definitions of $\hat{L}_x$, $\hat{L}_y$, and $\hat{L}_z$ given in equations 9.153–9.155, find the commutator of $\hat{L}_x$ with $\hat{L}_y$ and $\hat{L}_x$ with $\hat{L}_z$.

9.36 The $^{12}C^{16}O$ molecule has an equilibrium bond distance of 112.8 pm. Calculate (*a*) the reduced mass and (*b*) the moment of inertia. (*c*) Calculate the wavelength of the photon emitted when the molecule makes the transition from $\ell = 1$ to $\ell = 0$ using equation 9.144 for the energy levels.

9.37 (*a*) The distribution of wavelengths from a certain star peaks in the visible at $\lambda = 600$ nm. Assuming that the distribution obeys the Planck distribution law, use Wien's displacement law to estimate the temperature of the star. (*b*) A metal bar is heated to red heat so that its radiation peaks at $\lambda = 800$ nm. Estimate the temperature of the bar.

9.38 (*a*) Derive the value of the constant in the Wien displacement law (equation 9.186) in terms of h, c, and k. (*b*) If, from experiment, the values of h and c were measured to be 6.6×10^{-34} J s^{-1} and 3.0×10^8 m s^{-1}, and the value of the constant in the Wien displacement law was measured to be 2.9×10^{-3} K m^{-1}, find the value of k from (*a*). Since R is measured to be 8.3 J K^{-1} mol^{-1}, you can also calculate the Avogadro constant.

9.39 Calculate the number of photons emitted in 1 s from a 100-W red lamp, assuming for simplicity that all the photons have an average wavelength of 694 nm. Also, calculate the number of photons emitted from a ruby laser in a 5-ns pulse with 0.1 GW power ($\lambda = 694$ nm).

9.40 As is often said, quantum mechanics yields classical mechanics when $h \to 0$. What result is obtained classically for $\rho(\nu, T)$ when $h \to 0$? Why is this relation unsatisfactory? (*Hint:* Expand the exponential term in a power series.)

9.41 Using the Planck distribution law, equation 9.2, find the *frequency* of maximum emission as a function of temperature.

9.42 (*a*) Using the result of Problem 9.41, find the maximum emission frequency at 3000 K and at 10 000 K. (*b*) Using the Wien displacement law, equation 9.186, find the maximum emission wavelength at the same two temperatures.

9.43 Calculate the de Broglie wavelength (*a*) of an electron accelerated by a potential of 1000 V and (*b*) of a proton accelerated by a potential of 1000 V.

9.44 Calculate the de Broglie wavelength for thermal neutrons at a temperature of 100 °C.

9.45 What is the momentum of a photon with a wavelength of 500 nm? If this photon is absorbed by a $^{35}Cl_2$ molecule that is at rest, what will be the velocity of the Cl_2 molecule after absorbing the photon?

9.46 An atom makes a transition from an excited state with a lifetime of 10^{-9} s to the ground state and emits a photon with a wavelength 600 nm. What is the uncertainty in the energy of the excited state? What is the percentage uncertainty if the energy is measured from the ground state?

9.47 Since a wavefunction may be complex, it may be represented by

$$\psi = R + \mathrm{i}I$$

where $i = (-1)^{1/2}$. Show that $\psi^*\psi$ is always real.

9.48 (*a*) Show that the function $\psi = x\,\mathrm{e}^{-ax^2}$ is an eigenfunction of the operator $\mathrm{d}^2/\mathrm{d}x^2 - 4a^2x^2$. What is the eigenvalue? (*b*) Show that the function $\psi = K\,\mathrm{e}^{r/k}$ is an eigenfunction of the operator $\mathrm{d}/\mathrm{d}r$. What is the eigenvalue?

9.49 Derive the expression for the energy of a particle in a one-dimensional box using the de Broglie formula.

9.50 Calculate the first three energy levels, in kJ mol^{-1}, for an electron in a potential well 0.5 nm in width with infinitely high potential outside.

9.51 For a particle in a cubical box calculate $E(8ma^2/h^2)$ for the first 10 states. What is the degeneracy for each energy level?

9.52 Assume the form $\Psi_0 = N_0\,\mathrm{e}^{-ax^2}$ for the ground-state wavefunction of the harmonic oscillator, and substitute this into the Schrödinger equation. Find the value of a that makes this an eigenfunction.

(M) **9.53** Figure 9.6 shows that the quantum mechanical harmonic oscillator can be in regions forbidden to a *classical* harmonic oscillator. (*a*) Find an integral expression for the probability that the particle will be in the classically forbidden region for the ground state. (*b*) Using tables for Gaussian integrals or error functions (e.g., Abramowitz and Stegun, *Handbook of Mathematical Functions*, New York: Dover, 1964) or a numerical integration program on a microcomputer, compute the value of the integral in (*a*).

9.54 The expression for the energy levels of a quantum mechanical three-dimensional harmonic oscillator is obtained by expressing the potential energy by

$$V = \tfrac{1}{2}k_x x^2 + \tfrac{1}{2}k_y y^2 + \tfrac{1}{2}k_z z^2$$

where k_x, k_y, and k_z are the force constants in the three directions. What is the expression for the quantum mechanical energy levels, and what is the zero-point energy?

9.55 For the three-dimensional *isotropic* harmonic oscillator, $k_x = k_y = k_z$ (see Problem 9.54), write the formula for the energy levels. Give the energies and degeneracies of the first 10 states.

9.56 Using the definitions of $\hat{L}_z$ and $\hat{L}^2$ from equations 9.159 and 9.160 show that these two operators commute.

9.57 Sketch the possible orientation for angular momentum vectors for $\ell = \frac{3}{2}$.

9.58 Sketch a figure like Fig. 9.10 for $\ell = 3$.

9.59 What is the reduced mass of $^{14}N^{16}O$? What is its moment of inertia if $R_e = 115.1$ pm? Using equation 9.144, find the energies of the first three levels of rotational motion.

9.60 Check the normalization and orthogonality of Y_0^0 and Y_ℓ^m, $m = 0, 1, -1$, in Table 9.2.

9.61 Given that a particle is restricted to the region $-a < x < a$ and has a wavefunction Ψ proportional to $\cos(\pi x/2a)$, normalize the wavefunction.

Computer Problems

9.A (*a*) Plot the wavefunctions and probability densities for the first three levels of an electron in a box 0.1 nm in length. (*b*) Test the normalization and orthogonality of these wavefunctions. (*c*) Calculate the average value of x and the average value of x^2 for electrons in these energy levels. (*d*) Calculate the average value of p_x and the average value of p_x^2. (*e*) Check these values against the Heisenberg uncertainty principle.

9.B (*a*) Plot the first four wavefunctions for a harmonic oscillator to see the shapes of the wavefunctions. (*b*) Plot the corresponding probability densities for these four levels.

9.C In the wavefunction for the harmonic oscillator, α is taken as unity here because it is needed only to provide units in calculations. (*a*) Plot normalized wavefunctions and probability densities for vibrational quantum numbers of 0 and 1. (*b*) Check the normalization and orthogonality of these two wavefunctions. (*c*) Plot the wavefunction and probability density for $v = 30$.

9.D (*a*) Assuming that the $^{12}C^{16}O$ molecule is a harmonic oscillator with $\mu = 1.1385\times10^{-26}$ kg and $k = 1886$ N m^{-1}, write the expression for the vibrational wavefunction. (*b*) Plot the wavefunction versus the distance x for $v = 0$ and $v = 1$. (*c*) Calculate the average value of x, the average of the square of x, and the standard deviation of the internuclear distance for $v = 0$ and $v = 1$. Note that the mean internuclear distance is 113 pm.

9.E Calculate the probability that a harmonic oscillator is outside of its classical turning points when the vibrational wave number is 4.

9.F Calculate the radiant energy density for a blackbody at 1500, 1700, and 2000 K as a function of (*a*) wavelength in micrometers and (*b*) frequency in Hz.

9.G (*a*) The following differential equation is involved in the discussion of the particle in a box and the harmonic oscillator.

$$m = \frac{d^2x}{dt^2} = -kx$$

Use Mathematica to solve the differential equation and make some plots of solutions. (*b*) Since the potential energy of a classical harmonic oscillator is given by $V = kx^2/2$, plot the potential energy versus x for $k = 1$ and $k = 2$.

9.H Compare the wavefunctions and probability densities for the harmonic oscillator with the classical potential energy function for the harmonic oscillator. (*a*) Calculate the energy levels for a harmonic oscillator in units of $\hbar$ for a fundamental vibrational frequency of $1/2\pi s^{-1}$. (*b*) Assuming that $\alpha = 1$, write the expression for the vibrational wavefunction psi in Mathematica. (*c*) Plot the wavefunctions for the first five levels in comparison with the classical expression for the potential energy. (*d*) Plot the probability densities for the first five levels in comparison with the classical expression for the potential energy.

9.I The amplitude for a traveling wave is given by

$$A(x, t) = A_0 \cos(kx - \omega t)$$

where $k = 2\pi/\lambda$ and $\omega = 2\pi/\tau$ so that

$$A(x, t) = A_0 \cos(2\pi x/\lambda - 2\pi t/\tau) = A_0 \cos(2\pi x/\lambda - \omega t)$$

(*a*) Plot $A(x, t)/A_0$ versus x for constant t for $\lambda = 2$ m and 4 m. (*b*) Plot $A(x, t)/A_0$ versus t for constant x for $\tau = 0.1$ s and 0.2 s. (*c*) In general, for wave motion the phase velocity is $\nu\lambda$. For light the phase velocity is c so that at constant t,

$$A(x) = A_0 \cos(2\pi x\nu/c)$$

Plot $A(x)/A_0$ versus x for $\nu = 380 \times 10^{12}\ s^{-1}$, which is about the highest frequency the human eye can detect.

9.J The amplitude of traveling wave is given by

$$A = A_0 \cos(kx - \omega t)$$

Show that this equation for the amplitude satisfies the following partial differential equation and derive the expression for the phase velocity ν_p.

$$\frac{\partial^2 A}{\partial x^2} = \frac{1}{\nu_p^2}\frac{\partial^2 A}{\partial t^2}$$

How is the phase velocity related to ν and λ?

9.K As discussed in the text, when a quantum mechanical harmonic oscillator is excited with a very short pulse of radiation, which necessarily contains a distribution of wavelengths, a number of vibrational levels are excited coherently (that is, in phase). (*a*) Type in the expression for the stationary wavefunction for a single level as a function of the quantum number v and the bond distance x utilizing HermiteH. Plot this function for a couple of levels. (*b*) Add up the wavefunctions for $v = 14, 15, 16, 17, 18, 19, 20, 21$, and 22 with relative contributions of these wavefunctions to the superposition that correspond to a Gaussian distribution centered at $v = 18$. (*c*) Plot this wavefunction and the probability density function versus x to show that this wave packet is something like a particle. (*d*) Introduce weighting factors of $\exp(-ivt)$ into each of these terms so that the movement of the wave packet can be calculated. As a simplification, the energies of the levels are taken to be proportional to the vibrational quantum numbers v. To construct the probability density, this wavefunction has to be multiplied with the corresponding wavefunction with factors of $\exp(ivt)$. Plot the probability density function versus x at $t = 0$. (*e*) Increase t in steps to show that the system can be thought of as a particle oscillating back and forth. This behavior makes it possible to study the kinetics of ultrafast reactions. [More information is provided by J. S. Baskin and A. H. Zewail, *J. Chem. Educ.* 78:737 (2001).]

10

Atomic Structure

This chapter introduces the electronic structure of atoms. The electronic wavefunctions for an atom contain all the information about the electronic properties of the atom. The wavefunctions for the hydrogen atom and the one-electron atoms, such as He^+ and Li^{2+}, can be calculated exactly, but approximate methods have to be used with atoms having two or more electrons. Fortunately, these approximate methods can yield quite precise results, but at the cost of complicated calculations carried out on a computer.

We will find that an additional postulate has to be added to the nonrelativistic quantum theory discussed in the preceding chapter, namely, the Pauli exclusion principle.

We will consider the electronic structure of the hydrogenlike atoms in some detail because of the importance of the concept of orbitals that this introduces. The concept of the atomic orbital and the representation of many-electron systems by products of atomic orbitals will be useful for molecules as well as atoms. One of the great triumphs of quantum mechanics has been the insight it provides into the structure of the periodic table and therefore of the periodicity in the physical and chemical properties of the elements.

Only certain transitions between energy levels can occur in the absorption or emission of electromagnetic radiation. The rules governing these transitions are called selection rules, and we will discuss these restrictions on atomic spectra at the end of this chapter.

10.1 THE SCHRÖDINGER EQUATION FOR HYDROGENLIKE ATOMS

We have used the time-independent Schrödinger equation (equation 9.19) to calculate wavefunctions for a particle in a box, a harmonic oscillator, and a rigid rotor. Now we consider the two-particle system consisting of an electron (charge $-e$) and a nucleus having atomic number Z and charge Ze, that is, H, He^+, Li^{2+}, ..., U^{91+}, the hydrogenlike atoms. In such an atom, the electron interacts with the nuclear Coulomb potential, so the potential energy is

$$V(r) = \frac{-Ze^2}{4\pi\epsilon_0 r} \tag{10.1}$$

where ϵ_0 is the **permittivity of vacuum,** and r is the distance from the nucleus to the electron. Note that the potential energy of interaction between the two oppositely charged particles is taken to be zero when they are infinitely far apart and is increasingly negative as r decreases. This means that the force of attraction between oppositely charged particles increases as r gets smaller.

The expressions for the Hamiltonian operator and the Schrödinger equation can be written in terms of the x, y, z coordinates of the two particles, but to treat an isolated hydrogen atom, it is possible to use center-of-mass coordinates and write the Schrödinger equation for a particle of reduced mass μ moving around a fixed center at $x = y = z = 0$. This Schrödinger equation is

$$\left[\frac{-h^2}{8\pi^2\mu}\left(\frac{\partial^2}{\partial x^2}+\frac{\partial^2}{\partial y^2}+\frac{\partial^2}{\partial z^2}\right)-\frac{Ze^2}{4\pi\epsilon_0(x^2+y^2+z^2)^{1/2}}\right]\psi(x,y,z) = E\psi(x,y,z) \tag{10.2}$$

The reduced mass μ is given by

$$\mu = \frac{m_e m_N}{m_e + m_N} \tag{10.3}$$

where m_e is the mass of the electron and m_N is the mass of the nucleus.

Since a hydrogenlike atom is spherically symmetric, it is convenient to use spherical coordinates and write equation 10.2 as

$$\left[\frac{-h^2}{8\pi^2\mu}\nabla^2 - \frac{Ze^2}{4\pi\epsilon_0 r}\right]\psi(r,\theta,\phi) = E\psi(r,\theta,\phi) \tag{10.4}$$

where

$$\nabla^2 = \frac{1}{r^2}\frac{\partial}{\partial r}\left(r^2\frac{\partial}{\partial r}\right) + \frac{1}{r^2 \sin\theta}\frac{\partial}{\partial\theta}\left(\sin\theta\frac{\partial}{\partial\theta}\right) + \frac{1}{r^2\sin^2\theta}\frac{\partial^2}{\partial\phi^2} \tag{10.5}$$

as we saw earlier in equation 9.140. Note that the potential energy term $-Ze^2/4\pi\epsilon_0 r$ has no θ or ϕ dependence, but the wavefunctions are in general functions of θ and ϕ, which are introduced in the Laplacian operator.

We noted in Section 9.12 that the angle-dependent terms in ∇^2 can be represented as $(-1/\hbar^2 r^2)\hat{L}^2$ (equation 9.160), where $\hat{L}^2$ is the operator for the square of the angular momentum. Thus,

$$\nabla^2 = \frac{1}{r^2}\frac{\partial}{\partial r}\left(r^2\frac{\partial}{\partial r}\right) - \frac{1}{r^2}\frac{\hat{L}^2}{\hbar^2} \tag{10.6}$$

which, when substituted into the Schrödinger equation 10.4 and multiplied by $2\mu r^2$, gives

$$-\hbar^2\frac{\partial}{\partial r}\left(r^2\frac{\partial\psi}{\partial r}\right) + \hat{L}^2\psi - 2\mu r^2\left(\frac{Ze^2}{4\pi\epsilon_0 r} + E\right)\psi = 0 \tag{10.7}$$

Notice that the angular dependence of the operator is contained in $\hat{L}^2$, **whose eigenvalues and eigenfunctions we already know** (Section 9.12). This immediately suggests that we try to solve equation 10.7 by separation of variables by writing $\psi(r, \theta, \phi) = R_{n\ell}(r)Y_\ell^m(\theta, \phi)$, where $R_{n\ell}(r)$ is called the hydrogenlike radial wavefunction and satisfies the differential equation

$$-\left(\frac{\hbar^2}{2\mu}\right)\frac{1}{r^2}\frac{d}{dr}\left(r^2\frac{d}{dr}R_{n\ell}\right) + \left[\frac{\hbar^2\ell(\ell+1)}{2\mu r^2} - \frac{Ze^2}{4\pi\epsilon_0 r} - E\right]R_{n\ell} = 0 \tag{10.8}$$

A simpler equation can be found by making the substitution $R_{n\ell}(r) = S_{n\ell}(r)/r$. Then, after multiplying the entire equation by r, we find

$$-\left(\frac{\hbar^2}{2\mu}\right)\frac{d^2}{dr^2}S_{n\ell} + \left[\frac{\hbar^2\ell(\ell+1)}{2\mu r^2} - \frac{Ze^2}{4\pi\epsilon_0 r} - E\right]S_{n\ell} = 0 \tag{10.9}$$

which looks like the equations we have seen before. Notice that the term proportional to $\ell(\ell+1)/r^2$ (called the centrifugal potential) adds to the Coulomb potential to give an effective potential for $\ell > 0$, as shown in Fig. 10.1. Solutions to this equation can be found by expanding $S_{n\ell}(r)$ in a power series in r, and requiring that the wavefunction vanish as r gets large. These functions will represent the **bound** states of the electron in this atom and will have $E < 0$. The mathematical procedure is straightforward but lengthy. It is done carefully in a number of advanced books (see references at the end of the chapter, e.g., Pauling and Wilson).

The result of this calculation is that E can have only certain values given by

$$E_n = -\frac{\mu e^4 Z^2}{2(4\pi\epsilon_0)^2\hbar^2 n^2} \tag{10.10}$$

with the principal quantum number $n = 1, 2, 3, \ldots, \infty$. Note that E_n does not depend on ℓ; however, in solving the equation, it is found that $n > \ell$ so that for $n = 1$, ℓ can only be 0, while for $n = 2$, ℓ can be 0 and 1, and so on. The low-lying energy levels are shown in Fig. 10.2.

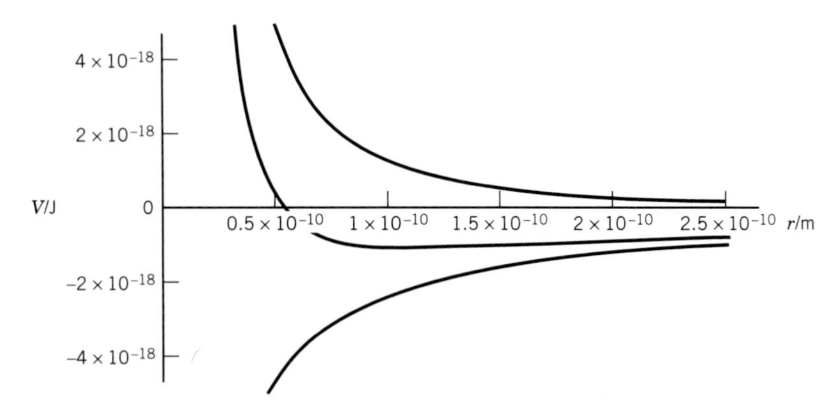

Figure 10.1 The lowest curve gives the Coulomb potential energy $(-e^2/4\pi\epsilon_0)$ for the hydrogen atom. The highest curve gives the centrifugal potential energy $(\hbar^2/\mu r^2)$ for $\ell = 1$. The middle curve gives the sum of these potential energies and shows that the minimum potential energy is at about 0.1 nm. (See Computer Problem 10.G.)

Example 10.1 *A solution to the Schrödinger equation for the hydrogen atom*

You may want to get more of a feel for the solutions of equation 10.8. A simple procedure is to guess a solution and see if it works. For example, pick $R = \mathrm{e}^{-\gamma r}$ and find the conditions to make this a solution of equation 10.9.

$$\frac{1}{r^2}\frac{\mathrm{d}}{\mathrm{d}r}\left(r^2\frac{\mathrm{d}\,\mathrm{e}^{-\gamma r}}{\mathrm{d}r}\right) = \frac{1}{r^2}\frac{\mathrm{d}}{\mathrm{d}r}(-\gamma r^2\,\mathrm{e}^{-\gamma r}) = \left(\frac{-2\gamma r + \gamma^2 r^2}{r^2}\right)\mathrm{e}^{-\gamma r}$$

Thus,

$$-\frac{\hbar^2}{2\mu}\left(\frac{-2\gamma r + \gamma^2 r^2}{r^2}\right)\mathrm{e}^{-\gamma r} + \left[\frac{\hbar^2\ell(\ell+1)}{2\mu r^2} - \frac{Ze^2}{4\pi\epsilon_0 r} - E\right]\mathrm{e}^{-\gamma r} = 0$$

For the terms multiplying $1/r^2$ and $1/r$ to vanish for all r, we must have $\ell = 0$ and

$$\frac{\gamma\hbar^2}{\mu} = \frac{Ze^2}{4\pi\epsilon_0}$$

and then

$$E = -\frac{\hbar^2\gamma^2}{2\mu} = -\frac{\mu Z^2 e^4}{2(4\pi\epsilon_0)^2\hbar^2}$$

This is the $n = 1$ eigenfunction and energy. Now try it with $R = r\,\mathrm{e}^{-\gamma r}$!

Note that E_n for a hydrogenlike atom depends on the mass of the nucleus, since μ depends on the mass of the nucleus. Since $m_\mathrm{N} \gg m_\mathrm{e}$, the reduced mass of a hydrogenlike atom is very close to the mass of the electron. You should convince yourself that the difference between μ and m_e in the hydrogen atom is 0.05%. We are especially interested in the energies and wavefunctions in the other hydrogenlike atoms and in molecules with heavier nuclei. Therefore, we will emphasize here the properties of the hydrogenlike atoms with nuclei of infinite mass (or fixed nucleus) so that μ is replaced with m_e. Replacing μ in equation 10.10 with m_e yields an expression for E_n, which can be written

$$E_n = -\frac{Z^2}{2n^2}\frac{e^2}{4\pi\epsilon_0 a_0} \tag{10.11}$$

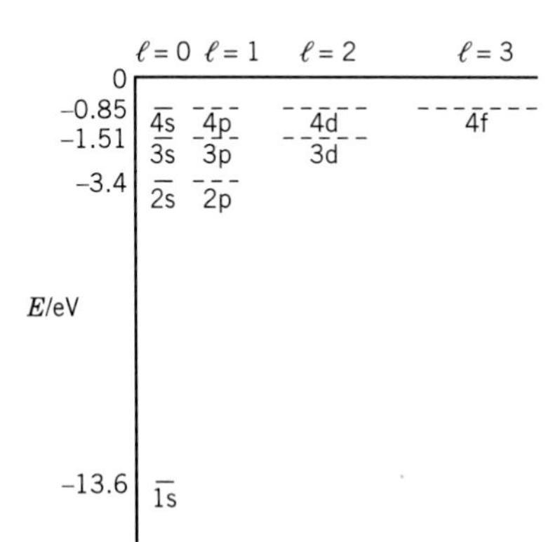

Figure 10.2 Electronic energy levels for the hydrogen atom. The dashes show the degeneracies of the levels. There are more states above those shown, and a continuum exists at positive energies.

where

$$a_0 = \frac{\hbar^2(4\pi\epsilon_0)}{m_e e^2} \tag{10.12}$$

is the **Bohr radius,** which we will find is the most probable distance from the nucleus to the electron in the ground state of a hydrogen atom with a fixed nucleus. We define

$$E_h = \frac{e^2}{4\pi\epsilon_0 a_0} \tag{10.13}$$

which is the potential energy of two electrons separated by a Bohr radius. This is a natural unit of energy to use in connection with atoms and molecules, and is called the **Hartree energy.** (Both Bohr and Hartree were atomic physicists of the early twentieth century.) Thus $E_n = -Z^2E_h/2n^2$.

Example 10.2 ***Values of the Bohr radius and the Hartree energy***

Calculate the values of a_0 and E_h.

$$a_0 = \frac{h^2(4\pi\epsilon_0)}{4\pi^2 m_e e^2} = \frac{h^2\epsilon_0}{\pi m_e e^2} = 52.917\ 724\ 9\ \text{pm}$$

$$E_h = \frac{e^2}{4\pi\epsilon_0 a_0} = \cdots = 4.359\ 748\ 2 \times 10^{-18}\ \text{J} = 27.211\ 396\ 1\ \text{eV}^*$$

(Note that the energy of the H atom in its ground state is $-E_h/2$.)

Example 10.3 ***Two other expressions for Hartree energy***

The Hartree energy was introduced in equation 10.13 as

$$E_h = \frac{e^2}{4\pi\epsilon_0 a_0}$$

It is convenient to express the Hartree energy in other ways. Use the expression for a_0 to express the Hartree energy in two other ways.

Since

$$a_0 = \frac{\hbar^2(4\pi\epsilon_0)}{m_e e^2} \quad \text{and} \quad \frac{e^2}{4\pi\epsilon_0} = \frac{\hbar^2}{m_e a_0}$$

replacing this ratio in the equation for the Hartree energy yields

$$E_h = \frac{\hbar^2}{m_e a_0^2}$$

Eliminating a_0 in the definition of the Hartree energy yields

$$E_h = \frac{m_e e^4}{(4\pi\epsilon_0)^2\hbar^2}$$

---

*1 eV is the work done in moving an electron through 1 V and is $1.602\ 177\ 33 \times 10^{-19}$ J.

10.2 THE SPECTRUM OF HYDROGEN ATOMS

In discussing spectroscopy it is often more convenient to use frequencies ν or wave numbers $\tilde{\nu}$ than wavelengths because they are proportional to energies. The energies of the various levels of hydrogen atoms are given by equation 10.10. The **wave number** $\tilde{\nu}$ of electromagnetic radiation is the reciprocal of the wavelength: $\tilde{\nu} = 1/\lambda$, so that $\tilde{\nu} = \nu/c$, where c is the speed of electromagnetic radiation. The SI unit of wave numbers is m^{-1}, but usually cm^{-1} is used.

The energies can be expressed in wave numbers by dividing by hc or $2\pi\hbar c$ since $\Delta E = h\nu = hc\tilde{\nu} = 2\pi\hbar c\tilde{\nu}$, so that $\Delta E/2\pi\hbar c = \tilde{\nu} = \tilde{E}_2 - \tilde{E}_1$. Dividing equation 10.10 by $2\pi c\hbar$ yields

$$\tilde{E}_n = -\frac{\mu e^4 Z^2}{4\pi c(4\pi\epsilon_0)^2\hbar^3 n^2} = -\frac{RZ^2}{n^2} \tag{10.14}$$

where the **Rydberg constant** is given by

$$R = \frac{\mu e^4}{4\pi c(4\pi\epsilon_0)^2\hbar^3} \tag{10.15}$$

The value of the Rydberg constant depends on the mass of the nucleus of the hydrogenlike atom. For the hydrogen atom, $R_H = 1.096\,775\,856 \times 10^7$ m^{-1} or $1.096\,775\,856 \times 10^5$ cm^{-1}. For a deuterium atom, $R_D = 1.097\,074\,275 \times 10^5$ cm^{-1}. When the mass of the nucleus is infinite, the reduced mass is equal to the mass of the electron m_e, and so $R_\infty = 1.097\,373\,153\,4 \times 10^5$ cm^{-1}. The value of the Rydberg constant can be determined very accurately because spectroscopic frequencies can be measured with high accuracy. The energy levels for the hydrogen atom are shown in Fig. 10.3. The lowest state ($n = 1$) is called the **ground**

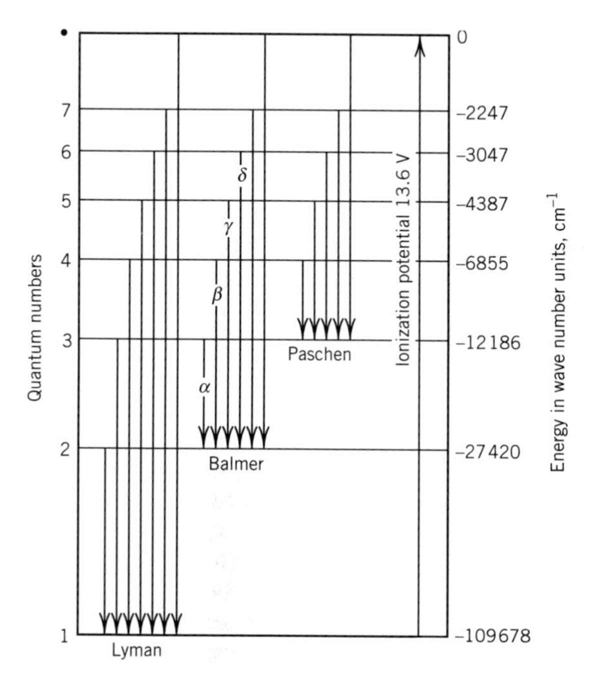

Figure 10.3 Energy levels for the hydrogen atom.

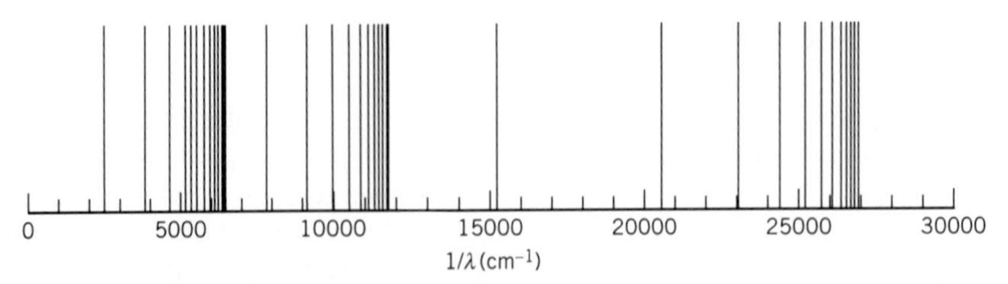

Figure 10.4 Hydrogen atom spectra for the Balmer series, Paschen series, and Brackett series. (See Computer Problem 10.C.)

state. The higher energy states ($n = 2, 3, \ldots$) are called **excited states**. An atom or molecule in an excited state can relax to a lower state with the emission of electromagnetic radiation. The energies given at the right in Fig. 10.3 are the wave numbers for radiation produced when an electron with no initial kinetic energy falls from an infinite distance into a given orbit, that is, the series limits.

Equation 10.14 shows that when a photon is emitted, the frequency of the radiation in wave numbers is given by

$$\tilde{\nu} = -\frac{RZ^2}{n_2^2} + \frac{RZ^2}{n_1^2} = RZ^2\left(\frac{1}{n_1^2} - \frac{1}{n_2^2}\right) \tag{10.16}$$

As shown in Fig. 10.3, electrons falling from excited levels into the ground state ($n = 1$) produce the lines in the Lyman series, those falling into the first excited state ($n = 2$) produce the Balmer series, etc. In each series, the α line results from the smallest change in energy in that series. As the quantum number of the higher state increases, the spectral lines get closer together and converge as $n_2 \to \infty$. The spectra for the Balmer, Paschen, and Brackett series are shown together in Fig. 10.4.

The **ionization energy** E_i is the energy required to take the electron in the ground state to $n_2 = \infty$. The energy of a hydrogenlike atom can be expressed in terms of electron volts by writing equation 10.10 as

$$E_n = -\frac{Z^2}{n^2}(13.605\,698 \text{ eV}) \tag{10.17}$$

where the Rydberg constant R_∞ is expressed in electron volts. This shows that hydrogenlike atoms with larger Z bind their electrons more strongly. Thus

$$E_i = (13.605\,698 \text{ eV})Z^2\left(\frac{1}{1^2} - \frac{1}{\infty}\right) = (13.605\,698 \text{ eV})Z^2 \tag{10.18}$$

Example 10.4 ***Ionization energies of hydrogenlike ions***

The ionization energy E_i for a hydrogenlike atom is the energy required to remove the electron from the atom in its ground state to a position very far from the nucleus, so that $E_i = (13.606 \text{ eV})Z^2$. Calculate the ionization energies of H, He^+, Li^{2+}, and Be^{3+}.

$$E_i(\text{H}) = 13.606 \text{ eV}$$

$$E_i(\text{He}^+) = 2^2 \times 13.606 \text{ eV} = 54.424 \text{ eV}$$

$$E_i(Li^{2+}) = 3^2 \times 13.606 \text{ eV} = 122.454 \text{ eV}$$
$$E_i(Be^{3+}) = 4^2 \times 13.606 \text{ eV} = 217.696 \text{ eV}$$

The wavefunctions of hydrogenlike systems have the form $R_{n\ell}(r)Y_{\ell}^{m}(\theta,\phi)$, indicating that for $n > 1$ there are a number of different wavefunctions with the same energy. For example if $n = 2$, we can have $\ell = 0$ (and $m = 0$) or $\ell = 1$ (and $m = +1, 0, -1$). In fact, as we saw in Section 9.12, the allowed values of m for a given ℓ are

$$m = -\ell, -\ell + 1, \ldots, \ell - 1, \ell \tag{10.19}$$

The value of ℓ determines the orbital angular momentum, and m determines the z component of angular momentum (equation 9.175). For historical reasons, we associate letter symbols with the values of ℓ:

$$\begin{array}{lcccc} \ell = & 0 & 1 & 2 & 3 \\ \text{Symbol} = & \text{s} & \text{p} & \text{d} & \text{f} \end{array} \tag{10.20}$$

In summary, there are three quantum numbers for each eigenfunction of a hydrogenlike atom which can take on the following values:

$$n = 1, 2, 3, \ldots \tag{10.21}$$

$$\ell = 0, 1, 2, \ldots, n - 1 \tag{10.22}$$

$$m = 0, \pm 1, \pm 2, \ldots, \pm \ell \tag{10.23}$$

Therefore, for a given value of n, there are n^2 degenerate eigenstates.

There is a fourth quantum number m_s that has not been introduced yet: the spin quantum number for the electron (see Section 10.5). That quantum number can be $+\frac{1}{2}$ or $-\frac{1}{2}$. The energy levels of the hydrogenlike atom do not depend on ℓ, m_ℓ, and m_s, and so the degeneracies of the level for $n = 1, 2$, and 3 are 2, 8, and 18, as shown in the following example.

Example 10.5 ***The total degeneracy (neglecting electron spin) of a hydrogen level is n^2***

Show that the total degeneracy g_{total} of the energy levels of the hydrogenlike atom is n^2 by writing out the possible quantum numbers for $n = 1, 2, 3$, and 4.

Possible values of quantum numbers n, ℓ, and m are

Orbital	n	ℓ	m	g	g_{total}
1s	1	0	0	1	1
2s	2	0	0	1	
2p		1	0, ±1	3	4
3s	3	0	0	1	
3p		1	0, ±1	3	
3d		2	0, ±1, ±2	5	9
4s	4	0	0	1	
4p		1	0, ±1	3	
4d		2	0, ±1, ±2	5	
4e		3	0, ±1, ±2, ±3	7	16

Inclusion of electron spin in the degeneracy calculation gives a total degeneracy of $2n^2$ (see Section 10.5).

The total wavefunction for the hydrogenlike atom is

$$\psi_{n\ell m_\ell}(r, \theta, \phi) = N_{n\ell}\rho^{\ell} \mathrm{e}^{-\rho/2} L_{n+\ell}^{2\ell+1}(\rho) Y_{\ell}^{m_\ell}(\theta, \phi) \tag{10.24}$$

where $\rho = 2Zr/na_0$ and $N_{n\ell}$ is a normalization constant given by

$$N_{n\ell} = -\left[\left(\frac{2Z}{na_0}\right)^3 \frac{(n-\ell-1)!}{2n\{(n+\ell)!\}^3}\right]^{1/2} \tag{10.25}$$

and $L_{n+\ell}^{2\ell+1}(\rho)$ is known as an associated Laguerre polynomial. $Y_{\ell}^{m_\ell}(\theta, \phi)$ represents the spherical harmonics we saw in Section 9.12.

In the next section, it will be convenient to discuss the hydrogenlike wavefunction in the form

$$\psi_{n\ell m_\ell}(r, \theta, \phi) = R_{n\ell}(r) Y_{\ell}^{m_\ell}(\theta, \phi) \tag{10.26}$$

mentioned earlier.

10.3 EIGENFUNCTIONS AND PROBABILITY DENSITIES FOR HYDROGENLIKE ATOMS

A wavefunction for a one-electron system is called an **orbital.** For an atomic system such as H, it is called an atomic orbital. These wavefunctions for the hydrogenlike atoms through $n = 3$ are given in Table 10.1. It is convenient to write these equations in terms of the Bohr radius a_0.

To visualize the nature of these functions it is helpful to consider the radial function and the spherical harmonics separately.

The **radial functions** $R_{n\ell}(r)$ for the hydrogenlike atoms depend on the principal quantum number n, the azimuthal quantum number ℓ, and the atomic number Z. The normalized radial functions for the hydrogen atom are shown in Fig. 10.5. The radial function always contains the factor e^{-Zr/na_0}, where n is the principal quantum number. As z is increased, the amplitude of the wavefunction falls off more rapidly with increasing r, indicating that the electron is attracted more closely to the higher positively charged nuclei. The radial functions have $n - \ell - 1$ zero values between $r = 0$ and $r = \infty$. These produce spherical nodal surfaces in ψ so that the electron density goes to zero at these surfaces. The existence of nodes is required so that, for example, the 1s and 2s and other orbitals will be orthogonal (Section 9.4); that is,

$$\int \psi_{1s}\psi_{2s}\, \mathrm{d}\tau = 0 \tag{10.27}$$

where $\mathrm{d}\tau$ represents the element of volume.

Probability densities are more useful for visualizing the electronic structure of atoms than are wavefunctions. It is obvious from Fig. 10.5 that $[R_{n\ell}(r)]^2$ is a maximum at the nucleus for s orbitals. However, if we are interested in the prob-

Table 10.1 Real Hydrogenlike Wavefunctions[a]

n	ℓ	m	*Wavefunction*
1	0	0	$\psi_{1s} = \dfrac{1}{\sqrt{\pi}}\left(\dfrac{Z}{a_0}\right)^{3/2} e^{-\sigma}$
2	0	0	$\psi_{2s} = \dfrac{1}{4\sqrt{2\pi}}\left(\dfrac{Z}{a_0}\right)^{3/2} (2-\sigma)\, e^{-\sigma/2}$
2	1	0	$\psi_{2p_z} = \dfrac{1}{4\sqrt{2\pi}}\left(\dfrac{Z}{a_0}\right)^{3/2} \sigma\, e^{-\sigma/2} \cos\theta$
2	1	± 1	$\psi_{2p_x} = \dfrac{1}{4\sqrt{2\pi}}\left(\dfrac{Z}{a_0}\right)^{3/2} \sigma\, e^{-\sigma/2} \sin\theta\cos\phi$
			$\psi_{2p_y} = \dfrac{1}{4\sqrt{2\pi}}\left(\dfrac{Z}{a_0}\right)^{3/2} \sigma\, e^{-\sigma/2} \sin\theta\sin\phi$
3	0	0	$\psi_{3s} = \dfrac{1}{81\sqrt{3\pi}}\left(\dfrac{Z}{a_0}\right)^{3/2} (27-18\sigma+2\sigma^2)\, e^{-\sigma/3}$
3	1	0	$\psi_{3p_z} = \dfrac{\sqrt{2}}{81\sqrt{\pi}}\left(\dfrac{Z}{a_0}\right)^{3/2} (6-\sigma)\sigma\, e^{-\sigma/3} \cos\theta$
3	1	± 1	$\psi_{3p_x} = \dfrac{\sqrt{2}}{81\sqrt{\pi}}\left(\dfrac{Z}{a_0}\right)^{3/2} (6-\sigma)\sigma\, e^{-\sigma/3} \sin\theta\cos\phi$
			$\psi_{3p_y} = \dfrac{\sqrt{2}}{81\sqrt{\pi}}\left(\dfrac{Z}{a_0}\right)^{3/2} (6-\sigma)\sigma\, e^{-\sigma/3} \sin\theta\sin\phi$
3	2	0	$\psi_{3d_{z^2}} = \dfrac{1}{81\sqrt{6\pi}}\left(\dfrac{Z}{a_0}\right)^{3/2} \sigma^2 e^{-\sigma/3}(3\cos^2\theta - 1)$
3	2	± 1	$\psi_{3d_{xz}} = \dfrac{\sqrt{2}}{81\sqrt{\pi}}\left(\dfrac{Z}{a_0}\right)^{3/2} \sigma^2 e^{-\sigma/3} \sin\theta\cos\theta\cos\phi$
			$\psi_{3d_{yz}} = \dfrac{\sqrt{2}}{81\sqrt{\pi}}\left(\dfrac{Z}{a_0}\right)^{3/2} \sigma^2 e^{-\sigma/3} \sin\theta\cos\theta\sin\phi$
3	2	± 2	$\psi_{3d_{x^2-y^2}} = \dfrac{1}{81\sqrt{2\pi}}\left(\dfrac{Z}{a_0}\right)^{3/2} \sigma^2 e^{-\sigma/3} \sin^2\theta\cos 2\phi$
			$\psi_{3d_{xy}} = \dfrac{1}{81\sqrt{2\pi}}\left(\dfrac{Z}{a_0}\right)^{3/2} \sigma^2 e^{-\sigma/3} \sin^2\theta\sin 2\phi$

[a] $\sigma = \dfrac{Z}{a_0} r$.

ability that the orbital electron is a certain distance from the nucleus, we need another approach. To calculate this probability density we need to take the product of $[R_{n\ell}(r)]^2$ and the volume of the spherical shell $4\pi r^2\, dr$ that has a radius of r. This is the **radial probability density** $p_{n\ell}(r)$, defined by

$$p_{n\ell}(r) = r^2 R_{n\ell}^2(r) \tag{10.28}$$

The radial probabilities $p(r)$ for a number of orbitals of the hydrogen atom are given in Fig. 10.5. For the hydrogen atom in a 1s orbital, the highest radial probability density occurs at $r = a_0$, the Bohr radius. Thus, a_0 is the most probable radius for a 1s hydrogen atom.

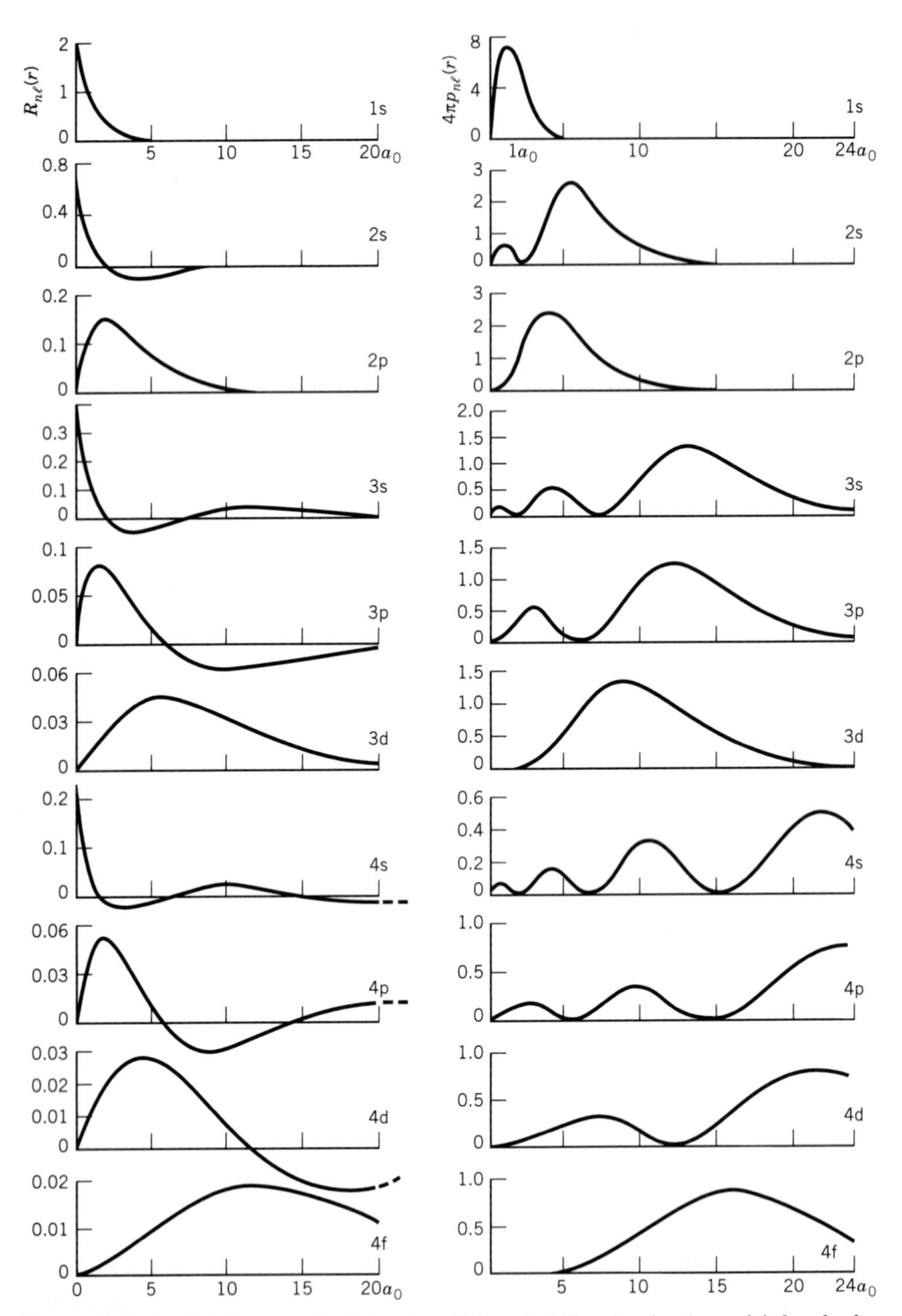

Figure 10.5 Radial function $R_{n\ell}(r)$ and radial probability density $4\pi p_{n\ell}(r)$ for the hydrogen atom. (From D. A. Davies, *Waves, Atoms and Solids.* London: Longman, 1978.)

Example 10.6 ***The radial probability density for the 1s state***

Find the maximum for the radial probability density $p(r)$ for a hydrogenlike atom in the 1s state, and thereby show that the most probable radial position, i.e., the radius with the highest radial probability density, is a_0/Z.

$$p(r) = r^2\left(\frac{1}{\pi}\right)\left(\frac{Z}{a_0}\right)^3 e^{-2Zr/a_0}$$

$$\frac{dp(r)}{dr} = \frac{2r}{\pi}\left(\frac{Z}{a_0}\right)^3 e^{-2Zr/a_0} - \frac{r^2}{\pi}\left(\frac{2Z}{a_0}\right)\left(\frac{Z}{a_0}\right)^3 e^{-2Zr/a_0} = 0$$

$$1 - \frac{Zr}{a_0} = 0$$

$$r = \frac{a_0}{Z}$$

As the principal quantum number increases, the electron moves out to greater distances from the nucleus. The average distance for an orbital electron may be expressed by the **expectation value** $\langle r \rangle$:

$$\langle r \rangle_{n\ell} = \int_0^\infty r p_{n\ell}(r)\, dr \tag{10.29}$$

$$= \frac{n^2 a_0}{Z}\left\{1 + \frac{1}{2}\left[1 - \frac{\ell(\ell+1)}{n^2}\right]\right\} \tag{10.30}$$

The expectation value for the radius for a hydrogenlike atom in the 1s state is $3a_0/2Z$. The most probable value of r and the expectation value of r are not equal.

Example 10.7 ***The average distance between the nucleus and the electron in the 1s state***

Use the wavefunction for an electron in a 1s orbital to derive the expression for the average distance between the electron and the nucleus in a hydrogenlike atom.

The average radius is obtained by multiplying the probability of finding the electron at r by r and integrating over all space:

$$\langle r \rangle = \int_0^\infty \psi_{1s} r \psi_{1s}\, d\tau = \int \psi_{1s}^2 r^3\, dr \sin\theta\, d\theta\, d\phi$$

The angular integration can be done immediately since ψ_{1s} is independent of θ and ϕ.

$$\langle r \rangle = \int \psi_{1s}^2 4\pi r^3\, dr$$

Integral tables show that

$$\int_0^\infty x^n e^{-kx}\, dx = \frac{n!}{k^{n+1}}$$

Thus the value of this integral with $n = 3$ is $6/k^4$. Substituting

$$\psi_{1s} = \frac{1}{\pi^{1/2}}\left(\frac{Z}{a_0}\right)^{3/2} e^{-Zr/a_0}$$

where z is the atomic number, yields

$$\langle r \rangle = \frac{3}{2}\frac{a_0}{Z}$$

The probability density for the electron in a hydrogenlike atom is given by

$$\psi^* \psi = |R(r)\Theta(\theta)\Phi(\phi)|^2 \tag{10.31}$$

where $Y_\ell^{m_\ell}(\theta, \phi)$ has been written $\Theta(\theta)\Phi(\phi)$ and where the symbols on the right represent the product of the wavefunction with its complex conjugate. The presentation of electron density as a function of r, θ, and ϕ would require four dimensions. One way to do this is to use the density of dots to represent the probability of finding an electron in a region of space and use stereo plots with a stereo viewer to see probability densities in three dimensions.* The method used most frequently is to depict surfaces that enclose some large percentage, say, 90%, of the electron density. Diagrams of this type are shown in Fig. 10.6. The probability densities in

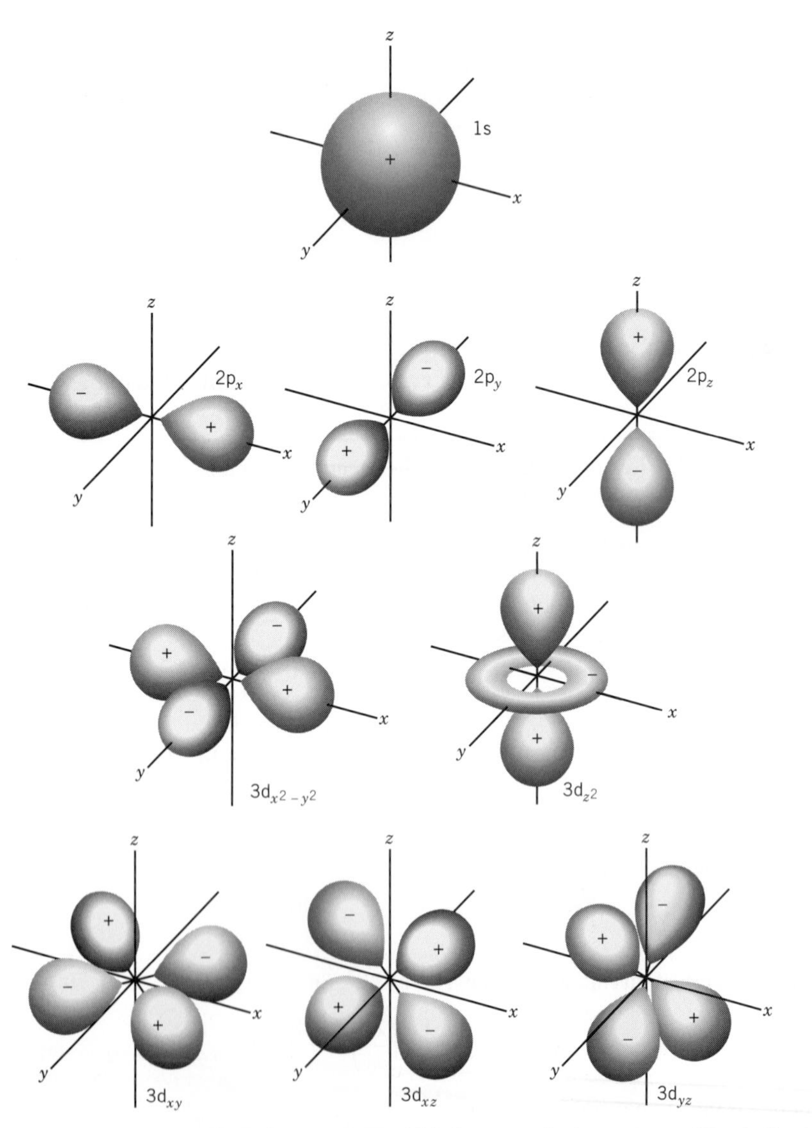

Figure 10.6 Contour surfaces for constant $\psi^*\psi$ for one-electron atoms. The indicated signs are those of the wavefunctions. These signs are indicated because they will be of interest later when we discuss molecular orbitals. The probability density is, of course, always positive.

*D. T. Cromer, *J. Chem. Educ.* **45**:626 (1968).

Fig. 10.6 are all positive because they are squares of the wavefunctions. The signs indicated are those of the wavefunctions themselves before squaring.

When $m \neq 0$, angular wavefunctions are more difficult to represent because they are complex, as we have seen in Table 9.2. For example, that table shows that Y_1^1 and Y_1^{-1} are both complex. This problem of representation is solved by using linear combinations $Y_1^1 + Y_1^{-1}$ and $Y_1^1 - Y_1^{-1}$, which are not complex, to form p_x and p_y orbitals. This is possible because Y_1^1 and Y_1^{-1} correspond to the same energy, and any linear combination of these wavefunctions is also an energy eigenfunction with the same energy. This has been done in writing the wavefunctions in Table 10.1. For example,

$$\psi_{2p_x} = \frac{1}{2^{1/2}}(Y_1^1 + Y_1^{-1})R_{21} \propto \sin\theta\cos\phi \tag{10.32}$$

$$\psi_{2p_y} = \frac{1}{2^{1/2}}(Y_1^1 - Y_1^{-1})R_{21} \propto \sin\theta\sin\phi \tag{10.33}$$

When $\ell = 2$, m can have values of 0, ± 1, and ± 2, and so there are five d orbitals. The possible linear combinations give the trigonometric forms in Table 10.1.

The orientations of the p orbitals can be calculated by considering the magnitudes and signs of the trigonometric functions at several angles. In the absence of an electric or magnetic field, electrons in p_x, p_y, and p_z orbitals all have the same energy; in fact, the energy depends only on the total quantum number n. For the hydrogen atom with quantum number $n = 2$ there are four degenerate states, all having the same energy in the absence of a magnetic or electric field.

The p orbitals do not have to point along the x, y, and z directions. Linear combinations of p_x, p_y, and p_z may be formed to point in any three mutually perpendicular directions. It will be seen in Chapter 11 that the directional character of certain chemical bonds results from the directed orientation of these and other orbitals. Note that opposite lobes of p orbitals have opposite signs.

There is another way to visualize the electron density associated with various atomic orbitals, and that is to use a three-dimensional plot to show the electron density in a plane through the nucleus. Figure 10.7 shows such plots of the electron density in the xy plane for atomic hydrogen in the 1s, 2s, $2p_x$, and $2p_y$ orbitals.

There are five independent d orbitals. The $3d_{z^2}$ orbital has two large regions of electron density above one axis, by convention the z axis, and a small donut-shaped orbital in the xy plane. The other four d orbitals have four equivalent lobes of electron density with two nodal planes separating them. Note that lobes that are opposite one another in these wavefunctions have the same sign.

One of the deficiencies in the diagrams in Fig. 10.6 for higher wavefunctions is that the nodal surfaces resulting from the radial functions $R(r)$ are not shown. Figure 10.5 shows that the number of radial nodes in a hydrogenlike wavefunction is $n - \ell - 1$. Figure 10.6 shows that the number of angular nodes (nodal planes) is ℓ. Thus the total number of nodes is $n - 1$.

10.4 ORBITAL ANGULAR MOMENTUM OF THE HYDROGENLIKE ATOM

A hydrogenlike atom may have orbital angular momentum, depending on its angular momentum quantum number ℓ. As we saw in equation 9.161, the square

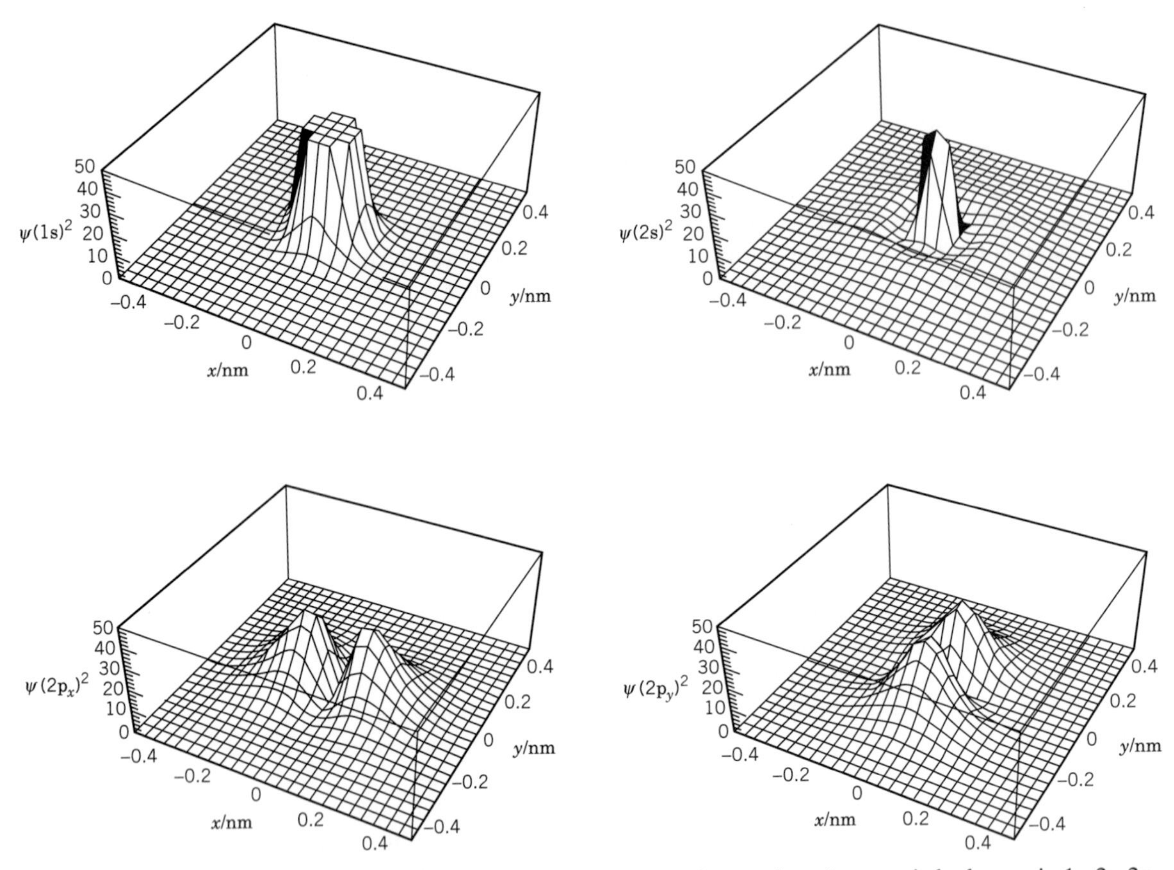

Figure 10.7 Plots of electron density in the xy plane for atomic hydrogen in 1s, 2s, $2p_x$, and $2p_y$ orbitals. (See Computer Problem 10.A.)

of the magnitude of the angular momentum is obtained by operating on the hydrogenlike wavefunction with the operator $\hat{L}^2$ for the square of the angular momentum. Since the wavefunction is $R(r)Y_\ell^m(\theta, \phi)$ and $\hat{L}^2$ operates only on θ and ϕ, we obtain equation 9.161, which was discussed earlier in connection with the rigid rotor. Thus, the angular momentum of a hydrogenlike atom can only have the values

$$L = \sqrt{\ell(\ell+1)}\hbar \qquad \ell = 0, 1, 2, \ldots \tag{10.34}$$

as given earlier in equation 9.162.

Operating on the spherical harmonics with the operator for the z component of the angular momentum $\hat{L}_z$, as we did in equation 9.163, yields the expression for the z component L_z of the angular momentum in terms of the magnetic quantum number m that we have seen before in equations 9.164 and 10.19:

$$L_z = m\hbar \qquad m = -\ell, -\ell+1, \ldots, \ell-1, \ell \tag{10.35}$$

The possible orientations of the angular momentum vectors of a hydrogenlike atom with $\ell = 1$ and $\ell = 2$ are given in Figs. 9.9 and 9.10. In the absence of a magnetic field, and *not including the spin of the electron,* which also is a kind of angular momentum, the energy of the hydrogenlike atom is independent of m. However, in the presence of a magnetic field, the energy depends on m, or the orientation of the angular momentum vector. In fact, the energies of the $2\ell + 1$ eigenstates with different values of m are all different. We say that the magnetic field has removed the $2\ell + 1$ degeneracy with respect to m. (Actually, the electron has an intrinsic angular momentum, called spin, in addition to the orbital angular momentum. For the present discussion, we will ignore the spin and treat the hydrogen atom as if the only angular momentum is from orbital motion.) The reason that the degeneracy (see the discussion on degeneracy in Section 9.7) is removed in a magnetic field is that when an atom has angular momentum L, the atom acts like a small magnet. We say that it has a **magnetic dipole moment** $\boldsymbol{\mu}$ given by

$$\boldsymbol{\mu} = \gamma_e \boldsymbol{L} \tag{10.36}$$

where γ_e is the **magnetogyric ratio** of the electron, equal to $-e/2m_e$. The z component of the dipole moment is then given by

$$\mu_z = -\frac{e}{2m_e} L_z \tag{10.37}$$

or in an eigenstate of L_z with eigenvalue $m\hbar$,

$$\mu_z = -\left(\frac{e\hbar}{2m_e}\right) m = -\mu_B m \tag{10.38}$$

where μ_B is called the **Bohr magneton:**

$$\mu_B = e\hbar/2m_e \tag{10.39}$$

which is the natural unit of magnetic dipole moment for electronic states.

When a magnetic dipole is placed in a magnetic field oriented along a given direction, the potential energy is given by the scalar product

$$E = -\boldsymbol{\mu} \cdot \boldsymbol{B} \tag{10.40}$$

where $\boldsymbol{B}$ is the **magnetic flux density.** Since we are free to choose the z direction any way we like, we can pick it to be along $\boldsymbol{B}$ so that $B_z = B$, the magnitude of $\boldsymbol{B}$. Then

$$E = -\mu_z B = \frac{eB}{2m_e} L_z \tag{10.41}$$

The Hamiltonian operator for the atom in a magnetic field is then found by adding this potential to $\hat{H}_0$, the Hamiltonian in the absence of the field:

$$\hat{H} = \hat{H}_0 + \frac{eB}{2m_e} \hat{L}_z \tag{10.42}$$

When this is applied to the eigenfunctions of a hydrogenlike atom (i.e., the eigenfunctions of $\hat{H}_0$), we find that these functions are eigenfunctions of $\hat{H}$ with eigenvalues

$$E_{n\ell m} = -\frac{m_e e^4 Z^2}{2(4\pi\epsilon_0)^2 n^2 \hbar^2} + \mu_B m B \tag{10.43}$$

with $n = 1, 2, \ldots$; $\ell = 0, 1, \ldots, n - 1$; and $m = \ell, \ell - 1, \ldots, -\ell$. Therefore, in the presence of a magnetic field the energy levels have been split into $2\ell + 1$ levels (see Fig. 10.8).

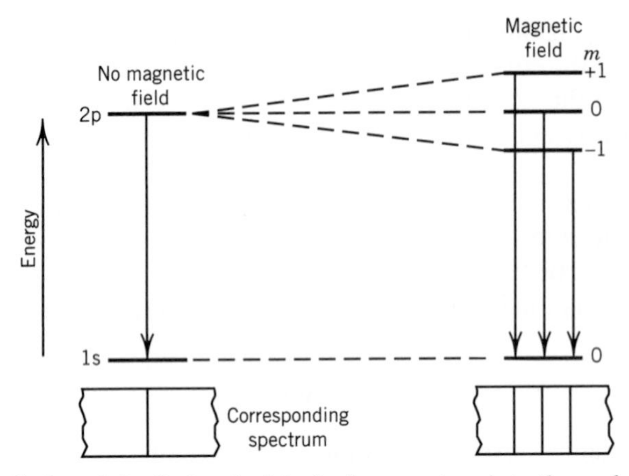

Figure 10.8 Splitting of the 2p level of the hydrogen atom into three closely spaced levels by a magnetic field, assuming for illustrative purposes that the electron spin is absent. The spectra for the transition from the 2p to the 1s level are shown in the absence and presence of a magnetic field. (From D. A. McQuarrie, *Quantum Chemistry.* Mill Valley, CA: University Science Books, 1983. Reprinted with permission from University Science Books.)

The splitting of energy levels in a magnetic field can be studied by measuring the magnetic susceptibility (see Section 22.5), by electron paramagnetic resonance (EPR, Section 15.9), and by optical spectroscopy in a magnetic field. The splitting of the spectral lines due to a magnetic field is called the **Zeeman effect.** When the splitting has a contribution due to the intrinsic (spin) magnetic moment of the electrons, to which we now turn, it is called the anomalous Zeeman effect (for historical reasons).

Example 10.8 ***Energy levels of hydrogenlike atoms in a magnetic field***

(*a*) Calculate the value of the Bohr magneton. (*b*) For a hydrogenlike atom with a 3d electron, what is the value of the orbital angular momentum? When the atom is placed in a magnetic field of 1 T, what are the relative values of the possible energy levels?

$$(a)\quad \mu_B = \frac{e\hbar}{2m_e} = \frac{(1.602\,177\times10^{-19}\ \text{C})(1.054\,572\times10^{-34}\ \text{J s})}{2(9.109\,390\times10^{-31}\ \text{kg})}$$
$$= 9.274\,01\times10^{-24}\ \text{J T}^{-1}$$

$$(b)\quad L = \sqrt{2(2+1)}\hbar = \sqrt{6}\hbar$$
$$L_z = -2\hbar, -\hbar, 0, \hbar, 2\hbar$$
$$E = -\mu_B m B_z$$
$$= -(9.274\times10^{-24}\ \text{J T}^{-1})(1\ \text{T})m$$

where $m = 0, \pm1, \pm2$.

10.5 ELECTRON SPIN

Because it is a nonrelativistic equation, the Schrödinger equation, by itself, does not account for all measurements on atoms and molecules. In 1928, Dirac developed the relativistic equation for a one-electron system and showed that it predicts the existence of electron spin, which Goudsmit and Uhlenbeck had proposed in 1925 to explain the splitting of certain spectroscopic lines. In nonrelativistic quantum theory, electron spin must be treated as an additional postulate. Since the spin angular momentum of an electron has no analogue in classical mechanics, we cannot construct spin angular momentum operators by first writing the classical Hamiltonian. However, it turns out that the treatment of spin angular momentum is closely analogous to the treatment of orbital angular momentum.

The **spin angular momentum** vector $\boldsymbol{S}$ has a magnitude $|\boldsymbol{S}| = S = [s(1+s)]^{1/2}\hbar$, where s is the **spin quantum number,** just as the orbital angular momentum $\boldsymbol{L}$ has a magnitude $|\boldsymbol{L}| = L = [\ell(\ell+1)]^{1/2}\hbar$, where ℓ is the orbital angular momentum quantum number. However, it is not correct to think of the spin angular momentum of an electron as being due to a spinning motion of the electron mass on its axis. Furthermore, the **component of the spin angular momentum in a particular direction,** arbitrarily referred to as the z direction, is given by $S_z = m_s\hbar$, where m_s is the **quantum number for the z component of the spin,** just as the component of the orbital angular momentum in a particular direction is given by $L_z = m_\ell\hbar$, where m_ℓ is the quantum number for the z component of the angular momentum. A remarkable fact about electrons is that their spin quantum number s, referred to simply as the **spin,** has the single value $\frac{1}{2}$. The magnitudes of the spin angular momentum S and its z component S_z are given by

$$S = \sqrt{s(s+1)}\hbar = \frac{\sqrt{3}}{2}\hbar \qquad \text{since } s = \frac{1}{2} \tag{10.44}$$

$$S_z = m_s\hbar \qquad m_s = \pm\frac{1}{2} \tag{10.45}$$

Although the spin s has a single value, the quantum number m_s for the z component has two possible eigenvalues, $\pm\frac{1}{2}$. The case where $m_s = +\frac{1}{2}$ is often referred to as "spin up," and $m_s = -\frac{1}{2}$ is referred to as "spin down."

We have introduced five new quantities in rapid succession. To keep them straight it will help to remember that these quantities and the operators $\hat{S}^2$ and $\hat{S}_z$ are comparable with quantities introduced for angular momentum. This comparison is shown in Table 10.2.

When the operators $\hat{S}^2$ and $\hat{S}_z$ are applied to spin functions, they yield eigenvalues. Since the spin eigenfunctions do not involve spatial coordinates, the two possible spin functions for an electron are represented by α and β:

$$\hat{S}^2\alpha = \tfrac{1}{2}(\tfrac{1}{2}+1)\hbar^2\alpha = \tfrac{3}{4}\hbar^2\alpha \tag{10.46}$$

$$\hat{S}^2\beta = \tfrac{1}{2}(\tfrac{1}{2}+1)\hbar^2\beta = \tfrac{3}{4}\hbar^2\beta \tag{10.47}$$

$$\hat{S}_z\alpha = +\tfrac{1}{2}\hbar\alpha \tag{10.48}$$

$$\hat{S}_z\beta = -\tfrac{1}{2}\hbar\beta \tag{10.49}$$

At this level of approximation, the operators $\hat{S}^2$ and $\hat{S}_z$ commute with the Hamiltonian operator $\hat{H}$, $\hat{L}^2$, and $\hat{L}_z$ so that the magnitude of the spin, the z component of the spin, the energy, the magnitude of the orbital angular momentum, and the z component of the orbital angular momentum can all have simultaneous eigenvalues (see Example 9.16).

Table 10.2 Quantities for Representing Orbital Angular Momentum and Spin Angular Momentum for a Single Electron

	Orbital Angular Momentum	*Spin Angular Momentum*
Angular momentum vector	$\boldsymbol{L}$	$\boldsymbol{S}$
Magnitude of above	$L = \sqrt{\ell(\ell+1)}\hbar$	$S = \sqrt{s(s+1)}\hbar$
Component of angular momentum vector in z direction	$L_z = m_\ell \hbar$	$S_z = m_s \hbar$
Operator for square of angular momentum	$\hat{L}^2$	$\hat{S}^2$
Operator for z component of angular momentum	$\hat{L}_z$	$\hat{S}_z$
Quantum number	$\ell\ (= 0, 1, 2, \ldots)$	$s\ (= \frac{1}{2})$
Quantum number for z component	$m_\ell\ (= -\ell, -\ell+1, \ldots, 0, \ldots, +\ell-1, +\ell)$	$m_s\ (= \pm\frac{1}{2})$
Magnetic dipole moment vector	$\boldsymbol{\mu}_\ell$	$\boldsymbol{\mu}_s$

The spin eigenfunctions, α and β, are orthonormal, as indicated by

$$\int \alpha^* \alpha \, d\sigma = \int \beta^* \beta \, d\sigma = 1 \tag{10.50}$$

$$\int \alpha^* \beta \, d\sigma = \int \alpha \beta^* \, d\sigma = 0 \tag{10.51}$$

where σ is called the spin variable, which has no classical analogue.

A *complete* wavefunction for a hydrogenlike atom must indicate the spin state of the electron. Since there are two spin functions, there are twice as many wavefunctions for the hydrogen atom as we indicated earlier: $\psi\alpha$ and $\psi\beta$ for each of the previous ψ's. Thus, a complete state specification for the hydrogen atom requires the four quantum numbers n, ℓ, m_ℓ, and m_s. This increases the degeneracy of the energy levels from n^2 to $2n^2$.

The two possible orientations for the spin angular momentum vector for an electron in a magnetic field are shown in Fig. 10.9. The vector has magnitude $S = [\frac{1}{2}(1+\frac{1}{2})]^{1/2}\hbar$. Its component S_z may be $+\frac{1}{2}\hbar$ or $-\frac{1}{2}\hbar$. The components S_x and S_y cannot be determined simultaneously with S_z, since S_x and S_y do not commute with S_z, just as L_x and L_y do not commute with L_z.

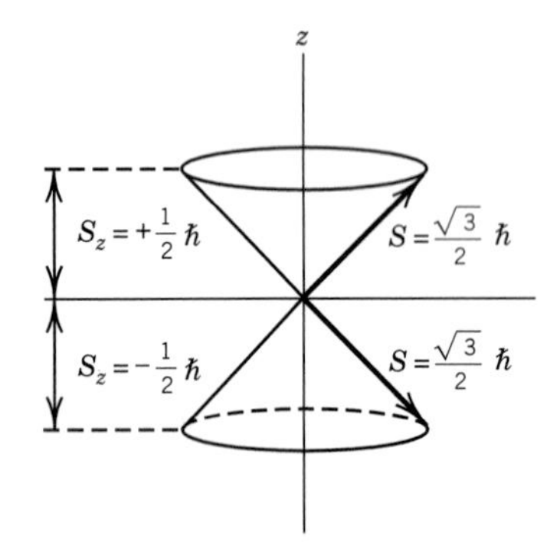

Figure 10.9 Possible orientations for the spin angular momentum $\boldsymbol{S}$ of the electron in a magnetic field. The magnitude S of the spin angular momentum is $[s(s+1)]^{1/2}\hbar$, where s is the spin quantum number. Thus, $S = (\sqrt{3}/2)\hbar$. The z component S_z of the spin angular momentum vector is $m_s\hbar$, where the spin quantum number m_s for the z is $\pm 1/2$.

Because of its charge and intrinsic spin angular momentum, an electron has a magnetic dipole moment $\boldsymbol{\mu}_s$. As with orbital angular momentum (equation 10.36), the magnetic moment of an electron is proportional to its spin angular momentum $\boldsymbol{S}$:

$$\boldsymbol{\mu}_s = -\frac{g_e e}{2m_e}\boldsymbol{S} \tag{10.52}$$

where g_e is the **electron** g **factor,** which is 2.002 322.

The component μ_z of the magnetic moment of the electron in the direction of the applied magnetic field is

$$\mu_z = -\frac{g_e e}{2m_e} S_z \tag{10.53}$$

where S_z is the component of the spin angular momentum in the direction of the field. Since S_z is given by $m_s \hbar$,

$$\begin{aligned} \mu_z &= -\frac{g_e e \hbar}{2m_e} m_s \\ &= -g_e \mu_B m_s \end{aligned} \tag{10.54}$$

The energy of the spin magnetic moment in a magnetic field B is (see equation 10.41)

$$E = g_e \mu_B m_s B \tag{10.55}$$

The electron spin has two energy states in a magnetic field: $E = +\frac{1}{2} g_e \mu_B B$ and $E = -\frac{1}{2} g_e \mu_B B$. The transition between these two levels is studied in electron spin resonance.

This energy is the eigenvalue of the operator $\hbar^{-1} g_e \mu_B B \, \hat{S}_z$ or $g_e(eB/2m_e)\,\hat{S}_z$. When this is added to the Hamiltonian of a hydrogenlike atom in a magnetic field, equation 10.42, we find that the total Hamiltonian is

$$\begin{aligned} \hat{H} &= \hat{H}_0 + \frac{eB}{2m_e}\hat{L}_z + \frac{g_e eB}{2m_e}\hat{S}_z \\ &= \hat{H}_0 + \frac{eB}{2m_e}(\hat{L}_z + g_e \hat{S}_z) \end{aligned} \tag{10.56}$$

The eigenvalues of $\hat{H}$ in a magnetic field are now given by

$$E_{n\ell m m_s} = -\frac{m_e e^4 Z^2}{2(4\pi\epsilon_0)^2 \hbar n^2} + \frac{eB\hbar}{2m_e}(m_\ell + g_e m_s) \tag{10.57}$$

In Fig. 10.10, we show the splitting of the 2s and 2p orbital states in a hydrogenlike system. Since g_e is very close to 2, the state with $m_\ell = 1$ and $m_s = -\frac{1}{2}$ and the state with $m_\ell = -1$ and $m_s = +\frac{1}{2}$ are approximately degenerate.

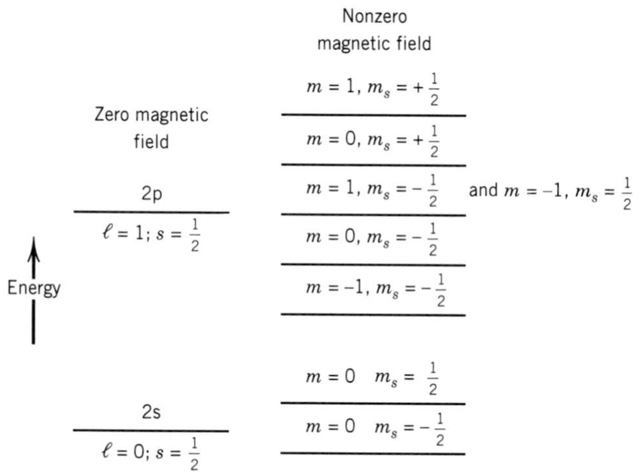

Nonzero magnetic field
Zero magnetic field
Energy
2p
$\ell = 1; s = \frac{1}{2}$
$m = 1,\ m_s = +\frac{1}{2}$
$m = 0,\ m_s = +\frac{1}{2}$
$m = 1,\ m_s = -\frac{1}{2}$ and $m = -1,\ m_s = \frac{1}{2}$
$m = 0,\ m_s = -\frac{1}{2}$
$m = -1,\ m_s = -\frac{1}{2}$
2s
$\ell = 0; s = \frac{1}{2}$
$m = 0\ \ m_s = \frac{1}{2}$
$m = 0\ \ m_s = -\frac{1}{2}$

Figure 10.10 The energies of 2s and 2p states of a hydrogenlike atom in a magnetic field including orbital and spin angular momentum.

In general, there is another term in the Hamiltonian that is due to the interaction between the spin and orbital parts of the angular momentum and is called the spin–orbit coupling. When this is important, we must define the total angular momentum of the atomic system, $\boldsymbol{J} = \boldsymbol{L} + \boldsymbol{S}$. We will discuss this briefly at the end of this chapter (Section 10.12).

10.6 VARIATIONAL METHOD

Although the Schrödinger equation can be solved for certain simple systems such as the particle in a box, the harmonic oscillator, the rigid rotor, and the hydrogen-like atom, it cannot be solved for many-electron atoms or molecules. It is therefore necessary to use approximation methods, of which the variational method is one of the most important because it allows us to calculate an upper bound for the energy eigenvalue.

The variational method is based on the theorem that if ψ is any normalized, well-behaved function of the coordinates, then

$$\int \psi^* \hat{H} \psi \, \mathrm{d}\tau \geq E_{\mathrm{gs}} \tag{10.58}$$

where $\hat{H}$ is the correct Hamiltonian and E_{gs} is the true ground-state energy. If ψ is the ground-state eigenfunction, the equality applies; however, the use of an approximate wavefunction always yields an energy that is higher than the ground-state energy. When the true eigenfunction is not known, a **trial wavefunction** can be devised by adding up functions ϕ_i that each obey the correct boundary conditions:

$$\psi = \sum c_i \phi_i \tag{10.59}$$

This wavefunction is used in equation 10.58, and the constants are varied to obtain the lowest possible energy E for this set of functions by use of the equation

$$E = \frac{\int \psi^* \hat{H} \psi \, \mathrm{d}\tau}{\int \psi^* \psi \, \mathrm{d}\tau} \tag{10.60}$$

The denominator is required to normalize the trial wavefunctions. The best values of the constants are obtained by solving the simultaneous equations $\partial E/\partial c_i = 0$. The use of equation 10.60 with these values of the constants in equation 10.59 will give an energy that is greater than E_{gs}. In principle, as one increases the flexibility of the trial function, one gets closer and closer to the true ground-state energy.

The variational method will be used in the next section to obtain an approximate energy for the helium atom. In Section 11.3, we will also see how the approximate energy for the ground state of the hydrogen molecule ion and its first excited state can be obtained by the variational method.

Example 10.9 ***The variational method applied to a particle in a box***

Use the variational method to obtain an upper bound to the ground-state energy of a particle in a one-dimensional box and compare the result with the true value given in equation 9.63.

A normalized trial variational function that satisfies the boundary conditions that $\psi = 0$ at $x = 0$ and $x = a$ is

$$\psi = \frac{\sqrt{30}}{a^{5/2}} x(a-x)$$

$$\hat{H} = -\frac{\hbar^2}{2m}\frac{d^2}{dx^2} \qquad 0 \le x \le a$$

$$\int_0^a \psi^* \hat{H} \psi \, d\tau = -\frac{30\hbar^2}{a^5 2m}\int_0^a (ax - x^2)\frac{d^2}{dx^2}(ax - x^2)\, dx$$

$$= -\frac{30\hbar^2}{a^5 m}\int_0^a (x^2 - ax)\, dx = \frac{5h^2}{4\pi^2 ma^2}$$

$$\frac{5h^2}{4\pi^2 ma^2} \ge E_0$$

The true value is $h^2/8ma^2$.

$$\% \text{ error} = \frac{(5/4\pi^2) - (1/8)}{(1/8)} \times 100 = 1.3\%$$

This is not a true variational calculation because there is no variable parameter, but it does provide an illustration that an approximate wavefunction always yields a higher energy than the true ground-state wavefunction. We will use the variational method in the next section and in Section 11.3.

10.7 HELIUM ATOM

The helium atom has two electrons, and the coordinates used in writing the Hamiltonian are shown in Fig. 10.11. Since we are interested only in the internal motions of the electrons with respect to the nucleus in a heliumlike atom, we will ignore the kinetic energy of the nucleus (i.e., fix the nucleus), so the Hamiltonian operator may be written

$$\hat{H} = -\frac{\hbar^2}{2m_e}(\nabla_1^2 + \nabla_2^2) - \frac{1}{4\pi\epsilon_0}\left(\frac{Ze^2}{r_1} + \frac{Ze^2}{r_2} - \frac{e^2}{r_{12}}\right) \qquad (10.61)$$

where $Z = 2$. The first term is the kinetic energy operator for the two electrons, the next two terms represent the potential energy of each of the electrons in the field of the nucleus, and the last term represents the interelectronic potential energy.

When this Hamiltonian operator is used in the Schrödinger equation $\hat{H}\psi = E\psi$, we can in principle obtain the eigenfunctions ψ and corresponding eigenvalues E. The wavefunctions ψ are functions of the coordinates of the two electrons $(x_1, y_1, z_1, x_2, y_2, z_2)$. The interelectronic repulsion term $e^2/4\pi\epsilon_0 r_{12}$ makes it impossible to obtain an analytic solution of the Schrödinger equation for a heliumlike atom; thus, it is necessary to use approximation methods to obtain the wavefunctions and energies. Fortunately, in the case of heliumlike atoms these approximation methods yield the wavefunctions and energies to any desired degree of accuracy.

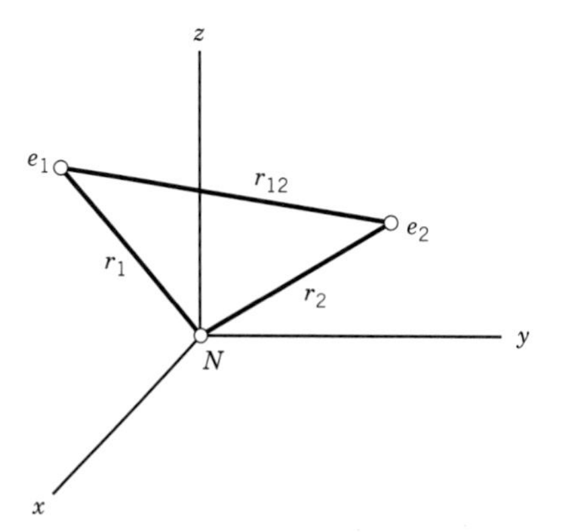

Figure 10.11 Coordinates of the electrons in the helium atom.

As a first approximation we will ignore the $e^2/4\pi\epsilon_0 r_{12}$ term. This amounts to treating the electrons as if they do not interact with one another. The Hamiltonian operator is then the sum of the Hamiltonian operators for the two independent particles:

$$\hat{H} = \hat{H}_1 + \hat{H}_2 \tag{10.62}$$

where

$$\hat{H}_1 = -\frac{\hbar^2}{2m}\nabla_1^2 - \frac{Ze^2}{4\pi\epsilon_0 r_1} \tag{10.63}$$

$$\hat{H}_2 = -\frac{\hbar^2}{2m}\nabla_2^2 - \frac{Ze^2}{4\pi\epsilon_0 r_2} \tag{10.64}$$

The approximate wavefunction can be written as the product of the wavefunctions ψ_1 and ψ_2 for the two independent particles because the Schrödinger equation can be divided into separate equations for the two electrons:

$$\hat{H}_1\psi_1 = E_1\psi_1 \tag{10.65}$$

$$\hat{H}_2\psi_2 = E_2\psi_2 \tag{10.66}$$

The energy of the whole system is given by the sum of the energies of the two electrons:

$$E = E_1 + E_2 \tag{10.67}$$

In this approximation the wavefunction of the ground state of the helium atom is given by the product of the two 1s orbitals:

$$\psi = \frac{1}{\pi^{1/2}}\left(\frac{Z}{a_0}\right)^{3/2} e^{-Zr_1/a_0} \frac{1}{\pi^{1/2}}\left(\frac{Z}{a_0}\right)^{3/2} e^{-Zr_2/a_0} \tag{10.68}$$

where $Z = 2$ for helium. This is generally abbreviated $\psi = 1s(1)1s(2)$, where (1) and (2) refer to electrons 1 and 2. We can now use this approximate wavefunction as a trial wavefunction.

The trial wavefunction in equation 10.68 can be tested by using it to calculate the approximate energy of the helium atom with respect to the nucleus and two electrons infinitely distant from one another. The exact energy for this is the negative of the experimental energy required to remove the two electrons from the helium atom, which is 79.0 eV. The approximate wavefunction (equation 10.68) is an eigenfunction of $\hat{H}_1 + \hat{H}_2$ with eigenvalue equal to twice the 1s orbital energy of a helium ion, or $2E_{1s}(Z = 2) = 8(-13.6) = -108.8$ eV. The repulsion energy of the two electrons in the approximate wavefunction (equation 10.68) is given by

$$\int d\upsilon_1 \int d\upsilon_2 \psi^* \left(\frac{e^2}{4\pi\epsilon_0 r_{12}}\right)\psi \tag{10.69}$$

When this is calculated it is found to have a value of 34.0 eV, so that the total approximate energy of the helium atom is -74.8 eV, compared with the exact value of -79.0 eV. Notice that the approximate value is larger than the true value, as is required by the variational theorem.

The variational method can be used to obtain a more accurate value for the ground-state energy of the helium atom. The nuclear charge Z in the 1s orbital wavefunctions of equation 10.68 can be used as a variational parameter, by replacing it with an effective nuclear charge Z'. The rationale for using Z' in equation 10.68 is that each electron "sees" only an effective nuclear charge $Z' < Z$ because the other electron "screens" the nucleus. If we calculate the average energy including the interaction term $e^2/4\pi\epsilon_0 r_{12}$ using an approximate wavefunction of the type 10.68, but with z replaced by Z', we find

$$E' = \left[(Z')^2 - 27\frac{Z'}{8}\right]\frac{e^2}{4\pi\epsilon_0 a_0} \tag{10.70}$$

We choose Z' to give the lowest energy by minimizing E' with respect to Z'. Differentiating equation 10.70 with respect to Z', we find

$$\frac{dE'}{dZ'} = \left(2Z' - \frac{27}{8}\right)\frac{e^2}{4\pi\epsilon_0 a_0} \tag{10.71}$$

so that $Z'_{\text{min}} = 27/16$ and $E'_{\text{min}} = -77.5$ eV, compared with the exact energy of -79.0 eV. We see from this how well we can do using the variational method with only one parameter. By adding further parameters and changing the form of the approximate wavefunction (for example, by adding small amounts of higher hydrogenlike orbitals, such as 2p, 3p, 3d, etc.), we can get an energy that will approach closer and closer to one that we can measure.

10.8 PAULI EXCLUSION PRINCIPLE

The wavefunction that we have just discussed for the ground state of the helium atom is incomplete in that it does not include the spin functions (α or β) for the two electrons. We can write the following four spin functions for the two electrons:

$$\alpha(1)\alpha(2) \qquad \beta(1)\beta(2) \qquad \alpha(1)\beta(2) \qquad \alpha(2)\beta(1) \tag{10.72}$$

where $\alpha(1)$ indicates that electron 1 has the spin function α. However, the last two spin functions cannot be used as they are because they imply that it is possible to distinguish between the two electrons. Electrons are identical to one another, and it is not possible to distinguish between them. The wavefunction for the electrons must reflect this, and since the electrons are in identical orbitals, the spin functions for the two electrons must be written in such a way that they do not distinguish between the two electrons. This may be done by writing the spin functions as follows:

$$\alpha(1)\alpha(2) \tag{10.73}$$

$$\beta(1)\beta(2) \tag{10.74}$$

$$2^{-1/2}[\alpha(1)\beta(2) + \alpha(2)\beta(1)] \tag{10.75}$$

$$2^{-1/2}[\alpha(1)\beta(2) - \alpha(2)\beta(1)] \tag{10.76}$$

The first three functions are unchanged when the two electrons are interchanged. These three functions are therefore **symmetric** with respect to electron interchange. The fourth function changes sign when the two electrons are interchanged and is therefore **antisymmetric** with respect to electron interchange.

It is found experimentally that wavefunctions (including both spatial and spin functions) of electrons must be antisymmetric with respect to the interchange of any two electrons. This fact was discovered by Pauli in 1926. The Pauli exclusion principle may be stated as follows: **The wavefunction for any system of electrons must be antisymmetric with respect to the interchange of any two electrons.** Another way of stating the Pauli principle in a way that is useful in chemistry is to say that no two electrons in an atom may have the same four quantum numbers n, ℓ, m_ℓ, and m_s. In the nonrelativistic quantum mechanics that we are using, the Pauli principle is an additional postulate.

Since we have described the ground state of the helium atom as 1s(1)1s(2), which is symmetric, we need to multiply it by an antisymmetric spin function so that the total wavefunction will be antisymmetric. Thus, the total wavefunction for the ground state of helium can be approximated as

$$\psi = 1s(1)1s(2)2^{-1/2}[\alpha(1)\beta(2) - \alpha(2)\beta(1)] \tag{10.77}$$

In 1929 Slater developed a mathematical method for constructing approximate wavefunctions satisfying the antisymmetry requirement by writing them as determinants. In a **Slater determinant** the elements in a given column involve the same spin orbital, while elements in the same row involve the same electron. The approximate helium wavefunction (equation 10.77) can be written in the form

$$\psi = \frac{1}{\sqrt{2}}\begin{vmatrix} 1s(1)\alpha(1) & 1s(1)\beta(1) \\ 1s(2)\alpha(2) & 1s(2)\beta(2) \end{vmatrix} \tag{10.78}$$

The requirement of the Pauli principle that no two electrons in an atom or molecule have all quantum numbers the same is provided for automatically by the Slater determinant. If two rows or two columns of a determinant are identical, the determinant vanishes. Another useful property of a determinant is that the interchange of two rows or columns changes the sign of the determinant, showing that the Pauli principle is satisfied automatically by a wavefunction expressed by a determinant.

The inclusion of spin does not alter the energy calculations of the preceding section because the Hamiltonian does not include spin. However, spin considerations are essential for the excited states of helium and for the lithium atom, as we will soon see. This is because of the Pauli principle, not because of explicitly spin-dependent terms occurring in the Hamiltonian.

Particles with half-integral spin ($s = \frac{1}{2}, \frac{3}{2}, \ldots$) all require antisymmetric wavefunctions and are referred to as **fermions** because they must obey a kind of statistics called Fermi–Dirac statistics. Particles with integral spin ($s = 0, 1, 2, \ldots$) all require symmetric wavefunctions and are referred to as **bosons** because they follow a different statistical law called Bose–Einstein statistics.

Example 10.10 ***The wavefunction in equation 10.78 is an eigenfunction of total electron spin***

Show that the Slater determinant given in equation 10.78 for the ground state for the helium atom is an eigenfunction of $\hat{S}_{z,\text{tot}} = \hat{S}_{z1} + \hat{S}_{z2}$. What is the eigenvalue?

We will use the facts that

$$\hat{S}_z \alpha = +\tfrac{1}{2}\hbar\alpha$$
$$\hat{S}_z \beta = -\tfrac{1}{2}\hbar\beta$$

The wavefunction is

$$\psi = 1s(1)1s(2)2^{-1/2}[\alpha(1)\beta(2) - \alpha(2)\beta(1)]$$
$$\hat{S}_{z1}\psi = 1s(1)1s(2)2^{-1/2}[\beta(2)\tfrac{1}{2}\hbar\alpha(1) + \alpha(2)\tfrac{1}{2}\hbar\beta(1)]$$
$$\hat{S}_{z2}\psi = 1s(1)1s(2)2^{-1/2}[-\alpha(1)\tfrac{1}{2}\hbar\beta(2) - \beta(1)\tfrac{1}{2}\hbar\alpha(2)]$$
$$\hat{S}_{z,\text{tot}}\psi = \hat{S}_{z1}\psi + \hat{S}_{z2}\psi = 0$$

Thus the eigenvalue of the z component of the spin is zero.

Comment:

We speak of the Pauli principle as being a postulate of quantum mechanics, and that raises questions as to what it takes to make a postulate and where this particular one comes from. We are going to see in the next several sections that the periodic table has the form it does because of the Pauli principle. If a system of electrons in an atom could be symmetric with respect to the interchange of any two electrons, there would be an entirely different set of elements. Since the Pauli principle cannot be derived from a nonrelativistic theory, it must be taken as a postulate of quantum mechanics.

We have seen that the wavefunction for the ground state of the helium atom may be approximated by the product of two 1s hydrogenlike wavefunctions (orbitals). We might expect that the wavefunction for the first excited state of helium may be approximated by the product of 1s and 2s hydrogenlike wavefunctions. Thus, we will examine $\psi = 1s(1)2s(2)$. This spatial wavefunction suffers from the problem that it is not possible to distinguish between two electrons. The actual wavefunction for the first excited state should be written in one of the two ways that do not distinguish between the two electrons:

$$\psi_a = 2^{-1/2}[1s(1)2s(2) + 1s(2)2s(1)] \tag{10.79}$$

$$\psi_b = 2^{-1/2}[1s(1)2s(2) - 1s(2)2s(1)] \tag{10.80}$$

The first function is symmetric and the second is antisymmetric. However, in either case the probability density ψ^2 is not altered by interchanging the electrons. Note that there is no way to make an antisymmetric function from 1s(1)1s(2). The two excited states represented by ψ_a and ψ_b have different energies, and ψ_b has the lower energy. The experimental value of the energy for the state approximated by ψ_b is -59.2 eV and that for ψ_a is -58.4 eV.

Now we want to incorporate spin into the spatial wavefunction (10.80) for the first excited state of the helium atom. Since this spatial wavefunction is antisymmetric, it must be multiplied by symmetric spin functions according to the

Pauli principle. There are three of these. Therefore, the wavefunctions of the first excited state of helium are

$$\psi_1 = 2^{-1/2}[1s(1)2s(2) - 1s(2)2s(1)]\alpha(1)\alpha(2) \tag{10.81}$$

$$\psi_2 = 2^{-1}[1s(1)2s(2) - 1s(2)2s(1)][\alpha(1)\beta(2) + \alpha(2)\beta(1)] \tag{10.82}$$

$$\psi_3 = 2^{-1/2}[1s(1)2s(2) - 1s(2)2s(1)]\beta(1)\beta(2) \tag{10.83}$$

Thus, the first excited state of helium is a **triplet** state because it has a degeneracy of 3 due to the spin. In the presence of a magnetic field the first excited state is split into three energy levels. Since the sum of the spin quantum numbers of the two electrons is 1, this net spin may be oriented in one of three ways in a magnetic field. The z components of the spin angular momentum may be $-\hbar$, 0, and $+\hbar$, and so the energy of the atom in a magnetic field may have three values. In the ground state for helium the electrons are paired, so the resultant electron spin is zero and the ground state is a singlet.

Since ψ_a is symmetric, it must be multiplied by an antisymmetric spin function to obtain an antisymmetric total wavefunction:

$$\psi = 2^{-1}[1s(1)2s(2) + 1s(2)2s(1)][\alpha(1)\beta(2) - \alpha(2)\beta(1)] \tag{10.84}$$

The second excited state is a singlet state. Since $\ell = 0$ and the total z component of spin is zero, this state is not split by application of a magnetic field.

The Pauli exclusion principle did not have an effect on our consideration of the hydrogen atom, and it had only a modest effect on our consideration of the helium atom, where we found the first excited state to be a triplet. But the Pauli principle has a major effect on the quantum mechanical treatment of the lithium atom. Since by the Pauli principle the 1s orbital can accommodate only two electrons, the third electron must be in a 2s orbital. Following the procedure of forming determinants, this wavefunction for the ground state of Li can be written as a Slater determinant:

$$\begin{aligned}\psi &= \frac{1}{\sqrt{6}}\begin{vmatrix} 1s(1)\alpha(1) & 1s(1)\beta(1) & 2s(1)\alpha(1) \\ 1s(2)\alpha(2) & 1s(2)\beta(2) & 2s(2)\alpha(2) \\ 1s(3)\alpha(3) & 1s(3)\beta(3) & 2s(3)\alpha(3) \end{vmatrix} \\ &= \frac{1}{\sqrt{6}}[1s(1)\alpha(1)1s(2)\beta(2)2s(3)\alpha(3) - 1s(1)\alpha(1)1s(3)\beta(3)2s(2)\alpha(2) \\ &\quad - 1s(1)\beta(1)1s(2)\alpha(2)2s(3)\alpha(3) + 1s(1)\beta(1)1s(3)\alpha(3)2s(2)\alpha(2) \\ &\quad + 2s(1)\alpha(1)1s(2)\alpha(2)1s(3)\beta(3) - 2s(1)\alpha(1)1s(3)\alpha(3)1s(2)\beta(2)]\end{aligned} \tag{10.85}$$

In contrast with the determinant for the first excited state of He, this wavefunction cannot be written as the product of the spatial function and a spin function. The last column of the determinant could have been written with a β instead of an α; thus, the ground state of the lithium atom is doubly degenerate.

In using equation 10.85 in a variational treatment the nuclear charge Z may be replaced by Z_1 in the 1s function and by Z_2 in the 2s function to allow for the fact that the electrons are partially screened from the nuclear charge of 3+. The variational treatment yields $Z_1 = 2.69$, $Z_2 = 1.78$. The variational energy is -201.2 eV, compared with the experimental ground-state energy of -203.48 eV.

This treatment is approximate in that the determinantal wavefunction is not an exact solution of the Schrödinger equation containing the interelectronic repulsion terms $(e^2/r_{12} + e^2/r_{13} + e^2/r_{23})/4\pi\epsilon_0$.

10.9 HARTREE–FOCK SELF-CONSISTENT FIELD METHOD

For atoms with more electrons, the methods for calculating wavefunctions that we have illustrated for helium and lithium rapidly become impractical. In 1928 Hartree introduced the self-consistent field (SCF) method. This method may be used to calculate the ground-state wavefunction and energy for any atom.

If interelectron repulsion terms in the Schrödinger equation are ignored, the Schrödinger equation for an n-electron atom may be separated into n one-electron hydrogenlike equations. The approximate wavefunction obtained in this way is the product of n one-electron functions that are hydrogenlike wavefunctions (orbitals). Hydrogenlike orbitals use the full nuclear charge Z, but we know that the outer electrons of an atom are shielded from the nuclear charge by the inner electrons so that the effective charge is less.

Hartree used a variational function ϕ, which is the product of n orbitals g_i that contain parameters to be evaluated by the variational method (e.g., effective nuclear charges):

$$\phi = g_1(r_1,\theta_1,\phi_1)g_2(r_2,\theta_2,\phi_2)\cdots g_n(r_n,\theta_n,\phi_n) \tag{10.86}$$

Each orbital in this variational function is taken to be the product of a radial factor $h_i(r_i)$ and a spherical harmonic $Y_{\ell_i}^{m_i}(\theta_i,\phi_i)$:

$$g_i = h_i(r_i)Y_{\ell_i}^{m_i}(\theta_i,\phi_i) \tag{10.87}$$

Hartree's procedure was to first estimate the form of the orbitals $g_1,\ldots,g_n$. Since the wavefunction of the atom is taken as the product of these orbitals, the Schrödinger equation for the atom may be separated into n equations of the type

$$\left[-\frac{\hbar^2}{2m}\nabla_i^2 + V(r_i)\right]g_i = \epsilon_i g_i \tag{10.88}$$

where ϵ_i is the energy of the orbital for electron i. These equations are solved by successive approximations. The potential energy function $V(r_i)$ for any one electron is obtained on the assumption that the electric charge of all of the other electrons is smeared out to form a spherically symmetric charge cloud. The orbital g_i of the first electron obtained in this way is used to improve the potential energy function $V(r_2)$ for use in the Schrödinger equation for the second electron to obtain an improved orbital g_2 for it. This process is continued for all n electrons, and then the process is started over with electron 1. The calculation of improved orbitals is continued in this way until there is no further change in the orbitals. The product of these orbitals gives the Hartree self-consistent field wavefunction for the atom.

Hartree provided for spin and the Pauli principle by putting no more than two electrons in each orbital, but his wavefunctions did not involve spin and were not made to be antisymmetric with respect to the interchange of electrons. In 1930 Fock (and Slater) pointed out that it is necessary to use spin orbitals and to take linear combinations of antisymmetric products of spin orbitals. A self-consistent field calculation carried out in this way is referred to as a Hartree–Fock calculation.

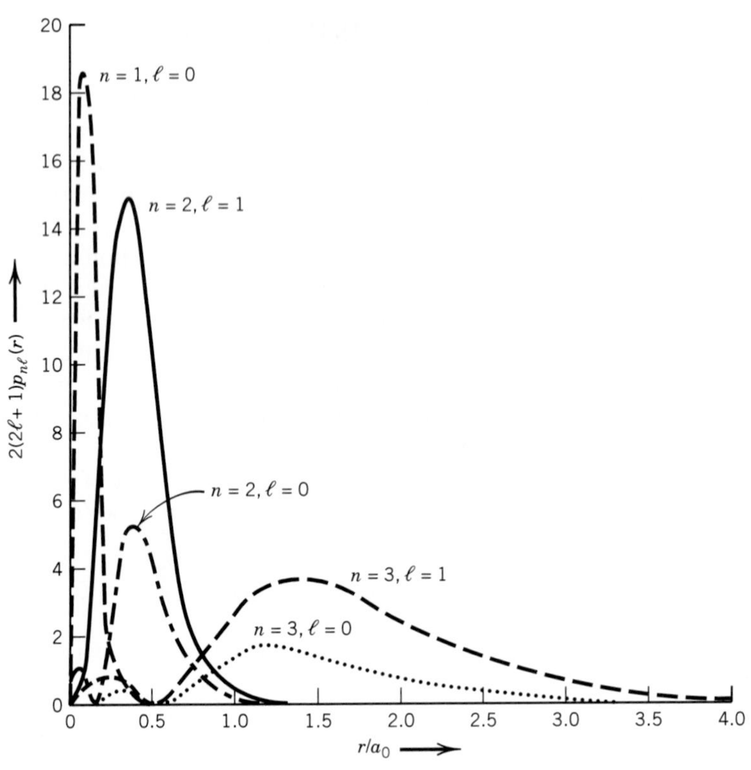

Figure 10.12 Radial probability densities for argon calculated by the Hartree–Fock method. (From R. Eisberg and R. Resnick, *Quantum Physics.* Hoboken, NJ Wiley, 1985.)

Figure 10.12 shows the radial electron density of argon orbitals calculated by the Hartree–Fock method. For each value of the principal quantum number n, the probability density is largely concentrated in a narrow range of radii. The set of orbitals having the same principal quantum number n is referred to as a **shell,** and the set of orbitals with the same n value and ℓ value is referred to as a **subshell.** Since the orbitals involve the same angular dependence as the hydrogenlike atomic wavefunctions, the subshells may be referred to as 1s, 2s, 2p, 3s, ... Note that the ns orbitals are more likely to be close to the nucleus than the np, nd, ... (penetration), so that the ns are less shielded from the nucleus than the np, nd, ... (see Fig. 10.13).

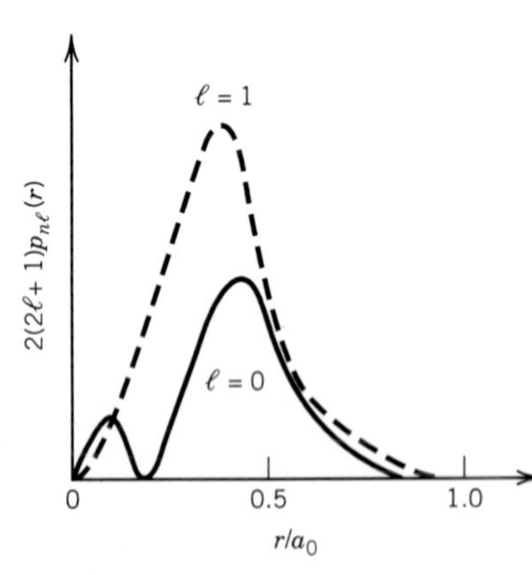

Figure 10.13 Blowup of the $n = 2$, $\ell = 1$ and $n = 2$, $\ell = 0$ radial probability densities for argon near the nucleus, showing the greater penetration of the $\ell = 0$ (s) orbital.

As the atomic number (nuclear charge) increases, the energies of the inner orbitals become more negative because of the increased attraction between the nucleus and the electrons. The p orbitals have higher (less negative) energies than the s orbitals because p electrons have lower probability densities in the neighborhood of the nucleus and feel the attraction of the nucleus less. At or very near to the nucleus an s orbital is not shielded by inner electrons; therefore, it has a low energy (i.e., a large negative potential).

The Hartree–Fock calculated energies for atoms usually agree with experimental values to about 1%. The method provides for the interactions between electrons in an average way, but it does not provide for their instantaneous interactions. Since electrons tend to stay away from one another, we may speak of correlation in the positions of electrons at any instant. The **correlation energy** is the difference between the exact energy and the Hartree–Fock energy. This energy, which is of the order of an electron volt or more, is large enough to be a serious problem in the calculation of differences in energies, as, for example, in calculating enthalpy changes for reaction.

The effects of instantaneous electron interaction may be provided for by including excited configurations in a trial wavefunction, using the variational approach. This method is referred to as configuration interaction. By including more configurations in the wavefunction used in the variational method, a representation of the true wavefunction may be approached more closely.

10.10 THE PERIODIC TABLE AND THE AUFBAU PRINCIPLE

The quantum theory of atoms provides an explanation of the structure of the periodic table. As we have seen, the electron subshells in atoms may be designated 1s, 2s, 2p, 3s, ... According to the Pauli exclusion principle, an s subshell may contain 2 electrons, a p subshell 6, and a d subshell 10. Thus, the subshells each contain $2(2\ell + 1)$ electrons.

To find the ground-state **electron configuration** of an atom, we add electrons to the subshells beginning with the lowest energy until we have added the correct number for that atom. This procedure is called the **aufbau** (building-up) **principle.**

The electron configurations for the first 36 elements found in this way are given in Table 10.3. The electron configurations of the other elements are given in Table C.6. The symbols [He], [Ne], and [Ar] represent the closed-shell electron configurations of these elements. In Section 10.13 we will consider the significance of the atomic term symbols given in the fourth column of Table 10.3.

The electron configurations of the elements account for the periodicity of physical and chemical properties of the elements. The periodic table can be arranged to show the electron configurations of the elements, as in Fig. 10.14.* This form of the periodic table makes it clear why the lanthanides (58 to 71) and actinides (90 to 103) have such similar chemical properties. Table 10.3 gives the orbital radii of isolated atoms; this is the radius of the maximum in radial probability density of the outermost orbital. Within a row of the periodic table, the atomic radius tends to decrease with increasing atomic number. Within a column in the periodic table, the atomic radius tends to increase with the atomic number. These general trends in atomic radius can be understood in terms of two factors that primarily determine the radius of the outermost orbital. The first factor is that the larger the principal quantum number, the larger is the radius of the orbital. The second factor is that increasing the effective nuclear charge reduces the size of the orbital. As we saw in the preceding section, the effective nuclear charge felt by an electron is equal to the nuclear charge less any shielding of this charge by intervening orbital electrons. In a given row of the periodic table the principal quantum number remains constant, but the effective nuclear charge increases so that the radius of the outermost orbital decreases as the atomic number increases. In a column in the periodic table the effective nuclear charge remains nearly constant, but the principal quantum number increases so that the atomic radius increases.

*H. C. Longuet-Higgins, *J. Chem. Educ.* **34**:30 (1957).

Table 10.3 Atomic Properties

Z	*Atom*	*Configuration*	*Ground-State Term Symbol*	*First Ionization Energy*, eV	*Orbital Radius*, pm
1	H	1s	$^2S_{1/2}$	13.505	52.9
2	He	$1s^2$	$^1S_0$	24.580	29.1
3	Li	[He]2s	$^2S_{1/2}$	5.390	158.6
4	Be	$[He]2s^2$	$^1S_0$	9.320	104.0
5	B	$[He]2s^22p$	$^2P_{1/2}$	8.296	77.6
6	C	$[He]2s^22p^2$	$^3P_0$	11.264	62.0
7	N	$[He]2s^22p^3$	$^4S_{3/2}$	14.54	52.1
8	O	$[He]2s^22p^4$	$^3P_2$	13.614	45.0
9	F	$[He]2s^22p^5$	$^2P_{3/2}$	17.42	39.6
10	Ne	$[He]2s^22p^6$	$^1S_0$	21.559	35.4
11	Na	[Ne]3s	$^2S_{1/2}$	5.138	171.3
12	Mg	$[Ne]3s^2$	$^1S_0$	7.644	127.9
13	Al	$[Ne]3s^23p$	$^2P_{1/2}$	5.984	131.2
14	Si	$[Ne]3s^23p^2$	$^3P_0$	8.149	106.8
15	P	$[Ne]3s^23p^3$	$^4S_{3/2}$	11.00	91.9
16	S	$[Ne]3s^23p^4$	$^3P_2$	10.357	81.0
17	Cl	$[Ne]3s^23p^5$	$^2P_{3/2}$	13.01	72.5
18	Ar	$[Ne]3s^23p^6$	$^1S_0$	15.755	65.9
19	K	[Ar]4s	$^2S_{1/2}$	4.339	216.2
20	Ca	$[Ar]4s^2$	$^1S_0$	6.111	169.0
21	Sc	$[Ar]4s^23d$	$^2D_{3/2}$	6.56	157.0
22	Ti	$[Ar]4s^23d^2$	$^3F_2$	6.83	147.7
23	V	$[Ar]4s^23d^3$	$^4F_{3/2}$	6.74	140.1
24	Cr	$[Ar]4s3d^5$	$^7S_3$	6.76	145.3
25	Mn	$[Ar]4s^23d^5$	$^6S_{5/2}$	7.432	127.8
26	Fe	$[Ar]4s^23d^6$	$^5D_4$	7.896	122.7
27	Co	$[Ar]4s^23d^7$	$^4F_{9/2}$	7.86	118.1
28	Ni	$[Ar]4s^23d^8$	$^3F_4$	7.633	113.9
29	Cu	$[Ar]4s3d^{10}$	$^2S_{1/2}$	7.723	119.1
30	Zn	$[Ar]4s^23d^{10}$	$^1S_0$	9.391	106.5
31	Ga	$[Ar]4s^23d^{10}4p$	$^2P_{1/2}$	6.00	125.4
32	Ge	$[Ar]4s^23d^{10}4p^2$	$^3P_0$	8.13	109.0
33	As	$[Ar]4s^23d^{10}4p^3$	$^4S_{3/2}$	10.00	100.1
34	Se	$[Ar]4s^23d^{10}4p^4$	$^3P_2$	9.750	91.8
35	Br	$[Ar]4s^23d^{10}4p^5$	$^2P_{3/2}$	11.84	85.1
36	Kr	$[Ar]4s^23d^{10}4p^6$	$^1S_0$	13.996	79.5

Source: J. T. Waber and D. T. Cromer, *J. Chem. Phys.* **42:**4116 (1965).

10.11 IONIZATION ENERGY AND ELECTRON AFFINITY

The **ionization energy** is the energy required to remove an electron completely from a gaseous atom, molecule, or ion. An atom has as many ionization energies as it has electrons. The first ionization energy corresponds to the reaction

$$A = A^+ + e^- \tag{10.89}$$

and the second corresponds to the reaction

$$A^+ = A^{2+} + e^- \tag{10.90}$$

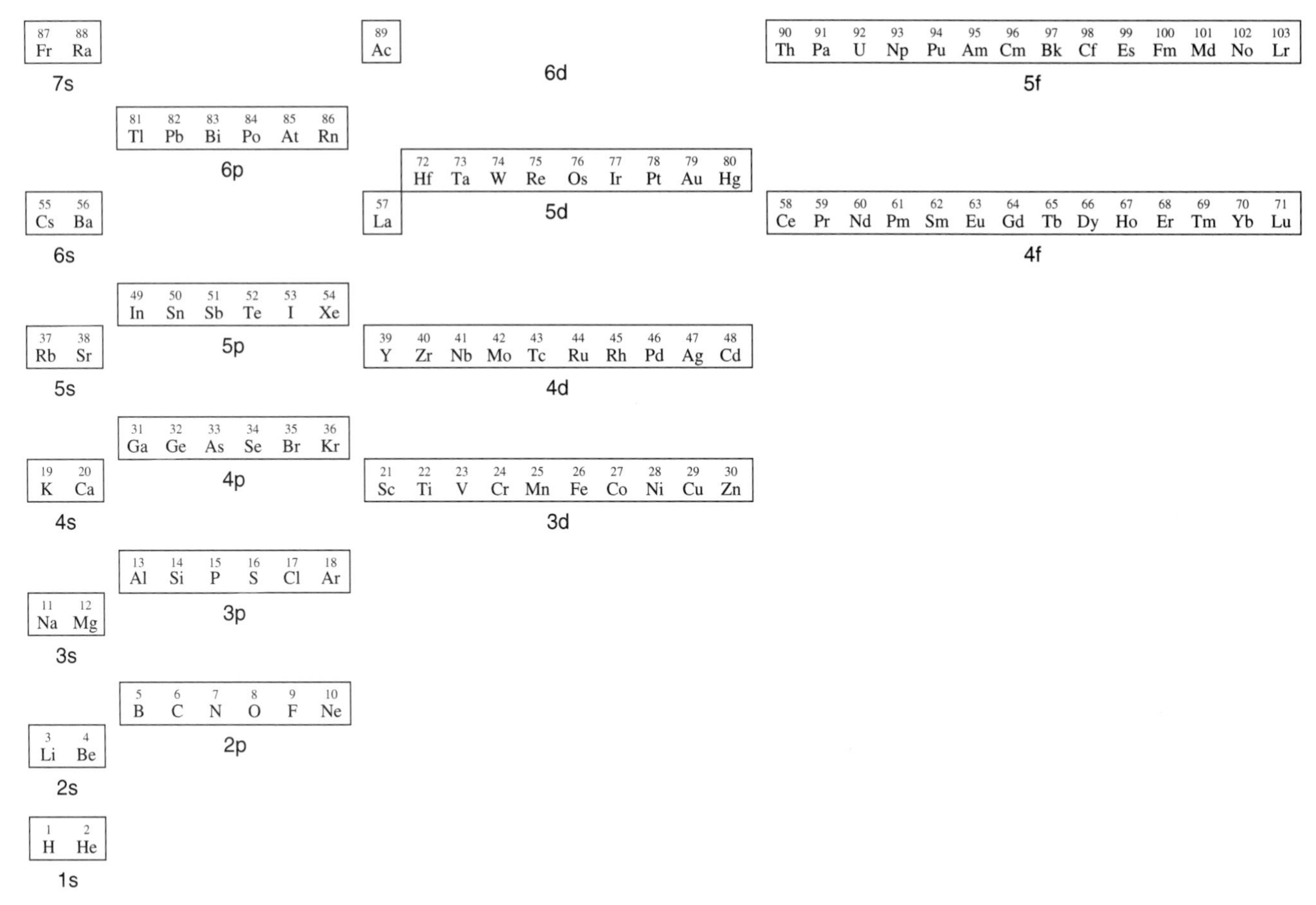

Figure 10.14 Periodic table arranged to show electron configurations of the atoms. The relative energies of the orbitals are shown by their vertical positions. We can think of the electrons being fed successively into higher and higher (i.e., less negative) energy orbitals.

The first ionization energies of the first 36 elements are given in Table 10.3, and the first ionization energies of the elements up to $Z = 89$ are plotted in Fig. 10.15. The most difficult atom to ionize is helium, and the ionization energies of the other inert gases decrease as the atomic radius increases. The alkali metals are the easiest atoms to ionize because the outer shell is occupied by a single electron. In the series lithium, sodium, potassium, rubidium, and cesium, the ionization energy decreases because of the increase in size of the outer orbital containing a single electron.

In contrast, the ionization energies of the halogens are almost as great as those of the inert gases. The electrons in the outer orbital of the halogen atoms are shielded from the nuclear charge mainly by the electrons in inner orbitals, since the electrons in the outer orbital are all approximately the same distance from the nucleus. A direct result of this incomplete shielding of the nuclear charge, as far as electrons in the outer orbital are concerned, is the fact that the halogen atoms readily take on an additional electron to form negative ions.

Within a period of the periodic table there is a general increase in ionization energy, but this increase is not regular. The decrease between beryllium and boron is due to the fact that boron has a single 2p electron outside the filled 2s orbital.

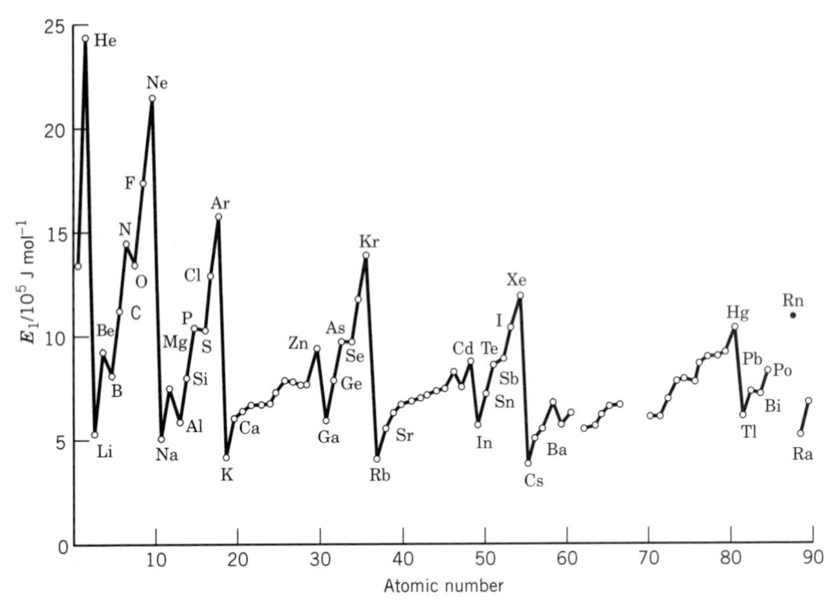

Figure 10.15 First ionization energies of the elements as a function of atomic number.

Ionization energies may be determined by irradiating atoms with light of very short wavelength (Section 14.10). All of the ionization energies of the first six elements are given in Table 10.4. The successive stages of ionization are indicated in the headings of the columns. Column I gives the ionization energy of the neutral atom, column II gives the ionization energy for the singly ionized atoms, and so on. After the first electron has been removed, it is more difficult to remove the second, and so on. There is an especially large increase in ionization energy after all the electrons in the outer shell have been removed; this point for each element is shown by the stairstep line in Table 10.4.

Table 10.4 Ionization Energies of the First Six Elements in Electron Volts

		Spectrum					
Atomic Number	*Element*	*I*	*II*	*III*	*IV*	*V*	*VI*
1	H	13.598					
2	He	24.587	54.416				
3	Li	5.392	75.638	122.451			
4	Be	9.322	18.211	153.893	217.713		
5	B	8.298	25.154	37.930	259.368	340.217	
6	C	11.260	24.383	47.887	64.492	392.077	489.981

Source: C. E. Moore, *National Standard Reference Data Series,* Vol. 34. Washington, DC: U.S. Government Printing Office, 1970.

The **electron affinity** E_{ea} of an atom is the energy released in the process of adding an electron. Since the halogen atoms lack a single electron to complete their outer shell, they have a high affinity for an electron, as illustrated by

$$\mathrm{Cl(g)} + e^- = \mathrm{Cl^-(g)} \qquad \Delta H°(0\ \mathrm{K}) = -347\ \mathrm{kJ\ mol^{-1}} = \frac{-347\ \mathrm{kJ\ mol^{-1}}}{96.485\ \mathrm{kJ\ mol^{-1}\ eV^{-1}}} = -3.60\ \mathrm{eV} \qquad (10.91)$$

Thus, the electron affinity of Cl(g) is 3.60 eV. As shown in Table 10.5 some electron affinities are negative, meaning that the negative ion is unstable with respect to the atom and an electron.

Table 10.5 Electron Affinities E_{ea} in Electron Volts[a]

H(g)	0.75415	Ne(g)	(−0.30)
He(g)	(−0.22)	Na(g)	0.548
Li(g)	0.602	Mg(g)	(−2.4)
Be(g)	(−2.5)	Al(g)	0.52
B(g)	0.86	Si(g)	1.24
C(g)	1.27	P(g)	0.77
N(g)	0	S(g)	2.077
O(g)	1.465	Cl(g)	3.614
F(g)	3.39		

[a] The values in parentheses are calculated values.

10.12 ANGULAR MOMENTUM OF MANY-ELECTRON ATOMS

An atom has total orbital angular momentum $\boldsymbol{L}$ and total spin angular momentum $\boldsymbol{S}$. These properties of an atom or molecule are important because they are **conserved;** that is, the angular momentum does not change unless an external force acts on the atom or molecule. A conserved quantity is called a **constant of motion,** and the quantum mechanical operator for a conserved quantity commutes with the Hamiltonian operator. Thus $\hat{L}^2$, $\hat{L}_z$, $\hat{S}^2$, and $\hat{S}_z$ commute with $\hat{H}$. These operators yield eigenvalues in the usual way:

$$\hat{L}^2\psi = \hbar^2 L(L+1)\psi \qquad (10.92)$$

$$\hat{L}_z\psi = \hbar M_L\psi \qquad (10.93)$$

$$\hat{S}^2\psi = \hbar^2 S(S+1)\psi \qquad (10.94)$$

$$\hat{S}_z\psi = \hbar M_S\psi \qquad (10.95)$$

The symbols used in this section are summarized in Table 10.6.

The orbital angular momentum and the spin angular momentum of an atom are each made up of contributions of individual electrons. Since angular momentum is a vector quantity, the contributions of the individual electrons add vectorially. Lowercase letters are used to represent angular momenta of individual electrons, and capital letters are used to represent angular momenta for atoms. Thus,

$$\boldsymbol{L} = \sum_i \boldsymbol{\ell}_i \qquad (10.96)$$

This vector addition is represented schematically in Fig. 10.16. Remember that the orbital angular momentum vectors of the individual electrons and the sum $\boldsymbol{L}$ have only certain orientations with respect to a given direction, which we label the z direction. The z component L_z of the total orbital angular momentum of an atom is the **scalar sum** of the z components for the individual electrons, as illustrated in Fig. 10.16.

$$L_z = \sum_i \ell_{zi} \qquad (10.97)$$

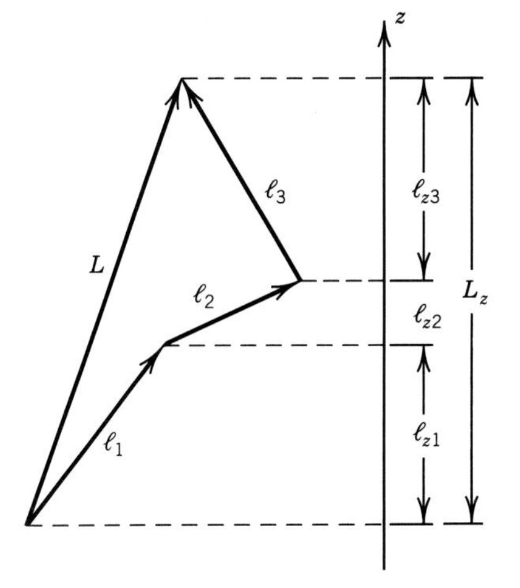

Figure 10.16 Schematic diagram for the addition of angular momentum vectors. Note that the angular momentum vectors add vectorially but the components in the z direction add as scalars. (From D. A. McQuarrie, *Quantum Chemistry.* Mill Valley, CA: University Science Books, 1983. Reprinted with permission from University Science Books.)

We saw earlier (Section 9.12) that the z component of the angular momentum of a single electron is directly proportional to the magnetic quantum number m: $L_z = m\hbar$. When this relation is applied to an atom containing several electrons,

Table 10.6 Angular Momentum of a Many-Electron Atom[a]

	Orbital Angular Momentum	
	Electron	*Atom*
Angular momentum vector	$\boldsymbol{\ell}_i$	$\boldsymbol{L} = \sum \boldsymbol{\ell}_i$
Component of angular momentum vector in z direction	ℓ_{zi}	$L_z = \sum \ell_{zi}$
Quantum number of angular momentum	ℓ_i (0, 1, 2, ...)	$L = \ell_1 + \ell_2, \ell_1 + \ell_2 - 1, \ldots, \lvert\ell_1 - \ell_2\rvert$
Quantum number for z component (magnetic quantum number)	m_i $(-\ell_i, \ldots, +\ell_i)$	$M_L = \sum m_i$ $(-L, \ldots, +L)$

[a]The four rows in the second half of the table are in the same order as in the first half of the table.

the magnetic quantum number for the orbital angular momentum of an atom is capitalized so that $L_z = M_L\hbar$. The z components of the orbital angular momenta of single electrons are represented by $\ell_{zi} = m_i\hbar$, where m_i is the magnetic quantum number for the ith electron. When these two relations are substituted in equation 10.97 we obtain

$$M_L = \sum_i m_i \tag{10.98}$$

The magnetic quantum numbers for an atom can range over $-L, \ldots, +L$, depending on the orientation in a magnetic field. Here L is the orbital angular momentum quantum number for the atom. For an atom containing two electrons, the maximum value of L is obtained when the two orbital angular momenta are lined up, and the minimum value is obtained when the two orbital angular momenta are opposed. When the quantum numbers for the orbital angular momenta of the two electrons are represented by ℓ_1 and ℓ_2, the orbital angular momentum quantum numbers for the atom can have the values

$$L = \ell_1 + \ell_2, \ell_1 + \ell_2 - 1, \ldots, |\ell_1 - \ell_2| \tag{10.99}$$

If the atom contains more than two electrons, this relation can be applied successively.

The total spin angular momentum for a light atom is the vector sum of the spin angular momenta $\boldsymbol{s}_i$ of the individual electrons:

$$\boldsymbol{S} = \sum_i \boldsymbol{s}_i \tag{10.100}$$

The z component S_z of the total spin angular momentum of an atom is the scalar sum of the z components for the individual electrons:

$$S_z = \sum_i s_{zi} \tag{10.101}$$

As shown in equation 10.45, the contribution of the ith electron to the total spin angular momentum in the z direction is directly proportional to the quantum number m_{si} for the ith electron: $S_{zi} = m_{si}\hbar$. The total spin angular momentum

Table 10.6 *(continued)*[a]

Spin Angular Momentum		Total Angular Momentum of Atom
Electron	*Atom*	
$\mathbf{s}_i$	$\mathbf{S} = \sum \mathbf{s}_i$	$\mathbf{J} = \mathbf{L} + \mathbf{S}$
s_{zi}	$S_z = \sum s_{zi}$	$J_z = L_z + S_z$
$s_i\left(\frac{1}{2}\right)$	$S = s_1 + s_2, s_1 + s_2 - 1, \ldots, \lvert s_1 - s_2 \rvert$	$J = L + S, L + S - 1, \ldots, \lvert L - S \rvert$
$m_{si}\left(\pm\frac{1}{2}\right)$	$M_S = \sum m_{si}$ $(-S, \ldots, +S)$	$M_J = M_L + M_S$ $(-J, \ldots, +J)$

[a]The four rows in the second half of the table are in the same order as in the first half of the table.

S_z in the z direction is directly proportional to the spin quantum number M_S for the z component for the whole atom. Thus, equation 10.101 becomes

$$M_S = \sum_i m_{si} \tag{10.102}$$

The value of M_S can range from $-S$ to $+S$, depending on the orientation of the total spin angular momentum for the atom in a magnetic field. For an atom containing two electrons, the spin quantum number S is a maximum if the spins are parallel and a minimum if the spins are opposed. If the spin quantum numbers of the individual electrons are represented by s_1 and s_2, the possible values of S are

$$S = s_1 + s_2, s_1 + s_2 - 1, \ldots, |s_1 - s_2| \tag{10.103}$$

For a two-electron atom, $S = 1, 0$.

The total angular momentum $\mathbf{J}$ of an atom is the vector sum of all the orbital $\boldsymbol{\ell}_i$ and spin $\mathbf{s}_i$ angular momenta of electrons in it. Like other angular momenta $\mathbf{J}$ is quantized, and its quantum number J can take on only integer and half-integer values. In principle these can range up to the sum of the orbital angular momentum quantum numbers and spin quantum numbers for the individual electrons in the atom. We are interested in the total angular momentum quantum number J because states with different angular momenta differ in energy. This is partly because the electrons repel each other electrostatically, and the strengths of the repulsions depend on the distribution of electric charge, which we know is connected with the orbital quantum numbers ℓ_i and m_i. This effect is given the shorthand name orbital–orbital, or $\ell\ell$, interaction. In addition, the atomic states must obey the Pauli exclusion principle, which, because it dictates which orbital states can be associated with which spin states, indirectly brings in a spin–spin, or ss, interaction determined by s_i and s_{zi}. In addition to these effects, there is also a direct spin–orbit, or ℓs, interaction.

The preceding paragraph describes a rather complicated situation, since the relative energies of different atomic states depend on the relative strengths of $\ell\ell$, ss, and ℓs interactions of the various electrons. In relatively light atoms ($Z <$ 40), however, it turns out that the ℓs interaction is distinctly weaker than the other two. This means that the electron orbits interact to give the total orbital angular momentum described in equation 10.96 by $\mathbf{L}$. At the same time, the spins interact to give a resultant spin angular momentum described in equation 10.100 by $\boldsymbol{S}$. Then the weaker ℓs interaction can be thought of as coupling $\boldsymbol{L}$ and $\boldsymbol{S}$ to give the total angular momentum $\boldsymbol{J}$:

$$\boldsymbol{J} = \boldsymbol{L} + \boldsymbol{S} \tag{10.104}$$

The $\boldsymbol{J}$ vector precesses in a magnetic field, and the z component is given by

$$J_z = L_z + S_z \tag{10.105}$$

The quantum number J for the total angular momentum of an atom has values given by

$$J = L + S, L + S - 1, \ldots, |L - S| \tag{10.106}$$

and the total magnetic quantum number M_J is given by

$$M_J = M_L + M_S \tag{10.107}$$

and can range from $-J$ to $+J$. This situation is referred to as LS coupling or Russell–Saunders coupling.

10.13 ATOMIC TERM SYMBOLS

The electron configurations used in describing the various elements in Table 10.3 are incomplete descriptions of the ground states because they do not fully specify how the spin and orbital angular momenta add vectorially. In general, several atomic states with different L, S, and J quantum numbers are represented by a single electron configuration. The energies of these states differ. However, atomic states can be classified according to L, S, and J, and the atomic term symbols used for this purpose have the form

$$^{2S+1}L_J \tag{10.108}$$

In a term symbol the total orbital angular momentum quantum number is not represented by a number, but by a capital letter, just as we represented the angular momentum quantum number for the electron in a hydrogenlike atom by s, p, d, f, ... for $\ell = 0, 1, 2, 3, \ldots$:

$$\begin{aligned} L &= 0\ 1\ 2\ 3\ 4 \ldots \\ \text{Symbol} &= \text{S P D F G} \ldots \end{aligned}$$

The superscript in the term symbol is the **spin multiplicity** and is referred to as a singlet, doublet, triplet, ... for 1, 2, 3, ...

In heavy atoms this hierarchy breaks down, and spin–orbit coupling eventually becomes much stronger than the other two. In this limit a different coupling scheme (jj coupling) is used in which individual electrons acquire total angular momenta $\boldsymbol{j}_i = \boldsymbol{\ell}_i + \boldsymbol{s}_i$. The total angular momentum is then obtained using $\boldsymbol{J} = \sum \boldsymbol{j}_i$. In atoms of intermediate Z, neither approximation is valid, and the situation is very complicated, but the eigenfunction of the Hamiltonian is still an eigenfunction of $\hat{J}^2$. Here we will consider only the LS coupling scheme, which is of great utility in unraveling the spectroscopy and chemistry of the atoms in the first two rows of the periodic table.

Example 10.11 ***The atomic term symbol for helium in its ground state***

What is the atomic term symbol for helium?

Since the electron configuration is $1s^2$, the magnetic quantum numbers m_i of both electrons are zero; and since the electrons are paired, the quantum number m_{si} for the z component of the spin is $+\frac{1}{2}$ for one electron and $-\frac{1}{2}$ for the other, as shown in the following summary:

m_1	m_{1s}	m_2	m_{2s}	M_L	M_S	M_J
0	$+\frac{1}{2}$	0	$-\frac{1}{2}$	0	0	0

Since $M_L = 0$ and there is only one possible state, the total orbital angular momentum quantum number L for the helium atom must be equal to zero. Similarly, since $M_S = 0$, the total spin angular momentum quantum number S for the helium atom must be equal to zero. Since $M_J = M_L + M_S$ is equal to zero, the total angular momentum quantum number J for the atom is 0, and the term symbol is $^1S_0$.

The conclusions from the preceding example may be generalized as follows. Although we have not proved it, closed shells do not contribute orbital or spin angular momentum to an atom because the individual angular momenta add vectorially to zero. Thus, it is the valence electrons that determine the term symbol. On the basis of the helium example other atoms with only ns^2 outside of closed shells will also have the atomic term symbol $^1S_0$. This is illustrated by Be, Mg, and Ca in Table 10.3. Since closed shells do not contribute, this is the term symbol for all the noble gases.

Example 10.12 ***Atomic term symbols for lithium and boron in their lowest states***

What are the atomic term symbols for lithium and boron in their lowest states?

Since lithium and boron each have a single electron outside of closed shells, their possible term symbols are readily identified. Since lithium has a 2s electron with $\ell = 0$, its magnetic quantum number m must also be zero.

m	m_s	M_L	M_S	M_J
0	$\pm\frac{1}{2}$	0	$\pm\frac{1}{2}$	$\pm\frac{1}{2}$

Thus, $L = 0$, $S = \frac{1}{2}$, and $J = \frac{1}{2}$, so the term symbol is $^2S_{1/2}$. Boron has a p electron, so that

m	m_s	M_L	M_S	M_J
0	$\pm\frac{1}{2}$	0	$\pm\frac{1}{2}$	$\pm\frac{1}{2}$
±1	$\pm\frac{1}{2}$	±1	$\pm\frac{1}{2}$	$\pm\frac{1}{2}, \pm\frac{3}{2}$

Thus, $L = 1$, $S = \frac{1}{2}$, and $J = \frac{3}{2}, \frac{1}{2}$ so there are two possible term symbols, $^2P_{3/2}$ and $^2P_{1/2}$.

As we have seen from the preceding example, an atom in a given configuration may exist in more than one state. The energies of atoms in levels corresponding to various terms can be calculated, but this is a demanding and time-consuming process. Fortunately, some patterns have emerged that usually make it possible to identify the lowest energy level for the ground-state configuration of an atom. The German spectroscopist Hund summarized these patterns with three empirical rules:

1. The term arising from the ground configuration with the maximum multiplicity $(2S + 1)$ lies lowest in energy.
2. For levels with the same multiplicity, the one with the maximum value of L lies lowest in energy.
3. For levels with the same S and L, the one with the lowest energy depends on the extent to which the subshell is filled.
   a. If the subshell is less than half-filled, the state with the smallest value of J is the most stable.
   b. If the subshell is more than half-filled, the state with the largest value of J is the most stable.

Thus, according to rule 3a, the ground state of boron discussed in Example 10.12 is $^2P_{1/2}$.

Example 10.13 ***Atomic term symbols for a two-electron atom with the electrons in different s orbitals***

What are the atomic term symbols for a two-electron atom with the electrons in different s orbitals?

Suppose an atom with ns^2 valence electrons is excited to a state with valence electrons $nsn's$, where n' is a higher principal quantum number. Now the electron spins do not have to be paired and so there are four possible states, which in general will have different energies:

m_1	m_{1s}	m_2	m_{2s}	M_L	M_S	M_J
0	$+\frac{1}{2}$	0	$+\frac{1}{2}$	0	1	1
0	$+\frac{1}{2}$	0	$-\frac{1}{2}$	0	0	0
0	$-\frac{1}{2}$	0	$+\frac{1}{2}$	0	0	0
0	$-\frac{1}{2}$	0	$-\frac{1}{2}$	0	-1	-1

In the first row the spins are parallel, so the z component of the spin quantum number M_S for the atom is 1 and the quantum number S for the spin angular momentum is zero. Since the valence electrons are both s electrons, $M = 0$ and $L = 0$. For all four possibilities, the quantum number M for the orbital angular momentum is zero. Therefore, the quantum number L for the total angular momentum is zero. Since the largest value of M_S is unity, the total spin quantum number S for the atom can have values of 1 and 0.

If $S = 1$ and $L = 0$, then $J = 1$ and the term symbol is $^3S_1$. Note that for this case $M_S = 1, 0, -1$, so that we have a triplet state. If $S = 0$ and $L = 0$, then $J = 0$ and the term symbol is $^1S_0$, a singlet state. As indicated by Hund's first rule, the triplet state is the most stable.

As the number of electrons increases, the number of different ways of assigning them increases. For example, there are 15 ways for carbon with two p electrons. Details of these assignments are given in more advanced texts.

Example 10.14 ***Spin–orbit coupling***

An atom with orbital angular momentum $\boldsymbol{L}$ and spin angular momentum $\boldsymbol{S}$ often has a term in its Hamiltonian of the form $A\boldsymbol{L}\cdot\boldsymbol{S}$ called the spin–orbit coupling term. Since the total angular momentum $J = L + S$ commutes with the Hamiltonian, the states of this atom can be labeled with the eigenvalues of J^2, J_z, L^2, and S^2. Notice that $\boldsymbol{L}\cdot\boldsymbol{S} = \frac{1}{2}(J^2 - L^2 - S^2)$ so that the eigenvalue of $\boldsymbol{L}\cdot\boldsymbol{S}$ is $\frac{1}{2}[J(J+1) - L(L+1) - S(S+1)]$. This can lead to observable energy level splittings. For example, the excited states of an alkali atom with the outermost electron excited from the s to the next higher p orbital will have $L = 1$ and $S = \frac{1}{2}$. This leads to two possible J values: $\frac{3}{2}$ and $\frac{1}{2}$. Calculate the energy level splitting due to spin–orbit coupling.

The energy term due to $A\boldsymbol{L}\cdot\boldsymbol{S}$ will be

$$A[J(J+1) - 1(1+1) - \tfrac{1}{2}(\tfrac{1}{2}+1)] = A[J(J+1) - \tfrac{11}{4}]$$

For $J = \frac{3}{2}$ the energy term equals A; for $J = \frac{1}{2}$ it equals $-2A$, giving a splitting of $3A$.

In Na the observed splitting of the intense yellow fluorescent lines is 17 cm^{-1}; thus, $A = 5.7$ cm^{-1} as illustrated in Fig. 10.17.

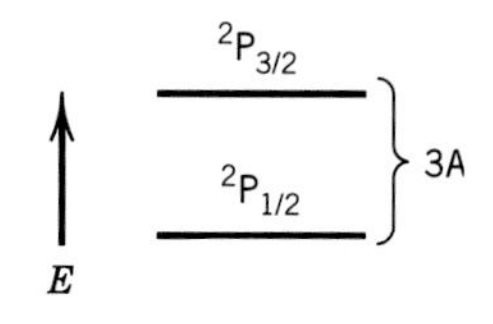

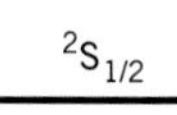

Figure 10.17 Low-lying energy levels in alkali atoms.

Comment:

Atomic term symbols provide a compact way to summarize the properties of an atom in a specific term, and therefore provide a name for the term. Atomic states that have the same electron configuration, same energy, same L value, and same S value constitute a term. In the atomic term symbol, the value of L is given by the code letter (i.e., S, P, ...), and the value of S is given by the left superscript. The value of J is given by the right subscript. The same basic procedure will be followed in the next chapter with molecular term symbols.

10.14 SPECIAL TOPIC: ATOMIC SPECTRA AND SELECTION RULES

We have seen that an atom can exist in a series of states, each of which is identified by a term symbol. Spectroscopic transitions between these states provide information on their energies and quantum numbers. The emission spectra of atoms may be excited in a gas discharge tube or flame. The absorption spectrum may be obtained by passing light through a gas of the atoms.

The energy levels of a hydrogenlike atom depend only on the principal quantum number and are given by equation 10.11, which provides an explanation for the Lyman ($n_1 = 1$), Balmer ($n_1 = 2$), Paschen ($n_1 = 3$), etc. series. However, not all possible transitions can occur in spectroscopic transitions, because photons have an intrinsic angular momentum equal to one unit, and angular momentum is conserved. For left-circularly polarized light $m = 1$, and for right-circularly polarized light $m = -1$. When a photon is absorbed by an atom, its angular momentum is transferred to the electrons of the atom. Since angular momentum has to be conserved in absorption, or emission, we conclude that $\Delta\ell = \pm 1$ for the atom. These plus and minus signs apply in either absorption or emission because the angular momentum of an electron can increase or decrease during absorption or emission, as illustrated in Fig. 10.18.

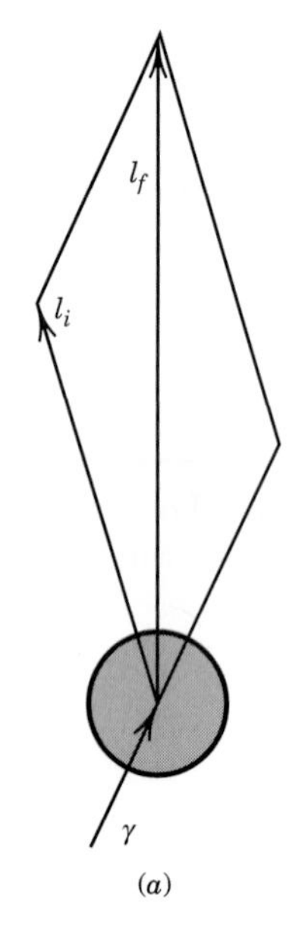

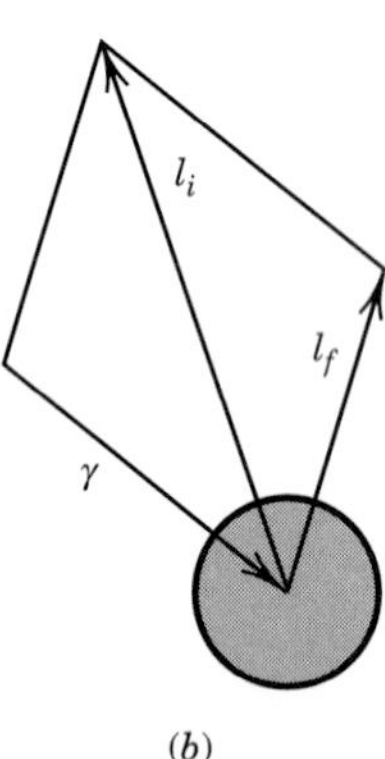

Figure 10.18 Conservation of orbital angular momentum on absorption of a photon (γ): (*a*) the angular momentum of the atom increases in the absorption process; (*b*) the angular momentum of the atom decreases in the absorption process. (From P. W. Atkins, *Molecular Quantum Mechanics.* New York: Oxford University Press, 1983.)

We will see in Chapter 13 that the most intense spectroscopic transitions involve the interaction of the electric vector of the radiation with instantaneous electric dipoles in the atom or molecule. These are called **electric dipole transitions.**

The spectroscopic transitions that occur for hydrogen atoms or hydrogenlike atoms are indicated in Fig. 10.19. This diagram, referred to as a Grotrian diagram, shows only lines for which $\Delta\ell = \pm 1$. The complete selection rule for electric dipole transitions of hydrogenlike atoms is

$$\Delta n, \text{ unrestricted} \qquad \Delta\ell = \pm 1 \qquad \Delta m = \pm 1, 0 \tag{10.109}$$

The spectrum of atomic hydrogen has a fine structure that we will not discuss because it is not important for chemistry. These further small splittings are explained by quantum electrodynamics.

Electric dipole transitions are due to the oscillating electric field component of light, and magnetic dipole transitions are due to the oscillating magnetic field component of light. Magnetic dipole transitions are generally about 10^5 times weaker (i.e., less probable) than electric dipole transitions. Selection rules were initially found experimentally, but they may be derived from the equation for the electric dipole transition moment that is discussed in Chapter 13.

As we have seen, the states of many-electron atoms depend on the coupling of angular and spin momenta and can be described in terms of L, J, and S. The electric dipole selection rules for atoms in general may be expressed in terms of these quantum numbers:

1. $\Delta L = 0, \pm 1$ except that the transition $L = 0$ to $L = 0$ does not occur. This is an extension of the selection rule $\Delta\ell = \pm 1$ for the hydrogen atom to transitions involving any number of electrons.

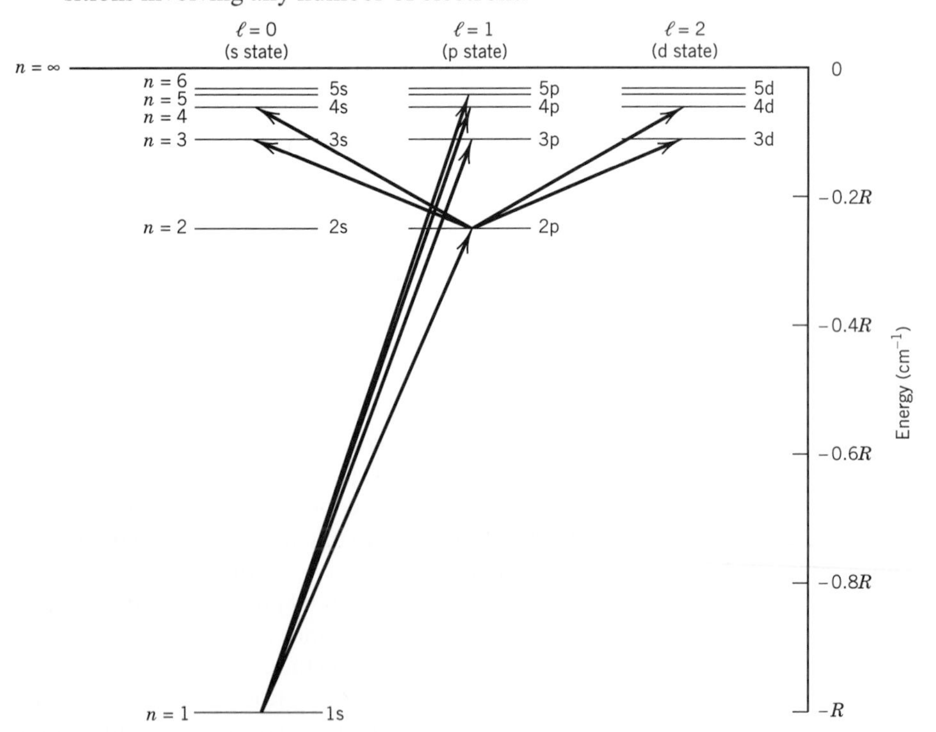

Figure 10.19 Grotrian diagram for some of the shortest wavelength transitions of hydrogenlike atoms, where R is the Rydberg constant.

2. For a transition to be dipole allowed, there must be a change in **parity.** An even function such as

$$\psi(-x) = +\psi(x)$$

is said to have even parity. For an odd function

$$\psi(-x) = -\psi(x)$$

the parity is odd. The parity of a many-electron atom is even if $\sum \ell_i$ is even and odd if $\sum \ell_i$ is odd, where the summation extends over all electrons. This selection rule is summarized by even $\nleftrightarrow$ even, odd $\nleftrightarrow$ odd, even $\leftrightarrow$ odd. This selection rule, which is referred to as the Laporte rule, is consistent with $\Delta\ell = \pm 1$ when only one electron is promoted from the ground configuration.

3. $\Delta J = 0, \pm 1$ except $J = 0 \nleftrightarrow J = 0$.

4. $\Delta S = 0$. This selection rule results from the fact that the electric component of the electromagnetic field has no effect on the total spin angular momentum of the electrons in an atom. This is illustrated by the radiative transitions shown by helium (Fig. 10.20). There are no transitions between singlet states and triplet states. This selection rule gives rise to the phenomenon of **metastable states.** The lowest triplet state of helium cannot emit a photon and

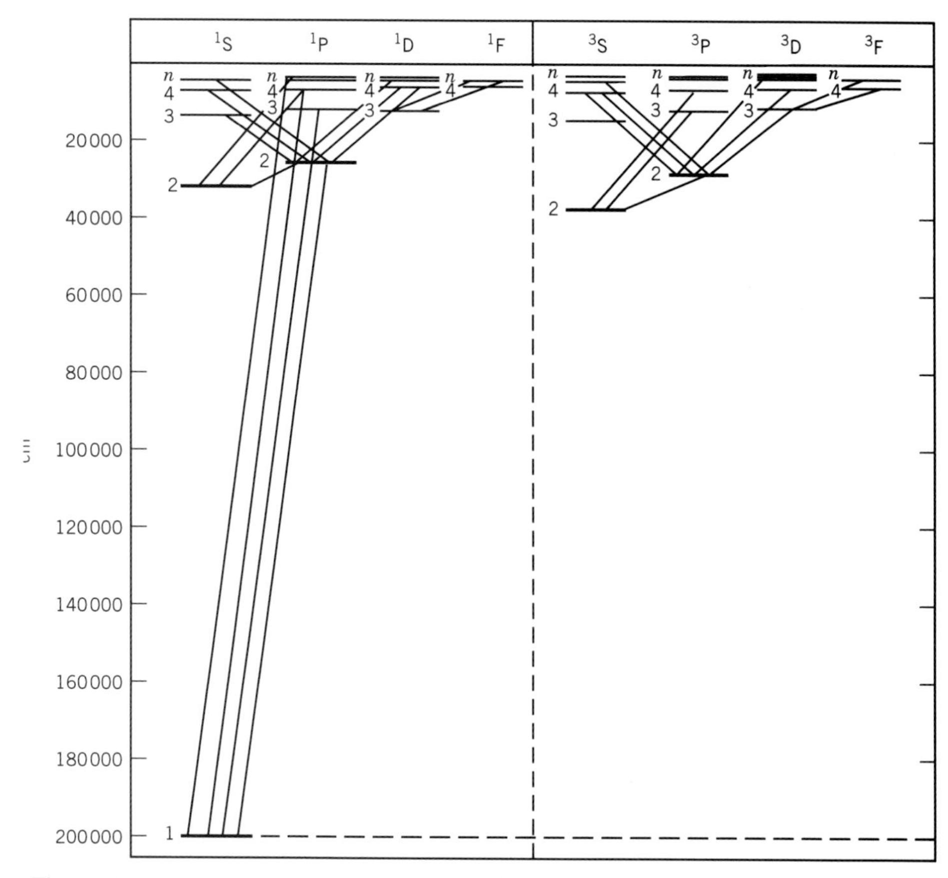

Figure 10.20 Grotrian diagram for the helium atom. [Reproduced from J. M. Hollas, *High Resolution Spectroscopy,* Fig. 6.17, p. 294. Boston: Butterworth, 1982; by permission of the publishers, Butterworth & Co. (Publishers) Ltd. ©.]

make a transition to the singlet ground state. The transition to the ground state can occur in a collision with another molecule, and so triplet helium has a lifetime that depends on pressure. Metastable atoms play an important role in photochemistry. This selection rule breaks down for atoms with higher atomic numbers.

Forbidden transitions (such as $\Delta S \neq 0$) may occur, but they usually occur infrequently and give weak lines. Their occurrence does not mean that quantum mechanics is wrong, but rather results from the approximations used in writing Hamiltonians. Various kinds of higher-order interactions are frequently omitted, and when they are included, correct (but considerably more complicated) conclusions are reached. When electric or magnetic fields are applied, there are changes in various energy levels, and further selection rules are involved. The Zeeman effect is the splitting of lines by the application of a magnetic field. The Stark effect is the splitting of lines by the application of an electric field.

10.15 SPECIAL TOPIC: ATOMIC UNITS

To make it easier to work with the Schrödinger equation, so-called atomic units are often used. These units are fundamental constants (and combinations of fundamental constants) that arise in quantum mechanical calculations and are conveniently treated like units. Atomic units form a coherent system of units based on the independent dimensions of length, mass, charge, and angular momentum. The base unit of length is the Bohr radius, $a_0 = 4\pi\epsilon_0\hbar^2/m_e e^2$. The base unit of mass is the electron rest mass m_e. The base unit of charge is the elementary charge e. The base unit of angular momentum is $\hbar = h/2\pi$. On the basis of these four units, it follows that the unit of energy is the hartree $E_h = m_e e^4/(4\pi\epsilon_0)^2\hbar^2$. The atomic units for time, velocity, force, etc. are readily derived from the basic four, and these expressions and the best current values in the SI are given in Table 10.7.

These units are important because the results of calculations in quantum mechanics are best expressed in atomic units. When this is done, the results of theoretical calculations can be converted later to SI units using the best values of $\hbar$, e, and m_e at that time. Since the values of atomic units change with

Table 10.7 Atomic Units and SI Equivalents

Physical Quantity	*Symbol for Unit*	*Value of Unit in SI (1996)*
Length	a_0	$5.291\ 772\ 49 \times 10^{-11}$ m
Mass	m_e	$9.109\ 389\ 7 \times 10^{-31}$ kg
Charge	e	$1.602\ 177\ 33 \times 10^{-19}$ C
Angular momentum	$\hbar$	$1.054\ 572\ 66 \times 10^{-34}$ J s
Energy	E_h	$4.359\ 748\ 2 \times 10^{-18}$ J
Time	$\hbar/E_h$	$2.418\ 884\ 334\ 1 \times 10^{-17}$ s
Velocity	$a_0E_h/\hbar$	$2.187\ 691\ 42 \times 10^{-6}$ m s^{-1}
Force	E_h/a_0	$8.238\ 729\ 5 \times 10^{-8}$ N
Electric dipole moment	ea_0	$8.478\ 357\ 9 \times 10^{-30}$ C m
Magnetic dipole moment	$e\hbar/m_e$	$1.854\ 803\ 08 \times 10^{-23}$ J T^{-1}

time as better experimental determinations are made, they are written in italic type, rather than roman type like the abbreviations for the meter, kilogram, and second.

In atomic units the Schrödinger equation for the hydrogenlike atom (see equation 10.4) is

$$\left(-\frac{1}{2}\nabla^2 - \frac{Z}{r}\right)\psi = E\psi \tag{10.110}$$

Note that the permittivity factor $4\pi\epsilon_0$ is replaced by unity in these units. The energy of the hydrogenlike atom is given by

$$E_n = -\frac{Z^2}{2n^2} \tag{10.111}$$

Ten Key Ideas in Chapter 10

1. The Schrödinger equation for hydrogenlike atoms can be solved exactly, and so their spectra and ionization energies can be predicted exactly. The spectra consist of series of lines that converge as the variable quantum number increases.
2. The total wavefunction for the hydrogenlike atom involves the product of an associated Laguerre polynomial and a spherical harmonic; three quantum numbers (n, ℓ, and m) are involved.
3. The radial wavefunction $R_{n\ell}(r)$ can be used to calculate the radial probability density and the expectation value for the radius of a hydrogenlike atom. Spherical harmonics can be used to calculate the shapes of the orbitals.
4. The angular momentum L of a hydrogenlike atom is proportional to the square root of $\ell(\ell + 1)$. The angular momentum in the z direction is given by $m\hbar$, where $m = -\ell, -\ell + 1, \ldots, \ell - 1, \ell$. This angular momentum leads to a splitting of spectral lines in a magnetic field that is referred to as the Zeeman effect.
5. Electrons have spin of $\frac{1}{2}$, so their spin angular momentum $\boldsymbol{S}$ is added to orbital angular momentum $\boldsymbol{L}$ to yield the total angular momentum of the atomic system $\boldsymbol{J} = \boldsymbol{L} + \boldsymbol{S}$.
6. In the variational method, a trial wavefunction is used to calculate an upper bound for the energy eigenvalue of a system.
7. The Pauli exclusion principle requires that the wavefunction for any system of electrons must be antisymmetric with respect to the interchange of any two electrons. This explains why the first excited state of helium is a triplet state.
8. The electron configurations of atoms explains the structure of the periodic table.
9. Atomic states can be classified by atomic term symbols that summarize L, S, and J for an atom.
10. Not all possible transitions can occur in spectroscopic transitions because photons have intrinsic angular momentum equal to one unit, and angular momentum is conserved.

REFERENCES

R. E. Christofferson, *Basic Principles and Techniques of Molecular Quantum Mechanics.* New York: Springer, 1989.

R. Eisberg and R. Resnick, *Quantum Physics,* 2nd ed. Hoboken, NJ: Wiley, 1985.

M. D. Fayer, *Elements of Quantum Mechanics*. Oxford: Oxford University Press, 2001.

C. F. Fischer, *The Hartree–Fock Method for Atoms.* Hoboken, NJ: Wiley, 1977.

G. Herzberg, *Atomic Spectra and Atomic Structure.* Englewood Cliffs, NJ: Prentice-Hall, 1937.

I. N. Levine, *Quantum Chemistry,* 5th ed. Upper Saddle River, NJ: Prentice-Hall, 1999.

J. P. Lowe, *Quantum Chemistry,* 2nd ed. Hoboken, NJ: Wiley, 1993.

J. L. McHale, *Molecular Spectroscopy.* Upper Saddle River, NJ: Prentice-Hall, 1999.

C. E. Moore, *Atomic Energy Levels,* Natl. Bur. Std. Circl. No. 467. Washington, DC: US Government Printing Office, 1949.

L. Pauling and E. B. Wilson, *Introduction to Quantum Mechanics.* New York: McGraw-Hill, 1935.

M. A. Ratner and G. C. Schatz, *Introduction to Quantum Mechanics in Chemistry*. Upper Saddle River, NJ: Prentice Hall, 2001.

G. C. Schatz and M. A. Ratner, *Quantum Mechanics in Chemistry*. Englewood Cliffs, NJ: Prentice-Hall, 1993.

J. Simons and J. Nichols, *Quantum Mechanics in Chemistry*. New York: Oxford University Press, 1997.

PROBLEMS

Ⓜ Problems marked with an icon may be more conveniently solved on a personal computer with a mathematical program.

10.1 Using data from Table C.3 at 0 K, what is the ionization energy of H(g)?

$$\mathrm{H(g)} = \mathrm{H^+(g)} + \mathrm{e^-}$$

10.2 How much energy in eV and kJ $\mathrm{mol^{-1}}$ is required to remove electrons from the following orbitals in an H atom: (*a*) 3d, (*b*) 4f, (*c*) 4p, (*d*) 6s?

10.3 Calculate the ground-state ionization potentials for He^+, Li^{2+}, Be^{3+}, and C^{5+}.

10.4 Since H and D have different reduced masses, they also have slightly different electronic energy levels. Calculate (*a*) the ionization potentials and (*b*) the wavelengths of the first line in the Balmer series for these atoms.

10.5 Since we know the expression for the wavefunction for the hydrogen atom in its ground state, show that

$$\hat{H}\psi_{100} = -\frac{\hbar^2}{2m_e a_0^2}\psi_{100}$$

and express the ground-state energy in hartrees.

10.6 A muon is an elementary particle with a negative charge equal to the charge of the electron and a mass approximately 200 times the electron mass. The muonium atom is formed from a proton and a muon. Calculate the reduced mass, the Rydberg constant, and the formula for the energy levels for this atom. What is the most probable radius of the 1s orbital for this atom?

10.7 A hydrogenlike atom has a series of spectral lines at λ = 26.2445, 19.4404, 17.3578, and 16.4028 nm. What is the nuclear charge on the atom? What is the formula for this spectral series (i.e., n_1 and n_2 in equation 10.16)?

10.8 Calculate the wavelengths in μm of the first three lines of the Paschen series for atomic hydrogen.

10.9 Calculate the wavelength of light emitted when an electron falls from the $n = 100$ orbit to the $n = 99$ orbit of the hydrogen atom. Such species are known as high Rydberg atoms. They are detected in astronomy and are more and more studied in the laboratory.

10.10 Calculate the Rydberg constant R_H for a hydrogen atom and the Rydberg constant R_D for a deuterium atom given the value for R_∞ given in Appendix B.

10.11 Check the normalization of the hydrogen atomic wavefunction ψ_{100}.

10.12 Show that ψ_{100} and ψ_{200} for the hydrogen atom are orthogonal.

10.13 In a hydrogenlike atom in the 1s state, there is a difference between the average distance $\langle r \rangle$ between the electron and the nucleus and the most probable distance between the electron and the nucleus. (*a*) Derive the expression for the average distance $\langle r \rangle$ between the electron and the nucleus. (*b*) Derive the expression for the most probable distance r_{mp}.

10.14 The spin functions α and β cannot be expressed in terms of spherical harmonics, but they can be expressed as column matrices:

$$\alpha = \begin{bmatrix} 1 \\ 0 \end{bmatrix} \quad \text{and} \quad \beta = \begin{bmatrix} 0 \\ 1 \end{bmatrix}$$

The spin operator can be represented by the following Pauli matrix:

$$\hat{S}_z = \frac{1}{2}\begin{bmatrix} 1 & 0 \\ 0 & -1 \end{bmatrix}$$

Show that $\hat{S}_z \alpha = \frac{1}{2}\alpha$ and $\hat{S}_z \beta = -\frac{1}{2}\beta$.

10.15 The spin parts of the wavefunctions for the first excited state of helium are $\psi_a = \alpha(1)\alpha(2)$, $\psi_b = \alpha(1)\beta(2) + \alpha(2)\beta(1)$, and $\psi_c = \beta(1)\beta(2)$. Since $\hat{S}_z \psi = \hbar M_S \psi$, what are the M_S values for the three wavefunctions?

10.16 What are the degeneracies of the following orbitals for hydrogenlike atoms: (*a*) $n = 1$, (*b*) $n = 2$, and (*c*) $n = 3$?

10.17 Show that for a 1s orbital of a hydrogenlike atom the most probable distance from proton to electron is a_0/Z. Find the numerical values for C^{5+} and B^{3+}.

10.18 Find the values of r for which the hydrogenlike atom 2s and 3s wavefunctions are equal to zero (these are the radial nodes). Compare to Fig. 10.1.

10.19 What is the average distance from an orbital electron to the nucleus for a 2s and 2p electron in (*a*) H and (*b*) Li^{2+}?

10.20 Calculate the expectation value $\langle r \rangle$ of the radius of a 2s orbital and a 2p orbital for a hydrogenlike atom. Is this the result that you expected?

10.21 In the laboratory there is a limit to the number of lines that can be observed in the spectrum of a hydrogenlike atom because of pressure broadening. As atoms are excited to higher quantum numbers their effective radii increase, and because of crowding, the excited atoms contact nearby atoms and do not act independently. However, in interstellar space, emissions from hydrogen atoms at extremely low pressures with very high quantum numbers can be detected because of the large volumes per atom. Radio astronomers have detected hydrogen atoms undergoing the transition $n = 253$ to $n = 252$ [D. B. Clark, *J. Chem. Educ.* **68**:454 (1991)]. At what frequency and wavelength was this observation made, and what is the expectation value for the radius of the emitting hydrogen atom?

10.22 Calculate the expectation value of the distance between the nucleus and the electron of a hydrogenlike atom in the $2p_z$ state using equation 9.35. Show that the same result is obtained using equation 10.30.

10.23 It can be shown that a linear combination of two eigenfunctions belonging to the same degenerate level is also an eigenfunction of the Hamiltonian with the same energy. In terms of mathematical formulas, if $\hat{H}\psi_1 = E_1\psi_1$ and $\hat{H}\psi_2 = E_1\psi_2$, then $\hat{H}(c_1\psi_1 + c_2\psi_2) = E_1(c_1\psi_1 + c_2\psi_2)$. The wavefunction $c_1\psi_1 + c_2\psi_2$ still needs to be normalized. Equation 10.8 yields the following expressions for the 2p eigenfunctions:

$$\psi_{2p_{+1}} = b\,e^{-Zr/2a}\,r\sin\theta\,e^{i\phi}$$

$$\psi_{2p_{-1}} = b\,e^{-Zr/2a}\,r\sin\theta\,e^{-i\phi}$$

$$\psi_{2p_0} = c\,e^{-Zr/2a}\,r\cos\theta$$

Use this information to find the real functions for the 2p orbitals that are given in Table 10.1. Given:

$$e^{i\phi} = \cos\phi + i\sin\phi$$

$$e^{-i\phi} = \cos\phi - i\sin\phi$$

10.24 For a hydrogenlike atom, what is the magnitude of the orbital angular momentum, and what are the possible values of L_z for electrons in the 2p and 3d orbitals?

10.25 What is the magnitude of the angular momentum for electrons in 3s, 3p, and 3d orbitals? How many radial and angular nodes are there for each of these orbitals?

10.26 How many angular, radial, and total nodes are there for the following hydrogenlike wavefunctions: (*a*) 1s, (*b*) 2s, (*c*) 2p, (*d*) 3p, and (*e*) 3d?

10.27 The antisymmetric spin function for two electrons is $N[\alpha(1)\beta(2) - \alpha(2)\beta(1)]$. Derive the value for the normalization constant N.

10.28 Using equation 10.55, calculate the difference in energy between the two spin angular momentum states of a hydrogen atom in the 1s orbital in a magnetic field of 1 T. What is the wavelength of radiation emitted when the electron spin "flips"? In what region of the electromagnetic spectrum is this?

10.29 For the wavefunction

$$\psi = \begin{vmatrix} \psi_A(1) & \psi_A(2) \\ \psi_B(1) & \psi_B(2) \end{vmatrix}$$

show that (*a*) the interchange of two columns changes the sign of the wavefunction, (*b*) the interchange of two rows changes the sign of the wavefunction, and (*c*) the two electrons cannot have the same spin orbital.

10.30 What are the electron configurations for H^-, Li^+, O^{2-}, F^-, Na^+, and Mg^{2+}?

10.31 How many electrons can enter the following sets of atomic orbitals: 1s, 2s, 2p, 3s, 3p, and 3d?

10.32 Calculate the frequency and the wavelength in nanometers for the line in the Paschen series of the hydrogen spectrum that is due to a transition from the sixth quantum level to the third.

10.33 There is a Brackett series in the hydrogen spectrum where $n_1 = 4$. Calculate the wavelengths in nm of the first two lines of this series.

10.34 Since the outer electron for Li is quite a bit farther out than the two 1s electrons, this atom is something like a hydrogen atom in the 2s state. The first ionization potential of Li is 5.39 eV. What ionization potential would be expected from this simple model of an Li atom? What effective nuclear charge Z' seen by the outer electron would give the correct first ionization potential?

10.35 The first ionization potentials of Na, K, and Rb are 5.138, 4.341, and 4.166 eV, respectively. Assume that the energy level of the outer electron can be represented by a hydrogenlike formula with an effective nuclear charge Z' and that the relevant orbitals are 3s, 4s, and 5s, respectively. Calculate Z' for these atoms.

10.36 What is the total electronic energy in hartrees of He, Li, and Be with respect to the nuclei and free electrons? Ionization potentials are given in Table 10.4.

10.37 Why is the ionization energy of boron less than beryllium and the ionization energy of oxygen less than nitrogen?

10.38 The enthalpy of formation of $H^-(g)$ at 0 K is given as 143.266 kJ mol^{-1} in Table C.3. What is the electron affinity of H(g)?

10.39 For a carbon atom in the configuration [He]$2s^2 2p^2$ the following term symbols are all possible: $^1S_0$, $^3P_0$, $^3P_1$, $^3P_2$, and $^1D_2$. According to Hund's rules, which is the most stable state?

10.40 Which of the following transitions are allowed in the electronic spectrum of a hydrogenlike atom: (*a*) 2s → 1s, (*b*) 2p → 1s, (*c*) 3d → 1s, and (*d*) 3d → 3p?

10.41 (*a*) When a hydrogenlike atom is in a 2s orbital, to which orbitals can the atom be excited by the absorption of radiation? (*b*) When a hydrogenlike atom is in a 3p orbital, to which orbitals can the atom be excited by the absorption of radiation? (*c*) When a hydrogenlike atom is in a 3d orbital, to which orbitals can the atom be excited by the absorption of radiation?

10.42 Show that the bound-state eigenfunctions of a hydrogenlike atom have the factor $\exp(-Zr/na_0)$ by examining the Schrödinger equation for large r.

10.43 Given that the 1s wavefunction for a hydrogenlike atom is proportional to $\exp(-Zr/a_0)$, calculate the normalization factor for this wavefunction.

10.44 Calculate the ionization potential of a hydrogen atom in its ground state from the Rydberg constant for the hydrogen atom.

10.45 If we use Ψ as a trial function for Hamiltonian $\hat{H}$, where

$$\Psi = \frac{(\phi_0 + a\phi_1)}{(1 + a^2)^{1/2}}$$

and ϕ_0 and ϕ_1 are the two *lowest* eigenfunctions with eigenvalues E_0 and E_1 (with $E_1 > E_0$), show that the variational method yields $a = 0$.

10.46 Calculate the classical energy of an electron and proton separated by 0.0529 nm, relative to the same system at infinite separation. Compare this with the quantum mechanical result for the energy.

10.47 For what value of n do adjacent energy levels in H have a separation of kT at room temperature?

10.48 What are the wavelengths of the first lines in the Balmer series for $^6Li^{2+}$ and $^7Li^{2+}$?

10.49 What is the expectation value for the radius of a hydrogen atom if $n = 50$ and $\ell = 0$?

10.50 Calculate $\langle r \rangle$ for a 2s electron in a hydrogen atom, given

$$\int_0^\infty x^n e^{-ax}\, dx = \frac{n!}{a^{n+1}} \qquad n > -1, a > 0$$

10.51 Positronium consists of an electron and positron (positive particle with the same mass as an electron). (*a*) Calculate the wavelength of the radiation emitted when the system goes from the $n = 2$ orbital to the $n = 1$ orbital. (*b*) Calculate the ionization energy.

10.52 What are the most probable positions for an electron in a $2p_z$ orbital in hydrogen?

10.53 A hydrogen atom is in a cubical box of 100 Å on a side. For what value of n (for an s state) does the expectation value of the radius equal one-half the box size?

10.54 Show that Ψ_{1s} and Ψ_{2s} are orthogonal for the H atom.

10.55 Using the integrals in Problem 10.50, show that the 2s and 3s wavefunctions are normalized.

10.56 What is the most probable distance for a 1s electron in Li^{2+}?

10.57 What is the average distance from the electron to the proton for the 3s, 3p, and 3d states of the hydrogen atom?

10.58 Calculate the splitting in kJ/mol and eV for an H atom in the 2p state in a 10-T magnetic field, neglecting the spin as in Fig. 10.8 and equation 10.43. Compare with kT at room temperature.

10.59 What is the magnitude of the angular momentum for the electrons in 4s, 4p, 4d, and 4f orbitals? How many radial and angular nodes are there for each of these orbitals?

10.60 Show that the following wavefunction for the hydrogen atom is antisymmetric to the interchange of the two electrons:

$$\Psi = \begin{vmatrix} 1s\alpha(1) & 1s\beta(1) \\ 1s\alpha(2) & 1s\beta(2) \end{vmatrix}$$

10.61 In a hydrogen atom, the 2s and 2p orbitals have the same energy. However, in a boron atom, the 2s orbital has a lower energy than the 2p. Explain this in terms of the shape of the orbitals.

10.62 Give the electronic configurations for the ground states of the first 18 electrons in the periodic table.

10.63 The first three ionization energies of scandium (Sc, atomic number 21) are as follows:

$$Sc = Sc^+ + e^- \qquad E_i = 6.54 \text{ eV}$$
$$Sc^+ = Sc^{2+} + e^- \qquad E_i = 12.8 \text{ eV}$$
$$Sc^{2+} = Sc^{3+} + e^- \qquad E_i = 24.75 \text{ eV}$$

What are the electron configurations of the three ions?

Computer Problems

10.A Make three-dimensional plots to show the electron density in the xy plane for atomic hydrogen in the 1s, 2s, $2p_x$, and

$2p_y$ orbitals. Restrict the plots to $x = -0.5$ to 0.5 nm and $y = -0.5$ to 0.5 nm.

10.B The Pauli spin operators for the three directions can be represented by matrices:

$$S_x = \begin{bmatrix} 0 & 0.5 \\ 0.5 & 0 \end{bmatrix}$$

$$S_y = \begin{bmatrix} 0 & -0.5i \\ 0.5i & 0 \end{bmatrix}$$

$$S_z = \begin{bmatrix} 0.5 & 0 \\ 0 & -0.5 \end{bmatrix}$$

(*a*) Calculate the commutator for S_x and S_y. (*b*) Calculate the sum of squares of the representations of the operators to obtain the spin matrix S^2. (*c*) Show that the square of the operator of spin angular momentum commutes with the operator for the x component.

10.C (*a*) Calculate the frequencies (in cm^{-1}) for the first 13 lines of the Balmer spectrum of hydrogen atoms. Plot the line spectrum and label the frequency axis. (*b*) Do the same for the Paschen spectrum. (*c*) Do the same for the Brackett spectrum. (*d*) Combine the spectra to show the experimental spectrum.

10.D Plot the angular parts of the wavefunctions for the atomic orbitals 2s, $2p_x$, $2p_y$, and $2p_z$ using computer software that provides spherical harmonics $Y(\ell, m)$. In order to plot $2p_x$ and $2p_y$ the sums and differences of $Y(1, 1)$ and $Y(1, -1)$ must be used as explained in the text.

10.E Plot the squares of the angular factors of the atomic orbitals $2p_x$, $2p_y$, and $2p_z$ in three dimensions and compare them with Fig. 10.6.

10.F Plot the squares of the angular factors $3d_{z^2}$, $3d_{xz}$, $3d_{yz}$, $3d_{xy}$, and $3d_{x^2-y^2}$ atomic orbitals in three dimensions. The angular factors are given in Table 10.1. Compare these plots with Fig. 10.6.

10.G Plot the Coulomb potential, the centrifugal potential, and the sum (effective potential) for a hydrogen atom with $\ell = 1$. Locate the minimum in the effective potential by taking the derivative and plotting it. Why doesn't the radius for the minimum potential correspond to the expectation value for r for the 1s orbital, which is $(3/2)a_0$?

11

Molecular Electronic Structure

Quantum mechanics has made it possible to understand the nature of chemical bonding and to predict the structures and properties of simple molecules. Our ideas about covalent bonds go back to 1916, when Lewis described the sharing of electron pairs between atoms. The pairs of electrons held jointly by two atoms were considered to be effective in completing a stable electronic configuration for each atom. This approach provided only a qualitative picture of chemical bonding. The first successful quantum mechanical explanation of a chemical bond, specifically that in molecular hydrogen, was made in 1927 by Heitler and London using the valence bond method. Since then the molecular orbital method has become the method of choice, and so it is emphasized in this chapter. Today the electronic structure, energy levels, bond angles, bond distances, dipole moments, and spectra of simple molecules may be calculated with a high degree of accuracy. For molecules with more than one electron, approximations have to be introduced. However, even approximate calculations are very helpful in understanding molecular structure, chemical properties, and molecular spectra.

11.1 THE BORN–OPPENHEIMER APPROXIMATION

Born and Oppenheimer pointed out that since nuclei are thousands of times more massive than electrons, they move much more slowly and can be treated as being stationary in considering the motions of the electrons in molecules. Thus a simpler electronic Schrödinger equation can be solved for each fixed internuclear distance R. The molecular vibrations can then be investigated by writing a Schrödinger equation for the motion of the nuclei in the average potential field produced by the electrons.

The hydrogen molecule ion H_2^+ is the simplest molecule, and so we will discuss it in some detail as an introduction to the treatment of more complicated molecules. The Schrödinger equation for the hydrogen molecule ion is

$$\hat{H}\,\psi(\boldsymbol{r}_1, \boldsymbol{R}_\mathrm{A}, \boldsymbol{R}_\mathrm{B}) = E\psi(\boldsymbol{r}_1, \boldsymbol{R}_\mathrm{A}, \boldsymbol{R}_\mathrm{B}) \tag{11.1}$$

where $\boldsymbol{r}_1$ is the vector locating the electron and $\boldsymbol{R}_\mathrm{A}$ and $\boldsymbol{R}_\mathrm{B}$ are vectors locating the two protons in a coordinate system. The Hamiltonian for the hydrogen molecule ion is given by

$$\hat{H} = -\frac{\hbar^2}{2M}(\nabla_\mathrm{A}^2 + \nabla_\mathrm{B}^2) - \frac{\hbar^2}{2m_\mathrm{e}}\nabla_\mathrm{e}^2 - \frac{e^2}{4\pi\epsilon_0 r_{1\mathrm{A}}} - \frac{e^2}{4\pi\epsilon_0 r_{1\mathrm{B}}} + \frac{e^2}{4\pi\epsilon_0 R} \tag{11.2}$$

where M is the mass of each nucleus, m_e is the mass of the electron, $r_{1\mathrm{A}}$ is the distance between the electron and the A nucleus, and $r_{1\mathrm{B}}$ is the distance between the electron and the B nucleus, as shown in Fig. 11.1. It is not practical to try to solve an equation like equation 11.1 directly, so the approximation introduced by Born and Oppenheimer is used. Since the electron moves much more rapidly than the nuclei, the Schrödinger equation for the electronic motion can be studied at a fixed distance R (not a vector) between the nuclei. By using the Born–Oppenheimer approximation, we assume that the wavefunction for the molecule is given by

$$\psi(\boldsymbol{r}_1, \boldsymbol{R}_\mathrm{A}, \boldsymbol{R}_\mathrm{B}) = \psi_\mathrm{e}(\boldsymbol{r}_1, R)\psi_\mathrm{n}(\boldsymbol{R}_\mathrm{A}, \boldsymbol{R}_\mathrm{B}) \tag{11.3}$$

where $\psi_\mathrm{n}(\boldsymbol{R}_\mathrm{A}, \boldsymbol{R}_\mathrm{B})$ is the wavefunction for nuclear motion. The Schrödinger equation for electronic motion is then

$$\hat{H}_\mathrm{e}\psi_\mathrm{e} = E_\mathrm{e}(R)\psi_\mathrm{e} \tag{11.4}$$

with the Hamiltonian for electronic motion given by

$$\hat{H}_\mathrm{e} = -\frac{\hbar}{2m_\mathrm{e}}\nabla_\mathrm{e}^2 - \frac{e^2}{4\pi\epsilon_0 r_{1\mathrm{A}}} - \frac{e^2}{4\pi\epsilon_0 r_{1\mathrm{B}}} + \frac{e^2}{4\pi\epsilon_0 R} \tag{11.5}$$

containing the electronic kinetic energy, the electrostatic attraction of the electrons to each nucleus, and the nuclear electrostatic repulsion. Notice that the latter term is a constant since the internuclear distance R is fixed. We can solve equation 11.4 for all R, giving us the electronic energy $E_\mathrm{e}(R)$ as a function of R.

The Born–Oppenheimer approximation shows that the nuclei move in the potential energy $E_\mathrm{e}(R)$ determined by the electronic motion, so the Schrödinger equation for nuclear motion becomes

$$\left[-\frac{\hbar^2}{2M}(\nabla_\mathrm{A}^2 + \nabla_\mathrm{B}^2) + E_\mathrm{e}(R)\right]\psi_\mathrm{n}(\boldsymbol{R}_\mathrm{A}, \boldsymbol{R}_\mathrm{B}) = E_\mathrm{n}\psi_\mathrm{n}(\boldsymbol{R}_\mathrm{A}, \boldsymbol{R}_\mathrm{B}) \tag{11.6}$$

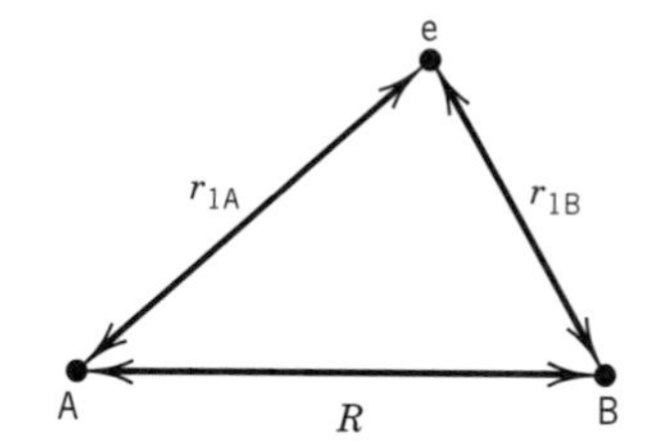

Figure 11.1 Coordinates for the hydrogen molecule ion H_2^+. The protons are labeled A and B.

where E_n is the energy of nuclear motion, which contains contributions for translational motion, rotational motion, and vibrational motion (as described in Section 13.3). The wavefunction for nuclear motion $\psi_n(\boldsymbol{R}_A, \boldsymbol{R}_B)$ is the product of wavefunctions for translational, rotational, and vibrational motions (Section 13.3).

11.2 THE HYDROGEN MOLECULE ION

The electronic Schrödinger equation for H_2^+ (equation 11.4) can be solved exactly, but the results are complicated. It is more useful to continue the approach we followed in Section 10.7 in treating the helium atom, and that is to use a trial wavefunction made up of hydrogenlike atomic orbitals. This approach is referred to as **molecular orbital theory.** In this approach, molecular wavefunctions for many-electron molecules are written in terms of determinants (Section 10.8) involving single electron wavefunctions (molecular orbitals). Molecular properties calculated in this way are approximate, but, as we will see later (Section 11.4), the treatment can be improved to any desired degree of accuracy. As trial wavefunctions $\psi_i(\boldsymbol{r}_A, \boldsymbol{r}_B, \boldsymbol{R})$ for the one-electron hydrogen molecule ion, we will use the linear combination

$$\psi_\pm = c_1 1s_A \pm c_2 1s_B \tag{11.7}$$

where $1s_A$ and $1s_B$ are atomic hydrogen orbitals on protons A and B, and c_1 and c_2 are constants. This type of function is referred to as a linear combination of atomic orbitals or an **LCAO molecular orbital.** Since the two nuclei are identical, $c_1 = c_2 = c$. For nuclei that are close, the 1s orbitals overlap as shown in Fig. 11.2.

In order to normalize the molecular orbital with the plus sign, the following integral must have the value unity:

$$\int \mathrm{d}\tau\, \psi_+^* \psi_+ = 1 \tag{11.8}$$

where $\mathrm{d}\tau$ is the volume element. Substituting the wavefunction yields

$$\begin{aligned} 1 &= c^2 \int \mathrm{d}\tau (1s_A^* + 1s_B^*)(1s_A + 1s_B) \\ &= c^2 \int \mathrm{d}\tau\, 1s_A^* 1s_A + c^2 \int \mathrm{d}\tau\, 1s_A^* 1s_B + c^2 \int \mathrm{d}\tau\, 1s_B^* 1s_A + c^2 \int \mathrm{d}\tau\, 1s_B^* 1s_B \end{aligned} \tag{11.9}$$

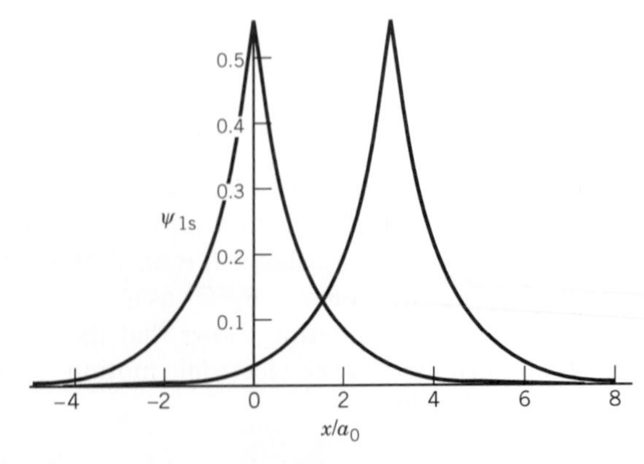

Figure 11.2 Magnitudes of the 1s wavefunctions on hydrogen atoms A and B when they are $3a_0$ apart, but not interacting. The abscissa is the distance from atom A along the internuclear axis in units of a_0. (See Computer Problem 11.F.)

The first and last integrals are each equal to unity because the 1s orbitals are normalized. Since the 1s wavefunction is real, the second and third integrals are equal. They are referred to as the **overlap integrals** S. Therefore, equation 11.9 can be written in terms of the overlap integral:

$$\begin{aligned} 1 &= c^2\left(2 + 2\int d\tau\, 1s_A 1s_B\right) \\ &= c^2(2 + 2S) \end{aligned} \tag{11.10}$$

so that $c = 1/[2(1 + S)]^{1/2}$. Thus the normalized wavefunction with a plus sign is given by

$$\psi_g = \frac{1}{[2(1+S)]^{1/2}}(1s_A + 1s_B) \tag{11.11}$$

In the same way it can be shown that the normalized wavefunction with a minus sign is

$$\psi_u = \frac{1}{[2(1-S)]^{1/2}}(1s_A - 1s_B) \tag{11.12}$$

When a molecule has a center of symmetry, the wavefunction may or may not change sign when it is inverted through the center of symmetry. If $\psi(x, y, z) = \psi(-x, -y, -z)$, the wavefunction is said to have **even parity** and is designated with a subscript g for *gerade* (German for even). If $\psi(x, y, z) = -\psi(-x, -y, -z)$, the wavefunction is said to have **odd parity** and is designated with a subscript u for *ungerade* (German for odd). This is the origin of the subscripts on ψ_g and ψ_u.

The overlap integral S can be evaluated analytically as a function of the **internuclear distance** R, but since this is difficult the result is simply given here:

$$S = e^{-R}\left(1 + R + \frac{R^2}{3}\right) \tag{11.13}$$

When $R = 0$, the two 1s orbitals overlap completely, and $S = 1$. As R is increased to infinity, the overlap integral decreases asymptotically to zero. The magnitude of the ψ_g molecular orbital along the axis through the two nuclei is given in Fig. 11.3*a* for $R = 3a_0$. The magnitude of the ψ_u molecular orbital along the axis through the two nuclei is given in Fig. 11.3*b* for $R = 3a_0$. The probability densities of the electron in the hydrogen molecule ion are given by the squares of the wavefunctions, as shown for the probability densities along the internuclear axis in Fig. 11.3*c* for ψ_g and in Fig. 11.3*d* for ψ_u.The electron density of ψ_g is large between the nuclei, which tends to pull them together. This buildup of electron density between the two nuclei for ψ_g causes bonding, and so ψ_g is referred to as a **bonding molecular orbital.** The electron density of ψ_u (Fig. 11.3*d*) has a nodal plane between the two nuclei where the electron density is zero. This lessening of the electron density between the two nuclei works against bonding; hence ψ_u is called an **antibonding molecular orbital.**

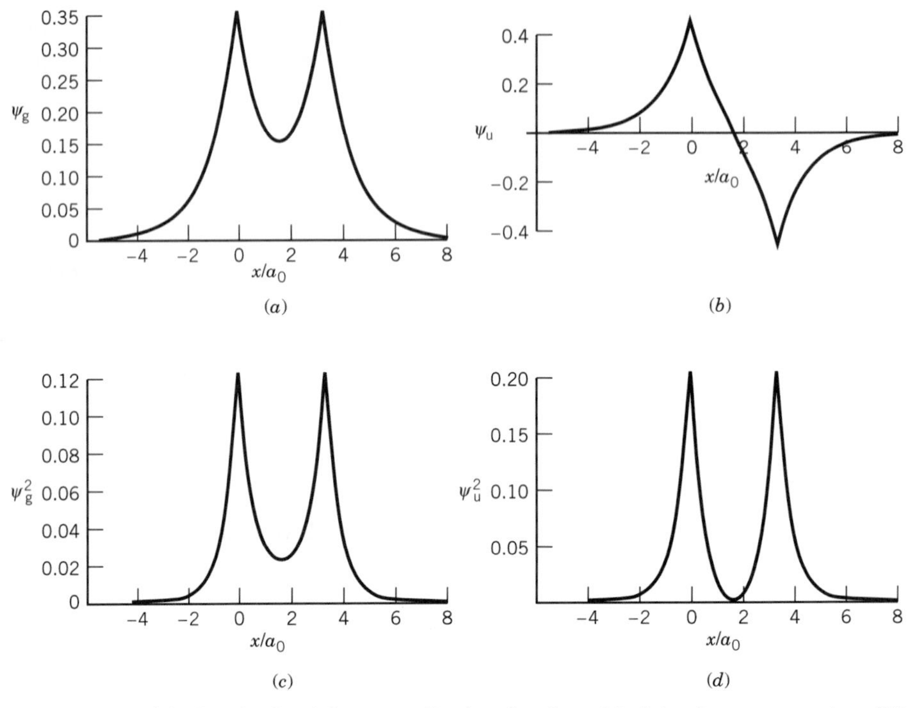

Figure 11.3 (*a*) Magnitude of the normalized molecular orbital ψ_g when protons A and B are separated by $3a_0$. (*b*) Magnitude of the normalized molecular orbital ψ_u. (*c*) Magnitude of ψ_g^2. (*d*) Magnitude of ψ_u^2. (See Computer Problem 11.G.)

11.3 CALCULATION OF THE ENERGY OF THE HYDROGEN MOLECULE ION

Using the molecular orbitals (ψ_g or ψ_u), we can calculate an upper bound on the energy of $H_2{}^+$ using the variation principle (Section 10.6) for every value of the internuclear distance R. The variational energy E for an orbital $\psi = c_1 1s_A + c_2 1s_B$ is given by

$$E = \frac{\int \psi^* \hat{H} \psi \, d\tau}{\int \psi^* \psi \, d\tau} = \frac{\int [c_1 1s_A + c_2 1s_B] \hat{H} [c_1 1s_A + c_2 1s_B] \, d\tau}{\int [c_1 1s_A + c_2 1s_B]^2 \, d\tau}$$

$$= \frac{c_1^2 H_{AA} + 2c_1c_2 H_{AB} + c_2^2 H_{BB}}{c_1^2 S_{AA} + 2c_1c_2 S_{AB} + c_2^2 S_{BB}} = \frac{c_1^2 H_{AA} + 2c_1c_2 H_{AB} + c_2^2 H_{BB}}{c_1^2 + 2c_1c_2 S + c_2^2} \tag{11.14}$$

where the following symbols have been used:

$$H_{AA} = \int 1s_A \hat{H} 1s_A \, d\tau = \int 1s_B \hat{H} 1s_B \, d\tau = H_{BB} \tag{11.15}$$

$$H_{AB} = \int 1s_A \hat{H} 1s_B \, d\tau = \int 1s_B \hat{H} 1s_A \, d\tau \tag{11.16}$$

$$S_{AA} = \int 1s_A 1s_A \, d\tau = \int 1s_B 1s_B \, d\tau = S_{BB} = 1 \tag{11.17}$$

$$S_{AB} \int 1s_A 1s_B \, d\tau = \int 1s_B 1s_A \, d\tau = S_{BA} = S \tag{11.18}$$

The integral H_{AA} is called the **Coulomb integral** because the difference between H_{AA} and the energy of a single hydrogen atom is just that of the Coulomb interaction of nucleus B with an electron centered on nucleus A ($H_{AA} = H_{BB}$ in this case). Since this interaction is attractive, its contribution to H_{AA} is negative. The energy of a single hydrogen atom is also negative; thus, H_{AA} is a negative number. The integral H_{AB} is referred to as the **resonance integral.**

We have already noted that the overlap integral is a function of the internuclear distance R (see equation 11.13). We have also used a symmetry argument to show that $c_1 = c_2 = c$, but now we are going to treat c_1 and c_2 as variational constants to be determined by minimizing the energy (see Section 10.6).

To find the minimum energy, we set the derivatives of E with respect to c_1 and c_2 equal to zero. To do this it is convenient to write equation 11.14 in the form

$$E(c_1^2 + 2c_1c_2S + c_2^2) = c_1^2 H_{AA} + 2c_1c_2 H_{AB} + c_2^2 H_{BB} \tag{11.19}$$

Differentiating this equation with respect to c_1 yields

$$E(2c_1 + 2c_2S) + \frac{\partial E}{\partial c_1}(c_1^2 + 2c_1c_2S + c_2^2) = 2c_1 H_{AA} + 2c_2 H_{AB} \tag{11.20}$$

and differentiating with respect to c_2 yields

$$E(2c_1S + 2c_2) + \frac{\partial E}{\partial c_2}(c_1^2 + 2c_1c_2S + c_2^2) = 2c_1 H_{AB} + 2c_2 H_{BB} \tag{11.21}$$

Since $\partial E/\partial c_1$ and $\partial E/\partial c_2$ are equal to zero for the minimum energy, these equations can be written as

$$c_1(H_{AA} - E) + c_2(H_{AB} - SE) = 0 \tag{11.22}$$

$$c_1(H_{AB} - SE) + c_2(H_{BB} - E) = 0 \tag{11.23}$$

There is a nontrivial solution to these equations only if the determinant of the coefficients is equal to zero:

$$\begin{vmatrix} H_{AA} - E & H_{AB} - SE \\ H_{AB} - SE & H_{BB} - E \end{vmatrix} = 0 \tag{11.24}$$

It can be shown that $H_{AA} = H_{BB} = E_{1s} + J$, where E_{1s} is the energy of a hydrogen atom in the 1s state and J is a function of the internuclear distance R. It can also be shown that $H_{AB} = E_{1s}S + K$, where K is also a function of R. These functions are given by

$$J = e^{-2R}\left(1 + \frac{1}{R}\right) \tag{11.25}$$

$$K = \frac{S}{R} - e^{-R}(1 + R) \tag{11.26}$$

When the expressions for H_{AA}, H_{BB}, and H_{AB} are substituted into the secular equation (11.24), we obtain

$$\begin{vmatrix} E_{1s} + J - E & E_{1s}S + K - SE \\ E_{1s}S + K - SE & E_{1s} + J - E \end{vmatrix} = (E_{1s} + J - E)^2 - (E_{1s}S + K - SE)^2 = 0 \tag{11.27}$$

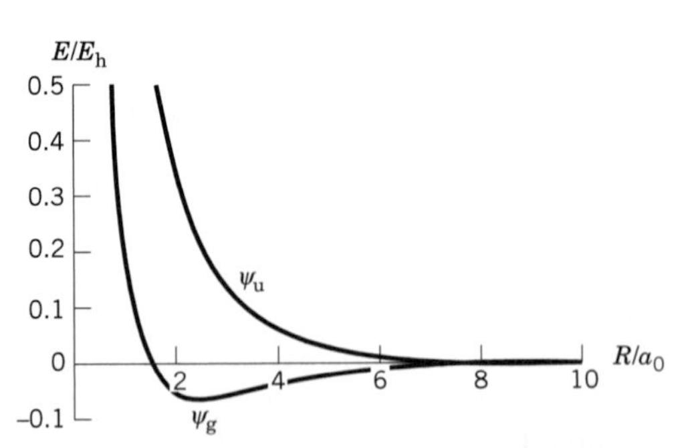

Figure 11.4 Plot of energies of the hydrogen molecule ion in bonding and antibonding molecular orbitals as a function of internuclear distance according to the approximate wavefunctions of equations 11.11 and 11.12. Energies are expressed in units of E_h and internuclear distances are expressed in terms of a_0. (See Computer Problem 11.B.)

This yields two solutions:

$$E_g = E_{1s} + \frac{J + K}{1 + S} \tag{11.28}$$

$$E_u = E_{1s} + \frac{J - K}{1 - S} \tag{11.29}$$

The energy ΔE_g of the hydrogen molecule ion in the bonding orbital ψ_g relative to the completely dissociated species H^+ and H is given by

$$\Delta E_g = E_g - E_{1s} = \frac{J + K}{1 + S} \tag{11.30}$$

The energy ΔE_u of the hydrogen molecule ion in the antibonding orbital ψ_u relative to the completely dissociated species H^+ and H is given by

$$\Delta E_u = E_u - E_{1s} = \frac{J - K}{1 - S} \tag{11.31}$$

These two energies are plotted in Fig. 11.4 versus internuclear distance R.

Figure 11.4 shows that for the bonding wavefunction ψ_g the minimum energy is at $R = 2.50a_0 = 132$ pm, in comparison with the experimental value $R_e = 2.00a_0 = 106$ pm. At $R = 132$ pm, the binding energy is $0.0648E_h = 170$ kJ mol^{-1}, in comparison with the experimental value of $0.102E_h = 258$ kJ mol^{-1}. The antibonding wavefunction ψ_u leads to repulsion at all internuclear distances. This is another reason it is called an antibonding orbital. This state is referred to as an **excited state.** The simple molecular orbital in equation 11.7 does not quantitatively explain the bonding in the hydrogen molecule ion, but this wavefunction can be improved by adding more terms. The next logical step is to add terms for 2s and 2p orbitals on the two nuclei. In the limit of adding more and more terms, the results obtained are equal to the exact solution of the Schrödinger equation.

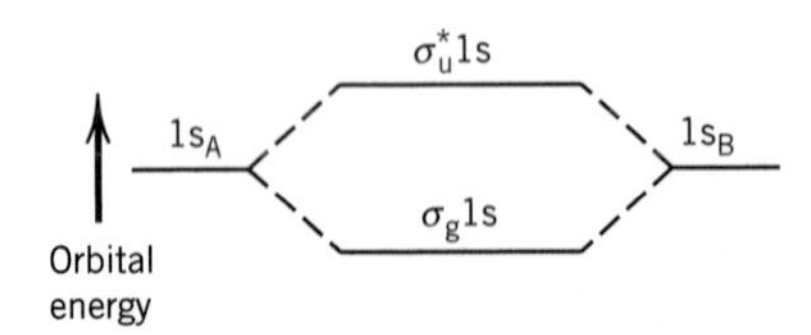

Figure 11.5 Correlation diagram for H_2^+. Note that the energy of the antibonding orbital is raised more than the energy of the bonding orbital is lowered, as indicated by equations 11.30 and 11.31.

It is useful to represent the energies of molecular orbitals by use of a correlation diagram, such as Fig. 11.5. Since H_{AA} and H_{AB} are negative and S is positive, E_g is more negative than E_u and is the energy of the more stable molecular orbital, as is seen in Fig. 11.4. Since ψ_g is symmetrical around the internuclear axis, it is referred to as a sigma (σ) orbital, and since it is even (*gerade*) and is made up of two 1s orbitals, it is designated $\sigma_g 1s$. The less stable orbital is represented by $\sigma_u^* 1s$. Note that two atomic orbitals give rise to two molecular orbitals.

To improve the result for the ground state, one uses the following atomic orbitals in the xy plane:

$$1s_A = k^{3/2}\pi^{-1/2}e^{-kr_A} \tag{11.32}$$

$$(2p_y)_A = \frac{\zeta^{5/2}}{4(2\pi)^{1/2}} y e^{-\zeta r_A/2} \tag{11.33}$$

as discussed by Levine.* The substitution of these atomic orbitals in equations 11.11 and 11.12 yields molecular orbitals proportional to $\psi(1s)$ and $\psi(2p)$ in the xy plane:

$$\psi(1s) = k^{3/2}\pi^{-1/2}(e^{-kr_A} \pm e^{-kr_B}) \tag{11.34}$$

$$\psi(2p) = \frac{\zeta^{5/2}}{4(2\pi)^{1/2}} y(e^{-\zeta r_A/2} \pm e^{-\zeta r_B/2}) \tag{11.35}$$

where the plus sign yields the bonding orbital and the minus sign yields the antibonding orbital. Perturbation theory using these wavefunctions leads to a minimum $U(R)$ at $2.01a_0$. At this internuclear distance $k = 1.246/a_0$ and $\zeta = 2.965/a_0$ when distances are expressed in bohrs. The calculated D_0 is 2.73 eV, compared with the true value of 2.79 eV. The electron densities in the xy plane for these molecular orbitals are shown in Fig. 11.6 for the 1s orbitals and in Fig. 11.7 for the 2p orbitals. In the 1s bonding orbital there is a buildup of electron density between the nuclei, and in the 2p bonding orbital there is a buildup of electron density between the lobes of the p orbitals.

We can use the same LCAO procedure to form other molecular orbitals from the atomic orbitals on the two protons. This is useful only if the molecular orbitals that are found have the bonding characteristics of the $\sigma_g 1s$ or $\sigma_u^* 1s$ orbitals we have discussed. Two conditions must be met for this to be true: The two atomic orbitals must have about the same energy, and they must have the same symmetry properties with respect to rotations about the internuclear axis. If the latter condition is not satisfied, then the integrals corresponding to H_{AB} and to S_{AB} are zero for all R; if the former condition is not satisfied, then the integrals corresponding to H_{AA} and H_{BB} will be very different. In either case the coefficients c_1 and c_2 will be very different (one approximately 0 and one approximately 1) so that little or no bonding takes place. It is quite easy to determine whether the overlap integral S_{AB} is zero by examining the positive and negative parts of the orbitals, as indicated in Fig. 11.8.

Figure 11.8*a* shows the situation when two 1s orbitals (or two *n*s orbitals) are brought together: The orbitals overlap and bonding can occur. Figure 11.8*b* shows what happens when a p orbital with axis perpendicular to the internuclear axis is brought up to an s orbital. Here the overlap is zero since the overlap of the positive lobe of the p orbital with the s orbital is exactly canceled by the overlap of the negative lobe of the p orbital, and thus no bonding can occur. If two parallel

*I. N. Levine, *Quantum Chemistry,* 5th ed. Upper Saddle River, NJ: Prentice-Hall, 1999, pp. 389–390.

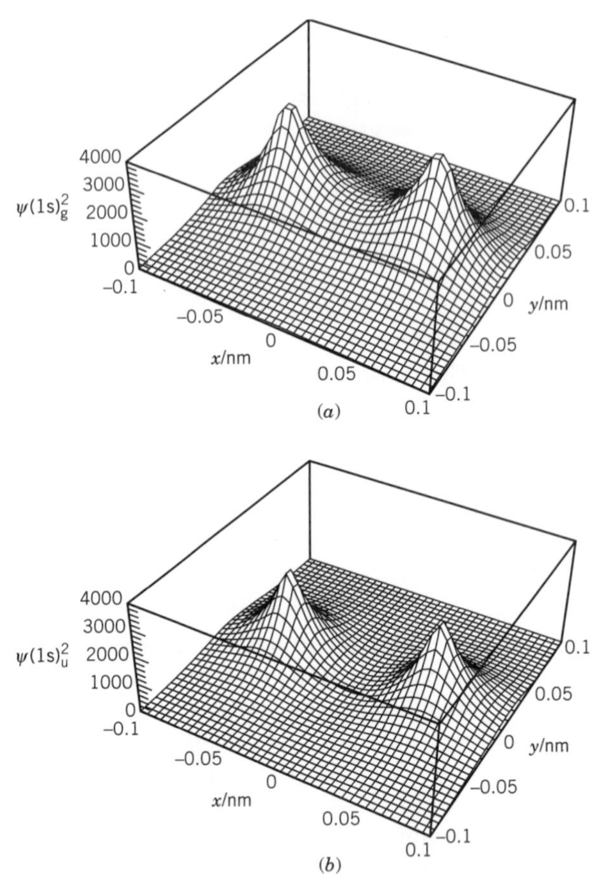

Figure 11.6 Plot of the electron density in the *xy* plane for the hydrogen molecule ion $H_2{}^+$ with (*a*) a 1s sigma bonding orbital and (*b*) a 1s sigma antibonding orbital. Note the buildup of electron density between the nuclei with the bonding orbital. Also note that with the antibonding orbital the electron density is zero along a line perpendicular to the line between the nuclei and halfway between the nuclei. The internuclear axis is along x. (See Computer Problem 11.A.)

p orbitals are brought together, as in Fig. 11.8*c*, *d*, bonding can occur. The orbitals formed in the last line of the figure are interesting because they have a nodal plane (the wavefunction is zero everywhere in the plane) perpendicular to the page and containing the internuclear axis. Such an orbital is called a π orbital. The bonding orbital changes sign on inversion and so is ungerade (π_u2p), while the antibonding orbital is gerade (π_g^* 2p). Note that there is another pair of orbitals, exactly like these, formed from the 2p orbitals perpendicular to the page. Therefore, the π orbitals are doubly degenerate.

This process of combining hydrogenlike orbitals can be continued to higher values of the orbital angular momentum quantum number ℓ. One-electron molecular orbitals are classified according to the quantum number λ for the angular momentum about the internuclear axis. In general, for an electron in a diatomic molecule the axial angular momentum is given by

$$L_z = \pm\lambda\hbar \tag{11.36}$$

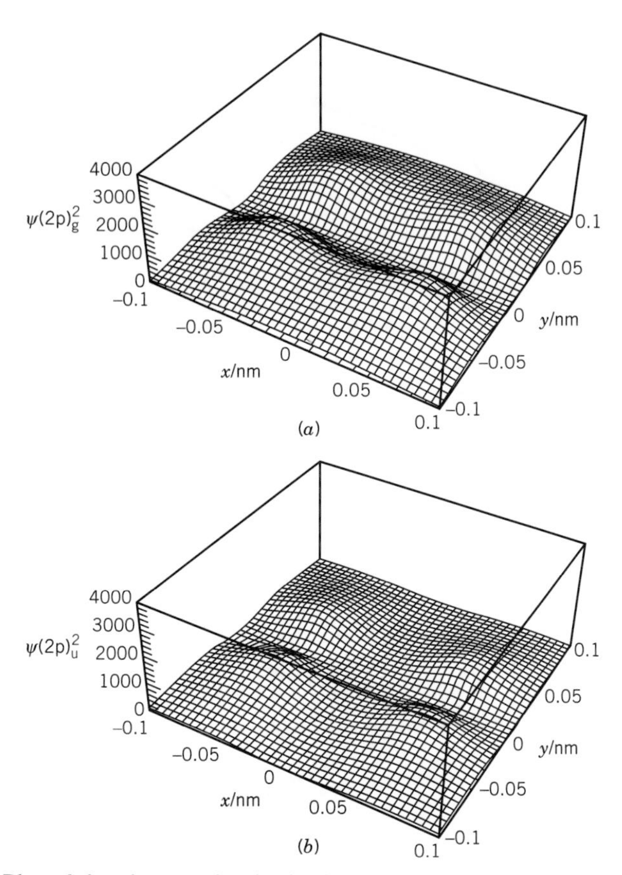

Figure 11.7 Plot of the electron density in the *xy* plane for the hydrogen molecule ion H_2^+ with (*a*) a 2p bonding orbital and (*b*) a 2p antibonding orbital. Note that with the 2p bonding orbital, the electron density is zero along the line joining the nuclei with the bonding orbital. Also note that with the antibonding orbital the electron density is also zero along a line perpendicular to the line between the nuclei and halfway between the nuclei. (See Computer Problem 11.A.)

where λ corresponds to the absolute value m for an atom. The value of λ for a one-electron orbital for a diatomic molecule is represented by a Greek letter according to

$$\begin{array}{lccccc} \lambda & = 0 & 1 & 2 & 3 \\ \text{Orbital} & = \sigma & \pi & \delta & \phi \end{array}$$

This is, of course, analogous to the classification of atomic orbitals as s, p, d, f, ... according to $\ell = 0, 1, 2, 3, \ldots$.

Excited states of H_2^+ are formed by exciting the electron to one of the higher levels, as shown by the potential energy curves in Fig. 11.9. From our analysis of the molecular orbitals of H_2^+, we can now go on to discuss other homonuclear diatomic molecules. We first must arrange the LCAO-MOs (linear combination of atomic orbitals–molecular orbitals) that we have discussed in order of increasing energy, so that electrons can be placed in the orbitals, two at a time, to account for the electronic structures of various diatomic molecules. This is the same aufbau process that was used with atoms (Section 10.10) and will be discussed in

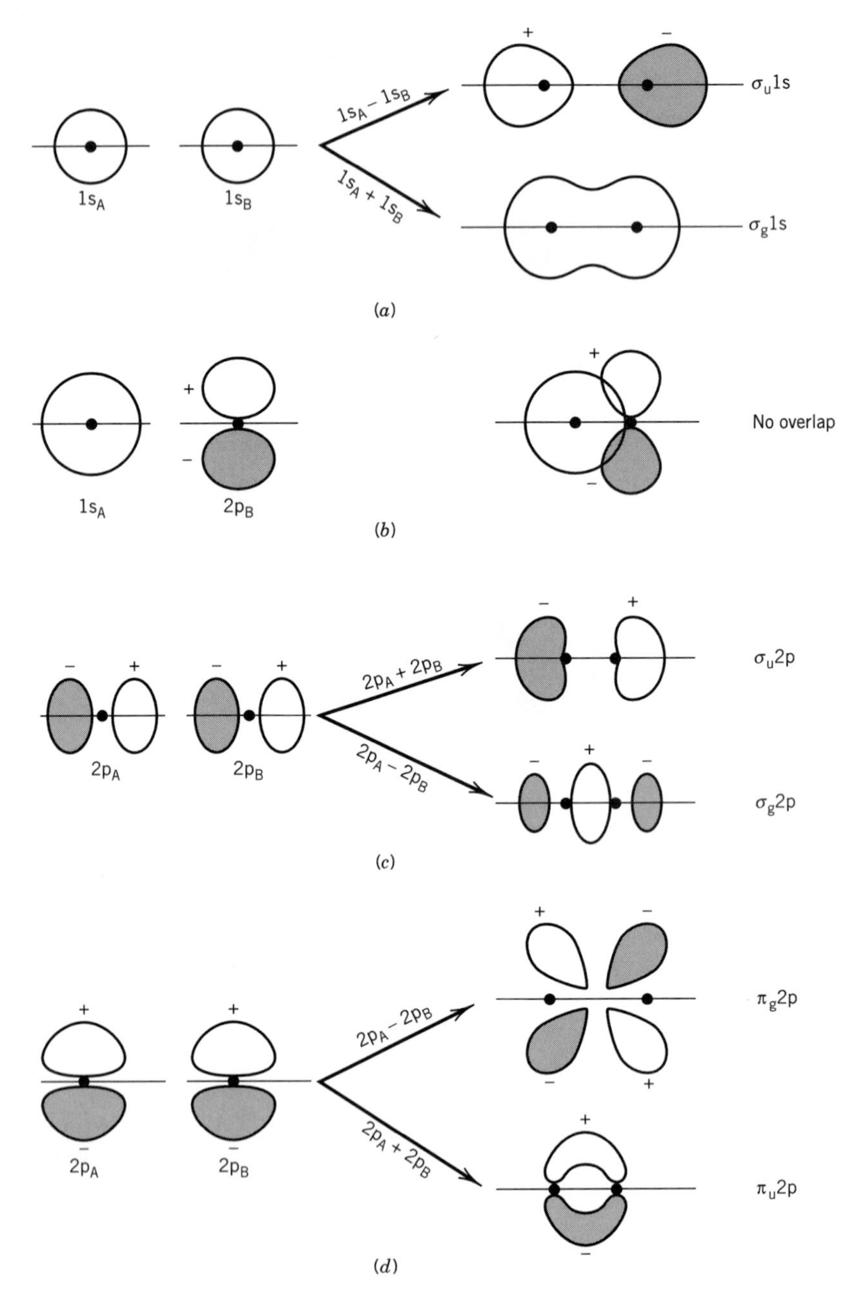

Figure 11.8 Formation of pairs of molecular orbitals from pairs of atomic orbitals. The solid points represent nuclei A and B.

Section 11.5. The sequence of homonuclear diatomic molecular orbitals is given approximately in Fig. 11.10. We have to say *approximately* because the order of the energy levels depends on the atomic number of the nuclei and the internuclear distance. As indicated in this diagram, two atomic orbitals combine to form two molecular orbitals, one with lower energy and the other with higher energy than the atomic orbitals from which they were formed.

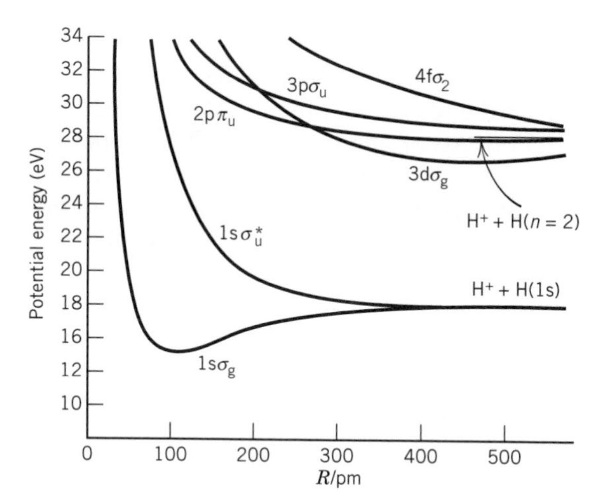

Figure 11.9 Lower excited states of H_2^+. The energy is measured relative to the ground state of the hydrogen molecule. (From T. Sharp, "Potential Energy Diagram for Molecular Hydrogen and Its Ions." In *Atomic Data,* Vol. 2, p. 119. New York: Academic, 1971.)

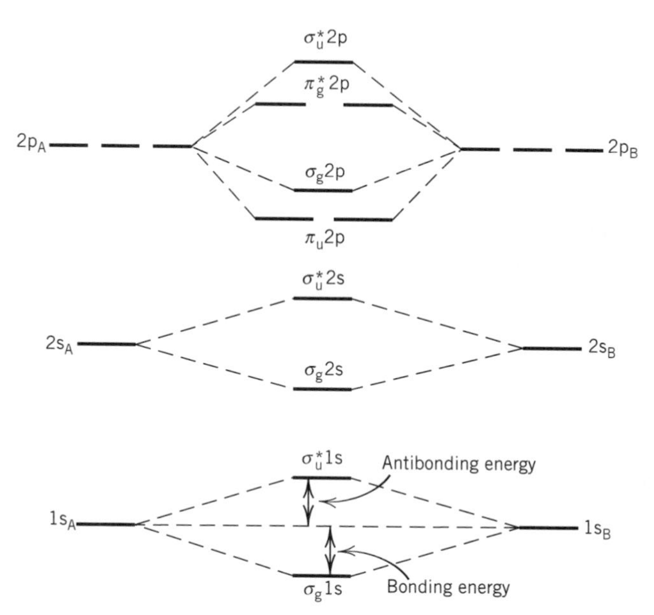

Figure 11.10 Schematic diagram for the lowest energy molecular orbitals of homonuclear diatomic molecules.

11.4 MOLECULAR ORBITAL DESCRIPTION OF THE HYDROGEN MOLECULE

Using the Born–Oppenheimer approximation, the electronic Hamiltonian for the hydrogen molecule may be written

$$\hat{H} = -\frac{\hbar^2}{2m}(\nabla_1^2 + \nabla_2^2) + \frac{e^2}{4\pi\epsilon_0}\left(-\frac{1}{r_{A1}} - \frac{1}{r_{A2}} - \frac{1}{r_{B1}} - \frac{1}{r_{B2}} + \frac{1}{r_{12}}\right) + \frac{e^2}{4\pi\epsilon_0 R} \tag{11.37}$$

where the coordinates are defined in Fig. 11.11.

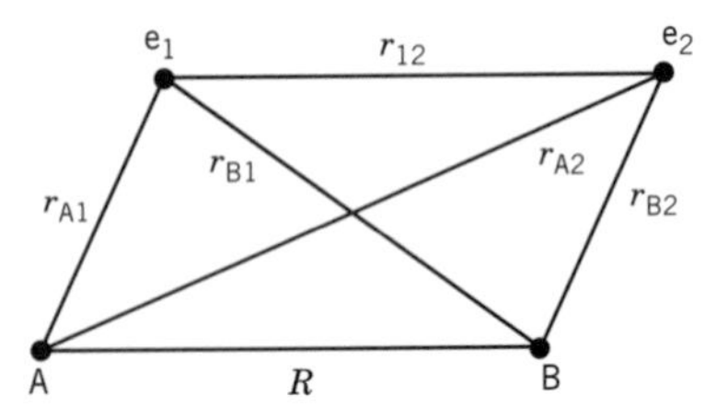

Figure 11.11 Electronic coordinates in the hydrogen molecule. The two protons are represented by A and B.

When the electronic Hamiltonian for the hydrogen molecule is used in equation 11.4, an exact solution cannot be obtained because of the $1/r_{12}$ term, just as in the atomic case. This is the reason that the LCAO-MO method is used to obtain an approximate solution. According to the LCAO-MO approach, molecular hydrogen is formed by putting two electrons with opposite spin in the $\sigma_g 1s$ orbital. That is, we assume that each electron can be assigned to an orbital and that the electronic wavefunction for the molecule is the product of the two wavefunctions for the two electrons:

$$\psi_{MO} = \psi_i(1)\psi_j(2) \tag{11.38}$$

where i and j designate the different orbitals and 1 and 2 designate the two electrons. According to the Pauli principle (Section 10.8), two electrons with opposite spin can be assigned to a given spatial orbital, and so as a first approximation we will assume that in the ground state of the hydrogen molecule the two electrons are placed in the $1\sigma_g$ orbital developed for H_2^+. Thus, the electronic configuration of H_2 will be described as $(1\sigma_g)^2$, just as we described the electronic configuration of He as $(1s)^2$.

The wavefunction for electron 1 in the $1\sigma_g$ molecular orbital given in equation 11.11 is represented by

$$1\sigma_g(1) = \frac{1}{[2(1+S)]^{1/2}}[1s_A(1) + 1s_B(1)] \tag{11.39}$$

In discussing the helium atom (Section 10.7) we found that the wavefunction satisfying the antisymmetry requirement is given by a Slater determinant. The same considerations apply here, and so the approximate wavefunction for the ground state of the hydrogen molecule is given by the following Slater determinant (remember that α and β are electron spin wavefunctions):

$$\psi_{MO}[(1\sigma_g)^2] = \frac{1}{\sqrt{2}}\begin{vmatrix} 1\sigma_g(1)\alpha(1) & 1\sigma_g(1)\beta(1) \\ 1\sigma_g(2)\alpha(2) & 1\sigma_g(2)\beta(2) \end{vmatrix} \tag{11.40}$$

This yields the following wavefunction for a hydrogen molecule in its ground state:

$$\begin{aligned} \psi_{MO}[(1\sigma_g)^2] &= [1\sigma_g(1)1\sigma_g(2)\alpha(1)\beta(2) - 1\sigma_g(1)1\sigma_g(2)\beta(1)\alpha(2)](1/2)^{1/2} \\ &= \frac{[1s_A(1) + 1s_B(1)][1s_A(2) + 1s_B(2)](1/2)^{1/2}[\alpha(1)\beta(2) - \beta(1)\alpha(2)]}{2(1+S_{AB})} \end{aligned} \tag{11.41}$$

The approximate energy of the hydrogen molecule is obtained by calculating the expectation value of the Hamiltonian (equation 11.37) using this wavefunction.

$$E = \int \psi^*_{MO}[(1\sigma_g)^2]\hat{H}_{el}\psi_{MO}[(1\sigma_g)^2]\,d\tau \tag{11.42}$$

This integration leads to a rather complicated equation for E, which we will simply write as

$$E = 2E_{1s} + \frac{e^2}{4\pi\epsilon_0 R} - \text{integrals} \tag{11.43}$$

The first term is the electronic energy of two hydrogen atoms at infinite distance. The second term is the energy of electrostatic repulsion of the two nuclei, and the last term is a series of integrals for the interactions of various charge distributions with one another.* These integrals can be evaluated at a series of internuclear distances to obtain the molecular potential energy curve. The minimum is at 84 pm, and the calculated dissociation energy is $D_e = 255$ kJ mol^{-1}. The experimental values are 74.1 pm and 458 kJ mol^{-1}.

Although simple molecular orbital theory does account for a large proportion of the binding energy of the hydrogen molecule, it has to be extended to yield accurate results. A detailed description of the improvements that are possible is beyond the scope of this book, but the general directions can be indicated.

We can see one deficiency in the approximate wavefunction by multiplying out the spatial part of equation 11.41 to obtain

$$1s_A(1)1s_A(2) + 1s_A(1)1s_B(2) + 1s_B(1)1s_A(2) + 1s_B(1)1s_B(2)$$

The first term and last term correspond to forms of the hydrogen molecule with ionic bonding, namely, $H_A^-H_B^+$ and $H_A^+H_B^-$, so this molecular orbital wavefunction describes a state at $R = \infty$ that is 50% H^+ and H^- and 50% H + H, which is clearly not correct. This problem can be reduced by introducing variable coefficients c_1 and c_2 in

$$\psi = c_1(R)\psi_{\text{covalent}} + c_2(R)\psi_{\text{ionic}} \tag{11.44}$$

where

$$\psi_{\text{covalent}} = 1s_A(1)1s_B(2) + 1s_A(2)1s_B(1) \tag{11.45}$$

$$\psi_{\text{ionic}} = 1s_A(1)1s_A(2) + 1s_B(1)1s_B(2) \tag{11.46}$$

Using the variational method, the values of c_1 and c_2 can be determined at each value of R. Since equation 11.44 yields $R_e = 74.9$ pm (experimental 74.1 pm) and $D_e = 386$ kJ mol^{-1} (experimental 458 kJ mol^{-1}), the inclusion of one variational parameter leads to considerable improvement.

Further improvements can be obtained by increasing the number of atomic orbitals used, that is, enlarging the basis set—for example, by adding 2s and $2p_z$ orbitals, and thereby introducing more variational parameters. The evaluation of parameters can be done in a systematic way by using the Hartree–Fock self-consistent field method (Section 10.9) that is used to obtain atomic orbitals. Equations of the form

$$\hat{H}_{\text{eff}}\psi_i = E_i\psi_i \tag{11.47}$$

where $\hat{H}_{\text{eff}}$ is the effective one-electron Hamiltonian, ψ_i is the molecular orbital for the ith electron, and E_i is the orbital energy for the ith electron, are solved by the iterative methods described in Section 10.9 for atoms to obtain self-consistent molecular orbitals. These Hartree–Fock wavefunctions do not adequately correlate the motion of electrons with unlike spins, and further improvements can be obtained by using a wavefunction that is a linear combination of functions representing different electronic configurations of the molecule. For example, the doubly excited configuration $(1\sigma_u)^2$ may be added to the molecular orbital because it

*These integrals are all given in P. W. Atkins, *Molecular Quantum Mechanics.* New York: Oxford University Press, 1983, Appendix 14.

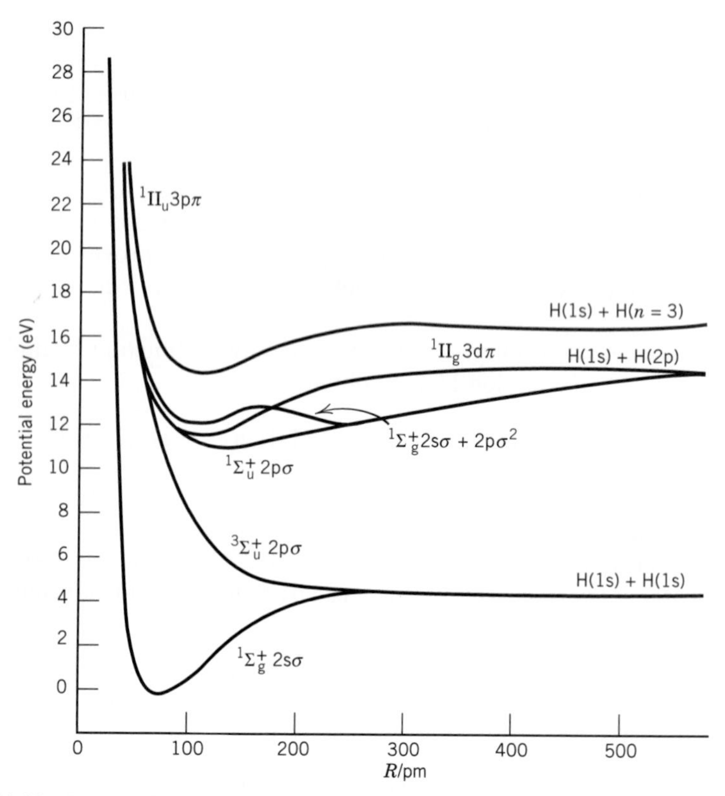

Figure 11.12 Potential energy curves for lower excited states of H_2. (From T. Sharp, "Potential Energy Diagram for Molecular Hydrogen and Its Ions." In *Atomic Data,* Vol. 2, p. 119. New York: Academic, 1971.)

has the same symmetry. By using 100-term wavefunctions, Kolos and Wolniewicz* have obtained $D_0 = 36\,117.8\ \text{cm}^{-1}$ in comparison with the observed† value of $36\,117.3 \pm 1.0\ \text{cm}^{-1}$. Their theoretical value of the internuclear distance R_e of H_2 is 74.140 pm, compared with the experimental value from spectroscopic measurements of 74.139 pm.

The hydrogen molecule has a large number of excited electronic states. The potential energy curves for some of the lowest are shown in Fig. 11.12. The states of the H_2 molecule (or any diatomic molecule) are given by molecular term symbols, analogous to the atomic term symbols of Section 10.13. In diatomic molecules the orbital angular momenta of the electrons couple to give a resultant orbital angular momentum L, and the electron spin momenta combine to give a resultant spin angular momentum S. The component of the orbital angular momentum along the axis of the molecule is given by

$$M_L = m_1 + m_2 + \cdots \tag{11.48}$$

*W. Kolos and L. Wolniewicz, *J. Chem. Phys.* **41**:3663 (1964); W. Kolos and L. Wolniewicz, *J. Chem. Phys.* **48**:3672 (1968); W. Kolos and L. Wolniewicz, *J. Chem. Phys.* **49**:404 (1968).

†G. Herzberg, *J. Mol. Spectrosc.* **33**:147 (1970); G. Herzberg, *Science* **177**:123 (1972).

where $m_i = 0$ for a σ orbital, $m_i = \pm 1$ for a π orbital, and so on. The quantum number Λ is defined as the absolute value of M_L and is represented by the following code letters:

$$\begin{array}{lllllll} \Lambda & = 0 & 1 & 2 & 3 & \ldots \\ \text{Symbol} & = \Sigma & \Pi & \Delta & \Phi & \ldots \end{array}$$

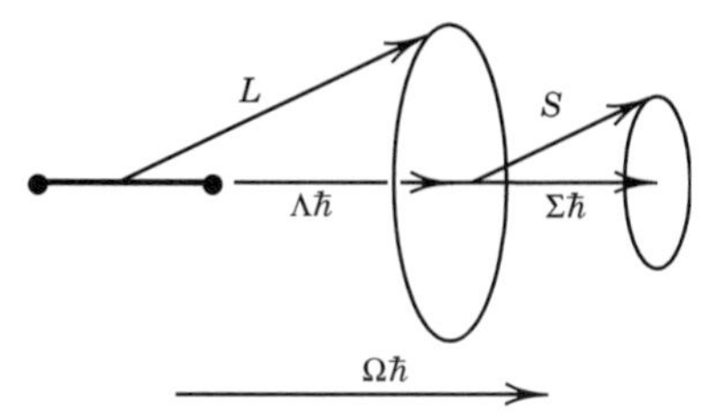

Figure 11.13 Orbital angular momentum L and spin angular momentum S for a diatomic molecule when the spin–orbit coupling is weak, as in Hund's case (a). [Reproduced from J. M. Hollas, *High Resolution Spectroscopy,* Fig. 6.31, p. 316. Boston: Butterworth, 1982; by permission of the publishers, Butterworth & Co. (Publishers) Ltd. ©.]

in analogy to atomic term symbols. The multiplicity of a state of a diatomic molecule is given by $2S + 1$, where S is the sum of the spins of the electrons in the molecule. The term symbol of a molecule is represented by

$$^{2S+1}\Lambda \tag{11.49}$$

For the hydrogen molecule in the ground state, there are two electrons in σ orbitals and so $m_1 = 0$, $m_2 = 0$, and $M_L = 0$. Since $\Lambda = 0$, the molecule is in a Σ state. Since the spins of the electrons are opposed, $S = 0$, and the molecular term symbol is $^1\Sigma$ (a singlet sigma state).

For Σ terms a superscript of plus or minus is added according to the behavior of the wavefunction on reflection in the plane containing the internuclear axis. A plus sign indicates that the wavefunction is invariant under this operation, and a negative sign indicates the wavefunction changes sign on reflection in this plane. If a diatomic molecule has a center of symmetry, a right subscript of g or u is attached to the term symbol to denote the parity of the orbital. As we have seen before (Section 11.3), the parity of an orbital is determined by observing the inversion symmetry. When a point on an orbital is inverted an equal distance through the center of the molecule, the orbital is gerade if it has the same sign at the two points. The parity of a multielectron molecule is obtained by noting g or u for every orbital and forming products using $g \times g = g$, $g \times u = u$, and $u \times u = g$.

There are four kinds of angular momenta in a diatomic molecule, of which we have discussed two: the electronic orbital angular momentum L and the electronic spin angular momentum S. The other two are the nuclear spin angular momentum I, which we will not consider, and the angular momentum N of the rotational motion of the nuclear framework. These angular momenta are coupled by small terms in the Hamiltonian. For low-mass diatomic molecules, the spin–orbit coupling is weak [called Hund's case (a) in the molecular spectroscopist's jargon], and the result of the coupling is illustrated in Fig. 11.13. The components of L ($\Lambda\hbar$) and S ($\Sigma\hbar$) along the internuclear axis are added to form a total angular momentum component along that axis labeled $\Omega\hbar$. Ω is given by $|\Lambda + \Sigma|$. These quantum numbers will come up again when we discuss selection rules in molecular spectroscopy.

11.5 ELECTRON CONFIGURATIONS OF HOMONUCLEAR DIATOMIC MOLECULES

As mentioned in Section 11.3, the order in which molecular orbitals of homonuclear diatomic molecules are filled depends on the nuclear charge and the internuclear distance. The variation in the sequence in energies is similar to that encountered with the relative energies of atomic orbitals of the elements, which depend on atomic number. To derive the possible sequences from the lowest energy up, it is useful to introduce a second way of bridging the gap between atomic and molecular systems. In addition to the separated atoms approach, there is the **united atom** approach. In this approach, the molecule is thought of first as the atom obtained from coalescing all the nuclei in the molecule. The united atom for H_2 is ^2He, and the united atom for N_2 is ^{28}Si. The electronic structure of the molecule is obtained by thinking about the changes in orbitals that occur when the nucleus of the united atom is pulled apart to form the nuclei of the molecule at their equilibrium distances. In the **correlation diagram** in Fig. 11.14, the possible energy levels of the united atom for a homonuclear diatomic molecule are given on the left and the sum of the energies of the two separated atoms are given on the right. An important principle for correlation diagrams is the noncrossing rule, which states that the lines for the energies of molecular orbitals with the same symmetry cannot cross. Symmetry in this context refers to whether the orbital is σ, π, δ, ... or whether it is g or u. The orbitals are connected in the order of increasing energy. The molecular orbitals coming out of the united atoms are designated $1s\sigma_g, 2s\sigma_g, 2p\sigma_u, \ldots$, and the resulting molecular orbitals are designated $1\sigma_g, 1\sigma_u, 2\sigma_g, \ldots$. The reason for dropping the s, p, d,... in the latter is that different angular momentum quantum numbers may be involved in the united atoms and separated atoms approaches; for example, $2p\sigma_u$ correlates with $\sigma_u 1s$.

The electronic structures of successive homonuclear diatomic molecules may be obtained from the correlation diagram by use of the aufbau principle; that is, electrons are added to orbitals in pairs, in order of increasing energy. Notice that the order of orbital energies varies with R. Two electrons may be placed in a σ level and four in a π or δ level (since $L_z = \pm\lambda\hbar$, making these doubly degenerate orbitals). The electron configurations of the ground states of homonuclear diatomic molecules from the first row of the periodic table are given in Table 11.1. In molecules with many electrons, the inner electrons tend to be concentrated around the nuclei and take little part in forming bonds. The fact that a configuration is listed does not mean that the molecule is stable. Orbitals that tend to build up electron density between the atoms are referred to as **bonding orbitals.** The bonding orbitals are $1\sigma_g, 2\sigma_g, 1\pi_u, 3\sigma_g, \ldots$. Orbitals with nodal planes between the nuclei are **antibonding orbitals** (Section 11.3). Thus, $1\sigma_u$, $2\sigma_u$, $1\pi_g$, and $3\sigma_u$ are antibonding orbitals. Notice that the orbitals of homonuclear diatomic molecules come in pairs, one bonding and one antibonding. The bonding energy of an electron in a bonding orbital is usually slightly less than the antibonding energy of an electron in an antibonding orbital.

It is useful to define a **bond order** that is roughly proportional to the strength of bonding. The bond order is equal to the number of bonding pairs of electrons minus the number of antibonding pairs. Bond orders of 1, 2, and 3 correspond with what are usually called single, double, and triple bonds.

We now discuss, one by one, the homonuclear diatomic molecules of the first row elements.

He2$^+$

The third electron in the molecular ion He_2^+ is in an antibonding orbital, but there is net bonding because there are two electrons in bonding orbitals. This ion is observed in electric arcs.

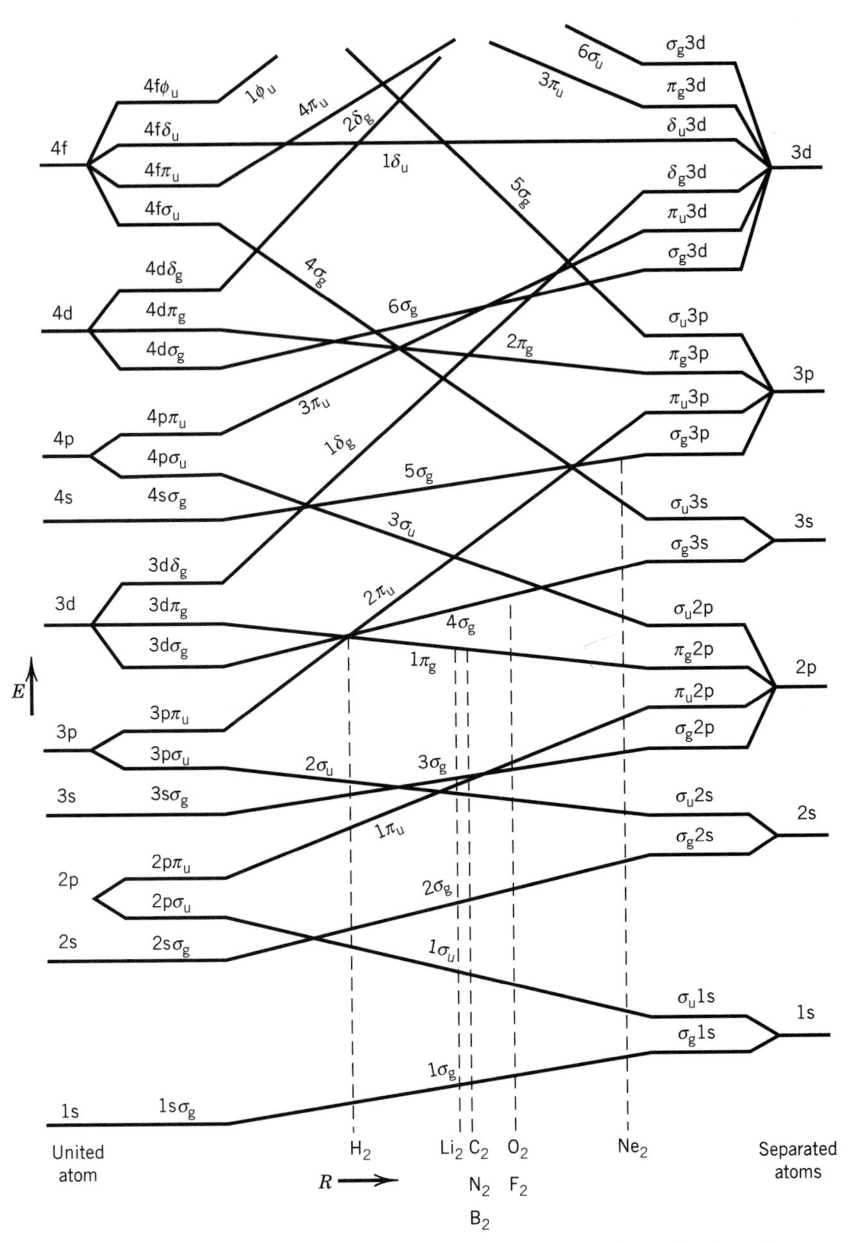

Figure 11.14 Correlation diagram for homonuclear diatomic molecules. The energy scale is schematic, but dashed lines show the correct sequence for filling of orbitals for the molecules indicated. (From R. S. Berry, S. A. Rice, and J. Ross, *Physical Chemistry*. Oxford University Press, 2000.)

He_2

According to simple molecular orbital theory, in He_2 a pair of electrons would occupy the $1\sigma_g$ orbital and a pair would occupy the $1\sigma_u$ orbital. Since both the bonding and antibonding orbitals are filled, there is no decrease in energy as compared with two isolated helium atoms, and a stable He_2 molecule is not formed.

Table 11.1 Ground States of Homonuclear Diatomic Molecules and Ions

Molecule	*Number of Electrons*	*Configuration*	*Term Symbol*	*Bond Order*	R_e/pm	D_e eV	D_e kJ mol^{-1}
H_2^+	1	$(1\sigma_g)$	$^2\Sigma_g$	$\frac{1}{2}$	106.0	2.793	269.483
H_2	2	$(1\sigma_g)^2$	$^1\Sigma_g$	1	74.12	4.7483	458.135
He_2^+	3	$(1\sigma_g)^2(1\sigma_u)$	$^2\Sigma_u$	$\frac{1}{2}$	108.0	2.5	238
He_2	4	$(1\sigma_g)^2(1\sigma_u)^2$		0	—	—	—
Li_2	6	$[He_2](2\sigma_g)^2$	$^1\Sigma_g$	1	267.3	1.14	110.0
Be_2	8	$[He_2](2\sigma_g)^2(2\sigma_u)^2$		0	—	—	—
B_2	10	$[Be_2](1\pi_u)^2$	$^3\Sigma_g$	1	158.9	~3.0	~290
C_2	12	$[Be_2](1\pi_u)^4$	$^1\Sigma_g$	2	124.2	6.36	613.8
N_2^+	13	$[Be_2](1\pi_u)^4(3\sigma_g)$	$^2\Sigma_g$	$2\frac{1}{2}$	111.6	8.86	854.8
N_2	14	$[Be_2](1\pi_u)^4(3\sigma_g)^2$	$^1\Sigma_g$	3	109.4	9.902	955.42
O_2^+	15	$[N_2](1\pi_g)$	$^2\Pi_g$	$2\frac{1}{2}$	112.27	6.77	653.1
O_2	16	$[N_2](1\pi_g)^2$	$^3\Sigma_g$	2	120.74	5.213	502.9
F_2	18	$[N_2](1\pi_g)^4$	$^1\Sigma_g$	1	143.5	1.34	118.8
Ne_2	20	$[N_2](1\pi_g)^4(3\sigma_u)^2$		0	—	—	—

Li_2

The additional pair of electrons in Li_2 beyond those for He_2 go into the $2\sigma_g$ orbital. Thus, there is a single bond between the two nuclei. The vapor of lithium metal is primarily monatomic because the Li_2 bond is quite weak and a high temperature is needed to vaporize the metal.

Be_2

The additional pair of electrons in Be_2 beyond Li_2 goes into the $2\sigma_u$ antibonding orbital. Thus, as in the case of He_2, there is no net stabilization as compared with isolated Be atoms.

B_2

According to Fig. 11.14 the additional pair of electrons in B_2 beyond Be_2 go into the $1\pi_u$ orbital. A single bond is formed, and the diatomic molecule is reasonably stable. Spectroscopic measurements show that its ground state is a triplet, so that the outer two electrons are in different $1\pi_u$ orbitals. If there are several orbitals with the same energy, the electrons spread themselves among several orbitals. Since there are two unpaired electrons, the ground state is a triplet.

C_2

The additional two electrons in C_2 beyond B_2 fill the two half-vacant $1\pi_u$ bonding orbitals. Thus, C_2 has four electrons in bonding orbitals that are not compensated by electrons in corresponding antibonding orbitals. Since C_2 has a bond order of 2, compared with 1 for B_2, we would expect C_2 to have tighter binding and a smaller internuclear distance than B_2.

Although C_2 exists in high-temperature gases rich in carbon, other species (C_3, C_4, ...) also exist, and carbon forms networks in the solid state (diamond and graphite) rather than condensing as C_2. In addition, molecular species such as C_{60} and C_{70} have been found to be stable. Both of these phenomena are due to the fact that the energies of the $1\pi_u$ and $3\sigma_g$ are so close that the molecule C_2 is easily raised into the excited configuration $(1\pi_u)^3(3\sigma_g)$. The unpaired electrons in this molecule can form bonds with further carbon atoms and thereby compensate for the energy of excitation.

N_2

The additional pair of electrons in N_2 beyond C_2 fills the $3\sigma_g$ bonding orbital. Thus, N_2 has a singlet ground state and a triple bond. Because of this strong bonding to form diatomic molecules, N_2 is a very stable molecule with a short internuclear distance. The first excited state of N_2 is 6.2 eV above the ground state.

Note from Fig. 11.14 that the order of the $1\pi_u$ and $3\sigma_g$ orbitals changes as we go from N_2 to O_2.

O_2

According to simple molecular orbital theory, the additional pair of electrons in O_2 beyond N_2 go into the $1\pi_g$ orbitals. Since there are two degenerate $1\pi_g$ orbitals, the two electrons can go into either the same or different orbitals. If they go into the same orbital, a singlet state is formed, and if they go into different orbitals, a triplet state is formed. As in the case of atoms, where Hund's rule predicts that the triplet state has the lower energy, in O_2 the electrons go into different orbitals and have parallel spins. Therefore, in the ground state O_2 has a spin of one and is paramagnetic (Section 22.6). This prediction was one of the early triumphs of MO theory.

F_2

The additional two electrons in F_2 beyond O_2 fill the $1\pi_g$ orbitals so that the ground state is a singlet. Since the electron pairs in two $1\pi_g$ antibonding orbitals approximately cancel the bonding due to two electron pairs in the bonding orbitals, the bonding is weaker than in O_2, and the internuclear distance is greater.

Ne_2

The $3\sigma_u$ orbital would be filled in Ne_2, and so the antibonding effects cancel the bonding effects; no stable molecule is formed.

Correlation diagrams can be constructed for heteronuclear diatomic molecules, but they have to take account of the fact that the energy levels in the two atoms may be quite different. For atomic orbitals in two atoms to be involved in bonding, they must have the same σ, π, ... properties and energies that are not too different. The electrons in heteronuclear diatomic molecules tend to localize more around one nucleus than the other, so that the molecule tends to have a dipole moment (Section 11.8). In the extreme case, ionic molecules (Section 11.9) are formed. The difference in affinity of different atoms in a molecule for electrons is discussed in terms of electronegativity (Section 11.8).

11.6 ELECTRONIC STRUCTURE OF POLYATOMIC MOLECULES: VALENCE BOND METHOD

The electronic structure of polyatomic molecules can also be described by the LCAO-MO method, but now the electronic energy will depend on many internuclear distances and angles. Therefore, finding the equilibrium geometry by minimizing the energy (calculated with the variational method) with respect to all these coordinates is a difficult and time-consuming process. Such calculations are becoming more routine as more powerful computers become available. The accuracy of the calculation depends on the number of atomic orbitals used. When a large number of atomic orbitals or basis set is used, the internuclear distances and bond angles of small molecules (fewer than ~6 atoms) can be calculated as accurately as they can be measured (distances to a few pm and bond angles to a degree or two). Today, the ground-state energies are often calculated accurately enough to be used in thermodynamic calculations.

Since chemical bond energies are the difference between large numbers (of the order of the energy of a 1s orbital in the hydrogen atom, or ~1300 kJ mol^{-1}), one must calculate these large numbers to very high accuracy to obtain chemical accuracy in the difference. It is therefore important to have simple ways of thinking about chemical bonding and molecular structure that, although not accurate, give a qualitative picture with little numerical work. One such method, introduced by Heitler and London for H_2 just after the advent of quantum mechanics, is called the valence bond method. We present a qualitative picture of this method in this section.

The valence bond method is based on the idea that a chemical bond is formed when there is good overlap between the atomic orbitals of the participating atoms. When the idea of **hybrid orbitals** is added to this, a rationalization of the bond angles in simple molecules is found. Hybrid orbitals are linear combinations of atomic orbitals on a single atom with definite angular relationships among them. We will illustrate these ideas by examining the molecules BeH_2, BH_3, CH_4, NH_3, and H_2O.

The H–Be–H bond angle is 180°. The ground-state electron configuration of the beryllium atom is $1s^2 2s^2$. To represent the directionality of the BeH_2 bonds and the fact that they are equivalent, two hybrid orbitals of Be are formed by taking a linear combination of the 2s orbital and one of the 2p orbitals. Energy is required to excite a 2s electron to a 2p orbital, but more than enough energy for this is obtained when a stable chemical compound with two bonds is formed. The two sp orbitals formed in this way are

$$\psi_{sp(i)} = 2^{-1/2}(2s + 2p_x) \tag{11.50}$$

$$\psi_{sp(ii)} = 2^{-1/2}(2s - 2p_x) \tag{11.51}$$

where these orbitals have been normalized in the usual way. There are two of these hybrid orbitals because two beryllium atomic orbitals have been used. When an s orbital is hybridized with a p_x orbital, it must be remembered that the p_x orbital is positive on one side of the nucleus and negative on the other. The node in the radial function for the 2s orbital, inside of which the orbital is negative and outside of which it is positive, may be neglected since the nodal surface is quite close to the nucleus and the contribution of this region to bonding is small. In $\psi_{sp(i)}$ the amplitudes tend to cancel on one side of the nucleus and add on the other as shown in Fig. 11.15. The $\psi_{sp(ii)}$ orbital is equivalent but points in the opposite direction. The lobes of these orbitals extend much farther along the x axis than the 2s and $2p_x$ orbitals, and so they provide more overlap with the 1s orbitals of the two hydrogen atoms in BeH_2.

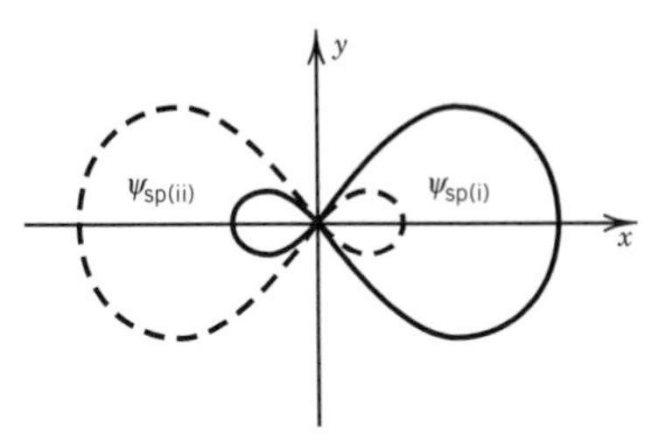

Figure 11.15 The two sp hybrids formed from an s orbital and a p orbital. One hybrid is shown by a solid line and the other is shown by a dashed line.

The two Be–H bonds are described by combining the $1s_A$ and $1s_B$ orbitals for the two protons with these hybrid orbitals to obtain the following two bond orbitals:

$$\psi = c_1 1s_A + c_2 \psi_{sp(i)} \tag{11.52}$$

$$\psi' = c_1' 1s_B + c_2' \psi_{sp(ii)} \tag{11.53}$$

According to valence bond theory BeH_2 is stabilized by the overlap of the two beryllium sp orbitals and the two hydrogen 1s orbitals. The $1s^2$ electrons of beryllium are not involved.

The three B–H bonds in BH_3 lie in a plane with H–B–H angles of 120°. The boron atom has the configuration $1s^2 2s^2 2p$. The following three hybrid orbitals of boron are constructed to account for three equivalent bonds in BH_3:

$$\psi_{sp^2(i)} = \frac{1}{\sqrt{3}} 2s + \sqrt{\frac{2}{3}} 2p_z \tag{11.54}$$

$$\psi_{sp^2(ii)} = \frac{1}{\sqrt{3}} 2s - \frac{1}{\sqrt{6}} 2p_z + \frac{1}{\sqrt{2}} 2p_x \tag{11.55}$$

$$\psi_{sp^2(iii)} = \frac{1}{\sqrt{3}} 2s - \frac{1}{\sqrt{6}} 2p_z - \frac{1}{\sqrt{2}} 2p_x \tag{11.56}$$

These wavefunctions have been normalized and are orthogonal. By substituting the expressions for the angular parts of the p_z and p_x orbitals, it is readily shown that the sp^2 orbitals lie in a plane with lobes pointed in directions separated by 120°, as shown in Fig. 11.16.

Carbon atoms have the electron configuration $1s^2 2s^2 2p^2$, and the outer four valence electrons may be used to form sp^3 hybrid orbitals:

$$\psi_{sp^3(i)} = \frac{1}{\sqrt{4}} (2s + 2p_x + 2p_y + 2p_z) \tag{11.57}$$

$$\psi_{sp^3(ii)} = \frac{1}{\sqrt{4}} (2s - 2p_x - 2p_y + 2p_z) \tag{11.58}$$

$$\psi_{sp^3(iii)} = \frac{1}{\sqrt{4}} (2s + 2p_x - 2p_y - 2p_z) \tag{11.59}$$

$$\psi_{sp^3(iv)} = \frac{1}{\sqrt{4}} (2s - 2p_x + 2p_y - 2p_z) \tag{11.60}$$

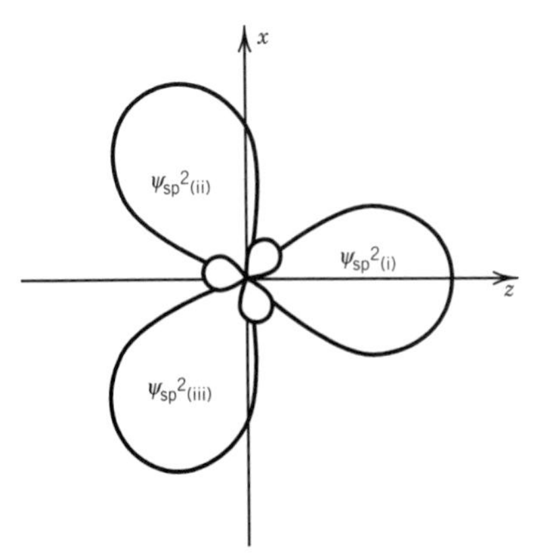

Figure 11.16 The three sp^2 hybrid orbitals formed from an s orbital and two p orbitals.

These orthonormal orbitals point in the directions shown in Fig. 11.17, in agreement with the tetrahedral structure of CH_4 and with the geometries of the alkanes. These four electrons in **hybrid sp³ orbitals** can form σ bonds with hydrogen or other elements. For other elements, further hybridization schemes

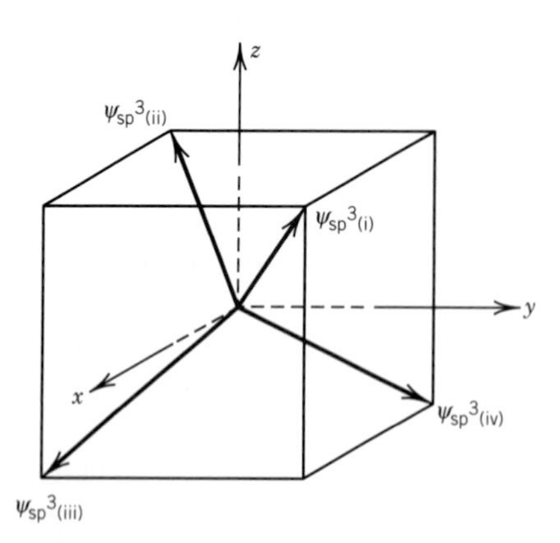

Figure 11.17 Direction of the sp^3 hybrids formed from an s orbital and three p orbitals.

involving d orbitals can be used to account for bipyramidal structures (coordination number 5) and octahedral structures (coordination number 6).

When we come to NH_3, we must introduce the idea of **lone-pair** electrons. We use the $2s^2 2p_x^1 2p_y^1 2p_z^1$ configuration on N to form four sp^3 hybrids just as in carbon. Now we must put two valence electrons of N in one of these orbitals, leaving three available to bond to the hydrogens. This leads to a tetrahedral structure of NH_3 (bond angle 109°) with one apex of the tetrahedron containing a lone electron pair, as shown in Fig. 11.18*a*. (The experimentally observed bond angle is 107°.) These lone-pair electrons are available for bonding to, for example, a proton, H^+, to form $NH_4{}^+$.

The water molecule can be treated in much the same manner, except that we now have two lone pairs, as shown in Fig. 11.18*b*. This predicts that the bond angle in H_2O is the tetrahedral angle 109° instead of the experimentally observed 104°. The small difference can be accounted for by adding more terms to the wavefunction. The lone pairs of H_2O are available to bond to other atoms. In particular, the interaction between hydrogen atoms on other water molecules and the lone pairs gives rise to the hydrogen bond (see Section 11.10) and the unusual properties of H_2O.

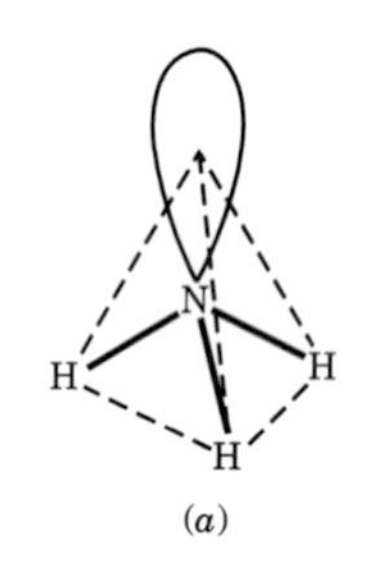

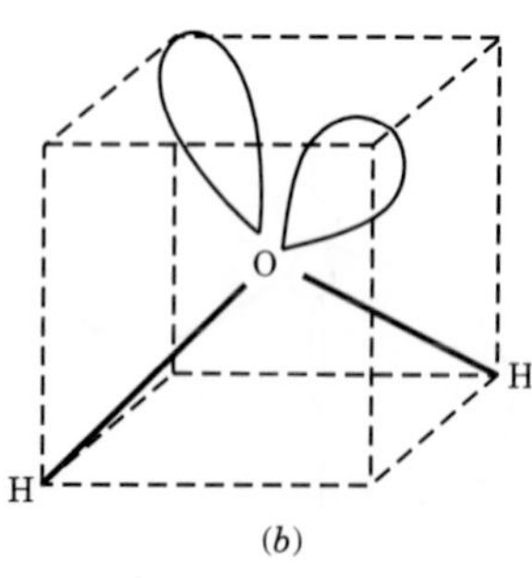

Figure 11.18 (*a*) Lone-pair electrons of NH_3. (*b*) Lone-pair electrons of H_2O.

As the speed of computer calculations and the size of computer memories have increased, it has become possible to make useful quantum mechanical calculations on larger and larger molecules. This is done by expressing molecular orbitals as linear combinations of atomic orbitals and then determining coefficients in the linear combinations by a self-consistent field calculation (LCAO-MO-SCF). The application of the Hartree–Fock self-consistent field method (Section 10.9) to the calculation of molecular properties was developed by Roothaan in the 1950s and is referred to as the Hartree–Fock-Roothaan method. When this method is applied to larger and larger molecules, integrals in the resulting secular determinant become very difficult to evaluate. More recently, computer programs have been written for these calculations using Gaussian functions, rather than Slater orbitals. The 1s Slater orbital (STO) and Gaussian function (GF) in each of these basis sets is

$$\phi_{1s}^{STO}(r,\xi) = \left(\frac{\xi^3}{\pi}\right)^{1/2} e^{-\xi r} \tag{11.61}$$

$$\phi_{1s}^{GF}(r,\alpha) = \left(\frac{2\alpha}{\pi}\right)^{3/4} e^{-\alpha r^2} \tag{11.62}$$

where ξ and α are orbital exponents. These two functions have different shapes, but the desired shape of the Slater orbital can be approximated by representing the 1s orbital by a sum of Gaussian functions such as equation 11.62.

$$\psi = \sum c_i \phi_i \tag{11.63}$$

The reason Gaussian functions are used is that all integrals (e.g., overlap integrals) that arise theoretically can be calculated analytically, rather than numerically.

Example 11.1 ***The product of Gaussian functions is a Gaussian function***

Show that the product of a one-dimensional Gaussian function centered at x_A

$$\phi_A = e^{-\alpha(x-x_A)^2}$$

and one centered at x_B

$$\phi_B = e^{-\beta(x-x_B)^2}$$

is a Gaussian function centered at

$$x_p = \frac{\alpha x_A + \beta x_B}{\alpha + \beta}$$

The product of these Gaussian functions is

$$\begin{aligned}\phi_A\phi_B &= \exp[-\alpha(x - x_A)^2 - \beta(x - x_B)^2] \\ &= \exp[-\alpha x^2 + 2\alpha x_A x - \alpha x_A^2 - \beta x^2 + 2\beta x_B x - \beta x_B^2]\end{aligned}$$

Now we want to rearrange this equation so as to find a term in $(x - x_p)^2$.

$$\begin{aligned}\phi_A\phi_B &= \exp[-(\alpha x_A^2 + \beta x_B^2)]\exp\left\{-(\alpha+\beta)\left[x^2 - 2x\frac{\alpha x_A + \beta x_B}{\alpha+\beta} + \left(\frac{\alpha x_A + \beta x_B}{\alpha+\beta}\right)^2\right]\right\} \\ &\quad \times \exp\left[(\alpha+\beta)\left(\frac{\alpha x_A + \beta x_B}{\alpha+\beta}\right)^2\right] \\ &= \exp[-(\alpha x_A^2 + \beta x_B^2)]\exp\{-(\alpha+\beta)[x^2 - 2xx_p + x_p^2]\} \times \exp[(\alpha+\beta)x_p^2] \\ &= \exp\left[-\frac{\alpha\beta(x_A - x_B)^2}{\alpha+\beta}\right]\exp[-(\alpha+\beta)(x - x_p)^2]\end{aligned}$$

Thus this product is a Gaussian function centered at x_p. This derivation can be extended to the three-dimensional Gaussian functions that are involved in molecules.

Programs for use on computers have been developed by Pople and coworkers,* and John Pople was awarded the Nobel Prize in chemistry for 1998.

11.7 HÜCKEL MOLECULAR ORBITAL THEORY

Molecules with extensive π bonding systems, such as benzene, are not described very well by valence bond theory because the π electrons are often not localized in

*W. J. Hehre, L. Random, P. V. R. Schleyer, and J. A. Pople, *Ab Initio Molecular Orbital Theory*. Hoboken, NJ: Wiley-Interscience, 1986.

a single bond. Their electronic structure can be described in a simple but approximate molecular orbital method developed by Hückel in 1930. The two types of bonds involved in these molecules are illustrated in Fig. 11.19 for ethylene (C_2H_4). Ethylene is a planar molecule, and the bonds between the carbon and hydrogen atoms in the plane are sp^2 hybrid orbitals of carbon and 1s orbitals of hydrogen. This forms the **σ-bond framework** shown in Fig. 11.19*a*. The $2p_z$ orbitals of the carbon atoms that are not involved in the σ framework overlap sideways, forming the π system, as shown in Fig. 11.19*b*. This simple picture explains the bond angles in ethylene and its planar structure.

Hückel molecular orbital theory assumes that the π electrons, which are responsible for the special properties of conjugated and aromatic hydrocarbons, do not interact with one another, and so the many-electron wavefunction is just a product of one-electron molecular orbitals. The molecular orbital of the delocalized π orbital of ethylene is represented by

$$\psi = c_1\phi_1 + c_2\phi_2 \tag{11.64}$$

where ϕ_1 and ϕ_2 are the $2p_z$ orbitals of the two carbon atoms. The corresponding secular determinant is

$$\begin{vmatrix} H_{11} - ES_{11} & H_{12} - ES_{12} \\ H_{21} - ES_{21} & H_{22} - ES_{22} \end{vmatrix} = 0 \tag{11.65}$$

where $H_{ij} = \int \phi_i^* \hat{H} \phi_j \, d\tau$ and $S_{ij} = \int \phi_i^* \phi_j \, d\tau$.

In Hückel molecular orbital theory, the secular equation is simplified by making the following assumptions:

1. Overlap integrals S_{ij} are set equal to zero unless $i = j$, when $S_{ii} = 1$.
2. All the diagonal elements in the secular equation are assumed to be the same; thus, the Coulomb integrals H_{ii} are all set equal to α.
3. The resonance integrals H_{ij} are set equal to zero, except for those on neighboring atoms, which are set equal to β.

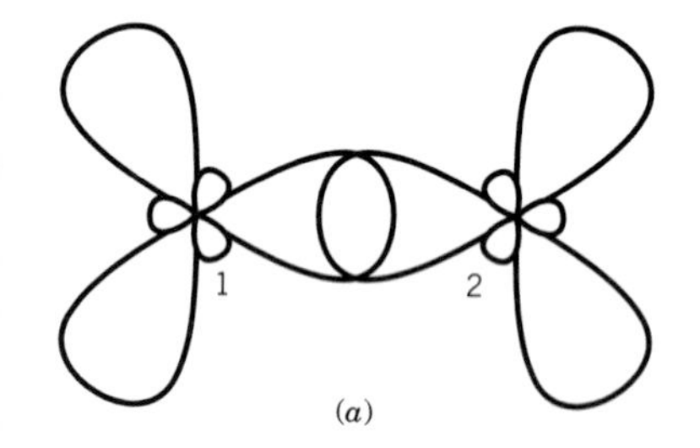

With these assumptions the secular equation for the π electrons in ethylene becomes

$$\begin{vmatrix} \alpha - E & \beta \\ \beta & \alpha - E \end{vmatrix} = 0 \tag{11.66}$$

In Hückel theory, the Coulomb integral α and the resonance integral β are regarded as empirical parameters to be evaluated from experimental data on the molecule. Thus, in Hückel theory it is unnecessary to specify the Hamiltonian operator.

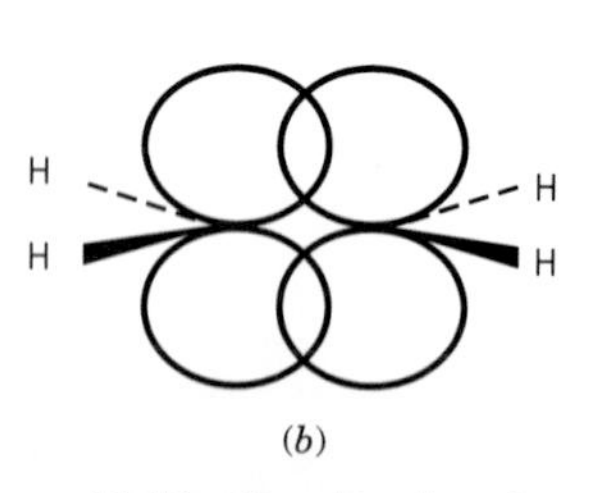

Figure 11.19 Bonding in ethylene. (*a*) In the plane of the nuclei: formation of σ bonds between carbon atoms 1 and 2 using sp^2 orbitals. (*b*) Perpendicular to the plane of the nuclei: formation of π bonds between p orbitals.

Equation 11.66 is readily solved for the energy by using the quadratic formula to obtain $E = \alpha \pm \beta$. Thus, there are bonding and antibonding orbitals, as shown in Fig. 11.20. The resonance integral is negative, and so the energy of the lowest level is $\alpha + \beta$. The bonding orbital is occupied by an electron pair, and so the π electronic energy of ethylene is $2\alpha + 2\beta$.

The wavefunctions for the bonding and antibonding orbitals are obtained by going back to the pair of linear algebraic equations that gave rise to the secular equation:

$$c_1(\alpha - E) + c_2\beta = 0 \tag{11.67}$$

$$c_1\beta + c_2(\alpha - E) = 0 \tag{11.68}$$

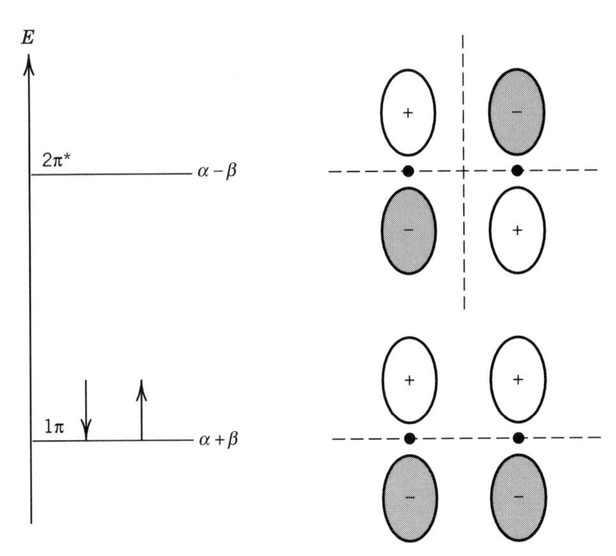

Figure 11.20 Hückel molecular orbitals for ethylene. The carbon nuclei are represented by dots, and the nodal planes for the molecular orbitals are represented by dashed lines.

Substituting $E = \alpha + \beta$ in either equation yields $c_1 = c_2$ so that the wavefunction for the bonding orbital is

$$\psi_1 = 2^{-1/2}(\phi_1 + \phi_2) \tag{11.69}$$

The bonding orbital is shown in Fig. 11.20 alongside the energy level. Substituting $E = \alpha - \beta$ into equation 11.67 or 11.68 yields $c_1 = -c_2$, so that the wavefunction for the antibonding orbital is

$$\psi_2 = 2^{-1/2}(\phi_1 - \phi_2) \tag{11.70}$$

The antibonding orbital is also shown in Fig. 11.20. Notice the resemblance to the LCAO orbitals of H_2^+ (Section 11.3).

The Hückel theory gives us an estimate of the excitation energy to the first excited state of ethylene; it is evident from Fig. 11.20 that this excitation energy is $2|\beta|$. This figure provides the opportunity to introduce some nomenclature that will be useful later. The **highest occupied molecular level** is referred to as HOMO, and the **lowest unoccupied molecular level** is referred to as LUMO.

The π electronic energy of planar 1,3-butadiene ($CH_2{=}CHCH{=}CH_2$) is readily calculated using Hückel theory. The secular determinantal equation is

$$\begin{vmatrix} \alpha - E & \beta & 0 & 0 \\ \beta & \alpha - E & \beta & 0 \\ 0 & \beta & \alpha - E & \beta \\ 0 & 0 & \beta & \alpha - E \end{vmatrix} = 0 \tag{11.71}$$

If β is factored from each column and $(\alpha - E)/\beta$ is replaced by x, we obtain

$$\begin{vmatrix} x & 1 & 0 & 0 \\ 1 & x & 1 & 0 \\ 0 & 1 & x & 1 \\ 0 & 0 & 1 & x \end{vmatrix} = 0 \tag{11.72}$$

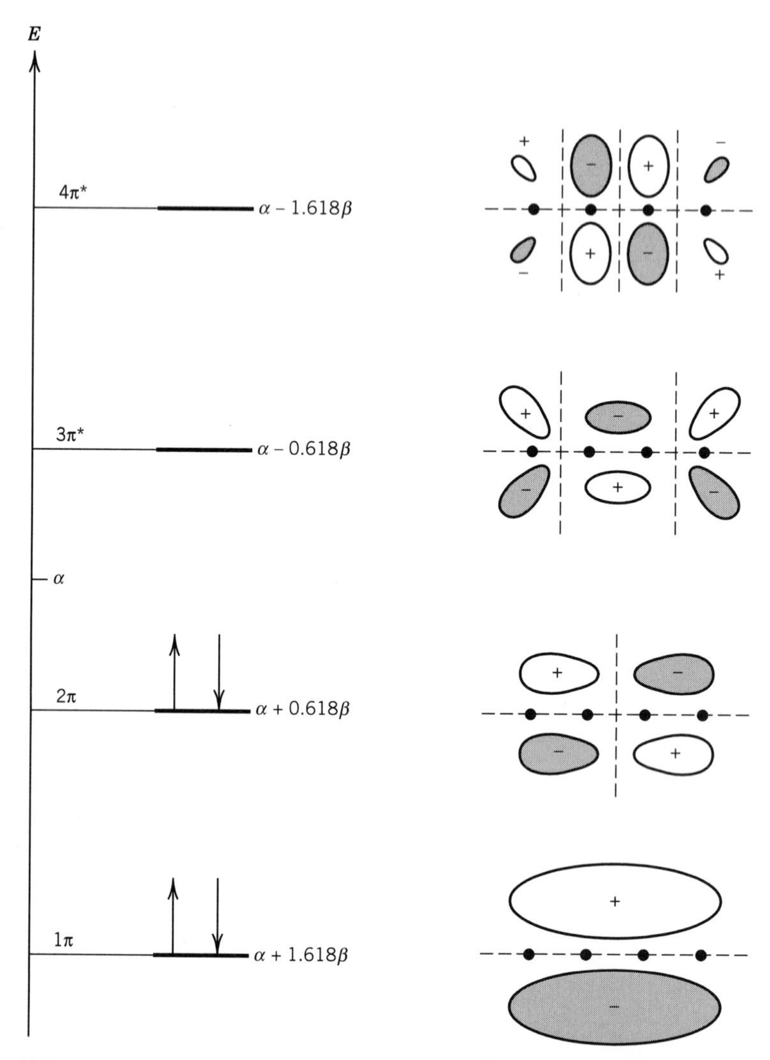

Figure 11.21 Hückel molecular orbitals for 1,3-butadiene. The carbon nuclei are represented by dots, and the nodal planes for the molecular orbitals are represented by dashed lines. (Adapted from I. N. Levine, *Quantum Chemistry,* 5/E © 2000, Fig. 16.1. Reprinted by permission of Prentice-Hall Inc., Upper Saddle River, NJ 07458.)

or $x^4 - 3x^2 + 1 = 0$, so that $x = \pm 0.618, \pm 1.618$. Thus, for 1,3-butadiene there are four possible energy levels, two bonding and two antibonding, as shown in Fig. 11.21:

$$\begin{aligned} E_1 &= \alpha + 1.618\beta \\ E_2 &= \alpha + 0.618\beta \\ E_3 &= \alpha - 0.618\beta \\ E_4 &= \alpha - 1.618\beta \end{aligned} \tag{11.73}$$

The four π electrons occupy the two bonding orbitals so that the π electronic energy is

$$\begin{aligned} E_\pi &= 2(\alpha + 1.618\beta) + 2(\alpha + 0.618\beta) \\ &= 4\alpha + 4.472\beta \end{aligned} \tag{11.74}$$

The lowest excitation energy is $1.236|\beta|$.

The 2π orbital of butadiene is the HOMO, a bonding orbital, and the $3\pi^*$ orbital is the LUMO, an antibonding orbital.

The four Hückel molecular orbitals for 1,3-butadiene are*

$$\begin{aligned}
\psi_1 &= 0.372\phi_1 + 0.602\phi_2 + 0.602\phi_3 + 0.372\phi_4 \\
\psi_2 &= 0.602\phi_1 + 0.372\phi_2 - 0.372\phi_3 - 0.602\phi_4 \\
\psi_3 &= 0.602\phi_1 - 0.372\phi_2 - 0.372\phi_3 + 0.602\phi_4 \\
\psi_4 &= 0.372\phi_1 - 0.602\phi_2 + 0.602\phi_3 - 0.372\phi_4
\end{aligned} \tag{11.75}$$

These four molecular orbitals are indicated in Fig. 11.21. Notice that the π orbitals extend the entire length of the molecule. (For the calculation of these coefficients, see Computer Problem 11.C.)

The Hückel secular determinant for benzene is

$$\begin{vmatrix}
\alpha - E & \beta & 0 & 0 & 0 & \beta \\
\beta & \alpha - E & \beta & 0 & 0 & 0 \\
0 & \beta & \alpha - E & \beta & 0 & 0 \\
0 & 0 & \beta & \alpha - E & \beta & 0 \\
0 & 0 & 0 & \beta & \alpha - E & \beta \\
\beta & 0 & 0 & 0 & \beta & \alpha - E
\end{vmatrix} = 0 \tag{11.76}$$

The six roots are

$$\begin{aligned}
E_1 &= \alpha + 2\beta \\
E_2 &= E_3 = \alpha + \beta \\
E_4 &= E_5 = \alpha - \beta \\
E_6 &= \alpha - 2\beta
\end{aligned} \tag{11.77}$$

Since benzene has six π electrons, pairs of electrons go in the three lowest energy orbitals (those with plus signs in E_i). Thus the π electronic energy in benzene is

$$\begin{aligned}
E_\pi &= 2(\alpha + 2\beta) + 4(\alpha + \beta) \\
&= 6\alpha + 8\beta
\end{aligned} \tag{11.78}$$

The equations for the six Hückel molecular orbitals for benzene are not given here, but the corresponding electron densities are shown in Fig. 11.22. Note that the π electronic energy in C_6H_6 is more negative than three times the value in C_2H_4, indicating that C_6H_6 does not contain three double bonds. The difference is called the delocalization energy.

The Hückel theory is an example of a semiempirical molecular orbital method. We have used a simple Hamiltonian (neglecting many terms) to find the orbitals and their energies. We now can use experimental quantities to fit α and β for ethylene. We then use the values to make predictions for butadiene, benzene, and so on. This method does not give quantitative results, but it does provide us with qualitative insights about larger systems for which the more computationally intensive methods are too costly or time-consuming, and it gives us insight into the excited electronic states of conjugated π electron molecules.

*The derivation of these wavefunctions is given in I. N. Levine, *Quantum Chemistry,* 5th ed., Upper Saddle River, NJ: Prentice-Hall, 1999.

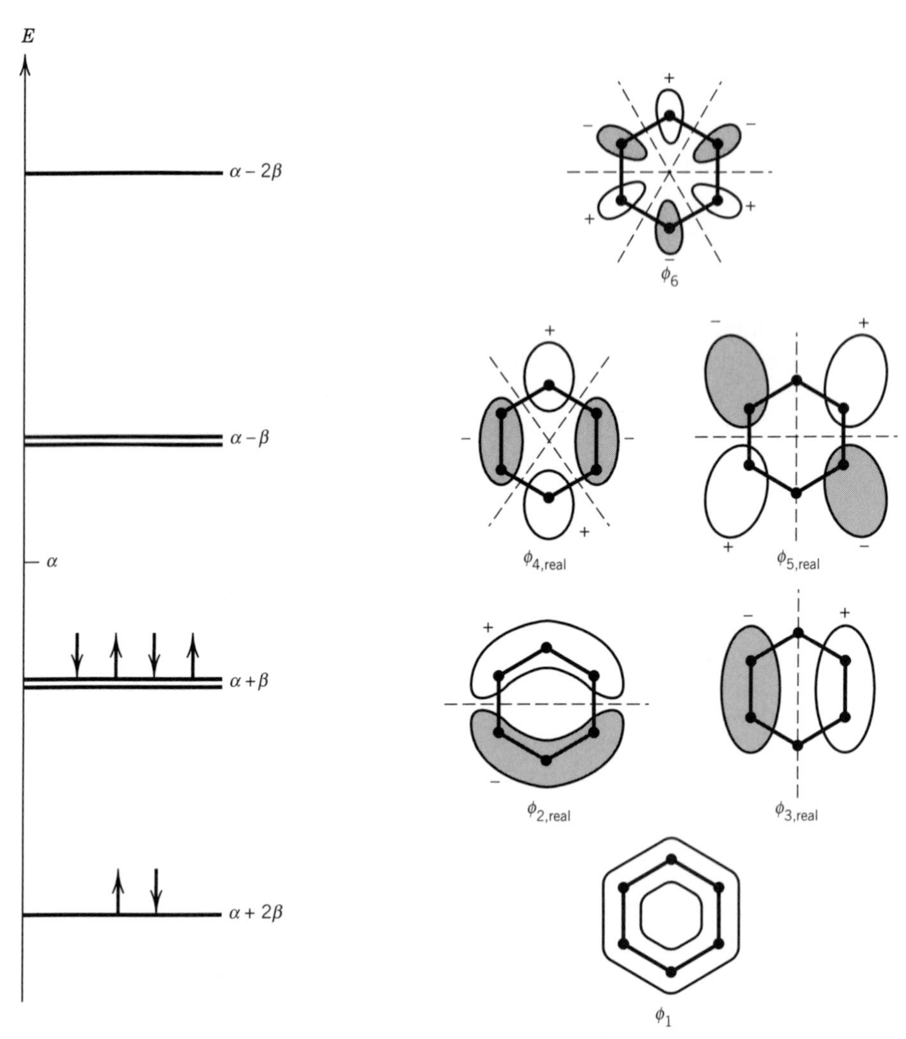

Figure 11.22 Hückel molecular orbitals for benzene. The carbon nuclei are represented by dots, and the nodal planes perpendicular to the molecular plane are represented by dashed lines. (Adapted from I. N. Levine, *Quantum Chemistry,* 5/E ©2000, Fig. 16.4. Reprinted by permission of Prentice-Hall Inc., Upper Saddle River, NJ 07458.)

Example 11.2 ***Hückel molecular orbitals for $CH_2CHCH_2\cdot$: Nonbonding orbitals***

Calculate the Hückel molecular orbital energies of the planar radical (i.e., one nonpaired electron) $CH_2CHCH_2\cdot$.

The secular determinantal equation is

$$\begin{vmatrix} \alpha - E & \beta & 0 \\ \beta & \alpha - E & \beta \\ 0 & \beta & \alpha - E \end{vmatrix} = 0$$

or

$$\begin{vmatrix} x & 1 & 0 \\ 1 & x & 1 \\ 0 & 1 & x \end{vmatrix} = 0 \qquad \text{with} \qquad x = \frac{\alpha - E}{\beta}$$

This yields

$$x^3 - 2x = 0$$

with roots $x = 0, -\sqrt{2}, +\sqrt{2}$. Therefore,

$$\begin{aligned} E_1 &= \alpha + \sqrt{2}\beta \\ E_2 &= \alpha \\ E_3 &= \alpha - \sqrt{2}\beta \end{aligned}$$

The orbital corresponding to E_2 is called a nonbonding orbital since its energy is unchanged from the atomic value $\alpha = H_{ii}$, which is the 2p atomic orbital energy in this system.

These ideas can be applied to conjugated polymeric molecules of indefinite length.

Comment:

The assumptions of Hückel theory are so drastic that we cannot expect to obtain very accurate results, but it is useful because it leads to simple visualizations of the electron distributions in molecules with conjugated double bonds. In particular, it emphasizes the delocalization of electrons in these systems. Much more complicated wavefunctions are required for accurate calculations of energy levels, bond distances, and bond angles.

11.8 DIPOLE MOMENTS AND IONIC BONDING

The electric dipole moment $\boldsymbol{\mu}$ of a neutral molecule containing charges Q_i is defined by

$$\boldsymbol{\mu} = \sum Q_i \boldsymbol{r}_i \tag{11.79}$$

where $\boldsymbol{r}_i$ is the position vector of Q_i. The dipole moment $\boldsymbol{\mu}$ is a vector because it points in a certain direction, but we will often use the magnitude μ of the vector. For two equal but opposite charges the magnitude of the dipole moment is given by $\mu = Qr$, where r is the distance between the charges. The dipole moment has the SI units C m, where C is the coulomb and m is the meter.*

*Dipole moments have often been expressed in debye units D after Peter Debye, who made many contributions to the understanding of polar molecules. The debye unit is 10^{-18} esu cm. Since the charge of a proton is $4.803\,21 \times 10^{-10}$ esu or $1.602\,177\,33 \times 10^{-19}$ C, the debye D may be expressed in SI units as follows:

$$\frac{(10^{-18}\text{ esu cm})(1.602\,177\,33 \times 10^{-19}\text{ C})(10^{-2}\text{ m cm}^{-1})}{(4.803\,21 \times 10^{-10}\text{ esu})} = 3.335\,64 \times 10^{-30}\text{ C m}$$

Example 11.3 ***Dipole moment of NaCl***

Calculate the dipole moment of NaCl in the gas phase assuming equal and opposite charges equal to the proton charge. The equilibrium internuclear distance is 236 pm.

$$\mu = Qr = (1.602 \times 10^{-19}\ \text{C})(236 \times 10^{-12}\ \text{m}) = 38 \times 10^{-30}\ \text{C m}$$

The dipole moment of a homonuclear diatomic molecule is zero because the electrons are shared equally by the two atoms. In general, a heteronuclear diatomic molecule will have a dipole moment, as expected from the difference in electronegativities. For a neutral molecule, the expectation value of the dipole moment in the nth electronic state $\langle\mu\rangle_n$ is given by

$$\langle\mu\rangle_n = \int \psi_n^* \hat{\mu} \psi_n \, d\tau \tag{11.80}$$

where $\hat{\mu}$ is the **electric dipole moment operator,** as given above by replacing $\boldsymbol{r}_i$ by $\hat{\boldsymbol{r}}_1$.

The determination of dipole moments and some uses are discussed in Chapter 22, along with magnetic dipole moments.

When atoms with nearly the same electronegativity form bonds, the molecular orbitals are spread more or less evenly over the two atoms, and covalent bonds are formed. When atoms have somewhat different electronegativities, the bonding electrons are not evenly shared, but are drawn toward the more electronegative atom. When the electronegativities are quite different, an electron moves completely to the more electronegative atom and an **ionic bond** is formed.

The interaction energy $E(R)$ of the atoms in an ionic bond can be calculated using Coulomb's law (equation 10.1). As two ions approach each other closely because of Coulomb attraction, an equilibrium position is reached at some point because of repulsive forces that balance the attraction. The short-range repulsion energy increases very rapidly when the charge clouds of the two ions begin to overlap. The repulsion energy may be represented by an empirical expression such as $b \exp(-aR)$. Thus, the potential energy $E(R)$ of a diatomic molecule is approximately

$$E(R) = \frac{Q_1 Q_2}{4\pi\epsilon_0 R} + b\, e^{-aR} \tag{11.81}$$

where the energy is measured with respect to the separated ions. Further effects may be taken into account; for example, the electric field of each ion polarizes (Section 22.2) the electron cloud of the other ion and produces a further energy of attraction. However, the short-range repulsion energy and polarization energy are so small and close to R_e that the Coulomb attractive term alone gives a good approximation to the dissociation energy of an ionic molecule.

Equation 11.81 applies to dissociation into separated ions. However, the ground state of the dissociated system consists of atoms rather than ions. The dissociation energy D_e into atoms is given by

$$D_e(\text{MX} \rightarrow \text{M} + \text{X}) = D_e(\text{MX} \rightarrow \text{M}^+ + \text{X}^-) - E_i(\text{M}) + E_{ea}(\text{X})$$

where $D_e(\text{MX} \rightarrow \text{M}^+ + \text{X}^-)$ is the dissociation energy into ions, $E_i(\text{M})$ is the ionization energy of metal atom, and $E_{ea}(\text{X})$ is the electron affinity of the nonmetal atom. Since the lowest ionization potentials are greater than the highest electron affinities, $D_e(\text{MX} \rightarrow \text{M} + \text{X})$ is smaller than $D_e(\text{MX} \rightarrow \text{M}^+ + \text{X}^-)$. What happens when MX dissociates to M + X is that at some large internuclear distance the system can decrease its energy by moving the electron from X^- to M^+.

Example 11.4 ***Dissociation of NaCl into ions***

The dissociation energy of NaCl(g) into atoms is 4.29 eV. What is the dissociation energy into ions, and how does this compare with what would be expected from Coulomb's law? The equilibrium internuclear distance in the gaseous NaCl molecule is 0.2361 nm.

$$\begin{aligned} D_e(\text{NaCl} \rightarrow \text{Na}^+ + \text{Cl}^-) &= D_e(\text{NaCl} \rightarrow \text{Na} + \text{Cl}) + E_i(\text{Na}) - E_{ea}(\text{Cl}) \\ &= 4.29 + 5.14 - 3.61 = 5.82 \text{ eV} \end{aligned}$$

If only the Coulomb term is used, then

$$\begin{aligned} D_e(\text{NaCl} \rightarrow \text{Na}^+ + \text{Cl}^-) &= \frac{Q_1 Q_2}{4\pi\epsilon_0 R_e} \\ &= \frac{(1.602 \times 10^{-19}\text{ C})^2}{4\pi(8.854 \times 10^{-12}\text{ C}^2\text{ N}^{-1}\text{ m}^{-2})(0.2361 \times 10^{-9}\text{ m})} \\ &= \frac{9.770 \times 10^{-19}\text{ J}}{1.602 \times 10^{-19}\text{ J eV}^{-1}} \\ &= 6.10 \text{ eV} \end{aligned}$$

Thus, the Coulomb term accounts for the observed dissociation energy within 0.3 eV.

11.9 INTERMOLECULAR FORCES

When two neutral molecules come close to one another, the various interactions between the electrons and nuclei of one molecule and the electrons and nuclei of the other produce a potential energy of interaction. At very small distances, the molecules repel each other strongly. At certain intermediate distances, the potential energy of interaction is negative; the molecules attract each other, but usually weakly. At very large distances, the potential energy approaches zero. Thus, if we plot the interaction energy versus intermolecular distance, we obtain curves such as Fig. 11.23.

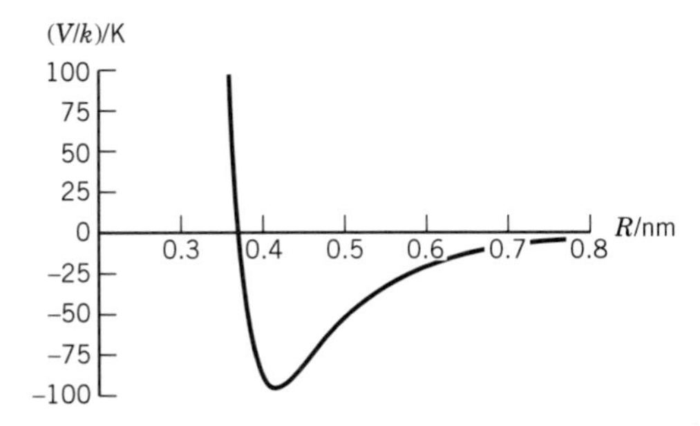

Figure 11.23 The energy of interaction of two nitrogen molecules as a function of distance according to the Lennard-Jones potential. As is customary, the energy is divided by the Boltzmann constant so that the ordinate is in kelvins. (See Computer Problem 11.H.)

Attractive Forces

The long-range attractive interactions between neutral molecules are called **van der Waals forces** and include a number of terms, all arising from the electrostatic interactions. For example, if the two molecules have permanent dipole moments (Section 11.8), then at fixed angles with respect to one another the dipole–dipole interaction between molecules varies as R^{-3}, where R is the intermolecular distance. In the gas or liquid phase, molecules will have random orientations, so the thermal average over the angular part of the interaction causes the first nonzero term in the interaction to be

$$\langle V(R)\rangle_{\text{dd}} = -\frac{2}{3kT}\left(\frac{\mu_{\text{A}}\mu_{\text{B}}}{4\pi\epsilon_0}\right)^2\frac{1}{R^6} \tag{11.82}$$

where μ_{A} and μ_{B} are the permanent dipole moments of the two molecules, T is the temperature, and dd means dipole–dipole. Notice that the interaction is negative (attractive) since there is a preference in the thermal (statistical mechanical) averaging for the molecules to orient in such a way as to attract each other and lower their mutual energy. As the temperature increases, this tendency is less important, so the attraction decreases.

Another term in the attractive potential energy arises from the **dipole–induced dipole interaction** between molecules. If molecule A has a dipole moment μ_{A}, this creates an electric field that polarizes the charges on molecule B, creating an induced dipole moment of magnitude $\alpha_{\text{B}}\mu_{\text{A}}$, where α_{B} is called the polarizability of molecule B. The dipole–induced dipole attractive energy is given by

$$\langle V(R)\rangle_{\text{ind}} = -\frac{\alpha_{\text{B}}\mu_{\text{A}}^2 + \alpha_{\text{A}}\mu_{\text{B}}^2}{(4\pi\epsilon_0)^2R^6} \tag{11.83}$$

Notice that it also varies as R^{-6}.

There is a third term varying as R^{-6} that was explained by London in 1930 and is sometimes called the **London force** or **dispersion force.** It occurs even when the molecules have no permanent dipole moment, and it can be visualized as the interaction between the fluctuating charge distribution on molecule A with that of molecule B. The rules of quantum mechanics say that even though molecule A has no permanent dipole moment, the average value of the square of the dipole moment operator is not zero. Therefore, we can say that at any instant the molecule has a dipole moment that polarizes molecule B (and vice versa), leading to a potential varying as R^{-6}. The exact form is complicated, but a good approximation is

$$\langle V(R)\rangle_{\text{disp}} = -\frac{3}{2}\left(\frac{E_{\text{A}}E_{\text{B}}}{E_{\text{A}} + E_{\text{B}}}\right)\frac{\alpha_{\text{A}}\alpha_{\text{B}}}{(4\pi\epsilon_0)^2}\frac{1}{R^6} \tag{11.84}$$

where E_{A} and E_{B} are approximately equal to the energies of the first electronic transitions of these molecules, α_{A} and α_{B} are their polarizabilities, and disp means dispersion.

These three terms add to give the total attractive energy between molecules A and B. In most cases, this energy is on the order of $\sim 2 \times 10^{-3}$ eV at a distance of 0.5 nm. Of course, the larger the dipole moments or the polarizabilities, the stronger is the attraction.

Intermolecular Potential Energy

At small distances, the molecules repel strongly, so when this force is added to the attractive forces we have discussed, we obtain the total intermolecular potential energy. To represent this as a simple function of R, we can take the repulsion to be proportional to R^{-n} or $e^{-\alpha R}$, where n and α have to be fit empirically. One very useful empirical function is named for J. E. Lennard-Jones and is given by

$$V = 4\epsilon\left[\left(\frac{\sigma}{R}\right)^{12} - \left(\frac{\sigma}{R}\right)^{6}\right] \tag{11.85}$$

In this "6–12" potential, the twelfth-power dependence for the repulsive energy is chosen for convenience and because it is a reasonable fit to the data. The minimum in this potential is at $R = (2)^{1/6}\sigma$, where $V = -\epsilon$. Note that $V = 0$ at $R = \sigma$, so the energy rises very steeply for small R (see Fig. 11.23). The values of ϵ and σ can be found by fitting experimental data on the second virial coefficient, gas viscosity, and molecular beam scattering cross sections. Typical values for the parameters ϵ and σ for interactions between like atoms or molecules are given in Table 11.2.

Table 11.2 Lennard-Jones Potential Parameters[a]

	$\frac{\epsilon}{k}$/K	σ/pm
Ar	120	341
Xe	221	410
H_2	37	293
N_2	95.1	370
O_2	118	358
Cl_2	256	440
CO_2	197	430
CH_4	148	382
C_6H_6	243	860

[a] Note that the depth of the potential well has been expressed as a temperature by dividing the energy ϵ by the Boltzmann constant k.

Example 11.5 ***Lennard-Jones interaction potential***

By differentiation of the expression for the Lennard-Jones potential, show that the distance R_m at the minimum where $dV/dR = 0$ is $R_m = 2^{1/6}\sigma$. Substituting this relationship into equation 11.85, show that the Lennard-Jones potential may also conveniently be given by

$$V = \epsilon\left[\left(\frac{R_m}{R}\right)^{12} - 2\left(\frac{R_m}{R}\right)^{6}\right]$$

Differentiating equation 11.85 and setting the derivative equal to zero, we obtain

$$\frac{dV}{dR} = 4\epsilon\left[-\frac{12\sigma^{12}}{R^{13}} + \frac{6\sigma^6}{R^7}\right] = 0$$

If we represent the value of R at the minimum by R_m, then

$$R_m = 2^{1/6}\sigma$$

Substituting this in equation 11.85 yields the desired expression for V.

Comment:

The intermolecular potential energy when atoms and molecules approach each other is so important that more time could be spent on this topic. More complicated equations for intermolecular potential energies are used because the twelfth-power dependence in the Lennard-Jones potential is arbitrary. Pair potentials can also be used to calculate molecule–molecule potentials by using the sum of atom–atom potentials. Calculations of this type are used in predicting minimum energy geometries of interacting macromolecules. Examples are the binding of an inhibitor by an enzyme and the folding of a peptide chain to form a globular protein.

$[F{-}H{-}F]^-$

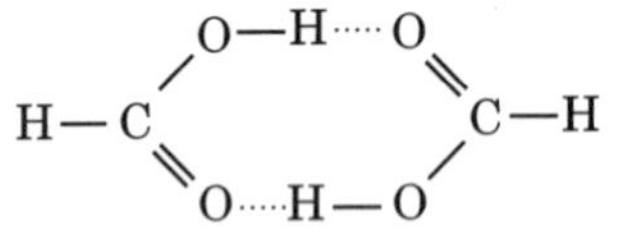

Formic acid dimer

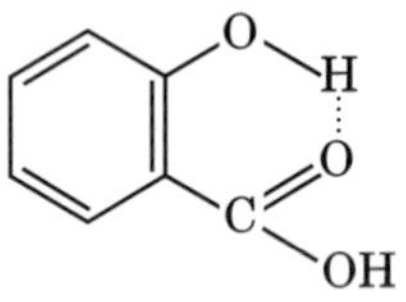

Salicylic acid

Figure 11.24 Examples of molecules exhibiting hydrogen bonding.

11.10 SPECIAL TOPICS: HYDROGEN BONDS, HYBRID ORBITALS, AND BAND THEORY OF SOLIDS

Hydrogen Bonds

A number of unusual structures such as HF_2^- and formic acid dimer in the gas phase (see Fig. 11.24) are evidence for the formation of hydrogen bonds. The unusually high acid dissociation constant of salicylic acid, as compared with the meta and para isomers, is also evidence for a hydrogen bond. A hydrogen bond results when a proton may be shared between two electronegative atoms, such as F, O, or N, that are the right distance apart. The proton of the hydrogen bond is attracted by the high concentration of negative charge in the vicinity of these electronegative atoms. Fluorine forms very strong hydrogen bonds; oxygen, weaker ones; and nitrogen, still weaker ones. The unusual properties of water are due to a large extent to the formation of hydrogen bonds involving the four lone-pair electrons on oxygen. In ice there is a tetrahedral arrangement, with each oxygen atom bonded to four hydrogen atoms. Hydrogen bonds are formed along the axis of each lone pair in ice, and their existence in liquid water is responsible for the high boiling point of water as compared with the boiling points of hydrides of other elements in the same column of the periodic table (H_2S, $-62\,°C$; H_2Se, $-42\,°C$; H_2Te, $-4\,°C$). When water is vaporized, these hydrogen bonds are broken, but in formic and acetic acids the hydrogen bonds are strong enough for dimers of the type illustrated in Fig. 11.24 to exist in the vapor. Hydrogen bonds between N and O are responsible for the stability of the α helix formed by polypeptides and found in protein molecules.

Hybrid Orbitals

We want to construct a hybrid orbital (made up of $2p_x$, $2p_y$, $2p_z$, and 2s orbitals) that points in a particular direction in space. To take a simple example, we will construct two equivalent orbitals in the xy plane, one pointing along a line making an angle χ with the x axis and the other pointing along a line making an angle $-\chi$ with the x axis. Since they are in the xy plane, we need only consider the $2p_x$, $2p_y$, and 2s orbitals. First we form linear combinations of the p orbitals that point in the specified directions:

$$\phi^{(1)} = (\cos\chi)\phi_{2p_x} + (\sin\chi)\phi_{2p_y} \qquad \phi^{(2)} = (\cos\chi)\phi_{2p_x} - (\sin\chi)\phi_{2p_y}$$

Notice that these orbitals are not orthogonal unless $2\chi = 90$:

$$\int d\tau\, \phi^{(1)}\phi^{(2)} = \cos^2\chi - \sin^2\chi = \cos 2\chi \tag{11.86}$$

We can now form linear combinations of $\phi^{(1)}$, $\phi^{(2)}$, and ϕ_{2s} that are orthogonal by using the following relation (note that the coefficients are equal so that the orbitals are equivalent):

$$\Phi_I = c_1\phi_{2s} + c_2\phi^{(1)} \qquad \Phi_{II} = c_1\phi_{2s} + c_2\phi^{(2)}$$

To find c_1 and c_2, Φ_I and Φ_{II} are normalized and made orthogonal:

$$\text{Normalization:} \qquad c_1^2 + c_2^2 = 1 \tag{11.87}$$

$$\text{Orthogonality:} \quad c_1^2 + c_2^2\cos 2\chi = 0 \tag{11.88}$$

Thus,

$$c_1^2 = \frac{\cos 2\chi}{\cos 2\chi - 1} \tag{11.89}$$

$$c_2^2 = \frac{1}{1 - \cos 2\chi} \tag{11.90}$$

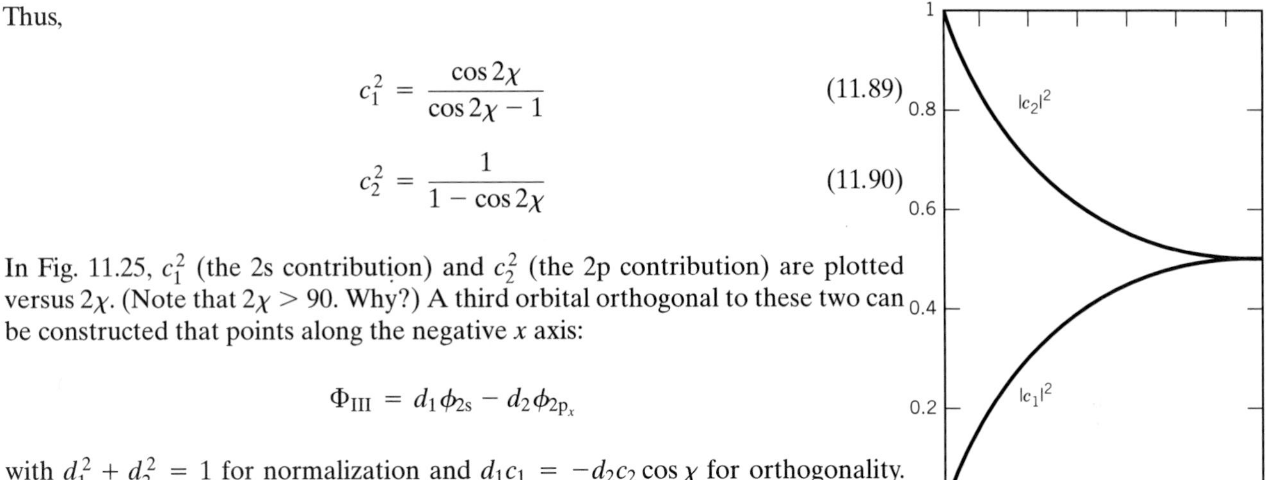

Figure 11.25 The 2s contribution (c_1^2) and the 2p contribution (c_2^2) to the hybrid orbitals as a function of angle 2χ.

In Fig. 11.25, c_1^2 (the 2s contribution) and c_2^2 (the 2p contribution) are plotted versus 2χ. (Note that $2\chi > 90$. Why?) A third orbital orthogonal to these two can be constructed that points along the negative x axis:

$$\Phi_{\rm III} = d_1\phi_{2s} - d_2\phi_{2p_x}$$

with $d_1^2 + d_2^2 = 1$ for normalization and $d_1c_1 = -d_2c_2\cos\chi$ for orthogonality. After squaring the last expression and using the results for c_1^2 and c_2^2, we find

$$d_1^2 = \frac{1 + \cos 2\chi}{1 - \cos 2\chi} \tag{11.91}$$

$$d_2^2 = \frac{2\cos 2\chi}{\cos 2\chi - 1} \tag{11.92}$$

These are plotted in Fig. 11.26. Note that $\Phi_{\rm III}$ is equivalent to $\Phi_{\rm I}$ and $\Phi_{\rm II}$ (i.e., $d_1^2 = c_1^2$) only when $2\chi = 120$. For this value of χ, $\Phi_{\rm I}$, $\Phi_{\rm II}$, and $\Phi_{\rm III}$ are the sp^2 hybrid orbitals of equations 11.54–11.56.

Band Theory of Solids

In Chapter 23 we will consider three-dimensional solids, but now we can consider the quantum theory of one-dimensional solids. In considering diatomic molecules in Fig. 11.8, we saw how combining of the 1s orbitals of the two atoms gives rise to two molecular orbitals. When more atoms are added in a line to form molecules with 3, 4, . . . , N atoms, there are N molecular orbitals, as shown in Fig. 11.27. When N is large, we can think of a continuous **band** of energy levels, although the band is really made up of discrete levels. The lowest level in the s band is fully bonding (i.e., the coefficients in the combination all have the same sign), and the highest level in the s band is fully antibonding (i.e., the coefficients alternate in sign). Above the band of s levels, there is a band of p levels, etc., as shown in Fig. 11.28. In general there is a **band gap** between bands, but sometimes the bands overlap.

In considering the electronic structure of a one-dimensional solid here, we will only consider the situation at $T = 0$ K so that we can avoid the effects of thermal excitation of electrons. Let us assume that the atoms have enough electrons to fill the orbitals in the s band (2 electrons per orbital), but only half-fill the p band. In this case electrons in the HOMO level of the p band have essentially the same energy as those in the LUMO level of the p band, and thus electrons can easily be excited. In this case, this hypothetical one-dimensional solid may be a conductor. Metallic conductors, semiconductors, and insulators and the effects of temperature are discussed in Chapter 23.

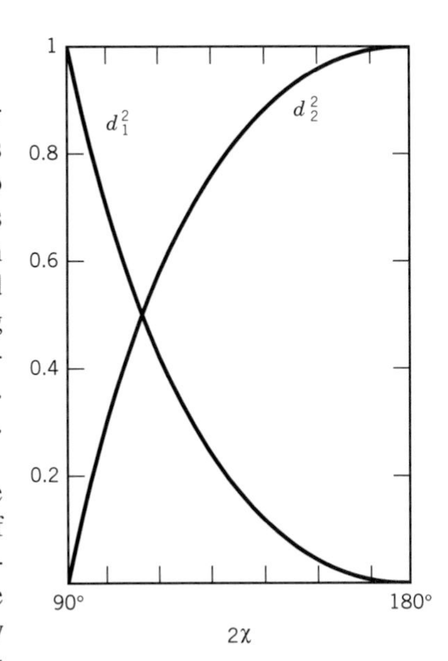

Figure 11.26 The 2s contribution (d_1^2) and the 2p contribution (d_2^2) to the third hybrid orbital as a function of 2χ.

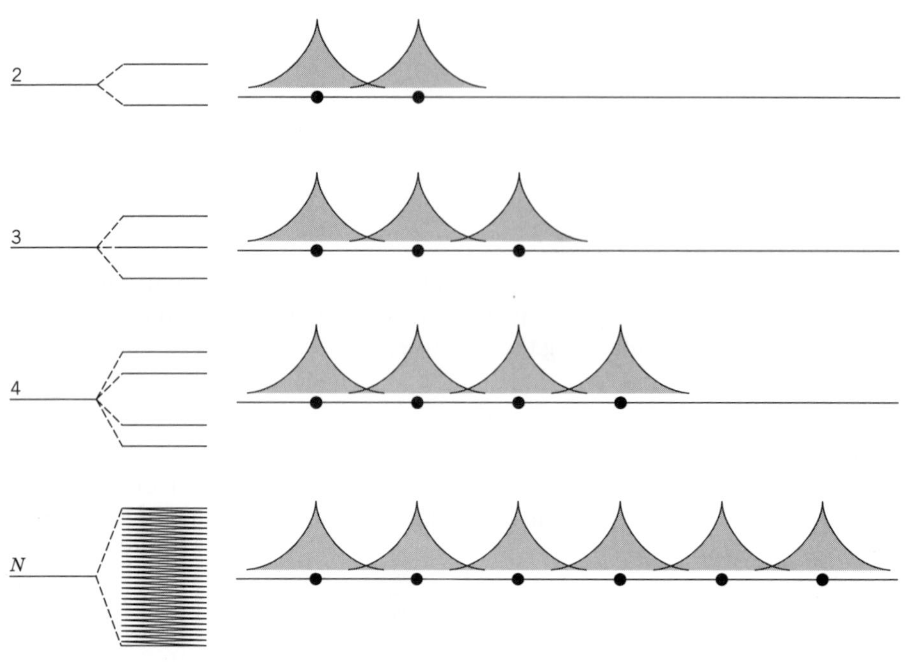

Figure 11.27 Formation of a band of very closely spaced energy levels as atoms overlap s levels in a one-dimensional solid.

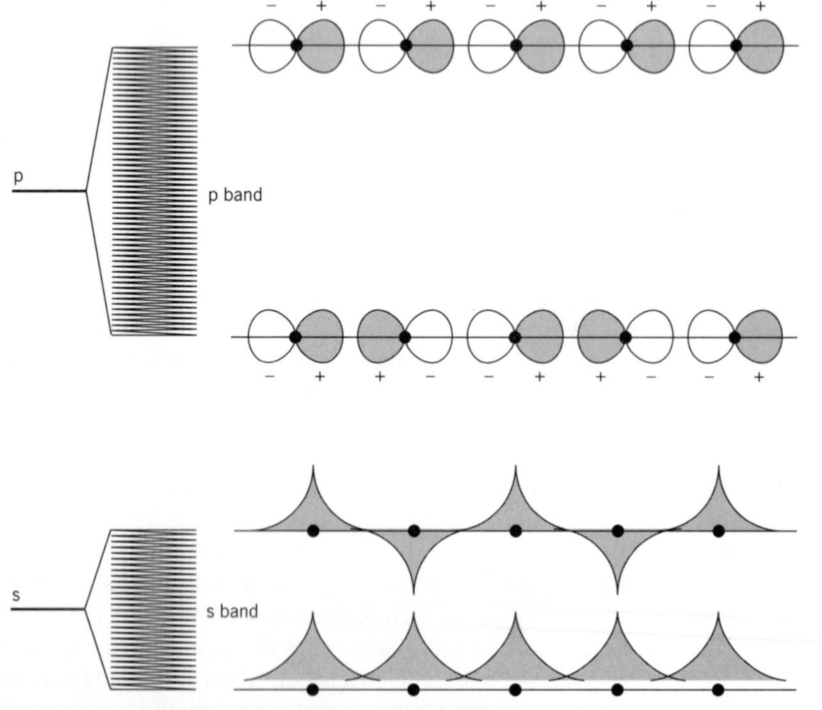

Figure 11.28 Formation of an s band and a p band in a one-dimensional solid. The lowest level in the s band is fully bonding, and the highest level in the s band is fully antibonding. This also applies to the p band. There is a gap between the s band and the p band in the solid that is smaller than the difference in the energies of the s level and the p level in the isolated atoms.

Eight Key Ideas in Chapter 11

1. Since nuclei are thousands of times more massive than electrons, they move more slowly and can be treated as being stationary in considering the motions of electrons in molecules. Therefore, the Hamiltonian for electronic motion takes internuclear distances as independent variables.
2. The Schrödinger equation for the hydrogen molecule ion can be solved exactly, but it is not discussed in this chapter because it is complicated. Instead, molecular orbitals are discussed because molecular orbitals formed as linear combinations of atomic orbitals (LCAO) are used with larger molecules.
3. The equation for the variation energy of the hydrogen molecule ion contains two constants and leads to two linear equations in these constants. This leads to a secular equation that can be solved for the energy of the ground state and the energy of the excited state.
4. One-electron molecular orbitals are labeled $\sigma, \pi, \delta, \ldots$ according to the angular momentum quantum number $\lambda = 0, 1, 2, \ldots$ in analogy with atomic term symbols.
5. In diatomic molecules, the orbital angular momenta of the electrons couple to give a resultant orbital angular momentum L, and the electron spin momenta combine to give a resultant spin angular momentum S. The absolute value of $M_L = m_1 + m_2 + \cdots$ is designated by Λ, which can have values of $0, 1, 2, \ldots$. The configurations with the different Λ are represented by the term symbols $\Sigma, \Pi, \Delta, \ldots$ in analogy with atomic term symbols.
6. The electron configurations of first-row diatomic molecules help us to understand their wide-ranging properties. Hybrid orbitals are very useful for understanding bond angles.
7. Hückel molecular orbital theory provides a qualitative understanding of energy levels and spectra of molecules with conjugated double bonds.
8. Molecules attract each other at short distances by dipole–dipole interaction, by dipole–induced dipole interaction, and by the dispersion force that occurs even when the molecules have no permanent dipole moments. The empirical Lennard-Jones 6–12 potential is often used as an intermolecular potential function.

REFERENCES

T. A. Albright and J. K. Burdett, *Problems in Molecular Orbital Theory.* New York: Oxford University Press, 1992.

W. H. Flygare, *Molecular Structure and Dynamics.* Englewood Cliffs, NJ: Prentice-Hall, 1978.

W. J. Hehre, L. Radom, P. V. R. Schleyer and J. A. Pople, *Ab Initio Molecular Orbital Theory.* Hoboken, NJ: Wiley-Interscience, 1986.

I. N. Levine, *Quantum Chemistry,* 5th ed. Upper Saddle River, NJ: Prentice-Hall, 1999.

J. P. Lowe, *Quantum Chemistry*, 2nd ed. Hoboken, NJ: Wiley, 1993.

J. L. McHale, *Molecular Spectroscopy*. Upper Saddle River, NJ: Prentice-Hall, 1999.

R. S. Mulliken and W. C. Ermler, *Diatomic Molecules.* New York: Academic, 1977.

L. Pauling, *The Nature of the Chemical Bond.* Ithaca, NY: Cornell University Press, 1960.

J. A. Pople and D. L. Beveridge, *Approximate Molecular Orbital Theory.* New York: McGraw-Hill, 1970.

M. A. Ratner and G. C. Schatz, *Introduction to Quantum Mechanics in Chemistry.* Upper Saddle River, NJ: Prentice Hall, 2001.

W. G. Richards and D. L. Cooper, *Ab Initio Molecular Orbital Calculations for Chemists.* Oxford, UK: Clarendon, 1983.

G. C. Schatz and M. A. Ratner, *Quantum Mechanics in Chemistry.* Englewood Cliffs, NJ: Prentice-Hall, 1993.

J. Simons and J. Nichols, *Quantum Mechanics in Chemistry.* New York: Oxford University Press, 1997.

PROBLEMS

(M) Problems marked with an icon may be more conveniently solved on a personal computer with a mathematical program.

11.1 Given that the equilibrium distance in H_2^+ is 106 pm and that of H_2 is 74.1 pm, calculate the internuclear repulsion energy in both cases at R_e. Using $D_e(H_2^+) = 2.79$ eV and $D_e(H_2) = 4.78$ eV, calculate E_{el} at R_e for both.

11.2 Given the equilibrium dissociation energy D_e for N_2 in Table 11.1 and the fundamental vibration frequency 2331 cm^{-1}, calculate the spectroscopic dissociation energy D_0 in kJ mol^{-1}.

11.3 Derive the values of the normalization constants given in equations 11.11 and 11.12.

11.4 Plot ψ_g and ψ_u versus distance along the internuclear axis for H_2^+ in the ground state without worrying about normalization ($R_e = 106$ pm).

11.5 The overlap integral S for the H_2^+ molecule can be evaluated as

$$S = \left(1 + \frac{R}{a_0} + \frac{R^2}{3a_0^2}\right) e^{-R/a_0}$$

Plot this as a function of R/a_0. At what value of R is it a maximum? At what value of R is it a minimum? What is the value of S at the equilibrium separation, $R = 106$ pm?

11.6 Express the four valence bond wavefunctions for H_2 as Slater determinants.

11.7 Show that the hybrid orbitals for tetravalent carbon given by equations 11.57 to 11.60 are orthogonal.

11.8 Using the hydrogenlike orbitals of Table 10.1, write out the sp^2 orbitals of equations 11.54–11.56. For $\psi_{sp^2(i)}$, write the wavefunction as a function of r for $\theta = 0$ (along the positive z axis), $\theta = 90°$ (in the xy plane), and $\theta = 180°$ (along the negative z axis).

11.9 Show that the $\psi_{sp^2(i)}$ orbital has been normalized.

11.10 The solutions of equation 11.24 were given in Section 11.3. Actually solve the secular determinant to obtain these values. For H_2^+ the secular determinant can be written more simply as

$$\begin{vmatrix} \alpha - E & \beta - ES \\ B - ES & \alpha - E \end{vmatrix} = 0$$

11.11 Bond order for a diatomic molecule can be defined by $\frac{1}{2}(N - N^*)$, where N is the number of electrons in bonding molecular orbitals and N^* is the number of electrons in antibonding orbitals. Calculate the bond orders of H_2^+, N_2^+, N_2, and O_2.

11.12 Discuss the electronic structure of the methyl radical and the location of the unpaired electron.

11.13 How many electrons are involved in σ and π bonding orbitals in the following molecules: (*a*) ethylene, (*b*) ethane, (*c*) butadiene, and (*d*) benzene?

11.14 The heat of hydrogenation of cyclohexene is -121 kJ mol^{-1}, and the heat of hydrogenation of benzene is -209 kJ mol^{-1}. What is the reduction in the energy due to the formation of the π bond system in benzene?

11.15 Consider the Hückel molecular orbitals for butadiene given in equation 11.75. Each ϕ_i is an atomic p$_z$ orbital on carbon atom i, so there is a nodal plane in the xy plane for each molecular orbital. There are other nodes in these orbitals as we move from atom 1 to atom 4 since the orbitals change sign. How many nodes are there for ψ_1, ψ_2, ψ_3, and ψ_4? Where are they? Compare with Fig. 11.21.

11.16 In Hückel theory, the contribution to the electronic energy of a single π bond (see the discussion regarding ethylene in Section 11.7) is 2β. For butadiene the contribution for two conjugated bonds is 4.472β, while for benzene (three conjugated bonds) it is 8β. What is the extra stabilization in butadiene and benzene due to the conjugation (in terms of β)? Using the data of Problem 11.14, calculate the value of β for benzene and predict the value of the extra stabilization for butadiene.

11.17 The delocalization energy of a conjugated molecule is the π electron energy minus the π electron energy for the corresponding amount of ethylene. Calculate the delocalization energies of 1,3-butadiene and benzene.

(M) **11.18** The Lennard-Jones parameters for nitrogen are $\epsilon/k = 95.1$ K and $\sigma = 0.37$ pm. Plot the potential energy (expressed as V/k in K) for the interaction of two molecules of nitrogen.

(M) **11.19** For Ne the parameters of the Lennard-Jones 6–12 potential are $\epsilon/k = 35.6$ K and $\sigma = 275$ pm. Plot V in J mol^{-1} versus r and calculate the distance r_m where $dV/dr = 0$.

11.20 For KF(g) the dissociation constant D_e is 5.18 eV, and the dipole moment is 28.7×10^{-30} C m. Estimate these values assuming that the bonding is entirely ionic. The ionization potential of K(g) is 4.34 eV, and the electron affinity of F(g) is 3.40 eV. The equilibrium internuclear distance in KF(g) is 0.217 nm.

11.21 The equilibrium internuclear distance for NaCl(g) is 236.1 pm. What dipole moment is expected? The actual value is 3.003×10^{-29} C m. How do you explain the difference?

11.22 Calculate the dipole moment that HCl would have if it consisted of a proton and a chloride ion (considered to be a point charge) separated by 127 pm (the internuclear distance obtained from the infrared spectrum). The experimental value is 3.44×10^{-30} C m. How do you explain the difference?

11.23 The equilibrium distances in HCl, HBr, and HI are 127, 141, and 161 pm, respectively. Given the dipole moments in Table 22.2, find the effective fractional charges on the H and X ions. Is this in accord with the order of the electronegativities of the halogens?

11.24 Show that the dipole moment defined by equation 11.79 is independent of the location of the arbitrary origin if the net charge on the molecule is zero.

11.25 The first excited states of He_2 are formed by exciting an electron from the antibonding $1\sigma_u$ molecular orbital to the bonding $2\sigma_g$ orbital. Write the electron configuration. What are the possible spin states? What is the bond order? Are the electronic states g or u?

11.26 Show that if $\hat{H}\psi = E\psi$, $(\hat{H} + c)\psi = (E + c)\psi$, where c is a constant. Thus if ψ is an eigenfunction of $\hat{H}$, the eigenvalue is $E + c$.

11.27 Cyclobutadiene, C_4H_4, is a four-carbon atom ring. Write the secular equation similar to equation 11.71 for the π molecular orbitals of this planar molecule. Find the energies of the orbitals. Predict the total π electronic energy of this compound. Is there extra π electron stabilization (as in butadiene or benzene) in this molecule?

11.28 An approximate description of the π electrons in conjugated polyene, $CH_2{=}CH(CH{=}CH)CH{=}CH_2$, is the free electron molecular orbital model. In this model, the π electrons are assumed to be noninteracting and to be in a one-dimensional box of length equal to one less than the number of carbons multiplied by the C–C distance of 150 pm. For butadiene and hexatriene, what are the electron configurations of the ground and first excited states in terms of particle-in-a-box eigenfunctions (see Section 9.7)? What is the excitation energy from the ground to the first excited state? What is the wavelength of this transition?

11.29 Using Fig. 11.14, give the electronic configurations and bondordersforthegroundstatesofLi_2^+,Be_2^+,B_2^+,andN_2^+found byremovinganelectronfromthehighestfilledmolecularorbital.

11.30 In H_2, what percentage is the equilibrium bond energy of the total electron energy?

11.31 In the cyclobutadiene molecule (see Problem 11.27), we can allow the bonds to become unequal by making one pair of opposite resonance integrals equal to β_1 and the other pair equal to β_2. This corresponds to unequal bond lengths. Write the secular determinant for this case and solve it as a function of β_2/β_1.

11.32 Consider the Hückel model description of butadiene given in Section 11.7. What is the electronic configuration of the positive ion? Of the negative ion? In electron paramagnetic resonance spectroscopy (EPR), only the unpaired spin densities are probed. For these two ions, what are the unpaired π electron densities on each carbon atom (i.e., the square of the coefficients in equation 11.75)?

11.33 lthough NaCl is an example of ionic bonding, it dissociates into atoms. The potential curve as a function of R (for R larger than the equilibrium distance) can be approximated by a purely electrostatic attraction, $V = -e^2/(4\pi\epsilon_0 R)$, until at some R this curve crosses the potential curve for the neutral atoms. Assume that the potential curve for the neutral atoms is independent of R and compute the value of R at the crossing in terms of the ionization potential of Na and the electron affinity of Cl. Given that $E_i(\text{Na}) = 5.14$ eV and $E_{ea}(\text{Cl}) = 3.60$ eV, compute this R.

11.34 Assume that the R^{-6} part of the Lennard-Jones potential for Ar is given by the dispersion force term, equation 11.84. If the polarizability (α) of Ar is 1.85×10^{-40} J^{-1} C^2 m, find the approximate energy of the first electronic excitation using the Lennard-Jones parameters in Table 11.2.

11.35 Calculate the spectroscopic dissociation energy D_0 of O_2 from the data in Tables 11.1 and 13.4 and compare it with the dissociation energy for ideal gas O_2 at absolute zero.

11.36 Show that the hybrid orbitals $\psi_{sp(i)}$ and $\psi_{sp(ii)}$ (equations 11.50 and 11.51) are normalized and orthogonal.

11.37 Show that the hybrid orbitals $\psi_{sp^2(i)}$ and $\psi_{sp^2(ii)}$ are orthogonal and normalized.

11.38 Calculate the lattice energy, which is the dissociation energy of NaCl(s) into gaseous ions. Given: $\Delta_f H°[\text{Na}^+(\text{g}), 298\text{ K}] = 609.358$ kJ mol^{-1}; $\Delta_f H°[\text{Cl}^-(\text{g}), 298\text{ K}] = -233.13$ kJ mol^{-1}.

11.39 The dipole moments of CH_3Cl, CH_3Br, and CH_3I are given in Table 22.2. Explain the relative magnitudes of these moments in terms of the electronegativity.

(M) **11.40** The Lennard-Jones parameters for argon are $\epsilon/k = 122$ K and $\sigma = 0.34$ nm. Plot the potential energy (expressed as V/k in K) for the interaction of two molecules of argon.

11.41 For methane the parameters for the Lennard-Jones 6–12 potential are $\epsilon/k = 148$ K and $\sigma = 0.382$ nm. Plot V in kJ mol^{-1} versus R and calculate R_m.

Computer Problems

11.A Make three-dimensional plots to show the electron density in the xy plane for the hydrogen molecule ion H_2^+. The parameters from a variational treatment (I. N. Levine, *Quantum Chemistry,* 5th ed., Upper Saddle River, NJ: Prentice-Hall, 1999, pp. 389–390) are $R_e = 2.01$ bohrs, $k = 1.246\,\text{bohr}^{-1}$, and $\zeta = 2.965\,\text{bohr}^{-1}$. Restrict the plots to $x = -0.1$ to 0.1 and $y = -0.1$ to 0.1.

11.B A trial wavefunction for the hydrogen molecule ion is

$$\psi_{\pm} = 1s_A \pm 1s_B$$

The energies in units of hartrees corresponding to these two wavefunctions can be calculated using

$$E_+ = \frac{J + K}{1 + S}$$

$$E_- = \frac{J - K}{1 - S}$$

where

$$S = e^{-R}\left(1 + R + \frac{R^2}{3}\right) \quad \text{(overlap integral)}$$

$$J = e^{-2R}\left(1 + \frac{1}{R}\right) \quad \text{(Coulomb integral)}$$

$$K = \frac{S}{R} - e^{-R}(1 + R) \quad \text{(exchange integral)}$$

where R is the distance between the protons in units of a_0. Plot S, J, K, E_+, and E_- versus R.

11.C Calculate the energy eigenvalues and eigenvectors for 1,3-butadiene using the Hückel method.

11.D Calculate the energy eigenvalues and eigenvectors for benzene using the Hückel method. (See Computer Problem 11.C.)

11.E Solve the Hückel secular equation for 1,3-butadiene and write the equations for the four energy levels.

11.F Plot the $1s_A$ and $1s_B$ wavefunctions with a separation of $3a_0$ between the protons in the hydrogen molecule ion.

11.G Plot the following normalized molecular orbitals of the hydrogen molecule ion with a separation of $3a_0$ between the protons in the hydrogen molecule ion.

$$\psi_g = \frac{1}{\sqrt{2(1 + S)}}(1s_A + 1s_B)$$

$$\psi_u = \frac{1}{\sqrt{2(1 - S)}}(1s_A - 1s_B)$$

11.H (*a*) Plot the Lennard-Jones potential for the interaction of two nitrogen molecules versus the distance between the molecules. When the energy is expressed as an equivalent temperature, $\epsilon/k = 98.1$ K and $\sigma = 0.37$ nm. (*b*) Calculate the value of R at the minimum by setting the derivative of the potential equal to zero. (*c*) Calculate the value of R at which the potential energy is equal to zero.

12 Symmetry

Ideas about symmetry are of great importance in connection with both theoretical and experimental studies of molecular structure. The basic principles of symmetry are applied in quantum mechanics, spectroscopy, and structural determinations by X-ray, neutron, and electron diffraction. Nature exhibits a great deal of symmetry, and this is especially evident when we examine molecules in their equilibrium configurations. By equilibrium configuration we refer to that with the atoms fixed in their mean positions.

In this chapter we confine our attention to isolated molecules and ions. The symmetry of an isolated molecule or ion is unaffected by molecules of solvent or adjacent molecules in a solid. When symmetry is present, certain calculations are simplified if the symmetry is taken into account. Aspects of symmetry also determine whether a molecule can be optically active or whether it may have a dipole moment. In this chapter we will see how symmetry may be treated quantitatively using group theory. Modern treatments of rotational, vibrational, and electronic spectroscopy of molecules all make extensive use of group theory.

12.1 SYMMETRY ELEMENTS AND SYMMETRY OPERATIONS

A **symmetry element** is an imaginary geometrical entity such as a line, a plane, or a point. The symmetry elements are C_n, σ, i, S_n, and E, which are defined in Table 12.1. For each symmetry element there are **symmetry operations** that move a molecule about a symmetry element so that the orientation and position of the molecule before and after the operation are indistinguishable. In other words, a symmetry operation brings every atom to an equivalent point or back to the identical point. The symbols for the symmetry operations are the same as those for the symmetry elements, but in the literature operations are given a caret like quantum mechanical operators because they can be applied to wavefunctions. The operations are referred to as rotation (sometimes called proper rotation) (C_n), reflection (σ), inversion (i), rotation–reflection (sometimes called improper rotation) (S_n), and identity (E). Every molecule or ion has the identity operation E. This operation leaves an object unchanged. This operation is often the result of carrying out an operation successively the number of times it takes to return the object to its initial position.

The symmetry of an object can be described by listing all the symmetry operations it possesses. The number of operations can be as few as one (the identity), or as many as infinity. All the corresponding symmetry elements for an object pass through a common point at the center of the object. Therefore, the symmetry of isolated molecules and ions is referred to as **point group symmetry.** In Section 12.8 we will discuss the conditions under which a collection of symmetry operations form a mathematical group. Symmetry operators often yield eigenvalues when they operate on wavefunctions. They may or may not commute with each other

Table 12.1 Symmetry Element Operations

Symbol for Element and Operation	*Element*	*Operation*
i	Center of symmetry (or inversion center)	Projection through the center of symmetry to an equal distance on the other side from the center
C_n	Proper rotation axis	Counterclockwise rotation about the C_n axis by $2\pi/n$ (or $360°/n$), where n is an integer
σ	Symmetry plane	Reflection across the plane of symmetry
S_n	Improper rotation axis (also referred to as a rotation–reflection axis or alternating axis)	Counterclockwise rotation about the S_n axis by $2\pi/n$ followed by reflection in a plane perpendicular to the axis (i.e., the combined operation of a C_n rotation followed by reflection across a σ_h mirror plane)
E	Identity element	The operation that leaves the system unchanged

and with other quantum mechanical operators. Remember, operators commute of the result of their successive applications is independent of the order in which they are applied.

In the next four sections we will consider the various symmetry operations, starting with rotations.

12.2 THE ROTATION OPERATION AND THE SYMMETRY AXIS

The operation of rotation is designated by C_n, where rotation about an axis by $2\pi/n$ radians ($360°/n$) brings the object back to an equivalent position. The value of n is referred to as the **order** of the rotation. The corresponding element is referred to as an n-fold rotation axis. Figure 12.1 shows the effects of successive fourfold clockwise rotations about an axis perpendicular to the plane of a planar MX_4 molecule. The four identical X atoms have been labeled X_A, X_B, X_C, and X_D so that we can see the results of each operation. Notice that carrying out two successive C_4 rotations about the same axis, which may be designated C_4^2, , has the same effect as a C_2 rotation. After a third C_4 rotation, the molecule is in a new equivalent configuration, which could also be obtained by a single fourfold counterclockwise rotation ($C_4^3 = C_4^{-1}$). This is generalized by writing $C_n^{n-1} = C_n^{-1}$. Also note $C_4^4 = E$, which can be generalized to $C_n^n = E$. A C_4 axis is

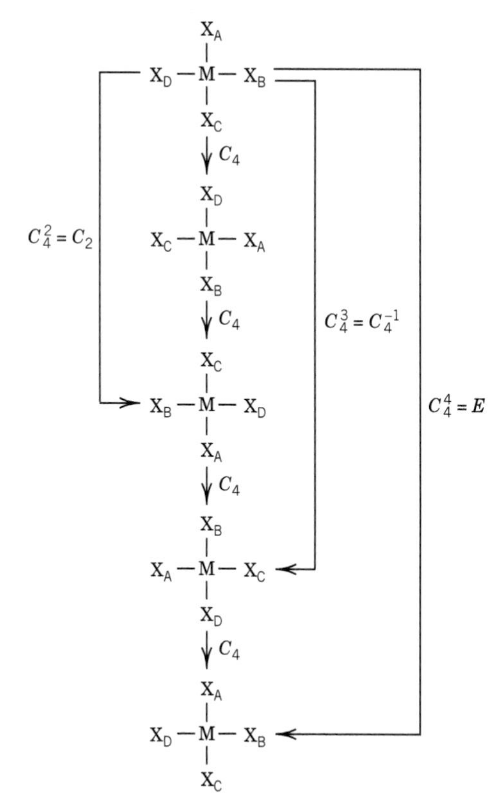

Figure 12.1 Successive C_4 clockwise rotations of a planar MX_4 molecule about an axis perpendicular to the plane of the molecule ($X_A = X_B = X_C = X_D$). (With permission from R. L. Carter, *Molecular Symmetry and Group Theory,* ©Wiley, Hoboken, NJ, 1998.)

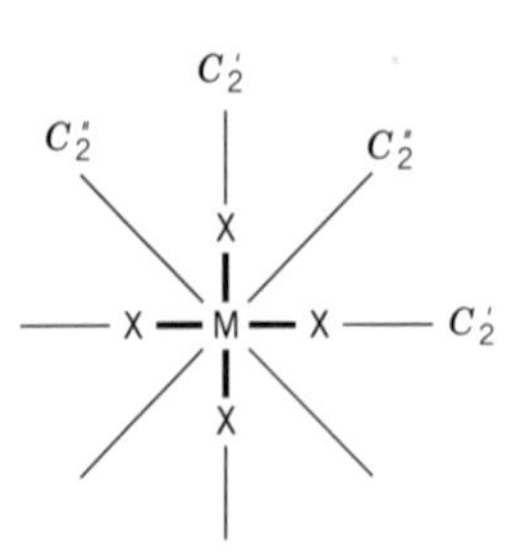

Figure 12.2 The C_2' and C_2'' axes of a planar MX_4 molecule. (With permission from R. L. Carter, *Molecular Symmetry and Group Theory*, ©Wiley, Hoboken, NJ, 1998.)

associated with three unique symmetry operations: C_4, $C_4^2 = C_2$, and $C_4^3 = C_4^{-1}$. C_4^4 is not unique because it is equal to E.

The C_4 and C_2 axes are referred to as being collinear. These two axes are not the only rotational axes of MX_4. There are four other C_2 axes in the plane of the molecule, as shown by Fig. 12.2. Note the single prime and double primes in these symbols. Only two notations are needed for the four axes because the two C_2' belong to the same class and the two C_2'' axes belong to a separate class. In listing the complete set of symmetry operations for a molecule, operations of the same class are designated by a single symbol preceded by the number of equivalent operations in the class. Thus for the planar MX_4 molecule, the rotational operations are $2C_4$, C_2, $2C_2'$, and $2C_2''$. The **principal axis** of rotation is the C_n axis with the highest n, so C_4 is the principal axis of rotation for planar MX_4.

Example 12.1 ***Distinct operations of a C_6 axis***

How many distinct operations (i.e., operations that cannot be represented in any other way) are implied by a C_6 axis?

$$\begin{array}{cccccc} C_6^1 & C_6^2 & C_6^3 & C_6^4 & C_6^5 & C_6^6 \\ & C_3^1 & C_2^1 & C_3^2 & & E \end{array}$$

Thus, two operations (C_6^1 and C_6^5) are characteristic only of a C_6 axis.

12.3 THE REFLECTION OPERATION AND THE SYMMETRY PLANE

The operation of reflection is represented by the lowercase sigma (σ), and the corresponding element is referred to as a **mirror plane.** When there is a mirror plane, the molecule is bisected by the mirror plane, which means that for any point a distance r along a normal to the mirror plane, there will be an equivalent point at a distance $-r$. Figure 12.3 shows the five mirror planes of MX_4. The first mirror plane, σ_h (horizontal mirror plane), is perpendicular to the principal axis of rotation. The centers of the atoms lie in the mirror plane, but if the atoms have a directional property perpendicular to the plane (like p_z orbitals), the operation σ_h transforms the property into the negative of itself. The σ_v planes (vertical mirror planes) go through three atoms, and the σ_d planes (dihedral mirror planes) go only through M. The mirror planes all go through a point at the center of the molecule.

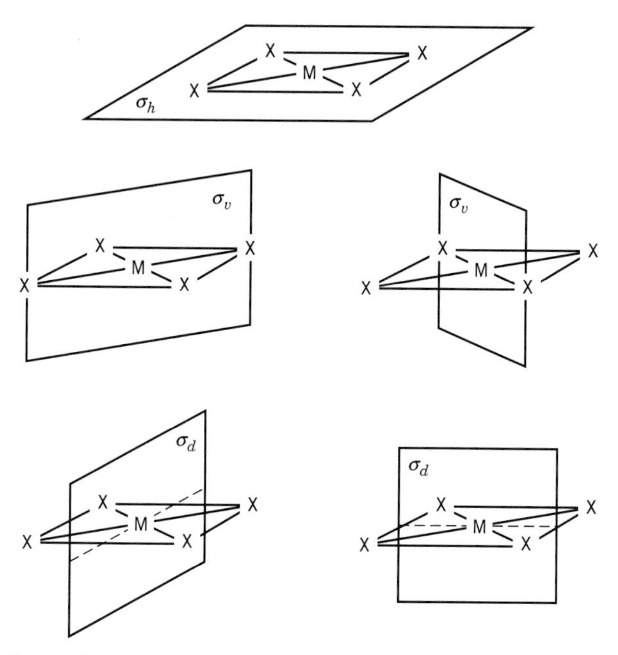

Figure 12.3 Mirror planes of square planar molecule MX_4. (With permission from R. L. Carter, *Molecular Symmetry and Group Theory,* ©Wiley, Hoboken, NJ, 1998.)

Note that performing two successive reflections about the same plane brings the molecule back to its original (identical) configuration, so that $\sigma_h \sigma_h = E$. Thus a reflection in a mirror plane is considered to be a single operation. Planar MX_4 has σ_h, σ_v, and σ_d.

12.4 THE INVERSION OPERATION AND THE CENTER OF SYMMETRY

The operation of inversion is relative to the central point in the molecule through which all the symmetry elements pass. There is an inversion operation and element i if for every point in the molecule (x, y, z) there is an equivalent point at coordinates $(-x, -y, -z)$. The central point (0, 0, 0) is referred to as a **center of symmetry.** Molecules that have inversion symmetry are referred to as being **centrosymmetric.** Since performing the inversion operation twice in succession brings every point back to itself, $ii = i^2 = E$. MX_4 is centrosymmetric, but a more general example is octahedral MX_6. The effect of the inversion operation on MX_6 is shown in Fig. 12.4.

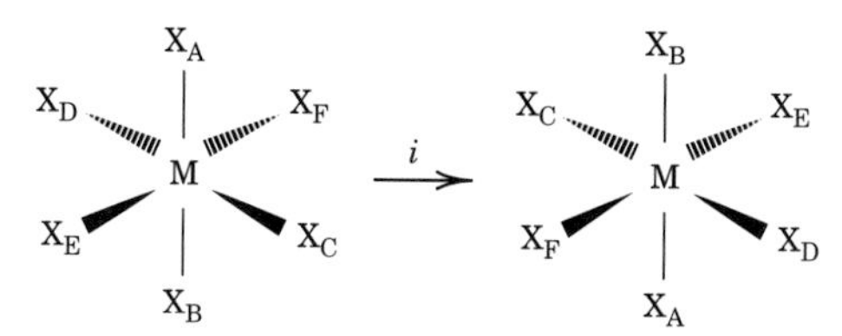

Fig. 12.4 The inversion operation on the octahedral molecule MX6. (With permission from R. L. Carter, *Molecular Symmetry and Group Theory,* ©Wiley, Hoboken, NJ, 1998.)

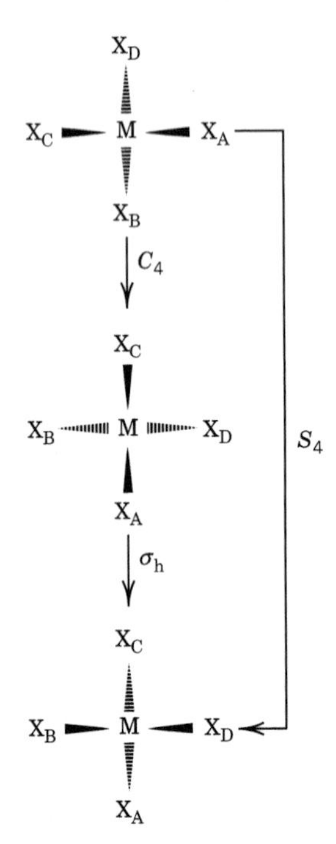

Figure 12.5 The improper rotation of the tetrahedral molecule MX4. The improper axis is perpendicular to the page. (With permission from R. L. Carter, Molecular *Symmetry and Group Theory*, ©Wiley, Hoboken, NJ, 1998.)

12.5 ROTATION-REFLECTION AND THE IMPROPER AXIS

Rotation–reflection is a compound operation that consists of a proper rotation followed by a reflection in a plane perpendicular to the axis of rotation. This axis is referred to as an **improper axis.** The compound operation is given the symbol S_n, where n refers to the initial rotation by $2(\pi/n$ radians ($360°/n$). The two parts of S_n may be the C_n and σ_h that the molecule has in its own right. For example, planar MX_4 has C_4 and σ_h, and so it also has S_4. However, it is not necessary for a molecule to have C_n and σ_h in order to have S_n. For example, the tetrahedral molecule MX_4 has an S_4 improper rotation, although it does not have either the C_4 or σ_h operation. This is illustrated in Fig. 12.5. By carrying out a second successive S4 operation to extend this figure, it can be shown that two S_4 operations are equivalent to the C_2 operation: $S_4^2 = C_2$. Carrying out two more S_4 operations shows that $S_4^4 = E$. There are only two S_4 operations (S_4 and $S_4^3 = S_4^{-1}$) because $S_4^2 = C_2$ and $S_4^4 = E$.

In discussing symmetry operations, it is convenient to orient the molecules in a right-hand Cartesian coordinate system. The thumb, index, and middle fingers of the right hand are pointed in three mutually perpendicular directions, and these are taken as the x, y, and z directions, respectively. The center of mass of the molecule under consideration is located at the origin of the Cartesian coordinate system, and its principal axis is aligned with the z axis.

Example 12.2 ***Effects of symmetry operations on a point (x,y,z) in a molecule***

Show the effects of i. C_2 (the axis of rotation is z), σ_{xz} (the symmetry plane is xz), and $\sigma_y z$ (the symmetry plane is yz) on a point represented by the column vector (x, y, z).

$$i \begin{bmatrix} x \\ y \\ z \end{bmatrix} = \begin{bmatrix} -x \\ -y \\ -z \end{bmatrix}$$

$$C_2 \begin{bmatrix} x \\ y \\ z \end{bmatrix} = \begin{bmatrix} -x \\ -y \\ z \end{bmatrix}$$

$$\sigma_{xz} \begin{bmatrix} x \\ y \\ z \end{bmatrix} = \begin{bmatrix} x \\ -y \\ z \end{bmatrix}$$

$$\sigma_{yz} \begin{bmatrix} x \\ y \\ z \end{bmatrix} = \begin{bmatrix} -x \\ y \\ z \end{bmatrix}$$

12.6 IDENTIFICATION OF POINT GROUPS OF MOLECULES

A given molecule can have a number of symmetry operations. The symmetry operations that do apply to a given molecule in its equilibrium configuration form a mathematical group. In order for a collection of operations to form a **group,** they must satisfy four requirements.

1. The operation that corresponds to the successive operation of two members of the group must be a member of the group. This also applies to the square of an operation.
2. The identity operation E must be a member of the group.
3. The operations must be associative, that is, $(AB)C = A(BC)$. They do not have to be commutative. Thus it is possible that $AB \neq BA$.
4. Each operation must have a unique inverse, that is, $AA^{-1} = A^{-1}A = E$. The **inverse operation** A^{-1} is that which returns the object to its original position.

The groups of operations for molecules were developed by Schoenflies and are referred to as **point groups** because one point in the molecules is left unchanged by any operation; this point is not necessarily occupied by a nucleus. It is useful to classify molecules according to their point groups, so there is a system of Schoenflies symbols for characterizing molecules. For example, H_2O belongs to the C_{2v} group.

The H_2O molecule has the operations C_2, σ_v, and σ_v', and, of course, it has the operation E. To make sure that these operations form a group, consider the multiplications in Table 12.2. All of the multiplications yield operations in the group, as required. Note that the operation of reflection in one vertical symmetry plane (σ_v) followed by the operation of reflection in the other vertical symmetry plane (σ_v') is equivalent to a twofold rotation; that is, $\sigma_v'\sigma_v = C_2$. Similarly, the successive operations of C_2 followed by σ_v yield the same result as the σ_v' operation (i.e., $\sigma_v C_2 = \sigma_v'$). It happens that for this particular point group each of the operations is its own inverse; thus $C_2C_2 = E$, $\sigma_v\sigma_v = E$, $\sigma_v'\sigma_v' = E$, and $E^2 = E$. For operations in certain other point groups, this is not the case (e.g., $C_3^1 C_3^2 = E$).

Table 12.2 Multiplication Table[a] for the Group C_{2v}

		Operation B			
		E	C_2^1	σ_v	σ_v'
Operation A	E	E	C_2^1	σ_v	σ_v'
	C_2^1	C_2^1	E	σ_v'	σ_v
	σ_v	σ_v	σ_v'	E	C_2^1
	σ_v'	σ_v'	σ_v	C_2^1	E

[a]The table contains the products AB for the indicated operations. Note that each column and each row has each symmetry operation represented only once.

The various types of point groups are defined below and are illustrated in Table 12.3. All molecules belong to one of the following point groups.

Point Group C_1

Molecules with no symmetry other than the identity are in point group C_1. An example is CHBrClF.

Point Group C_s

Point group C_s is the group for molecules that only have a reflection plane σ. An example is CH_2ClF.

Point Group C_i

Molecules, such as 1,2-dibromo-1,2-dichloroethane, that have only a center of symmetry i belong to point group C_i.

Point Groups C_n

Molecules possessing only an n-fold axis of rotation belong to a C_n point group.

Point Groups C_{nv}

Molecules with an n-fold axis of rotation and n vertical mirror planes (which are necessarily colinear with the n-fold axis) belong in one of the C_{nv} point groups.

Point Groups C_{nh}

Molecules with an n-fold axis and a plane of symmetry perpendicular to this axis belong to one of the C_{nh} point groups. Such a plane is referred to as a horizontal mirror plane. The C_{2h} point group necessarily involves a center of symmetry as well.

Point Groups D_n

Molecules with a C_n axis and a C_2 axis perpendicular to this axis are in the D_n point group.

Table 12.3 Common Schoenflies Point Groups with Examples

Schoenflies Symbol	*Symmetry Elements*	*Molecular Configuration*	*Schoenflies Symbol*	*Symmetry Elements*	*Molecular Configuration*
C_1	E	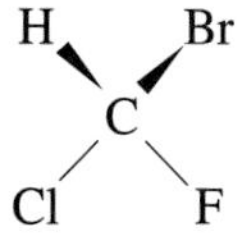 H, Br, C, Cl, F	D_{3h}	$E, 2C_3, 3C_2, \sigma_h, 2S_3, 3\sigma_v$	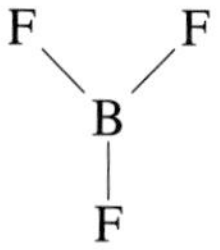 F, F, B, F
C_s	E, σ	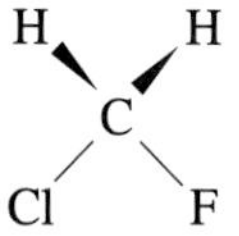 H, H, C, Cl, F	D_{4h}	$E, 2C_4, C_2, 2c'_2, 2C''_2, i, 2S_4, \sigma_h, 2\sigma_v, 2\sigma_d$	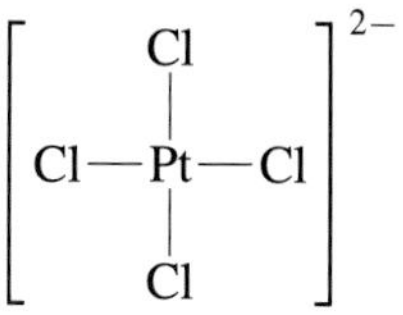 $[PtCl_4]^{2-}$
C_i	E, i	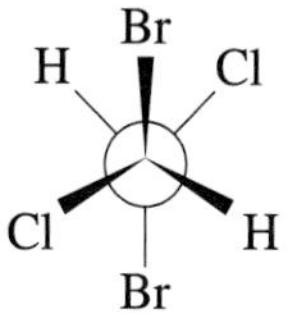 Br, H, Cl, Cl, H, Br	D_{5h}	$E, 2C_5, 2C_5^2, 5C_2, \sigma_h, 2S_5, 2S_5^2, 5\sigma_v$	 Ru
C_2	E, C_2	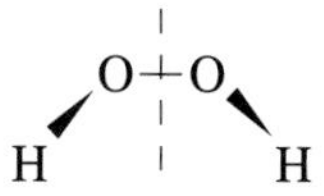 O—O, H, H	D_{6h}	$E, 2C_6, 2C_3, C_2, 3C'_2, 3C''_2, i, 2S_3, 2S_6, \sigma_h, 3\sigma_d, 3\sigma_v$	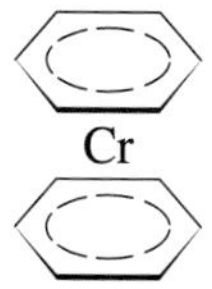 Cr
C_{2v}	$E, C_2, \sigma_v(xz), \sigma'_v(yz)$	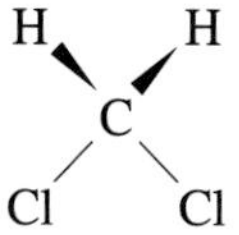 H, H, C, Cl, Cl	$D_{\infty h}$	$E, 2C_\infty, \infty\sigma_v, i, 2S_\infty, \infty C_2$	H—C≡C—H
C_{3v}	$E, 2C_3, 3\sigma_v$	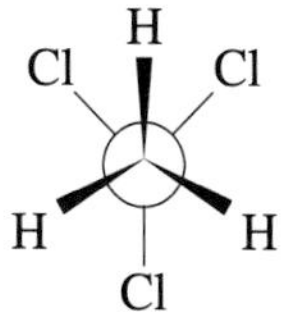 H, Cl, Cl, H, H, Cl	D_{2d}	$E, 2S_4, C_2, 2C'_2, 2\sigma_d$	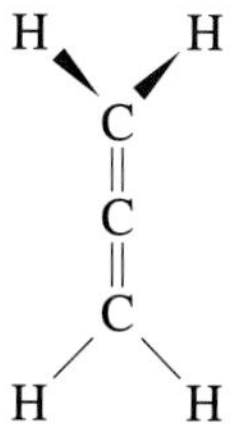 H, H, C=C=C, H, H
C_{4v}	$E, 2C_4, C_2, 2\sigma_v, 2\sigma_d$	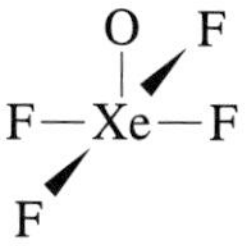 O, F, F—Xe—F, F	D_{3d}	$E, 2C_3, 3C_2, i, 2S_6, 3\sigma_d$	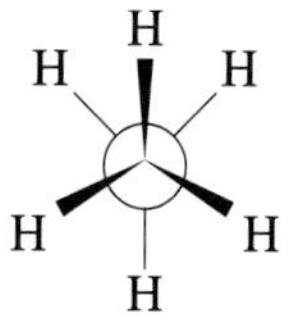 H, H, H, H, H, H
$C_{\infty v}$	$E, 2C_\infty, \infty\sigma_v$	 O≡C	D_{4d}	$E, 2S_8, 2C_4, 2S_8^3, C_2, 4C'_2, 4\sigma_d$	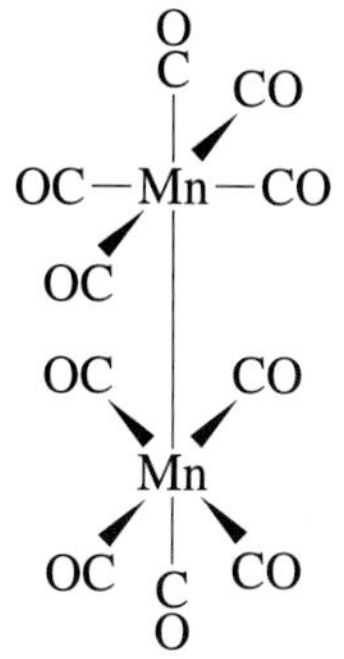 $(OC)_5Mn$—$Mn(CO)_5$

(*continued*)

Table 12.3 *(continued)*

Schoenflies Symbol	*Symmetry Elements*	*Molecular Configuration*	*Schoenflies Symbol*	*Symmetry Elements*	*Molecular Configuration*
C_{2h}	E, C_2, i, σ_h	H Cl C=C Cl H	D_{5d}	$E, 2C_5, 2C_5^2, 5C_2,$ $i, 2S_{10}^3, 2S_{10}, 5\sigma_d$	Fe
C_{3h}	$E, C_3, C_3^2, \sigma_h, S_3,$ S_3^5	H O O B H O H	T_d	$E, 8C_3, 3C_2, 6S_4,$ $6\sigma_d$	H C H H H
D_{2h}	$E, C_2(z), C_2(y),$ $C_2(x), i, \sigma_{xy}, \sigma_{xz},$ σ_{yz}	H H C=C H H	O_h	$E, 8C_3, 6C_2, 6C_4,$ $3C_2, i, 6S_4, 8S_6,$ $3\sigma_h, 6\sigma_d$	F F F—S—F F F

Point Groups D_{nd}

Molecules with a C_n axis, a perpendicular C_2 axis, and a dihedral mirror plane are in the D_{nd} point group. The dihedral mirror plane is colinear with the principal axis and bisects the two perpendicular C_2 axes.

Point Groups D_{nh}

As you may have guessed already, molecules in the D_{nh} point group have a horizontal mirror plane, that is, one perpendicular to the principal axis.

Point Groups S_n

To be in one of the S_n point groups a molecule has to have an n-fold improper rotation axis.

Special Point Groups

Linear molecules are either $C_{\infty v}$ or $D_{\infty h}$. Heteronuclear molecules, such as CO, are $C_{\infty v}$ because the molecular axis is an ∞-fold axis, and they have an infinite number of vertical mirror planes. Homonuclear diatomic molecules or polyatomics such as acetylene are $D_{\infty h}$ because the molecular axis is ∞-fold, and there is an infinite number of perpendicular C_2 axes since the molecule is symmetrical. Tetrahedral molecules are T_d. The T_h point group has all of the symmetry of a cube. Octahedral molecules, such as SF_6, are O_h. Molecules with the symmetry of an icosahedron or dodecahedron are I_h, and atoms with spherical symmetry are K_h.

Buckminsterfullerene,* is an example of I_h symmetry. The regular icosahedron and dodecahedron also belong to this point group. These molecules have the following symmetry operations: E, $12C_5$, $12C_5^2$, $20C_5$, and $15C_2$, i, $12S_{10}$, $12S_{10}^3$,

*H. W. Kroto, J. R. Heath, S. C. O'Brien, R. F. Curl, and R. E. Smalley, *Nature* **318;** 162–163 (1985).

$20S_6$, 15σ. The number of symmetry operations $h = 120$ is the largest likely to be encountered, except for $C_{\infty v}$ and $D_{\infty h}$, for which $h = \infty$. The C_2, C_3, and C_5 axes of C_{60} are shown in Fig. 12.6.

It is important to be able to identify the point group of a molecule so that group theory can be utilized in various applications to chemistry. Fortunately, it is not necessary to identify all the symmetry operations of a molecule to identify its point group. The most efficient way to proceed is to look for key symmetry elements in a prescribed sequence. This sequence is illustrated by the flow chart in Fig 12.7. To use this flow chart look sequentially for the symmetries indicated by the perpendicular lines. Then follow the right or left branch according to whether the particular kind of symmetry is present ("Yes") or absent ("No"). The special groups are discussed in the previous paragraph. If a molecule is not in one of these "special groups," look for a principal axis of rotation. If there is no axis of rotation, the molecule must belong to one of the low-symmetry nonrotational groups C_s C_i, or C_1. If a molecule has one or more rotational axes, it is necessary to identify the principal axis of rotation.

12.7 WHAT SYMMETRY TELLS US ABOUT DIPOLE MOMENTS AND OPTICAL ACTIVITY

The presence of symmetry in a molecule can be used to determine when certain molecular properties will be zero. For example, certain symmetry groups preclude

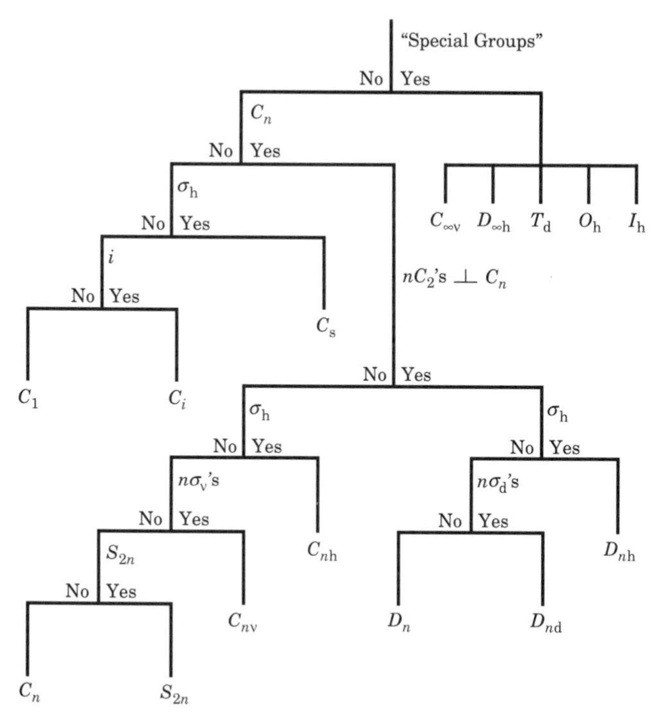

Fig. 12.7 Flow chart for systematically determining the point group of a molecule. (With permission from R. L. Carter, *Molecular Symmetry and Group Theory,* ©Wiley, Hoboken, NJ, 1998.)

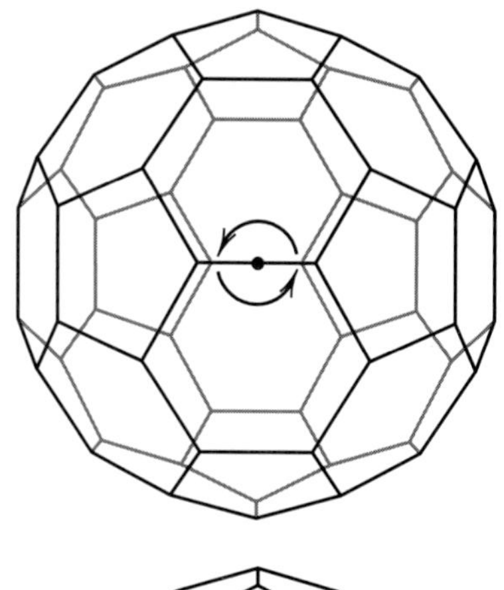

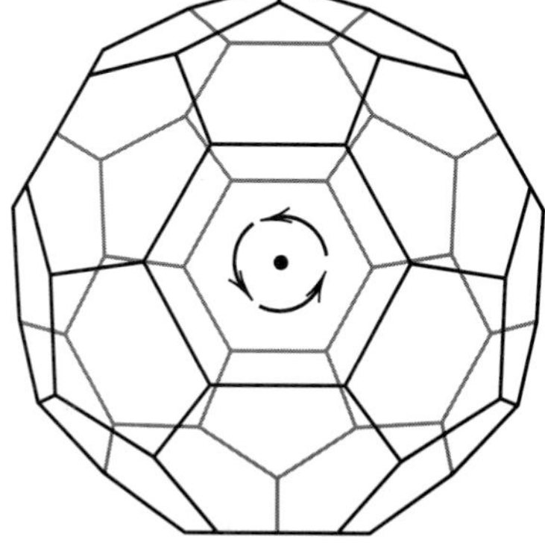

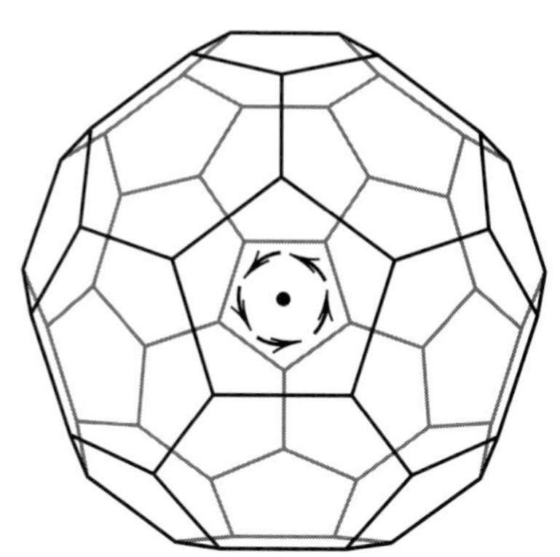

Figure 12.6 Twofold, threefold, and fivefold axes of C_{60}. [From F. Chung and S. Sternberg, *Am. Sci.* **81**:56 (1993).]

the possibility of a dipole moment or optical activity. We consider these two examples in this section.

The **dipole moment** (Section 11.8) is a vector quantity that is not affected either in direction or in magnitude by any symmetry operation of the molecule. Therefore, the dipole moment vector must be contained in each of the symmetry elements. Consequently, molecules that possess dipole moments belong only to the point groups C_n, C_s, and C_{nv}. The presence or absence of a dipole moment therefore tells something about the symmetry of a molecule. For example, carbon dioxide and water might have structures corresponding to a symmetrical linear molecule, to an unsymmetrical linear molecule, or to a bent molecule. The dipole moments recorded in Table 22.2 show that carbon dioxide in its ground electronic state has zero moment; therefore, the molecule must be symmetrical and linear. If it were unsymmetrical or bent, there would have been a permanent dipole moment. On the other hand, water in its ground electronic state has a pronounced dipole moment and cannot have the symmetrical linear structure. A molecule with a center of symmetry cannot have a dipole moment.

If a molecule and its mirror image cannot be superimposed, it is potentially **optically active.** Since a rotation followed by a reflection always converts a right-handed object to a left-handed object, an S_n axis guarantees that a molecule cannot exist in separate left- and right-handed forms.

All **improper** rotation axes (S_n), including a mirror plane ($\sigma = S_1$) and center of symmetry ($i = S_2$), convert a right-handed object into a left-handed object (i.e., produce a mirror image of the original object), whereas all **proper** rotation axes (C_n) leave a right-handed object unchanged in this respect. Hence, only molecules that have no improper symmetry elements can be optically active.

In a molecule in which internal rotation can take place (e.g., ethane or H_2O_2) it is possible to have optically active conformations, but in a gas or solution these conformers are so rapidly interconverted that optical isomers cannot be resolved.

Comment:

These discussions of the symmetry of molecules have added a new class of operators to the operators of quantum mechanics. The operators we encountered earlier operated on molecular wavefunctions. Symmetry operators operate on points in molecules, but they are related to the operators of quantum mechanics. Thus questions of commutability and noncommutability between these two types of operators arise. Unfortunately, we cannot follow up on this (but see the advanced textbooks in the reference list); it is important to know that the Hamiltonian operator for a molecule must be invariant under (commute with) all the symmetry operations of the molecule.

12.8 SPECIAL TOPIC: MATRIX REPRESENTATIONS

We have seen several examples of products of operations in connection with Table 12.2. These products and the effects of these operations on a point in a molecule can be given an actual algebraic significance by writing the symmetry operations as matrices (see Appendix D.8). For example, the effect of the inversion operation is to convert the point with coordinates x_1, y_1, and z_1 to point x_2, y_2, and z_2, as shown in Fig. 12.4. The new coordinates are given by the equations

$$x_2 = -x_1 + 0y_1 + 0z_1 \tag{12.1}$$

$$y_2 = 0x_1 - y_1 + 0z_1 \tag{12.2}$$

$$z_2 = 0x_1 + 0y_1 - z_1 \tag{12.3}$$

These relations are expressed in matrix notation by

$$\begin{bmatrix} x_2 \\ y_2 \\ z_2 \end{bmatrix} = \begin{bmatrix} -1 & 0 & 0 \\ 0 & -1 & 0 \\ 0 & 0 & -1 \end{bmatrix} \begin{bmatrix} x_1 \\ y_1 \\ z_1 \end{bmatrix} = \begin{bmatrix} -x_1 \\ -y_1 \\ -z_1 \end{bmatrix} \tag{12.4}$$

Thus, the transformation matrix of i is

$$R(i) = \begin{bmatrix} -1 & 0 & 0 \\ 0 & -1 & 0 \\ 0 & 0 & -1 \end{bmatrix} \tag{12.5}$$

The identity operation and the other three operations of the C_{2v} point group are represented by

$$R(E) = \begin{bmatrix} 1 & 0 & 0 \\ 0 & 1 & 0 \\ 0 & 0 & 1 \end{bmatrix} \tag{12.6}$$

$$R(C_2) = \begin{bmatrix} -1 & 0 & 0 \\ 0 & -1 & 0 \\ 0 & 0 & 1 \end{bmatrix} \tag{12.7}$$

$$R(\sigma_v) = \begin{bmatrix} 1 & 0 & 0 \\ 0 & -1 & 0 \\ 0 & 0 & 1 \end{bmatrix} \tag{12.8}$$

$$R(\sigma_v') = \begin{bmatrix} -1 & 0 & 0 \\ 0 & 1 & 0 \\ 0 & 0 & 1 \end{bmatrix} \tag{12.9}$$

$R(E)$ is the identity matrix (see Appendix D).

When these matrices are multiplied by each other the results are the same as when the operations are multiplied by each other, as shown in Table 12.2. Therefore, these matrices are referred to as **representatives** of their respective operations in the C_{2v} point group. The group multiplication table can be reproduced by matrix multiplications of the matrix representatives. The set of four matrices is referred to as a **representation** of the C_{2v} point group.

In Section 11.2, we used symmetry to classify wavefunctions of the hydrogen molecule ion. If $\psi(x, y, z) = \psi(-x, -y, -z)$, the wavefunction has even parity and is designated with the subscript g for *gerade*. Now we observe that $i\psi(x, y, z) = \psi(-x, -y, -z)$, so that the eigenvalue is $+1$. For an odd-parity wavefunction, the eigenvalue is -1 and the wavefunction is designated by subscript u for *ungerade*. A symmetry operation that applies to a molecule will commute with the electronic Hamiltonian operator, and the electronic wavefunction is an eigenfunction of this symmetry operation. The effect of the operations of the C_{2v} point group on the p_x orbital are shown in Fig. 12.8.

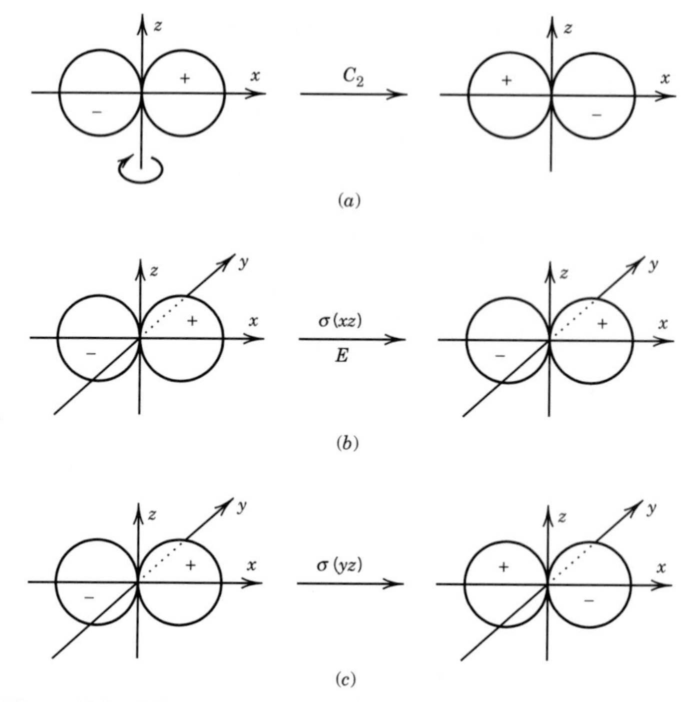

Figure 12.8 Effects of operations of the C_{2v} group on the p_x orbital.

Example 12.3 ***Multiplication of symmetry operations***

Multiply the transformation matrix for C_2 by the transformation matrix for σ_v and identify the operation that corresponds with the product. See Appendix D.

$$\begin{bmatrix} 1 & 0 & 0 \\ 0 & -1 & 0 \\ 0 & 0 & 1 \end{bmatrix} \begin{bmatrix} -1 & 0 & 0 \\ 0 & -1 & 0 \\ 0 & 0 & 1 \end{bmatrix} = \begin{bmatrix} -1 & 0 & 0 \\ 0 & 1 & 0 \\ 0 & 0 & 1 \end{bmatrix}$$

Thus, the product yields the transformation matrix for σ_v' so that

$$\sigma_v C_2 = \sigma_v'$$

as shown in Table 12.2.

Although we found this representation by considering the effect of the operations on a point in three-dimensional space, the notion of a representation is more general. Any set of numbers of matrices that have the same multiplication table as the operations in the group form a representation of the group. There are an infinite number of such representations of a group, but there are a finite number that are, in a mathematical sense, more fundamental than the others. These are called **irreducible** representations. The representation we found for the C_{2v} point group (equations 12.6–12.9) is not irreducible. It is, in fact, reducible to three different irreducible representations because the matrices are diagonal. In this case the diagonal elements themselves form an irreducible representation. For example, if we take the xx elements of 12.6–12.9 we have

$$R(E) = +1 \qquad R(C_2) = -1 \qquad R(\sigma_v) = +1 \qquad R(\sigma_v') = -1 \quad (12.10)$$

The yy elements give us

$$R(E) = +1 \qquad R(C_2) = -1 \qquad R(\sigma_v) = -1 \qquad R(\sigma_v') = +1 \quad (12.11)$$

and the zz elements

$$R(E) = +1 \qquad R(C_2) = +1 \qquad R(\sigma_v) = +1 \qquad R(\sigma_v') = +1 \quad (12.12)$$

These are representations because their multiplication table is identical to that of the C_{2v} group operations themselves. Another representation is

$$R(E) = +1 \qquad R(C_2) = +1 \qquad R(\sigma_v) = -1 \qquad R(\sigma_v') = -1 \quad (12.13)$$

These turn out to be all the irreducible representations of C_{2v}.

Example 12.4 ***The multiplication table of the representation is the same as that of the group***

Show that the representation given in equation 12.16 has the same multiplication table as the operations in the group C_{2v} (Table 12.2).

From the multiplication table of the R's of equation 12.16:

	$R(E)$	$R(C_2)$	$R(\sigma_v)$	$R(\sigma_v')$
$R(E)$	1	−1	1	−1
$R(C_2)$	−1	1	−1	1
$R(\sigma_v)$	1	−1	1	−1
$R(\sigma_v')$	−1	1	−1	1

If we compare these numbers with those we would obtain by replacing the operations in Table 12.2 by the R's of equation 12.10, we see that they are identical.

In the case of the C_{2v} group, as for all commutative groups, all the irreducible representations are one-dimensional (i.e., numbers). Many groups have higher-dimensional irreducible representations (e.g., D_{6h}, T_d), and then the matrices in the representation have that dimension.

Example 12.5 ***Matrix representation of a rotation***

To examine the effect of rotation in the xy plane, we rotate about the z axis by an angle θ. Then the x and y coordinates of a point change in the following way:

$$\begin{bmatrix} X \\ Y \end{bmatrix} = \begin{bmatrix} x\cos\theta - y\sin\theta \\ x\sin\theta + y\cos\theta \end{bmatrix} = \begin{bmatrix} \cos\theta & -\sin\theta \\ \sin\theta & \cos\theta \end{bmatrix} \begin{bmatrix} x \\ y \end{bmatrix}$$

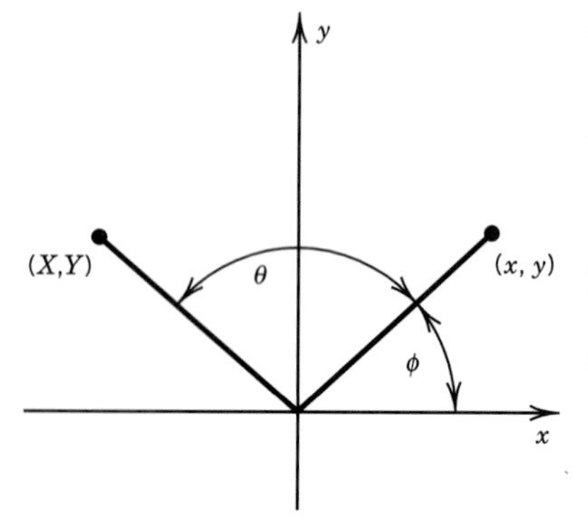

Figure 12.9 Effect of a rotation about the z axis on the coordinates of a point in the xy plane.

using matrix multiplication. This can be seen in Fig. 12.9 and using some simple trigonometry by the following argument. The initial x, y coordinates can be written as $x = r\cos\phi$, $y = r\sin\phi$. A rotation by θ brings these to

$$\begin{aligned} X &= r\cos(\phi+\theta) = r\cos\theta\cos\phi - r\sin\phi\sin\theta \\ Y &= r\sin(\phi+\theta) = r\cos\theta\sin\phi + r\sin\theta\cos\phi \end{aligned}$$

proving the formula above. What are the matrices representing operations C_3 and C_3^2?

The C_3 operation implies $\theta = 120°$ and C_3^2 implies $\theta = 240°$. Thus, the matrices for these are

$$C_3 = \begin{bmatrix} \frac{-1}{2} & \frac{-\sqrt{3}}{2} \\ \frac{\sqrt{3}}{2} & \frac{-1}{2} \end{bmatrix} \qquad C_3^2 = \begin{bmatrix} \frac{-1}{2} & \frac{\sqrt{3}}{2} \\ \frac{-\sqrt{3}}{2} & \frac{-1}{2} \end{bmatrix}$$

Note that when the matrix for C_3 is squared, the matrix for C_3^2 results, as it should. Finally, we note that since the z component of a point does not change for a rotation about the z axis, we can write the matrices in three-dimensional notation; for example,

$$C_3 = \begin{bmatrix} \frac{-1}{2} & \frac{-\sqrt{3}}{2} & 0 \\ \frac{\sqrt{3}}{2} & \frac{-1}{2} & 0 \\ 0 & 0 & 1 \end{bmatrix}$$

12.9 SPECIAL TOPIC: CHARACTER TABLES

Often we do not work with the matrix representations of a group themselves, but with the **character** of the representation, which is defined as the sum of the diagonal elements (trace) of the matrix representation. For one-dimensional representations, then, the character and the representation are identical. The character table for the C_{2v} group is given in Table 12.4. On the left is the label of the irreducible representation (A_1, A_2, B_1, B_2) and on the right are examples of functions of x, y, and z that transform like these representations under the operations of the C_{2v} group. For example, consider the function x. Operating on x with the elements of the C_{2v} group gives

$$Ex = (1)x \tag{12.14}$$
$$C_2x = (-1)x \tag{12.15}$$
$$\sigma_v x = (1)x \tag{12.16}$$
$$\sigma_v' x = (-1)x \tag{12.17}$$

By comparing this with Table 12.4, we see that x can be labeled B_1.

Table 12.4 Character Table for the C_{2v} Group

C_{2v}	E	C_2	$\sigma_v(xz)$	$\sigma_v'(yz)$	
A_1	1	1	1	1	z, z^2, x^2, y^2
A_2	1	1	−1	−1	xy
B_1	1	−1	1	−1	x, xz
B_2	1	−1	−1	1	y, yz

Example 12.6 ***Symmetry properties of the operations of C_{2v}***

Verify the symmetry properties of all the functions on the right-hand side of Table 12.4.

Consider the functions y and z. Under the operation of C_{2v}, they transform in the following way:

$$\begin{aligned} Ey &= (1)y & Ez &= (1)z \\ C_2y &= (-1)y & C_2z &= (1)z \\ \sigma_v y &= (-1)y & \sigma_v z &= (1)z \\ \sigma_v' y &= (+1)y & \sigma_v' z &= (1)z \end{aligned}$$

so that y transforms like B_2 and z like A_1. Composite functions such as yz can be found from the transformations of y and z alone, so that, for example,

$$\sigma_v(yz) = (-1)(+1)yz$$

Symmetry operations may also be applied to wavefunctions, and thus wavefunctions can be classified by their symmetry properties. For example, consider the effects of the operators of the C_{2v} group on p orbitals. The effect of the C_2 operation on a p_x orbital is illustrated in Fig. 12.8*a*. This is summarized by

$$C_2 p_x = -1 p_x \tag{12.18}$$

where -1 indicates sign reversal. Now what are the effects of the reflection operations $\sigma(xz)$ and $\sigma(yz)$? As shown in Fig. 12.8*b*, reflection in the xz plane has no effect, and so

$$\sigma(xz) p_x = 1 p_x \tag{12.19}$$

As shown in Fig. 12.8*c*, reflection in the yz plane causes a sign change so that

$$\sigma(yz) p_x = -1 p_x \tag{12.20}$$

The identity operation E has no effect, so the number that represents this operation is $+1$. Thus, the p_x orbital can be labeled B_1 in the C_{2v} point group.

Now consider the effects of the operators of the C_{2v} group on a p_y orbital.

$$E p_y = +1 p_y \tag{12.21}$$
$$C_2 p_y = -1 p_y \tag{12.22}$$
$$\sigma(xz) p_y = -1 p_y \tag{12.23}$$
$$\sigma(yz) p_y = +1 p_y \tag{12.24}$$

Therefore, p_y can be labeled B_2 in the C_{2v} group.

The importance of symmetry (or group theory) to chemical problems lies in the fact that if the symmetry of a molecule is that of a given point group, then the wavefunctions must transform like one of the irreducible representations of that group. Thus, the electronic wavefunctions for H_2O (in its C_{2v} equilibrium geometry) can be labeled A_1, A_2, B_1, or B_2 as in Table 12.4.

Furthermore, various operations, such as the Hamiltonian or the dipole moment operation, also transform like particular irreducible representations of the point group. From the transformation properties (i.e., symmetry labels) of the wavefunctions and operations, we can derive rules that tell us when certain integrals involving those operations equal zero. For example, the probability amplitude that a molecule in electronic eigenstate ϕ_i will absorb a photon and end up in state ψ_f depends on the integral

$$\int \psi_f^* \boldsymbol{\mu} \phi_i \, d\tau \tag{12.25}$$

(see Section 14.1). The vector operator $\boldsymbol{\mu}$ has three components, μ_x, μ_y, μ_z, which transform like x, y, and z in the molecular symmetry group. The above integral will be zero unless the integrand is unchanged when any of the symmetry operations of the group is applied to it. Suppose the molecule belongs to the C_{2v} group and ϕ_i transforms like A_1 (or z) and ψ_f transforms like B_2 (or y). Then the component of the integral with μ_x will vanish because the integrand will then transform as zyx (A_2) and therefore will change sign under σ_v (see Table 12.4). On the other hand, the y component of the integral will not necessarily vanish because the integrand then transforms as zy^2 or A_1, which is unchanged when any operation is applied to it.

Finding out which integrals vanish for symmetry reasons greatly decreases the amount of work we have to do in solving problems, and it yields general rules (selection rules) for spectroscopy and other areas of physical chemistry. Such chemical applications are the subject of many of the books listed at the end of the chapter. The programmed introduction to chemical applications by Vincent is especially recommended.

Example 12.7 ***Symmetry operations applied to molecular wavefunctions***

So far we have applied symmetry operations to points and atomic wavefunctions, but we can also apply them to molecular wavefunctions. (*a*) Apply the inversion operation i to H_2^+ in its ground state. What is the eigenvalue? (*b*) Apply the inversion operation i to H_2^+ in its first excited state. What is the eigenvalue?

(*a*) For ground-state H_2^+, $i\sigma_g 1s = (1)\sigma_g 1s$ because the molecule is symmetrical about its center. The eigenvalue is $+1$.

(*b*) For H_2^+ in the first excited state, $i\sigma_u 1s = (-1)\sigma_u 1s$ because the wavefunction has the opposite sign on the other side of the center of the molecule. The eigenvalue is -1.

Five Key Ideas in Chapter 12

1. The symmetry of a molecule can be described in terms of five types of symmetry elements and the corresponding operations. The symmetry operations operate on molecules and on their wavefunctions.
2. The operations for a molecule are associative, but they do not have to be commutative.
3. Molecules that possess dipole moments belong only to the point groups C_n, C_s, and C_{nv}.
4. An S_n axis guarantees that a molecule cannot exist in separate left- and right-handed forms.
5. Matrices provide representations of point groups; that is, matrices can be devised that have the same properties as the various symmetry operations.

REFERENCES

P. F. Bernath, *Spectra of Atoms and Molecules*. Oxford, UK: Oxford Press, 1995.
R. L. Carter, *Molecular Symmetry and Group Theory*. Hoboken, NJ: Wiley, 1998.
F. A. Cotton, *Chemical Applications of Group Theory*. Hoboken, NJ: Wiley, 1990.
B. E. Douglas and C. E. Hollingsworth, *Symmetry in Bonding and Spectra*. New York: Academic, 1985.
R. L. Flurry, *Symmetry Groups*. Englewood Cliffs, NJ: Prentice-Hall, 1980.
S. F. A. Kettle, *Symmetry and Structure*.Hoboken, NJ: Wiley, 1985.
A. Vincent, *Molecular Symmetry and Group Theory*. Hoboken, NJ: Wiley, 1977.
P. H. Wolton, *Beginning Group Theory for Chemistry*. Oxford, UK: Oxford University Press, 1998.

PROBLEMS

For problems 12.1–12.14, list the Schoenflies symbol and symmetry elements for each molecule.

12.1 H_2S

12.2 PCl_3

12.3 *trans*-$[CrBr_2(H_2O)_4]^+$ (ignore the H's)

12.4 *gauche*-CH_2ClCH_2Cl

12.5 $C_6H_3Br_3$ (1,3,5-tribromobenzene)

12.6 $CHClBr(CH_3)$

12.7 IF_5

12.8 C_6H_{12} (cyclohexane)

12.9 B_2H_6

12.10 $C_{10}H_8$ (naphthalene)

(planar)

12.11 C_5H_8 (spiropentane)

(triangles $\perp$ to each other)

12.12 C_4H_4S (thiophene)

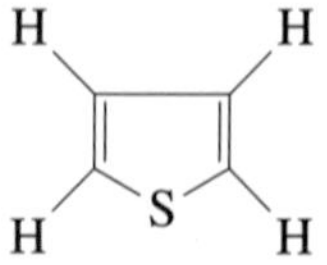

(planar)

12.13 $C_6H_4Cl_2$ (*p*-dichlorobenzene)

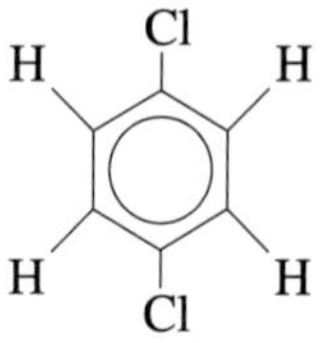

12.14 *trans*-CFClBrCFClBr

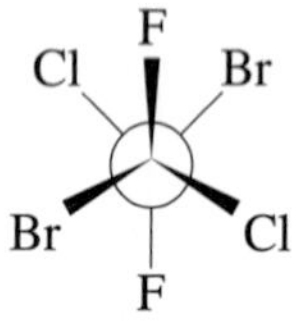

12.15 The symmetry elements for the staggered form of ethane are given in Table 12.3, and it is in the D_{3d} point group. What are the elements for the eclipsed form of ethane (this is the sterically hindered form), and what is the point group?

12.16 Construct the operation multiplication table for the point group C_{2h}.

12.17 List the operations associated with the S_6 elements and their equivalents, if any. How many distinct operations are produced?

12.18 Consider the three distinct isomers of dichloroethylene, $C_2H_2Cl_2$. To which symmetry group does each belong? Which can have a permanent dipole moment?

12.19 The first excited singlet state of ethylene is twisted so that the two hydrogens and carbon on one side are in a plane perpendicular to the plane containing the other three atoms. To which symmetry group does it belong? Does it have a dipole moment?

12.20 Which of the molecules in Problems 12.1–12.14 can have a permanent dipole moment?

12.21 Which of the molecules in Problems 12.1–12.14 can be optically active?

12.22 Consider the three distinct isomers of dichlorobenzene. To which symmetry group does each belong? Which can have a dipole moment?

12.23 There are 10 distinct isomers of dichloronaphthalene, $C_{10}H_6Cl_2$. Two of them do not have a dipole moment. List these two and find the symmetry group to which each belongs.

12.24 Some of the excited electronic states of acetylene are cis-bent and some are trans-bent. What is the symmetry group of these structures? (Cis-bent means that the hydrogens bend toward one another, while trans-bent means they bend away from one another.)

For the molecules in problems 12.25–12.39, give the Schoenflies symbol and symmetry elements.

12.25 C_8H_8

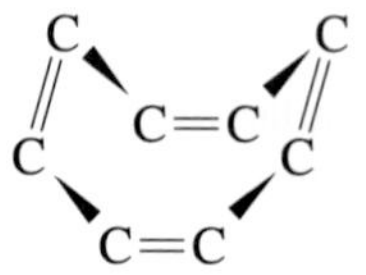

12.26 HCOOH (formic acid)

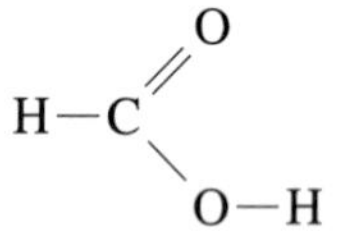

(planar)

12.27 $UO_2F_5^{3-}$

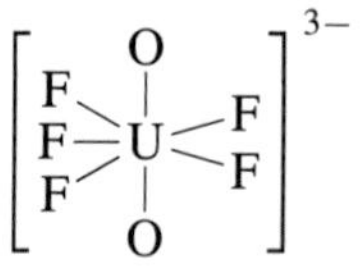

12.28 $C_{14}H_{10}$ (phenanthrene)

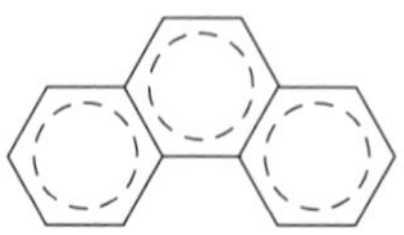

12.29 $Fe(CN)_6^{3-}$

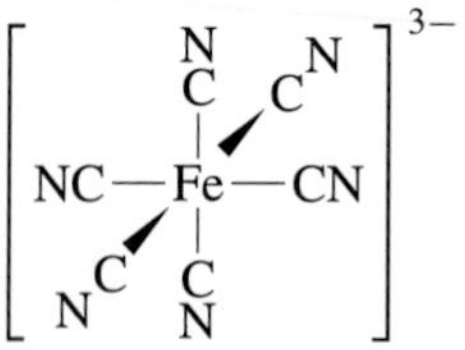

12.30 $(HNBCl)_3$

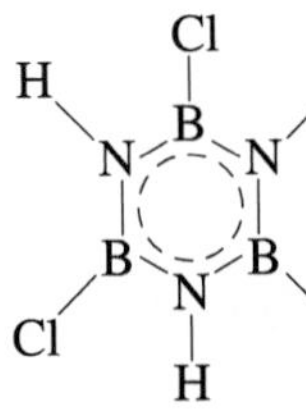

(planar)

12.31 $C_{10}H_{16}$ (adamantane)

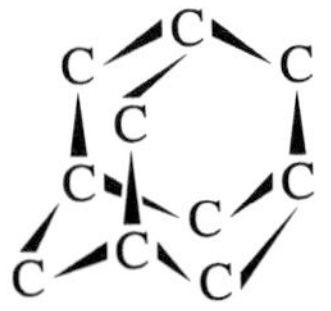

12.32 $C_6H_2O_2Cl_2$ (2,5-dichloroquinone)

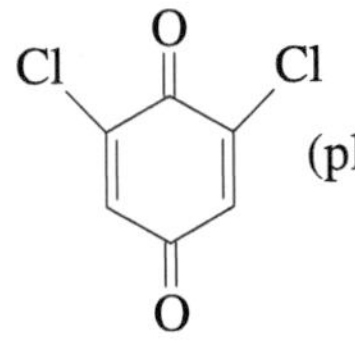

(planar)

12.33 HOCl

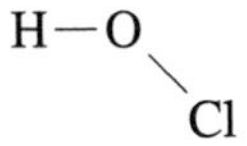

12.34 $(HNBH)_3$

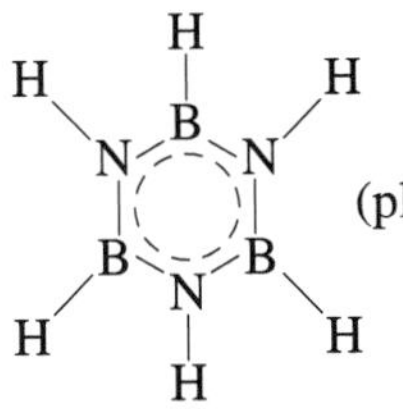

(planar)

12.35 $C_6H_3(C_6H_5)_3$ (1,3,5-triphenylbenzene)

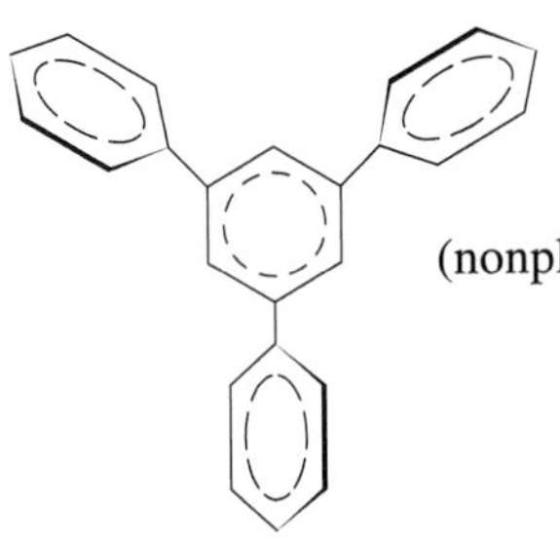

(nonplanar)

12.36 CH_3Cl (methyl chloride)

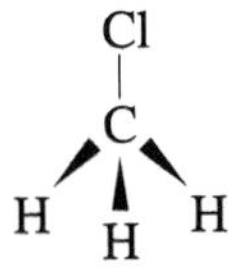

12.37 $Ni(CO)_4$

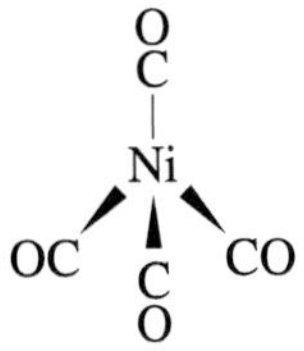

12.38 H_3CCH_2Br

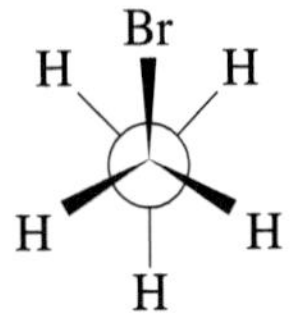

12.39 $Pt(Br)_4(NH_3)_2$ [tetrabromodiammineplatinum(IV)] (ignore the H's)

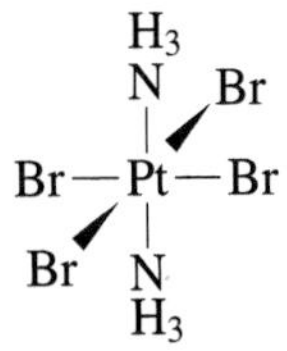

12.40 Construct the operation multiplication table for the point group C_{3v}.

12.41 In some of the excited states of benzene, the molecule is "stretched" so that the hexagon is elongated. What is the symmetry group of the molecule in such a state?

12.42 What is the symmetry group of HD? Can it have a dipole moment?

Computer Problems

12.A Show that the matrix product of the C_3 operation and the C_3^2 operation is equal to the identity operation E. A matrix for the C_3 operation is given in Example 12.5.

12.B The multiplication table for the group C_{2v} is given in Table 12.2, and the application of these operations to the water molecule is discussed in connection with this table. Since matrices for these operations are given in equations 12.6–12.9, verify that $\sigma_v'\sigma_v = C_2$, $\sigma_v C_2 = \sigma_v'$, and $C_2C_2 = E$, where these symbols refer to operations.

12.C Use *Mathematica* to show the shapes of a tetrahedron, an octahedron, a dodecahedron, an icosohedron, and a bucky ball.

12.D The following matrix transforms a point (x, y, z) through a cunterclocksise rotation about the z axis through an angle θ:

Apply this matrix to a point at (1, 1, 1) for the angles 0, $\pi/4$, $\pi/2$, $3\pi/4$, $3\pi/2$, and 2π.

13

Rotational and Vibrational Spectroscopy

Molecular spectroscopy is a powerful tool for learning about molecular structure and molecular energy levels. The study of rotational spectra gives us information about moments of inertia, interatomic distances, and angles. Vibrational spectra yield fundamental vibrational frequencies and force constants. Electronic spectra yield electronic energy levels and dissociation energies.

The types of spectroscopic transitions that can occur are limited by selection rules. As in the case of atoms, the principal interactions of molecules with electromagnetic radiation are of the electric dipole type, and so we will concentrate on them. Magnetic dipole transitions are about 10^5 times weaker than electric dipole transitions, and electric quadrupole transitions are about 10^8 times weaker. Although the selection rules limit the radiative transitions that can occur, molecular collisions can cause many additional kinds of transitions. Because of molecular collisions the populations of the various molecular energy levels are in thermal equilibrium.

13.1 THE BASIC IDEAS OF SPECTROSCOPY

When an isolated molecule undergoes a transition from one quantum eigenstate with energy E_1 to another with energy E_2, energy is conserved by the emission or absorption of a photon. The frequency ν of the photon is related to the difference in energies of the two states by Bohr's relation,

$$h\nu \equiv hc\tilde{\nu} = |E_1 - E_2| \tag{13.1}$$

where we have used the symbol $\tilde{\nu}$ ($= 1/\lambda$) introduced in Chapter 9 for the transition energy in **wave numbers** (SI unit m^{-1}, but usually cm^{-1} is used). The wave number $\tilde{\nu}$ is the number of waves per unit length. If $E_1 > E_2$, the process is photon emission; if $E_1 < E_2$, the process is photon absorption. The frequency range of photons, or the electromagnetic spectrum, is classified into different regions according to custom and experimental methods as outlined in Table 13.1. By measuring the frequency of the photon, we can learn about the eigenstates of the molecule being studied. This is called molecular spectroscopy.

The frequency of the photon in the absorption or emission process often tells us the kinds of molecular transitions that are involved. In the radio-frequency region (very low energy), transitions between nuclear spin states can occur (see Chapter 15). In the microwave region, transitions between electron spin states in molecules with unpaired electrons (Chapter 15) and, in addition, transitions between rotational states can take place. In the infrared region, transitions between vibrational states take place (with and without transitions between rotational states). In the visible and ultraviolet regions, the transitions occur between electronic states (accompanied by vibrational and rotational changes). Finally, in the far ultraviolet and X-ray regions, transitions occur that can ionize or dissociate molecules.

Table 13.1 Regions of the Electromagnetic Spectrum

	Wavelength in Vacuo, λ_0	*Wave Number in Vacuo,* $\tilde{\nu}$	*Frequency,* ν	*Photon Energy,* $h\nu$	*Molar Energy,* $N_A h\nu$
γ rays	10 pm	10^9 cm^{-1}	30.0 EHz	19.9×10^{-15} J	12.0 GJ/mol
X-rays	10 nm	10^6 cm^{-1}	30.0 PHz	19.9×10^{-18} J	12.0 MJ/mol
Vacuum UV	200 nm	50.0×10^3 cm^{-1}	1.50 PHz	993×10^{-21} J	598 kJ/mol
Near UV	380 nm	26.3×10^3 cm^{-1}	789 THz	523×10^{-21} J	315 kJ/mol
Visible	780 nm	12.8×10^3 cm^{-1}	384 THz	255×10^{-21} J	153 kJ/mol
Near IR	2.5 μm	4.00×10^3 cm^{-1}	120 THz	79.5×10^{-21} J	47.9 kJ/mol
Mid IR	50 μm	200 cm^{-1}	6.00 THz	3.98×10^{-21} J	2.40 kJ/mol
Far IR	1 mm	10 cm^{-1}	300 GHz	199×10^{-24} J	120 J/mol
Microwaves	100 mm	0.1 cm^{-1}	3.00 GHz	1.99×10^{-24} J	12.0 J/mol
Radio waves	1000 mm	0.01 cm^{-1}	300 MHz	0.199×10^{-24} J	1.2 J/mol

IR, infrared; UV, ultraviolet. The abbreviations for powers of 10 are given inside the back cover of the book. *Source:* IUPAC Report, "Names, Symbols, Definitions, and Units for Quantities in Optical Spectroscopy," 1984.

Example 13.1 ***Calculation of the energy of light***

Calculate the energy in joules per quantum, electron volts, and joules per mole of photons of wavelength 300 nm.

$$h\nu = \frac{hc}{\lambda} = \frac{(6.62 \times 10^{-34}\ \mathrm{J\ s})(3 \times 10^{8}\ \mathrm{m\ s^{-1}})}{(300 \times 10^{-9}\ \mathrm{m})} = 6.62 \times 10^{-19}\ \mathrm{J}$$
$$= (6.62 \times 10^{-19}\ \mathrm{J})/(1.602 \times 10^{-19}\ \mathrm{J\,eV^{-1}}) = 4.13\ \mathrm{eV}$$
$$N_A h\nu = (6.02 \times 10^{23}\ \mathrm{mol^{-1}})(6.62 \times 10^{-19}\ \mathrm{J}) = 398\ \mathrm{kJ\ mol^{-1}}$$

We shall see that the energy eigenvalues of a molecule can be written as

$$E = E_r + E_v + E_e \tag{13.2}$$

where E_r is the rotational energy, E_v the vibrational energy, and E_e the electronic energy. When the molecule undergoes a transition to another state with the emission or absorption of a single photon of frequency ν, then

$$h\nu = (E_r' - E_r'') + (E_v' - E_v'') + (E_e' - E_e'') \tag{13.3}$$

The primes refer to the state of higher energy and the double primes to the state of lower energy.

The classification of the various regions of the electromagnetic spectrum by the type of transition given above is possible because, in general,

$$E_r' - E_r'' \ll E_v' - E_v'' \ll E_e' - E_e'' \tag{13.4}$$

That is, electronic energy level differences are much greater than vibrational energy level differences, which are much greater than rotational energy level differences. Electronic transitions are often in the visible and ultraviolet part of the spectrum; vibrational transitions are in the infrared, and rotational transitions are in the far infrared and microwave regions.

13.2 EINSTEIN COEFFICIENTS AND SELECTION RULES

The spectrum of a molecule consists of a series of lines at the frequencies corresponding to all the possible transitions. Let us consider the transition from state 1 to state 2. The strength or intensity of a spectral line depends on the number of molecules per unit volume N_i that were in the initial state (the population density of that state) and the probability that the transition will take place. Einstein postulated that the rate of absorption of photons is proportional to the density of the electromagnetic radiation with the right frequency. The **radiant energy density** ρ is the radiant energy per unit volume, so it is expressed in J m^{-3}. (See Section 9.16.) The **spectral radiant energy density as a function of frequency** ρ_ν is the measure of the radiant energy of a particular frequency; it is given by

$$\rho_\nu = \mathrm{d}\rho/\mathrm{d}\nu \tag{13.5}$$

Thus, ρ_ν is expressed in J s m^{-3}. The energy density at the frequency required to excite atoms or molecules from E_1 to E_2 is represented by $\rho_\nu(\nu_{12})$. Thus Einstein's postulate about the **rate of absorption** of photons is summarized by the rate equation

$$\left(\frac{\mathrm{d}N_1}{\mathrm{d}t}\right)_{\mathrm{abs}} = -B_{12}\rho_\nu(\nu_{12})N_1 \tag{13.6}$$

where B_{12} is the **Einstein coefficient for stimulated absorption.** The SI unit for B_{12} is m kg^{-1}. (Note that N_1 can be taken as dimensionless or expressed in m^{-3}.) There is a minus sign because N_1 decreases when electromagnetic radiation is absorbed. Note that $\mathrm{d}N_1/\mathrm{d}t = -\mathrm{d}N_2/\mathrm{d}t$.

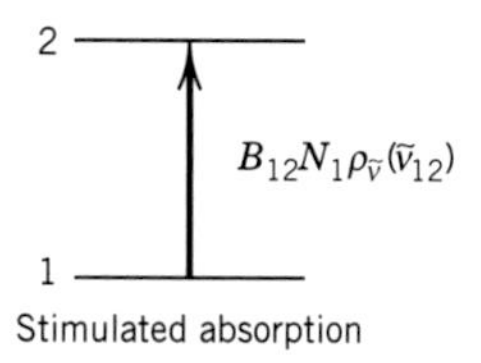

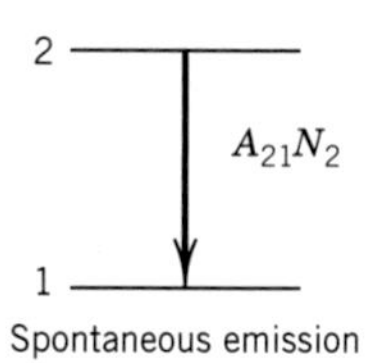

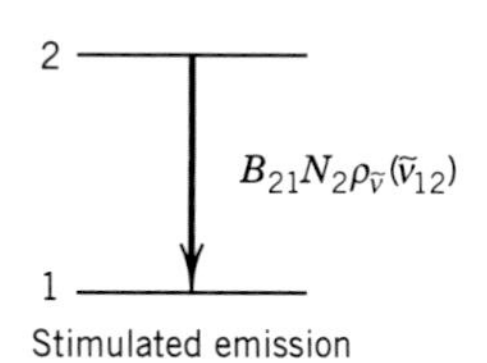

Figure 13.1 Definition of Einstein coefficients.

Excited atoms or molecules do not remain in excited states indefinitely, and Einstein postulated two processes for their return to the initial state, namely, spontaneous emission and stimulated emission, as illustrated in Fig. 13.1. The **rate of spontaneous emission** is given by (here N_2 is the population density of state 2)

$$\left(\frac{\mathrm{d}N_2}{\mathrm{d}t}\right)_{\text{spont}} = -A_{21}N_2 \tag{13.7}$$

where A_{21} is the **Einstein coefficient for spontaneous emission.** The SI unit for A_{21} is s^{-1}. The rate of spontaneous emission is independent of the radiation density, and the radiation is emitted in random directions with random phases.

Stimulated emission is quite different in that its rate is proportional to $\rho_\nu(\nu_{12})$, and the electromagnetic wave that is produced adds in phase and direction (i.e., coherently) to the stimulating wave. The **rate of stimulated emission** is indicated by the rate equation

$$\left(\frac{\mathrm{d}N_2}{\mathrm{d}t}\right)_{\text{stim}} = -B_{21}\rho_\nu(\nu_{12})N_2 \tag{13.8}$$

where B_{21} is the **Einstein coefficient for stimulated emission.** The interesting feature in stimulated emission is that it amplifies the radiation density. According to equation 13.8, incident light with frequency ν_{12} causes more radiation to be produced with exactly the same frequency and direction as long as there are molecules in state 2. As we will discuss later in more detail, this is the basis for a **laser,** which is the acronym for "light amplification by stimulated emission of radiation."

Equations 13.6–13.8 have been written for the three separate processes, but of course all three can occur in a system at the same time so that the whole rate equation is

$$\frac{\mathrm{d}N_1}{\mathrm{d}t} = -\frac{\mathrm{d}N_2}{\mathrm{d}t} = -B_{12}\rho_\nu(\nu_{12})N_1 + A_{21}N_2 + B_{21}\rho_\nu(\nu_{12})N_2 \tag{13.9}$$

This rate equation leads to several interesting conclusions. The first is that the three Einstein coefficients are related to each other. This can be seen by considering the equilibrium situation in which $\mathrm{d}N_1/\mathrm{d}t = -\mathrm{d}N_2/\mathrm{d}t = 0$. When the system is in equilibrium, equation 13.9 can be solved for the equilibrium spectral radiant energy density $\rho_\nu(\nu_{12})$ to obtain

$$\rho_\nu(\nu_{12}) = \frac{A_{21}}{(N_1/N_2)B_{12} - B_{21}} \tag{13.10}$$

When the system is in equilibrium, the ratio N_1/N_2 is given by the Boltzmann distribution (Section 16.1). When E_2 is the energy of the higher level and E_1 is the energy of the lower level, the **Boltzmann distribution** shows that

$$N_2 = N_1\,\mathrm{e}^{-(E_2-E_1)/kT} \tag{13.11}$$

Since $E_2 - E_1$ is positive, most of the atoms or molecules will be in the lower energy level at thermal equilibrium. If the system is exposed to electromagnetic radiation with frequency ν_{12}, where $h\nu_{12} = E_2 - E_1$, the equilibrium distribution can be written as

$$\frac{N_2}{N_1} = \exp(-h\nu_{12}/k_\mathrm{B}T) \tag{13.12}$$

Replacing N_1/N_2 in equation 13.10 with the Boltzmann distribution yields

$$\rho_\nu(\nu_{12}) = \frac{A_{21}}{B_{12}\, e^{h\nu_{12}/k_B T} - B_{21}} \tag{13.13}$$

This equation must be in agreement with **Planck's blackbody distribution law** (equation 9.2),

$$\rho_\nu(\nu_{12}) = \frac{8\pi h(\nu_{12}/c)^3}{e^{h\nu_{12}/kT} - 1} \tag{13.14}$$

because they both apply to a system at equilibrium. Comparison of equation 13.13 with equation 13.14 indicates that

$$B_{12} = B_{21} \tag{13.15}$$

and

$$A_{21} = \frac{8h\pi\nu_{12}^3}{c^3} B_{21} \tag{13.16}$$

Thus a measurement of any one of the three Einstein coefficients yields all three.

The second conclusion from equation 13.9 is that the time course of the irradiation can be calculated. Since $B_{12} = B_{21}$, these symbols can be replaced by B, and since there is no A_{12}, A_{21} can be replaced by A. N_1 can be replaced by $N_{\text{total}} - N_2$, where $N_{\text{total}} = N_1 + N_2$, and equation 13.9 can be integrated (see Problem 13.4) to obtain

$$\frac{N_2}{N_{\text{total}}} = \frac{B\rho_\nu(\nu_{12})}{A + 2B\rho_\nu(\nu_{12})}\left(1 - \exp\{-[A + 2B\rho_\nu(\nu_{12})]t\}\right) \tag{13.17}$$

At $t = 0$, there are no excited atoms or molecules. But if the radiation density is held constant, N_2/N_{total} rises to an asymptotic value of $B\rho_\nu(\nu_{12})/[A + 2B\rho_\nu(\nu_{12})]$. The interesting thing about this asymptotic value is that it is necessarily less than 1/2 because $A > 0$. **This means that irradiation of a two-level system can never put more atoms or molecules in the higher level than in the lower level.** This may be a surprise, but the significance of the conclusion is that laser action cannot be achieved with a two-level system. In order to obtain laser action, stimulated emission must be greater than the rate of absorption so that amplification of radiation of a particular frequency is obtained. This requires that

$$B_{21}\rho_\nu(\nu_{12})N_2 > B_{12}\rho_\nu(\nu_{12})N_1 \tag{13.18}$$

Since $B_{12} = B_{21}$, laser action can be obtained only when $N_2 > N_1$. This situation is referred to as a **population inversion.** The way population inversion can be achieved is discussed in the next chapter.

Quantum mechanics provides the means to calculate A_{nm} (and B_{nm}) between states n and m in terms of the transition dipole moment. A_{nm} (and B_{nm}) is proportional to the square of the **transition dipole moment** $\boldsymbol{\mu}_{nm}$, defined by

$$\boldsymbol{\mu}_{nm} = \int \psi_n^* \hat{\boldsymbol{\mu}} \psi_m \, d\tau \tag{13.19}$$

where $\hat{\boldsymbol{\mu}}$ is the **quantum mechanical dipole moment operator** for the molecule:

$$\hat{\boldsymbol{\mu}} = \sum_i q_i \boldsymbol{r}_i \tag{13.20}$$

where the sum is over all the electrons and nuclei of the molecule, q_i is the charge, and $\boldsymbol{r}_i$ is the position of the ith charged particle. To understand how the transition moment enters, we can think of the molecule interacting with the electric field of the radiation because of a transient or fluctuating dipole moment given by equation 13.19.

From equation 13.19, we see that if the transition dipole moment vanishes (usually because of symmetry), the spectral line has no intensity. The rules governing the nonvanishing of $\boldsymbol{\mu}_{nm}$ are called **selection rules,** and these allow us to make sense out of observed molecular spectra.

If the transition moment from state n to state m is nonzero and there is enough population in the initial state, then the spectral line will be seen in the spectrum. The quantum mechanical derivation of the relationship between the Einstein coefficients and the transition probability is too advanced for this book;* however, the final results are given here. When the ground state and excited states have degeneracies of g_1 and g_2, the Einstein coefficient A is given by

$$A = \frac{16\pi^3 \nu^3 g_1}{3\epsilon_0 h c^3 g_2}|\mu_{12}|^2 \tag{13.21}$$

This equation indicates that the rate of spontaneous emission, $A_{12}N_2$, increases rapidly with frequency; as a matter of fact, this rate is negligible in the microwave and infrared regions, and so only absorption spectra are measured. In the visible and ultraviolet regions spontaneous emission is significant, and both emission and absorption spectra are measured. The Einstein coefficient B is given by

$$B = \frac{2\pi^2 g_1}{3h^2 \epsilon_0 g_2}|\mu_{12}|^2 \tag{13.22}$$

If the rate of spontaneous emission is negligible, the net rate of absorption is given by

$$\text{rate}_{2\leftarrow 1} = B_{21}N_1\rho_{\tilde{\nu}}(\tilde{\nu}_{21}) - B_{12}N_2\rho_{\tilde{\nu}}(\tilde{\nu}_{21}) = (N_1 - N_2)B\rho_{\tilde{\nu}}(\tilde{\nu}_{21}) \tag{13.23}$$

This shows that if the populations of the two states are equal, there will be no net absorption of radiation.

We can also think of A_{12} as a measure of the lifetime of state 2. Consider molecules in (excited) state 2 with no radiation field present (and so no stimulated emission). The molecules will make a transition to state 1, emitting a photon frequency $\tilde{\nu}_{21}$, with a probability $A_{12}N_2$. Every time this occurs, N_2 decreases. After a time t, the number of molecules per unit volume in state 2 is given by

$$N_2(t) = N_2(0)\,\mathrm{e}^{-A_{12}t} = N_2(0)\,\mathrm{e}^{-t/\tau} \tag{13.24}$$

where we have defined the **lifetime** $\tau = A_{12}^{-1}$. Actually, if a molecule in state 2 can also make transitions to states 3, 4, ... (with photons of frequency $\tilde{\nu}_{23}, \tilde{\nu}_{24}, \ldots$), then the **total radiative lifetime** is given by

$$\frac{1}{\tau} = \sum_i A_{2i} \tag{13.25}$$

If other decay processes besides radiative transitions are possible (such as nonradiative transitions) we must add those rates to equation 13.25 to get the total decay rate (inverse lifetime).

*See J. Steinfeld, *Molecules and Radiation*. Cambridge, MA: MIT Press, 1985.

Example 13.2 ***Radiative lifetimes and transition moments***

The radiative lifetime of a hydrogen atom in its first excited level (2p) is 1.6×10^{-9} s. What is the magnitude of the electronic transition moment μ_{21} for this transition? The degeneracy g_2 of the 2p level is 3. [$\tilde{\nu} = (2.46 \times 10^{15}\ \text{s}^{-1})/(2.998 \times 10^8\ \text{m s}^{-1}) = 8.21 \times 10^6\ \text{m}^{-1}$.]

$$\mu_{21} = \left[\frac{3h\epsilon_0 g_2}{16\pi^3 \tilde{\nu}_{21}^3 \tau}\right]^{1/2}$$

$$= \left[\frac{(3)(6.626 \times 10^{-34}\ \text{J s})(8.854 \times 10^{-12}\ \text{C}^2\ \text{N}^{-1}\ \text{m}^{-2})(3)}{16\pi^3(8.21 \times 10^6\ \text{m}^{-1})^3(1.6 \times 10^{-9}\ \text{s})}\right]^{1/2}$$

$$= 10.9 \times 10^{-30}\ \text{C m}$$

A dipole moment of this magnitude corresponds to a distance from the proton to the electron of

$$r = \frac{10.9 \times 10^{-30}\ \text{C m}}{1.6 \times 10^{-19}\ \text{C}}$$

$$= 68.1\ \text{pm}$$

This transition dipole moment can be visualized as the movement of an electron 68.1 pm/52.9 pm = 1.29 Bohr radii.

13.3 SCHRÖDINGER EQUATION FOR NUCLEAR MOTION

We saw in Chapter 11 that the Schrödinger equation for a molecule can be treated in the Born–Oppenheimer approximation so that the **electronic Hamiltonian** is that for fixed nuclei, while the **Hamiltonian for nuclear motion** contains the kinetic energy operator of the nuclei and the electronic energy (as a function of the nuclear coordinates) as the potential energy operator:

$$\hat{H} = -\frac{\hbar^2}{2\mu}\nabla_{\boldsymbol{R}}^2 + E(\boldsymbol{R}) \tag{13.26}$$

In the absence of external fields (such as magnetic or electric fields), the potential energy term $E(\boldsymbol{R})$ can depend only on the relative positions of the nuclei, not on where the molecule is placed or on the orientation of the molecule in space.

The kinetic energy operator consists of the kinetic energy of the center of mass (leading to the translational energy of the molecule), the kinetic energy associated with rotational motion, and the kinetic energy of the vibrational motion. Thus, to a very good approximation, we may write

$$H = H_{\text{tr}} + H_{\text{rot}} + H_{\text{vib}} \tag{13.27}$$

where the translational and rotational Hamiltonians contain only kinetic energy terms, while the vibrational Hamiltonian contains $E(R)$, the potential energy depending on the internuclear distances. These internuclear distances are the **vibrational** coordinates of the molecule.

If the Hamiltonian is the sum of three terms, one for each kind of motion, then the wavefunction ψ can be written as a product of wavefunctions:

$$\psi = \psi_{\text{tr}}\psi_{\text{rot}}\psi_{\text{vib}} \tag{13.28}$$

The Schrödinger equations for the three terms are

$$\hat{H}_{tr}\psi_{tr} = E_{tr}\psi_{tr} \tag{13.29}$$

$$\hat{H}_{rot}\psi_{rot} = E_{rot}\psi_{rot} \tag{13.30}$$

$$\hat{H}_{vib}\psi_{vib} = E_{vib}\psi_{vib} \tag{13.31}$$

The translational wavefunction is that for a free particle (or particle in a very large box) with a mass equal to the mass of the molecule. The translational eigenvalues are very closely spaced and cannot be probed in molecular spectroscopy, so we will neglect them in our discussions.

To understand the number of coordinates required to describe a polyatomic molecule, consider the following. The total number of coordinates needed to describe the locations of the N atoms in a molecule is $3N$. However, to describe the internal motions in a molecule, we are not interested in its location in space, and so the three coordinates required to specify the position of the center of mass of the molecule can be subtracted, leaving $3N - 3$ coordinates. To describe the rotational motions of a molecule, we are interested in its orientation in a coordinate system. The orientation of a diatomic or linear molecule with respect to a coordinate system requires two angles, so this leaves $3N - 5$ coordinates to describe the internal motions. The orientation of a nonlinear polyatomic molecule with respect to a coordinate system requires three angles, so this leaves $3N - 6$ coordinates to describe the internal motions. These $3N - 5$ or $3N - 6$ internal motions are referred to as **vibrational degrees of freedom.**

To sum up, for a diatomic molecule, $\hat{H}_{rot}$ depends only on two angles, θ and ϕ (see equation 9.153); $\hat{H}_{vib}$ depends only on R, the internuclear separation. For polyatomic molecules, $\hat{H}_{vib}$ is more complex, depending on $3N - 6$ coordinates for nonlinear molecules and $3N - 5$ coordinates for linear molecules. We will now turn to a description of the rotational and vibrational eigenstates of both diatomic and polyatomic molecules.

13.4 ROTATIONAL SPECTRA OF DIATOMIC MOLECULES

To a first approximation the rotational spectrum of a diatomic molecule may be understood in terms of the Schrödinger equation for rotational motion of the rigid rotor (equation 9.142). The wavefunctions are the spherical harmonics $Y_J^M(\theta,\phi)$, and there are two quantum numbers J and M for molecular rotation. The energy eigenvalues are given by

$$E_r = \frac{\hbar^2}{2I}J(J+1) \qquad J = 0, 1, 2, \ldots$$
$$M = -J, \ldots, 0, \ldots, +J \tag{13.32}$$

where I is the moment of inertia (Section 9.11). Since the energy does not depend on M, the rotational levels are $(2J + 1)$-fold degenerate.

In spectroscopy it is standard to express the energies of various levels in wave numbers by dividing E by hc and referring to these values as **term values.** Term values are usually given in cm^{-1}, but the SI unit for a term value is m^{-1}. A tilde will be used to indicate the wave numbers in cm^{-1}. Rotational term values are represented by $\tilde{F}(J) = E_r/hc$, so that the rotational term values for a diatomic molecule are given by

$$\tilde{F}(J) = \frac{E_r}{hc} = \frac{J(J+1)h}{8\pi^2 Ic} = J(J+1)\tilde{B} \tag{13.33}$$

where the **rotational constant** is written

$$\tilde{B} = \frac{h}{8\pi^2 Ic} \tag{13.34}$$

where c is the speed of light, 2.998×10^{10} cm s^{-1}. The rotational energy levels for a rigid diatomic molecule are given in Fig. 13.2 in terms of the rotational constant.

According to the Born–Oppenheimer approximation (Section 11.1), the wavefunction for a molecule in the electronic state ψ_e, the vibrational state ψ_v, and having a particular set of rotational quantum numbers JM can be written as a product $\psi_e \psi_v \psi_{JM}$. The transition moment for an electric dipole transition from a rotational state JM to a rotational state $J'M'$ of the same electronic state is therefore given by

$$\int\int\int \psi_e^* \psi_v^* \psi_{J'M'}^* \hat{\boldsymbol{\mu}} \psi_e \psi_v \psi_{JM} \, d\tau_e \, d\tau_{rot} \, d\tau_{vib} \tag{13.35}$$

where $\hat{\boldsymbol{\mu}}$ is the dipole moment operator. Note that only the rotational function has changed in the transition. The permanent dipole moment $\boldsymbol{\mu}_0^{(e)}$ of a molecule in this electronic state is equal to the expectation value of the operator $\boldsymbol{\mu}$ over the wavefunction for the electronic state:

$$\boldsymbol{\mu}_0^{(e)} = \int \psi_e^* \hat{\boldsymbol{\mu}} \psi_e \, d\tau_e \tag{13.36}$$

Thus, equation 13.35 becomes

$$\int\int \psi_v^* \psi_{J'M'}^* \boldsymbol{\mu}_0^{(e)} \psi_v \psi_{JM} \, d\tau_{rot} \, d\tau_{vib} \tag{13.37}$$

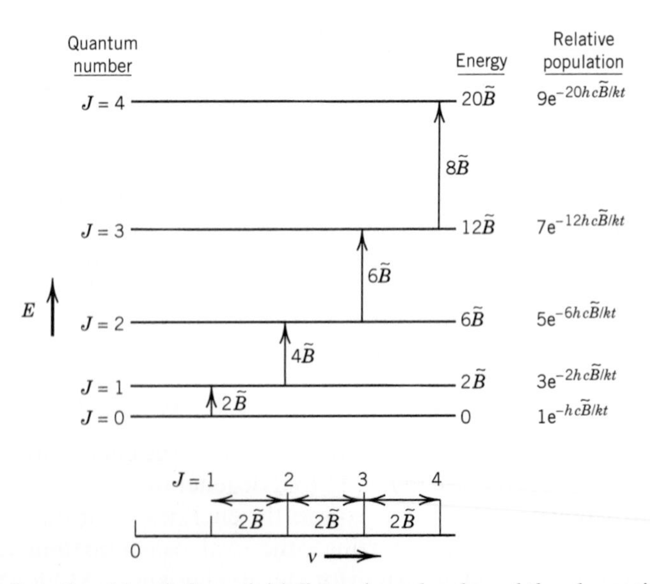

Figure 13.2 Rotational levels for a rigid diatomic molecule and the absorption spectrum that results from $\Delta J = 1$. The energies and relative populations of the two levels are indicated on the right. The transitions are labeled by the upper of the two J values involved. Note that the degeneracies of the levels have been taken into account in the population, and the intensities of the lines depend on the relative populations.

The integral over the vibrational coordinate yields the permanent dipole moment in that particular vibrational state. For simplicity, we will write it as $\boldsymbol{\mu}_0$, so that the final result for the integral is

$$\int \psi^*_{J'M'} \boldsymbol{\mu}_0 \psi_{JM} \, \mathrm{d}\tau_{\mathrm{rot}} \tag{13.38}$$

A molecule has a rotational spectrum only if this integral is nonzero. *Thus, the gross selection rule for rotational spectra is that a molecule must have a permanent dipole moment to emit or absorb radiation in making a transition between different states of rotation.* This is expected from the fact that a rotating dipole produces an oscillating electric field that can interact with the oscillating field of a light wave. A homonuclear diatomic molecule such as H_2 or O_2 does not have a dipole moment, so it does not show a pure rotational spectrum. Heteronuclear diatomic molecules do have dipole moments, so they do have rotational spectra. Polyatomic molecules are discussed in the next section. To find the specific selection rules we need to find the conditions on the quantum numbers that make the integral in equation 13.38 nonzero. For a linear molecule it can be shown that the transition moment is nonzero for

$$\Delta J = \pm 1 \qquad \Delta M = 0, \pm 1$$

This selection rule may be understood in the same way as that for atoms (Section 10.14). Since a photon has one unit of angular momentum, and angular momentum must be conserved in emission or absorption, the angular momentum of a molecule must change by a compensating amount.

The frequencies $\tilde{\nu}$ of the absorption lines due to $J \rightarrow J + 1$ are given by the difference between rotational term values (equation 13.33):

$$\begin{aligned} \tilde{\nu} &= \tilde{F}(J+1) - \tilde{F}(J) \\ &= [(J+1)(J+2) - J(J+1)]\tilde{B} \qquad J = 0, 1, 2, \ldots \\ &= 2\tilde{B}(J+1) \end{aligned} \tag{13.39}$$

As shown in Fig. 13.2, the frequencies of the successive lines in the rotational spectrum are given by $2\tilde{B}, 4\tilde{B}, 6\tilde{B}, \ldots$. Thus, there is a series of equally spaced lines with separations of $2\tilde{B}$. A separate series of lines is found for each isotopically different species of a given molecule, because the moments of inertia of isotopically substituted molecules are different.

We have been talking about diatomic molecules as if they are rigid rotors, but of course they are not. As the rotational motion increases, the chemical bond stretches due to centrifugal forces, the moment of inertia increases, and, consequently, the rotational energy levels come closer together. This may be taken into account by adding a term to equation 13.33:

$$\tilde{F}(J) = \frac{E_{\mathrm{r}}}{hc} = \tilde{B}J(J+1) - \tilde{D}J^2(J+1)^2 \tag{13.40}$$

The quantity $\tilde{D}$ is the **centrifugal distortion constant** in wave numbers. When centrifugal distortion is taken into account, the frequencies $\tilde{\nu}$ of the absorption lines due to $J \rightarrow J + 1$ are given by

$$\begin{aligned} \tilde{\nu} &= \tilde{F}(J+1) - \tilde{F}(J) \\ &= 2\tilde{B}(J+1) - 4\tilde{D}(J+1)^3 \qquad J = 0, 1, 2, \ldots \end{aligned} \tag{13.41}$$

The moment of inertia of a diatomic molecule also depends on its vibrational state because of the anharmonicity of vibrational motion. Since molecules are generally in their ground vibrational state at room temperature, we do not have to take this into account in considering pure rotational spectra; however, we will have to take it into account by an extension of equation 13.41 in discussing vibration–rotation spectra.

Example 13.3 ***Internuclear distance from rotational spectra***

In early measurements of the pure rotational spectrum of $H^{35}Cl$, Czerny found that the wave numbers of absorption lines are given by

$$\tilde{\nu} = (20.794\ \text{cm}^{-1})(J+1) - (0.000\,164\ \text{cm}^{-1})(J+1)^3$$

where J is the quantum number of the lower state. What is the internuclear distance in $H^{35}Cl$? What is the value of the centrifugal distortion constant?

From equation 13.41, $\tilde{B} = 10.397\ \text{cm}^{-1}$. Since

$$\tilde{B} = \frac{h}{8\pi^2 cI} = \frac{h}{8\pi^2 c\mu R_0^2}$$

we have

$$R_0 = \sqrt{\frac{h}{8\pi^2 c\mu\tilde{B}}}$$

$$= \sqrt{\frac{6.626\times10^{-34}\ \text{J s}}{8\pi^2(2.998\times10^{10}\ \text{cm s}^{-1})(1.626\,68\times10^{-27}\ \text{kg})(10.397\ \text{cm}^{-1})}}$$

$$= 129\ \text{pm}$$

(The reduced mass of $H^{35}Cl$ is given in Example 9.21.) The centrifugal distortion constant is given by

$$\tilde{D} = \tfrac{1}{4}(0.000\,164\ \text{cm}^{-1}) = 4.1\times10^{-5}\ \text{cm}^{-1}$$

We have discussed the selection rules that determine the transitions that can give rise to absorption or emission, but we already noted that there is another factor that determines the observed intensities, namely, the population of the initial state given by the Boltzmann distribution (equations 13.11 and 16.2). The fraction f_i of the molecules in the ith energy state is given by

$$f_i = \frac{e^{-\epsilon_i/kT}}{\sum_i e^{-\epsilon_i/kT}} = \frac{e^{-\epsilon_i/kT}}{q} \tag{13.42}$$

where q is the denominator. If the energy of a state is large compared with kT, the probability of finding a molecule in that state at equilibrium will be small. Because of degeneracy (Section 9.7), many states of a molecule may have the same energy, and these degenerate states make up the energy **level.** When energy levels are used, the Boltzmann distribution can be written

$$f_i = \frac{g_i\, e^{-\epsilon_i/kT}}{\sum_i g_i\, e^{-\epsilon_i/kT}} \tag{13.43}$$

where g_i is the degeneracy (Section 9.7) of the ith level. As discussed earlier, the component of the angular momentum in a particular direction is equal to $M_J\hbar$, where M_J may have values of $J, (J-1), \ldots, 0, \ldots, -J$, where J is the rotational quantum number. Thus, there are in all $2J+1$ different possible states with quantum number J. In the absence of an external electric or magnetic field the energies are identical for these various sublevels, and so the Jth energy level is said to have a degeneracy of $2J+1$. The rotational energy in the absence of an external electric or magnetic field, ignoring $\tilde{D}$ in equation 13.41, is given by $\epsilon_i = hcBJ(J+1)$ so that, using equation 13.42, the fraction of molecules in the Jth rotational level is given by

$$f_J = \frac{(2J+1)\,\mathrm{e}^{-[hcJ(J+1)B]/kT}}{q} \tag{13.44}$$

According to this equation, the number of molecules in level J increases with J at low J values, goes through a maximum, and then, because of the exponential term, decreases as J is further increased. The lines in the spectrum at the bottom of Fig. 13.2 have been labeled with the rotational quantum number J of the upper of the two states involved. The intensities of the lines are proportional to the populations in the lower state involved in the transition.

For molecules with larger moments of inertia I, the rotational energies are smaller, in fact, small compared with kT. The quantum numbers may become quite large before $\mathrm{e}^{-\epsilon_J/kT}$ becomes appreciably different from unity. For small quantum numbers populations are proportional to the degeneracies, since $\mathrm{e}^{-\epsilon_J/kT} \approx 1$ for $\epsilon_J \ll kT$.

There is a complication in rotational spectroscopy that we will not be able to discuss. The statistics of nuclear spin affect the number of degenerate states at each J level, and therefore the intensities of the rotational lines. The use of the Boltzmann distribution alone is an oversimplification.*

Although homonuclear diatomic molecules do not have permanent electric dipole moments and do not exhibit pure rotational spectra, they do show rotational Raman spectra (Section 13.9), and their electronic and vibrational spectra show rotational fine structure.

13.5 ROTATIONAL SPECTRA OF POLYATOMIC MOLECULES

For the treatment of its pure rotational spectrum we may consider a polyatomic molecule to be a rigid framework with fixed bond lengths and angles equal to their mean values. For a polyatomic molecule the **moment of inertia** about a particular axis that passes through the center of mass of the molecule is simply the sum of the moments due to the various nuclei about that axis:

$$I = \sum_i m_i R_i^2 \tag{13.45}$$

where R_i is the perpendicular distance of the nucleus mass m_i from the axis.

The rotation of a polyatomic molecule can be described in terms of moments of inertia taken relative to three mutually perpendicular axes. The moment about the z axis is

$$I_z = \sum_i m_i (x_i^2 + y_i^2) \tag{13.46}$$

*See the references at the end of the chapter, such as Herzberg.

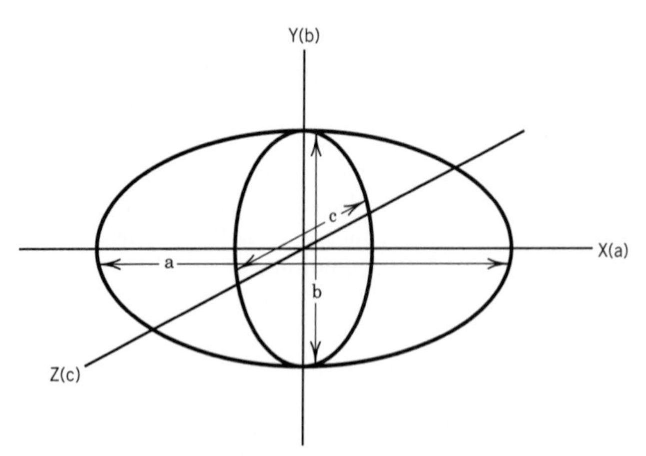

Figure 13.3 Momental ellipsoid with symmetry axes a, b, and c. The a, b, and c axes are fixed with respect to the molecule and rotate with it.

and I_x and I_y are defined similarly. In addition, there are three **products of inertia** that are defined like

$$I_{xy} = I_{yx} = \sum_i m_i x_i y_i \tag{13.47}$$

For any rigid molecule it is possible to choose a set of perpendicular axes that pass through the center of mass such that all products of inertia vanish. These three Cartesian axes, which are illustrated in Fig. 13.3, are called the **principal axes,** and the moments of inertia about these axes are called the **principal moments of inertia** I_a, I_b, and I_c. The axes are designated by a, b, and c and are fixed with respect to the molecule and rotate with it. The principal moments of inertia about these axes are always labeled so that $I_a \leq I_b \leq I_c$. The principal axes can often be assigned by inspecting the symmetry of the molecule. The momental ellipsoid is constructed as follows. Lines are drawn from the center of mass of the molecule in various directions with length proportional to $(I_\alpha)^{-1/2}$, where I_α is the moment of inertia about that line as an axis. Any symmetry operation of a molecule must apply to its momental ellipsoid.

The principal moments of inertia are used to classify molecules, as shown in Table 13.2. If all three principal moments of inertia are equal, the molecule is a **spherical** top. If two principal moments are equal, the molecule is a **symmetric** top. A molecule is a **prolate** top (cigar shaped) if the two larger moments are equal. The molecule is an **oblate** top (discus shaped) if the two smaller moments are equal. The molecule is an **asymmetric** top if all three principal moments are unequal.

The quantum mechanical Hamiltonian operator for the rotational motion of polyatomic molecules is found by first writing the classical mechanical energy in terms of angular momentum operators. Since we know how to convert classical angular momentum to its quantum mechanical form, we can then find the quantum Hamiltonian and solve the Schrödinger equation. The last part turns out to be straightforward for all the cases except the asymmetric top. We will not discuss the latter.

In classical mechanics the rotational energy of a rotor with one degree of freedom is

Table 13.2 Classification of Polyatomic Molecules According to Their Moments of Inertia

Moments of Inertia	*Type of Rotor*	*Examples*
$I_b = I_c, \quad I_a = 0$	Linear	HCN
$I_a = I_b = I_c$	Spherical top	CH_4, SF_6, UF_6
$I_a < I_b = I_c$	Prolate symmetric top	CH_3Cl
$I_a = I_b < I_c$	Oblate symmetric top	C_6H_6
$I_a \neq I_b \neq I_c$	Asymmetric top	CH_2Cl_2, H_2O

$$E_r = \tfrac{1}{2}I\omega^2 = \frac{(I\omega)^2}{2I} = \frac{L^2}{2I} \tag{13.48}$$

where ω is the angular velocity in radians per second, I is the moment of inertia, and L is the angular momentum. For an object that can rotate in three dimensions the classical expression for the rotational kinetic energy is

$$E_r = \tfrac{1}{2}I_{xx}\,\omega_x^2 + \tfrac{1}{2}I_{yy}\,\omega_y^2 + \tfrac{1}{2}I_{zz}\,\omega_z^2 \tag{13.49}$$

Since we will want to convert this to a quantum mechanical expression, it is more convenient to express it in terms of the angular momentum $L_q = I_{qq}\,\omega_q$, where q represents a direction,

$$E_r = \frac{L_x^2}{2I_{xx}} + \frac{L_y^2}{2I_{yy}} + \frac{L_z^2}{2I_{zz}} \tag{13.50}$$

in which the components of the total angular momentum about the three principal axes are given by

$$L_x = I_{xx}\,\omega_x \tag{13.51}$$

$$L_y = I_{yy}\,\omega_y \tag{13.52}$$

$$L_z = I_{zz}\,\omega_z \tag{13.53}$$

The total angular momentum is given by

$$L^2 = L_x^2 + L_y^2 + L_z^2 \tag{13.54}$$

The expressions for the energies of spherical tops, linear molecules, and symmetric tops are as follows.

Spherical Top

For a spherical top, $I_{xx} = I_{yy} = I_{zz} = I$, the momental ellipsoid is a sphere, and equation 13.48 becomes

$$E_r = \frac{(L_x^2 + L_y^2 + L_z^2)}{2I} = \frac{L^2}{2I} \tag{13.55}$$

where the second form has been obtained by introducing equation 13.54.

The quantum mechanical expression for the rotational energy is obtained by substituting the quantum mechanical expression for the eigenvalue of the square of the angular momentum, $J(J+1)\hbar^2$:

$$E = \frac{J(J+1)\hbar^2}{2I} \qquad J = 0, 1, 2, \ldots \tag{13.56}$$

However, spherical top molecules cannot have dipole moments for symmetry reasons. Only molecules belonging to point groups C_n, C_s, and C_{nv} can possess dipole moments. Therefore, spherical top molecules do not have pure rotational spectra. They do, however, have vibrational and electronic spectra with rotational fine structure. The moment of inertia for a symmetrical tetrahedral molecule, such as CH_4, is

$$I = \tfrac{8}{3}mR^2 \tag{13.57}$$

where R is the bond length and m is the mass of each of the four atoms arranged in a tetrahedral manner.

Linear Molecule

For a linear molecule, $I_{yy} = I_{xx}$ and $I_{zz} = 0$. Thus, L_z must be 0, and equation 13.48 becomes

$$E_r = \frac{L_y^2 + L_x^2}{2I_{xx}} = \frac{L^2}{2I_{xx}} \tag{13.58}$$

For a linear polyatomic molecule the equation for the rotational term $F(J)$ is the same as that given earlier for a diatomic molecule.

Symmetric Top

Examples of symmetric top molecules are NH_3, CH_3Cl, and the molecule shown later in Example 13.4. For these molecules $I_{xx} = I_{yy}$, but I_{zz} is different. We will use $I_\parallel$ for the moment of inertia parallel to the axis (I_{zz}) and $I_\perp$ for the moment perpendicular to the axis (I_{xx} and I_{yy}). Thus, the classical energy of rotation is

$$E_r = \frac{L_x^2 + L_y^2}{2I_\perp} + \frac{L_z^2}{2I_\parallel} \tag{13.59}$$

This can be written in terms of the magnitude of the angular momentum $L^2 = L_x^2 + L_y^2 + L_z^2$ as follows:

$$\begin{aligned} E_r &= \left(\frac{1}{2I_\perp}\right)(L_x^2 + L_y^2 + L_z^2) - \left(\frac{1}{2I_\perp}\right)L_z^2 + \left(\frac{1}{2I_\parallel}\right)L_z^2 \\ &= \left(\frac{1}{2I_\perp}\right)L^2 + \left[\left(\frac{1}{2I_\parallel}\right) - \left(\frac{1}{2I_\perp}\right)\right]L_z^2 \end{aligned} \tag{13.60}$$

The quantum mechanical expression for the energy is obtained by substituting $L^2 = J(J+1)\hbar^2$ (as we saw in connection with equation 9.162) and $L_z^2 = K^2\hbar^2$ (as we saw in connection with equation 9.164). This latter substitution comes from the fact that in quantum mechanics the component of angular momentum about any axis is restricted to the values of $K\hbar$, where $K = 0, \pm 1, \ldots, \pm J$:

$$E_r = \left(\frac{1}{2I_\perp}\right)J(J+1)\hbar^2 + \left[\left(\frac{1}{2I_\parallel}\right) - \left(\frac{1}{2I_\perp}\right)\right]K^2\hbar^2 \tag{13.61}$$

where $J = 0, 1, 2, \ldots$ and $K = 0, \pm 1, \pm 2, \ldots, \pm J$. This equation is generally used in the form

$$\frac{E_{JK}}{hc} = \tilde{B}J(J+1) + (\tilde{A} - \tilde{B})K^2 \tag{13.62}$$

where

$$\tilde{B} = \frac{\hbar}{4\pi c I_\perp} \quad \text{and} \quad \tilde{A} = \frac{\hbar}{4\pi c I_\parallel} \tag{13.63}$$

The quantum number K determines the component of the angular momentum along the axis of the symmetric top; this is the angular momentum of rotation about the symmetry axis. When $K = 0$ there is no rotation about the symmetry axis, and the rotation is about the axis perpendicular to the symmetry axis, that is, end-over-end rotation. When K has its maximum value ($+J$ or $-J$), most of the molecular rotation is about the symmetry axis (see Fig. 13.4).

The specific selection rules for rotational spectra of symmetric top molecules are $\Delta J = \pm 1$ and $\Delta K = 0$. The reason there cannot be a change in quantum number K is that the dipole vector of the molecule is oriented along the principal axis. The electromagnetic field of radiation can affect the rotation of the dipole, but it cannot affect the rotation of the molecule about its principal axis because there is no dipole moment perpendicular to the principal axis.

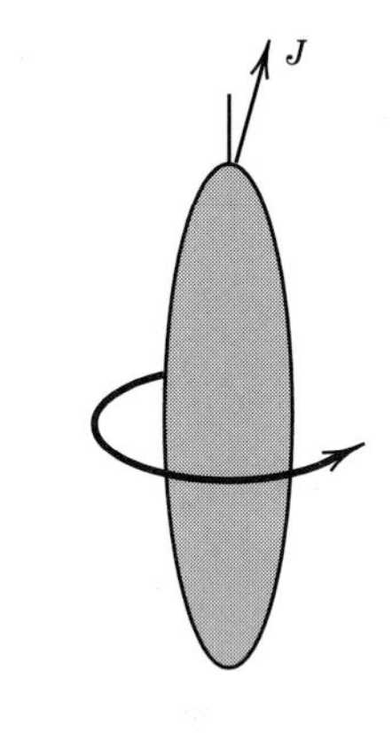

(a) $K \approx J$

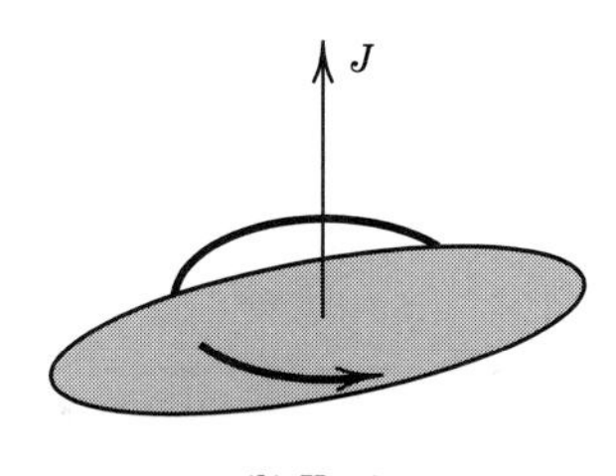

(b) $K = 0$

Figure 13.4 Meaning of the quantum number K.

Example 13.4 ***Moments of inertia of an octahedral symmetric top molecule***

Derive the expressions for the moments of inertia $I_\parallel$ and $I_\perp$ of the octahedral symmetric top molecule AB_2C_4 shown in the diagram.

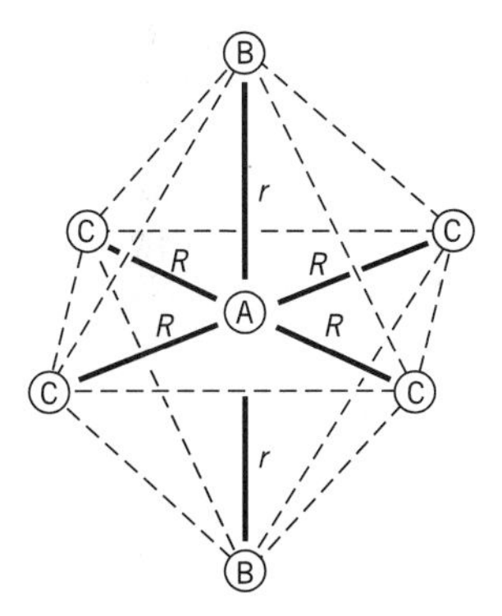

$$I_\parallel = 4m_C R^2$$

$$I_\perp = 2m_C R^2 + 2m_B r^2$$

The pure rotational spectroscopy of molecules has enabled the most precise evaluations of bond lengths and bond angles. The spectrum of a polyatomic molecule gives at most three principal moments of inertia; since usually more than three bond lengths and angles are involved, isotopically different molecules must be studied, and it must be assumed that isotopically different molecules have the same set of bond lengths and bond angles. In effect, a number of simultaneous equations are solved for the internuclear distances and angles.

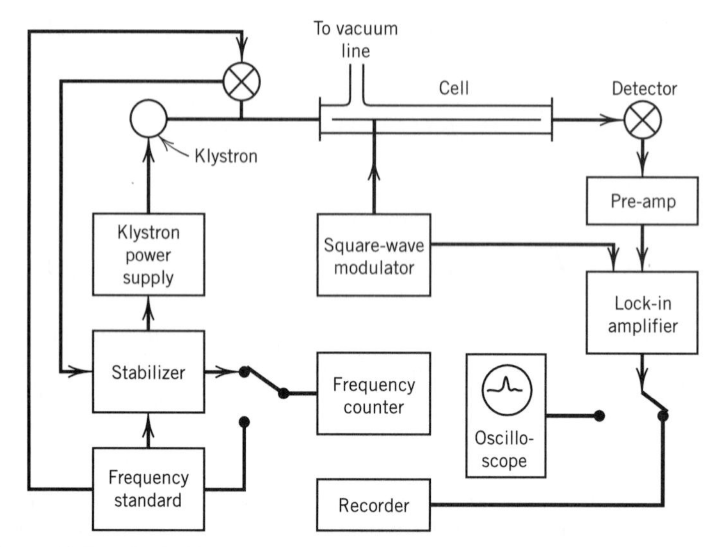

Figure 13.5 Block diagram of a Stark-modulated microwave spectrometer.

These spectra are in the microwave region. Microwave radiation is produced by special electronic oscillators called klystrons. Monochromatic radiation is produced, and the frequency may be varied continuously over wide ranges. The usual experimental arrangement is shown in Fig. 13.5. Microwave radiation is transmitted down in a waveguide that contains the gas being studied. The intensity of the radiation at the other end of the waveguide is measured by use of a crystal diode detector and amplifier. The oscillator frequency is swept over a range, and the transmitted intensity is presented on an oscilloscope or a recorder as a function of frequency.

According to the Heisenberg uncertainty principle, the accuracy with which an energy level may be determined is inversely proportional to the time the molecule is in this level. Hence, to obtain sharp rotational lines of a gas, the pressure must be maintained sufficiently low so that the average time between collisions is long compared with the period of a rotation. Usually it is necessary to determine microwave spectra at pressures below 10 Pa to reduce the line-broadening effects of collisions.

The lines in the microwave spectrum are split if the molecules being studied are in an electric field. This so-called Stark effect is due to the interaction of the dipole moment of the gaseous molecule and the electric field. Since the splitting is proportional to the permanent dipole moment, the magnitude of the dipole moment may be derived from the spectrum.

Comment:

Microwave spectroscopy of gases at low pressures can be used to determine rotational frequencies to one part per million since the lines are very sharp. Separate lines are obtained for molecules with different isotopic compositions. Since moments of inertia can be determined so accurately, bond lengths and bond angles can be determined with unprecedented precision.

13.6 VIBRATIONAL SPECTRA OF DIATOMIC MOLECULES

The harmonic oscillator was discussed in Sections 9.9 and 9.10, but in Chapter 12 we saw that the potential energy curves of diatomic molecules are not exactly parabolic. However, as shown in Fig. 13.6, the potential energy curve for a diatomic molecule is approximately parabolic in the vicinity of the equilibrium internuclear distance R_e. The potential energies indicated by the dashed line are given by the parabola

$$E(R) = \tfrac{1}{2}k(R - R_e)^2 \tag{13.64}$$

where k is the **force constant.** We have seen this earlier as equation 9.107.

It is difficult to solve the Schrödinger equation for the exact form of $E(R)$, but we can expand $E(R)$ in a **Taylor series** about the equilibrium separation R_e:

$$E(R) = E(R_e) + \left(\frac{dE}{dR}\right)_{R_e}(R - R_e) + \frac{1}{2}\left(\frac{d^2E}{dR^2}\right)_{R_e}(R - R_e)^2 + \frac{1}{3!}\left(\frac{d^3E}{dR^3}\right)_{R_e}(R - R_e)^3 + \cdots \tag{13.65}$$

The first term is simply a constant, the electronic energy at the equilibrium geometry, and the second term is zero since dE/dR is zero at the minimum of the potential energy curve. The third term is given by equation 13.64. If all higher terms are neglected as giving small corrections, then we have approximated the exact $E(R)$ by a harmonic potential, and we can solve the resulting Schrödinger equation. In Section 9.10, we discussed the solutions of the Schrödinger equation for the simple harmonic oscillator. There we saw that the energy levels are given by

$$E_v = (v + \tfrac{1}{2})h\nu \qquad v = 0, 1, 2, \ldots \tag{13.66}$$

where $\nu = (1/2\pi)(k/\mu)^{1/2}$ and μ is the red mass of the diatomic molecule (see Section 9.11). It is standard in spectroscopy to give the energy in terms of wave numbers, so we divide E_v by hc:

$$\tilde{G}(v) = \frac{E_v}{hc} = \tilde{\nu}(v + \tfrac{1}{2}) \tag{13.67}$$

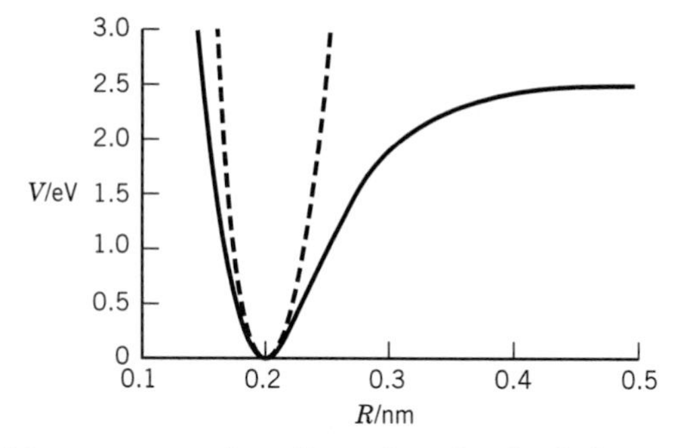

Figure 13.6 Potential energy curve for a diatomic molecule. At internuclear distances R in the neighborhood of the equilibrium distance R_e, the curve is nearly parabolic, as indicated by the dashed line. The parabolic approximation fails at higher excitation energies. (See Computer Problem 13.G.)

where $\tilde{G}(v)$ is referred to as the **vibrational term value** for the vth vibrational level. The tilde indicates that wave numbers (cm^{-1}) are used. In this approximation the energy levels are equally spaced. This is not a bad approximation for the lowest vibrational states of a diatomic molecule. For these levels, the neglect of higher terms in equation 13.65 is justified because the amplitude of vibrational motion is small.

The vibrational frequencies for many diatomics are of the order of 1000 cm^{-1}, with higher values for molecules with hydrogen atoms or strong bonds, and lower values for molecules with heavy atoms or weak bonds.

Not all diatomic molecules have an infrared (vibrational) absorption spectrum. To determine which transitions are possible in a vibrational spectrum, we must use equation 13.35 for the electric dipole transition moment. Since the dipole moment for a diatomic molecule, which is given by equation 13.37, depends on the internuclear distance, we expand this dipole moment in a Taylor series about $R = R_e$:

$$\boldsymbol{\mu}_0^{(e)} = \mu_e + \left(\frac{\partial \mu}{\partial R}\right)_{R_e} (R - R_e) + \frac{1}{2}\left(\frac{\partial^2 \mu}{\partial R^2}\right)_{R_e} (R - R_e)^2 + \cdots \qquad (13.68)$$

For a molecule in a given electronic state, the transition dipole moment for a vibrational transition is given by

$$\int \psi_{v''}^* \mu_0 \psi_{v'}\, d\tau = \mu_e \int \psi_{v''}^* \psi_{v'}\, d\tau + \left(\frac{\partial \mu}{\partial R}\right)_{R_e} \int \psi_{v''}^* (R - R_e)\psi_{v'}\, d\tau$$
$$+ \frac{1}{2}\left(\frac{\partial^2 \mu}{\partial R^2}\right)_{R_e} \int \psi_{v''}^* (R - R_e)^2 \psi_{v'}\, d\tau + \cdots \qquad (13.69)$$

The first term is equal to zero because the vibrational wavefunctions for different v are orthogonal. The second term is nonzero if the dipole moment depends on the internuclear distance R. *Thus, the selection rule for a diatomic molecule is that a molecule will show a vibrational spectrum only if the dipole moment changes with internuclear distance.*

Homonuclear diatomic molecules, such as H_2 and N_2, have zero dipole moment for all bond lengths and therefore do not show vibrational spectra. In general, heteronuclear diatomic molecules do have dipole moments that depend on internuclear distance, so they exhibit vibrational spectra.

The integral in the second term of equation 13.69 vanishes unless $v' = v'' \pm 1$ for harmonic oscillator wavefunctions. According to this specific selection rule, a harmonic oscillator would have a single vibrational absorption or emission frequency. In general, we would expect the second and higher derivatives of the dipole moment with respect to internuclear distance to be small; after all, if the dipole moment were due to fixed charges a variable distance apart, then $(\partial^2 \mu / \partial R^2)$ and higher derivatives would be equal to zero. Although these higher derivatives are small, they do give rise to overtone transitions with $\Delta v = \pm 2, \pm 3, \ldots$, with rapidly diminishing intensities.

These can be seen in the vibrational absorption spectrum of HCl represented schematically in Fig. 13.7. The strongest absorption band is at 3.46 μm; there is a much weaker band at 1.76 μm and a very much weaker one at 1.198 μm. These are the overtone transitions $v = 0$ to $v = 2$, and $v = 0$ to $v = 3$. The vibrational energy levels of $^{35}Cl_2$ are shown in Fig. 13.8.

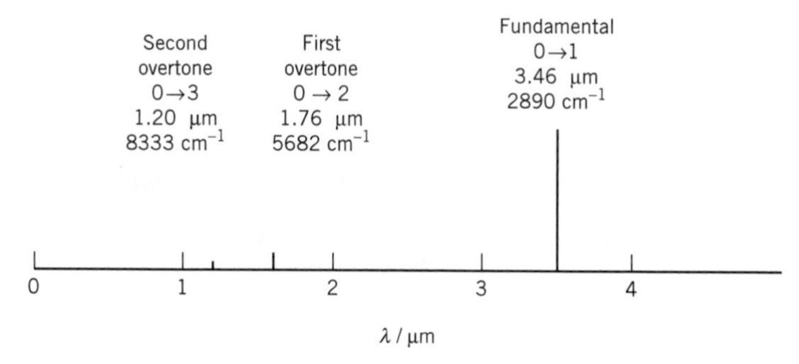

Figure 13.7 "Stick" representation of the vibrational absorption spectrum of $H^{35}Cl$. The relative intensities of the lines fall off five times as fast as indicated.

For a harmonic oscillator, equation 13.42 indicates that the fraction of the molecules in the vth energy level is given by (note that the levels are nondegenerate)

$$f_v = \frac{e^{-(v+1/2)h\nu/kT}}{\sum_{v=0}^{\infty} e^{-(v+1/2)h\nu/kT}}$$

$$= \frac{e^{-vh\nu/kT}}{\sum_{v=0}^{\infty} e^{-vh\nu/kT}} \tag{13.70}$$

The denominator is a geometric series with $x < 1$ for which the sum is given by

$$\sum_{v=0}^{\infty} x^v = \frac{1}{1-x} \tag{13.71}$$

so that

$$\sum_{v=0}^{\infty} e^{-vh\nu/kT} = \frac{1}{1 - e^{-h\nu/kT}} \tag{13.72}$$

Thus, the fraction of the molecules in the ith vibrational state is given by

$$f_v = (1 - e^{-h\nu/kT})\, e^{-vh\nu/kT} \tag{13.73}$$

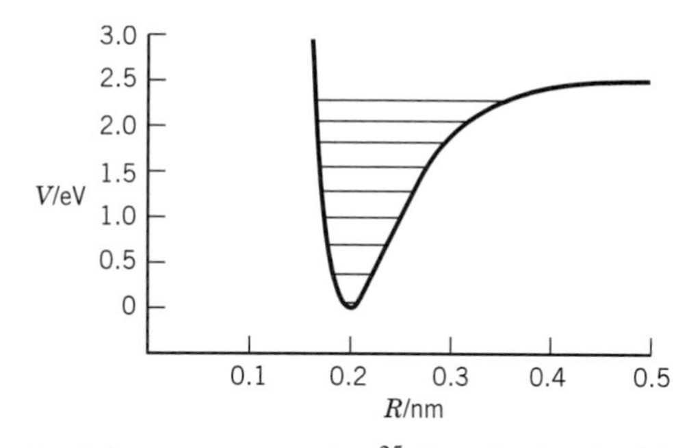

Figure 13.8 The potential energy curve for $^{35}Cl_2$ calculated with the Morse potential (equation 13.82) with every fifth vibrational level from $v = 0$ to $v = 40$. (See Computer Problem 13.B.)

At room temperature this relation predicts that the ratio of the population of $H^{35}Cl$ in $v = 1$ to that in $v = 0$ is 8.9×10^{-7}. Therefore, the molecules with $v = 1$ and higher do not contribute to the spectrum.

Example 13.5 ***Populations of vibrational states for different temperatures***

What fractions of $H^{35}Cl$ molecules are in the $v = 0, 1, 2$, and 3 states at (*a*) 1000 K and (*b*) 2000 K?

These fractions are given by equation 13.73 where, using Table 13.4,

$$\frac{hc\tilde{\nu}}{k} = \frac{(6.626 \times 10^{-34}\ \text{J s})(2.998 \times 10^{10}\ \text{cm s}^{-1})(2990.95\ \text{cm}^{-1})}{1.381 \times 10^{-23}\ \text{J K}^{-1}}$$
$$= 4302\ \text{K}$$

so that

$$f_v = (1 - \text{e}^{-4302/T})\text{e}^{-(4302/T)v}$$

(*a*) At 1000 K,

$$f_0 = 1 - \text{e}^{-4.302} = 0.9865$$
$$f_1 = 0.9865\,\text{e}^{-4.302} = 0.0133$$
$$f_2 = 0.9865\,\text{e}^{-4.302\times 2} = 0.0018$$
$$f_3 = 0.9865\,\text{e}^{-4.302\times 3} = 0.000002$$

(*b*) At 2000 K,

$$f_0 = 1 - \text{e}^{-2.151} = 0.8836$$
$$f_1 = 0.8836\,\text{e}^{-2.151} = 0.1028$$
$$f_2 = 0.8836\,\text{e}^{-2.151\times 2} = 0.0120$$
$$f_3 = 0.8836\,\text{e}^{-2.151\times 3} = 0.0014$$

Figure 13.6 shows that equation 13.67 is not sufficient to represent the energy levels of a diatomic molecule; if equation 13.67 did apply, the overtones would be at integral multiples of the fundamental. When the Schrödinger equation is solved for equation 13.65 truncated after the cubic term, it is found that the energy levels are given by an equation of the form

$$\tilde{G}(v) = \tilde{\nu}_e(v + \tfrac{1}{2}) - \tilde{\nu}_e x_e(v + \tfrac{1}{2})^2 + \tilde{\nu}_e y_e(v + \tfrac{1}{2})^3 \qquad (13.74)$$

where $\tilde{\nu}_e$ is the vibrational wave number, x_e and y_e are anharmonicity constants,* and $v = 0, 1, 2, \ldots$. When the third term in equation 13.74 can be ignored, the frequencies $\tilde{\nu}$ of absorption lines due to $v \to v + 1$ are given by

$$\tilde{\nu} = \tilde{G}(v + 1) - \tilde{G}(v) = \tilde{\nu}_e - 2\tilde{\nu}_e x_e(v + 1) \qquad (13.75)$$

Example 13.6 ***Calculation of vibrational absorption frequencies***

Calculate the vibrational frequencies in wave numbers for the fundamental absorption band of $H^{35}Cl$ and the first four overtones for (*a*) the harmonic oscillator approximation

*The anharmonicity constants are tabulated as $\tilde{\nu}_e x_e$ and $\tilde{\nu}_e y_e$ because early in the history of spectroscopy equation 13.74 was written $G(v) = \nu_e[(v + \frac{1}{2}) - x_e(v + \frac{1}{2})^2 + y_e(v + \frac{1}{2})^3]$.

and (*b*) the anharmonic oscillator approximation. The spectroscopic constants are given in Table 13.4.

(*a*) For the harmonic oscillator approximation, the frequencies in wave numbers are given by $\tilde{\nu}_e v$, where v is the vibrational quantum number in the higher level in $v = 0 \rightarrow 1, 2, 3, \ldots$.

(*b*) For the anharmonic oscillator approximation, the frequencies in wave numbers are given by $\tilde{\nu}_e v - \tilde{\nu}_e x_e v(v+1)$, where $v = 1, 2, 3, \ldots$. Since $\tilde{\nu}_e = 2990.95\ \text{cm}^{-1}$ and $\tilde{\nu}_e x_e = 52.819\ \text{cm}^{-1}$, the frequencies are given by the following table:

v (upper level)	1	2	3	4	5
Harmonic	2990.95	5981.9	8972.85	11 963.8	14 954.7
Anharmonic	2885.31	5664.99	8339.02	10 907.4	13 370.2

See Fig. 13.7 and Computer Problem 13.I.

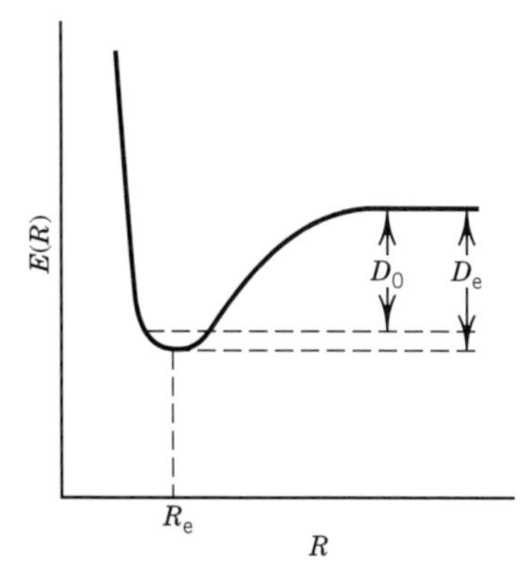

Figure 13.9 The potential energy of a diatomic molecule as a function of the internuclear distance. Only the $v = 0$ vibrational level is shown. The dissociation energy that we are primarily concerned with in this chapter is the spectroscopic dissociation energy D_0.

In Chapter 11 we dealt with the **equilibrium dissociation energy** D_e measured from the minimum in the potential energy curve. But now we will be dealing with the **spectroscopic dissociation energy** D_0 measured from the zeroth vibrational level. The relationship between these two dissociation energies is shown in Fig. 13.9.

The potential energy curves for H_2 and $H_2{}^+$ are shown in Fig. 13.10 along with their respective spectroscopic dissociation energies, $D_0(H_2)$ and $D_0(H_2{}^+)$.

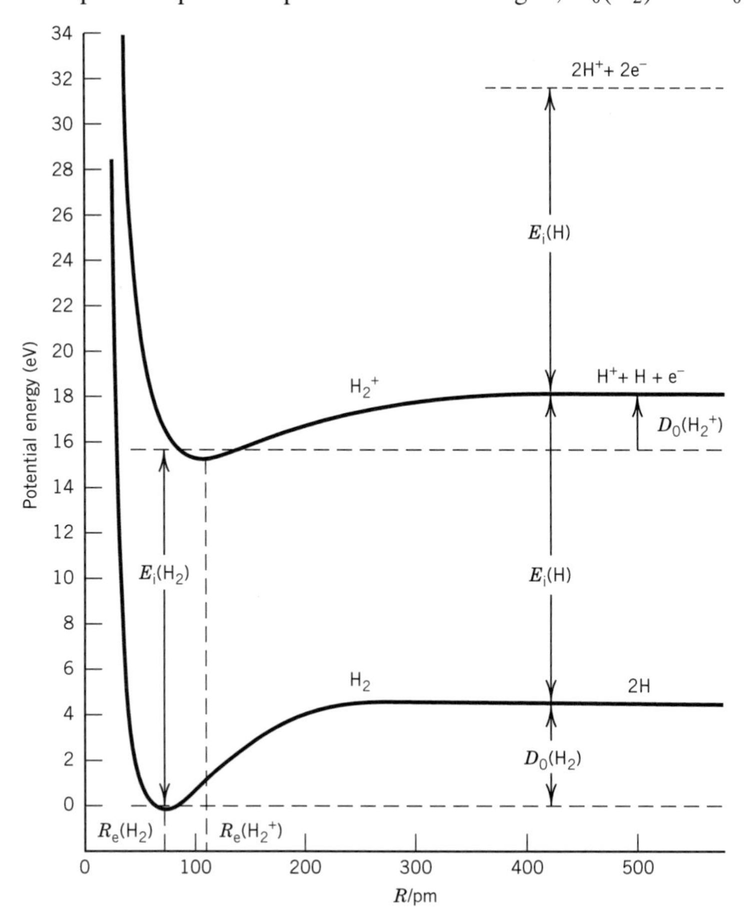

Figure 13.10 Potential energy curves for the ground electronic states of H_2 and $H_2{}^+$ with the zero-point vibrational levels shown.

The ionization energy $E_i(H_2)$ for hydrogen is the energy required to remove an electron to an infinite distance from H_2^+, and it has been measured accurately:

$$H_2(g) = H_2^+(g) + e^- \qquad E_i(H_2) = 15.4259 \text{ eV} \tag{13.76}$$

Thus, the zero-point levels of H_2 and H_2^+ are separated by 15.4259 eV, as shown in Fig. 13.10. The potential energy curves for H_2 and H_2^+ at infinite internuclear distance are separated by the ionization potential of a hydrogen atom in its ground state. The ionization potential calculated in Example 10.4 can be corrected for the finite mass of the nucleus:

$$H(g) = H^+(g) + e^- \qquad E_i(H) = 13.598\,396 \text{ eV} \tag{13.77}$$

As can be seen from Fig. 13.10,

$$E_i(H_2) + D_0(H_2^+) = D_0(H_2) + E_i(H) \tag{13.78}$$

It is very difficult to measure the spectroscopic dissociation energy of H_2^+ directly, so equation 13.78 is used to calculate $D_0(H_2^+)$.* The values of these dissociation energies and ionization energies are shown in Table 13.3 in eV, cm^{-1}, and kJ mol^{-1}.

The vibrational parameters for a number of diatomic molecules are given in Table 13.4. According to equation 13.74 the energy of the ground state of a diatomic molecule is given by

$$\tilde{G}(0) = \frac{\tilde{\nu}_e}{2} - \frac{\tilde{\nu}_e x_e}{4} + \frac{\tilde{\nu}_e y_e}{8} \tag{13.79}$$

Thus, the **equilibrium dissociation energy** is given by

$$\tilde{D}_e = \tilde{D}_0 + \frac{\tilde{\nu}_e}{2} - \frac{\tilde{\nu}_e x_e}{4} + \frac{\tilde{\nu}_e y_e}{8} \tag{13.80}$$

For $^1H^1H$, the values of $\tilde{\nu}_e$, $\tilde{\nu}_e x_e$, and $\tilde{\nu}_e y_e$ are 4401.21, 121.33, and 0.813 cm^{-1}. Therefore, the zero-point energy is $G(0) = 4401.21/2 - 121.33/4 + 0.813/8 = 2170\ cm^{-1}$. This is the value used in Chapter 11. Note, however, that H_2 does not have an infrared spectrum, so these values are determined by other means.

Table 13.3 Dissociation Energies for $H_2^+(g)$ and $H_2(g)$ and Ionization Potentials E_i for $H_2(g)$ and H(g)

	eV	cm^{-1}	kJ mol^{-1}
	H_2^+		
D_0	2.650 79	21 380	255.760
D_e	2.793	22 527	269.481
	H_2		
D_0	4.477 97	36 117	432.055[a]
D_e	4.748 3	38 297	458.135
$E_i(H_2)$	15.425 9	124 417	1488.361
$E_i(H)$	13.598 396	109 677.6	1312.035

[a]This spectroscopic dissociation energy of H_2 is in agreement with $\Delta H_0^\circ = 432.074$ kJ mol^{-1} calculated from Table C.3.

*G. Herzberg, *Science* **177:**123 (1972).

Table 13.4 Constants of Diatomic Molecules

	State	T_e/cm^{-1}	ν_e/cm^{-1}	$\nu_e x_e/\text{cm}^{-1}$	$\tilde{B}_e/\text{cm}^{-1}$	$\tilde{\alpha}_e/\text{cm}^{-1}$	R_e/pm	$\frac{N_A\mu}{10^{-3}\ \text{kg mol}^{-1}}$	D_0/eV	E_i/eV
$^{79}Br_2$	$^1\Sigma_g^+$	0	325.321	1.077	0.082107	3.187×10^{-4}	228.10	39.459 166	1.9707	10.52
$^{12}C_2$	$^1\Sigma_g^+$	0	1854.71	13.34	1.8198	0.0176	124.25	6.000 000	6.21	12.15
$^{12}C^1H$	$^2\Pi_r$	0	2858.5	63.0	14.457	0.534	111.99	0.929 741	3.46	10.64
$^{35}Cl_2$	$^1\Sigma_g^+$	0	559.7	2.67	0.2439	1.4×10^{-3}	198.8	17.484 427	2.47937	11.50
$^{12}C^{16}O$	$^1\Sigma^+$	0	2169.814	13.288	1.931281	0.017504	112.832	6.856 209	11.09	14.01
	$^1\Pi$	65 075.8	1518.2	19.40	1.6115	0.0233	123.53			
$^1H_2$	$^1\Sigma_g^+$	0	4401.21	121.34	60.853	3.062	74.144	0.503 913	4.4781	15.43
	$^1\Sigma_u^+$	91 700	1358.09	20.888	20.015	1.1845	129.28			
$^1H^{81}Br$	$^1\Sigma^+$	0	2648.98	45.218	8.46488	0.23328	141.443	0.995 427	3.758	11.67
$^1H^{35}Cl$	$^1\Sigma^+$	0	2990.95	52.819	10.5934	0.30718	127.455	0.979 593	4.434	12.75
$^1H^{127}I$	$^1\Sigma^+$	0	2309.01	39.644	6.4264	0.1689	160.916	0.999 884	3.054	10.38
$^{127}I_2$	$^1\Sigma_g^+$	0	214.50	0.614	0.03737	1.13×10^{-4}	266.6	63.452 238	1.54238	9.311
$^{39}K^{35}Cl$	$^1\Sigma^+$	0	281	1.30	0.128635	7.89×10^{-4}	266.665	18.429 176	4.34	8.44
$^{14}N_2$	$^1\Sigma_g^+$	0	2358.57	14.324	1.99824	0.017318	109.769	7.001 537	9.759	15.58
	$^3\Pi_g$	59 619	1733.39	14.122	1.6375	0.0179	121.26			
	$^3\Pi_u$	89 136	2047.18	28.445	1.8247	0.0187	114.87			
$^{23}Na^{35}Cl$	$^1\Sigma^+$	0	366	2.0	0.218063	1.62×10^{-3}	236.08	13.870 687	4.23	8.9
$^{14}N^{16}O$	$^2\Pi_r$ $\Omega = \frac{1}{2}$	119.82	1904.04	14.100	1.72	0.0182	115.077	7.466 433	6.496	9.26
	$\Omega = \frac{3}{2}$	0	1904.20	14.075	1.67	0.0171				
$^{16}O_2$	$^3\Sigma_g^-$	0	1580.19	11.98	1.44563	0.0159	120.752	7.997 458	5.115	12.07
	$^1\Delta_g$	7918.1	1483.5	12.9	1.4264	0.0171	121.56			
	$^3\Sigma_u^-$	49 793.3	709.31	10.65	0.8190	0.01206	160.43			
$^{16}O^1H$	$^2\Pi_i$	0	3737.76	84.811	18.911	0.7242	96.966	0.948 087	4.392	12.9

Source: K. P. Huber and G. Herzberg, *Molecular Spectra and Molecular Structure IV, Constants of Diatomic Molecules.* New York: Van Nostrand, 1979.

The observed absorption frequencies for $\upsilon = 0$ to $\upsilon = 1, 2, 3, \ldots$ are given by

$$\tilde{\nu} = \tilde{G}(\upsilon) - \tilde{G}(0) = \tilde{\nu}_e \upsilon - \tilde{\nu}_e x_e \upsilon(\upsilon + 1) \tag{13.81}$$

The Taylor series in equation 13.65 represents only the potential energy of a diatomic molecule in the neighborhood of the minimum. What is really needed is a potential energy function for the whole range of R values. The **Morse potential** is a simple function that provides an approximate potential energy V as a function of internuclear distance R in terms of the equilibrium dissociation energy D_e and other spectroscopic properties:

$$V(R) = D_e\{1 - \exp[-a(R - R_e)]\}^2 \tag{13.82}$$

When $R \to \infty$ the potential energy approaches the equilibrium dissociation energy, and the potential energy is zero at $R = R_e$. The Schrödinger equation can be solved for the Morse potential, and the corresponding term value expression is

$$\tilde{G}(\upsilon) = a\left(\frac{\hbar D_e}{\pi c \mu}\right)^{1/2}\left(\upsilon + \frac{1}{2}\right) - \left(\frac{\hbar a^2}{4\pi c \mu}\right)\left(\upsilon + \frac{1}{2}\right)^2 \tag{13.83}$$

By comparing this equation with equation 13.74, we find that

$$\tilde{\nu}_e = a\left(\frac{\hbar D_e}{\pi c \mu}\right)^{1/2} \tag{13.84}$$

$$\tilde{\nu}_e x_e = \frac{\hbar a^2}{4\pi c \mu} \tag{13.85}$$

Equations 13.84 and 13.85 provide two expressions for the parameter a. That indicates that the physical properties in the expressions for a are not all independent. When the two expressions are set equal, the following relation is obtained:

$$D_e = \frac{\tilde{\nu}_e}{4x_e} \tag{13.86}$$

Since actual potential energy curves differ from the Morse equation, this is not an exact relation, but it is useful when the dissociation energy of an excited molecule, for example, is not known.

Example 13.7 ***The Morse potential for $H^{35}Cl$***

Calculate the parameters in the equation for the Morse potential of $H^{35}Cl$ and plot the potential energy curve.

The spectroscopic properties are given in Table 13.4. Since various units are used in this table, it is convenient to make the calculation in SI units. The reduced mass in kilograms is given by

$$\mu = \frac{(1.007\,825)(34.968\,852)(1.660\,540 \times 10^{-27})}{1.007\,825 + 34.968\,852} = 1.626\,65 \times 10^{-27}\,\text{kg}$$

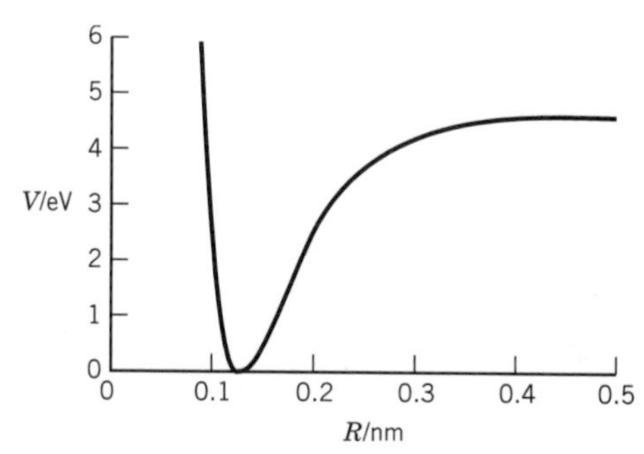

Figure 13.11 Plot of the potential energy of $H^{35}Cl$ versus internuclear distance according to the Morse equation. The actual potential energy curve has a slightly different shape. (See Computer Problem 13.H.)

The equilibrium dissociation energy in m^{-1} is given by

$$D_e = \frac{4.434 \text{ eV}}{1.239\,842\,4 \times 10^{-6} \text{ eV/m}^{-1}} + \tilde{\nu}_e/2 - \tilde{\nu}_e x_e/4$$

$$= \frac{4.434 \text{ eV}}{1.239\,842\,4 \times 10^{-6} \text{ eV/m}^{-1}} + \frac{299\,095 \text{ m}^{-1}}{2} - \frac{5281.9 \text{ m}^{-1}}{4}$$

$$= 3.724\,49 \times 10^6 \text{ m}^{-1}$$

The parameter a in the Morse equation is given by

$$a = \tilde{\nu}_e \left(\frac{\pi \mu c}{\hbar D_e} \right)^{1/2}$$

$$= 299\,095 \text{ m}^{-1} \left[\frac{\pi (1.626\,65 \times 10^{-27} \text{ kg})(2.997\,925 \times 10^8 \text{ m s}^{-1})}{(1.054\,57 \times 10^{-34} \text{ J s})(3.724\,49 \times 10^6 \text{ m}^{-1})} \right]^{1/2}$$

$$= 1.867\,97 \times 10^{10} \text{ m}^{-1}$$

The plot of the potential energy as a function of internuclear distance is given in Fig. 13.11.

13.7 VIBRATION–ROTATION SPECTRA OF DIATOMIC MOLECULES

At high resolution, each of the absorptions in the vibrational spectrum in Fig. 13.7 is found to have a complicated structure that results from simultaneous changes in rotational energy. Because of this structure, molecular spectra are often referred to as **band spectra.** The fundamental vibration band for HCl ($v = 0 \rightarrow 1$) is shown in Fig. 13.12.

When a molecule in a state with vibrational quantum number v and rotational quantum number J makes a spectral transition to another state, the vibrational quantum number changes to $v \pm 1$ (according to the harmonic oscillator selection rules), and the rotational quantum number can change to $J \pm 1$ or remain the same. The possible transitions are shown in Fig. 13.13. The transitions with $\Delta J = +1$ give rise to lines in the R branch of the spectrum, and the transitions with $\Delta J = -1$ give rise to lines in the P branch of the spectrum. The intensities of the lines in these branches reflect the thermal populations of the initial rotational states. The Q branch, when it occurs, consists of lines corresponding to $\Delta J = 0$.

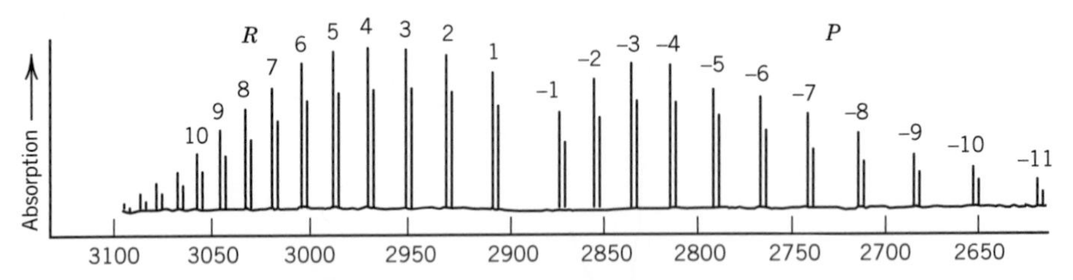

Figure 13.12 Fundamental vibrational band for HCl ($v = 0 \rightarrow 1$). The double peaks are due to the presence of $H^{35}Cl$ (75% abundance) and $H^{37}Cl$ (25% abundance). (Reprinted with permission from A. R. H. Cole, *Tables of Wavenumbers for the Calibration of Infrared Spectrometers,* 2nd ed. Copyright © 1977, Pergamon Press on behalf of IUPAC.)

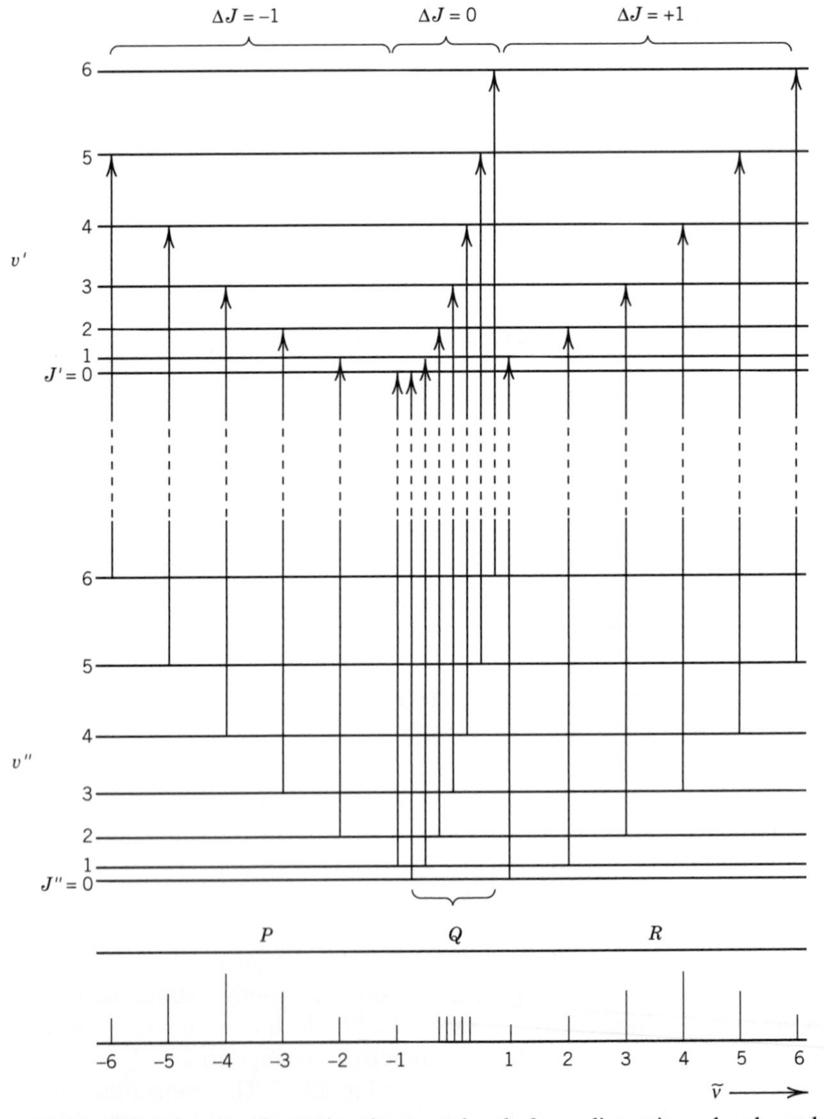

Figure 13.13 Vibrational and rotational energy levels for a diatomic molecule and the transitions observed in the vibration–rotation spectrum when the transition between v' and v'' is allowed. In the spectrum shown at the bottom, the relative heights of the spectral lines indicate relative intensities of absorption.

Generally, these transitions are forbidden, except for molecules such as NO, which have orbital angular momentum about their axes.

The energies of the levels in Fig. 13.13 are given to the accuracy we need here by the equation

$$\begin{aligned} E(v, J)/hc &= \tilde{G}(v) + \tilde{F}_v(J) \\ &= \tilde{\nu}_e(v + \tfrac{1}{2}) - \tilde{\nu}_e x_e(v + \tfrac{1}{2})^2 + \tilde{B}_v J(J + 1) \end{aligned} \quad (13.87)$$

which expresses the energy of a level as the sum of the first two terms of the vibrational term value (equation 13.74) and the first term of the rotational term value (equation 13.40). Now it is necessary to put a subscript v on $\tilde{B}$ since the rotational constant depends on the vibrational quantum number v. Since $\tilde{B}$ is inversely proportional to the moment of inertia I, it varies as R_e^{-2}, where R_e is the equilibrium internuclear distance. R_e varies with the vibrational state. R_e in $v = 1$ is slightly larger than in $v = 0$; therefore, $B_1 < B_0$.

The dependence of the rotational constant on the vibrational quantum number is generally represented by

$$\tilde{B}_v = \tilde{B}_e - \tilde{\alpha}_e(v + \tfrac{1}{2}) \quad (13.88)$$

where α_e is the **vibration–rotation coupling constant.**

Now let us consider a vibrational transition from $v = 0$ to $v = 1$. A molecule with $v = 0$ can have various J values, and in going to $v = 1$, the value of J can go to $J + 1$ or $J - 1$ because the selection rule is $\Delta J = \pm 1$. In the vibrational ground state, equation 13.87 indicates that the energy is given by

$$\tilde{E}(v = 0, J) = \tilde{\nu}_e/2 - \tilde{\nu}_e x_e/4 + \tilde{B}_0 J(J + 1) \quad (13.89)$$

where $\tilde{B}_0$ is the rotational constant when $v = 0$. When the molecule absorbs a photon and $v \to 1$ and $J \to J + 1$, the energy of the upper state is given by

$$\tilde{E}(v = 1, J + 1) = \tilde{\nu}_e 1.5 - \tilde{\nu}_e x_e 1.5^2 + \tilde{B}_1(J + 1)(J + 2) \quad (13.90)$$

where $\tilde{B}_1$ is the rotational constant when $v = 1$. These transitions lead to the R branch of the vibration–rotation spectrum, and the absorption frequencies are given by

$$\begin{aligned} \tilde{\nu}_R &= \tilde{E}(v = 1, J + 1) - \tilde{E}(v = 0, J) \\ &= \tilde{\nu}_0 + \tilde{B}_1(J + 1)(J + 2) - \tilde{B}_0 J(J + 1) \\ &= \tilde{\nu}_0 + 2\tilde{B}_1 + (3\tilde{B}_1 - \tilde{B}_0)J + (\tilde{B}_1 - \tilde{B}_0)J^2 \end{aligned} \quad (13.91)$$

where

$$\tilde{\nu}_0 = \tilde{\nu}_e - 2\tilde{\nu}_e x_e \quad (13.92)$$

is the center of the vibration–rotation band where there is no absorption because $\Delta J = 0$ is forbidden. If $B_1 = B_0$, then these frequencies are equally spaced.

When the molecule absorbs a photon and $v \to 1$ and $J \to J - 1$, the energy of the upper state is given by

$$\tilde{E}(v = 1, J - 1) = \tilde{\nu}_e 1.5 - \tilde{\nu}_e x_e 1.5^2 + \tilde{B}_1(J - 1)J \quad (13.93)$$

These transitions lead to the P branch of the vibration–rotation spectrum, and the absorption frequencies are given by

$$\begin{aligned}\tilde{\nu}_P &= \tilde{E}(v = 1, J - 1) - \tilde{E}(v = 0, J) \\ &= \tilde{\nu}_0 + \tilde{B}_1(J - 1)J - \tilde{B}_0 J(J + 1) \\ &= \tilde{\nu}_0 - (\tilde{B}_1 + \tilde{B}_0)J + (\tilde{B}_1 - \tilde{B}_0)J^2\end{aligned} \tag{13.94}$$

If $B_1 = B_0$, these frequencies are again equally spaced; however, since $\tilde{B}_1 < \tilde{B}_0$, the spacing between lines in the R branch decreases with increasing J, and the spacing in the P branch increases with increasing J. These features are evident in the fundamental infrared absorption spectrum for HCl shown in Fig. 13.12.

Table 13.4 gives the vibrational and rotational constants for a number of diatomic molecules and several of their electronic excited states. The electronic energy relative to the ground state is represented by $\tilde{T}_e$.

Example 13.8 ***Population of rotational states of $H^{35}Cl$ at 300 K***

Calculate the relative populations of the first five rotational levels of the ground vibrational state of $H^{35}Cl$ at 300 K.

According to equation 13.88 and Table 13.4, $\tilde{B}_v = 10.5934\ \text{cm}^{-1} - (0.3072\ \text{cm}^{-1})(0 + \frac{1}{2}) = 10.4398\ \text{cm}^{-1}$ for $v = 0$. We have

$$\frac{N_J}{N_0} = (2J + 1)\,e^{-hcJ(J+1)\tilde{B}_v/kT}$$

where N_0 is the number of molecules in the $J = 0$ state (i.e., we are assuming the Boltzmann distribution, equation 13.42). First we need to calculate the following factor:

$$\begin{aligned}\frac{hc\tilde{B}_v}{kT} &= \frac{(6.626 \times 10^{-34}\ \text{J s})(2.998 \times 10^8\ \text{m s}^{-1})(10.44\ \text{cm}^{-1})(10^2\ \text{cm m}^{-1})}{(1.3806 \times 10^{-23}\ \text{J K}^{-1})(300\ \text{K})} \\ &= 5.007 \times 10^{-2}\end{aligned}$$

For $J = 1$

$$\begin{aligned}\frac{N_1}{N_0} &= 3\,e^{-2(5.007\times10^{-2})} \\ &= 2.71\end{aligned}$$

The relative populations for $J = 0, 1, 2, 3, 4, 5$ are 1.00, 2.71, 3.70, 3.84, 3.31, and 2.45, in excellent agreement with Fig. 13.12.

13.8 VIBRATIONAL SPECTRA OF POLYATOMIC MOLECULES

In Section 13.3, we saw that $3N - 5$ coordinates are required to describe the internal motions of a diatomic or linear molecule and $3N - 6$ coordinates are required for a nonlinear polyatomic molecule. The different types of vibrational motion that are possible can be described in terms of **normal modes of vibration,** which are described below. For a diatomic molecule, $3N - 5 = 1$, and so there is a single degree of vibrational freedom and a single normal mode. For a linear triatomic molecule, such as CO_2, $3N - 5 = 4$, and so there are four normal modes. This

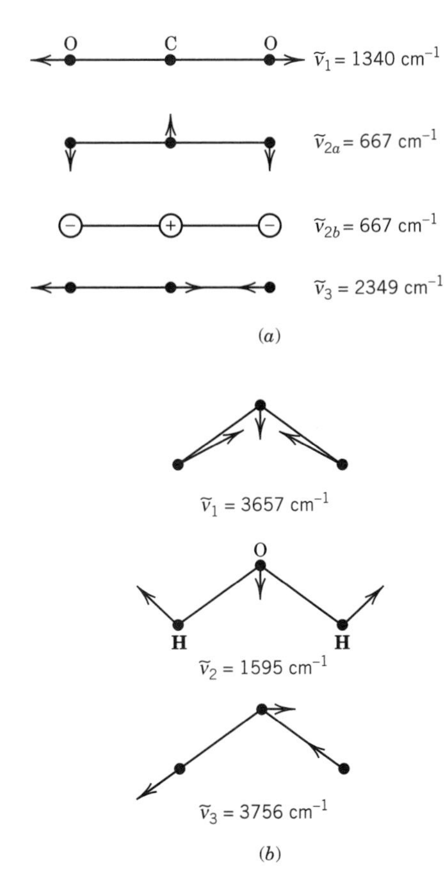

Figure 13.14 (*a*) Normal modes of vibration of the symmetrical linear triatomic molecule CO_2. (*b*) Normal modes of vibration of the nonlinear triatomic molecule H_2O. The vectors representing the magnitudes of the oxygen vibrations have been increased relative to those of hydrogen.

means that there are four types of vibrational motion. Figure 13.14*a* provides a schematic representation of four types of vibrational motion for a symmetrical linear triatomic molecule and gives the vibrational frequencies in wave numbers for CO_2. For a nonlinear triatomic molecule, such as H_2O, $3N - 6 = 3$, and so there are three normal modes (Fig. 13.14*b*). NH_3, CH_4, and N_2O_4 have 6, 9, and 12 normal modes of vibration.

To see what normal modes of vibration are, we first consider the vibration of polyatomic molecules from a classical mechanical viewpoint. The kinetic energy T of a polyatomic molecule is given by

$$T = \frac{1}{2}\sum_{k=1}^{N} m_k\left[\left(\frac{dx_k}{dt}\right)^2 + \left(\frac{dy_k}{dt}\right)^2 + \left(\frac{dz_k}{dt}\right)^2\right] \tag{13.95}$$

This equation can be simplified by introducing mass-weighted Cartesian displacement coordinates $q_1, \ldots, q_{3N}$.

$$q_1 = m_1^{1/2}(x_1 - x_{1e}) \qquad q_2 = m_1^{1/2}(y_1 - y_{1e}) \qquad q_3 = m_1^{1/2}(z_1 - z_{1e})$$
$$q_4 = m_2^{1/2}(x_2 - x_{2e}) \qquad \cdots \qquad q_{3N} = m_N^{1/2}(z_N - z_{Ne}) \tag{13.96}$$

where x_{ie} and so on are the values of the coordinates at the equilibrium geometry of the molecule. Since these are independent of time, the kinetic energy becomes

$$T = \frac{1}{2}\sum_{i=1}^{3N}\left(\frac{dq_i}{dt}\right)^2 \tag{13.97}$$

Since the potential energy V is a function of the coordinates $x_1, \ldots, z_N$, it is also a function of the mass-weighted coordinates. We will use a Taylor series expansion about the equilibrium position as we did in equation 13.65, this time for a function of several variables:

$$V = V_e + \sum_{i=1}^{3N}\left(\frac{\partial V}{\partial q_i}\right)_e q_i + \frac{1}{2}\sum_{i=1}^{3N}\sum_{k=1}^{3N}\left(\frac{\partial^2 V}{\partial q_i\,\partial q_k}\right)_e q_i q_k + \cdots \tag{13.98}$$

Since V_e is the potential energy at the equilibrium configuration, it is a constant which we can set equal to zero, and the terms in $(\partial V/\partial q_i)_e$ are all equal to zero because the potential energy is a minimum at the equilibrium configuration by definition. If we neglect terms higher than quadratic, equation 13.98 can be written

$$V = \frac{1}{2}\sum_{i=1}^{3N}\sum_{k=1}^{3N} K_{ik} q_i q_k \tag{13.99}$$

so that the total energy is given by

$$E = T + V = \frac{1}{2}\sum_{i=1}^{3N}\left(\frac{dq_i}{dt}\right)^2 + \frac{1}{2}\sum_{i=1}^{3N}\sum_{k=1}^{3N} K_{ik} q_i q_k \tag{13.100}$$

where K_{ik} is the second derivative of V with respect to q_i and q_k evaluated at the equilibrium configuration. The problem in using this expression is with the cross terms. Fortunately, it is possible to make a linear transformation of the mass-weighted coordinates q to new coordinates Q such that the quadratic term does not contain cross terms:

$$E = \frac{1}{2}\sum_{i=1}^{3N}\left(\frac{dQ_i}{dt}\right)^2 + \frac{1}{2}\sum_{i=1}^{3N-6 \text{ or } 3N-5} \kappa_i Q_i^2 \tag{13.101}$$

We have used the fact that translational and rotational motion have only kinetic energy so that there are $3N - 6$ vibrational coordinates for a nonlinear molecule and $3N - 5$ for a linear molecule. These $3N - 6$ or $3N - 5$ coordinates are referred to as **normal coordinates,** and the corresponding $3N - 6$ or $3N - 5$ vibrations are referred to as **normal modes of vibration.**

In a normal mode of vibration, the nuclei move in phase (i.e., the nuclei pass through the extremes of their motion simultaneously). The motions of the nuclei in a normal mode are such that the center of mass does not move, and the molecule as a whole does not rotate. This means that different atoms move different distances. Each normal mode has a characteristic vibration frequency. Sometimes several modes have identical vibration frequencies and are referred to as degenerate modes. It can be shown that any vibrational motion of a polyatomic molecule can be expressed as a linear combination of normal modes of vibrations.

Turning now to the quantum mechanical treatment of a molecule, the vibrational Hamiltonian obtained from equation 13.101 is simply a sum of terms, one for each coordinate:

$$\hat{H} = \sum_{i=1}^{\substack{3N-6 \\ \text{or } 3N-5}} \hat{H}_i \tag{13.102}$$

This indicates that the vibrational wavefunctions for the molecule can be written as the product of harmonic oscillator wavefunctions, one for each coordinate Q_i. We have seen in equation 9.116 that the eigenvalues for the harmonic oscillator are given by $E_i = (v_i + \frac{1}{2})hc\tilde{\nu}_i$ so that the total vibrational energy of a polyatomic molecule is

$$E = \sum_{i=1}^{\substack{3N-6 \\ \text{or } 3N-5}} (v_i + \tfrac{1}{2})hc\tilde{\nu}_i \tag{13.103}$$

The frequency of a normal mode depends both on the force constant k_Q for the mode and on the reduced mass μ_Q for the mode: $2\pi\nu_i = (k_Q/\mu_Q)^{1/2}$.

The four normal modes of CO_2 are shown in Fig. 13.14*a*. The first normal mode is a symmetrical stretching vibration in which the carbon atom remains fixed. The third normal mode is an asymmetrical stretching vibration. The other two normal modes are orthogonal bending vibrations. The lower vibration frequency for the bending vibrations indicates that it is generally easier to bend a molecule than to stretch it. Figure 13.14*b* shows the three normal modes of vibration of H_2O. As indicated in the diagrams, the displacements of various atoms in a normal mode are not equal, but depend on the masses and force constants.

For a polyatomic molecule, some normal modes of vibration are spectroscopically active and some are not. *The gross selection rule is still that the displacements of a normal mode must cause a change in dipole moment in order to be spectroscopically active in the infrared.*

Of the four normal-mode vibrations for CO_2 the symmetric stretch is not active in the infrared, but the other vibrations are. Since CO_2 is linear and symmetrical in its equilibrium state, it does not have a dipole moment, and the symmetrical stretching vibration does not create one. The asymmetric stretch and bending vibrations produce a changing dipole moment. The three normal modes of H_2O are all active in the infrared because the magnitude of the dipole moment changes in each type of vibration.

The specific selection rule for vibrational spectroscopy is that $\Delta v = \pm 1$ in the harmonic oscillator approximation. In addition, combination bands are formed in which two or more vibrational modes change simultaneously.

The frequencies, in cm^{-1}, of the strongest bands for H_2O vapor are summarized in Table 13.5. The weaker bands in the spectrum are the overtones and combinations shown in the table. As shown in Table 13.5, the vibrations are not harmonic, and so the overtones are not exact multiples, and the combinations are not exact sums.

One of the vibrational motions of a polyatomic molecule may be an internal rotation. If there is an appreciable potential energy barrier for an internal rotation about some bond, there will be an oscillation about the mean position. For example, in ethylene, $CH_2{=}CH_2$, there is a large potential energy barrier for internal rotation, so that there are only small oscillations about the C=C bond. In some

Table 13.5 Infrared Bands of H_2O Vapor

$\tilde{\nu}/\text{cm}^{-1}$	*Intensity*	*Interpretation*
1595.0	Very strong	$\tilde{\nu}_2$
3151.4	Medium	$2\tilde{\nu}_2$
3651.7	Strong	$\tilde{\nu}_1$
3755.8	Very strong	$\tilde{\nu}_3$
5332.0	Medium	$\tilde{\nu}_2 + \tilde{\nu}_3$
6874	Weak	$2\tilde{\nu}_2 + \tilde{\nu}_3$

cases, such as in $CH_3{-}CH_3$, the potential energy barrier is small enough that the internal rotation is said to be "free" at room temperature.

The vibrational spectra of polyatomic molecules are very useful for identification and serve also as criteria of purity. For such practical applications the infrared spectra for a large number of compounds have been cataloged and are used like fingerprints. Groups of atoms within the molecule have quite characteristic absorption bands. The wavelengths at which a certain group absorbs vary slightly, depending on the structure of the rest of the molecule.

The infrared spectrum of a molecule may be considered to be made up of several regions.

1. Hydrogen stretching vibrations, 3700–2500 cm^{-1}. These vibrations occur at high frequencies because of the low mass of the hydrogen atom. If an OH group is not involved in hydrogen bonding (Section 11.10), it usually has a frequency in the vicinity of 3600–3700 cm^{-1}. Hydrogen bonding causes this frequency to drop by 300–1000 cm^{-1} or more. The NH absorption falls in the 3300- to 3400-cm^{-1} range, and the CH absorption falls in the 2850- to 3000-cm^{-1} range. For SiH, PH, and SH, it is approximately 2200, 2400, and 2500 cm^{-1}.
2. Triple-bond region, 2500–2000 cm^{-1}. Triple bonds have high frequencies because of the large force constants. The $C{\equiv}C$ group usually causes absorption between 2050 and 2300 cm^{-1}, but this absorption may be weak or absent because of the symmetry of the molecule. The $C{\equiv}N$ group absorbs near 2200–2300 cm^{-1}.
3. Double-bond region, 2000–1600 cm^{-1}. Absorption bands of substituted aromatic compounds fall in the range 2000–1600 cm^{-1} and are a good indicator of the position of the substitution. Carbonyl groups, $C{=}O$, of ketones, aldehydes, acids, amides, and carbonates usually show strong absorption in the vicinity of 1700 cm^{-1}. Olefins, $C{=}C$, may show absorption in the vicinity of 1650 cm^{-1}. The bending of the $C{-}N{-}H$ bond also occurs in this region.
4. Single-bond stretch and bend region, 500–1700 cm^{-1}. The region 500–1700 cm^{-1} is not diagnostic for particular functional groups, but it is a useful "fingerprint" region, since it shows differences between similar molecules. Organic compounds usually show peaks in the region between 1300 and 1475 cm^{-1} because of the bending motions of bonds to hydrogen. Out-of-plane bending motions of olefinic and aromatic CH groups usually occur between 700 and 1000 cm^{-1}.

Comment:

Applying the selection rule that a normal mode will have a vibrational spectrum only if the dipole moment changes in the vibration may be difficult for the normal modes of a polyatomic molecule. Fortunately, that information is in the character table for the symmetry group. An example of a character table was given at the end of the preceding chapter (see Table 12.4). That character table for C_{2v} shows that for any molecule in this symmetry group some of the vibrational modes will be infrared active. Water is an example, as shown by Fig. 13.14b.

13.9 RAMAN SPECTRA

When a sample is irradiated with monochromatic light, the incident radiation may be absorbed, may stimulate emission, or may be scattered. A part of the scattered radiation is referred to as the **Raman spectrum.** It is found that some photons lose energy in scattering from a molecule in the sample and emerge with a lower frequency; these photons produce what are referred to as **Stokes lines** in the spectrum of the scattered radiation. A smaller fraction of the scattered photons gains energy in striking a molecule in the sample and emerges with a higher frequency; these photons produce what are referred to as **anti-Stokes lines** in the spectrum of the scattered radiation. Only a very small fraction (usually less than 1 part in 10^6) of the incident radiation is scattered, and the frequency shifts may be quite small; since lasers can produce very intense radiation that is highly monochromatic, they are used as the radiation source.

The interpretation of Raman spectra is based on the conservation of energy, which requires that when a photon of frequency ν is scattered by a molecule in a quantum state with energy E_{i} and the outgoing photon has a frequency ν', the molecule ends up in quantum state f with energy E_{f}:

$$h\nu + E_{\text{i}} = h\nu' + E_{\text{f}} \tag{13.104}$$

or

$$h(\nu' - \nu) = E_{\text{i}} - E_{\text{f}} = h\,\Delta\nu_{\text{R}} = hc\,\Delta\tilde{\nu}_{\text{R}} \tag{13.105}$$

where the shift in frequency is labeled $\Delta\nu_{\text{R}}$ and the shift in wave number is labeled $\Delta\tilde{\nu}_{\text{R}}$. Notice that Raman spectroscopy is different from absorption or emission spectroscopy in that the **incident** light need not coincide with a quantized energy difference in the molecule. Therefore, any frequency of light can be used. Since many final states are possible, of both higher and lower energy than the initial state, many Raman spectral lines can be observed. A typical experimental apparatus is shown in Fig. 13.15.

The frequency shifts seen in Raman experiments correspond to vibrational or rotational energy differences, so this kind of spectroscopy gives us information on the vibrational and rotational states of molecules.

The Raman effect arises from the induced polarization of scattering molecules that is caused by the electric vector of the electromagnetic radiation. Some aspects of the Raman effect can be understood classically. First we will consider an isotropic molecule, that is, one that has the same optical properties in all

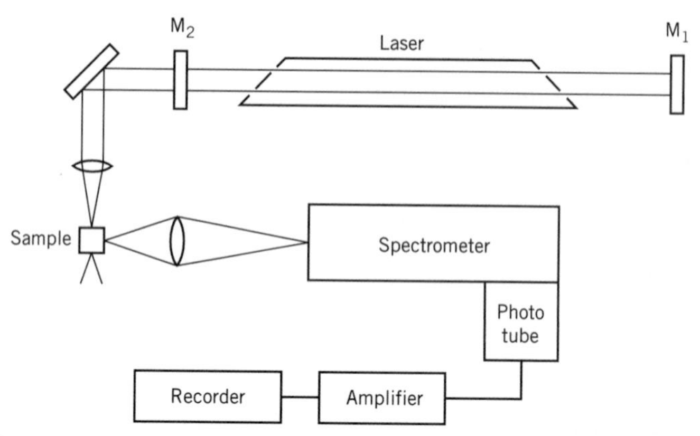

Figure 13.15 Apparatus for obtaining Raman spectra. Mirrors M_1 and M_2 are required to obtain laser action.

directions (CH_4 is an example). A dipole moment $\boldsymbol{\mu}$ is induced in the molecule by an electric field $\boldsymbol{E}$:

$$\boldsymbol{\mu} = \boldsymbol{\alpha E} \quad \text{or} \quad \mu = \alpha E \tag{13.106}$$

where α is the **polarizability.** The polarizability has units of dipole moment divided by electric field strength, that is, C m/V m^{-1} = C^2 m^2/J. For an isotropic molecule the vectors $\boldsymbol{\mu}$ and $\boldsymbol{E}$ point in the same direction, and the polarizability α is a scalar. The polarizability α of a molecule that is rotating or vibrating is not constant, but varies with some frequency ν_k (for example, a vibration or rotation frequency) according to

$$\alpha = \alpha_0 + (\Delta\alpha)\cos 2\pi\nu_k t \tag{13.107}$$

where α_0 is the equilibrium polarizability and $\Delta\alpha$ is its maximum variation. Since the electric field of the impinging electromagnetic radiation varies with time according to

$$E = E^\circ \cos 2\pi\nu_0 t \tag{13.108}$$

the induced dipole moment of the molecule is given by

$$\begin{aligned}\mu &= [\alpha_0 + (\Delta\alpha)\ \cos\ 2\pi\nu_k t]E^\circ \cos\ 2\pi\nu_0 t \\ &= \alpha_0 E^\circ \cos 2\pi\nu_0 t + \tfrac{1}{2}(\Delta\alpha)E^\circ[\cos 2\pi(\nu_0 + \nu_k)t + \cos 2\pi(\nu_0 - \nu_k)t]\end{aligned} \tag{13.109}$$

where the last form has been obtained using the relation $\cos a \cos b = \frac{1}{2}[\cos(a + b) + \cos(a - b)]$. The three terms in this equation provide the classical explanation for Rayleigh scattering (ν_0), anti-Stokes lines ($\nu_0 + \nu_k$), and Stokes lines ($\nu_0 - \nu_k$), respectively. However, the classical treatment incorrectly implies that the Stokes and anti-Stokes lines will occur with equal intensity. The anti-Stokes lines are, of course, weaker because they depend on the populations of excited levels.

In order for a molecular motion to be Raman active, the polarizability $\boldsymbol{\alpha}$ must change when that motion occurs (that is, $\Delta\alpha \neq 0$). *In order for a vibrational mode to be Raman active, the polarizability* $\boldsymbol{\alpha}$ *must change during the vibration, and for a rotation to be Raman active, the polarizability must change as the molecule rotates in an electric field.* The polarizability of both homonuclear and heteronuclear diatomic molecules changes as the distance between nuclei changes because this alters the electronic structure. The polarizability of an atom or a spherical rotor (Section 13.5) does not change in a rotation; indeed, we cannot even talk about the rotation of an atom. Thus, spherical rotors do not have a rotational Raman effect. All other molecules are anisotropically polarizable; that means that the polarization is dependent on the orientation of the molecule in the electric field.

When a molecule is anisotropic, the application of an electric field $\boldsymbol{E}$ in a particular direction induces a moment $\boldsymbol{\mu}$ in a different direction. In this case $\boldsymbol{\alpha}$ is a tensor, and the induced dipole moment is given by

$$\boldsymbol{\mu} = \boldsymbol{\alpha}\boldsymbol{E} \tag{13.110}$$

which is expressed by the following matrix equation (see Appendix D.8):

$$\begin{bmatrix} \mu_x \\ \mu_y \\ \mu_z \end{bmatrix} = \begin{bmatrix} \alpha_{xx} & \alpha_{xy} & \alpha_{xz} \\ \alpha_{yx} & \alpha_{yy} & \alpha_{yz} \\ \alpha_{zx} & \alpha_{zy} & \alpha_{zz} \end{bmatrix} \begin{bmatrix} E_x \\ E_y \\ E_z \end{bmatrix} \tag{13.111}$$

This is equivalent to the following set of algebraic equations:

$$\mu_x = \alpha_{xx} E_x + \alpha_{xy} E_y + \alpha_{xz} E_z \tag{13.112}$$

$$\mu_y = \alpha_{yx} E_x + \alpha_{yy} E_y + \alpha_{yz} E_z \tag{13.113}$$

$$\mu_z = \alpha_{zx} E_x + \alpha_{zy} E_y + \alpha_{zz} E_z \tag{13.114}$$

Thus, each component (μ_x, μ_y, μ_z) of the induced dipole moment $\boldsymbol{\mu}$ can depend on each component (E_x, E_y, E_z) of the electric field $\boldsymbol{E}$. Only six of the nine coefficients of the polarizability are independent, since it can be shown that $\alpha_{xy} = \alpha_{yx}$, $\alpha_{xz} = \alpha_{zx}$, and $\alpha_{yz} = \alpha_{zy}$.

The quantum mechanical theory for the selection rules for the Raman effect is more complicated than for pure rotational and vibrational spectra because Raman scattering is a kind of two-photon process: The incident photon is absorbed and the leaving photon is emitted by the molecule in a single quantum process.

The specific selection rules for rotational Raman transitions are as follows for linear and symmetric top molecules:

Linear molecules	$\Delta J = 0, \pm 2$	
Symmetric top	$\Delta J = 0, \pm 2, \Delta K = 0$	when $K = 0$
molecules	$\Delta J = 0, \pm 1, \pm 2, \Delta K = 0$	when $K \neq 0$

where K is the component of the angular momentum J along the principal symmetry axis. The $\Delta J = 0$ applies in vibration–rotation transitions. The fact that $\Delta J = \pm 2$ for linear molecules is a result of the fact that the polarizability of a molecule returns to its initial value twice in a 360° revolution, as shown in Fig. 13.16. The $\Delta K = 0$ is a result of the fact that the dipole of a symmetric top molecule is along the principal axis, so there cannot be a component of the dipole moment perpendicular to this axis.

The frequencies of the Stokes lines ($\Delta J = 2$) in the rotational Raman spectrum of a linear molecule are given by

$$\Delta\tilde{\nu}_{\mathrm{R}} = \tilde{B}J'(J'+1) - \tilde{B}J''(J''+1) \tag{13.115}$$

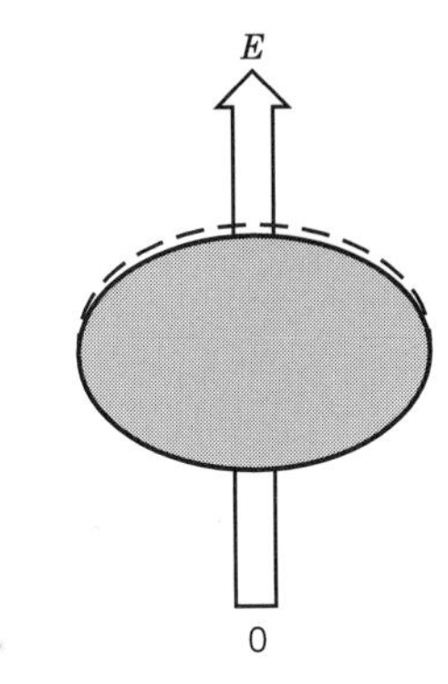

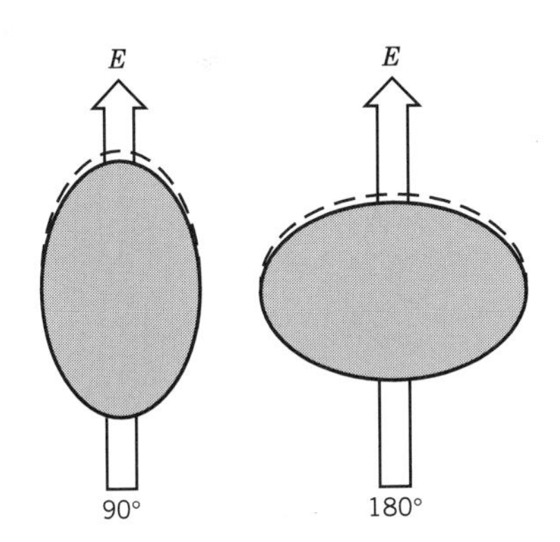

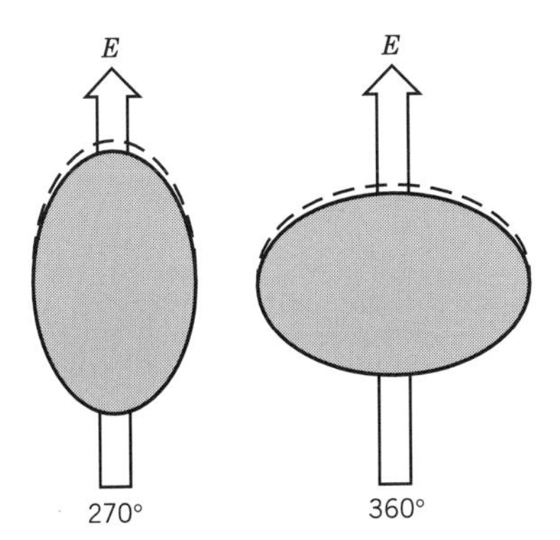

Figure 13.16 Polarizability ellipsoid for a molecule that returns to its initial value twice in a 360° revolution.

where J'' is the initial quantum number. Substituting $J' = J'' + 2$,

$$\begin{aligned} \Delta\tilde{\nu}_R &= \tilde{B}[(J'' + 2)(J'' + 3) - J''(J'' + 1)] \\ &= 2\tilde{B}(2J'' + 3) \end{aligned} \tag{13.116}$$

These lines appear at lower frequencies than the exciting line and are referred to as the S branch. The relative intensities of these lines are determined by the populations of the initial states, as we have discussed for the vibration–rotation spectra in the infrared.

The frequencies of the anti-Stokes lines ($\Delta J = -2$) in the rotational Raman spectrum are given by

$$\Delta\tilde{\nu}_R = -2\tilde{B}(2J'' - 1) \qquad \text{where } J'' \geq 2 \tag{13.117}$$

The lines appear at higher frequencies and are referred to as the O branch. In addition, there is a Q branch for $\Delta J = 0$. The S, Q, and O branches correspond to the P, Q, and R branches of infrared spectroscopy.

The pure rotational Raman spectrum of CO_2 is shown in Fig. 13.17. Notice the large number of initially populated rotational states, since the rotational splitting is small compared with kT.

As noted above, for a molecule to have a vibrational Raman spectrum it is necessary for the polarizability to change as the molecule vibrates. The polarizabilities of both homonuclear and heteronuclear diatomic molecules change as the molecule vibrates, so both types of molecules have vibrational Raman spectra in contrast to infrared vibrational spectra. The specific selection rule for the vibrational Raman effect is $\Delta v = \pm 1$. The vibrational transitions are accompanied by rotational Raman transitions with the specific selection rules $\Delta J = 0, \pm 2$, as

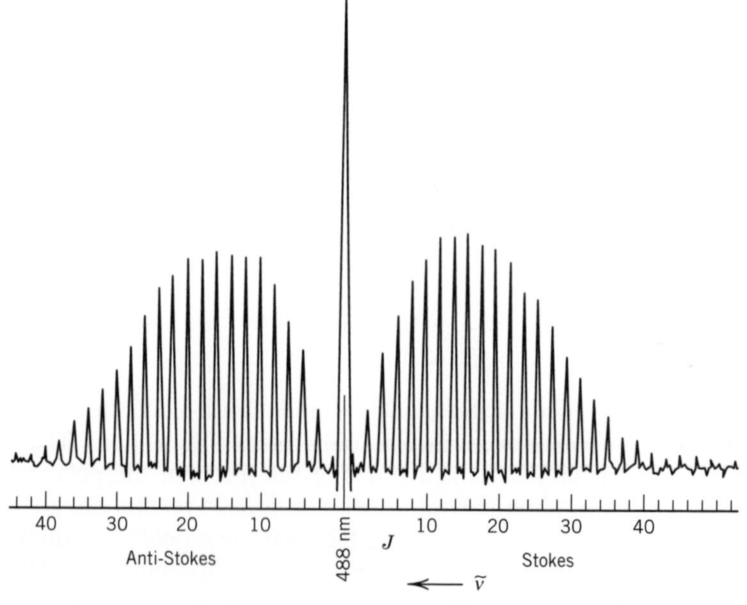

Figure 13.17 Pure rotational Raman spectrum of CO_2. The abscissa gives the quantum numbers for the lower quantum state. The intense peak at 488 nm is due to elastic scattering. (From B. P. Straughan and S. Walker, *Spectroscopy,* Vol. 2. London: Chapman & Hall, 1976.)

before. The vibrational Raman spectra of homonuclear diatomic molecules are of special interest because they yield force constants and rotational constants that are not available from infrared absorption spectroscopy.

Figure 13.18 shows the theoretical rotation–vibration Raman spectrum for $\Delta v = +1, 0, -1$ and $\Delta J = +2, 0, -1$ for a linear molecule. It is assumed that there is a small population in the first excited vibrational state. The center series of lines is the rotational Raman spectrum of the molecule. Since homonuclear diatomic molecules give spectra of this type, Raman spectroscopy provides the possibility, not available in microwave or infrared spectroscopy, of determining their internuclear distances and force constants.

For a polyatomic molecule, some normal modes will be Raman active and some will not, depending on what happens to the polarizability ellipsoid when the generalized coordinate for the normal mode has changed. This is most easily seen for a linear symmetric XY_2 molecule, such as CO_2, in which the principal axes of the polarizability ellipsoid are coincident with the symmetry axes of the molecule. In these molecules, only the symmetrical stretching normal mode is Raman active. This particular mode is not active in the infrared (Section 13.8). According to the **mutual exclusion rule** for molecules with a center of symmetry, fundamental transitions that are active in the infrared are forbidden in the Raman scattering, and vice versa. However, there are some vibrations that are forbidden in both spectra. The torsional vibration of ethylene is neither infrared nor Raman active; this is the vibration in which ethylene is twisted out of its planar equilibrium structure. Benzene has thirty normal modes of vibration, and eight of them are totally inactive in both infrared and Raman; these are referred to as spectroscopically dark vibrations.

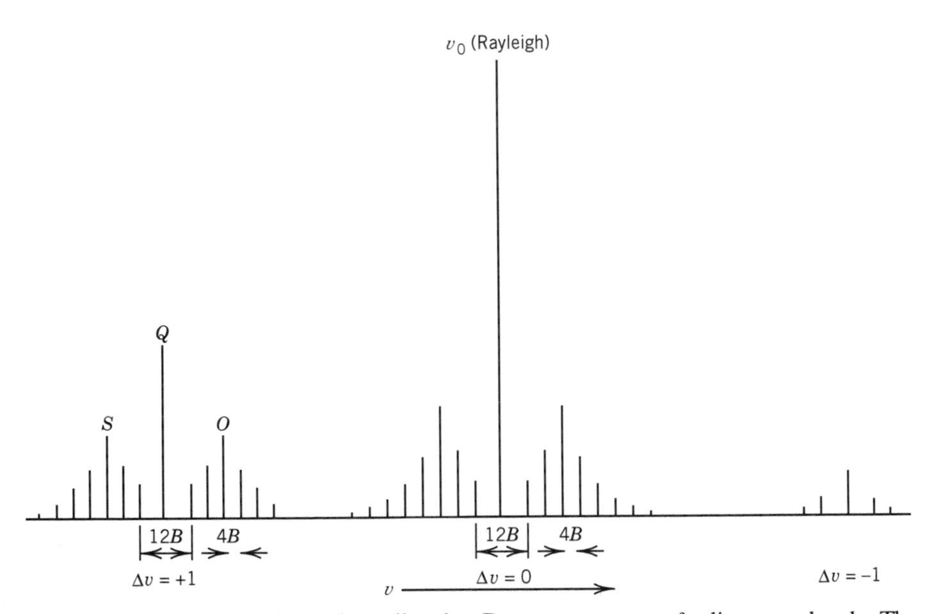

Figure 13.18 Theoretical rotation–vibration Raman spectrum of a linear molecule. The effects of nuclear spin statistics have been omitted from this illustration. (From W. A. Guillory, *Introduction to Molecular Structure and Spectroscopy.* Boston: Allyn & Bacon, 1977. Used by permission.)

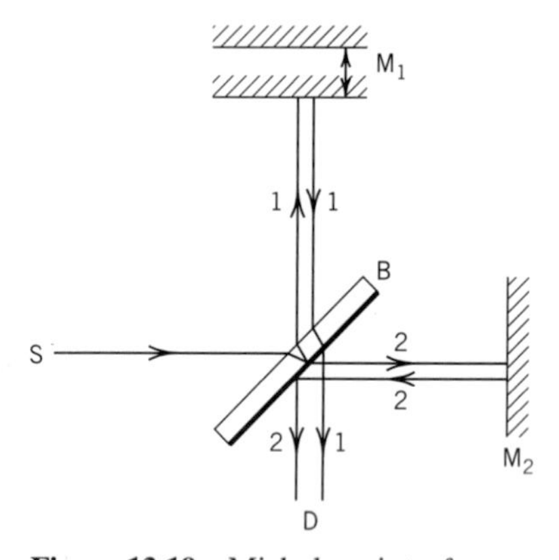

Figure 13.19 Michelson interferometer with source S, beam splitter B, movable mirror M_1, fixed mirror M_2, and detector D. Rays 1 and 2 are recombined at the detector, but they have a path difference δ, which causes interference.

The symmetry group of the molecule can be used to determine whether a particular vibration is Raman active or infrared active. Those normal modes whose symmetry is the same as the functions x, y, or z will be infrared active, while those whose symmetry is the same as x^2, y^2, or z^2 will be Raman active (see Chapter 12).

The intensity of Raman scattering is very much increased when the exciting frequency ν_0 coincides with an electronic absorption frequency (see Section 14.5); this is called **resonance Raman spectroscopy.**

13.10 SPECIAL TOPIC: FOURIER TRANSFORM INFRARED SPECTROSCOPY

The sensitivity of infrared absorption measurements can be greatly increased by using a method involving Fourier transforms. A Michelson interferometer is built into the spectrometer, along with a dedicated computer. The construction of the Michelson interferometer is shown in Fig. 13.19. The infrared radiation that has been transmitted by the sample, designated as the source S, is split into two rays by B, which is usually a very thin film of germanium supported on a potassium bromide substrate. The beam splitter transmits half of the infrared radiation from the sample and reflects half toward a movable mirror M_1. The transmitted ray is reflected from a stationary mirror M_2. When the two rays reach the detector D, there is interference because of the path difference δ. If the radiation from the sample is monochromatic, the intensity measured at the detector $I(\delta)$ is given by

$$I(\delta) = I(\tilde{\nu})(1 + \cos 2\pi\tilde{\nu}\delta) \tag{13.118}$$

where $I(\tilde{\nu})$ is the intensity of the beam from the sample. The dependence of intensity on path difference is shown in Fig. 13.20. The path difference δ is changed by moving mirror M_1. If the radiation transmitted by the sample is polychromatic, equation 13.118 is replaced by a summation or integral:

$$I(\delta) = \int_0^\infty I(\tilde{\nu})(1 + \cos 2\pi\tilde{\nu}\delta)\, \mathrm{d}\tilde{\nu} \tag{13.119}$$

This intensity can be measured at a series of path differences by moving M_1 continuously. The plot of $I(\delta)$ versus δ shows beats, as illustrated by Fig. 13.21 for radiation containing two frequencies. Much more complicated signals as a function of path difference are obtained from actual infrared spectra. These measurements can be used to calculate $I(\tilde{\nu})$ by use of a Fourier transformation, which gives

$$I(\tilde{\nu}) = \int_0^\infty \left[I(\delta) - \frac{1}{2}I(0)\right] \cos 2\pi\tilde{\nu}\delta\, \mathrm{d}\delta \tag{13.120}$$

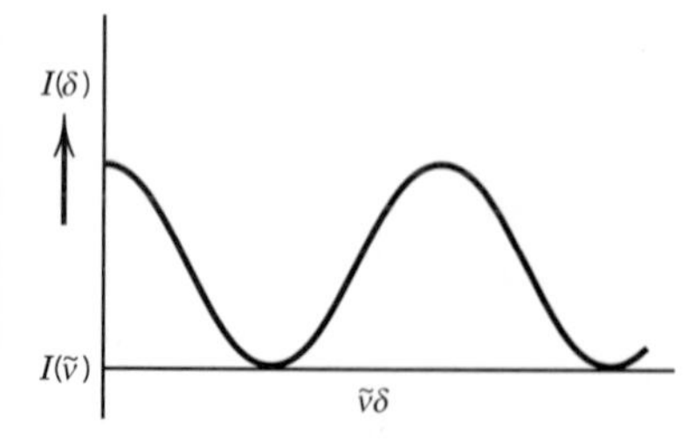

Figure 13.20 Intensity $I(\delta)$ measured at the detector as a function of path difference for monochromatic radiation from the source.

where $I(0)$ is the intensity for zero path difference. This integration is carried out by the dedicated computer in the spectrometer, and the spectrum $I(\tilde{\nu})$ is plotted out. To a first approximation, the resolution is inversely proportional to the distance moved by mirror M_1, but there are problems with making this distance greater than about 5 cm, which gives a resolution of about 0.1 cm^{-1}.

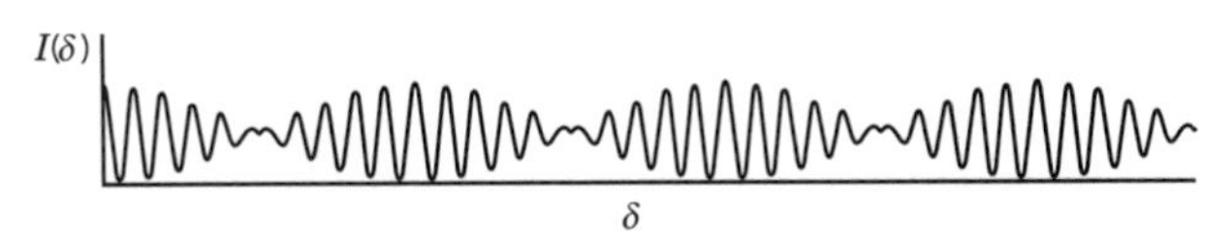

Figure 13.21 $I(\delta)$ versus δ when radiation from the source contains two frequencies.

The advantage of a Fourier transform spectrometer is that it makes use of the radiation at all wave numbers from the source for all of the time of recording. Fourier transforms are also used in nuclear magnetic resonance spectroscopy (Section 15.10) and in determining the structures of crystals by X-ray diffraction (Section 23.6).

Ten Key Ideas in Chapter 13

1. The electromagnetic spectrum is divided into regions by the modes of detection, but these regions are also characterized by the ranges of photon energy they involve. The absorption of a photon leads to different types of energetic changes in molecules in different regions of the spectrum.
2. Consideration of the equations for Einstein's stimulated absorption, spontaneous emission, and stimulated emission shows that irradiation of a two-level system can never put more atoms or molecules in the higher level than in the lower level.
3. The Hamiltonian of a molecule can, to a good approximation, be separated into translational, vibrational, and rotational contributions. When this is satisfactory, the wavefunction can be written as the product of translational, vibrational, and rotational wavefunctions.
4. The rotational lines for a rigid diatomic molecule are equally spaced, and the spacing yields the moment of inertia. To have a rotational spectrum, a molecule must have a permanent dipole moment.
5. Diatomic molecules are not harmonic oscillators, and the deviation from equal spacing of vibrational lines yields anharmonicity constants. To have a vibrational spectrum, a molecule must have a dipole moment that changes with internuclear distance.
6. Vibration–rotation spectra yield vibration–rotation coupling constants.
7. Vibrational spectra of polyatomic molecules with N atoms can be described in terms of $3N - 5$ coordinates for linear molecules and $3N - 6$ coordinates for nonlinear molecules. There is a Hamiltonian for each of these normal modes of vibration.
8. When light is scattered by molecules, part of the scattered light emerges with lower frequency (Stokes lines) and a smaller part emerges with a higher frequency (anti-Stokes lines). The incident light need not coincide with a quantized energy difference in the molecule. Laser light sources are used because of their brightness.
9. In order for a vibrational mode to be Raman active, the polarizability must change as the molecule vibrates, and for a rotation to be Raman active, the polarizability must change as the molecule rotates.
10. Spectroscopic measurements yield properties of gas molecules that have many scientific and practical applications.

REFERENCES

P. F. Bernath, *Spectra of Atoms and Molecules.* Oxford, UK: Oxford University Press, 1995.
G. Herzberg, *Infrared and Raman Spectra of Polyatomic Molecules.* Princeton, NJ: Van Nostrand, 1950.
G. Herzberg, *Spectra of Diatomic Molecules.* Princeton, NJ: Van Nostrand, 1950.
G. Herzberg, *The Spectra and Structures of Simple Free Radicals.* Ithaca, NY: Cornell University Press, 1971.
J. M. Hollas, *Modern Spectroscopy.* Hoboken, NJ: Wiley, 1996.
K. P. Huber and G. Herzberg, *Molecular Spectra and Molecular Structure IV, Constants of Diatomic Molecules.* New York: Van Nostrand–Reinhold, 1979.
J. L. McHale, *Molecular Spectroscopy.* Upper Saddle River, NJ: Prentice-Hall, 1999.
J. I. Steinfeld, *Molecules and Radiation.* Cambridge, MA: MIT Press, 1985.
B. P. Straughan and S. Walker, *Spectroscopy,* Vols. 1, 2, and 3. Hoboken, NJ: Wiley, 1976.
E. B. Wilson, J. C. Decius, and P. C. Cross, *Molecular Vibrations.* New York: McGraw-Hill, 1955.

PROBLEMS

Problems marked with an icon may be more conveniently solved on a personal computer with a mathematical program.

13.1 Since the energy of a molecular quantum state is divided by kT in the Boltzmann distribution, it is of interest to calculate the temperature at which kT is equal to the energy of photons of different wavelengths. Calculate the temperature at which kT is equal to the energy of photons of wavelength 10^3 cm, 10^{-1} cm, 10^{-3} cm, and 10^{-5} cm.

13.2 Most chemical reactions require activation energies ranging between 40 and 400 kJ mol^{-1}. What are the equivalents of 40 and 400 kJ mol^{-1} in terms of (*a*) nm, (*b*) wave numbers, and (*c*) electron volts?

13.3 (*a*) What vibrational frequency in wave numbers corresponds to a thermal energy of kT at 25 °C? (*b*) What is the wavelength of this radiation?

13.4 Show that equation 13.17 is a solution of equation 13.9 by differentiating equation 13.17 and substituting it into equation 13.9.

13.5 Calculate the reduced mass and the moment of inertia of $D^{35}Cl$, given that $R_e = 127.5$ pm.

13.6 The H—O—H bond angle for ^1H_2O is 104.5°, and the H—O bond length is 95.72 pm. What is the moment of inertia of H_2O about its C_2 axis?

13.7 Some of the following gas molecules have pure microwave absorption spectra and some do not: N_2, HBr, CCl_4, CH_3CH_3, CH_3CH_2OH, H_2O, CO_2, O_2. What is the gross selection rule for rotational spectra, and which molecules satisfy it?

13.8 Calculate the frequency in wave numbers and the wavelength in cm of the first rotational transition ($J = 0 \rightarrow 1$) for $D^{35}Cl$.

13.9 The pure rotational spectrum of $^{12}C^{16}O$ has transitions at 3.863 and 7.725 cm^{-1}. Calculate the internuclear distance in $^{12}C^{16}O$. Predict the positions, in cm^{-1}, of the next two lines.

13.10 Assume the bond distances in $^{13}C^{16}O$, $^{13}C^{17}O$, and $^{12}C^{17}O$ are the same as in $^{12}C^{16}O$. Calculate the position, in cm^{-1}, of the first rotational transitions in these four molecules. (Use the information in Problem 13.9.)

13.11 The far-infrared spectrum of HI consists of a series of equally spaced lines with $\Delta\tilde{\nu} = 12.8$ cm^{-1}. What is (*a*) the moment of inertia and (*b*) the internuclear distance?

13.12 For $H^{35}Cl$ calculate the relative populations of rotational levels, f_J/f_0, for the first three levels at 300 K and 1000 K.

13.13 Using equation 13.44, show that J for the maximally populated level is given by

$$J_{\max} = \sqrt{\frac{kT}{2hcB}} - \frac{1}{2}$$

13.14 Using the result of Problem 13.13, find the J nearest $J_{\max}$ at room temperature for $H^{35}Cl$ and $^{12}C^{16}O$. (*a*) What is the ratio of the population at that J to the population at $J = 0$? (*b*) What is the energy of that J relative to $J = 0$ in units of kT?

13.15 The moment of inertia of $^{16}O^{12}C^{16}O$ is 7.167×10^{-46} kg m^2. (*a*) Calculate the CO bond length, R_{CO}, in CO_2. (*b*) Assuming that isotopic substitution does not alter R_{CO}, calculate the moments of inertia of (1) $^{18}O^{12}C^{18}O$ and (2) $^{16}O^{13}C^{16}O$.

13.16 Derive the expression for the moment of inertia of a symmetrical tetrahedral molecule such as CH_4 in terms of the bond length R and the masses of the four tetrahedral atoms. The easiest way to derive the expression is to consider an axis along one CH bond. Show that the same result is obtained if the axis is taken perpendicular to the plane defined by one group of three atoms HCH.

13.17 What are the values of $\tilde{A}$ and $\tilde{B}$ (from equation 13.62) for the symmetric top NH_3 if $I_{\parallel} = 4.41 \times 10^{-47}$ kg m^2 and $I_{\perp} = 2.81 \times 10^{-47}$ kg m^2? What is the wavelength of the $J = 0$ to $J = 1$ transition? What are the wavelengths of the $J = 1$ to $J = 2$ transitions (remember the selection rules, $\Delta J = \pm 1$, $\Delta K = 0$, and find all allowed transitions)?

13.18 Consider a linear triatomic molecule, ABC. Find the center of mass (which by symmetry lies on the molecular axis). Show that the moment of inertia is given by

$$I = \frac{1}{M}[R_{AB}^2 m_A m_B + R_{BC}^2 m_B m_C + (R_{AB} + R_{BC})^2 m_A m_C]$$

where R_{AB} is the AB bond distance, R_{BC} is the BC bond distance, m_i are the masses of the atoms, and $M = m_A + m_B + m_C$. Show that if $R_{AB} = R_{BC}$ and $m_A = m_C$, then $I = 2m_A R_{AB}^2$.

13.19 The fundamental vibration frequency of $H^{35}Cl$ is 8.967×10^{13} s^{-1} and that of $D^{35}Cl$ is 6.428×10^{13} s^{-1}. What would the separation be between infrared absorption lines of $H^{35}Cl$ and $H^{37}Cl$ on one hand and those of $D^{35}Cl$ and $D^{37}Cl$ on the other, if the force constants of the bonds are assumed to be the same in each pair?

13.20 Find the force constants of the halogens $^{127}I_2$, $^{79}Br_2$, and $^{35}Cl_2$ using the data of Table 13.4. Is the order of these the same as the order of the bond energies?

13.21 Given the following fundamental frequencies of vibration, calculate $\Delta H°$ for the reaction

$$H^{35}Cl(v = 0) + {}^2D_2(v = 0) = {}^2D^{35}Cl(v = 0) + H^2D(v = 0)$$

$H^{35}Cl$: 2989 cm^{-1} H^2D: 3817 cm^{-1}

$^2D^{35}Cl$: 2144 cm^{-1} $^2D^2D$: 3119 cm^{-1}

(M) **13.22** If the fundamental vibration frequency of $^1H_2$ is 4401.21 cm^{-1}, compute the fundamental vibration frequency of $^2D_2$ and $^1H^2D$ assuming the same force constants. If D_0 for $^1H_2$ is 4.4781 eV, what is D_0 for $^2D_2$ and ^1H_2D? Neglect anharmonicities.

13.23 Using the values for $\tilde{\nu}_e$ and $\tilde{\nu}_e\tilde{x}_e$ in Table 13.4 for $^1H^{35}Cl$, estimate the dissociation energy assuming the Morse potential is applicable.

13.24 Apply the Taylor expansion to the potential energy given by the Morse equation $\tilde{V}(R) = D_e\{1 - \exp[-a(R - R_0)]\}^2$ to show that the force constant k is given by $k = 2D_e a^2$.

13.25 (*a*) What fraction of H_2(g) molecules are in the $v = 1$ state at room temperature? (*b*) What fractions of Br_2(g) molecules are in the $v = 1$, 2, and 3 states at room temperatures?

13.26 The first three lines in the R branch of the fundamental vibration–rotation band of $H^{35}Cl$ have the following frequencies in cm^{-1}: 2906.25 (0), 2925.78 (1), 2944.89 (2), where the numbers in parentheses are the J values for the initial level. What are the values of $\tilde{\nu}_0$, B_v', B_v'', B_e, and α?

13.27 In Table 13.3, D_e for H_2 is given as 4.7483 eV or 458.135 kJ mol^{-1}. Given the vibrational parameters for H_2 in Table 13.4, calculate the value you would expect for $\Delta_f H°$ for H(g) at 0 K.

13.28 Calculate the wavelengths in (*a*) wave numbers and (*b*) micrometers of the center two lines in the vibration spectrum of HBr for the fundamental vibration. The necessary data are to be found in Table 13.4.

13.29 How many normal modes of vibration are there for (*a*) SO_2 (bent), (*b*) H_2O_2 (bent), (*c*) HC≡CH (linear), and (*d*) C_6H_6?

13.30 List the numbers of translational, rotational, and vibrational degrees of freedom for (*a*) Ne, (*b*) N_2, (*c*) CO_2, and (*d*) CH_2O.

(M) **13.31** Acetylene is a symmetrical linear molecule. It has seven normal modes of vibration, two of which are doubly degenerate. These normal modes may be represented as follows:

$$\leftarrow \mathrm{H}-\overset{\rightarrow}{\mathrm{C}}\equiv\overset{\leftarrow}{\mathrm{C}}-\mathrm{H} \rightarrow \qquad \underset{\downarrow}{\mathrm{H}}-\overset{\uparrow}{\mathrm{C}}\equiv\underset{\downarrow}{\mathrm{C}}-\overset{\uparrow}{\mathrm{H}}$$

$$\tilde{\nu}_1 = 3374 \text{ cm}^{-1} \qquad \tilde{\nu}_4 = 612 \text{ cm}^{-1}$$

$$\mathrm{H} \rightarrow -\overset{\rightarrow}{\mathrm{C}}\equiv\overset{\leftarrow}{\mathrm{C}}- \leftarrow \mathrm{H} \qquad \overset{\uparrow}{\mathrm{H}}-\underset{\downarrow}{\mathrm{C}}\equiv\underset{\downarrow}{\mathrm{C}}-\overset{\uparrow}{\mathrm{H}}$$

$$\tilde{\nu}_2 = 1974 \text{ cm}^{-1} \qquad \tilde{\nu}_5 = 729 \text{ cm}^{-1}$$

$$\mathrm{H} \rightarrow \leftarrow \mathrm{C}\equiv\overset{\leftarrow}{\mathrm{C}}-\mathrm{H} \rightarrow$$

$$\tilde{\nu}_3 = 3287 \text{ cm}^{-1}$$

(*a*) Which are the doubly degenerate vibrations? (*b*) Which vibrations are infrared active? (*c*) Which vibrations are Raman active?

13.32 Calculate the wave number and wavelength of the pure fundamental ($v = 0 \rightarrow 1$) vibrational transitions for (*a*) $^{12}C^{16}O$ and (*b*) $^{39}K^{35}Cl$ using data in Table 13.4.

13.33 (*a*) Consider the four normal modes of vibration of a linear molecule AB_2 from the standpoint of changing dipole moment and changing polarizability. Which vibrational modes are infrared active, and which are Raman active? (Note the exclusion rule.) (*b*) Consider the three normal modes of a nonlinear molecule AB_2. Which vibrational modes are infrared active, and which are Raman active?

13.34 Calculate the fraction of Cl_2 molecules ($\tilde{\nu} = 559.7$ cm^{-1}) in the $i = 0, 1, 2, 3$ vibrational states at 1000 K.

13.35 When CCl_4 is irradiated with the 435.8-nm mercury line, Raman lines are obtained at 439.9, 441.8, 444.6, and 450.7 nm. Calculate the Raman frequencies of CCl_4 (expressed in wave numbers). Also calculate the wavelengths (expressed in μm) in the infrared at which absorption might be expected.

13.36 The first several Raman frequencies of $^{14}N_2$ are 19.908, 27.857, 35.812, 43.762, 51.721, and 59.662 cm^{-1}. These lines are due to pure rotational transitions with $J = 1, 2, 3, 4, 5$, and 6. The spacing between the lines is $4B_e$. What is the internuclear distance?

13.37 What Raman shifts are expected for the first four Stokes lines for CO_2?

13.38 Some of the following gas molecules have a pure rotational Raman spectrum and some do not: N_2, HBr, CCl_4, CH_3CH_3, CH_3CH_2OH, H_2O, CO_2, O_2. What is the gross selection rule for pure rotational Raman spectra, and which molecules satisfy it?

13.39 Calculate the factors for converting between eV and cm^{-1} and between eV and kJ mol^{-1}

13.40 Energies in electron volts (eV) may be expressed in terms of temperature by use of the relation $e\phi = kT$, where ϕ is the difference in potential in V. What temperature corresponds to 1 V? 100 V? 1000 V? What is the electron volt equivalent of room temperature?

13.41 The internuclear distance in CO is 112.82 pm. Calculate (*a*) the reduced mass and (*b*) the moment of inertia.

13.42 Calculate the frequencies in cm^{-1} and the wavelengths in μm for the pure rotational lines in the spectrum of $H^{35}Cl$ corresponding to the following changes in rotational quantum number: $0 \rightarrow 1, 1 \rightarrow 2, 2 \rightarrow 3$, and $8 \rightarrow 9$.

13.43 Assuming that the internuclear distance is 74.2 pm for (*a*) H_2, (*b*) HD, (*c*) HT, and (*d*) D_2, calculate the moments of inertia of these molecules.

13.44 Calculate the energy difference in cm^{-1} and kJ mol^{-1} between the $J = 0$ and $J = 1$ rotational levels of OH, using the data of Table 13.4. Assuming that OD has the same internuclear distance as OH, calculate the energy difference between $J = 0$ and $J = 1$ in OD.

13.45 In the pure rotational spectrum of $^{12}C^{16}O$, the lines are separated by 3.8626 cm^{-1}. What is the internuclear distance in the molecule?

13.46 Consider the molecular radicals ^{12}CH and ^{13}CH. Calculate their moments of inertia using R_e from Table 13.4 and assuming R_e is the same in both. Using the results of Problem 13.13, find the value of J closest to J_{max} at room temperature, and compute the difference in energy between this state and the next higher energy state.

13.47 The separation of the pure rotation lines in the spectrum of CO is 3.86 cm^{-1}. Calculate the equilibrium internuclear separation.

13.48 Show that for large J the frequency of radiation absorbed in exciting a rotational transition is approximately equal to the classical frequency of rotation of the molecule in its initial or final state.

13.49 For the rotational Raman effect, what are the displacements of the successive Stokes lines in terms of the rotational constant B? Is the answer the same for the anti-Stokes lines?

13.50 Show that the moments of inertia of a regular hexagonal molecule made up of six identical atoms of mass m are given by

$$I_{\parallel} = 6mr^2 \quad \text{and} \quad I_{\perp} = 3mr^2$$

where r is the bond distance.

13.51 What are the frequencies of the first three lines in the rotational spectrum of $^{16}O^{12}C^{32}S$ given that the O—C distance is 116.47 pm, the C—S distance is 155.76 pm, and the molecule is linear. Atomic masses of isotopes are given inside the back cover. The moment of inertia of a linear molecule ABC is given in Problem 13.18.

13.52 What are the rotational frequencies for the first three rotational lines in $^{16}O^{12}C^{34}S$, assuming the same bond lengths as in Problem 13.51?

13.53 Ammonia is a symmetric top with

$$I_{xx} = I_{yy} = I_{\perp} = 2.8003 \times 10^{-47} \text{ kg m}^2$$
$$I_{zz} = I_{\parallel} = 4.4300 \times 10^{-47} \text{ kg m}^2$$

Calculate the characteristic rotational temperatures Θ_r where

$$\Theta_r = \frac{h^2}{8\pi^2 Ik}$$

13.54 Using the Morse potential expression, equation 13.82, estimate D_e for HBr, HCl, and HI from the data in Table 13.4.

13.55 Calculate the values of D_e for HCl, HBr, and HI using the data of Table 13.4 and equation 13.80 (neglect y_e).

13.56 From the data of Table 13.4, calculate the vibrational force constants of HCl, HBr, and HI. Are these in the same order as the dissociation energies?

13.57 Using the Boltzmann distribution (equation 16.17), calculate the ratio of the population of the first vibrational excited state to the population of the ground state for $H^{35}Cl$ ($\tilde{\nu}_0 = 2990$ cm^{-1}) and $^{127}I_2$ ($\tilde{\nu}_0 = 213$ cm^{-1}) at 300 K.

13.58 Use the Morse potential to estimate the equilibrium dissociation energy for $^{79}Br_2$ using $\tilde{\nu}_e$ and $\tilde{\nu}_e x_e$ from Table 13.4.

13.59 The wave numbers of the first several lines in the R branch of the fundamental ($v = 0 \rightarrow 1$) vibrational band for $^2H^{35}Cl$ have the following frequencies in cm^{-1}: 2101.60 (0), 2111.94 (1), 2122.05 (2), where the numbers in parentheses are the J values for the initial level. What are the values of $\tilde{B}'_v$, $\tilde{B}''_v$, $\tilde{B}_e$, and α? How does the internuclear distance compare with that for $^1H^{35}Cl$?

13.60 Gaseous HBr has an absorption band centered at about 2645 cm^{-1} consisting of a series of lines approximately equally spaced with an interval of 16.9 cm^{-1}. For gaseous DBr estimate the frequency in wave numbers of the band center and the interval between lines.

13.61 Some of the following gas molecules have infrared absorption spectra and some do not: N_2, HBr, CCl_4, CH_3CH_3, CH_3CH_2OH, H_2O, CO_2, O_2. What is the gross selection rule for vibrational spectra, and which molecules satisfy it?

13.62 List the numbers of translational, rotational, and vibrational degrees of freedom of Cl_2, H_2O, and C_2H_2.

13.63 List the numbers of translational, rotational, and vibrational degrees of freedom of NNO (a linear molecule) and NH_3.

13.64 The rotational Raman spectrum of hydrogen gas is measured using a 488-nm laser. Stokes lines are observed at 355, 588, 815, and 1033 cm^{-1}. Since these transitions are of the type $J \rightarrow J+2$, it may be shown that the wave numbers of these lines are given by

$$\Delta\tilde{\nu}_R = 4\tilde{B}_e(J + \tfrac{3}{2})$$

where J is the rotational quantum number of the initial state (0, 1, 2, and 3, respectively, for the above lines) and $\tilde{B}_e$ is given by equation 13.34. What is R_e? [L. C. Hoskins, *J. Chem. Educ.* **54:**642 (1977).]

13.65 The rotational Raman spectrum of nitrogen gas shows Raman shifts of 19, 27, 34, 53, ... cm^{-1}, corresponding to rotational quantum numbers of the initial state of $J = 1, 2, 3, 4, \ldots$. Since the spacing is $4B_e$ ignoring centrifugal distortion, what is R_e? [L. C. Hoskins, *J. Chem. Educ.* **52:**568 (1975).]

13.66 Calculate $\Delta H°$(298 K) for the reaction

$$H_2 + D_2 = 2HD$$

assuming that the force constant is the same for all three molecules.

Computer Problems

13.A Plot the Morse potentials for molecular hydrogen in the ground state and the first excited state (see Table 13.4 for parameters), and put the plots on the same graph so that they can be compared with Fig. 11.12.

13.B Plot the Morse potential for molecular chlorine in the ground state (see Table 13.4 for parameters) and put in lines for every fifth vibrational level from $v = 0$ to $v = 40$.

13.C (*a*) Calculate the energies (in cm^{-1}) of the rotational levels of $H^{35}Cl$ for $J = 0$ to $J = 15$, taking into account the centrifugal distortion ($D = 4.4 \times 10^{-4}$ cm^{-1}). (*b*) Calculate the frequencies of absorption due to $J \rightarrow J+1$ transitions in cm^{-1}. (*c*) Calculate the fractions f of the molecules in each of these 15 levels using equation 13.43 at 300 K and plot f versus frequencies of absorption in wave numbers (cm^{-1}). (*d*) Calculate the fractions f of the molecules in each of these 15 levels using equation 13.43 at 500 K and plot these fractions on the same graph.

13.D Calculate the frequencies of lines in the vibration–rotation spectrum of HCl for $v = 0$ to $v = 1$, including the dependence of the rotational constant on the vibrational quantum number. Energies can be expressed in cm^{-1}. Plot the line spectra for $v = 0$ to $v = 2$ and $J = 0$ to $J = 10$. Make further plots to show the effect of a larger $\tilde{\nu}_e x_e$, a larger $\tilde{B}_e$, and a larger $\tilde{\alpha}_e$.

13.E (*a*) Given the spectroscopic constants for $H^{35}Cl$ in Table 13.4, calculate the energy levels (in cm^{-1}) for $J = 0, 1, 2, 3$, and 4 in the vibrational states $v = 0$ and $v = 1$. (*b*) Calculate the frequencies (in cm^{-1}) of the first two lines in the R branch of $v = 0 \rightarrow 1$ and the first two lines of the P branch of the absorption spectrum. Compare your calculations with Fig. 13.12.

13.F The observed vibrational frequencies for an anharmonic oscillator are given by

$$\tilde{\nu}_{obs} = \tilde{G}(v) - \tilde{G}(0) = \tilde{\nu}_e v - \tilde{\nu}_e x_e v(v+1)$$

Calculate the fundamental vibration frequency $\tilde{\nu}_e$ and the anharmonicity constant x_e for $H^{35}Cl$, for which the frequencies for the $0 \rightarrow v$ transitions are 2885.9, 5668.0, 8347.0, 10 923.1, and 13 396.5 cm^{-1} for $v = 1, 2, 3, 4$, and 5.

13.G Plot the Morse potential for the chlorine molecule and the parabolic curve $V = k(R - R_e)^2/2$ with a force constant k that corresponds to that for the Morse potential at the minimum of the potential energy plot.

13.H Plot the Morse potential for $H^{35}Cl$ using the parameters calculated in Example 13.7 and the fact that $R_e = 0.1275$ nm.

13.I Calculate the vibrational frequencies in wave numbers for the fundamental absorption band of $H^{35}Cl$ and the first four overtones for (*a*) the harmonic oscillator approximation and (*b*) the anharmonic oscillator. The spectroscopic constants are given in Table 13.4.

14

Electronic Spectroscopy of Molecules

Transitions between electronic levels of molecules lead to absorption and emission in the visible and ultraviolet parts of the spectrum. For some molecules the energy required to change electronic structure is so great that absorption occurs only in the high-energy vacuum ultraviolet parts of the spectrum. Electronic spectra contain many lines because electronic excitation is accompanied by change in vibrational and rotational states as well. When the lines can be resolved, electronic spectra are a rich source of information about molecular properties. At higher pressures in the gas phase the lines are so closely spaced that continuous absorption is obtained. The phenomena of fluorescence and phosphorescence involve electronic changes. Electronic spectra of molecules provide a means for learning about excited electronic states that are involved in photochemical reactions.

The development of lasers has revolutionized many areas of spectroscopy because of the extraordinary properties of laser radiation.

14.1 ELECTRONIC ENERGY LEVELS AND SELECTION RULES

In discussing H_2^+ in Chapter 11, we formed one-electron molecular orbitals from atomic orbitals and then used these to describe the electronic states of the H_2

molecule by putting the electrons into these orbitals in various ways. This provided us with an introduction to molecular term symbols that designate the symmetry, multiplicity, and angular momentum properties of the electronic state. **Electronic spectroscopy** is the excitation of electrons from low-energy orbitals to higher-energy orbitals by the absorption of light. Since the energies required to do this are generally much larger than vibrational and rotational energies, this form of spectroscopy uses light in the visible, ultraviolet, and even shorter-wavelength parts of the spectrum. In addition, vibrational and rotational energy changes almost always occur during an electronic transition, complicating the spectrum. The analysis of such spectra can give us information about dissociation energies, bond lengths, force constants, and potential energy curves.

We will begin by studying the electronic spectra of diatomic molecules, which are the most developed and best understood. In the gas phase at low pressure, the vast majority of diatomic molecules are in the ground electronic state and mainly in the lowest vibrational state. However, there is usually a broad distribution of rotational levels (since the rotational energy spacings are small compared with kT). When light is absorbed by these molecules, the electronic energy level is changed, the vibrational level may change, and the rotational level will change. The selection rules for vibrational level changes are not as stringent as in infrared (pure vibration–rotation) spectroscopy; therefore, there are many absorption lines in the electronic spectrum. At low pressures these are narrow, and we can often analyze the spectrum completely. At high pressures, collisions between molecules reduce the lifetimes of the initial and final states, thereby broadening the absorption spectra. In solution, collisions produce a spectrum that appears smooth and continuous, obscuring the very large number of lines making up the spectrum.

To analyze the spectra of diatomic molecules, we need to know the electric dipole selection rules just as we did for atoms (Section 10.14). For atoms, we saw that the levels were labeled by the angular momentum quantum numbers L, S, and J. In diatomic molecules, we have to consider in addition the angular momenta due to molecular rotation and due to nuclear spins. The selection rules for diatomic molecules become complicated and depend on the way the angular momenta are coupled in the molecule. The most common situation (especially for diatomics made up of light atoms in the first several rows of the periodic table) is called Hund's case (a). For this case the **selection rules** are as follows.

1. $\Delta\Lambda = 0, \pm 1$. Λ is the component of orbital angular momentum along the z axis; it can have values $0, 1, 2, \ldots, L$. Since electronic states are designated Σ, Π, Δ, Φ, Γ, ... corresponding to $\Lambda = 0, 1, 2, 3, 4, \ldots$, we see that Σ–Σ, Π–Σ, and Δ–Π transitions are allowed, but Δ–Σ and Φ–Π are not.
2. $\Delta S = 0$. Thus, as in atomic spectra, singlet–singlet and triplet–triplet transitions may occur, but singlet–triplet transitions are forbidden. This rule breaks down in molecules with nuclei with large atomic numbers.
3. $\Delta\Sigma = 0$. The quantum number Σ in a molecule is analogous to M_s in an atom and can take the values $S, S-1, \ldots, -S$. This quantum number, which determines the multiplicity $(2S+1)$ of a state, is reported as a presuperscript in a molecular term symbol.
4. $\Delta\Omega = 0, \pm 1$. The total angular momentum $\Omega\hbar$ along the internuclear axis, which is given by $|\Lambda + \Sigma|$, is sometimes reported as a postsubscript in the molecular term symbol.

5. Σ^+–Σ^+ and Σ^-–Σ^- transitions are allowed, but Σ^+–Σ^- are not. These post-superscripts refer to whether the wavefunction for the electronic state is symmetric (+) or antisymmetric (−) to reflection across any σ_v plane.

6. $g \leftrightarrow u$, $g \not\leftrightarrow g$, $u \not\leftrightarrow u$. Thus, the only allowed transitions are those involving a change in parity.

Forbidden transitions may occur, but they generally occur at rates several orders of magnitude slower than allowed transitions. The fact that so-called forbidden transitions do occur is not an indication that quantum mechanics is wrong, but that the approximations in the calculations are not satisfied for the real system. The treatment of a simpler model is often useful, even though the results are approximate.

To illustrate these selection rules we will consider some electronic transitions of the molecules N_2, O_2, and NO. Potential energy diagrams for the ground states and several of the excited electronic states of these molecules and their cations are given in Fig. 14.1. The molecular term symbols for these states are given, and the atomic term symbols are given for the dissociated atoms.

In addition to the term symbol, electronic states of molecules are also given letter symbols. The ground electronic state is labeled X, and excited states of the same multiplicity are labeled A, B, C,... in order of increasing energy. Excited states of different multiplicity are labeled with lowercase letters a, b, c,....

Before we can discuss these transitions we have to discuss the manner in which they occur.

Example 14.1 ***Electric dipole transitions in O_2 and NO***

For O_2 and NO, list possible electric dipole transitions from the ground state to excited states of these molecules that are shown.

O_2 $\quad X^3\Sigma_g^- \rightarrow B^3\Sigma_u^-$ (Schumann–Runge bands)

NO $\quad X^2\Pi \rightarrow A^2\Sigma^+$

$\quad X^2\Pi \rightarrow B^2\Pi, C^2\Pi$

Example 14.2 ***Reaction energies from spectroscopic transitions***

According to Fig. 14.1, what are the changes in internal energy for the following reactions?

(*a*) $O_2(X^3\Sigma_g^-) + N(^4S) = NO(X^2\Pi) + O(^3P)$

(*b*) $N_2(X^1\Sigma_g^+) + O(^3P) = NO(X^2\Pi) + N(^4S)$

Reaction (*a*) is the sum of the reactions

$$O_2(X^3\Sigma_g^-) = O(^3P) + O(^3P) \qquad \Delta E = 5.0 \text{ eV}$$

$$N(^4S) + O(^3P) = NO(X^2\Pi) \qquad \Delta E = -6.5 \text{ eV}$$

so that $\Delta E = -1.5 \text{ eV} = -145 \text{ kJ mol}^{-1}$.

Reaction (*b*) is the sum of the reactions

$$N_2(X^1\Sigma_g^+) = N(^4S) + N(^4S) \qquad \Delta E = 9.5 \text{ eV}$$

$$O(^3P) + N(^4S) = NO(X^2\Pi) \qquad \Delta E = -6.5 \text{ eV}$$

so that $\Delta E = 3.0 \text{ eV} = 290 \text{ kJ mol}^{-1}$.

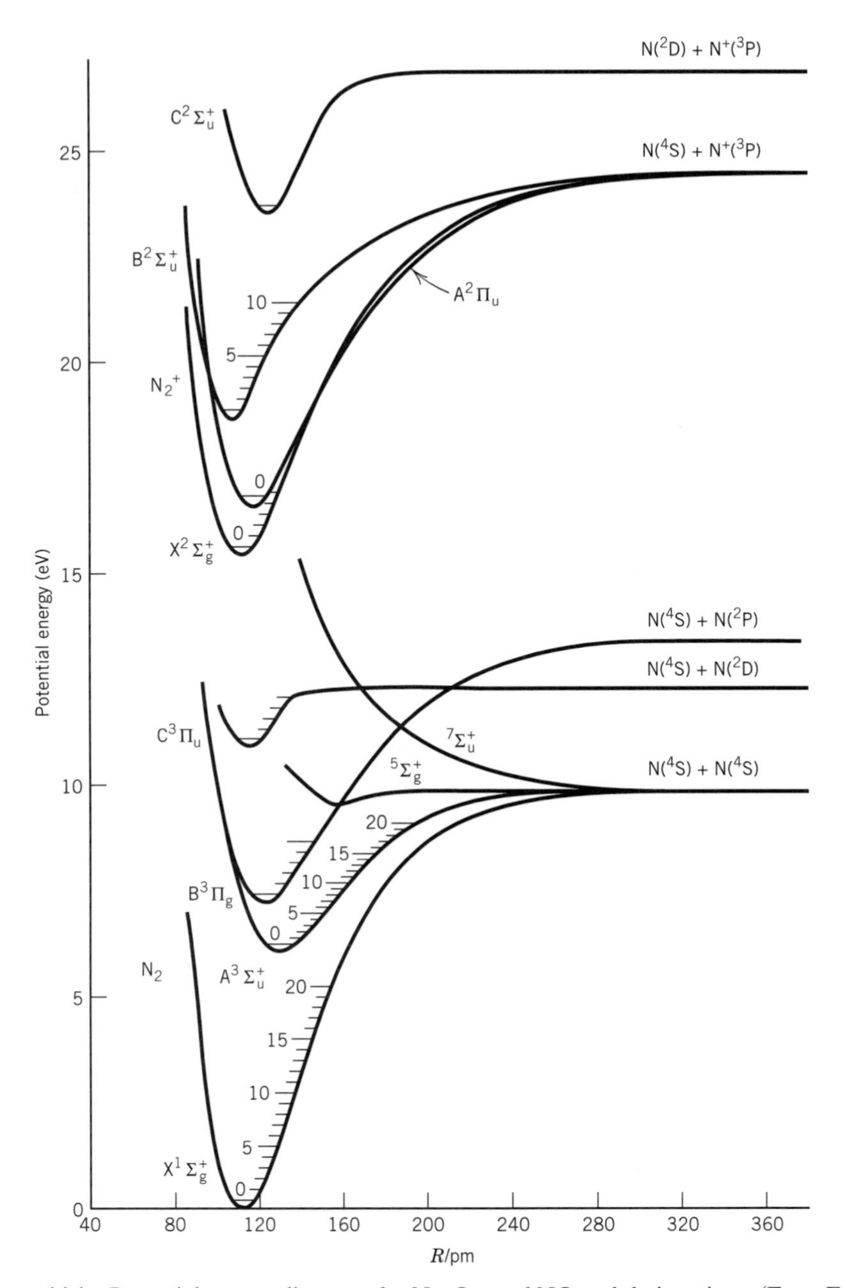

Figure 14.1 Potential energy diagrams for N_2, O_2, and NO and their cations. (From F. R. Gilmore, RAND Corporation Memorandum R-4034-PR, June 1964.)

14.2 ELECTRONIC ABSORPTION SPECTRA OF DIATOMIC MOLECULES AND THE FRANCK–CONDON PRINCIPLE

In the preceding chapter we discussed the vibrational spectra of diatomic molecules; we now consider vibrational contributions to the electronic spectra of diatomic molecules. As a simplification we will ignore the rotational changes that also accompany changes in electronic states because they involve smaller energy changes. When a diatomic molecule in its ground vibrational state absorbs a photon that raises it to an excited electronic state, the transitions are represented by

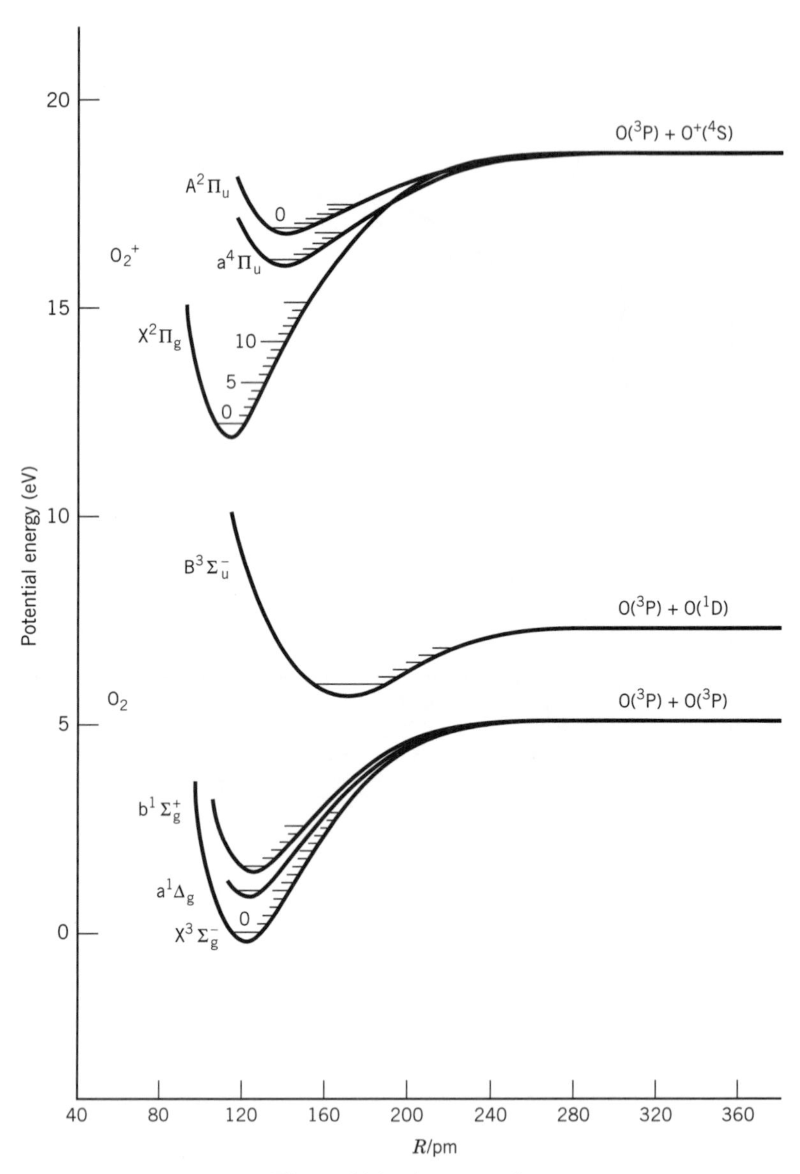

Figure 14.1 (*continued*)

Fig. 14.2, where the vibrational quantum number in the excited state is indicated by a single prime and the vibrational quantum number in the ground electronic state by a double prime. Note that the absorption frequencies for these transitions increase to a **convergence limit.** If higher vibrational levels in the ground electronic state are significantly populated, there will be other series, but we will concentrate on the transitions that start with the ground vibrational state of the ground electronic state.

The sum $\tilde{E}'$ of the electronic and vibrational energies of a diatomic molecule in one of the vibrational levels of the higher electronic state is given by (where we have allowed for anharmonicity)

$$\begin{aligned}\tilde{E}' &= \tilde{\nu}'_{\text{el}} + \tilde{\nu}'_{\text{e}}(n' + \tfrac{1}{2}) - \tilde{\nu}'_{\text{e}}x'_{\text{e}}(n' + \tfrac{1}{2})^2 \\ &= \tilde{\nu}'_{\text{el}} + \tilde{\nu}'_{\text{e}}(n' + \tfrac{1}{2}) - \tilde{\nu}'_{\text{e}}x'_{\text{e}}[(n')^2 + n' + \tfrac{1}{4}]\end{aligned} \tag{14.1}$$

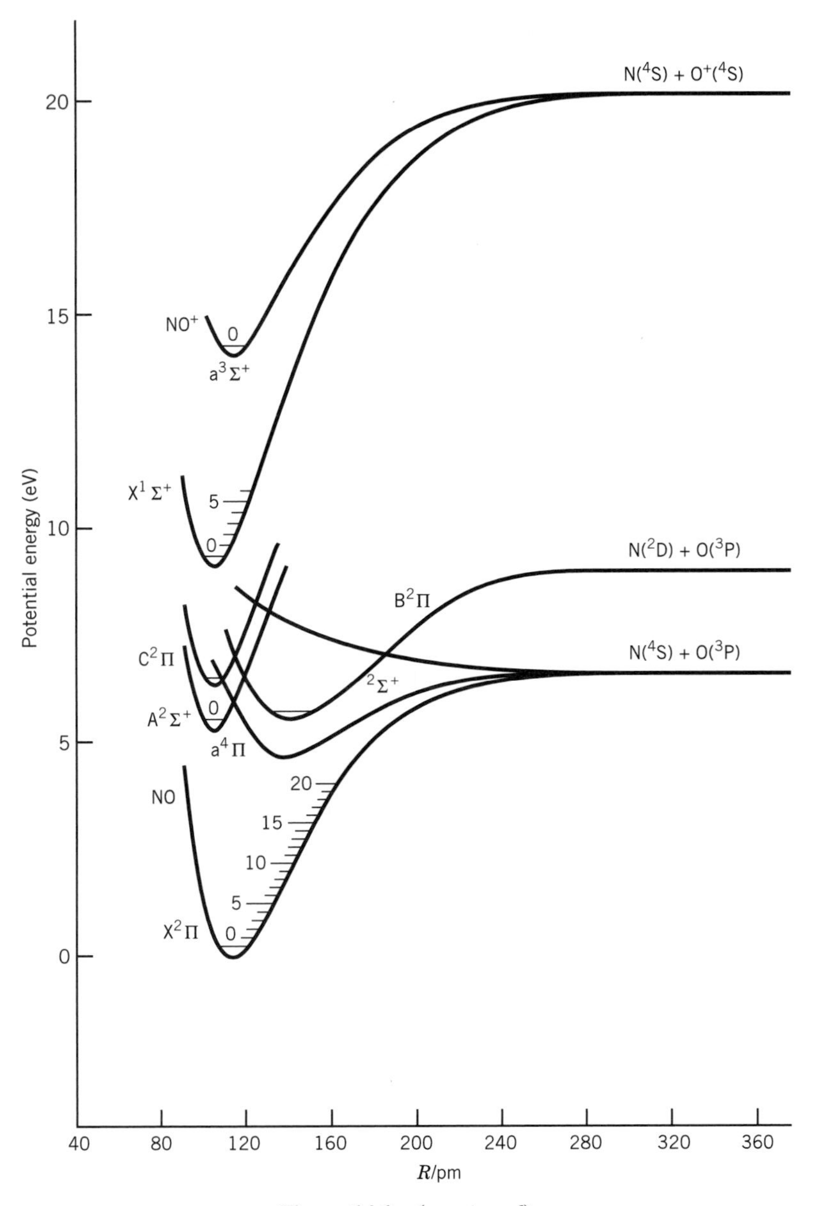

Figure 14.1 *(continued)*

where $\tilde{\nu}'_{\text{el}}$ is the electronic energy in wave numbers at the minimum of the electronic potential energy curve, $\tilde{\nu}'_{\text{e}}$ is the vibrational frequency of the excited state in wave numbers, and n' is the vibrational quantum number of the excited state. When absorption of a photon takes place, a diatomic molecule usually begins in its ground state ($n'' = 0$) because this (the zero-point level) is generally the only level populated at normal temperatures. The energy $\tilde{E}''$ of the diatomic molecule in the zero-point level of the lowest electronic state is given by

$$\tilde{E}''_{\text{total}} = \tilde{\nu}''_{\text{el}} + \frac{\tilde{\nu}''_{\text{e}}}{2} - \frac{\tilde{\nu}''_{\text{e}}x''_{\text{e}}}{4} \tag{14.2}$$

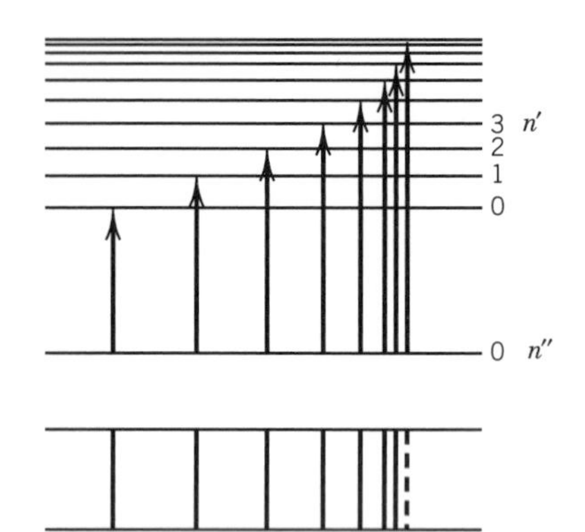

Figure 14.2 Electronic spectra for a diatomic molecule in its ground vibrational state, $n'' = 0$, being excited to various vibrational levels in the excited state with vibrational quantum numbers $n' = 0, 1, 2, 3, \ldots$. The dashed line in the spectrum under the energy levels shows the convergence limit for the progression.

The observed frequencies $\tilde{\nu}_{\text{obs}}$ in the electronic absorption spectrum are a function of the vibrational quantum numbers n' of the excited state and are given by

$$\tilde{\nu}_{\text{obs}} = \tilde{E}' - \tilde{E}'' = \tilde{\nu}_{0,0} + \tilde{\nu}'_{\text{e}}n' - \tilde{\nu}'_{\text{e}}x'_{\text{e}}n'(n' + 1) \tag{14.3}$$

where $\tilde{\nu}_{0,0}$ is the frequency in wave numbers of the $0 \rightarrow 0$ vibronic transition:

$$\tilde{\nu}_{0,0} = \tilde{T}_{\text{e}} + \left(\frac{\tilde{\nu}'_{\text{e}}}{2} - \frac{\tilde{\nu}'_{\text{e}}x'_{\text{e}}}{4}\right) - \left(\frac{\tilde{\nu}''_{\text{e}}}{2} - \frac{\tilde{\nu}''_{\text{e}}x''_{\text{e}}}{4}\right) \tag{14.4}$$

The difference $\tilde{T}_{\text{e}}$ in energies of the minima of the two electronic potential energy curves is given by

$$\tilde{T}_{\text{e}} = \tilde{\nu}'_{\text{el}} - \tilde{\nu}''_{\text{el}} \tag{14.5}$$

The series of transitions is referred to as a progression. The electronic absorption spectrum of a diatomic molecule has a characteristic shape since as n' in equation 14.3 increases, the separation between lines decreases. Consequently, the progression converges to a limit.

Now that we have the transition energies, we will consider the strength of the transitions due to absorption of a photon. Consider a diatomic molecule in its lowest electronic and vibrational state, in which the most probable internuclear separation is the equilibrium separation. The excited electronic state does *not*, in general, have the same equilibrium internuclear distance. **Franck and Condon recognized that the electronic transition occurs faster than the nuclei can adjust to their new equilibrium position,** so the most probable position for the nuclei in the excited state immediately after excitation is still the *ground*-state equilibrium position. Thus, an electronic transition can be represented, approximately, by a vertical line, as shown in Fig. 14.3. The vibrational wavefunctions of the upper electronic state will be (approximately) harmonic oscillator-like functions (see Section 9.9), so that the largest overlap of probabilities (for the ground- and excited-state vibrations) will occur for vibrational states with quantum numbers greater than 0. This can be seen mathematically from the transition dipole moment. In Chapter 13, we saw that the **transition moment** for a spectroscopic transition is given by

$$\boldsymbol{\mu}_{\text{gv}JM;\text{ev}'J'M'} = \int \psi^*_{\text{g}}\chi^*_{\text{gv}}\theta^*_{JM}\,\boldsymbol{\mu}\,\psi_{\text{e}}\chi_{\text{ev}'}\theta_{J'M'}\,d\tau_{\text{elec}}\,d\tau_{\text{vib}}\,d\tau_{\text{rot}} \tag{14.6}$$

where ψ_{g} and ψ_{e} are the electronic wavefunctions of ground and excited states, respectively; χ_{gv} and $\chi_{\text{ev}'}$ are the vibrational wavefunctions of initial and final states; and θ_{JM} and $\theta_{J'M'}$ are the corresponding rotational wavefunctions. We can do the integration over electronic coordinates first, obtaining $\boldsymbol{\mu}_{\text{ge}}$. This is nonzero for allowed electronic transitions, and can often be taken to be approximately independent of vibrational coordinates (Condon approximation), so that

$$\boldsymbol{\mu}_{\text{gv}JM;\text{ev}'J'M'} = \left(\int \theta^*_{JM}\,\boldsymbol{\mu}_{\text{ge}}\,\theta_{J'M'}\,d\tau_{\text{rot}}\right)\int \chi^*_{\text{gv}}\chi_{\text{ev}'}\,d\tau_{\text{vib}}$$

$$= \left(\int \theta^*_{JM}\,\boldsymbol{\mu}_{\text{ge}}\,\theta_{J'M'}\,d\tau_{\text{rot}}\right)S_{\text{vv}'} \tag{14.7}$$

The integral of the vibrational wavefunctions is called a **Franck–Condon overlap integral** $S_{\text{vv}'}$. The intensity of a transition is proportional to the square of the transition moment, and therefore to $S^2_{\text{vv}'}$. This leads to the fact that the

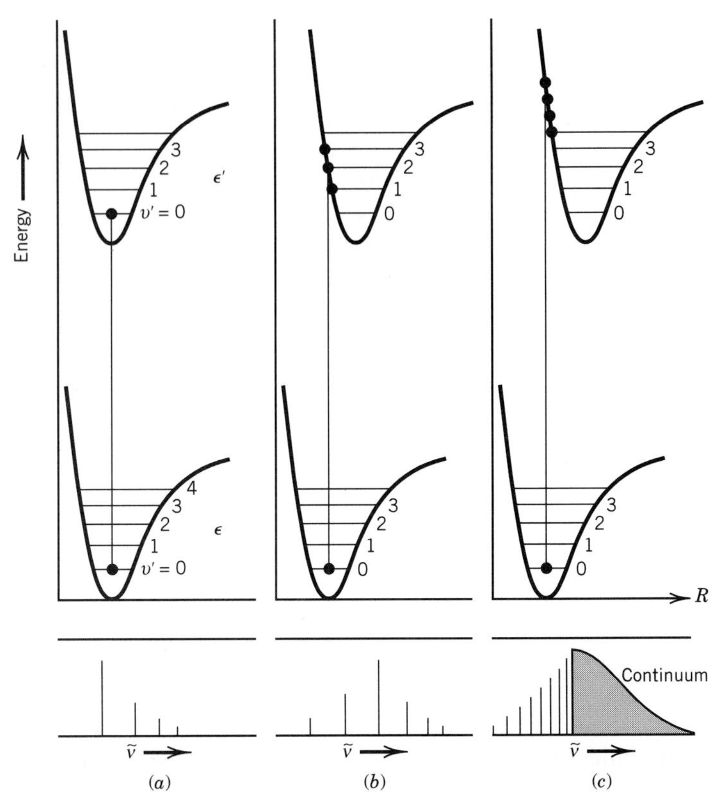

Figure 14.3 Potential energy curves and absorption spectra for three electronic transitions. (*a*) The minimum in the potential energy curve in the excited state coincides with that in the ground state; this is not the usual situation. (*b*) The minimum in the potential energy curve in the excited state is at a larger internuclear distance than in the ground state; this is the usual situation. (*c*) The absorption may raise the excited state to a higher energy than its dissociation energy so that the absorption is continuous.

absorption probability is largest into that vibrational state of the excited electronic states whose probability is largest directly above the equilibrium internuclear distance. In the three cases illustrated in Fig. 14.3, the potential energy curve for the excited electronic state lies directly above the ground state in (*a*) and is displaced to larger internuclear separations in (*b*) and (*c*). In Fig. 14.3*a*, the Franck–Condon overlap is largest between $v = 0$ and $v' = 0$. In Fig. 14.3*b*, the overlap is largest between $v = 0$ and $v' = 2$, so that the line is most intense. Note that this Franck–Condon overlap is greatest because the vibrational wavefunctions for $v > 1$ tend to pile up at the classical turning points (i.e., the potential walls), as shown in Fig. 14.4. There will be intensity in other transitions as well, since the overlap for these is not zero.

In Fig. 14.3*c* there is a significant probability of excitation to an energy above the dissociation energy of the excited molecule. Since all energies above this energy are allowed eigenvalues, continuous absorption occurs. The borderline between absorption lines and continuous absorption is referred to as the **convergence limit.** It, of course, gives the difference in energy between the vibrational ground state of the lower electronic state and the dissociation products of the upper electronic state. Cases (*b*) and (*c*) are more often encountered than case (*a*)

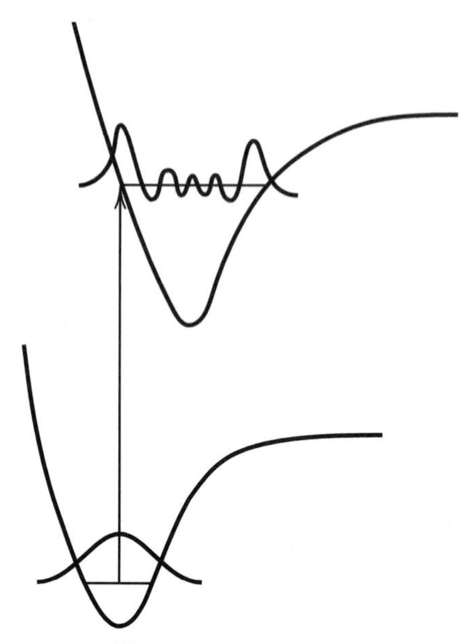

Figure 14.4 Franck–Condon principle. In an electronic transition, the overlap of the ground vibrational wavefunction in the lower electronic state and the various vibrational wavefunctions in the upper electronic state is greatest for the vibrational level whose classical turning point is at the equilibrium separation in the lower state.

because the bonding is generally weaker in the excited electronic state. Since the bonding is weaker, the equilibrium internuclear distance is greater in the excited state.

14.3 DETERMINATION OF DISSOCIATION ENERGIES

The determination of dissociation energies of diatomic molecules from vibrational spectra was referred to in Section 13.8. However, it is virtually impossible to obtain accurate values in this way because of the generally long extrapolation involved.

If the onset between discrete lines and the continuum in Fig. 14.3*c* is sharp, the dissociation energies D_0 and D_0' of the ground state and the excited state can be determined quite accurately, as shown in Fig. 14.5. This figure shows that

$$\tilde{\nu}_{\text{limit}} = D_0' + \tilde{\nu}_0 = D_0 + \Delta\tilde{\nu}_{\text{atomic}} \tag{14.8}$$

where $\tilde{\nu}_{\text{limit}}$ is the wave number of the onset of the continuum in the progression and where the dissociation energies are expressed in wave numbers. If the states of the atoms produced in the dissociation of the ground state and in the dissociation of the excited state are known, $\Delta\tilde{\nu}_{\text{atomic}}$ is known, and D_0 can be calculated from $\tilde{\nu}_{\text{limit}}$.

If the wave number of the 0–0 band can be obtained from the vibrational spectrum, the dissociation energy D_0' for the excited state can be calculated from $\tilde{\nu}_{\text{limit}}$ using equation 14.8. The dissociation energies relative to the minima in the potential energy curves are given by

$$D_{\text{e}} = D_0 + \frac{\tilde{\nu}_{\text{e}}}{2} - \frac{\tilde{\nu}_{\text{e}}x_{\text{e}}}{4} + \frac{\tilde{\nu}_{\text{e}}y_{\text{e}}}{8} + \cdots \tag{14.9}$$

as may be seen from equation 13.80.

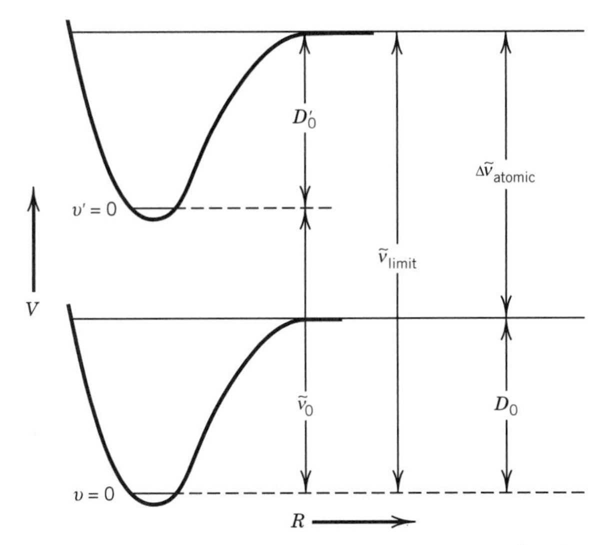

Figure 14.5 Determination of the dissociation energies D_0 and D_0' of the ground state and the excited state from $\tilde{\nu}_{limit}$, the wave number of the onset of a continuum in a progression.

If the potential energy curve for a diatomic molecule is crossed by the curve for a repulsive excited state, as shown in Fig. 14.6, a region of diffuseness is found in the spectrum. When the internuclear distance and energy of the molecule in the ground state are near the crossing point, there is a probability that the molecule will transfer from curve AB to curve CD and dissociate. This has the effect of broadening the vibrational and rotational levels in this region of the spectrum. This effect is referred to as **predissociation** because the molecule with the potential energy curve AB can dissociate at a lower energy when the repulsive curve CD crosses it.

14.4 SPECTROPHOTOMETERS AND THE BEER–LAMBERT LAW

In considering rotational and vibrational spectra we were concerned primarily with the frequencies of the lines and secondarily with their relative intensities. However, in considering electronic spectra we will be increasingly concerned with the intensity of absorption or emission. The intensity of absorption at a particular wavelength can be determined by passing a monochromatic beam of light through a sample of known thickness and concentration and measuring the intensity I of the transmitted light relative to the intensity I_0 that would be transmitted in the absence of the absorbing substance. **The intensity I of a beam of light is defined as the energy per unit area per unit time.**

The construction of a spectrophotometer is indicated schematically in Fig. 14.7. The principal parts are the source of electromagnetic radiation, the monochromator, the cell compartment, the photoelectric detector, and a device for indicating the output from the detector (electric meter, potentiometer, or recording potentiometer). The cell compartment contains an optical absorption cell filled with the solution to be studied and another absorption cell filled with a reference solution, usually pure solvent. The ratio of the intensity I of transmitted light for the solution to the intensity I_0 for the solvent is called the **transmittance.**

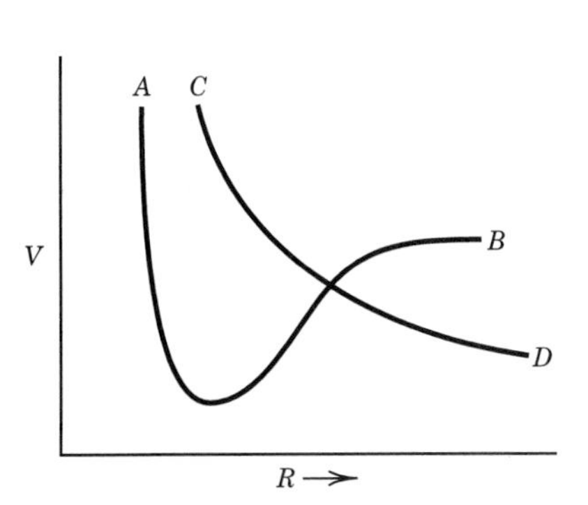

Figure 14.6 Predissociation can result when two potential curves cross in this way.

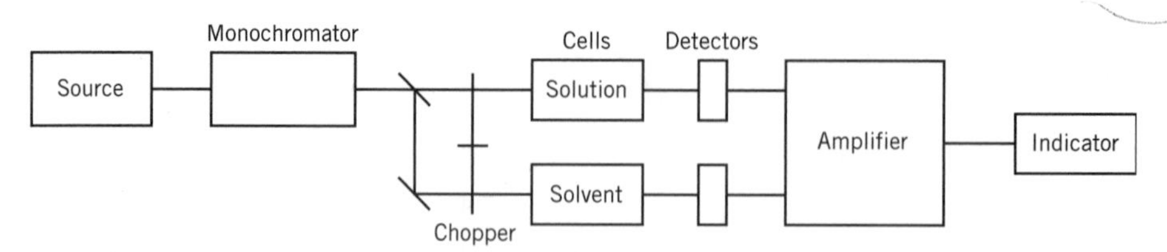

Figure 14.7 Schematic diagram for a simple UV-visible spectrometer. A laser can be used in place of the source and monochromator.

The transmittance I/I_0 can be determined at different wavelengths, and the absorption spectrum can be mapped. With some spectrophotometers such a plot is recorded automatically. The positions and intensities of the absorption bands and lines serve for identification and for criteria of purity; the transmittance serves for quantitative analysis of the concentration of material present.

Lambert developed the equation for the attenuation of a light beam as a function of the thickness of a homogeneous medium. Beer developed the equation for the effect of concentration. The Beer–Lambert law may be derived as follows.

The probability that a photon will be absorbed is usually directly proportional to the concentration of absorbing molecules and to the thickness of the sample for a very thin sample. This probability is expressed mathematically by the equation

$$\frac{\mathrm{d}I}{I} = -\kappa c \,\mathrm{d}x \tag{14.10}$$

where I is the intensity of light of a particular wavelength, that is, energy per unit area per unit time; $\mathrm{d}I$ is the change in light intensity produced by absorption in a thin layer of thickness $\mathrm{d}x$ and concentration c; and κ is the **Naperian molar absorption coefficient.** Distance x is measured through the cell in the direction of the beam of light that is being absorbed. The concentration c is usually expressed in mol L^{-1}.

The intensity of a beam of light after passing through length L of solution is related to the incident intensity I_0 by equation 14.12, which is obtained by integrating equation 14.10 between the limits I_0 when $x = 0$ and I when $x = L$:

$$\int_{I_0}^{I} \frac{\mathrm{d}I}{I} = -\kappa c \int_0^L \mathrm{d}x \tag{14.11}$$

$$\ln \frac{I}{I_0} = 2.303 \log \frac{I}{I_0} = -\kappa c L \tag{14.12}$$

Since it is convenient to use logarithms to the base 10, the Beer–Lambert law is used in the form

$$\log \frac{I_0}{I} = A = \epsilon c L \tag{14.13}$$

where the quantity $\log(I_0/I)$ is referred to as the **absorbance** A, and ϵ is referred to as the **molar absorption coefficient** or molar absorptivity. It can be seen from equation 14.13 that the absorbance is directly proportional to the concentration c and to the path length L. The proportionality constant is characteristic of the solute and depends on the wavelength of the light, the solvent, and the temperature. Since the molar absorption coefficient ϵ depends on wavelength, the absorbance A is wavelength dependent, and so if the radiation is not monochromatic, the

Beer–Lambert law in the form of equation 14.13 may not be obeyed. In addition, the Beer–Lambert law may not be obeyed for a substance that associates or dissociates in solution since A will change with concentration because of the changing ratio of concentrations of absorbing species.

Figure 14.8 shows the absorption spectra of benzene and *para*-xylene.*

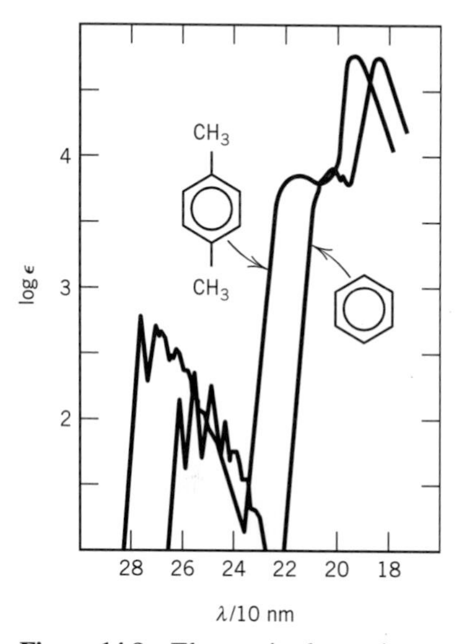

Figure 14.8 Electronic absorption spectra of benzene and *para*-xylene in solution.

Example 14.3 *Using Beer's law*

The percentage transmittance of an aqueous solution of disodium fumarate at 250 nm and 25 °C is 19.2% for a 5×10^{-4} mol L^{-1} solution in a 1-cm cell. (*a*) Calculate the absorbance A and the molar absorption coefficient ϵ. (*b*) What will be the percentage transmittance of a 1.75×10^{-5} mol L^{-1} solution in a 10-cm cell?

$$(a)\quad A = \log\frac{I_0}{I} = \log\frac{100}{19.2} = 0.717$$

$$\epsilon = \frac{A}{Lc} = \frac{0.717}{(1\text{ cm})(5\times10^{-4}\text{ mol L}^{-1})} = 1.43\times10^{3}\text{ L mol}^{-1}\text{ cm}^{-1}$$

$$(b)\quad \log\frac{I_0}{I} = (1.43\times10^{3}\text{ L mol}^{-1}\text{ cm}^{-1})(10\text{ cm})(1.75\times10^{-5}\text{ mol L}^{-1}) = 0.251$$

$$\frac{I_0}{I} = 1.782 \qquad \text{and} \qquad \frac{100I}{I_0} = 56.1\%$$

For mixtures of independently absorbing substances the absorbance is given by

$$\log\frac{I_0}{I} = A = (\epsilon_1 c_1 + \epsilon_2 c_2 + \cdots)L \tag{14.14}$$

where $c_1, c_2, \ldots$ are the concentrations of the substances having molar absorption coefficients of $\epsilon_1, \epsilon_2, \ldots$. A mixture of n components may be analyzed by measuring A at n wavelengths at which the molar absorption coefficients are known for each substance, provided that these coefficients are sufficiently different. The concentrations of the several substances may then be obtained by solving the n simultaneous linear equations.

Example 14.4 *Using Beer's law for absorbing mixtures*

An aqueous solution of A and B has an absorbance of 0.800 at λ_1 and 0.500 at λ_2. At λ_1 the molar absorption coefficient of A is 1.5×10^3 L mol^{-1} cm^{-1}, and the molar absorption coefficient of B is 4.0×10^3 L mol^{-1} cm^{-1}. At λ_2 the molar absorption coefficient of A is 3.0×10^3 L mol^{-1} cm^{-1}, and the molar absorption coefficient of B is 2.0×10^3 L mol^{-1} cm^{-1}. What is the composition of the solution?

When the concentrations are expressed in mM, the two equations are:

$$0.800 = 1.5c_A + 4.0c_B$$

$$0.500 = 3.0c_A + 2.0c_B$$

There are several ways to solve simultaneous linear equations. In this case the first equation can be multiplied by 0.500 and the second can be multiplied by 0.800, and the difference

*P. E. Stevenson, *J. Chem. Educ.* **41**:234 (1964).

between the two equations taken. This gives a relation between c_A and c_B that can be substituted into the above equations to obtain $c_A = 0.044$ mM and $c_B = 0.183$ mM. If you have a personal computer with a mathematical program, there are easier ways to solve even large systems (see Computer Problem 14.C).

When a sample is irradiated continuously at low intensity, its absorption coefficient remains constant in the absence of chemical reaction, and this indicates that excited molecules are continuously deactivated so that they do not accumulate. Usually the excitation energy is simply degraded to thermal energy in molecular collisions, but a chemical reaction may occur and change the composition and absorption spectrum of the sample (cf. Chapter 19). An excited molecule may also emit a quantum of radiation. Such emission is referred to as fluorescence or phosphorescence, depending on the difference in excited- and ground-state spin multiplicities (Section 14.8).

The Beer–Lambert law may be written in alternative ways. In a given situation one form may be more convenient to use than another. We will use all of the following forms:

$$I = I_0 10^{-\epsilon c L} \tag{14.15}$$

$$I = I_0 \, \mathrm{e}^{-\kappa c' L'} \tag{14.16}$$

$$I = I_0 \, \mathrm{e}^{-\sigma N x} \tag{14.17}$$

In the first equation, c is expressed in mol L^{-1} and L is expressed in cm, so that ϵ has the units L mol^{-1} cm^{-1}, as we have seen. In the second equation we will express c' in SI base units so that c' is in mol m^{-3}, L' is in m, and κ is in m^2 mol^{-1}. In the third equation, N is in m^{-3} and x is in m. Since σ has the units m^2_y it is referred to as the **absorption cross section.** Although σ has the units of area; it is not to be interpreted literally as the area of an absorbing molecule.

Example 14.5 ***Absorption cross section and absorbancy***

The molar absorbancy index ϵ of a solute is 44 000 L mol^{-1} cm^{-1}. What is the absorption cross section σ?

Comparison of equations 14.15 and 14.17 yields

$$\sigma = \frac{2.303 \epsilon c L}{N x}$$

Since N is the number of molecules per m^3 and c is the amount per liter,

$$c = \frac{N}{N_A (10^3 \text{ L m}^{-3})}$$

Since x is in meters and L is in centimeters,

$$L = x(10^2 \text{ cm m}^{-1})$$

so that

$$\sigma = \frac{2.303 \epsilon (10^2 \text{ cm m}^{-1})}{N_A (10^3 \text{ L m}^{-3})}$$

If $\epsilon = 44\,000$ L mol^{-1} cm^{-1}, then

$$\sigma = \frac{2.303(44\,000 \text{ L mol}^{-1} \text{ cm}^{-1})(10^2 \text{ cm m}^{-1})}{(6.022 \times 10^{23} \text{ mol}^{-1})(10^3 \text{ L m}^{-3})}$$
$$= 1.7 \times 10^{-20} \text{ m}^2$$

Since an electronic absorption band contains many lines that may not be resolved, the intensity is not as accurately measured by the maximum absorption (as represented by the maximum absorption coefficient ϵ_{max}) as by the integral over the entire band. The integrated absorption coefficient is defined by

$$\int_{\text{band}} \epsilon \, d\tilde{\nu} \tag{14.18}$$

Thus, it has the units L mol^{-1} cm^{-2}.

If the absorption band is Gaussian in shape, then the integrated intensity can be related to the molar absorbancy index ϵ_{max} and the width $\Delta\tilde{\nu}_{1/2}$ at half the maximum absorbancy index:

$$\int_{\text{band}} \epsilon \, d\tilde{\nu} = 1.06 \epsilon_{max} \Delta\tilde{\nu}_{1/2} \tag{14.19}$$

For relatively strong absorptions of molecules $\epsilon_{max} = 10^4$ to 10^5 L mol^{-1} cm^{-1}, and $\Delta\tilde{\nu}_{1/2}$ is of the order of 1000 to 5000 cm^{-1}. For weak absorptions $\epsilon_{max} =$ 10 L mol^{-1} cm^{-1}, and $\Delta\tilde{\nu}_{1/2}$ is of the order of 100 cm^{-1}. Extremely weak (forbidden) absorptions may have ϵ_{max} of the order of 10^{-3} to 10^{-4} L mol^{-1} cm^{-1}.

Example 14.6 ***Calculating the integrated absorption coefficient***

The maximum value of an observed absorbancy index is 44 000 L mol^{-1} cm^{-1} at 30 000 cm^{-1}. If the width of the band at half-maximum is 5000 cm^{-1}, what is the value of the integrated absorption coefficient?

Assuming that the band is Gaussian,

$$\int \epsilon \, d\tilde{\nu} = 1.06 \epsilon_{max} \Delta\tilde{\nu}_{1/2}$$
$$= 1.06(44\,000 \text{ L mol}^{-1} \text{ cm}^{-1})(5000 \text{ cm}^{-1})$$
$$= 2.33 \times 10^8 \text{ L mol}^{-1} \text{ cm}^{-2}$$

14.5 OSCILLATOR STRENGTH

The concept of oscillator strength f was developed to provide a theoretical reference for the intensity of a spectroscopic transition. **The oscillator strength is the ratio of the strength of a transition to the strength of a transition for an electron oscillating harmonically in three dimensions.** The oscillator strength for an actual transition may be calculated from the measured integrated absorption coefficient of the absorption band. Allowed electric dipole transitions yield oscillator strengths of approximately unity. Forbidden transitions have oscillator strengths much less than unity. Singlet–triplet transitions typically have oscillator strengths of the order of 10^{-5}.

The integrated absorption coefficient is given by

$$\int \kappa \, d\nu = \frac{8\pi^3 N_A \nu}{3hc(4\pi\epsilon_0)} |\boldsymbol{\mu}_{12}|^2 \tag{14.20}$$

where $\boldsymbol{\mu}_{12}$ is the transition dipole moment given by equation 13.19 and ν is the average frequency for the absorption band. In this equation we are using an integrated absorption coefficient expressed in terms of κ in $m^2\ mol^{-1}$ and frequency rather than $\int \epsilon \, d\tilde{\nu}$. It is more convenient to do derivations with quantities in SI base units and convert later, as we will.

The square of the transition moment for the three-dimensional harmonic oscillator for the transition from the ground state to the first excited state is given by

$$|\boldsymbol{\mu}_{12}|^2_{osc} = \frac{3he^2}{8\pi^2 m_e \nu} \tag{14.21}$$

We define the oscillator strength f of a transition with transition moment $\boldsymbol{\mu}_{12}$ as

$$f = \frac{|\boldsymbol{\mu}_{12}|^2}{|\boldsymbol{\mu}_{12}|^2_{osc}} \tag{14.22}$$

Thus $f = 1$ for an electronic three-dimensional harmonic oscillator. Substituting from equation 14.20 for $|\boldsymbol{\mu}_{12}|^2$ and from equation 14.21 for $|\boldsymbol{\mu}_{12}|^2_{osc}$, we find

$$f = \frac{4m_e c \epsilon_0}{N_A e^2} \int \kappa \, d\nu \tag{14.23}$$

This is the experimentally measured oscillator strength. To calculate the oscillator strength, which is dimensionless, all the quantities on the right-hand side of equation 14.23 need to be expressed in SI base units. However, the integrated absorption coefficient is more often given in $L\ mol^{-1}\ cm^{-2}$, as we have seen in Example 14.6, and so some conversion factors must be included in equation 14.23 in order to calculate the oscillator strength from the integrated absorption coefficient expressed in its usual units. If a single electron can undergo more than one transition, the sum of the oscillator strengths for all the transitions arising from any one level to all other levels is unity:

$$\sum_i f_i = 1 \tag{14.24}$$

Example 14.7 ***Expressing the oscillator strength in terms of ϵ***

Show that the oscillator strength is given by the following equation when the integrated absorption coefficient (equation 14.18) is expressed in $L\ mol^{-1}\ cm^{-2}$:

$$f = (4.32 \times 10^{-9}\ L^{-1}\ mol\ cm^2) \int \epsilon \, d\tilde{\nu}$$

To convert $\int \kappa \, d\nu$, expressed in SI base units, to $\int \epsilon \, d\tilde{\nu}$, expressed in SI base units, we need to multiply by $2.303c$, where c is the velocity of light. However, since ϵ is usually expressed in $L\ mol^{-1}\ cm^{-1}$, we need to multiply $\int \epsilon \, d\tilde{\nu}$ by $10^3\ cm^3\ L^{-1}$ to obtain the value in $cm\ mol^{-1}$. To complete the conversion to SI base units a further factor $10^{-2}\ m\ cm^{-1}$ is required. Therefore,

$$f = \frac{4m_e c^2 \epsilon_0}{N_A e^2} 2.303(10^3\ cm^3\ L^{-1})(10^{-2}\ m\ cm^{-1}) \int \epsilon \, d\tilde{\nu}$$

Example 14.8 ***Calculating f from the integrated absorption coefficient***

What is the oscillator strength f of the solute of Example 14.6?

$$\begin{aligned} f &= (4.32 \times 10^{-9}\ \text{L}^{-1}\ \text{mol cm}^2) \int \epsilon\, \mathrm{d}\tilde{\nu} \\ &= (4.32 \times 10^{-9}\ \text{L}^{-1}\ \text{mol cm}^2)(2.33 \times 10^8\ \text{L mol}^{-1}\ \text{cm}^{-2}) \\ &= 1.01 \end{aligned}$$

There is a relationship between the integrated absorption coefficient and the radiative lifetime τ for a transition. This relationship goes back to the relation between B_{21} and A_{21} (equation 13.16) since $A_{21} = 1/\tau$ for a single transition. Rather than giving the derivation here, we will simply refer to the fact that for the solute described in Example 14.8, the theoretical radiative lifetime is 1.7×10^{-9} s. The radiative lifetime is inversely proportional to the integrated absorption coefficient, and so for the conditions in Example 14.8, a maximum molar absorption coefficient that was tenfold smaller would yield a relaxation time tenfold larger.

14.6 ELECTRONIC SPECTRA OF POLYATOMIC MOLECULES

The absorption spectra of polyatomic molecules lie mainly in the visible and ultraviolet regions, involving excitation of electrons from the higher-energy filled orbitals to the lower-energy unfilled orbitals of the molecule. In many unsaturated organic molecules and in inorganic transition metal complexes, the difference in energy between occupied and unoccupied orbitals is small enough that the absorption of light occurs in the visible. The absorption spectra of saturated organic molecules (e.g., C_2H_6) lie in the vacuum ultraviolet, since it takes a large amount of energy to promote an electron from an occupied orbital (in a single bond) to an unoccupied orbital.

In discussing ethylene in terms of the Hückel molecular orbital theory in Section 11.7, we saw that ultraviolet light can be absorbed if it provides enough energy to raise an electron from the 1π bonding orbital to the π^* antibonding orbital. This is referred to as a $\pi^* \leftarrow \pi$ transition, and for ethylene it occurs at a wavelength of about 180 nm. We also saw that the excitation energy required for 1,3-butadiene is lower. In the next section, we will see that this trend continues.

In organic compounds with oxygen, nitrogen, or halogen atoms, there are often filled **nonbonding** molecular orbitals usually associated with the lone pairs. Since these do not participate in the bonding, they may lie relatively high in energy. Excitation of an electron from such an orbital to an unfilled (antibonding) π^* orbital leads to absorption in the ultraviolet and visible regions. Thus, groups such as C=O, —N=N—, and —N=O that cause such absorption at wavelengths longer than 180 nm are called **chromophores.** These groups have characteristic absorption wavelengths. This is illustrated by the ultraviolet absorption of formaldehyde, for which the ground-state configuration is $\ldots(1b_1)^2(2b_2)^2$. The nature of these orbitals is indicated by Fig. 14.9. The lowest energy excitation is provided by taking an electron from the $2b_2$ orbital (a nonbonding orbital represented by n) and promoting it to the $2b_1$ orbital (a π^* orbital) to give the excited-state configuration$\ldots(1b_1)^2(2b_2)^1(2b_1)^1$. This is referred to as a $\pi^* \leftarrow n$ transition, and it occurs at a wavelength of 290 nm.

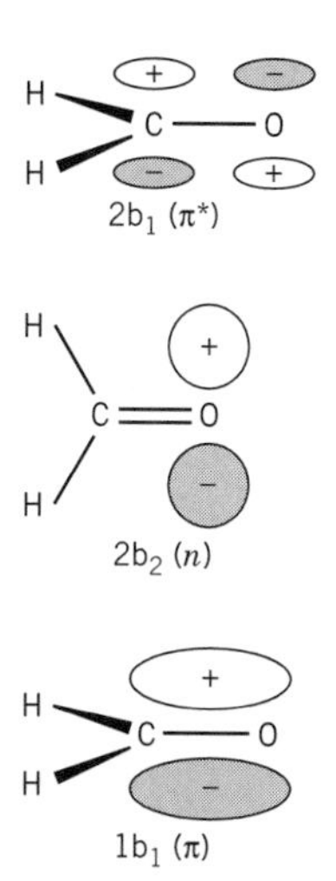

Figure 14.9 HOMO-1, HOMO, and LUMO orbitals of formaldehyde. (From J. M. Hollas, *Modern Spectroscopy*. © Wiley & Sons Ltd. Reproduced with permission, 1996.)

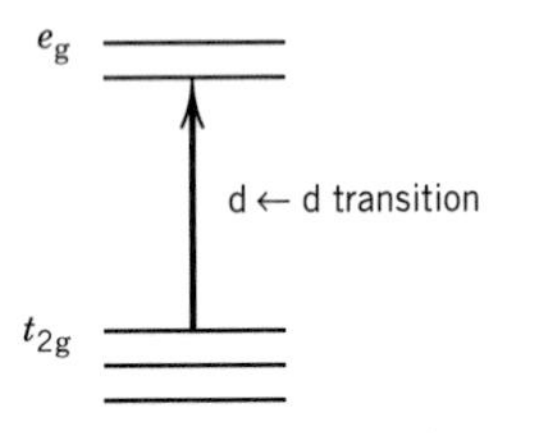

Figure 14.10 The d orbital states in an octahedral complex and the meaning of d ← d transitions.

Transitions of the $\pi^* \leftarrow n$ type are blue-shifted, that is, shifted to a shorter wavelength, in a hydrogen-bonding solvent such as ethanol. Such a solvent forms a weak bond with the n orbital and the 1s orbital of the hydrogen atom of the OH group of the solvent. This hydrogen bonding lowers the energy of the n orbital and increases the energy of the excitation, shifting it to the blue.

Other types of polyatomic molecules that have strong absorptions in the visible are transition metal complexes. These are known for their beautiful colors, which arise because of these electronic transitions. Here the low-lying unfilled orbitals are atomic d orbitals from the transition metal. The presence of the ligands splits the five d orbitals into two groups at different energies in octahedral or tetrahedral complexes, as shown in Fig. 14.10. Electrons in orbitals on the ligands can then be excited into an unfilled d orbital on the metal, giving a charge transfer transition that is usually very intense. In addition, electrons can be excited from one set of d orbitals to the other, also giving rise to absorption in the visible. These latter transitions are weaker than the charge transfer bands because d → d transitions are forbidden unless some perturbation occurs, such as a distortion of the octahedron (as shown in Fig. 14.11) to a lower symmetry with no inversion center of symmetry.

14.7 CONJUGATED MOLECULES: FREE-ELECTRON MODEL

For molecules with conjugated systems of double bonds [i.e., $R(CH{=}CH)_nR'$], it is found that the electronic absorption bands shift to longer wavelengths as the number of conjugated double bonds is increased.* Approximate quantitative calculations of the absorption frequencies may be made on the basis of the free-electron model for the π electrons of these molecules. The energy for the lowest electronic transition is that required to raise an electron from the highest filled level to the lowest unfilled level. In a system of conjugated double bonds each carbon atom has three σ bonds that lie in a plane, and each σ bond involves one outer electron of that carbon atom. Above and below this plane are the π orbital systems.

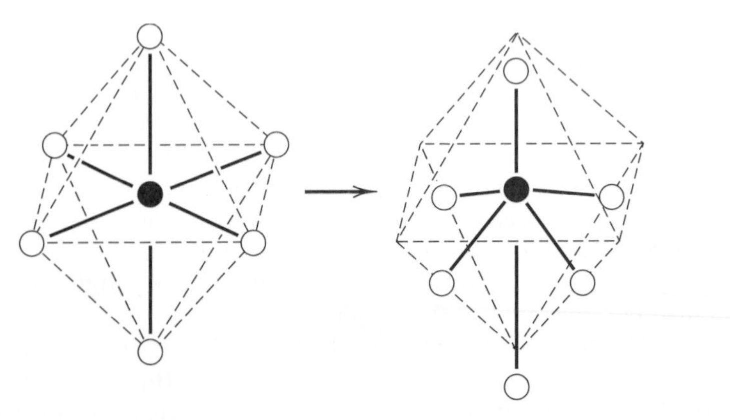

Figure 14.11 Vibrational motion or distortion in an octahedral complex that leads to the destruction of the center of symmetry, thereby making d ← d transitions allowed.

*In conjugated molecules double and single bonds alternate in the classical structure.

Each carbon atom contributes one electron to this π system, but these electrons are free to move the entire length of the series of π orbitals and are not localized at a given carbon atom. In the free-electron model it is assumed that the π system is a region of uniform potential and that the potential energy rises sharply to infinity at the ends of the system (i.e., a square-well potential). Thus, the energy levels E available to the π electrons would be expected to be those calculated for the particle in a one-dimensional box (Section 9.6):

$$E = \frac{n^2 h^2}{8 m_e a^2} \tag{14.25}$$

The length of the box a is usually taken to be the length of the chain between terminal carbon atoms plus a bond length or two.

The π electrons (one for each carbon atom) are assigned to orbitals so that there are two (one with spin $+\frac{1}{2}$ and the other with spin $-\frac{1}{2}$) in each level, starting with the lowest. For a completely conjugated hydrocarbon the number of π electrons is even, and the quantum number of the highest filled level will be $n = N/2$, where N is the number of π electrons (the number of carbon atoms involved). In absorption an electron from the highest filled level is excited to the next higher level with quantum number $n' = N/2 + 1$. The difference in energy of these two levels is

$$\Delta E = \frac{h^2}{8 m_e a^2}(n'^2 - n^2) = \frac{h^2}{8 m_e a^2}\left[\left(\frac{N}{2} + 1\right)^2 - \left(\frac{N}{2}\right)^2\right] = \frac{h^2}{8 m_e a^2}(N + 1) \tag{14.26}$$

The absorption frequency in wave numbers is given by

$$\tilde{\nu} = \frac{\Delta E}{hc} = \frac{h(N+1)}{8 c m_e a^2} \tag{14.27}$$

For linear molecules, the size of the system a is proportional to N, so the absorption frequency will vary as $1/N$ for large N.

Example 14.9 ***Absorption frequency for a conjugated molecule***

Calculate the lowest absorption frequency for octatetraene (C_8H_{10}), which contains a series of four conjugated double bonds. The length of the π bond system is about 0.95 nm.

$$\tilde{\nu} = \frac{h(N+1)}{8 c m_e a^2} = \frac{(6.62 \times 10^{-34}\ \text{J s})(9)(10^{-2}\ \text{m cm}^{-1})}{8(3 \times 10^8\ \text{m s}^{-1})(9.109 \times 10^{-31}\ \text{kg})(0.95 \times 10^{-9}\ \text{m})^2}$$

$$= 30\,200\ \text{cm}^{-1}$$

The observed absorption band is at 33 100 cm^{-1}.

14.8 FLUORESCENCE AND PHOSPHORESCENCE

In **fluorescence,** the radiation is emitted during a transition between electronic states of the same spin or multiplicity, whereas in **phosphorescence** the radiation is emitted in a transition between electronic states of different multiplicities, for example, between a triplet and a singlet state. Since the latter are (approximately) forbidden, the rate is low and therefore the lifetime of the lowest triplet state of a molecule (with singlet ground state) is long (10^{-4} to 100 s). On the other hand, the lifetime of excited singlet states is usually between 10^{-6} and 10^{-9} s.

When a ground-state (S_0) molecule is excited to the first excited singlet state, S_1 (see Fig. 14.12), a number of processes can occur. Since the molecule is generally in an excited vibrational state, collisions with other molecules in the gas or solution can remove vibrational energy from the excited molecule in a process called vibrational relaxation. Thus, the excited molecule ends up in the lowest vibrational state of S_1. Now, when the excited molecule radiates (fluoresces), the frequency is lower than that of the exciting radiation, as shown in Fig. 14.12.

Before a molecule can fluoresce, other processes can occur. The molecule can chemically react, as discussed in Chapter 19, or the molecule can lose its energy in a collision with another molecule, and it may make a transition to another excited electronic state. Such a nonradiative transition from one singlet state to another singlet state, or more generally between states of the same multiplicity, is called **internal conversion,** while nonradiative transitions between states of different multiplicities is called **intersystem crossing.** Intersystem crossing can occur where the potential energy curves for the states of different multiplicity cross. Since an electron spin flip is required, there must be a mechanism by which this can occur. This is provided by spin–orbit coupling, the interaction of the spin and the orbital

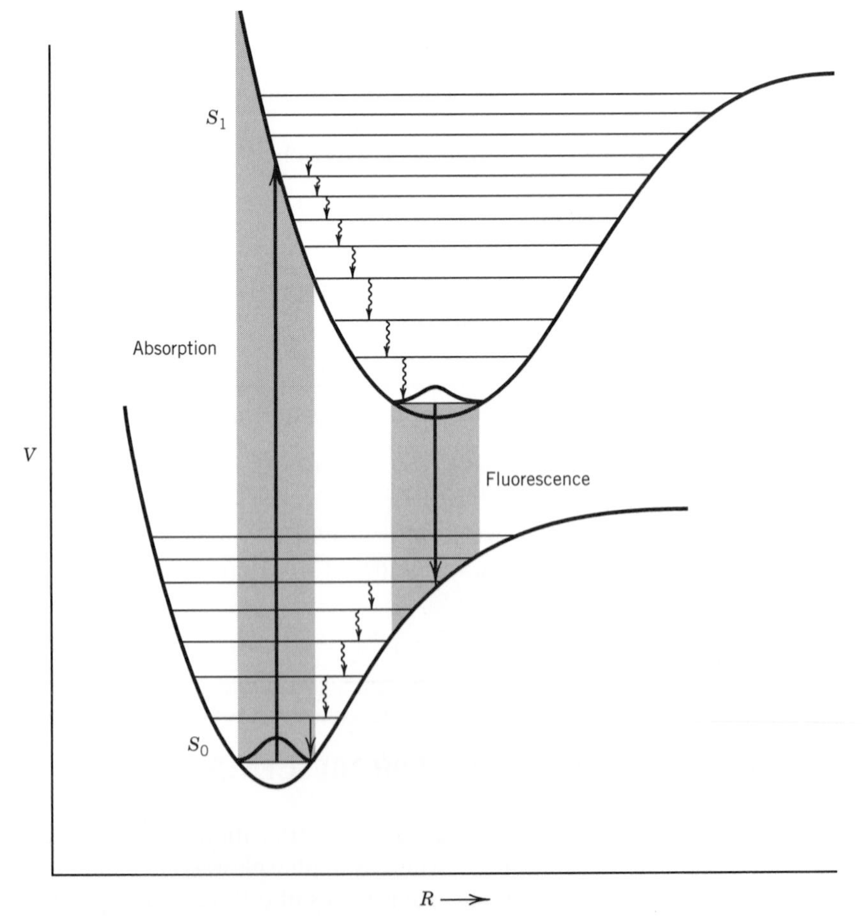

Figure 14.12 Absorption and fluorescence by a diatomic molecule. The shaded areas indicate the transitions to a range of vibrational quantum numbers.

angular momentum of an electron. Intersystem crossing is significantly slower (10^{-2} to 10^{-6}) than internal conversion. Various factors influence the rates of nonradiative transitions, including the difference in energies between the two electronic states involved.

If intersystem crossing occurs, the molecule can undergo further vibrational relaxation, as shown in Fig. 14.13, finally ending up in the lowest vibrational state of the excited triplet (T_1). The molecule can now undergo collisional loss of energy, or it can emit a photon in phosphorescence. Triplet-state molecules are especially likely to be involved in a chemical reaction because of their high energy and long lifetime.

The energy of a triplet state is usually lower than the energy of the excited singlet state with the same molecular orbital occupancy because in a triplet state the electrons having the same spin tend to avoid each other (because of the Pauli principle) by staying in different regions. Since the electrons are farther apart, there is a decrease in electronic repulsion and the energy of the molecule is lower.

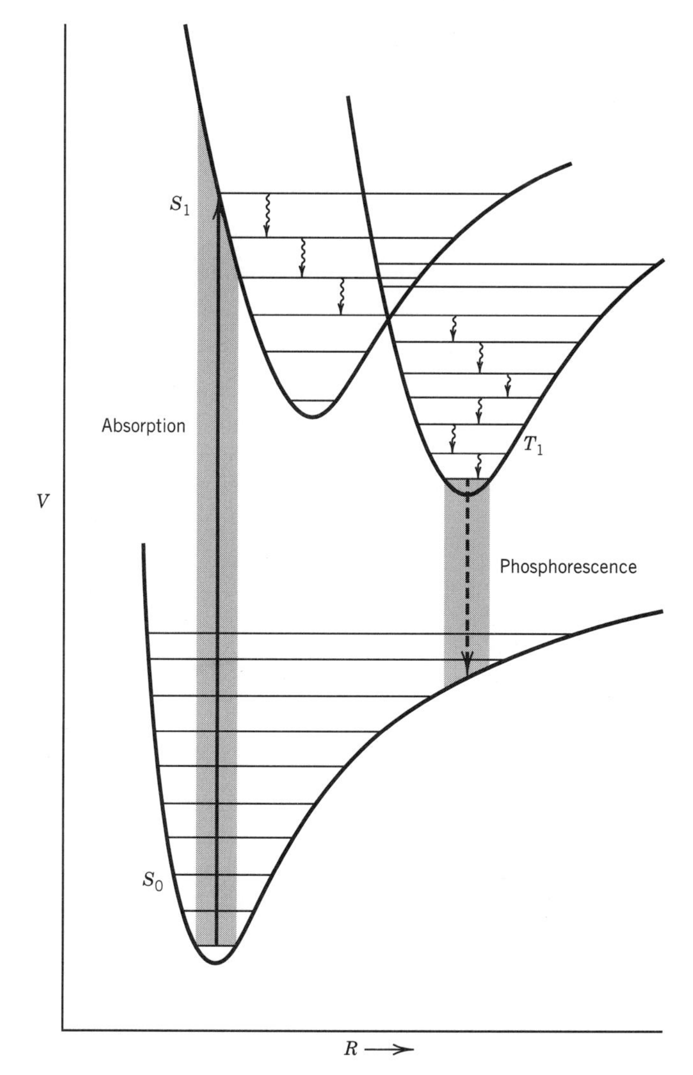

Figure 14.13 Absorption and phosphorescence by a diatomic molecule. The shaded areas indicate transitions to a range of vibrational quantum numbers.

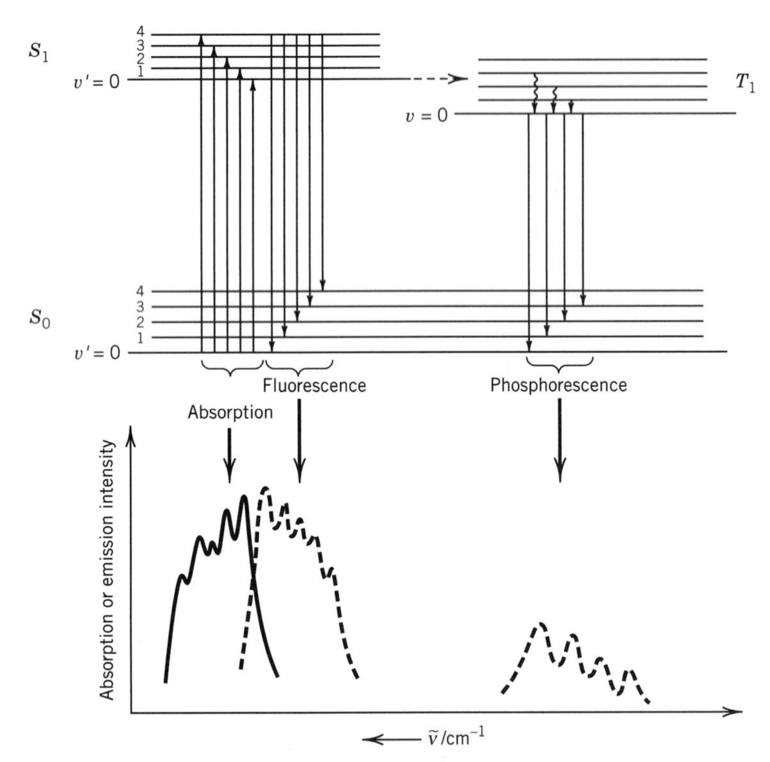

Figure 14.14 Schematic diagram showing absorption, fluorescence, and phosphorescence spectra. (From E. F. H. Brittain, W. O. George, and C. H. J. Wells, *Introduction to Molecular Spectroscopy.* New York: Academic, 1970.)

Consequently, phosphorescence usually occurs at a lower frequency than fluorescence, as shown in Fig. 14.14. This figure shows that the fluorescence spectrum has an approximate "mirror image" relationship to the absorption spectrum, if the spacings of the vibrational levels in the S_0 and S_1 states are similar.

The transition from the $v = 0$ vibrational level of the upper electronic state to the $v = 0$ vibrational level of the lower electronic state is called the 0–0 band. The small difference in the 0–0 bands for absorption and fluorescence is due to the difference in the solvation of the initial and final states in the two cases. As shown in Fig. 14.15, the excited state does not become equilibrated with the solvent in the absorption process, and the ground state does not become equilibrated with the solvent in the fluorescence process. As indicated in the diagram, the 0–0 fluorescence transition will be of lower energy than the 0–0 absorption transition.

The preceding discussion referred to a single excited singlet state and a single excited triplet state, but a molecule will have many singlets and triplets (and higher multiplicities). The higher excited states tend to relax by various processes very quickly (10^{-4}–10^{-12} s) to the lowest excited singlet or triplet, and so can be probed only by using ultrafast spectroscopic techniques.

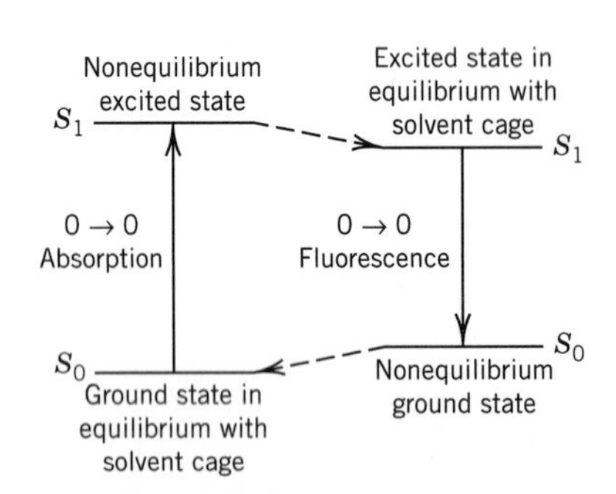

Figure 14.15 Effect of solute–solvent equilibrium in causing a difference in the energy of the 0–0 absorption transition and the 0–0 fluorescence transition. (From C. H. J. Wells, *Introduction to Molecular Photochemistry.* London: Chapman & Hall, 1972. © C. H. J. Wells, 1972.)

Example 14.10 ***Absorption wavelength and fluorescence***

Use Fig. 14.1 to calculate the longest wavelength of light that can be used to excite $NO(^2\Pi)$ to an electronically excited state that can fluoresce back to the ground state.

The transition

$$NO(^2\Pi) \rightarrow NO(^2\Sigma)$$

satisfies the selection rules for absorption. The excited state can fluoresce back to the ground state. The energy required for a vertical transition is 6.0 eV, which is equivalent to $(6.0 \text{ eV})(8066 \text{ cm}^{-1} \text{ eV}^{-1}) = 48 \times 10^3 \text{ cm}^{-1}$ or 2.08×10^{-5} cm or 208 nm.

Comment:

Another development is laser-induced fluorescence (LIF). In this type of experiment a frequency-tunable laser with a narrow line is directed at a gaseous sample and the frequency is slowly changed. When the laser frequency corresponds to an allowed molecular transition, excited states are formed and fluoresce. The fluorescence intensity is measured during the scanning process and is interpreted in terms of transitions from the initial vibration–rotation state of the molecule being irradiated.

14.9 LASERS

Normally a beam of light loses intensity as it passes through an absorbing material. However, if molecules are present in an excited state, stimulated emission (Section 13.2) can occur, and the light beam can gain intensity. A laser achieves this condition and produces an intense and coherent beam. By coherent we mean that the light waves are in phase. The name *laser* comes from **light amplification by stimulated emission of radiation.**

When a two-state system is irradiated, the rate of change in the population of state 2 is equal to the difference between the rate of absorption and the rate of emission (Fig. 13.1):

$$\frac{dN_2}{dt} = BN_1\rho_{\tilde{\nu}}(\tilde{\nu}_{12}) - A_{21}N_2 - BN_2\rho_{\tilde{\nu}}(\tilde{\nu}_{12})$$

$$= B\rho_{\tilde{\nu}}(\tilde{\nu}_{12})(N_1 - N_2) - A_{21}N_2 \qquad (14.28)$$

where the subscripts have been left off of the B's because of equation 13.15. Let us consider the case in which spontaneous emission is negligible so that we can ignore the last term in equation 14.28. We can see that as long as $N_1 > N_2$ there will be absorption. However, as irradiation is continued N_2 will approach N_1, and the rate of absorption of radiation will decrease to zero. The system will now be transparent because the net absorption of light of frequency $\tilde{\nu}_{12}$ is zero, and the transition is said to be **saturated.** If somehow $N_2 > N_1$, a situation known as **population inversion,** then dN_2/dt will be negative, which means that the number of emitted photons increases, and the intensity of radiation in the direction of the incident radiation will increase. In other words, there will be amplification. It is this amplification that creates the coherent beam of a laser.

Laser action requires a population inversion, but we have seen that we cannot get a population inversion by irradiating a two-level system. However, population inversion can be obtained in multilevel systems such as those in Fig. 14.16. In Fig. 14.16*a* the system is raised to level E_3 by the absorption of radiation (pump) and then undergoes a rapid nonradiative transition to level E_2. If the system can be pumped hard enough so that $N_2 > N_1$, then laser action can be obtained on $E_2 \rightarrow E_1$.

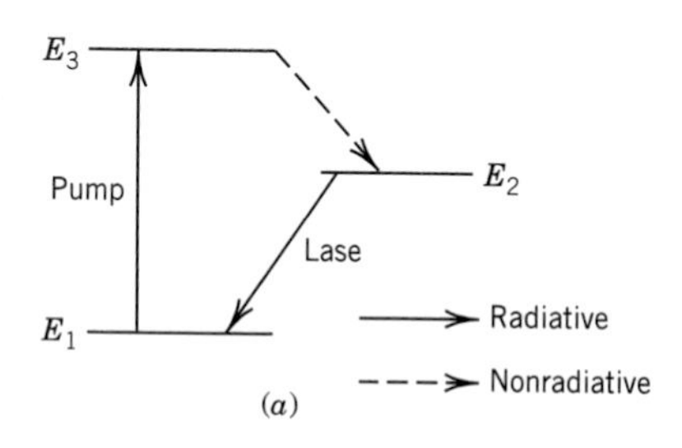

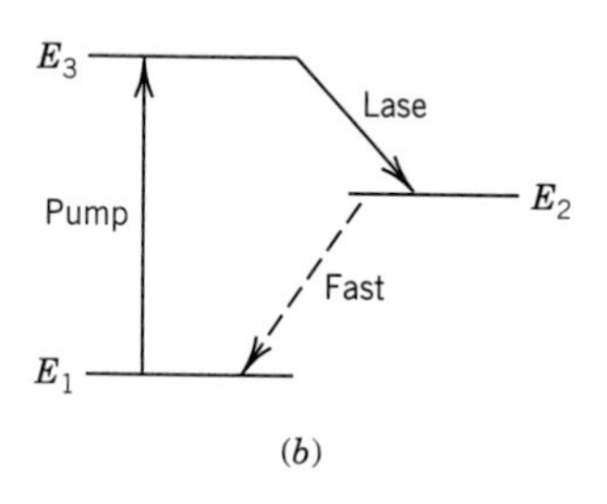

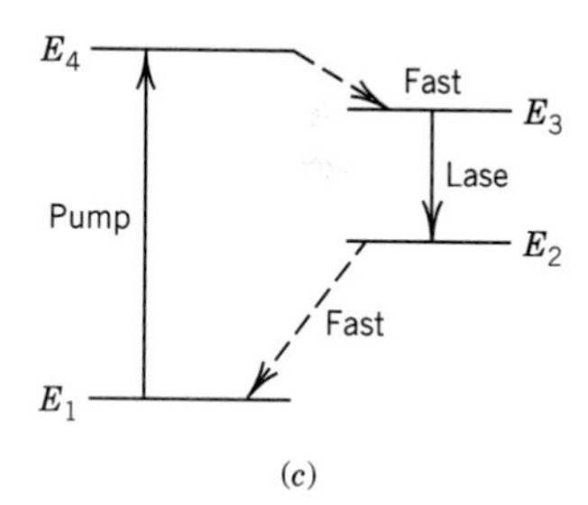

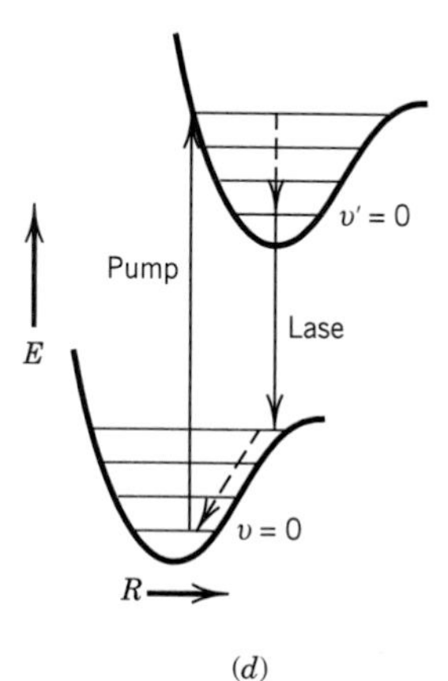

Figure 14.16 Multilevel lasing schemes. (*a, b*) Three-level systems and (*c, d*) four-level systems represented in different ways. [From W. F. Coleman, *J. Chem. Educ.* **59:**441 (1982).]

Figure 14.16*b* shows a more suitable system because laser action depends on $N_3 > N_2$. This population inversion is readily achieved because N_2 is essentially zero initially and will remain so if the $E_2 \rightarrow E_1$ nonradiative process is fast. A number of lasers depend on four-level systems such as that in Fig. 14.16*c*, *d*. Here the lasing action occurs between two levels, neither of which is involved in the pumping process.

Usually a laser cavity has mirrors at each end to increase the radiation density that stimulates emission. In Fig. 14.17 the two ends of the ruby rod are mirrors. The silver mirror at the back end is nearly 100% reflecting, and the silver mirror at the front end has a reflectivity that is less than 100%. The gain of the system is the amount of amplification, usually per round-trip through the cavity, for a given population inversion. For a given system some minimum gain is required to overcome the optical losses in the cavity and the energy emission of the laser. Figure 14.17 shows the construction of a ruby laser that is powered by a flashlamp. The output beam is emitted through the partially silvered end of the ruby crystal.

The length d of the resonant cavity created by the two mirrors is critical. Standing waves are formed in the cavity so that there is an integral number of waves in the cavity. The electric field of the standing wave in the cavity is zero at the surface of the two mirrors. Since $\lambda = 2d/n$, where n is an integer, the frequency of the standing waves is given by

$$\nu = \frac{nc}{2d} \tag{14.29}$$

where c is the velocity of light.

Lasers may be operated in a **continuous wave** (cw) mode or a **pulse mode.** To operate in cw mode a laser has to be pumped steadily by irradiation with a light source or by electrical discharge at a rate sufficient to supply the emergent laser radiation and overcome any losses. In the pulse mode the laser is excited with a pulsed electric discharge or pulsed lamp, and some kind of shutter is used to control the release of energy stored in the laser so that an intense pulse is obtained. The shutter may be mechanical or electro-optical, but the simplest method at visible wavelengths is to use a saturable dye. A cell containing a dye solution with a peak absorption at the laser wavelength is located in the laser cavity. Laser action cannot start until the gain exceeds the loss in the dye plus other losses. However, there is some laser action in the cavity, and if this can build up to a high enough value, then the dye begins to bleach because of saturation (equation 14.28), and the radiation density builds up rapidly. The time between pulses is determined by the concentration of the dye.

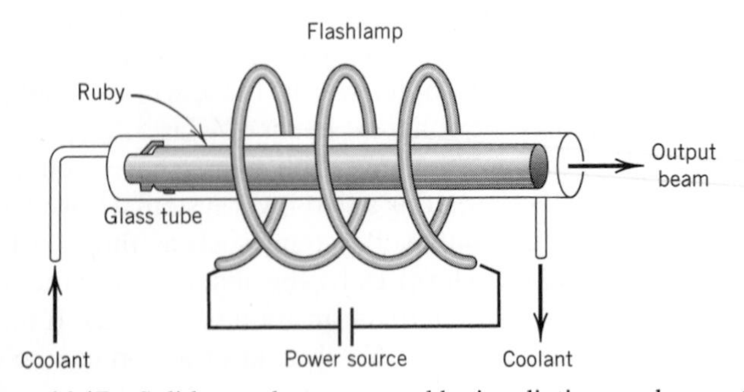

Figure 14.17 Solid-state laser pumped by irradiating a ruby crystal.

The initiation of pulses is often called Q switching because the Q factor is a measure of the energy stored to the energy discharged per cycle. Lasers have large Q values. Various types of shutters can be used in a laser cavity so that lasing does not occur while the population inversion is building. Then the shutter is opened, and a large fraction of the energy stored is dumped in a single pulse.

Example 14.11 ***The YAG solid-state laser***

A widely used solid-state laser is made of $Y_3Al_2O_{15}$ (yttrium aluminum garnet, commonly referred to as YAG) in which some of the Y^{3+} ions are replaced by Nd^{3+}. This laser is generally pumped with a xenon-filled flashlamp. A particular YAG laser yields 10 pulses per second at 1064 nm. Each pulse is 20 ns long and has an energy of 350 mJ. What are the peak and average powers? How many photons are produced per pulse and per minute?

The peak power is $0.35\ \text{J}/20 \times 10^{-9}\ \text{s} = 17.5$ MW. The average power is $(0.35\ \text{J})(10\ \text{s}^{-1}) = 3.5$ W. The energy per photon is $hc/\lambda = (6.626 \times 10^{-34}\ \text{J s})(2.998 \times 10^8\ \text{m s}^{-1})/(1064 \times 10^{-9}\ \text{m}) = 1.867 \times 10^{-19}$ J. The number of photons per pulse is $(0.35\ \text{J})/(1.867 \times 10^{-19}\ J) = 1.875 \times 10^{18}$. The number of photons per minute is $(1.875 \times 10^{18})(10)(60\ \text{min}^{-1}) = 1.125 \times 10^{21}\ \text{min}^{-1}$.

Gas lasers are generally pumped by passage of an electric current. The ions and electrons that are produced are accelerated, and the electrons cause excitation by collisions with the molecules of the gas. The first gas laser was the He–Ne laser, which can oscillate at three wavelengths, $\lambda_1 = 3.39\ \mu\text{m}$, $\lambda_2 = 1.15\ \mu\text{m}$, and $\lambda_3 = 0.633\ \mu\text{m}$ (the most widely used). The helium energy levels are involved in the pumping process, and the laser action occurs between energy levels of Ne.

Some gas lasers use transitions between vibration–rotation levels of a molecule. The CO_2 laser (which also contains N_2 and He) utilizes transitions between two vibrational levels. This laser is one of the most powerful, and it can be operated at 1 MW continuously. It is also one of the most efficient: 15 to 20% of the electrical power put into the electrical discharge is converted to laser radiation.

Figure 14.18 shows the lowest vibrational levels for CO_2 and N_2. Since CO_2 has three modes of vibration, the vibrational level may be described by giving the quantum number for each vibrational mode. The level is designated by giving the quantum numbers for (1) symmetric stretching mode, (2) bending mode, and (3) asymmetric stretching mode in that order. Thus, the state is specified by n_1, n_2, n_3. Since the bending mode is doubly degenerate, a bending vibration consists of a combination of the orthogonal bending vibrations. The superscript 0 or 1 on n_2 indicates whether the angular momentum about the axis of the molecule is 0 ($\ell = 0$) or $\hbar$ ($\ell = 1$). Thus, 01^10 represents the level with one quantum in the bending vibration with an angular momentum of unity, and 02^00 represents the level with two quanta in the bending vibration with an angular momentum of zero. The 00^01 level is populated by collisions of electrons with ground-state (00^00) molecules and by resonant energy transfer from N_2 molecules. Laser action is produced by the transition between the 00^01 and 10^00 levels ($\lambda \approx 10.6\ \mu\text{m}$). It is also possible to obtain laser action between the 00^01 and 02^00 levels ($\lambda \approx 9.6\ \mu\text{m}$). The transitions shown in Fig. 14.18 with wavy lines are radiationless transitions that occur rapidly. In this description we have neglected the rotational fine structure.

Some gas lasers may be tuned to several discrete frequencies using a grating or other device that rejects unwanted frequencies from the laser cavity, but dye lasers can provide a continuous range of wavelengths over approximately

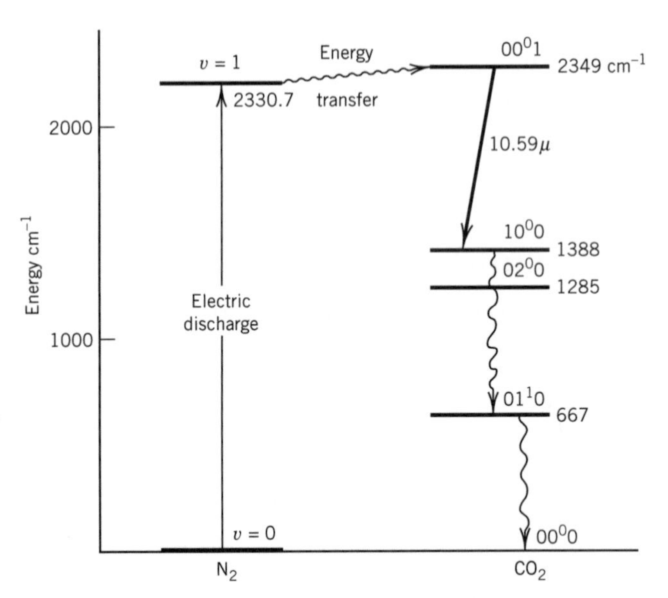

Figure 14.18 Low-lying vibrational energy levels of nitrogen and carbon dioxide molecules. (From J. I. Steinfeld, *Molecules and Radiation.* Cambridge, MA: MIT Press, 1985.)

40–50 nm. The active medium is a fluorescent organic molecule, and the solution of the dye is pumped with another laser or flashlamp. The optical cavity can be tuned over much of the fluorescence spectrum of the dye. By using a number of dyes, it is possible to cover the whole range from 300 to 1000 nm.

In chemical lasers the population inversion is produced by an exothermic chemical reaction. For example, a laser may be produced by mixing hydrogen and fluorine. The following reactions produce HF in higher vibrational levels:

$$F + H_2 \rightarrow HF^* + H \tag{14.30}$$

$$H + F_2 \rightarrow HF^* + F \tag{14.31}$$

The first reaction is exothermic by 132 kJ mol^{-1}. Therefore, HF* may represent molecules in the $v = 3, 2$, or 1 vibrational state. The second reaction is exothermic by 410 kJ mol^{-1}, so that HF* produced may be in a vibrational level as high as $v = 10$. However, the second reaction is not very efficient for pumping the HF laser. Laser action takes place between several vibrational levels. The reaction of hydrogen with fluorine takes place very slowly unless atomic fluorine is provided. Atomic fluorine can be provided by adding SF_6 and dissociating it by electrical means:

$$SF_6 + e^- \rightarrow SF_5 + F + e^- \tag{14.32}$$

Lasers are important in chemistry because they can be used to initiate photochemical reactions, and the short pulses permit the study of very fast reactions. Lasers have also revolutionized Raman spectroscopy because exposures are greatly reduced, and even weak lines may be detected in experiments of short duration. A laser may be used to selectively photodissociate one isotopic species of a molecule, owing to the extreme monochromaticity associated with the laser beam. The dissociated or ionized molecules may be allowed to react with another substance so that the isotopes may be separated by a chemical method.

Comment:

Solid-state and gas lasers have a small tuning range, but dye lasers can be tuned over a larger range because of the broad absorption spectrum resulting from the solvent broadening of the vibrational structure into bands. The dye solution is pumped through the laser cavity to avoid overheating by the laser used to produce excited molecules. A diffraction grating is used to vary the wavelength. By use of a series of different dyes, the range from the near infrared to the near ultraviolet can be covered continuously.

14.10 PHOTOELECTRON SPECTROSCOPY

Photoelectron spectroscopy involves the ejection of electrons from atoms or molecules by radiation with monochromatic ultraviolet or X-ray photons and the interpretation of the spectra in terms of orbital energies. The ejected electrons are called **photoelectrons.** The photoionization process is represented by

$$M + h\nu = M^+ + e^- \tag{14.33}$$

where molecule M is generally in its ground electronic and vibrational state. The molecule is ionized; some of the energy of the photon may be used to raise the resulting ion to a higher vibrational level or excited electronic level, and the remaining energy is converted to kinetic energy of the photoelectron. The kinetic energy of the emitted electron is given by

$$\frac{1}{2}mv^2 = h\nu - I = h\nu + E(\mathrm{M}) - E(\mathrm{M}^+) \tag{14.34}$$

where $E_i = E(\mathrm{M}^+) - E(\mathrm{M})$ is the ionization energy. $E(\mathrm{M})$ is the energy of M in its initial electronic and vibrational state (usually the ground state), and $E(\mathrm{M}^+)$ is the energy of the ion in the electronic and vibrational state in which it is produced. Thus, measurement of the kinetic energy of electrons emitted gives information about the different electronic and vibrational states of M^+.

Figure 14.19*a* shows the process for ultraviolet photoelectron spectroscopy (UPS); the electron is emitted from a valence orbital. When X-ray photons are used, the electron can be emitted from a core orbital, as indicated in Fig. 14.19*b*, and we speak of X-ray photoelectron spectroscopy (XPS). We will see later, in

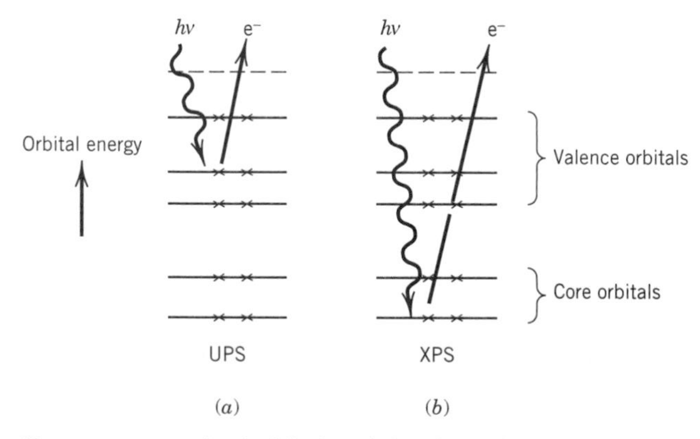

Figure 14.19 Processes occurring in (*a*) ultraviolet photoelectron spectroscopy (UPS) and (*b*) X-ray photoelectron spectroscopy (XPS).

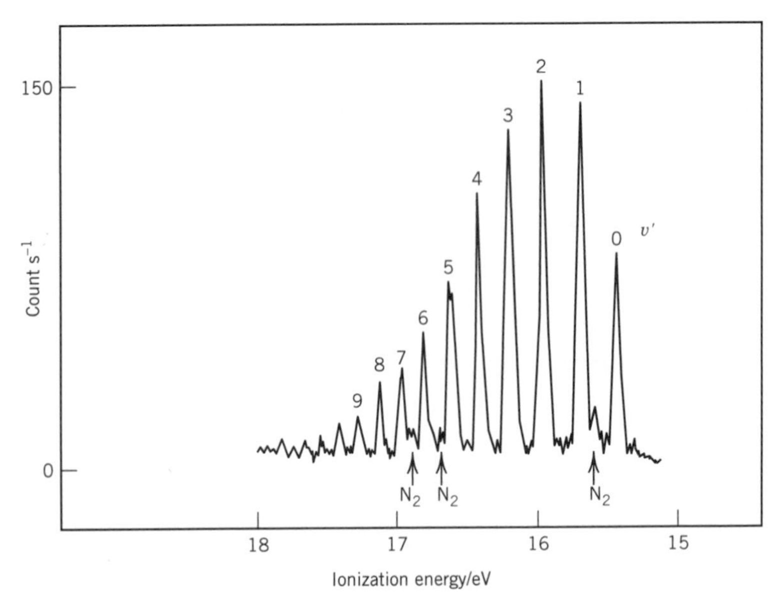

Figure 14.20 The He I UPS spectrum of H_2. The sample was contaminated with a little nitrogen. (From D. W. Turner, C. Baker, A. D. Baker, and C. R. Brundle, *Molecular Photoelectron Spectroscopy.* © John Wiley & Sons, Ltd. Reproduced with permission, 1970.)

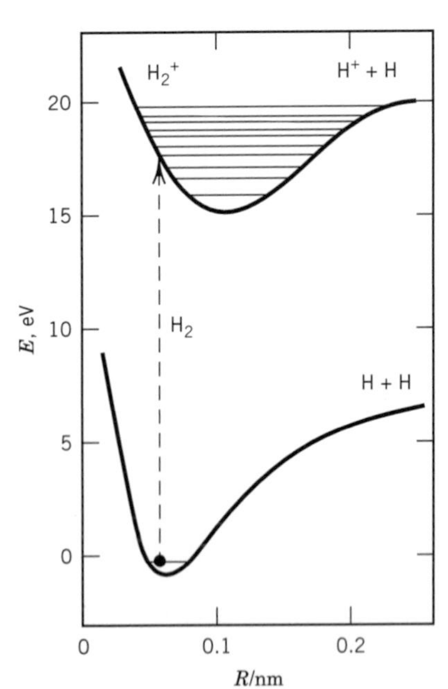

Figure 14.21 Potential energies of H_2 and $H_2{}^+$ with energy levels involved in the photoelectron spectrum.

Chapter 24, that UPS and XPS can also be used to determine the surface electronic structure of a solid.

The most commonly used source of ultraviolet photons is a helium gas discharge tube; the most intense line has a wavelength of 58.4 nm and an energy of 21.22 eV. The kinetic energy of emitted electrons may be measured with a focusing deflection analyzer, utilizing either magnetic or electrostatic fields.

Figure 14.20 shows the photoelectron spectrum of H_2. The count of electrons per second is plotted versus the ionization energy $I = E(M^+) - E(M)$, which is calculated from $h\nu - \frac{1}{2}mv^2$. The most energetic electrons emitted by the sample (line labeled 0) come from the production of hydrogen molecule ions in their ground state:

$$H_2(v = 0) + h\nu = H_2^+(v = 0) + e^- \tag{14.35}$$

The energy difference $E(M^+) - E(M)$ for the $v = 0 \rightarrow 0$ transition is $21.22 - 5.77 = 15.45$ eV, which is the adiabatic ionization potential for the hydrogen molecule (cf. Section 13.6). The other lines in the spectrum result when $H_2{}^+$ is produced in higher vibrational states. Since the minimum of the potential energy curve for $H_2{}^+$ is at a somewhat larger internuclear distance than for H_2 (Fig. 14.21), the Franck–Condon principle leads us to expect that the 0–0 transition will not be the most probable. The energy difference for the strongest band (in this case $v = 0 \rightarrow 2$) is referred to as the vertical ionization potential; this ionization potential is $21.22 - 5.24 = 15.98$ eV.

The photoelectron spectra of molecules with more electrons are, of course, considerably more complicated and yield information about the dissociation of inner electrons as well as valence electrons. Photoelectron spectroscopy offers the most direct method for determining the ionization potentials and provides a great deal of information about molecular electronic structure.

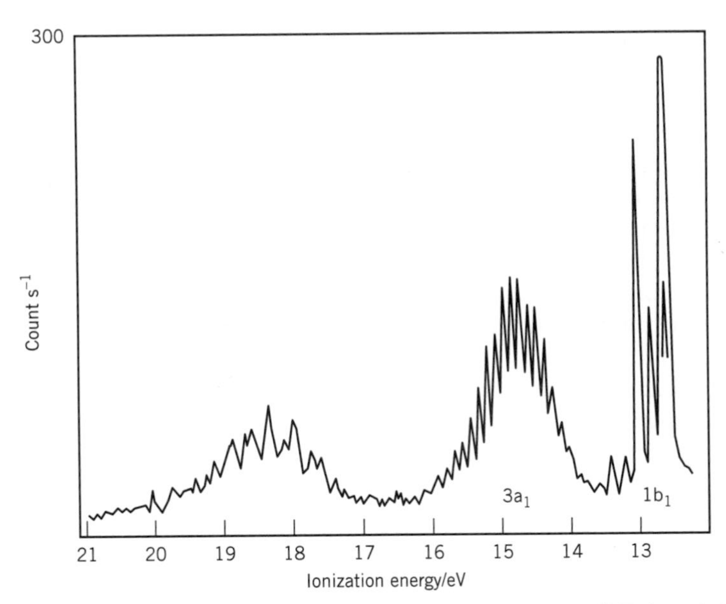

Figure 14.22 The He I UPS spectrum of H_2O. (From D. W. Turner, C. Baker, A. D. Baker, and C. R. Brundle, *Molecular Photoelectron Spectroscopy.* John Wiley & Sons, Ltd. Reproduced with permission, 1970.)

Figure 14.22 gives the UPS spectrum of H_2O, which can be interpreted in terms of the ground molecular orbital configuration$\ldots(2a_1)^2(1b_2)^2(3a_1)^2(1b_1)^2$. Removal of an electron from the $1b_1$ orbital shows short progressions for the ν_1 symmetric stretching and the ν_2 angle bending vibrations. Removal of an electron from $3a_1$ leads to a linear configuration and a long progression of the ν_2 bending vibration. The third band system is more complex and will not be discussed here.

14.11 SPECIAL TOPIC: OPTICAL ACTIVITY AND OPTICAL ROTATION*

Chiral molecules are those that exist as two nonsuperimposable structures that are mirror images. The stereoisomers of chiral molecules are referred to as **enantiomers.** These molecules are **optically active.** The term *optical activity* refers to the rotation of the plane of plane-polarized light when it passes through a substance or solution. One enantiomer will rotate light in one direction, while the other enantiomer will rotate light in the opposite direction. A closely related phenomenon is the formation of elliptically polarized light from plane-polarized light produced by an optically active medium in the vicinity of its absorption bands. These effects can be understood in terms of the differences in refractive index n and absorbancy index ϵ for left- and right-circularly polarized light.

*D. J. Caldwell and H. Eyring, *The Theory of Optical Activity.* Hoboken, NJ: Wiley, 1971; C. Djerassi, *Optical Rotatory Dispersion.* New York: McGraw-Hill, 1959; J. G. Foss, *J. Chem. Educ.* **40:**592 (1963); G. Snatzke, *Optical Rotatory Dispersion and Circular Dichroism in Organic Chemistry.* Philadelphia: Sadtler Research Labs., 1967; L. Velluz, M. Legrand, and M. Grosjean, *Optical Circular Dichroism.* New York: Academic, 1965; E. Charney, *The Molecular Basis of Optical Activity.* Hoboken, NJ: Wiley, 1979.

Circularly polarized light is light in which the electric vector rotates as the light beam advances. If the electric vector rotates clockwise as observed facing the light source, the light is said to be right-hand circularly polarized light. If it rotates counterclockwise, it is left-hand circularly polarized light. When left- and right-circularly polarized beams of equal intensity are combined, they yield plane-polarized light. In plane-polarized light the electric vector remains in a plane. A separated beam of circularly polarized light may be obtained by passing plane-polarized light through a quarter wave plate oriented at 45° to the direction of the electric vector of the polarized light. Since the quarter wave plate may be inclined to the right or the left, either right- or left-circularly polarized light may be obtained in this way.*

Figure 14.23 gives the variation of refractive index (dispersion curve) and the variation of the absorption coefficient (absorption curve) for an optically active material measured with left- and right-circularly polarized light. The difference in refractive index for the two components is referred to as **circular birefringence,** and the difference in absorption is referred to as **circular dichroism.** The difference curves are shown in the lower part of Fig. 14.23. The plot of Δn versus λ is referred to as the rotatory dispersion curve, and the plot of $\Delta\epsilon$ versus λ is referred to as the circular dichroism spectrum. When an absorption band causes the effects shown in Fig. 14.23, the whole phenomenon is referred to as a Cotton effect. In contrast with ordinary dispersion, a strong absorption band may or may not produce a large effect on the rotatory dispersion, and a weak absorption band may produce a large effect on the rotatory dispersion.

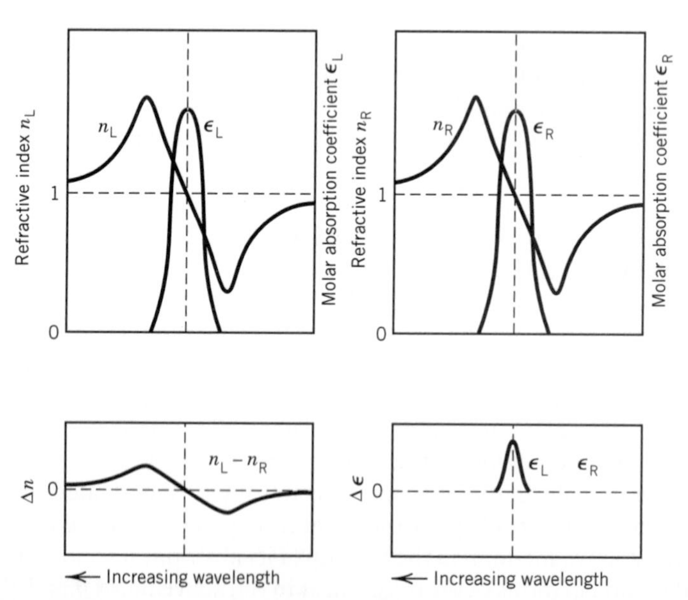

Figure 14.23 Variation of refractive index n and molar absorption coefficient ϵ in the neighborhood of a single absorption line of an optically active substance, as measured with left- and right-circularly polarized light. The difference curves are referred to as the rotary dispersion curve (Δn) and the circular dichroism spectrum ($\Delta\epsilon$). [From J. G. Foss, *J. Chem. Educ.* **40**:592 (1963).]

*D. Halliday and R. Resnick, *Physics,* Part 2, p. 1081. New York: Wiley, 1978.

Optically active substances can be divided into two classes: one in which optical activity is found only in the crystal form (for example, quartz) and one in which it is found in the gaseous, liquid, and certain nonsymmetric crystalline states of the pure substance or in solutions. Optical activity arises in the former group due to the right- or left-hand spiral structure in the crystal and disappears when this structure is melted. Substances in the latter category are optically active because of the asymmetry of the molecule itself. For a molecule whose mirror image is not superimposable on itself, left- and right-circularly polarized light have different refractive indices and correspondingly different absorption coefficients. This may happen for any molecule having only proper rotation elements of symmetry (Chapter 12). A molecule possessing any improper rotation axis (S_n), a mirror plane, or a center of symmetry cannot be optically active.

The rotation of plane-polarized light is measured with a polarimeter that consists of a light source, linear polarizer, sample, and analyzer (another linear polarizer). The rotation of the plane of polarization by the sample is measured by rotating the analyzer. If a substance rotates the plane of polarized light to the right, or clockwise, as viewed looking toward the light source, it is said to be dextrorotatory, and the rotation is given a positive sign. If the rotation is counterclockwise, the substance is levorotatory, and the rotation is given a negative sign. The magnitude of the rotation α is directly proportional to the length L of the sample and the concentration c of the optically active molecules, so it is convenient to calculate a **specific rotation** $[\alpha]$ that is defined by

$$[\alpha] = \frac{\alpha}{cL} \tag{14.36}$$ *

where L is the path length and c is the concentration in mass per unit volume. For a pure substance, c equals the density of the pure substance. The specific rotation varies with the wavelength, temperature, and solvent, so these variables must be specified.

The reason for the rotation of the plane of polarization may be understood by thinking of plane-polarized light as being formed by equal contributions of left- and right-circularly polarized light. When this light enters a medium in which the refractive indices n_L and n_R are different, then the speed of propagation of light in this medium is different for the left- and right-circularly polarized components. Thus, one moves ahead of the other, and they are now out of phase with one another. Since they are equal in amplitude, this produces a rotation of the net electric field vector, as shown in Fig. 14.24*a*, *b*. The magnitude of this rotation is linearly related to the path length of the light in the medium.

If the medium has greater absorption for left- over right-circularly polarized light, then (neglecting the effect of the rotation of the plane polarization) linearly polarized light will become elliptically polarized in an optically active medium. We can understand this by once again thinking of plane-polarized light as being composed of equal components of left- and right-polarized beams. If the absorption is greater for left polarization, that beam will be diminished in intensity relative to the right polarization. The net result is elliptic polarization of the resultant light, as shown in Fig. 14.24*c*. If both effects are present, the axes of the elliptically polarized light are rotated.

*In the past it has been customary to express c in g cm^{-3} and L in decimeters. Thus, most values of $[\alpha]$ in the literature are given in deg dm^{-1} cm^3 g^{-1}. In the SI system it is preferable to express c in kg m^{-3} and L in meters so that $[\alpha]$ is expressed in deg m^2 kg^{-1}.

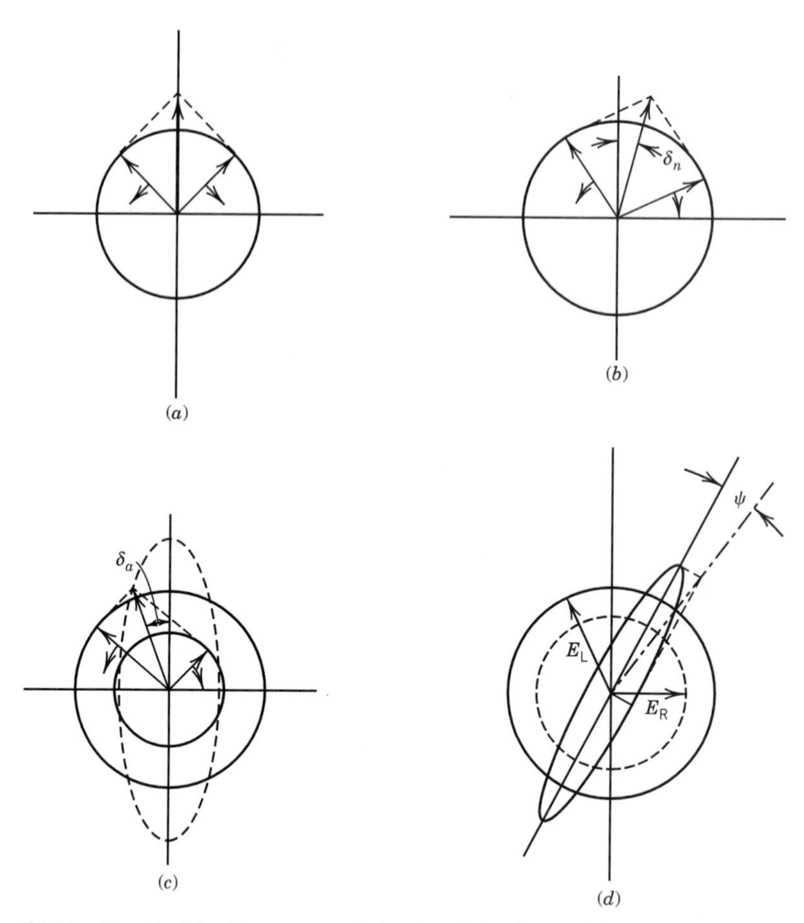

Figure 14.24 Circular birefringence and circular dichroism. (*a*) Representation of plane-polarized light as the sum of two components of circularly polarized light rotating in opposite directions. (*b*) Rotation of plane-polarized light by an angle δ_n due to different velocities of propagation of left- and right-circularly polarized light. (*c*) Production of elliptically polarized light from plane-polarized light by different absorption coefficients for left- and right-circularly polarized light. (*d*) Effects on plane-polarized light of differences in both refractive indices and absorption coefficients for the two circularly polarized components. In (*c*) and (*d*) the emerging beam is elliptically polarized.

The effects are wavelength dependent, because the closer the light beam is to an absorption, the more pronounced the effect can become. The change in optical rotation with wavelength is called **optical rotatory dispersion** (ORD). The measurements of the rotatory dispersion and circular dichroism can be used to determine the structure, configuration, and conformation of complex optically active molecules (such as proteins, synthetic polypeptides, and steroids).

Ten Key Ideas in Chapter 14

1. In electronic spectra of molecules, as in atomic spectra, singlet–singlet and triplet–triplet transitions may occur, but singlet–triplet transitions are forbidden.

2. Electronic absorption spectra of diatomic molecules generally start from the zeroth vibrational level of the ground electronic state and show progressions that have a convergence limit at the high-frequency end.
3. The intensities of lines in the vibronic spectrum are determined by Franck–Condon overlap integrals.
4. The Beer–Lambert law shows that the absorbance, $\log(I_0/I)$, is directly proportional to the concentration and the length of the optical path; thus, the intensity of a light beam in an absorbing medium decreases exponentially.
5. The absorption spectra of molecules with conjugated double bonds can be accounted for semiquantitatively by using the energy levels of an electron in a one-dimensional box the length of the conjugated system.
6. Fluorescence follows absorption when vibrational relaxation in the excited electronic state is followed by a singlet–singlet transition to the ground electronic state. Phosphorescence, which is generally slower, occurs when there is intersystem crossing to a triplet state, followed by a transition to the ground electronic state.
7. Laser action requires a three-level system as a minimum because a population inversion has to be created by irradiation or the passage of an electric current. Light emission is the result of stimulated emission, and as a consequence the laser beam is monochromatic and coherent. Lasers are also used to make very short pulses.
8. In photoelectron spectroscopy, irradiation of a gas by high-energy photons results in the emission of electrons with various kinetic energies, and determination of the spectrum of these energies provides information about the electronic energy levels in the molecule irradiated.
9. Enantiomers of chiral molecules rotate polarized light in opposite directions. When circularly polarized light is used, there is a difference in refractive index (circular birefringence) and a difference in absorption (circular dichroism).
10. The effect of wavelength on the absorption of circularly polarized light (optical rotatory dispersion) can be used to determine the structure, configuration, and conformation of complex optically active molecules such as proteins, synthetic polypeptides, and steroids.

REFERENCES

P. F. Bernath, *Spectra of Atoms and Molecules.* Oxford, UK: Oxford University Press, 1995.

G. Herzberg, *Spectra of Diatomic Molecules.* Princeton, NJ: Van Nostrand, 1950.

J. M. Hollas, *Modern Spectroscopy.* Hoboken, NJ: Wiley, 1996.

K. P. Huber and G. Herzberg, *Molecular Spectra and Molecular Structure IV, Constants of Diatomic Molecules.* New York: Van Nostrand–Reinhold, 1975.

H. Lefebre-Brion and R. W. Field, *Perturbations in the Spectra of Diatomic Molecules.* Orlando, FL: Academic, 1986.

J. L. McHale, *Molecular Spectroscopy.* Upper Saddle River, NJ: Prentice-Hall, 1999.

A. E. Siegman, *Lasers,* 2nd ed. Sausalito, CA: University Science Books, 1986.

J. I. Steinfeld, *Molecules and Radiation.* Cambridge, MA: MIT Press, 1985.

W. Struve, *Fundamentals of Molecular Spectroscopy.* Hoboken, NJ: Wiley Interscience, 1989.

PROBLEMS

(M) Problems marked with an icon may be more conveniently solved on a personal computer with a mathematical program.

14.1 The spectroscopic dissociation energy of $H_2(g)$ into ground-state hydrogen atoms is 4.4763 eV. What is the spectroscopic dissociation energy of $H_2(g)$ into one ground-state H and one H atom in the 2p state? If H_2 is dissociated with photons of energy 15 eV, what is the velocity of the H atoms coming off in the 1s and 2p states?

14.2 According to the hypothesis of Franck, the molecules of the halogens dissociate into one normal atom and one excited atom. The wavelength of the convergence limit in the spectrum of iodine is 499.5 nm. (*a*) What is the energy of dissociation in kJ mol^{-1} of iodine into one normal and one excited atom? (*b*) The thermochemical value of the heat of dissociation of I_2 into ground-state atoms can be found in Table C.3. Calculate the energy of the excited state of I that is formed from the spectroscopic dissociation in kJ mol^{-1} and eV.

14.3 The ultraviolet absorption of O_2 includes a series of lines (the Schumann–Runge bands) due to transitions from the $^3\Sigma_g^-$ ground state to the excited electronic state $^3\Sigma_u^-$, which are shown in Fig. 14.1. These lines converge to 175.9 nm, which corresponds to dissociation to one O atom in its ground state ^{3}P and one O atom in an excited state ^{1}D. What is D_0 for O_2? How does this compare with the enthalpy of formation at 0 K? Given: The ^{1}D state of O is 1.970 eV above the ground state ^{3}P.

14.4 The spectroscopic dissociation energy of $^{127}I_2$ is 1.542 38 eV according to Table 13.4. What wavelength of light would you use to dissociate ground-state molecules to ground-state atoms if you wanted the atoms to fly away with velocities of 10^3 m s^{-1}?

14.5 A solution of dye containing 0.1 mol L^{-1} transmits 80% of the light at 435.6 nm in a glass cell 1 cm thick. (*a*) What percentage of light will be absorbed by a solution containing 2 mol L^{-1} in a cell 1 cm thick? (*b*) What concentration will be required to absorb 50% of the light? (*c*) What percentage of the light will be transmitted by a solution of the dye containing 0.1 mol L^{-1} in a cell 5 cm thick? (*d*) What thickness should the cell be to absorb 90% of the light with solution of this concentration?

14.6 Derive equation 14.19 for the integrated intensity of a Gaussian absorption line. A Gaussian line has the form $\epsilon_m e^{-\sigma(\tilde{\nu}-\tilde{\nu}_m)^2}$, where $\tilde{\nu}_m$ is the frequency at the intensity maximum ϵ_m. [*Hint:* Relate σ to the width of the line at half-maximum intensity and use the integral $\int_{-\infty}^{+\infty} e^{-\sigma x^2}\,dx = (\pi/\sigma)^{1/2}$.]

14.7 According to equation 14.19, the integrated absorption coefficient $\int \kappa(\tilde{\nu})\,d\tilde{\nu}$ is equal to $1.06\kappa_{max}\Delta\tilde{\nu}_{1/2}$ when the absorption band is Gaussian in shape. The quantity $\Delta\tilde{\nu}_{1/2}$ is the width of the band when $\kappa(\tilde{\nu}) = \kappa_{max}/2$. For such a band, the Naperian absorption coefficient $\kappa(\tilde{\nu})$ is given by

$$\kappa(\tilde{\nu}) = \kappa_{max}\exp[-\alpha(\tilde{\nu} - \tilde{\nu}_{max})^2]$$

Derive equation 14.19.

14.8 The following absorption data are obtained for solutions of oxyhemoglobin in pH 7 buffer at 575 nm in a 1-cm cell:

g/cm^{-3}	3×10^{-4}	5×10^{-4}	10×10^{-4}
Transmission, %	53.5	35.1	12.3

The molar mass of hemoglobin is 64.0 kg mol^{-1}. (*a*) Is Beer's law obeyed? What is the molar absorption coefficient? (*b*) Calculate the percent transmission for a solution containing 10^{-4} g cm^{-3}.

(M) **14.9** The protein metmyoglobin and imidazole form a complex in solution. The molar absorption coefficients in L mol^{-1} cm^{-1} of the metmyoglobin (Mb) and the complex (C) are as follows:

λ/nm	500	630
$\epsilon_{Mb}/10^3$ L mol^{-1} cm^{-1}	9.42	6.88
$\epsilon_c/10^3$ L mol^{-1} cm^{-1}	9.42	1.30

An equilibrium mixture in a cell of 1-cm path length has an absorbance of 0.435 at 500 nm and 0.121 at 630 nm. What are the concentrations of metmyoglobin and complex?

14.10 The absorption spectrum for benzene in Fig. 14.8 shows maxima at about 180, 200, and 250 nm. Estimate the integrated absorption coefficients using ϵ_{max} and $\Delta\tilde{\nu}_{1/2}$ and assuming that the width at half-maximum is 5000 cm^{-1} in each case. What are the three oscillator strengths? (See Example 14.7.)

14.11 Relatively strong absorption bands have $\epsilon_{max} = 10^4$–10^5 L mol^{-1} cm^{-1} and $\Delta\tilde{\nu}_{1/2}$ of the order 1000–5000 cm^{-1}, while weak absorption bands have $\epsilon_{max} = 10$ L mol^{-1} cm^{-1} and $\Delta\tilde{\nu}_{1/2}$ of the order 100 cm^{-1}. Assuming that the absorption lines are Gaussian, compute the integrated absorption coefficient and the oscillator strengths for these bands.

14.12 The measured oscillator strength of a transition can be used to compute the transition moment, $|\boldsymbol{\mu}_{12}|^2$, by combining equations 14.20 and 14.23 to find

$$|\boldsymbol{\mu}_{12}|^2 = \frac{f3he^2}{8\pi^2 m_e \nu}$$

For strong transitions (for which $f \cong 1$), moderately weak transitions ($f \cong 10^{-3}$), and weak transitions ($f \cong 10^{-6}$), calculate $|\boldsymbol{\mu}_{12}|$ and $|\boldsymbol{R}_{12}| = |\mu_{12}|/e$, assuming a transition energy of 25 000 cm^{-1}.

14.13 In Chapter 11, the Hückel molecular orbital model was introduced to describe the electronic states of conjugated molecules. In this chapter, the free-electron model (FEMO) was introduced for the same systems. Consider the butadiene molecule in both descriptions. The Hückel model (equation 11.75) gives the energies and wavefunctions for four orbitals, while the FEMO model gives an infinite number of orbitals. Consider the lowest four in the FEMO model. Do they have the same number of nodes as the Hückel orbitals? Is there any way of choosing α and β in the Hückel model or a in the FEMO model to make the predictions for the energies of all four orbitals agree? Suppose we are content to make the lowest electronic absorption energy agree in both models; what is the formula for β in terms of a?

14.14 The lifetimes of vibrationally excited states of molecules of a liquid are limited by the collision rates in the liquid. If one in 10 collisions deactivates a vibrationally excited state, what is the broadening of vibrational lines if a molecule undergoes 10^{13} collisions per second?

14.15 Calculate the line width for (*a*) an electronic excited state with a lifetime of 10^{-8} s and (*b*) a rotational state with a lifetime of 10^3 s. In each case express the line width in cm^{-1} and MHz.

14.16 A laser is powered by a 100-W flashlamp that produces pulses with a repetition rate of 100 Hz. If the efficiency of the laser is 1%, how many photons will there be in a laser pulse with a wavelength of 500 nm?

14.17 What is the width of the frequency distribution for a 10-fs pulse and for a 100-fs pulse from a laser?

14.18 A laser operating at 700 nm produces 20 fs pulses with a repetition rate of 100 MHz. The average radiant power of the laser is 1 W. (*a*) What is the radiant power in each pulse? (*b*) How many photons are there in a pulse? (*c*) How many photons are emitted by the laser in 1 s?

14.19 A laser powered by a 100-W light source produces photons with 1000-nm wavelength at a repetition rate of 10 Hz. The actual number of photons emitted per pulse is 10^{17}. What is the efficiency of converting energy from the light source to laser output?

14.20 The first ionization potentials of Ar, Kr, and Xe are 15.755, 13.966, and 12.130 eV, respectively. Calculate the velocity of the emitted electrons when photons from a He discharge lamp with $\lambda = 58.43$ nm are used to record the photoelectron spectrum of these gases.

14.21 A sample of oxygen gas is irradiated with Mg $K_{\alpha_1 \alpha_2}$ radiation of 0.99 nm (1253.6 eV). A strong emission of electrons with velocities of 1.57×10^7 m s^{-1} is found. What is the binding energy of these electrons?

14.22 The photoelectron spectrum of molecules shows that similar atoms in different chemical environments have slightly different core orbital binding energies. For example, the 1s binding energy of carbon in CH_4 is 290 eV, while it is 293 eV in CH_3F. (*a*) Explain this shift on the basis of the electronegativity difference between carbon and fluorine. (*b*) In the molecule $F_3CCOOCH_2CH_3$, predict the order of the carbon 1s binding energies in the four carbon atoms.

14.23 When α-D-mannose ($[\alpha]_D^{20} = +29.3°$) is dissolved in water, the optical rotation decreases as β-D-mannose is formed until at equilibrium $[\alpha]_D^{20} = +14.2°$. This process is referred to as mutarotation. As expected, when β-D-mannose ($[\alpha]_D^{20} = -17.0°$) is dissolved in water, the optical rotation increases until $[\alpha]_D^{20} = +14.2°$ is obtained. Calculate the percentage of the α form in the equilibrium mixture.

14.24 The dissociation energies of HCl(g), H_2(g), and Cl_2(g) into normal atoms have been determined spectroscopically and are 4.431, 4.476, and 2.476 eV, respectively. Calculate the enthalpy of formation of HCl(g) at 0 K in kJ mol^{-1} from these data.

14.25 The limit of continuous absorption for Br_2 gas occurs at 19 750 cm^{-1}. The dissociation that occurs is

$$Br_2(\text{ground}) = Br(\text{ground}) + Br(\text{excited})$$

The transition of a ground bromine atom to an excited one corresponds to a wave number of 3685 cm^{-1}:

$$Br(\text{ground}) = Br(\text{excited})$$

Calculate the energy increase for the process

$$Br_2(\text{ground}) = 2Br(\text{ground})$$

in (*a*) cm^{-1} and (*b*) electron volts.

14.26 (*a*) Calculate the energy levels for $n = 1$ and $n = 2$ for an electron in a potential well of width 0.5 nm with infinite barriers on either side. The energies should be expressed in J and kJ mol^{-1}. (*b*) If an electron makes a transition from $n = 2$ to $n = 1$, what will be the wavelength of the radiation emitted?

14.27 The Schumann–Runge bands of O_2 are due to absorption from the ground state ($^3\Sigma_g^-$) to the B $^3\Sigma_u^-$ excited state. From Fig. 14.1, estimate the longest wavelength for this transition from the lowest vibrational state of ground-state O_2.

14.28 When a 1.9-cm absorption cell was used, the transmittance of 436-nm light by bromine in carbon tetrachloride solution was found to be as follows:

c/mol L^{-1}	0.005 46	0.003 50	0.002 10
I/I_0	0.010	0.050	0.160

c/mol L^{-1}	0.001 25	0.000 66
I/I_0	0.343	0.570

Calculate the molar absorption coefficient. What percentage of the incident light would be transmitted by 2 cm of solution containing 1.55×10^{-3} mol L^{-1} bromine in carbon tetrachloride?

14.29 The absorption band of a certain molecule in solution has a Gaussian shape with maximum molar absorption coefficient of 2×10^4 L mol^{-1} cm^{-1} and a full width at half-maximum of 4000 cm^{-1}. (*a*) What is the integrated absorption coefficient for this band? (*b*) What is the oscillator strength f for this transition?

14.30 The absorption coefficient α for a solid is defined by $I = I_0 e^{-\alpha x}$, where x is the thickness of the sample. The absorption coefficients for NaCl and KBr at a wavelength of 28 μm are 14 and 0.25 cm^{-1}. Calculate the percentage of this infrared radiation transmitted by 0.5 cm thicknesses of these crystals.

14.31 Commercial chlorine from electrolysis contains small amounts of chlorinated organic impurities. The concentrations of impurities may be calculated from infrared absorption spectra of liquid Cl_2. Calculate the concentration of $CHCl_3$ in g mL^{-1} in a sample of liquid Cl_2 if the transmittance at $\tilde{\nu} = 1216$ cm^{-1} is 45% for a 5-cm cell. At this wavelength liquid Cl_2 does not absorb, and the absorption coefficient for $CHCl_3$ dissolved in liquid Cl_2 is 900 ± 80 cm^{-1}(g cm^{-3})$^{-1}$.

14.32 To test the validity of Beer's law in the determination of vitamin A, solutions of known concentrations were prepared and treated by a standard procedure with antimony trichloride in chloroform to produce a blue color. The percent transmission of the incident filtered light for each concentration, expressed in μg mL^{-1}, was as follows:

Concentration, μg mL^{-1}	1.0	2.0	3.0	4.0	5.0
Transmission, %	66.8	44.7	29.2	19.9	13.3

Plot these data so as to test Beer's law. A solution, when treated in the standard manner with antimony trichloride, transmitted 35% of the incident light in the same cell. What was the concentration of vitamin A in the solution?

14.33 The protein metmyoglobin and the azide ion (N_3^-) form a complex. The molar absorption coefficients of the metmyoglobin (Mb) and of the complex (C) in a buffer are as follows:

λ	ϵ_{Mb}	ϵ_C
nm	10^4 L mol^{-1} cm^{-1}	10^4 L mol^{-1} cm^{-1}
490	0.850	0.744
540	0.586	1.028

An equilibrium mixture in a 1-cm cell gave an absorbance of 0.656 at 490 nm and 0.716 at 540 nm. (*a*) What are the concentrations of metmyoglobin and complex? (*b*) Since the total azide concentration is 1.048×10^{-4} mol L^{-1}, what is the equilibrium constant for the following reaction?

$$Mb + N_3^- = C$$

14.34 An acid–base indicator is a weak acid (Section 8.1) for which the acidic and basic forms have different absorption spectra and at least one of the forms absorbs strongly. For phenolphthalein, the basic form absorbs strongly and the acidic form absorbs so weakly that this absorption can be neglected. Plot the apparent molar absorbancy index versus pH for an indicator such as phenolphthalein, for which the acidic form does not absorb.

14.35 Acetone dissolved in water has a maximum absorption coefficient ϵ of 20 L mol^{-1} cm^{-1} at 38 000 cm^{-1}, and the width of the absorption at half-maximum is about 8000 cm^{-1}. What are the values of the integrated absorption coefficient and the oscillator strength?

14.36 The ionization potential of an atom may be determined by exposing it to high-energy monochromatic radiation and measuring the speed of ejected electrons. When krypton is irradiated with 58.4 nm light from a helium discharge lamp, ejected electrons have a velocity of 1.59×10^6 m s^{-1}. What is the ionization potential?

14.37 The most prominent line in the photoelectron spectrum of H_2 is due to the transition

$$H_2(v = 0) + h\nu \rightarrow H_2^+(v = 2) + e^-$$

If helium resonance radiation with an energy of 21.22 eV is used, what will be the electron kinetic energy, assuming that H_2^+ is a harmonic oscillator with a fundamental vibration frequency of 2297 cm^{-1}? The 0–0 ionization potential is 15.45 V.

14.38 When α-D-glucose ($[\alpha]_D^{20} = +112.2°$) is dissolved in water, the optical rotation decreases as β-D-glucose is formed until at equilibrium $[\alpha]_D^{20} = +52.7°$. As expected, when β-D-glucose ($[\alpha]_D^{20} = +18.7°$) is dissolved in water, the optical rotation increases until $[\alpha]_D^{20} = +52.7°$ is obtained. Calculate the percentage of the β form in the equilibrium mixture.

Computer Problems

14.A Rhodopsin (see Section 19.11) has a system of four conjugated double bonds, so we can think of it as having the structure H(CH=CH)$_4$H. Since each of the eight carbon atoms contributes one electron to the π bond that extends the length of the molecule, the $n = 1, 2, 3,$ and 4 levels in the free-electron model are filled, and absorption of a photon excites an electron from the $n = 4$ level to the $n = 5$ level. Assuming that the molecule is linear and that a C=C bond contributes 135 pm, a C—C bond contributes 154 pm, and a CH bond at each end contributes 77 pm, calculate the wavelength of maximum absorption.

14.B (*a*) Calculate the first several lines in the electronic absorption spectrum of $^{12}C^{16}O$, for which $\tilde{T}_e = 6.508\,043 \times 10^4$ cm^{-1}, $\tilde{\nu}_e' = 1514.10$ cm^{-1}, $\tilde{\nu}_e'' = 2169.81$ cm^{-1}, $\tilde{\nu}_e' x_e' = 17.40$ cm^{-1}, and $\tilde{\nu}_e'' x_e'' = 13.29$ cm^{-1}. (*b*) Make a plot of $\tilde{\nu}$ versus the vibrational quantum number v' of the excited state that goes to quantum numbers high enough to show that a convergence limit is reached.

14.C A solution of A and B has an absorbance of 0.800 at λ_1 and 0.500 at λ_2 in a 1-cm cell. At λ_1 the molar absorption coefficient of A is 1.5×10^3 L mol^{-1}cm^{-1} and the molar absorption coefficient of B is 4.0×10^3 L mol^{-1} cm^{-1}. At λ_2 the molar absorption coefficient of A is 3.0×10^3 L mol^{-1} cm^{-1} and the molar absorption coefficient of B is 2.0×10^3 L mol^{-1} cm^{-1}. What is the composition of the solution?

15

Magnetic Resonance Spectroscopy

Magnetic resonance spectroscopy differs from most other kinds of spectroscopy in that a magnetic field is used to provide the energy level separations probed by the radiation. For magnetic fields that can be routinely produced in the laboratory, the transitions between energy levels for nuclear magnetic dipoles occur in the radio-frequency range, and the transitions between energy levels for unpaired electron spins occur in the microwave range. Nuclear magnetic resonance (NMR) and electron paramagnetic resonance (EPR) yield such valuable structural information that they have become indispensable in chemistry.

Different methods for studying nuclear magnetic resonance were developed independently by Purcell and Bloch in 1946. Until about 1980 only continuous-wave (cw) NMR spectrometers were used in chemistry. Now Fourier transform spectrometers are used because of their greater sensitivity and the introduction of two-dimensional experiments. This has opened up the whole periodic table to NMR spectroscopy.

15.1 NUCLEAR MAGNETISM AND NUCLEAR MAGNETIC RESONANCE

The nuclei of certain isotopes of elements have nuclear spin because they are made up of protons and neutrons, each of which has spin angular momentum of $\hbar/2$. The spins of all the nucleons add, just like electron spins. In a nucleus with an even number of protons and an even number of neutrons, all spins are paired and the total **nuclear spin quantum number** I is equal to zero. Examples are ^{12}C and ^{16}O. The nuclear spin quantum number I can be integral or half-integral, as shown in Table 15.1.

For a nucleus, the **total spin angular momentum** is represented by $\boldsymbol{I}$, the spin quantum number by I, and the z **component of nuclear spin** by I_z. Since $\boldsymbol{I}$ is an angular momentum just like $\boldsymbol{S}$ for electrons, the eigenvalue of $\hat{I}^2$ is $I(I+1)\hbar^2$ and the magnitude of $\boldsymbol{I}$ is given by

$$|\boldsymbol{I}| = [I(I+1)]^{1/2}\hbar \tag{15.1}$$

Similarly, the eigenvalues of I_z are $m_I\hbar$ where

$$m_I = -I, -I+1, \ldots, I-1, I \tag{15.2}$$

so there are $2I+1$ values of m_I, each associated with an eigenstate of I^2 and I_z. In the absence of a magnetic field these states have the same energy, but in the presence of a magnetic field a nucleus with spin has $2I+1$ equally spaced energy levels. A proton has spin $\frac{1}{2}$, and so in the presence of a magnetic field it has two states, a state of low energy aligned with the field ($m_I = +\frac{1}{2}$) and a state of high energy opposed to the field ($m_I = -\frac{1}{2}$). The nucleus of a nitrogen atom in its most abundant isotope, ^{14}N, has three possible orientations in a magnetic field because $I = 1$, and so $m_I = -1, 0, 1$.

As we saw earlier in discussing electron spin in connection with spectroscopy, spin gives rise to a magnetic dipole moment vector $\boldsymbol{\mu}_S$.

Table 15.1 Magnetic Properties of Selected Nuclei

Nucleus	*% Abundance*	*Spin I*	g_N	*Magnetogyric Ratio* $\gamma/10^7\ T^{-1}\ s^{-1}$	*Larmor Frequency for* 1 T *in* MHz
^{1}H	99.99	$\frac{1}{2}$	5.585	26.7519	42.5759
^{2}D	0.01	1	0.857	4.1066	6.53566
^{7}Li	92.5	$\frac{3}{2}$	2.171	10.3975	16.546
^{13}C	1.11	$\frac{1}{2}$	1.405	6.7283	10.7054
^{14}N	99.6	1	0.403	1.9338	3.0756
^{15}N	0.4	$\frac{1}{2}$	−0.567	−2.712	4.3142
^{17}O	0.04	$\frac{5}{2}$	−0.757	−3.6279	5.772
^{19}F	100	$\frac{1}{2}$	5.257	25.181	40.0541
^{23}Na	100	$\frac{3}{2}$	1.478	7.08013	11.262
^{31}P	100	$\frac{1}{2}$	2.2634	10.841	17.238
^{33}S	0.74	$\frac{3}{2}$	0.4289	2.054	3.266

In equation 10.52 it was stated that the magnetic dipole moment vector $\boldsymbol{\mu}_s$ for electron spin is given by

$$\boldsymbol{\mu}_s = -\frac{g_e e}{2m_e}\boldsymbol{S} \tag{15.3}$$

where g_e is the electron g factor (2.002 322), m_e is the mass of the electron, and $\boldsymbol{S}$ is the angular momentum vector for electron spin. We also saw that the z component of the magnetic moment due to electron spin μ_{sz} is proportional to the magnitude of the spin angular momentum vector in the direction of the magnetic field; $S_z = m_s\hbar$, so the relation is given by

$$\mu_{sz} = -\frac{g_e e}{2m_e}S_z = -g_e\mu_B m_s \tag{15.4}$$

where μ_B is the Bohr magneton defined in equation 10.39 and m_s is the quantum number for the z component of spin.

The equations for the **nuclear magnetic moment** $\boldsymbol{\mu}$ and its z component μ_z are similar except that the g factor becomes the **nuclear *g* factor** g_N, the mass of the electron is replaced by the mass of the proton m_p, the Bohr magneton is replaced by the nuclear magneton μ_N, the spin quantum number m_s for the z component of the spin is replaced with the spin quantum number m_I for the z component of the nuclear spin, and the sign is changed:

$$\boldsymbol{\mu} = \frac{g_N e}{2m_p}\boldsymbol{I} = g_N\mu_N\boldsymbol{I}/\hbar \tag{15.5}$$

Since g_N can be positive or negative, and $\boldsymbol{\mu}$ and $\boldsymbol{I}$ are either parallel or antiparallel, we can write this equation in terms of the magnitudes of the two vectors as

$$\mu = \frac{g_N e}{2m_p}|\boldsymbol{I}| = g_N\mu_N|\boldsymbol{I}|/\hbar \tag{15.6}$$

The z component for the nuclear magnetic dipole moment is given by an equation similar to equation 15.5 since $I_z = m_I\hbar$:

$$\mu_z = \frac{g_N e}{2m_p}I_z = g_N\mu_N m_I \tag{15.7}$$

The basic unit of nuclear magnetism, the nuclear magneton μ_N, is defined by

$$\mu_N = \frac{e\hbar}{2m_p} \tag{15.8}$$

Example 15.1 ***Value of μ_N, the nuclear magneton***

What is the value of the nuclear magneton?

$$\mu_N = \frac{e\hbar}{2m_p} = \frac{(1.602\,177\,3\times10^{-19}\text{ C})(1.054\,573\times10^{-34}\text{ J s})}{2(1.672\,623\,1\times10^{-27}\text{ kg})}$$
$$= 5.050\,787\times10^{-27}\text{ J T}^{-1}$$

The nuclear magneton is $\frac{1}{1836}$ of the Bohr magneton ($\mu_N/\mu_B = m_e/m_p$). This tells us that nuclear spin levels have much smaller splittings than electron spin levels when a magnetic field is applied.

Usually nuclear magnetic moments are discussed in terms of their ***magnetogyric ratio*** γ, rather than in terms of nuclear g factors. The magnetogyric ratio was introduced in equation 10.36 as the ratio of the magnetic moment vector to the angular momentum vector. For nuclear magnetic moments,

$$\boldsymbol{\mu} = \gamma \boldsymbol{I} \qquad \text{or} \qquad \mu = \gamma|\boldsymbol{I}| \tag{15.9}$$

The magnetogyric ratio is important because we will see that the frequency of the radiation absorbed or emitted in a nuclear magnetic transition is proportional to the magnetogyric ratio. Equation 15.5 indicates that the magnetogyric ratio is given by

$$\gamma = g_N \mu_N / \hbar \tag{15.10}$$

Substituting equations 15.1 and 15.10 into equation 15.9 yields

$$\mu = g_N \mu_N [I(I+1)]^{1/2} = \gamma\hbar[I(I+1)]^{1/2} \tag{15.11}$$

Thus the magnetic moment of a nucleus in a magnetic field depends on its magnetogyric ratio and its nuclear spin quantum number I. The magnetic moment in the direction of the magnetic field is given by

$$\mu_z = g_N \mu_N m_I = \gamma\hbar m_I \tag{15.12}$$

15.2 ENERGY LEVELS IN NUCLEAR MAGNETIC RESONANCE

In the absence of a magnetic field,* the energy of an isolated nucleus is independent of the quantum number m_I. As mentioned earlier in discussing the energy of a hydrogenlike atom in a magnetic field $\boldsymbol{B}$, the energy is given by

$$E = -\boldsymbol{\mu} \cdot \boldsymbol{B} = -\mu_z B \tag{15.13}$$

where μ_z is the magnetic dipole moment in the direction of the field and B is the magnitude of the field (see equations 10.40 and 10.41). Because of equation 15.9, this equation can be written

$$E = -\gamma B I_z \tag{15.14}$$

Since the potential energy is the only energy involved, the spin Hamiltonian operator for an isolated nucleus is given by

$$\hat{H} = -\gamma B \hat{I}_z \tag{15.15}$$

where $\hat{I}_z$ is the operator for the z component of the angular momentum. The Schrödinger equation shows what happens when the spin Hamiltonian operates on a spin function ψ:

$$\hat{H}\psi = -\gamma B \hat{I}_z \psi = E\psi \tag{15.16}$$

*Strictly speaking, the magnetic field strength is represented by $\boldsymbol{H}$, and the magnetic flux density is represented by $\boldsymbol{B}$. It is the magnetic flux density $\boldsymbol{B}$ that determines the magnetic force on a moving charged particle; it is often referred to simply as the magnetic field, as it is in this chapter.

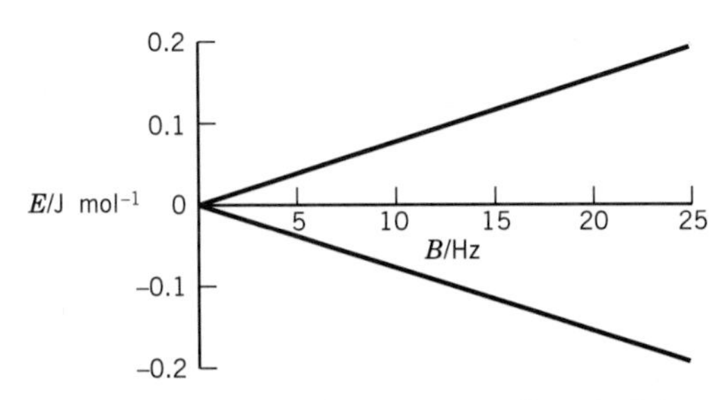

Figure 15.1 The energy of a mole of hydrogen atoms aligned with a magnetic field ($m_I = +\frac{1}{2}$, lower plot) and opposed to a magnetic field ($m_I = -\frac{1}{2}$, upper plot). (See Computer Problem 15.A.)

The spin functions are also eigenfunctions of the operator for the z component of the angular momentum:

$$\hat{I}_z \psi = \hbar m_I \psi \tag{15.17}$$

where $m_I = I, I-1, \ldots, -I$. Thus the possible energies of a magnetic dipole in a magnetic field are given by

$$E = -\hbar \gamma m_I B \tag{15.18}$$

When this is applied to protons, which can have $m_I = \frac{1}{2}$ or $-\frac{1}{2}$, the dependence of energy on magnetic field strength is shown in Fig. 15.1. The energy difference and the corresponding frequency are given by

$$\Delta E = E(m_I = -\tfrac{1}{2}) - E(m_I = \tfrac{1}{2}) = -\hbar\gamma B(-\tfrac{1}{2}) + \hbar\gamma B/2 = \hbar\gamma B = h\nu \tag{15.19}$$

Thus

$$\nu = \frac{\gamma B}{2\pi} \tag{15.20}$$

This is referred to as the **Larmor frequency.** The larger the value of γ, the easier it is to observe a nucleus in NMR. Since absorption at this frequency is determined by use of resonance methods, this is known as **nuclear magnetic resonance** (NMR) spectroscopy. The Larmor frequencies for protons are given in Fig. 15.2 as a function of the magnetic field strength, and Larmor frequencies for a number of nuclei at 1 T magnetic field are included in Table 15.1.

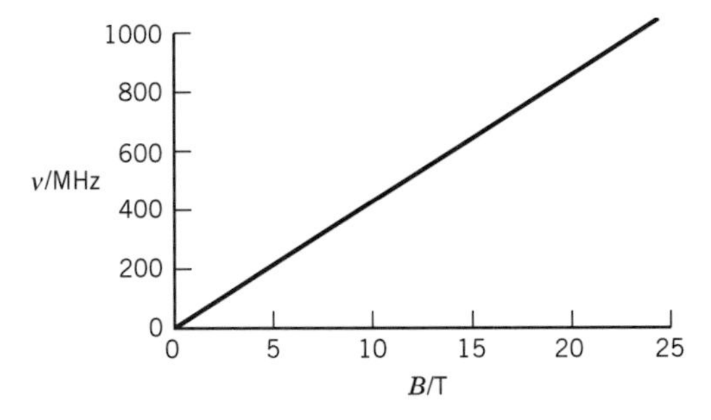

Figure 15.2 The NMR frequency for hydrogen atoms as a function of the strength of the magnetic field. (See Computer Problem 15.B.)

Since these transitions take place between magnetic states, they are induced by the oscillating magnetic field of radiation rather than the oscillating electric field, and are called **magnetic dipole transitions.** For magnetic field strengths that are practical for use in nuclear magnetic resonance, the frequencies for various nuclei are in the range of 1000 MHz to a few kilohertz. To achieve resonance, either the frequency of the electromagnetic radiation or the magnetic field strength can be adjusted.

Example 15.2 ***Calculating the magnetic field from the Larmor frequency***

At a certain magnetic field strength the frequency of radiation that is absorbed by a sample containing protons is 220 MHz. What is the magnetic field strength?

$$B = \frac{2\pi\nu}{\gamma} = \frac{2\pi(220 \times 10^6\ s^{-1})}{26.7519 \times 10^7\ \mathrm{T}^{-1}\,\mathrm{s}^{-1}} = 5.1671\ T$$

If we consider a single nuclear spin in a classical picture, its magnetic moment $\boldsymbol{\mu}$ can be considered to be rotating around the field direction $\boldsymbol{B}$, as shown in Fig. 15.3*a*. The magnetic moment of the nucleus does not line up with the field but precesses about it at an angle θ,

$$\cos\theta = \frac{m_I}{[I(I+1)]^{1/2}} \tag{15.21}$$

as shown in Fig. 15.3*a*. The angular velocity around the cone is $\gamma\boldsymbol{B}$. The frequency of complete rotations is the Larmor frequency, $\gamma\boldsymbol{B}2/\pi$. For an assembly of spin $\frac{1}{2}$

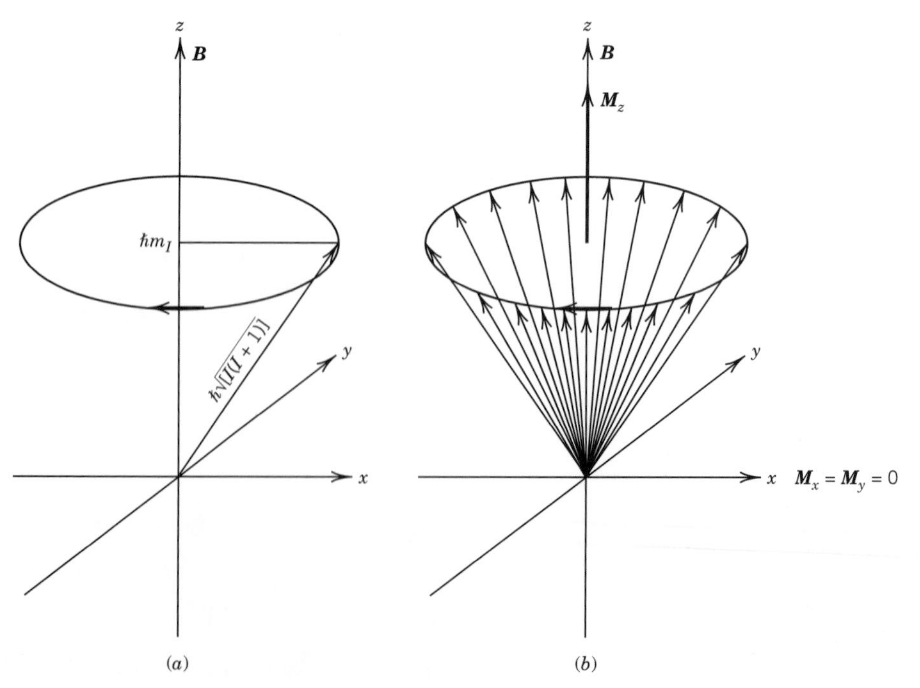

Figure 15.3 (*a*) A freely precessing nucleus in a magnetic field $\boldsymbol{B}$. (*b*) Precession of low-energy nuclei in a magnetic field $\boldsymbol{B}$. [From J. W. Akitt and B. E. Mann, *NMR and Chemistry,* Taylor and Francis Books Ltd. (Nelson Thornes), London, ©2000.]

nuclei, there are two precession cones, one for nuclei with $m = +\frac{1}{2}$ and one for $m = -\frac{1}{2}$. These cones point in opposite directions, but Fig. 15.3*b* shows only the *excess* lower energy ($m = \frac{1}{2}$) nuclei in the sample. The vectors all rotate with the same frequency and are distributed evenly over the conical surface so that $\boldsymbol{M}_x = \boldsymbol{M}_y = 0$. The vectors, however, add to give a net magnetization $\boldsymbol{M}_z$ along the z axis (the direction of $\boldsymbol{B}$).

In discussing transitions induced by electromagnetic radiation in Section 13.2, we pointed out that the radiation induces emission from higher to lower levels as well as absorption from lower to higher levels. This occurs in NMR as well. Since spontaneous emission is extremely small and can be neglected for NMR, the processes of absorption and stimulated emission are the only ones that are relevant. The rate of absorption is proportional to the number of nuclei in the lower state, while the rate of stimulated emission is proportional to the population (number) of nuclei in the upper state. Since the proportionality constant is the same for both these processes (see Sections 13.2 and 14.9), we see that the net loss of energy from the electromagnetic field is proportional to the *difference* in populations between the lower and upper levels. Let N_u be the population in the upper state and N_l be the population in the lower state in Fig. 15.1. At thermal equilibrium, these populations obey the Boltzmann equation (see Eq. 13.11)

$$\left(\frac{N_u}{N_l}\right)_{eq} = e^{-\Delta E/kT} = e^{-g_N \mu_N B/kT} \cong 1 - \frac{g_N \mu_N B}{kT} \tag{15.22}$$

where we have expanded the exponential term since at temperatures above a few kelvins for any nucleus, the exponent is very small.

If the intensity of the radiation field is increased, the rate of absorption and stimulated emission may become higher than the rate of **thermal equilibration** so that the populations of the two levels can become equal, leading to saturation, that is, zero net absorption.

Example 15.3 ***Ratio of proton spins in the two spin states***

What is the ratio of the number of proton spins in the lower state to the number in the higher state in a magnetic field of 1 T at room temperature?

$$\frac{N_l}{N_u} = 1 + \frac{g_N \mu_N B}{kT} = 1 + \frac{(5.585)(5.05 \times 10^{-27}\ \text{J T}^{-1})(1\ \text{T})}{(1.38 \times 10^{-23}\ \text{J K}^{-1})(298\ \text{K})}$$
$$= 1 + 6.86 \times 10^{-6}$$

Notice that the exponent, $g_N \mu_N B/kT$, is indeed very small, so the expansion in equation 15.22 is justified.

To detect a nuclear resonance, the system (with field $\boldsymbol{B}$ in the z direction) is perturbed by applying a sinusoidally oscillating magnetic field $\boldsymbol{B}_1$ in the xy plane with frequency $\gamma \boldsymbol{B}/2\pi$. This oscillating magnetic field stimulates both absorption and emission of energy by the spin system, leading to a net absorption of energy. The oscillating field is produced by passing a radio-frequency alternating current through the Helmholtz double coils on either side of the sample. As indicated in Fig. 15.4, the field $\boldsymbol{B}$ produces a magnetization $\boldsymbol{M}$ in the system in the z direction due to the nuclear spins. In addition, the nuclear spins will *tend to precess* about

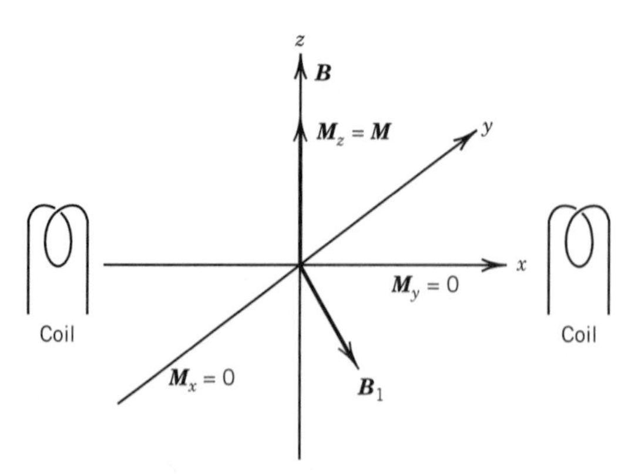

Figure 15.4 When a rotating magnetic field $\boldsymbol{B}_1$ with the same angular velocity as the nuclei is applied, the nuclei tend to precess around $\boldsymbol{B}_1$. This causes the rotating cone to tip. [From J. W. Akitt and B. E. Mann, *NMR and Chemistry,* Taylor and Francis Books Ltd. (Nelson Thornes), London, ©2000.] x, y, and z are laboratory coordinates.

$\boldsymbol{B}_1$, which is rotating with the Larmor frequency in the xy plane. This causes the cone of vectors to tip and contribute a rotating vector $\boldsymbol{M}_{xy}$ in the xy plane.

To follow changes in the magnetization in the xy plane, it is convenient to use a rotating xy coordinate system (frame) in which the x and y axes rotate around the z axis at the Larmor frequency. In Fig. 15.5, x' and y' are measured in the rotating frame, and the $\boldsymbol{B}_1$ vector is now stationary. A sufficiently long or powerful $\boldsymbol{B}_1$ along the x' axis will turn the magnetization $\boldsymbol{M}$ so that it lies along the y' axis. $\boldsymbol{M}_y$ is oriented along the y' axis, but in the lab frame it creates a signal that can be picked up by the Helmholtz coils. This is a description of a so-called 90° pulse. The signal that is picked up is referred to as a ***resonance*** signal.

So far we have assumed that the $\boldsymbol{B}_1$ pulse is monochromatic, but if it is a short pulse, it covers a band of frequencies, and all the nuclei with precession frequencies within this band are excited by the pulse. The effectiveness of $\boldsymbol{B}_1$ varies as $\sin(x)/x$, where x is the frequency, as shown in Fig 15.6. This figure shows that the nuclei within $\pm 1/4$PW Hz of the spectrometer frequency are almost equally

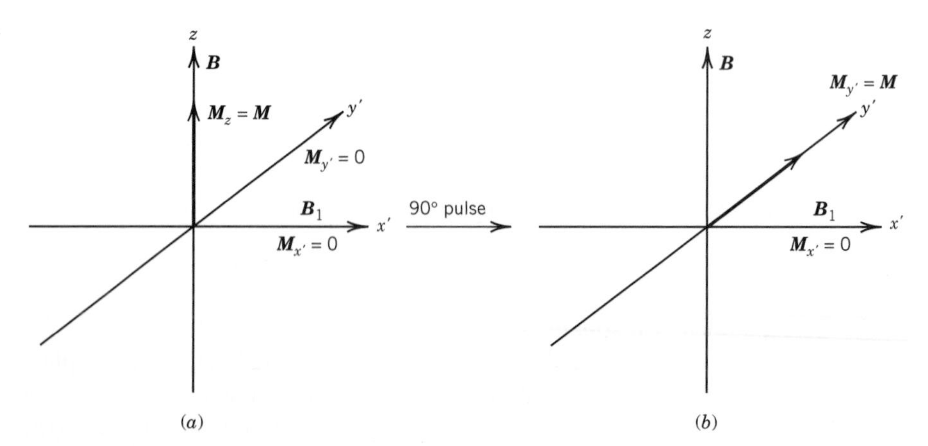

Figure 15.5 (a) The system before the pulse. (b) When a 90° pulse is applied along the x' axis the nuclear magnetism is rotated until it lies along the y' axis. [From J. W. Akitt and B. E. Mann, *NMR and Chemistry,* Taylor and Francis Books Ltd. (Nelson Thornes), London, ©2000.]

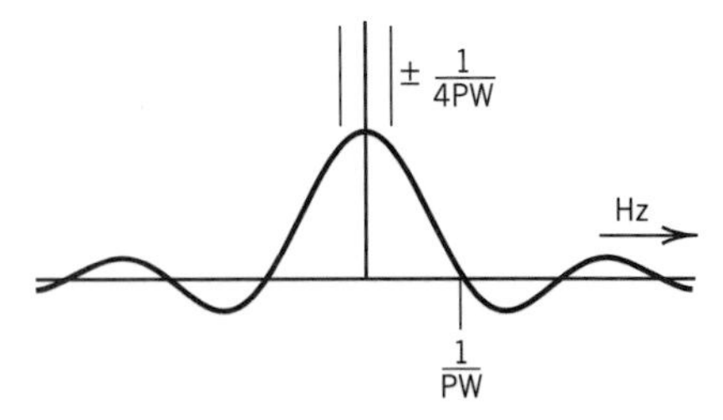

Figure 15.6 The effectiveness of a pulse of length PW versus frequency. The effectiveness is essentially constant in the range ±1/4PW. [From J. W. Akitt and B. E. Mann, *NMR and Chemistry,* Taylor and Francis Books Ltd. (Nelson Thornes), London, ©2000.]

affected, but that there are null points at $\pm n$/PW where the nuclei are not excited. For the nuclei with Larmor frequencies in the central region of the pulse, the magnetization in the laboratory frame, $\boldsymbol{M}_{xy}$, precesses around $\boldsymbol{B}$ with the Larmor frequency. After the pulse has terminated, this rotating magnetization continues and induces a signal in the Helmholtz coils that can be detected without interference from $\boldsymbol{B}_1$. As the system returns to equilibrium, the signal $\boldsymbol{M}_{xy}$ diminishes to zero, usually within less than 10 s.

15.3 FOURIER TRANSFORM NMR SPECTROMETER

The FT NMR spectrometer involves a strong, highly stable magnet, a pulse generator, receiver, and computer, as shown in Fig. 15.7. The sample is surrounded by transmitter/receiver coils (Helmholtz coils). The magnetic field, which is along the z axis, is created by a superconducting cylindrical solenoid whose windings are made of a niobium–tin alloy that is a superconductor at the temperature of liquid helium (4.2 K). This can provide a persistent magnetic field $\boldsymbol{B}$ of 12 T or greater. The liquid helium dewar is thermally isolated, by means of a vacuum chamber, from an outer dewar containing liquid nitrogen (77 K). After a superconducting magnet has been energized, no additional electric current is needed, but it does need to be kept cold.

In a FT NMR spectrometer, a short powerful $\boldsymbol{B}_1$ pulse is applied to the sample, centered at the Larmor frequency of the nuclear spin under study. As explained above, this pulse excites all the nuclear spins so that there is a rotating magnetization in the xy plane. After the pulse, the magnetization relaxes back to equilibrium. The signal is amplified and compared with the input pulse to obtain the time-dependent output signal (the free induction decay, or FID) as shown in Fig. 15.8.

A FID is a plot of signal versus time, and it is NMR in the "time domain." Since recording a FID takes a very short time, this process is repeated many times, and signal averaging is used to obtain a more accurate FID. Spectra are more readily interpreted as absorption as a function of frequency, and such spectra are in the "frequency domain." These two types of spectra are interconvertible through the Fourier transform, which we discuss in more detail in Section 15.10. A computer in the FT NMR spectrometer is used to calculate the Fourier transform of the FID signal. As the simplest possible example, Fig. 15.8 shows the relation between the time domain and the frequency domain for a single-resonance signal.

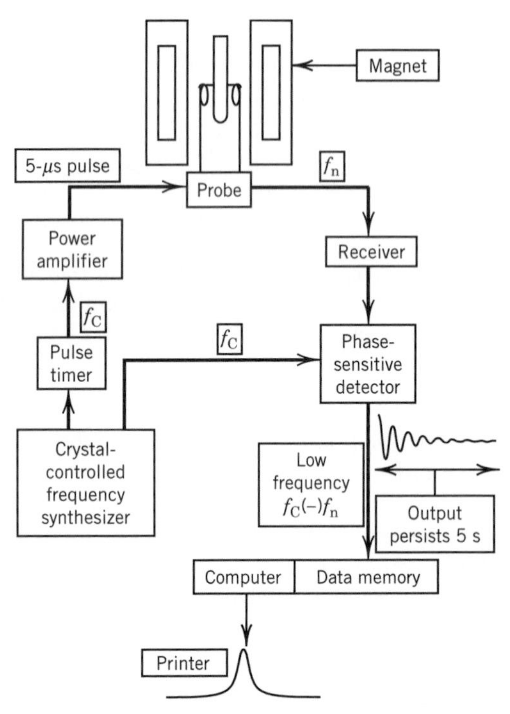

Figure 15.7 Fourier transform NMR spectrometer. The radio-frequency pulse tips the nuclei in the sample by 90° provided their frequency lies within the bandwidth of the pulse. The nuclear output frequency f_n will be close to the input frequency f_C and the difference $f_C - f_n$ is obtained at the output of the phase sensitive detector. The time-dependent output (FID) that is shown is Fourier transformed to obtain a plot of intensity versus frequency. [From J. W. Akitt and B. E. Mann, *NMR and Chemistry,* Taylor and Francis Books Ltd. (Nelson Thornes), London, ©2000.]

For now, we will simply assume that the frequency domain plot can be obtained and will discuss the significance of NMR spectra.

Comment:

We have seen that certain isotopes of a very large number of elements can be detected by NMR. However, some magnetic nuclei give stronger signals than others. It can be shown that the strength of the signal (receptivity) is proportional to the natural isotopic abundance times the cube of the magnetogyric ratio. This shows that the nuclei that give the strongest signals are ^{1}H, ^{19}F, *and* ^{31}P. *However, the sensitivities of Fourier transform NMR spectrometers are great enough that a wide variety of nuclei can be used.*

15.4 THE CHEMICAL SHIFT

A single isotope gives rise to a single nuclear magnetic resonance, and this would not be of much interest in chemistry except for the fact that the magnetic field at the nucleus is not equal to the applied magnetic field. The magnetic field that the nuclear spin feels is the vector sum of the applied field and the field due to the other nuclear spins and the electron spins in the molecule. In the liquid state,

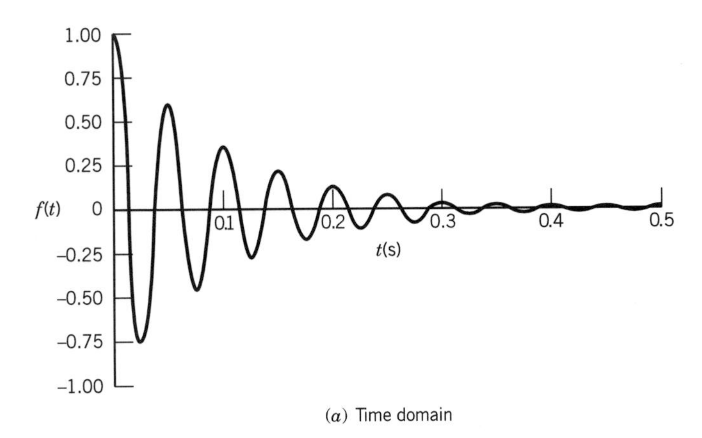

(a) Time domain

1.4
1.2
1.0
0.8
f(ν)
0.6
0.4
0.2
0
10 20 30 40 50 60
ν

(b) Frequency domain

Figure 15.8 (*a*) Free induction decay (FID) for hydrogen atoms with a Larmor frequency of 20 s^{-1} after a 90° pulse. The relaxation time T_2 is 0.10 s. (*b*) Fourier transform of the FID. This is the NMR spectrum and shows the frequency of 20 s^{-1} and a peak width that corresponds with the relaxation time T_2 (see Computer Problems 15.C and 15.D). See Section 15.7 for a discussion of the relaxation time T_2.

molecules rotate rapidly and randomly, so direct nuclear magnetic fields fluctuate and average to zero. However, a diamagnetic screening effect is produced by the orbital electrons in a molecule. The origin of this screening effect is illustrated in Fig. 15.9. When a molecule is placed in a magnetic field, the field induces motion of orbital electrons that set up an additional magnetic field normally in opposition to the applied field. The nature of this magnetic field is indicated by the dashed lines in the figure. The magnetic field at the nucleus is less than the applied field. The nucleus of interest is said to be **shielded** by electrons. This (and other) shielding effects can be taken into account by writing $B = B_0 - \sigma_i B_0$, where B_0 is the externally applied field, so that equation 15.20 becomes

$$\nu_i = \frac{\gamma B_0}{2\pi}(1 - \sigma_i) \tag{15.23}$$

where σ_i is the **shielding constant** for nucleus i.

It is important to note that the shielding constant σ_i depends on the local electronic or chemical structure around the nucleus. The shielding constant σ is less than 10^{-5} for protons and less than about 10^{-3} for most other nuclei. Since

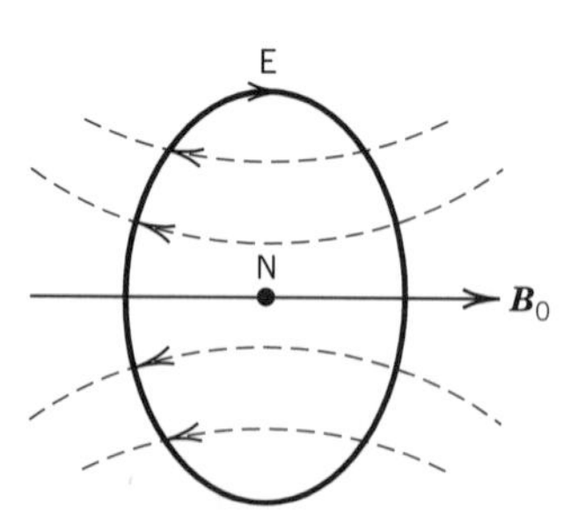

Figure 15.9 Motion of electron cloud E around nucleus N that creates a magnetic field at the nucleus in opposition to the applied field $\boldsymbol{B}_0$. [From J. W. Akitt and B. E. Mann, *NMR and Chemistry,* Taylor and Francis Books Ltd. (Nelson Thornes), London, ©2000.]

the magnitude of the shielding depends on the orientation of the molecule with respect to the applied field, σ is actually a second-rank tensor. However, for a liquid or a gas the directional part of σ averages out so that it may be treated as a scalar. Since it is difficult to measure the magnetic field strength B_0 at a nucleus accurately, it is not practical to measure σ_i, but it is possible to determine relative values by using the same nuclei in a **reference compound.** For ^{1}H, ^{13}C, and ^{29}Si, the reference compound is tetramethylsilane (TMS), $(CH_3)_4Si$.

The frequency for ^{1}H, for example, in the reference compound is given by

$$\nu_{\text{ref}} = \frac{\gamma B_0}{2\pi}(1 - \sigma_{\text{ref}}) \tag{15.24}$$

The difference in frequency for a ^{1}H nucleus in a certain position in molecule i and in the reference compound is obtained by subtracting equation 15.24 from equation 15.23 to obtain

$$\nu_i - \nu_{\text{ref}} = \frac{\gamma B_0}{2\pi}(\sigma_{\text{ref}} - \sigma_i) \tag{15.25}$$

This equation is then divided by equation 15.24 to obtain, to very high accuracy,

$$\frac{\nu_i - \nu_{\text{ref}}}{\nu_{\text{ref}}} = (\sigma_{\text{ref}} - \sigma_i) = 10^{-6}\delta_i \tag{15.26}$$

where δ_i is called the **chemical shift.** Note that we have used $1 - \sigma_{\text{ref}} \approx 1$. In other words, the chemical shift δ_i is the difference in shielding constants σ_i expressed in parts per million (ppm). The chemical shift can be calculated from the difference frequency between the chosen nuclei and the reference, expressed in Hz, divided by the spectrometer frequency, expressed in MHz.

NMR spectra are conventionally arranged so that the frequency increases to the left. This means that the chemical shifts increase to the left. Since it is convenient to have chemical shifts for most organic compounds be positive, tetramethylsilane was chosen as a reference because it has a low Larmor frequency. Tetramethylsilane has a larger shielding constant for ^{1}H than most organic substances because the electron density around ^{1}H is higher in TMS than in most organic compounds. Since frequency increases to the left in an NMR spectrum, we can also think of the magnetic field strength at the nuclei as increasing to the right. Thus chemical shifts for protons are calculated using

$$\delta_{\text{H}} = \frac{\Delta\nu}{\nu_{\text{spect}}} \times 10^6 \tag{15.27}$$

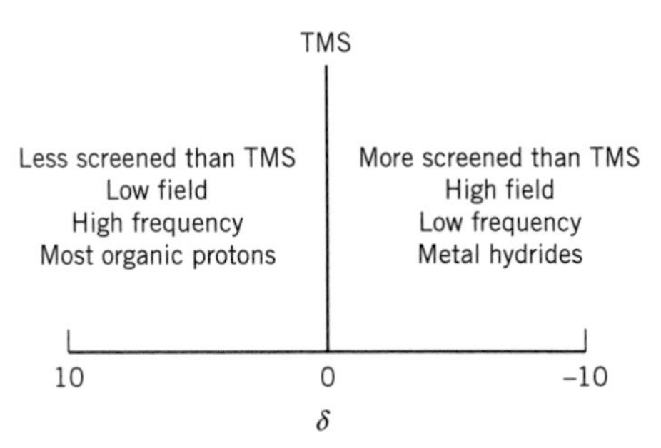

Figure 15.10 Terminology used in discussing proton chemical shifts. (From J. W. Akitt, *NMR and Chemistry,* 3rd ed. The Netherlands: Kluwer, 1992.)

where ν_{spect} is the spectrometer frequency and $\Delta\nu$ for a line with respect to TMS is taken as positive for an absorption at a frequency higher than TMS and as negative for a frequency lower than TMS. Figure 15.10 summarizes some of the terminology that is used in discussing ^1H chemical shifts.

Chemical shifts for protons in various molecules, expressed as δ values, are given in Table 15.2. Note that the protons of benzene are not shielded very much, and therefore resonate at lower magnetic field strengths than methyl protons.

The reason protons of benzene resonate at lower fields can be seen in Fig. 15.11. When a benzene ring is oriented perpendicular to the magnetic field, the circulation of electrons in the π orbitals induces a field that is in the same direction as the applied field at the protons. Therefore, aromatic protons resonate at a lower applied field than they otherwise would. This effect is reduced by molecular tumbling because when the benzene ring is oriented parallel to the field, there is no such effect.

Proton chemical shifts in organic compounds can be correlated with the electronegativities of neighboring groups, types of carbon bonding, and hydrogen bonding. In

$$\text{H}-\overset{|}{\underset{|}{\text{C}}}-\text{X}$$

the chemical shift for the proton depends on the electronegativity of X. The greater the electronegativity of X, the more it will draw electrons away from H.

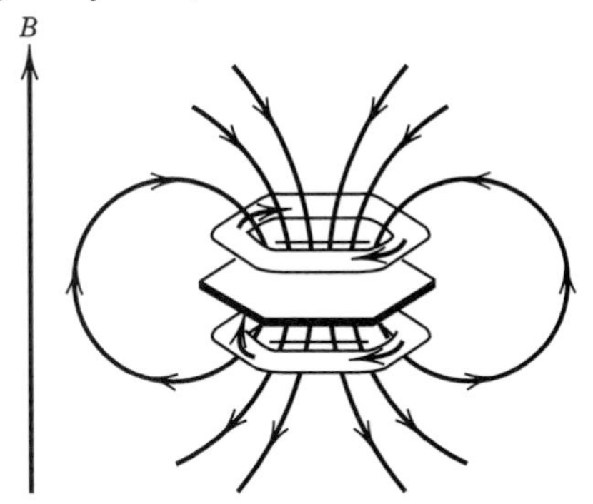

Figure 15.11 Magnetic flux density produced by electron circulation in the π orbitals of the benzene ring. At the protons, the induced field is in the same direction as the applied field.

Table 15.2 Chemical Shifts of ^1H in Hydrocarbons

	δ/ppm
CH_3R	−0.5 to 0.5
RCH_2R	0.8 to 1.9
$C{=}CCH_2R$	1.9 to 2.9
$C{=}CCHR_2$	1.9 to 3.7
C=CH	3.9 to 8.0
C≡CH	1.3 to 3.1
Ar-H	6.5 to 9.0
HCO_2R	7.8 to 8.8
RCHO	9.0 to 11.0

Ar = aromatic.

Source: J. H. Nelson, *Nuclear Magnetic Resonance Spectroscopy,* 2003 ©. Reprinted by permission of Pearson Education, Inc., Upper Saddle River, NJ.

The more electrons are drawn away, the lower the magnetic field required for resonance, and the larger the δ value. The electronegativities of the halogens are F > Cl > Br > I, and the proton chemical shifts δ for protons in the methyl halides are $CH_3F > CH_3Cl \gg CH_3Br > CH_3I$.

The protons of metal hydrides have negative chemical shifts on the δ scale; that is, they are more highly shielded than the protons of TMS.

The chemical shifts for other nuclei are larger than for protons because they are surrounded by a larger number of electrons. For ^{13}C the chemical shifts, measured with respect to TMS, range from 200 ppm for

$$R'-\overset{\overset{\displaystyle O}{\|}}{C}-R$$

to approximately zero for methyl groups. Chemical shifts for ^{14}N and ^{15}N, measured with respect to NH_3, range up to about 800 ppm. Phosphorus chemical shifts, measured with respect to P_4O_6, range from about 100 to −200 ppm.

15.5 INTERNUCLEAR SPIN–SPIN COUPLING

So far we have not taken account of the fact that neighboring magnetic dipoles in a sample interact with each other. This **spin–spin coupling** affects the magnetic field at the positions of nuclei being observed. Nuclear dipoles can interact directly through space, and this **direct spin–spin coupling** is important in solid-state NMR spectroscopy. But in high-resolution spectroscopy of solutions in low-viscosity liquids, this coupling is averaged to zero by molecular motions. In this section, we are concerned with **indirect spin–spin coupling** that takes place through bonds in a single molecule. This coupling via bonding electrons is called the Fermi contact interaction.

The usual presentation of a proton resonance spectrum is illustrated in Fig. 15.12. The multiplets arise from spin–spin splitting and provide information in addition to that arising from the chemical shift δ. The coupling pattern and the coupling constants provide information on connections between groups and make it possible to deduce the structure of a molecule. In Fig. 15.12, the CH_3 resonance is split into three lines because of coupling to the neighboring CH_2 group. The spins of the protons in the neighboring CH_2 group can be arranged in three ways as shown in Fig. 15.13. The CH_3 resonance is split into four resonances because the spins of the protons in the neighboring CH_3 group can be arranged in four different ways as shown in Fig. 15.13. In the absence of a nearby group containing protons, a CH_2 group would give a single absorption line because these two protons do not split each other. The protons in CH_2 are said to be isochronous and magnetically equivalent. The reason they give a single resonance line is clarified in the next section using quantum mechanics.

There is a simple rule for splitting in groups of $I = \frac{1}{2}$ nuclei: A neighboring group with n equivalent $I = \frac{1}{2}$ nuclei split a resonance into $n+1$ lines. The relative intensities of the lines are given by the binomial coefficients of $(1 + 1)^n$ or by Pascal's triangle (see Example 15.4).

The intervals between lines in a multiplet are equal, and this interval in cycles per second is referred to as the **coupling constant** J. In contrast to the separation between the absorptions of different groups on the frequency scale, the coupling

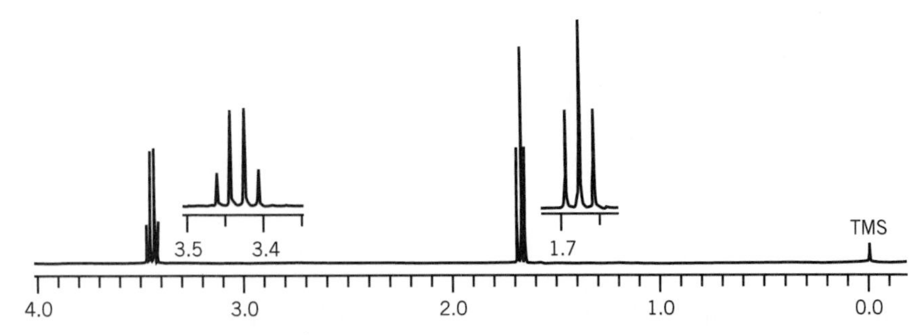

Figure 15.12 The 400 MHz ^{1}H NMR spectrum of CH_3CH_2Br in $CDCl_3$. The CH_3 resonance at about $\delta = 1.7$ and the CH_2 resonance at about $\delta = 3.4$ are also shown in expanded form. (From J. W. Akitt and B. E. Mann, *NMR and Chemistry,* Taylor and Francis Books Ltd. (Nelson Thornes), London, ©2000.)

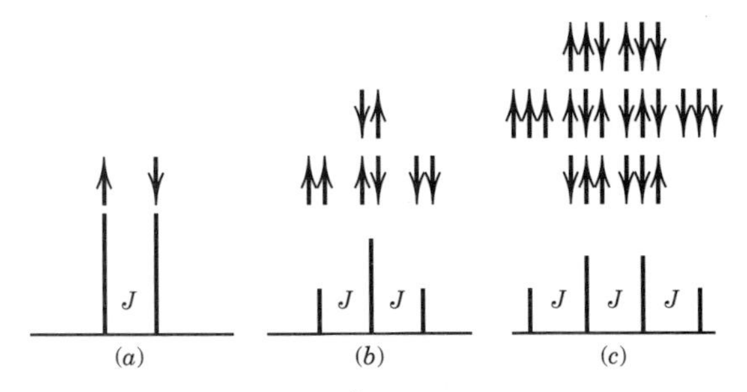

Figure 15.13 Splitting patterns with spin $\frac{1}{2}$ nuclei resulting for (a) one neighboring spin, (b) two neighboring spins, and (c) three neighboring spins.

constant is independent of the magnetic field strength. Note that chemical shifts are measured from the center of a multiplet. Molecular structure is of critical importance in determining the values of coupling constants, and that means that they are of great interest to chemists. Table 15.3 gives some values for proton spin–spin coupling constants, which indicates how they lead to structural information about organic molecules. When various conformations of a molecule are rapidly interconverted, the equivalence of nuclei should be determined on an average basis,

Table 15.3 Proton Spin–Spin Coupling Constants of ^{1}H

Structure	*J*/Hz	*Structure*	*J*/Hz
$>C(H)H$ (geminal H–C–H)	−20 to +6	H–C=C–H (trans)	12 to 9
$-C(H)-C(H)-$ (H H)	5.5 to 7.5	benzene ring	*o* 6 to 9 *m* 0.5 to 4 *p* 0 to 2.5
H–C=C–H (cis)	7 to 10	cyclohexane chair	ax, ax 9 to 14 ax, eq 2 to 4 eq, eq 2.5 to 4

rather than for one of the conformers. If the ethanol molecule CH_3CH_2OH were perfectly rigid, the methyl protons would not be equivalent. However, because of rapid rotation about the C–C bond, the electronic environments of the three protons are magnetically equivalent.

The number of lines in a multiplet is referred to as the multiplicity M, and in general

$$M = 2nI + 1 \tag{15.28}$$

where n is the number of equivalent neighbor nuclei and I is the nuclear spin. For nuclei with $I = \frac{1}{2}$, this yields the $n + 1$ rule.

Example 15.4 ***Binomial coefficients and Pascal's triangle***

Calculate the coefficients of the binomial expansion for the first several numbers n of protons in a molecule.

For small numbers this can be done by hand, but for larger numbers, it is convenient to use a computer with a mathematical program (see Computer Problem 15.I).

Pascal's triangle:

0 proton							1						
1 proton						1		1					
2					1		2		1				
3				1		3		3		1			
4			1		4		6		4		1		
5		1		5		10		10		5		1	
6	1		6		15		20		15		6		1

An important point to note is that within a group of equivalent nuclei (e.g., the protons of CH_3), each nucleus does not split the resonances of the others; the group of equivalent nuclei resonates as one collective system. Of course, the intensity of the resonance line is proportional to the number of nuclei in the group.

We will see in the next section that the $n + 1$ rule for protons is an approximation, but we will also see that this is an approximation that gets better as the magnetic field strength is increased.

15.6 SPIN–SPIN SPLITTING IN AX AND AB SYSTEMS

Since the previous treatment is qualitative, we now turn to quantum mechanics for a complete treatment of spin–spin splitting. As a simplification, we will consider two spin $\frac{1}{2}$ nuclei, and so we can refer to them as two protons in different environments. First, we will consider AX systems, where this NMR notation indicates that $J_{12} \ll \nu_0|\sigma_1 - \sigma_2|$. In other words, nuclei A and X have sufficiently different shielding constants σ_i that the spin–spin splitting J_{12} is small in comparison with the difference in chemical shifts. Then we will consider AB systems, where $J_{12} \approx \nu_0|\sigma_1 - \sigma_2|$. The spectrum of an AX system with spin–spin splitting is referred to as the **first-order spectrum** of a two-spin system, and that is what was discussed in the preceding section without equations. The spectrum of an AB system is referred to as the **second-order spectrum** of a two-spin system. The quantum mechanical treatment of a first-order spectrum is based on first-order perturbation theory, and the treatment of a second-order spectrum requires a variational treatment (Section 10.6).

In the absence of spin–spin coupling, the Hamiltonian operator for a two-spin system (see equation 15.15) is given by

$$\hat{H} = -\gamma B_0(1 - \sigma_1)\hat{I}_{z1} - \gamma B_0(1 - \sigma_2)\hat{I}_{z2} \tag{15.29}$$

This Hamiltonian leads to a two-line spectrum where one line is at the Larmor frequency for nucleus 1 and the other is at the Larmor frequency for nucleus 2.

When there is spin–spin interaction between A and X, the classical Hamiltonian for the system involves a term proportional to $\boldsymbol{\mu}_1 \cdot \boldsymbol{\mu}_2$, the dot product of the two magnetic dipole moments $\boldsymbol{\mu}_1$ and $\boldsymbol{\mu}_2$. In quantum mechanics, these two magnetic dipole moments are replaced by the corresponding spin operators $\hat{I}_{z1}$ and $\hat{I}_{z2}$ for the two magnetic dipoles. Therefore, the Hamiltonian operator for the coupled two-spin system is

$$\hat{H} = -\gamma B_0(1 - \sigma_1)\hat{I}_{z1} - \gamma B_0(1 - \sigma_2)\hat{I}_{z2} + \frac{hJ_{12}}{\hbar^2}\hat{I}_{z1}\hat{I}_{z2} \tag{15.30}$$

where J_{12} is the spin–spin coupling constant. The factor $h/\hbar^2$ ensures that the coupling constant has the unit s^{-1}.

This two-spin system has four possible wavefunctions:

$$\begin{aligned} \psi_1 &= \alpha(1)\alpha(2) \\ \psi_2 &= \beta(1)\alpha(2) \\ \psi_3 &= \alpha(1)\beta(2) \\ \psi_4 &= \beta(1)\beta(2) \end{aligned} \tag{15.31}$$

First-order perturbation theory can be used to show that the energies of the four energy levels for the AX system are given by

$$E_1 = -h\nu_0\left(1 - \frac{\sigma_1 + \sigma_2}{2}\right) + \frac{hJ_{12}}{4} \tag{15.32}$$

$$E_2 = -\frac{h\nu_0}{2}(\sigma_1 - \sigma_2) - \frac{hJ_{12}}{4} \tag{15.33}$$

$$E_3 = \frac{h\nu_0}{2}(\sigma_1 - \sigma_2) - \frac{hJ_{12}}{4} \tag{15.34}$$

$$E_4 = h\nu_0\left(1 - \frac{\sigma_1 + \sigma_2}{2}\right) + \frac{hJ_{12}}{4} \tag{15.35}$$

where $\nu_0 = \gamma B_0/2\pi$. The energy levels for an AX system are shown in Fig. 15.14. The only transitions that are allowed are $1 \rightarrow 2$, $1 \rightarrow 3$, $2 \rightarrow 4$, and $3 \rightarrow 4$ because the selection rule requires that only one type of nucleus at a time can

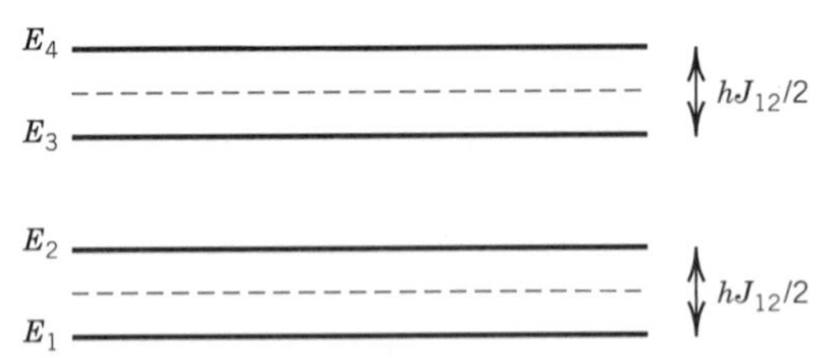

Figure 15.14 Four energy levels of an AX system according to first order perturbation theory.

undergo a transition. As shown in Problem 15.18, the frequencies for these four transitions are given by

$$\nu(1 \to 2) = \nu_0(1 - \sigma_1) - \frac{J_{12}}{2} \tag{15.36}$$

$$\nu(1 \to 3) = \nu_0(1 - \sigma_2) - \frac{J_{12}}{2} \tag{15.37}$$

$$\nu(2 \to 4) = \nu_0(1 - \sigma_2) + \frac{J_{12}}{2} \tag{15.38}$$

$$\nu(3 \to 4) = \nu_0(1 - \sigma_1) + \frac{J_{12}}{2} \tag{15.39}$$

The spectrum for an AX system is shown in Fig. 15.15 for two spectrometer frequencies. The spin–spin splitting is independent of the spectrometer frequency. This spectrum illustrates the $n + 1$ rule since the neighboring spin splits the line for a proton into each doublet.

Now we turn to a consideration of the spectrum of an AB system, that is, one where $J_{12} \approx \nu_0|\sigma_1 - \sigma_2|$. This case cannot be treated by perturbation theory because the perturbation is not small. The second-order treatment of the system with two spins requires a variational calculation (Section 10.6). This is done by using a linear combination of the four basic wavefunctions.

$$\psi^{(2)} = c_1\psi_1 + c_2\psi_2 + c_2\psi_3 + c_4\psi_4 \tag{15.40}$$

This is referred to as a **second-order treatment,** but actually it is exact in this case because equation 15.40 represents all the possible spin functions for two hydrogen atoms. The expansion of the 4×4 secular determinant (Section 11.7) yields the following four energy levels.

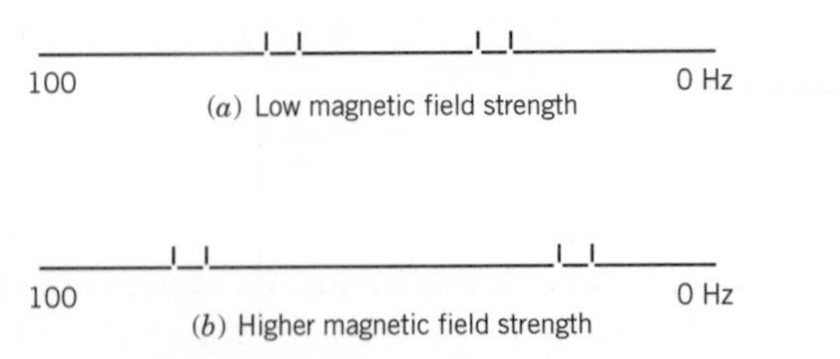

Figure 15.15 NMR spectrum of a molecule with two interacting hydrogen atoms. Note that the splitting is independent of the magnetic field strength.

$$E_1 = -h\nu_0\left(1 - \frac{\sigma_1 + \sigma_2}{2}\right) + \frac{hJ_{12}}{4} \tag{15.41}$$

$$E_2 = -\frac{h}{2}[\nu_0^2(\sigma_1 - \sigma_2)^2 + J_{12}^2]^{1/2} - \frac{hJ_{12}}{4} \tag{15.42}$$

$$E_3 = \frac{h}{2}[\nu_0^2(\sigma_1 - \sigma_2)^2 + J_{12}^2]^{1/2} - \frac{hJ_{12}}{4} \tag{15.43}$$

$$E_4 = h\nu_0\left(1 - \frac{\sigma_1 + \sigma_2}{2}\right) + \frac{hJ_{12}}{4} \tag{15.44}$$

where $\nu_0 = \gamma B_0/2\pi$. The equations for E_1 and E_4 are the same as for the AX system. These energy levels and the allowed transitions are shown in Fig. 15.16.

The frequencies of the four lines in the AB spectrum are given by

Frequency	Intensity	
$\nu(1 \to 2) = -\frac{1}{2}[\nu_0^2(\sigma_1 - \sigma_2)^2 + J_{12}^2]^{1/2} - \frac{J_{12}}{2}$	$\frac{(r-1)^2}{(r+1)^2}$	(15.45)
$\nu(1 \to 3) = \frac{1}{2}[\nu_0^2(\sigma_1 - \sigma_2)^2 + J_{12}^2]^{1/2} - \frac{J_{12}}{2}$	1	(15.46)
$\nu(2 \to 4) = \frac{1}{2}[\nu_0^2(\sigma_1 - \sigma_2)^2 + J_{12}^2]^{1/2} + \frac{J_{12}}{2}$	$\frac{(r-1)^2}{(r+1)^2}$	(15.47)
$\nu(3 \to 4) = -\frac{1}{2}[\nu_0^2(\sigma_1 - \sigma_2)^2 + J_{12}^2]^{1/2} + \frac{J_{12}}{2}$	1	(15.48)

where

$$r = \left[\frac{(\Delta^2 + J_{12}^2)^{1/2} + \Delta}{(\Delta^2 + J_{12}^2)^{1/2} - \Delta}\right]^{1/2} \quad \text{and} \quad \Delta = \nu_0(\sigma_1 - \sigma_2) \tag{15.49}$$

A term $(\nu_0/2)(2-\sigma_1-\sigma_2)$ has been omitted from each frequency because here we are interested only in relative frequencies. Note that the intensities of the four lines are not the same, as they were in the AX spectrum in Fig. 15.15. The spectra of AB systems are plotted in Fig. 15.17 for various ratios of $\nu_0(\sigma_1 - \sigma_2)/J_{12}$ and constant J_{12}. As the magnetic field strength or the chemical shift is increased, the

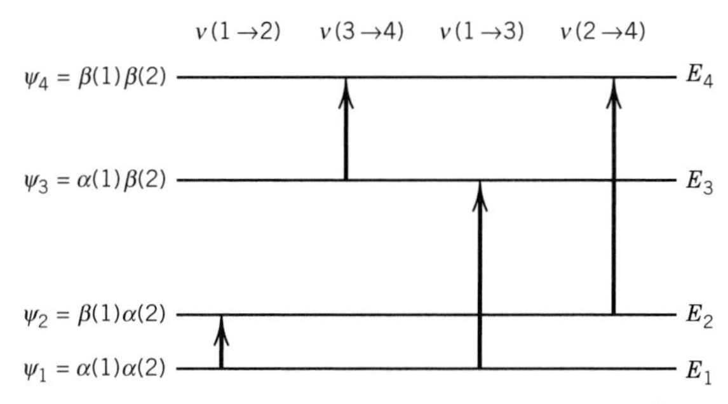

Figure 15.16 Energy levels of an AB molecule for which $J_{12} \approx \nu_0|\sigma_1 - \sigma_2|$. The allowed transitions in a magnetic field are shown by the vertical arrows.

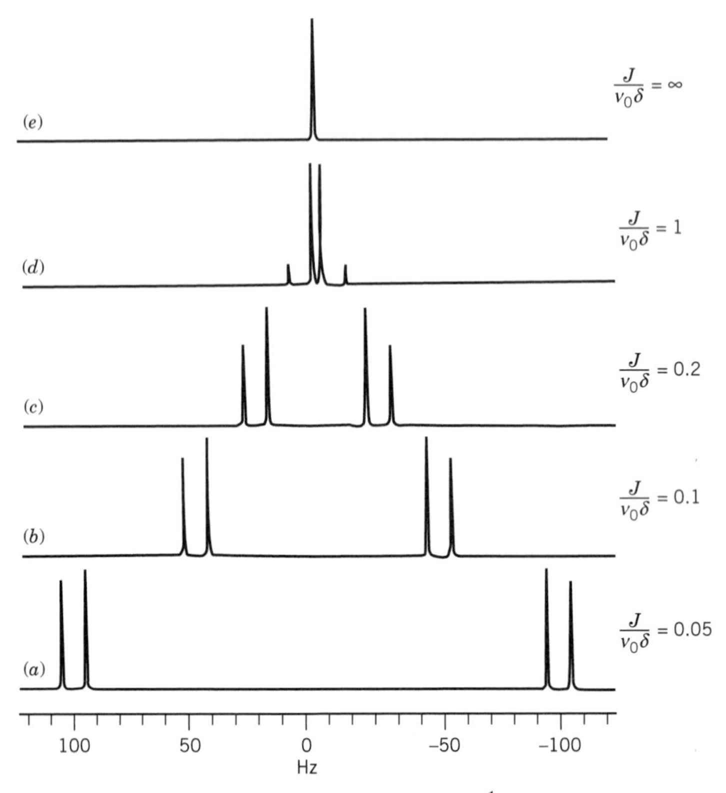

Figure 15.17 Spectra for an AB system with two spin $\frac{1}{2}$ nuclei for several values of the ratio $J/\nu_0\delta$. The coupling constant is kept at 10 Hz, and the chemical shift separations are 200, 100, 50, 10, and 0 Hz.

spectrum approaches the AX spectrum of Fig. 15.15. In the limit as the chemical shift decreases, there is a single line. This shows that for an AA system, where the two protons have the same chemical shift, the protons do not split each other's resonance. *In other words, there is no spin–spin splitting between chemically equivalent protons.*

For systems with larger numbers of protons, the second-order spectrum can be very complicated, but computer programs are available for the calculation of the second-order spectrum of systems with more protons. Figure 15.17 illustrates the advantage of achieving higher and higher magnetic field strengths: As the magnetic field strength increases, the NMR spectrum approaches the first-order spectrum.

15.7 NUCLEAR MAGNETIC RELAXATION

In this section we discuss the time-dependent NMR signals and FID in more detail. When a system at equilibrium is perturbed by a short pulse, its relaxation back to equilibrium usually follows the exponential equation

$$(n - n_{\text{eq}})_t = (n - n_{\text{eq}})_0 \exp(-t/T) \tag{15.50}$$

where T is the relaxation time. As described in Section 15.2, an assembly of $I = \frac{1}{2}$ nuclear spins in a magnetic field has two populations: one oriented in the direction of the field and the other opposed to the field with a small excess number in the lower energy state. These excess nuclei precess around the direction of the field with a net magnetization $\boldsymbol{M}_z$. with no detectable transverse magnetization in the xy plane (Fig. 15.3). We now consider two perturbations on this system: a 180° pulse and a 90° pulse.

The effect of a 180° pulse is to swing the magnetization to the opposite direction, as shown in Fig. 15.18. Note that after the pulse the spin system has more energy than before the pulse. This magnetization then decays with a relaxation time T_1, which is referred to as the **spin-lattice relaxation** time because it involves an exchange of energy between the spins and their environment. It is also referred to as a longitudinal relaxation time because it all takes place in the direction of the magnetic field.

The effect of a 90° pulse (Fig. 15.5) is to swing the magnetization to the direction of the y axis, as shown in Fig. 15.19. This figure utilizes the rotating frame introduced in Fig. 15.4. Although the magnetization is shown in the y direction, the Larmor frequency of each spin differs slightly from its companions, so the xy magnetization starts to lose coherence and therefore decreases in magnitude with relaxation time T_2, which is referred to as the **spin–spin relaxation time** (or transverse relaxation time) because it involves an exchange of energy within the spin system. Figure 15.19 also shows the longitudinal relaxation time (growth of $\boldsymbol{M}_z$) that occurs on the time scale of T_1. Usually $T_2 \ll T_1$ so that T_2 determines

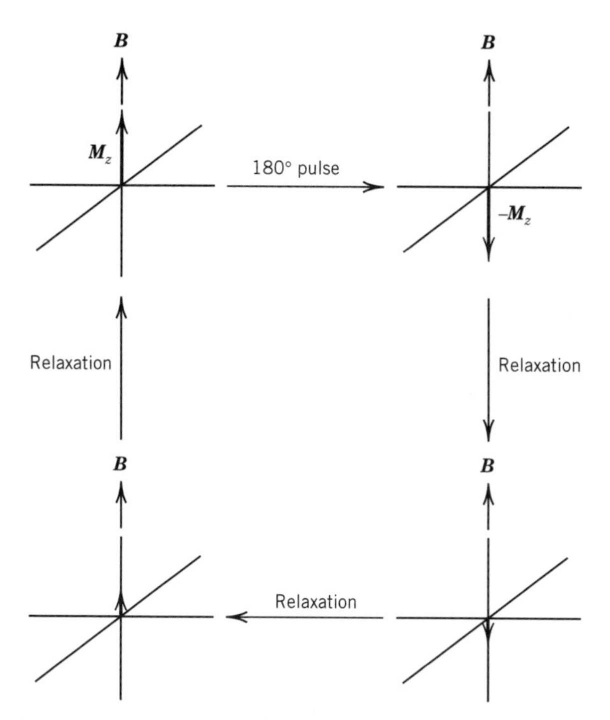

Figure 15.18 The relaxation after a 180° pulse yields T_1. After the pulse, the spin system returns to equilibrium by decaying to zero, and then increasing again in the $\boldsymbol{B}_0$ direction. This involves an exchange of energy between the spins and their environment. [From J. W. Akitt and B. E. Mann, *NMR and Chemistry,* Taylor and Francis Books Ltd. (Nelson Thornes), London, ©2000.]

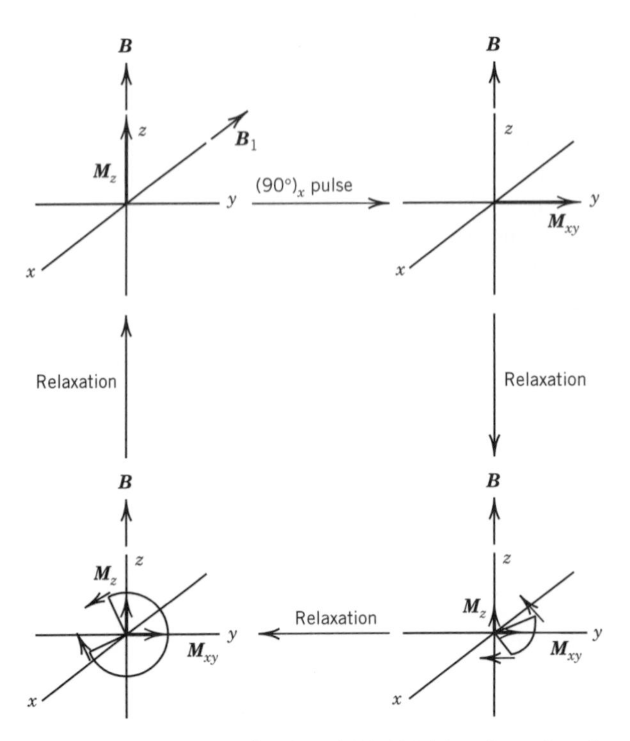

Figure 15.19 The relaxation after a 90° pulse yields T_2. After the pulse, the magnetization is shown in the y direction, as would be seen if the observer were rotating in the same direction at the Larmor frequency. There is a spreading out of the nuclear frequencies that reduces $\boldsymbol{M}_{xy}$. [From J. W. Akitt and B. E. Mann, *NMR and Chemistry,* Taylor and Francis Books Ltd. (Nelson Thornes), London, ©2000.]

the line width of the signal. Because of the T_2 relaxation, the xy component of the magnetization (which gives rise to the free induction decay (FID)) obeys the equation

$$M_{xy}(t) = M_{xy}(0)\cos(2\pi\nu t)\,\mathrm{e}^{-t/T_2} \tag{15.51}$$

where $M_{xy}(0)$ is the transverse magnetization immediately after the pulse, and ν is the Larmor frequency. If the sample contains several spins with different relaxation times, the equation for the FID is simply a sum of terms:

$$M_{xy}(t) = \sum a_i \cos(2\pi\nu_i t)\ \exp(-t/T_{2i}) \tag{15.52}$$

where a_i is the amplitude. Figure 15.20 shows the FID for a system with two spins with $\nu_1 = 40\ \mathrm{s}^{-1}$, $\nu_2 = 80\ \mathrm{s}^{-1}$, $T_{21} = 0.10$ s, $T_{22} = 0.30$ s, $a_1 = 0.50$, and $a_2 = 1$ and the Fourier transform $F(\nu)$ of the FID. The lines have different widths at half-height $\Delta\nu_{1/2}$ because they have different relaxation times T_2. The positions of the peaks are determined by their Larmor frequencies, and their line widths are determined by their respective T_2's.

The line shapes are Lorentzian, which means that the intensities $F(\nu)$ depend on frequency according to

$$F(\nu) = \frac{T_2}{1 + T_2^2(2\pi)^2(\nu - \nu_0)^2} \tag{15.53}$$

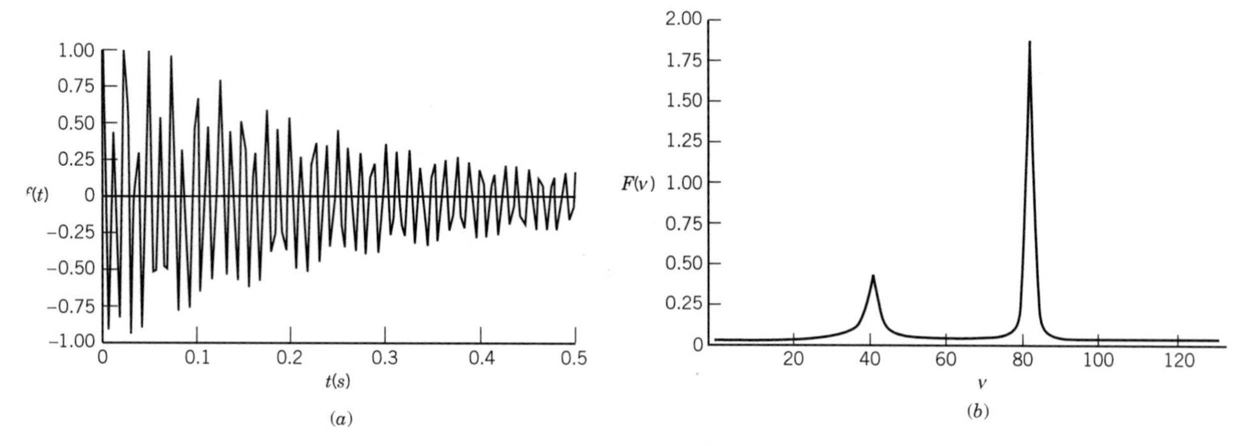

Figure 15.20 (*a*) FID for a system with two spins with $\nu_1 = 40\ \text{s}^{-1}$, $\nu_2 = 80\ \text{s}^{-1}$, $T_{21} = 0.10$ s, $T_{22} = 0.30$ s, $a_1 = 0.50$, and $a_2 = 1$. (*b*) The Fourier transform $F(\nu)$ of the FID. (See Computer Problem 15.G.)

It can be shown that for this (Lorentzian) absorption line, the width at half-maximum intensity $\Delta\nu_{1/2}$ is given by $1/\pi T_2$. The observed line widths are generally wider than that due to the true T_2 for the spin because of inhomogeneities in the strength of the magnetic field throughout the sample volume. There are now NMR methods that eliminate the effect of field inhomogeneity by a clever technique of reversing the transverse magnetization after a short time; these are so-called **spin–echo** techniques.

Example 15.5 ***Lorentzian function for absorption line shape***

Plot the Lorentzian function for absorption line shape for $\nu = 20\ \text{s}^{-1}$ and $T_2 = 0.10$ s, and check that the width at half-maximum $\Delta\nu_{1/2}$ is given by $1/\pi T_2$.

The plot of equation 15.53 for these values is given in Fig. 15.21. Note that the width at half-maximum height is given by

$$\Delta\nu_{1/2} = \frac{1}{\pi T_2} = \frac{1}{(3.14)(0.10\ \text{s})} = 3.2\,\text{s}^{-1}$$

(See Computer Problem 15.H.)

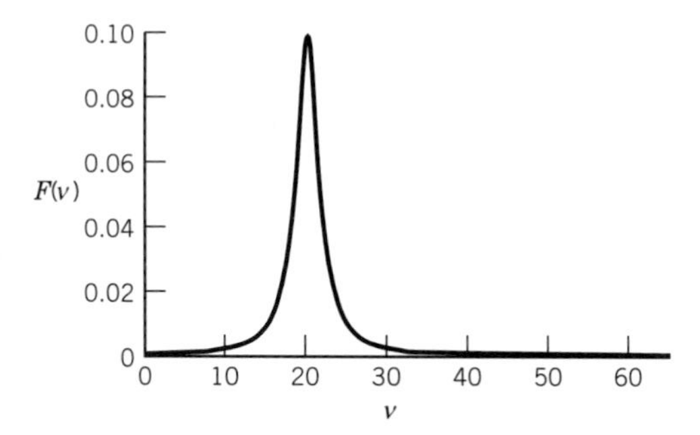

Figure 15.21 Plot of the Lorentzian function (equation 15.55) for absorption line shape for $\nu_0 = 20\ \text{s}^{-1}$ and $T_2 = 0.10$ s. (See Computer Problem 15.H.)

The NMR spectrum of a molecule A_nX_m in which the two types of nuclei are spin-coupled can be simplified by decoupling the spins in X. This is accomplished by applying a second strong radio-frequency field $\boldsymbol{B}_2$ at the Larmor frequency of X. The X nuclei precess around $\boldsymbol{B}_2$, and if this precession is fast enough their effective z magnetization at the coupled A nucleus averages to zero. Nucleus A is then said to be decoupled from X. In this manner, a decoupled spectrum for a complicated molecule may be much easier to interpret.

The usefulness of NMR in determining the structures of complicated molecules, including proteins, has been greatly increased by the development of pulse sequences. Many of these experiments can be described by the sequence: Relaxation → Preparation→ Evolution → Mixing → Acquisition. During the relaxation period, the nuclear spins are allowed to return to their equilibrium distribution. During the preparation period, the spins of the nuclei are subjected to a (possibly complicated) radio-frequency pulse. The system evolves under the influence of the magnetic field during the evolution period. It may be necessary to apply a 180° pulse in the mixing period, and more than one pulse may be applied. Then the FID signal is acquired. This is the description of a one-dimensional NMR experiment. A second time dimension can be introduced by varying the time after one of the pulses. This makes it possible to collect a series of FIDs, each at a different time. After Fourier transforming in both time variables, this is referred to as two-dimensional NMR spectroscopy. These one- and two-dimensional experiments provide information on the connectivity between nuclei of a molecule, and therefore its structure. Coupling between groups generally decreases with the number of intervening bonds, but coupling over as many as five bonds has been detected.

15.8 TWO-DIMENSIONAL NMR

In one-dimensional NMR the ordinate gives the intensity of the resonance and the abscissa gives the frequencies. In two-dimensional NMR both the ordinate and abscissa give frequencies. The intensities of the lines are indicated by contour lines, as in a topographic map. The acronym COSY (for *correlation spectroscopy*) is used when the nuclei that are studied are all the of the same type, and the term HETCOSY is used when two different types of nuclei are involved such as $^1H/^{13}C$. The pulse sequence for a COSY experiment is shown in Fig. 15.22. The two 90° pulses are separated by a time period t_1 that is varied. Figure 15.23 gives a schematic representation of a COSY spectrum for an AX system.

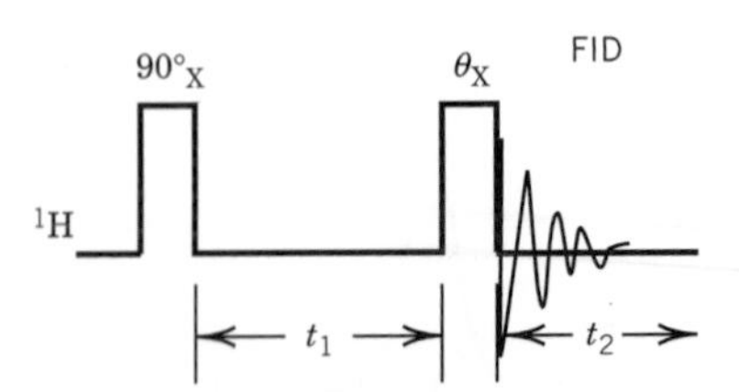

Figure 15.22 Pulse sequence for a COSY NMR experiment. The 90° pulse is followed by a time period t_1 that is varied. The second 90° pulse produces a FID that is Fourier transformed. (From J. H. Nelson, *Nuclear Magnetic Resonance Spectroscopy*, 2003 ©. Reprinted by permission of Pearson Education, Inc., Upper Saddle River, NJ.)

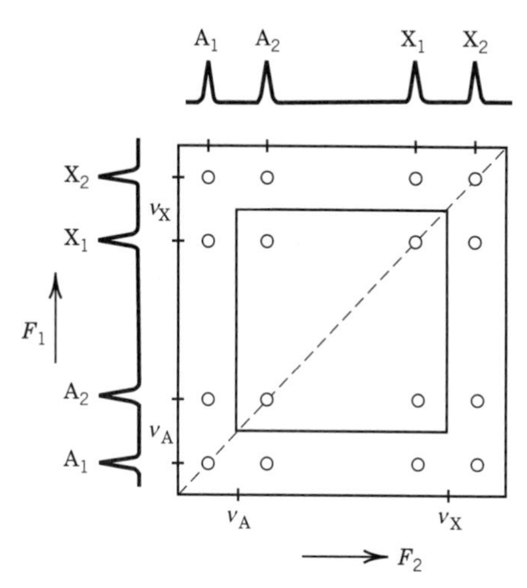

Figure 15.23 Schematic representation of a COSY spectrum for an AX spin system. The spectra along the ordinate and abscissa were obtained in a separate experiment. The 1D spectrum for AX also appears on the diagonal. (From J. H. Nelson, *Nuclear Magnetic Resonance Spectroscopy,* 2003 ©. Reprinted by permission of Pearson Education, Inc., Upper Saddle River, NJ, 2003.)

This pulse sequence introduces extra peaks off the diagonal for the protons that are spin coupled and none for those that are not. These extra peaks are referred to as cross peaks. If A and X were not coupled, the cross peaks would be absent.

The COSY experiment is equivalent to carrying out simultaneously a series of decoupling experiments at each multiplet in a system to find the part of a spectrum where a perturbation has occurred. COSY spectra are very useful for identifying the connectivity in complicated molecules. NMR spectra can be extended to more dimensions and many other pulse sequences. This has made it possible to determine amino acid sequences and folding in protein molecules in aqueous solution.

Another method of interest is called **magic-angle spinning** (MAS), which can average out the dipole–dipole spin interactions in solids. We have already pointed out that such interactions broaden the NMR spectrum in solids (and in large molecules whose rotation is slow). The dipole–dipole interaction depends on the angle that the spins make with $\boldsymbol{B}_0$ (conventionally, the z axis). By spinning the sample quickly at the angle 54.74° with respect to $\boldsymbol{B}_0$ (i.e., the angle at which $1-3\cos^2\theta = 0$), the dipole–dipole interaction averages to zero. The problem is that the frequency at which the sample is spun must be greater than the width of the spectrum due to dipole–dipole interactions. This requires spinning at 4–5 kHz. Once the dipolar spin coupling is removed from solid samples, only the chemical shift remains.

Comment:

Two-dimensional proton NMR spectra can be used to determine the three-dimensional structures of small protein molecules in solution. This method complements the determination of structures of protein molecules in crystals by use of X-ray diffraction (Chapter 23). In a crystal the configurations of side chains on the external surface of the protein are affected by neighboring protein molecules. In solution these effects are missing, and so the three-dimensional structure is slightly different. Since the inception of two-dimensional NMR in the late 1980s, NMR solution structures have been determined for more than 200 proteins.

15.9 ELECTRON SPIN RESONANCE

The basic equations for electron spin resonance (ESR) follow the same pattern as for nuclear magnetic resonance. The relation between the magnetic dipole moment $\boldsymbol{\mu}_s$ for the electron and its spin angular momentum vector $\boldsymbol{S}$ is given in equation 15.3. The expression for the magnetic energy of an electron in a magnetic field is

$$E = g_e \mu_B m_s B \tag{15.54}$$

where g_e is the g factor for the electron (2.002 322 for a free electron), μ_B is the Bohr magneton (Section 10.4), and m_s is the quantum number for the z component of the electron spin. This equation was derived earlier in the discussion of the effect of electron spin on optical spectra (see equation 10.55).

The two energy levels of a single electron in a magnetic field are shown in Fig. 15.24. Because of the negative charge of the electron, the magnetic moment μ_e of an electron is in the direction opposite to its spin angular momentum, and the electron spin quantum number is $-\frac{1}{2}$ in the lower level, in contrast to the situation with nuclei. For a transition from $m_s = -\frac{1}{2}$ to $m_s = +\frac{1}{2}$,

$$\Delta E = h\nu = g_e \mu_B B \tag{15.55}$$

In a molecule an unpaired electron is shielded to a greater or lesser extent by its environment in the radical. Thus equation 15.55 becomes

$$h\nu = g_e \mu_B (1 - \sigma) B \tag{15.56}$$

where σ is the shielding constant. This equation is usually written

$$h\nu = g \mu_B B \tag{15.57}$$

where the dimensionless g factor $g = g_e(1 - \sigma)$ is used to define the position of an ESR absorption.

Substances that show ESR spectra include free radicals, odd-electron molecules, triplet states of organic molecules, and paramagnetic transition metal ions

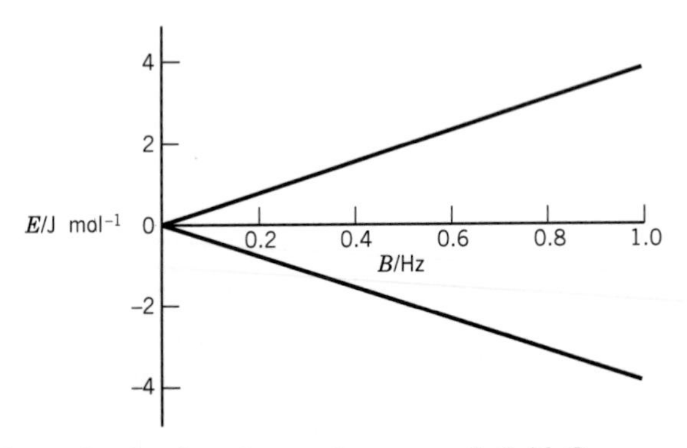

Figure 15.24 Energy levels of an electron in a magnetic field. Compare to Fig. 15.1. The upper line is for $m_s = +\frac{1}{2}$, and the lower line is for $m_s = -\frac{1}{2}$. (See Computer Problem 15.J.)

and their complexes. Any paramagnetic substance can be studied, but as we have seen, most substances are not paramagnetic because electron spins are usually paired.

An electron resonance experiment is similar to an NMR experiment, but since μ_B for the electron is about 10^3-fold larger than μ_N for nuclei, the frequencies required fall in the microwave range instead of the radio-frequency range when magnetic fields of convenient laboratory strength are used. Usually a frequency of about 10 GHz ($\lambda = 3$ cm) is used with a magnetic field of 0.3 to 0.4 T.

Example 15.6 ***Magnetic flux density for an ESR experiment***

Calculate the magnetic flux density required to give a precessional frequency of 9500 MHz for a free electron.

$$B = \frac{h\nu}{g_e\mu_B} = \frac{(6.6262 \times 10^{-34}\ \text{J s})(9500 \times 10^6\ \text{s}^{-1})}{(2.0023)(9.2741 \times 10^{-24}\ \text{J T}^{-1})} = 0.3390\ \text{T}$$

ESR spectroscopy has been very useful in determining the structure of organic and inorganic free radicals. Free radicals may be produced chemically, photochemically, or by use of high-energy radiation. If a radical has a very short life, a flow system or continuous radiation may have to be used to maintain a sufficiently high concentration for detection. Actually, a concentration of only about 10^{-10} mol L^{-1} is required to obtain a spectrum under favorable conditions.

A molecule in a triplet state has a total electron spin of 1 ($S = 1$). In this case there are three sublevels that have spin angular momentum S_z about a chosen axis of $+1$, 0, or -1. A triplet molecule has an even number of electrons, two of them unpaired, while a radical with spin $\frac{1}{2}$ has an odd number of electrons. For a molecule to have a triplet state, the unpaired electrons must interact; a molecule with two unpaired electrons a great distance apart is a diradical, not a triplet.

The splitting of lines in the ESR spectrum provides information of chemical interest. This splitting arises from the fact that unpaired electrons feel the effect of the applied magnetic field and the magnetic field produced by nuclei *within its orbital* that have spin (see Table 15.1). In ESR this splitting of lines is called hyperfine splitting instead of spin–spin splitting, as it is in NMR.

The simplest example of hyperfine splitting is the ESR spectrum of atomic hydrogen, which consists of two lines: The absorption due to the unpaired electron is split into a doublet by the spin of the proton. The splitting is caused by the magnetic field due to the magnetic moment of the proton. The field experienced by the unpaired electron is the sum of the applied magnetic field and the magnetic field of the proton. Since the proton can take on two orientations with respect to the applied field, the local field at the electron is $B_{\text{loc}} = B_{\text{appl}} \pm a/2$, where a is referred to as the **hyperfine splitting constant.** The general relation that applies to all nuclei with spin I is

$$B_{\text{loc}} = B_{\text{appl}} + am_I \tag{15.58}$$

where m_I is the quantum number for the z component of the nuclear angular momentum.

If several magnetic nuclei are present in the same radical, each contributes to the splitting with different hyperfine splitting constants. For two nuclei,

$$B_{\text{loc}} = B_{\text{appl}} + a_1 m_{I_1} + a_2 m_{I_2} \tag{15.59}$$

15.10 SPECIAL TOPIC: FOURIER TRANSFORMS

A periodic phenomenon can be considered in the frequency domain or the time domain. It is possible to go back and forth between these domains by use of Fourier transforms. If the periodic phenomenon in the time domain is represented by $f(t)$, the corresponding function $F(\omega)$ in the (circular) frequency domain is given by the Fourier equation

$$F(\omega) = \int_{-\infty}^{\infty} f(t) e^{-i\omega t} \, dt \tag{15.60}$$

where $i = (-1)^{1/2}$. This Fourier transform can be viewed as mapping a function $f(t)$ in t space into another function $F(\omega)$ in frequency space. When the data on $f(t)$ are made up of a list of length n of equally spaced values a_r, the Fourier transform b_s is given by

$$b_s = \frac{1}{n^{1/2}} \sum_{r=1}^{n} a_r \exp[-2\pi i (r-1)(s-1)/n] \tag{15.61}$$

The Fourier transform of the FID is made up of complex numbers; that is, they each consist of a real part and an imaginary part. The real part is the absorption mode signal, and that is the form of the spectrum that we are familiar with. But the imaginary part may also be useful, and it is referred to as the dispersion mode signal.

When the function of frequency $F(\omega)$ is known (that is, the spectrum), the function of time $f(t)$ can be calculated using the inverse Fourier transform:

$$f(t) = \frac{1}{2\pi} \int_{-\infty}^{\infty} F(\omega) e^{i\omega t} \, d\omega \tag{15.62}$$

When the data on $F(\omega)$ are made up of equally spaced values b_s, the inverse Fourier transform a_r is given by

$$a_r = \frac{1}{n^{1/2}} \sum_{s=1}^{n} b_s \exp[2\pi i (r-1)(s-1)/n] \tag{15.63}$$

These operations can be carried out with mathematical programs for personal computers.

Ten Key Ideas in Chapter 15

1. Since many nuclei have spin, the transitions between energy levels in a magnetic field can be detected at a frequency equal to their Larmor frequency at that magnetic field strength.
2. Most NMR spectrometers are based on irradiation of a sample by a 90° pulse, recording the free induction decay (FID), and converting it to an absorption spectrum using a Fourier transform.
3. The Larmor frequency for a given nucleus in a molecule depends on its electronic environment, so its chemical shift δ provides information about this environment.

4. The absorption line for a particular nucleus is split into a multiplet by neighboring spins, and the number of peaks in the multiplet and the spin–spin coupling constant J provide further information about structure.
5. The first-order NMR spectrum follows the $n + 1$ rule for multiplets, and it can be treated quantitatively by first-order perturbation theory, which shows that the spin–spin coupling is independent of magnetic field strength.
6. The actual spectrum (second-order spectrum), which is more complicated, can be treated by the variational method, which is exact; this treatment shows that the actual spectrum approaches the first-order spectrum as the magnetic field strength is increased.
7. When a spin is perturbed by a pulse of electromagnetic radiation, the longitudinal magnetization decays with a relaxation time T_1 (spin–lattice relaxation time), and the transverse magnetization decays with a relaxation time T_2 (spin–spin relaxation time), which determines peak width.
8. A Fourier transform converts a free induction decay (FID) to a spectrum (a plot of intensity of absorption versus frequency) in which the peak positions yield the Larmor frequencies and the widths of peaks at half-height yield the spin–spin relaxation times.
9. Two-dimensional NMR gives further information about interactions with neighboring spins and can be used, for example, to determine the sequence of amino acids in a protein.
10. Electron spin resonance of molecules with unpaired electrons yields information on coupling between the electron spin and the spins on neighboring nuclei.

REFERENCES

J. W. Akitt and B. E. Mann, *NMR and Chemistry*. London: Taylor and Francis Books Ltd. (Nelson Thornes), 2000.

D. Canet, *Nuclear Magnetic Resonance*. Hoboken, NJ: Wiley, 1996.

R. Ernst, G. Bodenhausen, and A. Wokaun, *Principles of Magnetic Resonance in One and Two Dimensions*. New York: Oxford University Press, 1987.

H. Friebolin, *Basic One- and Two-Dimensional NMR*. Heidelberg: VCH, 1991.

A. Günther, *Nuclear Magnetic Resonance Spectroscopy*. Hoboken, NJ: Wiley, 1995.

M. H. Levitt, *Spin Dynamics: Basics of Nuclear Magnetic Resonance*. Hoboken, NJ: Wiley, 2001.

J. H. Nelson, *Nuclear Magnetic Resonance Spectroscopy*. Upper Saddle River, NJ: Pearson Education, 2003.

PROBLEMS

(M) Problems marked with an icon may be more conveniently solved on a personal computer with a mathematical program.

15.1 NMR spectrometers usually have fixed frequencies between 60 and 750 MHz. Calculate the magnetic fields needed to give an NMR transition frequency for hydrogen equal to these frequencies.

15.2 For the frequencies of Problem 15.1, calculate the corresponding energies in kJ mol^{-1} and compare these with RT at 300 K.

15.3 (*a*) What are the energy levels for a ^{23}Na nucleus in a magnetic field of 2 T? (*b*) What is the absorption frequency?

15.4 The magnetogyric ratio γ_N for a nucleus is defined by

$$\mu_N = \gamma_N \hbar I$$

What is the value of γ_N for H?

15.5 What is the difference in fractional populations of ^{13}C spins between the upper and lower states in a magnetic field of 2 T at room temperature?

15.6 In a magnetic field of 2 T, what fraction of the protons have their spin lined up with the field at room temperature?

15.7 It is now possible to do NMR experiments at very low temperatures. Calculate the ratio of the number of protons in the upper spin state to that in the lower spin state in a magnetic field of 2 T at 1 and 10 mK.

15.8 (*a*) What is the value of the magnetogyric ratio for the proton? (*b*) What is the Larmor frequency for the proton at 10 T?

15.9 Calculate the magnetic fields required for resonance at 300 MHz for (*a*) ^{31}P and (*b*) ^{33}S.

15.10 Using information from Tables 15.2 and 15.3, sketch the spectrum you would expect for ethyl acetate ($CH_3CO_2CH_2CH_3$).

15.11 Chemical shifts δ are expressed in ppm, but they can also be expressed in Hz. What magnetic fields are necessary to produce frequency shifts of 100 and 500 Hz for protons with a $\delta = 1$?

15.12 What is the separation of the CH_3 and CH_2 proton resonances in ethanol at (*a*) 60 MHz and (*b*) 300 MHz? (See Table 15.2.)

15.13 At a magnetic field of 1.41 T, the frequency separation between protons in benzene and protons in tetramethylsilane is 436.2 Hz. What is the chemical shift?

15.14 Equation 15.26 indicates that the chemical shift δ measured with respect to a reference is a million times greater than the difference in shielding constants σ for the reference and the group of interest. Since the reference is arbitrary, we can also apply this equation to the difference between two groups. In the ethanol molecule, the chemical shift is 1.17 ppm for the protons in CH_3 and 3.59 ppm for the protons in CH_2. (*a*) What is the difference in shielding constants for these two types of protons? (*b*) What is the difference in the magnetic field at the protons in CH_3 and CH_2 when the applied field B_0 is 1 T and (*c*) 2 T?

15.15 (*a*) Using equation 15.26, show that

$$\nu_1 - \nu_2 = \nu_{\text{ref}} \times 10^{-6}(\delta_1 - \delta_2)$$

where ν_1 and ν_2 are the resonance frequencies for protons in groups 1 and 2. (*b*) What is the difference in resonance frequencies for protons in CH_3 and CH_2 at 60 MHz?

15.16 Sketch the proton resonance spectrum of $D_2CHCOCD_3$ (deuteroacetone containing a little hydrogen). Indicate the relative intensities of the lines.

(M) **15.17** The proton resonance pattern of 2,3-dibromothiophene shows an AB-type spectrum with lines at 405.22, 410.85, 425.07, and 430.84 Hz measured from tetramethylsilane at 1.41 T [K. F. Kuhlmann and C. L. Braun, *J. Chem. Educ.* **46**:750 (1969)]. (*a*) What is the coupling constant J? (*b*) What is the difference in the chemical shifts of the A and B hydrogens? (*c*) At what frequencies would the lines be found at 2 T? Use the results of Section 15.6.

(M) **15.18** Derive the equations for the frequencies for an AX system from the equations for the energy levels and plot the line spectrum for $\nu_0(\sigma_1 - \sigma_2) = 50\ \text{s}^{-1}$ and $J_{12} = 5.0\ \text{s}^{-1}$.

15.19 Calculate the relative frequencies and relative intensities for an AB system in which $\nu_0(\sigma_1 - \sigma_2) = 10\ \text{s}^{-1}$ and $J_{12} = 5.0\ \text{s}^{-1}$.

15.20 Verify the equation

$$E_2^{(0)} = -\hbar\gamma B_0\left(\frac{\sigma_1 - \sigma_2}{2}\right)$$

for the energy corresponding to $\psi_2 = \beta(1)\alpha(2)$ for a molecule containing two hydrogen atoms that do not interact.

15.21 Show that for a Lorentzian absorption line, the width at half-maximum intensity $\Delta\nu_{1/2}$ is given by $1/\pi T_2$.

15.22 At room temperature the chemical shift of cyclohexane protons is an average of the chemical shifts of the axial and equatorial protons. Explain.

15.23 The two lines in the proton magnetic resonance spectrum for the two methyl groups connected to nitrogen in *N,N*-dimethylacetamide coalesce when the temperature is raised. What is the rate constant for the cis–trans isomerization when the multiplet structure is just lost at 331 K? The difference in chemical shifts between the two peaks is 10.85 Hz.

15.24 Calculate the transition (Larmor) frequency of a free electron in a 3-T field. What energy in cm^{-1} does this correspond to?

15.25 Line separations in ESR may be expressed in G or MHz. Show how the conversion factor $1\ \text{T} = 2.80 \times 10^4$ MHz is obtained.

15.26 What is the magnitude of the magnetic moment of the proton?

15.27 Using data from Table 15.1, calculate the magnetic field strengths at which (*a*) ^{13}C and (*b*) ^{19}F will have splittings corresponding to 5 MHz.

15.28 With superconducting magnets, it is possible to reach magnetic fields of ~12 T. At this field strength, what is the splitting between spin levels for protons in MHz and cm^{-1}?

15.29 What is the frequency for proton resonance at a magnetic field strength of 10 T?

15.30 Calculate the magnetic field strength B necessary for resonance at 200 MHz for ^{19}F and ^{13}C.

15.31 A sample containing protons and fluorines (^{19}F) is placed in a magnetic field of 1 T at 25 °C. Calculate the difference in the fraction of spins lined up parallel to the field and those antiparallel for both nuclei.

15.32 Sketch the proton resonance spectrum for a compound, represented by AMX, with three protons with rather different chemical shifts ($\nu_0\delta_A = 100$, $\nu_0\delta_M = 200$, and $\nu_0\delta_X = 700$ Hz). The three protons are coupled with $J_{AM} = 9$, $J_{AX} = 7$, and $J_{MX} = 3$ Hz.

15.33 Describe the proton and deuteron NMR spectra of HD, neglecting the quadrupole magnetic moment of the deuteron.

15.34 On the basis of the spin–spin coupling constants in Table 15.3, describe the spectrum expected for

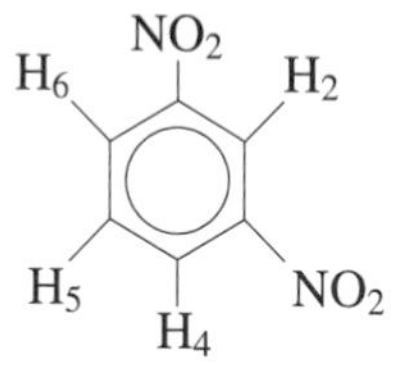

NO_2 H_6 H_2 H_5 NO_2 H_4

The protons are labeled to assist with labeling the spectrum.

15.35 Are the protons in 1,1-difluoroethylene and 1,1-difluoroallene magnetically equivalent or chemically equivalent? The only spin $\frac{1}{2}$ nuclei are fluorine and hydrogen.

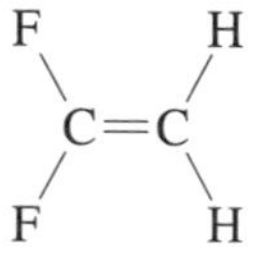

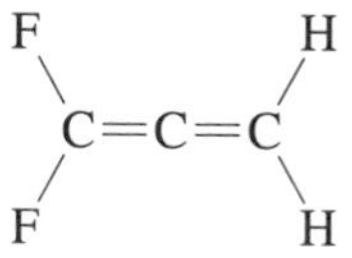

$F_2C{=}CH_2$ $F_2C{=}C{=}CH_2$

15.36 Calculate the resonance frequency for electrons at 0.33 T.

Computer Problems

15.A Plot the energy for a system containing 1 mol of hydrogen atoms (*a*) aligned with the magnetic field and (*b*) opposed to the magnetic field up to 25 T.

15.B Plot the frequency in MHz versus magnetic field strength up to 25 T for NMR with hydrogen atoms. This is essentially the range available with commercial spectrometers.

15.C When a system involves a single type of spin, the intensity of magnetization after a 90° pulse is a function of time that we will represent by $f(t)$. This signal, which is referred to as the free induction decay (FID), has the form

$$f(t) = a_1 \cos(2\pi\nu_1 t) \exp(-t/T_2)$$

where a_1 is the amplitude, ν_1 is the Larmor frequency, and T_2 is the transverse relaxation time for the spin. The signal $f(t)$ oscillates and damps down to zero at $t \gg T_2$. The Fourier transform of $f(t)$ is a plot of the corresponding spectrum $F(\nu)$ versus frequency. The position of the peak depends on the Larmor frequency ν, and the width of the peak corresponds to the transverse relaxation time T_2. A plot of $f(t)$ versus t is said to be in the time domain, and a plot of $F(\nu)$ versus ν is said to be in the frequency domain. Both are useful. A spectrum can be converted to a plot of $f(t)$ versus t with an inverse Fourier transform. When a system involves more than one type of spin, $f(t)$ is made up of a sum of terms of this type, each with its own amplitude, frequency, and transverse relaxation time. As shown in Fig. 15.5, the width of a spectral line at half the peak height is given by $\Delta\nu_{1/2} = 1/\pi T_2$. Perform the following calculations to show the relation between the FID and the spectrum for several simple cases:

(*a*) There is one spin with $\nu_1 = 20\ s^{-1}$ and $T_2 = 0.10$ s.

(*b*) There is one spin with $\nu_1 = 40\ s^{-1}$ and $T_2 = 0.10$ s.

(*c*) There are two spins with $\nu_1 = 40\ s^{-1}$, $\nu_2 = 80\ s^{-1}$, $T_{21} = 0.10$ s, and $T_{22} = 0.10$ s.

15.D (*a*) Write a program to calculate the free induction decay (FID) for a system involving one spin with $\nu_1 = 20\ s^{-1}$ and $T_2 = 0.10$ s. (*b*) Plot the FID for $t = 0$ to $t = 0.50$ s. (*c*) Make a Fourier transform of this FID to obtain the NMR spectrum and plot the intensity $F(\nu)$ from $\nu = 0$ to $\nu = 64\ s^{-1}$.

15.E This problem is like Computer Problem 15.C, part (*b*), except that the transverse relaxation time is longer. (*a*) Calculate the free induction decay (FID) for a system involving one spin with $\nu_1 = 40\ s^{-1}$ and $T_2 = 0.30$ s. (*b*) Plot the FID for $t = 0$ to $t = 0.50$ s. (*c*) Make a Fourier transform of this FID to obtain the NMR spectrum and plot the intensity $F(\nu)$ from $\nu = 0$ to $\nu = 64\ s^{-1}$. Is the spectrum broader or narrower? Why?

15.F (*a*) Write a program to calculate the free induction decay (FID) for a system involving two spins with $a_1 = 0.50$, $a_2 = 1$, $\nu_1 = 40\ s^{-1}$, $\nu_2 = 80\ s^{-1}$, $T_{21} = 0.10$ s, and $T_{22} = 0.10$ s. (*b*) Plot the FID for $t = 0$ to $t = 0.50$ s. (*c*) Make a Fourier transform of this FID to obtain the NMR spectrum and plot the intensity $F(\nu)$ from $\nu = 0$ to $\nu = 128\ s^{-1}$.

15.G (*a*) Write a program to calculate the free induction decay (FID) for a system involving two spins with $a_1 = 0.50$, $a_2 = 1$, $\nu_1 = 40\ s^{-1}$, $\nu_2 = 80\ s^{-1}$, $T_{21} = 0.10$ s, and $T_{22} = 0.30$ s. (*b*) Plot the FID for $t = 0$ to $t = 0.50$ s. (*c*) Make a Fourier transform of this FID to obtain the NMR spectrum and plot the intensity $F(\nu)$ from $\nu = 0$ to $\nu = 128\ s^{-1}$.

15.H Plot the Lorentzian line shape for absorption when $\nu_0 = \omega/2\pi = 20\ s^{-1}$ and (*a*) $T_2 = 0.10$ s and (*b*) $T_2 = 0.30$ s.

15.I Calculate Pascal's triangle for up to 10 protons.

15.J Plot the energy levels of an electron in a magnetic field in J/mol up to 1.0 T. (Compare with Fig. 15.1.)

16

Statistical Mechanics

The equilibrium properties of matter may be considered from two points of view: the macroscopic and the microscopic. Thermodynamics is a macroscopic view that describes the behavior of large numbers of molecules in terms of pressure, volume, composition, and exchanges of heat and work. The quantitative relationships between various measured properties are not based on any model of the microscopic structure of matter.

On the other hand, quantum mechanics provides a microscopic description for the structure and interactions of molecules. Ideally, we would like to be able to predict the thermodynamic behavior of substances using our knowledge about individual molecules obtained from spectroscopic measurements and from theoretical calculations of wavefunctions.

Statistical mechanics provides the needed bridge between microscopic mechanics (classical and quantum) and macroscopic thermodynamics. The classical aspects of the science were developed during the latter part of the nineteenth century by Boltzmann in Austria, Maxwell in England, and Gibbs in the United States. From their work, we can now calculate the thermodynamic properties for ideal gases from information on single molecules. The molecular information required includes vibrational frequencies, moments of inertia, and energies of dissociation. For simple molecules the values of thermodynamic properties so obtained are often more accurate than the ones measured directly. For more complicated systems, especially those involving the strong interactions between molecules, the use of statistical mechanics is much more difficult and is the subject of current research. Statistical mechanics also helps us to understand the properties of real gases, solids, polymers, and biomacromolecules.

Statistical mechanics provides insight into the laws of thermodynamics, and through it we will see heat, work, temperature, irreversible processes, and state functions in a new light.

16.1 THE BOLTZMANN DISTRIBUTION

We want to apply our knowledge of quantum mechanics to a macroscopic system, i.e., a system consisting of large numbers of molecules, but we cannot expect to be able to deal with all details of the mechanical motions and electronic excitations of such a system. As a consequence, it is necessary to use averages over dynamic states, the calculations of which are referred to as statistical mechanics. We have already dealt with averages over dynamic states when we used the Boltzmann distribution in equation 13.13 on Einstein coefficients, equation 13.70 on vibrational spectroscopy, and equation 15.22 on NMR. The barometric equation 1.46 is also an example of a Boltzmann distribution. In the next chapter we will find that the Maxwell velocity distribution is another example of the Boltzmann distribution. According to the Boltzmann distribution, if the energy associated with some state of a macroscopic system is E, then the probability of occurrence of that state is proportional to $\exp(-E/kT)$, where k is the Boltzmann constant (1.38×10^{-23} J K^{-1}). In statistical mechanics, we will regularly use k rather than R, but we will use the usual symbols for thermodynamic properties. Remember that $N_A k = R$, where N_A is the Avogadro constant. Widom (*Statistical Mechanics*, Cambridge University Press, 2002) has emphasized the usefulness of the Boltzmann distribution in introducing statistical mechanics.

This exponential dependence is a consequence of the product law for probabilities of independent events. When the simultaneous occurrence of two events is viewed as a single event, the energy associated with the single event is $E_1 + E_2$, and the probability of the single event is the product of the separate probabilities. Notice that the exponential function is the unique function with the property $f(E_1 + E_2) = f(E_1)f(E_2)$. The probability of the simultaneous occurrence of the two events viewed as a single event is proportional to

$$e^{-E_1/kT} e^{-E_2/kT} = e^{-(E_1+E_2)/kT} \tag{16.1}$$

If the energy of a macroscopic system in a particular microscopic state i is E_i, then according to the Boltzmann distribution, the probability p_i of a microscopic state with energy E_i is given by

$$p_i = \frac{e^{-E_i/kT}}{\sum_i e^{-E_i/kT}} \tag{16.2}$$

In this formula, the sum is over all possible microscopic states of the system. For a given energy E, there may be a very large number of microscopic states with that energy. The summation in the denominator is required to normalize p_i so that

$$\sum p_i = 1 \tag{16.3}$$

We will use equation 16.2 in the form

$$p_i = \frac{e^{-E_i/kT}}{Q} \tag{16.4}$$

where Q is the **canonical partition function** for the system:

$$Q = \sum_i e^{-E_i/kT} \tag{16.5}$$

This partition function is the key to statistical mechanics because it can be used to calculate all the thermodynamic properties of the system. We will illustrate this first for the internal energy U.

The internal energy was introduced by the first law as the sum of all the kinetic and potential energies of all the molecules that make up the system. Now we are going to consider a macroscopic system in contact with a heat reservoir. The energy of such a system can fluctuate, but these fluctuations are minute compared with the total energy. Therefore, the thermodynamic energy U of a system at fixed temperature can be identified with the mean energy calculated as

$$U = \sum p_i E_i = \frac{\sum E_i e^{-E_i/kT}}{Q} \tag{16.6}$$

where the E_i are all the possible energy levels of the system. From our earlier discussion of the internal energy, we know that it can be considered a function of $T, V, N_1, N_2, \ldots$, where numbers of molecules of species are used rather than the corresponding amounts $n_1, n_2, \ldots$.

There is another way to relate the internal energy to the canonical partition function that is based on the observation that the numerator of equation 16.6 is related to a derivative of $\ln \sum \exp(-E_i/kT) = \ln Q$. From $\mathrm{d} \ln x/\mathrm{d}x = 1/x$ and the chain rule for differentiation, we can see that

$$U(T, V, N_1, N_2, \ldots) = -\left(\frac{\partial \ln \sum e^{-E_i/kT}}{\partial(1/kT)}\right)_{V,N_1,N_2,\ldots} = -k\left[\frac{\partial \ln Q}{\partial(1/T)}\right]_{V,T,N_1,N_2,\ldots} \tag{16.7}$$

This expression for the internal energy in terms of the canonical partition function Q can be compared with the Gibbs–Helmholtz equation for the internal energy in terms of the Helmholtz energy A (see equation 4.62).

$$U = \left(\frac{\partial(A/T)}{\partial(1/T)}\right)_V \tag{16.8}$$

Comparison of equations 16.7 and 16.8 suggests that

$$A = -kT \ln Q \tag{16.9}$$

Strictly speaking, there could be an additional term of the form $Tf(V)$ in equation 16.9, but later calculations show that this term is not necessary.

As we have seen earlier, in Chapter 4, if we know a thermodynamic potential for a system as a function of its natural variables, all the other thermodynamic properties of the system can be calculated by taking derivatives with respect to the natural variables. Thus if the Helmholtz energy of a system can be obtained from statistical mechanics as a function of T, V, and $\{N_i\}$, the entropy, pressure, and chemical potentials can be calculated by use of

$$S = -\left(\frac{\partial A}{\partial T}\right)_{V,\{N_i\}} \tag{16.10}$$

$$P = -\left(\frac{\partial A}{\partial V}\right)_{T,\{N_i\}} \tag{16.11}$$

$$\mu_i = -\left(\frac{\partial A}{\partial N_i}\right)_{T,V,\{N_j \neq i\}} \tag{16.12}$$

Example 16.1 ***The Gibbs equation for the entropy S***

Using equation 16.9 for A and equation 16.6 for U, we can find a new equation for the entropy. Remember that $A = U - TS$, so

$$\frac{S}{k} = \frac{U - A}{kT} = \frac{\sum_i E_i \mathrm{e}^{-E_i/kT}}{kTQ} + \ln Q$$

Since $E_i = -kT \ln \mathrm{e}^{-E_i/kT}$, we can rewrite the first term to find

$$\frac{S}{k} = -\sum_i \frac{\mathrm{e}^{-E_i/kT}}{Q}(\ln \mathrm{e}^{-E_i/kT}) + \ln Q$$

We can then multiply the second term by $\sum_i \mathrm{e}^{-E_i/kT}/Q = 1$ and combine the terms to get

$$\frac{S}{k} = -\sum_i \frac{\mathrm{e}^{-E_i/kT}}{Q} \ln\left(\frac{e^{-E_i/kT}}{Q}\right)$$

Using equation 16.4, we have

$$S = -k \sum_i p_i \ln p_i$$

This is called the Gibbs equation for entropy. Note that if we are considering an *isolated* system, then the energy E is fixed. Therefore, all the possible microscopic states have the same energy E and the same probability $p = 1/\Omega$, where Ω is the number of microscopic states with energy E. Substituting into the Gibbs equation, we find

$$S = k \ln \Omega \qquad \text{(for an isolated system)}$$

This equation was first written down by Boltzmann.

When A, S, P, and μ_i have been calculated in this way, U, H, and G can be calculated from their definitions. This is the sequence of calculations we now want to make for ideal gases.

We can use equation 16.5 to calculate the partition function Q for a system of independent molecules, such as an ideal gas. First, we must specify the states i of the system and their energies E_i. In a system containing N molecules, we will let n_1 represent all of the quantum numbers (translational, vibrational, rotational, and electronic) of molecule number 1. In order to specify a microscopic state of the whole system, it is necessary to specify the state of each molecule, that is, n_1, $n_2, \ldots, n_N$. The energy E of that state would then be the sum of the energies of the individual molecules: $E = \epsilon_{n_1} + \epsilon_{n_2} + \cdots + \epsilon_{n_N}$, each in its specified state. The sum in equation 16.5 over the states i of the whole system would then be the product of the summations for the individual molecules:

$$Q = \sum_{n_1}\sum_{n_2}\cdots\sum_{n_N} e^{-(\epsilon_{n_1}+\epsilon_{n_2}+\cdots+\epsilon_{n_N})/kT} \tag{16.13}$$

Since the exponential of a sum is the product of exponentials, equation 16.13 can be written as

$$Q = \left(\sum_{n_1} e^{-\epsilon_{n_1}/kT}\right)\left(\sum_{n_2} e^{-\epsilon_{n_2}/kT}\right)\cdots\left(\sum_{n_N} e^{-\epsilon_{n_N}/kT}\right) \tag{16.14}$$

where each term is for a single molecule. Note that the sums are equal, and we call each sum the **single-molecule partition function** q:

$$q = \sum e^{-\epsilon_i/kT} \tag{16.15}$$

where the summation is over all possible states of a single molecule. Therefore, equation 16.14 can be written as

$$Q = q^N \tag{16.16}$$

This equation is correct only if the molecules are distinguishable so that we can differentiate between cases such as (a) molecule 1 in state j, molecule 2 in state k and (b) molecule 1 in state k, molecule 2 in state j. This would be true if the molecules are localized in space, but not if they are moving around as in a gas or a liquid.

If the molecules are indistinguishable, as in an ideal gas, then we have over-counted the number of possible macroscopic states; for example, we have counted the two cases mentioned above as two states instead of one. For indistinguishable molecules, we have to correct equation 16.16 by dividing by the number of distinct ways of arranging N distinguishable molecules into a microscopic state, or $N!$.

The number of ways of arranging N molecules into a macroscopic state is $N!$ only if the number of microscopic states is large compared with N. Fortunately, this is the case for an ideal gas (except at very low temperature) because of the large number of translational energy states (see Example 9.15). In order to treat the general case correctly, we have to consider the symmetry of the wave functions. This leads to corrections to the Boltzmann distribution known as the Fermi–Dirac and Einstein–Bose distributions (Section 16.11).

Therefore, the partition function for a system of N identical noninteracting molecules is

$$Q_{\text{dist}} = q^N \tag{16.17a}$$

$$Q_{\text{ind}} = q^N/N! \tag{16.17b}$$

depending on whether the molecules are distinguishable or indistinguishable.

If we have a mixture of noninteracting molecules, N_1 of type 1, N_2 of type 2, ..., the partition function for the mixture is

$$Q_{\text{dist}} = q_1^{N_1} q_2^{N_2} \cdots \tag{16.17c}$$

$$Q_{\text{dist}} = \frac{q_1^{N_1} q_2^{N_2} \cdots}{N_1! N_2! \cdots} \tag{16.17d}$$

Let us consider a single-species ideal gas and derive the thermodynamic functions of the system.

$$Q = \frac{1}{N!} q^N \tag{16.18}$$

Equation 16.9 shows how the Helmholtz energy for an ideal gas containing a single species can be calculated.

$$A = -kT \ln Q = -kT \ln \frac{q^N}{N!} = -NkT \ln q + kT \ln N! \tag{16.19}$$

Fortunately, $N!$ and $\ln N!$ can be represented in a simple way if N is a large number. According to **Stirling's approximation,**

$$\ln N! = N \ln N - N \text{ or } N! = \mathrm{e}^{-N} N^N \tag{16.20}$$

If you try this out with some small numbers, you will see that the approximation becomes better and better as N increases. We will be dealing with such large numbers (on the order of 10^{23}) that we will not have to worry about the fact that this is only an approximation.

Thus equation 16.19 can be written

$$A = -NkT \ln q + NkT \ln N - NkT = -NkT \ln \frac{q\mathrm{e}}{N} \tag{16.21}$$

The entropy, pressure, and chemical potential of the ideal gas can be calculated by using equations 16.10 to 16.12.

$$S = Nk \left[\ln \frac{q\mathrm{e}}{N} + T \left(\frac{\partial \ln q}{\partial T} \right)_V \right] \tag{16.22}$$

$$P = NkT \left(\frac{\partial \ln q}{\partial V} \right)_T \tag{16.23}$$

$$\mu = -kT \ln \left(\frac{q}{N} \right) \tag{16.24}$$

The expressions for U, H, and G can be obtained from their definitions.

$$U = A + TS = NkT^2 \left(\frac{\partial \ln q}{\partial T} \right)_V \tag{16.25}$$

$$H = U + PV = NkT \left[T \left(\frac{\partial \ln q}{\partial T} \right)_V + 1 \right] \tag{16.26}$$

$$G = A + PV = -NkT \ln \left(\frac{q}{N} \right) \tag{16.27}$$

This shows that all of the thermodynamic properties of an ideal gas can be calculated from its single-molecule partition function. In this chapter, the expression for the Gibbs energy in terms of q will be especially useful.

Comment:

The canonical partition function plays a fundamental role in statistical mechanics. It is a number that depends on the temperature and that can be used to calculate the various thermodynamic properties of an ideal gas. This limitation arises from the fact that the derivation of the equation for the molecular partition function is based on a system of particles with negligible interactions. It is also based on the assumption that the occupation numbers are considerably smaller than the degeneracy. This is the so-called Boltzmann limit of the Bose–Einstein and Fermi–Dirac distributions. This chapter is primarily concerned with the calculation of the thermodynamic properties of ideal gases, but Section 16.11 goes beyond this to gases with intermolecular interactions.

16.2 SINGLE-MOLECULE PARTITION FUNCTION FOR AN IDEAL GAS

An independent particle can have several different types of energy. An atom can have translational energy, electronic energy, and nuclear energy. A molecule can have, in addition, vibrational energy and rotational energy. As a first approximation these various kinds of energy can be considered to be independent. In that case the energy of a particle is given by

$$\epsilon = \epsilon_{\mathrm{t}} + \epsilon_{\mathrm{v}} + \epsilon_{\mathrm{r}} + \epsilon_{\mathrm{e}} + \epsilon_{\mathrm{n}} \tag{16.28}$$

where the subscripts t, v, r, e, and n indicate translational, vibrational, rotational, electronic, and nuclear energy. Since transitions in nuclear energy levels are not involved in chemical reactions, we will simply ignore nuclear energies in this treatment. If the various modes of motion can be treated as independent, the degeneracy of an energy level is equal to the product of the degeneracies for the various modes, and the molecular partition function q can be written as

$$\begin{aligned} q &= \sum_{ijkl} \left[g_{i\mathrm{t}} g_{j\mathrm{v}} g_{k\mathrm{r}} g_{l\mathrm{e}} \exp\left(-\frac{\epsilon_{i\mathrm{t}} + \epsilon_{j\mathrm{v}} + \epsilon_{k\mathrm{r}} + \epsilon_{l\mathrm{e}}}{kT} \right) \right] \\ &= \sum_{ijkl} \left[g_{i\mathrm{t}} \exp\left(-\frac{\epsilon_{i\mathrm{t}}}{kT} \right) \right] \left[g_{j\mathrm{v}} \exp\left(-\frac{\epsilon_{j\mathrm{v}}}{kT} \right) \right] \\ &\quad \times \left[g_{k\mathrm{r}} \exp\left(-\frac{\epsilon_{k\mathrm{r}}}{kT} \right) \right] \left[g_{l\mathrm{e}} \exp\left(-\frac{\epsilon_{l\mathrm{e}}}{kT} \right) \right] \end{aligned} \tag{16.29}$$

which is a sum of products. A sum of independent products can always be rewritten as a product of sums. This is illustrated by $(1 + 2)(3 + 4) = 21$ and the sum of all possible products, which is $3 + 4 + 6 + 8 = 21$. Thus, the preceding equation can be rewritten as

$$\begin{aligned} q &= \sum_{i} \left[g_{i\mathrm{t}} \exp\left(-\frac{\epsilon_{i\mathrm{t}}}{kT} \right) \right] \sum_{j} \left[g_{j\mathrm{v}} \exp\left(-\frac{\epsilon_{j\mathrm{v}}}{kT} \right) \right] \\ &\quad \times \sum_{k} \left[g_{k\mathrm{r}} \exp\left(-\frac{\epsilon_{k\mathrm{r}}}{kT} \right) \right] \sum_{l} \left[g_{l\mathrm{e}} \exp\left(-\frac{\epsilon_{l\mathrm{e}}}{kT} \right) \right] \end{aligned} \tag{16.30}$$

which is a product of sums. For molecules this equation applies only to the ground electronic state because higher electronic states have different vibrational frequencies, different moments of inertia, and, of course, different electronic energies.

Since each of the sums in equation 16.30 has the form of a partition function, the partition function of a molecule in its ground state can be written as

$$q = q_t q_v q_r q_e \tag{16.31}$$

where

$$q_t = \sum_i g_{it} \exp\left(-\frac{\epsilon_{it}}{kT}\right), \text{ etc.} \tag{16.32}$$

Therefore, the canonical partition function Q for an ideal gas is given by

$$Q = \frac{(q_t q_v q_r q_e)^N}{N!} \tag{16.33}$$

Since the Helmholtz energy is proportional to the logarithm of Q, it is made up of a sum of terms for these types of energy:

$$\begin{aligned} A = -NkT \ln Q &= -NkT \ln \frac{q_t e}{N} - NkT \ln q_v - NkT \ln q_r - NkT \ln q_e \\ &= A_t + A_v + A_r + A_e \end{aligned} \tag{16.34}$$

where

$$A_t = -NkT \ln \frac{q_t e}{N},\ A_v = -NkT \ln q_v,\ A_r = -NkT \ln q_r,\ A_e = -NkT \ln q_e \tag{16.35}$$

Associating $N!$ with the translational contribution has the advantage of making A_t the Helmholtz energy of an ideal monatomic gas without electronic excitation. In addition, since q_t is proportional to V (see the next section), q_t/N depends only on the molar volume and is therefore an intensive quantity. The separate calculation of vibrational, rotational, and electronic contributions is a useful approximation, but for more accurate calculations it is necessary to take into account the interactions between these types of energy, which were discussed earlier.

Example 16.2 ***Showing that the partition function is a product of partition functions***

Consider a molecule that has only two translational energy levels and two vibrational energy levels. Assume that the translational partition function is given by

$$q_t = g_{1t}\, e^{-\epsilon_{1t}/kT} + g_{2t}\, e^{-\epsilon_{2t}/kT}$$

Assume that the vibrational partition function is given by

$$q_v = g_{1v}\, e^{-\epsilon_{1v}/kT} + g_{2v}\, e^{-\epsilon_{2v}/kT}$$

Show that the molecular partition function can be written as a sum of products, that is, in the form of equation 16.29.

Equation 16.31 for the molecular partition function is

$$q = q_t q_v$$

Substituting the expressions given above and multiplying yields

$$q = (g_{1t}\,e^{-\epsilon_{1t}/kT})(g_{1v}\,e^{-\epsilon_{1v}/kT}) + (g_{1t}\,e^{-\epsilon_{1t}/kT})(g_{2v}\,e^{-\epsilon_{2v}/kT}) + (g_{2t}\,e^{-\epsilon_{2t}/kT})(g_{1v}\,e^{-\epsilon_{1v}/kT}) + (g_{2t}\,e^{-\epsilon_{2t}/kT})(g_{2v}\,e^{-\epsilon_{2v}/kT})$$

This is the sum of products that corresponds to equation 16.29.

16.3 TRANSLATIONAL CONTRIBUTIONS TO THE THERMODYNAMIC PROPERTIES OF IDEAL GASES

In Chapter 9 we saw that a particle in a cubic box of volume V can have only certain energies specified by the quantum numbers n_x, n_y, and n_z. Using equation 9.88 in equation 16.15 yields the molecular partition function for translational motion:

$$q_t = \sum_{n_x}\sum_{n_y}\sum_{n_z} \exp\left[-\frac{h^2(n_x^2+n_y^2+n_z^2)}{8mV^{2/3}kT}\right] \tag{16.36}$$

Replacing the sum of products with a product of sums yields

$$\begin{aligned} q_t &= \sum_{n_x}\exp\left(-\frac{h^2n_x^2}{8mV^{2/3}kT}\right)\sum_{n_y}\exp\left(-\frac{h^2n_y^2}{8mV^{2/3}kT}\right)\sum_{n_z}\exp\left(-\frac{h^2n_z^2}{8mV^{2/3}kT}\right) \\ &= \left[\sum_{n=1}^{\infty}\exp\left(-\frac{h^2n^2}{8mV^{2/3}kT}\right)\right]^3 \end{aligned} \tag{16.37}$$

To evaluate this sum, we note that for macroscopic containers the exponent is very small, unless T is very small. Since the successive terms in the summation differ by only small amounts, we may replace the summation by integration, considering the quantum number n as a continuous variable that is not restricted to integer values:

$$\int_0^{\infty} e^{-h^2n^2/8mV^{2/3}kT}\,dn = \frac{(2\pi mkT)^{1/2}V^{1/3}}{h} \tag{16.38}$$

The value of the definite integral involved here is given in Table 17.1. Thus, the **translational partition function** is given by

$$q_t = \left[\frac{(2\pi mkT)^{1/2}V^{1/3}}{h}\right]^3 = \left(\frac{2\pi mkT}{h^2}\right)^{3/2} V \tag{16.39}$$

Note that the translational partition function for a molecule is proportional to the volume V in which it moves. Thus, the translational partition function is an extensive variable. The quantity $(h^2/2\pi mkT)^{1/2}$ is called the **thermal wavelength** and is represented by Λ. Thus, the expression for q_t in equation 16.39 can be written

$$q_t = \frac{V}{\Lambda^3} \tag{16.40}$$

The condition for the applicability of Boltzmann statistics is that the thermal wavelength must be small compared with the mean distance between molecules.

Example 16.3 ***Translational partition function for a hydrogen atom***

What is the molecular partition function for translational motion of a hydrogen atom at 3000 K in a volume of 0.2494 m^3? (This is the molar volume of an ideal gas at this temperature and 1 bar pressure.) What is the thermal wavelength?

Using equation 16.39,

$$q_t = \left[\frac{2\pi(1.0080 \times 10^{-3}\ \text{kg mol}^{-1})(1.3807 \times 10^{-23}\ \text{J K}^{-1})(3000\ \text{K})}{(6.022 \times 10^{23}\ \text{mol}^{-1})(6.626 \times 10^{-34}\ \text{J s})^2}\right]^{3/2} \times (0.2494\ \text{m}^3)$$

$$= 7.791 \times 10^{30}$$

Note that the molecular partition function is a dimensionless quantity and may be interpreted approximately as the number of energy levels accessible to the single hydrogen atom in this volume. To calculate Λ, we rearrange equation 16.40 to obtain

$$\Lambda = \left(\frac{V}{q_t}\right)^{1/3} = \left(\frac{0.2494\ \text{m}^3}{7.791 \times 10^{30}}\right)^{1/3} = 3.175 \times 10^{-11}\ \text{m}$$

Note that the thermal wavelength is much smaller than the length of the side of the container, as required by the derivation of equation 16.39.

The single-molecule partition function in equation 16.39 applies to a monatomic ideal gas with no internal degrees of freedom. Substituting equation 16.39 in equation 16.18 yields

$$Q = \frac{1}{N!}\left(\frac{2\pi mkT}{h^2}\right)^{3N/2} V^N \tag{16.41}$$

The Helmholtz energy of the monatomic ideal gas is obtained by substituting this in equation 16.9 or by using equation 16.35.

$$A = -kT\left\{N\ln\left[\left(\frac{2\pi mkT}{h^2}\right)^{3/2} V\right] - \ln N!\right\} = -NkT\ln\left[\left(\frac{2\pi mkT}{h^2}\right)^{3/2}\frac{V\text{e}}{N}\right] \tag{16.42}$$

Now that we have the Helmholtz energy for a monatomic ideal gas without electronic excitation as a function of its natural variables T, V, and N, we can calculate S, V, and μ for an ideal monatomic gas by using equations 16.10 to 16.12.

$$S_t = -\left(\frac{\partial A}{\partial T}\right)_{V,N} = Nk\ln\left[\left(\frac{2\pi mkT}{h^2}\right)^{3/2}\frac{V\text{e}^{5/2}}{N}\right] \tag{16.43}$$

$$P = -\left(\frac{\partial A}{\partial V}\right)_{T,N} = \frac{NkT}{V} \tag{16.44}$$

$$\mu_t = \left(\frac{\partial A}{\partial N}\right)_{T,V} = -kT\ln\left[\left(\frac{2\pi mkT}{h^2}\right)^{3/2}\frac{V}{N}\right] \tag{16.45}$$

Knowledge of the translational contributions A_t, S_t, P, and μ_t for an ideal gas makes it possible to calculate the translational contributions to U, H, G, and C_V.

$$U_t = A_t + TS_t = (3/2)NkT \tag{16.46}$$

$$H_t = U_t + PV = (5/2)NkT \tag{16.47}$$

$$G_t = H_t - TS_t = N\mu_t = -NkT\ln\left[\left(\frac{2\pi mkT}{h^2}\right)^{3/2}\frac{V}{N}\right] \tag{16.48}$$

$$C_{Vt} = \left(\frac{\partial U_t}{\partial T}\right)_{V,N} = \frac{3}{2}Nk \tag{16.49}$$

Example 16.4 ***Thermodynamic properties as functions of temperature and pressure***

What are the expressions for the translational contributions A_t, S_t, μ_t, and G_t for ideal gases as functions of temperature and pressure?

The following equations are obtained by substituting $V = NkT/P$ into equations 16.35, 16.43, 16.45, and 16.48:

$$A_t = -NkT\ln\left[\left(\frac{2\pi mkT}{h^2}\right)^{3/2}\frac{kT\,\mathrm{e}}{P}\right] \tag{16.50}$$

$$S_t = Nk\ln\left[\left(\frac{2\pi mkT}{h^2}\right)^{3/2}\frac{kT\,\mathrm{e}^{5/2}}{P}\right] \tag{16.51}$$

$$\mu_t = -kT\ln\left[\left(\frac{2\pi mkT}{h^2}\right)^{3/2}\frac{kT}{P}\right] \tag{16.52}$$

$$G_t = -NkT\ln\left[\left(\frac{2\pi mkT}{h^2}\right)^{3/2}\frac{kT}{P}\right] \tag{16.53}$$

We now have equations to calculate all the thermodynamic properties of an ideal monatomic gas without electronic excitation at specified T and V or T and P. If the pressure is equal to the standard state pressure $P°$, the standard properties are obtained. Remember that the pressure must be expressed in pascals if SI units are used. In Chapter 2, we referred to the fact that for an ideal monatomic gas, $\overline{U} = \frac{3}{2}RT$, $\overline{H} = \frac{5}{2}RT$, $\overline{C_V} = \frac{3}{2}R$, and $\overline{C_P} = \frac{5}{2}R$. In Chapter 3 we referred to the Sackur–Tetrode equation for the molar entropy of an ideal monatomic gas. We have now seen how statistical mechanics explains the form of this equation (and more).

Example 16.5 ***The Sackur–Tetrode equation***

Express equation 16.51 in a convenient form for calculating the molar entropy of an ideal monatomic gas, which does not have electronic excitation, at temperature T in kelvins and pressure P in bars.

To do this we express the mass of the atom by $A_r m_u$, where A_r is the relative atomic weight and m_u is the atomic mass constant [$m(^{12}C)/12$]. We express the pressure in equa-

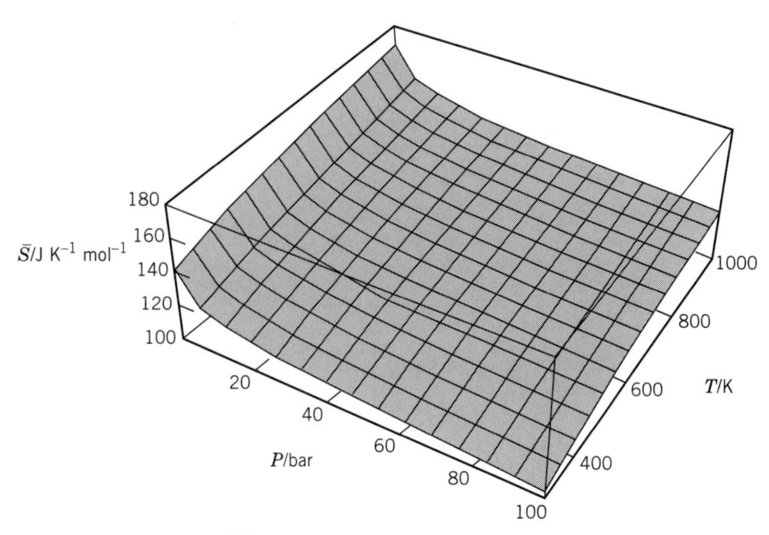

Figure 16.1 Molar entropy $\overline{S}$ of O(g) from 298 K to 1000 K and 1 bar to 100 bar using the Sackur–Tetrode equation, which does not include the contribution from the electronic energy. (See Computer Problem 16.K.)

tion 16.51 by $(P/P^\circ)P^\circ$, where P° is the standard state pressure, and the temperature in equation 16.51 by $(T/\text{K})\text{K}$, where T_1 is 1 K. This leads to

$$\overline{S} = R\left\{\frac{5}{2} + \ln\left[\left(\frac{2\pi m_u k T_1}{h^2}\right)^{3/2}\frac{kT_1}{P^\circ}\right] + \frac{3}{2}\ln A_r - \ln\frac{P}{P^\circ} + \frac{5}{2}\ln\frac{T}{\text{K}}\right\}$$

$$= R\left(-1.151\,693 + \frac{3}{2}\ln A_r - \ln\frac{P}{P^\circ} + \frac{5}{2}\ln\frac{T}{\text{K}}\right)$$

This is a convenient form of the **Sackur–Tetrode equation** for calculating $\overline{S}$ at T and P.

Figure 16.1 shows a three-dimensional plot of the molar entropy of O(g) over a range of temperature and pressure. The values shown are smaller than those in Table C.3 because the Sackur–Tetrode equation does not include the contribution from the electronic energy. (See Computer Problem 16.K.)

The methods used in deriving these equations become inapplicable as $T \to 0$. As $T \to 0$ the thermal wavelength becomes so large that the wavefunctions of the particles overlap significantly, and the Boltzmann distribution cannot be used.

Example 16.6 ***Translational contributions for oxygen atoms***

What are the translational contributions to $\overline{C}_P^\circ$, $\overline{S}^\circ$, $\overline{H}^\circ$, and $\overline{G}^\circ$ for O(g) at 298.15 K?

$$m = A_r m_u = (15.9994)(1.660\,540\,2 \times 10^{-27}\ \text{kg})$$
$$= 2.656\,77 \times 10^{-26}\ \text{kg}$$

$$(\overline{C}_P^\circ)_t = \tfrac{5}{2}R = \tfrac{5}{2}(8.314\,51\ \text{J K}^{-1}\ \text{mol}^{-1})$$
$$= 20.786\ \text{J K}^{-1}\ \text{mol}^{-1}$$

$$\overline{H}_t^\circ = \tfrac{5}{2}RT = \tfrac{5}{2}(8.314\,51\ \text{J K}^{-1}\ \text{mol}^{-1})(298.15\ \text{K})$$
$$= 6.197\ \text{kJ mol}^{-1}$$

$$\left(\frac{2\pi mkT}{h^2}\right)^{3/2}\frac{kT}{P^\circ} = \left[\frac{2\pi(2.656\,77\times10^{-26}\text{ kg})(1.380\,658\times10^{-23}\text{ J K}^{-1})(298.15\text{ K})}{(6.626\,075\,5\times10^{-34}\text{ J s})^2}\right]^{3/2}$$

$$\times\frac{(1.380\,658\times10^{-23}\text{ J K}^{-1})(298.15\text{ K})}{10^5\text{ N m}^{-2}} = 2.548\,789\times10^6$$

$$\overline{S}_t^\circ = (8.314\,51\text{ J K}^{-1}\text{ mol}^{-1})(2.5+\ln 2.548\,789\times10^6)$$

$$= 143.435\text{ J K}^{-1}\text{ mol}^{-1}$$

Equation 16.53 shows that

$$\overline{G}_t^\circ = -RT\ln\frac{q_t}{N_A} = -RT\ln\left[\left(\frac{2\pi mkT}{h^2}\right)^{3/2}\frac{RT}{N_AP^\circ}\right]$$

$$= -(8.314\,51\text{ J K}^{-1}\text{ mol}^{-1})(298.15\text{ K})\ln 2.548\,789\times10^6$$

$$= -36.568\text{ kJ mol}^{-1}$$

Example 16.7 ***Translational contributions for oxygen molecules***

What are the translational contributions to $\overline{C}_P^\circ$, $\overline{H}^\circ$, $\overline{S}^\circ$, and $\overline{G}^\circ$ for O_2(g) at 298.15 K?

The translational contributions to $\overline{C}_P^\circ$ and $\overline{H}^\circ$ are the same as for O(g), or any other ideal monatomic gas. Since the mass of the molecule is twice that of the atom, the translational entropy is larger by $R\ln 2^{3/2} = 8.645\text{ J K}^{-1}\text{ mol}^{-1}$, and the translational Gibbs energy is more negative by $RT\ln 2^{3/2} = -2.577\text{ kJ mol}^{-1}$. Thus, at 298.15 K

$$(\overline{C}_P^\circ)_t = 20.786\text{ J K}^{-1}\text{ mol}^{-1}$$

$$\overline{S}_t^\circ = 152.080\text{ J K}^{-1}\text{ mol}^{-1}$$

$$\overline{H}_t^\circ = 6.197\text{ kJ mol}^{-1}$$

$$\overline{G}_t^\circ = -39.145\text{ kJ mol}^{-1}$$

The values of $(\overline{C}_P^\circ)_t$ and $\overline{S}_t^\circ$ are slightly less than $\overline{C}_P^\circ$ and $\overline{S}^\circ$ in Table C.2 because of additional contributions due to rotation, vibration, and electronic excitation. The values of $\overline{H}_t^\circ$ and $\overline{G}_t^\circ$ cannot be compared with $\Delta_f H^\circ$ and $\Delta_f G^\circ$ in Table C.2 because of the difference in reference states, and there are also rotational, vibrational, and electronic contributions.

16.4 VIBRATIONAL CONTRIBUTIONS TO THE THERMODYNAMIC PROPERTIES OF IDEAL GASES

The vibrational levels of diatomic molecules have been discussed in Section 13.6 on spectroscopy. If there is excitation to very high vibrational levels, the anharmonicity has to be taken into account in calculating the partition function, but for many purposes, it is sufficient to use the harmonic oscillator approximation because only lower levels are occupied. In calculating the partition function, vibrational energy is usually measured from the ground state ($v = 0$) rather than from the bottom of the potential energy curve. This simplifies the equations because the **zero-point energy** ($h\nu/2$, where ν is the vibration frequency) does not appear.

First we will consider a diatomic molecule at lower vibrational quantum numbers, where it is effectively a harmonic oscillator. The energies with respect to the ground state are given by $nh\nu$ with $n = 0, 1, 2, \ldots$. The single-molecule partition function for a harmonic oscillator is given by

$$q_{\mathrm{v}} = \sum_{n=0}^{\infty} \mathrm{e}^{-nh\nu/kT} = 1 + x + x^2 + x^3 + \cdots \tag{16.54}$$

where $x = \exp(-h\nu/kT)$ When $x < 1$, the infinite series can be replaced with

$$\frac{1}{1-x} = 1 + x + x^2 + x^3 + \cdots \tag{16.55}$$

Thus if $\exp(-h\nu/kT) < 1$, the single-molecule partition function is given by

$$q_{\mathrm{v}} = \frac{1}{1 - \exp(-h\nu/kT)} = \frac{1}{1 - \exp(-\Theta_{\mathrm{v}}/T)} \tag{16.56}$$

where the **characteristic vibrational temperature** Θ_{v} is equal to $h\nu/k$. The characteristic temperature Θ_{v} of a vibration is the temperature at which the energy of vibration is equal to kT. The vibrational partition function for a normal mode is generally not much greater than unity, except at high temperatures, since the vibrational frequencies of simple molecules are usually large compared with kT_{room}.

The vibrational contribution to the Helmholtz energy is given by

$$A_{\mathrm{v}} = -NkT \ln q_{\mathrm{v}} \tag{16.57}$$

Substituting equation 16.56 yields

$$A_{\mathrm{v}} = NkT \ln [1 - \exp(-\Theta_{\mathrm{v}}/T)] \tag{16.58}$$

The vibrational contributions to the entropy, pressure, and chemical potential are given by equations 16.59 to 16.61:

$$S_{\mathrm{v}} = -\left(\frac{\partial A_{\mathrm{v}}}{\partial T}\right)_{V,N} = Nk\left[-\ln(1 - \exp(-\Theta_{\mathrm{v}}/T) + \frac{\Theta_{\mathrm{v}}/T}{\exp(\Theta_{\mathrm{v}}/T) - 1}\right] \tag{16.59}$$

$$P = -\left(\frac{\partial A_{\mathrm{v}}}{\partial V}\right)_{T,N} = 0 \tag{16.60}$$

$$\mu_{\mathrm{v}} = \left(\frac{\partial A_{\mathrm{v}}}{\partial N}\right)_{T,V} = kT \ln[1 - \exp(-\Theta_{\mathrm{v}}/T)] \tag{16.61}$$

The vibrational contributions to the other thermodynamic properties can be obtained by use of partial derivatives and definitions, as shown for the translational contributions. These contributions are shown later in Table 16.1.

When a molecule has several normal modes of vibration, q_{v} is the product of the vibrational partition functions for the various normal modes, as shown by

$$q_{\mathrm{v}} = q_{\mathrm{v1}} q_{\mathrm{v2}} q_{\mathrm{v3}} \cdots \tag{16.62}$$

Therefore, the vibrational contributions to a thermodynamic property are the sum of contributions from each normal mode. For polyatomic molecules, there are $3N - 5$ normal modes of vibration for linear molecules and $3N - 6$ normal modes for nonlinear molecules, as we have seen in Section 13.8.

Table 16.1 Contributions to Molar Thermodynamic Properties of Ideal Gases

	Translation	Vibration	Rotation: Diatomic and Polyatomic Linear	Rotation: Nonlinear	Electronic[a]
$\frac{\overline{U}}{R}$	$\frac{3}{2}T$	$\sum_i \frac{\Theta_{vi}}{\exp(\Theta_{vi}/T)-1}$	T	$\frac{3}{2}T$	0
$\frac{\overline{H}}{R}$	$\frac{5}{2}T$	$\sum_i \frac{\Theta_{vi}}{\exp(\Theta_{vi}/T)-1}$	T	$\frac{3}{2}T$	0
$\frac{\overline{C}_V}{R}$	$\frac{3}{2}$	$\sum_i \frac{(\Theta_{vi}/T)^2\exp(\Theta_{vi}/T)}{[\exp(\Theta_{vi}/T)-1]^2}$	1	$\frac{3}{2}$	0
$\frac{\overline{C}_P}{R}$	$\frac{5}{2}$	$\sum_i \frac{(\Theta_{vi}/T)^2\exp(\Theta_{vi}/T)}{[\exp(\Theta_{vi}/T)-1]^2}$	1	$\frac{3}{2}$	0
$\frac{\overline{S}}{R}$	$\ln\left[\left(\frac{2\pi mkT}{h^2}\right)^{3/2}\frac{kT}{P}\right]+\frac{5}{2}$	$\sum_i\left\{\frac{\Theta_{vi}/T}{[\exp(\Theta_{vi}/T)-1]} - \ln\left[1-\exp\left(\frac{-\Theta_{vi}}{T}\right)\right]\right\}$	$\ln\left(\frac{T}{\sigma\Theta_r}\right)+1$	$\ln\left[\frac{(T^3\pi/\Theta_a\Theta_b\Theta_c)^{1/2}}{\sigma}\right]+\frac{3}{2}$	$\ln g_0$
$\frac{\overline{A}}{R}$	$-T\ln\left[\left(\frac{2\pi mkT}{h^2}\right)^{3/2}\frac{kT}{P}\right]-T$	$\sum_i T\ln\left[1-\exp\left(-\frac{\Theta_{vi}}{T}\right)\right]$	$-T\ln\frac{T}{\sigma\Theta_r}$	$-T\ln\frac{(T^3\pi/\Theta_a\Theta_b\Theta_c)^{1/2}}{\sigma}$	$-T\ln g_0$
$\frac{\overline{G}}{R}=\frac{\mu}{R}$	$-T\ln\left[\left(\frac{2\pi mkT}{h^2}\right)^{3/2}\frac{kT}{P}\right]$	$\sum_i T\ln\left[1-\exp\left(-\frac{\Theta_{vi}}{T}\right)\right]$	$-T\ln\frac{T}{\sigma\Theta_r}$	$-T\ln\frac{(T^3\pi/\Theta_a\Theta_b\Theta_c)^{1/2}}{\sigma}$	$-T\ln g_0$

[a] As a simplification, the temperature dependence of the electronic partition function has been neglected.

Example 16.8 ***Vibrational contributions for oxygen molecules***

What are the vibrational contributions to $\overline{C}_P^\circ$, $\overline{S}^\circ$, $\overline{H}^\circ$, and $\overline{G}^\circ$ for $O_2(g)$ at 298.15 K? The vibrational frequency is 1580.246 cm^{-1}.

$$\Theta_v = \frac{hc\tilde{\nu}}{k} = \frac{(6.626\,076\times10^{-34}\text{ J s})(2.997\,925\times10^{8}\text{ m s}^{-1})(1.580\,246\times10^{5}\text{ m}^{-1})}{1.380\,658\times10^{-23}\text{ J K}^{-1}}$$

$$= 2273.73\text{ K}$$

$$(\overline{C}_P^\circ)_v = (8.314\,51\text{ J K}^{-1}\text{ mol}^{-1})\left(\frac{2273.73}{298.15}\right)^2\frac{e^{2273.73/298.15}}{(e^{2273.73/298.15}-1)^2}$$

$$= 0.236\text{ J K}^{-1}\text{ mol}^{-1}$$

$$\overline{S}_v^\circ = (8.314\,51\text{ J K}^{-1}\text{ mol}^{-1})\left[\frac{(2273.73)/(298.15)}{(e^{2273.73/298.15}-1)} - \ln(1-e^{-2273.73/298.15})\right]$$

$$= 0.035\text{ J K}^{-1}\text{ mol}^{-1}$$

$$\overline{H}_v^\circ = (8.314\,51\text{ J K}^{-1}\text{ mol}^{-1})(298.15\text{ K})\frac{(2273.73/298.15)}{e^{2273.73/298.15}-1}$$

$$= 0.009\text{ kJ mol}^{-1}$$

$$\overline{G}^\circ_{\text{v}} = (8.314\,51 \text{ J K}^{-1} \text{ mol}^{-1})(298.15 \text{ K}) \ln(1 - \text{e}^{-2273.73/298.15})$$
$$= -0.001 \text{ kJ mol}^{-1}$$

The small value of the vibrational contribution to $\overline{C}^\circ_P$ at 298.15 K indicates that $O_2(g)$ has very little vibrational energy at this temperature. The classical vibrational heat capacity is $R = 8.314\,51$ J K^{-1} mol^{-1} (Section 16.9), and this value of $(\overline{C}^\circ_P)_{\text{v}}$ is approached at 3000 K.

16.5 ROTATIONAL CONTRIBUTIONS TO THE THERMODYNAMIC PROPERTIES OF IDEAL GASES

The expression for the rotational energy of a rigid diatomic molecule was derived in Section 13.4:

$$\epsilon_{\text{r}} = \frac{J(J+1)h^2}{8\pi^2 I} \tag{16.63}$$

Here J is the rotational quantum number, and I is the moment of inertia of the diatomic molecule. Although the rotational energy depends only on J, the state of a rigid rotor is specified by the quantum number J and an additional quantum number M, where M can have integral values between $-J$ and $+J$. Thus, there are $2J + 1$ values of M for each value of J; in other words, the rotational levels are $(2J + 1)$-fold degenerate. The rotational partition function is thus

$$q_{\text{r}} = \sum_{J=0}^{\infty} (2J+1)\,\text{e}^{-J(J+1)h^2/8\pi^2 IkT} \tag{16.64}$$

For molecules with large moments of inertia, the energy levels are so close together that the summation may be replaced by an integration when T is reasonably high (above 10 to 100 K):

$$q_{\text{r}} = \int_0^{\infty} (2J+1)\,\text{e}^{-J(J+1)h^2/8\pi^2 IkT}\,\text{d}J \tag{16.65}$$

Since $(2J+1)\,\text{d}J$ is the differential of $J(J+1) = J^2 + J$, this equation is integrated by substituting

$$u = \frac{h^2}{8\pi^2 IkT}(J^2 + J) \tag{16.66}$$

so that

$$\text{d}u = \frac{h^2}{8\pi^2 IkT}(2J+1)\,\text{d}J \tag{16.67}$$

Thus equation 16.65 can be written

$$q_{\text{r}} = \frac{8\pi^2 IkT}{h^2}\int_0^{\infty} \text{e}^{-u}\,\text{d}u = \frac{8\pi^2 IkT}{h^2} = \frac{T}{\Theta_{\text{r}}} \tag{16.68}$$

where Θ_{r} is the **characteristic rotational temperature** defined by

$$\Theta_{\text{r}} = \frac{h^2}{8\pi^2 Ik} \tag{16.69}$$

The calculation of the rotational partition function for a homonuclear diatomic molecule (such as Cl_2) has to take into account the fact that rotation by 180° interchanges two equivalent nuclei. Since the new orientation is indistinguishable from the first, we have to divide by 2 so that the indistinguishable orientations are counted only once. For a heteronuclear diatomic molecule (such as HCl), a 180° rotation produces a distinguishable orientation. The effect of symmetry is taken into account by introducing a **symmetry number** σ that has the value 2 for a homonuclear molecule and the value 1 for a heteronuclear molecule. In general, then,

$$q_r = \frac{T}{\sigma \Theta_r} \tag{16.70*}$$

Example 16.9 ***Characteristic rotational temperature for hydrogen molecules***

What is the characteristic rotational temperature Θ_r for $H_2(g)$? What is the value of the molecular partition function for rotation at 3000 K? The moment of inertia is 4.6052×10^{-48} kg m^2.

$$\Theta_r = \frac{h^2}{8\pi^2 Ik} = \frac{(6.626\,08 \times 10^{-34}\ \text{J s})^2}{8\pi^2(4.6052 \times 10^{-48}\ \text{kg m}^2)(1.380\,66 \times 10^{-23}\ \text{J K}^{-1})} = 87.544\ \text{K}$$

$$q_r = \frac{T}{\sigma\Theta_r} = \frac{3000\ \text{K}}{(2)(87.544\ \text{K})} = 17.134$$

The rotational contributions for diatomic molecules obtained by substituting equation 16.70 into the expressions for the various thermodynamic properties are given later in Table 16.1.

Example 16.10 ***Rotational contributions for oxygen molecules***

What are the rotational contributions to $\overline{C}_P^\circ$, $\overline{S}^\circ$, $\overline{H}^\circ$, and $\overline{G}^\circ$ for $O_2(g)$ at 298.15 K? The moment of inertia for $O_2(g)$ is $I = 1.9373 \times 10^{-46}$ kg m^2.

*The behavior of H_2, D_2, and T_2 at low temperatures is complicated by the quantum mechanical symmetry requirements applicable to molecules containing identical nuclei. Since the proton has spin $\frac{1}{2}$, the wavefunction for H_2 must be antisymmetric to exchange of the nuclei. The wavefunctions for electronic and vibrational motion are both symmetric. Therefore, either the rotational or nuclear spin wavefunction must be antisymmetric. The rotational wavefunction for a symmetric rotor is symmetric with respect to inversion of the coordinates for even values of the rotational quantum number J, and antisymmetric for odd values of J. Therefore, for even values of J the nuclear wavefunction must be antisymmetric, and these levels have a degeneracy of 1. For odd values of J the nuclear wavefunction must be symmetric, and these levels have a degeneracy of 3. Thus, H_2 is a mixture of two molecular species. Hydrogen with a symmetric nuclear wavefunction and J odd is called **orthohydrogen.** Hydrogen with an antisymmetric nuclear wavefunction and J even is called **parahydrogen.** The rotational partition function for the equilibrium mixture is

$$q_r = \sum_{J_{\text{even}}}(2J+1)\exp\left[\frac{J(J+1)\Theta_r}{T}\right] + 3\sum_{J_{\text{odd}}}(2J+1)\exp\left[\frac{J(J+1)\Theta_r}{T}\right]$$

If hydrogen is prepared at room temperature or higher, where the ortho–para ratio is 3, it will retain this composition at lower temperatures if no catalyst (such as activated charcoal) is present, even though this is a nonequilibrium composition. If a catalyst is present, equilibrium results and a different plot of C_V versus T is obtained at low temperatures.

$$\Theta_r = \frac{h^2}{8\pi^2 Ik} = \frac{(6.626\,08 \times 10^{-34} \text{ J s})^2}{8\pi^2(1.9373 \times 10^{-46} \text{ kg m}^2)(1.380\,66 \times 10^{-23} \text{ J K}^{-1})} = 2.079 \text{ K}$$

$$(\overline{C}_P^\circ)_r = R = 8.314\,51 \text{ J K}^{-1} \text{ mol}^{-1}$$

$$\overline{S}_r^\circ = R \ln \frac{eT}{\sigma\Theta_r} = (8.314\,51 \text{ J K}^{-1} \text{ mol}^{-1}) \ln \frac{(2.7183)(298.15 \text{ K})}{(2)(2.079 \text{ K})}$$

$$= 43.838 \text{ J K}^{-1} \text{ mol}^{-1}$$

$$\overline{H}_r^\circ = RT = (8.314\,51 \text{ J K}^{-1} \text{ mol}^{-1})(298.15 \text{ K}) = 2.479 \text{ kJ mol}^{-1}$$

$$\overline{G}_r^\circ = -RT \ln \frac{T}{\sigma\Theta_r} = -(8.314\,51 \text{ J K}^{-1} \text{ mol}^{-1})(298.15 \text{ K}) \ln \frac{298.15 \text{ K}}{2 \times 2.079 \text{ K}}$$

$$= -10.591 \text{ kJ mol}^{-1}$$

The rotational contributions for diatomic molecules can also be used for linear polyatomic molecules such as CO_2, C_2H_2, N_2O, and HCN. The symmetry number is 2 for symmetrical linear molecules such as CO_2 and C_2H_2, and it is 1 for nonsymmetrical linear molecules such as N_2O and HCN.

A linear molecule with a small moment of inertia I has a small rotational partition function and a small rotational contribution to S° and G°. Molecules with larger moments of inertia have larger rotational contributions to S° and G°. The reason for this is that when the moment of inertia is large, the rotational energy levels are close together and more states are populated.

For polyatomic molecules in general the calculation of rotational contributions has to take into account the fact that the molecule may have three different moments of inertia. As shown in Section 13.5, the three moments of inertia for a spherical top molecule, such as CH_4, are equal: $I_a = I_b = I_c$. Two moments are equal for a symmetric top: $I_a < I_b = I_c$ for a prolate top such as CH_3Cl, and $I_a = I_b < I_c$ for an oblate top such as C_6H_6. For an asymmetric top, such as H_2O, the three moments of inertia are all different.

At temperatures well above the characteristic rotational temperatures, it may be shown* that the rotational partition function for a nonlinear polyatomic molecule is given by

$$q_r = \frac{\pi^{1/2}}{\sigma}\left(\frac{T^3}{\Theta_a\Theta_b\Theta_c}\right)^{1/2} \tag{16.71}$$

where the characteristic rotational temperatures are given by

$$\Theta_a = \frac{h^2}{8\pi^2 I_a k} \tag{16.72}$$

$$\Theta_b = \frac{h^2}{8\pi^2 I_b k} \tag{16.73}$$

$$\Theta_c = \frac{h^2}{8\pi^2 I_c k} \tag{16.74}$$

*D. McQuarrie, *Statistical Mechanics,* p. 136. Sausalito, CA: University Science Books, 2000.

and σ is the symmetry number. The symmetry number of a molecule is equal to the number of distinct proper rotational operations plus the identity operation (see Chapter 12). For H_2O, $\sigma = 2$. For NH_3, $\sigma = 3$. For ethylene, $\sigma = 4$. For CH_4, $\sigma = 12$. For C_6H_6, $\sigma = 12$. Each freely rotating methyl group in a molecule contributes a factor of 3 to the symmetry number.

The rotational contributions for nonlinear polyatomic molecules are given later in Table 16.1.

Nonlinear polyatomic molecules have three degrees of rotational freedom rather than the two for linear molecules; therefore, the rotational contribution to $\overline{C}_P^\circ$ is $3R/2$, and the enthalpy is $3RT/2$ relative to the hypothetical ideal gas at absolute zero, as suggested by the principle of equipartition.

Example 16.11 ***Rotational contributions for water molecules***

For $H_2O(g)$, $I_aI_bI_c = 5.7658 \times 10^{-141}$ kg^3 m^6 and $\sigma = 2$. What is the value of the rotational partition function for $H_2O(g)$, and what are the rotational contributions to $\overline{C}_P^\circ$, $\overline{S}^\circ$, $\overline{H}^\circ$, and $\overline{G}^\circ$ at 3000 K?

$$\Theta_a\Theta_b\Theta_c = \left(\frac{h^2}{8\pi^2 k}\right)^3 \frac{1}{I_aI_bI_c} = \left[\frac{(6.626\,076 \times 10^{-34}\ \text{J s})^2}{8\pi^2(1.380\,661 \times 10^{-23}\ \text{J K}^{-1})}\right]^3$$

$$\times \frac{1}{5.7658 \times 10^{-141}\ \text{kg}^3\ \text{m}^6} = 1.133\,05 \times 10^4\ \text{K}^3$$

$$q_r = \frac{\pi^{1/2}}{2}\left(\frac{3000^3}{1.133\,05 \times 10^4}\right)^{1/2} = 1368.0$$

$$(\overline{C}_P^\circ)_r = \tfrac{3}{2}R = \tfrac{3}{2}(8.314\,51\ \text{J K}^{-1}\ \text{mol}^{-1}) = 12.472\ \text{J K}^{-1}\ \text{mol}^{-1}$$

$$\overline{S}_r^\circ = (8.314\,51\ \text{J K}^{-1}\ \text{mol}^{-1})\left\{\ln\left[\left(\frac{1}{2}\right)\left(\frac{3000^3\,\pi}{1.133\,05 \times 10^4}\right)^{1/2}\right] + \frac{3}{2}\right\}$$

$$= 72.511\ \text{J K}^{-1}\ \text{mol}^{-1}$$

$$\overline{H}_r^\circ = \tfrac{3}{2}RT = \tfrac{3}{2}(8.314\,51\ \text{J K}^{-1}\ \text{mol}^{-1})(3000\ \text{K}) = 37.415\ \text{kJ mol}^{-1}$$

$$\overline{G}_r^\circ = -(8.314\,51\ \text{J K}^{-1}\ \text{mol}^{-1})(3000\ \text{K})\ln\left[\frac{\pi^{1/2}}{2}\left(\frac{3000^3}{1.133\,05 \times 10^4}\right)^{1/2}\right]$$

$$= -180.118\ \text{kJ mol}^{-1}$$

16.6 ELECTRONIC CONTRIBUTIONS TO THE THERMODYNAMIC PROPERTIES OF IDEAL GASES

The electronic state of an atom or of a diatomic molecule is represented by three quantum numbers:

1. An orbital quantum number L for an atom or Λ for a diatomic molecule that can have the values 0, 1, 2, ...
2. A spin quantum number S that determines the multiplicity $(2S + 1)$ of the state.

3. A total angular momentum quantum number J for an atom or Ω for a diatomic molecule.

The characteristic temperature Θ_e for an electronic transition is defined by

$$\Theta_e = \frac{\epsilon_e}{k} \tag{16.75}$$

where ϵ_e is the energy of the excited electronic energy level relative to the ground state. If energies are measured with respect to the electronic ground state, the electronic molecular partition function for an atom or molecule is given by

$$q_e = g_0 + g_1 \exp\left(-\frac{\Theta_{e1}}{T}\right) + g_2 \exp\left(-\frac{\Theta_{e2}}{T}\right) + \cdots \tag{16.76*}$$

where the degeneracy g_i is given by

$$g_i = 2J_i + 1 \quad \text{or} \quad g_i = 2\Omega_i + 1 \tag{16.77}$$

However, for a molecule, only the thermodynamic properties of the ground state can be calculated by combining the first term of equation 16.76 with the partition functions for vibration and rotation **for the ground electronic state.** For the excited electronic levels represented by the second and third terms in equation 16.76, the vibrational frequencies and rotational frequencies are different from those in the ground state. Often excited electronic states of molecules have very much higher energies, so they affect the calculations of thermodynamic properties only at very high temperatures.

Example 16.12 ***Electronic contributions for oxygen atoms***

What are the electronic contributions to $\overline{S}°$ and $\overline{G}°$ for O(g) at 298.15 K to the degree of completeness we have used here? The degeneracy in the ground state is 5, the first excited state has an energy of 158.2 cm^{-1} with respect to the ground state and has a degeneracy of 3, and the second excited state has an energy of 226.5 cm^{-1} with no degeneracy.

$$\Theta_{e1} = \frac{hc\tilde{\nu}}{k} = \frac{(6.626\,076 \times 10^{-34}\ \text{J s})(2.997\,925 \times 10^8\ \text{m s}^{-1})(1.582 \times 10^4\ \text{m}^{-1})}{1.380\,658 \times 10^{-23}\ \text{J K}^{-1}}$$

$$= 227.6\ \text{K}$$

$$\Theta_{e2} = 325.9\ \text{K}$$

$$q_e = 5 + 3\,e^{-227.6/298.15} + e^{-325.9/298.15} = 6.7335$$

*S. J. Strickler [*J. Chem. Educ.* **43**:364 (1966)] discusses the fact that the electronic partition function q_e for the hydrogen atom at 25 °C calculated from

$$q_e = \sum_{n=1}^{\infty} n^2 e^{-\epsilon_n/kT}$$

is infinite. In this equation n^2 is the degeneracy, and ϵ_n is the energy of the orbital with quantum number n, which is proportional to n^{-2}. This is contrary to experience, since it would require the population of the ground state to be zero. The problem is that there are an infinite number of terms in the partition function with energy less than 13.60 eV. The paradox is resolved by considering the size of the orbital of a hydrogen atom for a very large quantum number n. For a finite container the maximum n is finite, and the value of q_e is not detectably different from unity. In fact, the cutoff probably comes from neighboring molecules that limit the radius of a hydrogen atom to about $(V/N)^{1/3}$.

$$\overline{S}_e^\circ = R \ln q_e = (8.314\,51 \text{ J K}^{-1}\text{ mol}^{-1}) \ln 6.7335 = 15.853 \text{ J K}^{-1}\text{ mol}^{-1}$$

$$\overline{G}_e^\circ = -T\overline{S}_e^\circ = -(298.15 \text{ K})(15.853 \text{ J K}^{-1}\text{ mol}^{-1}) = -4.727 \text{ kJ mol}^{-1}$$

In calculating the electronic partition function for a diatomic molecule, it is convenient to measure energy with respect to the atoms in an ideal gas state at absolute zero. When this is done the calculated thermodynamic properties of the molecule are relative to its constituent atoms in their ground states. On this basis the energy of the electronic ground state of the molecule is $-D_0$, where D_0 is the spectroscopic dissociation energy (Section 13.2). Equation 16.76 may therefore be written as

$$q_e = g_0\,e^{D_0/kT} \tag{16.78}$$

when T is low enough that the second and higher terms can be neglected.

The contributions to $(\overline{C}_P^\circ)_e$, $\overline{H}_e^\circ$, $\overline{S}_e^\circ$, and $\overline{G}_e^\circ$ for a molecule are therefore

$$(\overline{C}_P^\circ)_e = 0 \tag{16.79}$$

$$\overline{H}_e^\circ = -D_0 \tag{16.80}$$

$$\overline{S}_e^\circ = R \ln g_0 \tag{16.81}$$

$$\overline{G}_e^\circ = -D_0 - RT \ln g_0 \tag{16.82}$$

Example 16.13 ***Electronic contributions for oxygen molecules***

What are the electronic contributions to $\overline{C}_P^\circ$, $\overline{H}^\circ$, $\overline{S}^\circ$, and $\overline{G}^\circ$ for O_2(g) at 298.15 K? The electronic ground state is a triplet, and the spectroscopic dissociation energy D_0 is 491.888 kJ mol^{-1}. The energy of the first excited electronic state is so high that it does not have to be considered.

$$(\overline{C}_P^\circ)_e = 0$$

$$\overline{H}_e^\circ = -D_0 = -491.888 \text{ kJ mol}^{-1}$$

$$\overline{S}_e^\circ = (8.314\,51 \text{ J K}^{-1}\text{ mol}^{-1}) \ln 3 = 9.134 \text{ J K}^{-1}\text{ mol}^{-1}$$

$$\overline{G}_e^\circ = -491.888 \text{ kJ mol}^{-1} - (8.314\,51 \times 10^{-3} \text{ kJ K}^{-1}\text{ mol}^{-1})(298.15 \text{ K}) \ln 3$$

$$= -494.611 \text{ kJ mol}^{-1}$$

16.7 THERMODYNAMIC PROPERTIES OF IDEAL GASES

The formulas for calculating the molar thermodynamic properties of ideal gases are summarized in Table 16.1 on page 582.

Now that we have shown how to calculate the translational, rotational, vibrational, and electronic contributions to the thermodynamic properties, we can obtain the properties by summing up the contributions, as shown, for example, in equation 16.34.

Example 16.14 ***Molar thermodynamic properties for oxygen atoms***

What are the values of $\overline{C}_P^\circ$, $\overline{S}^\circ$, $\overline{H}^\circ$, and $\overline{G}^\circ$ for O(g) at 298.15 K?

Atoms have only translational and electronic contributions, and therefore we sum the contributions calculated in Examples 16.6 and 16.12:

$$\overline{C}_P^\circ = (\overline{C}_P^\circ)_t + (\overline{C}_P^\circ)_e = 20.786 \text{ J K}^{-1} \text{ mol}^{-1} \qquad \text{since } (\overline{C}_P^\circ)_e \text{ is taken here as zero}$$

$$\overline{S}^\circ = \overline{S}_t^\circ + \overline{S}_e^\circ = (143.435 + 15.853) \text{ J K}^{-1} \text{ mol}^{-1} = 159.288 \text{ J K}^{-1} \text{ mol}^{-1}$$

$$\overline{H}^\circ = \overline{H}_t^\circ + \overline{H}_e^\circ = 6.197 \text{ kJ mol}^{-1} \qquad \text{since } \overline{H}_e^\circ \text{ is taken here as zero}$$

$$\overline{G}^\circ = \overline{G}_t^\circ + \overline{G}_e^\circ = (-36.568 - 4.727) \text{ kJ mol}^{-1} = -41.295 \text{ kJ mol}^{-1}$$

The values of $\overline{C}_P^\circ$ and $\overline{S}^\circ$ are in pretty good agreement with Table C.2, but the values of $\overline{H}^\circ$ and $\overline{G}^\circ$ need to be adjusted to the proper reference state.

Example 16.15 ***Molar thermodynamic properties for oxygen molecules***

What are the values of $\overline{C}_P^\circ$, $\overline{S}^\circ$, $\overline{H}^\circ$, and $\overline{G}^\circ$ for O_2(g) at 298.15 K?

These values are obtained by adding the translational, vibrational, rotational, and electronic contributions calculated in Examples 16.7, 16.8. 16.10, and 16.13.

$$\begin{aligned}\overline{C}_P^\circ &= (\overline{C}_P^\circ)_t + (\overline{C}_P^\circ)_v + (\overline{C}_P^\circ)_r + (\overline{C}_P^\circ)_e \\ &= (20.786 + 0.236 + 8.314 + 0) \text{ J K}^{-1} \text{ mol}^{-1} \\ &= 29.336 \text{ J K}^{-1} \text{ mol}^{-1}\end{aligned}$$

$$\begin{aligned}\overline{S}^\circ &= \overline{S}_t^\circ + \overline{S}_v^\circ + \overline{S}_r^\circ + \overline{S}_e^\circ = (152.080 + 0.035 + 43.838 + 9.134) \text{ J K}^{-1} \text{ mol}^{-1} \\ &= 205.088 \text{ J K}^{-1} \text{ mol}^{-1}\end{aligned}$$

$$\begin{aligned}\overline{H}^\circ &= \overline{H}_t^\circ + \overline{H}_v^\circ + \overline{H}_r^\circ + \overline{H}_e^\circ = (6.197 + 0.009 + 2.479 - 491.888) \text{ kJ mol}^{-1} \\ &= -483.203 \text{ kJ mol}^{-1}\end{aligned}$$

$$\begin{aligned}\overline{G}^\circ &= \overline{G}_t^\circ + \overline{G}_v^\circ + \overline{G}_r^\circ + \overline{G}_e^\circ = (-39.145 - 0.001 - 10.591 - 494.611) \text{ kJ mol}^{-1} \\ &= -544.349 \text{ kJ mol}^{-1}\end{aligned}$$

The values of $\overline{C}_P^\circ$ and $\overline{S}^\circ$ are in pretty good agreement with Table C.2, but the values of $\overline{H}^\circ$ and $\overline{G}^\circ$ need to be adjusted to the proper reference state, as shown later.

Since the value of $\overline{C}_P^\circ$ calculated here for O_2(g) is in good agreement with the value in Table C.2 (29.355 J K^{-1} mol^{-1}), the electronic contribution that we ignored is small at this temperature. The value of $\overline{S}^\circ$ is also in good agreement with the value in Table C.2 (205.138 J K^{-1} mol^{-1}). The values of $\overline{H}^\circ$ and $\overline{G}^\circ$ calculated here cannot be compared with $\Delta_f H^\circ$ and $\Delta_f G^\circ$ because of the difference in reference states, but we will calculate $\Delta_f H^\circ$ and $\Delta_f G^\circ$ for O(g) in Example 16.16.

Molecular parameters of a number of gases are given in Table 16.2. The dissociation energies D_0 are with respect to the constituent atoms in the ground state. For diatomic molecules these dissociation energies have been obtained from spectra. For polyatomic molecules the values of D_0 have been obtained from thermochemical measurements and theoretical calculations of heat capacities of gases. In Table 16.2 values of Θ_e greater than 10 000 K are not given.

Table 16.2 Molecular Parameters of Gases

Gas	M/g mol^{-1}	D_0/kJ mol^{-1}	σ	Θ_r/K[a]	Θ_v/K[b]	g_0	Θ_e/K[b]
H	1.008 0	0				2	
C	12.001	0				1	23.6(3)
							62.6(5)
N	14.007	0				4	
O	15.994	0				5	228.1(3)
							325.9
Cl	35.453	0				4	1 269.53(2)
I	126.904 5	0				4	10 939.3(2)
H_2	2.016	432.073	2	87.547	6338.2	1	
N_2	28.013 4	941.4	2	2.875 05	3392.01	1	
O_2	31.998 8	491.888	2	2.079	2273.64	3	
Cl_2	70.906	239.216	2	0.345 6	807.3	1	
I_2	253.82	148.81	2	0.053 76	308.65	1	
HCl	36.465	427.772	1	15.234 4	4301.38	1	
HI	127.918	294.67	1	9.369	3322.24	1	
CO	28.010 55	1070.11	1	2.777 1	3121.48	1	
NO	30.008	627.7	1	2.452 0	2738.87	2	174.2(2)
CO_2	44.009 95	1596.23	2	0.561 67	960.10(2)	1	
					1932.09		
					3380.14		
NO_2[c]	46.008	928.3	2	4.243 01	1088.9	2	
					1953.6		
					2396.3		
H_2O	18.016	917.773	2	11 331.5	2294.27	1	
					5261.71		
					5403.78		
NH_3	17.036 1	1157.77	3	1 876.0	1367	1	
					2341(2)		
					4800		
					4955(2)		
CH_4	16.043	1640.57	12	435.6	1957(3)	1	
					2207.1(2)		
					4196.2		
					4343.3(3)		
N_2O_4	92.016	1909.82	4	6.5793×10^{-3}	72	1	
					374		
					554		
					619		
					691		
					971		
					1079		
					1184		
					1814		
					1975		
					2460		
					2515		

Source: M. Chase et al., JANAF Thermochemical Tables. *J. Phys. Chem. Ref. Data* **14** (1985), Supplement 1.

[a] The values of Θ_r for nonlinear polyatomic molecules are values of $\Theta_a\Theta_b\Theta_c$.

[b] The values in parentheses are degeneracies. Otherwise, the levels have a degeneracy of unity.

[c] NO_2 is a bent molecule with a bond angle of 134° 15′.

The values of $\overline{C}_P^\circ$ and $\overline{S}^\circ$ calculated with equations given in this chapter may be compared with the values from the JANAF Thermochemical Tables (Table C.2), which are more accurate because they take more effects into account (e.g., deviations from the rigid rotor–harmonic oscillator approximation). For many simple molecules, spectroscopic measurements have yielded the values of the anharmonicity constant x_e, the centrifugal distortion constant D, and the rotation–vibration coupling constant α_e.

The values of $\overline{H}^\circ$ and $\overline{G}^\circ$ are with respect to ideal gases of the constituent atoms at absolute zero. However, they can be used to calculate $\Delta_r H^\circ$ and $\Delta_r G^\circ$ for chemical reactions because $\overline{H}^\circ$ and $\overline{G}^\circ$ for reactants and products have all been calculated for the same reference state.

Example 16.16 ***Reaction properties for $\frac{1}{2}O_2$ (g) = O(g)***

What are the values of $\Delta_r C_P^\circ$, $\Delta_r S^\circ$, $\Delta_r H^\circ$, and $\Delta_r G^\circ$ for $\frac{1}{2}O_2(g) = O(g)$ calculated statistically mechanically, and what is the equilibrium constant at 298.15 K?

Using values calculated in Examples 16.14 and 16.15,

$$\Delta_r C_P^\circ = 20.786 - \tfrac{1}{2}(29.336) = 6.118 \text{ J K}^{-1} \text{ mol}^{-1}$$

$$\Delta_r S^\circ = 159.288 - \tfrac{1}{2}(205.088) = 56.744 \text{ J K}^{-1} \text{ mol}^{-1}$$

$$\Delta_r H^\circ = 6.197 - \tfrac{1}{2}(-483.203) = 247.799 \text{ kJ mol}^{-1}$$

$$\Delta_r G^\circ = -41.295 - \tfrac{1}{2}(-544.349) = 230.880 \text{ kJ mol}^{-1}$$

$$K = e^{-\Delta_r G^\circ/RT} = e^{-230.880/(8.31451)(298.15)} = 3.55 \times 10^{-41} = \frac{P_O/P^\circ}{(P_{O_2}/P^\circ)^{1/2}}$$

The values of $\Delta_r H^\circ$ and $\Delta_r G^\circ$ should be equal to $\Delta_f H^\circ$ and $\Delta_f G^\circ$ for O(g). The small differences from Table C.2 are due to the neglect of higher-order terms.

Table 16.3 gives $\Delta_r C_P^\circ$, $\Delta_r S^\circ$, $\Delta_r H^\circ$, and $\Delta_r G^\circ$ for two endothermic reactions and two exothermic reactions, calculated with the molecular parameters in Table 16.2.

The reaction for the formation of NH_3 from its elements is an example of a reaction that is thermodynamically spontaneous at room temperature, but does not occur at an appreciable rate. As the temperature is raised to increase the rate of formation of NH_3, the equilibrium constant becomes less favorable. Therefore, it has been important to find catalysts for $\frac{1}{2}N_2(g) + \frac{3}{2}H_2(g) = NH_3(g)$ so that the reaction can be carried out at lower temperatures.

The reaction forming methane from carbon monoxide is another reaction that is spontaneous at low temperatures, but it has such a low rate that higher temperatures and a catalyst must be used to obtain a reasonable yield.

16.8 DIRECT CALCULATION OF EQUILIBRIUM CONSTANTS FOR REACTIONS OF IDEAL GASES

The equilibrium constant can be calculated from thermodynamic properties calculated from spectroscopic properties, as we have seen above, but it is convenient to express the equilibrium constant directly in terms of the partition

Table 16.3 Changes in Standard Thermodynamic Properties for Chemical Reactions of Gases

T/K	$\Delta_r C_P^\circ$ / J K^{-1} mol^{-1}	$\Delta_r S^\circ$ / J K^{-1} mol^{-1}	$\Delta_r H^\circ$ / kJ mol^{-1}	$\Delta_r G^\circ$ / kJ mol^{-1}	K
		$\frac{1}{2}O_2(g) = O(g)$			
298.15	6.118	56.744	247.799	230.880	3.55×10^{-41}
500.00	5.305	60.551	248.961	218.685	1.43×10^{-23}
1000.00	3.487	64.321	251.095	186.774	1.75×10^{-10}
2000.00	2.499	66.755	253.950	120.441	7.16×10^{-4}
3000.00	2.272	67.862	256.314	52.728	1.21×10^{-1}
		$\frac{1}{2}N_2(g) + \frac{1}{2}O_2(g) = NO(g)$			
298.15	−0.052	10.638	88.942	85.770	9.41×10^{-16}
500.00	−0.096	11.346	88.922	83.249	2.01×10^{-9}
1000.00	0.136	11.994	88.942	76.948	9.56×10^{-5}
2000.00	0.102	12.426	89.075	64.223	2.10×10^{-2}
3000.00	0.055	12.574	89.151	51.429	1.27×10^{-1}
		$\frac{1}{2}N_2(g) + \frac{3}{2}H_2(g) = NH_3(g)$			
298.15	−22.729	−98.497	−46.195	−16.763	8.65×10^{2}
500.00	−16.977	−108.930	−50.218	4.357	3.50×10^{-1}
1000.00	−5.165	−116.749	−55.516	61.451	6.17×10^{-4}
2000.00	4.347	−116.620	−54.785	178.893	2.13×10^{-5}
3000.00	6.648	−114.328	−49.054	294.584	7.43×10^{-6}
		$CO(g) + 3H_2(g) = CH_4(g) + H_2O(g)$			
298.15	−47.608	−213.781	−206.788	−143.048	1.15×10^{25}
500.00	−36.260	−235.941	−215.340	−97.369	1.49×10^{10}
1000.00	−9.704	−252.178	−226.280	25.899	4.44×10^{-2}
2000.00	9.773	−251.154	−223.691	278.618	5.29×10^{-8}
3000.00	13.941	−246.219	−211.383	527.276	6.59×10^{-10}

function for the gaseous species involved. To do that we will use the fact that $K = \exp(-\Delta_r G^\circ/RT)$ and that, for indistinguishable molecules, $G_i = -N_i kT \ln(q_i/N_i)$ (equation 16.27) for species i. Replacing $N_i k$ in this equation with $n_i R$ and dividing by n_i yields the molar Gibbs energy:

$$\overline{G_i} = \frac{G_i}{n_i} = -RT \ln \frac{q_i}{N_i} \tag{16.83}$$

This is the molar Gibbs energy of species i with respect to the ground state of species i. In making statistical mechanical calculations on chemical reactions it is necessary to use the molar Gibbs energies of the reactants and products with respect to the same ground state for all species involved in the reaction. In thermodynamic tables the standard properties are with respect to the elements in their specified reference states, but in statistical mechanical calculations on gas reactions it is more convenient to use the gaseous atoms in their ground states as the reference states. This means that the electronic contribution to the molecular partition function for an atom in its ground state is $q_{ei} = g_{0i}$, the degeneracy of the ground state of the atom. The electronic contribution to the molecular partition function for a molecule is $q_{ei} = g_{0i}\, e^{D_{0i}/kT}$, where D_{0i} is the spectroscopic dissociation energy of the molecule, that is, the energy of dissociation of the molecule in its ground state to atoms in their ground states. Since dissociation energies are often tabulated in kJ mol^{-1}, this relation can also be written $q_{ei} = g_{0i}\, e^{D_{0i}/RT}$.

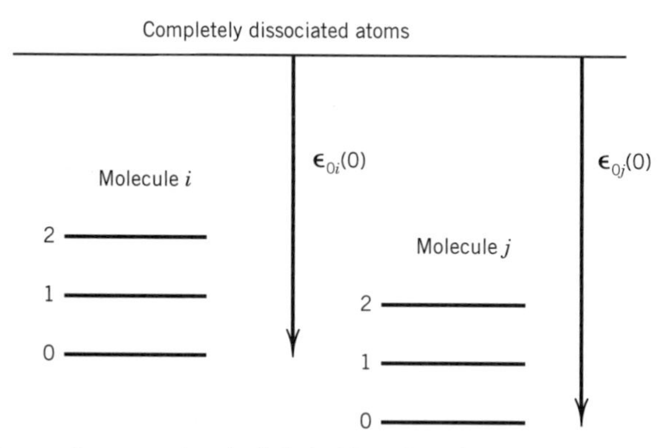

Figure 16.2 Electronic energy levels (labeled 0, 1, 2, ...) of molecules i and j that contain the same atoms, where $\epsilon_{0i}(0)$ is the negative of the dissociation energy ($-D_{0i}$).

Example 16.17 ***The molar Gibbs energy with respect to atoms at absolute zero***

Show that the molar Gibbs energy of species i is increased by $N_A\epsilon_{0i}(0)$ when the reference state is changed from the ground state of the species to the ground states of the atoms involved.

The energies of electronic states in a species contribute to the electronic partition function q_{ei}. For example, equation 16.76 can be written

$$q_{ei} = g_{0i} + g_{1i}\exp\left(-\frac{\epsilon_{1i}}{kT}\right) + \cdots$$

If we want to express the energies of all of the molecules in a chemical reaction on the basis of the atoms they contain, as shown in Fig. 16.2, we have to add the energy $\epsilon_{0i}(0)$ of the ground state of i with respect to the atoms in their ground states. Thus the equation becomes

$$q_{ei} = g_{0i}\exp\left[-\frac{\epsilon_{0i}(0)}{kT}\right] + g_{1i}\exp\left[-\frac{\epsilon_{0i}(0)+\epsilon_{1i}}{kT}\right] + \cdots$$

This is equivalent to multiplying each term by $\exp[-\epsilon_{0i}(0)/kT]$. Note that $\epsilon_{0i}(0)$ is the negative of the dissociation energy of the species from its ground state. Multiplying q_i in equation 16.83 by $\exp[-\epsilon_{0i}(0)/kT]$ yields

$$\begin{aligned}\overline{G_i} &= -RT\ln\frac{q_i\exp[-\epsilon_{0i}(0)/kT]}{N_i}\\ &= -RT\ln\exp[-\epsilon_{0i}(0)/kT] - RT\ln\frac{q_i}{N_i}\\ &= N_A\epsilon_{0i}(0) - RT\ln\frac{q_i}{N_i}\end{aligned}$$

Now we want to rearrange equation 16.83 for the molar Gibbs energy to the form

$$\overline{G_i} = \overline{G_i^\circ} + RT\ln\frac{P_i}{P^\circ} \tag{16.84}$$

so that we can calculate $\overline{G}_i^\circ$ and use it in calculating the equilibrium constant using $\sum \nu_i \overline{G}_i^\circ = -RT \ln K$. When gases are ideal, $P_i V = (N_i/N_A)RT$, and so equation 16.83 can be written

$$\overline{G}_i = -RT \ln \frac{q_i RT}{VP_i N_A} \tag{16.85}$$

In order to be able to take out a term in P_i/P°, both the numerator and denominator are multiplied by P°. Then this equation can be rearranged to

$$\overline{G}_i = -RT \ln \frac{q_i RT}{VP^\circ N_A} + RT \ln \frac{P_i}{P^\circ} \tag{16.86}$$

which shows that the standard molar Gibbs energy of species i is given by

$$\overline{G}_i^\circ = -RT \ln \frac{q_i RT}{VP^\circ N_A} = -RT \ln \frac{q_i}{V} - RT \ln \frac{RT}{P^\circ N_A} \tag{16.87}$$

Substituting equation 16.87 into the equation for the equilibrium constant yields

$$K = \left(\frac{RT}{N_A P^\circ}\right)^{\sum \nu_i} \prod \left(\frac{q_i}{V}\right)^{\nu_i} = \left(\frac{kT}{P^\circ}\right)^{\sum \nu_i} \prod \left(\frac{q_i}{V}\right)^{\nu_i} \tag{16.88}$$

where $P^\circ = 10^5$ Pa. Note that the equilibrium constant is dimensionless.

Since the equilibrium constant for a reaction of ideal gases can be calculated using q/V for the reacting species, it is useful to review these expressions for monatomic and diatomic gases. For a monatomic gas,

$$\frac{q}{V} = g_0 \left(\frac{2\pi mkT}{h^2}\right)^{3/2} \tag{16.89}$$

where m is the mass of the atom and g_0 is the degeneracy of its ground electronic state. For a diatomic gas,

$$\frac{q}{V} = \left(\frac{2\pi mkT}{h^2}\right)^{3/2} \frac{T}{\sigma\Theta_r} \frac{1}{1 - \exp(-\Theta_v/T)} g_0 \exp\left(\frac{D_0}{kT}\right) \tag{16.90}$$

where σ is the symmetry number, Θ_r is the characteristic rotational temperature, Θ_v is the characteristic vibrational temperature, and D_0 is the spectroscopic dissociation energy. When it is convenient to express D_0 in kJ mol^{-1}, the last term becomes $\exp(D_0/RT)$, where R is in kJ K^{-1}mol^{-1}. We have assumed that only the ground electronic state need be considered.

Example 16.18 ***Calculation of the equilibrium constant for $H_2(g) = 2H(g)$***

Calculate the equilibrium constant for the dissociation of molecular hydrogen to atoms at 1000 K, assuming the gases are ideal and using the harmonic oscillator–rigid rotor approximation.

$$H_2(g) = 2H(g) \qquad K = \frac{[P(H)/P^\circ]^2}{P(H_2)/P^\circ}$$

The molecular parameters are given in Table 16.2.

The mass of a hydrogen atom $m_H = 1.007\,825 \times 10^{-3}$ kg mol$^{-1}/N_A = 1.673\,53 \times 10^{-27}$ kg and the degeneracy g_{0H} of the ground electronic state is 2, so that

$$\frac{q_H}{V} = g_{0H}\left(\frac{2\pi m_H kT}{h^2}\right)^{3/2}$$

$$= 2\left[\frac{2\pi(1.673\,53 \times 10^{-27}\ \text{kg})(1.3807 \times 10^{-23}\ \text{J K}^{-1})(1000\ \text{K})}{(6.626 \times 10^{-34}\ \text{J s})^2}\right]^{3/2} = 1.203 \times 10^{31}\ \text{m}^{-3}$$

The symmetry number σ for molecular hydrogen is 2, the characteristic rotational temperature Θ_r is 87.547 K, the characteristic vibrational temperature Θ_v is 6338.3 K, the degeneracy $g_0(H_2)$ of the ground state is unity, and the spectroscopic dissociation energy D_{0H_2} (from the ground state molecule to atoms in their ground states) is 432.073 kJ mol^{-1}, so that

$$\left(\frac{2\pi m_{H_2} kT}{h^2}\right)^{3/2} = 1.701 \times 10^{31}\ \text{m}^{-3}$$

$$\frac{T}{\sigma\Theta_r} = \frac{1000\ \text{K}}{2(87.547\ \text{K})} = 5.771$$

$$\frac{1}{1-\exp(-\Theta_v/T)} = \frac{1}{1-\exp\left(\frac{-6338.3\ \text{K}}{1000\ \text{K}}\right)} = 1.002$$

$$g_{0H_2}\exp\left(\frac{D_0}{kT}\right) = 2\exp\left[\frac{432.073\ \text{kJ mol}^{-1}}{(8.314\,51 \times 10^{-3}\ \text{kJ K}^{-1}\ \text{mol}^{-1})(1000\ \text{K})}\right] = 3.703 \times 10^{22}$$

so that

$$\frac{q_{H_2}}{V} = \left(\frac{2\pi m_{H_2} kT}{h^2}\right)^{3/2} \frac{T}{\sigma\Theta_r} \frac{1}{1-\exp(-\Theta_v/T)} g_{0H_2}\exp\left(\frac{D_0}{kT}\right) = 3.604 \times 10^{54}\ \text{m}^{-3}$$

$$\frac{kT}{P^\circ} = \frac{(1.3807 \times 10^{-23}\ \text{J K}^{-1})(1000\ \text{K})}{10^5\ \text{Pa}} = 1.381 \times 10^{-25}\ \text{m}^3$$

Equation 16.88 shows that the equilibrium constant is given by

$$K = \left(\frac{kT}{P^\circ}\right)\frac{(q_H/V)^2}{q_{H_2}/V} = \frac{(1.381 \times 10^{-25}\ \text{m}^3)(1.203 \times 10^{31}\ \text{m}^{-3})^2}{(3.604 \times 10^{54}\ \text{m}^{-3})} = 5.54 \times 10^{-18}$$

The value calculated in Chapter 5 using the JANAF tables is 5.16×10^{-18}, which is more accurate because further vibrational and rotational terms have been taken into account.

This is an example of a problem that is much easier to solve using a computer with a mathematical program. This makes it easier to check the input and calculate K at a series of temperatures. Also, parts of the program can be used to calculate partition functions for other molecules. This is illustrated in the *Solutions Manual* using Mathematica. (See Computer Problems 16.G–16.J.)

The equilibrium constants in Table 16.4 calculated from equations 16.88–16.90 using Mathematica given in the *Solutions Manual* are compared with values calculated using JANAF tables in Chapter 5. The JANAF tables are more accurate because they are not based on the harmonic oscillator–rigid rotor approximation. Deviations from this approximation are more serious at higher temperatures.

The equilibrium constant for a gas reaction can be calculated from spectroscopic information only for diatomic molecules or small polyatomic molecules for

Table 16.4 Equilibrium Constants Calculated Using Equations 16.88 to 16.90 and Using the JANAF Tables

Reaction	Computer Problem	T/K 500	1000	2000
$H_2(g) = 2(H)g$	16.G	5.30×10^{-41}	5.54×10^{-18}	2.89×10^{-6}
	5.D	4.84×10^{-41}	5.16×10^{-18}	2.65×10^{-6}
$I_2(g) = 2I(g)$	16.H	3.32×10^{-11}	3.21×10^{-3}	37.6
	5.D	3.24×10^{-11}	3.08×10^{-3}	34.4
$HI(g) = H(g) + I(g)$	16.I	5.58×10^{-27}	2.41×10^{-11}	2.84×10^{-3}
	5.D	3.50×10^{-27}	2.33×10^{-11}	6.05×10^{-1}
$2HI(g) = H_2(g) + I_2(g)$	16.J	7.28×10^{-3}	3.25×10^{-2}	7.40×10^{-2}
	5.D	7.80×10^{-3}	3.44×10^{-2}	7.99×10^{-2}

which dissociation energies may be determined spectroscopically. For larger polyatomic molecules the statistical mechanical calculation of K requires enthalpy of formation data determined by chemical methods, since it is not practical to determine dissociation energies for these larger molecules from spectroscopic measurements.

Comment:

In the development of large rockets in the 1950s, severe difficulties were encountered in performance calculations on propellant systems. Thermodynamic data were needed at temperatures higher than can ordinarily be reached in the laboratory, and so statistical mechanics was used to calculate the standard thermodynamic properties up to 6000 K. The JANAF (Joint Army, Navy, Air Force) tables were started, and the current edition was produced in 1985. The JANAF tables list the properties of about 1800 substances. Excerpts from these tables are given in Table C.3.

16.9 EQUIPARTITION

According to the classical principle of equipartition, each squared term in the classical energy expression for a gas molecule contributes $R/2$ to the molar heat capacity at constant volume. This principle is not exact because the heat capacity decreases with decreasing temperature (quantum mechanics explains why), but the classical calculation does give the limit that the heat capacity approaches at high temperature. Statistical mechanics shows why this is so, and it yields the temperature range over which each contribution increases to its classical value.

The full translational contribution is obtained for a gas at very low temperatures. Table 16.5 shows that $\overline{U}_t = \frac{3}{2}RT$ and $\overline{C}_V = \frac{3}{2}R$, independent of the mass

Table 16.5 Contributions to $\overline{C}_V/R$ for Ideal Gases at High Temperatures[a]

Molecule	*Translational*	*Vibrational*	*Rotational*
Monatomic	$\frac{3}{2}$	0	0
Diatomic	$\frac{3}{2}$	1	1
Linear polyatomic	$\frac{3}{2}$	$3N - 5$	1
Nonlinear polyatomic	$\frac{3}{2}$	$3N - 6$	$\frac{3}{2}$

[a] N is the number of atoms in the molecule.

or structure of a molecule of an ideal gas. In classical mechanics the translational energy of a gas molecule is given by

$$U_t = \frac{mv_x^2}{2} + \frac{mv_y^2}{2} + \frac{mv_z^2}{2} \tag{16.91}$$

This suggests that each quadratic term in the classical energy expression contributes $RT/2$ to $\overline{U}_t$ and $R/2$ to $\overline{C}_V$. This is in agreement with the fact that, for monatomic gases, $\overline{C}_V = \frac{3}{2}R$ at pressures where they behave as ideal gases and at temperatures low enough that there is negligible electronic excitation.

In Section 16.7 we saw that each normal mode of vibration of a polyatomic molecule contributes RT to the internal energy and R to $\overline{C}_V$ in the high-temperature limit. For a one-dimensional harmonic oscillator, classical mechanics yields

$$U_v = \frac{mv_x^2}{2} + \frac{kx^2}{2} \tag{16.92}$$

Thus, we see again that each squared term contributes $R/2$ to the molar heat capacity in the limit as the temperature is increased.

Well above the characteristic rotational temperature, the contribution of rotation to the molar heat capacity of a diatomic molecule or a linear polyatomic molecule is R. The classical expression for the rotational energy of a diatomic or linear polyatomic molecule is

$$U_r = \frac{I\omega_x^2}{2} + \frac{I\omega_y^2}{2} \tag{16.93}$$

Since there are two squared energy terms, each contributes $R/2$ to the molar heat capacity at constant volume. There is one more term for a nonlinear polyatomic molecule, and so the contribution of rotation to the molar heat capacity is $\frac{3}{2}R$, as expected.

The high-temperature contributions to the molar heat capacities at constant volume are summarized in Table 16.5.

16.10 ENSEMBLES

We began our study of statistical mechanics by assuming the Boltzmann distribution for a macroscopic system at temperature T, volume V, and number of particles N. There is a different approach, invented by Gibbs, that provides a more general way of thinking about statistical mechanics, called the **ensemble method.** An ensemble is an imaginary collection of a large number of isolated systems having certain macroscopic properties in common that we average over to obtain the thermodynamic properties of a system. A **microcanonical ensemble** is a collection of systems having the same number of particles, the same volume, and the same energy (N, V, U), that contains all possible microstates of the system. Averaging over these systems gives the thermodynamic properties.

A **canonical ensemble** is made up of a very large number η of systems, each of which contains N particles in volume V. The ensemble is prepared by putting systems in contact with a heat reservoir at temperature T, and then removing them to form the ensemble. The systems in the ensemble are isolated from each other, as illustrated in Fig. 16.3. Thus, all the systems in the canonical ensemble are characterized by the same fixed N and V values. The ensemble of systems is also isolated, and contains η systems, which are replicas of the system of interest. The total isolated ensemble contains ηN particles, and has a volume of ηV and a fixed energy U. The energy of a system can have any possible value less than U. The average energy of a system is calculated using $\langle U \rangle = \sum p_i U_i$, where p_i is the probability that a system has energy U_i. This is the ensemble that leads to the Boltzmann distribution of Section 16.1.

The distribution of systems between the various possible amounts of energy is described by

$$
\begin{aligned}
&\eta_0 \text{ of the systems have energy } U_0 \\
&\eta_1 \text{ of the systems have energy } U_1 \\
&\quad\vdots \\
&\eta_i \text{ of the systems have energy } U_i \\
&\quad\vdots
\end{aligned}
$$

Other types of ensembles are also useful. The ensemble that yields the Gibbs energy most directly is the isothermal–isobaric ensemble with partition function $\Delta(T, P, N)$. In a grand canonical ensemble, the systems in the ensemble have constant T, V, and μ. The systems in the ensemble have the same chemical potential because they are each placed in contact with a reservoir of particles through a

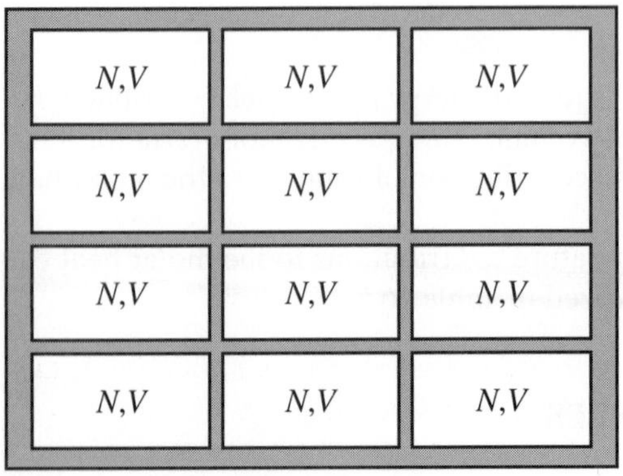

N,V	N,V	N,V
N,V	N,V	N,V
N,V	N,V	N,V
N,V	N,V	N,V

Figure 16.3 Canonical ensemble of systems each having the same N and V. The canonical ensemble contains η systems, isolated from one another after having been in contact with a heat reservoir at temperature T.

semipermeable membrane before being added to the grand canonical ensemble. The number of particles in a system fluctuates, but $\langle N \rangle$ can be calculated.

16.11 NONIDEAL GASES

In earlier sections of this chapter we have ignored intermolecular interactions and have therefore obtained thermodynamic properties of ideal gases. To calculate thermodynamic properties of nonideal gases it is necessary to take into account the potential energy V of intermolecular interactions. This cannot be done by using the molecular partition function. For nonideal gases it is necessary to use the canonical ensemble partition function, which is discussed in the preceding section. Using the canonical ensemble, it is possible to show that the second virial coefficient $B(T)$ (Section 1.5) is related to the potential energy of interaction $V(R)$ of two molecules (Section 11.9) by

$$B(T) = 2\pi N_A \int_0^\infty [1 - e^{-V(R)/kT}] R^2 \, dR \tag{16.94}$$

The second virial coefficient depends on pairwise interactions, and the third virial coefficient depends on interactions between triplets of molecules. For **an ideal** gas the potential energy $V(R)$ of interaction is zero and so $B(T) = 0$. We will apply equation 16.94 to the calculation of the second virial coefficient of a gas made up of hard spheres and a gas obeying the Lennard-Jones potential.

Hard-Sphere Potential

The simplest type of potential energy of interaction of two molecules is that due to short-range repulsive interactions. If molecules can be represented by hard spheres, their potential energy of interaction is given by

$$V(R) = \infty \qquad \text{for } 0 \le R \le d \tag{16.95}$$

$$V(R) = 0 \qquad \text{for } R > d \tag{16.96}$$

Therefore,

$$\begin{aligned} B(T) &= 2\pi N_A \int_0^d R^2 \, dR + 2\pi N_A \int_d^\infty (1-1) R^2 \, dR \\ &= 2\pi N_A \left(\frac{R^3}{3} \right)_0^d = \frac{2}{3}\pi N_A d^3 = 4 N_A b_0 \end{aligned} \tag{16.97}$$

Since the radius of the hard-sphere molecule is half the distance d of closest approach, the volume of a hard-sphere molecule b_0 is equal to $\frac{4}{3}\pi(d/2)^3 = \frac{1}{6}\pi d^3$. Thus, $B(T)$ is equal to four times the volume of a mole of hard-sphere molecules. Alternatively, we can point out that $B(T)$ for a gas of hard spheres is one-half the volume excluded to the center of one hard sphere of diameter d by another of diameter d.

Lennard-Jones Potential

Equation 16.94 for $B(T)$ with the Lennard-Jones potential (Section 11.9) must be evaluated numerically. In using this equation it is convenient to express the result in terms of reduced variables so that it may be readily applied to gases with different Lennard-Jones parameters. In making the numerical integration, distance is expressed as a multiple of the Lennard-Jones σ, and the temperature is expressed in terms of a reduced temperature $T^* = kT/\epsilon$, where ϵ is the depth of the potential well. The results of the numerical integration are shown in Fig. 16.4, which gives the reduced second virial coefficient $B^*(T^*) = B(T)/B_0$, where B_0 is the hard-sphere second virial coefficient ($B_0 = 2\pi N_A \sigma^3/3$). At low temperatures the second virial coefficient reflects the attractive interactions between molecules, and at high temperatures it reflects the repulsive interactions.

Example 16.19 ***Calculating the second virial coefficient from the Lennard-Jones potential***

Using the Lennard-Jones parameters for argon in Table 11.2, estimate the second virial coefficient B at 300 and 1000 K using Fig. 16.4. Compare these values with values calculated from Fig. 1.9. From Table 11.2, $\sigma = 0.341$ nm and $\epsilon/k = 120$ K.

At 300 K, $T^* = 300\text{ K}/120\text{ K} = 2.50$. From Fig. 16.4, $B^*(T^*) = -0.31$.

$$
\begin{aligned}
B &= \frac{2\pi}{3} N_A \sigma^3 B^* \\
&= \frac{2\pi}{3}(6.02 \times 10^{23}\text{ mol}^{-1})(0.341 \times 10^{-9}\text{ m})^3(-0.31) \\
&= (-1.55 \times 10^{-5}\text{ m}^3\text{ mol}^{-1})(10^2\text{ cm m}^{-1})^3 \\
&= -15.5\text{ cm}^3\text{ mol}^{-1}
\end{aligned}
$$

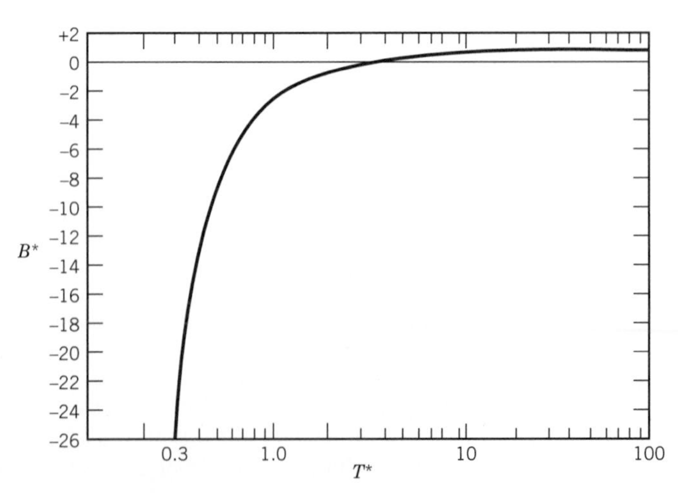

Figure 16.4 Reduced second virial coefficient B^* for the Lennard-Jones potential as a function of reduced temperature T^*. (From J. O. Hirschfelder, C. F. Curtiss, and R. B. Bird, *Molecular Theory of Gases and Liquids.* Hoboken, NJ: Wiley, 1954.)

Figure 1.9 indicates that $B = -20\ \text{cm}^3\ \text{mol}^{-1}$. At 1000 K, $T^* = 1000\ \text{K}/120\ \text{K} = 833$. From Fig. 16.4, $B^* = 0.4$.

$$B = \frac{2\pi}{3} N_A \sigma^3 B^*$$

$$= \frac{2\pi}{3}(6.02 \times 10^{23}\ \text{mol}^{-1})(0.341 \times 10^{-9}\ \text{m})^3(0.4)(10^2\ \text{cm m}^{-1})^3$$

$$= 20\ \text{cm}^3\ \text{mol}^{-1}$$

Figure 1.9 indicates $B = 20\ \text{cm}^3\ \text{mol}^{-1}$.

It is important to be able to calculate the equation of state for a nonideal gas from molecular parameters because, once this has been done, other thermodynamic properties may be calculated using the equation of state.

Figure 16.4 is an example of the law of corresponding states. To derive this particular form it was assumed that the intermolecular interactions are spherically symmetric. This is obviously not the case for some molecules, but the law can be extended.

The law of corresponding states provides a means for estimating Lennard-Jones parameters. Since the reduced temperature T_c^* of the critical point calculated from kT_c/ϵ for a number of gases is approximately 1.3, the interaction energy ϵ can be estimated from the critical temperature T_c using $\epsilon/k = T_c/1.3$. Since the reduced molecular volume at the critical point $V_c/N_A\sigma^3$ is approximately 2.7 for a number of gases, the distance parameter σ may be estimated from $\sigma = (V_c/2.7N_A)^{1/3}$.

Even for ideal gases, equation 16.18 for Q becomes seriously in error when the number of levels is not much greater that the number of molecules because of multiple occupancy of levels. In this case it is necessary to distinguish between bosons and fermions. **Bosons** have integer spin (0, 1, 2, ...) and are not restricted in their occupancy of states. **Fermions** have half-integer spin (1/2, 3/2, ...) and are subject to the Pauli principle (Section 10.8). Protons, neutrons, and electrons are fermions. ^{4}He atoms are bosons. Thus electrons in metals and ^{4}He atoms at low temperatures do not follow the Boltzmann distribution.

16.12 HEAT CAPACITIES OF SOLIDS

To simplify the calculation of the heat capacity of a monatomic solid, we can imagine that each atom oscillates about its equilibrium lattice point with a small amplitude. As mentioned in Section 16.9, according to classical theory each mole of atoms would contribute R for each of its three vibrational degrees of freedom, so that the molar heat capacity at constant volume would be $3R = 25\ \text{J K}^{-1}\ \text{mol}^{-1}$. This is observed at high enough temperatures for all atomic solids. However, classical theory was not able to explain the decrease of $\overline{C}_V$ to zero as absolute zero is approached.

To calculate thermodynamic properties at a particular temperature for a crystal made up of atoms, we can regard the crystal as one gigantic molecule with $3N - 6$ internal degrees of freedom. For an atomic crystal these degrees of freedom all correspond to lattice (center-of-mass) vibrations, but for a molecular crystal there are some internal degrees of freedom that correspond to rotations, internal vibrations, or torsions. In principle, we can imagine a normal mode analysis being carried out on any crystal.

In Einstein's first statistical mechanical calculation of the heat capacity of an idealized monatomic crystal in 1907, he assumed that all the vibrational frequencies of the solid have one frequency that we will represent by ν_E. Thus, according to Table 16.1, the molar heat capacity at constant volume is given by $3N$ terms of the type indicated there:

$$\overline{C_V} = \frac{3N_A k(\Theta_E/T)^2 \exp(\Theta_E/T)}{[\exp(\Theta_E/T) - 1]^2} \tag{16.98}$$

where $\Theta_E = h\nu_E/k$ is the **Einstein temperature.** This theoretical result was of great importance because it explained why the heat capacity decreased from the classical result ($3R$) at high temperatures to zero as $T \to 0$. It also helped explain why $\overline{C_V}$ rises more slowly with increasing temperature for diamond and graphite than for silver and copper. High values of ν_E correspond to crystals where the force constant (equation 9.101) is large and/or the reduced mass is small. However, the Einstein theory predicted too rapid a decrease in heat capacity in the neighborhood of absolute zero.

In 1912 Debye introduced the idea of a spectrum of vibrational frequencies for an atomic crystal and was thereby able to derive an expression for the heat capacity that more accurately represented the experimental data at very low temperatures. The theory of sound waves in a continuous medium shows that the probability density (the number of modes between ν and $\nu + d\nu$) $f(\nu)$ for vibrations is proportional to the square of the frequency:

$$f(\nu)\,d\nu = \frac{12\pi\nu^2 V}{\overline{v}^3}\,d\nu \tag{16.99}$$

where $\overline{v}$ is the average speed of the sound waves and V is the volume. The low-frequency vibrations in an atomic crystal have wavelengths that extend over hundreds or thousands of atoms. These vibrations are nearly independent of the atomic-scale structure involved and are characteristic of a continuous medium with given elastic constants. The form of the distribution function used by Debye is shown in Fig. 16.5. He used the probability density given in equation 16.99 up to a frequency ν_D, chosen so that the total number of frequencies would be $3N$:

$$\int_0^{\nu_D} f(\nu)\,d\nu = 3N \tag{16.100}$$

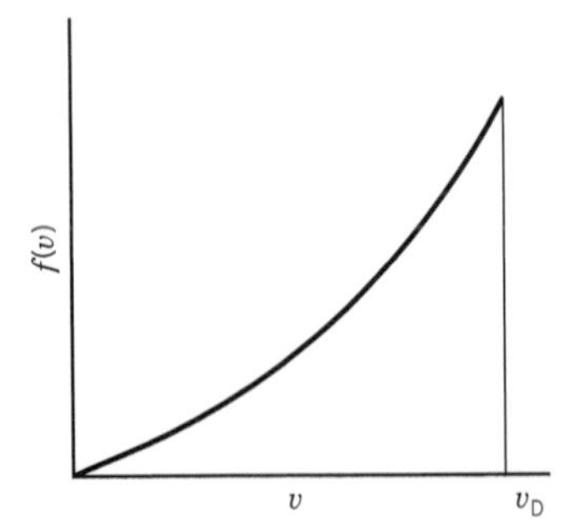

Figure 16.5 Frequency spectrum for an isotropic crystal. The highest frequency permitted is the Debye frequency ν_D.

The **Debye wavelength** $\lambda_D = \overline{v}/\nu_D$ corresponds to a minimum wavelength of the order of the interatomic distance in the crystal. According to the theory of elasticity of an isotropic medium (same properties in each direction), two different velocities of sound have to be taken into account: the longitudinal velocity v_l and the transverse velocity v_t with two polarization directions perpendicular to the direction of propagation. Since these two velocities of sound are different, equation 16.99 becomes

$$f(\nu)\,d\nu = 4\pi\nu^2 V\left(\frac{1}{v_l^3} + \frac{2}{v_t^3}\right) \tag{16.101}$$

When this expression is substituted in equation 16.100 and the integration is carried out, we can use the resulting expression to derive

$$f(\nu) = \frac{9N\nu^2}{\nu_D^3} \tag{16.102}$$

The expression for the energy of the Debye crystal can be derived using this expression, and the resulting expression for the heat capacity of a crystal is

$$C_V = 9Nk\left(\frac{T}{\Theta_D}\right)^3 \int_0^{\Theta_D/T} \frac{x^4\,e^x}{(e^x-1)^2}\,dx \tag{16.103}$$

where the **Debye temperature** $\Theta_D \equiv h\nu_D/kT$ and x is a dimensionless variable equal to $h\nu/kT$. As the temperature approaches absolute zero, this equation reduces to

$$C_V = \frac{12\pi^4}{5}Nk\left(\frac{T}{\Theta_D}\right)^3 \tag{16.104}$$

which represents the experimental results rather well for many atomic crystals. If a crystal is anisotropic—for example, if it has strong interactions in a plane like graphite does—the low-temperature behavior is more complicated.

The Debye equation is an example of a law of corresponding states (Section 16.11). It indicates that the heat capacities of all atomic solids should lie on the same curve when $\overline{C}_V$ is plotted versus T/Θ_D. Actually, there are various kinds of deviations from the Debye law because the actual frequency spectrum is more complicated than that shown in Fig. 16.5.

16.13 SPECIAL TOPIC: FLUCTUATIONS OF THERMODYNAMIC QUANTITIES

Given the Boltzmann form for the probability of a particular microstate, equation 16.4, we can infer that the macroscopic properties of a macroscopic system have fluctuations, i.e., that the macroscopic property has a range of values. For example, we can calculate the mean square deviation of the energy, as we did in Chapter 9 for quantum mechanical properties.

Using equation 16.4, the mean square energy at constant volume is given by

$$\langle E_i^2\rangle = \frac{1}{Q}\sum E_i^2\,e^{-E_i/kT} = \frac{1}{Q}\left(\frac{\partial^2 Q}{\partial(1/kT)^2}\right)_V \tag{16.105}$$

and the square of the mean energy is given by

$$\langle E_i\rangle^2 = \left(\frac{1}{Q}\sum E_i\,e^{-E_i/kT}\right)^2 = \left(\frac{1}{Q}\frac{\partial Q}{\partial(1/kT)}\right)_V^2 \tag{16.106}$$

From these two equations we find the mean square deviation in the energy, σ_u^2, to be

$$\sigma_u^2 = \frac{1}{Q}\left(\frac{\partial^2 Q}{\partial(1/kT)^2}\right)_V - \left(\frac{1}{Q}\frac{\partial Q}{\partial(1/kT)}\right)_V^2 \tag{16.107}$$

For the case of one mole of an ideal gas, $q = q_n/N_A!$, where q is given by equation 16.39 and N_A is Avogadro's constant. Substituting, we find

$$\sigma_u^2 = N_A(N_A - 1)\frac{1}{q^2}\left(\frac{\partial q}{\partial(1/kT)}\right)_V^2 + \frac{N_A}{q}\left(\frac{\partial^2 q}{\partial(1/kT)^2}\right)_V - \left(\frac{N_A}{q}\frac{\partial Q}{\partial(1/kT)}\right)_V^2$$

$$= \frac{N_A}{q}\left(\frac{\partial^2 q}{\partial(1/kT)^2}\right)_V - \left(\frac{N_A}{q^2}\frac{\partial q}{\partial(1/kT)}\right)_V^2 \tag{16.108}$$

Substituting for q from equation 16.39, we have

$$\sigma_u^2 = \frac{2}{3}N_A(kT)^2 = C_V kT^2 \tag{16.109}$$

The root-mean-square deviation of the energy σ_u^2 is then proportional to $N^{1/2}$. Since N is so large, this seems to be a huge deviation, but compared with the average energy itself, which is proportional to N, it is quite small: $\sigma_u/U \sim N^{-1/2}$. From this we see that fluctuations are normally quite small and can be neglected when N is very large. In certain circumstances, however, fluctuations can become observable even in macroscopic systems. For example, at the critical point, the density or concentration fluctuations can become large enough to be seen in light scattering (see Chapter 21).

Eight Key Ideas in Chapter 16

1. The Boltzmann distribution gives the distribution of molecules over energy levels, and it introduces the canonical partition function Q that is the link between the properties of individual molecules in an ideal gas and its macroscopic properties.
2. The Helmholtz energy A for an ideal gas is given by $-kT \ln Q$, which is a function of the natural variables $(T, V, N_1, N_2, \ldots)$ of A. Therefore, all the thermodynamic properties of an ideal gas can be calculated from Q.
3. The canonical ensemble partition function Q for an ideal gas is related to the single-molecule partition function q by $Q = q^N/N!$.
4. As a first approximation, the translational, vibrational, rotational, and electronic energies of gas molecules can be considered to be independent, and this leads to the conclusion that the one-molecule partition function for a gas molecule is equal to the product of the partition functions for these various types of energy.
5. The translational partition function can be expressed in terms of a thermal wavelength, and the condition for the applicability of Boltzmann statistics is that the thermal wavelength is small compared with the mean distance between molecules.
6. The vibrational, rotational, and electronic contributions to the single-molecule partition function can be calculated from molecular properties calculated from spectroscopic measurements. The thermodynamic properties for the electronic ground state are readily calculated, but calculations for excited electronic states are more complicated because excited states have different vibrational and rotational properties.

7. Equilibrium constants for gas reactions can be calculated from the molecular partition functions for the reactants and products. This has been especially useful for reactions at high temperatures, as in rocket propellant systems.
8. According to the classical principle of equipartition, each squared term in the classical expression for the energy of a gas molecule contributes $R/2$ to the molar heat capacity at constant volume.

REFERENCES

H. B. Callen, *Thermodynamics and an Introduction to Thermostatistics.* Hoboken, NJ: Wiley, 1985.

A, H. Carter, *Classical and Statistical Thermodynamics.* Upper Saddle River, NJ: Prentice-Hall, 2001.

D. Chandler, *Introduction to Modern Statistical Mechanics.* New York: Oxford University Press, 1987.

C. Garrod, *Statistical Mechanics and Thermodynamics.* New York: Oxford University Press, 1995.

C. E. Hecht, *Statistical Thermodynamics and Kinetic Theory.* New York: Freeman, 1990.

D. A. McQuarrie, *Statistical Mechanics.* Sausalito, CA: University Science Books, 2000.

R. K. Pathria, *Statistical Mechanics.* Oxford: Pergamon Press, 1972.

B. Widom, *Statistical Mechanics: A Concise Introduction for Chemists.* Cambridge, UK: Cambridge University Press, 2002.

R. E. Wilde and S. Singh, *Statistical Mechanics—Fundamentals and Modern Applications.* Hoboken, NJ: Wiley, 1998.

PROBLEMS

(M) Problems marked with an icon may be more conveniently solved on a personal computer with a mathematical program.

16.1 Using the Boltzmann distribution, calculate the ratio of populations at 25 °C of energy levels separated by (*a*) 1000 cm^{-1} and (*b*) 10 kJ mol^{-1}.

16.2 Calculate the ratio of populations at 25 °C of energy levels separated by (*a*) 1 eV and (*b*) 10 eV. (*c*) Calculate the ratios at 1000 °C.

16.3 Starting with the definition of the molecular partition function (equation 16.32) and $U = \sum_i N_i \epsilon_i$, derive equation 16.25 for U.

16.4 Show that the energy E of a system of N independent protons in a magnetic field B is $-N\hbar\gamma B/2$ in the limit as $T \to 0$ and is equal to 0 in the limit as $T \to \infty$. (See equation 15.18.) How do you interpret these results?

16.5 Use the relation between pressure P and the molecular partition function q_t to derive the equation of state of an ideal gas starting with equation 16.39 for q_t.

16.6 Derive the expression for the translational partition function for a molecule that is moving along a line, rather than in three-dimensional space. This is of interest in connection with transition-state theory (Section 19.4).

16.7 Show that the same expression is obtained for the chemical potential from A (equation 16.21) as from G (equation 16.27) for an ideal gas.

16.8 What is the ratio of the thermal wavelength to the length of one side of the container for (*a*) a hydrogen atom in a cube 1 nm on a side at 2 K and (*b*) an oxygen molecule in 0.25 m^3 at 300 K?

16.9 Calculate the translational partition function for $H_2(g)$ at 1000 K and 1 bar.

16.10 What are the translational partition functions of hydrogen atoms and hydrogen molecules at 500 K in a volume of 4.157×10^{-2} m^3? (This is the molar volume of an ideal gas at this temperature and a pressure of 1 bar.)

16.11 Calculate the molar entropy of gaseous H atoms at 1000 K and (*a*) 1 bar and (*b*) 1000 bar.

16.12 Calculate the molar entropy of neon at 25 °C and 1 bar.

16.13 Write out the summation

$$S = \sum_{i=1}^{3} \sum_{j=0}^{1} x^i y^j$$

and show that the summation can also be written as

$$S = \sum_{i=1}^{3} x^i \sum_{j=0}^{1} y^j$$

16.14 Write out the summation

$$S = \sum_{i=1}^{2} \sum_{j=1}^{2} x^{i+j}$$

and show that the summation can also be written as

$$S = \sum_{i=1}^{2} x^i \sum_{j=1}^{2} x^j$$

16.15 Derive the expression for the vibrational contribution to the internal energy

$$U_{\mathrm{v}} = \frac{RTx}{e^x - 1}$$

where $x = h\nu/kT$. What is the limit of the vibrational contribution to the internal energy at high temperature?

16.16 By use of series expansions show that the vibrational contribution to $\overline{C}_V^\circ$ for a diatomic molecule approaches R as $T \to \infty$.

16.17 According to Fig. 13.14, the normal-mode vibrational frequencies of H_2O are 3657, 1595, and 3756 cm^{-1}. What is the value of the vibrational partition function of H_2O at 2000 K?

16.18 What are the rotational contributions to $\overline{C}_P^\circ$, $\overline{S}^\circ$, $\overline{H}^\circ$, and $\overline{G}^\circ$ for NH_3(g) at 25 °C?

16.19 What fraction of HCl molecules is in the state $v = 2$, $J = 7$ at 500 °C? The characteristic vibrational and rotational temperatures are given in Table 16.2.

16.20 Calculate the translational partition functions for H, H_2, and H_3 at 1000 K and 1 bar. What are the rotational partition functions of H_2 and H_3 (linear) at 1000 K? The internuclear distances in H_3 are 94 pm.

16.21 What are the symmetry numbers of the following organic molecules, assuming free rotation of methyl groups: (*a*) ethane, (*b*) propane, (*c*) 2-methylpropane, and (*d*) 2,2-dimethylpropane?

16.22 Calculate the symmetry numbers of methane (CH_4) and ethylene (C_2H_4) by adding up the number of distinct property rotational operations in Table 12.3 plus the identity operation.

16.23 (*a*) Calculate the symmetry number for the ethane structure shown in Table 12.3. This is the symmetry number of the rigid structure. (*b*) Since there is essentially free rotation about the C–C bond, there are three equivalent positions of the second CH_3 group with respect to the first. What is the symmetry number of a freely rotating ethane molecule?

16.24 What are the electronic contributions to $\overline{S}^\circ$ and $\overline{G}^\circ$ for I(g) at 298.15 K and 3000 K?

16.25 What is the electronic partition function for C(g) at 1000 K, according to the data of Table 16.2? What are the relative populations of these levels?

16.26 Calculate the electronic contribution to the standard molar entropy and standard molar Gibbs energy of C(g) at 1000 K, to the degree of completeness we have used here.

16.27 Calculate the molar entropies of H(g) and N(g) at 25 °C and 1 bar. The degeneracies of the ground states are 2 and 4, respectively. Compare these values with those in Table C.2.

16.28 A molecule has a ground state and two excited electronic energy levels, all of which are nondegenerate: $\epsilon_0 = 0$, $\epsilon_1 = 1 \times 10^{-20}$ J, and $\epsilon_2 = 3 \times 10^{-20}$ J. What fraction of each level is occupied at 298 and 1000 K?

16.29 The ground state of Cl(g) is fourfold degenerate. The first excited state is 875.4 cm^{-1} higher in energy and is twofold degenerate. What is the value of the electronic partition function at 25 °C? At 1000 K?

16.30 What is the partition function for oxygen atoms at 1000 K according to the data in Table 16.2? What are the relative populations of these levels at equilibrium?

16.31 Derive the expression for the electronic internal energy of an atom or molecule. What is the electronic energy per mole for a chlorine atom at 298 K and 1000 K? (See Problem 16.29.)

16.32 Calculate the fraction of hydrogen atoms that at equilibrium at 1000 °C would have $n = 2$.

16.33 A quantum mechanical system has two energy levels, ϵ_1 and ϵ_2. Derive equations for the probability p_1 that the system will be in state 1 and the probability p_2 that the system will be in state 2. What are these probabilities at $T/\mathrm{K} = 0$ and ∞? What are the values of p_1 and p_2 at $\Delta\epsilon = kT$?

16.34 Calculate $\overline{C}_P^\circ$ for NH_3(g) at 1000 K. The characteristic vibrational temperatures for the six normal modes are given in Table 16.2.

16.35 Calculate the molar entropy of nitrogen gas at 25 °C and 1 bar pressure. The equilibrium separation of atoms is 109.5 pm, and the vibrational wave number is 2330.7 cm^{-1}.

16.36 Calculate $\overline{C}_P^\circ$ for CO_2 at 1000 K. Compare the actual contributions to $\overline{C}_P^\circ$ from the various normal modes with the classical expectations.

16.37 Calculate the equilibrium constant for the isotope exchange reaction $D + H_2 = H + DH$ at 25 °C. Assume that the equilibrium distance and force constants of H_2 and DH are the same.

16.38 Calculate the equilibrium constant at 25 °C for the reaction $H_2 + D_2 = 2HD$. It may be assumed that the equilibrium distance and force constant k are the same for all three molecular species, so that the additional vibrational frequencies required may be calculated from $2\pi\nu = (k/\mu)^{1/2}$. Because of the zero-point vibration, $\Delta\epsilon_0$ for this reaction is given by

$$\Delta\epsilon_0 = \tfrac{1}{2}N_A h(2\nu_{HD} - \nu_{H_2} - \nu_{D_2})$$

16.39 Express the equilibrium constant for the reaction $H_2 + I_2 = 2HI$ in terms of molecular properties.

16.40 The classical limits of heat capacities of molecules of ideal gases are readily calculated using the principle of equipartition. Calculate $\overline{C}_V^\circ/R$ and $\overline{C}_P^\circ/R$ for Ar, O_2, CO_2, and CH_4 and compare $\overline{C}_P^\circ/R$ with the values in Table C.3 at 3000 K.

16.41 Considering H_2O to be a rigid nonlinear molecule, what value of $\overline{C}_P^\circ$ for the gas would be expected classically? If vibration is taken into account, what value is expected? Compare these values of $\overline{C}_P^\circ$ with the actual values at 298 and 3000 K in Table C.3.

16.42 Show how $P = kT(\partial \ln Q/\partial V)_T$ leads to $PV = nRT$.

16.43 Show that the statistical mechanical expressions for $\overline{H}$, $\overline{S}$, and $\overline{G}$ of a monatomic gas without electronic excitation are consistent with $\overline{G} = \overline{H} - T\overline{S}$.

16.44 The average energy and average square of the energy of a macroscopic system are given by

$$\langle E\rangle = \frac{\sum E_j e^{-\beta E_j}}{Q} \quad \text{and} \quad \langle E^2\rangle = \frac{\sum E_j^2 e^{-\beta E_j}}{Q}$$

Show that the square of the standard deviation of the energy is given by

$$\sigma_E^2 = \langle E^2\rangle - \langle E\rangle^2 = k_B T^2 C_V$$

Calculate $\sigma_E/\langle E\rangle$ for a monatomic ideal gas for which $\langle E\rangle = \frac{3}{2}Nk_BT$ and $C_V = \frac{3}{2}Nk_B$. What do you think of the chances of observing a fluctuation in the energy of a macroscopic system?

16.45 The canonical ensemble partition function Q for a mixture of two monatomic ideal gases is given by

$$Q = \frac{q_1^{N_1}\, q_2^{N_2}}{N_1!\, N_2!}$$

Show that

$$U = \tfrac{3}{2}(n_1 + n_2)RT$$

and

$$PV = (n_1 + n_2)RT$$

16.46 Since the Helmholtz energy A of an ideal gas is given by $-NkT\ln(qe/N)$, all the other thermodynamic properties can be calculated if $q = q_t q_v q_r q_e$ can be expressed as a function of T, V, and N. Derive these relations, which are given in equations 16.22 to 16.27.

16.47 Consider a molecule that has two energy levels separated by ϵ, where the ground state has a degeneracy of 2 and the excited state has a degeneracy of 3. (*a*) What is the expression for the partition function at temperature T? (*b*) What are the fractional populations of the two states at temperature T? (*c*) What is the internal energy per particle at temperature T?

16.48 The energies of the $n = 2$ and $n = 1$ orbitals of the hydrogen atom are 27 420 and 109 678 cm^{-1}, respectively. What are the relative populations in these levels at (*a*) 25 °C and (*b*) 2000 °C?

16.49 A helium atom is in a volume of 10^{-9} m^3. What are the values of its translational partition function at 298, 1000, and 5000 K?

16.50 The thermal wavelength defined in connection with equation 16.40 is a little different from the de Broglie wavelength. (*a*) What is the de Broglie wavelength for hydrogen atoms at 3000 K, using the root-mean-square average momentum as p? (*b*) How does it compare with the thermal wavelength calculated in Example 16.3 (*c*) How does this thermal wavelength compare with the mean distance between hydrogen atoms in a gas of hydrogen atoms at 3000 K and 1 bar?

16.51 Calculate $\overline{S}^\circ$ and $\overline{C}_P^\circ$ for argon ($M = 39.948$ g mol^{-1}) at 25 °C and 1 bar.

16.52 Calculate the molar entropy of helium in the ideal gas state at 25 °C and 1 bar pressure.

16.53 Compare the translational partition function of I(g) at 1000 K and 1 bar with that for H(g) calculated in Problem 16.16.

16.54 (*a*) Calculate the thermal wavelength for an O_2 molecule at 1 K and 298 K. For Boltzmann statistics to be applicable, the thermal wavelength must be small compared with the mean distance between molecules. (*b*) Calculate the mean distance between gas molecules at 1 bar at these temperatures assuming each molecule is in the center of a cube. (*c*) Are Boltzmann statistics applicable at both temperatures?

16.55 What are the most probable populations of the first several vibrational levels of O_2(g) at 1000 K? The characteristic vibrational temperature is 2274 K.

16.56 What are the characteristic vibrational temperatures for oscillators with frequencies of 10^9 s^{-1} (radio waves), 10^{12} s^{-1} (far infrared), 10^{15} s^{-1} (near ultraviolet), and 10^{18} s^{-1} (X-rays)?

16.57 What are the rotational contributions to $\overline{C}_P^\circ$ and $\overline{S}^\circ$ of CH_4 at 298.15 K?

16.58 What are the symmetry numbers of the following organic compounds, assuming free rotation of methyl groups: (*a*) ethylene, (*b*) 1-methylethylene, (*c*) 1,1-dimethylethylene, (*d*) 1,1,2-trimethylethylene, and (*e*) 1,1,2,2-tetramethylethylene?

16.59 (*a*) The water molecule belongs to the C_{2v} point group, which includes the symmetry elements C_2 and E. What is the symmetry number of a water molecule?

16.60 Calculate the ratio of the number of HBr molecules in state $v = 2$, $J = 5$ to the number in state $v = 1$, $J = 2$ at 1000 K. Assume that all of the molecules are in their electronic ground states. ($\Theta_v = 3700$ K, $\Theta_r = 12.1$ K.)

16.61 Show that at high temperatures, $q_v = kT/hc\tilde{\nu}$.

16.62 Calculate the temperature at which 10% of the molecules in a system will be in the first excited electronic state if this state is 400 kJ mol^{-1} above the ground state.

16.63 (*a*) In Problem 16.28, what is the electronic energy of the molecule at 298 and 1000 K? (*b*) Since the internal energy is given in terms of the molecular partition function by

$$U = NkT^2\left(\frac{\partial \ln q}{\partial T}\right)_V$$

calculate the electronic energy of the molecule at 298 and 1000 K using this equation.

16.64 A molecule exists in singlet and triplet forms with the singlet having the higher energy by 4.11×10^{-21} J per molecule. The singlet level has a degeneracy of 1, and the triplet level has a degeneracy of 3. (*a*) Ignoring higher levels, what is the electronic partition function? (*b*) What is the ratio of the concentration of triplets to singlet molecules at 298 K?

16.65 What fraction of hydrogen atoms have $n = 2$ at room temperature according to the Boltzmann distribution? At 3000 K?

16.66 The Sackur–Tetrode equation

$$\overline{S}(V, T) = \overline{S}_0' + R \ln V + \overline{C}_V \ln T$$

seems to predict that $\overline{S} \rightarrow -\infty$ when $T \rightarrow 0$. Explain why this is not in conflict with the third law.

16.67 For actual calculation of the molar entropy of a monatomic gas, the Sackur–Tetrode equation may be written in the form

$$\overline{S} = \overline{S}' + \frac{3}{2}R \ln A_r - R \ln \frac{P}{P^\circ} + \frac{5}{2}R \ln \frac{T}{\text{K}}$$

Show that for $P^\circ = 1$ bar, $\overline{S}'/R = -1.151\,693$.

16.68 Calculate $\overline{C}_P^\circ$ for hydrogen gas at 298.15 and 2000 K. This calculation is discussed in some detail by C. Marzzacco and M. Waldman, *J. Chem. Educ.* **50**:444 (1973).

16.69 Calculate the molar entropy for chlorine gas at 25 °C and 1 bar pressure.

16.70 Calculate the statistical mechanical values of $\overline{C}_P^\circ, \overline{S}^\circ, \overline{H}^\circ$, and $\overline{G}^\circ$ for H(g) at 3000 K.

16.71 Calculate the statistical mechanical values of $\overline{C}_P^\circ, \overline{H}^\circ, \overline{S}^\circ$, and $\overline{G}^\circ$ for H_2(g) at 3000 K.

16.72 What are the values of $\Delta_r C_P^\circ$, $\Delta_r H^\circ$, $\Delta_r S^\circ$, and $\Delta_r G^\circ$ for $H_2(g) = 2H(g)$ at 3000 K calculated in the preceding two problems? What is the value of K? What is the degree of dissociation at 1 bar?

16.73 Calculate the values of D_0 in Table 16.2 from data in Table C.3 for H_2(g), O_2(g), Cl_2(g), HCl(g), and CO(g).

16.74 Derive the statistical mechanical expression for the equilibrium constant for the reaction

$$A_2(g) + B_2(g) = 2AB(g)$$

where A and B are isotopes. The contribution of the vibrational partition function may be ignored because it is so close to unity.

16.75 Tabulate the translational, vibrational, and rotational contributions to $\overline{C}_V^\circ/R$ for H, H_2, H_2O, and NH_3 in the ideal gas state that are expected classically. Calculate the classical limits for $\overline{C}_P^\circ$ and compare them with the values in Table C.3 at 3000 K.

16.76 Starting with $\langle U\rangle = kT^2(\partial \ln Q/\partial T)_{N,V}$, show that

$$\langle\epsilon\rangle = \frac{\sum_i \epsilon_i \,\mathrm{e}^{-\epsilon_i/kT}}{\sum_i \mathrm{e}^{-\epsilon_i/kT}}$$

Computer Problems

16.A Consider a molecule that has two nondegenerate energy levels separated by ϵ. (*a*) Plot the partition function versus kT/ϵ. (*b*) Plot the fractional populations of the two states versus kT/ϵ. (*c*) Plot the ratio of the internal energy U to the energy if all the molecules were in the excited state versus kT/ϵ.

16.B Calculate the fractional populations of $^{16}O_2$ gas molecules in vibrational levels up to $v = 6$ at (*a*) 1000 K and (*b*) 2000 K.

16.C Calculate the molecular partition function for translational motion of a hydrogen atom at 300 K in a volume of 0.2494 m^3. Calculate the thermal wavelength.

16.D Plot the molar heat capacity at constant pressure of $^{14}N_2$(g) versus T from 298.15 to 2000 K. The rigid rotor–harmonic oscillator approximation can be used. Plot values from Table C.3 on the same graph. What do you think is responsible for the differences?

16.E Calculate the relative fractions N_J/N_0 of $^{12}C^{16}O$ molecules in rotational levels up to $J = 20$ at 300 and 500 K. The rotational constant is 2.7771 K.

16.F Calculate the relative fractions N_J/N_0 of $^1H^{35}Cl$ molecules in rotational levels up to $J = 20$ at 300 and 1000 K. The rotational constant is 15.2344 K.

16.G Calculate the equilibrium constant for the reaction $H_2(g) = 2H(g)$ at 500, 1000, and 2000 K using the harmonic oscillator–rigid rotor approximation. Compare these results with values calculated using Table C.3.

16.H Calculate the equilibrium constant for the reaction $I_2(g) = 2I(g)$ at 500, 1000, and 2000 K using the harmonic oscillator–rigid rotor approximation. Compare these results with values calculated using Table C.3 in Computer Problem 5.D.

16.I Calculate the equilibrium constant for the reaction $HI(g) = H(g) + I(g)$ at 500, 1000, and 2000 K using the harmonic oscillator–rigid rotor approximation. Compare these results with values calculated using Table C.3.

16.J Calculate the equilibrium constant for the reaction $2HI(g) = H_2(g) + I_2(g)$ at 500, 1000, and 2000 K using the harmonic oscillator–rigid rotor approximation. Compare these results with values calculated using Table C.3.

16.K (*a*) Make a three-dimensional plot of the molar entropy of O(g) from 298 to 1000 K and 1 to 100 bar using the Sackur–Tetrode equation. (*b*) Make a two-dimensional plot of the molar entropy at 1 bar for 298 to 1000 K.

PART

THREE

Kinetics

Kinetic theory introduces the calculation of the rates of certain processes by use of a simple model of atoms and molecules in the gas phase. The probabilities of molecular speeds and the values of average speeds depend on the molecular mass and temperature for noninteracting gas molecules. The frequency of collisions and the transport properties (viscosity, diffusion, and heat conduction) for gases of rigid spherical molecules can be calculated. However, the behavior of real gases is more complicated, again because of intermolecular interactions.

The prediction of rates of chemical reactions is much more difficult, so we will first consider the experimental aspects of gas reactions and the use of this information to obtain mechanisms of reactions. Then we turn to chemical dynamics to learn about the role of the transition state and to photochemistry to learn about the various processes that can occur after a molecule has absorbed a photon.

The last chapter in this part of the book deals with the kinetics of reactions in the liquid state. The study of viscosity, diffusion, and electrical transport of ions provides information that is useful in understanding the rates of reactions in liquids. Relaxation methods are useful for studying very fast reactions in the liquid phase, and the theory of diffusion-controlled reactions yields an upper limit for the rate constants of bimolecular reactions. This will help us to better understand the acid–base catalysis, enzyme catalysis, and the rates of electrochemical reactions.

17

Kinetic Theory of Gases

The kinetic theory of gases is concerned with the properties of idealized models of molecules. We will calculate the distribution of molecular speeds, the pressure of an ideal gas, and the rate of collision with a surface assuming point molecules. Then we will calculate the rates of molecular collisions and the mean free path assuming that the molecules are tiny hard spheres. These calculations will help us interpret the rates of chemical reactions. This simple model can also be used to calculate rates of mixing of gases by diffusion, the rate of conduction of heat, and viscosity.

17.1 PROBABILITY DENSITY FOR MOLECULAR SPEEDS OF GAS MOLECULES

In beginning our consideration of elementary kinetic theory, we will assume that molecules are represented by points in space that move in straight lines. In other words, the molecules are assumed not to have volume or cross-sectional area, and they are assumed to move in straight lines because they do not interact with each other except in collisions.

Earlier we used the position vector $\boldsymbol{r}$ to specify the position of a particle in three-dimensional space in terms of the unit vectors $\boldsymbol{i}$, $\boldsymbol{j}$, and $\boldsymbol{k}$ in the direction of the x, y, and z axes:

$$\boldsymbol{r} = \boldsymbol{i}x + \boldsymbol{j}y + \boldsymbol{k}z \tag{17.1}$$

Taking the derivative of $\boldsymbol{r}$ with respect to time yields the **velocity vector** for that particle:

$$\boldsymbol{v} = \boldsymbol{i}v_x + \boldsymbol{j}v_y + \boldsymbol{k}v_z \tag{17.2}$$

where $v_x = \mathrm{d}x/\mathrm{d}t$. Since the velocity of a gas molecule is represented by a vector, the velocity has a magnitude and a direction. The velocity vector for a molecule can be plotted in velocity space as shown in Fig. 17.1. The component velocities, v_x, v_y, and v_z, of a molecule have signs, but we are often more interested in the magnitude v of the velocity vector $\boldsymbol{v}$ than its direction. The magnitude v of the velocity vector is referred to as the **speed** of the particle. As shown by Fig. 17.1, the speed v can be calculated from the components of the velocity vector by using the Pythagorean theorem:

$$v = |\boldsymbol{v}| = (v_x^2 + v_y^2 + v_z^2)^{1/2} \tag{17.3}$$

This quantity is also referred to as the absolute value of the velocity vector.

At a given instant, the velocity vectors for molecules in a gas can be represented by points at the ends of the vectors, as shown in Fig. 17.2. To describe the distribution of velocities in three dimensions, we represent the probability of finding a molecule with a velocity in the range v_x to $v_x + \mathrm{d}v_x$, v_y to $v_y + \mathrm{d}v_y$, and v_z to $v_z + \mathrm{d}v_z$ by $f(v_x, v_y, v_z)\,\mathrm{d}v_x\,\mathrm{d}v_y\,\mathrm{d}v_z$, where $\mathrm{d}v_x\,\mathrm{d}v_y\,\mathrm{d}v_z$ is the infinitesimal volume in velocity space. This element of volume is illustrated in Fig. 17.3. Thus the **probability density** $f(v_x, v_y, v_z)$ is the probability per unit volume at a point in velocity space. The probability for all of velocity space is unity:

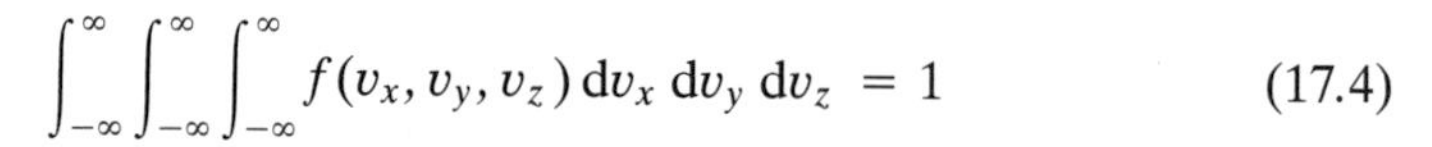

$$\int_{-\infty}^{\infty}\int_{-\infty}^{\infty}\int_{-\infty}^{\infty} f(v_x, v_y, v_z)\,\mathrm{d}v_x\,\mathrm{d}v_y\,\mathrm{d}v_z = 1 \tag{17.4}$$

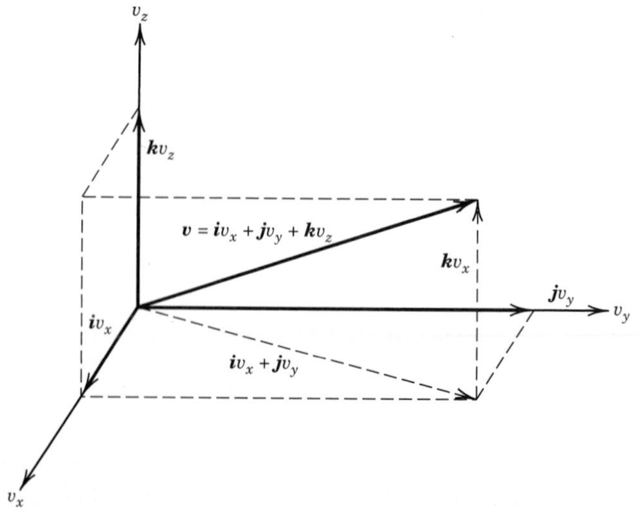

Figure 17.1 Velocity vector of a particle in velocity space. The length v of the vector $\boldsymbol{v}$ that represents the speed and direction of a particle can be calculated from the components v_x, v_y, and v_z by use of the Pythagorean theorem, given in equation 17.3.

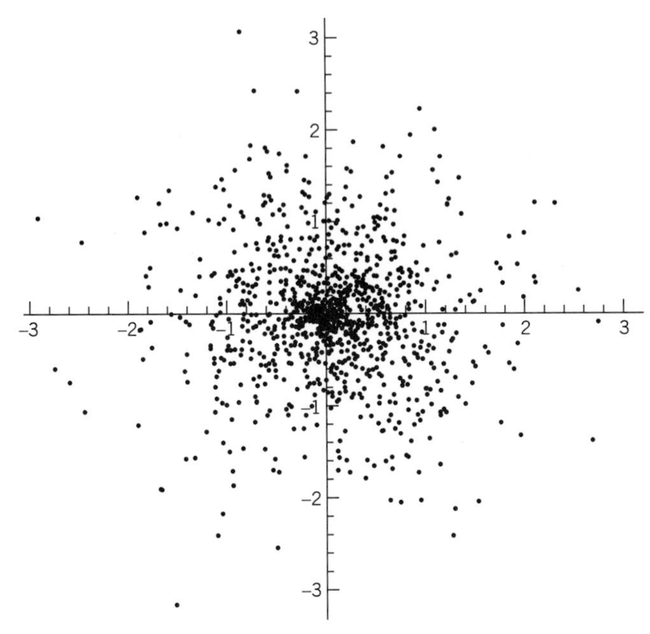

Figure 17.2 Points representing the heads of velocity vectors for the molecules in a plane. Note that very few molecules have very high speeds, that is, large absolute values of the velocity vector, and that the distribution is isotropic, that is, the same in each direction. (See Computer Problem 17.G.)

Alternatively, this equation can be written

$$\int_{-\infty}^{\infty} f(\boldsymbol{v})\,\mathrm{d}\boldsymbol{v} = 1 \tag{17.5}$$

where $\boldsymbol{v}$ is the velocity vector.

The probability density $f(v_x, v_y, v_z) = f(\boldsymbol{v})$ is called a joint probability density because three things have to occur: The component velocities must be in the range v_x to $v_x + \mathrm{d}v_x$, v_y to $v_y + \mathrm{d}v_y$, and v_z to $v_z + \mathrm{d}v_z$. In the case of a gas, the three velocity components are independent. Therefore, the probability density for the velocity vector is the product of the probability densities in the three directions:

$$f(v_x, v_y, v_z) = f(v_x)f(v_y)f(v_z) = f(\boldsymbol{v}) \tag{17.6}$$

The probability density in the x direction is represented by $f(v_x)$, and $f(v_x)\,\mathrm{d}v_x$ is the probability that a molecule has a velocity in the x direction between v_x and $v_x + \mathrm{d}v_x$.

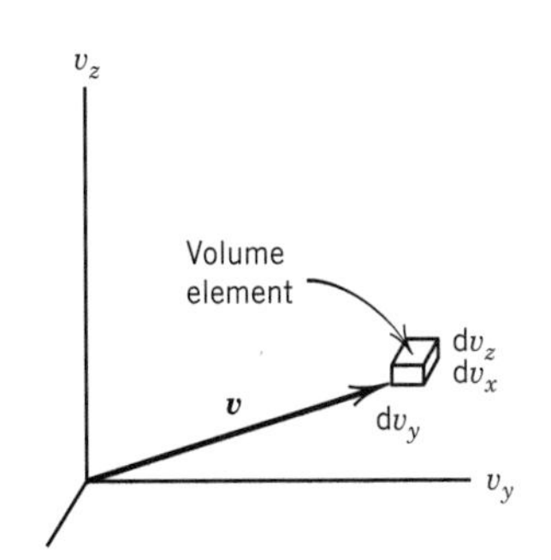

Figure 17.3 The volume element in velocity space is infinitesimal in size, and it has the density of points at the end of a specific velocity vector $\boldsymbol{v}$.

17.2 VELOCITY DISTRIBUTION IN ONE DIRECTION

The energy of a molecule of mass m moving in the x direction with velocity v_x is $mv_x^2/2$, and so the Boltzmann distribution (equation 16.2) indicates that the probability density $f(v_x)$ that a molecule has velocity v_x is given by

$$f(v_x) = \text{const}\,\mathrm{e}^{-mv_x^2/2kT} \tag{17.7}$$

Table 17.1 Definite Integrals Occurring in the Kinetic Theory of Gases

Integral	n = 0	1	2	3	4	5
$\int_0^\infty x^n \exp(-ax^2)\, dx$	$\frac{1}{2}\left(\frac{\pi}{a}\right)^{1/2}$	$\frac{1}{2a}$	$\frac{1}{4}\left(\frac{\pi}{a^3}\right)^{1/2}$	$\frac{1}{2a^2}$	$\frac{3}{8}\left(\frac{\pi}{a^5}\right)^{1/2}$	$\frac{1}{a^3}$
$\int_{-\infty}^{+\infty} x^n \exp(-ax^2)\, dx$	$\left(\frac{\pi}{a}\right)^{1/2}$	0	$\frac{1}{2}\left(\frac{\pi}{a^3}\right)^{1/2}$	0	$\frac{3}{4}\left(\frac{\pi}{a^5}\right)^{1/2}$	0

The value of the integration constant can be determined by integrating from $-\infty$ to ∞:

$$\int_{-\infty}^{\infty} f(v_x)\, dv_x = 1 = \text{const} \int_{-\infty}^{\infty} e^{-mv_x^2/2kT}\, dv_x \tag{17.8}$$

Using an integral from Table 17.1 shows that the constant in equation 17.8 is equal to $(m/2\pi kT)^{1/2}$, so that the **Maxwell–Boltzmann distribution** of molecular velocities is given by

$$f(v_x) = \left(\frac{m}{2\pi kT}\right)^{1/2} e^{-mv_x^2/2kT} \tag{17.9}$$

This probability density has the form of the Gaussian error function (Appendix D.7), as shown in Fig. 17.4. The most probable velocity in the x direction is zero because of the form of equation 17.9. This can be shown by integrating the velocity in the x direction times its probability over all values of v_x:

$$\langle v_x \rangle = \int_{-\infty}^{\infty} v_x f(v_x)\, dv_x = 0 \tag{17.10}$$

This integral is readily evaluated by noting that $f(v_x)$ is symmetrical and v_x is an odd function. When the temperature is raised or the mass of the particle is decreased, the distribution becomes broader, but the area under the curve remains constant because $f(v_x)$ is normalized.

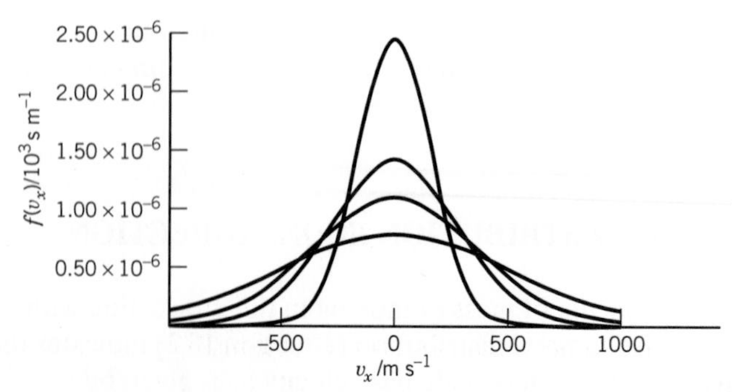

Figure 17.4 Probability density for the velocity of oxygen molecules in an arbitrarily chosen direction at 100, 300, 500, and 1000 K. (See Computer Problem 17.C.)

Equation 17.8 can be used to obtain the following expression for the **average kinetic energy in the x direction** (see Example 17.2):

$$\epsilon_x = \tfrac{1}{2}m\langle v_x^2\rangle = \tfrac{1}{2}kT \tag{17.11}$$

Of course, similar expressions apply to the y and z directions. This is an example of the principle of equipartition of energy (Section 16.9).

Example 17.1 ***Probability density in a specific direction***

Calculate the probability density for v_x of O_2 molecules at 300 K at 0, 300, and 600 m s^{-1}.

At 300 m s^{-1},

$$f(v_x) = \left(\frac{M}{2\pi RT}\right)^{1/2} \exp\left(-\frac{Mv_x^2}{2RT}\right)$$

$$= \left[\frac{0.032 \text{ kg mol}^{-1}}{2\pi(8.3145 \text{ J K}^{-1} \text{ mol}^{-1})(300 \text{ K})}\right]^{1/2} \exp\left[-\frac{(0.032 \text{ kg mol}^{-1})(300 \text{ m s}^{-1})^2}{2(8.3145 \text{ J K}^{-1} \text{ mol}^{-1})(300 \text{ K})}\right]$$

$$= 8.022 \times 10^{-4} \text{ s m}^{-1}$$

The probability densities at 0 and 600 m s^{-1} are 1.429×10^{-3} s m^{-1} and 1.419×10^{-4} s m^{-1}, in agreement with Fig. 17.4.

Example 17.2 ***Average square of the velocity in a specific direction***

Using the distribution function for velocities in the x direction, show that

$$\langle v_x^2\rangle = \frac{kT}{m}$$

The average is the integral of v_x^2 multiplied by the distribution function:

$$\langle v_x^2\rangle = \int_{-\infty}^{\infty} v_x^2 f(v_x)\,dv_x$$

$$= \left(\frac{m}{2\pi kT}\right)^{1/2} \int_{-\infty}^{\infty} e^{-mv_x^2/2kT} v_x^2\,dv_x$$

Using the value of the definite integral given in Table 17.1,

$$\langle v_x^2\rangle = \left(\frac{m}{2\pi kT}\right)^{1/2} \frac{\pi^{1/2}}{2(m/2kT)^{3/2}}$$

$$= \frac{kT}{m}$$

17.3 MAXWELL DISTRIBUTION OF SPEEDS

Since no direction in space is favored, the same result is obtained for $f(v_y)$ and $f(v_z)$. Thus, the probability density in three dimensions is obtained by substituting

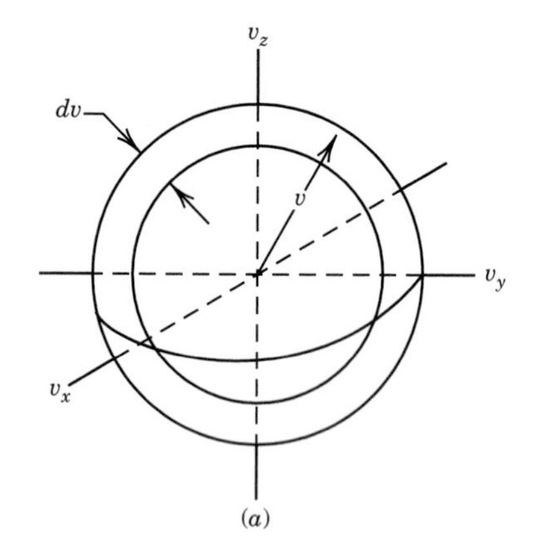

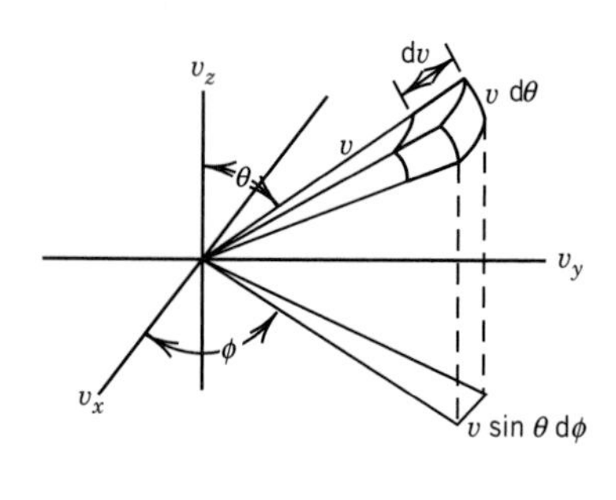

Figure 17.5 (*a*) Calculation of the probability of velocity components in the range v to $v + dv$. (*b*) Volume element in spherical coordinates.

equation 17.9 and the corresponding equations for $f(v_y)$ and $f(v_z)$ in equation 17.6:

$$f(v_x, v_y, v_z) = \left(\frac{m}{2\pi kT}\right)^{3/2} \exp\left[-\frac{m}{2kT}(v_x^2 + v_y^2 + v_z^2)\right] \tag{17.12}$$

However, usually we are more interested in the distribution of speeds than in the distribution of component velocities. The **speed** v of a molecule is related to its component velocities by equation 17.3. The speed is represented by the distance in velocity space of a point from the origin in Fig. 17.1. Therefore, the probability $F(v)\,dv$ that a molecule has a speed between v and $v + dv$ is given by the probable number of points in a spherical shell of thickness dv, as shown in Fig. 17.5*a*. The required integration is most conveniently carried out by converting to spherical coordinates, which are shown in Fig. 17.5*b*, using

$$v_x = v \sin\theta \cos\phi \tag{17.13}$$

$$v_y = v \sin\theta \sin\phi \tag{17.14}$$

$$v_z = v \cos\theta \tag{17.15}$$

The differential volume element $dv_x\, dv_y\, dv_z$ can be written in spherical coordinates as

$$dv_x\, dv_y\, dv_z = v^2\, dv \sin\theta\, d\theta\, d\phi \tag{17.16}$$

The probability $F(v)\,dv$ can now be found by integration of $f(v_x, v_y, v_z)$ $dv_x\, dv_y\, dv_z$ over the angles θ and ϕ:

$$F(v)\,dv = \int_0^{\pi} d\theta \int_0^{2\pi} d\phi\, f(v_x, v_y, v_z) \sin\theta\, v^2\, dv \tag{17.17}$$

Substituting equation 17.12 into this expression and using equation 17.3, we find

$$F(v)\,dv = 4\pi v^2 \left(\frac{m}{2\pi kT}\right)^{3/2} \exp\left(-\frac{mv^2}{2kT}\right) dv \tag{17.18}$$

Thus, the probability density $F(v)$, for the **Maxwell distribution** of speeds, is

$$F(v) = 4\pi v^2 \left(\frac{m}{2\pi kT}\right)^{3/2} \exp\left(-\frac{mv^2}{2kT}\right) \tag{17.19}$$

As a result, the probability density at a speed of 0 is zero. The probability density increases with the speed up to a maximum and then declines.

A plot of $F(v)$ versus the molecular speed v is shown in Fig. 17.6 for oxygen at 100, 300, 500, and 1000 K. The probability that a molecule has a speed between any two values is given by the area under the curve between these two values of the speed. The plot of $F(v)$ versus v is approximately quadratic near the origin. At higher speeds the probability decreases toward zero because the exponential term decreases much more rapidly than v^2 increases. Thus, very few molecules have very high or very low speeds. The fraction of the molecules having speeds greater than 10 times the most probable speed (defined in the next section) is 9×10^{-42} at any temperature. The Avogadro constant times this fraction is so much less than 1 that we can say that no molecule has a velocity this high.

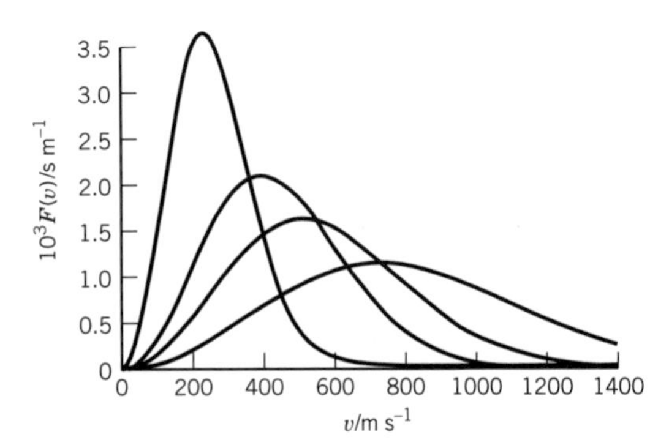

Figure 17.6 Probability density of various speeds v for oxygen molecules at 100, 300, 500, and 1000 K calculated using equation 17.19. (See Computer Problem 17.D.)

Notice that the maximum of $F(v)$ moves to higher v as the temperature is raised. The speed at the maximum is the most probable speed, for which we will soon derive a formula. Notice also that the width of the curve becomes larger as T increases (and, since the area under the curve is always unity, the maximum gets lower). Thus, as T increases or m decreases, the most probable speed increases and so do the numbers of molecules at high speeds.

Sometimes it is more useful to know the probability density as a function of molecular translational energy than in terms of the probability density for molecular speed. The probability $F(\epsilon)\,d\epsilon$ that the molecular energy is in the range ϵ to $\epsilon + d\epsilon$ can be calculated from the probability of molecular speeds $F(v)\,dv$ (equation 17.18) as follows. Since the kinetic energy of a molecule is given by $\epsilon = mv^2/2$, the speed is given by $v = (2\epsilon/m)^{1/2}$ and the differential of the speed is given by $dv = d\epsilon/(2m\epsilon)^{1/2}$. Substituting these relations into equation 17.18 to change the variable from v to ϵ yields

$$\begin{aligned} F(\epsilon)\,d\epsilon &= 4\pi\left(\frac{m}{2\pi kT}\right)^{3/2}\left(\frac{2\epsilon}{m}\right)e^{-\epsilon/kT}\frac{d\epsilon}{(2m\epsilon)^{1/2}} \\ &= \frac{2\pi}{(\pi kT)^{3/2}}\epsilon^{1/2}e^{-\epsilon/kT}\,d\epsilon \end{aligned} \tag{17.20}$$

Note that the probability that a molecule has a certain translational energy is independent of its mass. Equation 17.20 can be used to calculate the average kinetic energy of an ideal gas molecule:

$$\begin{aligned} \langle\epsilon\rangle &= \int_0^\infty \epsilon F(\epsilon)\,d\epsilon = \frac{2\pi}{(\pi kT)^{3/2}}\int_0^\infty \epsilon^{3/2}e^{-\epsilon/kT}\,d\epsilon \\ &= \frac{2\pi}{(\pi kT)^{3/2}}3\left(\frac{kT}{2}\right)^2(\pi kT)^{1/2} = \frac{3}{2}kT \end{aligned} \tag{17.21}$$

as obtained earlier in Section 16.9.

Figure 17.7 shows the probability density $F(\epsilon)$ as a function of the translational energy of an ideal gas molecule at 300 K.

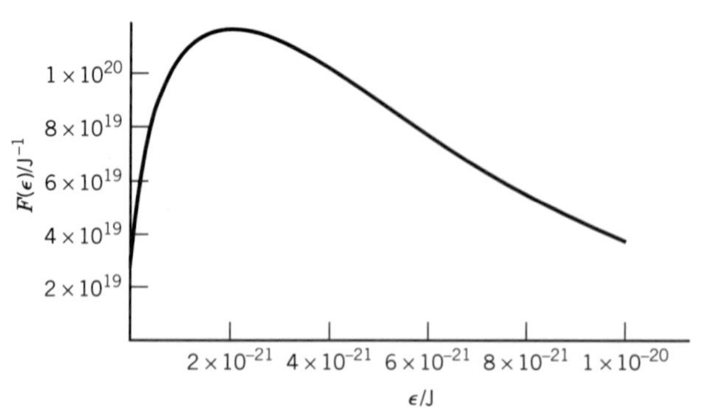

Figure 17.7 Plot of the probability density $F(\epsilon)$ versus ϵ for the translational energy of an ideal gas molecule at 300 K. (See Computer Problem 17.F.)

17.4 TYPES OF AVERAGE SPEEDS

Since there is a distribution of molecular speeds, there are different measures of the average speed. We will discuss the most probable speed v_{mp}, the mean speed $\langle v \rangle$, and the root-mean-square speed $\langle v^2 \rangle^{1/2}$.

The **most probable speed** v_{mp} is the speed at the maximum of $F(v)$. Setting dF/dv equal to zero, we find

$$\frac{dF(v)}{dv} = \left(\frac{m}{2\pi kT}\right)^{3/2} e^{-mv^2/2kT} \left[8\pi v + 4\pi v^2 \left(-\frac{mv}{kT}\right)\right] = 0 \qquad (17.22)$$

or

$$v_{mp} = \left(\frac{2kT}{m}\right)^{1/2} = \left(\frac{2RT}{M}\right)^{1/2} \qquad (17.23)$$

The **mean speed** $\langle v \rangle$ is calculated as the average of v using the probability distribution $F(v)$:

$$\langle v \rangle = \int_0^\infty v F(v)\, dv \qquad (17.24)$$

Substituting equation 17.19 and performing the integration with the help of Table 17.1, we find

$$\langle v \rangle = 4\pi \left(\frac{m}{2\pi kT}\right)^{3/2} \int_0^\infty \exp\left(-\frac{mv^2}{kT}\right) v^3\, dv \qquad (17.25)$$

$$\langle v \rangle = \left(\frac{8kT}{\pi m}\right)^{1/2} = \left(\frac{8RT}{\pi M}\right)^{1/2} \qquad (17.26)$$

The last speed we consider is the **root-mean-square speed,** which is defined as the square root of $\langle v^2 \rangle$:

$$\langle v^2 \rangle^{1/2} = \left[\int_0^\infty v^2 F(v)\, dv\right]^{1/2} \qquad (17.27)$$

Table 17.2 Various Types of Average Speeds of Gas Molecules at 298 K

Gas	$\langle v^2 \rangle^{1/2}$/m s^{-1}	$\langle v \rangle$/m s^{-1}	v_{mp}/m s^{-1}
H_2	1920	1769	1568
O_2	482	444	394
CO_2	411	379	336
CH_4	681	627	556

Substituting equation 17.19 and using Table 17.1 again, we find

$$\langle v^2 \rangle^{1/2} = \left(\frac{3kT}{m}\right)^{1/2} = \left(\frac{3RT}{M}\right)^{1/2} \tag{17.28}$$

From these three calculations, we can see that at any temperature,

$$\langle v^2 \rangle^{1/2} > \langle v \rangle > v_{mp} \tag{17.29}$$

Each of these measures of the probability distribution is proportional to $(T/M)^{1/2}$, so that each increases with temperature and decreases with molar mass. Lighter molecules therefore move faster than heavier molecules on average, as shown in Table 17.2.

The **speed of sound** in a gas is, not surprisingly, also about the same magnitude as the average speeds. It can be shown* that sound waves in a gas are longitudinal contractions and rarefactions that are adiabatic and reversible, and that travel at the speed v_s given by the thermodynamic quantity:

$$v_s^2 = -\frac{V}{\rho\left(\frac{\partial V}{\partial P}\right)_S} \tag{17.30}$$

Here V is the volume, S is the entropy, P is the pressure, and ρ is the density of the gas. Since for an ideal gas undergoing a reversible adiabatic expansion or contraction, PV^{γ} = constant (see equation 2.85), we have

$$\left(\frac{\partial V}{\partial P}\right)_S = -\frac{V}{\gamma P} \tag{17.31}$$

Substituting this into equation 17.30 and using the ideal gas law, we find that

$$v_s^2 = \frac{\gamma P}{\rho} = \frac{\gamma RT}{M} \tag{17.32}$$

Thus,

$$v_s = \left(\frac{\gamma RT}{M}\right)^{1/2} \tag{17.33}$$

For monatomic gases $\gamma = \frac{5}{3}$, so that v_s is just smaller than v_{mp}. For real gases, the velocity of sound depends slightly on pressure.

*M. Zemansky and R. Dittman, *Heat and Thermodynamics*. New York: McGraw-Hill, 1981.

Example 17.3 ***Various speeds for hydrogen molecules***

Calculate the most probable speed v_{mp}, the mean speed $\langle v \rangle$, and the root-mean-square speed $\langle v^2 \rangle^{1/2}$ for hydrogen molecules at 0 °C.

$$v_{mp} = \left(\frac{2RT}{M}\right)^{1/2} = \left[\frac{(2)(8.3145\ \mathrm{J\ K^{-1}\ mol^{-1}})(273.15\ \mathrm{K})}{(2.016 \times 10^{-3}\ \mathrm{kg\ mol^{-1}})}\right]^{1/2}$$

$$= 1.50 \times 10^3\ \mathrm{m\ s^{-1}}$$

$$\langle v \rangle = \left(\frac{8RT}{\pi M}\right)^{1/2} = \left[\frac{(8)(8.3145\ \mathrm{J\ K^{-1}\ mol^{-1}})(273.15\ \mathrm{K})}{(3.1416)(2.016 \times 10^{-3}\ \mathrm{kg\ mol^{-1}})}\right]^{1/2}$$

$$= 1.69 \times 10^3\ \mathrm{m\ s^{-1}}$$

$$\langle v^2 \rangle^{1/2} = \left(\frac{3RT}{M}\right)^{1/2} = \left[\frac{3(8.3145\ \mathrm{J\ K^{-1}\ mol^{-1}})(273.15\ \mathrm{K})}{2.016 \times 10^{-3}\ \mathrm{kg\ mol^{-1}}}\right]^{1/2}$$

$$= 1.84 \times 10^3\ \mathrm{m\ s^{-1}}$$

The root-mean-square speed of a hydrogen molecule at 0 °C is 6620 km h^{-1}, but at ordinary pressures a molecule travels only an exceedingly short distance before colliding with another molecule and changing direction.

In discussing spectroscopy, we have always tacitly assumed that the emitting atom or molecule is at rest, but since atoms and molecules in a gas are in motion there is a Doppler broadening of spectral lines. If the frequency that would be emitted if the atom or molecule were at rest is ν_0, then the frequency measured by a stationary observer is given by the following approximation at molecular velocities considerably less than the velocity of light:

$$\nu \approx \nu_0 \left(1 + \frac{v_x}{c}\right) \tag{17.34}$$

where v_x is the velocity with which the emitting atom or molecule is moving toward the observer and c is the speed of light. At temperature T the spectral line will be spread out by the Maxwell distribution by the emitting species. Equation 17.34 indicates that the velocity in the x direction is given by $v_x = c(\nu - \nu_0)/\nu_0$, so the distribution of molecular velocities given by equation 17.7 is proportional to

$$\mathrm{e}^{-mv_x^2/2k_\mathrm{B}T} = \mathrm{e}^{-mc^2(\nu-\nu_0)^2/2\nu_0^2 kT} \tag{17.35}$$

The equation for a Gaussian distribution of frequencies about ν_0 can be written

$$p(\nu)\,\mathrm{d}\nu = (2\pi\sigma_\nu^2)^{-1/2}\,\mathrm{e}^{-(\nu-\nu_0)^2/2\sigma_\nu^2}\,\mathrm{d}\nu \tag{17.36}$$

The standard deviation σ_ν for the frequency distribution is given by

$$\sigma_\nu = \left(\frac{\nu_0^2 k_\mathrm{B} T}{mc^2}\right)^{1/2} \tag{17.37}$$

For sodium atoms, $\nu_0 = 5 \times 10^8$ Hz, and the molar mass is 0.022 99 kg mol^{-1}. Computer Problem 17.E shows that when the temperature is 500 K, the standard deviation for the frequency distribution due to Doppler broadening is 708.7 Hz.

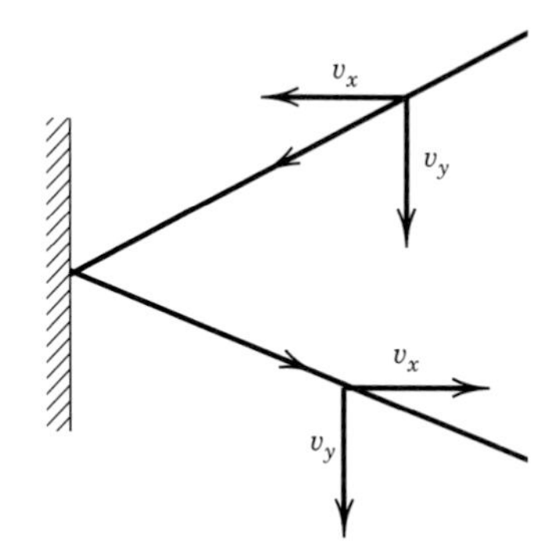

Figure 17.8 Specular reflection of a molecule with a wall in the yz plane. The x component of the velocity has its sign reversed by the collision, but v_y and v_z are unchanged if the impinging molecule is in the xy plane.

17.5 PRESSURE OF AN IDEAL GAS

The pressure of an ideal gas can be calculated by assuming that the walls of the container are flat, and that the collisions of the molecules with the walls are elastic. That means that the molecules do not lose kinetic energy in their collisions with the walls. The collisions with a flat wall are specular; that means that the angle of incidence is equal to the angle of reflection, as shown in Fig. 17.8. This figure shows that when a molecule collides with the wall in the yz plane, v_y and v_z are not changed, but the sign of v_x is reversed. Thus, $v_x^2 + v_y^2 + v_z^2$ is not changed by a collision with the wall.

The pressure is the average force per unit area that the wall must exert on the molecules to hold them at constant volume. The average force $\langle F_x \rangle$ in the x direction is equal to the time rate of change in the momentum in the x direction of the molecules that strike the wall:

$$\langle F_x \rangle = ma_x = m\frac{\mathrm{d}v_x}{\mathrm{d}t} = \frac{\mathrm{d}(mv_x)}{\mathrm{d}t} \tag{17.38}$$

Consider the molecules striking area A of the yz plane, as shown in Fig. 17.9. Since the momenta in the y and z directions do not change, the change in momentum of the molecule is that in the x direction, namely, $-2mv_x$, where $v_x \geq 0$ prior to collision with the wall. In a time $\mathrm{d}t$, a molecule with v_x will hit the surface if that molecule is within a volume $v_x\,\mathrm{d}t\,A$ of the surface ($v_x \geq 0$). Since we assume that the molecules are randomly distributed throughout the volume of the gas, the probable number of molecules having a velocity in the range of v_x to $v_x + \mathrm{d}v_x$ and within the necessary distance to hit the surface in time $\mathrm{d}t$ is

$$Nf(v_x)\,\mathrm{d}v_x\left(v_x\,\mathrm{d}t\frac{A}{V}\right) \tag{17.39}$$

The factor in parentheses is the volume in which molecules will hit the surface in a time $\mathrm{d}t$ divided by the total volume. This is the probability of finding the molecule in the correct volume to hit the surface for a random spatial distribution. To find the force on the surface exerted by the gas, we must multiply expression 17.39 by the negative of the momentum change of the molecules or $2mv_x$, divide by $\mathrm{d}t$, and integrate over all positive v_x. Therefore, the average force in the x direction is

$$\langle F_x \rangle = N\int_0^\infty \mathrm{d}v_x\,(2mv_x)\,f(v_x)\,v_x\,\frac{A}{V} \tag{17.40}$$

or, by substituting for $f(v_x)$,

$$\begin{aligned}\langle F_x \rangle &= \frac{NA}{V}\,2m\left(\frac{m}{2\pi kT}\right)^{1/2}\int_0^\infty \mathrm{d}v_x v_x^2\,\mathrm{e}^{-mv_x^2/2kT} \\ &= \frac{NA}{V}\,2m\left(\frac{m}{2\pi kT}\right)^{1/2}\left(\frac{2kT}{m}\right)^{3/2}\int_0^\infty \mathrm{d}x\,x^2\,\mathrm{e}^{-x^2} \\ &= \frac{NA}{V}\,kT\end{aligned} \tag{17.41}$$

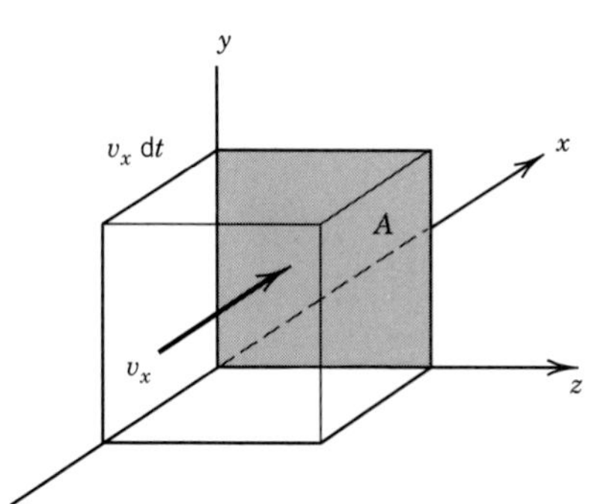

Figure 17.9 Molecule with velocity v_x striking a wall of area A. All molecules in the volume $Av_x\,\mathrm{d}t$ with $v_x \geq 0$ will strike this surface in time $\mathrm{d}t$.

The pressure is the force per unit area; therefore,

$$P = \frac{NkT}{V} \tag{17.42}$$

which is the ideal gas equation of state.

Comment:

Historically, this derivation was made before statistical mechanics was developed, and it showed that a thermodynamic property of a gas could be calculated with a very simple model. This model also provides a molecular explanation for Dalton's law that $P = \sum P_i$. In Section 16.3 we saw that the ideal gas law could also be derived using the microcanonical partition function. The calculation of the equation of state of a nonideal gas requires the use of a canonical ensemble partition function and the potential of molecular interaction, as discussed in Section 16.11.

17.6 COLLISIONS WITH A SURFACE AND EFFUSION

In studying the reaction of a gas with a solid it is necessary to calculate the number of gas molecules that hit the plane surface per unit time. In addition, it is often necessary to compute the rate at which molecules pass through a small opening into an evacuated vessel (effusion). For small holes the rate is small enough *not* to upset the equilibrium speed distribution in the bulk gas. In addition, the mean free path is assumed to be large compared with the diameter of the hole, so that collisions in the neighborhood of the hole can be neglected.

In the preceding section, we saw that the number of collisions with a wall of area A in time dt is given by $Nf(v_x)\,dv_x(v_x\,dt\,A/V)$, where N is the number of molecules in volume V. It is convenient to discuss the number of collisions with a wall or the number of molecules passing through a small opening in terms of the **flux** J_N, which is the number of particles striking the wall or passing through an imaginary surface per unit area per unit time. Thus the flux is given by

$$J_N = \int_0^\infty (N/V) f(v_x)\, v_x\, dv_x = \rho \int_0^\infty f(v_x)\, v_x\, dv_x \tag{17.43}$$

where ρ is the number density (N/V). The use of an integral in Table 17.1 yields

$$J_N = \rho \left(\frac{kT}{2\pi m} \right)^{1/2} \tag{17.44}$$

The use of the expression for the mean velocity (equation 17.26) leads to a simple equation for the flux:

$$J_N = \frac{\rho \langle v \rangle}{4} \tag{17.45}$$

Since we are assuming that the gas is ideal, the number density ρ may be eliminated from equation 17.44 by use of the ideal gas law $P = \rho kT$. Thus,

$$J_N = \frac{P}{(2\pi mkT)^{1/2}} \tag{17.46}$$

For a pure substance, the measurement of the rate of escape J_N through a small hole can be used to calculate the pressure. This is the basis of the Knudsen method for measuring the vapor pressure of a solid or a liquid. The solid or liquid is placed in a container with a small hole. This container is placed in an evacuated chamber, and the loss in mass Δw of the container and sample is measured after time t. If the area of the hole is A, the flux is given by

$$J_N = \frac{\Delta w}{mtA} \tag{17.47}$$

A sufficiently large surface area of the solid or liquid must be exposed to maintain the saturation vapor pressure. These simple equations cannot be used if the gaseous sample has molecules with several different masses. For example, the vapor in equilibrium with graphite at high temperatures contains C_1, C_2, C_3, C_4, ... It was not possible to obtain a precise value for $\Delta H°$ for the reaction C(graphite) = C(g) until the composition of the vapor at a series of temperatures had been obtained by mass spectrometry.

Example 17.4 ***Vapor pressure from an effusion measurement***

The vapor pressure of solid beryllium was measured by R. B. Holden, R. Speiser, and H. L. Johnston [*J. Am. Chem. Soc.* **70**:3897 (1948)] using a Knudsen cell. The effusion hole was 0.318 cm in diameter, and they found a mass loss of 9.54 mg in 60.1 min at a temperature of 1457 K. What is the vapor pressure?

$$\begin{aligned} J_N &= \frac{\Delta w}{mtA} = \frac{\Delta w N_A}{MtA} \\ &= \frac{(9.54 \times 10^{-6}\ \text{kg})(6.022 \times 10^{23}\ \text{mol}^{-1})}{(9.012 \times 10^{-3}\ \text{kg mol}^{-1})(60 \times 60.1\ \text{s})\pi(0.159 \times 10^{-2}\ \text{m})^2} \\ &= 2.23 \times 10^{22}\ \text{m}^{-2}\ \text{s}^{-1} \end{aligned}$$

$$\begin{aligned} P &= J_N(2\pi mkT)^{1/2} \\ &= (2.23 \times 10^{22}\ \text{m}^{-2}\ \text{s}^{-1}) \\ &\quad \times \left[\frac{2\pi(9.012 \times 10^{-3}\ \text{kg mol}^{-1})(1.381 \times 10^{-23}\ \text{J K}^{-1})(1457\ \text{K})}{(6.022 \times 10^{23}\ \text{mol}^{-1})}\right]^{1/2} \\ &= 0.968\ \text{Pa} = 0.968 \times 10^{-5}\ \text{bar} \end{aligned}$$

Since the flux is inversely proportional to the square root of the mass, effusion through a porous barrier can be used to separate different isotopic species of gas molecules.

Comment:

So far we have treated properties that can be discussed in terms of point molecules, but now we come to properties that depend on the cross-sectional areas of molecules in collisions with each other. As a simplification we are going to consider only hard spherical molecules that do not interact with each other except by collision. The discussion of collision rates will increase our understanding of rates of chemical reactions, mean free paths, diffusion, thermal conductivity, and viscosity.

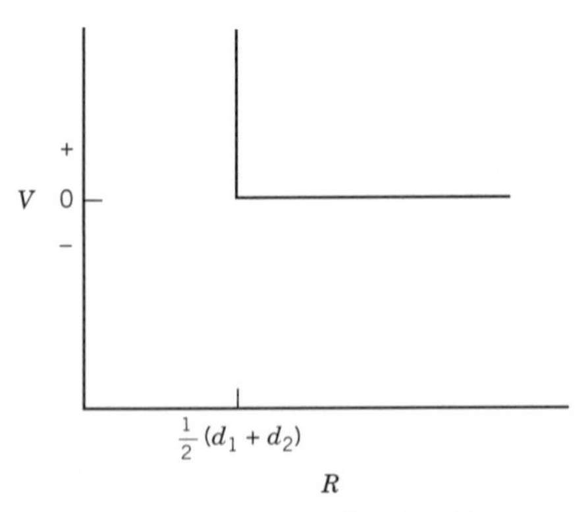

Figure 17.10 Potential energy as a function of distance R between centers for two hard-sphere molecules with diameters d_1 and d_2.

17.7 COLLISIONS OF HARD-SPHERE MOLECULES

The interactions of molecules in the gas phase are very complicated because of the shape of the intermolecular potential (Section 11.9). In this section we will use a very simple molecular model—the **hard sphere.** This is equivalent to assuming that the intermolecular potential is zero at distances between centers greater than $\frac{1}{2}(d_1 + d_2)$, where d_1 and d_2 are the diameters of the two molecules. The potential energy is infinite at shorter distances, as illustrated in Fig. 17.10. Thus, hard-sphere molecules 1 and 2 do not interact unless the distance between their centers is $\frac{1}{2}(d_1 + d_2)$, and then they bounce like idealized billiard balls.

As shown in Fig. 17.11 hard spherical molecules collide with each other if their centers come within distance d equal to their diameters if the molecules are alike, or distance $d_{12} = \frac{1}{2}(d_1 + d_2)$ if they are different. The distance d_{12} is called the **collision diameter.**

Let us consider collisions of molecules of type 1 with molecules of type 2. If molecules of type 2 are stationary, a molecule of type 1 will collide in unit time with all molecules of type 2 that have their centers in a cylinder of volume $\pi d_{12}^2 v_1$. According to this simple calculation a molecule of type 1 would undergo $\pi d_{12}^2 v_1 \rho_2$

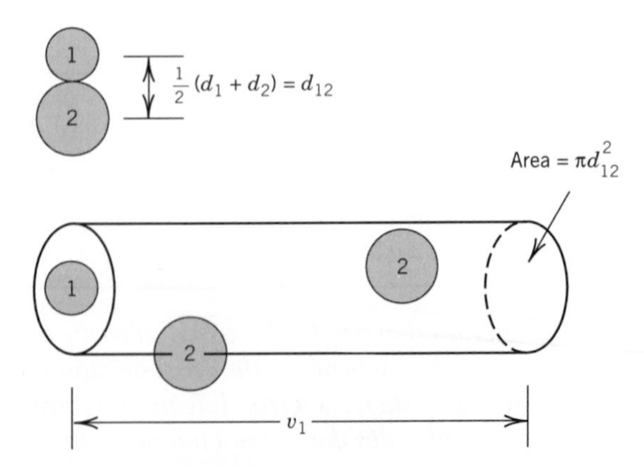

Figure 17.11 Collisions of hard-sphere molecules. If molecules of type 2 are stationary, a molecule of type 1 will collide in unit time with all molecules of type 2 that have their centers in a cylinder of volume $\pi d_{12}^2 v_1$.

collisions per unit time, where ρ_2 is the number of type 2 molecules per unit volume. However, molecules of type 2 are not stationary, so we need to use the relative speed v_{12} in calculating the rate of collisions z_{12} of a molecule of type 1 with molecules of type 2. Thus,

$$z_{12} = \rho_2 \pi d_{12}^2 \int f(v_1) f(v_2) v_{12} \, dv_1 \, dv_2 \tag{17.48}$$

where we are averaging over the product of distribution functions for each molecule. The quantity z_{12} is referred to as the **collision frequency** of molecules of type 1 with molecules of type 2 because it has the unit s^{-1}.

Note that the product $f(v_1)f(v_2)$ contains the sum of kinetic energies of particle 1 and particle 2 in the exponential term. We have already seen (Section 9.9) that we can convert from the sum of kinetic energies of particle 1 and particle 2 to the kinetic energy of the center of mass plus the relative kinetic energy. In addition, the volume element $dv_1 \, dv_2$ becomes the volume element $dv_1 \, dv_{CM}$. The integral over the velocity of the center of mass yields unity; after the integrations over angles have been done, we have

$$z_{12} = \rho_2 \pi d_{12}^2 \int f(v_{12}) \, v_{12} \, dv_{12} \tag{17.49}$$

where

$$f(v_{12}) = 4\pi \left(\frac{\mu}{2\pi kT} \right)^{3/2} v_{12}^2 \, e^{-\mu v_{12}^2/2kT} \tag{17.50}$$

and μ is the reduced mass equal to $m_1 m_2/(m_1 + m_2)$. The integration in equation 17.49 can now be done using Table 17.1 to obtain the collision frequency z_{12} of molecules of type 1 with molecules of type 2:

$$z_{12} = \rho_2 \pi d_{12}^2 \left(\frac{8kT}{\pi\mu} \right)^{1/2} = \rho_2 \pi d_{12}^2 \langle v_{12} \rangle \tag{17.51}$$

Since the density ρ_2 of molecules of type 2 has the SI unit m^{-3}, the collision diameter d_{12} has the unit m, the mean relative speed $\langle v_{12} \rangle$ has the units m s^{-1}, and the collision frequency has the unit s^{-1}.

Equation 17.51 introduces a new type of molecular speed, the **mean relative speed** $\langle v_{12} \rangle$:

$$\langle v_{12} \rangle = \left(\frac{8kT}{\pi\mu} \right)^{1/2} \tag{17.52}$$

Let us take a minute to consider why it has the form it does. If we square both sides of equation 17.52 and introduce the definition of the reduced mass μ, we obtain

$$\begin{aligned} \langle v_{12} \rangle^2 &= \left(\frac{8kT}{\pi} \right) \left(\frac{1}{m_1} + \frac{1}{m_2} \right) \\ &= \langle v_1 \rangle^2 + \langle v_2 \rangle^2 \end{aligned} \tag{17.53}$$

As shown by Fig. 17.12, we can use the Pythagorean theorem to interpret the mean relative speed in terms of the mean speeds of molecules 1 and 2. Molecules 1 and 2 can collide with each other with any angle between 0° and 180° between their

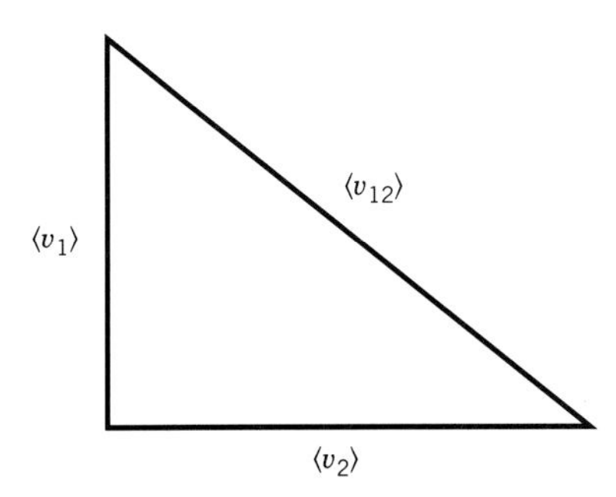

Figure 17.12 The mean relative speed $\langle v_{12} \rangle$ of molecules 1 and 2 can be calculated from a right triangle involving the mean speeds of molecules of types 1 and 2. According to the Pythagorean theorem, $\langle v_{12} \rangle^2 = \langle v_1 \rangle^2 + \langle v_2 \rangle^2$.

paths, but equation 17.53 shows that the average collision is at 90°. For collisions of identical particles, $\langle v_{11}\rangle^2 = 2\langle v_1\rangle^2$, so that $\langle v_{11}\rangle = 2^{1/2}\langle v_1\rangle = 2^{1/2}\langle v\rangle$.

Example 17.5 ***Mean relative speed of two different molecules***

What is the mean relative speed of hydrogen molecules with respect to oxygen molecules (or oxygen molecules with respect to hydrogen molecules) at 298 K?

The molecular masses are

$$\begin{aligned} m_1 &= \frac{2.016 \times 10^{-3}\ \text{kg mol}^{-1}}{6.022 \times 10^{23}\ \text{mol}^{-1}} \\ &= 3.348 \times 10^{-27}\ \text{kg} \\ m_2 &= \frac{32.000 \times 10^{-3}\ \text{kg mol}^{-1}}{6.022 \times 10^{23}\ \text{mol}^{-1}} \\ &= 5.314 \times 10^{-26}\ \text{kg} \\ \mu &= [(3.348 \times 10^{-27}\ \text{kg})^{-1} + (5.314 \times 10^{-26}\ \text{kg})^{-1}]^{-1} \\ &= 3.150 \times 10^{-27}\ \text{kg} \\ \langle v_{12}\rangle &= \left(\frac{8kT}{\pi\mu}\right)^{1/2} \\ &= \left[\frac{8(1.381 \times 10^{-23}\ \text{J K}^{-1})(298\ \text{K})}{\pi(3.150 \times 10^{-27}\ \text{kg})}\right]^{1/2} \\ &= 1824\ \text{m s}^{-1} \end{aligned}$$

Note that the mean relative speed is closer to the mean speed of molecular hydrogen (1920 m s^{-1}) than to that of molecular oxygen (482 m s^{-1}).

If the molecule of type 1 is moving through molecules of type 1 rather than molecules of type 2, equation 17.51 becomes

$$z_{11} = 2^{1/2}\rho\pi d^2\langle v\rangle \tag{17.54}$$

since $(8kT/\pi\mu)^{1/2}$ becomes $2^{1/2}\langle v\rangle$ because $1/\mu = 1/m + 1/m = 2/m$. The collision frequency z_{11} is the rate of collisions of molecules of type 1 with molecules of type 1.

In connection with chemical kinetics, we will also be interested in the number of collisions per unit time per unit volume. This quantity is referred to as the **collision density,** and it is represented by Z. To calculate the number of collisions of molecules of type 1 with molecules of type 2 per unit time per unit volume of gas Z_{12}, we simply multiply z_{12} by the number density ρ_1, so that

$$Z_{12} = \rho_1\rho_2\pi d_{12}^2\langle v_{12}\rangle \tag{17.55}$$

If we are interested in the number of collisions of molecules of type 1 with other molecules of type 1 per unit time per unit volume of gas Z_{11}, equation 17.55 reduces to

$$\begin{aligned} Z_{11} &= \tfrac{1}{2}\rho^2\pi d^2\langle v_{11}\rangle \\ &= 2^{-1/2}\rho^2\pi d^2\langle v\rangle \end{aligned} \tag{17.56}$$

Table 17.3 Collision Frequencies z_{11} and Collision Densities Z_{11} for Four Gases at 298 K

	z_{11}/s^{-1}		$Z_{11}/\text{mol L}^{-1}\text{ s}^{-1}$	
Gas	1 bar	10^{-6} bar	1 bar	10^{-6}bar
H_2	14.13×10^9	14.13×10^3	2.85×10^8	2.85×10^{-4}
O_2	6.24×10^9	6.24×10^3	1.26×10^8	1.26×10^{-4}
CO_2	8.81×10^9	8.81×10^3	1.58×10^8	1.58×10^{-4}
CH_4	11.60×10^9	11.60×10^3	2.08×10^8	2.08×10^{-4}

where a divisor of 2 has been introduced so that each collision is not counted twice, and $\langle v_{11} \rangle$ has been replaced by $2^{1/2}\langle v \rangle$ by means of the reduced mass of like particles. The collision density is readily expressed in mol m^{-3} s^{-1} simply by dividing by the Avogadro constant.

The collision density is of interest because it sets an upper limit on the rate with which two gas molecules can react (see Section 19.1). Actual chemical reaction rates are usually much smaller than the collision rates, indicating that not every collision leads to reaction.

Collision frequencies z_{12} and collision densities Z_{11} for four gases are given in Table 17.3 at 25 °C. The collision densities are expressed in mol L^{-1} s^{-1} because it is easier to think about chemical reactions in these units.

Example 17.6 ***Collision frequency and collision density***

For molecular oxygen at 25 °C, calculate the collision frequency z_{11} and the collision density Z_{11} at a pressure of 1 bar.

The collision diameter of oxygen is 0.361 nm or 3.61×10^{-10} m, as determined in a manner to be described shortly (Section 17.10):

$$\langle v \rangle = \left(\frac{8RT}{\pi M} \right)^{1/2} = \left[\frac{(8)(8.3145 \text{ J K}^{-1} \text{ mol}^{-1})(298 \text{ K})}{\pi(32 \times 10^{-3} \text{ kg mol}^{-1})} \right]^{1/2} = 444 \text{ m s}^{-1}$$

The number density is given by

$$\rho = \frac{N}{V} = \frac{PN_A}{RT} = \frac{(1 \text{ bar})(6.022 \times 10^{23} \text{ mol}^{-1})(10^3 \text{ L m}^{-3})}{(0.083\,145 \text{ L bar K}^{-1} \text{ mol}^{-1})(298 \text{ K})} = 2.43 \times 10^{25} \text{ m}^{-3}$$

The collision frequency is given by

$$\begin{aligned} z_{11} &= \sqrt{2}\rho\pi d^2 \langle v \rangle \\ &= (1.414)(2.43 \times 10^{25} \text{ m}^{-3})\pi(3.61 \times 10^{-10} \text{ m})^2(444 \text{ m s}^{-1}) \\ &= 6.24 \times 10^9 \text{ s}^{-1} \end{aligned}$$

The collision density is given by

$$\begin{aligned} Z_{11} &= \frac{1}{2^{1/2}}\rho^2 \pi d^2 \langle v \rangle \\ &= (0.707)(2.43 \times 10^{25} \text{ m}^{-3})^2 \pi(3.61 \times 10^{-10} \text{ m})^2(444 \text{ m s}^{-1}) \\ &= 7.58 \times 10^{34} \text{ m}^{-3} \text{ s}^{-1} = \frac{(7.58 \times 10^{34} \text{ m}^{-3} \text{ s}^{-1})(10^{-3} \text{ m}^3 \text{ L}^{-1})}{6.022 \times 10^{23} \text{ mol}^{-1}} \\ &= 1.26 \times 10^8 \text{ mol L}^{-1} \text{ s}^{-1} \end{aligned}$$

Example 17.7 ***Relation between collision frequencies and collision densities***

Above we have explicit expressions for the collision frequency z_{12} between molecules of type 1 and type 2, z_{11} between molecules of the same type, collision density Z_{12} between molecules of type 1 and type 2, and Z_{11} between molecules of the same type. What are the relations between z_{12} and Z_{12} and between z_{11} and Z_{11}?

Comparing the equations in the text, we see that

$$Z_{12} = \rho_1 z_{12}$$

so that the number of collisions per unit volume per unit time between molecules of types 1 and 2 is equal to the density ρ_1 of molecules of type 1 times the frequency of collisions between molecules of types 1 and 2. Comparing equations in the text, we also see that

$$Z_{11} = \rho z_{11}/2$$

so that the number of collisions per unit volume per unit time between molecules of the same type is equal to the density ρ of the molecules times the collision frequency, divided by 2 to avoid double counting.

The mean free path λ is the average distance traveled between collisions. Although it is not a directly measurable quantity, it is a very useful concept, as we shall see. It can be computed by dividing the average distance traveled per unit time by the collision frequency. For a molecule moving through like molecules,

$$\lambda = \frac{\langle v \rangle}{z_{11}} = \frac{1}{2^{1/2}\rho\pi d^2} \tag{17.57}$$

Assuming that the collision diameter d is independent of temperature, the temperature and pressure dependence of the mean free path may be obtained by substituting the ideal gas law in the form $\rho = P/kT$:

$$\lambda = \frac{kT}{2^{1/2}\pi d^2 P} \tag{17.58}$$

Thus, at constant temperature, the mean free path is inversely proportional to the pressure.

Example 17.8 ***Calculating the mean free path***

For oxygen at 25 °C the collision diameter is 0.361 nm. What are the mean free paths in meters and molecular diameters at (*a*) 1 bar pressure and (*b*) 0.1 Pa pressure?

(*a*) From Example 17.7, $\rho = 2.43 \times 10^{25}\ \text{m}^{-3}$ at 1 bar, and using equation 17.57,

$$\lambda = \frac{1}{2^{1/2}\rho\pi d^2}$$

$$\lambda = [(1.414)(2.43 \times 10^{25}\ \text{m}^{-3})\pi(3.61 \times 10^{-10}\ \text{m})^2]^{-1} = 7.11 \times 10^{-8}\ \text{m}$$

and

$$(7.11 \times 10^{-8}\ \text{m})/(3.61 \times 10^{-10}\ \text{m}) = 197 \text{ molecular diameters}$$

$$(b)\quad \rho = \frac{PN_A}{RT} = \frac{(0.1\ \text{Pa})(6.022 \times 10^{23}\ \text{mol}^{-1})}{(8.3145\ \text{J K}^{-1}\ \text{mol}^{-1})(298\ \text{K})} = 2.43 \times 10^{19}\ \text{m}^{-3}$$

$$\lambda = [(1.414)(3.14)(3.61 \times 10^{-10}\ \text{m})^2(2.43 \times 10^{19}\ \text{m}^{-3})]^{-1} = 0.071\ \text{m} = 7.1\ \text{cm}$$

and

$$(7.11 \times 10^{-2}\ \text{m})/(3.61 \times 10^{-10}\ \text{m} = 1.97 \times 10^{8}\ \text{molecular diameters}$$

At pressures so low that the mean free path becomes comparable with the dimensions of the containing vessel, the flow properties of the gas become markedly different from those at higher pressures.

Comment:

The discussion in this section and the preceding section marks the first time we have considered the rates of processes. Thus these two sections begin to lay the foundations for considerations of chemical kinetics. In these discussions it has been necessary to assume that molecules are rigid spheres, but we know from quantum mechanics and the experimental evidence for intermolecular potentials that they are not. Because of these intermolecular interactions it is actually hard to define a collision, and in more complete discussions it is necessary to use scattering theory.

17.8 EFFECTS OF MOLECULAR INTERACTIONS ON COLLISIONS

Collisions between gas molecules are more complicated than indicated in the preceding sections because of intermolecular attractive and repulsive forces. In connection with the discussion of the **Lennard-Jones potential** (Section 11.9), we saw that as molecules approach each other there is first intermolecular attraction and then, at shorter distances, repulsion. This is shown by the paths of two colliding molecules illustrated in Fig. 17.13. This figure has been drawn so that the center of mass of the two molecules is stationary and the motion is confined to the plane of the paper. The numbers 1, 2, 3, ... indicate the successive positions of the two molecules. As the molecules approach, they first attract each other so that their paths are drawn together. As the molecules approach each other more closely, they repel each other, and their paths begin to diverge. After the interaction, the paths of the molecules make an angle χ with the directions of the initial paths.

The initial parameters of the collision are the relative kinetic energy ($\frac{1}{2}\mu v_{12}^2$) and the **impact parameter** b. The impact parameter is the minimum distance at which the molecules would pass each other if there were no molecular interactions. If b is large, the angle of deflection χ will be small.

The trajectories for collisions at various values of the impact parameters and for two values of the kinetic energy of approach are shown in Fig. 17.14. In this figure one of the molecules approaches from the top of the diagrams with various values of the impact parameter b, and the other molecule approaches from below in a symmetrical fashion. In Fig. 17.14*a* the low-energy collisions lead to a very complicated pattern. For large values of the impact parameter the molecules

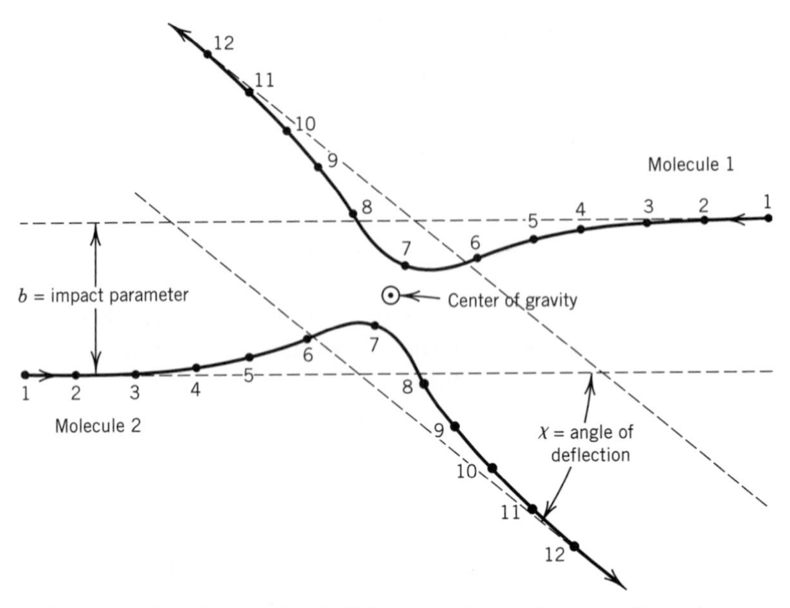

Figure 17.13 Collision of two spherically symmetrical molecules with an impact parameter b. The diagram is drawn so that the center of gravity of the system does not shift during the collision. The angle of deflection is χ. (From W. Kauzmann, *Thermal Properties of Matter, Vol. 1, Kinetic Theory of Gases,* © 1966. Benjamin/Cummings, Menlo Park, CA. Reprinted with permission.)

attract each other along the whole trajectory, and the deflection is negative by definition (although it is not possible experimentally to distinguish positive from negative deflections). As the value of the impact parameter decreases, the deflection becomes more and more negative, as shown in the diagram, until the repulsive force begins to be felt. As the impact parameter is further reduced, the repulsive force becomes dominant, and there are large positive deflections. For a head-on collision ($b = 0$) the deflection is 180°. In Fig. 17.14*b* the high-energy collisions give results that are close to, but not identical with, what would be expected for collisions of rigid spheres. The scattering angle χ can be calculated from the parameters for the molecular interaction, the impact parameter, and the relative kinetic energy of the two molecules. However, this calculation cannot be made analytically, and so we will not pursue it here.

When collisions of molecules interacting according to a Lennard-Jones potential are considered classically, we encounter the rather unsatisfactory situation that the cross section is infinite; notice in Fig. 17.14 that even "collisions" with large impact parameters b have some deflection. This problem is resolved by quantum mechanics, but we will not be able to go into quantum mechanical scattering theory, which shows that the cross section is similar in size to the hard-sphere result, but is energy dependent.

17.9 SPECIAL TOPIC: TRANSPORT PHENOMENA IN GASES

If a gas is not uniform with respect to composition, temperature, and velocity, transport processes occur until the gas does become uniform. The transport

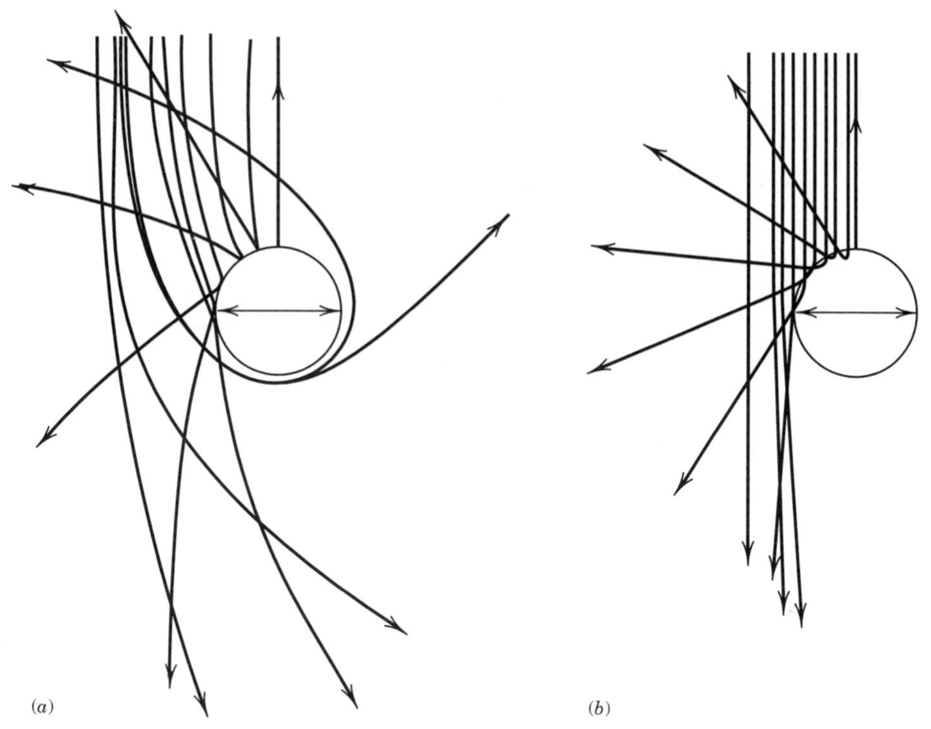

Figure 17.14 Collision trajectories for a pair of molecules interacting by a Lennard-Jones 6–12 potential. The center of gravity is stationary at the center of the circle, which has a diameter equal to the Lennard-Jones constant σ. (*a*) Trajectories for molecules that approach at energies equal to 0.1ϵ, where ϵ is the depth of the potential well in the Lennard-Jones potential. (*b*) Trajectories for molecules that approach at energies equal to 50ϵ. (From W. Kauzmann, *Thermal Properties of Matter, Vol. 1, Kinetic Theory of Gases,* © 1966. Benjamin/Cummings, Menlo Park, CA. Reprinted with permission.)

of matter in the absence of bulk flow is referred to as **diffusion.** The transport of heat from regions of high temperature to regions of lower temperature without convection is referred to as **thermal** conduction, and the transfer of momentum from a region of higher velocity to a region of lower velocity gives rise to the phenomenon of **viscous** flow. In each case the rate of flow is proportional to the rate of change of some property with distance, a so-called gradient.

The flux of component i in the z direction due to diffusion is proportional to the concentration gradient $\mathrm{d}c_i/\mathrm{d}z$, according to Fick's law:

$$J_{iz} = -D\frac{\mathrm{d}c_i}{\mathrm{d}z} \tag{17.59}$$

The proportionality constant is the **diffusion coefficient** D. The flux J_{iz} is expressed in terms of quantity per unit area per unit time. If SI units are used, J_{iz} has the units mol $\mathrm{m^{-2}\,s^{-1}}$, $\mathrm{d}c_i/\mathrm{d}z$ has the units of mol $\mathrm{m^{-4}}$, and D has the units of $\mathrm{m^2\,s^{-1}}$. The negative sign comes from the fact that if c_i increases in the positive z direction, $\mathrm{d}c_i/\mathrm{d}z$ is positive, but the flux is in the negative z direction because the flow is in the direction of lower concentrations.

The diffusion coefficient for the diffusion of one gas into another may be determined by use of a cell such as that shown schematically in Fig. 17.15*a*. The

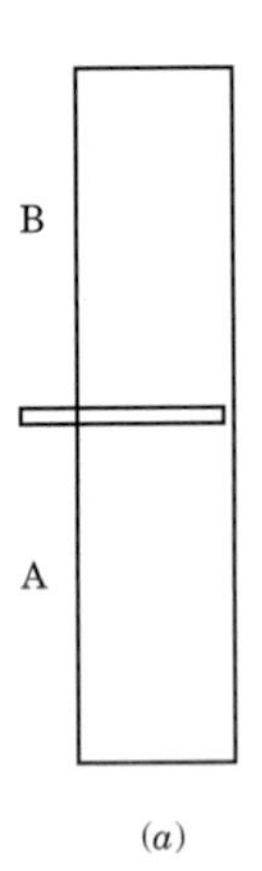

(a)

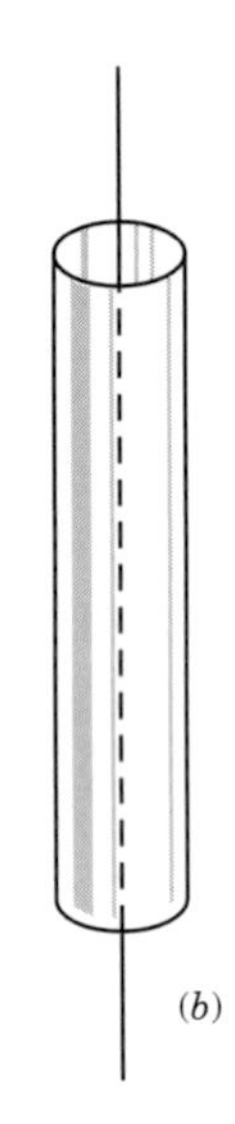

(b)

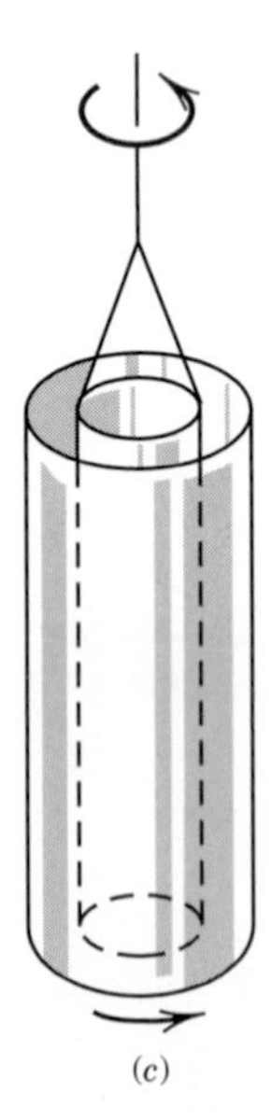

(c)

Figure 17.15 Schematic diagrams of apparatus for measurements of irreversible properties. (*a*) In the measurement of the diffusion coefficient D, the sliding separator is withdrawn so that a substance in chamber A can diffuse into B. (*b*) In the measurement of the thermal conductivity κ, the rate of heat transfer from an axial hot wire is measured. (*c*) In the measurement of the viscosity η of a gas, the outer cylinder is rotated and the torsion on the inner cylinder is determined from the twist in the suspension wire.

heavier gas is placed in chamber A and the lighter in chamber B. The sliding partition is withdrawn for a definite interval of time. From the average composition of one chamber or the other, after a time interval, D may be calculated.

The transport of heat is due to a gradient in temperature. Thus, the flux of energy q_z in the z direction due to the temperature gradient in that direction is given by

$$q_z = -\kappa \frac{dT}{dz} \tag{17.60}$$

where the proportionality constant κ is the **thermal conductivity.** When q_z has the units of J m^{-2} s^{-1} and dT/dz has the units of K m^{-1}, κ has the units of J m^{-1} s^{-1} K^{-1}. The negative sign in equation 17.60 indicates that if dT/dz is positive, the flow of heat is in the negative z direction, which is the direction toward lower temperature.

The determination of the thermal conductivity by the hot-wire method is illustrated schematically in Fig. 17.15*b*. The outer cylinder is kept at a constant temperature by a controlled bath. The tube is filled with the gas under investigation, and the fine wire at the axis of the tube is heated electrically. When a steady state is achieved, the temperature of the wire is measured by determining its electrical resistance. The thermal conductivity is calculated from the temperature of wire and wall, the heat dissipation, and the dimensions of the apparatus.

Thermal diffusion is the flux of material due to a temperature gradient of dT/dz. The fact that the thermal diffusion coefficients depend on mass makes it possible to separate isotopes by use of this effect.

Viscosity is a measure of the resistance that a fluid offers to an applied shearing force. Consider what happens to the fluid between parallel planes, illustrated in Fig. 17.16, when the top plane is moved in the y direction at a constant speed relative to the bottom plane while a constant distance between the planes (coordinate z) is maintained. The planes are considered to be very large, so that edge effects may be ignored. The layer of fluid immediately adjacent to the moving plane moves with the velocity of this plane. The layer next to the stationary plane is stationary; in between the velocity usually changes linearly with distance, as shown. The velocity gradient (i.e., the rate of change of velocity with respect to distance measured *perpendicular* to the direction of flow) is represented by dv_y/dz. The **viscosity** η is defined by the equation

$$F = -\eta \frac{dv_y}{dz} \tag{17.61}$$

Here F is the force per unit area required to move one plane relative to the other. The negative sign comes from the fact that if F is in the $+y$ direction, the velocity v_y decreases in successive layers away from the moving plane and dv_y/dz is negative. If F has the units of kg m s^{-2}/m^2 and dv_y/dz has the units of m s^{-1}/m, then the viscosity η has the units of kg m^{-1} s^{-1}. The SI unit of viscosity is the pascal-second. Since 1 N = 1 kg m s^{-2}, 1 Pa s = 1 kg m^{-1} s^{-1}. A fluid has a viscosity of 1 Pa s if a force of 1 N is required to move a plane of 1 m^2 at a velocity of 1 m s^{-1} with respect to a plane surface a meter away and parallel with it.

Although the viscosity is conveniently defined in terms of this hypothetical experiment, it is easier to measure it by determining the rate of flow through a tube, the torque on a disk that is rotated in the fluid, or other experimental arrangement. In the experimental arrangement illustrated in Fig. 17.15*c*, the outer cylinder is rotated at a constant velocity by an electric motor. The inner coaxial cylinder is suspended on a torsion wire. A torque is transmitted to the inner cylinder by the fluid, and this torque is calculated from the angular twist of the torsion wire.

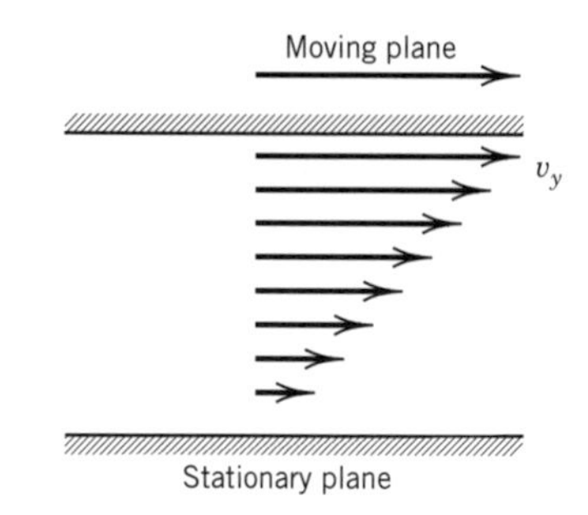

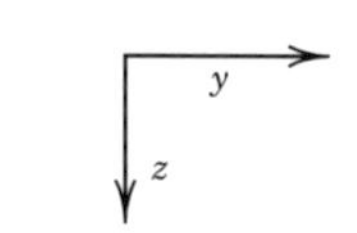

Figure 17.16 Velocity gradient in a fluid due to a shearing action.

17.10 SPECIAL TOPIC: CALCULATION OF TRANSPORT COEFFICIENTS

To calculate the transport coefficients introduced in the last section (D, κ, and η), even for hard-sphere molecules, we would need to consider how the Maxwell–Boltzmann distribution is disturbed by a gradient of concentration, temperature, or velocity. This calculation, which is too advanced for this book, can be found in some of the references listed at the end of the chapter (e.g., Hirschfelder et al.).

In spite of this, we can get a good qualitative understanding by a highly simplified discussion. Consider the diffusion of molecules in a concentration gradient in the z direction. Imagine that we are at $z = 0$ and we construct planes parallel to the xy plane at $z = \pm\lambda$, where λ is the mean free path (see Fig. 17.17). We choose planes at the mean free path because molecules from more distant points will, on average, have suffered collisions before reaching $z = 0$. Now let us calculate the flux of particles (see Section 17.6) across $z = 0$ due to the molecules above ($z > 0$) and below ($z < 0$). The flux across $z = 0$ from above is

$$J_+ = \left[\rho_0 + \lambda\left(\frac{d\rho}{dz}\right)\right]\frac{\langle v\rangle}{4} \tag{17.62}$$

where ρ_0 is the number density of particles in the plane at $z = 0$. We have used equation 17.45, and the density of particles at $z = +\lambda$ is given by the term in brackets. Similarly, the flux across $z = 0$ due to the molecules below $z = 0$ is

$$J_- = \left[\rho_0 - \lambda\left(\frac{d\rho}{dz}\right)\right]\frac{\langle v\rangle}{4} \tag{17.63}$$

The net flux of particles across the plane $z = 0$ is then

$$J = -\frac{1}{2}\langle v\rangle\lambda\frac{d\rho}{dz} \tag{17.64}$$

This equation can be compared with equation 17.59 to obtain

$$D_a = \frac{1}{2}\langle v\rangle\lambda = \left(\frac{kT}{\pi m}\right)^{1/2}\frac{1}{\rho\pi d^2} \tag{17.65}$$

where the subscript a indicates "approximate."

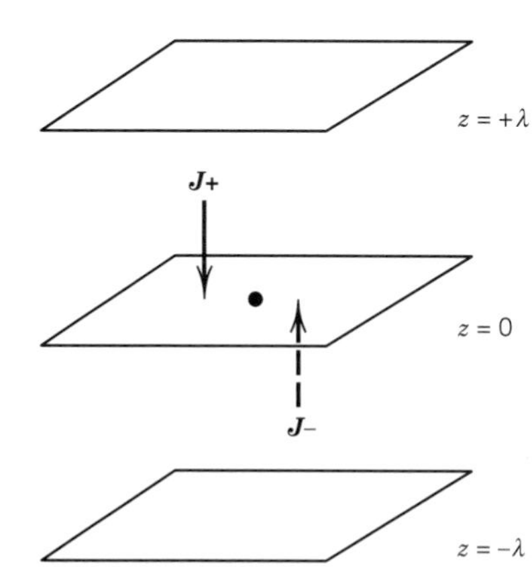

Figure 17.17 Planes constructed at distance $\pm\lambda$ (the mean free path) from the origin. The concentration gradient is in the z direction.

The exact theoretical expression for the diffusion coefficient of hard spheres is

$$D = \frac{3\pi}{8}\left(\frac{kT}{\pi m}\right)^{1/2}\frac{1}{\rho\pi d^2} \tag{17.66}$$

so that our highly simplified model yields a good qualitative result.

A similar simplified model for thermal conductivity of hard spheres yields the approximate value

$$\kappa_a = \frac{1}{3}\frac{\overline{C}_V}{N_A}\lambda\langle v\rangle\rho = \frac{2}{3}\frac{\overline{C}_V}{N_A}\left(\frac{kT}{\pi m}\right)^{1/2}\frac{1}{\pi d^2} \tag{17.67}$$

The exact expression for hard spheres is

$$\kappa = \frac{25\pi\overline{C}_V}{32\,N_A}\left(\frac{kT}{\pi m}\right)^{1/2}\frac{1}{\pi d^2} \tag{17.68}$$

Finally, the approximate model for the viscosity of hard spheres yields

$$\eta_a = \frac{1}{3}\rho\langle v\rangle m\lambda = \frac{2}{3}\left(\frac{kT}{\pi m}\right)^{1/2}\frac{m}{\pi d^2} \tag{17.69}$$

whereas the exact expression for hard spheres is

$$\eta = \frac{5\pi}{16}\left(\frac{kT}{\pi m}\right)^{1/2}\frac{m}{\pi d^2} \tag{17.70}$$

Note that although the approximate theory yields results that are too low, the dependences on T, m, and ρ agree with the exact theory. The exact expressions can be used to calculate molecular diameters d from experimental transport coefficients. Note that this does not imply that real molecules are hard spheres; in fact, we are forcing a model on the experiment. Nevertheless, the results in Table 17.4 show that a consistent set of molecular diameters results from this analysis of the data.

The equations discussed above as exact are first approximations, and higher approximations lead to coefficients in the equations for D, κ, and η that cannot be expressed in terms of π and integers.* The expressions above are adequate for

Table 17.4 Viscosity and Thermal Conductivity of Gases at 273.2 K and 1 bar and Calculated Molecular Diameters

	η	κ	*Molecular Diameter, d*/nm	
Gas	10^{-5} kg m^{-1} s^{-1}	10^{-2} J K^{-1}m^{-1} s^{-1}	*From* η	*From* κ
He	1.85	14.3	0.218	0.218
Ne	2.97	4.60	0.258	0.258
Ar	2.11	1.63	0.364	0.365
H_2	0.845	16.7	0.272	0.269
O_2	1.92	2.42	0.360	0.358
CO_2	1.36	1.48	0.464	0.458
CH_4	1.03	3.04	0.414	0.405

*B. L. Earl, *J. Chem. Educ.* **66**:147 (1989).

the calculation of collision diameters, but more accurate values of the coefficients are known.

Example 17.9 ***Viscosity of a gas***

Calculate the viscosity of molecular oxygen at 273.2 K and 1 bar. The molecular diameter is 0.360 nm.

Using the exact equation for hard spheres, we find

$$m = \frac{32.00 \times 10^{-3}\ \mathrm{kg\ mol^{-1}}}{6.022 \times 10^{23}\ \mathrm{mol^{-1}}} = 5.314 \times 10^{-26}\ \mathrm{kg}$$

$$\eta = \frac{5\pi}{16}\left(\frac{kT}{\pi m}\right)^{1/2}\frac{m}{\pi d^2}$$

$$= \frac{5\pi}{16}\left[\frac{(1.381 \times 10^{-23}\ \mathrm{J\ K^{-1}})(273.2\ \mathrm{K})}{\pi(5.314 \times 10^{-26}\ \mathrm{kg})}\right]^{1/2}\frac{5.314 \times 10^{-26}\ \mathrm{kg}}{\pi(0.360 \times 10^{-9}\ \mathrm{m})^2}$$

$$= 1.926 \times 10^{-5}\ \mathrm{kg\ m^{-1}\ s^{-1}}$$

Seven Key Ideas in Chapter 17

1. The probability density $f(v_x, v_y, v_z)$ is the joint probability that the component velocities are in the range v_x to $v_x + dv_x$, v_y to $v_y + dv_y$, and v_z to $v_z + dv_z$, and therefore it is the product of the probability densities $f(v_x)$, $f(v_y)$, and $f(v_z)$ in the three directions.
2. The Boltzmann distribution indicates that the probability density for a component velocity is proportional to the exponential of $-mv_x^2/2kT$, and the average energy in that direction is equal to $kT/2$.
3. The probability density for the Maxwell distribution of speeds is proportional to $v^2 \exp(-mv^2/2kT)$, and so it goes through a maximum at the most probable speed. It is important to distinguish between the most probable speed, the mean speed, the root-mean-square speed, and the speed of sound.
4. The flux of molecules through a small opening is proportional to the pressure, and so measurements of flux can yield the vapor pressure of a solid.
5. Calculations of collision frequencies z_{12} and collision densities Z_{12} are of interest in connection with the interpretation of the rates of gas reactions.
6. For ideal gases, the rates of irreversible processes can also be calculated using kinetic theory; these include the diffusion coefficient D, the thermal conductivity κ, and the viscosity η.
7. Since all of these calculations are concerned with ideal gases of hard spherical molecules, they are necessarily only approximate for real gases, but they are the predecessors of more advanced calculations.

REFERENCES

R. B. Bird, W. E. Steward, and E. N. Lightfoot, *Transport Phenomena*. Hoboken, NJ: Wiley, 2001.

S. Chapman and T. G. Cowling, *The Mathematical Theory of Non-Uniform Gases,* 3rd ed. New York: Cambridge University Press, 1970.

C. E. Hecht, *Statistical Thermodynamics and Kinetic Theory.* New York: Freeman, 1990.

J. O. Hirschfelder, C. F. Curtiss, and R. B. Bird, *The Molecular Theory of Gases and Liquids.* New York: Wiley, 1954.
P. C. Jordan, *Chemical Kinetics and Transport.* New York: Plenum, 1979.
D. A. McQuarrie, *Statistical Mechanics,* Chapter 16. Sausalito, CA: University Science Books, 2000.

PROBLEMS

Problems marked with an icon may be more conveniently solved on a personal computer with a mathematical program.

17.1 If the diameter of a gas molecule is 0.4 nm and each is imagined to be in a separate cube, what is the length of the side of the cube in molecular diameters at 0 °C and pressures of (*a*) 1 bar and (*b*) 1 Pa?

17.2 Plot the probability density $f(v)$ of molecular speeds versus speed for oxygen at 25 °C.

17.3 What is the ratio of the probability that gas molecules have two times the mean speed to the probability that they have the mean speed?

17.4 (*a*) Use equation 17.21 to calculate the average value of ϵ^2 for a molecule of an ideal gas. (*b*) Calculate the standard deviation σ_ϵ of the molecular energy from (*a*) and from equation 9.58. (*c*) Calculate the ratio of the standard deviation of the translational energy to the average translational energy.

17.5 Calculate the mean speed and the root-mean-square speed for the following set of molecules: 10 molecules moving 5×10^2 m s^{-1}, 20 molecules moving 10×10^2 m s^{-1}, and 5 molecules moving 15×10^2 m s^{-1}.

17.6 Calculate the most probable, mean, and root-mean-square speeds for oxygen molecules at 25 °C.

17.7 The mean speed of H_2 at 298 K is 1769 m s^{-1}. What is the mean speed of a hydrogen molecule relative to another hydrogen molecule? How do you rationalize your calculation?

17.8 What fraction of oxygen molecules at 300 K have velocities (*a*) between 400 and 410 m s^{-1} and (*b*) between 800 and 810 m s^{-1}? You can assume that $F(v)$ is independent of v in each of these intervals.

17.9 The standard deviation σ of a distribution is given by

$$\sigma = [\langle x^2 \rangle - \langle x \rangle^2]^{1/2}$$

What is the standard deviation of the distribution of speeds v of hydrogen molecules at 298.15 K?

17.10 Derive equations for $\overline{U}$ and $\overline{C}_V$ for any monatomic gas from kinetic theory.

17.11 Calculate the velocity of sound in nitrogen gas at 25 °C. (See Section 17.4.)

17.12 Calculate the speed of sound at 25 °C in (*a*) $H_2O(g)$ and (*b*) $CO_2(g)$. The molar heat capacities at constant pressure are given in Table C.2 and $\overline{C}_P - \overline{C}_V = R$.

17.13 (*a*) Calculate the collision frequency for a nitrogen molecule in nitrogen at 1 bar pressure and 25 °C. (*b*) What is the collision density? What is the effect on the collision density (*c*) of doubling the absolute temperature at constant pressure and (*d*) of doubling the pressure at constant temperature?

17.14 (*a*) Calculate the mean free path for hydrogen gas (d = 0.247 nm) at 1 bar and 0.1 Pa at 25 °C. (*b*) Repeat the calculation for chlorine gas (d = 0.496 nm).

17.15 The pressure in interplanetary space is estimated to be of the order of 10^{-14} Pa. Calculate (*a*) the average number of molecules per cubic centimeter, (*b*) the collision frequency, and (*c*) the mean free path in miles. Assume that only hydrogen atoms are present and that the temperature is 1000 K. Assume that d = 0.2 nm.

17.16 Calculate the collision frequency z_{11} and the collision density Z_{11} for molecular chlorine at 25 °C and 1 bar. The collision diameter is 0.544×10^{-9} m.

17.17 A gas mixture contains H_2 at 0.666 bar and O_2 at 0.333 bar at 25 °C. (*a*) What is the collision frequency z_{12} of a hydrogen molecule with an oxygen molecule? (*b*) What is the collision frequency z_{21} of an oxygen molecule with a hydrogen molecule? (*c*) What is the collision density Z_{12} between hydrogen molecules and oxygen molecules in mol L^{-1} s^{-1}? The collision diameters of H_2 and O_2 are 0.272 nm and 0.360 nm, respectively.

17.18 For O_2(g), d = 0.361 nm, Θ_r = 2.079 K, Θ_v = 2273.64 K, and M = 31.9988 g mol^{-1}. At 1 bar and 25 °C what is the average time between collisions? How many vibrational oscillations will have occurred during this time?

17.19 (*a*) How many molecules of H_2 strike the wall per unit area per unit time at 1 bar at 298 K? 1000 K? (*b*) How many molecules of O_2 strike the wall per unit area per unit time at 1 bar at 298 K? 1000 K?

17.20 In Section 17.6, we derived the expression for the flux J_N of gas molecules through a surface in terms of velocities v_x of molecules in the direction perpendicular to the surface. The derivation can be made in a more general way by considering that the molecules can approach the surface with velocity v at angles θ and ϕ in the system of spherical coordinates. In this case, the differential of the flux is given by

$$\mathrm{d}J_N = \frac{\rho}{4\pi} v\, F(v)\, \mathrm{d}v \cos\theta \sin\theta\, \mathrm{d}\theta\, \mathrm{d}\phi$$

Integrate this equation to obtain the flux J_N.

17.21 A Knudsen cell containing crystalline benzoic acid ($M = 122$ g mol^{-1}) is carefully weighed and placed in an evacuated chamber thermostated at 70 °C for 1 h. The circular hole through which effusion occurs is 0.60 mm in diameter. Calculate the sublimation pressure of benzoic acid at 70 °C in Pa from the fact that the weight loss is 56.7 mg.

17.22 R. B. Holden, R. Speiser, and H. L. Johnston [*J. Am. Chem. Soc.* **70**:3897 (1948)] found the rate of loss of weight of a Knudsen effusion cell containing finely divided beryllium to be 19.8×10^{-7} g cm^{-2} s^{-1} at 1320 K and 1210×10^{-7} g cm^{-2} s^{-1} at 1537 K. Calculate $\Delta_{sub}H$ for this temperature range.

17.23 A 5-mL container with a hole 10 μm in diameter is filled with hydrogen. This container is placed in an evacuated chamber at 0 °C. How long will it take for 90% of the hydrogen to effuse out?

17.24 The vapor pressure of naphthalene ($M = 128.16$ g mol^{-1}) is 17.7 Pa at 30 °C. Calculate the weight loss in a period of 2 h of a Knudsen cell filled with naphthalene and having a round hole 0.50 mm in diameter.

17.25 Atoms and molecules can escape from the uppermost layer of the earth's atmosphere only if they have the escape velocity. The minimum escape velocity is the velocity in the direction perpendicular to the surface of the earth. (*a*) Show that the minimum escape velocity v_e is given by

$$v_e = \left(\frac{2GM_{earth}}{R_{earth}}\right)^{1/2}$$

where M_{earth} is the mass of the earth, R_{earth} is the radius of the earth (radius of the uppermost layer of the atmosphere), and G is the gravitational constant, which is defined by

$$V(r) = -\frac{Gm_1m_2}{r}$$

Given: $G = 6.67 \times 10^{-11}$ J m kg^{-1}, $M_{earth} = 5.98 \times 10^{24}$ kg, and $R_{earth} = 6.36 \times 10^6$ m. An atom or molecule with mass m can escape if its kinetic energy is equal and opposite to the potential energy between the atom or molecule and the earth. (*b*) Calculate the escape velocity of an atom or molecule.

(M) **17.26** The viscosity of helium is 1.88×10^{-5} Pa s at 0 °C. Calculate (*a*) the collision diameter and (*b*) the diffusion coefficient at 1 bar.

17.27 What is the self-diffusion coefficient of radioactive CO_2 in ordinary CO_2 at 1 bar and 25 °C? The collision diameter is 0.40 nm.

17.28 The probability $F(\epsilon)$ that the molecular translational energy is in the range $\epsilon + d\epsilon$ is given by equation 17.20. Use this equation to calculate the most probable translational energy.

17.29 What is the speed of a molecule with $v_x = 100$ m s^{-1}, $v_y = 200$ m s^{-1}, and $v_z = 300$ m s^{-1}?

17.30 In the text, we derive $U_t = \frac{3}{2}RT$ by using the ideal gas law. Show that the same result may be obtained by averaging over the Maxwell speed distribution to obtain the kinetic energy of an average molecule.

$$U = \int_0^\infty \tfrac{1}{2}mv^2 F(v)\,dv$$

17.31 What is the ratio of the number of molecules having twice the most probable speed to the number having the most probable speed?

17.32 Suppose that a gas contains 10 molecules having an instantaneous speed of 2×10^2 m s^{-1}, 30 molecules with a speed of 4×10^2 m s^{-1}, and 15 molecules with a speed of 6×10^2 m s^{-1}. Calculate $\langle v \rangle$, v_{mp}, and $\langle v^2 \rangle^{1/2}$.

17.33 Calculate the root-mean-square speed of oxygen molecules having a kinetic energy of 10 kJ mol^{-1}. At what temperature would this be the root-mean-square speed?

17.34 What is the root-mean-square speed of a hexane molecule at 0 °C?

17.35 Calculate the velocity of sound in (*a*) He and (*b*) N_2 at 25 °C.

17.36 The speed of sound in argon at the triple point of water (273.16 K) has been measured at a series of low pressures and extrapolated to zero pressure to obtain

$$v_s^2 = 94\,756.75 \text{ m}^2 \text{ s}^{-2}$$

What is the value of the gas constant R if the molar mass of argon is 39.947 753 g mol^{-1} and $\gamma = \frac{5}{3}$?

17.37 What is the average time between collisions of an oxygen molecule in oxygen at 298 K and (*a*) 1 bar, (*b*) 10^{-6} bar, and (*c*) 10^{-12} bar? ($d = 0.36 \times 10^{-9}$ m.)

17.38 What is the mean free path of nitrogen at 1 bar and 25 °C? What is the average time between collisions?

17.39 Oxygen ($d = 0.361$ nm) is contained in a vessel at 250 Pa pressure and 25 °C. Calculate (*a*) the number of collisions between molecules per second per cubic meter and (*b*) the mean free path.

17.40 For O_2 at 10^{-3} bar at 25 °C, (*a*) what is the collision frequency z_{11}? (*b*) What is the collision density Z_{11}? (*c*) What is the average time between collisions of a single molecule?

17.41 An equal number of moles of H_2 and Cl_2 are mixed and held at 298 K and a total pressure of 1 bar. (*a*) Calculate the collision frequencies z_{12} and z_{21}, where hydrogen is component 1 and chlorine is component 2. (*b*) Calculate the collision density Z_{12}. Given: $d_1 = 0.272$ nm and $d_2 = 0.544$ nm.

17.42 Ultrahigh vacuum is defined as about 10^{-3} Pa. What is the mean free path and average number of collisions per second for a single O_2 molecule ($d = 0.361$ nm) at this pressure and 25 °C?

17.43 Calculate the number of collisions per square centimeter per second of oxygen molecules with a wall at a pressure of 1 bar and 25 °C.

17.44 Large vacuum chambers have been built for testing space vehicles at 10^{-6} Pa. Calculate (*a*) the mean free path of nitrogen ($d = 0.375$ nm) at this pressure and (*b*) the number of molecular impacts per square meter of wall per second at 25 °C.

17.45 The vapor pressure of water at 25 °C is 3160 Pa. (*a*) If every water molecule that strikes the surface of liquid water sticks, what is the rate of evaporation of molecules from a square centimeter of surface? (*b*) Using this result, find the rate of evaporation in g cm^{-2} min^{-1} of water into perfectly dry air.

17.46 A substance of $M = 200$ g mol^{-1} has a vapor pressure of 10^{-5} Pa at 25 °C. What mass of the substance will effuse from a Knudsen cell in 2 h through a hole 0.1 cm in diameter?

Computer Problems

17.A Plot the probability density for the molecular speed of nitrogen molecules at 200, 500, and 800 K.

17.B Plot the probability density of the x component of the velocity of nitrogen molecules at 200, 500, and 800 K.

17.C Plot the probability density for the velocity of oxygen molecules in an arbitrary direction at 100, 300, 500, and 1000 K.

17.D Calculate the probability density of various speeds v for molecular oxygen at 100, 300, 500, and 1000 K

17.E A sodium atom emits a frequency of 5×10^8 MHz from an emission cell at 500 K. (*a*) What is the standard deviation of the spectral line due to Doppler broadening? (*b*) Plot the shape of the spectral lines at 500 and 1000 K.

17.F (*a*) Calculate the probability density for molecular energy of a molecule of ideal gas at 100, 300, and 500 K. Note that the probability density is independent of the molar mass. (*b*) Calculate the average energies of an ideal gas molecule at these temperatures. If it bothers you that the energy at the maximum of the probability density for molecular energy differs from the average energy, see B. A. Morrow and D. F. Tessler, *J. Chem. Educ.* **59**:193 (1982).

17.G Plot points at the ends of random vectors with a normal distribution in a plane.

17.H Plot the most probable speed, the mean speed, and the root-mean-square speed of ideal gas molecules in m s^{-1} at 273.15 K versus molar mass M in g mol^{-1}.

18

Experimental Kinetics and Gas Reactions

So far this book has emphasized thermodynamics and equilibrium states, but now we move on to a more difficult subject: the rates of chemical reactions. You have a substance, or several substances, in the gas or liquid phase under a certain set of conditions, and you find that when you add another substance, add a catalyst, irradiate the system, or change the temperature or pressure, chemical changes take place in the system. From a purely experimental point of view there is a question as to how rapidly these changes take place and how their rates depend on independent variables such as concentrations of reactants and catalysts, temperature, and pressure. Sometimes it is found that the changes in concentrations with time follow rather simple mathematical relations all the way to equilibrium. However, chemists are not satisfied with simply representing experimental results, but want to understand what is going on in molecular terms. Even when there is a plausible mechanism for the chemical changes that occur, there is the question as to why the steps in the mechanism have the rates they do and how the time course of reaction could have been predicted in advance. Thus chemical kinetics is a challenging field, and we have delayed its consideration until thermodynamics, quantum mechanics, statistical mechanics, and kinetic theory can all be applied to its elucidation.

18.1 RATE OF REACTION

In discussing the rate of a reaction, the first thing to be clear about is its stoichiometry because the rates of consumption of reactants and rates of production of products are in the ratios of their stoichiometric numbers. As we have seen, a chemical reaction can be represented by

$$0 = \sum \nu_i B_i \tag{18.1}$$

where ν_i is the stoichiometric number of reactant B_i. The extent of reaction ξ (Section 2.11) is the same for each reactant and product, and $n_i = n_{i0} + \nu_i \xi$, where n_{i0} is the initial amount of reactant i. Taking the time derivative of n_i yields

$$\frac{dn_i}{dt} = \nu_i \frac{d\xi}{dt} \tag{18.2}$$

The quantity $d\xi/dt$ is called the **rate of conversion.** By convention the **rate of reaction** v is defined in terms of the rate of change of the concentration of a reactant or product so that

$$v = \frac{1}{V}\frac{d\xi}{dt} = \frac{1}{\nu_i V}\frac{dn_i}{dt} = \frac{1}{\nu_i}\frac{d[B_i]}{dt} \tag{18.3}$$

Therefore, the rate of the reaction

$$A + 2B = X \tag{18.4}$$

is

$$v = \frac{1}{-1}\frac{d[A]}{dt} = \frac{1}{-2}\frac{d[B]}{dt} = \frac{d[X]}{dt} \tag{18.5}$$

This has the advantage that the same rate of reaction v is obtained, no matter which reactant or product is studied. However, the rate of reaction does depend on how the stoichiometric equation is written; in discussing kinetics, stoichiometric equations with fractional stoichiometric numbers are avoided, even though they are permissible in thermodynamics.

If the reaction goes in the forward direction, the rate of reaction v is positive. If the reaction goes in the backward direction, the rate of reaction is negative. If the reaction is at equilibrium, the rate of reaction v is zero.

We will generally be concerned with reactions occurring at constant volume, but that is not always the case. If the volume changes during the reaction, equation 18.3 can be written

$$v = \frac{1}{\nu_i V}\frac{d([B_i]V)}{dt} = \frac{1}{\nu_i}\frac{d[B_i]}{dt} + \frac{[B_i]}{\nu_i V}\frac{dV}{dt} \tag{18.6}$$

Example 18.1 ***Rate of conversion and rate of reaction***

The reaction $H_2 + Br_2 = 2HBr$ is carried out in a 0.250-L reaction vessel. The change in the amount of Br_2 in 0.01 s is −0.001 mol. (*a*) What is the rate of conversion $d\xi/dt$? (*b*) What is the rate of reaction v? (*c*) What are the values of $d[H_2]/dt$, $d[Br_2]/dt$, and $d[HBr]/dt$?

(*a*) $$\frac{d\xi}{dt} = \frac{0.001\text{ mol}}{0.01\text{ s}} = 0.1\text{ mol s}^{-1}$$

(*b*) $$v = \left(\frac{1}{V}\right)\left(\frac{d\xi}{dt}\right) = \frac{0.1\text{ mol s}^{-1}}{0.25\text{ L}} = 0.40\text{ mol L}^{-1}\text{ s}^{-1}$$

(*c*) $$\frac{d[H_2]}{dt} = -0.40\text{ mol L}^{-1}\text{ s}^{-1}$$

$$\frac{d[Br_2]}{dt} = -0.40\text{ mol L}^{-1}\text{ s}^{-1}$$

$$\frac{d[HBr]}{dt} = 0.80\text{ mol L}^{-1}\text{ s}^{-1}$$

Example 18.2 ***Dependence of reaction rate on the chemical equation***

In discussing the rate v of a chemical reaction, it is important to know how the stoichiometric equation is written because the rate may depend on that. Show that different rates are obtained when the reaction is written as

$$2A + B = 2C$$

and

$$A + \tfrac{1}{2}B = C$$

According to the first stoichiometric equation,

$$v = \frac{1}{-2}\frac{d[A]}{dt} = -\frac{d[B]}{dt} = \frac{1}{2}\frac{d[C]}{dt}$$

According to the second stoichiometric equation,

$$v = \frac{1}{-1}\frac{d[A]}{dt} = -2\frac{d[B]}{dt} = \frac{d[C]}{dt}$$

This problem is largely avoided by not using fractional stoichiometric numbers in stoichiometric equations used to interpret rates of reactions.

The rates of chemical reactions are obtained from measurements of concentration as a function of time. Chemical analytical methods may be used when the reaction can be stopped suddenly. This may be done by rapid cooling for high-temperature reactions, or by catalyst inactivation for a catalyzed reaction. **Physical methods** are especially useful for determining the rate of a chemical reaction because they offer the possibility of continuous measurement of the extent of reaction. A wide variety of physical methods have been used, but spectroscopic methods are the most generally useful. Although the focus in chemical kinetics is on rates, we will often use integrated rate equations so that concentrations measured at various times can be used directly in the quantitative representation of kinetic data, as shown in the next section.

An important characteristic of any measurement method is its response time. The measuring device obviously must respond more rapidly than the concentration is changing. Pulsed lasers have opened many opportunities for studying very fast reactions. Reactions occurring in picoseconds (10^{-12} s) may be studied in this way. Special mixing methods have been developed for studying very fast reactions; their use for solution reactions is discussed in Section 20.6.

To study certain gas reactions at high temperatures it is necessary to heat the gas to the higher temperature very quickly because the reaction occurs rapidly. This may be accomplished by means of a **shock tube,** in which a shock wave is used to heat the gas suddenly. The tube is divided into two sections separated by a diaphragm that can be ruptured. The gas to be studied is placed on one side of the diaphragm, and a driver gas at higher pressure on the other. When the diaphragm is ruptured, a shock wave passes through the reacting gas, heating it suddenly to a higher temperature. In some reactions the extent of reaction may be determined as a function of time after passage of the shock wave by measuring the absorption of a beam of light passing perpendicularly across the tube.

Photochemical reactions may be initiated rapidly by a light pulse from a flash lamp or a laser. In the **flash photolysis** method, a reaction vessel is exposed to a very-high-intensity flash of visible or ultraviolet radiation. The flash dissociates and excites molecules in the sample, and the concentrations of these species are then determined over a period of time using subsequent flashes at a much lower intensity.

18.2 ORDER OF REACTION

At constant temperature the rate of reaction v depends on the concentrations of reactants and products; it may also depend on catalysts and inhibitors, but we will neglect that for now. For example, if reaction 18.4 goes essentially to completion to the right, the rate of reaction may be experimentally found to be

$$v = k[\mathrm{A}]^{\alpha}[\mathrm{B}]^{\beta} \tag{18.7}$$

In this **rate equation,** k is the **rate constant** (or rate coefficient), and α and β are independent of concentration and time. Note that α and β are *not* the stoichiometric numbers in the balanced chemical equation, but have to be obtained from rate experiments. The exponent α is referred to as the **order** of the reaction with respect to reactant A. The order is not necessarily an integer. When the rate law has this general form, the sum of the orders for the reactants is referred to as the **overall order** of the reaction. In this case, the overall order is $\alpha + \beta$.

Rates of reaction are often discussed as if the reactions go to completion to the right. In many cases this is a good approximation, but if the equilibrium constant for the reaction is of the order of unity, it is not. If a reaction does not go to completion, the rate law is of the form $v = v_{\mathrm{f}} - v_{\mathrm{b}}$, where v_{f} is the rate of the forward reaction and v_{b} is the rate of the backward reaction. Determination of the initial rate (see Fig. 18.3 later) yields the rate equation for v_{f}. The rate equation for v_{b} can be obtained by studying the backward reaction. At chemical equilibrium, the net rate v is equal to zero, and the equilibrium expression obtained from $v_f = v_b$ should be in agreement with thermodynamics.

The rate equation for a reaction is frequently more complicated than equation 18.7. For example, the rate of reaction may have additional terms such as $1+k[A]^n$ in the denominator when the mechanism is complicated. Also, the rate of reaction may be affected by products, even if the reaction goes essentially to completion. The concentration of a catalyst or inhibitor may also have to be included in the rate law. If a reaction can go by two paths, for example, a catalyzed path and an uncatalyzed path, the rate equation will consist of two additive terms, one for each path (see Section 20.7).

When rate laws are simple they can be integrated to give the concentration of a reactant as a function of time. We will derive integrated forms for first-order reactions, second-order reactions, zero-order reactions, and higher-order reactions for certain special cases. All of the equations derived in this section apply only at constant temperature and volume for reactions that go to completion.

The rate equation for a **first-order reaction** A $\rightarrow$ products,

$$-\frac{d[A]}{dt} = k[A] \tag{18.8}$$

may be integrated after it is written in the form

$$-\frac{d[A]}{[A]} = k\,dt \tag{18.9}$$

If the concentration of A is $[A]_1$ at t_1 and $[A]_2$ at t_2, then

$$-\int_{[A]_1}^{[A]_2} \frac{d[A]}{[A]} = k\int_{t_1}^{t_2} dt$$

$$\ln\frac{[A]_1}{[A]_2} = k(t_2 - t_1) \tag{18.10}$$

An especially useful form of this equation is obtained if t_1 is taken to be zero and the initial concentration is represented by $[A]_0$:

$$\ln\frac{[A]_0}{[A]} = kt \tag{18.11}$$

or

$$[A] = [A]_0\,e^{-kt} \tag{18.12}$$

$$\ln[A] = \ln[A]_0 - kt \tag{18.13}$$

The last form indicates that the rate constant k may be calculated from a plot of $\ln[A]$ versus t; the slope of the line in such a plot is $-k$. The rate constant for a first-order reaction has units of reciprocal time.

For a first-order reaction aA $\rightarrow$ products,

$$v = -\frac{1}{a}\frac{d[A]}{dt} = k[A] \tag{18.14}$$

so that

$$-\frac{d[A]}{dt} = ak[A] = k_A[A] \tag{18.15}$$

where $k_A = ak$.

It is evident from equation 18.10 that to determine the rate constant for a first-order reaction it is necessary only to determine the **ratio** of the concentrations at two times. Physical quantities proportional to concentration may be substituted for concentrations in these equations, since the proportionality constants cancel.

The **half-life** $t_{1/2}$ of a reaction is the time required for half of the reactant to disappear. For the first-order reaction A → products, equation 18.11 leads to

$$k = \frac{1}{t_{1/2}} \ln \frac{1}{1/2} = \frac{0.693}{t_{1/2}} \quad \text{or} \quad t_{1/2} = \frac{0.693}{k} \tag{18.16}$$

For the first-order reaction aA → products, equation 18.15 leads to $t_{1/2} = 0.693/k_A = 0.693/ak$. For a first-order reaction the half-life is independent of the initial concentration. Thus, 50% of the substance remains after one half-life, 25% remains after two half-lives, 12.5% after three, and so on (see Fig. 18.1).

The **relaxation time** τ for a first-order reaction is equal to the reciprocal of the first-order rate constant: $\tau = 1/k$. Thus, equation 18.12 may be written

$$[A] = [A]_0 \, e^{-t/\tau} \tag{18.17}$$

A reaction is **second order** if the rate is proportional to the square of the concentration of one reactant or is proportional to the product of the concentrations of two reactants. If the rate is proportional to the square of the concentration of A in the reaction A → products, the rate law

$$-\frac{d[A]}{dt} = k[A]^2 \tag{18.18}$$

may be integrated after arranging it in the form

$$-\frac{d[A]}{[A]^2} = k\,dt \tag{18.19}$$

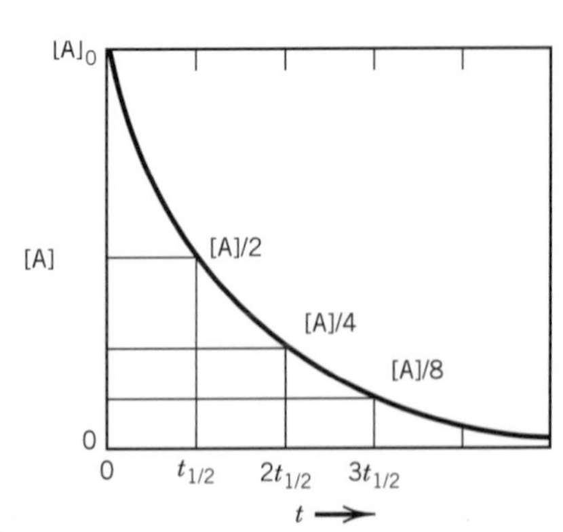

Figure 18.1 For a first-order reaction, one-half of the reactant disappears in $t_{1/2}$ independent of the initial concentration.

If the concentration is $[A]_0$ at $t = 0$ and [A] at time t, integration yields

$$kt = \frac{1}{[A]} - \frac{1}{[A]_0} \tag{18.20}$$

Thus, a plot of 1/[A] versus t is linear for such a second-order reaction, and the slope is equal to the second-order rate constant. This integrated rate equation also applies if the rate is given by k[A][B], the stoichiometry is represented by A + B = products, and A and B are initially at the same concentration. As may be seen from equation 18.20, the half-life for such a second-order reaction is given by

$$t_{1/2} = \frac{1}{k[A]_0} \tag{18.21}$$

Thus, the half-life is inversely proportional to the initial concentration (see Fig. 18.2).

If the reaction $aA \rightarrow$ products is second order, then

$$v = -\frac{1}{a}\frac{d[A]}{dt} = k[A]^2 \tag{18.22}$$

so that equation 18.20 becomes

$$akt = \frac{1}{[A]} - \frac{1}{[A]_0} \tag{18.23}$$

The half-life for this reaction is given by

$$t_{1/2} = \frac{1}{ak[A]_0} \tag{18.24}$$

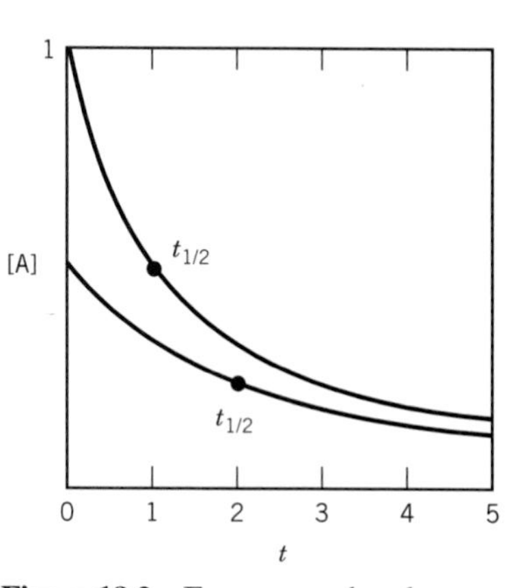

Figure 18.2 For a second-order reaction $A + B \rightarrow$ products, with $[A]_0 = [B]_0$, the half-life is inversely proportional to $[A]_0$; that is, $t_{1/2} = 1/k[A]_0$. Thus, when $[A]_0$ is reduced by a factor of 2, the half-life doubles.

A different integrated rate law for a second-order reaction is obtained if the rate is given by $k[A][B]$ and the stoichiometry is given by $aA + bB \rightarrow$ products. The rate constant is defined by

$$v = -\frac{1}{a}\frac{d[A]}{dt} = -\frac{1}{b}\frac{d[B]}{dt} = k[A][B] \tag{18.25}$$

If the initial reactants are not in stoichiometric proportions (i.e., $b[A]_0 \neq a[B]_0$), then the integrated rate equation is

$$kt = \frac{1}{b[A]_0 - a[B]_0} \ln \frac{[A][B]_0}{[A]_0[B]} \tag{18.26}$$

If the stoichiometric numbers of A and B are unity, equation 18.26 becomes

$$kt = \frac{1}{[A]_0 - [B]_0} \ln \frac{[A][B]_0}{[A]_0[B]} \tag{18.27}$$

Thus, for a second-order reaction in which the initial reactants are not in stoichiometric proportions, a plot of $\ln([A]/[B])$ versus t is linear.

The rate constant for a second-order reaction has the units $(\text{concentration})^{-1}$ s^{-1}. If the concentrations are expressed in mol L^{-1}, the second-order rate constant has the units L mol^{-1} s^{-1}. If concentrations are expressed in mol m^{-3}, the second-order rate constant has the units m^3 mol^{-1} s^{-1}. If concentrations are expressed in molecules per cubic centimeter, the second-order rate constant has the units cm^3 s^{-1}.

A reaction is **zero order** if the rate is independent of the concentration of the reactant:

$$-\frac{d[A]}{dt} = k \tag{18.28}$$

This can occur if the rate is limited by the concentration of a catalyst; in this case k may be proportional to the concentration of the catalyst. This can also occur in a photochemical reaction if the rate is determined by the light intensity; in this case k may be proportional to the light intensity. Integration of equation 18.28 yields

$$[A]_0 - [A] = kt \tag{18.29}$$

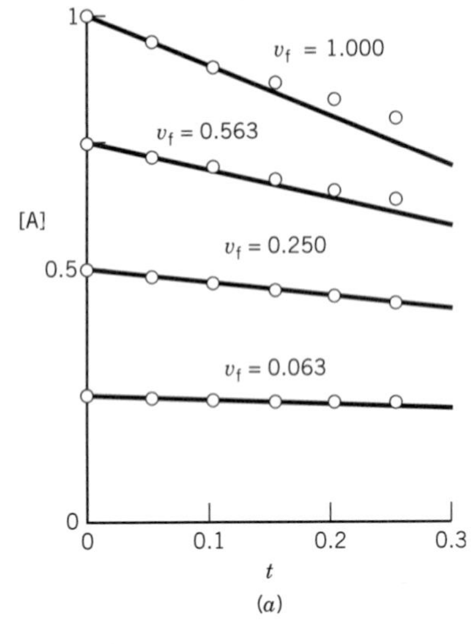

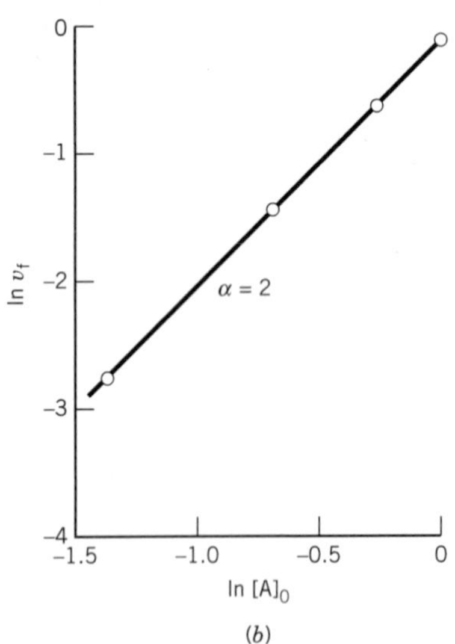

Figure 18.3 (*a*) Determination of the initial rate of a second-order reaction A → X at four initial concentrations. The initial slopes b determined by least squaring $[X]/t = b + ct$ are equal to the initial velocities v_f. The initial velocities v_f divided by L mol^{-1} s^{-1} are indicated. (*b*) Determination of the order of the reaction with respect to A by plotting $\ln v_f$ versus $\ln[A]_0$.

The zero-order rate constant has the units mol L^{-1} s^{-1}, or m^{-3} s^{-1} when concentrations are expressed in molecules per cubic meter.

The rate equation for a reaction A → products may also be integrated if it is of the form $k[A]^\alpha$, where $\alpha = \frac{1}{2}, \frac{3}{2}, 2, 3, \ldots$. All of these rate equations yield

$$\frac{1}{[A]^{\alpha-1}} - \frac{1}{[A]_0^{\alpha-1}} = (\alpha - 1)kt \tag{18.30}$$

In Section 18.11 we will see that the rate of reaction of H_2 with Br_2 is proportional to $[Br_2]^{1/2}$ in the absence of HBr and proportional to $[Br_2]^{3/2}$ if sufficient HBr is added initially. The rate of the formation of phosgene ($COCl_2$) from CO and Cl_2 is given by $k[Cl_2]^{3/2}[CO]$.

The first objective in studying the kinetics of a chemical reaction is to determine the rate equation. As pointed out earlier, the rate equation for a reaction A + 2B = X may be of the form $v = k[A]^\alpha[B]^\beta$, but it may be more complicated (see Sections 18.3 and 18.4). In particular, the concentration of the product X may also occur in the rate equation for the forward reaction. To avoid this complication, it is desirable to determine the **initial reaction rate,** that is, the rate at the initial concentrations of the reactants in the absence of product.

Figure 18.3*a* shows the concentrations of A at various times determined for the reaction

$$A = X \tag{18.31}$$

by use of a physical method, such as light absorption. To determine the initial rate at each of the four initial concentrations, the rates may be estimated from successive experimental points using $\Delta c/\Delta t$ and extrapolated to $t = 0$. However, it is better to fit the concentration of a product to the power series $[X] = bt + ct^2$ using the method of least squares. If data on only the first 10% of reaction are used, two terms are sufficient for most analyses.* Linear regression may be used if the equation is written $[X]/t = b + ct$. The parameter b is the initial (forward) velocity v_f of equation 18.31. Then a plot of $\ln v_f$ versus $\ln[A]_0$ may be used to calculate the order with respect to A, as shown in Fig. 18.3*b*. The advantage of this method is that it avoids complications due to products.

In determining initial velocities it is advantageous to use high concentrations of all the reactants but one, so that only one concentration changes significantly during the kinetics experiment. Under these conditions the reaction will have the order for the reactant at the lowest concentration. This is referred to as the **method of isolation.** If the reaction is first order with respect to the substance at low concentration the overall reaction under these conditions is said to be pseudo–first order. The pseudo–first-order rate constant k' is directly proportional to the concentration of the reactant at the higher concentration if the rate is also first order with respect to that substance:

$$v = k[A][B]_0 = (k[B]_0)[A] = k'[A] \tag{18.32}$$

since $[B]_0 \gg [A]_0$.

The concentration of A is plotted versus time for zero-, half-, first-, and second-order reactions in Fig 18.4*a*. In each case the initial concentration was

*K. J. Hall, T. I. Quickenden, and D. W. Watts, *J. Chem. Educ.* **53**:493 (1976).

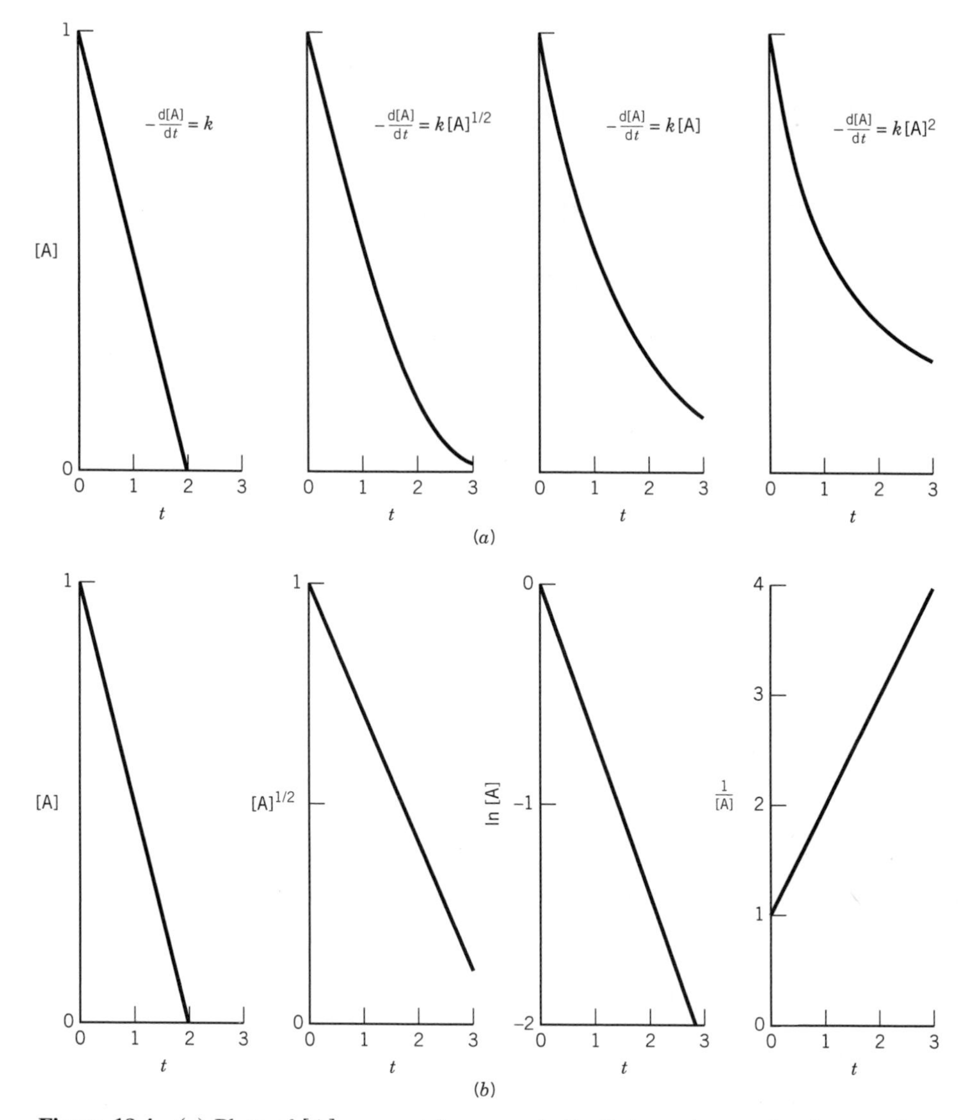

Figure 18.4 (*a*) Plots of [A] versus t for zero-, half-, first-, and second-order reactions with $[A]_0 = 1$ mol L^{-1}, each having a half-life of 1 min. (*b*) Linear plots for the zero-, half-, first-, and second-order reactions.

1 mol L^{-1}, and the volume was constant. The rate constant was adjusted so that the half-life in each case was 1 min. It is apparent that the change in concentration with time is not very different for these different orders during the first part of the reaction. Thus, if a reaction is followed only during the first half-life, it takes very accurate analytical data to determine the order. After longer times the differences are greater, and the plots shown in Fig. 18.4*b* can be used to distinguish the order and to determine the value of the rate constant.

The study of initial rates may not reveal the full rate law. For example, in Sections 18.7 and 18.11, we will see that products may be inhibitory. The effect of a product on the reaction rate may be determined by adding it initially. A reaction is said to be autocatalytic if a product of the reaction causes it to go faster.

So far we have considered reactions that go essentially to completion. Now we will consider first-order reactions that do not go to completion, parallel first-order reactions, and consecutive first-order reactions.

18.3 REVERSIBLE FIRST-ORDER REACTIONS

When we consider a system in which both the forward and backward reactions are important, the net rate of reaction can generally be expressed as the difference between the rate in the forward direction and that in the backward direction. As a simple example we will consider the reversible reaction

$$A \underset{k_2}{\overset{k_1}{\rightleftharpoons}} B \tag{18.33}$$

The rate law for this reversible* reaction is

$$\frac{d[A]}{dt} = -k_1[A] + k_2[B] \tag{18.34}$$

If initially only A is present, then $[B] = [A]_0 - [A]$, and

$$\begin{aligned}\frac{d[A]}{dt} &= -k_1[A] + k_2([A]_0 - [A]) = k_2[A]_0 - (k_1 + k_2)[A] \\ &= -(k_1 + k_2)\left([A] - \frac{k_2}{k_1 + k_2}[A]_0\right) = -(k_1 + k_2)([A] - [A]_{eq})\end{aligned} \tag{18.35}$$

where the expression for $[A]_{eq}$ is obtained as follows:

$$\frac{[B]_{eq}}{[A]_{eq}} = \frac{[A]_0 - [A]_{eq}}{[A]_{eq}} = \frac{k_1}{k_2} = K \tag{18.36}$$

where K is the equilibrium constant for reaction 18.33. This equation can be solved for $[A]_{eq}$:

$$[A]_{eq} = \frac{k_2}{k_1 + k_2}[A]_0 \tag{18.37}$$

The expression for the equilibrium constant in terms of the forward and backward rate constants can be used to eliminate k_2 from equation 18.34:

$$\begin{aligned}\frac{d[A]}{dt} &= -k_1[A] + \frac{k_1}{K}[B] \\ &= -k_1[A](1 - [B]/[A]K)\end{aligned} \tag{18.38}$$

Thus, we can see that the reaction has an initial rate of $k_1[A]$ and that it slows down as B accumulates. When $[B]/[A] = K$, the reaction is at equilibrium and the rate is zero. Similar equations can be derived for more complicated mechanisms.

*Note that "reversible" here means something different than in thermodynamics. In thermodynamics a reversible process goes through equilibrium states. In kinetics this term is used to indicate that the backward reaction is significant.

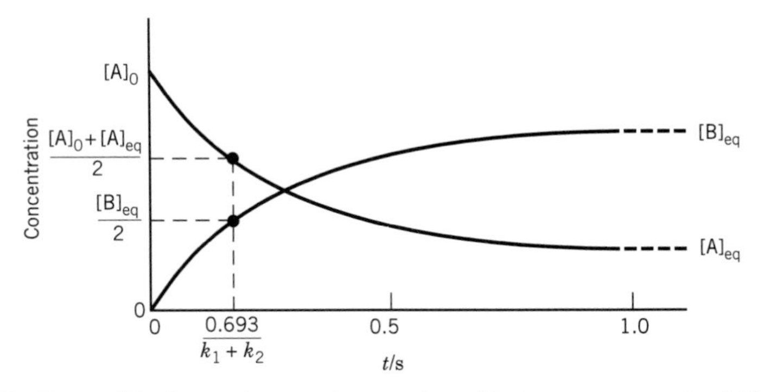

Figure 18.5 Reversible first-order reaction starting with A at concentration $[A]_0$. The values of the rate constants are $k_1 = 3\ s^{-1}$ and $k_2 = 1\ s^{-1}$.

Integrating equation 18.35 yields

$$-\int_{[A]_0}^{[A]} \frac{d[A]}{[A] - [A]_{eq}} = (k_1 + k_2)\int_0^t dt \tag{18.39}$$

$$\ln \frac{[A]_0 - [A]_{eq}}{[A] - [A]_{eq}} = (k_1 + k_2)t \tag{18.40}$$

For such a reaction the concentrations of A and B as functions of time are illustrated in Fig. 18.5. The concentrations of A and B will be halfway to their equilibrium values in a time of $0.693/(k_1 + k_2)$.

Thus, a plot of $-\ln([A] - [A]_{eq})$ versus time is linear, and $(k_1 + k_2)$ may be calculated from the slope. It should be especially noted that the rate of approach to equilibrium in this reaction is determined by the sum of the rate constants of the forward and reverse reactions, not by the rate constant for the forward reaction. Since the ratio k_1/k_2 may be calculated from the equilibrium concentrations by use of equation 18.36, the values of k_1 and k_2 may be obtained.

For some purposes it is more convenient to have equations for [A] and [B] in exponential form. Equation 18.40 may be written as

$$[A] = \frac{k_2[A]_0}{k_1 + k_2}\left[1 + \frac{k_1}{k_2} e^{-(k_1+k_2)t}\right] \tag{18.41}$$

Since $[B] = [A]_0 - [A]$,

$$[B] = \frac{k_1[A]_0}{k_1 + k_2}\left[1 - e^{-(k_1+k_2)t}\right] \tag{18.42}$$

18.4 CONSECUTIVE FIRST-ORDER REACTIONS

Two consecutive irreversible first-order reactions can be represented by

$$A_1 \xrightarrow{k_1} A_2 \xrightarrow{k_2} A_3 \tag{18.43}$$

To determine the way in which the concentrations of the substances in such a mechanism depend on time, the rate equations are first written down for each substance. It is then necessary to obtain the solution of these simultaneous differential equations. For the foregoing reactions the rate equations are as follows:

$$\frac{d[A_1]}{dt} = -k_1[A_1] \tag{18.44}$$

$$\frac{d[A_2]}{dt} = k_1[A_1] - k_2[A_2] \tag{18.45}$$

$$\frac{d[A_3]}{dt} = k_2[A_2] \tag{18.46}$$

It will be assumed that, at $t = 0$, $[A_1] = [A_1]_0$, $[A_2] = 0$, and $[A_3] = 0$. The rate equation for A_1 is readily integrated to obtain

$$[A_1] = [A_1]_0\, e^{-k_1 t} \tag{18.47}$$

Substitution of this expression into equation 18.45 yields

$$\frac{d[A_2]}{dt} = k_1[A_1]_0\, e^{-k_1 t} - k_2[A_2] \tag{18.48}$$

which may be integrated to obtain

$$[A_2] = \frac{k_1[A_1]_0}{k_2 - k_1}(e^{-k_1 t} - e^{-k_2 t}) \tag{18.49}$$

Because of conservation of the number of moles, $[A_1]_0 = [A_1] + [A_2] + [A_3]$ at any time, so the concentration of A_3 is given by

$$[A_3] = [A_1]_0 - [A_1] - [A_2] = [A_1]_0\left[1 + \frac{1}{k_1 - k_2}(k_2\, e^{-k_1 t} - k_1\, e^{-k_2 t})\right] \tag{18.50}$$

Figure 18.6 shows the concentrations of A_1, A_2, and A_3 as a function of time when $k_1 = 1\ s^{-1}$ and $k_2 = 1, 5$, and $25\ s^{-1}$. In Fig. 18.6*a* note the induction period in the appearance of A_3; it is not formed initially because A_2 has to be formed first. As k_2 is increased, this induction period becomes less important. Also note that as k_2 becomes larger than k_1, less A_2 is formed and that after the induction period, $d[A_2]/dt \approx 0$. This is the basis for the **steady-state approximation** that is often useful in deriving rate equations for systems of reactions. When $k_2 \gg k_1$, this approximation can be used to treat the kinetics of the system of reactions in equation 18.43. In the steady state, the rate of change of $[A_2]$ is zero, so that

$$\frac{d[A_2]}{dt} = -k_2[A_2] + k_1[A_1] = 0 \tag{18.51}$$

The concentration of A_2 in the steady state is given by $[A_2]_{ss} = (k_1/k_2)[A_1]$, and since $[A_1]$ is given by

$$[A_1] = [A_1]_0 e^{-k_1 t} \tag{18.52}$$

the steady-state concentration of A_2 is given by

$$[A_2]_{ss} = \frac{k_1}{k_2}[A_1]_0 e^{-k_1 t} \tag{18.53}$$

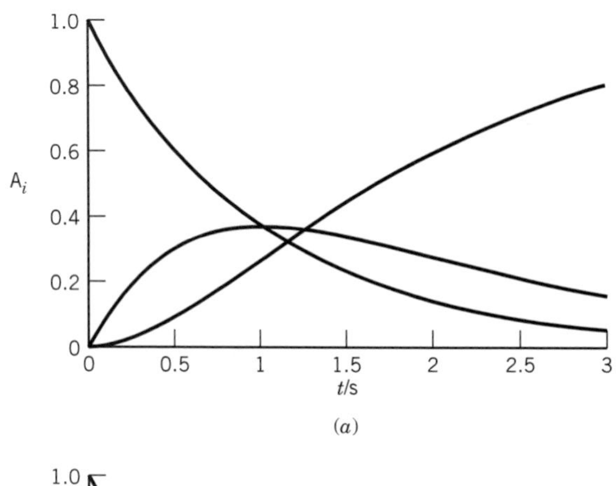

(a)

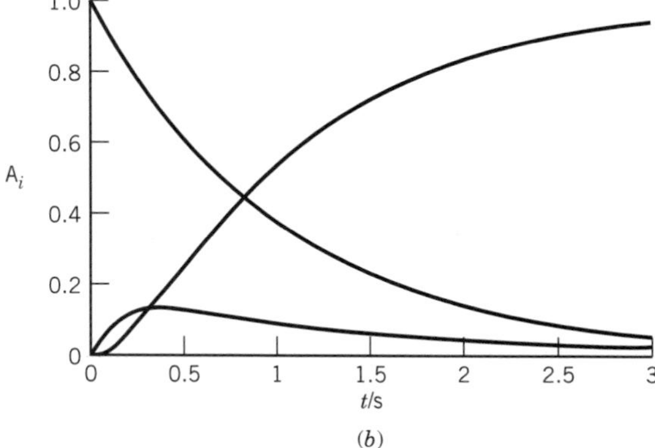

(b)

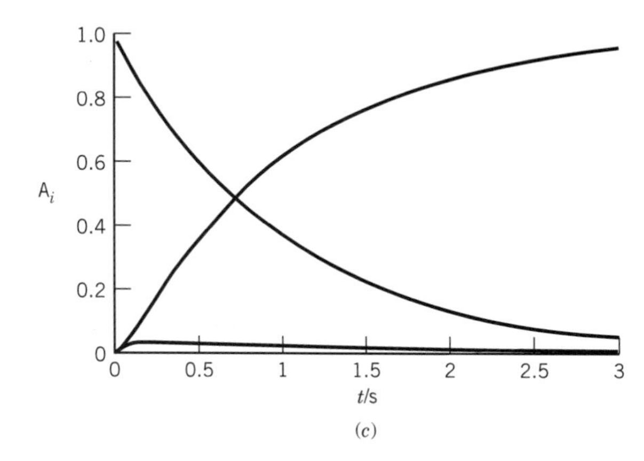

(c)

Figure 18.6 Plots of concentrations of reactants in $A_1 \rightarrow A_2 \rightarrow A_3$ when $k_1 = 1\ s^{-1}$ and k_2 is (*a*) 1, (*b*) 5, and (*c*) 25 s^{-1}. These plots could be calculated using equations 18.47, 18.49, and 18.50, but they were actually calculated by solving the three simultaneous differential equations that describe the system (see Computer Problem 18.A).

The steady-state concentration of A_3 is given by

$$[A_3]_{ss} = [A_1]_0 - [A_1] - [A_2]_{ss} = [A_1]_0\left[1 - \left(1 + \frac{k_1}{k_2}\right)e^{-k_1 t}\right] \tag{18.54}$$

This agrees with equation 18.50 when the term $k_1 e^{-k_2 t}$ is neglected. Thus as k_2 becomes larger and larger with respect to k_1, the behavior of this system approaches that of $A_1 \rightarrow A_3$, as can be seen from Fig. 18.6.

Since all reactions are reversible to some extent, it is more realistic to consider

$$A_1 \underset{k_2}{\overset{k_1}{\rightleftarrows}} A_2 \underset{k_4}{\overset{k_3}{\rightleftarrows}} A_3 \tag{18.55}$$

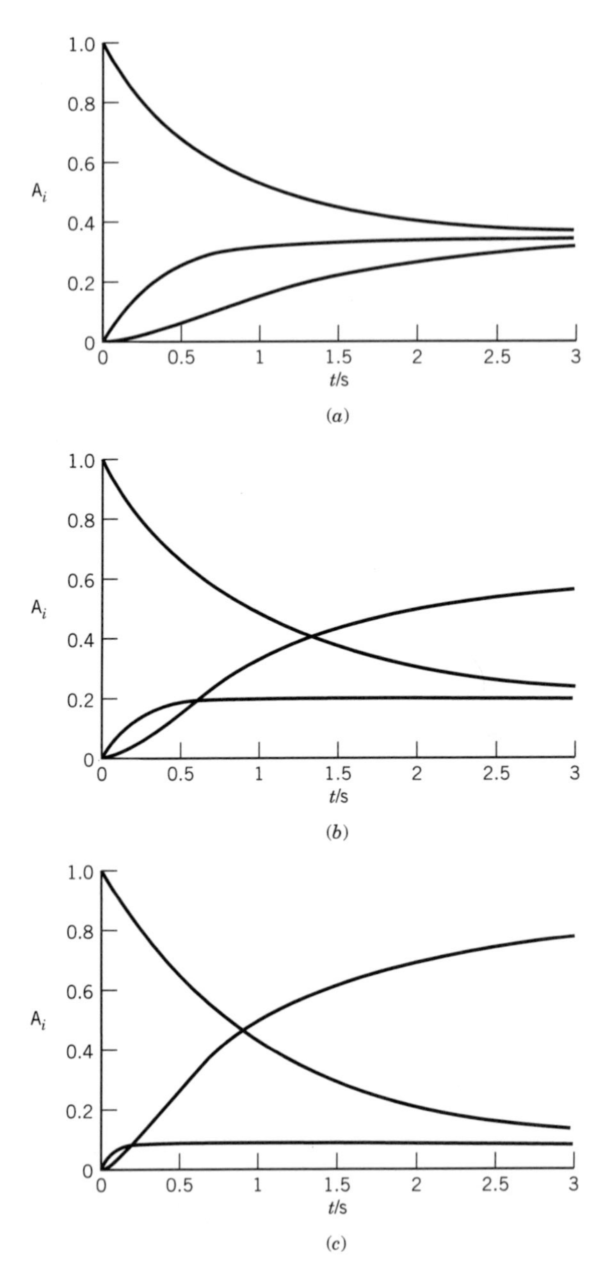

Figure 18.7 Plots of concentrations of reactants in

$$A_1 \underset{k_2}{\overset{k_1}{\rightleftarrows}} A_2 \underset{k_4}{\overset{k_3}{\rightleftarrows}} A_3$$

when $k_1 = k_2 = k_4 = 1\ s^{-1}$ and k_3 is (*a*) 1, (*b*) 3, and (*c*) 9 s^{-1}. These plots were calculated by solving the three simultaneous differential equations that describe the system (see Computer Problem 18.B).

Figure 18.7*a* shows the concentrations of these three reactants as a function of time when all the rate constants are 1 s^{-1}; in Fig. 18.7*b* $k_3 = 3\ s^{-1}$, and in Fig. 18.7*c* $k_3 = 9\ s^{-1}$. When all the rate constants are the same there is an induction period in the formation of A_3, but as k_3 is increased there is less of an induction period and A_2 shows a sustained steady state. As reaction systems become more complicated, it becomes impractical to derive general equations, such as equations 18.47, 18.49, and 18.50, but concentrations of reactants can be calculated by solving the system of rate equations with a computer. When a system involves a wide range of rate constants, special computer methods are needed: When concentrations change rapidly, small steps are required in numerical integration, but when the concentrations change slowly, larger steps can be used. Sets of differential equations of this type are said to be "stiff," in analogy with the equations for the vibration of a stiff rod.

Example 18.3 ***Parallel reactions***

Derive equations for the concentrations of B and C as functions of time as they are produced in the parallel reactions

$$\mathrm{A} \xrightarrow{k_1} \mathrm{B} \qquad \mathrm{A} \xrightarrow{k_2} \mathrm{C}$$

The rate equation for A is

$$-\frac{\mathrm{d}[\mathrm{A}]}{\mathrm{d}t} = k_1[\mathrm{A}] + k_2[\mathrm{A}] = (k_1 + k_2)[\mathrm{A}]$$

Thus, the disappearance of A will be first order, and on the basis of the earlier discussion of first-order reactions we can write

$$[\mathrm{A}] = [\mathrm{A}]_0\,\mathrm{e}^{-(k_1+k_2)t}$$

The rate equation for B is

$$\frac{\mathrm{d}[\mathrm{B}]}{\mathrm{d}t} = k_1[\mathrm{A}] = k_1[\mathrm{A}]_0\,\mathrm{e}^{-(k_1+k_2)t}$$

Integration yields

$$[\mathrm{B}] = \frac{-k_1[\mathrm{A}]_0}{(k_1+k_2)}\,\mathrm{e}^{-(k_1+k_2)t} + \text{constant}$$

If $[\mathrm{B}] = 0$ at $t = 0$, the constant is $k_1[\mathrm{A}]_0/(k_1 + k_2)$ and

$$[\mathrm{B}] = \frac{k_1[\mathrm{A}]_0}{(k_1+k_2)}[1 - \mathrm{e}^{-(k_1+k_2)t}]$$

Thus, the fraction of A that is converted to B at infinite time is $k_1/(k_1 + k_2)$. At any time the sum of [A], [B], and [C] must be equal to the total concentration of A at the beginning, $[\mathrm{A}]_0$. Consequently, if $[\mathrm{C}]_0 = 0$, then

$$[\mathrm{C}] = \frac{k_2[\mathrm{A}]_0}{(k_1+k_2)}[1 - \mathrm{e}^{-(k_1+k_2)t}]$$

It is apparent from these equations that the ratio of the concentrations of B and C is always given by k_1/k_2, which is referred to as the **branching ratio.**

18.5 MICROSCOPIC REVERSIBILITY AND DETAILED BALANCE*

The principle of microscopic reversibility is a consequence of the invariance of the equations of classical mechanics when time is reversed. Consider a particle that

*L. Onsager, *Phys. Rev.* **37:**405 (1931); B. H. Mahan, *J. Chem. Educ.* **52:**299 (1975).

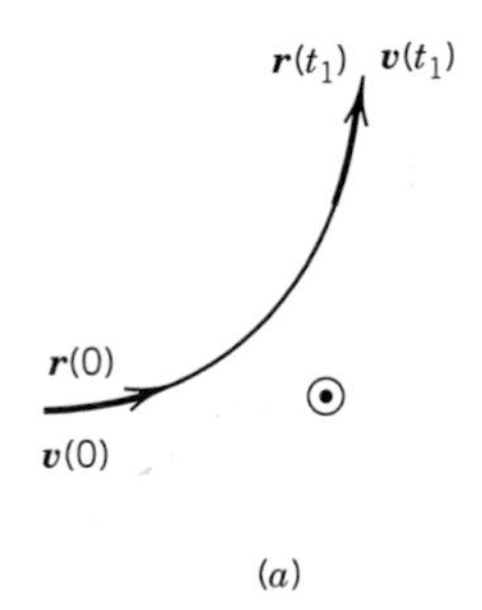

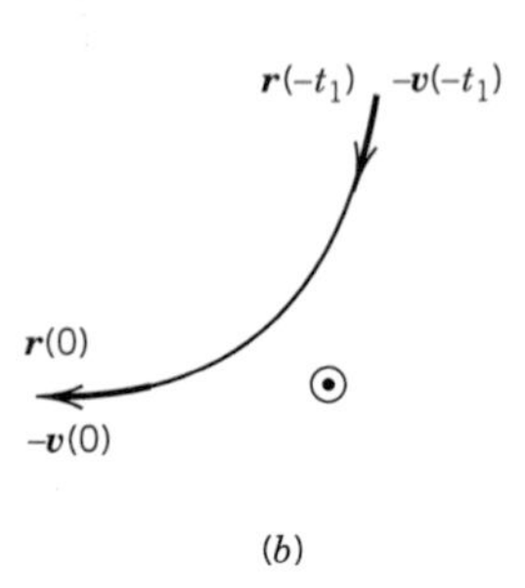

Figure 18.8 (*a*) A particle with coordinates $\boldsymbol{r}(0)$ and velocity vector $\boldsymbol{v}(0)$ moves under the action of a force to position $\boldsymbol{r}(t_1)$, where it has the velocity vector $\boldsymbol{v}(t_1)$. (*b*) We imagine that the direction of motion is reversed and the clock is set to read $-t_1$. At $t = 0$, the particle will have returned to its original position, but its velocity components are reversed. [From B. H. Mahan, *J. Chem. Educ.* **52**:299 (1975). Copyright ©1975 *Journal of Chemical Education.*]

moves under the action of a force that is a function of position only. The particle moves from position $\boldsymbol{r}(0)$ at $t = 0$ to $\boldsymbol{r}(t_1)$ at time t_1, as shown in Fig. 18.8*a*. The initial velocity of the particle is $\boldsymbol{v}(0)$, and its velocity at t_1 is $\boldsymbol{v}(t_1)$. Now suppose that at time t_1 we could instantaneously reverse all of the components of the velocity $\boldsymbol{v}(t_1)$ and allow the particle to move for another time period t_1. The particle would retrace its path to its initial position $\boldsymbol{r}(0)$, but the velocity components would be reversed. As shown in Fig. 18.8*b*, the reversed trajectory can be thought of as beginning at a time $-t_1$ and evolving as time goes forward to $t = 0$. A system is said to be invariant under time reversal if the equation is invariant under the following transformation:

$$\begin{aligned} t &= -t \\ \boldsymbol{r}(t) &= \boldsymbol{r}(-t) \\ \boldsymbol{v}(t) &= -\boldsymbol{v}(-t) \end{aligned} \tag{18.56}$$

Quantum mechanics is also time-reversal invariant. The principle of microscopic reversibility illustrated here for a classical particle can be extended to transition probabilities and cross reactions.

Starting with the principle of microscopic reversibility, it can be shown that, for an elementary reaction (Section 18.7), the ratio of the forward and backward rate constants is equal to the equilibrium constant obtained from statistical mechanics. For an elementary reaction

$$\mathrm{A + B = C + D} \tag{18.57}$$

at equilibrium,

$$\frac{[\mathrm{C}]_{\mathrm{eq}}[\mathrm{D}]_{\mathrm{eq}}}{[\mathrm{A}]_{\mathrm{eq}}[\mathrm{B}]_{\mathrm{eq}}} = \frac{k_{\mathrm{f}}}{k_{\mathrm{b}}} \tag{18.58}$$

so that

$$k_{\mathrm{f}}[\mathrm{A}]_{\mathrm{eq}}[\mathrm{B}]_{\mathrm{eq}} = k_{\mathrm{b}}[\mathrm{C}]_{\mathrm{eq}}[\mathrm{D}]_{\mathrm{eq}} \tag{18.59}$$

Thus, for an elementary reaction at equilibrium, the rate of the forward reaction is equal to the rate of the backward reaction. This is an example of what is referred to as the **principle of detailed balance.** In a system with many reactions this principle applies to each reaction individually.

As an application of this principle, consider the suggestion that the isomers A, B, and C can be interconverted by the following mechanism:

$$\begin{array}{c} \mathrm{A} \longrightarrow \mathrm{B} \\ \nwarrow \quad \swarrow \\ \mathrm{C} \end{array} \tag{18.60}$$

Although this mechanism at first sight appears reasonable, no actual process follows this mechanism because it violates the principle of detailed balance. According to this principle, at equilibrium the forward rate of *each* step is equal to the backward rate of that step. The mechanism has to be written as

$$\begin{array}{c} \mathrm{A} \underset{k_2}{\overset{k_1}{\rightleftharpoons}} \mathrm{B} \\ k_5 \searrow\!\!\nwarrow k_6 \qquad k_4 \swarrow\!\!\nearrow k_3 \\ \mathrm{C} \end{array} \tag{18.61}$$

with

$$\frac{[\mathrm{B}]_{\mathrm{eq}}}{[\mathrm{A}]_{\mathrm{eq}}} = \frac{k_1}{k_2} \tag{18.62}$$

$$\frac{[\mathrm{C}]_{\mathrm{eq}}}{[\mathrm{B}]_{\mathrm{eq}}} = \frac{k_3}{k_4} \tag{18.63}$$

$$\frac{[\mathrm{A}]_{\mathrm{eq}}}{[\mathrm{C}]_{\mathrm{eq}}} = \frac{k_5}{k_6} \tag{18.64}$$

Multiplying the left sides of equations 18.62, 18.63, and 18.64 yields unity, so that

$$\frac{k_1 k_3 k_5}{k_2 k_4 k_6} = 1 \tag{18.65}$$

and the six rate constants of mechanism 18.61 are not independent.

Comment:

The existence of a backward reaction is often ignored in discussions of rate equations, and that is certainly justified from an experimental point of view when the equilibrium constant is very large. It is important to remember, however, that an equilibrium is reached in which the transition of molecules from products to reactants occurs at the same rate as the transition from reactants to products, and that this applies individually to each path for the reaction.

18.6 EFFECT OF TEMPERATURE

The dependence of rate constants on temperature over a limited range can usually be represented by an empirical equation proposed by Arrhenius in 1889:

$$k = A\,\mathrm{e}^{-E_{\mathrm{a}}/RT} \tag{18.66}$$

where A is the **pre-exponential factor** and E_{a} is the **activation energy.** The pre-exponential factor A has the same units as the rate constant. Equation 18.66 may be written in logarithmic form:

$$\ln k = \ln A - \frac{E_{\mathrm{a}}}{RT} \tag{18.67}$$

According to this equation, a straight line should be obtained when the logarithm of the rate constant is plotted against the reciprocal of the absolute temperature. This is often called an Arrhenius plot. Differentiating equation 18.67 with respect to temperature yields

$$E_{\mathrm{a}} = RT^2 \frac{\mathrm{d}\ln k}{\mathrm{d}T} \tag{18.68}$$

This equation may be regarded as the definition of the activation energy. When this equation is integrated we obtain

$$\ln \frac{k_2}{k_1} = \frac{E_{\mathrm{a}}}{R}\left(\frac{T_2 - T_1}{T_1 T_2}\right) \tag{18.69}$$

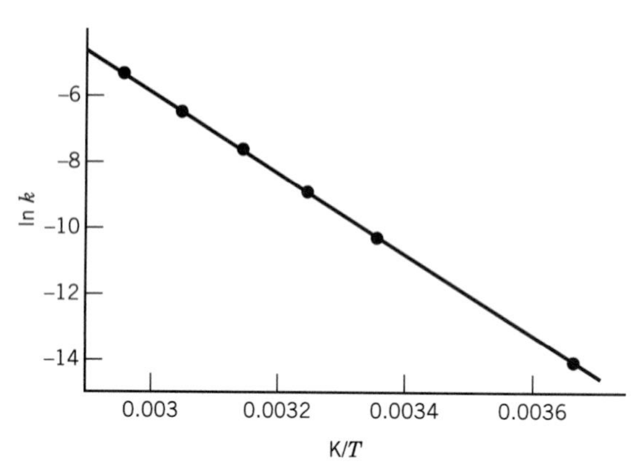

Figure 18.9 Plot of $\ln k$ versus $1/T$ for the decomposition of N_2O_5 from which the Arrhenius activation energy E_a may be calculated. (See Computer Problem 18.F.)

Example 18.4 ***Determination of the activation energy***

The rate constants for the first-order gas reaction $N_2O_5 = 2NO_2 + \frac{1}{2}O_2$ are as follows:

T/K	273	298	308	318	328	338
$k/10^{-5}\ \text{s}^{-1}$	0.0787	3.46	13.5	49.8	150	487

What are the values of the activation energy and the pre-exponential factor?

The plot of ln k versus $1/T$ is given in Fig. 18.9. The plot of the points was fit by the least squares method, which yielded a slope of 12 375 K^{-1} and an intercept of 31.27. Therefore, the activation energy $E_a = (12\,375\ \text{K}^{-1})(8.314\ \text{J K}^{-1}\ \text{mol}^{-1}) = 103 \times 10^3\ \text{J mol}^{-1}$. The pre-exponential factor is given by $\exp(31.27) = 3.96 \times 10^{13}\ \text{s}^{-1}$. Thus, equation 18.66 becomes

$$k = (3.96 \times 10^{13}\ \text{s}^{-1})\exp\left[\frac{-103 \times 10^3\ \text{J mol}^{-1}}{(8.3145\ \text{J K}^{-1}\ \text{mol}^{-1})T}\right]$$

It should be realized that for any given pre-exponential factor there is a fairly narrow range of activation energies that will give reaction rates in the range measurable by conventional techniques, that is, with half-lives from 1 min to 10 days. For example, if $A = 10^{13}\ \text{s}^{-1}$ and the temperature is 298 K, reactions with activation energies less than about 80 kJ mol^{-1} will be too fast to study with ordinary methods, and reactions with activation energies greater than about 100 kJ mol^{-1} will be too slow.

The Arrhenius plots for some reactions are curved, so the activation energy is a function of temperature. As we will see in the next chapter there are theoretical reasons for expecting k for some reactions to vary with temperature according to

$$k = aT^m \mathrm{e}^{-E_0/RT} \tag{18.70}$$

When m is known, E_0 may be calculated from the slope of a plot of $\ln(k/T^m)$ versus $1/T$. If m is not provided by a theory, it is difficult to determine its value experimentally because the exponential dependence on $1/T$ is usually much stronger than the temperature dependence of T^m.

Some reactions actually proceed more slowly at higher temperatures. For them the activation energy is negative, but for such reactions other equations may be more suitable in representing the rate constant as a function of temperature.

Example 18.5 ***Curved Arrhenius plots***

For a reaction that follows equation 18.70, what are the Arrhenius parameters at temperature T'?

The Arrhenius activation energy is defined by equation 18.68, so it may be obtained as follows:

$$\ln k = \ln a + m \ln T - \frac{E_0}{RT}$$

$$E_a = RT^2 \frac{d \ln k}{dT} = RT^2 \left(\frac{m}{T} + \frac{E_0}{RT^2} \right)$$

$$= E_0 + mRT$$

Thus, E_0 is a hypothetical activation energy at absolute zero. Substitution of this relation in equation 18.70 yields

$$k = aT^m e^m e^{-E_a/RT} \tag{18.71}$$

Thus, the pre-exponential factor A at temperature T' is equal to $a(T')^m e^m$. When the Arrhenius plot is curved and E_a is calculated at two temperatures, E_0 and m can be calculated. Then the factor a can be calculated from the rate constant at either temperature.

18.7 MECHANISMS OF CHEMICAL REACTIONS

Many reactions that follow simple rate laws such as we have been discussing actually occur through a series of steps. These steps are called **elementary** reactions because they cannot be broken down further into simpler chemical reactions. However, when we look at elementary reactions in more detail in Sections 18.9 and 18.10, we will see that they may involve identifiable **physical** steps. The sequence of elementary reactions that add up to give the overall reaction is called the **mechanism** of the reaction. A mechanism is a hypothesis about the elementary steps through which chemical change occurs. Sometimes the elementary steps can be studied in isolation. However, the evidence for a mechanism is often indirect, and there is always the possibility that a different mechanism is also in accord with all the facts about the kinetics of the reaction and is in better accord with other knowledge about the reactants and intermediates involved. A valid mechanism must of course explain the rate law of the backward reaction as well as the forward reaction.

Often it is possible to devise several mechanisms that are consistent with an experimentally determined rate law. Sometimes these mechanisms can be distinguished by use of nonkinetic data. For example, optical and mass spectroscopy may be used to detect intermediates. Isotopically labeled reactants can be used to trace the paths of atoms in a reaction.

In discussing elementary reactions we refer to their **molecularity.** The molecularity is the number of reactant molecules in an elementary step. Thus, elementary reactions are referred to as **unimolecular, bimolecular,** and **trimolecular,** depending on whether one, two, or three molecules are involved as reactants. In contrast to what we said earlier about the lack of relationships between the stoichiometric equation and reaction order, the rate law of an elementary step can be obtained directly from its chemical equation. A unimolecular reaction is first order, a bimolecular reaction is second order, and a trimolecular reaction is third order. Most elementary reactions are unimolecular or bimolecular. Trimolecular reactions are uncommon and, as we will see later, really involve two bimolecular steps.

The mechanism of a reaction leads directly to a set of differential equations that completely describes the kinetic behavior of that mechanism. A differential equation may be written for each molecular species in the mechanism by writing positive terms for each reaction by which the species is formed and negative terms for each reaction in which the species disappears. However, these differential equations are not all independent. There are conservation equations that must also be satisfied, and so the number of independent rate equations is reduced by this number. The concentrations of various species as functions of time may be obtained by solving the boundary value problem; for a mechanism with a number of steps this can only be done numerically using a computer. This of course requires numerical values for all the rate constants. However, there are two approximation methods that yield expressions for the rate of the overall reaction in terms of the concentrations of the various reactants and products and the rate constants. These methods are the steady-state method and the rapid equilibrium method.

The **steady-state method** is based on the fact that the concentrations of intermediates may not change much during a reaction after an initial buildup. (See Section 18.4.) Therefore, rate equations are written for intermediates, and these rates are set equal to zero. This yields a set of algebraic equations that can be solved for the concentrations of intermediates. These equations can then be used to eliminate these concentrations from the rate equations for the formation of product. As we will see, this provides the relations between the experimentally determined rate parameters and the rate constants for the elementary steps in the mechanism.

A reaction $A \rightleftharpoons B$ may go through an unstable intermediate I; thus, there are two elementary reactions as represented by the following mechanism:

$$A \underset{k_{-1}}{\overset{k_1}{\rightleftharpoons}} I \underset{k_{-2}}{\overset{k_2}{\rightleftharpoons}} B \qquad K = \frac{[B]_{eq}}{[A]_{eq}} = \frac{k_1 k_2}{k_{-1} k_{-2}} \tag{18.72}$$

Since $[A] + [I] + [B] = [A]_0$ if we start with A, there are only two independent rate equations, and they can be written as

$$\frac{d[I]}{dt} = k_1[A] + k_{-2}[B] - (k_{-1} + k_2)[I] \tag{18.73}$$

$$\frac{d[B]}{dt} = k_2[I] - k_{-2}[B] \tag{18.74}$$

If I is an unstable intermediate, its concentration rises to a low value that remains rather constant after an initial increase. For this reaction, the steady-state approximation is $d[I]/dt = 0$. By use of this approximation, [I] can be eliminated from equation 18.74 to obtain

$$\frac{d[B]}{dt} = \frac{k_1 k_2[A] - k_{-1} k_{-2}[B]}{k_{-1} + k_2}$$
$$= k_f[A] - k_b[B] \tag{18.75}$$

Thus, the reaction behaves like a reversible first-order reaction (Section 18.3) with $k_f = k_1k_2/(k_{-1} + k_2)$ and $k_b = k_{-1}k_{-2}/(k_{-1} + k_2)$. Note that $k_f/k_b = K$. **If the concentration of the intermediate is very small during the entire reaction, it may be very difficult to distinguish mechanism 18.72 from the mechanism $A \rightleftharpoons B$.** The Arrhenius plots for k_f and k_b will probably be curved since they are each a composite of three rate constants.

If the reaction A = B goes to completion because $k_{-2} = 0$ and, in addition, $k_{-1} \gg k_2$, the unstable intermediate will essentially be in equilibrium with A and

$$\frac{d[B]}{dt} = \frac{k_1k_2}{k_{-1}}[A]$$
$$= k_2K_1[A] = k_f[A] \quad (18.76)$$

where $K_1 = k_1/k_{-1}$.

As an example of these ideas, consider the decomposition of ozone, which is represented by the reaction $2O_3 = 3O_2$. Thus, the reaction rate is defined by

$$v = -\frac{1}{2}\frac{d[O_3]}{dt} = \frac{1}{3}\frac{d[O_2]}{dt} \quad (18.77)$$

The rate law for this decomposition in the presence of relatively high concentrations of a chemically inert gas M is

$$v = \frac{k[O_3]^2[M]}{k'[O_2][M] + [O_3]} \quad (18.78)^*$$

Thus,

$$-\frac{d[O_3]}{dt} = \frac{2k[O_3]^2[M]}{k'[O_2][M] + [O_3]} \quad (18.79)$$

This and other information has led to the following mechanism:

$$1.\ O_3 + M \underset{k_{-1}}{\overset{k_1}{\rightleftharpoons}} O_2 + O + M \quad (18.80)$$

$$2.\ O + O_3 \xrightarrow{k_2} 2O_2 \quad (18.81)$$

$$\text{Overall reaction: } 2O_3 = 3O_2 \quad (18.82)$$

Two independent rate equations for this mechanism are 18.83 and 18.84; we do not have to write a separate equation for O_2 since at constant volume $3[O_3] + 2[O_2] + [O] =$ constant:

$$\frac{d[O]}{dt} = k_1[O_3][M] - k_{-1}[O_2][O][M] - k_2[O][O_3] \quad (18.83)$$

$$-\frac{d[O_3]}{dt} = k_1[O_3][M] - k_{-1}[O_2][O][M] + k_2[O][O_3] \quad (18.84)$$

*S. W. Benson and A. E. Axworthy, Jr., *J. Chem. Phys.* **29**:1718 (1957).

Since oxygen atoms are not produced in the overall reaction and since their concentration is always low, it is a good approximation to set $d[O]/dt = 0$. Thus, in the steady state

$$[O] = \frac{k_1[O_3][M]}{k_{-1}[O_2][M] + k_2[O_3]} \tag{18.85}$$

This relation may be substituted into equation 18.84 for the decomposition of O_3 to obtain

$$-\frac{d[O_3]}{dt} = \frac{2k_1k_2[O_3]^2[M]}{k_{-1}[O_2][M] + k_2[O_3]} \tag{18.86}$$

This rate equation is in accord with the empirical rate equation (18.78). Early in the reaction starting with pure ozone $k_{-1}[O_2][M] \ll k_2[O_3]$, and this rate equation reduces to

$$-\frac{d[O_3]}{dt} = 2k_1[O_3][M] \tag{18.87}$$

Under these conditions the **first step is rate determining,** and two molecules of O_3 are decomposed each time step 1 occurs because the oxygen atom produced causes the destruction of a second molecule of O_3. On the other hand, if the **second step is rate determining,** that is, $k_{-1}[O_2][M] \gg k_2[O_3]$, then

$$-\frac{d[O_3]}{dt} = \frac{2k_2K_1[O_3]^2}{[O_2]} \tag{18.88}$$

where $K_1 = k_1/k_{-1}$ is the equilibrium constant for the first step. Oxygen (O_2) inhibits the forward reaction because it reduces the concentration of oxygen atoms in equilibrium with O_3 and O_2 (step 1). [Note that an incorrect rate equation is obtained by starting with the assumption that reaction 1 remains in equilibrium because this step, as well as the second step, destroys O_3; D. C. Tardy and E. D. Cater, *J. Chem. Educ.* **60:**109 (1983).] This reaction is discussed again in the next chapter, on photochemistry, because ozone is formed in the upper atmosphere by a photochemical reaction. The photochemical formation of ozone and the various reactions by which it can be converted to oxygen are extremely important because ozone in the upper atmosphere shields the surface of the earth from harmful ultraviolet radiation that would otherwise be transmitted through the atmosphere.

18.8 RELATION BETWEEN RATE CONSTANTS FOR THE FORWARD AND BACKWARD REACTIONS

In discussing the principle of detailed balance (Section 18.5) we referred to the fact that at equilibrium the forward rate of each step is equal to the backward rate of that step. For elementary reactions this leads to a relationship between the rate constants for the forward and backward reactions that is not necessarily obeyed for the overall reaction.

For an elementary reaction there is an exact correspondence between the chemical equation and the rate equation. For the elementary reaction

$$\mathrm{AB} \underset{k_\mathrm{b}}{\overset{k_\mathrm{f}}{\rightleftharpoons}} \mathrm{A} + \mathrm{B} \tag{18.89}$$

the complete rate equation is

$$-\frac{\mathrm{d[AB]}}{\mathrm{d}t} = k_\mathrm{f}[\mathrm{AB}] - k_\mathrm{b}[\mathrm{A}][\mathrm{B}] \tag{18.90}$$

At equilibrium, $\mathrm{d[AB]}/\mathrm{d}t = 0$ and so

$$\frac{[\mathrm{A}]_\mathrm{eq}[\mathrm{B}]_\mathrm{eq}}{[\mathrm{AB}]_\mathrm{eq}} = \frac{k_\mathrm{f}}{k_\mathrm{b}} \tag{18.91}$$

It is easy to say that the ratio of rate constants is equal to the equilibrium constant K_c, as we did earlier in Sections 18.3 and 18.4, but here we have to be careful because K_c has been defined (Section 5.7) as a dimensionless quantity. In general,

$$K_c = (c^\circ)^{-\sum \nu_i} \prod [\mathrm{B}_i]_\mathrm{eq}^{\nu_i} \tag{18.92}$$

For reaction 18.89,

$$K_c = \frac{[\mathrm{A}]_\mathrm{eq}[\mathrm{B}]_\mathrm{eq}}{[\mathrm{AB}]_\mathrm{eq} c^\circ} \tag{18.93}$$

In general,

$$\frac{k_\mathrm{f}}{k_\mathrm{b}} = K_c (c^\circ)^{\sum \nu_i} \tag{18.94}$$

For AB = A + B,

$$\frac{k_\mathrm{f}}{k_\mathrm{b}} = \frac{[\mathrm{A}]_\mathrm{eq}[\mathrm{B}]_\mathrm{eq}}{[\mathrm{AB}]_\mathrm{eq}} = K_c c^\circ \tag{18.95}$$

Here, as elsewhere in this chapter, we assume that the gas mixtures and solutions are ideal.

Thus, if the rate constant for the forward reaction is known, the rate constant for the backward reaction can be calculated using the equilibrium constant for the elementary reaction. Since rate equations are generally written in terms of concentrations, it is necessary to use K_c for a gas reaction. The value of K_c can be calculated using

$$K_P = \left(\frac{c^\circ RT}{P^\circ}\right)^{\sum \nu_i} K_c \tag{18.96}$$

which was derived in Section 5.7. Standard Gibbs energies of formation of gases may be used to calculate K_P, and equation 18.96 may then be used to calculate K_c.

Example 18.6 ***Rate constant for the backward reaction***

The rate constant for the elementary reaction

$$C_2H_6 \rightarrow 2CH_3$$

is $1.57 \times 10^{-3}\ s^{-1}$ at 1000 K. What is the rate constant for the backward reaction at this temperature? For $CH_3(g)$, $\Delta_f G^\circ = 159.74\ kJ\ mol^{-1}$ at 1000 K.

Using the value of $\Delta_f G^\circ$ for C_2H_6 at 1000 K from Table C.3 yields $K_P = 1.083 \times 10^{-11}$. Using equation 18.96,

$$K_P = 1.083 \times 10^{-11} = \frac{(1\ mol\ L^{-1})(0.083\,145\ L\ bar\ K^{-1}\ mol^{-1})(1000\ K)}{(1\ bar)} K_c$$

$$K_c = 1.302 \times 10^{-13} = \frac{([CH_3]/c^\circ)^2}{[C_2H_6]/c^\circ}$$

Now we can use equation 18.94:

$$\frac{k_f}{k_b} = K_c (c^\circ)^{\sum \nu_i}$$

Since $\sum \nu_i = 1$,

$$\frac{k_f}{k_b} = (1.302 \times 10^{-13})(1\ mol\ L^{-1}) = \frac{1.57 \times 10^{-3}\ s^{-1}}{k_b}$$

$$k_b = 1.21 \times 10^{10}\ L\ mol^{-1}\ s^{-1}$$

If a rate equation has a sum of terms for the forward reaction, indicating multiple paths, the principle of detailed balancing requires that *each* term for the forward reaction be balanced by a thermodynamically appropriate term for the backward reaction at equilibrium.

Example 18.7 ***Rate law for the backward reaction***

Suppose a reaction A = B, with equilibrium constant $K = [B]_{eq}/[A]_{eq}$, goes to the right by an uncatalyzed pathway with rate $k_{0f}[A]$ and by a proton-catalyzed pathway with a rate $k_{Hf}[H^+][A]$. Thus, the rate law for the forward reaction involves the sum of two terms:

$$v_f = (k_{0f} + k_{Hf}[H^+])[A] = k_f[A]$$

What is the corresponding rate law for the backward reaction, and how are the rate constants related to k_{0f} and k_{Hf}?

The rate constants k_f and k_b for the forward and backward reactions are related by

$$K = \frac{[B]_{eq}}{[A]_{eq}} = \frac{k_f}{k_b} = \frac{k_{0f} + k_{Hf}[H^+]}{k_b}$$

Thus

$$k_b = \frac{k_{0f}}{K} + \frac{k_{Hf}}{K} = k_{0b} + k_{Hb}[H^+]$$

where

$$k_{0b} = k0f/k$$

$$k_{Hb} = k_{Hf}/k$$

The rate law for the backward reaction is

$$v_b = (k_{0b} + k_{Hb}[H^+])[B]$$

Now we need to take a closer look at elementary reactions. The most common type of reaction is bimolecular.

18.9 BIMOLECULAR REACTIONS

When two uncharged, nonpolar molecules approach each other at the short distances required for chemical reaction, the predominant force is a strong repulsion, as we have seen in connection with the Lennard-Jones intermolecular potential (Section 11.9). If these molecules can react, the activation energy is a measure of the energy required to deform the electron clouds of the reactants so that the reaction can occur. When two molecules with closed electronic shells react, the activation energy is generally found to be in the range 80–200 kJ mol^{-1}. As a result, such reactions are rare because there are frequently other reaction paths involving free radicals that are faster, and are therefore the ones by which reaction occurs.

When one of the reactants is a radical, the activation energy is generally in the range 0–60 kJ mol^{-1}, as indicated in Table 18.1. By the term *radical* we refer to molecules with at least one orbital vacancy in their valence shells and to atoms, except for the rare gases. Since reactions of radicals with closed-shell molecules are faster than reactions of closed-shell molecules with each other, many reactions observed in the gas phase involve radicals. The reactions of radicals with each other are generally even faster because they have activation energies of about zero, as indicated in Table 18.1. In fact, the activation energies for the recombination of radicals may be negative so that the reaction goes more slowly at a higher temperature.

A common type of bimolecular reaction is a metathesis reaction involving an atom or a radical. The enthalpy change for such a reaction may be positive or negative, depending on whether the newly formed bond has a higher or lower dissociation energy than the bond broken. The Arrhenius parameters for some bimolecular reactions are given in Table 18.2. The pre-exponential factors for metathesis reactions involving atoms are all in the range of about $10^{10.5}$–$10^{11.5}$ L mol^{-1} s^{-1}, but the activation energies differ appreciably. The pre-exponential factors for metathesis reactions not involving atoms are smaller, as are those for association reactions of radicals. The activation energies of association reactions of radicals are all about zero. We will see in the next chapter that the pre-exponential factors of such reactions can be estimated quite well using collision theory for rigid spherical molecules.

Table 18.1 Activation Energies for Exothermic Bimolecular Reactions

Reaction Type	*Electronic Structures*	E_a/kJ mol^{-1}
Molecule + molecule	Two closed shells	80–200
Radical + molecule	One closed shell/one open shell	0–60
Radical + radical	Two open shells	~ 0

Source: S. W. Benson, *Thermochemical Kinetics.* Hoboken, NJ: Wiley, 1976.

Table 18.2 Arrhenius Parameters for Bimolecular Reactions

Reaction	$\log\left(\frac{A}{\text{L mol}^{-1}\text{ s}^{-1}}\right)$	$\frac{E_a}{\text{kJ mol}^{-1}}$
	Metathesis reactions involving atoms	
$Br + H_2 \rightarrow HBr + H$	10.8	76.2
$I + H_2 \rightarrow HI + H$	11.4	143
$Cl + H_2 \rightarrow HCl + H$	10.9	23.0
$O + O_3 \rightarrow 2O_2$	10.5	23.9
$O + NO_2 \rightarrow O_2 + NO$	10.3	4.2
$O + H_2 \rightarrow OH + H$	10.5	42.7
$N + O_2 \rightarrow NO + O$	9.3	26.4
$N + NO \rightarrow N_2 + O$	10.2	0
$O + OH \rightarrow O_2 + H$	10.3	0
	Metathesis reactions not involving atoms	
$CH_3 + C_2H_6 \rightarrow CH_4 + C_2H_5$	8.5	45.2
$2C_2H_5 \rightarrow C_2H_4 + C_2H_6$	9.6	0
$C_6H_5 + CH_4 \rightarrow C_6H_6 + CH_3$	8.6	46.4
	Association reactions of radicals	
$2CH_3 \rightarrow C_2H_6$	10.5	0
$2C_2H_5 \rightarrow C_4H_{10}$	10.4	0
$CH_3 + NO \rightarrow CH_3NO$	8.8	0

Some bimolecular reactions are more complex than simple two-body reactions. In some cases the bimolecular rate constant depends on the pressure, and this is an indication that the reaction leads to a **collision complex.** A collision complex is a weakly bound molecule that survives for a time that is longer than the characteristic periods of its vibrations and rotations. If a weakly bound molecule is formed the reaction may be written

$$A + B \rightleftharpoons AB \rightarrow \text{products} \tag{18.97}$$

When a weakly bound intermediate is formed, the reaction is no longer really a bimolecular reaction, but the dividing line is not sharp. If the pressure is sufficiently low so that the unimolecular dissociation of AB is in the falloff region (see the next section), the overall reaction will approach third order, and may appear to be a trimolecular reaction. Under these conditions the distinction between bimolecular and trimolecular becomes less meaningful. The reaction

$$ClO + NO \rightleftharpoons ClONO \rightarrow Cl + NO_2 \tag{18.98}$$

is an example of such a reaction. Its rate constant is given by $(6.2 \times 10^{-12}\text{ cm}^3\text{ s}^{-1})\exp(294\text{ K}/T)$. A further example of the complications that may be encountered when a reaction is studied in detail is the reaction

$$HO_2 + HO_2 \rightarrow H_2O_2 + O_2 \tag{18.99}$$

This reaction goes by two paths: one is a bimolecular path with a negative activation energy, and the other is a trimolecular reaction that also has a negative activation energy.

If the molecule formed in an association reaction of radicals has enough bonds, the energy of the exothermic reaction can be absorbed by various molecular vibrations, without causing dissociation. When this is the situation the forward reaction is bimolecular and the backward reaction is unimolecular. However, when we go from radicals to atoms the energized product of the association reaction has to be deactivated by collision or it will dissociate.

Example 18.8 ***Use of the equilibrium assumption***

A weak complex AB is formed in the reaction

$$\mathrm{A + B} \underset{k_{-1}}{\overset{k_1}{\rightleftharpoons}} \mathrm{AB} \xrightarrow{k_2} \mathrm{C}$$

Assuming that $k_{-1} \gg k_2$, derive the rate equation for the formation of C.

Since the first step remains in equilibrium, the rate equation is

$$\frac{\mathrm{d[C]}}{\mathrm{d}t} = k_2[\mathrm{AB}]$$

Equation 18.94 yields

$$\frac{k_1}{k_{-1}} = K_c(c^\circ)^{\sum \nu_i} = K_c/c^\circ$$

where

$$K_c = \frac{[\mathrm{AB}]c^\circ}{[\mathrm{A}][\mathrm{B}]}$$

Thus

$$\frac{k_1}{k_{-1}} = \frac{[\mathrm{AB}]}{[\mathrm{A}][\mathrm{B}]}$$

Using this equation to eliminate [AB] from the rate equation yields

$$\frac{\mathrm{d[C]}}{\mathrm{d}t} = \frac{k_1 k_2}{k_{-1}}[\mathrm{A}][\mathrm{B}] = k_2(K_c/c^\circ)[\mathrm{A}][\mathrm{B}]$$

18.10 UNIMOLECULAR AND TRIMOLECULAR REACTIONS

It may appear unusual to discuss unimolecular reactions and trimolecular reactions together, but these two types of reactions are related through the forward and backward reactions in association reactions. Unimolecular reactions are either isomerizations or dissociations, as illustrated by

$$\mathrm{CH_3NC} \rightarrow \mathrm{CH_3CN} \tag{18.100}$$

$$\mathrm{C_2H_6} \rightarrow 2\mathrm{CH_3} \tag{18.101}$$

The reverse of reaction 18.101 is an association reaction. Since unimolecular reactions have more complicated kinetics than bimolecular reactions, the kinetics of association reactions, such as the reverse of equation 18.101, involve the same complications. First, we will discuss the forward reactions of reactions such as

18.100 and 18.101, and second, we will discuss the reverse of a reaction such as 18.101. The following discussion applies only when reactants and products are polyatomic so that there are many vibrational modes. The recombination of atoms to make diatomic molecules is discussed at the end of this section.

Isomerizations and dissociations in the gas phase are first order at pressures of 1 bar and higher, but they become second order at lower pressures. Both of these observations are puzzling. Why should an isolated molecule in a gas suddenly isomerize or dissociate? Why should the reaction become second order when the pressure is lowered? The apparent rate constant k_{uni} for a unimolecular reaction $A \rightarrow$ products is defined by

$$\frac{d[A]}{dt} = -k_{uni}[A] \tag{18.102}$$

Figure 18.10 shows how k_{uni} varies with pressure from the isomerization of methyl isocyanide to methyl cyanide (reaction 18.100). The kinetics are said to be in the **falloff region** when the pressure is lowered below about 1 bar. Unimolecular reactions generally have large activation energies, as shown in Table 18.3. This raises another interesting question: How is this large activation energy supplied?

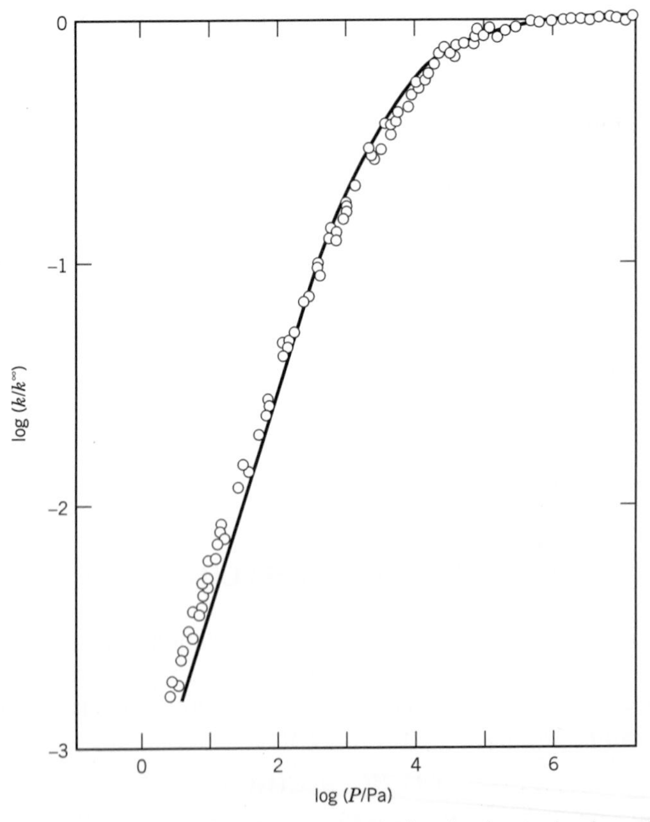

Figure 18.10 Apparent first-order rate constants for the isomerization of CH_3NC at 503.5 K. The curve shows the results of the RRKM (Rice, Ramsberger, Kassel, and Marcus) theory. At high pressure the first-order rate constant is independent of pressure and is designated as k^∞. As the pressure is reduced, the rate constant decreases and is said to be in the falloff region. [Reprinted with permission from F. W. Schneider and B. S. Rabinovitch, *J. Am. Chem. Soc.* **84**:4215 (1962). Copyright © 1962 American Chemical Society.]

Table 18.3 Arrhenius Parameters for Unimolecular Gas Reactions at Atmospheric Pressure (k^{∞})

Reaction	$\log\left(\frac{A^{\infty}}{s^{-1}}\right)$	$\frac{E_a}{\text{kJ mol}^{-1}}$	*Temperature Range* / K
Isomerizations			
$CH_3NC \rightarrow CH_3CN$	13.6	160.5	470–530
cyclo = $C_3H_6 \rightarrow CH_3CH{=}CH_2$	15.45	274	700–800
Dissociation to stable molecules			
$C_2H_5Cl \rightarrow C_2H_4 + HCl$	14.6	254	670–770
cyclo = $C_4H_8 \rightarrow 2C_2H_4$	15.6	262	690–740
Dissociation to free radicals			
$C_2H_6 \rightarrow 2CH_3$	16.0	360	820–1000
$HNO_3 \rightarrow OH + NO_2$	15.3	205	890–1200

Source: I. W. M. Smith, *Kinetics and Dynamics of Elementary Gas Reactions.* Boston: Butterworths, 1980, Fig. 4.2, page 116, by permission of the publishers, Butterworth and Co. (Publishers) Ltd, ©.

A first approximation of the answers to these questions was provided by Lindemann in 1922. He pointed out that when a molecule is excited to a higher energy state by a bimolecular collision, there is a time lag before decomposition or isomerization. During this time lag the **excited molecule** may lose its extra energy in a second bimolecular collision. Since such reactions are usually studied by diluting the reacting gas A with an excess of inert gas M, the excitation step and de-excitation step usually involve collisions of A and M, as indicated in the following mechanism:

$$A + M \xrightarrow{k_1} A^* + M \tag{18.103}$$

$$A^* + M \xrightarrow{k_2} A + M \tag{18.104}$$

$$A^* \xrightarrow{k_3} B + C \tag{18.105}$$

Here A^* is an A molecule with enough vibrational energy to isomerize or decompose.* In other words, part of the kinetic energy of a bimolecular collision has been used to raise an A molecule to a higher vibrational level. Although A^* has enough energy to react, it does not do so immediately because the energy has to become distributed within the molecule in such a way that the reaction can occur. Any collision can raise A to a higher vibrational level; thus, M in this mechanism might be another A molecule, or a product molecule, or a molecule of the added inert gas. If A^* collides with another molecule before it undergoes a unimolecular reaction, it will almost certainly lose its high level of vibrational energy. This possibility is represented by the step with rate constant k_2. The other possibility is that A^* undergoes a unimolecular reaction with rate constant k_3.

Only molecules with three or more atoms undergo unimolecular reactions. A diatomic molecule cannot dissociate in this way because it has a single mode of

*In the next chapter we will discuss A molecules in the transition state, which are represented by $A^{\ddagger}$. The symbol $A^{\ddagger}$ indicates that A is in the middle of the process of chemical change. Excited molecules A^* are not in the process of chemical change, but are simply A molecules with additional internal energy.

vibrational freedom. If this mode is excited by an amount equal to the dissociation energy, the molecule dissociates in about 10^{-12} s.

Since A^* is never present at a very high concentration, we can use the steady-state approximation to obtain an expression for the rate of reaction in terms of [A] and [M]:

$$\frac{d[A^*]}{dt} = k_1[A][M] - (k_2[M] + k_3)[A^*] = 0 \tag{18.106}$$

In the steady state, the rate of disappearance of A is equal to the rate of appearance of B (or C) so that $-d[a]/dt = k_3[A^*]$. Solving equation 18.106 for $[A^*]$ yields

$$-\frac{d[A]}{dt} = \frac{k_1 k_3[A][M]}{k_2[M] + k_3} = k_{uni}[A] \tag{18.107}$$

so that the unimolecular rate constant k_{uni} is given by

$$k_{uni} = \frac{k_1 k_3[M]}{k_2[M] + k_3} \tag{18.108}$$

Rate equation 18.107 has two limiting forms:

1. In the high-pressure limit, $k_2[M] \gg k_3$ so that

$$-\frac{d[A]}{dt} = \frac{k_1 k_3}{k_2}[A] = k^\infty[A] \tag{18.109}$$

Thus, at high pressures the reaction is first order, as observed. The first-order rate constant at high pressures is referred to as k^∞, which is equal to $k_1 k_3/k_2$, and the values of the pre-exponential factor A^∞ and E_a are given in Table 18.3.

2. In the low-pressure limit, $k_2[M] \ll k_3$ so that

$$-\frac{d[A]}{dt} = k_1[A][M] \tag{18.110}$$

Under these conditions all A^* isomerize or dissociate, so the rate of reaction is determined by the number of collisions of A with M that are sufficiently energetic. At very low pressures the reaction is found to be second order when the partial pressure of M is changed. However, in a single kinetic experiment [M] is held constant and the reaction is pseudo–first order. In a unimolecular dissociation, [M] increases during the experiment, but since the products are less efficient in energizing A, $k_1[M]$ remains approximately constant, and the reaction is again pseudo–first order.

Equation 18.108 provides a qualitative description of the change from first order at high pressures to second order at low pressures, but it is not satisfactory quantitatively because it predicts a steeper falloff in k_{uni} than is actually observed. The RRKM (Rice, Ramsberger, Kassel, and Marcus) theory has been developed to deal with this, but that will be left to more advanced books.

Now we introduce **trimolecular reactions** by considering a reaction like 18.101 and the reverse reaction. To discuss this, we introduce a mechanism that activates the dimer A_2 by collision with an inert atom (or molecule) M:

$$A_2 + M \underset{k_2}{\overset{k_1}{\rightleftharpoons}} A_2^* + M \tag{18.111}$$

$$A_2^* \underset{k_4}{\overset{k_3}{\rightleftharpoons}} 2A \tag{18.112}$$

The assumption that A_2^* is in a steady state yields

$$\frac{d[A_2^*]}{dt} = k_1[A_2][M] + k_4[A]^2 - (k_2[M] + k_3)[A_2^*] = 0 \tag{18.113}$$

so that the steady-state concentration of excited molecules is given by

$$[A_2^*] = \frac{k_1[A_2][M] + k_4[A]^2}{k_2[M] + k_3} \tag{18.114}$$

Substituting this in

$$\frac{d[A_2]}{dt} = -k_1[A_2][M] + k_2[A_2^*][M] \tag{18.115}$$

yields

$$\frac{d[A_2]}{dt} = -\frac{k_1k_3[A_2][M]}{k_2[M] + k_3} + \frac{k_2k_4[M][A]^2}{k_2[M] + k_3} \tag{18.116}$$

This shows that the rate of change in the concentration of dimer molecules is equal to the difference between the rate of the forward reaction

$$\text{forward rate} = \frac{k_1k_3[A_2][M]}{k_2[M] + k_3} \tag{18.117}$$

and the rate of the backward reaction

$$\text{backward rate} = \frac{k_2k_4[M][A]^2}{k_2[M] + k_3} \tag{18.118}$$

As we found in equation 18.108, the rate of the forward reaction at high [M] is first order, $(k_1k_3/k_2)[A_2]$, and at low [M] is second order, $k_1[A_2][M]$. In contrast, the rate of the backward reaction at high [M] is second order, $k_4[A]^2$, and at low [M] is third order, $(k_2k_4/k_3)[M][A]^2$.

The direction in which the reaction will go depends on the equilibrium constant expression

$$K_c = \frac{[A]^2}{[A_2]c^\circ} = \frac{k_1k_3}{k_2k_4} \tag{18.119}$$

This derivation shows that the reverse of a dissociation reaction (in this case, $2A = A_2$) is necessarily trimolecular at low concentrations. Since this is so, we have to conclude that equation 18.116 provides a qualitative description of the change from second order at high pressures to third order at low pressures for an association reaction. However, equation 18.116 is not satisfactory quantitatively, and the RRKM theory is required for a quantitative description.

Note that the mechanisms discussed here (18.103 to 18.105 and 18.111 to 18.112) are different from other mechanisms discussed in this chapter because energy transfer processes are included, even though they occur in any system. They are written out here to help us understand the formation and deactivation of excited molecules.

Two molecules can combine in a bimolecular reaction in the gas phase to form a dimer, as in reaction 18.112, because the energy released by bond formation can go into vibrational degrees of freedom. However, two atoms cannot combine in a bimolecular reaction in the gas phase to form a diatomic molecule because the energy release due to bond formation causes the molecule to dissociate. The reaction can occur if a third atom or molecule is involved and carries away the energy produced. Atom recombination reactions are third order in the gas phase, and so the elementary reactions are represented by

$$2A + M \rightarrow A_2 + M \tag{18.120}$$

$$A + B + M \rightarrow AB + M \tag{18.121}$$

The corresponding rate equations are

$$\frac{d[A_2]}{dt} = k[A]^2[M] \tag{18.122}$$

$$\frac{d[AB]}{dt} = k[A][B][M] \tag{18.123}$$

The values of the trimolecular rate constants for the recombination of atoms in Table 18.4 show that they are all about $10^{9.5\pm0.5}$ L^2 mol^{-2} s^{-1} at 300 K and about a power of 10 smaller at 2000 K. This corresponds to an activation energy of about -6 kJ mol^{-1}, but the data on trimolecular rate constants are actually better represented by $k = (\text{const})/T$ than by the Arrhenius equation. As indicated in Table 18.4, various third bodies M have different efficiencies in trimolecular reactions. As shown by several entries at the bottom of the table, reactions of atoms with diatomic molecules may be trimolecular.

There is another way of looking at trimolecular reactions that avoids nearly simultaneous three-body collisions and also provides an explanation of the negative activation energies. If the third body M is a polyatomic molecule capable of forming a complex with one of the recombining atoms, the mechanism can be written as

$$A + M \underset{k_{-1}}{\overset{k_1}{\rightleftharpoons}} AM \tag{18.124}$$

$$AM + A \xrightarrow{k_2} A_2 + M \tag{18.125}$$

Table 18.4 Rate Constants for Trimolecular Reactions

Reaction	T/K	$\log\left(\frac{k}{\text{L}^2\ \text{mol}^{-2}\ \text{s}^{-1}}\right)$
$H + H + M \rightarrow H_2 + M$	300	10.0 (H_2)
	1072	9.5 (H_2)
$O + O + M \rightarrow O_2 + M$	300	8.9 (O_2)
	2000	7.4 (Ar)
$I + I + M \rightarrow I_2 + M$	300	9.3 (Ne)
	300	10.8 (n-C_5H_{12})
$O + O_2 + M \rightarrow O_3 + M$	380	8.1 (O_2)
$O + NO + M \rightarrow NO_2 + M$	300	10.46 (O_2)
$H + NO + M \rightarrow HNO + M$	300	10.17 (H_2)

The third body is the molecule listed in parentheses.
Source: S. W. Benson, *Thermochemical Kinetics.* Hoboken, NJ: Wiley, 1976.

The complex AM can dissociate with rate constant k_{-1} to regenerate A and M, or it can react with an A molecule with rate constant k_2 to form A_2. The steady-state rate equation is

$$\frac{d[A_2]}{dt} = \frac{k_1 k_2 [A]^2 [M]}{k_{-1} + k_2 [A]} \tag{18.126}$$

If the intermediate is very short-lived, $k_{-1} \gg k_2[A]$, and equation 18.126 becomes

$$\frac{d[A_2]}{dt} = (k_1/k_{-1}) k_2 [A]^2 [M] \tag{18.127}$$

Since

$$\frac{k_1}{k_{-1}} = K_c (c^\circ)^{\sum \nu_i}$$

where

$$K_c = \frac{[AM] c^\circ}{[A][M]} \tag{18.128}$$

equation 18.126 can be written

$$\frac{d[A_2]}{dt} = (K_1/c^\circ) k_2 [A]^2 [M] \tag{18.129}$$

If k_2 follows the Arrhenius equation with activation energy E_2, the activation energy for the trimolecular reaction is given by

$$E_a = RT^2 \frac{d \ln K_1}{dT} + RT^2 \frac{d \ln k_2}{dT} = \Delta H_1^\circ + E_2 \tag{18.130}$$

If the first step is exothermic and the activation energy for the second step is not too high, the activation energy for the trimolecular reaction can be negative.

If the combination of an atom with another atom or diatomic molecule is trimolecular, the reverse reaction is bimolecular, and the Arrhenius equation for the reverse reaction may be calculated using K_c (Section 18.8). If the activation energy for the combination of two atoms is negative, the activation energy for the dissociation of the diatomic molecule is *less* than the dissociation energy for the diatomic molecule.

Comment:

If we think about molecules colliding in a gas, we can see that there really is no such thing as a simultaneous collision of three molecules. However, as we have seen in this section, third-order reactions can result from the deactivation of polyatomic excited molecules by collision or from the formation of a weak complex molecule between an atom and a polyatomic molecule. Chemists do not have to be concerned about molecularities of 4, 5, ... This restriction does not apply to the order of a reaction, which, as mentioned earlier, can have any integer or noninteger value from $-\infty$ to $+\infty$.

18.11 UNBRANCHED CHAIN REACTIONS

As discussed in Section 18.9, reactions of molecules with closed shells generally have high activation energies even when the change in Gibbs energy for the overall reaction is favorable. Faster reaction paths are frequently provided by radical–molecule reactions. As a consequence, many chemical reactions occur through a sequence of elementary reactions involving radicals. These may be nonchain reactions, or they may be chain reactions. Since a radical has an unpaired electron, its reaction with a molecule having paired electrons gives rise to another radical. In this way the reactive center is maintained and can give rise to a **chain of reactions.** We may ask why such a reaction ever stops. Sometimes, as a matter of fact, the chain reaction does not stop until all the material is consumed. At other times, however, the chain is broken when one of the radicals reacts at the wall of the containing vessel or with another radical to form a spin-paired molecule. The length of the chain (i.e., the number of molecules reacting per molecule activated) is determined by the relative rates of the chain-propagating and the chain-breaking reactions.

Mechanisms involving radicals may be **nonchain, straight chain,** and **branched chain.** Branched chain reactions produce explosions if they are highly exothermic; they are discussed in Section 18.12. The rate parameters for several types of elementary radical reactions are given in Table 18.2. Although the rate constants for radical fission reactions may have very large pre-exponential factors, they may be slow, except at very high temperatures, because of high activation energies. The activation energy of a radical fission reaction may be lower when rearrangement to a molecular product compensates for some of the energy required to break a bond.

As an example, let us consider the pyrolysis of ethane to ethylene:

$$C_2H_6 = C_2H_4 + H_2 \tag{18.131}$$

At temperatures of 700 to 900 K and pressures above about 0.2 bar, this reaction is first order in its early stages. Later in the reaction methane and propylene are formed, but we will consider only the early stages. There are various kinds of evidence that the reaction has the following mechanism:

Initiation
$$C_2H_6 \xrightarrow{k_1} 2CH_3 \tag{18.132}$$

Chain transfer
$$CH_3 + C_2H_6 \xrightarrow{k_2} CH_4 + C_2H_5 \tag{18.133}$$

Propagation
$$C_2H_5 \xrightarrow{k_3} C_2H_4 + H \tag{18.134}$$

$$H + C_2H_6 \xrightarrow{k_4} H_2 + C_2H_5 \tag{18.135}$$

Termination
$$H + C_2H_5 \xrightarrow{k_5} C_2H_6 \tag{18.136}$$

To the extent that reaction 18.133 occurs, the net chemical change is not given by equation 18.131.

The rate of reaction of ethane is

$$\frac{d[C_2H_6]}{dt} = -(k_1 + k_2[CH_3] + k_4[H])[C_2H_6] \tag{18.137}$$

The rate of formation of ethane in the last step is ignored because termination occurs only after long chains producing the products. In the steady state the rates of change of the concentrations of the radicals may be taken equal to zero:

$$\frac{d[CH_3]}{dt} = 2k_1[C_2H_6] - k_2[CH_3][C_2H_6] = 0 \tag{18.138}$$

$$\frac{d[C_2H_5]}{dt} = (k_2[CH_3] + k_4[H])[C_2H_6] - (k_3 + k_5[H])[C_2H_5] = 0 \tag{18.139}$$

$$\frac{d[H]}{dt} = k_3[C_2H_5] - k_4[H][C_2H_6] - k_5[H][C_2H_5] = 0 \tag{18.140}$$

The steady-state concentrations of the free radicals CH_3, C_2H_5, and H may be obtained from these three simultaneous equations. We find

$$[CH_3] = \frac{2k_1}{k_2} \tag{18.141}$$

$$[C_2H_5] = \frac{2k_1 + k_4[H]}{k_3 + k_5[H]}[C_2H_6] \tag{18.142}$$

$$[H] = \frac{2k_1k_5 \pm \sqrt{(2k_1k_5)^2 + 16k_1k_3k_4k_5}}{-4k_4k_5} \tag{18.143}$$

In general, we can expect that k_1 is small so that

$$[H] = \left(\frac{k_1k_3}{k_4k_5}\right)^{1/2} \tag{18.144}$$

Substituting equations 18.141 and 18.144 in equation 18.137 yields the rate of reaction of ethane

$$\frac{d[C_2H_6]}{dt} = -\left[3k_1 + k_4\left(\frac{k_1k_3}{k_4k_5}\right)^{1/2}\right][C_2H_6] \tag{18.145}$$

so that the reaction is first order in spite of its complicated mechanism.

According to this mechanism the termination reaction involves two different kinds of radicals colliding with each other. Other possibilities exist; for example, H radicals may collide with each other in the presence of a third body (Section 18.10). Depending on the orders of the initiation and termination reactions, the overall order of a pyrolysis reaction may be 0, $\frac{1}{2}$, 1, $\frac{3}{2}$, or 2.*

18.12 BRANCHED CHAIN REACTIONS

If a propagation step in a chain reaction produces two or more radicals from one, there is a possibility of a rapid increase in rate and, for an exothermic reaction, an explosion. The reaction of hydrogen with oxygen can be explosive above about 700 K. The ranges of temperature and pressure within which there are spontaneous thermal explosions are shown in Fig. 18.11. For example, at about 550 °C stoichiometric hydrogen–oxygen mixtures react very slowly at pressures below

*M. F. R. Mulcahy, *Gas Kinetics,* p. 89. Hoboken, NJ: Wiley, 1973.

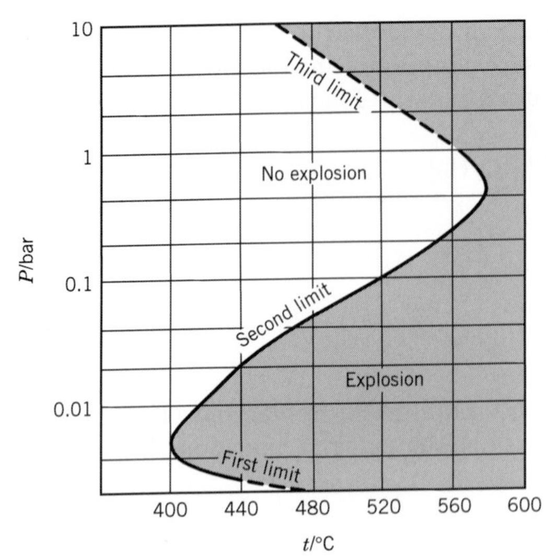

Figure 18.11 Explosion limits of a stoichiometric oxygen–hydrogen mixture.

10^{-3} bar. As the pressure is increased, the reaction rate increases slowly, but at a pressure of about 10^{-3} bar, depending on the volume of the vessel, there is a sudden explosion. On the other hand, if the gases are at a considerably higher pressure, the rate is again quite low. Hinshelwood found that if hydrogen at 0.26 bar and oxygen at 0.13 bar are placed in a 300-cm^3 quartz vessel at 550 °C, the rate of reaction is quite slow and becomes slower if the pressure is further reduced to 0.20 bar. If the pressure is reduced to 0.19 bar, however, an explosion occurs. Finally, as the total pressure is increased above the explosion zone, the reaction rate increases until it becomes so fast that the reaction mixture may be said to explode. The fact that the exact limits depend on the vessel surface and the vessel diameter indicates that radical chains may be terminated by reaction at the wall. If the vessel surface is coated with potassium chloride, radicals disappear when they strike the wall; if the vessel surface is coated with boric oxide, radicals are not destroyed so rapidly by collisions with the wall.

The first explosion limit can be understood in terms of the following mechanism:

$$\text{Initiation} \qquad H_2 + O_2 \xrightarrow{\text{wall}} 2OH \tag{18.146}$$

$$\text{Propagation} \qquad OH + H_2 \xrightarrow{k_2} H_2O + H \tag{18.147}$$

$$\text{Branching} \qquad H + O_2 \xrightarrow{k_3} OH + O \tag{18.148}$$

$$O + H_2 \xrightarrow{k_4} OH + H \tag{18.149}$$

$$\text{Termination} \qquad H + \text{wall} \xrightarrow{k_5} \tag{18.150}$$

The propagation reaction is exothermic and fast. The third and fourth reactions are called branching reactions because two radicals are formed from one. If the rate of branching is greater than the rate of termination, the number of radicals increases exponentially with time, and an explosion results. Reaction 18.150 is endothermic and slow below 700 K. The conditions of the first explosion limit are governed by the relative rates for branching $2k_3[H][O_2]$ and termination $k_5[H]$.

As the concentration of oxygen is increased, the rate of branching becomes greater than the rate of termination and an explosion occurs.

To explain the second limit, above which there is no explosion, it is necessary to invoke a new termination step to prevent the exponential increase in the number of radicals. For a new termination step to become more important as the pressure is increased, the new step must be higher order than the branching reaction. Thus, to explain the second limit, the following reaction must be added to the previous reactions:

$$\text{Termination} \qquad H + O_2 + M \xrightarrow{k_6} HO_2 + M \qquad (18.151)$$

In a stoichiometric mixture of oxygen and hydrogen, M may be hydrogen or oxygen, but these two gases have different efficiencies in this reaction. The HO_2 radical is relatively unreactive and does not produce another radical before it is quenched on the wall.

The third explosion limit results from the fact that the following reaction diminishes termination:

$$\text{Propagation} \qquad HO_2 + H_2 \xrightarrow{k_7} H_2O + OH \qquad (18.152)$$

Ten Key Ideas in Chapter 18

1. The rate of a reaction does depend on how the stoichiometric equation is written, so fractional stoichiometric numbers are avoided, even though they are permissible in thermodynamics.
2. The order of a reaction has to be determined experimentally since it cannot be deduced from the stoichiometric equation. The overall order is the sum of the orders with respect to the various reactants. The order with respect to a particular reactant can be determined by using high concentrations of all of the other reactants.
3. For irreversible reactions with simple orders the rate equations can be integrated, but for some reactions the rate laws are more complicated. The rate law for a simple reversible first-order reaction can be integrated, and the concentrations of the reactants can be expressed in terms of $\exp[-(k_1 + k_2)t]$.
4. The rate equations for consecutive irreversible reactions can be integrated. When the reactions are reversible, the expressions for the concentrations become much more complicated, and it becomes convenient to use a computer to integrate the rate equations. Under certain conditions the steady-state approximation can be used in deriving rate equations.
5. According to the principle of detailed balance, at equilibrium the forward rate for each step of a mechanism is equal to the backward rate of that step.
6. The Arrhenius equation is often useful for representing the dependence of a rate constant on temperature, but for some reactions more complicated equations have to be used.
7. Many chemical reactions are described by a mechanism, which is a series of steps of elementary reactions, that cannot be broken down into simpler chemical reactions. Elementary reactions are unimolecular, bimolecular, and trimolecular. Various steps in a mechanism do not necessarily all run at the same rate in the steady state, so it is necessary to use the stoichiometric numbers $s_1, s_2, \ldots$ of the steps.

8. There is a relationship between the rate constants for the forward and backward reactions and the equilibrium constant. When the equilibrium constant is known, the rate law for the reverse reaction can be calculated from the rate law for the forward reaction.

9. At an earlier time unimolecular gas reactions were puzzling because they were not second order and because their first-order rate constants were pressure dependent.

10. Chain reactions may have complicated rate laws and show surprising effects, such as explosion limits for branched chain reactions.

REFERENCES

S. W. Benson, *Thermochemical Kinetics.* Hoboken, NJ: Wiley, 1976.

W. C. Gardiner, Jr. (ed.), *Combustion Chemistry.* New York: Springer, 1984.

G. G. Hammes, *Principles of Chemical Kinetics.* New York: Academic, 1978.

P. L. Houston, *Chemical Kinetics and Reaction Dynamics.* New York: McGraw-Hill, 2001.

H. S. Johnston, *Gas Phase Reaction Theory.* New York: Ronald, 1966.

K. J. Laidler, *Chemical Kinetics,* 3rd ed. New York: Harper & Row, 1987.

W. L. Mallard, et al. *NIST Chemical Kinetics Database Version 4.0.* Gaithersburg, MD: National Institute of Science and Technology, date. (Covers gas phase kinetics for 7400 reactions.)

J. W. Moore and R. G. Pearson, *Kinetics and Mechanism.* Hoboken, NJ: Wiley, 1981.

M. J. Pilling and P. W. Seakins, *Reaction Kinetics.* New York: Oxford University Press, 1995.

A. V. Sapre and F. J. Krambeck, *Chemical Reactions in Complex Mixtures.* New York: Van Nostrand–Reinhold, 1991.

I. W. M. Smith, *Kinetics and Dynamics of Elementary Gas Reactions.* Boston: Butterworths, 1980.

J. I. Steinfeld, J. S. Francisco, and W. L. Hase, *Chemical Kinetics and Dynamics.* Englewood Cliffs, NJ: Prentice-Hall, 1999.

PROBLEMS

(M) Problems marked with an icon may be more conveniently solved on a personal computer with a mathematical program.

18.1 Nitrogen pentoxide (N_2O_5) gas decomposes according to the reaction

$$2N_2O_5 = 4NO_2 + O_2$$

At 328 K, the rate of reaction v under certain conditions is 0.75×10^{-4} mol L^{-1} s^{-1}. Assuming that none of the intermediates have appreciable concentrations, what are the values of $d[N_2O_5]/dt$, $d[N_2]/dt$, and $d[O_2]/dt$?

18.2 In studying the decomposition of ozone

$$2O_3(g) = 3O_2(g)$$

in a 2-L reaction vessel, it is found that $d[O_3]/dt = -1.5 \times 10^{-2}$ mol L^{-1} s^{-1}. (*a*) What is the rate of reaction v? (*b*) What is the rate of conversion $d\xi/dt$? (*c*) What is the value of $d[O_2]/dt$?

18.3 The decomposition of N_2O_5

$$2N_2O_5 = 4NO_2 + O_2$$

is studied by measuring the concentration of oxygen as a function of time, and it is found that

$$\frac{d[O_2]}{dt} = (1.5 \times 10^{-4}\ s^{-1})[N_2O_5]$$

at constant temperature and pressure. Under these conditions the reaction goes to completion to the right. What is the half-life of the reaction under these conditions?

18.4 The following data were obtained on the rate of hydrolysis of 17% sucrose in 0.099 mol L^{-1} HCl aqueous solution at 35 °C:

t/min	9.82	59.60	93.18
Sucrose remaining, %	96.5	80.3	71.0

t/min	142.9	294.8	589.4
Sucrose remaining, %	59.1	32.8	11.1

What is the order of the reaction with respect to sucrose, and what is the value of the rate constant k?

18.5 Methyl acetate is hydrolyzed in approximately 1 mol L^{-1} HCl at 25 °C. Aliquots of equal volume are removed at intervals and titrated with a solution of NaOH. Calculate the first-order rate constant from the following volumes of NaOH required to neutralize the aliquot:

t/s	339	1242	2745	4546	∞
V/cm^3	26.34	27.80	29.70	31.81	39.81

18.6 Prove that in a first-order reaction, where $dn/dt = -kn$, the average life, that is, the average life expectancy of the molecules, is equal to $1/k$.

18.7 The hydrolysis of 1-chloro-1-methylcycloundecane in 80% ethanol has been studied at 25 °C. The extent of hydrolysis was measured by titrating the acid formed after measured intervals of time with a solution of NaOH. The data are as follows on the volumes of NaOH required.

t/h	0	1.0	3.0
V/cm^3	0.035	0.295	0.715

t/h	5.0	9.0	12	∞
V/cm^3	1.055	1.505	1.725	2.197

(*a*) What is the order of the reaction? (*b*) What is the value of the rate constant? (*c*) What fraction of the 1-chloro-1-methylcycloundecane will be left unhydrolyzed after 8 h?

18.8 The following values of percent transmission are obtained with a spectrophotometer at a series of times during the decomposition of a substance absorbing light at a particular wavelength. Calculate k, $t_{1/2}$, and τ assuming the reaction is first order.

t	Percent Transmission
5 min	14.1
10 min	57.1
∞	100.0

Beer's law: $\log 100/T = abc$, where T = percent transmission, a = absorbancy index, b = cell thickness, and c = concentration.

18.9 Since radioactive decay is a first-order process, the decay rate for a particular nuclide is commonly given as the half-life. Given that potassium contains 0.0118% ^{40}K, which has a half-life of 1.27×10^9 years, how many disintegrations per second are there in a gram of KCl?

18.10 The decomposition of HI to $H_2 + I_2$ at 508 °C has a half-life of 135 min when the initial pressure of HI is 0.1 atm and 13.5 min when the pressure is 1 atm. (*a*) Show that this proves that the reaction is second order. (*b*) What is the value of the rate constant in L mol^{-1} s^{-1}? (*c*) What is the value of the rate constant in bar^{-1} s^{-1}? (*d*) What is the value of the rate constant in cm^3 s^{-1}?

18.11 The reaction between propionaldehyde and hydrocyanic acid has been studied at 25 °C. In a certain aqueous solution at 25 °C the concentrations at various times were as follows:

t/min	2.78	5.33	8.17
[HCN]/mol L^{-1}	0.0990	0.0906	0.0830
[C_3H_7CHO]/mol L^{-1}	0.0566	0.0482	0.0406

t/min	15.13	19.80	∞
[HCN]/mol L^{-1}	0.0706	0.0653	0.0424
[C_3H_7CHO]/mol L^{-1}	0.0282	0.0229	0.0000

What is the order of the reaction, and what is the value of the rate constant k?

18.12 Hydrogen peroxide reacts with thiosulfate ion in slightly acidic solution as follows:

$$H_2O_2 + 2S_2O_3^{2-} + 2H^+ \rightarrow 2H_2O + S_4O_6^{2-}$$

This reaction rate is independent of the hydrogen ion concentration in the pH range 4 to 6. The following data were obtained at 25 °C and pH 5.0. Initial concentrations: $[H_2O_2] = 0.036$ mol L^{-1}; $[S_2O_3^{2-}] = 0.020\,40$ mol L^{-1}.

t/min	16	36	43	52
$[S_2O_3^{2-}]/10^{-3}$ mol L^{-1}	10.30	5.18	4.16	3.13

(*a*) What is the order of the reaction? (*b*) What is the rate constant?

18.13 The reaction A = B is nth order (where $n = \frac{1}{2}, \frac{3}{2}, 2, 3, \ldots$) and goes to completion to the right. Derive the expression for the half-life in terms of k, n, and $[A]_0$.

18.14 A gas reaction 2A = B is second order in A and goes to completion in a reaction vessel of constant volume and temperature with a half-life of 1 h. If the initial pressure of A is 1 bar, what are the partial pressures of A and B, and what is the total pressure at 1 h, at 2 h, and at equilibrium?

18.15 The rate constant for the reaction

$$I + I + Ar \rightarrow I_2 + Ar$$

is 0.59×10^{16} cm^6 mol^{-2} s^{-1} at 293 K. What is the half-life of I if $[I]_0 = 2 \times 10^{-5}$ mol L^{-1} and $[Ar] = 5 \times 10^{-3}$ mol L^{-1}?

18.16 A solution of A is mixed with an equal volume of a solution of B containing the same number of moles, and the reaction

A + B = C occurs. At the end of 1 h, A is 75% reacted. How much of A will be left unreacted at the end of 2 h if the reaction is (*a*) first order in A and zero in B, (*b*) first order in both A and B, and (*c*) zero order in both A and B?

18.17 Show that for a first-order reaction R → P the concentration of product can be represented as a function of time by

$$[P] = a + bt + ct^2 + \cdots$$

and express a, b, and c in terms of $[R]_0$ and k.

18.18 For a reaction A → X, the following concentrations of A were found in a single kinetics experiment:

$[A]/\text{mol L}^{-1}$	1.000	0.952	0.909	0.870	0.833	0.800
t/h	0	0.05	0.10	0.15	0.20	0.25

What is the rate v of this reaction at $[A] = 1.000 \text{ mol L}^{-1}$?

18.19 The following table gives kinetic data for the following reaction at 25 °C:

$$OCl^- + I^- = OI^- + Cl^-$$

$[OCl^-]$ / mol L^{-1}	$[I^-]$ / mol L^{-1}	$[OH^-]$ / mol L^{-1}	$d[IO^-]/dt$ / $10^{-4}\ \text{mol L}^{-1}\ \text{s}^{-1}$
0.0017	0.0017	1.00	1.75
0.0034	0.0017	1.00	3.50
0.0017	0.0034	1.00	3.50
0.0017	0.0017	0.5	3.50

What is the rate law for the reaction, and what is the value of the rate constant?

18.20 For a reversible first-order reaction

$$A \underset{k_2}{\overset{k_1}{\rightleftharpoons}} B$$

$k_1 = 10^{-2}\ \text{s}^{-1}$ and $[B]_{eq}/[A]_{eq} = 4$. If $[A]_0 = 0.01 \text{ mol L}^{-1}$ and $[B]_0 = 0$, what will be the concentration of B after 30 s?

18.21 The first three steps in the decay of ^{238}U are

$$^{238}U \xrightarrow[4.5\times 10^9 y]{\alpha} {}^{234}Th \xrightarrow[24.1 d]{\beta} {}^{234}Pa \xrightarrow[1.14 m]{\beta} {}^{234}U$$

If we start with pure ^{238}U, what fraction will be ^{234}Th after 10, 20, 40, and 80 days?

18.22 Equation 18.27 for a second-order reaction becomes indeterminant when $[A]_0 = [B]_0$, but the text states that when the initial concentrations of A and B in a reaction A + B = X are equal,

$$\frac{1}{[A]} = \frac{1}{[A]_0} + kt \quad \text{and} \quad \frac{1}{[B]} = \frac{1}{[B]_0} + kt$$

Show that this is correct by using l'Hôpital's rule. According to l'Hôpital's rule, if a function of a variable x is indeterminant as $x \to 0$ because the function becomes 0/0, then the limit can be found by taking the limit of the derivative of the numerator divided by the derivative of the denominator.

18.23 Suppose the transformation of A to B occurs by both a reversible first-order reaction and a reversible second-order reaction involving hydrogen ion:

$$A \underset{k_2}{\overset{k_1}{\rightleftharpoons}} B \qquad A + H^+ \underset{k_4}{\overset{k_3}{\rightleftharpoons}} B + H^+$$

What is the relationship between these four rate constants?

18.24 Use the rapid equilibrium approximation to derive the rate law for the mechanism

$$\begin{array}{ccc} A^- & \xrightarrow{k_A} & B^- \\ K_{HA} \updownarrow & & \updownarrow K_{HB} \\ HA & \xrightarrow[k_B]{} & HB \end{array}$$

The acid dissociation reactions are rapid in comparison with the isomerization reactions.

18.25 Suppose that

$$A \xrightarrow{k_1} B \xrightarrow{k_2} C \longrightarrow \cdots$$

and you are interested in isolating the largest possible amount of B. Given the values of k_1 and k_2, derive an equation for the time that the concentration of B goes through a maximum. Now consider two cases: (*a*) A reacts more rapidly than B and (*b*) A reacts less rapidly than B. For a given value of k_2, in which case would you wait the longer time for B to go through its maximum?

18.26 The hydrolysis of

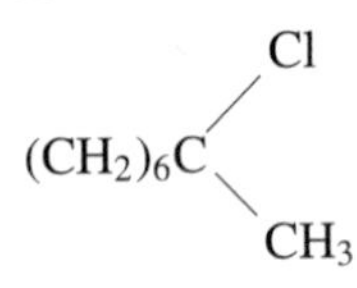

in 80% ethanol follows the first-order rate equation. The values of the specific reaction rate constants are as follows:

t/°C	0	25	35	45
k/s^{-1}	1.06×10^{-5}	3.19×10^{-4}	9.86×10^{-4}	2.92×10^{-3}

(*a*) Plot log k against $1/T$. (*b*) Calculate the activation energy. (*c*) Calculate the pre-exponential factor.

18.27 If a first-order reaction has an activation energy of 104 600 J mol^{-1} and a pre-exponential factor A of $5 \times 10^{13}\ \text{s}^{-1}$, at what temperature will the reaction have a half-life of (*a*) 1 min and (*b*) 30 days?

18.28 Isopropenyl allyl ether in the vapor state isomerizes to allyl acetone according to a first-order rate equation. The following equation gives the influence of temperature on the rate constant (in s^{-1}):

$$k = 5.4 \times 10^{11}\ e^{-123\ 000/RT}$$

where the activation energy is expressed in J mol^{-1}. At 150 °C, how long will it take to build up a partial pressure of 0.395 bar of allyl acetone, starting with 1 bar of isopropenyl allyl ether?

18.29 The pre-exponential factor for the trimolecular reaction

$$2NO + O_2 \rightarrow 2NO_2$$

is 10^9 cm^6 mol^{-2} s^{-1}. What is the value in L^2 mol^{-2} s^{-1} and in cm^6 s^{-1}?

18.30 A reaction A + B + C → D follows the mechanism

$$A + B \rightleftharpoons AB \qquad AB + C \rightarrow D$$

in which the first step remains essentially in equilibrium. Show that the dependence of rate on temperature is given by

$$k = A\,e^{-(E_a + \Delta H)/RT}$$

where ΔH is the enthalpy change for the first reaction.

18.31 Consider the following mechanism:

$$A + B \underset{k_2}{\overset{k_1}{\rightleftharpoons}} C \qquad C \xrightarrow{k_3} D$$

(*a*) Derive the rate law using the steady-state approximation to eliminate the concentration of C. (*b*) Assuming that $k_3 \ll k_2$, express the pre-exponential factor A and E_a for the apparent second-order rate constant in terms of A_1, A_2, and A_3 and E_{a1}, E_{a2}, and E_{a3} for the three steps.

18.32 For the two parallel reactions $A \xrightarrow{k_1} B$ and $A \xrightarrow{k_2} C$, show that the activation energy E for the disappearance of A is given in terms of activation energies E_1 and E_2 for the two paths by

$$E = \frac{k_1 E_1 + k_2 E_2}{k_1 + k_2}$$

18.33 Set up the rate expressions for the following mechanism:

$$A \underset{k_2}{\overset{k_1}{\rightleftharpoons}} B \qquad B + C \xrightarrow{k_3} D$$

If the concentration of B is small compared with the concentrations of A, C, and D, the steady-state approximation may be used to derive the rate law. Show that this reaction may follow the first-order equation at high pressures and the second-order equation at low pressures.

18.34 A dimerization $2A \rightarrow A_2$ is found to be first order, with a half-life of 666 s. This somewhat surprising result is explained by postulating the following mechanism:

$$A \xrightarrow{k_1} A^* \qquad A^* + A \xrightarrow{k_2} A_2$$

where $k_2 \gg k_1$. (*a*) What is the value for the rate constant k_1? (*b*) If the initial concentration of A is 0.05 M, how much time is required to reach [A] = 0.0125 M?

18.35 The reaction $NO_2Cl = NO_2 + \frac{1}{2}Cl_2$ is first order and appears to follow the mechanism

$$NO_2Cl \xrightarrow{k_1} NO_2 + Cl \qquad NO_2Cl + Cl \xrightarrow{k_2} NO_2 + Cl_2$$

(*a*) Assuming a steady state for the chlorine atom concentration, show that the empirical first-order rate constant can be identified with $2k_1$. (*b*) The following data were obtained at 180 °C. In a single experiment the reaction is first order, and the empirical rate constant is represented by k. Show that the reaction is second order at these low gas pressures and calculate the second-order rate constant.

$c/10^{-8}$ mol cm^{-3}	5	10	15	20
$k/10^{-4}$ s^{-1}	1.7	3.4	5.2	6.9

18.36 The reaction

$$2SO_2 + O_2 = 2SO_4$$

is catalyzed by the mechanism

$$2NO + O_2 \underset{k_{-1}}{\overset{k_1}{\rightleftharpoons}} 2NO_2$$

$$NO_2 + SO_2 \underset{k_{-2}}{\overset{k_2}{\rightleftharpoons}} NO + SO_3$$

To obtain the overall reaction from this mechanism, the second step has to be taken twice, and so the stoichiometric number s_2 of the second step is said to be 2. The equilibrium constant K_c for an overall reaction is related to the rate constants for the individual steps k_i and k_{-i} by

$$K_c = \prod_{i=1}^{S} \left(\frac{k_i}{k_{-i}}\right)^{s_i}$$

where s_i is the stoichiometric number of the ith step and S is the number of steps. Verify this relation for the above mechanism.

18.37 What is the rate constant for the following reaction at 500 K?

$$H + HCl \rightarrow Cl + H_2$$

The data required appear in Table 18.2 and Table C.3.

18.38 For the gas reaction

$$O + O_2 + M \underset{k'}{\overset{k}{\rightleftharpoons}} O_3 + M$$

where M = O_2, the rate constant is given by

$$k = (6.0 \times 10^7\ \mathrm{L^2\ mol^{-2}\ s^{-1}})\,e^{2.5/RT}$$

where the activation energy is in kJ mol^{-1}. Calculate the values of the parameters in the Arrhenius equation for the reverse reaction assuming $\Delta H°$ and $\Delta S°$ are independent of temperature.

18.39 (*a*) Write the steady-state equations for A and B in reaction 18.61. (*b*) Use these rate equatons to derive the equilibrium expressions for [B]/[A] and [C]/[A] by use of the principle of detailed balance, which requires that $k_1 k_3 k_6 = k_2 k_4 k_6$.

18.40 The mechanism of the pyrolysis of acetaldehyde at 520 °C and 0.2 bar is

$$CH_3CHO \xrightarrow{k_1} CH_3 + CHO$$

$$CH_3 + CH_3CHO \xrightarrow{k_2} CH_4 + CH_3CO$$

$$CH_3CO \xrightarrow{k_3} CO + CH_3$$

$$CH_3 + CH_3 \xrightarrow{k_4} C_2H_6$$

What is the rate law for the reaction of acetaldehyde, using the usual assumptions? (As a simplification further reactions of the radical CHO have been omitted and its rate equation may be ignored.)

18.41 For the reaction

$$H_2(g) + Br_2(g) = 2HBr(g)$$

spectroscopic measurements show that $d[Br_2]/dt = -1.2 \times 10^{-3}$ mol L^{-1} s^{-1}. (*a*) What is the rate of reaction v? (*b*) What is the value of $d[HBr]/dt$? (*c*) What is the rate of conversion $d\xi/dt$ if the reaction occurs in a 3-L vessel? (*d*) What amount of HBr is produced per second in the 3-L vessel under these conditions?

18.42 Under certain conditions, it is found that ammonia is formed from its elements at a rate of 0.10 mol L^{-1} s^{-1}.

$$N_2(g) + 3H_2(g) = 2NH_3(g)$$

(*a*) What is the rate of reaction v? (*b*) What is the value of $d[N_2]/dt$? (*c*) What is the value of $d[H_2]/dt$?

18.43 The rate of the gas reaction $H_2 + Br_2 = 2HBr$ doubles when the concentration of hydrogen is doubled, and it increases by a factor of 1.4 when the concentration of bromine is doubled. What is the order with respect to hydrogen, the order with respect to bromine, and the overall order?

18.44 The half-life of a first-order chemical reaction $A \longrightarrow B$ is 10 min. What percentage of A remains after 1 h?

18.45 A reaction is carried out with 1-cyclohexenyl allyl malonitrile at 135.7 °C. Calculate the first-order rate constant from the data on the first 5 min and the second 5 min.

t/min	0	5	10
% reaction	19.8	34.2	46.7

18.46 The reaction

$$SO_2Cl_2 = SO_2 + Cl_2$$

is first order with a rate constant of 2.2×10^{-5} s^{-1} at 320 °C. What percentage of SO_2Cl_2 is decomposed after being heated at 320 °C for 2 h?

18.47 The kinetics of the hydrolysis of an ester is studied by titrating the acid produced. A sample is withdrawn and titrated with alkali. The volumes required at various times are

t/min	0	27	60	∞
V/mL	0	18.1	26.0	29.7

(*a*) Prove that this reaction is first order. (*b*) Calculate the half-life.

18.48 A gas reaction A = 2B is first order in A and goes to completion in a reaction vessel of constant volume and temperature with a half-life of 10 min. If the initial pressure of A is 1 bar, what are the partial pressures of A and B at 10 min, at 20 min, and at equilibrium?

18.49 Modern carbon is radioactive because ^{14}C is produced by cosmic rays by the reaction $^{14}N(n, p)^{14}C$. This is the physicists' way of indicating that a neutron goes into the ^{14}N nucleus and a proton comes out. This nuclide of carbon has a half-life of 5720 years. Carbon recently incorporated into growing plants has a specific activity of 16 disintegrations per minute per gram. (*a*) What percentage of the carbon in growing plants is ^{14}C? (*b*) How many grams of modern carbon does it take to provide 0.05 microcuries of ^{14}C? A curie is 3.7×10^{10} nuclear transformations per second.

18.50 Living trees incorporate $^{14}C(t_{1/2} = 5720$ y) into their wood because there is ^{14}C in CO_2 due to cosmic rays and the nuclear reaction $^{14}N(n, p)^{14}C$. When a tree dies, this radioactivity of the wood slowly disappears. An archeological sample of wood has 42% of the ^{14}C found in living trees. Assuming the level of cosmic rays has been constant, what is the age of the archeological sample?

18.51 The reaction $2NO + O_2 \rightarrow 2NO_2$ is third order and $d[NO_2]/dt = k[NO]^2[O_2]$. The rate constant k has a value of 7.1×10^3 L^2 mol^{-2} s^{-1} at 25 °C. Air blown through a certain hot chamber and cooled quickly at 25 °C and 1 bar contains 1% by volume of nitric oxide, NO, and 20% of oxygen. How long will it take for 90% of this NO to be converted to nitrogen dioxide, NO_2 (or N_2O_4)?

18.52 The second-order rate constant for an alkaline hydrolysis of ethyl formate in 85% ethanol (aqueous) at 29.86 °C is 4.53 L mol^{-1} s^{-1}. (*a*) If the reactants are both present at 0.001 mol L^{-1}, what will be the half-life of the reaction? (*b*) If the concentration of one of the reactants is doubled and that of the other is cut in half, how long will it take for one-half the reactant present at the lower concentration to react?

18.53 The reaction

$$CH_3CH_2NO_2 + OH^- \rightarrow H_2O + CH_3CHNO_2^-$$

is second order, and k at 0 °C is 39.1 mol^{-1} min^{-1}. An aqueous solution is 0.004 molar in nitroethane and 0.005 molar in NaOH. How long will it take for 90% of the nitroethane to react?

18.54 The second-order rate constant for the reaction of ClO and NO is 6.2×10^{-12} cm^3 s^{-1}. What is its value in L mol^{-1} s^{-1}?

18.55 A solution of ethyl acetate and sodium hydroxide was prepared that contained (at $t = 0$) 5×10^{-3} mol L^{-1} ethyl acetate and 8×10^{-3} mol L^{-1} sodium hydroxide. After 400 s at 25 °C a 25-mL aliquot was found to neutralize 33.3 mL of 5×10^{-3} mol L^{-1} hydrochloric acid. (*a*) Calculate the rate constant for this second-order reaction. (*b*) At what time would you expect 20.0 mL of hydrochloric acid to be required?

18.56 A dimer is formed in the solution reaction $2A \rightarrow A_2$. The rate law is $v = k[A]^2$, where $k = 0.015\ M^{-1}\ s^{-1}$. What is the half-life of A when $[A]_0 = 0.05$ M?

18.57 In the preceding dimerization problem, how much time is required for [A] to reach a concentration of 0.0125 M?

18.58 The rate constant for a second-order reaction is $10^{8.2}\ L\ mol^{-1}\ s^{-1}$. What is the rate constant in $cm^3\ s^{-1}$?

18.59 Derive the integrated rate equation for a reaction of order $\frac{1}{2}$. Derive the expression for the half-life of such a reaction.

18.60 The gas-phase formation of phosgene, $CO + Cl_2 \rightarrow COCl_2$, is $\frac{3}{2}$ order with respect to CO. Derive the integrated rate equation for a $\frac{3}{2}$-order reaction. Derive the expression for the half-life.

18.61 Equal molar quantities of A and B are added to a liter of a suitable solvent. At the end of 500 s one-half of A has reacted according to the reaction A + B = C. How much of A will be reacted at the end of 800 s if the reaction is (*a*) zero order with respect to both A and B, (*b*) first order with respect to A and zero order with respect to B, and (*c*) first order with respect to both A and B?

18.62 For the reaction A + B → products, equation 18.27 applies. Show that when $[B]_0 \gg [A]_0$,

$$\ln \frac{[A]}{[A]_0} = -k[B]_0 t$$

This is referred to as a pseudo–first-order reaction.

18.63 When an optically active substance is isomerized, the optical rotation decreases from that of the original isomer to zero in a first-order manner. In a given case the half-time for this process is found to be 10 min. Calculate the rate constant for the conversion of one isomer to another.

18.64 The equations for $[A_2]$ and $[A_3]$ in Section 18.4 give an indeterminate result if $k_1 = k_2$. Rederive the equations, giving $[A_2]$ and $[A_3]$ as functions of time for the special case that

$$A_1 \xrightarrow{k_1} A_2 \xrightarrow{k_1} A_3$$

18.65 For the reaction 2A = B + C the rate law for the forward reaction is

$$-\frac{d[A]}{dt} = k[A]$$

Give two possible rate laws for the backward reaction.

18.66 The following rate constants were obtained for the first-order decomposition of acetone dicarboxylic acid in aqueous solution:

$t/°C$	0	20	40	60
$k/10^{-5}\ s^{-1}$	2.46	47.5	576	5480

(*a*) Calculate the energy of activation. (*b*) Calculate the pre-exponential factor *A*. (*c*) What is the half-life of this reaction at 80 °C?

18.67 Although the thermal decomposition of ethyl bromide is complex, the overall rate is first order, and the rate constant is given by the expression $k = (3.8 \times 10^{14}\ s^{-1})e^{-229\,000/RT}$, where the activation energy is in $J\ mol^{-1}$. Estimate the temperature at which (*a*) ethyl bromide decomposes at the rate of 1% per second and (*b*) the decomposition is 70% complete in 1 h.

18.68 Given that the first-order rate constant for the overall decomposition of N_2O_5 is $k = (4.3 \times 10^{13}\ s^{-1})e^{-103\,000/RT}$, calculate (*a*) the half-life at −10 °C and (*b*) the time required for 90% reaction at 50 °C. The activation energy is in $J\ mol^{-1}$.

18.69 Suppose that a substance X decomposes into A and B in parallel paths with rate constants given by

$$k_A = (10^{15}\ s^{-1})e^{-126\,000/RT} \qquad k_B = (10^{13}\ s^{-1})e^{-83\,700/RT}$$

where the activation energies are given in $J\ mol^{-1}$. (*a*) At what temperature will the two products be formed at the same rate? (*b*) At what temperature will A be formed 0.1 times as fast as B? (*c*) State a generalization concerning the effect of temperature on the relative rates of reactions with different activation energies.

18.70 For the reaction

$$O + NO + M \rightarrow NO_2 + M$$

$k_{300\ K} = 6 \times 10^9\ L^2\ mol^{-2}\ s^{-1}$ and $k_{1000\ K} = 3 \times 10^{10}\ L^2\ mol^{-2}\ s^{-1}$. Calculate the parameters in the Arrhenius equation.

18.71 (*a*) The viscosity of water changes about 2% per degree at room temperature. What is the activation energy for this process? (*b*) The activation energy for a reaction is 62.8 kJ mol^{-1}. Calculate $k_{35\,°C}/k_{25\,°C}$.

18.72 The reaction $2NO + O_2 \rightarrow 2NO_2$ is third order. Assuming that a small amount of NO_3 exists in rapid reversible equilibrium with NO and O_2 and that the rate-determining step is the slow bimolecular reaction $NO_3 + NO \rightarrow 2NO_2$, derive the rate equation for this mechanism.

18.73 The apparent activation energy for the recombination of iodine atoms in argon is −5.9 kJ mol^{-1}. This negative temperature coefficient may result from the following mechanism:

$$I + M = IM \qquad K = \frac{[IM]}{[I][M]} \qquad IM + I \underset{k_{-1}}{\overset{k_1}{\rightleftharpoons}} I_2 + M$$

Assuming that the first step remains at equilibrium, derive the rate equation that includes both the forward and backward reacnl tions. Show that the backward reaction is bimolecular and that the equilibrium constant expression for the dissociation of iodine is independent of the concentration of the third body.

18.74 Derive the steady-state rate equation for the following mechanism for a trimolecular reaction:

$$A + A \underset{k_{-1}}{\overset{k_1}{\rightleftharpoons}} A_2^* \qquad A_2^* + M \xrightarrow{k_2} A_2 + M$$

18.75 Show that the interconversion of ortho- and parahydrogen will be $\frac{3}{2}$ order, as obtained experimentally in the range 600 to 750 °C, if the rate-determining step is that between atoms and molecules of hydrogen:

$$\underset{\uparrow\downarrow}{\mathrm{H + para\text{-}H_2}} \underset{k_2}{\overset{k_1}{\rightleftharpoons}} \underset{\uparrow\uparrow}{\mathrm{ortho\text{-}H_2 + H}}$$

where the arrows represent the directions of the nuclear spins (cf. Section 15.1).

18.76 What are the Arrhenius parameters for the following elementary reaction?

$$\mathrm{H + HCl \longrightarrow Cl + H_2}$$

The data required are found in Table 18.2 and Table C.3.

18.77 The Arrhenius parameters for the reaction

$$\mathrm{Cl + H_2 \longrightarrow HCl + H}$$

are given in Table 18.2. What is the rate constant for the reverse reaction at 1000 K? The thermodynamic parameters for these substances are given in Table C.3.

18.78 Ozone is decomposed by the catalytic chain

$$\mathrm{NO + O_3 \xrightarrow{k_1} NO_2 + O_2} \qquad \mathrm{NO_2 + O \xrightarrow{k_2} NO + O_2}$$

What is the steady-state rate law for the formation of O_2?

18.79 The formation of phosgene by the reaction

$$\mathrm{CO + Cl_2 = COCl_2}$$

appears to follow the mechanism

$$\mathrm{Cl_2} \underset{k_{-1}}{\overset{k_1}{\rightleftharpoons}} \mathrm{2Cl}$$

$$\mathrm{Cl + CO} \underset{k_{-2}}{\overset{k_2}{\rightleftharpoons}} \mathrm{COCl}$$

$$\mathrm{COCl + Cl_2 \xrightarrow{k_3} COCl_2 + Cl}$$

Assuming that the intermediates Cl and COCl are in a steady state, what is the rate law for this reaction?

Computer Problems

18.A Two consecutive first-order reactions are represented by

$$\mathrm{A_1 \xrightarrow{k_1} A_2 \xrightarrow{k_2} A_3}$$

for which we have seen the general solution in equations 18.47–18.50. Use a mathematical application for solving differential equations to calculate the concentrations of A_1, A_2, andA_3 as a function of time without using equations 18.47–18.50. Assume that initially A_1 is at unit concentration and treat three cases: (*a*) $k_1 = k_2 = 1\ \mathrm{s^{-1}}$; (*b*) $k_1 = 1\ \mathrm{s^{-1}}, k_2 = 5\ \mathrm{s^{-1}}, k_3 = 3\ \mathrm{s^{-1}}$; (c) $\mathrm{k_1} = 1\ \mathrm{s^{-1}}, \mathrm{k_2} = 25\ \mathrm{s^{-1}}$.

18.B Two consecutive reversible first-order reactions are represented by

$$\mathrm{A_1} \underset{k_2}{\overset{k_1}{\rightleftharpoons}} \mathrm{A_2} \underset{k_4}{\overset{k_3}{\rightleftharpoons}} \mathrm{A_3}$$

Use a mathematical application for solving differential equations to calculate the concentrations of A_1, A_2, andA_3 as a function of time. Assume that initially A_1 is at unit concentration and treat three cases: (*a*) $k_1 = k_2 = k_3 = k_4 = 1\ \mathrm{s^{-1}}$; (*b*) $k_1 = k_2 = k_4 = 1\ \mathrm{s^{-1}}, k_3 = 3\ \mathrm{s^{-1}}$; (*c*) $k_1 = k_2 = k_4 = 1\ \mathrm{s^{-1}}$, $k_3 = 9\ \mathrm{s^{-1}}$.

18.C A reaction A + B = C + D is reversible and has an equilibrium constant equal to 2. (*a*) If the initial concentrations of A and B are 1 and 0.5 mol $\mathrm{L^{-1}}$, respectively, plot the four concentrations as a function of time, assuming the rate constant for the forward reaction is 1 L $\mathrm{mol^{-1}s^{-1}}$. (*b*) To confirm the equilibrium concentrations of C and D, calculate these concentrations using the equilibrium constant expression.

18.D In the case of a reaction like that in the preceding problem, there may be a question as to whether there is an intermediate X:

$$\mathrm{A + B} \underset{10\ \mathrm{s^{-1}}}{\overset{1\ \mathrm{L\ mol^{-1}\ s^{-1}}}{\rightleftharpoons}} \mathrm{X} \underset{0.5\ \mathrm{L\ mol^{-1}\ s^{-1}}}{\overset{10\ \mathrm{s^{-1}}}{\rightleftharpoons}} \mathrm{C + D}$$

For the indicated values of the rate constants, explore the effects of intermediate X on the plots of concentration versus time.

18.E The simplest example of an autocatalytic reaction is A + B → 2B. Assuming that the rate constant is unity and that $[\mathrm{A}]_0 = 1$, plot [A] and [B] versus time and explore the effect of varying the initial concentration of B from 0.01 to 0.2. Note that [B] levels off at $[\mathrm{A}]_0 + [\mathrm{B}]_0$.

18.F Calculate the activation energy and the pre-exponential factor for the gas reaction

$$\mathrm{N_2O_5 = 2NO_2 + \tfrac{1}{2}O_2}$$

T/K	273	298	308	318	328	338
$k \times 10^5/\mathrm{s^{-1}}$	0.0787	3.46	13.5	49.8	150	487

18.G (*a*) Calculate the activation energy and the pre-exponential factor for the hydrolysis of 2-chlorooctane. (*b*) Calculate k at 50 °C.

t/°C	0	25	35	45
$k/\mathrm{s^{-1}}$	1.06×10^{-5}	3.19×10^{-4}	9.86×10^{-4}	2.92×10^{-3}

18.H (*a*) Plot k_{uni} versus [M] according to the Lindemann equation with $k_1 = 1, k_2 = 3$, and $k_3 = 1$. (*b*) Plot $\log_{10}(k_{\mathrm{uni}}/k^{\infty})$ versus $\log_{10}(P)$, where the pressure is expressed in pascals. Compare this plot with Fig. 18.10.

18.I For the monomolecular triangle reaction (see equations 18.60 and 18.61), calculate concentrations as a function of time for the following three cases and discuss the results in terms of the principle of detailed balance. (*a*) The rate constants in mechanism 18.60 are all unity. (*b*) The rate constants in mechanism 18.61 are all unity. (*c*) The rate constants in mechanism 18.61 are all unity except for $k_1 = 1.1$. The initial concentration of A can be taken as 1 M.

18.J Assume that a reaction $A = B$ goes through an intermediate I, but that $k_2 \ll k_{-1}$.

$$A \underset{k_{-1}}{\overset{k_1}{\rightleftharpoons}} I \xrightarrow{k_2} B$$

Under these conditions, the first step remains at equilibrium.

$$\frac{d[B]}{dt} = k[A] = \frac{k_1 k_2}{k_{-1}}[A] = Kk_2[A]$$

Assume that for the first step $K = \exp(-\Delta H/RT)$, and $k_2 = s^{-1}$ at 298 K and $E_a = 20$ kJ mol^{-1}. Plot $\ln k$ versus $1/T$ for $\Delta H = 0, -10, -20$, and -30 kJ mol^{-1}.

18.K Sometimes Arrhenius plots are curved and can be represented by

$$k = aT^n e^{-E/RT}$$

(*a*) Plot $\log k$ versus $1/T$ for $n = -2, -1, 0, 1, 2$ when $a = 10^8$ and $E = 10$ kJ mol^{-1}. Use the temperature range 300 K to 1000 K. (*b*) Calculate the activation energies at temperatures 333, 500, and 1000 K and for $n = -2, -1, 0, 1, 2$ and make a table.

19

Chemical Dynamics and Photochemistry

The preceding chapter was concerned with macroscopic kinetics; this one is concerned with microscopic kinetics, that is, elementary reactions at the molecular level. The calculation of rate constants from properties of individual atoms and molecules is challenging because reactions occur as a result of collisions with a variety of energies, angles of approach, and states of reactants and products. Simple collision theory of bimolecular reaction is based on consideration of collisions of rigid spherical molecules. To go further and take electronic structure into account, it is necessary to use the concept of the potential energy surface for a reaction.

Transition-state theory attempts to simplify the problem by making a "dynamic bottleneck assumption." Transition-state theory is not exact, but is based on a series of approximations. However, it has been useful since its inception in 1935.

Since the absorption of light produces excited states of atoms and molecules, photochemistry is really the study of the chemistry of excited states. As pointed out in Section 13.1, electromagnetic radiation in the visible and ultraviolet is generally required to produce chemical reactions because changes in electronic energy levels are required. More recently it has been found that the absorption of many infrared photons from a high-intensity laser can also cause reaction.

19.1 SIMPLE COLLISION THEORY OF BIMOLECULAR REACTIONS

As we saw in the preceding chapter, many reactions of atoms or radicals with small molecules in the gas phase have pre-exponential factors in the range $10^{10.5}$–$10^{11.5}$ L mol^{-1} s^{-1}. Since bimolecular reactions of small radicals have zero activation energies, their actual rate constants may be in this range. If the activation energy is zero, we might expect molecules to react on their first collision so that the bimolecular rate constant can be estimated from the collision density Z_{12} between molecules of type 1 and type 2, as calculated for rigid spheres. The collision density Z_{12}, given by equation 17.55, is the number of collisions between molecules of type 1 and type 2 per unit volume per unit time.

If reaction occurs with each collision, we can obtain the reaction rate in moles of collisions between molecules of type 1 and type 2 per unit volume per unit time by dividing by the Avogadro constant to obtain

$$\frac{-\mathrm{d}[\mathrm{B}_1]}{\mathrm{d}t} = \frac{Z_{12}}{N_\mathrm{A}} = \frac{\pi d_{12}^2 \langle v_{12} \rangle}{N_\mathrm{A}} \rho_1 \rho_2 \tag{19.1}$$

In discussing chemical kinetics, concentrations are used rather than number densities, so we need to replace the number densities by using

$$\rho_i = \frac{N_i}{V} = \frac{n_i N_\mathrm{A}}{V} = [\mathrm{B}_i] N_\mathrm{A} \tag{19.2}$$

Substituting this and the expression for the mean relative velocity $\langle v_{12} \rangle$ yields

$$-\frac{\mathrm{d}[\mathrm{B}_1]}{\mathrm{d}t} = N_\mathrm{A} \pi d_{12}^2 \left(\frac{8 k_\mathrm{B} T}{\pi \mu} \right)^{1/2} [\mathrm{B}_1][\mathrm{B}_2]$$

$$= k[\mathrm{B}_1][\mathrm{B}_2] \tag{19.3}*$$

where the second-order rate constant is given by

$$k = N_\mathrm{A} \pi d_{12}^2 \left(\frac{8 k_\mathrm{B} T}{\pi \mu} \right)^{1/2} \tag{19.4}$$

When SI units are used, k has the units of m^3 mol^{-1} s^{-1} and has to be multiplied by 10^3 L m^{-3} to obtain the value in the usual units L mol^{-1} s^{-1}. If B_1 and B_2 are identical, a factor of 2 must be included on the right-hand side of equation 19.4. Equation 19.4 is only approximate even for bimolecular reactions that occur on the first collision because molecules are not hard spheres that interact only when they touch.

Example 19.1 ***Calculation of the rate constant for the reaction of two small radicals***

Calculate the bimolecular rate constant at 298 K for the reaction of two different "average" small radicals with a reduced mass μ of 30×10^{-3} kg mol^{-1}/N_A = 4.98×10^{-26} kg and a collision diameter d_{12} of 500 pm.

*Throughout this chapter we will write the Boltzmann constant as k_B to distinguish it from the rate constant k.

Using equation 19.4, we find

$$k = (6.022 \times 10^{23}\ \text{mol}^{-1})(3.14)(500 \times 10^{-12}\ \text{m})^2 \left[\frac{8(1.38 \times 10^{-23}\ \text{J K}^{-1})(298\ \text{K})}{(3.14)(4.98 \times 10^{-26}\ \text{kg})}\right]^{1/2}$$

$$= 2.17 \times 10^8\ \text{m}^3\ \text{mol}^{-1}\ \text{s}^{-1}$$

$$= (2.17 \times 10^8\ \text{m}^3\ \text{mol}^{-1}\ \text{s}^{-1})(10^3\ \text{L m}^{-3})$$

$$= 2.17 \times 10^{11}\ \text{L mol}^{-1}\ \text{s}^{-1}$$

In equation 19.4, the area πd_{12}^2 is referred to as the collision cross section, but we should really use the **reaction cross section** $\sigma(\epsilon_r)$, which is a function of the relative energy ϵ_r of the collision. Then the rate constant, which is also a function of the relative energy of collision, is given by

$$k(\epsilon_r) = \sigma(\epsilon_r)v_r \tag{19.5}$$

where v_r is the relative magnitude of the velocity of the reactants (same as the mean relative speed in Fig. 17.12). Now we have to think about what is meant by the relative energy of collision ϵ_r. If the spherical molecules collide head on, the relative energy of collision is equal to $\frac{1}{2}\mu v_r^2$, where μ is the reduced mass of the system. However, if the molecules barely touch when they pass (that is, the impact parameter b is only slightly less than d_{12}), the relative energy of collision is essentially zero. Figure 19.1 can be used to obtain the energy along the line of centers of the two molecules, $\epsilon_k = \mu v_k^2$, where v_k is the collision velocity along the line of centers.

The figure shows that the velocity along the line of centers is $v_k = v_r \cos\alpha$, where α is the angle indicated in the figure. Note that the right side of the figure shows that $\sin\alpha = b/b_{\text{max}}$, where b_{bmax} is the distance along the line of centers when the spheres touch. Thus the energy along the line of centers is given by

$$\epsilon_k = \frac{1}{2}\mu v_k^2 = \frac{1}{2}\mu v_r^2 \cos^2\alpha \tag{19.6}$$

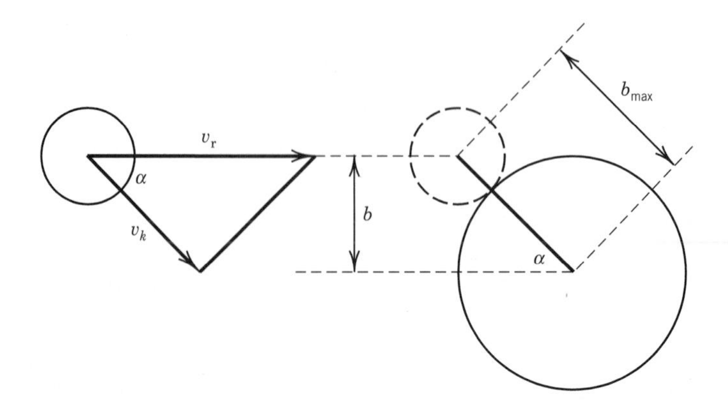

Figure 19.1 Collision of two spherical molecules with impact parameter b and relative velocity v_r. Note that $\cos\alpha = v_k/v_r$ and $\sin\alpha = b/b_{\text{max}}$. [With permission from P. L. Houston, *Chemical Kinetics and Reaction Dynamics,* McGraw-Hill, New York (2001).]

Since $\sin^2 \alpha + \cos^2 \alpha = 1$,

$$\epsilon_k = \frac{1}{2}\mu v_r^2(1 - \sin^2 \alpha) = \epsilon_r\left[1 - \left(\frac{b}{b_{max}}\right)^2\right] \tag{19.7}$$

Further steps in the derivation are based on the simplifying approximation that the probability of reaction is unity when $\epsilon_k > \epsilon^*$ and zero otherwise. Houston* shows that equation 19.7 yields a reaction cross section given by

$$\sigma(\epsilon_r) = \pi b_{max}^2\left(1 - \frac{\epsilon^*}{\epsilon_r}\right) \tag{19.8}$$

provided $\epsilon_r > \epsilon^*$ and $\sigma(\epsilon_r) = 0$ if $\epsilon_r < \epsilon^*$. The dependence of $\sigma(\epsilon)$ on ϵ is given in Fig. 19.2.

Given the functional form for the reaction energy, the value of the reaction cross section has to be obtained by an integration over impact parameters b, and the value of $k(T)$ has to be obtained by averaging over the Boltzmann energy distribution. These steps, which are described in detail by Houston, lead to

$$k(T) = \pi b_{max}^2 v_r \exp\left(-\frac{\epsilon^*}{kT}\right) \tag{19.9}$$

where the average relative velocity is $v_r = (8kT/\pi\mu)^{1/2}$ and ϵ^* is the minimum collision energy along the line of centers. The coefficient of the exponential term is simply the rate constant for hard-sphere collisions. The Boltzmann factor is the fraction of collisions that provide energy greater than ϵ^*. Note that equation 19.9 differs from the Arrhenius equation because of the factor $T^{1/2}$ in v_r, but that effect of temperature is not very significant in comparison with the exponential dependence on $1/T$.

According to collision theory, a collision that is sufficiently energetic to supply the activation energy may still fail to produce a reaction if the colliding molecules

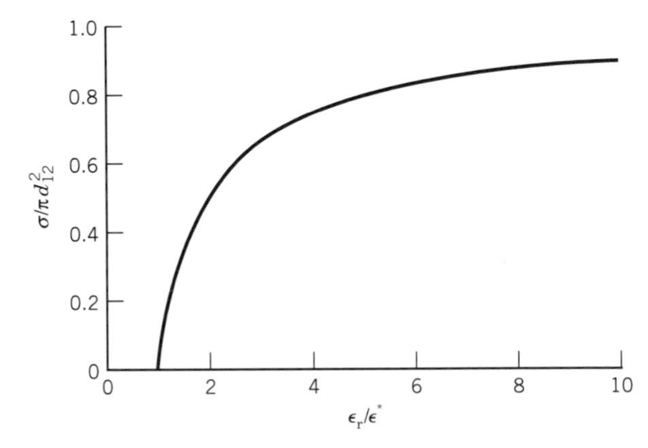

Figure 19.2 Dependence of the reaction cross section $\sigma(\epsilon)$ on the energy ϵ of collision along the line of centers. The lowest energy at which reaction occurs is ϵ^*. (See Computer Problem 19.C.)

*P. L. Houston, *Chemical Kinetics and Reaction Dynamics.* New York: McGraw-Hill (2001).

are not oriented in such a way that they can react with each other. This can be corrected by introducing a steric factor into equation 19.9. However, this is really not very useful because there is no simple theory for calculating its value. Since there are also very drastic assumptions underlying equation 19.9, we cannot expect it to yield very accurate results. For example, it ignores the changes in electronic structure that occur in any chemical reaction, and how these changes influence the reaction cross section. To correct this we must look at chemical reactions from the viewpoint of quantum mechanics.

19.2 POTENTIAL ENERGY SURFACES

Kinetic theory is helpful in that it tells us about molecular collisions, but it does not deal with the changes that take place on a molecular level when reactants are converted to products. When two molecules are very close to each other, they cannot be considered separately because their wavefunctions overlap. Thus, from the time the reactant molecules are close to each other until the products are well separated, the system is a kind of **supermolecule.** This supermolecule is different from an ordinary molecule because it is in the process of change, but it is a molecule in the sense that its energy and electron distribution can be calculated for each nuclear configuration by use of quantum mechanics. According to the Born–Oppenheimer approximation (Section 11.1), the electrons move much more rapidly than the nuclei, so the molecular electronic energy and wavefunction can be calculated for a given nuclear configuration by use of the electronic Schrödinger equation. This approximation was used earlier to calculate the electronic potential energy function so that the Schrödinger equation could be used to obtain molecular vibrational energy levels.

If a reaction involves N nuclei, there are $3N$ nuclear coordinates, but the group of nuclei has three translational coordinates of the center of mass and two or three rotational coordinates (about the center of mass) that do not affect the potential energy. Thus, the potential energy is a function of $3N - 5$ nuclear coordinates if the nuclei are constrained to a straight line and $3N - 6$ nuclear coordinates in general. For the simplest type of reaction,

$$A + BC \rightarrow AB + C \tag{19.10}$$

where A, B, and C are atoms, three coordinates are required. It is not possible to plot the potential energy as a function of three coordinates, but if the angle θ of approach of A to BC is fixed, the potential energy of the system can be plotted as a function of R_{AB} and R_{BC}, where R is intermolecular distance. Such a plot is shown in Fig. 19.3. If R_{AB} is rather large, as on the left face of the diagram, the potential energy is essentially that of the BC molecule. Similarly, the right face gives the potential energy of AB. Thus, Fig. 19.3 omits two monotonous valleys to the left and right that extend indefinite distances. Initially, the R_{AB} distance is very large. As A approaches BC, the lowest-energy path is given by the dashed line from reactants R to products P. This dashed line gives the minimum energy path, which is sometimes referred to as the reaction coordinate. We will soon see that the configuration of the system does not actually move along the reaction coordinate in the reaction, but the reaction coordinate does help us visualize the surface. The highest point along the reaction coordinate is a saddle point. At the **saddle point,** the potential energy is a maximum along the reaction coordinate,

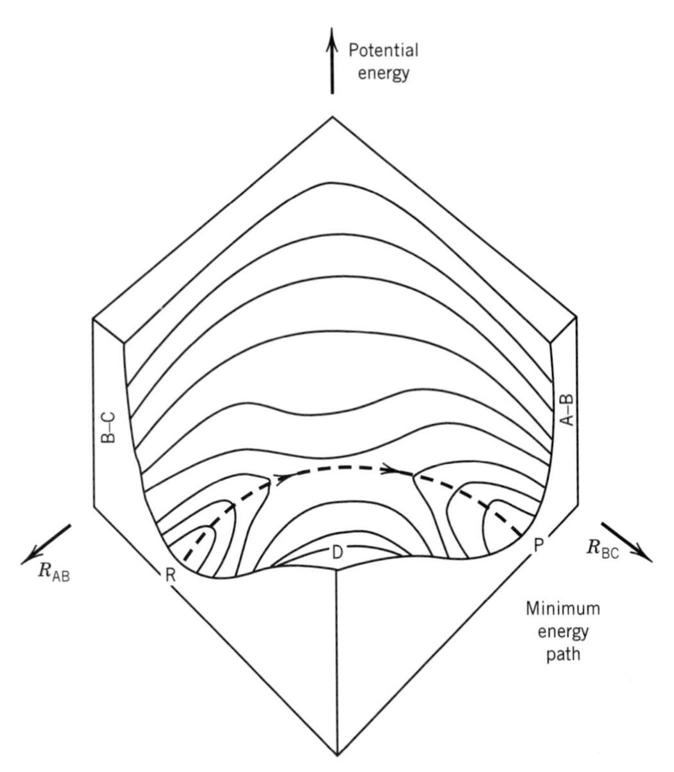

Figure 19.3 Potential energy surface for the reaction A + BC → AB + C with the nuclei constrained to a line.

but it is a minimum in the direction perpendicular to the reaction coordinate. The reaction system at this point is said to be in the **transition state.** In Fig. 19.3, D is a high plateau giving the potential energy of three atoms well separated from each other.

As a first simple example, consider what happens when A approaches a nonvibrating BC molecule along the internuclear axis. The point representing the configuration of the system moves along the minimum energy path, the dashed line in Fig. 19.3. As R_{AB} decreases, kinetic energy is converted to potential energy as the point representing the system of three nuclei moves up the valley from the left. If there is initially enough kinetic energy for the system to go over the saddle point, AB and C are formed and gain energy as the system goes down the valley to the right. If the kinetic energy is too low, the system returns down the valley to the left, and we would say that the reactants bounced off each other.

Figure 19.3 applies only when the nuclei are constrained to a line, and the potential energy surface will be different if there is a different angle θ of approach. The quantum mechanical calculation of an accurate potential surface for a reaction such as 19.10 is a difficult process, and surfaces have been calculated only for a few reactions.

A great deal of attention has been focused on the reaction of a hydrogen atom with a hydrogen molecule:

$$H_A + H_BH_C \rightarrow H_AH_B + H_C \tag{19.11}$$

The potential energy surface for this reaction for $\theta = 180°$ is described by means of the contour diagram in Fig. 19.4. This surface has been calculated using ab

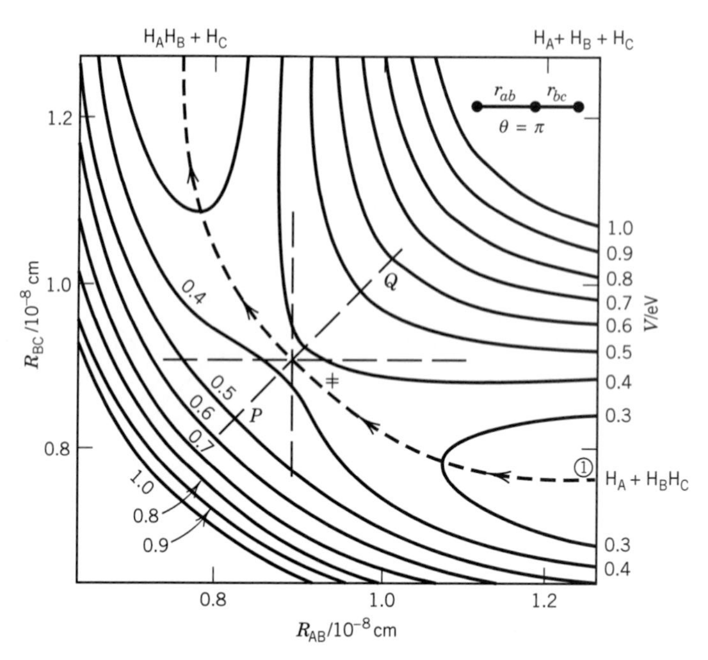

Figure 19.4 Potential energy surface for the reaction $H_A + H_BH_C \rightarrow H_AH_B + H_C$ for a linear approach and departure. [Based on R. N. Porter and M. Karplus, *J. Chem. Phys.* **40:**1105 (1964).]

initio methods with configuration interaction (as discussed in Section 11.4), and the error at any point on the surface is believed to be less than 0.03 eV (2.9 kJ mol^{-1}). As H_A approaches H_BH_C, along the minimum energy path, the potential energy of the system increases until the saddle point is reached at ‡. At this point $R_{AB} = R_{BC} = 93$ pm, and the potential energy of the system is 0.37 eV (42 kJ mol^{-1}), the highest along the dashed line. Since the saddle point is 0.37 eV higher than the potential of H_A and H_BH_C at an infinite distance, this energy must be supplied from relative kinetic energy or vibrational energy in order for the reaction to occur. In the upper right-hand corner of Fig. 19.4, there is a high plateau with energy of 432 kJ mol^{-1}. This is the energy of three hydrogen atoms infinitely far apart, with respect to separated reactants or products.

19.3 THEORETICAL CALCULATION OF A RATE CONSTANT

Once the potential energy curve has been obtained for various approach angles θ, the probability of a reaction for certain initial conditions (relative kinetic energy, vibrational energy, and θ) can, in principle, be calculated using the time-dependent Schrödinger equation. However, this is a very difficult calculation, and so classical mechanics is ordinarily used. **The force on a particular nucleus is given by the gradient in the potential energy.** For example, the component of the force on nucleus i in the x direction is given by (see Section 9.9)

$$F_{x,i} = -\frac{\partial V}{\partial x_i} \tag{19.12}$$

where V is the potential energy and x_i is the x coordinate of the nucleus. At each instant the system is represented by a point on the surface, and **Newton's law** $F = ma$ is integrated numerically to obtain the coordinates of the system as a function of time. Calculations can also be made when H_BH_C initially has vibrational motion, and as an approximation this is also treated classically. Rotational energy is not important in these calculations. These trajectory calculations yield a reaction probability of 0 for certain initial conditions and of 1 for others. Figure 19.5 shows the results of two calculations of this type for collinear collisions. In Fig. 19.5*a* the H_BH_C molecule is vibrating and H_A approaches with a certain initial velocity, but reaction does not occur. Note that in this nonreactive, inelastic collision, translational energy is converted to vibrational energy in H_BH_C. In Fig. 19.5*b* reaction does occur.

To calculate a rate constant in this way, it is necessary to make a very large number of trajectory calculations with initial states chosen to give a statistically representative sample of possible initial states at the chosen temperature. The initial conditions can be chosen by a Monte Carlo procedure to ensure that the distribution of each initial parameter approaches the correct distribution as the number of calculated trajectories increases. The relative kinetic energies of H_A and H_BH_C are given by the Boltzmann equation. All angles θ of approach have to be included, but most reactive collisions for reaction 19.11 occur at angles near 180°.

To see how a rate constant can be calculated from a series of trajectory calculations, we need to consider the reaction probability $P(b)$, often referred to as the opacity function, and the way it is used to calculate a reaction cross section $\sigma(v)$. The **reaction probability** $P(b)$ is simply the fraction of the total number of trajectories at a selected reactant relative velocity and impact parameter b (Section 17.8) that result in reaction. The reaction probability $P(b)$ for the $H + H_2$ reaction for a relative velocity of 1.17×10^6 cm s^{-1} is shown in Fig. 19.6 as a function of the impact parameter b. The reaction probability is greatest for an impact parameter of zero, and it decreases to zero at some finite impact parameter.

The contribution to the reaction cross section $\sigma(v)$ of collisions with an impact parameter b is $2\pi bP(b)\,db$, and so the cross section is given by

$$\sigma(v) = \int_0^{b_{\max}} 2\pi bP(b)\,db \tag{19.13}$$

When a chemical reaction occurs in bulk, molecules collide at all possible relative velocities, and so the rate constant $k(T)$ at temperature T is made up of a sum of terms for all possible relative velocities, with each weighted by the fraction f_i of collisions with that relative velocity:

$$\begin{aligned} k(T) &= f_1k(v_1) + f_2k(v_2) + \cdots \\ &= f_1\sigma(v_1)v_1 + f_2\sigma(v_2)v_2 + \cdots \end{aligned} \tag{19.14}$$

or

$$k(T) = \int_0^{\infty} f(v, T)v\sigma(v)\,dv \tag{19.15}$$

where $f(v, T)$ is the Maxwell–Boltzmann distribution for relative velocity v at temperature T. In 1965 Karplus, Porter, and Sharma* calculated a very large

*M. Karplus, R. N. Porter, and R. D. Sharma, *J. Chem. Phys.* **43:**3259 (1965).

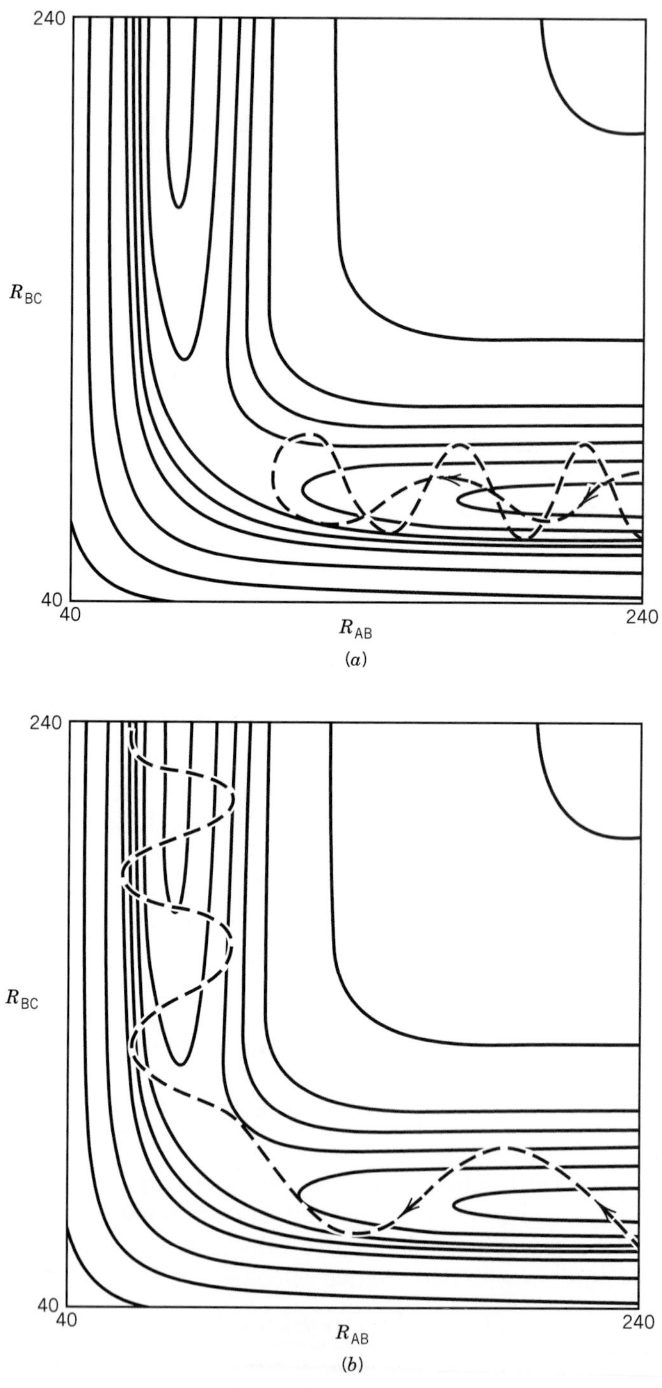

Figure 19.5 (*a*) Trajectory for a nonreactive, inelastic collision of a hydrogen atom and a hydrogen molecule. (*b*) Trajectory for a reactive collision. [From P. Siegbahn and B. Liu, *J. Chem. Phys.* **68:**2457 (1978), and C. J. Horowitz, *J. Chem. Phys.* **68:**2466 (1978).]

number of trajectories for the $H + H_2$ reaction and found that their results could be expressed by the following Arrhenius equation:

$$k = (4.3 \times 10^{13} \text{ mol}^{-1} \text{ cm}^3 \text{ s}^{-1}) \exp\left[-\frac{31\,000 \text{ J mol}^{-1}}{(8.3145 \text{ J K}^{-1} \text{ mol}^{-1})T}\right] \quad (19.16)$$

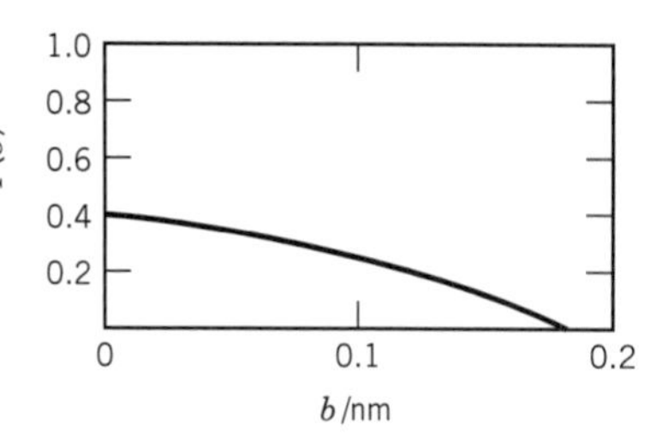

Figure 19.6 Reaction probability $P(b)$ for the reaction $H + H_2$ for a relative velocity of 1.17×10^6 cm s^{-1} as a function of impact parameter b. (From J. Nicholas, *Chemical Kinetics,* Hoboken, NJ: Wiley, 1976. Copyright ©1978 J. Nicholas. Reprinted with permission.)

This result is in pretty good agreement with experimental results obtained by studying the reaction

$$D + H_2 \rightarrow DH + H \quad (19.17)$$

or

$$H + \text{para-}H_2 \rightarrow \text{ortho-}H_2 + H \quad (19.18)$$

and with the value calculated using transition-state theory, which we will discuss in the next section. The transition-state rate constant over the same range of temperature is given by

$$k = (7.4 \times 10^{13} \text{ mol}^{-1} \text{ cm}^3 \text{ s}^{-1}) \exp\left[-\frac{34\,440 \text{ J mol}^{-1}}{(8.3145 \text{ J K}^{-1} \text{ mol}^{-1})T}\right] \quad (19.19)$$

Classical calculations have been made on a number of simple reactions, and the general conclusion is that classical calculations can provide an adequate description of the collision dynamics for some purposes. There are quantum mechanical effects in reaction kinetics, and penetration into classically forbidden regions ("tunneling") may be important, especially at lower temperatures.

The biggest difficulty in the quantum mechanical calculation of a reaction rate is calculating the potential energy surface with sufficient accuracy. The shape of this surface is extremely important in determining the role of the vibrational energy of a reactant in affecting the likelihood of reaction.

The potential energy diagram is symmetrical for the reaction $H_A + H_BH_C = H_AH_B + H_C$, but this is not true in general. The shape of the potential energy surface determines whether translational motion or vibrational motion will be most effective in causing reaction. Note that the reaction begins at the right-hand side of each diagram in Fig. 19.7 and proceeds from right to left. It is useful to distinguish between potential energy surfaces with early barriers and late barriers. Figure 19.7*a* shows a potential energy surface with an **early barrier** and a reactant with sufficient translational energy to cross the saddle point into the trough for products. The energy released as the system passes down from the saddle point results in vibrational energy for the products. Figure 19.7*b* shows a potential energy surface with an early barrier and a reactant with vibrational energy. Even though the total energy may be the same as in (a), the reactant may be reflected back from the barrier. Thus we conclude that an early barrier favors a reactant with translational energy and produces vibrationally excited products.

Figure 19.7*c* shows a potential energy surface with a **late barrier** and a reactant with vibrational energy. The molecule may bounce off of the wall of the valley it is in and cross the saddle point into the reactant valley, where it does not have much vibrational energy. Figure 19.7*d* shows that if the reactant does not have vibrational energy, it may bounce back into the reactant valley. Thus a late barrier favors a reactant with vibrational energy and produces a product with less vibrational energy. Note that a potential energy surface that has an early barrier in one direction has a late barrier in the other direction.

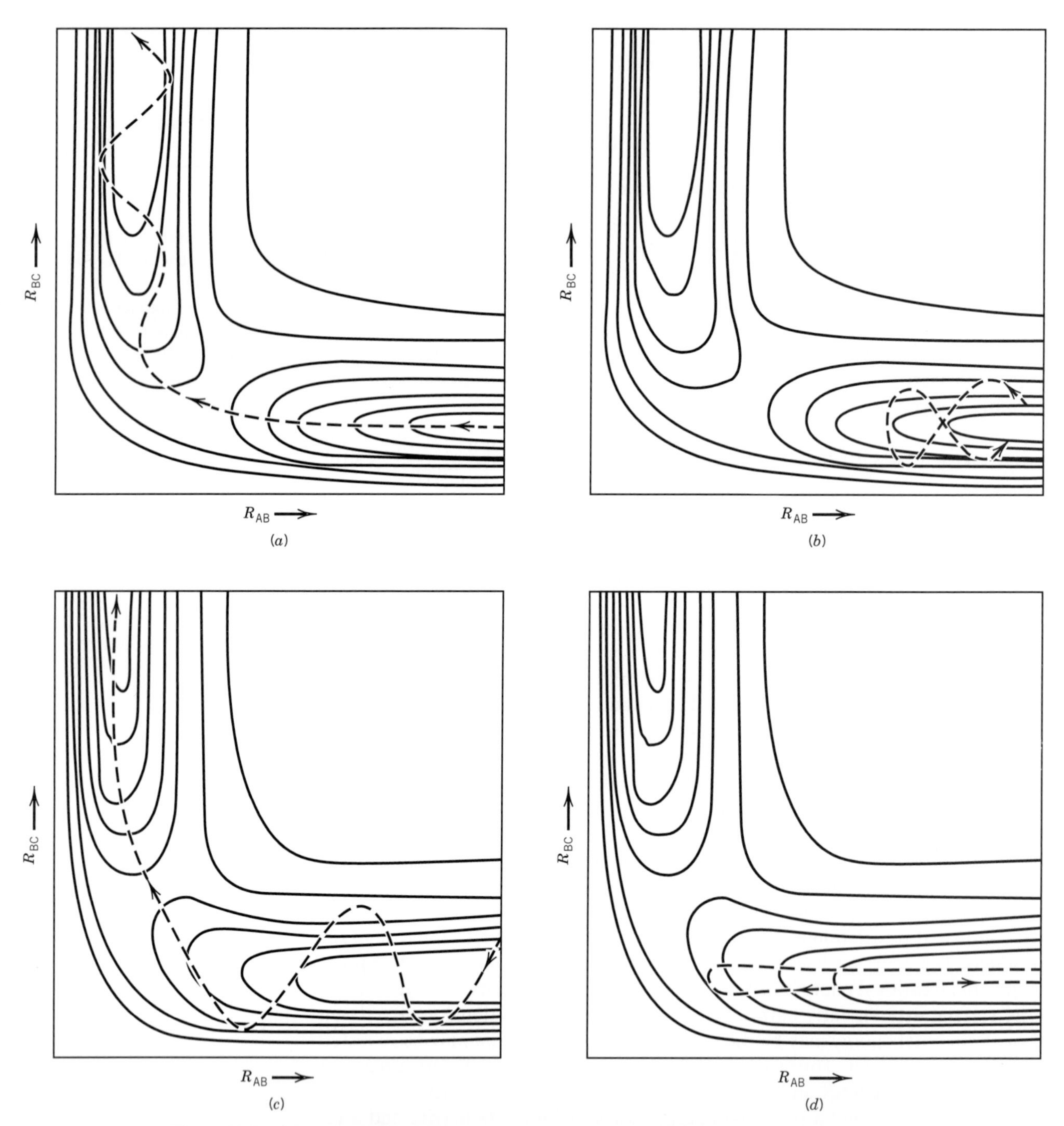

Figure 19.7 (*a*) A potential energy surface with an early barrier and a reactant with sufficient translational energy to cross the saddle point. (*b*) A potential energy surface with an early barrier and a vibrationally excited reactant. (*c*) A potential energy surface with a late barrier and a vibrationally excited reactant. (*d*) A potential energy surface with a late barrier and a reactant that has sufficient translational energy to reach the saddle point, but does not.

19.4 TRANSITION-STATE THEORY

Transition-state theory was developed before much was known about potential energy surfaces, and it, in effect, bypasses the problem of the dynamics of a reactive collision. Nevertheless, it is very useful because quantitative calculations can be made with estimated properties of the transition state for a reaction. The development of transition-state theory goes back to Eyring and Evans and Polanyi* in 1935. The basis of transition-state theory is that it is possible to define a surface in coordinate space and to calculate the flux of trajectories that pass through this surface from the reactant side to the product side *without turning back*. This flux is identified as the reactive flux. If the potential energy barrier is high, there is no difficulty in locating this surface; it is placed at the top of the energy barrier perpendicular to the reaction coordinate. In addition, transition-state theory is based on the Born–Oppenheimer approximation and the assumption that molecules are distributed among their states according to the Boltzmann distribution.

To derive an expression for the rate constant, we will focus our attention on what happens at the top of the potential energy barrier shown in Fig. 19.8 for the reaction

$$A + B \rightarrow C \tag{19.20}$$

The dashed lines show the quantum mechanical zero point energies for reactants, products, and the **activated complex** at the top of the potential energy barrier. The rate of this elementary gas reaction is given by

$$v = k[A][B] \tag{19.21}$$

According to transition-state theory the reaction proceeds through an activated complex $C^{\ddagger}$ that produces C at a rate $k^{\ddagger}[C^{\ddagger}]$. It is important to remember that $C^{\ddagger}$ is not an intermediate compound in the reaction, but rather is a structure that is in the process of change in the direction of the products. The rate of reaction can

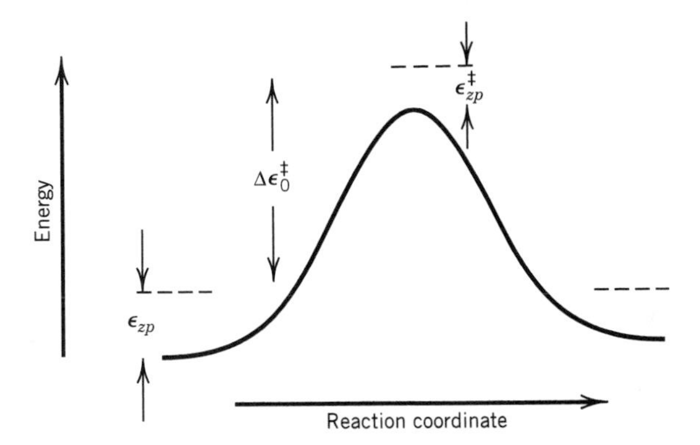

Figure 19.8 Potential energy diagram showing the relationship between the height of the barrier in the potential energy surface and the activation energy $\Delta\epsilon_0^{\ddagger}$. [Reproduced from I. W. M. Smith, *Kinetics and Dynamics of Elementary Gas Reactions*, 1980, Figure 4.2, p. 116, by permission of the publishers, Butterworth & Co. (Publishers) Ltd. ©.]

*H. Eyring, *J. Chem. Phys.* **3:**107 (1935); M. G. Evans and M. Polanyi, *Trans. Faraday Soc.* **31:**875 (1935).

be expressed in two ways, $k[\mathrm{A}][\mathrm{B}] = k^{\ddagger}[\mathrm{C}^{\ddagger}]$, so the second-order rate constant is given by

$$k = \frac{k^{\ddagger}[\mathrm{C}^{\ddagger}]}{[\mathrm{A}][\mathrm{B}]} \tag{19.22}$$

The concentration terms look like the expression for the equilibrium constant of the reaction $\mathrm{A} + \mathrm{B} = \mathrm{C}^{\ddagger}$, so transition-state theory assumes this is a rapid equilibrium and replaces the concentration terms in equation 19.22 with

$$\frac{K_c}{c^{\circ}} = \frac{[\mathrm{C}^{\ddagger}]}{[\mathrm{A}][\mathrm{B}]} \tag{19.23}$$

where the standard state concentration c° is required because the equilibrium constant K_c is dimensionless. Thus equation 19.22 can be written

$$k = \frac{k^{\ddagger}K_c}{c^{\circ}} \tag{19.24}$$

This shows that if we can derive expressions for $k^{\ddagger}$ and K_c theoretically, we can predict the rate of reaction 19.20.

We are going to use statistical mechanics to calculate K_c for the gas reaction $\mathrm{A} + \mathrm{B} = \mathrm{C}^{\ddagger}$, but first we must make a purely thermodynamic adjustment from a dimensionless K_P (obtained from statistical mechanics) to a dimensionless K_c (required for equation 19.24):

$$K_c = \left(\frac{c^{\circ}RT}{P^{\circ}}\right)^{-\sum \nu_i} K_P \tag{19.25}$$

where P° is the standard state pressure. This adjustment was discussed in Section 5.7. For the reaction under consideration, $\sum \nu_i = -1$, so this adjustment is given by

$$K_c = \left(\frac{c^{\circ}RT}{P^{\circ}}\right)K_P \tag{19.26}$$

Both K_c and K_P are dimensionless. Substituting equation 19.26 in equation 19.24 yields

$$k = k^{\ddagger}\left(\frac{RT}{P^{\circ}}\right)K_P \tag{19.27}$$

The calculation of $k^{\ddagger}$ is simple because $\mathrm{C}^{\ddagger}$ is going through the transition state and becoming C. $\mathrm{C}^{\ddagger}$ may have vibrations that are perpendicular to the reaction coordinate, and they have normal frequencies because their potential functions have the usual parabolic shape. However, the motion of $\mathrm{C}^{\ddagger}$ in the direction of the reaction coordinate has an inverted potential curve since the transition state is at a **saddle point.** Therefore, this motion represents advancement along the reaction coordinate through the transition state. Thus the reaction rate is proportional to the inverse time (or "frequency" ν) spent at the transition state. We replace $k^{\ddagger}$ with ν to obtain

$$k = \nu\left(\frac{RT}{P^{\circ}}\right)K_P \tag{19.28}$$

Sometimes a transmission coefficient κ is included in this equation to allow for the possibility that some of the activated complexes may return to reactants instead of going on to C. However, because it is difficult to calculate κ, we will assume that it has a value of unity so that the rate constant k, which is calculated by transition-state theory, is a maximum value for the reaction.

Statistical mechanics can be used to calculate the equilibrium constant for the formation of $C^{\ddagger}$. If this were an ordinary gas reaction, equation 16.88 shows that this equilibrium constant would have the value

$$K_P = \left(\frac{k_B T}{P^\circ}\right)^{-1} \frac{q(C^{\ddagger})/V}{[q(A)/V][q(B)/V]} e^{-\Delta\epsilon_0^{\ddagger}/k_B T} \tag{19.29}$$

where the change in energy at absolute zero is $\Delta\epsilon_0^{\ddagger} = \epsilon_0(C^{\ddagger}) - \epsilon_0(A) - \epsilon_0(B)$. In writing equation 19.29, we have assumed that $A + B = C^{\ddagger}$ is an ordinary reaction, but it is not because the activated complex $C^{\ddagger}$ is falling apart. The partition function for the motion of $C^{\ddagger}$ along the reaction coordinate can be written as if the motion were a vibration with frequency ν:

$$q_v = \frac{1}{1 - e^{-h\nu/k_B T}} \tag{19.30}$$

This frequency is lower than the frequencies of the other vibrations because the potential energy curve is relatively flat near the transition state. Since $h\nu/k_B T$ is small, $e^{-x} = 1-x$, so the denominator of equation 19.30 is close to $h\nu/k_B T$. Thus

$$q_v \approx \frac{k_B T}{h\nu} \tag{19.31}$$

Therefore, we write the partition function for $C^{\ddagger}$ as $(k_B T/h\nu)q'(C^{\ddagger})$, where $q'(C^{\ddagger})$ is the partition function for $C^{\ddagger}$, omitting the partition function for the vibration along the reaction coordinate. Making this substitution in equation 19.29 and substituting equation 19.29 in equation 19.28 yields

$$k = \left(\frac{k_B T}{h}\right)\left(\frac{RT}{P^\circ}\right)\left(\frac{k_B T}{P^\circ}\right)^{-1} \frac{q'(C^{\ddagger})/V}{[q(A)/V][q(B)/V]} e^{-\Delta\epsilon_0^{\ddagger}/k_B T} \tag{19.32}$$

Note that the frequency ν of passage over the potential energy barrier has canceled. The last three terms in equation 19.32 are the equilibrium constant $K^{\ddagger}$ for the formation of the activated complex from the reactants, omitting the partition function for the vibration along the reaction coordinate. Thus equation 19.32 can be written

$$k = \left(\frac{k_B T}{h}\right)\left(\frac{RT}{P^\circ}\right)K^{\ddagger} \tag{19.33}$$

where

$$K^{\ddagger} = \left(\frac{k_B T}{P^\circ}\right)^{-1} \frac{q'(C^{\ddagger})/V}{[q(A)/V][q(B)/V]} e^{-\Delta\epsilon_0^{\ddagger}/k_B T} \tag{19.34}$$

We have derived equation 19.33 for a bimolecular reaction, but it can be generalized to

$$k = \left(\frac{k_B T}{h}\right)\left(\frac{RT}{P^\circ}\right)^{m-1} K^\ddagger \tag{19.35}$$

where m is the **order of the reaction.**

For a unimolecular reaction, the translational partition functions for the activated complex and the reactant cancel, so the unimolecular rate constant is given by

$$k = \left(\frac{k_B T}{h}\right)\frac{q'_{\text{int}}(\text{C}^\ddagger)}{q_{\text{int}}(\text{A})} e^{-\Delta\epsilon_0^\ddagger/k_B T} \tag{19.36}$$

where the subscript int indicates that only the internal coordinates are involved. This yields a rate constant in s^{-1}, as it must. This is the rate constant in the high-pressure limit (see Section 18.10), where there are enough collisions to maintain the Boltzmann distribution of A^*. However, in the falloff region the rate of formation of vibrationally excited A^* molecules is too slow to maintain the Boltzmann distribution. To calculate rate constants in the falloff region it is necessary to calculate the rate constant for the activation process and its reverse.

The classical theory for unimolecular reactions was developed by Rice, Ramsberger, and Kassel. This theory has been improved in several aspects by Marcus, and it is now referred to as the RRKM theory. Since phenomena in the falloff region are quite complicated, we will not be able to discuss them here.

The pre-exponential factors for a number of bimolecular reactions are given in Table 19.1. The pre-exponential factors may be compared with 10^{13}–10^{15} $cm^3\ mol^{-1}\ s^{-1}$ calculated from simple collision theory (Section 19.1). The last column of Table 19.1 gives pre-exponential factors calculated from transition-state theory using *estimated* frequencies and geometrical parameters for the activated complex.

The calculation of the internal partition function $q^\ddagger_{\text{int}}$ for the transition state is a problem because in general we do not know the structure of the transition state. However, various hypotheses about the structure of the transition state may be made, or the structure can be calculated by ab initio methods.

Table 19.1 Pre-exponential Factors for Some Bimolecular Reactions Calculated by Transition-State Theory Compared with Experimental Values

Reaction	A_{exp} $cm^3\ mol^{-1}\ s^{-1}$	A_{calc} $cm^3\ mol^{-1}\ s^{-1}$
$H + H_2 \rightarrow H_2 + H$	5.4×10^{13}	3.5×10^{13}
$Br + H_2 \rightarrow HBr + H$	3×10^{13}	1×10^{14}
$H + CH_4 \rightarrow H_2 + CH_3$	1×10^{13}	2×10^{13}
$H + C_2H_6 \rightarrow H_2 + C_2H_5$	3×10^{12}	1×10^{13}
$CH_3 + H_2 \rightarrow CH_4 + H$	2×10^{12}	1×10^{12}
$CH_3 + CH_3COCH_3 \rightarrow CH_4 + CH_3COCH_2$	4×10^{11}	1×10^{11}
$CD_3 + CH_4 \rightarrow CD_3H + CH_3$	1×10^{11}	2×10^{11}
$2ClO \rightarrow Cl_2 + O_2$	6×10^{10}	1×10^{11}

Source: J. Nicholas, *Chemical Kinetics.* © Wiley, Hoboken, NJ, 1978. Reprinted with permission.

Example 19.2 ***Calculation of the pre-exponential factor for the rate constant of the reaction of a hydrogen atom with a hydrogen molecule***

Using transition-state theory, calculate the pre-exponential factor for the rate constant for the reaction

$$\mathrm{H} + \mathrm{H_2} \rightarrow \mathrm{H_2} + \mathrm{H}$$

at 500 K. Assume a linear activated complex with the nuclei each separated by 0.94×10^{-10} m. The vibrational partition functions may be ignored because their values are so close to unity. (The experimental value is 5.4×10^{13} cm^3 mol^{-1} s^{-1}.)

According to equation 19.35, the pre-exponential factor is given by

$$A = \left(\frac{RT}{h}\right)\frac{q'^{\ddagger}}{(q_{\mathrm{H}}/V)(q_{\mathrm{H_2}}/V)}$$

where $q'^{\ddagger} = q_{\mathrm{t}}^{\ddagger} q_{\mathrm{int}}^{\ddagger}$, $q_{\mathrm{H}} = q_{\mathrm{tH}} q_{\mathrm{int\,H}}$, and $q_{\mathrm{H_2}} = q_{\mathrm{tH_2}} q_{\mathrm{int\,H_2}}$. Since $q_{\mathrm{t}}^{\ddagger}/V$ and $q_{\mathrm{tH_2}}/V$ cancel,

$$A = \left(\frac{RT}{h}\right)\frac{q_{\mathrm{int}}^{\ddagger}/V}{(q_{\mathrm{tH}}/V) q_{\mathrm{int\,H}} q_{\mathrm{int\,H_2}}}$$

The internal partition function $q_{\mathrm{int}}^{\ddagger}$ for the transition state is the rotational partition function for a symmetrical and linear arrangement of three hydrogen atoms. Since the central hydrogen atom is on the axis of rotation, it does not contribute to the moment of inertia:

$$I = \mu R^2 = \frac{m_1 m_2}{m_1 + m_2} R^2 = \frac{(1.0078 \times 10^{-3}\ \mathrm{kg\ mol^{-1}})^2 (1.88 \times 10^{-10}\ \mathrm{m})^2}{(2.0156 \times 10^{-3}\ \mathrm{kg\ mol^{-1}})(6.022 \times 10^{23}\ \mathrm{mol^{-1}})}$$

$$= 2.96 \times 10^{-47}\ \mathrm{kg\ m^2}$$

The internal partition function for the transition state is therefore given by

$$q_{\mathrm{int}}^{\ddagger} = \frac{8\pi^2 I k_{\mathrm{B}} T}{2h^2} = \frac{8\pi^2 (2.96 \times 10^{-47}\ \mathrm{kg\ m^2})(1.381 \times 10^{-23}\ \mathrm{J\ K^{-1}})(500\ \mathrm{K})}{2(6.626 \times 10^{-34}\ \mathrm{J\ s})^2} = 18.4$$

The translational partition function for the relative motion of the reactants is given by

$$\frac{q_{\mathrm{t}}}{V} = \frac{(2\pi\mu k_{\mathrm{B}} T)^{3/2}}{h^3}$$

$$= \frac{[2\pi(1.116 \times 10^{-27}\ \mathrm{kg})(1.381 \times 10^{-23}\ \mathrm{J\ K^{-1}})(500\ \mathrm{K})]^{3/2}}{(6.626 \times 10^{-34}\ \mathrm{J\ s})^3}$$

$$= 1.157 \times 10^{30}\ \mathrm{m^{-3}}$$

where μ is the reduced mass of a hydrogen atom and a hydrogen molecule.

The rotational partition function for H_2 at 3000 K was calculated in Example 16.9 to be 17.13. At 500 K the rotational partition function is

$$q_{\mathrm{rH_2}} = 17.13 \frac{500}{3000} = 2.86$$

Thus, the pre-exponential factor is

$$A = \frac{(8.3145\ \mathrm{J\ K^{-1}\ mol^{-1}})(500\ \mathrm{K})(18.4)}{(6.626 \times 10^{-34}\ \mathrm{J\ s})(1.157 \times 10^{30}\ \mathrm{m^{-3}})(2.86)}$$

$$= 3.49 \times 10^7\ \mathrm{m^3\ mol^{-1}\ s^{-1}}$$

Converting to units used in the statement of the problem,

$$A = (3.49 \times 10^7 \text{ m}^3 \text{ mol}^{-1} \text{ s}^{-1})(100 \text{ cm m}^{-1})^3 = 3.49 \times 10^{13} \text{ cm}^3 \text{ mol}^{-1} \text{ s}^{-1}$$

which is about two-thirds of the experimental value of 5.4×10^{13} cm^3 mol^{-1} s^{-1}.

Comment:

Transition-state theory has been very useful in predicting the rates of chemical reactions, but it does depend on experimental data on vibrational frequencies and internuclear distances. The theory has been most useful for small molecules. In some cases the rate calculations are complicated by tunneling through the potential energy barrier.

19.5 MOLECULAR BEAM EXPERIMENTS

In gas kinetics experiments reactant molecules approach each other on trajectories with random angles and impact parameters and with relative energies that range around those having the highest probabilities at the reaction temperature. Colliding molecules may also have different amounts of vibrational, rotational, and electronic energies. Reaction rate constants are averages over these various angles of approach, relative energies, and the like.

In the 1960s it became possible to study the dynamics of elementary reactions by the use of crossed molecular beams. This makes it possible to control the energies of reacting atoms and molecules, to study the effect of molecular orientation, and to detect reaction intermediates and study their decay dynamics. This is done by measuring the velocity and angular distributions of the products. This has led to the discovery of bound states that tell more about the potential energy surface for a reaction. For example, in the reaction $F + CH_3I \rightarrow IF + CH_3$, it was found that CH_3IF was formed. Thus the use of crossed beams helps bridge the gap between the basic laws of mechanics and chemical reactions.

To obtain a microscopic view of a bimolecular gas reaction, crossed beams can be used in which reactant molecules are in known quantum states, and the quantum states of product molecules are identified. In the simplest type of apparatus, illustrated schematically in Fig. 19.9, A and B are sources of beams of the two reactants that collide in region C. These collisions occur in a chamber evacuated with a high-speed pump so that the only collisions are between the molecules from A and B. Product molecules and scattered reactant molecules are detected at D. The effect of changing the angle of approach may be studied by moving A or B, and the effect of the relative velocity of the reactants may be studied by use of velocity selectors on the beams as they leave A and B. Sometimes the molecules in molecular beams are put in selected electronic, vibrational, and rotational states so that the effects of these quantum numbers on the observed cross sections can be determined. The term *scattering* includes three types of phenomena:

1. Elastic scattering, in which there are no changes in rotational, vibrational, or electronic quantum numbers in the collision
2. Inelastic scattering, in which there are changes in quantum numbers, but no reaction
3. Reactive scattering, in which new products are formed

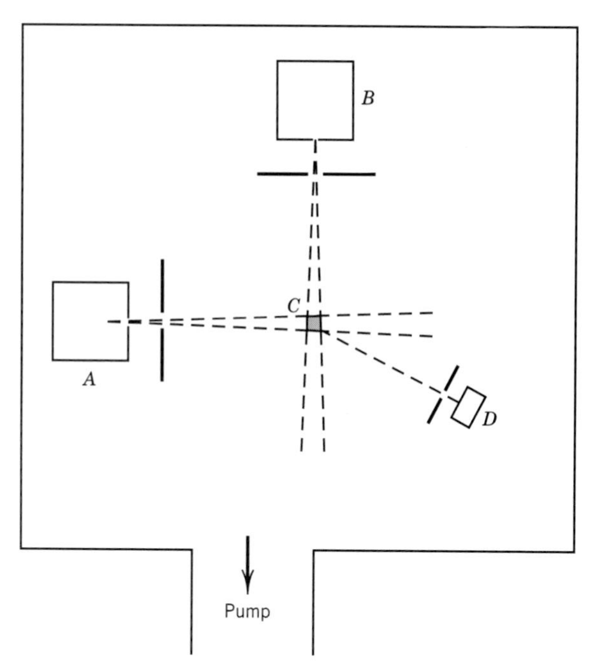

Figure 19.9 Schematic diagram for a molecular beam apparatus for studying the reaction of molecules from source A with molecules from source B. Products are detected at D.

To study the effects of the quantum numbers of molecules in colliding beams, the molecules must be prepared in a certain state, and there must not be collisions in the beams that would change these quantum numbers. Molecules can be put into desired vibrational and rotational states by laser excitation. Molecules with dipole moments can be oriented in an inhomogeneous electric field so that the effects of molecular orientation in collision can be studied:

$$\mathrm{Na} + \mathrm{Cl_2} \rightarrow \mathrm{NaCl} + \mathrm{Cl}$$

This reaction and some similar reactions have quite large cross sections (10^6 pm^2), which indicates that reaction occurs at distances of the order of 500 pm. The sodium atom loses an electron to the chlorine at a distance of this magnitude so that an ionic intermediate is formed:

$$\mathrm{Na} + \mathrm{Cl_2} \rightarrow (\mathrm{Na^+} \cdots \mathrm{Cl^-} \cdots \mathrm{Cl}) \rightarrow \mathrm{NaCl} + \mathrm{Cl}$$

This is referred to as a "harpooning" mechanism because the electron from the sodium atom can be thought of as a harpoon, and the sodium and chlorine ions formed can be thought of as being drawn together by electrostatic attraction.

In experiments with molecular beams, it may be useful to use a supersonic beam source to obtain reactant molecules with very high velocities. The supersonic molecular beam is generated by allowing a high-pressure, dilute mixture of the reactant of interest in an inert carrier gas to escape into the vacuum chamber through a pinhole. The reactant molecules in the beam move more rapidly than the speed of sound (see equation 17.33), and it can be shown that the peak speed of reactant molecules in the beam is approximated by

$$v_{\text{peak}} = \left(\frac{2RT}{M}\right)^{1/2} \left(\frac{\gamma}{\gamma - 1}\right)^{1/2} \tag{19.37}$$

where $\gamma = \overline{C}_P/\overline{C}_V$ for the carrier gas, M is the molar mass of the carrier gas, and T is the temperature of the source chamber. It can also be shown t40hat the reactant molecules have a narrower velocity distribution than the Maxwell–Boltzmann distribution.

19.6 PRINCIPLES OF PHOTOCHEMISTRY

In studying the electronic spectroscopy of molecules we have seen how the absorption of a photon by a molecule can raise it to a higher energy level, which may have a minimum in the potential energy curve, or may not. According to the Franck–Condon principle (Section 14.2), the internuclear distance in the molecule is not initially changed by the absorption, and the electric dipole transition moment (Section 13.2) is proportional to the overlap integral of the initial and final vibrational states. In general, selection rules are obeyed, but we must remember that the actual system may be more complicated than the ones for which the selection rules were developed. We will soon see that the selection rule $\Delta S = 0$ (Section 14.2), which prohibits transition between states of different multiplicities (i.e., singlet–triplet), is quite important in photochemistry. When a molecule absorbs a photon it may lose energy through fluorescence or phosphorescence, but now we will be interested in the fact that chemical reactions may ensue. As pointed out in Section 13.1, electromagnetic radiation in the visible and ultraviolet is generally required to produce chemical reactions because changes in electronic energy levels are required. More recently it has been found that the absorption of many infrared photons from a high-intensity laser can also cause reaction.

The **first principle** of photochemistry, which was stated by Grotthus in 1817 and Draper in 1843, is that only light that is absorbed can produce photochemical change. The **second principle,** which was proposed by Stark and Einstein from 1908 to 1912, is that a molecule absorbs a single quantum of light in becoming excited:

$$\mathrm{A} + h\nu \rightarrow \mathrm{A}^* \tag{19.38}$$

Thus, a mole of photons can excite a mole of molecules. If the electromagnetic radiation is extremely intense, as in a laser beam, two photons may be absorbed essentially simultaneously:

$$\mathrm{A} + 2h\nu \rightarrow \mathrm{A}^{**} \tag{19.39}$$

In discussing the intensity of light in connection with the Beer–Lambert law (Section 14.4), we used I to represent the intensity of light of a particular wavelength in terms of energy per unit area per unit time. Because of the two principles in the preceding paragraph, we will find it convenient in discussing photochemistry to use I_a to represent the **intensity of light absorbed** by a system, expressed as the amount of photons per unit volume per unit time, where, of course, the amount is expressed in moles. A mole of photons is a convenient unit in photochemistry, and it is frequently referred to as an **einstein.** The intensity I_a is calculated from the radiant energy of a specific wavelength absorbed per unit volume per unit time by dividing by $N_A h\nu$.

Example 19.3 ***Calculation of the intensity of light absorbed***

Monochromatic radiation at 400 nm, produced by a laser, is completely absorbed by a reaction mixture with a volume of 0.5 L. If the intensity of the radiation is 50 W, what amount of photons is absorbed in 10 min? What is the value of I_a?

$$\frac{E\lambda}{N_A hc} = \frac{(50 \text{ J s}^{-1})(600 \text{ s})(400 \times 10^{-9} \text{ m})}{(6.022 \times 10^{23} \text{ mol}^{-1})(6.626 \times 10^{-34} \text{ J s})(2.998 \times 10^{8} \text{ m s}^{-1})}$$

$$= 0.100 \text{ mol} = 0.100 \text{ einstein}$$

$$I_a = \frac{0.100 \text{ mol}}{(10 \text{ min})(60 \text{ s min}^{-1})(0.5 \text{ L})} = 3.33 \times 10^{-4} \text{ mol L}^{-1} \text{ s}^{-1}$$

If the gas or solution strongly absorbs the light, the reaction will occur only near the surface where the light enters. If the gas or solution is weakly absorbing, reaction will occur throughout the volume, but only a fraction of the light will be absorbed (see Problem 19.9).

The rate of a photochemical reaction is proportional to the intensity of light absorbed I_a, and the proportionality constant is the **quantum yield** ϕ. Thus for the reaction

$$A + 2B = C \tag{19.40}$$

the rate of reaction is given by

$$v = -\frac{d[A]}{dt} = -\frac{1}{2}\frac{d[B]}{dt} = \frac{d[C]}{dt} = \phi I_a \tag{19.41}$$

so that the quantum yield for a reaction is independent of the reactant studied. Thus the quantum yield is equal to the reaction rate v divided by the **intensity of light absorbed** I_a:

$$\phi = \frac{v}{I_a} \tag{19.42}$$

From a molecular point of view, the quantum yield is the ratio of the number of molecules undergoing some sort of change divided by the number of photons absorbed:

$$\phi = \frac{\text{molecules changed in a particular way}}{\text{photons absorbed}} \tag{19.43}$$

The quantum yield is also useful for expressing the rate of fluorescence and phosphorescence emission, as we will see later. When a photon is absorbed a number of things can happen to the energy, so the quantum yield may be less than unity.

The quantum yield for fluorescence or phosphorescence is necessarily less than unity, and usually much less. The quantum yield for a chemical reaction, however, may be a very large number if the absorption of light produces a radical that starts a chain reaction of a thermodynamically spontaneous reaction. The quantum yield for the first step of a chemical reaction, the so-called primary process, is equal to unity or less.

An electronically excited state of a molecule has a different electron distribution and nuclear configuration than the ground state. An electronically excited state of a molecule may be converted spontaneously to more possible products than the ground state because of the additional energy it has.

Before considering chemical reactions we must consider the physical processes that result from the absorption of electromagnetic radiation. The energy absorbed may produce electronically excited molecules that can react chemically, but often the energy is rapidly dissipated as heat.

Example 19.4 ***Calculation of the quantum yield for the photobromination of cinnamic acid***

In the photobromination of cinnamic acid, radiation at 435.8 nm with an intensity of 1.4×10^{-3} J s^{-1} was 80.1% absorbed in a liter of solution during an exposure of 1105 s. The concentration of Br_2 decreased by 7.5×10^{-5} mol L^{-1} during this period. What is the quantum yield?

$$E = \frac{N_A hc}{\lambda} = \frac{(6.02 \times 10^{23}\ \text{mol}^{-1})(6.62 \times 10^{-34}\ \text{J s})(3 \times 10^8\ \text{m s}^{-1})}{(435.8 \times 10^{-9}\ \text{m})}$$

$$= 2.74 \times 10^5\ \text{J mol}^{-1}$$

$$I_a = \frac{(1.4 \times 10^{-3}\ \text{J s}^{-1})(0.801)}{(2.74 \times 10^5\ \text{J mol}^{-1})(1\ \text{L})} = 4.09 \times 10^{-9}\ \text{mol L}^{-1}\ \text{s}^{-1}$$

$$v = \frac{7.5 \times 10^{-5}\ \text{mol L}^{-1}}{1105\ \text{s}} = 6.79 \times 10^{-8}\ \text{mol L}^{-1}\ \text{s}^{-1}$$

$$\phi = \frac{v}{I_a} = \frac{6.79 \times 10^{-8}\ \text{mol L}^{-1}\ \text{s}^{-1}}{4.09 \times 10^{-9}\ \text{mol L}^{-1}\ \text{s}^{-1}} = 16.6$$

Example 19.5 ***Calculation of the steady-state concentration of chlorine atoms***

A photochemical reaction may reach a steady state because the products undergo chemical changes. For example, the photochemical dissociation of molecular chlorine reaches a steady state as a result of the recombination of atoms.

$$Cl_2 + h\nu \underset{k_{-1}}{\overset{I_a}{\rightleftharpoons}} 2Cl$$

Derive the equation for the steady-state concentration of chlorine atoms.

The rate equation is

$$-\frac{d[Cl_2]}{dt} = \frac{1}{2}\frac{d[Cl]}{dt} = I_a - k_{-1}[Cl]^2$$

Note that each term in this equation has the units mol L^{-1} s^{-1}. In the steady state

$$[Cl] = \left(\frac{I_a}{k_{-1}}\right)^{1/2}$$

19.7 RATES OF INTRAMOLECULAR PROCESSES AND INTERMOLECULAR ENERGY TRANSFER

The electronic excitation of a molecule was discussed in Chapter 14, and in Section 14.8 we saw that the excitation energy may be dissipated by internal conversion (IC), intersystem crossing (ISC), fluorescence (F), or phosphorescence (P). An

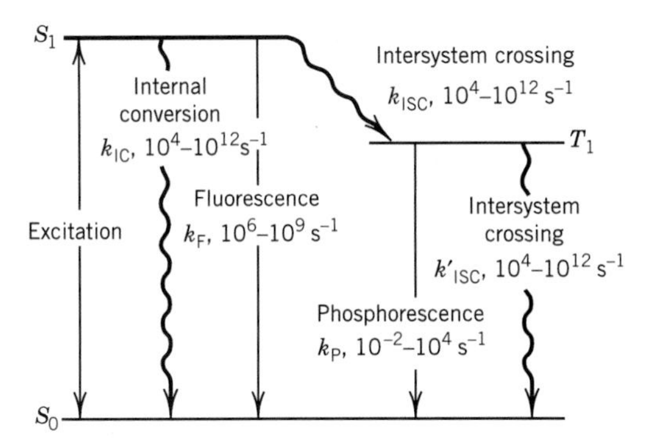

Figure 19.10 First-order rate constants for the various intramolecular processes. The wavy lines represent radiationless transitions. (From C. H. J. Wells, *Introduction to Molecular Photochemistry,* London: Chapman & Hall, 1972. © C. H. J. Wells, 1972.)

important characteristic of each of these processes is the **average lifetime of an excited molecule** before it undergoes the process. In this chapter we will emphasize the rate constant k, that is, the reciprocal of the lifetime τ, because we will be comparing the rates of these physical processes with the rates of chemical reactions. The approximate first-order rate constants for the various intramolecular processes are summarized in Fig. 19.10. Since the absorbing molecule is usually in a singlet ground state S_0, it is excited to a singlet excited state; this may be higher than state S_1, but we will assume that excitation is to S_1 for simplicity.

When a substance is illuminated with constant intensity, a steady state is reached in which the rates of formation of intermediates are equal to their rates of disappearance. If the absorbed intensity I_a is expressed in moles of photons absorbed per unit volume per unit time, the steady-state concentration of S_1 is given by

$$I_a = k_{IC}[S_1] + k_F[S_1] + k_{ISC}[S_1] \tag{19.44}$$

which may be written

$$[S_1] = \frac{I_a}{k_{IC} + k_F + k_{ISC}} \tag{19.45}$$

The steady-state rate equation for T_1 is (assuming T_1 is populated by ICS from S_1)

$$k_{ISC}[S_1] = k'_{ISC}[T_1] + k_P[T_1] \tag{19.46}$$

Solving for $[T_1]$ yields

$$[T_1] = \frac{k_{ISC}[S_1]}{k'_{ISC} + k_P} \tag{19.47}$$

Substituting equation 19.44 yields

$$[T_1] = \frac{k_{ISC}I_a}{(k'_{ISC} + k_P)(k_{IC} + k_F + k_{ISC})} \tag{19.48}$$

The rate constant k_F for fluorescence is equal to the Einstein probability A_{12} for spontaneous emission (Section 13.2) if there is a single lower state and excitation energy is not lost in various radiationless processes. We saw earlier

(Section 14.5) that the radiative lifetime $\tau_0 = 1/k_F$ is equal to $1/A_{12}$, and is approximately inversely proportional to the molar absorbancy index ϵ:

$$\tau_0 \approx \frac{10^{-4}\ \text{L mol}^{-1}\ \text{cm}^{-1}\ \text{s}}{\epsilon} \tag{19.49}$$

In other words, states that are populated readily are also depopulated readily. A strongly absorbing compound with $\epsilon = 10^5$ L mol^{-1} cm^{-1} would be expected to have a natural radiative lifetime of about 10^{-9} s, and a weakly absorbing compound with $\epsilon = 10^{-2}$ L mol^{-1} cm^{-1} would be expected to have a natural radiative lifetime of about 10^{-2} s.

The observed lifetime τ_S of an excited singlet state is less than the radiative lifetime τ_0 because there are other deactivation processes. According to Fig. 19.10, the rate of decay of S_1 is given by

$$-\frac{d[S_1]}{dt} = (k_{IC} + k_F + k_{ISC})[S_1] \tag{19.50}$$

so that the singlet lifetime is given by

$$\tau_S = \frac{1}{k_{IC} + k_F + k_{ISC}} \tag{19.51}$$

according to equation 18.17.

The singlet lifetime τ_S may be measured in the laboratory by observing the decay of the intensity of the fluorescence after a short ($\ll \tau_S$) pulse of excitation because the intensity of fluorescence is proportional to the excited state concentration. The singlet lifetimes for some organic molecules in solution at room temperature are given in Table 19.2.

The quantum yield for fluorescence ϕ_F is equal to the ratio of fluorescence $k_F[S_1]$ to the total rate of deactivation of the S_1 state, which is given by $(k_{IC} + k_F + k_{ISC})[S_1]$. Thus,

$$\phi_F = \frac{k_F}{k_{IC} + k_F + k_{ISC}} = \tau_S k_F = \tau_S/\tau_0 \tag{19.52}$$

This is the quantum yield for fluorescence in the absence of any quenching or chemical reaction. It is evident that the singlet lifetime is given by

$$\tau_S = \tau_0 \phi_F \tag{19.53}$$

Table 19.2 Fluorescence Lifetimes and Quantum Yields of Some Molecules in Solution at 25 °C

Molecule	*Solvent*	τ_S/ns	ϕ_F	τ_0/ns
Benzene	Hexane	26	0.070	370
Naphthalene	Hexane	106	0.380	280
Anthracene	Benzene	4	0.240	17
Chlorophyll *a*	Methanol	6.9	0.280	25
Chlorophyll *b*	Methanol	5.9	0.080	74
Eosin	Water	4.7	0.150	31

Source: R. B. Cundall and A. Gilbert, *Photochemistry.* London: Nelson, 1970.

Thus, measurement of the singlet lifetime τ_S and the quantum yield for fluorescence makes it possible to calculate the radiative lifetime τ_0. The radiative lifetimes for several molecules calculated in this way are given in Table 19.2. The quantum yield for internal conversion ϕ_{IC} and the quantum yield for intersystem crossing ϕ_{ISC} are given by

$$\phi_{IC} = \frac{k_{IC}}{k_{IC} + k_F + k_{ISC}} \tag{19.54}$$

$$\phi_{ISC} = \frac{k_{ISC}}{k_{IC} + k_F + k_{ISC}} \tag{19.55}$$

Thus,

$$\phi_F + \phi_{IC} + \phi_{ISC} = 1 \tag{19.56}$$

Returning to Fig. 19.10, we can see that if the rate constant k_{ISC} for the intersystem crossing S_1 to T_1 is fast enough, T_1 will be present at an appreciable concentration. If this happens, it is very significant for photochemistry, since triplet state molecules may have long lifetimes compared with singlet state molecules, and therefore have a higher probability of undergoing chemical reaction.

For the T_1 state of Fig. 19.10 the phosphorescence lifetime will be

$$\tau_{T_1} = \frac{1}{k_P + k'_{ISC}} \tag{19.57}$$

and the quantum yield for phosphorescence is given by

$$\phi_P = \frac{\text{rate of phosphorescence emission}}{\text{rate of absorption of radiation}} = \frac{k_P[T_1]}{I_a}$$

$$= \frac{k_P k_{ISC}}{(k'_{ISC} + k_P)(k_{IC} + k_F + k_{ISC})} = k_P k_{ISC} \tau_S \tau_{T_1} \tag{19.58}$$

where the second form has been obtained by substituting the expression for the steady-state value of $[T_1]$ given in equation 19.47 and equation 19.43.

Excited molecules may undergo several different types of reactions. Generally chemical reaction of one excited molecule with another molecule is accompanied by deactivation of the excited molecule, and the excited state is said to be quenched. Quenching an excited molecule (donor D) with a second molecule (acceptor A) may result in the electronic excitation of A with concomitant deactivation of D. **This quenching process is referred to as electronic energy transfer.** Since minute traces of impurities (quenchers) can rapidly deactivate excited molecules, substances and solvents used in photochemical studies must be carefully purified. For example, molecular oxygen reacts rapidly with excited molecules ($k \approx 10^9$–10^{10} L mol^{-1} s^{-1}), and it is therefore often important to deoxygenate the solutions under study.

We will represent quenchers by Q. The quenching of phosphorescence provides a means of determining the rate constant for the reaction of the triplet state molecules and the quencher. The steady-state concentration of triplet molecules in the absence of a quencher is given by equation 19.47. In the presence of a quencher this equation becomes

$$[T_1] = \frac{k_{ISC}[S_1]}{k_P + k'_{ISC} + k_Q[Q]} \tag{19.59}$$

because of the additional pathway for reaction of the triplet state molecules with the quencher. The quantum yield for phosphorescence in the absence of a quencher is given by equation 19.58 and will be represented by ϕ_P° in this section. If the steady-state concentration of T_1 in the presence of a quencher is given by equation 19.59, the ratio of the quantum yields in the absence and presence of the quencher is given by

$$\frac{\phi_P^\circ}{\phi_P} = \frac{k_P + k'_{ISC} + k_Q[Q]}{k_P + k'_{ISC}}$$

$$= 1 + k_Q\tau_{T_1}[Q] \tag{19.60}$$

where τ_{T_1}, the phosphorescence lifetime, is defined by equation 19.57. The ratio of the intensity I_P° of phosphorescence in the absence of quencher to the intensity I_P of phosphorescence in the presence of quencher is proportional to the ratio of the quantum yields, so the preceding equation may be written

$$\frac{I_P^\circ}{I_P} = 1 + k_Q\tau_{T_1}[Q] \tag{19.61}$$

This equation is referred to as the **Stern–Volmer equation.**

Electronic energy transfer processes may be classified as radiative transfer, short-range transfer, or long-range resonance transfer. In radiative transfer the donor D emits radiation which is absorbed by the acceptor A:

$$D^* \rightarrow D + h\nu \tag{19.62}$$

$$A + h\nu \rightarrow A^* \tag{19.63}$$

Short-range energy transfer can occur if the distance between donor and acceptor molecules approaches the collision diameter. A collision is not necessarily required, since the energy transfer can occur at distances slightly greater than the collision diameter.

In long-range energy transfer the donor and acceptor molecules are separated by a distance much greater than the collision diameter. The efficiency of the energy transfer depends on the extent of the overlap of the emission spectrum of the donor and the absorption spectrum of the acceptor. The energy transfer is not efficient unless the decay process $D^* \rightarrow D$ and the excitation process $A \rightarrow A^*$ are allowed electronic transitions. Under favorable circumstances energy transfer can occur over distances of 5 to 10 nm with rate constants of 10^{10}–10^{11} L mol^{-1} s^{-1}.

19.8 PHOTOCHEMICAL REACTIONS AND THEIR QUANTUM YIELDS

When an excited species undergoes a chemical reaction, we can add another step to the simplified mechanism that we have been discussing. If the excited species (here assumed to be T_1) reacts with reactant R with a bimolecular rate constant k_R, it may be possible to evaluate the rate constant:

$$T_1 + R \rightarrow \text{stable products} \qquad v = k_R[T_1][R] \tag{19.64}$$

In the steady state

$$[T_1] = \frac{k_{ISC}I_a}{(k_P + k'_{ISC} + k_R[R])(k_{IC} + k_F + k_{ISC})} \tag{19.65}$$

The quantum yield for the production of stable products is

$$\phi = \frac{k_R[T_1][R]}{I_a} \tag{19.66}$$

Substituting equation 19.65 and rearranging yields

$$\frac{1}{\phi} = \frac{k_F + k_{IC} + k_{ISC}}{k_{ISC}}\left(1 + \frac{k_P + k'_{ISC}}{k_R[R]}\right) \tag{19.67}$$

Even for this simplified mechanism there are so many rate constants to be determined before k_R can be obtained that this is not generally practical, so in further discussions, we will emphasize quantum yields. Note that from equation 19.43 the quantum yield can be determined by measuring the number of quanta absorbed and the number of molecules reacted. This requires no knowledge of rate constants.

To determine a quantum yield it is necessary to measure the intensity of the light. This may be done by use of a thermopile, which is a series of thermocouples with one set of junctions blackened to absorb all the radiation, which is then converted to heat. The other set of junctions is protected from radiation.

The amount of radiation may also be measured with a chemical **actinometer,** in which the amount of chemical change is determined. The yield of the photochemical reaction in the actinometer was determined originally by use of a thermopile. The quantum yields of a few photochemical reactions are summarized in Table 19.3.

Reaction 1 in Table 19.3 has the same value of ϕ from 280 to 300 nm, at low pressures and high pressures, in the liquid state or in solution in hexane. The primary process $HI + h\nu = H + I$ is followed by the reactions $H + HI = H_2 + I$ and $I + I = I_2$, thus giving two molecules of HI decomposed for each photon absorbed. Reaction 2, the dimerization of anthracene, has a quantum yield

Table 19.3 Quantum Yields in Photochemical Reactions at 25 °C

Reaction	*Approximate Wavelength Region,* nm	*Approximate* ϕ
1. $2HI \rightarrow H_2 + I_2$	300–280	2
2. $C_{14}H_{10} \rightleftharpoons \frac{1}{2}(C_{14}H_{10})_2$	<360	0–1
3. $CH_3CHO \rightarrow CO + CH_4(+C_2H_6 + H_2)$	310	0.5
	253.7	1
4. $(CH_3)_2CO \rightarrow CO + C_2H_6(+CH_4)$	<330	0.2
5. $H_2C_2O_4(+UO_2^{2+}) \rightarrow CO + CO_2 + H_2O(+UO_2^{2+})$	430–250	0.5–0.6
6. $Cl_2 + H_2 \rightarrow 2HCl$	400	10^5

of unity initially, but the reverse thermal reaction can occur as the product accumulates.

Reaction 3 in Table 19.3 is interesting because at 300 °C ϕ has a value of more than 300, indicating that the free radicals that are first produced by the absorption of light are able to propagate a chain reaction at the higher temperatures. At room temperature the reactions involved in the chain do not go fast enough to be detected. The products given in parentheses are present also but in small amounts.

The experimental determination of the quantum yield constitutes an excellent method for detecting **chain reactions** (Section 18.11). If several molecules of products are formed for each photon of light absorbed, the reaction is obviously a chain reaction in which the products of the reaction are able to promote reaction of other molecules.

Reaction 4 in Table 19.3 is an example of the fact that the absorption of light in a particular bond does not necessarily cause the rupture of that bond. Acetone, like other aliphatic ketones, absorbs ultraviolet light at about 280 nm. The C=O bond, which we designate as the chromophore, is very strong and does not break to give atomic oxygen. Instead, the absorption energy leads to the cleavage of an adjacent C—C bond that is weaker; thus,

$$(CH_3)_2C{=}O + h\nu \rightarrow CH_3\cdot + CH_3\dot{C}{=}O \qquad (19.68)$$

giving a methyl radical and an acetyl radical. The acetyl radical can then decarbonylate, giving CO and $CH_3\cdot$, or it can react with $CH_3\cdot$ to give back acetone. The methyl radicals can couple to form ethane.

Reaction 5 in Table 19.3 illustrates a photosensitized reaction. The photodecomposition of oxalic acid, sensitized by uranyl ion, is so reproducible that it is suitable for use as a chemical actinometer. In the uranyl oxalate actinometer the light is absorbed by the colored uranyl ion, and the energy is transferred to the colorless oxalic acid, which then decomposes. The uranyl ion remains unchanged and can be used indefinitely as a sensitizer. The fact that the molar absorption coefficient of uranyl ion is increased by the addition of colorless oxalic acid indicates the formation of a complex. The formation of a chemical complex is often necessary for photosensitization.

Reaction 6 in Table 19.3 is the best-known example of a chain reaction. About 10^5 molecules react for each quantum absorbed. The molecules of hydrogen chloride formed undergo further reaction with the hydrogen and chlorine atoms produced (Section 18.11). The measurement of the number of molecules per photon gives a measure of the average number of molecules involved in the chain. Initiation of the reaction with a flash of light can result in an explosively fast reaction.

19.9 THE OZONE LAYER IN THE STRATOSPHERE

The formation of ozone in the stratosphere is an example of a photochemical stationary state. The ozone formed in this way is important because it absorbs ultraviolet radiation that would otherwise cause damage to life at the surface of the earth. A simplified mechanism for the formation and destruction of ozone in the stratosphere is the following:

$$O_2 \xrightarrow[J_{O_2}]{h\nu} 2O \tag{19.69}$$

$$O + O_2 + M \xrightarrow{k_2} O_3 + M \tag{19.70}$$

$$O_3 \xrightarrow[J_{O_3}]{h\nu} O_2 + O \tag{19.71}$$

$$O_3 + O \xrightarrow{k_4} 2O_2 \tag{19.72}$$

where the J's are photodissociation constants defined below. The photolysis reaction in the first step occurs only at wavelengths less than 242 nm. Ozone is formed in the three-body reaction in step 2. The protective role of ozone is due to the third step, in which radiation in the 190- to 300-nm range dissociates ozone. The fourth step is a slow reaction. The recombination of oxygen atoms with a third body to form O_2 is very slow in the stratosphere and can be neglected. If the absorption of solar radiation in the first and third steps is steady, the concentration of ozone rises to a steady level that can be calculated as follows. The rate of change of ozone concentration is given by

$$\frac{d[O_3]}{dt} = k_2[O][O_2][M] - J_{O_3}[O_3] - k_4[O][O_3] \tag{19.73}$$

where J_{O_3} is the **photodissociation coefficient** for ozone. The photodissociation coefficient is the probability of dissociation of a molecule per second by light absorption. This coefficient is calculated using

$$J = \int_0^\infty \phi_{\tilde{\nu}} I_{\tilde{\nu}} \sigma_{\tilde{\nu}} \, d\tilde{\nu} \tag{19.74}$$

where $\phi_{\tilde{\nu}}$ is the quantum yield of dissociation of the molecule at wave number $\tilde{\nu}$, $I_{\tilde{\nu}}$ is the intensity of sunlight in quanta per unit area per unit time per wave number, and $\sigma_{\tilde{\nu}}$ is the absorption cross section of the molecule at wave number $\tilde{\nu}$. Thus, $I_{\tilde{\nu}}$ has SI units of $m^{-1}\,s^{-1}$, and J has SI units of s^{-1}.

The rate of change in the concentration of oxygen atoms is assumed to be zero in the steady state since they are present at very low concentrations:

$$\frac{d[O]}{dt} = 2J_{O_2}[O_2] - k_2[O][O_2][M] + J_{O_3}[O_3] - k_4[O][O_3] = 0 \tag{19.75}$$

Adding equations 19.73 and 19.75 yields

$$\frac{d[O_3]}{dt} = 2J_{O_2}[O_2] - 2k_4[O][O_3] \tag{19.76}$$

From experimental measurements it is known that $J_{O_3}[O_3] \gg J_{O_2}[O_2]$ and $k_2[O][O_2][M] \gg k_4[O][O_3]$, so equation 19.75 becomes

$$J_{O_3}[O_3] \cong k_2[O][O_2][M] \tag{19.77}$$

Substituting this for [O] in equation 19.76 yields

$$\frac{d[O_3]}{dt} = 2J_{O_2}[O_2] - \frac{2k_4 J_{O_3}[O_3]^2}{k_2[O_2][M]} \tag{19.78}$$

Therefore, the steady-state concentration of O_3 is given by

$$[O_3]_{ss} = [O_2]\sqrt{k_2 J_{O_2}[M]/k_4 J_{O_3}} \tag{19.79}$$

Since J_{O_2} increases with altitude and $[O_2]$ decreases with altitude, there is a maximum steady-state concentration in the stratosphere at an altitude of about 20 km.

In the late 1960s it was found that nitrogen oxides can catalyze reaction 19.72:

$$\begin{aligned} NO + O_3 &\rightarrow NO_2 + O_2 \\ NO_2 + O &\rightarrow NO + O_2 \\ \hline O_3 + O &\rightarrow 2O_2 \end{aligned} \tag{19.80}$$

In the mid-1970s it was discovered that chlorine atoms from the photolysis of chlorofluorocarbons ($CFCl_3 + h\nu \rightarrow CFCl_2 + Cl$ and $CF_2Cl_2 + h\nu \rightarrow CF_2Cl + Cl$) at the level of the ozone layer can also catalyze the decomposition of ozone:

$$\begin{aligned} Cl + O_3 &\rightarrow ClO + O_2 \\ ClO + O &\rightarrow Cl + O_2 \\ \hline O_3 + O &\rightarrow 2O_2 \end{aligned} \tag{19.81}$$

Comment:

The destruction of ozone by chlorine atoms in the stratosphere has become a serious issue because of the "ozone hole" in the Antarctic region, which can be surveyed by satellite. This has led to international controls on the manufacture of $CFCl_3$ and CF_2Cl_2. Several cycles of the type of 19.81 are involved, and there has been intense interest in quantitative calculations of the lifetimes of various chlorofluorocarbons in the stratosphere. The Nobel Prize in chemistry for 1995 was awarded to Paul Crutzen, Mario J. Molina, and F. Sherwood Rowland for their research on this topic.

19.10 FEMTOSECOND TRANSITION-STATE SPECTROSCOPY*†

In a chemical reaction a molecule is in the transition state for only about 10 femtoseconds (10^{-14}s, yet with modern spectroscopic methods it is possible to study dynamic processes that occur this rapidly. In Section 9.14 we saw how the irradiation of molecular iodine by an ultrashort laser pulse, referred to as a pump pulse, can produce a coherent (that is, in-phase) superposition of vibrational states. This showed how the wavefunctions for the coherent superposition interfere constructively and destructively to give a resultant wave packet that has a large amplitude only in a limited part of the classical region at a given time. After the pump pulse, this wave packet oscillates back and forward in the parabolic potential energy well much like a classical particle. Under some conditions, the particle may oscillate back and forth a number of times and then dissociate to iodine atoms. Under other conditions, the dissociation may occur immediately.

*A. H. Zewail, *Science* **242:**1645 (1988).

†J. C. Polanyi and A. H. Zewail, *Acc. Chem. Res.* **28:**119 (1995).

Since this phenomenon cannot be observed with a single molecule, it is important to understand that all the iodine molecules in the sample have very nearly the same internuclear distance before the pulse because in the ground state, the highest probability density is at this distance. Therefore, the wave packets are launched from the same internuclear distance for the entire sample of molecules. Thus a single-molecule trajectory is observed.

Probe pulses are then used to study the details of what happens. These probe pulses can be used to follow the particle-like oscillations and the decreases in intensity of the wave packet signal that indicate chemical reaction. Since the wavelength distribution in the probe pulse can be altered by the experimenter, the formation of product atoms or molecules can also be studied on a femtosecond scale.

Consider a photolysis reaction in which the excited molecule ABC* goes through a transition state $[A\cdots BC]^{\ddagger *}$ and dissociates into products. Note that $[A\cdots BC]^{\ddagger *}$ is different from an unstable intermediate in a reaction because it is in the process of flying apart:

$$ABC^* \rightarrow [A\cdots BC]^{\ddagger *} \rightarrow A + BC \tag{19.82}$$

The potential energy curves for the molecule ABC(V_0), for the first dissociative state (V_1) yielding A + BC, and the second dissociative state (V_2) yielding A + BC* are given in Fig. 19.11. The potential energy curves for A + BC and A + BC* do not have minima, and therefore lead to immediate dissociation. An initial pulse (the "pump" pulse), represented by λ_1, is absorbed at $t = 0$ by the ABC molecule at an internuclear distance close to the potential energy minimum.

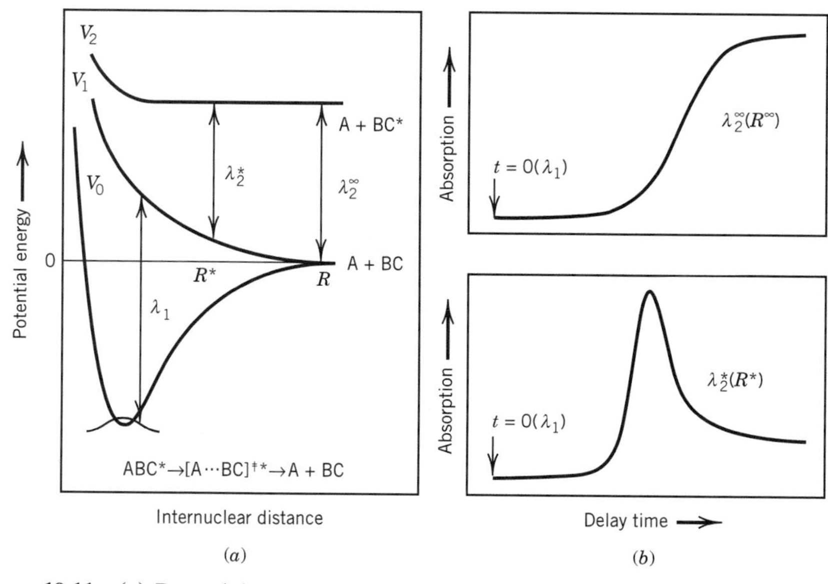

Figure 19.11 (*a*) Potential versus internuclear distance R for ABC(V_0), A + BC(V_1), and A + BC* (V_2). (*b*) Optical absorption as a function of time in femtoseconds after a pulse at λ_1. In the upper plot the absorption is for a probe pulse with wavelength λ_2^∞ that is absorbed by the completely dissociated products ($R = \infty$). In the lower plot the absorption is for a probe pulse λ_2^* that is absorbed by $[A\cdots BC]^{\ddagger *}$ with a particular internuclear distance. [Reprinted with permission from A. H. Zewail, *Science* **242,** 1645 (1988). Copyright ©1988 American Association for the Advancement of Science.]

The absorption of a photon by a molecule involves "instantaneous" vertical transition of the molecule from the ground-state potential energy V_0 to an excited-state potential energy V_1. For there to be appreciable absorption, the wavelength of the pump pulse must satisfy the relation $V_1(R_0) - V_0(R_0) \approx hc/\lambda_1$, where R_0 is the internuclear distance at the potential minimum. Classically the ABC molecules dissociate along V_1, with $R \to \infty$ as $t \to \infty$. After a time delay τ, the gas is irradiated with a probe pulse with wavelength λ_2.

In the simplest type of experiment, the wavelength λ_2^∞ of the probe pulse is tuned to the absorption wavelength of one of the free fragments. If the probe pulse follows the initial pulse closely in time, the absorption is negligible initially, as shown by the upper plot of optical absorption in Fig. 19.11*b*. The absorption of the probe pulse λ_2^∞ becomes substantial only when the fragments achieve large internuclear distance and are no longer interacting. The buildup of the signal from the λ_2^∞ pulse to half of its maximum takes about 200 fs, as shown by Fig. 19.11*b*.

The molecules in transition from the initial excited state can be detected by tuning λ_2 away from λ_2^∞. In this second type of experiment, the probe pulse, which is sent at time τ, will be absorbed significantly only if the transient-state configuration at time τ is such that $V_2(R^*) - V_1(R^*) = hc/\lambda_2^*$, where R^* is the internuclear distance at $t = \tau$. The lower absorption plot in Fig. 19.11*b* shows that the probe pulse λ_2^* is absorbed when the transition state has the internuclear distance R^* and that the absorption of λ_2^* subsequently falls as the internuclear distance increases and separated products are finally formed, after about 300 fs.

One particular reaction studied by Zewail and co-workers was

$$ICN^* \to [I\cdots CN]^{\ddagger *} \to I + CN \tag{19.83}$$

Thus, it is possible to measure absorption spectra of $[I\cdots CN]^{\ddagger *}$ at various times, and distances, as it flies apart. Since the recoil velocity of the fragments is about 1 km s^{-1}, the distance spanned in 100 fs is 100 pm. Zewail received the Nobel Prize in chemistry in 1999.

This method cannot be used directly for bimolecular reactions such as A + BC $\to$ AB + C because the bimolecular collisions occur at random times and cannot be controlled. However, it can be done for a class of bimolecular reactions in which a van der Waals "precursor molecule" contains the potential reagent molecules in close proximity. The van der Waals complex IH$\cdots$OCO was formed in a free jet expansion in an excess of helium carrier gas. A femtosecond pulse was used to dissociate the HI in the van der Waals complex. This caused a hot H atom to be ejected in the direction of the nearest-neighbor O atom in CO_2, thus initiating the reaction

$$H + OCO \to [HOCO]^\ddagger \to OH + CO \tag{19.84}$$

By use of a probe pulse suitable for the detection of OH, it was possible to measure the lifetime of the $[HOCO]^\ddagger$ collision complex. The lifetime depends on the translational energy, and is shorter at higher collision energies.

The range of applications of transition-state time domain spectroscopy has enlarged since 1988. Reactions studied include dissociation of NaI, isomerization of stilbene, and the exchange reaction Br + I_2, among many others. In some

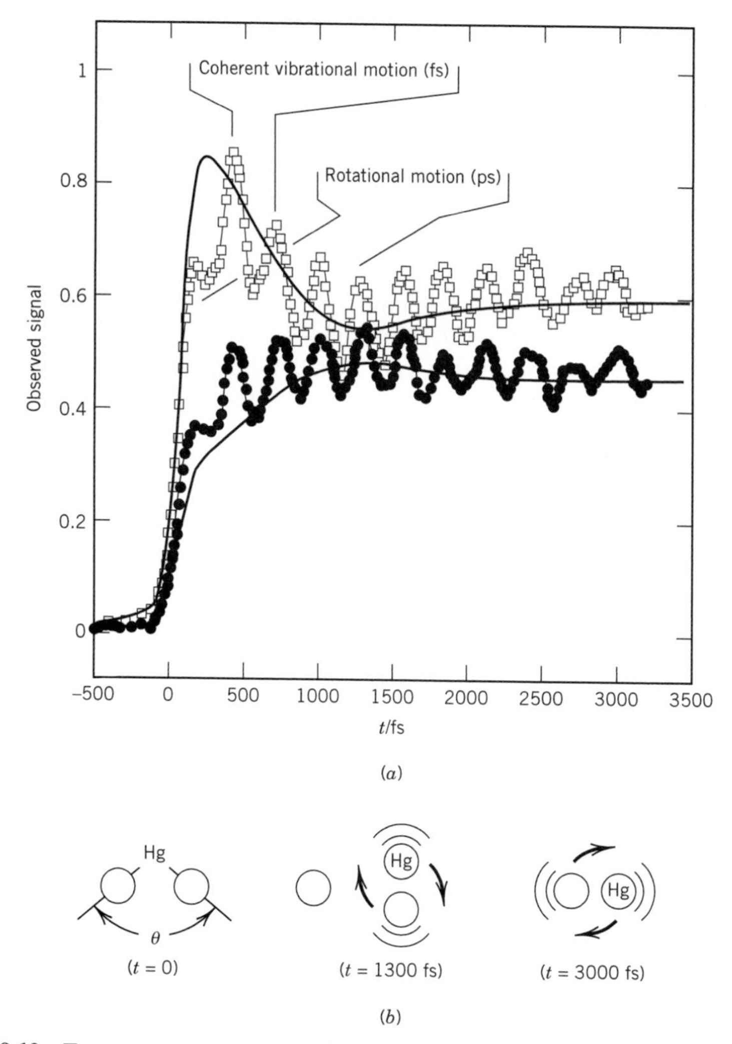

Figure 19.12 Femtosecond dynamics of barrier reaction. (*a*) Experimental observation of the vibrational (femtosecond) and rotational (picosecond) motions for the barrier (saddle-point transition state) descent, $[IHgI]^{*\ddagger} \rightarrow$ HgI(vib, rot) + I. The vibrational coherence in the reaction trajectories (oscillations) is observed in both parallel (squares) and perpendicular (circles) polarizations of femtosecond transition-state spectra (FTS). The rotational orientation (solid line) can be seen in the decay of FTS (parallel) and buildup of FTS (perpendicular) as the HgI rotates during bond breakage (*b*). [Reprinted with permission from J. C. Polanyi and A. H. Zewail, *Acc. Chem. Res.* **28,** 119–132 (1995). Copyright ©1995 American Chemical Society.]

of these examples, the vibrational and rotational motions are evident in the observed signal as a function of time. For example, in Fig. 19.12, the reaction of the excited transition-state complex $[IHgI]^{*\ddagger}$ is monitored by femtosecond laser pulses polarized either parallel or perpendicular to the initial laser pulse. Thus, it is now possible, in an increasing number of cases, to view bond breaking or formation in real time and study molecular dynamics in the region of the transition state.

19.11 SPECIAL TOPIC: APPLICATIONS OF PHOTOCHEMISTRY

In **photosynthesis** CO_2 and H_2O are converted to the glucose moiety as starch. The overall process can be represented by

$$CO_2 + H_2O \xrightarrow[\text{photons}]{\text{eight}} (CH_2O) + O_2 \tag{19.85}$$

where (CH_2O) represents one-sixth of a glucose moiety. Photosynthesis is carried out by plants, algae, and certain kinds of bacteria that have chloroplasts. The chloroplasts contain chlorophyll a, chlorophyll b, carotenes, electron carriers, and enzymes, and they have internal membranes that keep reactants separated. The first step in photosynthesis is the absorption of light by a chlorophyll molecule. Chlorophylls a and b contain networks of alternating single and double bonds and have strong absorption bands in the visible part of the spectrum, with molar absorbancy indices greater than 10^5 L mol^{-1} cm^{-1}. The energy of the absorbed photon is transferred from one chlorophyll molecule to another until it reaches a site called a reaction center. Experiments show that about eight photons of visible light have to be absorbed per O_2 liberated.

Example 19.6 ***Calculation of the efficiency of photosynthesis by white light***

Given that the energy change for reaction 19.88 is 477 kJ mol^{-1} and that the energy input between 400 and 700 nm at the earth's surface is equivalent to that at 575 nm, what is the theoretical maximum energy efficiency of photosynthesis by white light?

$$E = N_A h\nu = \frac{(6.02 \times 10^{23}\ \text{mol}^{-1})(6.62 \times 10^{-34}\ \text{J s})(3 \times 10^{8}\ \text{m s}^{-1})}{(575 \times 10^{-9}\ \text{m})(10^{3}\ \text{J kJ}^{-1})}$$

$$= 208\ \text{kJ mol}^{-1}$$

$$\text{Efficiency} = \frac{477\ \text{kJ mol}^{-1}}{(8)(208\ \text{kJ mol}^{-1})} = 0.29$$

The efficiency calculated on the basis of total solar radiation at the earth's surface that a plant could actually absorb leads to a maximum efficiency during the peak growing season of 0.066. Efficiencies of 0.032 have been achieved with corn.

In **vision** *cis*-retinal is converted to *trans*-retinal, as shown in Fig. 19.13. *cis*-Retinal forms a complex with the protein opsin, which is referred to as rhodopsin. The binding of *cis*-retinal to the protein changes its light absorption so that the peak absorption is at about 500 nm, which nicely matches sunlight. This is an example of absorption by a conjugated system, as discussed in Section 11.7. When a photon is absorbed by rhodopsin, the change in the configuration of retinal weakens its binding by opsin, and the complex dissociates. The conversion of *trans*-retinal back to *cis*-retinal is catalyzed by the enzyme retinal isomerase. Thus, the absorption of a single photon can trigger a nerve response.

Ten Key Ideas in Chapter 19

1. Since bimolecular reactions of small radicals have zero activation energies, we might expect them to react on each collision. However, their rate

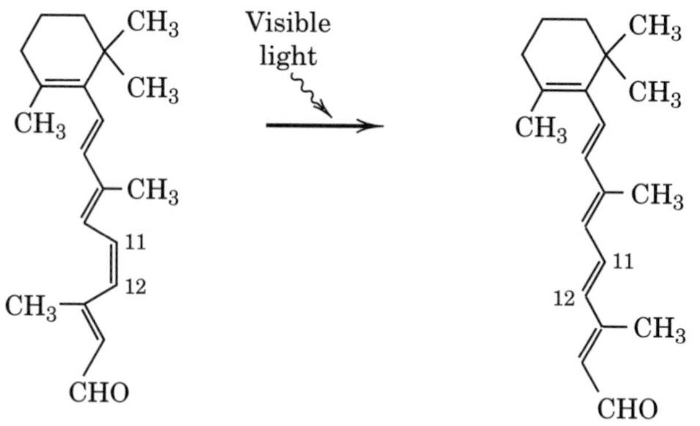

Rhodopsin (11-*cis*-retinal, bound to the protein opsin)

Activated rhodopsin (11-*trans*-retinal, released from the protein opsin)

Figure 19.13 Conversion of *cis*-retinal to *trans*-retinal by the absorption of a photon of visible light. *cis*-Retinal is the more stable form, and it is bound by the protein opsin.

constants are only approximately equal to the collision frequencies of hard spheres because molecules interact with each other over short distances and are not hard spheres.

2. When two molecules are very close to each other, they cannot be considered separately because their wavefunctions overlap. Molecules close to each other can be considered as being on a potential energy surface that gives the energy for given coordinates of the nuclei.
3. The highest point along the reaction coordinate is a saddle point, and in the neighborhood of this point the reaction system is said to be in the transition state. Classical calculations of large numbers of trajectories on a potential surface can yield pre-exponential factors and activation energies in pretty good agreement with experiment for simple systems.
4. Transition-state theory applies ideas of statistical mechanics to obtain estimates of rate constants for more complicated reactions.
5. The first principle of photochemistry is that only light that is absorbed can produce photochemical change. The second principle is that a molecule absorbs a single quantum of light in becoming excited, but with the intense radiation of lasers, successive photons can be absorbed.
6. An excited molecule can lose its additional energy in radiationless processes or in fluorescence or phosphorescence. Fluorescence can occur from various vibrational levels of a singlet excited state, but a molecule has to undergo intersystem crossing to a triplet state before phosphorescence occurs.
7. The quantum yield for fluorescence in the absence of any quenching or chemical reaction is equal to the ratio of the singlet lifetime to the radiative lifetime. The expression for the quantum yield for phosphorescence is much more complicated because it depends on the rate constant for intersystem crossing.
8. An excited molecule can also undergo chemical reaction or transfer its excess energy to a quencher. Short-range energy transfer can occur if the molecules are close, but a collision is not necessarily required. Long-range energy transfer can occur at much greater distance when the emission spectrum of the donor overlaps the absorption of the acceptor.

9. The formation of ozone in the stratosphere is an example of a photochemical stationary state. The conversion of ozone to molecular oxygen is catalyzed by nitrogen oxides and by chlorine atoms from the photolysis of chlorofluorocarbons.
10. The development of dye lasers capable of delivering pulses as short as 6×10^{-15} s have made it possible to observe the transition region of a chemical reaction.

REFERENCES

S. W. Benson, *Thermochemical Kinetics.* Hoboken, NJ: Wiley, 1976.

C. F. Bernasconi (ed.), *Investigation of Rates and Mechanisms of Reactions,* 4th ed. Hoboken, NJ: Wiley, 1985.

R. B. Bernstein, *Chemical Dynamics via Molecular Beam and Laser Techniques.* New York: Oxford University Press, 1982.

G. D. Billing and K. V. Mikkelsen, *Molecular Dynamics and Chemical Kinetics.* Hoboken, NJ: Wiley, 1996.

R. G. Gilbert and S. C. Smith, *Theory of Unimolecular and Recombination Reactions.* Boston: Blackwell, 1990.

G. G. Hammes, *Principles of Chemical Kinetics.* New York: Academic, 1978.

P. L. Houston, *Chemical Kinetics and Reaction Dynamics.* New York: McGraw-Hill, 2001.

R. D. Levine and R. B. Bernstein, *Molecular Reaction Dynamics and Chemical Reactivity.* New York: Oxford University Press, 1987.

J. W. Moore and R. G. Pearson, *Kinetics and Mechanisms.* Hoboken, NJ: Wiley, 1981.

I. W. M. Smith, *Kinetics and Dynamics of Elementary Gas Reactions.* Boston: Butterworths, 1980.

J. I. Steinfeld, J. S. Francisco, and W. L. Hase, *Chemical Kinetics and Dynamics.* Upper Saddle River, NJ: Prentice-Hall, 1999.

D. G. Truhlar, *Potential Energy Surfaces and Dynamics Calculations.* New York: Plenum, 1981.

N. J. Turro, *Modern Molecular Photochemistry.* Sausalito, CA: University Science Books, 1991.

PROBLEMS

19.1 Use the pre-exponential factor $A = 3 \times 10^{13}$ cm^3 mol^{-1} s^{-1} for the reaction $Br + H_2 \rightarrow HBr + H$ to calculate the reaction cross section and collision diameter for this reaction at 400 K.

19.2 (*a*) Calculate the second-order rate constant for collisions of dimethyl ether molecules with each other at 777 K. It is assumed that the molecules are spherical and have a radius of 0.25 nm. If every collision were effective in producing decomposition, what would be the half-life of the reaction (*b*) at 1 bar pressure and (*c*) at a pressure of 0.13 Pa?

19.3 Show that transition-state theory yields the simple collision theory result when it is applied to the reaction of two rigid spherical molecules.

19.4 The rate constant for the elementary reaction

$$K + Br_2 \rightarrow KBr + Br$$

is 1.0×10^{12} L mol^{-1} s^{-1} independent of temperature. Calculate the rate constant expected from collision theory at 298 K. The fact that the rate constant is greater than would be expected from collision theory is explained by the harpoon mechanism. According to this mechanism an electron jumps from K to Br_2 when these two molecules come within a certain distance that is greater than the collision diameter d_{12}, which is 400 pm.

19.5 A certain photochemical reaction requires an excitation energy of 126 kJ mol^{-1}. To what values does this correspond in the following units: (*a*) frequency of light, (*b*) wave number, (*c*) wavelength in nanometers, and (*d*) electron volts?

19.6 How many moles of photons does a laser with an intensity of 0.1 W at 560 nm produce in one hour?

19.7 A sample of gaseous acetone is irradiated with monochromatic light having a wavelength of 313 nm. Light of this wavelength decomposes the acetone according to the equation

$$(CH_3)_2CO \rightarrow C_2H_6 + CO$$

The reaction cell used has a volume of 59 cm^3. The acetone vapor absorbs 91.5% of the incident energy. During the experiment the following data are obtained:

$$\text{Temperature of reaction} = 56.7\,^\circ\text{C}$$
$$\text{Initial pressure} = 102.16 \text{ kPa}$$
$$\text{Final pressure} = 104.42 \text{ kPa}$$
$$\text{Time of radiation} = 7 \text{ h}$$
$$\text{Incident energy} = 48.1 \times 10^{-4} \text{ J s}^{-1}$$

What is the quantum yield?

19.8 A 100-cm^3 vessel containing hydrogen and chlorine was irradiated with light of 400 nm. Measurements with a thermopile showed that 11×10^{-7} J of light energy was absorbed by the chlorine per second. During an irradiation of 1 min the partial pressure of chlorine, as determined by the absorption of light and the application of Beer's law, decreased from 27.3 to 20.8 kPa (corrected to 0 °C). What is the quantum yield?

19.9 Show that if a solute follows the Beer–Lambert law, the intensity of absorbed radiation I_a in moles of photons per unit volume per second is given by

$$I_a = \frac{I_0}{l N_A h\nu}(1 - e^{-\kappa c l})$$

where l is the length of the cell in the direction of the incident monochromatic radiation and I_0 is in energy per unit area per unit time.

19.10 When CH_3I molecules in the vapor state absorb 253.7 nm light, they dissociate into methyl radicals and iodine atoms. The energy required to rupture the C—I bond is 209 kJ mol^{-1}. What are the velocities of the iodine atom and the methyl radical, assuming all of the excess energy goes into translational motion?

19.11 The phosphorescence of butyrophenone in acetonitrile is quenched by 1,3-pentadiene (P). The following quantum yields were measured at 25 °C:

$[P]/10^{-3}$ mol L^{-1}	0	1.0	2.0
ϕ/ϕ_0	1	0.61	0.43

Assuming that the quenching reaction is diffusion controlled and the rate constant has a value of 10^{10} L mol^{-1} s^{-1}, what is the lifetime of the triplet state?

19.12 Biacetyl triplets have a quantum yield of 0.25 for phosphorescence and a measured lifetime of the triplet state of 10^{-3} s. If its phosphorescence is quenched by a compound Q with a diffusion-controlled rate (10^{10} L mol^{-1} s^{-1}), what concentration of Q is required to cut the phosphorescence yield in half? (See Section 20.5.)

19.13 For 900 s, light of 426 nm was passed into a carbon tetrachloride solution containing bromine and cinnamic acid. The average power absorbed was 19.2×10^{-4} J s^{-1}. Some of the bromine reacted to give cinnamic acid dibromide, and in this experiment the total bromine content decreased by 3.83×10^{19} molecules. (*a*) What was the quantum yield? (*b*) State whether or not a chain reaction was involved.

19.14 The following calculations are made on a uranyl oxalate actinometer, on the assumption that the energy of all wavelengths between 254 and 435 nm is completely absorbed. The actinometer contains 20 cm^3 of 0.05 mol L^{-1} oxalic acid, which also is 0.01 mol L^{-1} with respect to uranyl sulfate. After 2 h of exposure to ultraviolet light, the solution required 34 cm^3 of potassium permanganate, $KMnO_4$, solution to titrate the undecomposed oxalic acid. The same volume, 20 cm^3, of unilluminated solution required 40 cm^3 of the $KMnO_4$ solution. If the average energy of the quanta in this range may be taken as corresponding to a wavelength of 350 nm, how many joules were absorbed per second in this experiment? ($\phi = 0.57$.)

19.15 A solution of a dye is irradiated with 400 nm light to produce a steady concentration of triplet state molecules. If the triplet state yield is 0.9 and the triplet state lifetime is 20×10^{-6} s, what light intensity, expressed in watts, is required to maintain a steady triplet concentration of 5×10^{-6} mol L^{-1} in a liter of solution? Assume that all of the light is absorbed.

19.16 The photochemical chlorination of chloroform,

$$CHCl_3 + Cl_2 = CCl_4 + HCl$$

is believed to proceed by the following mechanism:

$$Cl_2 + h\nu \xrightarrow{I_a} 2Cl$$
$$Cl + CHCl_3 \xrightarrow{k_1} CCl_3 + HCl$$
$$CCl_3 + Cl_2 \xrightarrow{k_2} CCl_4 + Cl$$
$$2CCl_3 + Cl_2 \xrightarrow{k_3} 2CCl_4$$

Derive the steady-state rate law for the production of carbon tetrachloride.

19.17 The mechanism for quenching fluorescence is

$$A + h\nu \rightarrow A^* \qquad I_a$$
$$A^* + Q \rightarrow A + Q \qquad k_q$$
$$A^* \rightarrow A + h\nu_f \qquad k_f = I_f/[A^*]$$

where I_a is the amount of exciting radiation absorbed per liter of solution per second, k_q is the rate constant for quenching, k_f is the rate constant for fluorescence, and I_f is the amount of fluorescence radiation per liter per second. Assuming a steady state is reached, derive the equation for the intensity of fluorescence radiation I_f as a function of [Q]. Describe how the data should be plotted to determine the rate constant for quenching k_q.

19.18 When a solution of anthracene in benzene is exposed to ultraviolet light, anthracene molecules are excited and form dimers with unexcited anthracene molecules. If the excited anthracene molecules fluoresce before they react with unexcited anthracene molecules to form dimers, they do not undergo dimerization. In concentrated solutions of anthracene, the quantum yield ϕ for the formation of dimers is high, but in dilute solutions it is low because the excitation is lost in fluorescence. Formulate a mechanism to represent these facts, and derive the quantum yield as a function of the concentration of anthracene. It is useful to assume that the excited anthracene molecules are in a steady state.

19.19 Professor Mario Molina (MIT) made an important contribution to the understanding of the role of chlorine atoms in the stratosphere by suggesting that the decomposition of ozone is catalyzed by the reactions

$$2ClO + M \rightarrow ClOOCl + M$$
$$ClOOCl + h\nu \rightarrow Cl + ClOO$$
$$ClOO + M \rightarrow Cl + O_2 + M$$
$$Cl + O_3 \rightarrow ClO + O_2$$

The steps in this mechanism add up to $2O_3 + h\nu \rightarrow 3O_2$, but what is the stoichiometric number of the last step, if the stoichiometric numbers of the first three steps are taken as unity?

19.20 Sunlight between 290 and 313 nm can produce sunburn (erythema) in 30 min. The intensity of radiation between these wavelengths in summer and at 45° latitude is about 50 $\mu W\ cm^{-2}$. Assuming that one photon produces chemical change in one molecule, how many molecules in a square centimeter of human skin must be photochemically affected to produce evidence of sunburn?

19.21 The pre-exponential factor for the reaction

$$H_2 + I_2 = 2HI$$

is 10^{11} L mol^{-1} s^{-1}, and the activation energy is 165 kJ mol^{-1} in the range 300 to 500 °C. If the collision diameter is 320 pm, what value of the pre-exponential factor is expected from collision theory at 600 K, and what is the value of the steric factor p? (At higher temperatures this reaction goes by the unbranched chain mechanism described in Section 18.11.)

19.22 Estimate the pre-exponential factor for the reaction

$$2CH_3 \rightarrow C_2H_6$$

using collision theory. The molecular diameter of CH_4 obtained from gas viscosity measurements at 0 °C is 0.414 nm (Table 17.4). The experimental value of A in Table 18.2 is $10^{10.5}$ L mol^{-1} s^{-1}.

19.23 Estimate the pre-exponential factor in the neighborhood of 500 K for the reaction

$$Cl + H_2 \rightarrow HCl + H$$

assuming the activated complex is linear. The H—H and H—Cl bond distances in the activated complex may be assumed to be 0.092 and 0.145 nm, respectively, and the vibrational partition functions may be taken as unity. The experimental value from Table 18.2 is $10^{10.9}$ L mol^{-1} s^{-1}.

19.24 Kinetic theory shows that the rate constant for a bimolecular gas reaction would be given by equation 19.9 for spherical molecules. Calculate the expression for the activation energy.

19.25 The equilibrium constant K for the following reaction of ideal gases at a certain temperature

$$A + B = C$$

is 100 when the standard state pressure $P°$ is 1 bar. What is the value of the equilibrium constant K^* when the standard state pressure is 1 Pa? Derive a general expression for making this calculation.

19.26 A solution absorbs 300 nm radiation at the rate of 1 W. What does this correspond to in amount of photons absorbed per second?

19.27 What intensities of light in J s^{-1} are required to produce 10^{-6} mol s^{-1} at (*a*) 700 nm and (*b*) 300 nm? (*c, d*) What are these powers in watts? (*Note*: 1 J s^{-1} = 1 W.)

19.28 Discuss the economic possibilities of using photochemical reactions to produce valuable products with electricity at 5 cents per kilowatt-hour. Assume that 5% of the electric energy consumed by a quartz–mercury vapor lamp goes into light, and 30% of this is photochemically effective. (*a*) How much will it cost to produce 1 lb (453.6 g) of an organic compound having a molar mass of 100 g mol^{-1}, if the average effective wavelength is assumed to be 400 nm and the reaction has a quantum yield of 0.8 molecule per photon? (*b*) How much will it cost if the reaction involves a chain reaction with a quantum yield of 100?

19.29 The quantum yield is 1 for the photolysis of gaseous HI to $H_2 + I_2$ by light of 253.7 nm wavelength. Calculate the amount of HI that will be decomposed if 300 J of light of this wavelength is absorbed.

19.30 In the stratosphere molecular oxygen absorbs solar radiation in the 185–220 nm wavelength region:

$$O_2 + h\nu = O + O$$

If the absorption cross section is 1.1×10^{-23} cm^2, what thickness of a layer of O_2 at 298 K and 1 bar is required to absorb half of the radiation?

19.31 The quantum yield for the photolysis of acetone

$$(CH_3)_2CO = C_2H_6 + CO$$

at 300 nm is 0.2. How many moles per second of CO are formed if the intensity of the 300-nm radiation absorbed is 10^{-2} J s^{-1}?

19.32 The fluorescence quantum yield for benzene at 25 °C is 0.070. The lifetime of the excited state is 26 nm. What is the radiative lifetime τ_0?

19.33 If the lifetime of a triplet is 1 s and the bimolecular rate constant for the quenching of the triplet by O_2 is

10^9 L mol^{-1} s^{-1}, what concentration of O_2 in a solution will reduce the intensity of fluorescence to 10%?

19.34 A cold high-voltage mercury lamp is to be used for a certain photochemical reaction that responds to ultraviolet light of 253.7 nm. The chemical analysis of the product is sensitive to only 10^{-4} mol. The lamp consumes 150 W and converts 5% of the electric energy into radiation, of which 80% is at 253.7 nm. The amount of the light that gets into the monochromator and passes out the exit slit is only 5% of the total radiation of the lamp. Fifty percent of this 253.7 nm radiation from the monochromator is absorbed in the reacting system. The quantum yield is 0.4 molecule of product per quantum of light absorbed. How long an exposure must be given in this experiment if it is desired to measure the photochemical change with an accuracy of 1%?

19.35 The quantum yield is unity for the dissociation of acetone vapor using 254-nm radiation at 150 °C. How long will it take to dissociate 10^{-2} mol using a 100-W laser producing 254-nm radiation?

19.36 A uranyl oxalate actinometer is exposed to light of wavelength 390 nm for 1980 s, and it is found that 24.6 cm^3 of 0.004 30 mol L^{-1} potassium permanganate is required to titrate an aliquot of the uranyl oxalate solution after illumination, in comparison with 41.8 cm^3 before illumination. Using the known quantum yield of 0.57, calculate the number of joules absorbed per second. The chemical reaction for the titration is

$$2MnO_4{}^- + 5H_2C_2O_4 + 6H^+ = 2Mn^{2+} + 10CO_2 + 8H_2O$$

19.37 A photochemical reaction of biological importance is the production of vitamin D, which prevents rickets and brings about the normal deposition of calcium in growing bones. Steenbock found that rickets could be prevented by subjecting the food as well as the patient to ultraviolet light below 310 nm. When ergosterol is irradiated with ultraviolet light below 310 nm, vitamin D is produced. When irradiated ergosterol was included in a diet otherwise devoid of vitamin D, it was found that absorbed radiant energy of about 7.5×10^{-5} J was necessary to prevent rickets in a rat when fed over a period of 2 weeks. The light used has a wavelength of 265 nm. (*a*) How many quanta are necessary to give 7.5×10^{-5} J? (*b*) If vitamin D has a molar mass of the same order of magnitude as ergosterol (382 g mol^{-1}), how many grams of vitamin D per day are necessary to prevent rickets in a rat? It is assumed that the quantum yield is unity.

19.38 Given that the intensity of solar radiation is 4.2 J cm^{-2} min^{-1}, how much carbon has to be burned to obtain the same amount of heat as the solar radiation on 1 m^2 in an 8-h day?

19.39 Given that solar radiation at noon at a certain place on the earth's surface is 4.2 J cm^{-2} min^{-1}, what is the maximum power output in W m^{-2}?

19.40 If a good agricultural crop yields about 2 tons acre^{-1} of dry organic material per year with a heat of combustion of about 16.7 kJ g^{-1}, what fraction of a year's solar energy is stored in an agricultural crop if the solar energy is about 4184 J min^{-1} ft^{-2} and the sun shines about 500 min day^{-1} on the average? One acre = 43 560 ft^2 and 1 ton = 907 000 g.

Computer Problems

19.A In molecular beam experiments, reactant molecules can be accelerated to supersonic velocities by allowing a dilute mixture of the reactant in an inert carrier gas to expand through a pinhole into a vacuum. (*a*) Use equation 19.37 to calculate the peak velocity of ethane molecules in a carrier gas of helium that expands from a source chamber at 298.15 K. (*b*) Calculate the temperature at which ethane molecules have this root-mean-square velocity.

19.B The simplest equation that gives a saddle-shaped surface is $z = y^2 - x^2$. Plot this surface in three dimensions and think about the path that a vibrating molecule would take across it with the minimum energy. Consider different angles of approach.

19.C Plot the ratio of the reaction cross section $\sigma(\epsilon)$ to the collision cross section πd_{12}^2 as a function of the ratio of the collision energy ϵ to the minimum energy ϵ^* along the line of centers to cause reaction.

19.D If the average energy of a probe photon in a femtosecond transition-state experiment is 5362 cm^{-1}, what is the average speed $\langle v \rangle$ of the wave packet in the dissociation of NaI?

20

Kinetics in the Liquid Phase

When chemical reactions occur in solution, the dynamics are different than for gas-phase reactions, for which kinetic theory has been so useful. We can learn about the motion of molecules in a liquid by studying rate processes—viscosity, diffusion, and electrical conductivity. These processes can be treated by transition-state theory, but here we will emphasize their phenomenological aspects and their relation to the kinetics of reactions in solution. Measurements of viscosity and diffusion provide a means for learning about the rates with which reactants can come together in solution. For ions there is an additional possibility of studying their motion in a solution that is subjected to an electric field. The interpretation of rates in the liquid phase is necessarily more complicated from a molecular viewpoint because of the much greater interaction between molecules. However, bimolecular reactions in solutions cannot occur more rapidly than the reactant molecules can diffuse together, and this rate may be calculated from measured diffusion coefficients of the reactants.

Acids and bases catalyze many reactions, and we will discuss two types of mechanisms for acid catalysis. Enzymes catalyze the reactions in living things and provide the mechanisms by which rates of these reactions are controlled. These reactions provide interesting examples of solution kinetics.

20.1 VISCOSITY OF A LIQUID

Viscosity was defined in the discussion of the kinetic theory of gases in Section 17.9. This definition applies to laminar flow, that is, flow in which one layer (lamina) slides smoothly relative to another. When this flow velocity is great enough, turbulence develops. The viscosity of a liquid may be measured by a number of methods, including the determination of the rate of flow through a capillary, the rate of settling of a sphere in a liquid, and the force required to turn one of two concentric cylinders at a certain angular velocity (Fig. 17.15*c*).

When a force F is applied to a particle in solution—as in an electric field, if the particle is charged, or in a centrifugal field—the particle will be accelerated. As the velocity of the particle increases, it experiences an increasing frictional force. For low velocities the frictional force is given by vf, where v is the velocity and f is the **frictional coefficient** of the particle. When the velocity is sufficiently high for the frictional force to be equal to the applied force,

$$vf = F \tag{20.1}$$

and the particle will move with constant velocity.

The frictional coefficient f is of interest because it provides some information about the size and shape of the particle. For spherical particles Stokes showed that for nonturbulent flow,

$$f = 6\pi\eta r \tag{20.2}$$

where η is the viscosity and r is the radius of the spherical particle. The frictional coefficients of prolate and oblate ellipsoids and long rods may be expressed in terms of the radius of a sphere of equal volume and a factor depending on the ratio of the major axis to the minor axis.

The **viscosity** η of a liquid may be determined by measuring the rate of settling of a sphere of known density. The force causing the sphere to settle in the fluid is equal to its effective mass times the acceleration of gravity; the effective mass is the mass of the sphere minus the mass of the fluid it displaces. If the sphere has a density ρ and the density of the medium is ρ_0, the force causing motion is $\frac{4}{3}\pi r^3(\rho - \rho_0)g$, where g is the acceleration of gravity (9.807 m s^{-2}). When the rate of settling of the sphere in the liquid is constant, the retarding force is equal to the force due to gravity, and so

$$\frac{4}{3}\pi r^3(\rho - \rho_0)g = 6\pi\eta r\left(\frac{\mathrm{d}x}{\mathrm{d}t}\right) \tag{20.3}$$

$$\frac{\mathrm{d}x}{\mathrm{d}t} = \frac{2r^2(\rho - \rho_0)g}{9\eta} \tag{20.4}$$

Thus, by measuring the velocity $\mathrm{d}x/\mathrm{d}t$ of settling of a sphere of known r and ρ in a liquid of known density ρ_0, the viscosity η may be obtained. This method is especially valuable for solutions of high viscosity, such as concentrated solutions of high polymers. Conversely, a determination of the rate of settling of colloidal particles of known density in a liquid of known viscosity provides a means for determining the effective particle radius.

The viscosity η can be determined by passing a liquid through a capillary tube and making use of the **Poiseuille equation:**

$$\eta = \frac{P\pi r^4 t}{8VL} \tag{20.5}$$

Table 20.1 Viscosity of Water in Pa s (kg m^{-1} s^{-1})

t/°C	($\eta \times 10^4$)/Pa s
0	1.786 5
10	1.303 7
20	1.001 9
25	0.890 9
40	0.654 0
60	0.467 4
80	0.355 4
100	0.282 9
125	0.220
150	0.183

where t is the time required for volume V of liquid to flow through a capillary tube of length L and radius r under an applied pressure P. The viscosities of most liquids decrease with increasing temperature. According to the "hole theory," there are vacancies in a liquid, and molecules are continually moving into these vacancies so that the vacancies move around. This process permits flow but requires energy because a molecule must surmount an activation barrier to move into a vacancy.

The viscosity of water at a series of temperatures is given in Table 20.1. The variation of the viscosity with temperature may be represented quite well by

$$\frac{1}{\eta} = A\,\mathrm{e}^{-E_a/RT} \tag{20.6}$$

where E_a is the activation energy for the **fluidity,** $1/\eta$.

The viscosity of a liquid increases as the pressure is increased because the number of holes is reduced, and it is therefore more difficult for molecules to move around each other.

In contrast with liquids, the viscosity of a gas increases as the temperature increases. The viscosity of an ideal gas is independent of pressure (Sections 17.9 and 17.10).

20.2 DIFFUSION

Fick's first law of diffusion was introduced in Section 17.9 as an illustration that a flux is proportional to the gradient of something, in this case the concentration. In this section we want to emphasize that diffusion occurs as a result of a gradient in the chemical potential μ. If the chemical potential for a species is uniform in a system, there is no driving force for diffusion.

A force F, equal to the negative gradient of the chemical potential, causes a substance to move from a region where its chemical potential is high to a region where it is low. For an ideal solution $\mu = \mu^\circ + RT \ln c$, and so the force F causing diffusion in the x direction in an ideal solution is given by

$$F = -\frac{\mathrm{d}\mu}{\mathrm{d}x} = -\frac{RT}{c}\frac{\mathrm{d}c}{\mathrm{d}x} \tag{20.7}$$

The force opposing the diffusion of a molecule or ion is the frictional coefficient f times the velocity v. Setting these forces on a molecule or ion equal to each other yields

$$fv = -\frac{RT}{N_A c}\frac{dc}{dx} \tag{20.8}$$

or

$$vc = -\frac{RT}{N_A f}\frac{dc}{dx} \tag{20.9}$$

This corresponds to **Fick's first law** (equation 17.59):

$$J = -D\frac{dc}{dx} \tag{20.10}$$

where J is the flux and D is the **diffusion coefficient.** Thus, the diffusion coefficient for ideal solutions is given by

$$D = \frac{RT}{N_A f} \tag{20.11}$$

which shows that the diffusion coefficient is inversely proportional to the frictional coefficient. This relation was first derived by Einstein.

In the preceding section we saw that $f = 6\pi\eta r$ for spherical particles, so the diffusion coefficient for a spherical particle is given by

$$D = \frac{RT}{N_A 6\pi\eta r} \tag{20.12}$$

In the next chapter (Section 21.5) we will see that this relation can be used to determine the molar mass of a protein molecule that is nearly spherical.

In studying transport processes the flux J is seldom measured directly. What is measured is the change in concentration with time at various points. The continuity equation is the expression of the fact that, since mass is conserved, the change in concentration in a region is due to the difference between the flows into and out of the region. To relate the flux to the change in concentration, consider the situation illustrated in Fig. 20.1. Flow of solute is occurring in the x direction in a cell of uniform cross section A. We want to calculate the change in concentration in a thin slab of thickness δx. The quantity of material crossing the plane at x in

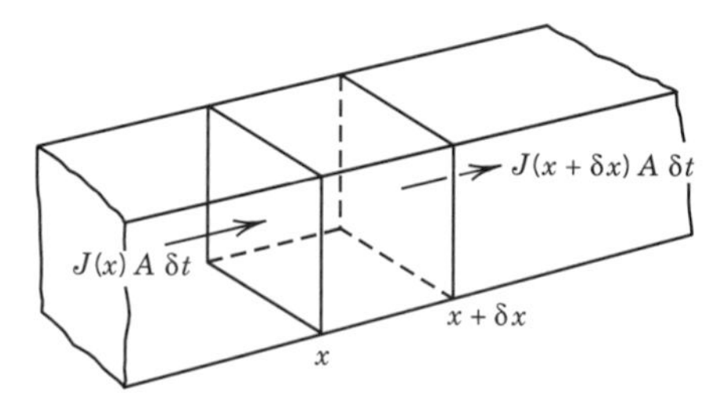

Figure 20.1 Cell of uniform cross section A in which there is transport by diffusion, sedimentation, or electrical migration.

time δt is $J(x)A\,\delta t$, whereas the quantity leaving through the plane at $x + \delta x$ in the same time is $J(x + \delta x)A\,\delta t$, which may be written

$$\left[J(x) + \frac{\partial J}{\partial x}\,\delta x\right]A\,\delta t \tag{20.13}$$

The net gain in the quantity of material between these hypothetical planes may be expressed in terms of the change of concentration in the volume $A\,\delta x$ or in terms of the difference between these two quantities of material transported:

$$A\,\delta c\,\delta x = JA\,\delta t - \left(J + \frac{\partial J}{\partial x}\,\delta x\right)A\,\delta t = -\frac{\partial J}{\partial x}\,A\,\delta x\,\delta t \tag{20.14}$$

In the limit, as the distances and times are made smaller,

$$\frac{\partial c}{\partial t} = -\frac{\partial J}{\partial x} \tag{20.15}$$

This is referred to as the **equation of continuity.** It relates the change in concentration at a given value of x in the cell to the rate of change of flux with distance.

Fick's first law (equation 20.10) is substituted in the equation of continuity (equation 20.15) to obtain Fick's second law:

$$\frac{\partial c}{\partial t} = \frac{\partial}{\partial x}D\,\frac{\partial c}{\partial x} \tag{20.16}$$

If the diffusion coefficient D is independent of the concentration and therefore of distance, then

$$\frac{\partial c}{\partial t} = D\,\frac{\partial^2 c}{\partial x^2} \tag{20.17}$$

which is known as **Fick's second law.**

Diffusion coefficients in solution are usually measured by first forming a sharp boundary between a solution and the solvent, as shown in Fig. 20.2. At a later time the boundary is diffuse.

To derive the expression for concentration as a function of distance and time for the experiment illustrated in Fig. 20.2, equation 20.17 is integrated with the following boundary conditions: When $t = 0$, $c = c_0$ for $x > 0$, and $c = 0$ for $x < 0$; and when $t > 0$, c approaches c_0 as x approaches ∞ and c approaches 0 as x approaches $-\infty$. The result is

$$c = \frac{c_0}{2}\left(1 - \frac{2}{\sqrt{\pi}}\int_0^{x/2\sqrt{Dt}} e^{-\beta^2}\,d\beta\right) \tag{20.18}$$

where the second term in the parentheses is referred to as the **error function.***

*The error function is

$$\operatorname{erf}(z) = \frac{2}{\pi^{1/2}}\int_0^z e^{-t^2}\,dt$$

Thus $\operatorname{erf}(0) = 0$ and $\operatorname{erf}(\infty) = 1$. Since the error function cannot be expressed in closed form, a table is given in Appendix D.2.

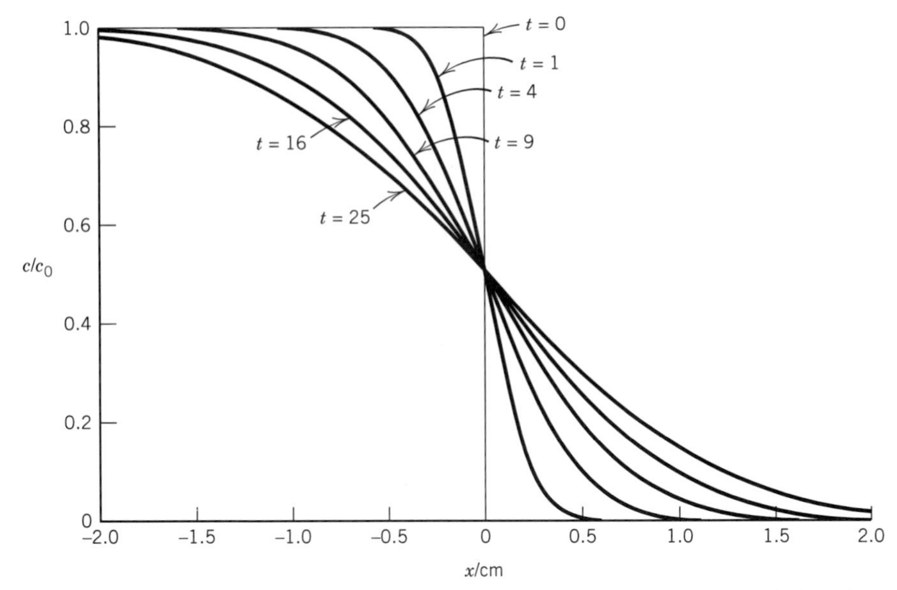

Figure 20.2 Diffusion of sucrose from an initially homogeneous aqueous solution of concentration c_0 to the left of $x = 0$ into water at 25 °C. The concentration of sucrose relative to its concentration in the homogeneous solution is given for $t = 0, 1, 4, 9, 16$, and 25 hours. In the actual experiment, x is the vertical axis. (See Computer Problem 20.B.)

The concentration of sucrose diffusing in water from an initially sharp boundary at 25 °C at various times is shown in Fig. 20.2.

The derivative of equation 20.18 with respect to distance is given by

$$\frac{\partial c}{\partial x} = \frac{c_0}{2\sqrt{\pi D t}} \mathrm{e}^{-x^2/4Dt} \tag{20.19}$$

This bell-shaped curve is referred to as a **Gaussian curve.** The concentration gradient curves for the diffusion of sucrose in water are given in Fig. 20.3.

The square of the standard deviation of the experimental bell-shaped curve is

$$\sigma^2 = \frac{\int_{-\infty}^{\infty} x^2(\partial c/\partial x)\,\mathrm{d}x}{\int_{-\infty}^{\infty}(\partial c/\partial x)\,\mathrm{d}x} \tag{20.20}$$

Substituting equation 20.19,

$$\sigma^2 = \frac{\int_{-\infty}^{\infty} x^2\,\mathrm{e}^{-x^2/4Dt}\,\mathrm{d}x}{\int_{-\infty}^{\infty} \mathrm{e}^{-x^2/4Dt}\,\mathrm{d}x} = 2Dt \tag{20.21}$$

where the last form is obtained by using the values of the definite integrals (see Table 17.1). Since the **standard deviation** σ of a Gaussian curve is the half-width at the inflection point, and the inflection points are at a height of 0.606 of the maximum ordinate, σ is readily obtained from the experimental curve, and D may be calculated using equation 20.21. The diffusion coefficients of a number of protein molecules are given in Table 21.3.

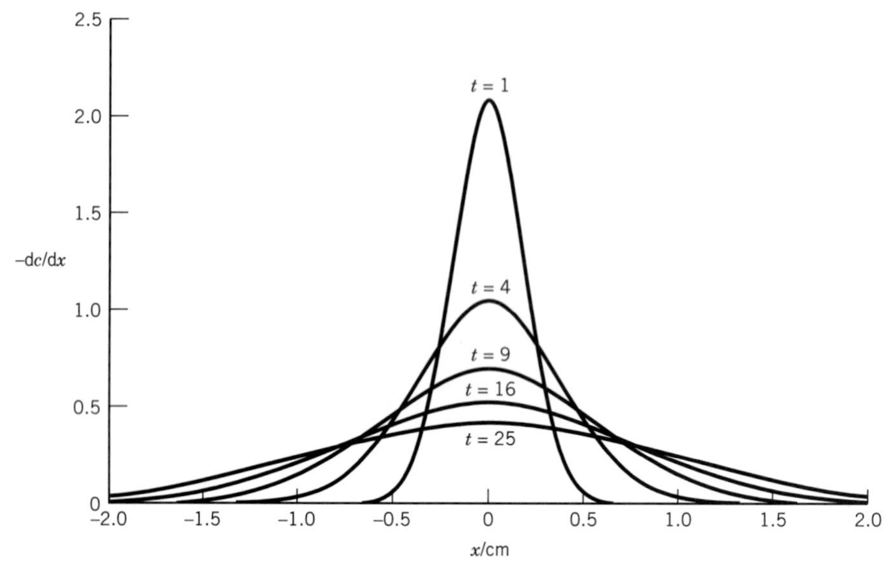

Figure 20.3 Concentration gradient curves for the diffusion of sucrose from an initially homogeneous aqueous solution to the left of $x = 0$. The plots are for $t = 1, 4, 9, 16$, and 25 hours. Note that the standard deviations σ are increasing linearly with the square root of time. (See Computer Problem 20.B.)

Example 20.1 ***Diffusion of sucrose in water***

(*a*) The diffusion coefficient of sucrose in water at 25 °C is 5.1×10^{-10} m^2 s^{-1}. What is the standard deviation of an initially sharp boundary (see Fig. 20.2) after 1 h? After 24 h? (*b*) The diffusion coefficient of hemoglobin in water at 25 °C is 6.9×10^{-11} m^2 s^{-1}. What is the standard deviation of an initially sharp boundary after 1 h? After 24 h?

(*a*) After 1 h,

$$\begin{aligned}\sigma &= (2Dt)^{1/2}\\ &= [2(5.1 \times 10^{-10}\ \mathrm{m^2\ s^{-1}})(1\ \mathrm{h})(60\ \mathrm{min\ h^{-1}})(60\ \mathrm{s\ min^{-1}})]^{1/2}\\ &= 0.00192\ \mathrm{m} = 1.92\ \mathrm{mm}\end{aligned}$$

After 24 h,

$$\sigma = (1.92\ \mathrm{mm})24^{1/2} = 11.8\ \mathrm{mm}$$

(*b*) After 1 h,

$$\begin{aligned}\sigma &= [2(6.9 \times 10^{-11}\ \mathrm{m^2\ s^{-1}})(1\ \mathrm{h})(60\ \mathrm{min\ h^{-1}})(60\ \mathrm{s\ min^{-1}})]^{1/2}\\ &= 0.705\ \mathrm{mm}\end{aligned}$$

After 24 h,

$$\sigma = (0.705\ \mathrm{mm})24^{1/2} = 3.45\ \mathrm{mm}$$

20.3 MOBILITY OF AN ION

A solution containing ions conducts an electric current because the ions move under the influence of an electric field. To measure the electric resistance R of a solution containing ions, the solution is placed in a cell with two electrodes that have a coating of platinum black, and an alternating current is used. When an alternating current is used, the electrolysis that occurs when the current passes in one direction is reversed when the current passes in the other direction, and the formation of a nonconducting gas film is prevented.

The **electric resistance** R of a uniform conductor is directly proportional to its length L and inversely proportional to its cross-sectional area A:

$$R = \frac{rL}{A} = \frac{L}{\kappa A} \tag{20.22}$$

where the proportionality constant r is called the **resistivity** and the proportionality constant $\kappa = 1/r$ is called the **electric conductivity.** In the SI system the electric conductivity κ has the units $\Omega^{-1}\ \text{m}^{-1}$, where the ohm is represented by Ω. Since the potential difference ϕ between two points on a uniform conductor in which a current I is flowing is $\phi = IR$, the unit of electric potential difference ϕ, the volt V, is equal to the product of the unit of electric current, the ampere A, and the unit of resistance, the ohm Ω: $1\ \text{V} = (1\ \text{A})(1\ \Omega)$.

Electric conductivities range from $10^8\ \Omega^{-1}\ \text{m}^{-1}$ for a metallic conductor at room temperature to $10^{-15}\ \Omega^{-1}\ \text{m}^{-1}$ for an insulator such as SiO_2. The electric conductivity κ of an electrolyte solution is made up of contributions from each of the types of ions present.

The **electric mobility** u of an ion is its drift velocity in the direction of the electric field, divided by the electric field strength E:

$$u = \frac{\mathrm{d}x/\mathrm{d}t}{E} \tag{20.23}$$

(Strictly speaking, the electric field strength is a vector and should be represented by $\boldsymbol{E}$, but here we will write equations in terms of the magnitude E of the electric field strength, which is a scalar.) The drift velocity of an ion is the average velocity in the direction of the field. Because of Brownian motion, an ion undergoes random displacements so that it does not move in a straight line over macroscopic distances. The electric field strength E is the negative gradient of the electric potential ϕ. When the electric potential varies only in the x direction, then

$$E = -\frac{\mathrm{d}\phi}{\mathrm{d}x} \tag{20.24}$$

For a uniform conductor the difference in the potential per unit distance may be calculated using Ohm's law. For a conductor of unit cross section the difference in potential between two points is equal to the current density I/A, where I is the current and A the area, multiplied by the resistivity $1/\kappa$:

$$E = \frac{I}{A\kappa} \tag{20.25}$$

Since the electric field strength is expressed in V m^{-1} in the SI system, the electric mobility u has the units $\text{m}^2\ \text{V}^{-1}\ \text{s}^{-1}$.

The drift velocities of ions can be determined in various ways, and the electric mobilities of a number of small ions at infinite dilution in water at 25 °C are given in Table 20.2.

Table 20.2 Electric Mobilities at 25 °C in Water at Infinite Dilution

Ion	$u/10^{-8}$ m^2 V^{-1} s^{-1}	*Ion*	$u/10^{-8}$ m^2 V^{-1} s^{-1}
H^+	36.25	OH^-	20.64
Li^+	4.01	F^-	5.74
Na^+	5.192	Cl^-	7.913
K^+	7.617	NO_3^-	7.406
NH_4^+	7.62	ClO_3^-	6.70
$N(CH_3)_4^+$	4.66	$CH_3CO_2^-$	4.24
Mg^{2+}	5.50	$C_6H_5CO_2^-$	3.36
Ca^{2+}	6.17	SO_4^{2-}	8.29
Pb^{2+}	7.20	CO_3^{2-}	7.18

The remarkably high electric mobility of the hydrogen ion is due to the fact that a proton may be transferred along a series of hydrogen-bonded water molecules by rearrangement of the hydrogen bonds (Section 11.10). In the following representation (*a*) shows the initial bonding in a group of oriented water molecules and (*b*) shows the final bonding:

```
    +                                                               +
H—O—H···O—H···O—H···O—H            H—O···H—O···H—O···H—O—H
  |     |     |     |                |      |      |      |
  H     H     H     H                H      H      H      H
         (a)                                   (b)
```

The net effect is rapid long-range proton mobility even though no single proton moves a long distance in a short time. For another hydrogen ion to be transferred to the right through this group of water molecules, molecular rotations must occur to produce again a favorable orientation for charge transfer.

This model for hydrogen ion mobility helps us to understand the remarkable fact that hydrogen ions move about 50 times more rapidly through ice than through liquid water. In ice each oxygen atom is surrounded by four oxygen atoms at a distance of 276 pm in a tetrahedral arrangement. Each hydrogen is near the line through the centers of the oxygen atoms and is about 100 pm from one oxygen and 176 pm from the other. Hydrogen ions may be conducted rapidly through this structure by the above mechanism when the water molecules are oriented correctly. A similar mechanism can be written for the transfer of hydroxyl ions in the opposite direction.

The separation of different macromolecular ions (such as proteins and nucleic acids) according to their electric mobilities is referred to as **electrophoresis.**

The electric conductivity κ of an electrolyte solution is the sum of the contributions of all the ionic species in the electrolyte. The electric current contributed by an ion depends on its charge number z_i as well as on its electric mobility.

If the ions of type i have a concentration c_i moles per unit volume, and all move with a velocity u_i in a field of 1 V/m, the transport of electric charge through a plane perpendicular to the direction of motion is $F|z_i|c_i u_i$, where F is the Faraday constant. Thus, for the solution as a whole, the electric conductivity κ is given by

$$\kappa = F \sum_i |z_i| c_i u_i = Fc \sum_i |z_i| \nu_i u_i \tag{20.26}$$

The second form gives the conductivity of a solution of a single electrolyte of concentration c with ν_i ions of type i.

Mobilities increase with the temperature, and the temperature coefficients are very nearly the same for all ions in a given solvent and are approximately equal to the temperature coefficient of the viscosity; for water this is 2% per degree in the neighborhood of 25 °C.

Example 20.2 ***Using conductivity to calculate ion concentrations***

The electric conductivity κ of pure water is $5.5 \times 10^{-6}\ \Omega^{-1}\ \text{m}^{-1}$ at 25 °C. What is the value of the ion product $K_w = [H^+][OH^-]$?

The concentrations of hydrogen and hydroxyl ions are, of course, equal and may be calculated using equation 20.26 and the values of the limiting ion mobilities in Table 20.2.

$$\kappa = Fc(u_{H^+} + u_{OH^-})$$

$$c = \frac{\kappa}{F(u_{H^+} + u_{OH^-})} = \frac{5.5 \times 10^{-6}\ \Omega^{-1}\ \text{m}^{-1}}{(96\,485\ \text{C mol}^{-1})(5.689 \times 10^{-7}\ \text{m}^2\ \text{V}^{-1}\ \text{s}^{-1})}$$

$$= 1.00 \times 10^{-4}\ \text{mol m}^{-3} = 1.00 \times 10^{-7}\ \text{mol L}^{-1}$$

Thus, $K_w = (1.00 \times 10^{-7})^2 = 1.00 \times 10^{-14}$ at 25 °C.

Example 20.3 ***Conductivity and transport of ions in water***

(*a*) What is the electrical conductivity of a solution of 0.01 mol L^{-1} sodium chloride in water at 25 °C, assuming that the electric mobilities of the ions are not significantly different than at infinite dilution? (*b*) Suppose a milliampere is passed through a 1-cm^3 cube of solution between opposite faces. How far will the sodium and chloride ions move in 10 min?

(*a*) Using equation 20.26,

$$\kappa = (96\,485\ \text{C mol}^{-1})(0.01\ \text{mol L}^{-1})(10^3\ \text{L m}^{-3})[(5.192 + 7.913) \times 10^{-8}\ \text{m}^2\ \text{V}^{-1}\ \text{s}^{-1}]$$

$$= 0.1264\ \text{C s}^{-1}\ \text{m}^{-1}\ \text{V}^{-1} = 0.1264\ \Omega^{-1}\ \text{m}^{-1}$$

$$(b) \quad E = \frac{I}{A\kappa} = \frac{0.001\ \text{A}}{(0.01\ \text{m})^2(0.1264\ \Omega^{-1}\ \text{m}^{-1})}$$

$$= 79.11\ \text{V m}^{-1}$$

For Na^+,

$$\Delta x = Eu\,\Delta t = (79.11\ \text{V m}^{-1})(5.192 \times 10^{-8}\ \text{m}^2\ \text{V}^{-1}\ \text{s}^{-1})(10\ \text{min})(60\ \text{s min}^{-1})$$

$$= 2.46\ \text{mm}$$

For Cl^-,

$$\Delta x = Eu\,\Delta t = (79.11\ \text{V m}^{-1})(7.913 \times 10^{-8}\ \text{m}^2\ \text{V}^{-1}\ \text{s}^{-1})(10\ \text{min})(60\ \text{s min}^{-1})$$

$$= 3.75\ \text{mm}$$

20.4 ENCOUNTER PAIRS AND SOLVENT CAGE

We have seen that in a gas, molecules may move many, many molecular diameters between collisions (Section 17.7). In a liquid, however, a molecule cannot move very far before it collides with a neighbor. This gives rise to the concept that a reactant molecule in a liquid is surrounded by a solvent **cage.** As a result, a given molecule has many collisions with its immediate neighbors before it moves to a new cage. This concept can be applied to two reactant molecules, say, A and B. If they do diffuse together, they will be surrounded by a solvent cage that will tend to keep them together until one or the other escapes from the cage. Therefore, collisions between reactant molecules will have a very different time sequence in a liquid than in a gas.

In a gas the collision frequency is independent of time, but in a liquid the collisions occur in groups, as shown in Fig. 20.4. The groups of collisions are referred to as **encounters.** At room temperature, an encounter may involve 10 to 10^5 collisions. If there is a significant probability that A and B will react when they collide, it is evident that there is a high probability that they will react during an encounter. When this is true, the rate of the reaction will be controlled by the rate with which A and B can diffuse together to form an encounter pair. Reaction under these conditions is said to be **diffusion controlled.**

We can distinguish between two types of bimolecular reactions in solution by consideration of the following simple mechanism:

$$\mathrm{A} + \mathrm{B} \underset{k_{-1}}{\overset{k_1}{\rightleftharpoons}} \{\mathrm{AB}\} \xrightarrow{k_2} \text{products} \tag{20.27}$$

where {AB} is the encounter pair. By use of the steady-state approximation (Section 18.7) it is readily shown that

$$-\frac{\mathrm{d}[\mathrm{A}]}{\mathrm{d}t} = \frac{k_1 k_2}{k_{-1} + k_2}[\mathrm{A}][\mathrm{B}] \tag{20.28}$$

If $k_2 \gg k_{-1}$, the reaction rate is determined by the rate $k_1[\mathrm{A}][\mathrm{B}]$, which is the rate of reactants diffusing together, and we refer to such a reaction as a **diffusion-controlled reaction.** In a diffusion-controlled reaction, reactant molecules within the same solvent cage collide enough times that reaction is highly likely before they can diffuse away from each other. For aqueous solutions it has been estimated that the cage lifetime for a pair of noninteracting molecules is of the order 10^{-12} to 10^{-8} s, during which time they may undergo 10 to 10^5 collisions with each other.

If $k_2 \ll k_{-1}$, the reaction rate is

$$-\frac{\mathrm{d}[\mathrm{A}]}{\mathrm{d}t} = k_2 K_{\mathrm{AB}}[\mathrm{A}][\mathrm{B}] \tag{20.29}$$

where $K_{\mathrm{AB}} = k_1/k_{-1}$ is the equilibrium constant of the formation of the encounter pair. This is an **activation-controlled reaction** because the reaction is largely determined by the activation energy for k_2.

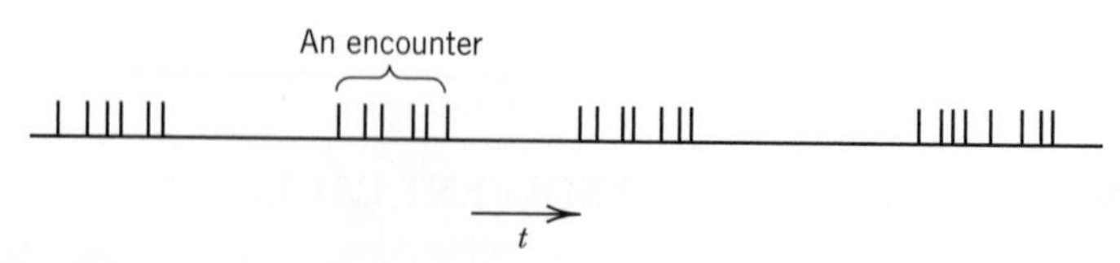

Figure 20.4 Time sequence of collisions between reactants A and B in a liquid. A group of collisions is referred to as an encounter.

20.5 DIFFUSION-CONTROLLED REACTIONS IN LIQUIDS

The maximum rate with which reactants can diffuse together in liquids may be calculated using the macroscopic theory of diffusion and experimentally determined diffusion coefficients of the reactants. The elementary theory of diffusion-controlled reactions was developed in 1917 by Smoluchowski in connection with his theoretical study of the coagulation of colloidal gold.

The diffusion coefficient D is defined in terms of Fick's first law, which is given in Sections 17.9 and 20.2. The diffusion coefficients of low-molar-mass solutes in aqueous solution at 25 °C are of the order of 10^{-9} m^2 s^{-1}. Experimental methods for determining diffusion coefficients in dilute aqueous solutions are discussed in Section 20.2. The diffusion coefficients for ions may be calculated from their ionic mobilities (Section 20.3).

Smoluchowski considered spherical particles with radii R_1 and R_2 that could be considered to react when they diffused within a distance $R_{12} = R_1 + R_2$ of each other. We may imagine one reactant molecule stationary and serving as a sink. Since the concentration is zero at distance R_{12}, a spherically symmetrical concentration gradient is set up. The flux through this concentration gradient is calculated and is expressed as a **second-order rate constant** k_a for association by

$$k_a = 4\pi N_A(D_1 + D_2)R_{12}f \tag{20.30}$$

where D_1 and D_2 are the diffusion coefficients of the reactants and f is an electrostatic factor. The **electrostatic factor** f is different from unity if the reactants are ions. It is larger than unity if the reactants have opposite charges and attract each other, and it is smaller than unity if the reactants have the same charge and repel each other. One ion can be visualized as moving in the electric field created by the other ion.

Because of the electrostatic factor f, the effective reaction radius ($R_{eff} = R_{12}f$) is substantially increased for the reaction of oppositely charged ions and substantially decreased for the reaction of two ions with the same sign. If the ionic strength is so low that ion atmospheres may be neglected, f is given by

$$f = \frac{z_1 z_2 e^2}{4\pi\epsilon_0\epsilon_r kTR_{12}}\left[\exp\left(\frac{z_1 z_2 e^2}{4\pi\epsilon_0\epsilon_r kTR_{12}}\right) - 1\right]^{-1} \tag{20.31}$$

where the charges on the ions are $z_1 e$ and $z_2 e$, ϵ_r is the relative permittivity, and ϵ_0 is the permittivity of free space. More details on the derivation of the equations for diffusion-controlled reactions are given by Hammes.*

The temperature coefficients for diffusion-controlled reactions in water are small because they correspond to the temperature coefficient of the viscosity of liquid water ($E_a = 17.4$ kJ mol^{-1} at 25 °C).

Example 20.4 ***Rate constant for a diffusion-controlled reaction***

A typical diffusion coefficient for a small molecule in aqueous solution at 25 °C is 5×10^{-9} m^2 s^{-1}. If the reaction radius is 0.4 nm, what value is expected for the second-order rate constant for a diffusion-controlled reaction of neutral molecules?

*G. G. Hammes, *Principles of Chemical Kinetics.* New York: Academic, 1978.

$$\begin{aligned} k_a &= 4\pi N_A(D_1 + D_2)R_{12} = 4\pi(6.022 \times 10^{23}\ \text{mol}^{-1})(10^{-8}\ \text{m}^2\ \text{s}^{-1})(0.4 \times 10^{-9}\ \text{m}) \\ &= 3.0 \times 10^7\ \text{m}^3\ \text{mol}^{-1}\ \text{s}^{-1} = (3.0 \times 10^7\ \text{m}^3\ \text{mol}^{-1}\ \text{s}^{-1})(10^3\ \text{L m}^{-3}) \\ &= 3.0 \times 10^{10}\ \text{L mol}^{-1}\ \text{s}^{-1} \end{aligned}$$

Example 20.5 ***Calculation of the electrostatic factor***

What is the electrostatic factor f in water at 25 °C if the reaction radius R_{12} is 0.2 nm for opposite unit charges? For like unit charges? The relative permittivity ϵ_r of water is 78.3.

For opposite charges,

$$\frac{z_1 z_2 e^2}{4\pi\epsilon_0 \kappa k T R_{12}} = \frac{(-1)(1)(1.602 \times 10^{-19}\ \text{C})^2(8.988 \times 10^9\ \text{N C}^{-2}\ \text{m}^2)}{78.3(1.3807 \times 10^{-23}\ \text{J K}^{-1})(298.15\ \text{K})(0.2 \times 10^{-9}\ \text{m})} = -3.58$$

$$f = -3.58[\text{e}^{-3.58} - 1]^{-1} = 3.68$$

Thus, a diffusion-controlled reaction is expected to be 3.68 times faster in this case than for uncharged particles. For ions of the same charge,

$$f = 3.58[\text{e}^{3.58} - 1]^{-1} = 0.103$$

Thus, a diffusion-controlled reaction is expected to be 0.103 times as fast for single charged particles with the same sign as for neutral particles.

Example 20.6 ***An example of a diffusion-controlled reaction***

The renaturation of DNA (see Section 8.9) involves the formation of AT and GC hydrogen bonds. This type of reaction can be studied in a simple model system using organic derivatives or uracil and adenine in an organic solvent. The formation of a hydrogen-bonded dimer between 1-cyclohexyluracil and 9-ethyladenine has been studied by G. G. Hammes and A. C. Park [*J. Am. Chem. Soc.* **91**:956 (1969)]. The second-order rate constant at 30 °C was found to be $2.8 \times 10^9\ \text{M}^{-1}\ \text{s}^{-1}$. This reaction can be interpreted in terms of a two-step rection in which the reactants form a dimer that is not hydrogen bonded and then form hydrogen bonds.

$$\text{A} + \text{B} \underset{k_{-1}}{\overset{k_1}{\rightleftharpoons}} \text{A, B} \underset{k_{-2}}{\overset{k_2}{\rightleftharpoons}} \text{A—B}$$

Assuming that the non–hydrogen-bonded complex is in a steady state, derive the expression for the steady-state rate and discuss the interpretation of the experimental second-order rate constant.

Since the complex is in a steady state,

$$\frac{\text{d[A, B]}}{\text{d}t} = k_1[\text{A}][\text{B}] + k_{-2}[\text{A—B}] - (k_2 + k_{-1})[\text{A, B}] = 0$$

so that in the steady state,

$$[\text{A, B}] = \frac{k_1[\text{A}][\text{B}] + k_{-2}[\text{A—B}]}{k_1 + K_2}$$

The rate of formation of the hydrogen-bonded dimer is given by

$$\frac{\text{d[A—B]}}{\text{d}t} = k_2[\text{A, B}] - k_{-2}[\text{A—B}]$$

Substituting the steady-state concentration of the non-hydrogen-bonded complex yields

$$\frac{\mathrm{d[A{-}B]}}{\mathrm{d}t} = k_\mathrm{f}[\mathrm{A}][\mathrm{B}] - k_\mathrm{r}[\mathrm{A{-}B}]$$

where

$$k_\mathrm{f} = \frac{k_1 k_2}{k_{-1} + k_2}$$

and

$$k_\mathrm{f} = \frac{k_{-1} k_{-2}}{k_{-1} + k_2}$$

If the reaction is diffusion controlled $k_2 \gg k_1$, so we would expect $k_\mathrm{f} = k_1$, where k_1 is the diffusion-controlled rate constant; thus the experimental rate constant k_f is expected to be of the order of k_1.

If a reaction occurs so rapidly that appreciable reaction occurs during the process of mixing the reactants conventionally, a flow method may be used. An early example was the study of the reaction of hemoglobin and oxygen.* A hemoglobin solution was forced into one arm of a Y-mixer and a solution of oxygen in a buffer into the other. In this way it is possible to mix liquids in about 10^{-3} s. In the stopped-flow method, reagents are forced into the mixer, the flow is brought to a sudden stop, and observations are made of the extent of reaction. In the continuous flow method the solutions are mixed and forced down the tube at a steady rate; the extent of reaction is constant at any given distance down the tube, but it increases with distance from the mixing chamber.

Some reactions occur in much less than 10^{-3} s, so their kinetics may not be studied by mixing methods. The time range has been extended down to about 10^{-9} s by the use of **relaxation methods** developed by M. Eigen and co-workers† in Göttingen, Germany. A solution in equilibrium is perturbed by rapidly changing one of the independent variables (usually temperature or pressure) on which the equilibrium depends. The change of the system to the new equilibrium is then followed by use of a rapidly responding physical method, for example, light absorption or electrical conductivity.

Equilibria may be shifted by changing the temperature (if $\Delta H \neq 0$) or by changing the pressure (if $\Delta V \neq 0$). A solution may be heated in a microsecond by use of a pulsed laser or by discharging a large electrical capacitor through a special conductivity cell containing the sample. Equilibria may also be shifted by reducing the pressure suddenly by allowing high-pressure gas to escape through a rupture disk. Figure 20.5 is a schematic diagram of a temperature-jump apparatus in which an increase in temperature in a small volume of solution is produced by passing a large current for about 1 μs. If there is a single reaction and if the displacement from equilibrium is small, the return to equilibrium at the new, higher temperature is represented by

$$\Delta c = \Delta c_0 \, \mathrm{e}^{-t/\tau} \tag{20.32}$$

*H. Hartridge and F. J. W. Roughton, *Proc. R. Soc. London Ser. A* **104:**395 (1923).

†M. Eigen and L. De Maeyer, in G. G. Hammes (ed.), *Investigation of Rates and Mechanisms of Reactions,* 3rd ed., p. 63. Hoboken, NJ: Wiley, 1974; G. G. Hammes, *Principles of Chemical Kinetics.* New York: Academic, 1978; D. N. Hague, *Fast Reactions.* Hoboken, NJ: Wiley-Interscience, 1971.

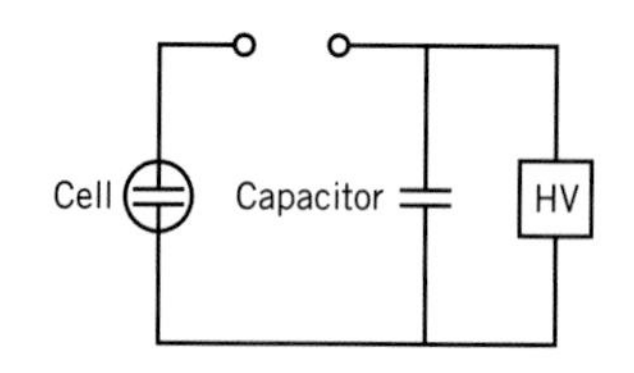

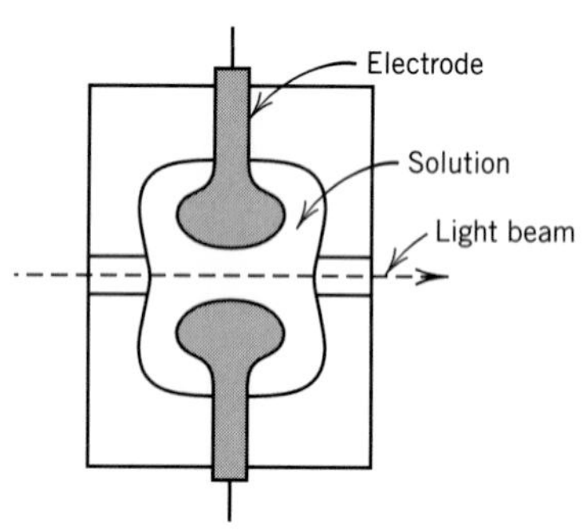

Figure 20.5 Schematic diagram of a temperature-jump apparatus.

where τ is the relaxation time (equation 18.17) and Δc_0 is the difference of the concentration of one of the reactants from its equilibrium value at $t = 0$. If several reactions are involved in the return to equilibrium, Δc is expressed by a sum of exponential terms with different relaxation times.

Comment:

The rates of chemical reactions range from those that are too slow to measure in the laboratory to processes that correspond to the time for an atom to move a fraction of a bond length. It is convenient to think about reaction rates in terms of relaxation times. Very different experimental methods have to be used to measure relaxation times over this broad range. The development of flow methods made it possible to measure relaxation times of solution reactions down to milliseconds. The development of pulse methods extended this down to nanoseconds (10^{-9} s), *and the development of lasers with very short pulses made it possible to work down to picoseconds* (10^{-12} s). *More recently, the experimental range has been extended to femtoseconds* (10^{-15} s) *(see, for example, Section 19.10).*

20.6 RELAXATION TIME FOR A ONE-STEP REACTION

To derive the relationship between the relaxation time τ and the rate constants for a one-step reaction, consider the reaction

$$\mathrm{A} + \mathrm{B} \underset{k_{-1}}{\overset{k_1}{\rightleftharpoons}} \mathrm{C} \tag{20.33}$$

for which the rate equation is

$$\frac{\mathrm{d}[\mathrm{C}]}{\mathrm{d}t} = k_1[\mathrm{A}][\mathrm{B}] - k_{-1}[\mathrm{C}] \tag{20.34}$$

At equilibrium

$$0 = k_1[\mathrm{A}]_{\mathrm{eq}}[\mathrm{B}]_{\mathrm{eq}} - k_{-1}[\mathrm{C}]_{\mathrm{eq}} \tag{20.35}$$

Equation 20.34 may be written in terms of the difference $\Delta[\mathrm{C}]$ from the final equilibrium concentrations by introducing

$$[\mathrm{A}] = [\mathrm{A}]_{\mathrm{eq}} - \Delta[\mathrm{C}] \tag{20.36}$$

$$[\mathrm{B}] = [\mathrm{B}]_{\mathrm{eq}} - \Delta[\mathrm{C}] \tag{20.37}$$

$$[\mathrm{C}] = [\mathrm{C}]_{\mathrm{eq}} + \Delta[\mathrm{C}] \tag{20.38}$$

The fact that the differences from equilibrium are the same for the three reactants, except for sign, comes from the stoichiometry of reaction 20.33. Substituting equations 20.36 to 20.38 into equation 20.34 yields

$$\frac{\mathrm{d}\,\Delta[\mathrm{C}]}{\mathrm{d}t} = k_1([\mathrm{A}]_{\mathrm{eq}} - \Delta[\mathrm{C}])([\mathrm{B}]_{\mathrm{eq}} - \Delta[\mathrm{C}]) - k_{-1}([\mathrm{C}]_{\mathrm{eq}} + \Delta[\mathrm{C}]) \tag{20.39}$$

If the displacement from equilibrium $\Delta[\mathrm{C}]$ is small,

$$\frac{\mathrm{d}\,\Delta[\mathrm{C}]}{\mathrm{d}t} = -\{k_1([\mathrm{A}]_{\mathrm{eq}} + [\mathrm{B}]_{\mathrm{eq}}) + k_{-1}\}\Delta[\mathrm{C}] = -\frac{\Delta[\mathrm{C}]}{\tau} \tag{20.40}$$

where equation 20.35 has been used and the term in $(\Delta[C])^2$ has been neglected because $\Delta[C]$ is small. Thus, the rate of approach to equilibrium is proportional to the displacement from equilibrium $\Delta[C]$. It is customary to use the relaxation time τ (equation 20.32) to characterize the rate of return to equilibrium. From equation 20.40 we see that for this example

$$\tau = \{k_{-1} + k_1([A]_{eq} + [B]_{eq})\}^{-1} \tag{20.41}$$

Thus, k_1 and k_{-1} may be obtained as slope and intercept of a plot of τ^{-1} versus $[A]_{eq} + [B]_{eq}$.

Example 20.7 ***Rate of ionization of water***

When a sample of pure water in a small conductivity cell is heated suddenly with a pulse of microwave radiation, equilibrium in the water dissociation reaction does not exist at the new higher temperature until additional dissociation occurs. It is found that the relaxation time for the return to equilibrium at 25 °C is 36 μs. Calculate k_1 and k_{-1}.

$$H^+ + OH^- \underset{k_{-1}}{\overset{k_1}{\rightleftharpoons}} H_2O$$

$$\tau = \frac{1}{k_{-1} + k_1([H^+] + [OH^-])}$$

$$K = \frac{[H^+][OH^-]}{[H_2O]c^\circ} = \frac{k_{-1}}{k_1 c^\circ} = \frac{10^{-14}}{55.5} = 1.8 \times 10^{-16}$$

Eliminating k_{-1}, we have

$$\tau = \frac{1}{k_1(Kc^\circ + [H^+] + [OH^-])} = \frac{1}{k_1[c^\circ(1.8 \times 10^{-16}) + (2 \times 10^{-7})]} = 36 \times 10^{-6}\ \mathrm{s}$$

$$k_1 = 1.4 \times 10^{11}\ \mathrm{L\ mol^{-1}\ s^{-1}}$$

$$k_{-1} = Kk_1c^\circ = (1.8 \times 10^{-16}\ \mathrm{mol\ L^{-1}})(1.4 \times 10^{11}\ \mathrm{L\ mol^{-1}\ s^{-1}}) = 2.5 \times 10^{-5}\ \mathrm{s^{-1}}$$

The relaxation time for the formation of a dimer

$$2A \underset{k_{-1}}{\overset{k_1}{\rightleftharpoons}} A_2 \tag{20.42}$$

is given by

$$\tau = \frac{1}{4k_1[A]_{eq} + k_{-1}} \tag{20.43}$$

If the return to equilibrium involves two steps, there will be two independent rate equations. If the reactions are both near equilibrium, these equations may be linearized, and the two linear differential equations will yield two relaxation times. The return to equilibrium will then be given by the sum of two exponential terms. In general, the number of exponential terms is equal to the number of independent reactions.*

*G. G. Hammes, *Principles of Chemical Kinetics,* p. 192. New York: Academic, 1978.

The rate constants for the forward and backward reactions are related to the equilibrium constant by

$$K = \frac{k_f}{k_b} \tag{20.44}$$

Taking the temperature derivative of ln K yields

$$\frac{d \ln K}{dT} = \frac{d \ln k_f}{dT} - \frac{d \ln k_b}{dT} = \frac{\Delta_r H}{RT^2} \tag{20.45}$$

Substituting the Arrhenius equation (equation 18.66) yields

$$\Delta_r H = E_{af} - E_{ab} \tag{20.46}$$

For solution reactions the difference between the internal energy and the enthalpy is usually negligible, so this relationship between the activation energies of the forward and backward reactions and the enthalpy of reaction is useful.

The fact that rates of ionic reactions are altered by changing the ionic strength indicates that perhaps rate equations should be written in terms of activities rather than concentrations. Since a complete rate equation has to yield the expression for the equilibrium constant, rate equations for reactions in which activity coefficients may deviate significantly from unity should be written in terms of activities. However, the general practice is to include the effect of ionic strength in the rate constant, as described later in Section 20.8.

20.7 ACID AND BASE CATALYSIS

Acids and bases catalyze many reactions. Suppose the rate of disappearance of a substance S (often called the substrate in a catalytic reaction) is first order in S: $-d[S]/dt = k[S]$. The first-order rate constant k for the reaction in a buffer solution may be a linear function of $[H^+]$, $[OH^-]$, $[HA]$, and $[A^-]$, where HA is the weak acid in the buffer and A^- is the corresponding conjugate base:

$$k = k_0 + k_{H^+}[H^+] + k_{OH^-}[OH^-] + k_{HA}[HA] + k_{A^-}[A^-] \tag{20.47}$$

In this expression k_0 is the first-order constant for the uncatalyzed reaction. The so-called catalytic coefficients k_{H^+}, k_{OH^-}, k_{HA}, and k_{A^-} may be evaluated from experiments with different concentrations of these species. If only the term $k_{H^+}[H^+]$ is important, the reaction is said to be subject to **specific hydrogen ion catalysis.** If the term $k_{HA}[HA]$ is important, the reaction is said to be subject to **general acid catalysis,** and if the term $k_{A^-}[A^-]$ is important, the reaction is said to be subject to **general base catalysis.**

By considering two types of catalytic mechanisms, we can see how different types of terms arise in equation 20.47. In the first mechanism a proton is transferred from an acid AH^+ to the substrate S, and then the acid form of the substrate reacts with a water molecule to form the product P:

$$S + AH^+ \underset{k_{-1}}{\overset{k_1}{\rightleftharpoons}} SH^+ + A$$

$$SH^+ + H_2O \xrightarrow{k_2} P + H_3O^+$$

$$H_3O^+ + A \underset{k_{-3}}{\overset{k_3}{\rightleftharpoons}} AH^+ + H_2O \tag{20.48}$$

The net reaction is S = P. Assuming that SH^+ is in a steady state, then

$$\frac{d[SH^+]}{dt} = 0 = k_1[S][AH^+] - (k_{-1}[A] + k_2)[SH^+] \qquad (20.49)^*$$

The rate of appearance of product is given by

$$\frac{d[P]}{dt} = k_2[SH^+] = \frac{k_1 k_2 [S][AH^+]}{k_{-1}[A] + k_2} \qquad (20.50)$$

where the second form is obtained by solving equation 20.49 for $[SH^+]$. If $k_2 \gg k_{-1}[A]$, then

$$\frac{d[P]}{dt} = k_1[S][AH^+] \qquad (20.51)$$

and the reaction is said to be general acid catalyzed. However, if $k_2 \ll k_{-1}[A]$, then

$$\frac{d[P]}{dt} = \frac{k_1 k_2 [S][AH^+]}{k_{-1}[A]} = \frac{k_1 k_2}{k_{-1} K}[S][H^+] \qquad (20.52)$$

where the second form is obtained by inserting $K = [A][H^+]/[AH^+]$. In this case the reaction is specifically hydrogen ion catalyzed.

In the second mechanism the acid form of the substrate reacts with a base A instead of a water molecule:

$$S + AH^+ \underset{k_{-1}}{\overset{k_1}{\rightleftharpoons}} SH^+ + A \qquad SH^+ + A \xrightarrow{k_2} P + AH^+ \qquad (20.53)$$

The steady-state treatment of this mechanism leads to

$$\frac{d[P]}{dt} = k_2[SH^+][A] = \frac{k_1 k_2 [S][AH^+]}{k_{-1} + k_2} \qquad (20.54)$$

which is an example of general acid catalysis.

For mechanisms of the type of 20.48 and 20.53 we might expect a relationship between the rate constant k_a for the acid-catalyzed reaction or k_b for the base-catalyzed reaction to depend on the strength of the acid or base. Indeed, Brønsted found that the rate constant k_a for acid catalysis or k_b for base catalysis is proportional to the ionization constant K_a for the acid or K_b for the base raised to some power:

$$k_a = C_A K_a^{\alpha} \qquad (20.55)$$

$$k_b = C_B K_b^{\beta} \qquad (20.56)$$

The exponents α and β are positive and have values between 0 and 1. The constants C_A, C_B, α, and β apply to a single reaction at a particular temperature catalyzed by different acids and bases. In the Brønsted equation (equations 20.55 and 20.56), low values of α and β indicate a low sensitivity of the catalytic constant to the strength of the catalyzing acid or base.

*Note that $[H_2O]$ is not written after k_2 because in dilute aqueous solutions $[H_2O]$ cannot be appreciably changed, so k_2 represents a first-order rate constant.

20.8 PRIMARY KINETIC SALT EFFECT

Brønsted and Bjerrum investigated the effect of electrolyte concentration on the rate constants of a number of reactions involving ions in aqueous solutions and obtained an unexpected result. We can show that the result they found can be derived from transition-state theory and the Debye–Hückel theory. The rate constant k for reactions of the type

$$A^{z_A} + B^{z_B} \rightarrow \text{products} \tag{20.57}$$

is defined by

$$-\frac{d[A]}{dt} = k[A][B] \tag{20.58}$$

Brønsted and Bjerrum found that the rate constant varied with ionic strength I according to

$$k = k^\circ 10^{2Az_Az_BI^{1/2}} \tag{20.59}$$

in the region of low ionic strength where the Debye–Hückel theory is obeyed. In equation 20.59, A is the Debye–Hückel constant (0.509 $kg^{1/2}$ $mol^{-1/2}$ at 25 °C), k° is the rate constant at zero ionic strength, and z_A and z_B are the charges on ions A and B, with signs. The magnitude of the effect is indicated by Fig. 20.6.

According to **transition-state theory** (Section 19.4), the rate of a reaction is given by

$$v = k^\ddagger[C^\ddagger] \tag{20.60}$$

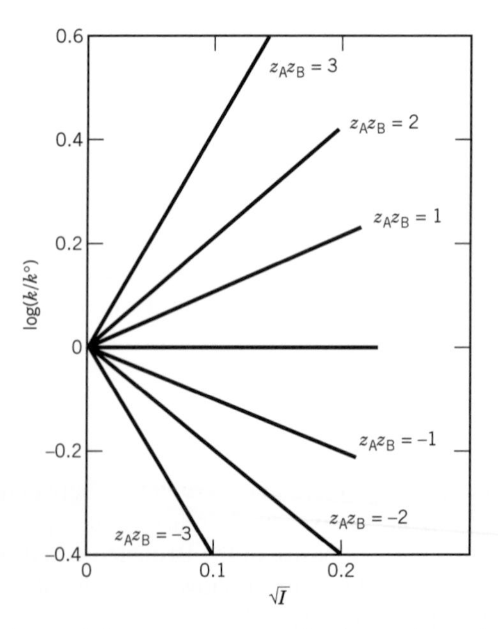

Figure 20.6 Dependence of the rate constants of ionic reactions of various charge types on ionic strength at 25 °C in aqueous solutions. The reactants studied can all be represented by $A^{z_A} + B^{z_B} \rightarrow$ products.

where $C^\ddagger$ is the activated complex. We assume that $C^\ddagger$ is in equilibrium with the reactants, as, for example, in $A^{z_A} + B^{z_B} = C^\ddagger$ with equilibrium constant expression

$$K_c = \frac{[C^\ddagger]\gamma_C}{[A][B]\gamma_A\gamma_B} \tag{20.61}$$

Eliminating $[C^\ddagger]$ between equations 20.60 and 20.61 yields

$$v = \frac{k^\ddagger K_c \gamma_A \gamma_B [A][B]}{\gamma_C} = k[A][B] \tag{20.62}$$

where the experimental rate constant k is given by

$$k = \frac{k^\ddagger K_c \gamma_A \gamma_B}{\gamma_C} = \frac{k^\circ \gamma_A \gamma_B}{\gamma_C} \tag{20.63}$$

It is of interest to use the symbol k° for the experimental rate constant at zero ionic strength; it is given by $k^\circ = k^\ddagger K_c$.

At low ionic strengths, the activity coefficients of the ions are given by

$$\log \gamma_i = -Az_i^2 I^{1/2} \tag{20.64}$$

We do not know much about the structure of the activated complex $C^\ddagger$, but we do know that its charge is $z_A + z_B$. When equation 20.63 is written in logarithmic form, and equation 20.64 is used, we obtain

$$\begin{aligned} \log k &= \log k^\circ + \log \gamma_A + \log \gamma_B - \log \gamma_C \\ &= \log k^\circ - AI^{1/2}[z_A^2 + z_B^2 - (z_A + z_B)^2] \\ &= \log k^\circ + 2z_A z_B A I^{1/2} \end{aligned} \tag{20.65}$$

If the reacting ions are oppositely charged, raising the ionic strength reduces the effective rate constant because the ions are shielded from each other to a greater extent.

Example 20.8 ***Dependence of rate constant on ionic strength***

The rate constant of a reaction

$$A^{1+} + B^{2-} \rightarrow \text{products}$$

is measured at 0.001 ionic strength and at 0.01 ionic strength at 25 °C in water. What is the expected ratio of the rate constants?

To derive a general relation, consider

$$A^{z_A} + B^{z_B} \rightarrow \text{products}$$

$$\log k_{0.001} = \log k^\circ + 2z_A z_B (0.509)(0.001)^{1/2}$$

$$\log k_{0.01} = \log k^\circ + 2z_A z_B (0.509)(0.01)^{1/2}$$

Therefore,

$$\begin{aligned} \log \frac{k_{0.001}}{k_{0.01}} &= 2z_A z_B (0.509)(0.001^{1/2} - 0.01^{1/2}) \\ &= 2(1)(-2)(0.509)(0.001^{1/2} - 0.01^{1/2}) \end{aligned}$$

$$= 0.183$$

$$k_{0.001} = 1.52 k_{0.01}$$

20.9 RATES OF ELECTRON TRANSFER REACTIONS

The transfer of an electron from donor D to an acceptor A in liquid solution can be represented by

$$D + A \rightleftharpoons (DA) \rightleftharpoons (D^+A^-) \rightleftharpoons D^+ + A^- \tag{20.66}$$

where (DA) and (D^+A^-) are the contact pair before and after the electron is transferred. The following discussion of a theory by Marcus* applies when (DA) $\rightarrow$ (D^+A^-) is the slow step in the reaction. When the electron is on the donor molecule its energy is represented by the parabolic plot on the left in Fig. 20.7, and when the electron is on the acceptor molecule its energy is represented by the parabolic curve on the right. These are plots for solvated molecules because the solvent molecules have to be taken into account since their orientations change with the changes in electric charge. The abscissa for this plot is the reaction coordinate (see Section 19.4). The change in Gibbs energy for the reaction $\Delta G°$ is negative, so the reaction will occur under these circumstances; but we

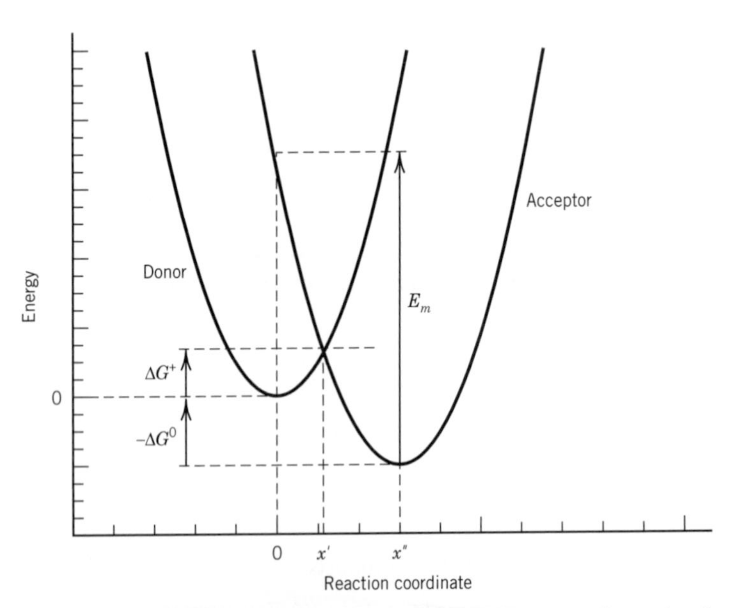

Figure 20.7 Energy dependence as a function of the reaction coordinate for the electron on donor or acceptor. (From P. L. Houston, *Chemical Kinetics and Reaction Dynamics*, McGraw-Hill, New York, 2001, with permission from McGraw-Hill.)

*R. Marcus, 1992 Nobel Prize in chemistry.

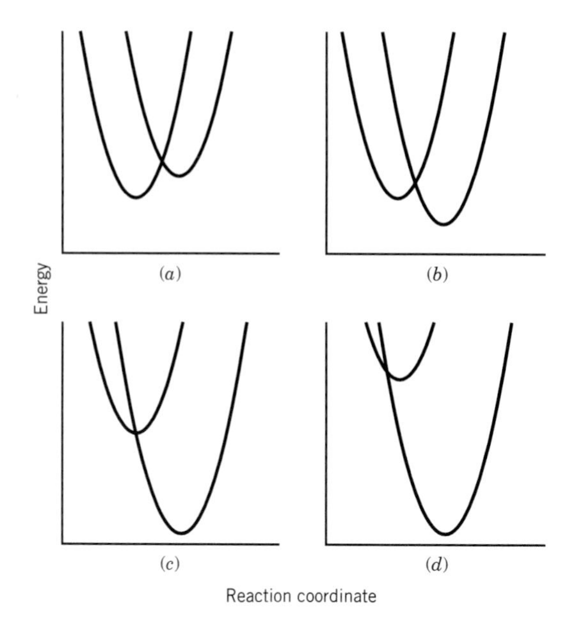

Figure 20.8 Donor and acceptor parabolas with the same reorganization energy E_m, but increasing ΔG° going from A to D. Note that the activation energy is large in A, smaller in B, zero in C, and larger in D. (From P. L. Houston, *Chemical Kinetics and Reaction Dynamics*, McGraw-Hill, New York, 2001, with permission from McGraw-Hill.)

are interested in the rate of electron transfer k_{ET}, which is given by

$$k_{\mathrm{ET}} = (kT/h)\exp\left(-\frac{(E_m + \Delta G^\circ)^2}{4E_m kT}\right) \tag{20.67}$$

The energy E_m, called reorganization energy, is the energy needed to reorganize the configuration of the acceptor and its surrounding solvent molecules into that of the donor and its surroundings, when back transfer of the electron is not allowed.*

Figure 20.8 shows the intersecting donor and acceptor parabolas for increasingly negative values of ΔG° but the same value of E_m. Equation 20.67 shows that k_{ET} increases with ΔG°, so we might expect that the rate of electron transfer would increase in going from A to D in the figure. The activation energy decreases in going from A to C, which reinforces the change in ΔG°, but the activation energy increases in going from C to D. This rather counterintuitive prediction of Marcus's equation has been confirmed by several researchers.

20.10 ENZYME CATALYSIS

The most amazing catalysts are the enzymes, which catalyze the multitudinous reactions in living organisms and also provide the means for controlling reaction rates so that the rates of various steps in a series of reactions are compatible.

*The derivation of this equation is described in P. L. Houston, *Chemical Kinetics and Reaction Dynamics*, New York: McGraw-Hill, 2001, and other, more advanced books on kinetics.

Enzymes are proteins, copolymers of amino acids with specific amino acid sequences and definite three-dimensional structures. Proteins provide various functional groups at the catalytic site that can interact with a substrate molecule and thereby catalyze a reaction. Some enzymes catalyze a single reaction. An example is fumarase, which catalyzes the hydration of fumarate to L-malate:

$$\text{H}(\text{CO}_2^-)\text{C}{=}\text{C}(\text{H})(^-\text{O}_2\text{C}) + \text{H}_2\text{O} = \text{H}(\text{H})(\text{CO}_2^-)\text{C}{-}\text{C}(\text{OH})(\text{H})(^-\text{O}_2\text{C}) \tag{20.68}$$

The reactants in an enzyme reaction are generally referred to as substrates. This reaction may be represented by S = P, since the concentration of water is constant. Other enzymes catalyze a class of reactions of a given type such as ester hydrolysis. Some enzymes require particular metal ions or coenzymes to operate.

Since enzymes are very effective catalysts, they are usually used in laboratory experiments at concentrations much lower than the concentration of the substrate. Generally, the reaction rate that is measured in the laboratory is a steady-state rate. In studies of enzyme kinetics it is advantageous to use initial steady-state rates because the product may be inhibitory.

The initial rate of an enzyme-catalyzed reaction of the type S = P is generally found to be directly proportional to the enzyme concentration. When the substrate concentration is varied, the initial rate is first order with respect to substrate at low [S] and zero order at high [S]. In 1913 **Michaelis and Menten** pointed out that these observations can be explained with the mechanism

$$\text{E} + \text{S} \underset{k_2}{\overset{k_1}{\rightleftharpoons}} \text{X} \xrightarrow{k_3} \text{E} + \text{P} \tag{20.69}$$

where X is an E–S complex. Since $[\text{E}] + [\text{X}] = [\text{E}]_0$ and $[\text{S}] + [\text{P}] = [\text{S}]_0$, there are two independent rate equations for this mechanism. These can be taken to be

$$\frac{\text{d}[\text{X}]}{\text{d}t} = k_1[\text{E}][\text{S}] - (k_2 + k_3)[\text{X}] \tag{20.70}$$

$$\frac{\text{d}[\text{P}]}{\text{d}t} = k_3[\text{X}] \tag{20.71}$$

These two rate equations cannot be solved to obtain analytic expressions for [E], [S], [X], and [P] as functions of time, but these concentrations may be calculated by use of a computer for specific values of the three rate constants.

Since enzymatic reactions are generally studied with enzyme concentrations (strictly speaking, molar concentrations of enzymatic sites) much lower than the concentrations of substrates, it is a good approximation to assume that the enzymatic reaction is in a steady state in which $\text{d}[\text{X}]/\text{d}t = 0$. By introducing the equation for the conservation of enzyme, $[\text{E}] = [\text{E}]_0 - [\text{X}]$, in equation 20.70, we obtain

$$[\text{X}] = \frac{k_1[\text{E}]_0[\text{S}]}{k_1[\text{S}] + k_2 + k_3} \tag{20.72}$$

Substituting this expression in equation 20.71 yields

$$\frac{d[P]}{dt} = \frac{k_3[E]_0}{1 + (k_2 + k_3)/k_1[S]} \tag{20.73}$$

This steady-state rate equation for the overall reaction is frequently written

$$v = \frac{k_{cat}[E]_0}{1 + K_M/[S]} \tag{20.74}$$

where k_{cat} is the **turnover number,** in this case k_3, and K_M is the **Michaelis constant,** in this case $(k_2 + k_3)/k_1$.

The initial steady-state velocity v is plotted as a function of the substrate concentration in Fig. 20.9*a*. At low substrate concentrations the reaction is first order with respect to substrate; however, as the substrate concentration is increased, the velocity asymptotically approaches a maximum of $k_{cat}[E]_0$. The quantity k_{cat} is called the turnover number because it is the number of product molecules produced per enzyme molecule (strictly, per catalytic site) per second. The turnover number is about $10^6\ s^{-1}$ for catalase, which catalyzes the decomposition of H_2O_2 to $H_2O + \frac{1}{2}O_2$, and about 100 s^{-1} for chymotrypsin, which catalyzes the hydrolysis of a number of esters and amides. It is evident from equation 20.74 that K_M is equal to the concentration of substrate required to give one-half the maximum velocity.

To obtain the best values of k_{cat} and K_M, it is useful to be able to plot the kinetic data as a straight line. There are three ways that this can be done, but since it is desirable to show the full accessible range of experiments along both axes, the Eadie–Hofstee plot is best. This method uses the Michaelis–Menten equation (20.74) written in the following form:

$$\frac{v}{[E]_0[S]} = \frac{k_{cat}}{K_M} - \frac{v}{K_M[E]_0} \tag{20.75}$$

As shown in Fig. 20.9*b*, the Michaelis constant can be obtained from the slope, and the turnover number can be obtained directly from the intercept on the abscissa.

The reversibility of the overall reaction can be provided for by including the reverse reaction for the second step:

$$E + S \underset{k_2}{\overset{k_1}{\rightleftharpoons}} X \underset{k_4}{\overset{k_3}{\rightleftharpoons}} E + P \tag{20.76}$$

The rate equations for this mechanism are

$$\frac{d[X]}{dt} = k_1[E][S] - (k_2 + k_3)[X] + k_4[E][P] \tag{20.77}$$

$$\frac{d[P]}{dt} = k_3[X] - k_4[E][P] \tag{20.78}$$

Substituting $[E]_0 = [E] + [X]$ in equation 20.77 and assuming that $d[X]/dt = 0$ make it possible to solve equation 20.77 for [X]. When this expression and the expression for the conservation of enzymatic sites are substituted in equation 20.78, it is found that the steady-state rate is given by

$$\frac{d[P]}{dt} = v = \frac{k_1k_3[S][E]_0 - k_2k_4[P][E]_0}{k_2 + k_3 + k_1[S] + k_4[P]} \tag{20.79}$$

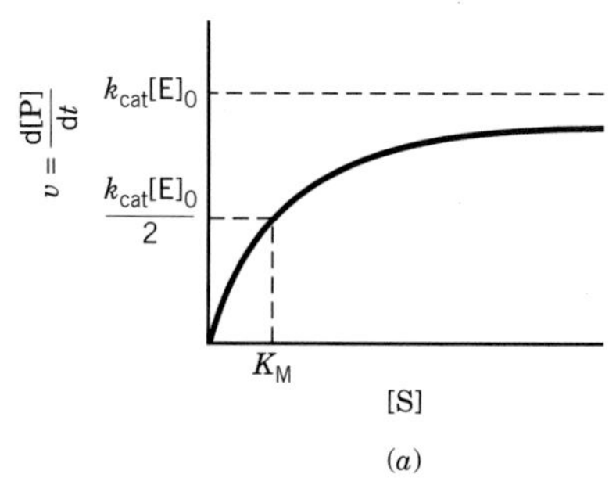

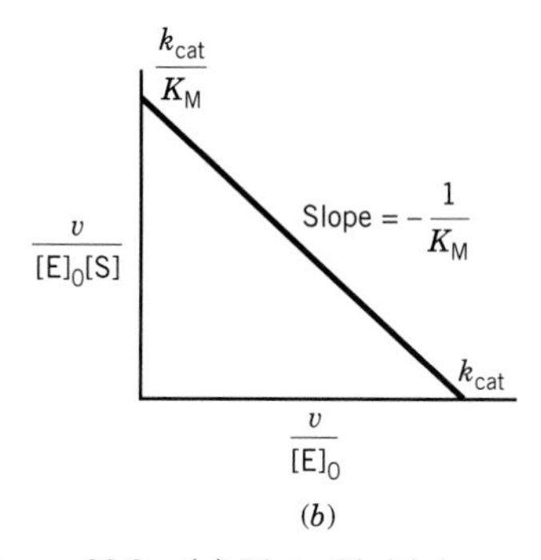

Figure 20.9 (*a*) Plot of initial steady-state velocity v of appearance of product P at different initial concentrations of substrate S. (*b*) Eadie–Hofstee plot of enzyme kinetic data to obtain the parameters K_M and k_{cat}.

It is convenient to rearrange this rate equation by introducing expressions for the Michaelis constants for the substrate and product:

$$K_S = \frac{k_2 + k_3}{k_1} \qquad K_P = \frac{k_2 + k_3}{k_4}$$

to obtain

$$\begin{aligned} v &= \frac{d[P]}{dt} = \frac{(k_3/K_S)[S] - (k_2/K_P)[P]}{1 + [S]/K_S + [P]/K_P}[E]_0 \\ &= \frac{k_3[E]_0(1 - [P]/[S]K)}{1 + K_S/[S] + K_S[P]/K_P[S]} \end{aligned} \tag{20.80}$$

Thus, the equilibrium constant for the overall reaction is given by

$$K = \frac{[P]_{eq}}{[S]_{eq}} = \frac{k_3 K_P}{k_2 K_S} = \frac{k_1 k_3}{k_2 k_4} \tag{20.81}$$

Equation 20.79 gives the steady-state velocity for any mixture of substrate and product. It reduces to equation 20.73 if product is not added initially and the reaction goes essentially to completion. The addition of product has two effects: The [P] term in the numerator of equation 20.79 results from the reverse reaction, and the [P] term in the denominator is due to product inhibition. It is evident from equation 20.81 that the four rate constants in the mechanism are not independent; they are related through the equilibrium constant.

Example 20.9 ***Using the Michaelis constant to calculate rates***

At 25 °C and pH 8 the turnover numbers and Michaelis constants for the fumarase reaction fumarate + H_2O = L-malate are

$$k_3 = 0.20 \times 10^3\ \text{s}^{-1} \qquad k_2 = 0.60 \times 10^3\ \text{s}^{-1}$$

$$K_S = 7.0 \times 10^{-6}\ \text{mol L}^{-1} \qquad K_P = 100 \times 10^{-6}\ \text{mol L}^{-1}$$

What are the values of k_1 and k_4 and the equilibrium constant $[P]_{eq}/[S]_{eq}$?

$$k_1 = \frac{k_2 + k_3}{K_S} = \frac{0.80 \times 10^3\ \text{s}^{-1}}{7.0 \times 10^{-6}\ \text{mol L}^{-1}} = 1.14 \times 10^8\ \text{L mol}^{-1}\ \text{s}^{-1}$$

$$k_4 = \frac{k_2 + k_3}{K_P} = \frac{0.80 \times 10^3\ \text{s}^{-1}}{100 \times 10^{-6}\ \text{mol L}^{-1}} = 8.0 \times 10^6\ \text{L mol}^{-1}\ \text{s}^{-1}$$

$$\begin{aligned} K_{eq} &= \frac{[P]_{eq}}{[S]_{eq}} = \frac{k_1 k_3}{k_2 k_4} \\ &= 4.8 \end{aligned}$$

Compounds that are structurally related to the substrate or product often combine with the catalytic site of the enzyme and cause **inhibition;** that is, the enzyme-catalyzed reaction is slowed down by the inhibitor. Since the substrate and the inhibitor compete for the same site, the effect of the inhibitor may be reduced by raising the substrate concentration. This type of inhibition, called competitive inhibition, may be represented by the mechanism

$$E + S \rightleftharpoons ES \longrightarrow E + P \qquad E + I \rightleftharpoons EI \qquad K_I = \frac{[E][I]}{[EI]} \tag{20.82}$$

where I is the inhibitor. The steady-state rate law is

$$v = \frac{V_S}{1 + (K_S/[S])(1 + [I]/K_I)} \tag{20.83}$$

where K_I is the dissociation constant of EI into E and I and $V_S = k_{cat}[E]_0$.

Substances that bind to the enzyme, although not at the active site, may not interfere with the binding of the substrate at the active site but may alter K_S and V_S. Such inhibitors are referred to as noncompetitive. The binding of substances that are not directly involved in an enzymatic reaction can also increase the reaction rate, and such substances are called **activators.**

The way in which V_S and K_S depend on pH, salt concentration, coenzyme concentration, and so on, gives further information about the enzymatic mechanism. In general, an enzymatic reaction has an optimum pH; the maximum velocity V_S decreases as the pH is raised or lowered from the optimum pH. In the neutral pH range the effects are generally reversible, but proteins are irreversibly denatured at extreme pH values. Reversible effects of pH on V_S may be attributable to the ionization of the enzyme–substrate complex. If the enzyme–substrate complex exists in three states with different numbers of protons, and if only the intermediate form breaks down to give product, the expression for the effect of pH on the maximum velocity may be derived from

$$\begin{array}{l} \text{ES} \\ K_{bES} \Big\Updownarrow \\ \text{HES} \xrightarrow[k_{-2}]{k} \text{enzyme} + \text{product} \\ K_{aES} \Big\Updownarrow \\ \text{H}_2\text{ES} \end{array} \tag{20.84}$$

where K_{aES} and K_{bES} are acid dissociation constants, and k is the rate constant for the rate-determining step. Since

$$\begin{aligned} [E]_0 &= [ES] + [HES] + [H_2ES] \\ &= [HES]\left(1 + \frac{[H^+]}{K_{aES}} + \frac{K_{bES}}{[H^+]}\right) \end{aligned} \tag{20.85}$$

then

$$\begin{aligned} V_S = k[HES] &= \frac{k[E]_0}{1 + [H^+]/K_{aES} + K_{bES}/[H^+]} \\ &= k'[E]_0 \end{aligned} \tag{20.86}$$

where k' is the **apparent rate constant** given by

$$k' = \frac{k}{1 + [H^+]/K_{aES} + K_{bES}/[H^+]} \tag{20.87}$$

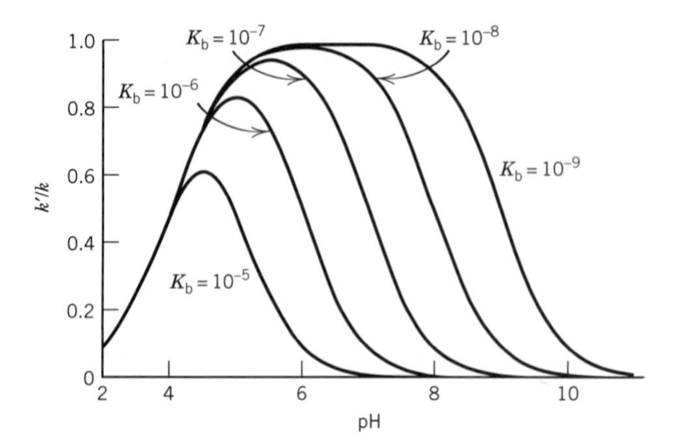

Figure 20.10 Plot of k'/k versus pH according to equation 20.87 for an enzyme with $K_a = 10^{-4}$ and $K_b = 10^{-5}, 10^{-6}, 10^{-7}, 10^{-8}$, and 10^{-9}. (See Computer Problem 20.D.)

Thus in discussing the kinetics of reactions involving reactants that are weak acids, there may be advantages in using apparent rate constants that are functions of $[H^+]$. If mechanism 20.84 is correct, then a plot of V_S versus pH is a symmetrical bell-shaped curve, as illustrated in Fig. 20.10.

Since a protein has many dissociable acid groups, it is perhaps surprising that the experimental results are sometimes represented by an equation like 20.84. There is a simple explanation of why two acid groups in the catalytic site might have the total effect on the kinetics that is represented by equation 20.86. If the catalytic function involves both an acidic function and a basic function, we would expect H_2ES to be inactive because the basic site is occupied by a proton and ES to be inactive because it cannot donate a proton to the substrate. Only HES can yield product because it has one group that can donate a proton and another that can accept a proton.

Example 20.10 ***Apparent rate constants***

In discussing the thermodynamics of biochemical reactions (Section 8.3) we found it convenient to use apparent equilibrium constants K' that are a function of pH. These apparent equilibrium constants are equal to the ratio of the apparent rate constants for the forward and backward reactions. As a simple example, consider the interconversion

$$A \underset{k_2'}{\overset{k_1'}{\rightleftharpoons}} B \tag{1}$$

where the symbols A and B represent sums of concentrations of species that are in rapid equilibrium by acid–base reactions. The apparent rate constants defined by the rate equation for reaction 1

$$\frac{d[A]}{dt} = -k_1'[A] + k_2'[B] \tag{2}$$

are functions of pH. It is evident that the apparent equilibrium constant is given by

$$K' = \frac{[B]}{[A]} = \frac{k_1'}{k_2'} \tag{3}$$

If HA and HB are monoprotic weak acids, the mechanism might be

$$\begin{array}{ccc} A^- & \underset{k_2}{\overset{k_1}{\rightleftharpoons}} & B^- \\ \Big\Updownarrow K_{HA} & & \Big\Updownarrow K_{HB} \\ HA & & HB \end{array} \tag{4}$$

where k_1 and k_2 are the rate constants for the interconversion of specific species. If the acid dissociations are equilibrated much more rapidly than the interconversion of the two anions, derive the expressions for the dependence of k_1' and k_2' on $[H^+]$ and show that they are consistent with the dependence of K' on $[H^+]$.

Since the acid and base forms of the reactants are in rapid equilibrium, we cannot write the rate equation for $d[A^-]/dt$, but we can write the rate equation for mechanism 4 as

$$\frac{d[A]}{dt} = -k_1[A^-] + k_2[B^-] \tag{5}$$

since there is only one path for the reaction. Since the species of A are in equilibrium,

$$[A^-] = \frac{[A]}{1 + [H^+]/K_{HA}} \tag{6}$$

and since the species of B are in equilibrium,

$$[B^-] = \frac{[B]}{1 + [H^+]/K_{HB}} \tag{7}$$

When equations 6 and 7 are substituted into equation 5, we obtain

$$\frac{d[A]}{dt} = -\frac{k_1[A]}{1 + [H^+]/K_{HA}} + \frac{k_2[B]}{1 + [H^+]/K_{HB}} \tag{8}$$

When we compare equation 8 with equation 2, it is clear that

$$k_1' = \frac{k_1}{1 + [H^+]/K_{HA}} \tag{9}$$

and

$$k_2' = \frac{k_2}{1 + [H^+]/K_{HB}} \tag{10}$$

When equations 9 and 10 are substituted into equation 3, we find that the apparent equilibrium constant is given by

$$K' = \frac{k_1(1 + [H^+]/K_{HB})}{k_2(1 + [H^+]/K_{HA})} = K\frac{(1 + [H^+]/K_{HB})}{(1 + [H^+]/K_{HA})} \tag{11}$$

where $K = k_1/k_2$. This derivation gives the same dependence of the apparent equilibrium constant on pH that we would have derived with thermodynamics in Chapter 8.

Mechanisms like the ones we have been discussing can give a variety of complicated effects; that is, the rate law may have a very complicated form. In general, however, the steady-state rate is somewhere between zero order and first order in substrate concentration. However, there are some enzymes for which the steady-state rate varies with a higher power of the substrate concentration. In other words, curves analogous to the sigmoid oxygen-binding curve for hemoglobin

(Section 8.6) are obtained. This has been found to be especially true of enzymes of importance in the regulation of metabolic pathways. These cooperative effects are encountered with multisite enzymes, not single-site enzymes, because the cooperative effect involves an increased affinity of a second site for a substrate when a first site is occupied. As in the case of hemoglobin, this interaction involves a structural change. According to the Monod–Wyman–Changeux model,* the multisite enzyme can exist in at least two states. In each of the two states the conformations of all the subunits are assumed to be the same. The binding of substrate shifts the equilibrium toward one or the other of these two states. If the effector drives the equilibrium in the direction that produces an enhanced rate of reaction, the effector is called an activator. If it causes a reduction in rate, it is called an inhibitor. As we have seen in the case of hemoglobin, the effect is multiplied by the fact that one effector molecule affects several catalytic sites on the molecule. The fact that enzymatic activities may be affected by various substances present in the cell provides a mechanism for the control of the rates of reactions in living things so that metabolic intermediates do not accumulate.

Comment:

More than 3500 enzymes are known, and they catalyze a very wide variety of reactions. In addition, there are RNA (ribonucleic acid) enzymes, catalytic antibodies, and synthetic enzymes (sometimes called ribozymes, abzymes, and synzymes, respectively) that follow many of the same principles. Some enzymes are very efficient catalysts and operate at diffusion-controlled rates. In contrast with the usual laboratory catalysts, they have the additional important feature that their catalytic activity is affected by other small molecules in the cell, which are also bound by the protein. In addition, enzymes may aggregate with other enzymes so that intermediates are not lost by diffusion into the medium before the next catalytic step.

20.11 OSCILLATING CHEMICAL REACTIONS

So far we have always described the approach to equilibrium in terms of a monotonic decrease in reactants and increase in products, but this is not necessarily the case. Some reactions involve autocatalysis, that is, catalysis by a product of the reaction. The simplest reaction system that exhibits oscillatory behavior was described by Lotka in 1925.

$$\mathrm{A} \xrightarrow{k_1} \mathrm{X} \tag{20.88}$$

$$\mathrm{X} + \mathrm{Y} \xrightarrow{k_2} 2\mathrm{Y} \tag{20.89}$$

$$\mathrm{Y} \xrightarrow{k_3} \mathrm{P} \tag{20.90}$$

Reactant A is continuously supplied to the system so that its concentration is always [A(0)]. The second reaction is autocatalytic because twice as much Y is produced as is consumed. The rate equations for the system have the following form:

$$\frac{\mathrm{d}[\mathrm{X}]}{\mathrm{d}t} = k_1[\mathrm{A}(0)] - k_2[\mathrm{X}][\mathrm{Y}] \tag{20.91}$$

*J. Wyman and S. J. Gill, *Binding and Linkage.* Mill Valley, CA: University Science Books, 1990.

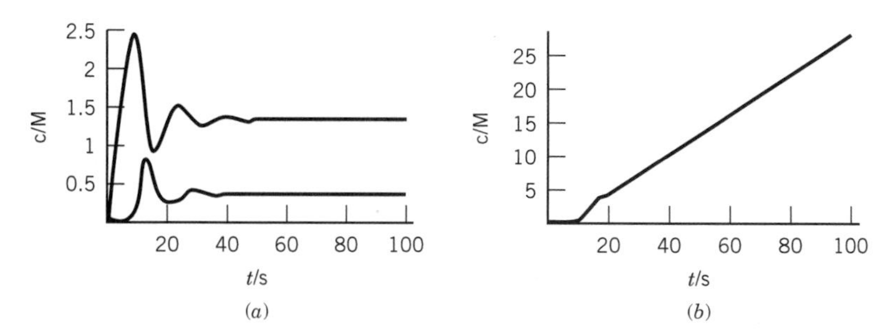

Figure 20.11 Concentration as a function of time for the three reactants in the reaction system defined by reactions 20.88, 20.89, and 20.90 for a constant concentration of A of 1 M. The initial concentrations are [X] = 0.2 M, [Y] = 0.1 M, and [P] = 0 M. The rate constants are $k_1 = 0.3\ \mathrm{s}^{-1}$, $k_2 = 0.6\ \mathrm{M}^{-1}\ \mathrm{s}^{-1}$, and $k_3 = 0.8\ \mathrm{s}^{-1}$. The concentration of X is given by the top curve in (*a*), and the concentration of Y is given by the lower curve in (*a*). The concentration of P is given in (*b*). (See Computer Problem 20.G.)

$$\frac{\mathrm{d}[\mathrm{Y}]}{\mathrm{d}t} = k_2[\mathrm{X}][\mathrm{Y}] - k_3[\mathrm{Y}] \tag{20.92}$$

$$\frac{\mathrm{d}[\mathrm{P}]}{\mathrm{d}t} = k_3[\mathrm{Y}] \tag{20.93}$$

For a particular set of rate constants and initial concentrations of X and Y, the concentrations of the three reactants change with time as shown in Fig. 20.11.

The best-known chemical oscillator is the Belousov-Zhabotinskii reaction:

$$2\mathrm{H}^+ + 2\mathrm{BrO_3}^- + 3\mathrm{CH_2(CO_2H)_2} = 2\mathrm{BrCH(CO_2H)_2} + 3\mathrm{CO_2} + 4\mathrm{H_2O} \tag{20.94}$$

When this reaction is catalyzed with Ce^{3+}, the concentration of the intermediate Br^- and the ratio $[Ce^{4+}]/[Ce^{3+}]$ oscillate as shown in Fig. 20.12. The catalyzed reaction involves about 18 steps and 21 different chemical species.

The underlying theory of oscillating reactions is becoming better understood and is of interest in connection with oscillations observed in biological systems. Chemical systems that produce oscillating reactions can also form spatial structures in initially homogeneous systems.

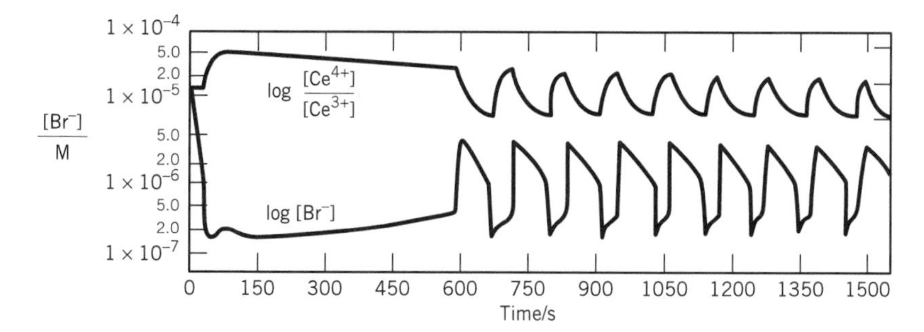

Figure 20.12 Oscillation in $[Br^-]$ and $[Ce^{4+}]$ in reaction 20.94 catalyzed by cerium ion. Note that Br^-, Ce^{4+}, and Ce^{3+} are not reactants or products in net reaction. [Reprinted with permission from R. J. Field, E. Körös, and R. M. Noyes, *J. Am. Chem. Soc.* **94**:8649 (1972). Copyright ©1972 American Chemical Society.]

Eleven Key Ideas in Chapter 20

1. The frictional coefficient of a particle in a liquid is of interest because it provides information about the size and shape of the particle. The force required to move a particle in a liquid is given by the product of the velocity of the particle and its frictional coefficient.
2. Diffusion in a liquid is a consequence of the gradient of the chemical potential. The flux is proportional to the concentration gradient, and the proportionality constant is the diffusion coefficient.
3. When a single solute diffuses from an initially sharp boundary between solution and solvent, the concentration profile at a later time can be calculated using the error function. The concentration gradient at a later time is given by a Gaussian curve.
4. The mobility of an ion in an electric field is equal to its velocity divided by the potential gradient. The electric conductivity of an electrolyte solution is the sum of the contributions of all of the ionic species in the electrolyte.
5. The patterns of collisions between reactants in a liquid and a gas are very different because reactant molecules are surrounded by a solvent cage. In a diffusion-controlled reaction, reactant molecules within the same solvent cage collide enough times that reaction is highly likely before they can diffuse away from each other.
6. Very fast reactions can be studied by perturbing a solution that is at equilibrium with a very rapid change in temperature or pressure, and then following the relaxation back to equilibrium spectroscopically. The relaxation is first order, but rate constants for bimolecular reactions can be determined.
7. The rate constant for the reaction of H^+ with OH^- is faster than would be expected for a diffusion-controlled reaction because the proton undergoes quantum mechanical tunneling to the hydroxyl ion once it is close.
8. The rate constant for a reaction between ions at low ionic strength depends on the ionic strength in the way expected from the Debye–Hückel theory. We know the charge of the activated complex, even though we do not know its structure.
9. Rates of electron transfer reactions may increase and then decrease with the change in the Gibbs energy of the reaction.
10. Enzyme-catalyzed reactions involve intermediates (enzyme–substrate complexes), so that they approach maximum velocities as substrate concentrations are increased. Since all reactions are reversible, the kinetic constants for the forward and backward reactions are related to the apparent equilibrium constant for the reaction that is catalyzed.
11. For certain complicated reactions, the concentrations of intermediates may oscillate rather than simply going through maxima and then leveling off to the equilibrium value.

REFERENCES

A. Cornish-Bowden, *Principles of Enzyme Kinetics.* London: Portland Press, 1995.

J. H. Espenson, *Chemical Kinetics and Reaction Mechanisms* New York: McGraw-Hill, 1995.

A. Fersht, *Structure and Mechanism in Protein Science: A Guide to Enzyme Catalysis and Protein Folding.* New York: W. H. Freeman, 1999.
G. G. Hammes, *Thermodynamics and Kinetics for the Biological Sciences*. Hoboken, NJ: Wiley, 2000.
P. L. Houston, *Chemical Kinetics and Reaction Dynamics*. New York: McGraw-Hill, 2001.
E. Zeffren and P. L. Hall, *The Study of Enzyme Mechanisms*. Hoboken, NJ: Wiley-Interscience, 1973.

PROBLEMS

20.1 A steel ball ($\rho = 7.86$ g cm^{-3}) 0.2 cm in diameter falls 10 cm through a viscous liquid ($\rho_0 = 1.50$ g cm^{-3}) in 25 s. What is the viscosity at this temperature?

20.2 Estimate the rate of sedimentation of water droplets of 1 μm diameter in air at 20 °C. The viscosity of air at this temperature is 1.808×10^{-5} Pa s.

20.3 The viscosity of mercury is 1.661×10^{-3} Pa s at 0 °C and 1.476×10^{-3} Pa s at 35 °C. What is the activation energy, and what viscosity is expected at 50 °C?

20.4 The viscosity of a liquid can be determined by measuring the falling velocity of a sphere of known density in the liquid. The force of gravity on the sphere is given by the apparent mass of the sphere (its mass minus the mass of liquid displaced) times the acceleration of gravity, $g = 9.807$ m s^{-2}. The retarding force is given by the frictional coefficient of the sphere times the velocity of fall. Derive the equation for the velocity of fall.

20.5 How fast does a bubble of air rise in water at 25 °C if its diameter is 1 mm?

20.6 A sharp boundary is formed between a dilute aqueous solution of sucrose and water at 25 °C. After 5 h the standard deviation of the concentration gradient is 0.434 cm. (*a*) What is the diffusion coefficient for sucrose under these conditions? (*b*) What will be the standard deviation after 10 h?

20.7 (*a*) Calculate the time required for the half-width of a freely diffusing boundary of dilute potassium chloride in water to become 0.5 cm at 25 °C ($D = 1.99 \times 10^{-9}$ m^2 s^{-1}). (*b*) Calculate the corresponding time for serum albumin ($D = 6.15 \times 10^{-11}$ m^2 s^{-1}).

20.8 The standard deviation σ of a freely diffusing boundary between dilute salt solution and water at 25 °C is 3.8 mm after 1 h. What is the diffusion coefficient of the salt in water? What will the standard deviation be after 2 h?

20.9 Using a table of the probability integral, calculate enough points on a plot of c versus x (like Fig. 20.2) to draw in the smooth curve for diffusion of 0.1 mol L^{-1} sucrose into water at 25 °C after 4 h and 29.83 min ($D = 5.23 \times 10^{-10}$ m^2 s^{-1}).

20.10 Calculate the conductivity of 0.001 mol L^{-1} HCl at 25 °C. The limiting ion mobilities may be used for this problem.

20.11 One hundred grams of sodium chloride are dissolved in 10 000 L of water at 25 °C, giving a solution that may be regarded in these calculations as infinitely dilute. (*a*) What is the conductivity of the solution? (*b*) This dilute solution is placed in a glass tube of 4 cm diameter provided with electrodes filling the tube and placed 20 cm apart. How much current will flow if the potential drop between the electrodes is 80 V?

20.12 It is desired to use a conductance apparatus to measure the concentration of dilute solutions of sodium chloride. If the electrodes in the cell are each 1 cm^2 in area and are 0.2 cm apart, calculate the resistance that will be obtained for 1, 10, and 100 ppm NaCl at 25 °C.

20.13 Derive the expression for log k for the reaction $A^{z_A} + B^{z_B} + C^{z_C} \longrightarrow$ products as a function of ionic strength.

20.14 Derive the expression for the relaxation times τ for the following two reactions:

(*a*) $A \underset{k_2}{\overset{k_1}{\rightleftharpoons}} B$

(*b*) $A + B \overset{k_1}{\rightleftharpoons} C + D$

20.15 Show that if A and B can be represented by spheres of the same radius that react when they touch, the second-order rate constant is given by

$$k_a = \frac{8 \times 10^3 RT}{3\eta} \text{ L mol}^{-1} \text{ s}^{-1}$$

where R is in J K^{-1} mol^{-1}. To obtain this result the diffusion coefficient is expressed in terms of the radius of a spherical particle by use of equation 20.12. For water at 25 °C, $\eta = 8.91 \times 10^{-4}$ kg m^{-1} s^{-1}. Calculate k_a at 25 °C.

20.16 What is the reaction radius for the reaction

$$H^+ + OH^- \xrightarrow{1.4 \times 10^{11} \text{ L mol}^{-1} \text{ s}^{-1}} H_2O$$

at 25 °C, given that the diffusion coefficients of H^+ and OH^- at this temperature are 9.1×10^{-9} m^2 s^{-1} and 5.2×10^{-9} m^2 s^{-1}?

20.17 For acetic acid in dilute aqueous solution at 25 °C, $K = 1.73 \times 10^{-5}$ and the relaxation time is 8.5×10^{-9} s for a 0.1 M solution. Calculate k_a and k_d in

$$CH_3CO_2H \underset{k_a}{\overset{k_d}{\rightleftharpoons}} CH_3CO_2^- + H^+$$

20.18 Derive the relation between the relaxation time τ and the rate constants for the reaction $A + B \underset{k_2}{\overset{k_1}{\rightleftharpoons}} C + D$, which is subjected to a small displacement from equilibrium.

20.19 Derive the relation between the relaxation time τ and the rate constants for the mechanism

$$\begin{array}{ccc} A & \underset{k_{-1}}{\overset{k_1}{\rightleftharpoons}} & B \\ K_A \Updownarrow & & \Updownarrow K_B \\ A' & \underset{k'_{-1}}{\overset{k'_1}{\rightleftharpoons}} & B' \end{array}$$

which is subjected to a small displacement from equilibrium. It is assumed that the equilibria, $A \rightleftharpoons A'$, $K_A = [A']/[A]$, and $B \rightleftharpoons B'$, $K_B = [B']/[B]$, are adjusted very rapidly so that these steps remain in equilibrium.

20.20 Calculate the first-order rate constants for the dissociation of the following weak acids: acetic acid, acid form of imidazole $C_3N_2H_5^+$, and NH_4^+. The corresponding acid dissociation constants are 1.75×10^{-5}, 1.2×10^{-7}, and 5.71×10^{-10}, respectively. The second-order rate constants for the formation of the acid forms from a proton plus the base are 4.5×10^{10}, 1.5×10^{10}, and 4.3×10^{10} L mol^{-1} s^{-1}, respectively.

20.21 The hydrolysis of pyrophosphate ($P_2O_7^{4-}$) at pH 7 at 25 °C by the enzyme pyrophosphatase occurs with an apparent first-order rate constant of $k' = 0.001$ s^{-1}. The reaction is first-order because the concentration of pyrophosphate is much lower than the Michaelis constant. Calculate the apparent first-order rate constant at pH 6 and pH 8 assuming that the mechanism is

$$\begin{array}{l} P_2O_7^{4-} + H_2O \underset{k_{-1}}{\overset{k}{\longrightarrow}} 2HPO_4^{2-} \\ \Updownarrow K_{HA} = 10^{-8.95} \\ HP_2O_7^{3-} \\ \Updownarrow K_{H_2A} = 10^{-6.12} \\ H_2P_2O_7^{2-} \end{array}$$

and that the acid dissociations are fast compared with the hydrolysis. The reaction goes so far to the right that we do not have to be concerned with the reverse reaction.

20.22 The solution reaction

$$I^- + OCl^- = OI^- + Cl^-$$

is believed to go by the mechanism

$$OCl^- + H_2O \underset{k_{-1}}{\overset{K_1}{\rightleftharpoons}} HOCl + OH^- \quad \text{(fast)}$$

$$I^- + HOCl \underset{k_{-1}}{\overset{k}{\longrightarrow}} HOI + Cl^- \quad \text{(slow)}$$

$$HOI + OH^- \underset{k_{-1}}{\overset{K_2}{\rightleftharpoons}} H_2O + OI^- \quad \text{(fast)}$$

Derive the rate equation for the forward rate of this reaction that shows the effect of the concentration of OH^-.

20.23 The mutarotation of glucose is first order in glucose concentration and is catalyzed by acids (A) and bases (B). The first-order rate constant may be expressed by an equation of the type that is encountered in reactions with parallel paths:

$$k = k_0 + k_{H^+}[H^+] + k_A[A] + k_B[B]$$

where k_0 is the first-order rate constant in the absence of acids and bases other than water. The following data were obtained by J. N. Brønsted and E. A. Guggenheim [*J. Am. Chem. Soc.* **49**:2554 (1927)] at 18 °C in a medium containing 0.02 mol L^{-1} sodium acetate and various concentrations of acetic acid:

$[CH_3CO_2H]$/mol L^{-1}	0.020	0.105	0.199
$k/10^{-4}$ min^{-1}	1.36	1.40	1.46

Calculate k_0 and k_A. The term involving k_{H^+} is negligible under these conditions.

20.24 The rate of a reaction between oppositely charged ions is measured at an ionic strength of 0.01 mol L^{-1}. How will the rate be affected if the ionic strength is raised to 0.05 mol L^{-1} if the reaction is (*a*) $A^+ + B^-$ or (*b*) $A^{2+} + B^{2+}$?

20.25 Suppose that an enzyme has a turnover number of 10^4 min^{-1} and a molar mass of 60 000 g mol^{-1}. How many moles of substrate can be turned over per hour per gram of enzyme if the substrate concentration is twice the Michaelis constant? It is assumed that the substrate concentration is maintained constant by a preceding enzymatic reaction and that products do not accumulate and inhibit the reaction.

20.26 The kinetics of the fumarase reaction

$$\text{fumarate} + H_2O = \text{L-malate}$$

is studied at 25 °C using a 0.01 ionic strength buffer of pH 7. The rate of the reaction is obtained using a recording ultraviolet spectrometer to measure the fumarate concentration [F]. The following rates of the forward reaction are obtained using a fumarase concentration of 5×10^{-10} mol L^{-1}:

[F]/10^{-6} mol L^{-1}	$v_F/10^{-7}$ mol L^{-1} s^{-1}
2	2.2
40	5.9

The following rates of the reverse reaction are obtained using a fumarase concentration of 5×10^{-10} mol L^{-1}:

[M]/10^{-6} mol L^{-1}	$v_M/10^{-7}$ mol L^{-1} s^{-1}
5	1.3
100	3.6

(*a*) Calculate the Michaelis constants and turnover numbers for the two substrates. In practice many more concentrations would be studied. (*b*) Calculate the four rate constants in the mechanism

$$E + F \underset{k_{-1}}{\overset{k_1}{\rightleftharpoons}} EX \underset{k_{-2}}{\overset{k_2}{\rightleftharpoons}} E + M$$

where E represents the catalytic site. There are four catalytic sites per fumarase molecule. (*c*) Calculate K_{eq} for the reaction catalyzed. The concentration of H_2O is omitted in the expression for the equilibrium constant because its concentration cannot be varied in dilute aqueous solutions.

20.27 Derive the steady-state rate equation for the mechanism

$$E + S \underset{k_2}{\overset{k_1}{\rightleftharpoons}} X \xrightarrow{k_3} E + P \qquad E + I \underset{k_5}{\overset{k_4}{\rightleftharpoons}} EI$$

for the case where $[S] \gg [E]_0$ and $[I] \gg [E]_0$.

20.28 The following initial velocities were determined spectrophotometrically for solutions of sodium succinate to which a constant amount of succinoxidase was added. The velocities are given as the change in absorbancy at 250 nm in 10 s. Calculate V, K_M, and K_I for malonate.

	$\dfrac{A \times 10^3}{10\ \text{s}}$	
[Succinate] 10^{-3} mol L^{-1}	No Inhibitor	15×10^{-6} mol L^{-1} Malonate
10.0	16.7	14.9
2.0	14.2	10.0
1.0	11.3	7.7
0.5	8.8	4.9
0.33	7.1	—

20.29 In the Eadie–Hofstee method for determining k_{cat} and K_M for an enzymatic reaction, $v/[E]_0[S]$ is plotted versus $v/[E]_0$. How are the kinetic parameters obtained from this plot?

20.30 The maximum initial velocities ($V = k_{cat}[E]_0$) for an enzymatic reaction are determined at a series of pH values:

pH	6.0	6.4	7.0	7.5	8.0	8.5	9.0
V	11	30	74	129	147	108	53

Calculate the values of the parameters V', K_a, and K_b in

$$V = \frac{V'}{1 + [H^+]/K_a + K_b/[H^+]}$$

See problem 20.31.

20.31 Use equation 20.86 to show that

$$V_S = \frac{V_{max}(1 + 2(K_b/K_a)^{1/2})}{1 + [H^+]/K_a + K_b/[H^+]}$$

where V_{max} is the maximum initial velocity in the plot of V_S versus pH. Further show that

$$K_a K_b = [H^+]_a [H^+]_b$$

where $[H^+]_a$ is the hydrogen ion concentration at which $v/V_{max} = \frac{1}{2}$ on the acidic side of the plot and $[H^+]_b$ is the hydrogen ion concentration at which $v/V_{max} = \frac{1}{2}$ on the basic side of the plot. Further show that the value of K_a can be calculated using

$$K_a = [H^+]_a + [H^+]_b - 4([H^+]_a[H^+]_b)^{1/2}$$

20.32 For the reaction

$$CO_2(+H_2O) \underset{k_2}{\overset{k_1}{\rightleftharpoons}} H_2CO_3$$

where the parentheses indicate that H_2O is not included in the equilibrium constant expression or in the rate equation, the following data were obtained: $\Delta_r H^\circ = 4730$ J mol^{-1} and $\Delta_r S^\circ = -33.5$ J K^{-1} mol^{-1}. At 25 °C, $k_1 = 0.0375$ s^{-1}, and at 0 °C, $k_1 = 0.0021$ s^{-1}. Assuming that $\Delta_r H^\circ$ and $\Delta_r S^\circ$ are independent of temperature in this range, (*a*) calculate the equilibrium constant and k_2 values at 25 °C and 0 °C, and (*b*) calculate the activation energies for the forward and backward reactions.

20.33 Calculate the time necessary for a quartz particle 10 μm in diameter to sediment 50 cm in distilled water at 25 °C. The density of quartz is 2.6 g cm^{-3}. The coefficient of viscosity of water is 8.91×10^{-4} kg m^{-1} s^{-1}.

20.34 How long will it take a spherical air bubble 0.5 mm in diameter to rise 10 cm through water at 25 °C?

20.35 Using data in Table 20.1 and equation 20.6, estimate the activation energy for water molecules to move into a vacancy at 25 °C.

20.36 A sharp boundary is formed between a solution of hemoglobin in a buffer and the buffer solution at 25 °C. After 10 h the half-width of the concentration gradient curve at the inflection point is 0.226 cm. What is the diffusion coefficient of hemoglobin under these conditions?

20.37 A sharp boundary is formed between a dilute buffered solution of hemoglobin ($D = 6.9 \times 10^{-11}$ m^2 s^{-1}) and the buffer at 20 °C. What is the half-width of the boundary after 1 and 4 h?

20.38 Since σ varies as $\sqrt{2Dt}$, the gradient curve has a certain width after time t, and it will be twice as wide after time $4t$ and three times as wide after time $9t$. Sketch $\partial c/\partial x$ versus x for sucrose in water at 25 °C after 1, 4, and 9 h for $c_0 = 0.1$ mol L^{-1}. Given: $D = 4.65 \times 10^{-10}$ m^2 s^{-1}.

20.39 For an electrolyte such as HCl it can be shown that the diffusion coefficient in a dilute solution in water is given by

$$D = \frac{2u_1 u_2 RT}{(u_1 + u_2)F}$$

where u_1 and u_2 are the electric mobilities of the two ions. What is the diffusion coefficient of dilute HCl in water at 25 °C? The electric mobilities are given in Table 20.2.

20.40 Using a table of the normal probability function, calculate enough points on a plot of dc/dx versus x (like Fig. 20.3) to draw in the smooth curve for diffusion of 0.01 mol L^{-1} sucrose into water at 25 °C after 3 h.

20.41 Calculate the conductivity at 25 °C of a solution containing 0.001 mol L^{-1} hydrochloric acid and 0.005 mol L^{-1} sodium chloride. The limiting ionic mobilities at infinite dilution may be used to obtain a sufficiently good approximation.

20.42 Estimate the conductivity at 25 °C of water that contains 70 ppm by weight of magnesium sulfate.

20.43 It may be shown that the diffusion coefficient at infinite dilution of an electrolyte with two univalent ions is given by

$$D = \frac{2u_1u_2RT}{(u_1 + u_2)F}$$

where u_1 and u_2 are the limiting values of the mobilities of the two ions. What is the diffusion coefficient of potassium chloride in water at 25 °C?

20.44 A study of conductivities at high electric field strengths reveals that the conductivity increases slightly with increasing electric field strength. A microsecond pulse at 10^7 V m^{-1} may be used. Approximately how far will a sodium ion move during such a pulse at room temperature?

20.45 The diffusion coefficient D of an ion is related to its ionic mobility u by

$$D = \frac{uRT}{zF}$$

The ionic mobilities of H^+ and OH^- are 3.63×10^{-7} m^2 V^{-1} s^{-1} and 2.06×10^{-7} m^2 V^{-1} s^{-1} at 25 °C. What is the rate constant for the following reaction?

$$H^+ + OH^- \rightarrow H_2O$$

The reaction radius is 0.75 nm, because once the proton is this close the reaction can proceed very rapidly by quantum mechanical tunneling. The electrostatic factor f is 1.70.

20.46 Pure solutions of the α and β chains of hemoglobin α_2/β_2 can be prepared. Assuming that α and β exist only as monomers in these solutions, and that they react on the first collision, estimate the half-life for the reaction

$$\alpha + \beta \rightarrow \alpha\beta$$

in water at 25 °C. The viscosity of water at this temperature is 8.95×10^{-4} kg m^{-1} s^{-1}. Calculate the half-life if equal volumes of 10^{-6} mol L^{-1} solutions of α and β are mixed. (See Problem 20.15.)

20.47 Derive the relation between the relaxation time τ and the rate constants for the reaction

$$A \underset{k_2}{\overset{k_1}{\rightleftharpoons}} B$$

which is subjected to a small displacement from equilibrium.

20.48 An imidazole buffer of pH 7 containing 0.05 mol L^{-1} imidazole has a relaxation time of 2.9×10^{-9} s at 25 °C. What are the values of the rate constants for the following reactions?

$$C_3N_2H_4 + H^+ \underset{k_{-1}}{\overset{k_1}{\rightleftharpoons}} C_3N_2H_5^+$$

The pK for imidazole at this temperature is 7.21.

20.49 Calculate the first-order rate constants for the following reactions at 25 °C.

H^+ Production	OH^- Production
$HOAc \rightarrow H^+ + OAc^-$	$OAc^- + H_2O \rightarrow HOAc + OH^-$
$ImH^+ \rightarrow H^+ + Im$	$Im + H_2O \rightarrow ImH^+ + OH^-$
$NH_4^+ \rightarrow H^+ + NH_3$	$NH_3 + H_2O \rightarrow NH_4^+ + OH^-$

where HOAc is acetic acid and Im is imidazole ($C_3N_2H_4$). The reverse reactions given above may all be assumed to be diffusion controlled with $k = 10^{10}$ L mol^{-1} s^{-1}. Acid dissociation constants at 25 °C are

HOAc	1.75×10^{-5}
ImH^+	1.2×10^{-7}
NH_4^+	5.71×10^{-10}

Which conjugate acid–base pair can play both H^+ and OH^- production roles about equally effectively?

20.50 The mutarotation of glucose is catalyzed by acids and bases and is first order in the concentration of glucose. When perchloric acid is used as a catalyst, the concentration of hydrogen ions may be taken to be equal to the concentration of perchloric acid, and the catalysis by perchlorate ion may be ignored since it is such a weak base. The following first-order constants were obtained at 18 °C:

$[HClO_4]$/mol L^{-1}	0.0010	0.0048	0.0099
$k/10^{-4}$ min^{-1}	1.25	1.38	1.53
$[HClO_4]$/mol L^{-1}	0.0192	0.0300	0.0400
$k/10^{-4}$ min^{-1}	1.90	2.15	2.59

Calculate the values of the constants in the equation $k = k_0 + k_{H^+}[H^+]$.

20.51 The initial rate v of oxidation of sodium succinate to form sodium fumarate by dissolved oxygen in the presence of the enzyme succinoxidase may be represented by equation 20.74. Calculate k_{cat}, $[E]_0$, and K_M from the following data:

$[S]/10^{-3}$ mol L^{-1}	10	2	1	0.5	0.33
$v/10^{-6}$ mol L^{-1} s^{-1}	1.17	0.99	0.79	0.62	0.50

20.52 For the fumarase reaction

$$\text{fumarate} + H_2O = \text{L-malate}$$

at pH 7, 25 °C, and 0.01 ionic strength, the Michaelis–Menten parameters have the following values:

$$V_F = (1.3 \times 10^3 \text{ s}^{-1})[E]_0 \qquad V_M = (0.8 \times 10^3 \text{ s}^{-1})[E]_0$$
$$K_F = 4 \times 10^{-6} \text{ mol L}^{-1} \qquad K_M = 10 \times 10^{-6} \text{ mol L}^{-1}$$

where $[E]_0$ is the molar concentration of the enzyme, which has four catalytic sites per molecule. Calculate (a) the four rate constants in the mechanism

$$E + F \underset{k_{-1}}{\overset{k_1}{\rightleftharpoons}} EX \underset{k_{-2}}{\overset{k_2}{\rightleftharpoons}} E + M$$

and (b) $\Delta G^{\circ\prime}$ for the overall reaction.

20.53 At 25 °C and pH 7.8, the following values are obtained for the Michaelis constant and maximum initial velocity for the forward reaction catalyzed by fumarase:

$$\mathrm{F} + \mathrm{H_2O} = \mathrm{M} \qquad K = \frac{[\mathrm{M}]_{\mathrm{eq}}}{[\mathrm{F}]_{\mathrm{eq}}} = 4.4$$

$$V_{\mathrm{F}} = (0.8 \times 10^3\ \mathrm{s^{-1}})[\mathrm{E}]_0$$

$$K_{\mathrm{F}} = 7 \times 10^{-6}\ \mathrm{mol\ L^{-1}}$$

where the enzyme concentration is in moles of enzyme per liter. The enzyme has four catalytic sites per molecule. In some experiments L-malate was added and was found to be inhibitory with a constant

$$K_{\mathrm{M}} = 100 \times 10^{-6}\ \mathrm{mol\ L^{-1}}$$

Calculate the values of the four rate constants in the mechanism

$$\mathrm{E} + \mathrm{F} \underset{k_{-1}}{\overset{k_1}{\rightleftharpoons}} \mathrm{EX} \underset{k_{-2}}{\overset{k_2}{\rightleftharpoons}} \mathrm{E} + \mathrm{M}$$

where E represents an enzymatic site.

20.54 The Michaelis constant for succinate being oxidized by succinoxidase is 0.5×10^{-3} mol $\mathrm{L^{-1}}$, and the competitive inhibition constant for malonate is 10×10^{-3} mol $\mathrm{L^{-1}}$. In an experiment with 10^{-3} mol $\mathrm{L^{-1}}$ succinate and 15×10^{-3} mol $\mathrm{L^{-1}}$ malonate, what is the percent inhibition?

20.55 Show how to plot rate data on an enzymatic reaction that is inhibited competitively by an inhibitor I at $[\mathrm{I}]_1$ to obtain the value of K_{I} by using an Eadie–Hofstee plot.

20.56 Derive the steady-state rate equation for the mechanism

$$\begin{array}{ccc} \mathrm{E} + \mathrm{S} & \underset{k_{-1}}{\overset{k_1}{\rightleftharpoons}} \mathrm{ES} & \xrightarrow{k_2} \mathrm{E} + \mathrm{P} \\ \Big\Updownarrow K_{\mathrm{EH}} & \Big\Updownarrow K_{\mathrm{EHS}} & \\ \mathrm{EH} & \mathrm{EHS} & \end{array}$$

Sketch the shape of the plots of V_{S} and K_{S} versus pH.

Computer Problems

20.A In Example 20.9 the following four rate constants were deduced from steady-state rate measurements at 25 °C and pH 8:

$$\mathrm{E} + \text{fumarate} \underset{0.60\times10^3}{\overset{1.14\times10^8}{\rightleftharpoons}} \mathrm{X} \underset{8.0\times10^6}{\overset{0.20\times10^3}{\rightleftharpoons}} \mathrm{E} + \text{L-malate}$$

where the unimolecular rate constants are in $\mathrm{s^{-1}}$ and the bimolecular constants are in mol $\mathrm{L^{-1}\ s^{-1}}$. Plot the concentrations of the four reactants when the initial concentrations of fumarate and free enzymatic sites, [E], are 10^{-4} M and 10^{-8} M, respectively.

20.B The diffusion coefficient of sucrose at 25 °C in dilute aqueous solution is $5.1 \times 10^{-6}\ \mathrm{cm^2\ s^{-1}}$. (*a*) Plot the ratio c/c_0, where c_0 is the initial concentration of sucrose, versus distance at 1, 4, 9, 16, and 25 hours. (*b*) Plot the concentration gradient $-dc/dx$ versus distance at these times.

20.C The diffusion coefficient of hemoglobin at 25 °C in dilute aqueous solution is $6.9 \times 10^{-7}\ \mathrm{cm^2\ s^{-1}}$. (*a*) Plot the ratio c/c_0 of sucrose, where c_0 is the initial concentration, versus distance at 1, 4, 9, 16, and 25 hours. (*b*) Plot the concentration gradient $-dc/dx$ versus distance at these times.

20.D As described in equations 20.84 to 20.87, the rate of an enzyme-catalyzed reaction may go through a maximum as the pH is varied. Plot the apparent rate constant divided by the rate constant for the rate-determining step versus pH for an enzyme–substrate complex that has $K_a = 10^{-4}$ and $K_b = 10^{-5}, 10^{-6}, 10^{-7}, 10^{-8}$, and 10^{-9}.

20.E Calculate the Michaelis constant K_{M} and maximum velocity V_{M} for fumarase when L-malate is the substrate using the Eadie–Hofstee method and make a plot to show how well the Michaelis equation fits the data.

$[\mathrm{M}]/10^{-6}$ M	0.1	0.333	1.0	3.33	10	33.3	100
v	1.9	4.2	6.1	6.5	7.2	7.4	6.9

20.F Plot the electrostatic factor f for the second-order rate constant at 25 °C in water for the reaction of two ions as a function of R_{12} when there are (*a*) opposite unit charges and (*b*) like unit charges.

20.G Solve the simultaneous rate equations for the Lotka mechanism for an autocatalytic reaction described in Section 20.11. Assume that the concentration of A is held at 1 M and the rate constants are $k_1 = 0.3\ \mathrm{s^{-1}}$, $k_2 = 0.6\ \mathrm{M^{-1}\ s^{-1}}$, and $k_3 = 0.8\ \mathrm{s^{-1}}$. The initial concentrations of W, Y, and P are to be taken as 0.2 M, 0.1 M, and 0 M, respectively. (*a*) Plot the concentrations of X, Y, and P at short times and discuss why the plots have these shapes. (*b*) Plot the concentrations of X, Y, and P at longer times and discuss what happens. [The Mathematica programs for making these plots are given in Ferreira et al., *J. Chem. Educ.* **76**:861 (1999).]

20.H The Lotka–Volterra mechanism for an autocatalytic reaction gives more striking results than the Lokta mechanism.

$$\mathrm{A} + \mathrm{X} \xrightarrow{k_1} 2\mathrm{X}$$

$$\mathrm{X} + \mathrm{Y} \xrightarrow{k_2} 2\mathrm{Y}$$

$$\mathrm{Y} \xrightarrow{k_3} \mathrm{P}$$

Assume that the concentration of A is held at 1 M and the rate constants are $k_1 = 1\ \mathrm{s^{-1}}$, $k_2 = 1.7\ \mathrm{M^{-1}\ s^{-1}}$, and $k_3 = 1.6\ \mathrm{s^{-1}}$. The initial concentrations of W, Y, and P are to be taken as 0.2 M, 0.1 M, and 0 M respectively. (*a*) Plot the concentrations of X, Y, and P at short times. (*b*) Plot the concentration product P as a function of time. Try varying the rate constants and initial concentrations to see what happens. [Mathematica programs for making these plots are given in Ferreira et al., *J. Chem. Educ.* **76**:861 (1999).]

PART FOUR

Macroscopic and Microscopic Structures

The second and third parts of the book were primarily concerned with the properties and dynamics of small molecules. Now we turn our attention to macromolecules, electric and magnetic properties of molecules, solid-state chemistry, and surface dynamics.

The chapter on macromolecules is concerned with high polymers, proteins, nucleic acids, and other macromolecules that have molar masses ranging from about 10^4 to 10^6 g mol^{-1}, or higher. These substances form viscous solutions, so measurements of viscosity provide information about the size and shape of these molecules in solution. More information can be obtained from measurements of diffusion, ultracentrifugation, and light scattering.

We have seen that the electric and magnetic fields involved in electromagnetic radiation are responsible for interactions that provide so much information about molecules from spectroscopy. But now we explore the effects of these fields on bulk properties. The effects of electric and magnetic fields are treated in similar ways, but there are significant differences.

Structural information about solids is primarily obtained by use of X-ray diffraction. This process is simplified by the recognition of the various types of symmetry that the internal structure of a crystal may have. Since the location of individual atoms is obtained, this is a powerful method for determining interatomic distances and angles.

The last chapter is concerned with the equilibrium and dynamics of processes that occur at the interface between a solid and a gas. When a molecule strikes a solid surface, it may rebound elastically or inelastically, or it may be adsorbed. An adsorbed molecule may dissociate on the surface or react with other species on the surface, or it may desorb from the surface. The catalysis of reactions by the surfaces of solids is of tremendous practical importance. The development of a number of "surface-sensitive" experimental methods, such as low-energy electron diffraction, electron emission from surfaces, and scanning tunneling microscopy, have made it possible to learn about processes at the interface.

21 Macromolecules

Although macromolecules (or polymers), both naturally occurring and man-made, play an immensely important role in our lives, the study of these systems is a relatively new activity. In fact, it was not until the 1920s that scientists became convinced that macromolecules could exist, based largely on the work of Staudinger in Germany and Carothers in the United States. Now it is commonplace to say that life is basically the biochemistry of macromolecules, and that the major part of the chemical industry is devoted to the production of macromolecules.

21.1 SIZE AND SHAPE OF MACROMOLECULES

The term *macromolecule* covers a very wide range of types from synthetic high polymers, which exist in solution in chains of variable length, to proteins, which have a unique structure and are folded in a unique way. Nevertheless, there are certain methods and concepts that apply quite generally to macromolecules; these are the principal subjects of this chapter.

Synthetic high polymers consist of long chains of atoms held together by covalent bonds. Such a chain is formed through the process of polymerization, in which monomer molecules chemically react to form linear chains, branched chains, or three-dimensional networks (see Fig. 21.1). The properties of polymers depend on both their chemical structure and their physical structure. Thus, plastics are usually linear or branched polymers that can be melted at reasonable temperatures and formed into various shapes, while rubbers are lightly cross-linked networks. These cross-links give rubbers their elastic properties.

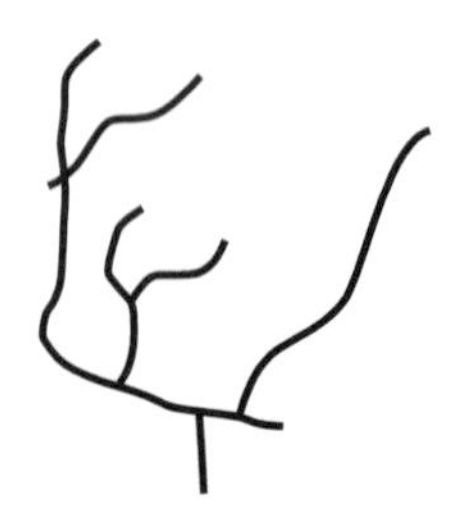

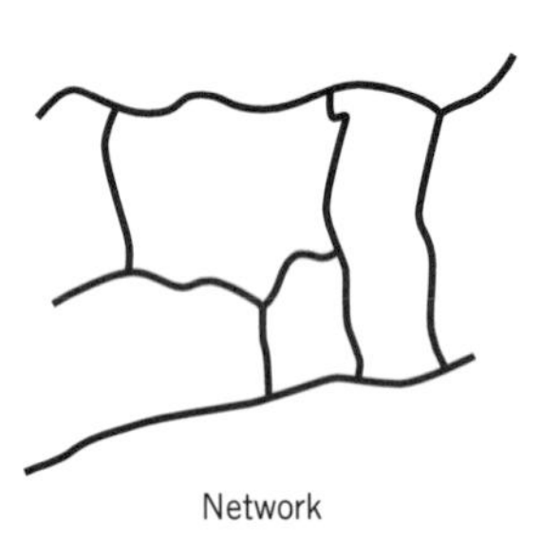

Figure 21.1 Structures of polymers. (From R. J. Young, *Introduction to Polymers,* p. 3. London: Chapman & Hall, 1983. Reprinted with permission.)

In discussing osmotic pressure in Section 6.7, it was pointed out that osmotic pressure measurements yield the number average molar mass. In discussing synthetic high polymers, it will be useful to distinguish between two types of average molar mass because the polymer almost always contains molecules with different masses—in other words, the polymer is polydisperse. The **number average molar mass** $\overline{M}_\text{n}$ is defined by

$$\overline{M}_\text{n} = \frac{\sum N_i M_i}{\sum N_i} = \frac{\sum n_i M_i}{\sum n_i} \tag{21.1}$$

where N_i is the number of polymer molecules with molar mass M_i and n_i is the amount of polymer with molar mass M_i. The **mass average molar mass** $\overline{M}_\text{m}$ is defined by

$$\overline{M}_\text{m} = \frac{\sum N_i M_i^2}{\sum N_i M_i} = \frac{\sum n_i M_i^2}{\sum n_i M_i} = \frac{\sum m_i M_i}{\sum m_i} \tag{21.2}$$

where $m_i = n_i M_i$ is the mass of polymer molecules with molar mass M_i. If the sample contains molecules of a single mass, the average molar mass is equal to the mass average molar mass; in a sample with dispersity, however, they are different, so the ratio $\overline{M}_\text{m}/\overline{M}_\text{n}$ is a measure of the dispersity of the sample. The mass average molar mass can be determined by use of light scattering, as described in a later section.

The size and dispersity of most synthetic macromolecules make the determination of their structure a very difficult task. In most cases, it is impossible to form a single crystal, so that X-ray diffraction methods are not useful. For certain systems (such as the nucleic acids and the proteins myoglobin and cytochrome c) single crystals are formed and the structure can be determined by X-ray methods. Even if single crystals can be formed, we are often more interested in the structure of the macromolecule in solution than in the crystalline state, and it is always difficult to know whether the structure is affected by crystallinity. Recently, NMR methods have been devised to determine the structure of some proteins. A number of methods have been devised to study the size and shape of macromolecules; they give less detailed information than X-ray diffraction or NMR, but they are less time-consuming and more easily applicable to most polymers. In the following, we discuss the measurements of osmotic pressure, viscosity, sedimentation, and light scattering.

21.2 OSMOTIC PRESSURE OF POLYMER SOLUTIONS

In Section 6.7, on colligative properties, we introduced the **osmotic pressure** Π of a solution. The osmotic pressure Π is related to the concentration c (mass per unit volume) of the polymer in solution by

$$\Pi = RT\frac{c}{M} + RT\,b(T)\frac{c^2}{M^2} + \cdots \tag{21.3}$$

where the first term is the ideal solution result and the succeeding terms are the corrections due to nonideality. This equation is the result for a monodisperse polymer in solution. As we have already indicated, most polymer solutions are polydisperse, so equation 21.3 must be changed. The term c/M in equation 21.3 is proportional to the number of polymer molecules N,

$$\frac{c}{M} = \frac{N}{N_A V} \tag{21.4}$$

where N_A is Avogadro's constant and V is the volume of the solution. For a polydisperse solution, Π is given by

$$\Pi = RT\frac{N}{N_A V} + RT\, b(T)\left(\frac{N}{N_A V}\right)^2 + \cdots \tag{21.5}$$

where N is the total number of molecules. The concentration of a solution of a polydisperse polymer is given by

$$c = \sum_i \frac{N_i M_i}{N_A V} = \overline{M}_n \frac{N}{N_A V} \tag{21.6}$$

and therefore

$$\Pi = \frac{RTc}{\overline{M}_n} + RT\, b(T)\frac{c^2}{\overline{M}_n^2} + \cdots \tag{21.7}$$

By taking the limit of Π/c as $c \to 0$, we can find the number average molar mass $\overline{M}_n$ of the polymer. The expansion of Π in powers of concentration is called a virial expansion just as in the equation of state of gases (Section 1.5).

Example 21.1 ***Molar mass from osmotic pressure measurements***

Osmotic pressures were measured for two dilute solutions of a sample of nitrocellulose in methanol at 25 °C. Use the following experimental values of Π/RTc to calculate the number average molar mass $\overline{M}_n$ and the second virial coefficient $b(T)$.

c/g cm^{-3}	2.5×10^{-3}	7.0×10^{-3}
(Π/RTc)/mol kg^{-1}	0.0102	0.0145

These concentrations can be expressed in SI base units by multiplying them by $(10^{-3}\text{ kg g}^{-1})(10^2\text{ cm m}^{-1})^3$.

The two data points allow us to write equation 21.7 in two ways.

$$0.0102\text{ mol kg}^{-1} = \frac{1}{\overline{M}_n} + \frac{b(T)(2.5\text{ kg m}^{-3})}{\overline{M}_n^2}$$

$$0.0145\text{ mol kg}^{-1} = \frac{1}{\overline{M}_n} + \frac{b(T)(7.0\text{ kg m}^{-3})}{\overline{M}_n^2}$$

Solving these equations simultaneously yields

$$\overline{M}_n = 98.1\text{ kg mol}^{-1}$$
$$b(T) = 9.2\text{ m}^3\text{ mol}^{-1}$$

In discussing osmotic pressure measurements and some other properties, it is useful to introduce the concepts of "good" and "poor" solvents. In a good solvent, each solute molecule is surrounded by a shell of solvent molecules, and this shell

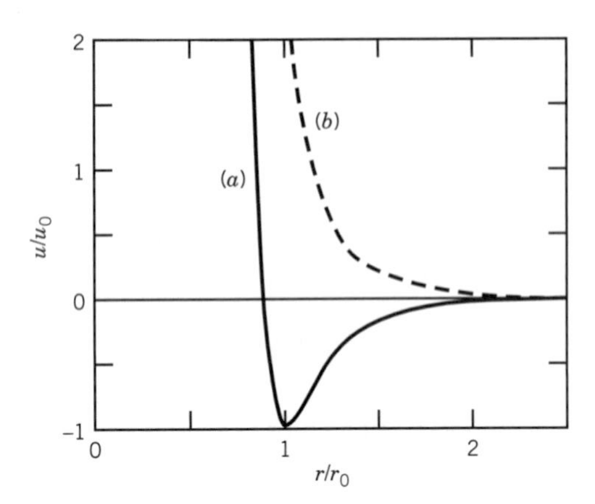

Figure 21.2 Pair interaction potential for two monomers in a poor solvent (a) and in a good solvent (b). The potential (a) is also representative for a van der Waals gas. (From G. R. Strobl, *The Physics of Polymers*, New York: Springer, 1997.)

has to be destroyed when two solute molecules approach each other. This means that the potential energy of interaction is repulsive at all distances between solute molecules, as shown in Fig. 21.2b. In a poor solvent, there is preference for solute–solute contacts. This means that solute molecules attract each other as they approach and repel each other at short distances for the same reason that real gas molecules repel each other at very short distances. This type of potential energy curve is shown in Fig. 21.2a.

Thus the interactions of polymer molecules with other polymer molecules and interactions between parts of a single molecule are like the interactions between gas molecules discussed in connection with the van der Waals equation and the explanation of the Boyle temperature. At high temperatures, the repulsive interactions predominate (the good solvent effect), and the second virial coefficient is positive. At low temperatures, the attractive interactions predominate (poor solvent effect), and the second virial coefficient is negative. At some intermediate temperature, the second virial coefficient is equal to zero, which means that the plot of Π/c versus c is horizontal at low concentrations. In polymer solutions this temperature is referred to as the **theta temperature,** and the solvent at this temperature is referred to as a theta solvent. These effects are illustrated in Fig. 21.3. Number average molar masses of polymers are often determined in theta solvents because Π/c is nearly independent of c over a range of low concentrations. Note that the theta temperature is analogous to the Boyle temperature for gases (Section 1.5). When the temperature is lowered below the theta temperature, the attractive forces between parts of a polymer molecule may become large enough that the solution becomes turbid because of the aggregation.

These changes in the average configuration of a polymer molecule in solution can be confirmed by any method that yields the root-mean-square end-to-end distance (Section 21.3). Thus, the contribution of the polymer to the viscosity of a solution is greater above the theta temperature than below the theta temperature.

The solvent can play an important role in determining the size of a polymer. For example, in a good solvent the polymer–solvent interaction results in a swelling of the polymer. The Gibbs energy is lowered significantly by this. In a poor solvent, the polymer lowers its Gibbs energy by having less contact with the solvent (contracting), and its complete solution may be impossible.

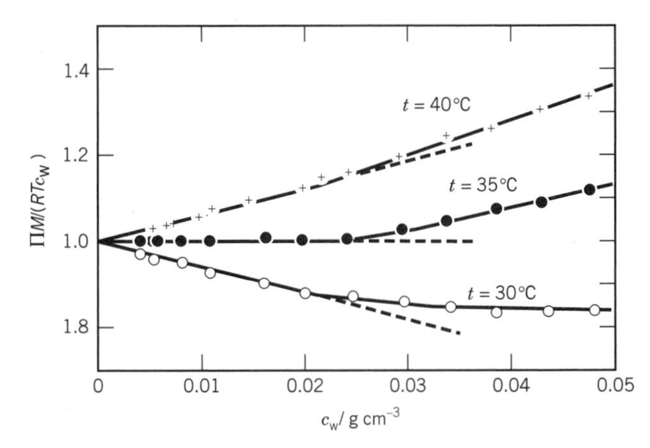

Figure 21.3 Plot of $\Pi M/RTc$ versus c for polystyrene ($\overline{M}_n = 1.3 \times 10^5$ g mol^{-1}) in cyclohexane at three temperatures. (From G. R. Strobl, *The Physics of Polymers*. New York: Springer, 1997.)

In solutions of proteins or other colloidal electrolytes it is necessary to distinguish between the **total osmotic pressure,** which would be obtained with a membrane impermeable to both salt and protein, and the **colloid osmotic pressure,** which is obtained with a membrane permeable to salt ions but not to protein. The latter type of membrane is always used when it is desired to obtain the molar mass of the protein or other colloidal electrolyte.

For colloidal electrolytes in solutions with low concentrations of electrolytes, the measured osmotic pressure is greater than that expected for the colloidal ions alone. This is a result of the fact that, although the salt ions may pass through the membrane, they will not be distributed equally at equilibrium. Donnan showed that because of the high molar mass ion on one side of the membrane, the concentration of the small ion of the same sign as the macro-ion is lower on that side of the membrane than in the salt solution, and that this is compensated by an increased concentration of the small ion of opposite charge. The Donnan effect may be reduced by increasing the salt concentration and, if possible, adjusting the pH to the pH where the net charge of the colloidal electrolyte is zero.

21.3 SPATIAL CONFIGURATION OF POLYMER CHAINS

Polymers are differentiated at the molecular level from other substances by their long chains of covalently bonded atoms. The useful mechanical properties of polymers are a consequence of the special attributes of these long chains. Since chains can have many spatial configurations, statistical mechanics is required to obtain a quantitative understanding of these molecular properties. The bond lengths and bond angles in polymer chains are the same as in substances of lower molecular mass within the limits of experimental measurements. Even with these restrictions on bond angles, however, the number of possible configurations of a chain several thousand bonds in length is prodigious. In solution the conformation of a particular chain undergoes continual change due to thermal agitation.

In this section we are interested in the average distance between ends of a chain containing a certain number of bonds. To make this calculation we will first consider an idealized model of the simplest possible kind. We will assume that

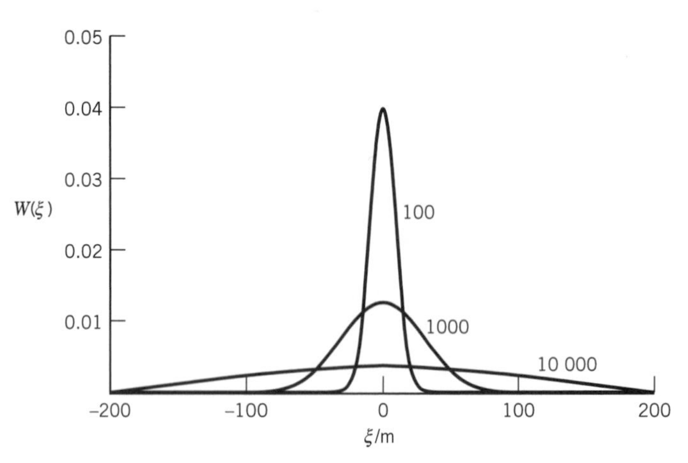

Figure 21.4 Probability density $W(\xi)$ for a random walker along a line after 100, 1000, and 10 000 steps of one meter in length. (See Computer Problem 21.A.)

the chain has N links of length l and that the direction in space of any link with respect to the preceding link is entirely random. This **freely jointed chain model** is an oversimplification because (1) angles of successive bonds in real molecules are limited to certain angles and (2) real chains cannot double back on themselves or occupy space filled with another part of the chain. This is known as the excluded volume effect. The mathematical treatment of this simple model involves an idea that is encountered in other areas of physical science, the idea of random walk. The simplest example of random walk is Brownian motion in one dimension.

Let us consider a walker who starts at the origin and takes a step of length h along a line to the right or left at successive intervals of time with equal probability ($p = \frac{1}{2}$). If the number N of steps is large, it is readily shown* that the probability $W(\xi)\,\mathrm{d}\xi$ that the walker is at a distance ξ to $\xi + \mathrm{d}\xi$ from the origin is given by

$$W(\xi)\,\mathrm{d}\xi = \frac{1}{(2\pi N h^2)^{1/2}}\,\mathrm{e}^{-\xi^2/2Nh^2}\,\mathrm{d}\xi \tag{21.8}$$

where h is the length of each step. Figure 21.4 shows the probability density for a random walker along a line after 100, 1000, and 10 000 steps. If the steps are not the same length, equation 21.8 becomes

$$W(\xi)\,\mathrm{d}\xi = \frac{1}{(2\pi N \overline{h^2})^{1/2}}\,e^{-\xi^2/2N\overline{h^2}}\,\mathrm{d}\xi \tag{21.9}$$

where $\overline{h^2}$ is the average of the square of the step length. This equation may be used to obtain one component of the displacement after N steps in random walk in three dimensions with a constant step length of l. The probability $W(x)\,\mathrm{d}x$ that there is a displacement in the range x to $x + \mathrm{d}x$ after N steps of length l in three dimensions is obtained by replacing ξ by x and h by $l\cos\theta$, where θ is the angle between the random walk vector and the x axis. The angle θ may take on all values with equal probabilities, and hence $\overline{h^2}$ in equation 21.9 is given by $l^2\overline{\cos^2\theta} = l^2/3$:

$$W(x)\,\mathrm{d}x = \left(\frac{3}{2\pi N l^2}\right)^{1/2}\mathrm{e}^{-3x^2/2Nl^2}\,\mathrm{d}x \tag{21.10}$$

*D. A. McQuarrie, *Statistical Thermodynamics,* Chapter 14. New York: Harper & Row, 1973.

The probability $W(x, y, z)\,dx\,dy\,dz$ that the coordinates after a random walk of N steps are between x and $x + dx$, y and $y + dy$, and z and $z + dz$ is given by the product of three probabilities of the type given by equation 21.10:

$$W(x, y, z)\,dx\,dy\,dz = \left(\frac{3}{2\pi N l^2}\right)^{3/2} e^{-3(x^2+y^2+z^2)/2Nl^2}\,dx\,dy\,dz \qquad (21.11)$$

This equation shows that the most probable coordinates after a random walk process are $x = 0$, $y = 0$, and $z = 0$. However, we are more interested in another question, namely, what is the root-mean-square value of the end-to-end distance after a three-dimensional random walk? Equation 21.11 can be converted to spherical coordinates to obtain

$$W(r)4\pi r^2\,dr = 4\pi\left(\frac{3}{2\pi N l^2}\right)^{3/2} e^{-3r^2/2Nl^2}\,r^2\,dr \qquad (21.12)$$

by use of $r^2 = x^2 + y^2 + z^2$. The **mean-square end-to-end distance** is obtained from

$$\overline{r^2} = \int_0^\infty W(r)4\pi r^2\,dr = Nl^2 \qquad (21.13)$$

so that the root-mean-square end-to-end distance for a **freely jointed chain** is

$$(\overline{r^2})^{1/2} = N^{1/2} l \qquad (21.14)$$

Thus, a very simple result has been obtained. For a freely jointed linear polymer with a fully extended length of Nl, the root-mean-square end-to-end distance is proportional to $N^{1/2}$ and, therefore, to the square root of the molar mass. Thus quadrupling the molar mass simply doubles $(\overline{r^2})^{1/2}$.

When a computer is used to calculate r^2 by having a random number generator calculate the orientation of each successive bond, a different value of r^2 will be obtained each time. Equation 21.14 gives the average result for $(r^2)^{1/2}$ for a large number of these calculations, and equation 21.18 gives the distribution of values of r^2 found in these calculations.

The freely jointed chain model discussed in the preceding three paragraphs is not a model for any real polymer because the bond angle θ, shown in Fig. 21.5, is determined by the structure of the monomer. In the idealized chain shown in the figure, C_1, C_2, and C_3 define a plane. Atom C_4 can occur any place on the circle that is the base of the cone described by rotation of bond 3. For polymethylene, θ is the tetrahedral angle 109° 28′. When θ is constant and ϕ varies randomly, the model is referred to as the **freely rotating chain model.** For this model, it can be shown* that for long chains

$$\overline{r^2} = Nl^2\left(\frac{1 - \cos\theta}{1 + \cos\theta}\right) \qquad (21.15)$$

For polymethylene $\theta = 109°28'$, $\cos\theta = -\frac{1}{3}$, and so $\overline{r^2} = 2Nl^2$.

Equation 21.15 illustrates the general conclusion that the root-mean-square end-to-end length is proportional to the square root of the number of bonds or links it contains, even if there are restrictions on the bond angles.

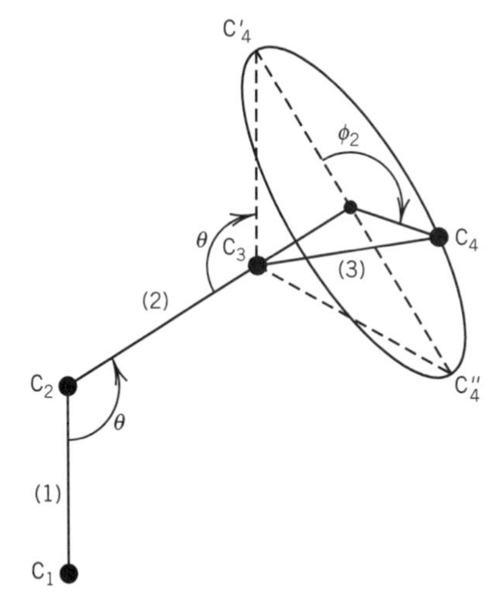

Figure 21.5 Three successive bonds in a singly bonded carbon chain. The first two bonds are in the plane of the page.

*P. J. Flory, *Polymer Chemistry*. Ithaca, NY: Cornell University Press, 1953.

Example 21.2 ***Difference between two models of polymer structure***

Illustrate the difference in the predictions of the root-mean-square end-to-end distance using the freely jointed chain model and the freely rotating chain model by considering a polymethylene chain with $M = 10^5\ \mathrm{g\,mol^{-1}}$. The length of a carbon–carbon bond is 155 pm. The number N of units in the chain is $(10^5\ \mathrm{g\,mol^{-1}})/(14\ \mathrm{g\,mol^{-1}}) = 7.14 \times 10^3$.

For the freely jointed chain model,

$$(\overline{r^2})^{1/2} = N^{1/2}l = (7.14 \times 10^3)^{1/2}(155\ \mathrm{pm}) = 13.1\ \mathrm{nm}$$

For the freely rotating chain model,

$$(\overline{r^2})^{1/2} = (2N)^{1/2}l = (2 \times 7.14 \times 10^3)^{1/2}(155\ \mathrm{pm}) = 18.5\ \mathrm{nm}$$

The length of the chain is $(7.14 \times 10^3)(155\ \mathrm{pm}) = 1107\ \mathrm{nm}$.

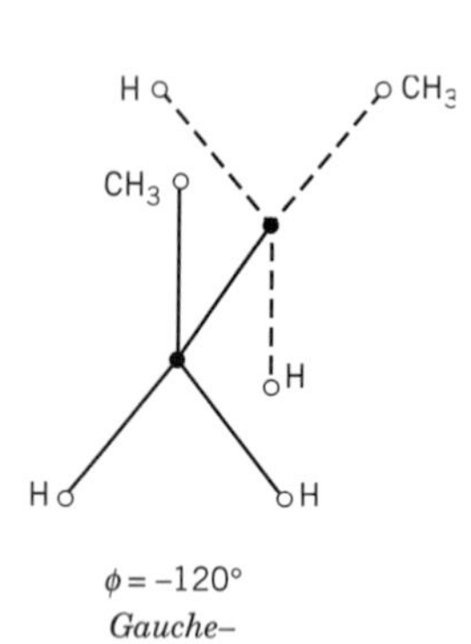

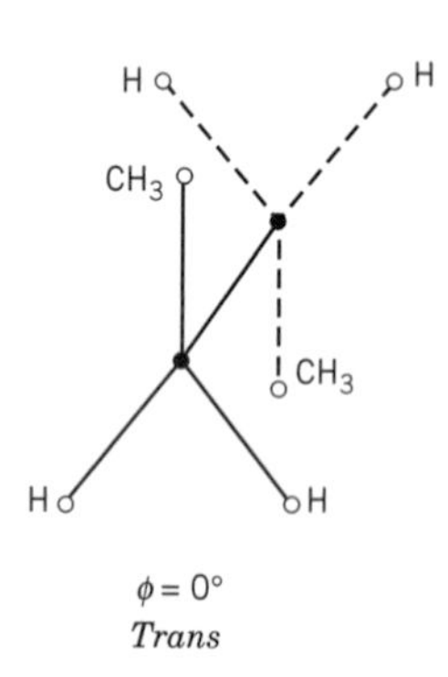

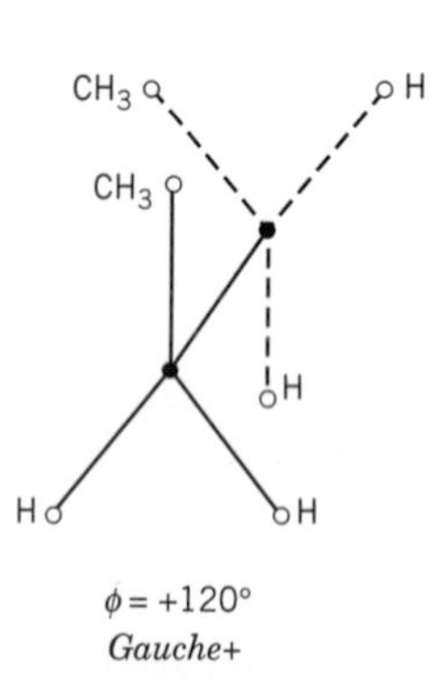

Figure 21.6 Conformations of *n*-butane.

So far in this discussion we have assumed that there is free rotation around each bond; that is, angle ϕ_2 in Fig. 21.5 can have any value. However, this rotation is not free because there is a potential energy associated with such internal rotation.

The simplest example of the restrictions on bond rotations is provided by *n*-butane. The three conformations with the lowest energies are shown in Fig. 21.6, and the corresponding potential energy curve is shown in Fig. 21.7. The eclipsed form ($\phi = 180°$) is not shown because it has a high energy relative to the other three forms. The gauche minima lie about 2.1 kJ mol^{-1} above the energy of the trans conformation. The gauche forms are each about half as probable as the trans form; $p = e^{-E/RT}$. When the potential energy of internal rotation in a singly bonded carbon chain is taken into account, equation 21.15 becomes

$$\overline{r^2} = Nl^2\left(\frac{1-\cos\theta}{1+\cos\theta}\right)\left(\frac{1+\overline{\cos\phi}}{1-\overline{\cos\phi}}\right) \tag{21.16}$$

where

$$\overline{\cos\phi} = \frac{\int_0^{2\pi}\cos\phi\, e^{-E(\phi)/RT}\,d\phi}{\int_0^{2\pi} e^{-E(\phi)/RT}\,d\phi} \tag{21.17}$$

where $E(\phi)$ is the potential energy of internal rotation given in Fig. 21.7.

Note that in both equations 21.15 and 21.16, $\overline{r^2}$ is still of the form $\beta^2 N$. Thus, the effect of restricting the bond angle and the free rotation about bonds is to increase the "effective bond length" β. Therefore, the form of equation 21.16 is not affected by bringing in these restrictions. These theoretical results are of great importance for an understanding of rubberlike elasticity and of hydrodynamic and thermodynamic properties of dilute polymer solutions.

In the model of a polymer we have considered, there are no interactions between different segments of the polymer. In a real polymer in solution, however, there will be interactions, of which the most important is the excluded volume interaction, which is due to the finite size of the segments. Even parts of the chain that are far apart *along the chain* can produce this effect (see Fig. 21.8). The real interactions in polymers are very complicated; however, as far as the long length scale properties such as end-to-end distance are concerned, the *details* of the

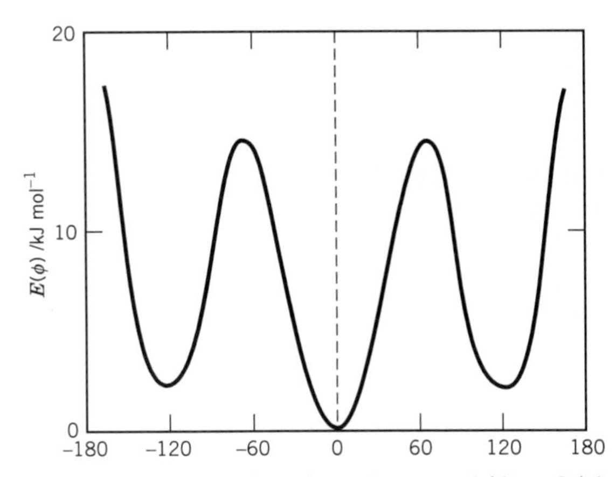

Figure 21.7 Potential energy of rotation about the central (C_2—C_3) bond of n-butane.

interaction are not important. Flory worked out a simple form of the theory including excluded volume, which gave as the root-mean-square end-to-end distance

$$\langle r^2 \rangle^{1/2} \approx N^{3/5} \tag{21.18}$$

instead of $\langle r^2 \rangle^{1/2} \approx N^{1/2}$. Thus, the excluded volume interactions tend to swell the chain from the ideal random chain. Since the intersegment interactions that give rise to the excluded volume effect are mediated by the solvent in which the polymer is dissolved, the form of $\langle r^2 \rangle$ and the probability distribution function for r^2 depend on solvent. In a good solvent, the polymer is surrounded by solvent molecules and is therefore swollen. In a poor solvent, the polymer prefers its own segment–segment interactions to the segment–solvent interactions, so the polymer tends to be small. In between these limiting cases, there is a point (the theta point) where the interactions are more or less equal, and the excluded volume effects are negligible.

There is another measure of the physical size of a chain molecule, and that is the **radius of gyration** $\langle R_G^2 \rangle^{1/2}$. The square of the radius of gyration R_G^2 of a chain molecule in a particular configuration is calculated by summing the squares of the distances R_G from each residue to the chain's center of mass and averaging by dividing by the number of residues:

$$R_G^2 = \frac{\sum R_{Gi}^2}{N} \tag{21.19}$$

To obtain the radius of gyration, the average of R_G^2 is calculated for all of the possible configurations; this yields $\langle R_G^2 \rangle$, which it turns out is directly proportional to the mean-square end-to-end distance $\langle r^2 \rangle$:

$$\langle R_G^2 \rangle = \frac{\langle r^2 \rangle}{6} \tag{21.20}$$

Thus, substituting the expression for $\langle r^2 \rangle^{1/2}$ (equation 21.14) and taking the square root yields

$$\langle R_G^2 \rangle^{1/2} = \frac{N^{1/2} l}{6^{1/2}} \tag{21.21}$$

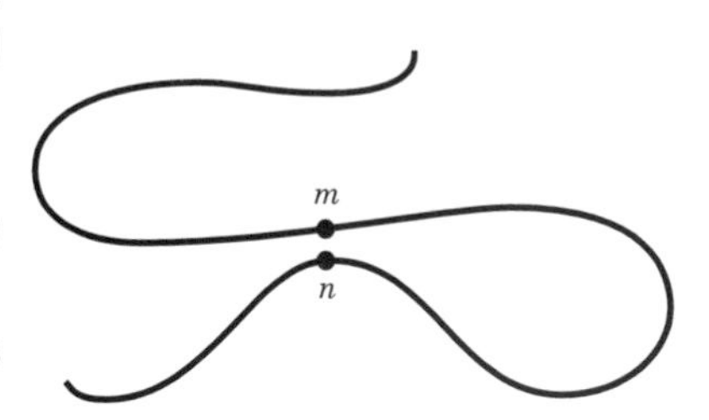

Figure 21.8 Excluded volume interaction between segments m and n.

The radius of gyration gives a better idea of the physical size of the chain molecule than the root-mean-square end-to-end distance.

21.4 MOLAR MASS DISTRIBUTIONS OF STEP-GROWTH POLYMERS

Synthetic high polymers are produced by the reaction of monomer molecules to produce linear or branched chain molecules. Polymerization reactions can be classified according to the mechanism of the reaction. There are basically two types of polymerization reactions: **step-growth polymerization** and **free-radical polymerization.** Their kinetics are quite different. In step-growth polymerization the chains grow in a slow, stepwise manner. In free-radical polymerization individual chains grow rapidly to their final length.

In step-growth polymerization, a linear polymer is produced by the stepwise condensation or addition of the reactive groups in bifunctional monomers. Common functional groups for this type of polymerization are $-OH$, $-CO_2H$, and $-NH_2$. For example, a polyester can be formed by the reaction of diacids with dialcohols via a polycondensation reaction:

$$x\,\mathrm{HOOC{-}R{-}COOH} + x\,\mathrm{HO{-}R'{-}OH} \rightarrow \mathrm{H}\!\left[\mathrm{O{-}\underset{\overset{\|}{O}}{C}{-}R{-}\underset{\overset{\|}{O}}{C}{-}O{-}R'}\right]_x\!\mathrm{OH} + (2x-1)\mathrm{H_2O} \tag{21.22}$$

Alternatively, a polyester can be formed by polymerization of a hydroxy acid:

$$x\,\mathrm{HO{-}R{-}COOH} \rightarrow \mathrm{H}\!\left[\mathrm{O{-}R{-}\underset{\overset{\|}{O}}{C}}\right]_x\!\mathrm{OH} + (x-1)\mathrm{H_2O} \tag{21.23}$$

In the production of polyesters or polyamides, the reactions are reversible, so that water must be removed as the reaction progresses. In the production of polyurethanes there is no elimination of a small molecule. Two molecules react to form a dimer. The dimer then reacts with another monomer to form a trimer, and so on. Therefore, the average molar mass of the product increases as the reaction proceeds.

During a step-growth polymerization the average molar mass or degree of polymerization increases steadily. The **number average degree of polymerization** $\overline{X}_n$ is equal to the average number of monomer units in the polymer molecules. If the initial number of monomer molecules in the reaction mixture is n_0 and the number present at time t is n, all the rest are in polymer molecules, so that the extent of p of reaction is

$$p = \frac{n_0 - n}{n_0} = 1 - \frac{n}{n_0} \tag{21.24}$$

The degree of polymerization is

$$\overline{X}_n = \frac{n}{n_0} = \frac{1}{1-p} \tag{21.25}$$

where the second form has been obtained by use of equation 21.24. For the average number of monomer units in the polymer molecules to be 100, it is evident that the extent of reaction p will have to be equal to 0.99.

Now we are going to consider the polymerization of a hydroxy acid HO—R—COOH, which is represented as AB. The number of monomer units in a given chain i is called the **degree of polymerization.** It is therefore equal to M/M_0, where M is the molar mass of the chain and M_0 the molar mass of the monomer.

To calculate the molar mass distribution in solution, we consider the **probability** π_i of forming a chain of size i. This probability must be equal to the probability of having $i-1$ elementary steps. Since the probability of having any one step is proportional to the fraction p of monomers reacted, then π_i must be proportional to p^{i-1}:

$$\pi_i = cp^{i-1} \tag{21.26}$$

To find c, we note that the sum of π_i over all i must equal 1, since the sum of all probabilities is unity:

$$1 = \sum_{i=1}^{\infty} cp^{i-1} = c(1 + p + p^2 + \cdots) = \frac{c}{1-p} \tag{21.27}$$

Therefore, $c = 1 - p$. This makes sense since then the probability of finding a chain with zero bonds, that is, a monomer, is just $1 - p$, the probability of finding an unreacted monomer. We therefore find that the probability of finding a chain of length i (or equivalently the mole fraction of chains of length i) is

$$\pi_i = (1-p)p^{i-1} \tag{21.28}$$

We can now calculate the number and mass average molar masses. The number average molar mass $\overline{M}_n$ is given by

$$\overline{M}_n = M_0 \sum_{i=1}^{\infty} i\pi_i = M_0(1-p)\sum_{i=1}^{\infty} ip^{i-1} \tag{21.29}$$

The sum $1 + 2p + 3p^2 + \cdots = (1-p)^{-2}$, so that

$$\overline{M}_n = \frac{M_0}{1-p} = \overline{X}_n M_0 \tag{21.30}$$

We can calculate the mass average molar mass by noting that

$$\overline{M}_m = \frac{M_0(\sum_i i^2\pi_i)}{\sum_i i\pi_i} \tag{21.31}$$

from the definition of $\overline{M}_m$ (see equation 21.2). Equivalently, we can consider the probabilities w_i defined on the basis of mass rather than number:

$$w_i = \frac{i\pi_i}{\sum i\pi_i} = i\pi_i(1-p) \tag{21.32}$$

Then the preceding definition of mass average molar mass follows directly. From this, we find

$$\overline{M}_m = M_0\left(\frac{1+p}{1-p}\right) \tag{21.33}$$

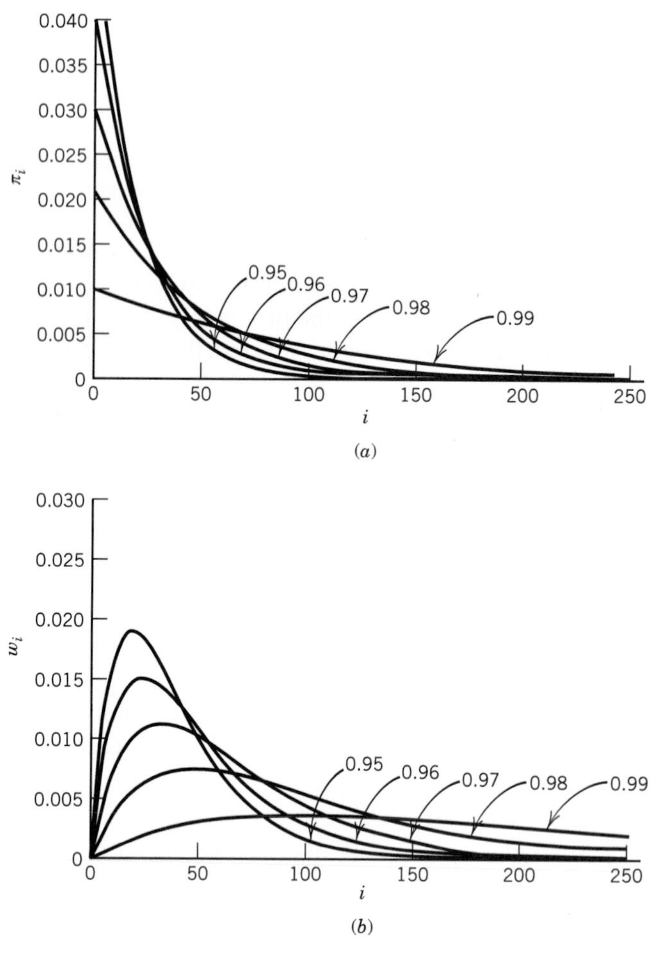

Figure 21.9 (*a*) Mole fraction distribution of condensation polymer for extents of reaction p of 0.95, 0.96, 0.97, 0.98, and 0.99. (*b*) Weight fraction distribution of condensation polymer for extents of reaction p of 0.95, 0.96, 0.97, 0.98, and 0.99. Note that the extent of reaction must be very close to unity to obtain the high polymer. (See Computer Problem 21.C.)

since

$$\sum_{i=1}^{\infty} i^2 p^{i-1} = (1+p)(1-p)^{-3} \tag{21.34}$$

The number and mass fractions or probabilities for different values of p are plotted in Fig. 21.9. Note that the mass fractions w_i go through a maximum while the number fractions π_i do not. These results have been confirmed by experiments in which polymer samples are fractionated (by solubility methods) into narrow ranges of polymer size and the various fractions analyzed.

If we multiply w_i by M_0 we have the probability of finding a chain of mass $M = iM_0$ in the solution. Therefore, the graph of the probability density for molar mass M, $P(M_0)$, is the same as the graph for w_i versus i (Fig. 21.9*b*), and is shown in Fig. 21.10. It can be shown that for p close to 1, the maximum of the curve

$P(M)$ occurs for $M \cong \overline{M}_n$. From the definitions of $\overline{M}_m$ and $\overline{M}_n$, it is clear that $\overline{M}_m \geq \overline{M}_n$. The equal sign can occur only if there is only one size of polymer in the sample (monodisperse), so that the ratio $\overline{M}_m\overline{M}_n^{-1}$ is a good measure of the polydispersity of the sample. From above, we see that

$$\frac{\overline{M}_m}{\overline{M}_n} = 1 + p \tag{21.35}$$

We have already seen that p must be close to 1 for the high polymer to be formed, so that for such systems $\overline{M}_m$ is twice $\overline{M}_n$.

Example 21.3 ***Mole fractions and weight fractions of polymers***

For the condensation polymerization of $HO{-}CH_2CH_2{-}CO_2H$, calculate the mole fractions π_i and weight fractions w_i of polymers with 10, 20, 30, and 40 monomer units when the fraction p of monomers reacted is 0.95. Also calculate the number average molar mass and mass average molar mass of this polymer.

$$\pi_i = (1-p)p^{i-1} = (0.05)0.95^{i-1}$$
$$w_i = i\,\pi_i(1-p) = i\,\pi_i(0.05)$$

i	10	20	30	40
π_i	3.16×10^{-2}	1.89×10^{-2}	1.13×10^{-2}	6.76×10^{-3}
w_i	1.57×10^{-2}	1.89×10^{-2}	1.69×10^{-2}	1.35×10^{-2}

$$\overline{M}_n = \frac{M_0}{1-p} = \frac{78\text{ g/mol}}{0.05} = 1560\text{ g/mol}$$

$$\overline{M}_m = M_0\frac{1+p}{1-p} = 78\text{ g/mol}\left(\frac{1.95}{0.05}\right) = 3042\text{ g/mol}$$

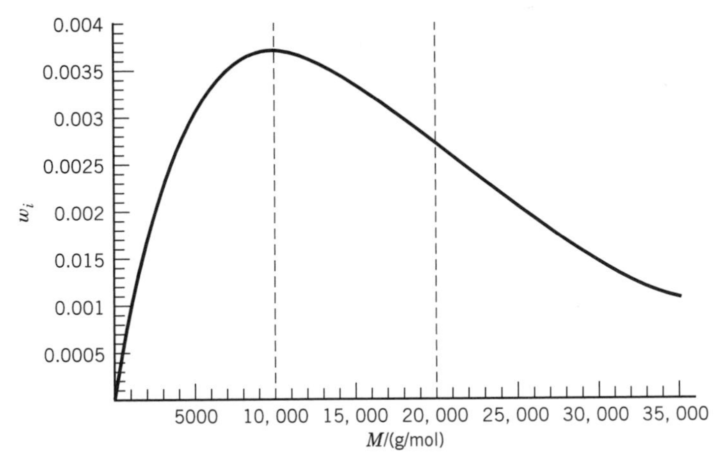

Figure 21.10 Plot of the weight fraction w_i versus the molar mass M for a step-growth polymer with $M_0 = 100$ g/mol and extent of reaction $p = 0.99$. The first vertical line is the number average molar mass and the second vertical line is the mass average molar mass. (See Computer Problem 21E.)

This discussion has been concerned with linear condensation polymers that arise from bifunctional polymers. When the monomer is trifunctional, polymer that is branched and cross-linked can be obtained. This introduces the possibility of forming an infinite network. This process is referred to as gelation, and the point of transition from a soluble polymer to an insoluble polymer is referred to as the gel point.

Polymers can also be formed from unsaturated organic compounds by free radical chain reactions. These polymerizations can be initiated with organic peroxides, azo compounds, or ultraviolet light or X-rays. When polymer is formed in a chain reaction, the growing chain bears an unpaired electron, and the electron is transferred to the new chain end in each addition step. This is very different from the condensation polymer polymerization described above because the chains grow to their final lengths very quickly. Chain growth is terminated by reaction of the free radical of another chain to form either one or two molecules of polymer.

21.5 DETERMINATION OF MOLAR MASSES USING VISCOSITY, SEDIMENTATION, AND LIGHT SCATTERING

The determination of the number average molar mass of a polymer was discussed in Section 21.2, but now we will discuss three other methods for determining molar masses of macromolecules. A suspension of particles or a solution of macromolecules has a viscosity η (Section 20.1) that is higher than the viscosity η_0 of the solvent. Einstein showed in 1906 that the viscosity of a dilute suspension of spheres is given by

$$\frac{\eta}{\eta_0} = 1 + \frac{5}{2}\phi \qquad \text{or} \qquad \frac{\eta}{\eta_0} - 1 = \eta_{sp} = \frac{5}{2}\phi \tag{21.36}$$

where ϕ is the volume fraction occupied by the spheres. The ratio η/η_0 is referred to as the **relative viscosity,** and $\eta/\eta_0 - 1 = \eta_{sp}$ is referred to as the **specific viscosity,** although neither of these quantities is a viscosity. It is of interest to note that the viscosity is independent of the size of the spherical particles. However, the coefficient of ϕ increases with the axial ratio for prolate and oblate ellipsoids of revolution. The volume for a polymer in solution may be written as cv_2, where c is the concentration of the polymer in mass per unit volume and v_2 is the partial specific volume of the polymer. The partial specific volume v_2 is equal to the partial molar volume V_i (Section 1.10) divided by the molar mass. Thus we might expect that equation 21.36 for nonspherical particles might be written as $\eta_{sp}/c = (\text{const})v_2$. The limit of η_{sp}/c as c approaches zero is referred to as the **intrinsic viscosity** $[\eta]$:

$$\lim_{c \to 0} \frac{\eta_{sp}}{c} = [\eta] \tag{21.37}$$

Intrinsic viscosities give strong indications of the axial ratios of molecules in solution. For a spherical, unhydrated protein with $v = 0.75\ \text{cm}^3\ \text{g}^{-1}$, we would expect $[\eta] = 1.9\ \text{cm}^3\ \text{g}^{-1}$. As shown in Table 21.1, the intrinsic viscosity of ribonuclease is not very much greater than this value. Bushy stunt virus is very nearly spherical, even though it has a molar mass of $1.07 \times 10^7\ \text{g mol}^{-1}$. The elongated molecules of fibrinogen provide the structure for blood clots, and the even

Table 21.1 Intrinsic Viscosities of Macromolecules in Water at 25 °C

Molecule	M/g mol^{-1}	$[\eta]$/cm^3 g^{-1}
Ribonuclease	13 683	2.3
Bovine serum albumin	66 500	3.7
Bushy stunt virus	10 700 000	3.4
Fibrinogen	330 000	27
Tobacco mosaic virus	40 000 000	37
Myosin	493 000	217
DNA	6 000 000	5000

Source: C. Tanford, *Physical Chemistry of Macromolecules.* New York: Wiley, 1961. Reprinted with permission.

more elongated molecules of myosin are the contractile part of muscle. The very high intrinsic viscosity of double-stranded deoxyribonucleic acid (DNA) shows that the molecules are very long and thin. High-molar-mass DNA may be degraded (reduced in molar mass) by the shear gradients encountered in pipetting solutions. To measure the viscosities of DNA solutions at very low shear, it has been necessary to use rotating cylinder, or Couette, viscometers.

In Section 21.3 we found that the root-mean-square end-to-end distance of a linear polymer in a theta solvent is proportional to the square root of its molar mass. On the basis of this result we can derive the form of the relation between the intrinsic viscosity and the molar mass.

The polymer molecule in solution may be visualized as a spherical cloud of segments, with the cloud getting thinner as we go out from the center. The size of this spherical cloud is measured by the root-mean-square end-to-end distance $(\overline{r^2})^{1/2}$. As an approximation of the spherical cloud, we may replace it with a rigid sphere with a radius proportional to $(\overline{r^2})^{1/2}$ and a volume proportional to $(\overline{r^2})^{3/2}$. Since the intrinsic viscosity of a sphere is proportional to the volume per unit mass of the macromolecule, the intrinsic viscosity of a random coil is expected to be proportional to $(\overline{r^2})^{3/2}/M_r$, where M_r is the relative molar mass defined by m/m_u, where m is the mass of a molecule and $m_u = m(^{12}C)/12$ is the atomic mass constant. The reason for using the dimensionless relative molar mass M_r here is that this avoids units when M_r is raised to various powers.

Since $(\overline{r^2})$ is proportional to the number of segments in the chain (see equation 21.14), or to M_r, the intrinsic viscosity of a random coil is expected to be proportional to $M_r^{1/2}$:

$$[\eta] = KM_r^{1/2} \tag{21.38}$$

This is exactly what is found for polystyrene in cyclohexane at 34.5 °C. In general, the intrinsic viscosity is related to the relative molar mass of the polymer by the Mark–Houwink equation, which is

$$[\eta] = KM_r^a \tag{21.39}$$

where a is an empirical constant that usually has a value in the range 0.5 to 0.8. The higher values of a are obtained in good solvents where the polymer chain is more extended than a random coil. The values of K and a for a number of polymer–solvent systems are shown in Table 21.2.

Table 21.2 Parameters in $[\eta] = KM^a$ for Polymer–Solvent Systems

Polymer	*Solvent*	t/°C	$K/10^{-2}$ cm^3 g^{-1}	a
Natural rubber	Toluene	25	5.0	0.67
Polymethyl methacrylate	Acetone	25	0.75	0.70
	Chloroform	25	0.48	0.80
	Methyl ethyl ketone	25	0.68	0.72
Polystyrene	Benzene	20	1.23	0.72
	Methyl ethyl ketone	20–40	3.82	0.58
	Toluene	20–30	1.05	0.72
Polyvinyl alcohol	Water	25	30.0	0.50

Source: H. R. Allcock and F. W. Lampe, *Contemporary Polymer Chemistry.* Englewood Cliffs, NJ: Prentice-Hall, 1981.

Since K and a have to be determined experimentally, this is a secondary method for the determination of the average molar mass of a high polymer. In the case of a heterogeneous polymer the molar mass determined using viscosity measurements is closer to a mass average molar mass than a number average molar mass (Section 21.1).

Example 21.4 ***Viscosity of DNA solutions***

Given that the intrinsic viscosity of a sample of DNA ($M = 6 \times 10^6$ g mol^{-1}) is 5.0×10^3 cm^3 g^{-1}, approximately what concentration of DNA in water would have a relative viscosity of 1.1?

$$\frac{(\eta/\eta_0) - 1}{c} = \frac{0.1}{c} = 5.0 \times 10^3 \text{ cm}^3 \text{ g}^{-1}$$

$$c = 2.0 \times 10^{-5} \text{ g cm}^{-3} = 2.0 \times 10^{-2} \text{ g L}^{-1}$$

Since the molar mass of DNA per nm measured along the axis of the helix is 1920 g mol^{-1}, the length of a molecule of DNA with $M = 6 \times 10^6$ g mol^{-1} is 3100 nm or 3.1 μm.

Comment:

The properties of solutions of small DNA molecules can be determined, as indicated by Table 21.1 where the intrinsic viscosity is given for a DNA molecule with molar mass 6×10^6 g mol^{-1}. *However, DNA molecules can be much bigger. Electron microscopy measurements show that the DNA molecule from a single human chromosome is about* 4 cm *long and has a molar mass of the order of* 10^{12} g mol^{-1}. *Such long molecules are broken by the shear stresses of pipetting and viscosity determinations.*

In Section 20.1 we saw that the radius r of a spherical particle can be determined by measuring the velocity with which it settles in a liquid by use of $f = 6\pi\eta r$, where f is equal to the ratio of the force on the particle to its velocity.

Dissolved molecules tend to sediment in the earth's gravitational field or to float upward, depending on their density relative to that of the solvent, but this tendency is counteracted by the random translational motion of the molecules. However, sufficiently powerful **ultracentrifuges** have been built to cause even molecules as small as sucrose to sediment at measurable rates. Svedberg* was the leader in the development of ultracentrifuges, which he defined as centrifuges adapted for quantitative measurements of convection-free and vibration-free sedimentation. There are two distinct types of ultracentrifuge experiments: (1) those where the velocity of sedimentation of a component of the solution is measured (sedimentation velocity) and (2) those where the redistribution of molecules is determined at equilibrium (sedimentation equilibrium).

The acceleration a of a centrifugal field is equal to $\omega^2 r$, where ω is the angular velocity of the centrifuge in radians per second (i.e., 2π times the number of revolutions per second) and r is the distance from the axis of rotation. Ultracentrifuges in which r is about 6 cm are commonly operated at 60 000 rpm or 1000 rps, and so the acceleration is

$$a = \omega^2 r = (2\pi 1000\ \mathrm{s}^{-1})^2(0.06\ \mathrm{m}) = 2.36 \times 10^6\ \mathrm{m\ s}^{-2} \tag{21.40}$$

Since the acceleration of the earth's gravitational field is 9.80 m s^{-2}, the acceleration is 240 000 times greater than in the earth's field.

When a solution containing a polymer is in a cell in an ultracentrifuge, the movement of molecules away from the axis of the centrifuge produces a boundary in the cell that moves away from the axis of rotation if the molecules are denser than the solvent. The direction of "sedimentation" depends on the sign of the Archimedes factor $(1 - v\rho)$, where ρ is the density of the solvent. For a complex mixture, such as blood plasma, there will be several boundaries corresponding to different protein molecules.

The ratio of the velocity to the centrifugal acceleration is called the **sedimentation coefficient** S:

$$S = \frac{1}{\omega^2 r}\frac{\mathrm{d}r}{\mathrm{d}t} \tag{21.41}$$

Since ω^2 has the units s^{-2}, the sedimentation coefficient has the unit seconds. The sedimentation coefficients of proteins fall in the range 10^{-13} s to 200 $\times 10^{-13}$ s, and the unit 10^{-13} s is called a **svedberg.**

In a centrifugal field a solute molecule is accelerated until the frictional force resisting its motion is equal to the acceleration of the centrifugal field times the effective mass $m(1 - v\rho)$, where m is the mass of the molecule. The frictional force is the product of the velocity $\mathrm{d}r/\mathrm{d}t$ and the frictional coefficient f (Section 20.1). Thus, when the steady-state velocity $\mathrm{d}r/\mathrm{d}t$ is reached,

$$f\frac{\mathrm{d}r}{\mathrm{d}t} = m(1 - v\rho)\omega^2 r = \frac{M(1 - v\rho)\omega^2 r}{N_\mathrm{A}} \tag{21.42}$$

where M is the molar mass.

Substituting equation 21.42 into equation 21.41 yields

$$S = \frac{M(1 - v\rho)}{N_\mathrm{A} f} \tag{21.43}$$

*T. Svedberg and K. O. Pedersen, *The Ultracentrifuge.* Oxford, UK: Oxford University Press, 1940.

Table 21.3 Physical Constants of Proteins at 20 °C in Water

Protein	$S/10^{-13}$ s	$D/10^{-11}$ m^2 s^{-1}	v/cm^3 g^{-1}	M/g mol^{-1}
Beef insulin	1.7	15.0	0.72	12 000
Lactalbumin	1.9	10.6	0.75	17 400
Myoglobin	2.06	12.4	0.749	16 000
Ovalbumin	3.6	7.8	0.75	44 000
Serum albumin	4.3	6.15	0.735	64 000
Hemoglobin	4.6	6.9	0.749	64 400
Serum globulin	7.1	4.0	0.75	167 000
Urease	18.6	3.4	0.73	490 000
Tobacco mosaic virus	185.0	0.53	0.72	40 000 000

The sedimentation coefficient by itself cannot be used to determine the molar mass of the sedimenting component unless the molecules are spherical. If the molecules are spherical, $f = 6\pi\eta r$, and equation 21.43 may be used to calculate the molar mass. Since the velocity of sedimentation is so low that there is no appreciable orientation of the molecules, the frictional coefficient involved in sedimentation is taken to be the same as that involved in diffusion. Introduction of equation 20.11 into equation 21.43 yields

$$M = \frac{RTS}{D(1 - v\rho)} \tag{21.44}$$

To calculate the molar mass from measured values of S and D, it is necessary to correct sedimentation and diffusion coefficients to the same temperature, usually 20 °C, and if S and D depend appreciably on concentration, to zero concentration. Equation 21.44 has probably been the most widely used in the calculation of molar masses of proteins, and the wide range of molar masses that can be obtained by this method is indicated by Table 21.3.

Example 21.5 ***Calculation of the molar mass of hemoglobin***

Using the data of Table 21.3, what is the molar mass of hemoglobin? The density of water at 20 °C is 0.9982×10^3 kg m^{-3}.

$$\begin{aligned} M &= \frac{RTS}{D(1 - v\rho)} \\ &= \frac{(8.31 \text{ J K}^{-1} \text{ mol}^{-1})(293 \text{ K})(4.6 \times 10^{-13} \text{ s})}{(6.9 \times 10^{-11} \text{ m}^2 \text{ s}^{-1})[1 - (0.749 \times 10^{-3} \text{ m}^3 \text{ kg}^{-1})(0.9982 \times 10^3 \text{ kg m}^{-3})]} \\ &= 64.4 \text{ kg mol}^{-1} \\ &= 64\,400 \text{ g mol}^{-1} \end{aligned}$$

For spherical macromolecules or colloidal particles, the particle radius r and molar mass M may be calculated from the experimental value of the diffusion coefficient by using equation 20.12. The molar mass of a spherical particle is given by

$$M = \frac{4\pi r^3 N_A}{3v} \tag{21.45}$$

where v is the specific volume (that is, the reciprocal of the density). If equation 21.45 is substituted in equation 20.12, we obtain

$$D = \frac{RT}{N_A 6\pi\eta}\left(\frac{4\pi N_A}{3Mv}\right)^{1/3} \tag{21.46}$$

For spherical particles, then, the diffusion coefficient is inversely proportional to the cube root of the molar mass. Of course, if the particles or molecules are not spherical, the value of the molar mass calculated from equation 21.46 will not be correct. This equation, however, does give the maximum molar mass that is consistent with a given D and v. For a nonspherical particle the frictional coefficient is larger than $6\pi\eta r$ and the molar mass is smaller than that calculated from equation 21.46.

Light scattering can also be used to measure the size of polymer molecules in solution. When the oscillating electromagnetic field of light impinges on a molecule, the molecule is polarized by the electric field vector and at the frequency of the light. This causes an oscillating dipole to be set up in the molecule, which then acts to radiate a new electromagnetic field. Since the field of a dipole radiates in all directions, we can say that the incident light has been scattered by the molecule. The theory of light scattering was developed by Rayleigh, who then explained the blue color of the sky and the red color at sunset as being due to the preferential scattering of blue light by the molecules and particles of the atmosphere.

If every molecule scatters light, then it might seem as if all liquids and solids would scatter light powerfully. However, if the concentration of molecules is the same everywhere in the system, then the summation of the scattered electromagnetic fields from all the molecules gives zero intensity except in the forward direction. Thus, as Rayleigh showed, it is the **fluctuations** in the concentration that give rise to fluctuations in the refractive index of the system, which then cause net scattering of light. This is why liquids at the critical point scatter light and become cloudy: at this point the concentration fluctuations are very large, and thus so are the refractive index fluctuations.

We can use classical electromagnetic theory to calculate the scattering of light from a single molecule. An oscillating dipole in the molecule produces an oscillating electric field at a distance r and angle ϕ with respect to the polarization of the dipole (see Fig. 21.11).

A straightforward experiment is to measure the transmitted intensity of a light beam through a polymer solution. If there is little absorption of light, the loss of intensity is due to scattering. The total scattered intensity is the integral of i over all angles, so that the transmitted intensity is given by

$$I = I_0 \, e^{-\tau x} \tag{21.47}$$

where τ is the **turbidity** and x is the thickness of the cell in the direction of the incident beam. This is of the form of Beer's law for light absorption (Section 14.4). It can be shown that the turbidity is proportional to the molar mass of the polymer and proportional to its concentration:

$$\tau = \frac{32\pi^3 n_0^2 (\mathrm{d}n/\mathrm{d}c)^2 Mc}{3N_A \lambda^4} = HMc \tag{21.48}$$

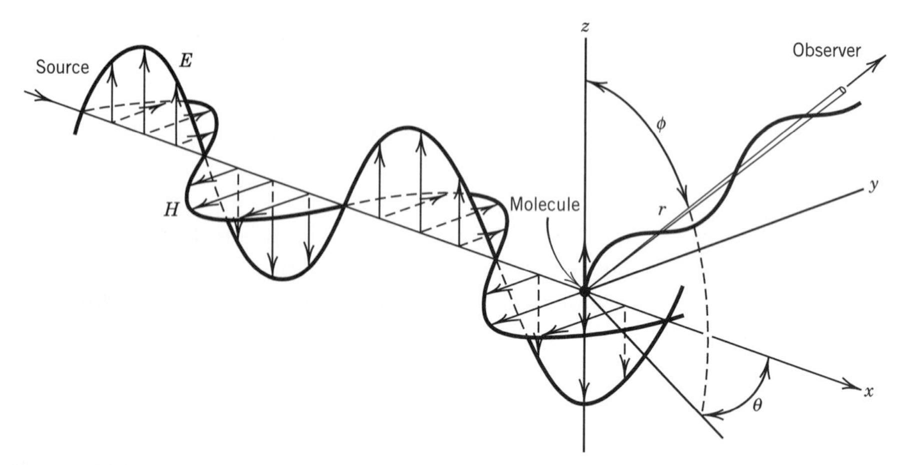

Figure 21.11 Scattering of plane-polarized light by a single molecule. For an isotropic molecule the induced dipole moment is along the z axis. Therefore, no light is scattered in the z direction.

where

$$H = \frac{32\pi^3 n_0^2 (\mathrm{d}n/\mathrm{d}c)^2}{3N_A\lambda^4} \tag{21.49}$$

and n_0 is the refractive index of the solvent, $\mathrm{d}n/\mathrm{d}c$ is the rate of change of the refractive index of the solution with increasing concentration, and λ is the wavelength of the light that is scattered. Since the turbidity τ is proportional to the molar mass of the polymer, this method is more useful as the molar mass increases, in contrast with osmotic pressure, which is inversely proportional to molar mass.

If one of the dimensions of the polymer is comparable to or greater than the wavelength of light, then there will be interference of the scattered waves from different parts of the same molecule. It is found that large molecules tend to scatter more light in the forward direction, and so the angular dependence of light scattering can also be used to obtain information about polymer shape.

Seven Key Ideas in Chapter 21

1. Synthetic polymers and the naturally occurring polymers from living organisms (proteins, nucleic acids, and polysaccharides) pose tremendous challenges to understanding their structures and the relation between structure and function.
2. The osmotic pressure provides a way to determine number average molar mass. The theta temperature of a polymer–solvent system is the analogue of the Boyle temperature of nonideal gases.
3. The spatial configurations of chains of synthetic high polymers can be discussed in terms of the freely jointed chain model or the freely rotating chain model. These calculations lead to mean-square end-to-end distances and radii of gyration.
4. The mechanisms and product distributions of step-growth polymerization and free-radical polymerization are very different. In step-growth polymerization, high molecular masses are obtained only in the last several percent of reaction. In free-radical polymerization, the chains grow rapidly until the free radical of a growing chain reacts with the free radical of another chain.

5. The intrinsic viscosity of a high polymer solution increases with some fractional power of the relative molar mass, where that fraction is in the range 0.5 to 0.8. The intrinsic viscosity of DNA is extremely high.
6. Sedimentation constants determined with an ultracentrifuge and diffusion constants can be used to calculate molar masses of proteins from the smallest to the largest molecules.
7. Measurements of light scattering from a solution can be used to determine molar masses of macromolecules because fluctuations in local concentration give rise to fluctuations in refractive index. The intensity of light scattered is proportional to the molar mass.

REFERENCES

R. B. Bird, W. E. Stewart, and E. N. Lightfoot, *Transport Phenomena.* Hoboken, NJ: Wiley, 2001.

F. A. Bovey and F. H. Winslow, *Macromolecules.* New York: Academic, 1979.

C. R. Cantor and P. R. Schimmel, *Biophysical Chemistry.* San Francisco: Freeman, 1980.

P. J. Flory, *Principles of Polymer Chemistry.* Ithaca, NY: Cornell University Press, 1953.

R. S. Seymour and C. E. Carraher, *Polymer Chemistry.* New York: Dekker, 1988.

G. Strobl, *The Physics of Polymers.* New York: Springer, 1997.

S. F. Sun, *Physical Chemistry of Macromolecules.* Hoboken, NJ: Wiley-Interscience, 1994.

J. W. Williams, *Ultracentrifugation of Macromolecules: Modern Topics.* New York: Academic, 1973.

PROBLEMS

(M) Problems marked with an icon may be more conveniently solved on a personal computer with a mathematical program.

21.1 A polymer solution contains 250 molecules of molar mass 75 000 g mol^{-1}, 500 molecules of molar mass 100 000 g mol^{-1}, and 250 molecules of molar mass 125 000 g mol^{-1}. Calculate $\overline{M}_n$, $\overline{M}_m$, and the ratio $\overline{M}_m/\overline{M}_n$ (the polydispersity).

(M) **21.2** Plot the probability density $W(r)$ for random walk in three dimensions after 1000 steps with a step length of unity. Indicate the root-mean-square end-to-end distance on this plot.

21.3 In polyethene $H(CH_2{-}CH_2)_nH$ the bond length l is 0.15 nm. What is the root-mean-square end-to-end distance for a molecule with universal joints with a molar mass of 10^5 g mol^{-1}? Taking into account the fact that carbon forms tetrahedral bonds, what is $\langle r^2\rangle^{1/2}$?

21.4 Derive the expression for the mean separation $\langle r\rangle$ for the ends of a freely jointed chain of N bonds of length l. See the definite integrals in Table 17.1 or Appendix D.3.

21.5 Derive the expression for the root-mean-square separation $\langle r^2\rangle^{1/2}$ for the ends of a freely jointed chain of N bonds of length l. See the definite integrals in Table 17.1 or Appendix D.3.

21.6 For a condensation polymerization of a hydroxy acid in which 99% of the acid groups are used up, calculate (*a*) the average number of monomer units in the polymer molecules, (*b*) the probability that a given molecule will have the number of residues given by this value, and (*c*) the weight fraction having this particular number of monomer units.

21.7 A hydroxy acid $HO{-}(CH_2)_5{-}CO_2H$ is polymerized, and it is found that the product has a number average molar mass of 20 000 g mol^{-1}. (*a*) What is the extent of reaction p? (*b*) What is the degree of polymerization $\overline{X}_n$? (*c*) What is the mass average molar mass?

21.8 A general polymerization reaction in the liquid phase can be written

$$M = \frac{1}{n}P_n$$

The values of $\Delta_r H°$ and $\Delta_r G°$ have been determined for some polymerization reactions. This makes it possible to calculate $\Delta_r S°$, and some values at 25 °C are shown in the following table:

Monomer	$\dfrac{-\Delta_r H°}{\text{kJ mol}^{-1}}$	$\dfrac{-\Delta_r S°}{\text{J K}^{-1}\text{ mol}^{-1}}$	$\dfrac{-\Delta_r G°}{\text{kJ mol}^{-1}}$
Styrene	69.9	104	38.5
α-Methylstyrene	35.2	104	4.2
Tetrafluoroethylene	154.8	112	121

If we assume that $\Delta_r H°$ and $\Delta_r S°$ are independent of temperature, we can calculate the temperature at which the equilibrium constant for the polymerization reaction is unity. At this temperature, depolymerization occurs, and that temperature is called the ceiling temperature. Calculate the ceiling temperatures of these three polymers, and interpret these temperatures.

21.9 Show that the intrinsic viscosity can also be defined by

$$[\eta] = \lim_{c \to 0} \left(\frac{1}{c}\right) \ln\left(\frac{\eta}{\eta_0}\right)$$

[*Hint:* $\ln(1+x) \approx x$ if $x \ll 1$.]

Ⓜ **21.10** The relative viscosities of a series of solutions of a sample of polystyrene in toluene were determined with an Ostwald viscometer at 25 °C:

$c/10^{-2}$ g cm^{-3}	0.249	0.499	0.999	1.998
η/η_0	1.355	1.782	2.879	6.090

The ratio η_{sp}/c was plotted against c and extrapolated to zero concentration to obtain the intrinsic viscosity. If the constants in equation 21.65 are $K = 3.7 \times 10^{-2}$ and $a = 0.62$ for this polymer, when concentrations are expressed in g/cm^3, calculate the molar mass.

21.11 At 34 °C the intrinsic viscosity of a sample of polystyrene in toluene is 84 cm^3 g^{-1}. The empirical relation between the intrinsic viscosity of polystyrene in toluene and molar mass is

$$[\eta] = 1.15 \times 10^{-2} M^{0.72}$$

What is the molar mass of this sample?

21.12 Given that the intrinsic viscosity of myosin is 217 cm^3 g^{-1}, approximately what concentration of myosin in water would have a relative viscosity of 1.5?

21.13 The sedimentation coefficient of myoglobin at 20 °C is 2.06×10^{-13} s. What molar mass would it have if the molecules were spherical? Given: $v = 0.749 \times 10^{-3}$ m^3 kg^{-1}, $\rho = 0.9982 \times 10^3$ kg m^{-3}, and $\eta = 0.001\,005$ Pa s.

21.14 The sedimentation and diffusion coefficients for hemoglobin corrected to 20 °C in water are 4.41×10^{-13} s and 6.3×10^{-11} m^2 s^{-1}, respectively. If $v = 0.749$ cm^3 g^{-1} and $\rho_{H_2O} = 0.998$ g cm^{-3} at this temperature, calculate the molar mass of the protein. If there is 1 mol of iron per 17 000 g of protein, how many atoms of iron are there per hemoglobin molecule?

21.15 Given the diffusion coefficient for sucrose at 20 °C in water ($D = 45.4 \times 10^{-11}$ m^2 s^{-1}), calculate its sedimentation coefficient. The partial specific volume v is 0.630 cm^3 g^{-1}.

21.16 A beam of sodium D light (589 nm) is passed through 100 cm of an aqueous solution of sucrose containing 10 g sucrose per 100 cm^3. Calculate I/I_0, where I_0 is the intensity that would have been obtained with pure water, given that $M = 342.30$ g mol^{-1} and $dn/dc = 0.15$ g^{-1} cm^3 for sucrose. The refractive index of water at 20 °C is 1.333 for the sodium D line.

21.17 (*a*) Integrate equation 21.41 to obtain S as a function of t_1, t_2, r_1, and r_2. (*b*) In an ultracentrifuge experiment, the sedimenting boundary of a protein is 5.949 cm from the axis at t_1 and 6.731 cm from the axis at $t_1 + 70$ minutes. If the speed of the rotor is 50 400 rpm, what is the sedimentation coefficient of the protein?

21.18 A sample of polymer contains 0.50 mole fraction with molar mass 100 000 g mol^{-1} and 0.50 mole fraction with molar mass 200 000 g mol^{-1}. Calculate (*a*) M_n and (*b*) M_m.

21.19 Calculate (*a*) number average and (*b*) mass average molar masses for the following mixture of high polymer fractions: 1 g of $M = 20$ kg mol^{-1}, 2 g of $M = 50$ kg mol^{-1}, and 0.5 g of $M = 100$ kg mol^{-1}.

21.20 Human blood plasma contains approximately 40 g of albumin ($M = 69\,000$ g mol^{-1}) and 20 g of globulin ($M = 160\,000$ g mol^{-1}) per liter. Calculate the colloid osmotic pressure at 37 °C, ignoring the Donnan effect.

21.21 For a condensation polymerization of a hydroxy acid in which 95% of the acid groups are used up, calculate (*a*) the average number of monomer units in the polymer molecules, (*b*) the probability that a molecule chosen at random will have this number of residues, and (*c*) the weight fraction having this particular number of monomer units.

21.22 For the polymer described in Problem 21.21, calculate the number average and mass average molar mass.

21.23 In the condensation polymerization of a hydroxy acid with a residue mass of 200, it is found that 99% of the acid groups are used up. Calculate (*a*) the number average molar mass and (*b*) the mass average molar mass.

21.24 The relation between M and $[\eta]$ for double-stranded linear DNA is $0.665 \ln M = 1.987 + \log([\eta] + 500)$ when $[\eta]$ is expressed in cm^3 g^{-1}. What is the molar mass of DNA that has an intrinsic viscosity of 5000 cm^3 g^{-1}?

Ⓜ **21.25** A sample of polystyrene was dissolved in toluene, and the following flow times in an Ostwald viscometer at 25 °C were obtained for different concentrations:

$c/10^{-2}$ g cm^{-3}	0	0.1	0.3	0.6	0.9
t/s	86.0	99.5	132	194	301

If the constants in equation 21.65 are $K = 3.7 \times 10^{-2}$ and $a = 0.62$ for this polymer, calculate the molar mass.

21.26 Given that the viscosity of a suspension of spheres is given by

$$\eta = \eta_0\left(1 + \tfrac{5}{2}\phi\right)$$

where η_0 is the viscosity of the solvent and ϕ is the volume fraction of the spheres, calculate the intrinsic viscosity of a solution of a spherical protein molecule with a partial specific volume v of 0.75 cm^3 g^{-1}. Assume that the molecules are not hydrated. Note that the intrinsic viscosity is independent of the radius of the spheres.

21.27 When a solution of a protein is in a cell in the rotor of an ultracentrifuge that is run for a long time, an equilibrium is finally reached in which the concentration c_2 at the bottom (r_2) of the cell is higher than the concentration c_1 at the top of the cell (r_1) because of the difference in centrifugal potential (note $r_2 > r_1$). Thermodynamics shows that at equilibrium

$$\ln \frac{c_2}{c_1} = \frac{M(1 - v\rho)\omega^2(r_2^2 - r_1^2)}{2RT}$$

where ω is the angular velocity in radians per second. An experiment is to be carried out with myoglobin ($M = 16\,000$ g mol^{-1}) in an ultracentrifuge operating at 15 000 rpm. The bottom of the cell is 6.93 cm from the axis of rotation, and the meniscus is 6.67 cm from the axis of rotation. What ratio of concentrations is expected at 20 °C if the partial specific volume of the protein is $v = 0.75 \times 10^{-3}$ m^3 kg^{-1} and the density of the solvent is $\rho = 1.00 \times 10^3$ kg m^{-3}?

21.28 Calculate the sedimentation coefficient of tobacco mosaic virus from the fact that the boundary moves with a velocity of 0.454 cm h^{-1} in an ultracentrifuge at a speed of 10 000 rpm at a distance of 6.5 cm from the axis of the centrifuge rotor.

21.29 The sedimentation coefficient of gamma-globulin at 20 °C is 7.1×10^{-13} s. Calculate how far the protein boundary will sediment in $\frac{1}{2}$ h if the speed of the centrifuge is 60 000 rpm and the initial boundary is 6.50 cm from the axis of rotation.

21.30 Using data from Table 21.3, calculate the molar mass of serum albumin.

21.31 The diffusion coefficient for serum globulin at 20 °C in a dilute aqueous salt solution is 4.0×10^{-11} m^2 s^{-1}. If the molecules are assumed to be spherical, calculate their molar mass. Given: $\eta_{H_2O} = 0.001\,005$ Pa s at 20 °C and $v = 0.75$ cm^3 g^{-1} for the protein.

21.32 The diffusion coefficient of hemoglobin at 20 °C is 6.9×10^{-11} m^2 s^{-1}. Assuming its molecules are spherical, what is the molar mass? Given: $v = 0.749 \times 10^{-3}$ m^3 kg^{-1} and $\eta = 0.001\,005$ J m^{-3} s.

21.33 The diffusion coefficient of a certain virus having spherical particles is 0.50×10^{-11} m^2 s^{-1} at 0 °C in a solution with a viscosity of 0.001 80 Pa s. Calculate the molar mass of this virus, assuming that the density of the virus is 1 g cm^{-3}.

21.34 The diffusion coefficient of myoglobin is 12.4×10^{-11} m^2 s^{-1} at 20 °C, and the viscosity of water is 0.001 005 Pa s. The partial specific volume is 0.749 cm^3 g^{-1}. What is the molar mass, assuming the protein is not hydrated and is spherical?

21.35 A solution of a high polymer in benzene has a concentration of 1 g/100 cm^3 and a refractive index for the sodium D line (589 nm) of 1.5021. The refractive index of benzene under these conditions is 1.5011. The turbidity τ is 2×10^{-4} cm^{-1}. What is the molar mass of the polymer? What is I/I_0 for a 10-cm cell?

Computer Problems

21.A (*a*) Plot the probability density $W(\xi)$ that a random walker along a line is at the distance ξ to $\xi + d\xi$ from the origin after 100 steps, 1000 steps, and 10 000 steps. Assume that the step length is $h = 1$ m. (*b*) Check that the standard deviation in each case is given by $\sigma = n^{1/2}h$.

21.B For polyethylene, $H(CH_2CH_2)_nH$, (*a*) calculate the root-mean-square end-to-end distance for a freely jointed chain and a freely rotating chain for molar masses of 10^3, 10^4, and 10^5 g mol^{-1} and make a table. (*b*) Calculate the radius of gyration for a freely jointed chain and a freely rotating chain for molar masses of 10^3, 10^4, and 10^5 g mol^{-1} and make a table. The bond length is 0.15 nm, and there are tetrahedral bond angles (109°).

21.C (*a*) Plot the mole fraction distribution of condensation polymer versus chain length i for extents of reaction from 0.95 to 0.99. (*b*) Plot the weight fraction distribution of condensation polymer versus chain length i for extents of reaction from 0.95 to 0.99.

21.D Plot the viscosity data in Problem 21.10 as $(1/c)\ln(\eta/\eta_0)$ to see whether the same intrinsic viscosity is obtained as in Problem 21.10.

21.E A hydroxy acid with $M_0 = 100$ g/mol is polymerized to the point that 99% of the monomers have reacted. (*a*) Plot the weight fraction w_i versus the chain size i. (*b*) Plot the weight fraction w_i versus the molar mass of the polymer. (*c*) Calculate the number average molar mass and the mass average molar mass, and indicate them on the plot in (*b*).

22

Electric and Magnetic Properties of Molecules

Much of our knowledge of molecular systems has been obtained by studying the ways in which they respond to electric and magnetic fields, especially their response to electromagnetic radiation in spectroscopic experiments. Dipole moments were defined in connection with molecular electronic structure in Section 11.8. We have also considered electric dipole moments in connection with transition moments (Section 13.2), intermolecular interactions (Section 11.9), and the Raman effect (Section 13.9). Nuclear magnetic moments have been discussed in connection with nuclear magnetic resonance. We have completed our discussions of spectroscopy, but now we will be concerned with electric and magnetic properties of bulk matter. Electric dipole moments are involved in dielectric and optical properties of bulk matter, and magnetic dipole moments are involved in diamagnetism, paramagnetism, and ferromagnetism. Liquids with dipolar molecules are called polar solvents. Polar properties are especially important for the ability of a solvent to dissolve electrolytes. In a solution of an electrolyte in a polar solvent the ions are surrounded by dipolar solvent molecules oriented around ions. The energy of this interaction makes the solution more stable.

22.1 POLARIZATION OF A DIELECTRIC

The capacitance C of a capacitor is the ratio of the charge Q on the conducting plates to the potential difference $\Delta\phi$ between the plates:

$$C = \frac{Q}{\Delta\phi} \tag{22.1}$$

Here we will consider only parallel-plate capacitors, such as that shown in Fig. 22.1. If the capacitor has vacuum between the plates, the capacitance C_0 is given by

$$C_0 = \frac{\epsilon_0 A}{d} \tag{22.2}$$

where A is the area of a plate, d is the distance between plates, and ϵ_0 is the **permittivity of vacuum** [$\epsilon_0 = 1/\mu_0 c^2 = 8.854\,187\,817\ldots \times 10^{-12}$ C^2 N^{-1} m^{-2} (exactly), where $\mu_0 = 4\pi \times 10^{-7}$ N A^{-2} (exact) is called the permeability of vacuum]. If a nonconductor (dielectric) is inserted between the plates, the capacitance is increased. To understand why the capacitance is increased, we need to consider the polarization of the dielectric that is illustrated in Fig. 22.1.

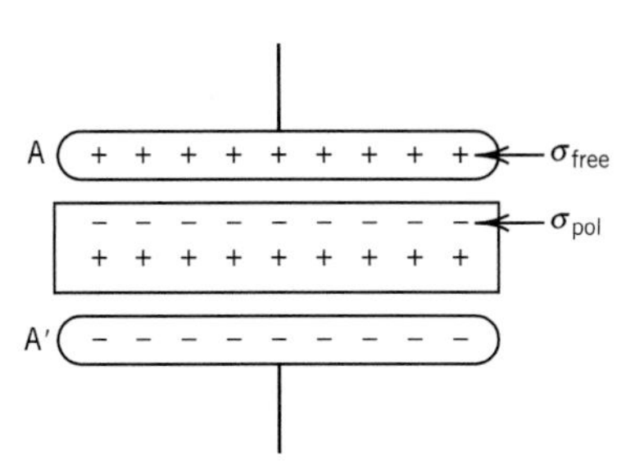

Figure 22.1 Parallel-plate capacitor filled with a dielectric. The plates are labeled A and A′; the dielectric is between the plates.

When a dielectric is placed in an electric field, the electrons in the molecules are pulled toward the positive plate and the positively charged nuclei are pulled toward the negative plate. The charges remain bound in their atoms or molecules, but the center of negative charge in a molecule is shifted with respect to the center of positive charge. The dielectric remains electrically neutral in the bulk because the positive and negative charges remain bound, but this displacement produces a layer of negative charges on the surface of the dielectric close to the positive plate and a layer of positive charges on the surface close to the negative plate, as shown in Fig. 22.1. These surface charges reduce the electric field strength within the dielectric because they produce an electric field in the opposite direction from that of the capacitor plates.

The dielectric in a capacitor is said to be polarized by the application of the electric field, and the **polarization** $\boldsymbol{P}$ of the dielectric is defined as the vector sum of dipole moments $\boldsymbol{\mu}_i$ of individual molecules per unit volume ($\boldsymbol{P} = \sum \boldsymbol{\mu}_i/V$). The polarization $\boldsymbol{P}$ is a vector pointing in the direction of the dipoles, but we will simply use its magnitude P. To have a simple model of the dielectric, we will assume that each molecule has a charge $+q$ that is separated from a charge $-q$ by a distance δ at a particular electric field strength so that each of the N molecules per unit volume contributes a dipole moment of $q\delta$. Thus the magnitude of the polarization is given by

$$P = Nq\delta \tag{22.3}$$

The dipole moment per unit volume is equal to the surface charge per unit area σ_{pol}, which we will refer to as the surface charge density on the dielectric: $P = \sigma_{\text{pol}}$.

We want to calculate the capacitance of the capacitor with a dielectric between the plates. The charge density on the plates is represented by σ_{free} because these charges can move through the wires connected to the plates; this charge density is unchanged by inserting the dielectric. When a dielectric is placed between the plates, Gauss's law shows that the electric field strength E in the dielectric is given by

$$E = \frac{\sigma_{\text{free}} - \sigma_{\text{pol}}}{\epsilon_0} \tag{22.4}$$

If there were no polarization, the electric field would be $\sigma_{\text{free}}/\epsilon_0$. Since σ_{pol} is equal to the polarization, this equation can be written

$$E = \frac{\sigma_{\text{free}} - P}{\epsilon_0} \tag{22.5}$$

The polarization is produced by the electric field, and so we cannot use this equation to calculate the electric field unless we know how P depends on E. At electric fields that are not too high, the polarization is proportional to the electric field strength, and this proportionality can be written

$$P = \chi_{el}\epsilon_0 E \tag{22.6}$$

where χ_{el} is the **electric susceptibility** of the dielectric. The susceptibility of vacuum is zero. Substituting equation 22.6 in 22.5 yields

$$E = \frac{\sigma_{free}}{\epsilon_0(1+\chi_{el})} \tag{22.7}$$

Thus the electric field strength in the dielectric is reduced by a factor of $1/(1+\chi_{el})$ by the charge on the surface of the dielectric.

Now we can calculate the capacitance of a capacitor filled with a dielectric because the potential difference $\Delta\phi$ between the plates is given by the product of the electric field strength E and the distance d between the plates.

$$\Delta\phi = Ed = \frac{\sigma_{free} d}{\epsilon_0(1+\chi_{el})} \tag{22.8}$$

Substituting this into the expression $C = \sigma_{free}A/\Delta\phi$ (equation 22.1) for the capacitance yields

$$C = \frac{\epsilon_0(1+\chi_{el})A}{d} = \frac{\epsilon_r\epsilon_0 A}{d} \tag{22.9}$$

where $\epsilon_r = 1+\chi_{el}$ is the **relative permittivity** (dielectric constant). Comparison of this equation with equation 22.2 shows that the relative permittivity is equal to the ratio of the capacitance C of the capacitor filled with dielectric to the capacitance C_0 with vacuum between the plates:

$$\epsilon_r = \frac{C}{C_0} \tag{22.10}$$

The relative permittivities of a number of substances are given in Table 22.1.

Table 22.1 Relative Permittivities ϵ_r of Gases and Liquids

Gas (1 atm)	ϵ_r *at* 0 °C	*Liquid*	ϵ_r *at* 20 °C
Hydrogen	1.000 272	Hexane	1.874
Argon	1.000 545	Cyclohexane	2.023
Air (CO_2 free)	1.000 567	Carbon tetrachloride	2.238
Carbon monoxide	1.000 70	Benzene	2.283
Methyl chloride	1.000 94	Toluene	2.387
Methane	1.000 944	Chlorobenzene	5.708
Carbon dioxide	1.000 985	Acetic acid	6.15
Ethane	1.001 50	Ammonia	15.5
Hydrogen iodide	1.002 34	Acetone	21.4
Hydrogen chloride	1.004 6	Methanol	33.6
Ammonia	1.007 2	Nitrobenzene	35.74
Water (steam at 110 °C)	1.012 6	Water	80.37

Since the polarization is given by $P = \chi_{el}\epsilon_0 E$ (equation 22.6),

$$\epsilon_r - 1 = \frac{P}{\epsilon_0 E} \tag{22.11}$$

and measurements of the relative permittivity yield information on the polarization of the dielectric. There are two kinds of polarization, **distortion polarization** and **orientation polarization.** All molecules undergo distortion polarization when they are placed in an electric field, and those with dipole moments become partially oriented in an electric field in addition if they are free to rotate, as they are in gases and liquids. First we will discuss distortion polarization and then orientation polarization.

22.2 POLARIZABILITY OF A DIELECTRIC

Dipole moments can be induced in all substances by application of an electric field. If the electric field strength $\boldsymbol{E}$ is not too high, the dipole moment due to the distortion polarization is given by

$$\boldsymbol{\mu} = \boldsymbol{\alpha E} \tag{22.12}$$

where $\boldsymbol{\alpha}$ is the molecular **polarizability.** In general, the polarizability is a matrix, and we have seen this matrix in equation 13.111. Equation 22.12 provides for the fact that the dipole moment vector is not necessarily parallel to the electric field strength vector. If $\alpha_{xx} = \alpha_{yy} = \alpha_{zz}$ and the off-diagonal components are equal to zero, the induced dipole moment is parallel to the electric field vector, and the substance is said to be **isotropic.** Since the dipole moment has units C m, and the electric field strength has units of V m^{-1} = J C^{-1} m^{-1}, the polarizability has the units C^2 m^2 J^{-1}. Values of the mean electric polarizability, that is, $(\alpha_{xx} + \alpha_{yy} + \alpha_{zz})/3$, are given for a number of molecules in the gas phase in Table 22.2 in the next section. Mean polarizabilities tend to be approximately proportional to molecular volumes.*

The polarization of a macroscopic sample is the number of molecules per unit volume (N/V) multiplied by the dipole moment αE:

$$P = \frac{N}{V}\alpha E \tag{22.13}$$

Note that this is a relation between a bulk (macroscopic) property (P) and a molecular property (α).

22.3 ORIENTATION POLARIZATION OF A DIELECTRIC

In the absence of an electric field, the individual molecular dipoles point in random directions, and so the net dipole moment per unit volume is zero. When an electric field is applied, the dipoles tend to line up and produce a net electric polarization. If all the dipoles were to line up, a very large polarization would be

*Electric polarizabilities are often given in the literature as $\alpha' = \alpha/4\pi\epsilon_0$, where the permittivity of vacuum ϵ_0 is $8.854\,19 \times 10^{-12}$ J^{-1} C^2 m^{-1}. α' has the units of volume and may be expressed in units of a_0^3, where a_0 is the Bohr radius.

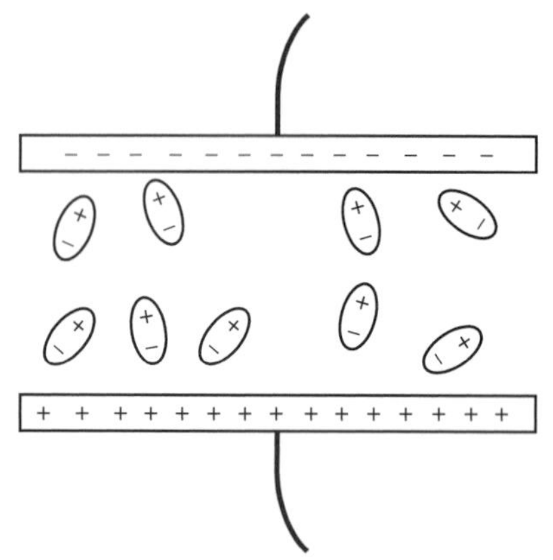

Figure 22.2 Representation of a capacitor containing dipolar molecules. The molecules are shown in their instantaneous positions indicating the partial orientation of the dipoles.

obtained. Because of thermal motion, the orientation is only partial, as illustrated in Fig. 22.2. Statistical mechanics can be used to calculate the polarization as a function of temperature. The energy of a dipole in an electric field is given by

$$U = -\boldsymbol{\mu} \cdot \boldsymbol{E} = -\mu E \cos\theta \tag{22.14}$$

where θ is the angle between $\boldsymbol{\mu}$ and $\boldsymbol{E}$. At equilibrium the relative number of molecules with potential energy U is proportional to the Boltzmann factor $\exp(-U/kT)$. The number of molecules $N(\theta)$ per unit solid angle at θ, N, is given by

$$N(\theta) = N_0\, e^{\mu E \cos\theta/kT} \tag{22.15}$$

When the exponent is small, this can be approximated with

$$N(\theta) = N_0 \left(1 + \frac{\mu E \cos\theta}{kT}\right) \tag{22.16}$$

Each molecule contributes a term $\mu \cos\theta$ to the net dipole moment in the direction of the electric field. Multiplying $\mu \cos\theta$ by $N(\theta)$ and integrating over the angular distribution shows that the polarization due to orientation of dipoles is given by

$$P = \frac{N}{V} \frac{\mu^2 E}{3kT} \tag{22.17}$$

The magnitude of the polarization of a dielectric in an electric field is given by the sum of the distortion polarization (equation 22.13) and the orientation polarization (equation 22.17):

$$P = \frac{N}{V}\left(\alpha + \frac{\mu^2}{3kT}\right) E_i \tag{22.18}$$

where E_i is the average local electric field. The average local field is not the same as the applied field because the molecular dipoles also produce a field. If the concentration of polar molecules is low, it can be shown that $E_i = E(\epsilon_r + 2)/3$. Substituting this expression for E_i in equation 22.18 and eliminating the polarization between equations 22.11 and 22.18 yields

$$\frac{\epsilon_r - 1}{\epsilon_r + 2} = \frac{N}{3V\epsilon_0}\left(\alpha + \frac{\mu^2}{3kT}\right) \tag{22.19}$$

Replacing N/V with $N_A\rho/M$, where M is the molar mass of the substance studied and ρ is the density, yields

$$P_m = \frac{\epsilon_r - 1}{\epsilon_r + 2} \frac{M}{\rho} = \frac{N_A}{3\epsilon_0}\left(\alpha + \frac{\mu^2}{3kT}\right) \tag{22.20}$$

where P_m is the **molar polarization.** Note that this connects the macroscopic property P_m to the molecular properties α and μ. The molar polarization has the units of a molar volume (m^3 mol^{-1}). Measurement of the molar polarization at a series of temperatures makes it possible to determine the polarizability α and the dipole moment μ. A plot of the molar polarization of fluorobenzene versus $1/T$

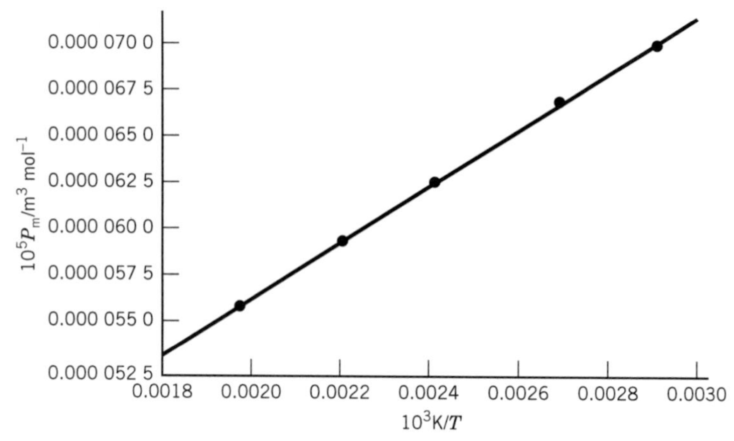

Figure 22.3 Dependence of the molar polarization of fluorobenzene on temperature. (See Computer Problem 22.A.)

is given in Fig. 22.3. The intercept yields the value of the molecular polarizability $\alpha = 1.146 \times 10^{-39}$ C^2 m^2 J^{-1}. The slope yields the dipole moment 5.26×10^{-30} C m. Dipole moments of molecules are often expressed in terms of debye units, represented by D, which are equal to 3.336×10^{-30} C m. Thus the dipole moment of fluorobenzene is 1.57 D. The dipole moments and polarizabilities of some gaseous molecules are given in Table 22.2.

Table 22.2 Dipole Moments and Polarizabilities of Gaseous Molecules

Gas (1 atm)	$\mu/10^{-30}$ C m	$\alpha/10^{-40}$ C^2 m^2 J^{-1}
Ar	0	1.85
He	0	0.22
N_2	0	1.97
H_2	0	0.911
CH_4	0	2.89
C_6H_6	0	11.6
CCl_4	0	11.7
CO_2	0	2.93
CO	0.390	2.20
HF	6.37	0.57
HCl	3.60	2.93
HBr	2.67	4.01
HI	1.40	6.06
$CHCl_3$	3.37	9.46
CH_2Cl_2	5.24	7.57
CH_3Cl	6.24	5.04
CH_3OH	5.70	3.59
H_2O	6.17	1.65
NH_3	4.90	2.47

Source: C. J. F. Böttcher and P. Bordewijk, *Theory of Electric Polarization.* Copyright (1978) with permission from Elsevier Science.

Example 22.1 ***The relative permittivity of $H^{35}Cl(g)$***

Given the polarizability α and dipole moment of HCl(g) in Table 22.2, calculate the relative permittivity ϵ_r at 1 atm and 273 K.

The relation between these quantities is given by equation 22.19. If we treat HCl(g) as an ideal gas,

$$\frac{N}{V} = \frac{N_A P}{RT}$$

Thus equation 22.19 can be written as

$$\frac{\epsilon_r - 1}{\epsilon_r + 2} = \frac{N_A P}{RT}\frac{1}{3\epsilon_0}\left(\alpha + \frac{\mu^2}{3kT}\right)$$

First we will calculate the factor

$$\frac{N_A P}{RT}\frac{1}{3\epsilon_0} = \frac{(6.022 \times 10^{23}\ \text{mol}^{-1})(1.013 \times 10^5\ \text{N m}^{-2})}{(8.3145\ \text{J K}^{-1}\ \text{mol}^{-1})(273\ \text{K})(3)(8.854 \times 10^{-12}\ \text{C}^2\ \text{N}^{-1}\ \text{m}^{-2})}$$
$$= 1.0119 \times 10^{36}\ \text{C}^{-2}\ \text{m}^{-2}\ \text{J}$$

The contribution of the dipole moment is

$$\frac{\mu^2}{3kT} = \frac{(3.60 \times 10^{-30}\ \text{C m})^2}{3(1.381 \times 10^{-23}\ \text{J K}^{-1})(273\ \text{K})}$$
$$= 1.146 \times 10^{-39}\ \text{C}^2\ \text{m}^2\ \text{J}^{-1}$$

Adding the contribution from the polarizability α gives

$$\frac{\epsilon_r - 1}{\epsilon_r + 2} = (1.0119 \times 10^{36}\ \text{C}^{-2}\ \text{m}^{-2}\ \text{J})(0.293 + 1.146) \times 10^{-39}\ \text{C}^2\ \text{m}^2\ \text{J}^{-1}$$
$$= 1.456 \times 10^{-3}$$

Thus

$$\epsilon_r = \frac{1 + 2(1.456 \times 10^{-3})}{1 - 1.456 \times 10^{-3}} = 1.0041$$

Table 22.1 gives a value of 1.0046 for the relative permittivity of HCl(g) under these conditions.

22.4 REFRACTIVE INDEX

The **refractive index** n is the ratio of the speed of light in a vacuum to the speed v of light in a medium. Electromagnetic theory shows that the speed of light in a vacuum is given by

$$c = \frac{1}{(\epsilon_0 \mu_0)^{1/2}} \tag{22.21}$$

This is the relation between the permittivity of vacuum ϵ_0 and the permeability of vacuum μ_0. The permeability of vacuum is given exactly by

$$\mu_0 = 4\pi \times 10^{-7}\ \text{N A}^{-2} \tag{22.22}$$

When light travels through a medium, its speed is given by

$$v = \frac{1}{(\epsilon_r \epsilon_0 \mu_r \mu_0)^{1/2}} = \frac{c}{n} \tag{22.23}$$

where ϵ_r is the **relative permittivity** and μ_r is the **relative permeability** (see Section 22.5). Using equation 22.21 to eliminate $\epsilon_0 \mu_0$ shows that the refractive index is given by

$$n = (\epsilon_r \mu_r)^{1/2} \tag{22.24}$$

If the relative permeability μ_r is taken as unity, which is a pretty good approximation for all materials that are not ferromagnetic, we see that

$$n = \epsilon_r^{1/2} \tag{22.25}$$

The refractive index depends on the relative permittivity because the electromagnetic wave induces polarization or orients electric dipoles and these dipoles interact with electromagnetic radiation. Optical properties and the dielectric properties of bulk matter can be treated together, and both can be interpreted in terms of molecular properties.

At optical frequencies, equation 22.19 becomes

$$\frac{\epsilon_r - 1}{\epsilon_r + 2} = \frac{N\alpha}{3V\epsilon_0} \tag{22.26}$$

because there is not enough time in a cycle to reorient permanent dipoles. Substituting equation 22.25 yields

$$\frac{n^2 - 1}{n^2 + 2} = \frac{N\alpha}{3\epsilon_0 V} \tag{22.27}$$

Thus the refractive index provides information on the polarizability α.

Since both the refractive index and the polarizability depend on frequency, we want to consider that briefly. First, let us consider a molecule without a permanent dipole. It can be shown that in the absence of damping, the polarizability α varies with frequency according to $1/(\omega_0^2 - \omega^2)$, where ω_0 is a transition frequency of the molecule and ω is the frequency of the electromagnetic radiation. This leads to the frequency dependence of the refractive index that is shown in Fig. 22.4. As ω approaches ω_0 from lower frequencies, the refractive index increases to $+\infty$. As the frequency increases above ω_0, the refractive index increases from $-\infty$ to unity. However, because of radiative damping of the transition moment, the dashed line is followed. This is the behavior of n at infrared frequencies and above.

Second, we consider a molecule with a permanent dipole moment, where orientation polarization is important. Since it takes time to orient a physical dipole, the radiation from the dipole is delayed and the refractive index is increased. Consequently, the relative permittivity ϵ_r at low frequency (where there is time to orient the dipoles) is greater than at high frequency where there is not enough time in a cycle. At low frequency, the dependence of the relative permittivity on frequency ω can be represented by

$$\epsilon_r = \epsilon_h + \frac{\epsilon_l - \epsilon_h}{1 + i\omega\tau} \tag{22.28}$$

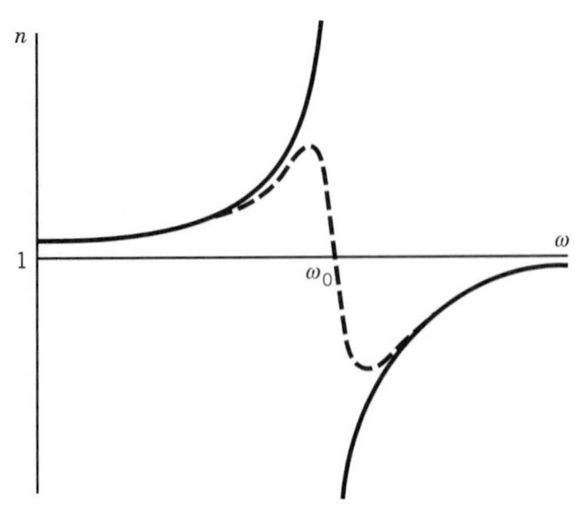

Figure 22.4 Frequency dependence of the refractive index for a harmonic oscillator of frequency ω_0. Since the oscillation is damped, the refractive index actually follows the dashed line.

where ϵ_h is the relative permittivity at high frequencies and ϵ_l is the relative permittivity at low frequencies, $i = (-1)^{1/2}$, and τ is the relaxation time for the orientation of a dipole. The relative permittivity ϵ_r is complex and can be written

$$\epsilon_r = \epsilon' - i\epsilon'' \qquad (22.29)$$

where the real part is given by

$$\epsilon' = \epsilon_h + \frac{\epsilon_l - \epsilon_h}{1 + \omega^2\tau^2} \qquad (22.30)$$

and the imaginary part is given by

$$\epsilon'' = \frac{(\epsilon_l - \epsilon_h)\omega\tau}{1 + \omega^2\tau^2} \qquad (22.31)$$

You can verify this by substituting these relations in equation 22.29 to obtain equation 22.28. The frequency dependencies of ϵ' and ϵ'' are shown in Fig. 22.5. The real part ϵ' changes from the value ϵ_l to ϵ_h in the region where $\omega\tau \approx 0.1$ to 10. There is a corresponding change in the refractive index, and this curve is called

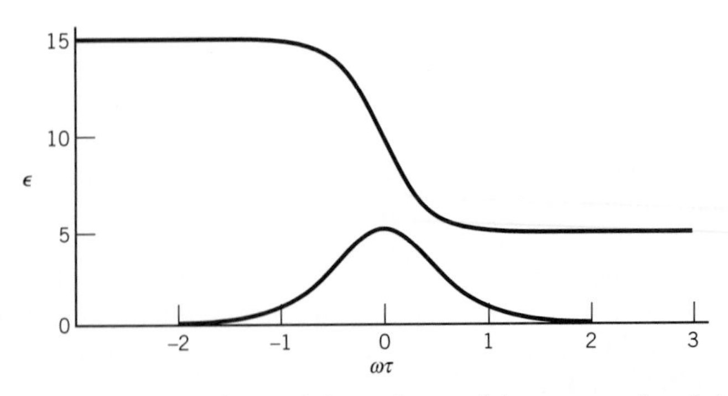

Figure 22.5 Frequency dependence of the real part ϵ' (upper curve) and the imaginary part ϵ'' (lower curve) of the relative permittivity for a process with a single relaxation time τ. This plot is for $\epsilon_l = 15$ and $\epsilon_h = 5$. (See Computer Problem 22.B.)

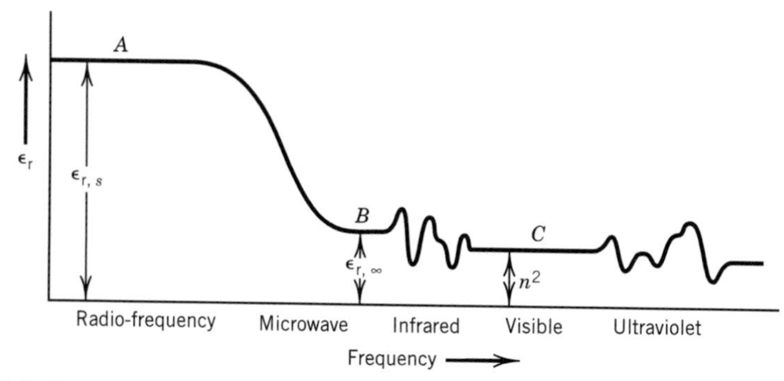

Figure 22.6 Variation of relative permittivity with frequency for a typical polar substance.

a **dispersion curve.** The imaginary part ϵ'' has significant values only in the range $\omega\tau \approx 0.1$ to 10, and it is a measure of the loss of useful energy into heat. This bell-shaped plot is called an **absorption curve.**

These ideas are also illustrated by Fig. 22.6, which shows the variation of relative permittivity with frequency for a typical polar substance. As the frequency is increased, the relative permittivity decreases in several steps. As we pointed out above, in the optical region the frequency is so high that permanent dipoles are not oriented and the refractive index varies with frequency as expected from Fig. 22.4. At low frequencies, all the terms on the right-hand side of equation 22.18 contribute and the value of ϵ_r is approximately constant, equal to $\epsilon_{r,s}$ ($= \epsilon_l$ in equation 22.28), the zero-frequency value. As the frequency is increased above the radio-frequency range, the relative permittivity decreases and the orientation polarization eventually becomes negligible because there is insufficient time for molecular orientation to occur. At these frequencies, ϵ_r reaches a plateau equal to $\epsilon_{r,\infty}$ ($= \epsilon_h$ in equation 22.28), the high frequency value. The relative permittivity shows dispersion in the neighborhood of absorption lines in the infrared, visible, and ultraviolet.

22.5 MAGNETIZATION

The treatment of the magnetization of a sample of matter by a magnetic field is similar to the polarization of a dielectric by an electric field, but there are some very significant differences. As we saw in the case of nuclear magnetic resonance, the principal measure of the effect of a magnetic field is the magnetic flux density $\boldsymbol{B}$, which we will refer to as the **magnetic field.** (In older literature, a different quantity $\boldsymbol{H}$ is referred to as the magnetic field, but we will not need to use that symbol.) Since magnetic fields are produced by electric currents, it is possible to connect the strength of the field of any magnet, including a single magnetic dipole, with basic mechanical and electrical units. The SI* unit of a magnetic field is the **tesla** (T), which is defined as $\mathrm{T} = \mathrm{N\ A^{-1}\ m^{-1}} = \mathrm{kg\ s^{-2}\ A^{-1}}$, where A is the ampere.

In discussing quantum mechanics, we have used $\boldsymbol{\mu}$ for the **magnetic dipole moment vector,** but in the following discussion of bulk magnetic properties, we will use $\boldsymbol{m}$ to avoid using the same symbol as for the electric dipole moment. The

*The gauss (G) is the unit for the magnetic field in the cgs Gaussian system of units; $1\ \mathrm{T} = 10^4\ \mathrm{G}$.

magnitude m of the magnetic dipole moment of a flat coil enclosing an area A_s and carrying a current I is $m = IA_s$. Thus the units of the magnetic dipole moment m are A m^2. Magnetic dipole moments are more commonly expressed in J T^{-1}, which can be readily shown to be the same as A m^2 by the use of the units of the tesla given above. The direction of the $\boldsymbol{m}$ vector is perpendicular to the plane of the flat coil pointing in the same sense as the linear motion of a right-hand screw turned in the same direction as the current.

Magnetic dipole moments $\boldsymbol{m}$ of atoms and molecules were discussed in Section 10.4 on the orbital angular momentum of the hydrogenlike atom and Section 10.5 on electron spin. Then we discussed nuclear magnetic moments in connection with NMR spectroscopy. In this chapter, we will be concerned with magnetic properties of matter in bulk. In a sample containing permanent magnetic dipole moments, thermal motion generally leads to random orientations so that the net magnetic moment of the sample is zero. However, if the sample is placed in a magnetic field, the magnetic dipoles tend to line up with the field, and this increases the magnetic field in the medium. The sample is said to be magnetized, and the **magnetization $\boldsymbol{M}$** is defined as the vector sum of the dipoles in the sample per unit volume ($\boldsymbol{M} = \sum \boldsymbol{m}/V$); the magnetization has the SI units A m^{-1}.

Magnetic moments can also be induced by the field. In the case illustrated in Fig. 22.7, the applied field $\boldsymbol{B}_0$ and the field $\mu_0\boldsymbol{M}$ due to the magnetization add to give a net field $\boldsymbol{B}$:

$$\boldsymbol{B} = \boldsymbol{B}_0 + \mu_0\boldsymbol{M} \tag{22.32}$$

Here $\boldsymbol{B}_0$ is the applied magnetic field and μ_0 is the **permeability of vacuum** ($4\pi \times 10^{-7}$ N A^{-2} exactly). The magnetic field $\boldsymbol{B}$ is expressed in teslas (T = N A^{-1} m^{-1}), and so $\mu_0\boldsymbol{M}$ has the same units.

For magnetic fields that are not too large, the magnetization is proportional to the applied magnetic field, so

$$\boldsymbol{B} = \mu_r\boldsymbol{B}_0 \tag{22.33}$$

where μ_r is the **relative permeability,** which is dimensionless. The relative permeability of vacuum is unity, and the relative permeabilities are close to unity for everything except for ferromagnetic and ferrimagnetic substances. Combining equations 22.32 and 22.33 yields

$$\mu_0\boldsymbol{M} = (1 + \mu_r)\boldsymbol{B}_0 = \chi_{mag}\boldsymbol{B}_0 \tag{22.34}$$

The **magnetic susceptibility** χ_{mag} is defined by

$$\chi_{mag} = \frac{\mu_0\boldsymbol{M}}{\boldsymbol{B}_0} \tag{22.35}$$

Note that $\chi_{mag} = 1 + \mu_r$, so χ_{mag} is dimensionless. Equation 22.34 is analogous to equation 22.6 for electric polarization. In considering dielectric materials, we emphasized the relative permittivity ϵ_r rather than the electric susceptibility χ_{el}, but in considering magnetic materials, magnetic susceptibility χ_{mag} is used more often than the relative permeability μ_r. The magnetic susceptibility of a sample can be determined by measuring the force exerted on the sample by an inhomogeneous magnetic field* or by using a number of other methods.

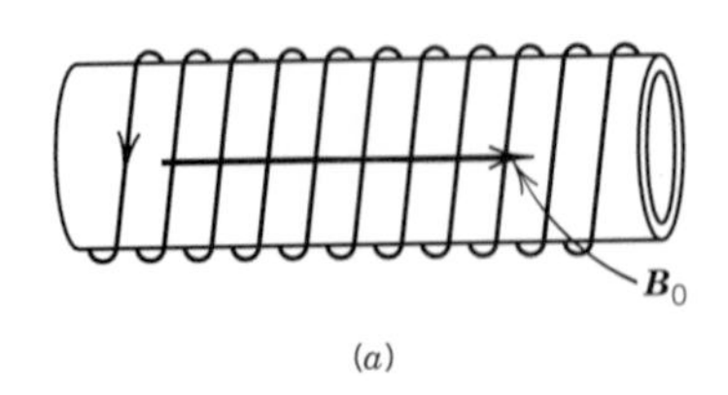

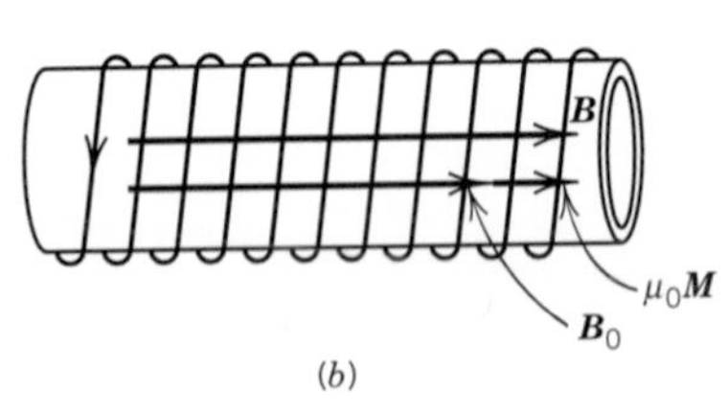

Figure 22.7 (a) The solenoid produces a magnetic field $\boldsymbol{B}_0$. (b) The solenoid is filled with magnetic material that contributes $\mu_0\boldsymbol{M}$ to produce a magnetic field $\boldsymbol{B} = \boldsymbol{B}_0 + \mu_0\boldsymbol{M}$.

*C. W. Garland, J. W. Nibler, and D. P. Shoemaker, *Experiments in Physical Chemistry.* New York: McGraw-Hill, 2003.

22.6 TYPES OF MAGNETIC MATERIALS

Paramagnetism

Paramagnetism results from the orientation of permanent magnetic dipoles in a substance. These permanent magnetic dipoles are due to the spins of unpaired electrons or to the angular momentum of electrons in orbitals of atoms or molecules. Electrons in orbitals with $\ell = 1, 2, 3, \ldots$ have angular momentum and therefore produce a magnetic dipole moment. In most substances the magnetic effects of electron spin and electron orbital motions cancel because electrons are paired in filled shells. Many rare earth and transition metal ions are paramagnetic because they have unpaired electrons. Free radicals have an odd number of electrons and are therefore paramagnetic. For a solid to be paramagnetic, it is necessary that the individual magnetic moments have some degree of isolation; that is, they must be only weakly interacting. As we saw in studying molecular electronic structure, the magnetic moment of a molecule also includes contributions from **orbital angular momentum.** Nuclei with magnetic moments produce a paramagnetic effect about a million times smaller than that due to orbital electrons and unpaired electrons.

In the absence of an applied magnetic field, the magnetic dipoles have random orientations, so the magnetization is zero. When a magnetic field is applied, the magnetic dipoles tend to line up, but this is opposed to some extent by thermal motions. Since the magnetic field in the sample is larger than the applied field, the magnetic susceptibility χ_{mag} is positive. The magnetic susceptibilities of some common paramagnetic substances are given in Table 22.3. A paramagnetic sample is attracted into the stronger part of an inhomogeneous magnetic field. The fact that molecular oxygen is paramagnetic may be a surprise, but the fact that it has a spin of one in its ground electronic state was mentioned in Section 11.5.

Since thermal motion tends to disturb the alignment of magnetic dipoles, the magnetization M of a paramagnetic material decreases with increasing temperature, obeying Curie's law,

$$M = \frac{CB_0}{T} \tag{22.36}$$

where C is a constant. This applies at low magnetic field strengths. As the applied magnetic field is increased, the magnetization saturates because when all of the dipoles are parallel, M has its maximum value $(N/V)m$, where N is the number of magnetic dipoles in volume V.

Table 22.3 Magnetic Susceptibilities and Molar Magnetic Susceptibilities of Paramagnetic Substances at 293 K and 1 bar

Substance	χ_{mag}	$\dfrac{M}{\text{kg mol}^{-1}}$	$\dfrac{\rho}{\text{kg m}^{-3}}$	$\dfrac{\chi_{\text{mag,m}}}{\text{m}^3\ \text{mol}^{-1}}$
O_2	$+1.9 \times 10^{-6}$	32.0×10^{-3}	1.32	4.6×10^{-8}
NO	$+0.8 \times 10^{-6}$	30.0×10^{-3}	1.23	2.0×10^{-8}
Chromium	$+3.3 \times 10^{-4}$	52.0×10^{-3}	7.2×10^{3}	0.24×10^{-8}
Tungsten	$+6.8 \times 10^{-5}$	183.9×10^{-3}	19.3×10^{3}	0.065×10^{-8}
$CuSO_4 \cdot 5H_2O$	$+1.76 \times 10^{-4}$	249.7×10^{-3}	2.28×10^{3}	1.93×10^{-8}
$MnSO_4 \cdot 4H_2O$	$+1.71 \times 10^{-3}$	223.1×10^{-3}	2.11×10^{3}	18.1×10^{-8}

The calculation of the contribution of the orientation of magnetic dipoles in a magnetic field to the magnetization M is similar to the calculation of the contribution of the orientation of electric dipoles in an electric field to the polarization P (see equation 22.17), so that the classical result for the magnitude M of the magnetization is

$$M = \frac{N}{V}\frac{m^2 B_0}{3kT} \tag{22.37}$$

where N is the number of magnetic dipoles and V is the volume. This equation is correct only if $mB_0/kT \ll 1$. For paramagnetic materials, the magnetic moment depends on the spin quantum number S (see Table 10.2), and quantum mechanics shows that equation 22.37 is replaced by

$$M = \frac{NS(S+1)g_e^2 \mu_B^2 B_0}{3kTV} \tag{22.38}$$

where S is the total spin quantum number, g_e is the g factor for a free electron (2.002 322), and μ_B is the Bohr magneton. This result is the explanation of Curie's law and provides a means for calculating the total spin quantum number S from measurements of the magnetic susceptibility χ_{mag}, which is given by equation 22.35,

$$\chi_{mag} = \frac{\mu_0 NS(S+1)g_e^2 \mu_B^2}{3kTV} \qquad (22.39)*$$

Example 22.2 ***The units of magnetic susceptibility***

Show that equation 22.39 yields the correct units for the magnetic susceptibility.

Since μ_0 has the units N A^{-2}, N/V has the units m^{-3}, μ_B has the units J T^{-1}, k has the units J K^{-1}, T has the units K, and the tesla T has the units N A^{-1} m^{-1},

$$\chi_{mag} = \frac{(\text{N A}^{-2})(\text{m}^{-3})(\text{J T}^{-1})^2}{(\text{J K}^{-1})(\text{K})} = 1$$

so that the magnetic susceptibility is dimensionless.

Example 22.3 ***The total spin quantum number of $CuSO_4 \cdot 5\,H_2O$***

The magnetic susceptibility of $CuSO_4 \cdot 5H_2O$ is 1.76×10^{-4} at 293 K. Calculate the apparent total spin quantum number S. Given: The molar mass of $CuSO_4 \cdot 5H_2O$ is 249.68 g mol^{-1} and its density is 2.284 g cm^{-3}.

Equation 22.39 can be rearranged to the form

$$S(S+1) = \frac{3kT\chi_{mag}}{\mu_0 N g_e^2 \mu_B^2}$$

*Sometimes in analyzing data, $\chi_{mag} = \mu_0 N m_{eff}^2 / 3kT$ is used to calculate an effective magnetic moment m_{eff}. In simple cases, $m_{eff}^2 = S(S+1)g_e^2\mu_B^2$.

where N is the number of molecules per unit volume. The number of molecules per unit volume is

$$N = \frac{(6.022 \times 10^{23}\ \mathrm{mol^{-1}})(2.284\ \mathrm{g\ cm^{-3}})(10^{6}\ \mathrm{cm^{3}\ m^{-3}})}{249.68\ \mathrm{g\ mol^{-1}}}$$

$$= 5.51 \times 10^{27}\ \mathrm{m^{-3}}$$

The factor $3kT/\mu_0$ is given by

$$\frac{3(1.381 \times 10^{-23}\ \mathrm{J\ K^{-1}})(293\ \mathrm{K})}{4\pi \times 10^{-7}\ \mathrm{N\ A^{-2}}} = 9.66 \times 10^{-15}\ \mathrm{m\ A^{2}}$$

The factor $g_e^2 \mu_B^2$ is

$$(2.002)^2(9.274 \times 10^{-24}\ \mathrm{J\ T^{-1}})^2 = 3.45 \times 10^{-46}\ \mathrm{J^2\ T^{-2}}$$

Thus

$$S(S+1) = \frac{(9.66 \times 10^{-15}\ \mathrm{m\ A^{2}})(1.76 \times 10^{-4})}{(5.51 \times 10^{27}\ \mathrm{m^{-3}})(3.45 \times 10^{-46}\ \mathrm{J^2\ T^{-2}})} = 0.895$$

The solution to the quadratic equation is

$$S = \frac{-1 + [1 + 4(0.895)]^{1/2}}{2} = 0.570$$

This suggests one unpaired spin that would contribute $\frac{1}{2}$. The spin calculated using equation 22.39 will not be a multiple of $\frac{1}{2}$ when orbital angular momentum is nonzero.

In discussing paramagnetic susceptibility in terms of spins, there are advantages in defining and using a molar magnetic susceptibility $\chi_{\mathrm{mag,m}}$, which is a property of a mole of a substance. For solids and liquids, the number of magnetic dipoles in a system is given by $N = mN_A/M$, where m is the mass of substance and M is the molar mass. Substituting this in equation 22.39 yields

$$\chi_{\mathrm{mag}} = \frac{m}{VM}\frac{N_A \mu_0 S(S+1) g_e^2 \mu_B^2}{3kT} \tag{22.40}$$

Since $m/V = \rho$ (density), the **molar magnetic susceptibility** $\chi_{\mathrm{mag,m}}$ is defined by

$$\chi_{\mathrm{mag,m}} = \chi_{\mathrm{mag}}\frac{M}{\rho} = \frac{N_A \mu_0 S(S+1) g_e^2 \mu_B^2}{3kT} \tag{22.41}$$

For a gas, $N/V = nN_A/V = N_A/\overline{V}$, and so the molar magnetic susceptibility is given by

$$\chi_{\mathrm{mag,m}} = \chi_{\mathrm{mag}}\overline{V} = \chi_{\mathrm{mag}}\frac{RT}{P} = \frac{N_A \mu_0 S(S+1) g_e^2 \mu_B^2}{3kT} \tag{22.42}$$

where the second expression for the molar magnetic susceptibility applies to an ideal gas. Thus for solids, liquids, and gases, the molar magnetic susceptibility for paramagnets is the Avogadro constant times the contribution of a single molecule.

Example 22.4 ***The total spin quantum number of molecular oxygen***

The molar magnetic susceptibility of O_2(g) is $4.6 \times 10^{-8}\ \mathrm{m^3\ mol^{-1}}$ at 293 K and 1 bar. What is the total spin quantum number S for molecular oxygen?

Equation 22.42 can be written

$$S(S+1) = \chi_{\text{mag,m}} \frac{3kT}{N_A \mu_0 g_e^2 \mu_B^2}$$

Since $3kT/\mu_0 = 9.66 \times 10^{-15}$ m A^2 and $g_e^2 \mu_B^2 = 3.45 \times 10^{-46}$ J^2 T^{-2} from the preceding example,

$$S(S+1) = (4.6 \times 10^{-8} \text{ m}^3 \text{ mol}^{-1}) \frac{9.66 \times 10^{-15} \text{ m A}^2}{(6.02 \times 10^{23} \text{ mol}^{-1})(3.45 \times 10^{-46} \text{ J}^2 \text{ T}^{-2})}$$

$$= 2.14$$

The solution to the quadratic equation $S^2 + S - 2.14 = 0$ is

$$S = \frac{-1 + [1 + 4(2.14)]^{1/2}}{2} = 1.05$$

This is just what we expect from the discussion of the electronic structure of molecular oxygen, which has two unpaired electrons. The left-hand superscript of the molecular term symbol for O_2 in Table 11.1 indicates that $2S + 1 = 3$, so $S = 1$. This shows how the molecular property of spin shows up in a bulk property.

Diamagnetism

Diamagnetism is the result of the induction of magnetic moments by a magnetic field. Even atoms with closed shells show diamagnetism. The magnetic susceptibility of a paramagnetic substance has a diamagnetic contribution, but the paramagnetic effect is usually considerably larger. Orbiting electrons can be considered as current loops, and when a magnetic field is applied the motion of the electrons must change in such a way that the induced magnetic field opposes the applied field. Since the induced magnetic moments oppose the applied field, diamagnetic substances have negative magnetic susceptibilities, as illustrated in Table 22.4. Since the induced magnetic dipoles oppose the field, a diamagnetic sample experiences a force away from the increasing magnetic field. This is the opposite of the effect of an inhomogeneous field on a paramagnetic sample.

Table 22.4 Magnetic Susceptibilities of Diamagnetic Substances at 293 K and 1 bar

Substance	χ_{mag}
He	-1.1×10^{-9}
H_2	-2.2×10^{-9}
N_2	-6.8×10^{-9}
Mercury	-3.3×10^{-5}
Copper	-9.7×10^{-6}
Ethyl alcohol	-1.3×10^{-5}

Ferromagnetism

The magnetic field in a ferromagnetic substance may be up to 10^7 times the applied magnetic field. Ferromagnetic substances, like paramagnetic substances, have permanent magnetic dipole moments, but they are distinguished by the fact that there is a strong interaction between the magnetic moments that keeps them aligned even after the magnetic field has been removed. Familiar ferromagnetic elements are iron, cobalt, and nickel, but many compounds and oxides can also be ferromagnetic. Chromium is not ferromagnetic at room temperature, but chromium oxide (CrO_2) is ferromagnetic and is used in magnetic recording media. When the temperature of a ferromagnetic substance is raised, the susceptibility decreases; the temperature at which the ferromagnetic property is lost is called the **Curie temperature.** The Curie temperature of iron is 770 °C, and iron is paramagnetic above this temperature.

The strong interaction (called exchange interaction) between spins in a ferromagnetic substance leads to ordering and minimization of the various energies involved and results in **magnetic domains** within a crystal in which the magnetic moments are nearly perfectly aligned.

Five Key Ideas in Chapter 22

1. When a dielectric is placed in an electric field, it becomes polarized both by distortion of the electron clouds and by the orientation of electric dipoles. The molar polarization is therefore made up of two terms, and they can be determined separately by studying the effect of temperature because there is less orientation of dipoles at high temperatures.
2. At low frequencies there is time for electric dipoles to become oriented in an oscillating electric field, and the relative permittivity is high. As the frequency of the electromagnetic radiation is increased, there is not enough time to orient the dipoles in a cycle and the relative permittivity decreases. The relative permittivity is complex; the plot of the real part is the dispersion curve, and the plot of the imaginary part is the absorption curve.
3. When a sample is placed in a magnetic field, the magnetic dipoles tend to line up with the field. Paramagnetism results from the orientation of permanent magnetic dipoles of the substance. Diamagnetism is the result of the induction of magnetic moments by the magnetic field; even atoms with closed shells show diamagnetism.
4. Measurements of magnetic susceptibilities provide the means for determining the total spin quantum number for a substance.
5. In ferromagnetic substances, there is a strong interaction between the magnetic moments that keeps them aligned even after the magnetic field has been removed. This interaction results in magnetic domains in small regions in a crystal. In antiferromagnetic substances, the magnetic moments of nearest neighbors are opposed.

REFERENCES

C. W. Garland, J. W. Nibler and D. P. Shoemaker, *Experiments in Physical Chemistry.* New York: McGraw-Hill, 2003

A. Hinchliffe and R. W. Munn, *Molecular Electromagnetism.* Hoboken, NJ: Wiley, 1985.

O. Kahn, *Molecular Magnetism.* New York: VCH, 1993.

M. Rigby, E. B. Smith, W. A. Wakeham, and G. C. Maitland, *The Forces between Molecules.* Oxford, UK: Oxford University Press, 1986.

S. Vulfson, *Molecular Magnetochemistry.* Neward, NJ: Gordon and Beach, 1998.

PROBLEMS

(M) Problems marked with an icon may be more conveniently solved on a personal computer with a mathematical program.

22.1 If a molecule has two groups with dipole moments μ_1 and μ_2, the square of the dipole moment of the molecule is given by

$$\mu^2 = \mu_1^2 + \mu_2^2 + 2\mu_1\mu_2 \cos\theta$$

where θ is the angle between the vectors. Show that when the dipole moments of the groups are equal,

$$\mu = 2\mu_1 \cos(\theta/2)$$

Given: There is a trigonometric identity $\cos 2x = 2\cos^2 x - 1$.

22.2 Calculate the SI units of the electric susceptibility χ_{el} from its definition in equation 22.6.

22.3 The relative permittivity of HI(g) at 1 atm and 273 K is 1.002 34. Given that its dipole moment is 1.40×10^{-30} C m, calculate its polarizability.

22.4 Given that the mean electric polarizability α of CH_4 is 2.90×10^{-40} J^{-1} C^2 m^2, express α' in units of a_0^3, where a_0 is the Bohr radius. Compare the volume α' with the volume corresponding to the molecular diameter of CH_4 obtained from kinetic theory (0.414 nm in Table 17.4).

22.5 The magnetic susceptibility of molecular oxygen at 1 bar and 300 K is 1.9×10^{-6}. What does this tell us about the spin of an oxygen molecule?

(M) **22.6** Given that the molar magnetic susceptibility of NO(g) is 2.0×10^{-8} m^3 mol^{-1} at 293 K and 1 bar, calculate the total spin quantum number S.

(M) **22.7** Use the molar magnetic susceptibility of $MnSO_4 \cdot 4H_2O$ to calculate the total spin quantum number at 293 K.

22.8 The relative permittivity of H_2(g) at 273 K and 1 atm is 1.000 272. Calculate the polarizability of H_2(g) under these conditions.

22.9 Show that the SI units are the same on the two sides of equation 22.20.

22.10 A gaseous paramagnetic atom with a magnetic dipole moment m equal to the Bohr magneton μ_B is placed in a magnetic field of 10 T. At what temperature will its magnetic energy be equal to its mean translational kinetic energy?

22.11 The left superscript of the atomic term symbol for molecular oxygen (Table 11.1) shows that $2S + 1 = 3$, so that the total spin quantum number $S = 1$. Calculate the magnetic susceptibility χ_{mag} of O_2(g) at 200 K and 1000 K at 1 bar.

Computer Problems

22.A Calculate the dipole moment of fluorobenzene, using a plot of P_m versus $1/T$, from the following molar polarizations for gaseous samples.

T/K	343.6	371.4	414.1	453.2	507.0
P_m/cm^3 mol^{-1}	69.9	66.8	62.5	59.3	55.8

22.B Plot the frequency dependence of the real part ϵ' and the imaginary part ϵ'' of the relative permittivity for a process with a single relaxation time τ with $\epsilon_l = 15$ and $\epsilon_h = 5$, as shown in Fig. 22.5. (*a*) Do this by using equation 22.29 with equations 22.30 and 22.31. (*b*) Do this using equation 22.28. Obtaining the same result in two ways will confirm these different ways of expressing the relative permittivity.

23

Solid-State Chemistry

In this chapter we first consider crystal geometry, then X-ray diffraction and the structure of specific crystals, and finally synthetic high polymers. The ideas of point-group symmetry developed in Chapter 12 are basic to the understanding of crystals, but the introduction of translations produces new kinds of symmetry and requires space groups (Section 23.3) rather than point groups.

In 1912 Laue suggested that the wavelength of X-rays might be about the same as the distance between atoms in a crystal, so that a crystal could serve as a diffraction grating for X-rays. This experiment was carried out by Frederick and Knipping, who observed the expected diffraction. Almost immediately afterward, W. L. Bragg (1913) improved on the Laue experiment, mainly by substituting monochromatic for polychromatic radiation and by providing a more physical interpretation to the Laue theory of the scattering experiment. Bragg also determined the structures of a number of simple crystals, including those of NaCl, CsCl,

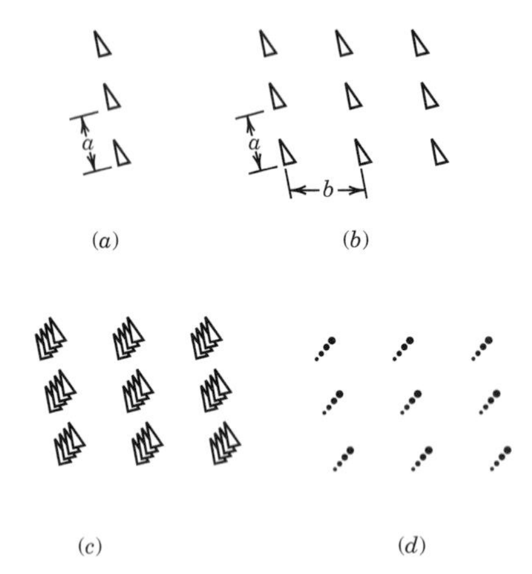

Figure 23.1 (*a*) One-dimensional pattern with vector ***a***. (*b*) Two-dimensional pattern with vectors ***a*** and ***b***. (*c*) Three-dimensional pattern with vectors ***a***, ***b***, and ***c***. (*d*) Lattice of pattern (*c*).

and ZnS. Since that time, single-crystal X-ray diffraction has developed into the most powerful method known for obtaining the atomic arrangement in the solid state. Since the 1950s, with the advent of high-speed computers capable of handling X-ray data, it has been possible to determine the structures of compounds as complex as proteins.

The chapter ends with special topics on superconductivity and quantum confined semiconductor structures.

23.1 CLASSIFICATION OF CRYSTAL STRUCTURES

A crystal may be described as a three-dimensional pattern in which a structural motif is repeated in such a way that the environment of every motif is the same throughout the crystal. The motif may be an atom or a molecule, or it may be a group of atoms or a group of molecules.

A linear pattern may be described by saying that there is a set of parallel motifs at the end of a vector ***a*** and its multiples. These vectors are given by

$$\boldsymbol{T} = u\boldsymbol{a} \tag{23.1}$$

where u is an integer. An example of a linear pattern is shown in Fig. 23.1*a*. The linear pattern formed in this way can be repeated in a second direction represented by the vector ***b*** to form a two-dimensional pattern in which every structural pattern has the same environment as every other, as shown in Fig. 23.1*b*. Finally, the whole two-dimensional pattern can be repeated in a third direction ***c*** to produce a translationally ordered pattern in three dimensions, as shown in Fig. 23.1*c*. The three vectors ***a***, ***b***, and ***c*** are called primitive vectors of the crystal. The **structural pattern** is repeated at the end of every vector of the form

$$\boldsymbol{T} = u\boldsymbol{a} + v\boldsymbol{b} + w\boldsymbol{c} \qquad \text{where } u, v, w \text{ are integers} \tag{23.2}$$

For many purposes it is convenient to concentrate on the geometry of the repetition and replace the structural pattern by a point, as shown in Fig. 23.1*d*, to obtain a **lattice.** The lattice may be generated from a single starting point by the infinite repetition of a set of fundamental translations that characterize the lattice. Any three noncoplanar vectors ***a***, ***b***, and ***c*** describe a lattice, but a given lattice can be described by an infinite number of sets of three vectors. This is illustrated in two dimensions in Fig. 23.2. The ***a*** vector together with any one of the ***b*** vectors may be chosen to generate the pattern.

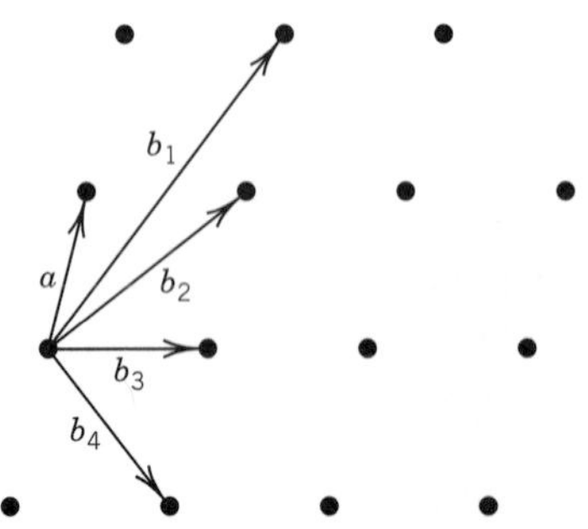

Figure 23.2 Alternate choices for the second translational vector in a two-dimensional lattice.

The space occupied by a lattice may be divided into **unit cells.** The repetition of a cell (with everything in it) in three dimensions generates the entire pattern of a crystal. A given lattice can be blocked out in cells in different ways, as shown in Fig. 23.3. If the corners of the cells include all of the lattice points in the crystal, the cell is called a **primitive unit cell.** Primitive cells have one lattice point per cell because each of the corner lattice points is shared by eight cells. A lattice can also be blocked out in cells that do not include all lattice points as corners. This is illustrated by one of the cells in Fig. 23.3. Such cells, referred to as **multiple unit cells,** are useful in simplifying the geometry of crystals for which the primitive unit cell is oblique, but the multiple unit cell has two or more edges that are at 90°.

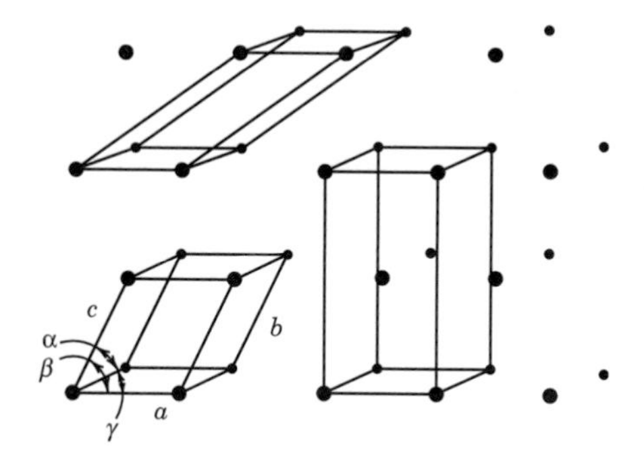

Figure 23.3 Cells in a lattice.

The classification of crystals is based on their **symmetry,** rather than on the dimensions of their unit cells. It is not satisfactory to use the equality of all edges, for example, because they may become unequal when the temperature changes. The use of symmetry avoids this difficulty because, if two directions in a crystal are equivalent by symmetry, they will necessarily have the same thermal expansion coefficient. The recognition of the types of symmetry we discussed in Chapter 12 reduces the number of coordinates of atoms that have to be specified to describe the structure of a unit cell because of the relationship between the coordinates of symmetry-related atoms.

We saw in Chapter 12 that a molecule can have various types of rotational symmetry. For a molecule with n-fold symmetry, rotation through an angle of $360°/n$ brings it into an equivalent position, and a molecule may have $n = 1, 2, 3, \ldots, \infty$. However, in contrast with individual molecules, crystals and crystal lattices can only have $n = 1, 2, 3, 4$, or 6. This statement can be proved by use of Fig. 23.4. The lattice shown in this figure has an axis of n-fold symmetry perpendicular to the page. The lattice points A_1, A_2, A_3, and A_4 are separated by distance a. Because of the assumed symmetry, rotation of the lattice about any lattice point through an angle $\alpha = 360°/n$ will produce a lattice indistinguishable from the original. Therefore, clockwise rotation by α about A_3 and counterclockwise rotation by α about A_2 requires that there be lattice points at B_1 and B_2. Since the line B_1B_2 is parallel to A_1A_4, B_1 and B_2 must be separated by an integer multiple of a, represented by ma. Thus,

$$a + 2a\cos\alpha = ma \qquad \cos\alpha = \frac{N}{2}$$

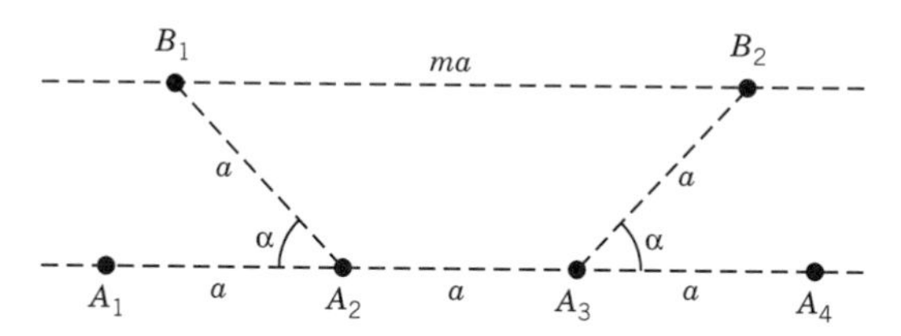

Figure 23.4 Restriction on rotational order in a crystal. There are n-fold rotational axes at lattice points A_1, A_2, A_3, A_4, B_1, and B_2.

Table 23.1 Crystallographic Point Groups

Crystal System	*Schoenflies Symbol*	*Hermann–Mauguin Symbol*
Triclinic	C_1	1
	C_i	$\bar{1}$
Monoclinic	C_2	2
	C_s	m
	C_{2h}	$2/m$
Orthorhombic	D_2	222
	C_{2v}	$mm2$
	D_{2h}	mmm
Tetragonal	C_4	4
	S_4	$\bar{4}$
	C_{4h}	$4/m$
	D_4	422
	C_{4v}	$4mm$
	D_{2d}	$\bar{4}2m$
	D_{4h}	$4/mmm$
Trigonal	C_3	3
	C_{3i}	$\bar{3}$
	D_3	32
	C_{3v}	$3m$
	D_{3d}	$\bar{3}m$
Hexagonal	C_6	6
	C_{3h}	$\bar{6}$
	C_{6h}	$6/m$
	D_6	622
	C_{6v}	$6mm$
	D_{3h}	$\bar{6}m2$
	D_{6h}	$6/mmm$
Cubic	T	23
	T_h	$m3$
	O	432
	T_d	$\bar{4}3m$
	O_h	$m3m$

where $N = m - 1$. The only values of α that satisfy this equation are 0°, 60°, 90°, 120°, 180°, 240°, and 360°, which means that **the rotational symmetry of the lattice can be only one-, two-, three-, four-, or sixfold.**

As a result of this restriction on rotational symmetry, all crystals can be classified as belonging to one of only 32 **crystallographic point groups.** The Schoenflies symbols of these 32 crystallographic point groups are shown in Table 23.1, where they are divided into seven crystal systems. The seven crystal systems can also be described in terms of unit cell axes and angles (see Fig. 23.5), but the classification according to symmetry elements is more fundamental.

The symmetries of the 32 crystallographic point groups can also be represented by the **Hermann–Mauguin symbols*** in Table 23.1. Crystallographers

*N. F. M. Henry and K. Lonsdale (eds.), *International Tables for X-Ray Crystallography, Vol. 1, Symmetry Groups.* Birmingham, UK: Kynoch, 1952.

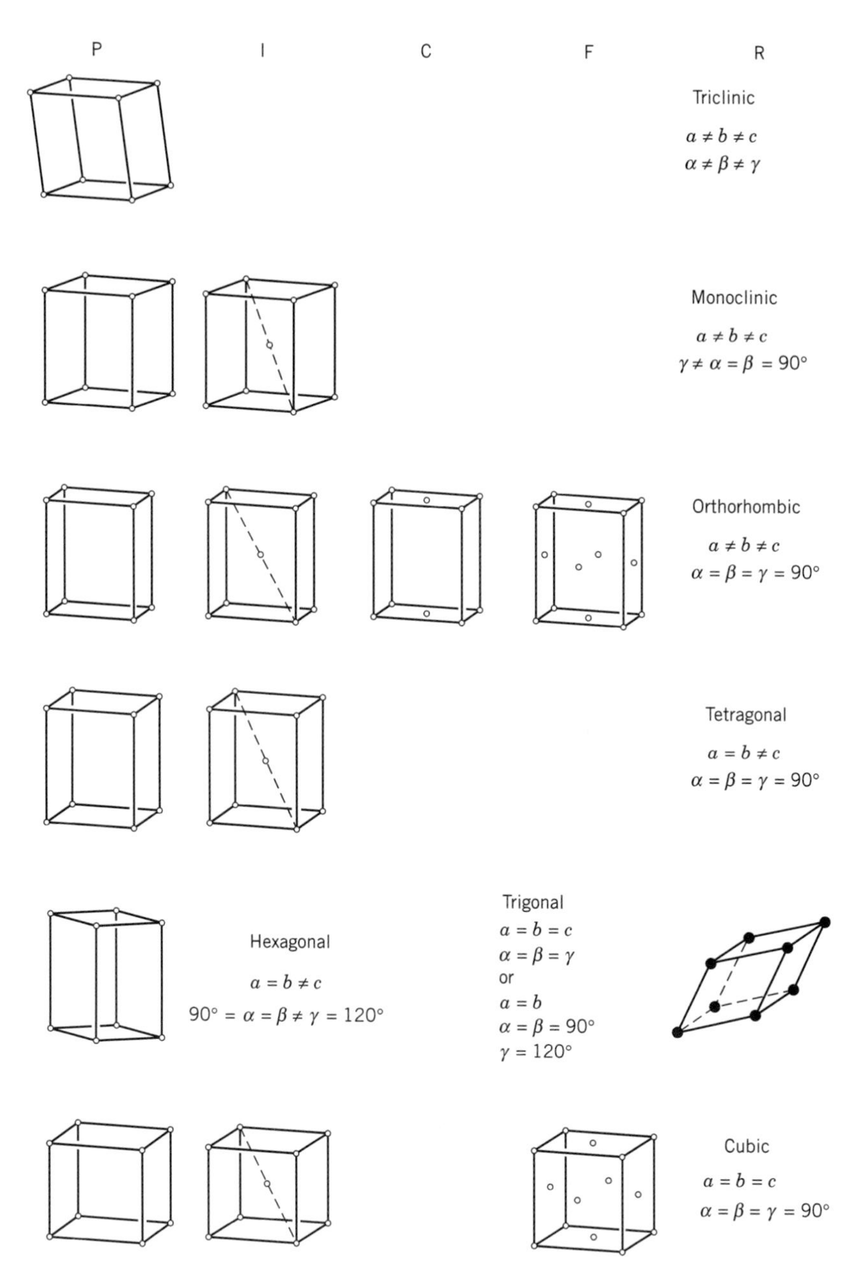

Figure 23.5 The 14 space lattices (often called Bravais lattices) and the seven crystal systems. P refers to primitive, I to body centered, C to end centered, F to face centered, and R to rhombohedral.

prefer the Hermann–Mauguin symbols because they can be extended to include translational symmetry, as we will soon see. Rotation axes are represented by a number; for example, a fourfold rotation axis is represented by 4. Crystallographers use rotary–inversion axes, while spectroscopists use rotary–reflection axes. In a rotary inversion, a rotation by $360°/n$ is followed by inversion through the center of symmetry. Rotary–inversion axes are represented by a number with an overbar; for example, a fourfold rotary–inversion axis is represented by $\bar{4}$. An inversion center is just a onefold rotary–inversion axis, or $\bar{1}$. A mirror plane is represented by the letter m. Each component of a symbol refers to a different direction; for example, $4/m$ indicates that there is a mirror plane perpendicular to a fourfold rotation axis. If we label axes x, y, z, the symbol $mm2$ indicates that mirror planes are perpendicular to x and y, and a twofold rotation axis is parallel to z. In Chapter 12 a decision tree was given for assigning Schoenflies symbols; a similar decision tree for assigning Hermann–Mauguin symbols to the 32 crystallographic point groups is given by Breneman.*

In addition to the primitive (P) unit cells we have been talking about, unit cells can also have lattice points that are **body centered** (I), **face centered** (F), or **end centered** (C), and still have the same crystallographic point group symmetry. This leads to the 14 Bravais lattices shown in Fig. 23.5.

To complete the classification of crystal structures, we must consider translational symmetry. The lattice translations discussed earlier also satisfy our definition of symmetry operations since the translation of the whole crystal in this way leads to an identical configuration. The Hermann–Mauguin system uses two other types of symmetry elements that result from combining the motions of rotations or reflections with the translational symmetry of the lattice. The operation corresponding to **screw axis,** for which the symbol is n_p, where n and p are integers, is a rotation of $360°/n$ followed by a translation of p/n in the direction of the axis. For example, a 3_1 screw axis involves rotation by 120° followed by translation by one-third of a unit cell parallel to the axis. A 3_2 screw axis implies a rotation of 120° and a translation of two-thirds. The possible screw axes are 2_1, 3_1, 3_2, 4_1, 4_2, 4_3, 6_1, 6_2, 6_3, 6_4, and 6_5.

The operation corresponding to a **glide plane** is a reflection in a plane followed by a translation. If the glide is parallel to the a axis, the symbol for the glide plane is simply a, and the operation is reflection in the plane and translation by $a/2$. There are diagonal and other glides that we will not go into.

Space groups are groups whose elements include the point symmetry elements, the translations, and screw axes and glide planes. There are 230 space groups, and all possible crystal structures fall into one of these space groups. The symbol for a space group always starts with P, I, C, F, or R to indicate the type of Bravais lattice (see Fig. 23.5).

The fact that there are 230 ways in which these symmetry operations may be combined in the three-dimensional patterns of crystals was derived independently by three men: Federow, a Russian crystallographer, in 1890; Schoenflies, a German mathematician, in 1891; and Barlow, a British amateur, in 1895. The actual determination of space groups of crystals did not become possible until diffraction techniques allowed the determination of the internal symmetry of crystals. Knowledge of the space group of a crystal simplifies the determination of the structure because only the asymmetric portion of the unit cell needs to be studied; the rest of the contents may be obtained from symmetry operations.

23.2 DESIGNATION OF CRYSTAL PLANES

The position of an atom in a unit cell is designated by giving its coordinates as fractions x, y, z of the unit cell edges a, b, c. The point at (x, y, z) is located by

*G. L. Breneman, *J. Chem. Educ.* **64:**216 (1987).

starting at the origin (0, 0, 0) and moving a distance xa along the a axis, then a distance yb parallel to the b axis, and finally a distance zc parallel to the c axis. For example, the **fractional coordinates** for the lattice points of a face-centered cubic structure are 0, 0, 0; $\frac{1}{2}, \frac{1}{2}$, 0; $\frac{1}{2}$, 0, $\frac{1}{2}$; and 0, $\frac{1}{2}, \frac{1}{2}$. The remaining lattice points may be obtained by adding unity to each of these coordinates. In expressing the locations of atoms in a unit cell the set of coordinates (000) stands for the locations of all eight corners, that is, (100), (111), (101), (110), (001), (011), (010), and (000). In a crystal a lattice point is not necessarily occupied by an atom or molecule, but it does represent a collection of atoms that is repeated in three dimensions.

The distance l between the points x_1, y_1, z_1 and x_2, y_2, z_2 is

$$\begin{aligned} l = {} & [(x_1 - x_2)^2 a^2 + (y_1 - y_2)^2 b^2 + (z_1 - z_2)^2 c^2 \\ & + 2(x_1 - x_2)(y_1 - y_2)ab \cos \gamma + 2(y_1 - y_2)(z_1 - z_2)bc \cos \alpha \\ & + 2(z_1 - z_2)(x_1 - x_2)ca \cos \beta]^{1/2} \end{aligned} \tag{23.3}$$

where the angles are defined in Fig. 23.3. For a cubic crystal this equation simplifies to

$$l = a[(x_2 - x_1)^2 + (y_2 - y_1)^2 + (z_2 - z_1)^2]^{1/2} \tag{23.4}$$

The volume of a unit cell is given by

$$V = abc(1 - \cos^2 \alpha - \cos^2 \beta - \cos^2 \gamma + 2 \cos \alpha \cos \beta \cos \gamma)^{1/2} \tag{23.5}$$

If the unit cell is cubic, orthorhombic, or tetragonal, this equation reduces to $V = abc$.

The planes through the lattice points of crystals are important because they represent possible crystal faces and because they help us to understand X-ray diffraction phenomena. Figure 23.6 shows the lattice points in one plane of a crystal. The c axis is taken as perpendicular to the page. Various sets of planes that are parallel to the c axis are indicated in the figure. Actually, there is an infinite number of such sets of planes.

The orientation of a set of planes of a crystal lattice may be specified by means of the intercepts of one of the planes on the three axes a, b, and c of the unit cell. Suppose that a plane intercepts the a axis at a/h, measured from the origin, the b axis at b/k, and the c axis at c/l. **This plane is referred to by the indices** hkl.* The indices of a set of planes through a lattice may be obtained by counting the number of planes crossed in moving one lattice space in the $\boldsymbol{a}$, $\boldsymbol{b}$, and $\boldsymbol{c}$ directions, respectively. For the set of planes in the lower left-hand corner of Fig. 23.6, two planes are crossed in going one lattice distance in the direction of the a axis and one plane is crossed in going one lattice direction horizontally in the direction of the b axis, whereas no plane would be crossed in going one lattice distance into the paper, since the planes are parallel to the c axis. Thus, this set of planes is

*The indices are often called "Miller indices" because this designation, invented by Whewell in 1825 and Grossman in 1829, was popularized by Miller's textbook of crystallography in 1829. They showed that faces of crystals could be designated by three integers, although nothing was then known about the internal structures of crystals.

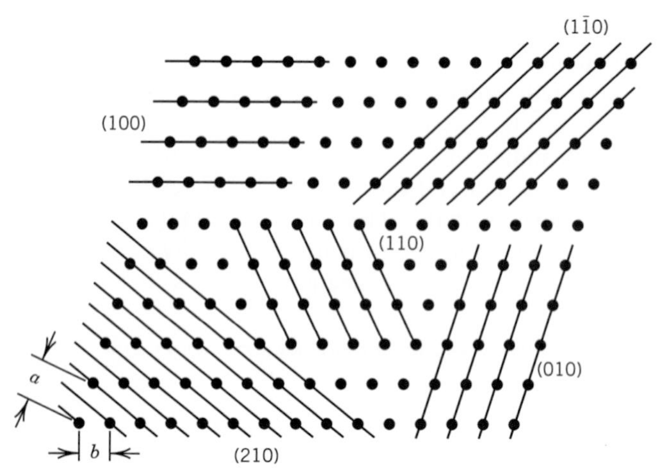

Figure 23.6 Sets of planes through lattice points as seen along the c axis of a crystal.

designated by the indices 210. The indices of the set of planes in the upper right-hand corner are $1\bar{1}0$. The overbar indicates that if a particular plane is intercepted by going in the positive direction along a, it is necessary to go in the negative direction along b in order to intercept the same plane. This representation for the exterior faces of a crystal and for the internal planes within the crystal will specify the orientation but not the spatial position of a plane. The faces of crystals are usually planes with high densities of atoms or molecules, and so they are planes with low indices.

The perpendicular distance d between adjacent planes of a set is given by

$$\begin{aligned} d = V[&h^2b^2c^2 \sin^2\alpha + k^2a^2c^2 \sin^2\beta + l^2a^2b^2 \sin^2\gamma \\ &+ 2hlab^2c(\cos\alpha\cos\gamma - \cos\beta) + 2hkabc^2(\cos\alpha\cos\beta - \cos\gamma) \\ &+ 2kla^2bc(\cos\beta\cos\gamma - \cos\alpha)]^{-1/2} \end{aligned} \tag{23.6}$$

where V is the unit cell volume given by equation 23.5. For the special case that the unit cell axes are mutually perpendicular (i.e., for orthorhombic, tetragonal, and cubic unit cells),

$$\frac{1}{d} = \left(\frac{h^2}{a^2} + \frac{k^2}{b^2} + \frac{l^2}{c^2}\right)^{1/2} \tag{23.7}$$

Some of the planes through cubic lattices are shown in Fig. 23.7 with their Miller indices.

In a cubic crystal $a = b = c$, so equation 23.7 for the perpendicular distance d between adjacent planes of a set may be written

$$d_{hkl} = \frac{a}{\sqrt{h^2 + k^2 + l^2}} \tag{23.8}$$

where a is the length of the side of the unit cell. The perpendicular distances between adjacent planes in a cubic crystal are obtained by substituting 0, 1, 2, 3, ...

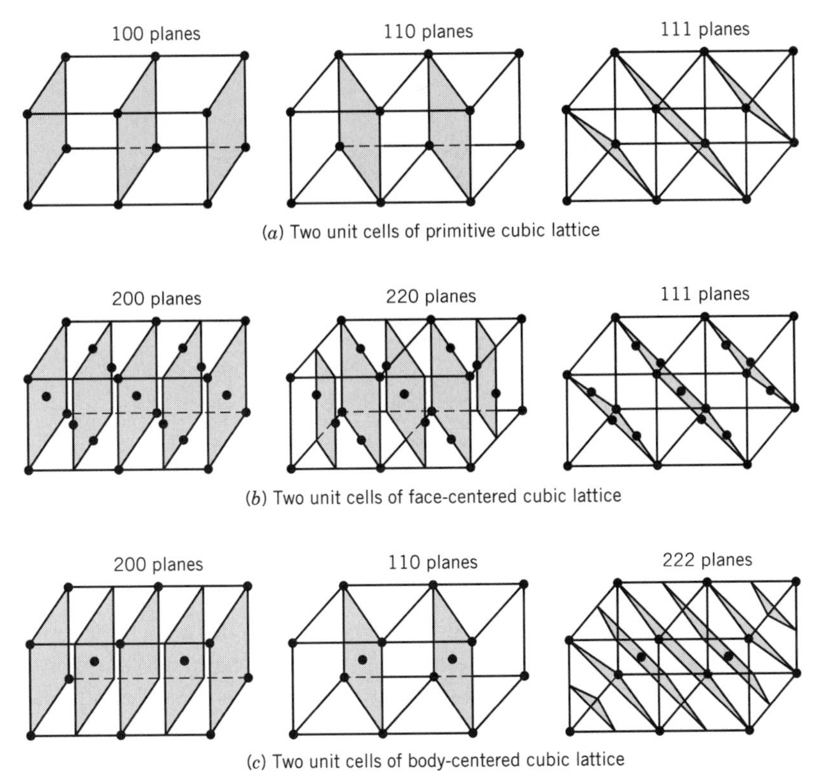

Figure 23.7 Planes through cubic lattices.

for h, k, and l in this equation. The distance between 100 planes is a, and the distance between 200 planes is $a/2$. The distance between 111 planes is $a/3^{1/2}$, and the distance between 222 planes is half as great. Diffraction methods make it possible to determine the distances between planes of atoms, and so the length of the side of a unit cell can be obtained indirectly by use of equation 23.7.

23.3 DIFFRACTION METHODS

In a crystal it is the electrons that scatter X-rays. Bragg pointed out that it is convenient to consider that the X-rays are "reflected" from a stack of planes in the crystal. For a given stack of planes (hkl) the reflected monochromatic radiation occurs only at certain angles that are determined by the wavelength of the X-rays and the perpendicular distance between adjacent planes. The relationship between these three variables is the **Bragg equation,** which may be derived by referring to Fig. 23.8. The horizontal lines represent planes in the crystal separated by the distance d. The plane ABC is perpendicular to the incident beam of parallel monochromatic X-rays, and the plane LMN is perpendicular to the reflected beam. As the angle of incidence θ is changed, a reflection will be obtained only when the waves are in phase at plane LMN, that is, when the difference in distance between planes ABC and LMN, measured along rays reflected from different planes, is a whole-number multiple of the wavelength. This occurs when

$$FS + SG = n\lambda \tag{23.9}$$

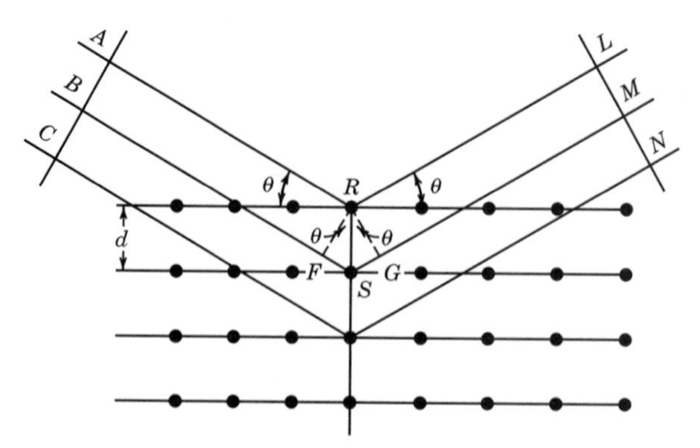

Figure 23.8 Diagram used in proving that $n\lambda = 2d \sin \theta$.

Since $\sin \theta = FS/d = SG/d$,

$$2d \sin \theta = n\lambda \tag{23.10}$$

This is the Bragg equation, θ is called the Bragg angle, and n is the order of a Bragg reflection.

This important equation gives the relationship of the distance between planes in a crystal and the angle at which the reflected radiation has a maximum intensity for a given wavelength λ; that is, all the scattered X-ray waves are in phase. If λ is longer than $2d$, there is no solution for n and no diffraction. Thus, light waves pass through crystals without being diffracted by the planes of scattering centers. If $\lambda \ll d$, the X-rays are diffracted through inconveniently small angles. The Bragg equation does not indicate the intensities of the various diffracted beams. The intensities depend on the nature and arrangement of the atoms within each unit cell.

The reflection corresponding to $n = 1$ for a given family of planes is called the first-order reflection, the reflection corresponding to $n = 2$ is the second-order reflection, and so on. Each successive order exhibits a larger Bragg angle. In discussing X-ray reflections it is customary to set $n = 1$ in equation 23.10 and consider that the second-order reflection is from a parallel stack of planes separated by half the lattice distance, and so on. Equation 23.10 may be written

$$\lambda = 2\frac{d}{n} \sin \theta = 2d_{nh,nk,nl} \sin \theta \tag{23.11}$$

where $d_{nh,nk,nl}$ is the perpendicular distance between adjacent planes having the indices nh, nk, nl. The planes nh, nk, nl are parallel to the hkl planes, and the perpendicular interplanar distance is $d_{nh,nk,nl} = d/n$.

To determine the angles at which X-rays are diffracted, an oriented single crystal may be rotated in an X-ray beam and the intensity of X-rays at the reflection angle determined with a counter. Various types of X-ray cameras have been developed in which the photographic film is moved as the crystal is rotated.

Example 23.1 ***Calculation of a Bragg angle***

At what angles θ will X rays of wavelength 1.542×10^{-10} m be reflected by planes separated by 3.5×10^{-10} m? What is an alternative interpretation of these reflections?

$$\theta = \sin^{-1} \frac{n\lambda}{2d}$$

For $n = 1$,

$$\theta = \sin^{-1} \frac{(1.542 \times 10^{-10}\ \text{m})}{2(3.5 \times 10^{-10}\ \text{m})}$$

$$= 12.73^\circ$$

For $n = 2$, $\theta = 26.14^\circ$, and for $n = 3$, $\theta = 41.37^\circ$. These reflections can also be interpreted as being due to first-order reflections from (100), (200), and (300) planes, which have interplanar spacings of d_{hkl} of 3.5×10^{-10}m, 1.75×10^{-10} m, and 1.17×10^{-10}m.

Neutrons can also be diffracted from the planes in a crystal. The **average de Broglie wavelength** (Section 9.1) of thermal neutrons is 252 pm at room temperature. An essentially monochromatic beam may be obtained by diffraction from a crystal monochromator that selects a small band of wavelengths from the beam of **thermal neutrons** from a nuclear reactor. The diffraction of neutrons from a crystal is different from that of X-rays because neutrons are scattered primarily by the nuclei in the crystal, while the X-rays are scattered by electrons. This means that neutron diffraction, in contrast to X-ray diffraction, is especially useful for accurately locating hydrogen atoms in a crystalline structure. For example, in a compound such as uranium hydride, X-ray diffraction can be utilized to determine the uranium coordinates, and neutron diffraction the hydrogen coordinates. Hydrogen atoms scatter X-rays weakly because they have only one electron.

Since neutrons possess a magnetic moment by virtue of having a spin of $\frac{1}{2}$, there is an additional scattering if the compound contains paramagnetic atoms or ions with unpaired electrons. Thus, neutron diffraction has been widely utilized to investigate structures of magnetic materials such as MnO and Fe_3O_4 in order to determine the arrangement of the atomic magnetic moments in the solids.

Example 23.2 ***The wavelength of thermal neutrons***

What is the wavelength of neutrons moving in a particular direction if they are in thermal equilibrium with their surroundings at 25 °C?

The kinetic energy of the neutrons moving in a particular direction can be expressed in terms of their temperature $kT/2$ (because of equipartition) or in terms of their momentum $p^2/2m$, where p is the momentum and m is the mass of a neutron:

$$\frac{kT}{2} = \frac{p^2}{2m}$$

$$p = (mkT)^{1/2}$$

The de Broglie wavelength is given by

$$\lambda = \frac{h}{p} = \frac{h}{(mkT)^{1/2}}$$

Thus, at 25 °C,

$$\lambda = \frac{6.626 \times 10^{-34}\ \mathrm{J\ s}}{[(1.6749 \times 10^{-27}\ \mathrm{kg})(1.38 \times 10^{-23}\ \mathrm{J\ K^{-1}})(298\ \mathrm{K})]^{1/2}}$$
$$= 252\ \mathrm{pm}$$

23.4 CUBIC LATTICES

The three types of cubic lattices are illustrated in Fig. 23.7. The smallest angles of incidence of X-ray diffraction for a given type of crystal are for the planes that are the farthest apart. For the **primitive cubic lattice** the 100 planes are the farthest apart, and the 110 planes and 111 planes are closer together. The distances between planes can be calculated with equation 23.8, and they are a, $a(2)^{-1/2}$, $a(3)^{-1/2}$, $a(4)^{-1/2}$, $a(5)^{-1/2}$, $a(6)^{-1/2}$, $a(8)^{-1/2}$, and so on. Note that $a(7)^{-1/2}$ is missing because 7 cannot be obtained from $h^2 + k^2 + l^2$, where h, k, and l are integers. The primitive cubic lattice has one lattice point per unit cell because the eight lattice points at the corners are each shared with eight other unit cells.

In the **face-centered cubic lattice,** there are lattice points at the center of each face of the unit cell in addition to the lattice points at the corners. Figure 23.7 shows that the planes that are the farthest apart are the 111 planes. It can be shown that the reflections from a face-centered cubic lattice have Miller indices that are all odd or all even; this statement also applies to face-centered lattices in other crystal systems. Thus only reflections 111, 200, 220, 311, 222, 400, 331, 420, etc., are found. The distances between planes in decreasing order are $a(3)^{-1/2}$, $a(4)^{-1/2}$, $a(8)^{-1/2}$, $a(11)^{-1/2}$, $a(12)^{-1/2}$, $a(16)^{-1/2}$, $a(19)^{-1/2}$, $a(20)^{-1/2}$, and so on. The eight face-centered lattice points are shared with adjoining unit cells, so there are $1 + 3 = 4$ lattice points per unit cell.

In the **body-centered cubic lattice,** there is a lattice point in the middle of the unit cell, and so there are two lattice points per unit cell. It can be shown that for body-centered cubic lattices, the hkl reflections for which the sum $h + k + l$ is odd are not observed. Accordingly, the distances between planes are $a(2)^{-1/2}$, $a(4)^{-1/2}$, $a(6)^{-1/2}$, $a(8)^{-1/2}$, and so on. It can be seen that the various types of cubic crystals may be distinguished by their diffraction patterns, since the patterns are qualitatively different.

Example 23.3 ***Smallest Bragg angles for a primitive cubic crystal***

For a primitive cubic crystal with $a = 3 \times 10^{-10}$ m, what are the smallest diffraction angles θ for (*a*) 100, (*b*) 110, and (*c*) 111 planes for $\lambda = 1.50 \times 10^{-10}$ m?

$$\theta = \sin^{-1} \frac{\lambda}{2d_{hkl}} = \sin^{-1} \frac{\lambda(h^2 + k^2 + l^2)^{1/2}}{2a}$$

(*a*) $\theta = \sin^{-1} \dfrac{1.5 \times 10^{-10}\ \mathrm{m}}{6 \times 10^{-10}\ \mathrm{m}}$

$= 15.48°$

(*b*) $\theta = \sin^{-1} \dfrac{(1.5 \times 10^{-10}\ \mathrm{m})2^{1/2}}{6 \times 10^{-10}\ \mathrm{m}}$

$= 20.70°$

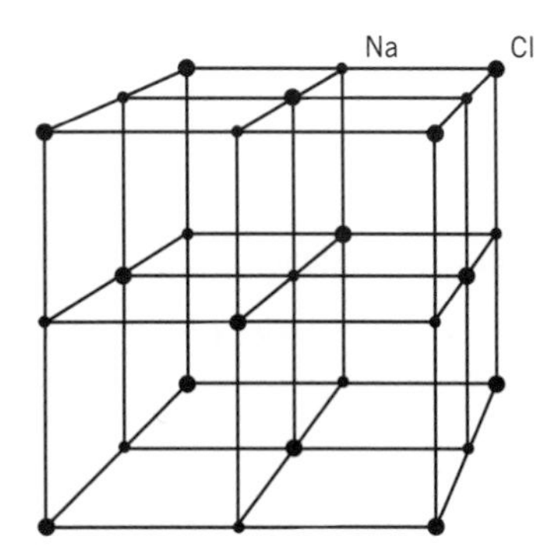

Figure 23.9 Unit cell of sodium chloride. Notice that the chloride ions form a face-centered cubic lattice and that the sodium and chloride lattices are interpenetrating.

$$(c) \quad \theta = \sin^{-1} \frac{(1.5 \times 10^{-10}\ \text{m})3^{1/2}}{6 \times 10^{-10}\ \text{m}}$$

$$= 25.66^\circ$$

Sodium chloride forms an interpenetrating face-centered cubic lattice as shown in Fig. 23.9. Eight of the chloride ions are at the corners of the cube, so they are each shared with eight other unit cells. Six of the chloride ions are in the faces of the cube, so they are each shared with another unit cell. Thus this unit cell contains 8/8 + 6/2 = 4 chloride ions. One sodium ion is at the center of the unit cell, and 12 sodium ions are on edges where they are each shared with four other unit cells. Thus this unit cell contains 1 + 12/4 = 4 sodium ions. Note that each ion is surrounded by six nearest neighbors with opposite charges.

23.5 ION RADII AND ATOM RADII

For cubic crystals, the length a of the side of the unit cell can be obtained from equation 23.8. This length must, of course, be consistent with the density ρ of the crystal and the number z of molecules per unit cell. The mass of the contents of a unit cell is zM/N_A, where z is the number of molecules of molar mass M in a unit cell. Thus, the density ρ for a perfect crystal is

$$\rho = \frac{zM}{N_A V} \tag{23.12}$$

where V is the volume of the unit cell.

Example 23.4 ***Using X-ray scattering and density to compute the number of ions in a unit cell***

The density of sodium chloride at 25 °C is 2.163×10^3 kg m^{-3}. When X rays from a palladium target having a wavelength of 58.1 pm are used, the 200 reflection of sodium chloride occurs at an angle of 5.91°. How many sodium and chloride ions are there in a unit cell?

According to Bragg's law,

$$d_{200} = \frac{\lambda}{2\sin\theta} = \frac{58.1\ \text{pm}}{2\sin 5.9°} = 282\ \text{pm}$$

Equation 23.8 yields $a = 564$ pm. Using equation 23.12, we find

$$2.163 \times 10^3\ \text{kg m}^{-3} = \frac{z(58.443 \times 10^{-3}\ \text{kg mol}^{-1})}{(6.022\,14 \times 10^{23}\ \text{mol}^{-1})(564 \times 10^{-12}\ \text{m})^3}$$

$$z = 3.999$$

Thus, as expected, the unit cell contains four sodium ions and four chloride ions.

Example 23.5 ***The closest distance between atoms***

Potassium crystallizes with a body-centered cubic lattice and has a density of 0.856×10^3 kg m^{-3}. What is the length of the side of the unit cell a, and what is the distance between (200), (110), and (222) planes? What is the closest distance between atoms, and what is the potassium atom radius r?

$$0.856 \times 10^3\ \text{kg m}^{-3} = \frac{2(39.098 \times 10^{-3}\ \text{kg mol}^{-1})}{(6.022\,137 \times 10^{23}\ \text{mol}^{-1})a^3}$$

$$a = 533.3 \times 10^{-12}\ \text{m} = 533.3\ \text{pm}$$

Using equation 23.8, we obtain

$$\text{For (200) planes, } d_{200} = 533.3/\sqrt{4} = 266.7\ \text{pm}$$

$$\text{For (110) planes, } d_{110} = 533.3/\sqrt{2} = 377.1\ \text{pm}$$

$$\text{For (222) planes, } d_{222} = 533.3/\sqrt{12} = 154.0\ \text{pm}$$

$$(2r)^2 = \left(\frac{a}{2}\right)^2 + \left(\frac{a}{2}\right)^2 + \left(\frac{a}{2}\right)^2 \qquad 2r = 461.9\ \text{pm} \qquad r = 231.0\ \text{pm}$$

Diamond has a face-centered cubic lattice with atoms at 000 and $\frac{1}{4}\frac{1}{4}\frac{1}{4}$ associated with each lattice point. This structure is represented in two different ways in Fig. 23.10. Since there are two atoms per lattice point, there are eight atoms per unit cell. The unit cell distance for diamond is 356.7 pm. Silicon, germanium, and gray tin also have this structure with unit cell distances of 543.1, 565.7, and 649.1 pm.

Example 23.6 ***The diamond structure***

Calculate the C—C bond distance in diamond and the C—C—C angle.

Using equation 23.4 for the points 000 and $\frac{1}{4}\frac{1}{4}\frac{1}{4}$,

$$l = a[(x_2 - x_1)^2 + (y_2 - y_1)^2 + (z_2 - z_1)^2]^{1/2}$$

$$= (356.7\ \text{pm})(\tfrac{3}{16})^{1/2} = 154.5\ \text{pm}$$

which is the C—C distance. The distance l between a carbon atom at a corner of the unit cell and one at a face-centered position is given by

$154.5 \text{ pm} = (356.7 \text{ pm})\left(\frac{3}{16}\right)^{1/2}$

$\frac{\sqrt{2}}{4}(356.7 \text{ pm})$

θ

$$l^2 = \left(\frac{356.7 \text{ pm}}{2}\right)^2 + \left(\frac{356.7 \text{ pm}}{2}\right)^2$$

$$l = \frac{\sqrt{2}}{2}(356.7 \text{ pm})$$

$$\sin\theta = \frac{2^{1/2}(356.7 \text{ pm})}{4(356.7 \text{ pm})\left(\frac{3}{16}\right)^{1/2}}$$

$$\theta = 54.736°$$

The C—C—C bond angle is 2θ or 109.472°, which is referred to as the tetrahedral angle.

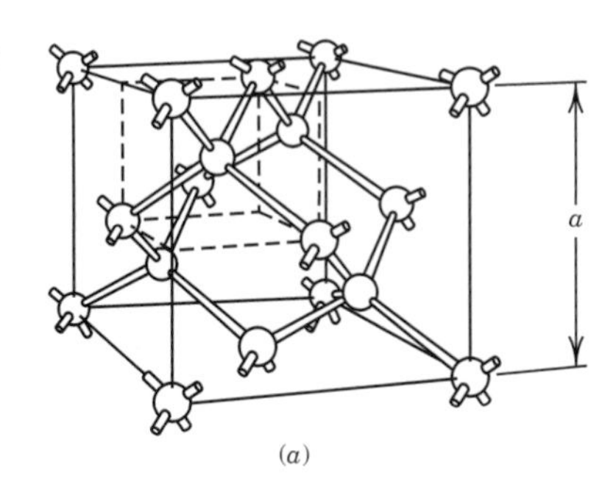

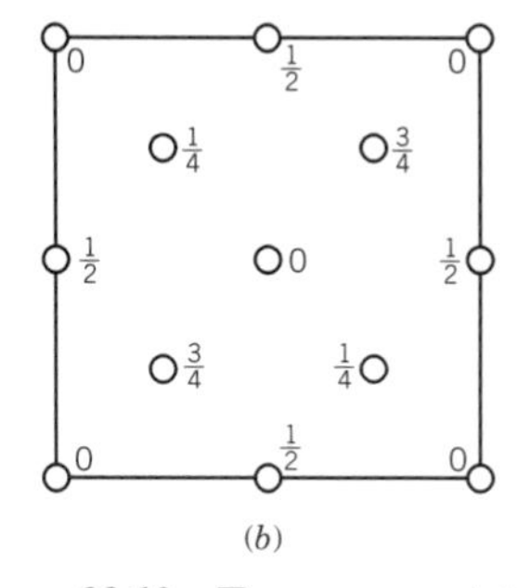

Figure 23.10 Two representations of the diamond structure. (*a*) Space model showing tetrahedral bonds. (*b*) Projection showing fractional coordinates. It is instructive to draw lines showing the bonds in this projection.

The value of the Avogadro constant N_A can be calculated from the density of a crystal, the relative atomic mass, and the unit cell length. To be useful for this purpose, a crystal must be free of defects. Very accurate values of these quantities for silicon have been measured at the National Institute for Standards and Technology.* The value of the Avogadro constant determined in this way $[6.022\,097\,6(63) \times 10^{23} \text{ mol}^{-1}]$ is tied to measurements of other fundamental constants through a least-squares adjustment to obtain the best values given in Appendix B.

There is a condition on the ratio of ion radii R that must be satisfied for an ionic substance MX to have the NaCl structure. Since the ions are in contact along a cell edge,

$$a = 2(R_+ + R_-) \tag{23.13}$$

In addition, ions cannot overlap along the diagonal of the face of the unit cell. Therefore,

$$(4R_-)^2 \leq 2a^2 \tag{23.14}$$

$$(4R_+)^2 \leq 2a^2 \tag{23.15}$$

Thus,

$$a \geq 2\sqrt{2}R_- \quad \text{and} \quad a \geq 2\sqrt{2}R_+ \tag{23.16}$$

Using equation 23.13,

$$2(R_+ + R_-) \geq 2\sqrt{2}R_- \tag{23.17}$$

$$\frac{R_+}{R_-} \geq \sqrt{2} - 1 = 0.414 \tag{23.18}$$

Atom and ion radii for a number of elements are summarized in Table 23.2. It is seen that in each column of the periodic table the ionic radius increases with the principal quantum number of the valence orbital electrons. The radius

*R. D. Deslattes, A. Hemins, R. M. Schoonover, C. L. Carroll, and H. A. Bowman, *Phys. Rev. Lett.* **36:**898 (1976).

Table 23.2 Crystal Structure Data

Atomic Number	*Element*	*Structure*[a]	*Atom Radius*/pm	*Ion*	*Ion Radius*/pm[b]
3	Li	b.c.c.	152	Li^{+}	60
4	Be	c.p.h.	112	Be^{2+}	31
6	C	Cubic (diamond)	77		
8	O			O^{2+}	140
9	F			F^{-}	136
11	Na	b.c.c.	186	Na^{+}	95
12	Mg	c.p.h.	161	Mg^{2+}	65
14	Si	Cubic (diamond)	118	Si^{4+}	41
17	Cl			Cl^{-}	181
19	K	b.c.c.	232	K^{+}	133
20	Ca	f.c.c.	197	Ca^{2+}	99
26	Fe	b.c.c.	124	Fe^{2+}	80
				Fe^{3+}	64
29	Cu	f.c.c.	128	Cu^{+}	96
30	Zn	c.p.h.	133	Zn^{2+}	74
32	Ge	Cubic (diamond)	128	Ge^{4+}	53
35	Br			Br^{-}	195
37	Rb	b.c.c.	245	Rb^{+}	148
53	I	Orthorhombic	136	I^{-}	216
55	Cs	b.c.c.	263	Cs^{+}	169
78	Pt	f.c.c.	139	Pt^{4+}	65

[a] b.c.c., body-centered cubic; c.p.h., close-packed hexagonal; f.c.c., face-centered cubic.
[b] The radius of an ion depends on the number of neighboring ions of opposite sign, or, in other words, the coordination number. The values in the table apply to an ion with a coordination number of 6. In structures where the coordination number is 4 the radius should be decreased by about 7%; with coordination number 8 the radius should be increased by 3%, and with coordination number 12 the radius should be increased by 6%.
Source: A. Kelly and G. W. Groves, *Crystallography and Crystal Defects.* Reading, MA: Addison-Wesley, 1970.

of an ion is nearly the same in different crystals because the repulsive force increases very sharply as the internuclear distance becomes smaller than a certain value.

It has been found that the distance between two kinds of atoms connected by a covalent bond of a given type (single, double, etc.) is nearly the same in different molecules. The distance between two atoms is taken to be equal to the sum of the bond radii of the two atoms. Since the C—C bond distance is 154 pm in many compounds, the radius for a carbon single bond is taken to be 77 pm. Since the C≡C distance in acetylene is 120 pm, the radius for a carbon triple bond is taken to be 60 pm. By consideration of the bond distances in many compounds it has been possible to build up tables of bond radii, such as Table 23.3, which are useful in predicting the structures of molecules. It must be realized, however, that the effective radius of an atom depends in part also on its environment and on the nature of the bonds in the molecule under consideration.

Table 23.3 Covalent Radii for Atoms in pm

	H	C	N	O	F
Single-bond radius	30	77.2	70	66	64
Double-bond radius		66.7	60	56	
Triple-bond radius		60.3			
		Si	P	S	Cl
Single-bond radius		117	110	104	99
Double-bond radius		107	100	94	89
Triple-bond radius		100	93	87	
		Ge	As	Se	Br
Single-bond radius		122	121	117	114
Double-bond radius		112	111	107	104
		Sn	Sb	Te	I
Single-bond radius		140	141	137	133
Double-bond radius		130	131	127	123

Source: L. Pauling, *The Nature of the Chemical Bond.* Ithaca, NY: Cornell University Press, 1960, which should be consulted for details concerning the source and constancy of these radii.

Comment:

The determination of the density of a crystal, the relative atomic masses of the atoms, the isotopic abundance, and the size of the unit cell has proved to be one of the most accurate ways to determine the value of the Avogadro constant. These quantities have been determined very accurately for silicon crystals at the National Institute of Standards and Technology by Deslattes and co-workers. Silicon has been found to be the most suitable for this purpose because crystals can be grown with very few defects. The value of the Avogadro constant determined in this way has been tied to measurements of other fundamental constants through least squares adjustment to obtain the value $6.022\,136\,7 \times 10^{23}$ mol^{-1} *that is recommended in Appendix B.*

23.6 SCATTERING OF X-RAYS FROM A UNIT CELL

X-rays are scattered by electrons, and the greater the electron density, $\rho(r)$, the greater the scattering. Heavy atoms scatter X-rays to a greater extent because they have higher electron densities in their vicinity. Note that the electron density of an atom is spherically symmetric. The scattering factor for an atom depends on the electron distribution $\rho(r)$ in the atom and on the angle θ of scattering. Specifically, the scattering factor is given by

$$f = 4\pi \int_0^\infty \rho(r) \frac{\sin kr}{kr} r^2 \, \mathrm{d}r \tag{23.19}$$

where $k = (4\pi/\lambda)\sin\theta$, where θ is the Bragg angle. The scattered intensity is greatest in the forward direction, for which $\theta = 0$. The atomic scattering factor

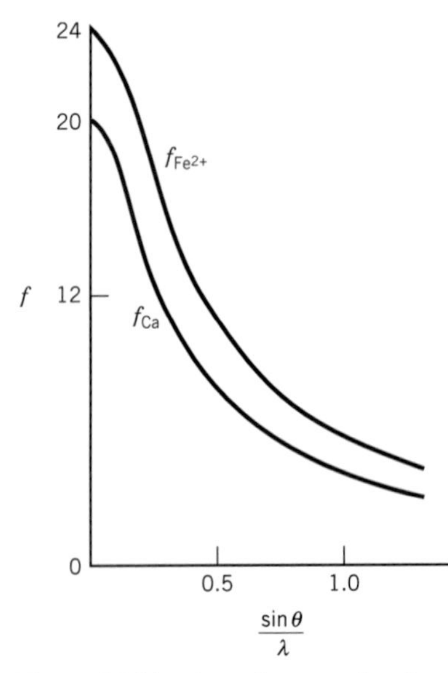

Figure 23.11 Atomic scattering factors f for Fe^{2+} and Ca. Note that as the Bragg angle goes to zero, f becomes equal to the number of electrons. [From D. E. Sands, *Introduction to Crystallography.* New York: Dover Publications, Inc., 1993.]

in the forward direction is equal to the number of electrons in the atom or ion. In the forward direction, $\theta = 0$ and $k = 0$, and so equation 23.19 becomes*

$$f = 4\pi \int_0^\infty \rho(r) r^2 \, dr \tag{23.20}$$

The integration of the electron density times the spherical shell $4\pi r^2 \, dr$ yields the total number of electrons in the atom. Figure 23.11 shows plots of the atomic scattering factors for Fe^{2+} and Ca.

We can discuss the scattering of X-rays from a crystal by considering a single unit cell because when the scattering from two atoms in a unit cell is in phase for certain Miller indices, it will be in phase with scattering from corresponding atoms in the other unit cells in the crystal. Consider a unit cell containing N atoms, or ions, with **fractional coordinates** x_i, y_i, and z_i. The path difference δ_i associated with waves of wavelength λ scattered by atom i can be shown to be given by

$$\delta_i = \lambda(hx_i + ky_i + lz_i) \tag{23.21}$$

where h, k, and l are Miller indices. The corresponding **phase difference** (angular measure) is given by

$$\phi_i = \frac{2\pi}{\lambda}\delta_i = 2\pi(hx_i + ky_i + lz_i) \tag{23.22}$$

Reflections from two atoms reinforce if the phase difference is an integer times 2π and interfere destructively if the phase difference is an odd integer times π (180°).

For a unit cell the amplitude of the scattering of X-rays is referred to as the **structure factor** $F(hkl)$, which is made up of additive contributions from all of the atoms or ions in the unit cell:

$$F(hkl) = \sum f_i \, e^{i\phi_i} \tag{23.23}$$

Substituting equation 23.22 in equation 23.23 yields

$$F(hkl) = \sum f_i \, e^{i2\pi(hx_i + ky_i + lz_i)} \tag{23.24}$$

The intensity $I(hkl)$ of the (hkl) reflection is proportional to the square of the amplitude; that is, $I(hkl) \propto |F(hkl)|^2$.

Example 23.7 ***Structure factor*** $F(h, k, l)$ ***for a primitive cubic unit cell***

Derive the structure factor for a primitive cubic unit cell of identical atoms. The relative coordinates of the lattice points are given by (0,0,0), (1,0,0), (0,1,0), (0,0,1), (1,1,0), (1,0,1), (0,1,1), and (1,1,1).

Since these atoms are each shared with eight other unit cells, the sum in equation 23.24 should be multiplied by $\frac{1}{8}$ so that it applies to one atom.

$$F(hkl) = \frac{1}{8} f [e^{2\pi i(0)} + e^{2\pi i(h)} + e^{2\pi i(k)} + e^{2\pi i(l)} + e^{2\pi i(h+k)} + e^{2\pi i(k+l)} + e^{2\pi i(h+l)} + e^{2\pi i(h+k+l)}]$$

*Since kr occurs in the numerator and the denominator of the integrand, at $\theta = 0$ the ratio $(\sin kr)/kr$ is replaced by its limit when $kr \to 0$. When $k = 0$, $(\sin kr)/kr$ in equation 23.19 becomes unity since $\sin kr = kr + \cdots$ for small angles.

Since $e^{2\pi i} = \cos 2\pi + i \sin 2\pi = 1$,

$$F(hkl) = \tfrac{1}{8}f[1^0 + 1^h + 1^k + 1^l + 1^{h+k} + 1^{k+l} + 1^{h+l} + 1^{h+k+l}] = f$$

This shows that there will be reflections for a primitive unit cell for all integer values of h, k, and l.

According to Fourier's theorem, a continuous, single-valued periodic function can be represented by a series composed of sine and cosine terms. This theorem can be applied to the electron density $\rho(x, y, z)$ in a crystal, but as a simplification, we will first consider a one-dimensional crystal. The electron density $\rho(x)$ in a one-dimensional crystal with a repeat distance a is shown in Fig. 23.12:

$$\rho(x) = \frac{1}{a}\sum_{n=-\infty}^{\infty}\left(A_n \cos\frac{2\pi nx}{a} + B_n \sin\frac{2\pi nx}{a}\right) \tag{23.25}$$

Since

$$\cos\frac{2\pi nx}{a} = (e^{i2\pi nx/a} + e^{-i2\pi nx/a})/2 \tag{23.26}$$

$$\sin\frac{2\pi nx}{a} = (e^{i2\pi nx/a} - e^{-i2\pi nx/a})/2i \tag{23.27}$$

equation 23.25 can be written

$$\rho(x) = \frac{1}{a}\sum_{n=-\infty}^{\infty} F(n)\, e^{-i2\pi nx/a} \tag{23.28}$$

Be sure not to confuse the fractional coordinate x_i of atom i in equation 23.24 with the distance x measured in the one-dimensional crystal. Fourier showed that when an equation has the form of equation 23.28, the function $F(n)$ can be obtained from the following integration (Fourier transform). The Fourier transform of equation 23.28 is

$$F(n) = \int_{-a/2}^{a/2} \rho(x)\, e^{i2\pi nx/a}\, dx \tag{23.29}$$

Equations 23.28 and 23.29 are said to be Fourier transforms of each other. If $F(n)$ is known for all values of n, we can calculate $\rho(x)$. This is the situation when we discussed a unit cell of known structure. If $\rho(x)$ is known from $-a/2$ to $+a/2$, we can calculate $F(n)$.

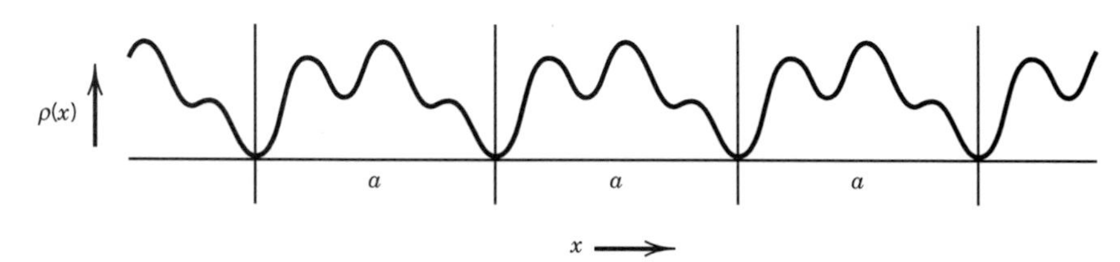

Figure 23.12 Electron density as a function of distance x in a one-dimensional crystal with repeat distance a.

Equation 23.28 can be generalized to three dimensions, where it becomes

$$\rho(x, y, z) = \frac{1}{V} \sum_{h=-\infty}^{\infty} \sum_{k=-\infty}^{\infty} \sum_{l=-\infty}^{\infty} F(hkl) \exp[-2\pi i(hx + ky + lz)] \quad (23.30)$$

where V is the volume of the unit cell. Alternatively, the electron density can be written as

$$\rho(x, y, z) = \frac{1}{V} \sum_{h} \sum_{k} \sum_{l} |F(hkl)| e^{i\phi(hkl)} e^{-i2\pi(hx+ky+lz)} \quad (23.31)$$

where $\phi(hkl)$ is the phase difference (see equation 23.22).

Since the structure factor $F(hkl)$ is a complex function, this equation appears to imply that the electron density is a complex function, but the electron density $\rho(x, y, z)$ is a real function. If chemical bonding effects are neglected, $\rho(x, y, z)$ is the linear superposition of spherical electron densities of the individual atoms. It can be shown that this function can be written

$$\rho(x, y, z) = \frac{1}{V} \sum_{h} \sum_{k} \sum_{l} |F(hkl)| \cos[2\pi(hx + ky + lz) - \phi(hkl)] \quad (23.32)$$

The summations over h, k, and l go from $-\infty$ to $+\infty$, but in practice the electron density can be represented quite well with a finite number of terms. Higher resolution is obtained by including more reflections (i.e., terms). Equation 23.32 shows that it is necessary to know the phases in order to calculate the electron densities.

The Fourier transform of equation 23.30 can be written

$$F(hkl) = \int\int\int \rho(x, y, z) e^{-2\pi i(hx+ky+lz)} \, dx \, dy \, dz \quad (23.33)$$

where the volume has been ignored. This is known as the Fourier transform of the function $\rho(x, y, z)$. If we knew $\rho(x, y, z)$ at every point in a unit cell, we could calculate $F(hkl)$ by use of this equation. The actual situation in determining a crystal structure is that $|F(hkl)|$ is known from the intensities of the X-ray reflections, but there is no information on the phase because film and other radiation detectors measure only energy. Phase information is needed to calculate the positions of the atoms. This is the **phase problem.** Various methods for coping with this problem are discussed at the end of this section.

To illustrate the usefulness of the structure factor, let us consider its dependence on the reflection indices hkl for several lattices. In primitive lattices with atoms at the lattice sites, x_j, y_j, and z_j are zero, so $F(hkl) = \sum f_j$. Thus, the structure factor $F(hkl)$ has the same value for all values of h, k, and l, and there will be reflections for all integer values of h, k, and l.

Example 23.8 ***Absent reflections***

Which reflections will be absent in the diffraction pattern for a body-centered cubic unit cell?

The eight corner atoms are each only $\frac{1}{8}$th in the unit cell, and their contribution is equivalent to one atom at $x = y = z = 0$. The atom at the center of the unit cell is at $x = y = z = \frac{1}{2}$, so the structure factor is given by

$$\begin{aligned} F(hkl) &= f\,\mathrm{e}^{2\pi\mathrm{i}(0+0+0)} + f\,\mathrm{e}^{2\pi\mathrm{i}(h/2+k/2+l/2)} \\ &= f[1 + \mathrm{e}^{\mathrm{i}\pi(h+k+l)}] \end{aligned}$$

where f is the atomic scattering factor. Since $\mathrm{e}^{\mathrm{i}\pi} = -1$,

$$F(hkl) = f[1 + (-1)^{h+k+l}]$$

If $h + k + l$ is even, then $F(hkl) = 2f$, but if $h + k + l$ is odd, the reflection will be absent. Thus, the reflections (100), (111), (210), (300), ... will be absent.

Example 23.9 ***The structure factor expression***

What is the structure factor expression for a face-centered cubic lattice? Derive a rule for absences in reflections in terms of the values of h, k, l.

The four atoms in the unit cell may be assigned the positions (000), $(\frac{1}{2}\frac{1}{2}0)$, $(\frac{1}{2}0\frac{1}{2})$, and $(0\frac{1}{2}\frac{1}{2})$:

$$\begin{aligned} F(hkl) &= f\,\mathrm{e}^{2\pi\mathrm{i}(h\cdot 0+k\cdot 0+l\cdot 0)} + f\,\mathrm{e}^{2\pi\mathrm{i}(h/2+k/2+0)} \\ &\quad + f\,\mathrm{e}^{2\pi\mathrm{i}(h/2+0+l/2)} + f\,\mathrm{e}^{2\pi\mathrm{i}(0+k/2+l/2)} \\ &= f[1 + \mathrm{e}^{\pi\mathrm{i}(h+k)} + \mathrm{e}^{\pi\mathrm{i}(h+l)} + \mathrm{e}^{\pi\mathrm{i}(k+l)}] \\ &= f[1 + (-1)^{h+k} + (-1)^{h+l} + (-1)^{k+l}] \end{aligned}$$

If h, k, l are all even or all odd, then $F(hkl) = 4f$ and there are reflections. If one is even and the other two are odd, or the reverse, the reflections are absent. This is illustrated by the powder pattern for NaCl.

These examples show that the structure factors contain all the information about all the atoms in a unit cell. For a known structure they may be used to calculate the electron density throughout the unit cell using equation 23.32. But the X-ray diffractionist faces a much more difficult problem. The structure factors $F(hkl)$ are related to the intensities $I(hkl)$ of radiation reflected from the planes (hkl) by

$$I(hkl) \propto |F(hkl)|^2 \tag{23.34}$$

Measurements of the densities of spots on photographic film or of counts recorded by a Geiger counter for each reflection may be subjected to routine corrections to obtain $I(hkl)$ values. Thus, a set of $|F(hkl)|^2$ values may be obtained. Unfortunately, what is needed to calculate $\rho(x, y, z)$ using equation 23.32 are values of $F(hkl)$ instead of $|F(hkl)|^2$. Since $F(hkl)$ is a complex number, we can write

$$F(hkl) = A(hkl) + \mathrm{i}B(hkl) \tag{23.35}$$

so that

$$\begin{aligned} |F(hkl)|^2 &= [A(hkl) + \mathrm{i}B(hkl)][A(hkl) - \mathrm{i}B(hkl)] \\ &= [A(hkl)]^2 + [B(hkl)]^2 \end{aligned} \tag{23.36}$$

Since values of $A(hkl)$ and $B(hkl)$ are not obtained directly, indirect methods must be used to obtain these quantities and the electron density $\rho(x, y, z)$. This is another statement of the **phase problem.** Fortunately, the number of parameters needed to describe a crystal structure is far smaller than the number of reflections, so that the problem is greatly overdetermined. Several methods are used to get around the phase problem. If heavy atoms are present in the unit cell, they may by themselves determine enough phases so that a Fourier map of the electron density may reveal the position of some of the lighter atoms.

The method of **isomorphous replacement** is especially useful in determining the three-dimensional structure of protein molecules in crystals. X-ray diffraction data are determined on the protein crystal, and these are called the native data. Then a heavy atom that is bound at specific locations is introduced into the crystal, and the X-ray diffraction data determined on this altered but isomorphous crystal are referred to as the heavy atom data. A difference data set is obtained by subtraction. The difference data are used to determine the positions and phases of the heavy atoms in the unit cell. The phases determined in this way are used to solve for the structure of the protein.*

X-ray analyses of thousands of crystal structures have led to detailed knowledge of the geometrical properties of different groups of atoms, including well-established values of bond lengths and angles. Modern crystallographic analyses using data-collecting diffractometers and high-speed computers have made it possible to determine the molecular structure of proteins. X-ray diffraction data were used in the determination of the structure of deoxyribonucleic acid, and this led to an understanding of the hydrogen bonding that makes that structure stable.

In certain cases X-ray diffraction may be used to determine the absolute configuration of an optically active substance. In 1951 Bijvoet, Peerdeman, and van Bommel studied sodium rubidium (+)-tartaric acid by X-ray diffraction and found that the absolute configuration was the one arbitrarily chosen from the two possible enantiomorphic structures by Fischer 60 years earlier.

Comment:

In 1985 Hauptman and Karle received the Nobel Prize for their development of a direct method for determining phases. For unit cells that are not too large, they showed that use of prior structural knowledge makes it possible to determine the positions of the atoms from the scattered intensities alone. However, this method can be used only for unit cells containing up to about 100 atoms.

23.7 BINDING FORCES AND PACKING IN CRYSTALS

A number of different types of binding forces are involved in holding crystals together. The physical properties of a crystal are very dependent on the type of bonding.

Ionic crystals are held together by the strong coulomb attractions of the oppositely charged ions. The lattice energy determined from heat of formation

*More detailed information is provided by K. E. van Holde, W. C. Johnson, and P. S. Ho, *Principles of Physical Biochemistry,* Upper Saddle River, NJ: Prentice-Hall, 1998.

and heat of vaporization measurements agrees with that calculated on the assumption that the units of the crystal are ions held together by electrostatic forces.

In ionic crystals there is no fixed directed force of attraction. Although the ionic crystals are strong, they are likely to be brittle. They have very little elasticity and cannot be easily bent or worked. The melting points of ionic crystals are generally high (NaCl, 800 °C; KCl, 790 °C). In ionic crystals some of the atoms may be held together by covalent bonds to form ions having definite positions and orientations in the crystal lattice. For example, in calcium carbonate a carbonate ion does not "belong" to a given calcium ion, but three particular oxygen atoms are bonded to a given carbon atom.

Covalent crystals, which are held together by covalent bonds in three dimensions, are strong and hard and have high melting points. Diamond is an example of this type of crystal.

The great difference between graphite and diamond can be understood in terms of the crystal lattice. Graphite has hexagonal networks in sheets like benzene rings. The distance between atoms in the plane is 142 pm, but the distance between these atomic layer planes is 335 pm. In two directions, then, the carbon atoms are tightly held as in the diamond, but in the third direction the force of attraction is much less. As a result, one layer can slip over another. The crystals are flaky, and yet the material is not wholly disintegrated by a shearing action. This planar structure is part of the explanation of the lubricating action of graphite, but this action also depends on absorbed gases, and the coefficient of friction is much higher in a vacuum.

Covalently bonded crystals are insulators because the bonding orbitals are fully occupied. They become conductors only if electrons are excited to unoccupied levels. Thus, they become photoconductors when they are irradiated at a wavelength sufficiently short to raise electrons to excited levels.

Molecular crystals are held together by van der Waals forces (Section 11.9). Examples are provided by crystals of neutral organic compounds and rare gases. Since van der Waals forces are weak, such molecular crystals have low melting points and low cohesive strengths.

Hydrogen-bonded crystals are held together by the sharing of protons between electronegative atoms. Hydrogen bonds are involved in many organic and inorganic crystals and in the structure of ice and water. They are comparatively weak bonds, but they play an extremely important role in determining the atomic arrangement in hydrogen-bonded substances such as proteins and polynucleotides.

Metallic bonds exist only between large aggregates of atoms. This type of bonding gives metals their characteristic properties: opacity, luster, malleability, and conductivity of electricity and heat. Metallic bonding is due to the outer, or valence, electrons. The overlapping of the wavefunctions for the valence electrons in metals results in orbitals that extend over the entire crystal. The electrons pass throughout the volume of the crystal, and for certain purposes, we may consider that there is an electron gas—except that we will see that it is fundamentally different from other gases.

There is a gradual transition between metallic and nonmetallic properties. Atoms with fewer and more loosely held electrons form metals with the most prominent metallic properties. Examples are sodium, copper, and gold. As the number of valence electrons increases and they are held more tightly, there is a

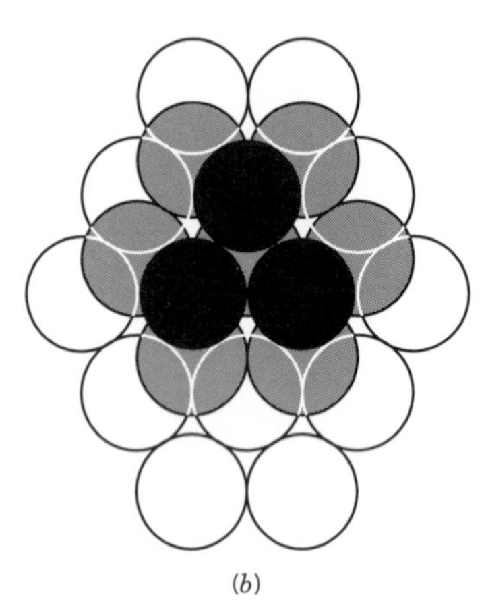

Figure 23.13 (*a*) Cubic close packing (face-centered cubic). (*b*) Hexagonal close packing.

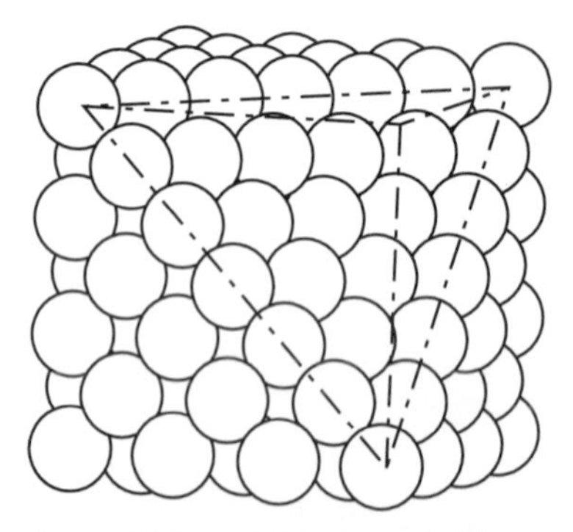

Figure 23.14 Cubic close packing (face-centered cubic). Some atoms have been omitted to show that the close-packed planes are (111) planes.

transition to covalent properties. The close-packed structures are often found in metals because the binding energy per unit volume is maximized.

When the bonding of atoms is not highly directional, it is often found that the lowest energy structure is that in which each atom is surrounded by the greatest possible number of neighbors. It is therefore of interest to consider the ways in which uniform spheres can be stacked to form close-packed structures. When spheres are packed in a plane, they arrange themselves so that each sphere is surrounded hexagonally by six others. When the second layer is formed by placing spheres in the hollows on top of the first layer, it is evident that all of the hollows in the first layer are not occupied, as may be seen from Fig. 23.13. When a third layer is added, there is a choice as to whether the spheres in this layer are stacked so that they are not above the spheres in the first layer, as in Fig. 23.13*a*, or are, as in Fig. 23.13*b*. If the spheres in the third layer are not directly above the spheres in the first layer, as shown in Fig. 23.13*a*, the structure has cubic symmetry and the cubic unit cell is face centered. The fact that **cubic close packing** is really face-centered cubic may be seen from Fig. 23.14. Since this is a close-packed structure, it is of interest to calculate the fraction of the volume occupied by spheres. Since the length of the diagonal of the face of a unit cell is $\sqrt{2}a$, the radii of the spheres that just touch are given by $(\sqrt{2}/4)a$. Since there are four spheres per unit cell, the fraction of the volume occupied by spheres is

$$\frac{4(\frac{4}{3}\pi)[(\sqrt{2}/4)a]^3}{a^3} = 0.7405 \tag{23.37}$$

In cubic close packing each sphere has 12 nearest neighbors: 6 within its own layer, 3 in the layer above, and 3 in the layer below. Metals and rare gases often form cubic close-packed structures.

If the spheres in the third layer are placed over the spheres in the first layer, as shown in Fig. 23.13*b*, and the spheres in the fourth layer are placed over those in the second layer, and so on, the unit cell is hexagonal and the packing is referred to as **hexagonal close packing.** As in the case of cubic close packing, each sphere has 12 nearest neighbors, and the fraction of the volume occupied by spheres is again 0.7405. The coordinates of the atoms in the unit cell are (000) and $(\frac{1}{3}\frac{1}{3}\frac{1}{2})$. There are two atoms in the unit cell.

The unit cell dimensions in terms of the radius of a sphere are $a = b = 2r$, $c = 4\sqrt{2}r/\sqrt{3}$, and $c/a = 2\sqrt{2}/\sqrt{3} = 1.633$. A number of metals have hexagonal close-packed structures, but c/a usually deviates a little from the ideal ratio of 1.633. This indicates that the atoms are not exactly spherical in shape.

The layers of hexagonal close packing may be described as ABABAB.... The layers of cubic close packing may be described as ABCABC....

Hexagonal close packing and cubic close packing are the only two ways of close-packing identical spheres so that the environment of each sphere is identical to the environment of all the other spheres, but there are other ways of close-packing spheres so that the environment of each sphere is not identical, for example, ABCABABCAB.... In principle, there are an infinite number of these other ways.

In a number of crystal structures containing two types of atoms, one type of atoms forms a close-packed structure and the other type occupies interstices between the close-packed spheres. There are two types of interstices between close-packed spheres: tetrahedral sites and octahedral sites. When one sphere rests on three others, the centers of the four spheres lie at the apices of a regular tetrahedron, and the space at the center of this tetrahedron is called a tetrahedral site. Since in any close-packed structure each sphere is in contact with three spheres in the layer below it and three spheres in the layer above it, there are two tetrahedral sites per sphere. Thus, if X atoms form a close-packed structure, and Y atoms are small enough to fit in the tetrahedral sites, this might be a convenient structure for a compound XY_2. Half of the sites would be occupied in a compound XY. For the smaller sphere to occupy a tetrahedral site without disturbing the closest-packed lattice, the radius of the smaller spheres should be no greater than 0.225 of that of the larger spheres.

The other type of interstitial site in a close-packed structure is the octahedral site that is surrounded by six spheres whose centers lie at the apices of a regular octahedron. There is one octahedral site for every sphere in a close-packed structure so that a compound of the type XY can be accommodated. For Y atoms to fit into octahedral sites, their radii must be less than 0.414 of the radii of the larger spheres.

Although the majority of metallic elements crystallize with hexagonal close packing or cubic close packing, some crystallize with the body-centered cubic arrangement, which is not a close-packed structure. In this structure, which is illustrated in Fig. 23.15, each atom has eight nearest neighbors and six other next-nearest neighbors slightly farther away at the body-centered positions of neighboring cells. By use of the Pythagorean theorem it is readily shown that the distance from the body-centered point to one of the corners of the cubic unit cell is $(\sqrt{3}/2)a$. If the structure is made up of spheres that touch, they must have a radius of $(\sqrt{3}/4)a$. The fraction of the volume of the unit cell (and hence of the entire crystal) occupied by spheres is

$$\frac{2(\frac{4}{3}\pi)[(\sqrt{3}/4)a]^3}{a^3} = 0.6802 \tag{23.38}$$

The alkali metals and tungsten crystallize in a body-centered cubic structure.

The characteristics of cubic lattices for spheres are summarized in Table 23.4. It is very rare to find spheres packed in a simple cubic lattice.

Example 23.10 ***The magnesium atom radius***

Magnesium forms hexagonal close-packed crystals with $a = 320.9$ pm at 25 °C. What is the density of the metal, and what is the magnesium atom radius?

$$V = a^2c(1 - \cos^2\gamma)^{1/2} = a^2c\sin\gamma$$

Table 23.4 Characteristics of Cubic Lattices

	Simple	*Body-Centered*	*Face-Centered*
Volume of unit cell	a^3	a^3	a^3
Lattice points per cell	1	2	4
Nearest neighbors	6	8	12
Distance to nearest neighbor	a	$(\sqrt{3}/2)\,a$	$(1/\sqrt{2})\,a$
Fraction of volume occupied	0.524	0.680	0.740

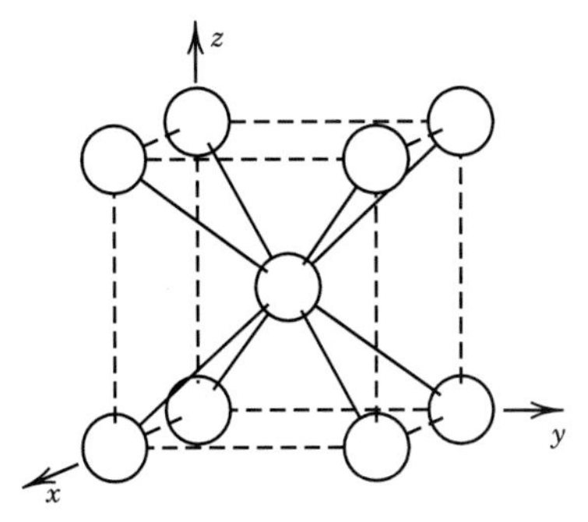

Figure 23.15 Body-centered cubic structure.

Since $c = 1.633a$,

$$V = 1.633(320.9 \times 10^{-12}\ \text{m})^3 \sin 120° = 4.673 \times 10^{-29}\ \text{m}^3$$

$$\rho = \frac{2(24.305 \times 10^{-3}\ \text{kg mol}^{-1})}{(6.022\,137 \times 10^{23}\ \text{mol}^{-1})(4.673 \times 10^{-29}\ \text{m}^3)} = 1.727 \times 10^3\ \text{kg m}^{-3}$$

$$r = \tfrac{1}{2}a = \tfrac{1}{2}(320.9\ \text{pm}) = 160.5\ \text{pm}$$

23.8 STRUCTURE OF LIQUIDS*

In a perfect crystal the atoms, ions, or molecules occur at definite distances from any individual atom, ion, or molecule that is taken as the origin of a coordinate system. In a gas the molecules have random positions at a given time. Liquids are intermediate between crystals and gases in that the molecules are not arranged in a definite lattice, but there is some local order. By a detailed analysis of the intensity of the scattered X-rays, it is possible to calculate the distribution of atoms or molecules in a liquid.

When X-rays are scattered by an amorphous phase the intensity $I(\theta)$ of scattered radiation is a function only of the scattering angle θ, which is defined as the angle between scattered and incident directions. (Note that this angle is twice the Bragg angle used earlier.) The intensity tends to fall off as θ increases. It may be shown that the intensity is given by

$$I(\theta) \propto \int_0^\infty \mathcal{P}(R) \frac{\sin kR}{kR}\, \mathrm{d}R \tag{23.39}$$

where

$$k = \frac{4\pi}{\lambda} \sin \frac{\theta}{2}$$

and $\mathcal{P}(R)\,\mathrm{d}R$ is the probability of finding a particle between R and $R + \mathrm{d}R$, R being measured from one particle at the origin. For a uniform medium $\mathcal{P}(R)$ would be proportional to $4\pi R^2$, and so it is convenient to introduce a **pair correlation function** $g_2(R)$ defined by

$$\mathcal{P}(R) = 4\pi R^2 \rho g_2(R) \tag{23.40}$$

where ρ is the number density for the medium. If the medium were continuous and homogeneous, the pair correlation function $g_2(R)$ would be constant. As shown in Fig. 23.16, the pair correlation function for a liquid is zero for small values of R, has a maximum at the most probable nearest-neighbor distance, has diminishing maxima that correspond to second-nearest-neighbor distance, and so on. The first maximum in Fig. 23.16 is due to the 8 or 12 nearest neighbors that surround each molecule. The pair correlation function $g_2(R)$ is essentially zero at distances less than one molecular diameter because of the strong short-range intermolecular repulsions (Section 11.9). The existence of a shell of nearest neighbors means that there will be a relatively high probability of finding molecules one molecular diameter away, as well as two, three, and four molecular diameters removed. Since the liquid has no long-range order, $g_2(R)$ is essentially constant after a few

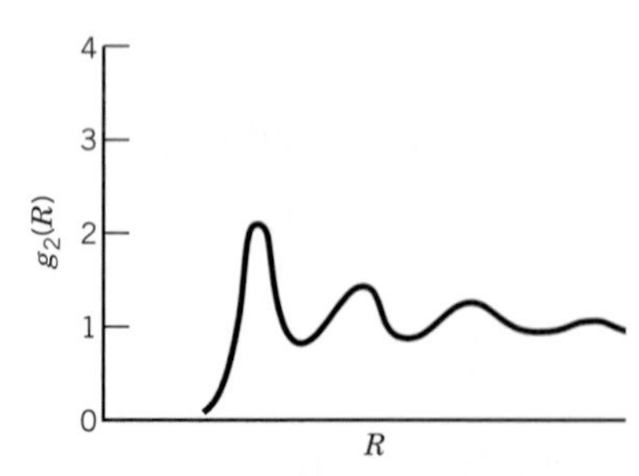

Figure 23.16 Pair correlation function for a liquid.

*Y. Marcus, *Introduction to Liquid State Chemistry.* Hoboken, NJ: 1977.

molecular diameters. As the temperature of the liquid is raised, the maxima and the minima in the pair correlation function become less pronounced.

23.9 LIQUID CRYSTALS

In certain liquids new phases, which are intermediate between liquid and solid phases, appear on cooling. These phases often have a translucent or cloudy appearance and are called liquid crystals.

In a liquid of asymmetric molecules the molecular axes are arranged at random, but in liquid crystals there is some kind of alignment. As shown in Fig. 23.17, there are three types of liquid crystals. In **nematic** liquid crystals the long axes of the molecules are aligned parallel to each other, but the molecules are not arranged in layers. The word *nematic* was coined from the Greek root for thread to describe the appearance of this particular type of liquid crystal under a microscope. Nematic liquid crystals have a translucent appearance because they scatter light strongly.

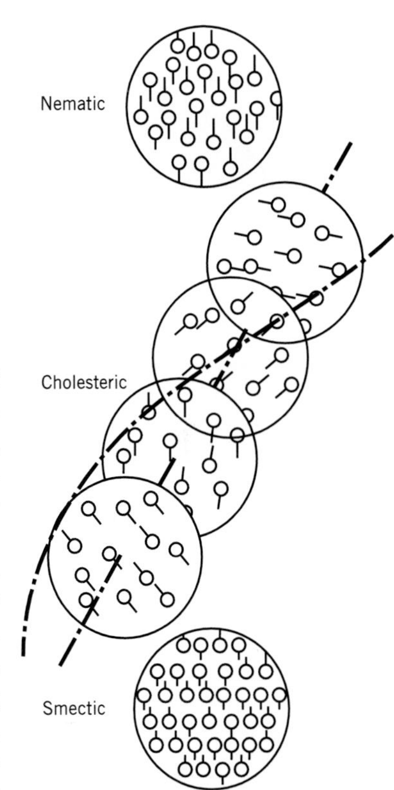

Figure 23.17 Structures of liquid crystals.

In **cholesteric** liquid crystals the molecular axes are aligned, and the molecules are arranged in layers in which the orientation of the axes shifts in a regular way in going from one layer to the next, as shown in Fig. 23.17. The distance measured perpendicular to the layers through which the direction of alignments shifts 360° is of the order of the wavelength of visible light. As a result of the strong Bragg reflection of light, cholesteric liquid crystals have vivid iridescent colors. The pitch of the spiral and the reflected color depends sensitively on the temperature, and so these liquid crystals have been used to measure skin and other surface temperatures. The name *cholesteric* comes from the fact that many derivatives of cholesterol (but not cholesterol itself) form this type of liquid crystal.

The third type of liquid crystals, **smectic,** are formed by certain molecules with chemically dissimilar parts. The chemically similar parts attract each other, and there is a tendency to form layers as well as to have the molecules aligned in one direction, as illustrated in Fig. 23.17. Smectic phases are soaplike in feel and structure and may have some relationships with cell membranes.

23.10 THEORETICAL TREATMENT OF THE ELECTRON DISTRIBUTION IN SOLIDS

The high electrical conductivity of metals is a result of the ease with which electrons in the metal can move under the influence of a static or low-frequency electric field. In the free-electron model of a metal each valence electron is treated as a particle in a three-dimensional box the size of the metal crystal. In this oversimplified theory the energy levels of the electron are given by equation 9.63 for a particle in a box. For each eigenstate there are actually two possible states of the electron, corresponding to the two values of the spin. It can be shown that the density-of-state function $g(\epsilon)$ is given by

$$g(\epsilon) = C\epsilon^{1/2} \tag{23.41}$$

The number of one-electron states with energy between ϵ and $\epsilon + d\epsilon$ is given by $g(\epsilon)\,d\epsilon$.

The density of states $g(\epsilon)$ is shown as a function of energy in Fig. 23.18*a*. As electrons are added at 0 K, the energy levels are filled up to some maximum energy

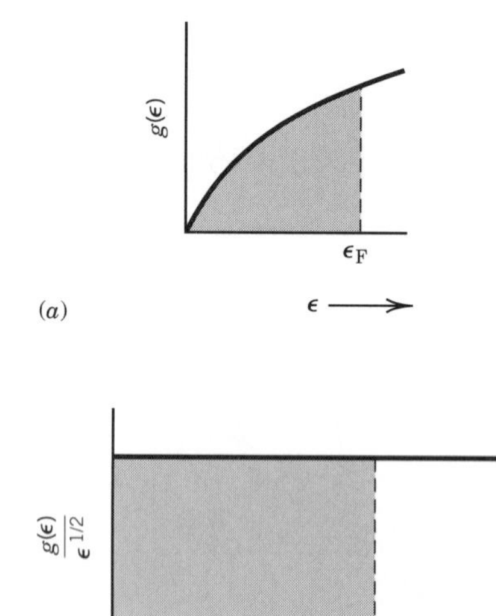

Figure 23.18 (*a*) Density of states for a free electron gas as a function of energy. (*b*) Plot of $g(\epsilon)/\epsilon^{1/2}$ versus energy for a free electron gas. The shading indicates the levels occupied at 0 K.

ϵ_F determined by the number of electrons. This is indicated by the shading in the figure. This maximum energy is called the **Fermi energy.** The Fermi energy is the chemical potential of the electrons and determines their tendency to move at an interface. It is sometimes convenient to present the same information by plotting $g(\epsilon)/\epsilon^{1/2}$ versus ϵ, as shown in Fig. 23.18*b*. In contrast to gas molecules, where Maxwell–Boltzmann statistics allow any number of particles to have exactly the same energy, electrons follow Fermi–Dirac statistics, which means that only one particle is allowed in each state of the system. At a temperature above absolute zero the number of occupied states (per unit volume) in the energy range ϵ to $\epsilon + d\epsilon$ is given by $d(N/V)$:

$$d\left(\frac{N}{V}\right) = f(\epsilon,T)g(\epsilon)\,d\epsilon \qquad (23.42)$$

where $f(\epsilon,T)$ is the Fermi–Dirac distribution function,

$$f(\epsilon,T) = \frac{1}{e^{(\epsilon-\mu)/kT} + 1} \qquad (23.43)$$

The quantity μ is constant at any given temperature. The value of μ can be obtained by integrating equation 23.42 over all energies since this integration must yield the number of electrons per unit volume. Since $f(\epsilon, T) = \frac{1}{2}$ when $\epsilon = \mu$, the quantity μ is equal to the energy at which $f(\epsilon, T)$ has half its maximum value. At 0 K, $f(\epsilon, 0) = 1$ for energies ϵ less than the Fermi energy ϵ_F, and $f(\epsilon, 0) = 0$ for energies ϵ greater than the Fermi energy ϵ_F. As $T \to 0$, $\mu \to \epsilon_F$. It is a good approximation to take $\mu \approx \epsilon_F$ at other temperatures provided that $\epsilon_F \gg kT$ ($\epsilon_F \approx 5$ eV for most metals). Combining the preceding two equations and using this approximation yield

$$d\left(\frac{N}{V}\right) = \frac{C\epsilon^{1/2}\,d\epsilon}{e^{(\epsilon-\epsilon_F)/kT} + 1} \qquad (23.44)$$

Figure 23.19 shows the distribution of electron energies at temperatures T_1 and T_2 where $0 < T_1 < T_2$. At room temperature the distribution of electron energies differs only slightly from that at 0 K for $\epsilon_F \approx 5$ eV. A small fraction of the electrons have energies greater than the Fermi energy ϵ_F, and they leave behind holes (unoccupied states) at $\epsilon < \epsilon_F$. The excited electrons and holes both contribute to the electrical conductivity.

Figure 23.19 provides the explanation for the very small contribution of electrons to the heat capacity for a metal. When the temperature of a metal is raised a small amount, only a small fraction of the electrons have their energies raised. Since the energies of most electrons are not affected by raising the temperature, the electronic heat capacity is negligible compared with the vibrational heat capacity $3R$ predicted by the Einstein and Debye theories (Section 16.12).

The free electron theory of metals is only an approximation. The electronic states of solids (metals, insulators, and semiconductors) can be studied by more exact quantum mechanical methods.

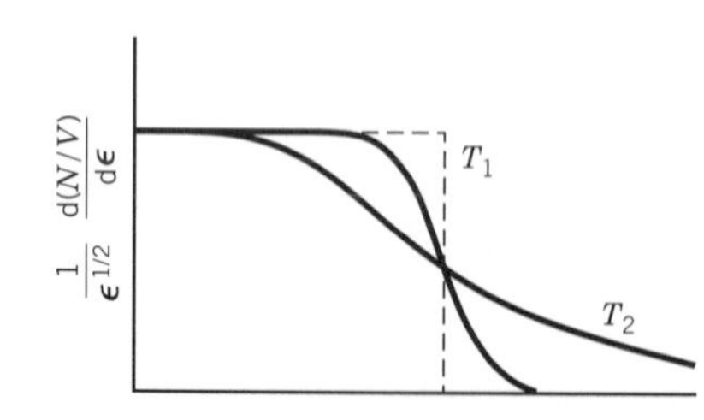

Figure 23.19 Distribution of electron energies in a metal at 0 K (dashed line) and two higher temperatures $T_2 > T_1$.

23.11 SPECIAL TOPIC: SUPERCONDUCTIVITY

In 1911, Kamerlingh Onnes, a Dutch physicist at the University of Leiden, discovered that the resistivity of mercury suddenly drops to zero as the temperature is

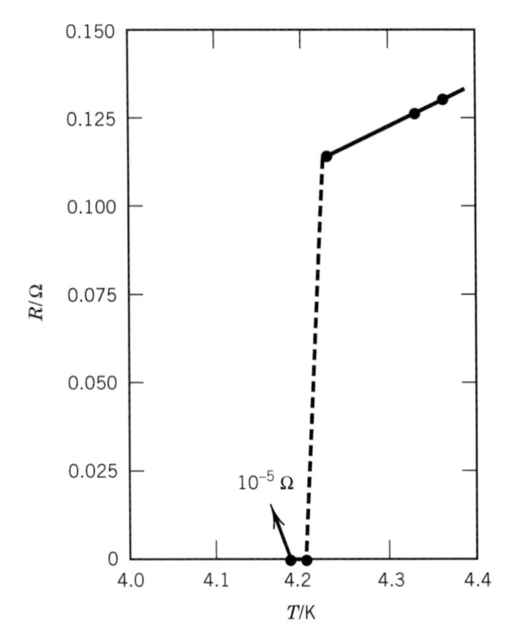

Figure 23.20 Resistance in ohms of a specimen of mercury versus temperature. This is the original plot by Kamerlingh Onnes showing his discovery of superconductivity. (From C. Kittel, *Introduction to Solid State Physics.* Hoboken, NJ: Wiley, 1976.)

lowered below 4.2 K (see Fig. 23.20). The same phenomenon occurs (at different characteristic temperatures) for many other metals, alloys, and compounds. We describe this by saying that at a critical temperature T_c the sample undergoes a phase transition from a normal to a superconducting state. In addition to zero resistivity, superconductors have unusual magnetic properties. When a sample in a weak magnetic field is cooled through T_c, the magnetic flux originally in the sample is ejected. This is called the **Meissner effect.** On the other hand, a superconducting state can be destroyed by a strong enough magnetic field (called the critical field H_c). The thermodynamic properties of superconductors are also interesting. For example, the entropy of a superconductor decreases considerably on cooling, indicating that the superconducting state is more ordered than the normal state. Detailed studies suggest that the superconducting state consists of two kinds of electrons: "ordered" pairs and "normal" electrons. As the temperature decreases, the number of ordered pairs increases. The temperature dependence of the populations suggests that the normal electrons are in energy states separated from the energy states of the ordered pairs by a gap Δ, which is on the order of $\sim$3–5$k_B T_c$.

The critical temperature T_c has been found to vary with isotopic substitution, suggesting that the motion of atoms is intimately connected with superconductivity. The first successful theory of superconductivity was given by Bardeen, Cooper, and Schrieffer in 1957 and is called the BCS theory after these physicists. The theory suggests that the attractive interaction between the electrons and the lattice of metal ions can be large enough that it overcomes the repulsion between electrons, leading to pairing of electrons (with opposite spins). These pairs then interact very weakly with lattice vibrations and so experience no friction (or scattering) as they move through the sample. The BCS theory explained the energy

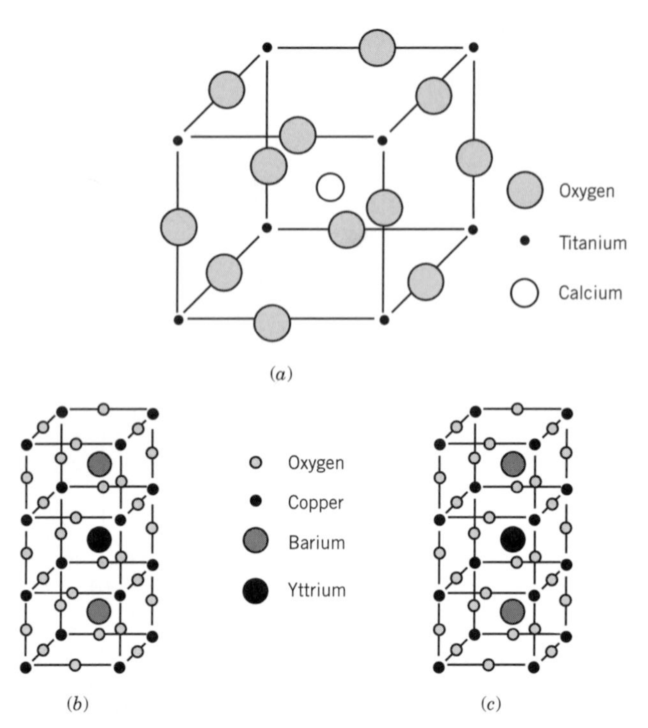

Figure 23.21 (*a*) The perovskite structure of $CaTiO_3$ has a cubic unit cell. (*b*) Idealized unit cell of the hypothetical $YBa_2Cu_3O_9$ based on a perovskite substructure. (*c*) Structure of the 1-2-3 oxide compound $YBa_2Cu_3O_{7-x}$, obtained from X-ray diffraction analyses. Ionic sizes are not drawn to scale.

gap, the thermodynamic properties, and many of the electromagnetic properties, and it also gave a formula for T_c in terms of parameters in the Hamiltonian of the system.

The usefulness of superconducting materials for practical purposes is decreased because of the very low temperatures at which most materials become superconducting. Until 1986, the highest known T_c was about 23 K. In that year, Bednorz and Müller, at the IBM lab in Zurich, discovered that an oxide of La, Ba, and Cu became superconducting at about 30 K. The next year another oxide was discovered to be superconducting at ~90 K, above the boiling point of liquid N_2 (77 K), which is a plentiful and inexpensive refrigerant. Suddenly, the possibility of using superconductors at reasonable temperatures became real.

The most studied of the so-called high-T_c superconductors is $YBa_2Cu_3O_{7-x}$ ($x \le 0.1$). The nonstoichiometry seems to be necessary for superconductivity. The solid-state structure of this compound is in the perovskite family (Fig. 23.21). The nonstoichiometry means that there are oxygen vacancies in the lattice, which undoubtedly play a role in the properties of the material. Speculation has focused on the sheets and chains of Cu and O atoms in the structure as being necessary for the superconducting state. Through 2003, no agreement had been reached on whether these new materials require a new theoretical model or whether they are examples of a BCS-type theory.

23.12 SPECIAL TOPIC: QUANTUM CONFINED SEMICONDUCTOR STRUCTURES

One class of solid-state material that has had tremendous impact in the last 50 years is the semiconductor. The modern world of computing was born when Bardeen, Brattain, and Shockley developed the first transistor in 1947 at Bell Laboratories in Murray Hill, New Jersey. Semiconductors have a ground-state electronic structure that consists of an energy band filled with electrons, called the **valence band;** an energy band with no electrons in it, called the **conduction band;** and an **energy gap,** denoted by E_g, that separates the valence and conduction bands. An electron near the bottom of the conduction band is well approximated as a free particle with an effective mass m_e^*. This mass is generally much smaller than that of an electron in free space. As a result, the wavelength of an electron in a semiconductor can be many nanometers. The absence of an electron in the valence band is called a **hole.** The hole behaves as if it were a positively charged particle with its own effective mass m_h^*, also generally smaller than that of an electron in free space. Defining the zero of energy at the top of the valence band, the energy E_e of an electron near the bottom of the conduction band and the energy E_h of a hole near the top of the valence band are well approximated as

$$E_e = \frac{\hbar^2 k_e^2}{2m_e^*} + E_g \qquad \text{and} \qquad E_h = -\frac{\hbar^2 k_h^2}{2m_h^*} \tag{23.45}$$

where k_e and k_h are the wave vectors for the electron and hole.

The energy of a photon absorbed by the semiconductor to promote an electron from the valence band to the conduction band, leaving a hole behind, is

$$h\nu = E_g + \frac{\hbar^2}{2}\left[\frac{k_e^2}{m_e^*} + \frac{k_h^2}{m_h^*}\right] \tag{23.46}$$

giving rise to an absorption spectrum that consists of a continuum starting at the band gap energy.

The development of sophisticated semiconductor growth technologies in the 1970s led to the emergence of classes of semiconductor structures that have one or more dimensions restricted to the nanometer length scale, smaller than the wavelength of the electron in the semiconductor. These are called **quantum confined** structures where the motion of conduction band electrons or valence band holes is best described quantum mechanically. Structures known as **quantum wells** confine the electron and hole in one dimension. Quantum wells consist of a semiconductor layer that is only ~10 nm thick surrounded by a semiconductor or an insulator that has a band gap larger than that in the layer, as illustrated schematically in Fig. 23.22*a*. The electron is confined to a one-dimensional box with finite potential walls in one direction, while it is free in the other two directions. This leads to a set of discrete, particle-in-a-box electronic states in the small dimension, illustrated in Fig. 23.23, and a continuum in the plane of the layer. The optical absorption spectrum consists of a staircase structure, compared with the bulk continuum, shown schematically in Fig. 23.22*b*, with an absorption onset that is at a higher energy than the band gap, and where each step corresponds to accessing a new discrete transition in the small direction. The onset of the first step in the absorption spectrum of a quantum well is size dependent, scaling as the inverse of

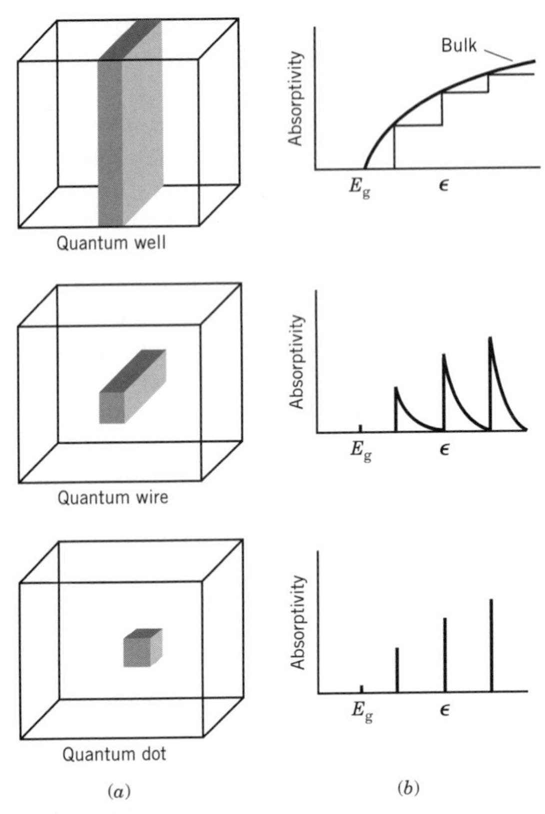

Figure 23.22 (*a*) Schematic structures for a semiconductor quantum well, wire, and dot. (*b*) Idealized absorption spectra for quantum well, wire, and dot structures, with electrons confined by the smallest dimension of the system.

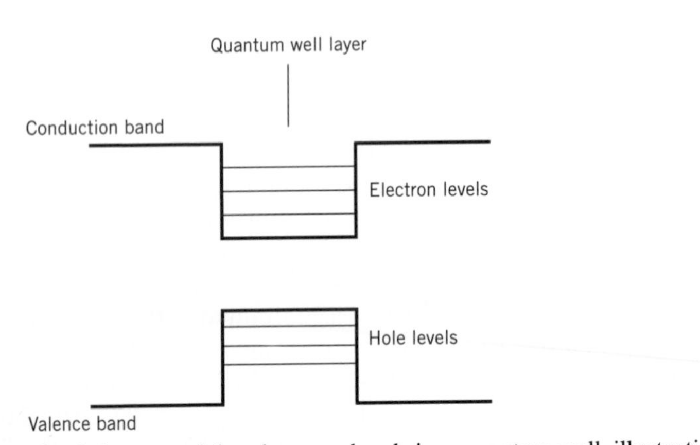

Figure 23.23 Schematic of the potential and energy levels in a quantum well, illustrating quantum confinement along the smallest direction.

the thickness of the well. The flat part of the step corresponds to increasing the kinetic energy of the electron and hole in the other two directions, where they behave as free particles. The optical and electronic effects that result from quantum confinement in a quantum well are directly responsible for the creation of optoelectronic devices that have become ubiquitous, including new and efficient lasers that are widely used in sophisticated telecommunications optical networks as well as in inexpensive consumer electronics, photodetectors, electro-optic modulators, and a slew of other technologically important devices.

Increasing the confinement to two directions leads to **quantum wires,** and confinement in all three directions results in **quantum dots,** with structures schematically rendered in Fig. 23.22*a*. The staircase pattern of the quantum well absorption spectrum becomes a sawtooth pattern for the quantum wire, and then a set of sharp, discrete absorption features for the idealized quantum dot, as illustrated in Fig. 23.22*b*. The quantum dot absorption features correspond to transitions between discrete, three-dimensional particle-in-a-box states of the electron and the hole, both confined to the same nanometer-size box. These discrete transitions are reminiscent of atomic spectra and have resulted in quantum dots also being called **artificial atoms.**

There have been many approaches to the fabrication of quantum dot structures, ranging from sophisticated growth methods that use expensive semiconductor fabrication facilities, to simple solution-based chemical methods. One chemical approach is the nucleation and growth of nanometer-size crystals of semiconductors in solution. These **nanocrystals** consist of a small inorganic core that is generally <10 nm in its smallest dimension, surrounded by a shell of organic groups that are loosely bound to the core and that regulate its growth and stabilize the nanocrystals in solvents. This structure is illustrated in Fig. 23.24. Solutions of nanocrystal quantum dots look like solutions of organic dye molecules. The nanocrystals are well approximated electronically as quantum dots with a spherical shape. Their electronic structure can be idealized using a three-dimensional spherical quantum mechanical box with wavefunctions that are like those of atomic orbitals. The lowest energy transition of a spherical nanocrystal with radius R within the spherical particle-in-a-spherical-box model corresponds to a photon with energy

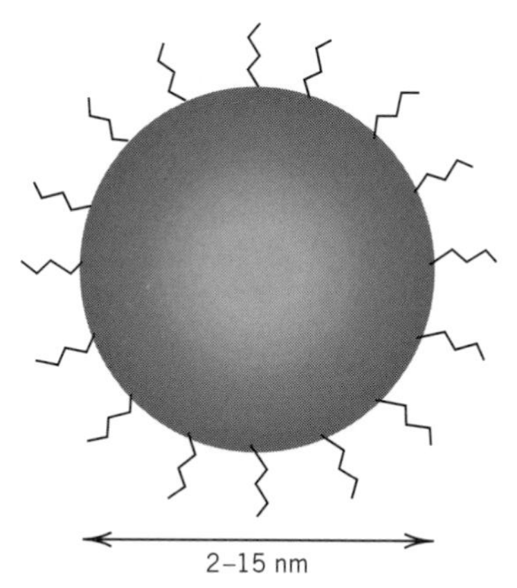

Figure 23.24 Illustration of the structure of a nanocrystal quantum dot grown in solution. A nanometer-size crystal of semiconductor has its surface functionalized by organic molecules.

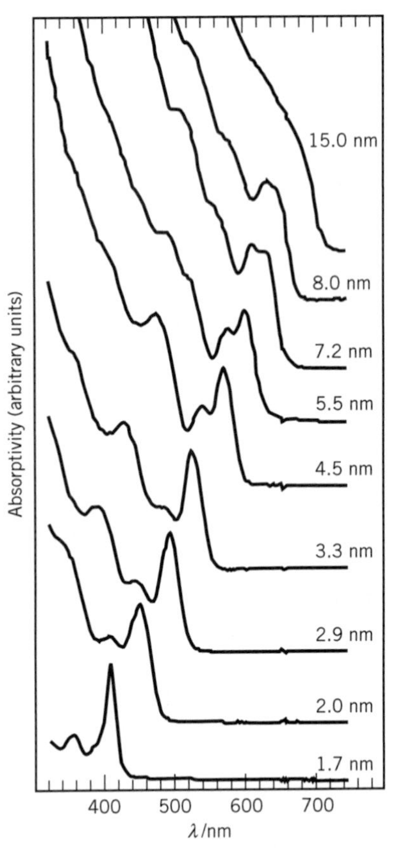

Figure 23.25 Absorption spectra from a set of solutions of CdSe nanocrystal quantum dots ranging in size from 1.7 nm to 15.0 nm, illustrating quantum confinement and artificial atom concepts. [From C. B. Murray, C. R. Kagan, and M. G. Bawendi, *Ann Rev. Mat. Sci.* **30:** 545–610 (2000). Copyright © 2000 Annual Reviews. Reprinted with permission.]

$$h\nu = E_g + \frac{\hbar^2 \pi^2}{2R^2}\left(\frac{1}{m_e^*} + \frac{1}{m_h^*}\right) \tag{23.47}$$

This transition results in a state with the electron and hole fully delocalized inside the nanocrystal in the equivalent of 1s atomic orbitals. The energy of this lowest transition is strongly sensitive to the size of the nanocrystal, scaling as $1/R^2$. Although chemically synthesized quantum dots do not have the infinitely sharp absorption spectra idealized in Fig. 23.22*b*, partly due to imperfect samples and finite size distributions, the strong size dependence of the lowest transition and "artificial atom" absorption bands are clearly observed as shown in Fig. 23.25. Nanocrystal quantum dots can also exhibit high-quantum-efficiency fluorescence from the recombination of the lowest-energy electron and hole. The strong size dependence of this fluorescence is such that a semiconductor such as CdSe that has a band gap in the red portion of the visible spectrum can have its fluorescence tuned with size throughout the visible spectrum with vials of nanocrystal quantum dots fluorescing in a rainbow of colors, from the blue edge of the visible spectrum for particles with $R \sim 1$ nm to the red edge of the visible spectrum for particles with $R \sim 4$ nm. This size-dependent fluorescence presents an appealing demonstration of the quantum mechanical particle-in-a-box model, and a direct example of some of the new properties that emerge when materials have dimensions in the nanometer size range. This size-dependent fluorescence has also led to the initially unexpected commercial application of nanocrystal quantum dots as fluorescent markers in biological systems.

Eight Key Ideas in Chapter 23

1. The space occupied by a crystal lattice can be divided into unit cells. The repetition of the unit cell in three dimensions generates the entire pattern of the crystal.
2. The classification of crystals is based on their symmetry, rather than the dimensions of their unit cells. In contrast with individual molecules, crystals can have rotational symmetry of only one-, two-, three-, four-, or six-fold. As a result of this restriction, all crystals can be classified in one of 32 crystallographic point groups.
3. The planes through lattice points are important because they represent possible crystal faces and because they help us understand X-ray diffraction phenomena. There is an infinite number of sets of planes in a crystal, and they are referred to by their Miller indices.
4. The Bragg equation gives the angles at which reflection occurs from a stack of planes. The reflections can be referred to as first-order, second-order, . . . reflections, or the perpendicular interplanar distance can be taken to be d/n, where n is the order of the reflection, in which case the Miller indices are multiplied by n.
5. One of the most accurate measurements of the Avogadro constant is based on the determination of the density, relative atomic mass, and unit cell length for silicon.
6. The phase problem complicates the determination of crystal structures by X-ray diffraction because the intensities of reflections give only the absolute value of the structure factor from which the electron density function can be calculated. One of the ways for getting around this problem, which has been very useful in determining the structures of proteins, is isomorphous replacement.
7. When the bonding of atoms in crystals is not highly directional, it is often found that the lowest-energy structure is that in which each atom is surrounded by the greatest number of possible neighbors. Hexagonal close packing and cubic close packing (face-centered cubic) are the only two ways of close-packing identical spheres so that the environment of each sphere is identical to that of the others, but there is an infinite number of other ways of close-packing where the environment of each sphere is not identical.
8. The free electron model of metals can explain many properties, including the low-temperature specific heat.

REFERENCES

J. Drenth, *Principles of Protein X-Ray Crystallography,* 2nd ed. Springer, 1999.

J. P. Glusker and K. N. Trueblood, *Crystal Structure Analysis, A Primer.* London: Oxford University Press, 1985.

T. Hahn, *International Tables for Crystallography; Brief Teaching Edition of Volume A, Space-Group Symmetry.* Dordrecht: Reidel, 1985.
C. Hammond, *The Basis of Crystallography and Diffraction*. New York: Oxford University Press, 2001.
C. Kittel, *Introduction to Solid State Physics,* 6th ed. Hoboken, NJ: Wiley, 1986.
M. F. C. Ladd and R. A. Palmer, *Structure Determination by X-Ray Crystallography,* 2nd ed. New York: Plenum, 1985.
K. E. van Holde, W. C. Johnson, and P. S. Ho, *Principles of Physical Biochemistry*. Upper Saddle River, NJ: Prentice-Hall, 1998.
M. A. White, *Properties of Materials*. Oxford University Press, 2000.

PROBLEMS

(M) Problems marked with an icon may be more conveniently solved on a personal computer with a mathematical program.

23.1 What is the equation for the distances between 110 planes for a crystal with mutually perpendicular axes?

23.2 Calculate the angles at which the first-, second-, and third-order reflections are obtained from planes 500 pm apart, using X-rays with a wavelength of 100 pm.

23.3 Calculate the structure factor for a cubic unit cell of AB in which the B atoms occupy the body-centered position. Which reflections will be strong and which weak?

23.4 The crystal unit cell of magnesium oxide is a cube 420 pm on an edge. The structure is interpenetrating face centered. What is the density of crystalline MgO?

23.5 Platinum forms face-centered cubic crystals. If the radius of a platinum atom is 139 pm, what is the length of the side of the unit cell? What is the density of the crystal?

23.6 Tungsten forms body-centered cubic crystals. From the fact that the density of tungsten is 19.3 g cm^{-3}, calculate (*a*) the length of the side of this unit cell and (*b*) d_{200}, d_{110}, and d_{222}.

23.7 (*a*) Metallic iron at 20 °C is studied by the Bragg method, in which the crystal is oriented so that a reflection is obtained from the planes parallel to the sides of the cubic crystal, then from planes cutting diagonally through opposite edges, and finally from planes cutting diagonally through opposite corners. Reflections are first obtained at $\theta = 11° 36'$, $8° 3'$, and $20° 26'$, respectively. What type of cubic lattice does iron have at 20 °C? (*b*) Metallic iron also forms cubic crystals at 1100 °C, but the reflections determined as described in (*a*) occur at $\theta = 9° 8'$, $12° 57'$, and $7° 55'$, respectively. What type of cubic lattice does iron have at 1100 °C? (*c*) The density of iron at 20 °C is 7.86 g cm^{-3}. What is the length of the side of the unit cell at 20 °C? (*d*) What is the wavelength of the X-rays used? (*e*) What is the density of iron at 1100 °C?

23.8 Cesium chloride, bromide, and iodide form interpenetrating simple cubic crystals instead of interpenetrating face-centered cubic crystals like the other alkali halides. The length of the side of the unit cell of CsCl is 412.1 pm. (*a*) What is the density? (*b*) Calculate the ion radius of Cs^+, assuming that the ions touch along a diagonal through the unit cell and that the ion radius of Cl^- is 181 pm.

23.9 Deslattes et al. [*Phys. Rev. Lett.* **33**:463 (1974)] found the following values for a single crystal of very pure silicon at 25 °C: $\rho = 2.328\,992$ g cm^{-3}, $a = 543.1066$ pm. Silicon has a face-centered cubic lattice like diamond. The atomic mass is 28.085 41 g mol^{-1}. What value of Avogadro's constant is obtained from these values?

23.10 Insulin forms crystals of the orthorhombic type with unit-cell dimensions of $13.0 \times 7.48 \times 3.09$ nm. If the density of the crystal is 1.315 g cm^{-3} and there are six insulin molecules per unit cell, what is the molar mass of the protein insulin?

23.11 Molybdenum forms body-centered cubic crystals, and at 20 °C, the density is 10.3 g cm^{-3}. Calculate the distance between the centers of the nearest molybdenum atoms.

23.12 Silicon has a face-centered cubic structure with two atoms per lattice point, just like diamond. At 25 °C, $a =$ 543.1 pm. What is the density of silicon?

23.13 The common form of ice has a tetrahedral structure with protons located on the lines between oxygen atoms. A given proton is closer to one oxygen atom than the other and is said to belong to the closer oxygen atom. How many different orientations of a water molecule in space are possible in this lattice?

23.14 The diamond has a face-centered cubic crystal lattice, and there are eight atoms in a unit cell. Its density is 3.51 g cm^{-3}. Calculate the first six angles at which reflections would be obtained using an X-ray beam of wavelength 7.12 pm.

23.15 Derive the structure factor for a body-centered cubic unit cell of identical atoms. The relative coordinates of the lattice points are given by (0,0,0), (1,0,0), (0,1,0), (0,0,1), (1,1,0), (1,0,1), (0,1,1), (1,1,1), and $(\frac{1}{2}, \frac{1}{2}, \frac{1}{2})$.

23.16 Calculate the ratio of the radii of small and large spheres for which the small spheres will just fit into octahedral sites in a close-packed structure of the large spheres.

23.17 A close-packed structure of uniform spheres has a cubic unit cell with a side of 800 pm. What is the radius of the spherical molecule?

23.18 Titanium forms hexagonal close-packed crystals. Given the atomic radius of 146 pm, what are the unit cell dimensions, and what is the density of the crystals?

23.19 What neutron energy in electron volts is required for a wavelength of 100 pm?

23.20 The only metal that crystallizes in a primitive cubic lattice is polonium, which has a unit cell side of 334.5 pm. What are the perpendicular distances between planes with indices (110), (111), (210), and (211)?

23.21 Calculate the highest-order diffraction line that can be observed for the 100 planes of NaCl using an X-ray tube with a copper target ($\lambda = 154$ pm).

23.22 For a C lattice (Fig. 23.5), what is the expression for the structure factor? Derive the rule for absent reflections.

23.23 The density of platinum is 21.45 g cm^{-3} at 20 °C. Given the fact that the crystal is face-centered cubic, calculate the length of the side of the unit cell.

23.24 A substance forms face-centered cubic crystals. Its density is 1.984 g cm^{-3}, and the length of the edge of the unit is 630 pm. Calculate the molar mass.

23.25 Potassium bromide has a face-centered cubic lattice, and the edge of the unit cell is 654 pm. What is the density of the crystal?

23.26 Calculate the density of a diamond from the fact that it has a face-centered cubic structure with two atoms per lattice point and a unit cell edge of 356.7 pm.

23.27 Tantalum crystallizes with a body-centered cubic lattice. Its density is 17.00 g cm^{-3}. (*a*) How many atoms of tantalum are there in a unit cell? (*b*) What is the length of a unit cell? (*c*) What is the distance between (200) planes? (*d*) What is the distance between (110) planes? (*e*) What is the distance between (222) planes?

23.28 Iron crystallizes in body-centered cubic packing at room temperature. Since the density is 7.88 g cm^{-3} at 25 °C, what is the length of the side of the unit cell at this temperature? What is the radius of an iron atom in this crystalline form?

23.29 Cobalt has a hexagonal close-packed structure with a = 250.7 pm. What is its density?

23.30 Aluminum forms face-centered cubic crystals, and the length of the side of the unit cell is 405 pm at 25 °C. Calculate (*a*) the density of aluminum at this temperature and (*b*) the distances between (200), (220), and (111) planes.

23.31 From the fact that the length of the side of a unit cell for lithium is 351 pm, calculate the atomic radius of Li. Lithium forms body-centered cubic crystals.

23.32 Platinum forms face-centered cubic crystals, and the length of the side of the unit cell is 393 pm. What is the number of atoms per cm^2 of surface on the (100), (110), and (111) planes?

23.33 If spherical molecules of 500 pm radius are packed in cubic close packing and in body-centered cubic crystals, what are the lengths of the side of the cubic unit cells in the two cases?

23.34 What is the de Broglie wavelength of thermal neutrons at 200 °C?

23.35 Metallic sodium forms a body-centered cubic unit cell with a = 424 pm. What is the sodium atom radius?

Computer Problems

23.A Calculate the angles θ for the first-order Bragg reflections for 100, 110, and 111 planes of an orthorhombic unit cell with a = 488.2 pm, b = 665.7 pm, and c = 831.6 pm. The wavelength of the monochromatic X-rays is 154.433 pm.

23.B Copper forms face-centered cubic crystals with a 361.6-pm unit cell at 25 °C. Calculate the first five Bragg angles obtained with 154.05-pm X-rays.

23.C Calculate the Fourier transform of the function $\sin 2\pi\nu_1 t + 2\sin 2\pi\nu_2 t$, where $\nu_1 = 30\ s^{-1}$ and $\nu_2 = 20\ s^{-1}$. (*a*) Plot the function in the time domain over a period of one second. (*b*) Plot the corresponding spectrum in the frequency domain.

23.D Calculate the angles for the first 11 first-order Bragg reflections for a face-centered cubic crystal with $\lambda/a = 0.289$.

24

Surface Dynamics

Previous chapters have been largely concerned with the properties of matter under conditions where surface effects are negligible. Now we shall consider the equilibrium and dynamics of processes that occur at the interface between a solid and a gas. When a molecule strikes a solid surface, it may rebound elastically or inelastically, undergo a reaction, or be adsorbed. If it is adsorbed, it may diffuse around on the surface, remain fixed, or dissolve in the bulk phase, but we will concentrate on the processes that occur on the surface. An adsorbed molecule may dissociate on the surface or react with another molecule on the surface. If a chemical reaction occurs on the surface, the products may desorb into the gas phase. The use of solid surfaces as catalysts in chemical technology is of tremendous practical importance.

24.1 PHYSISORPTION AND CHEMISORPTION

In the preceding chapter, we have seen that a crystal can have a number of faces with different Miller indices. Faces with low Miller indices and high densities of atoms or molecules are more likely to be observed because in general they are more stable than surfaces with high Miller indices. The various faces of a crystal

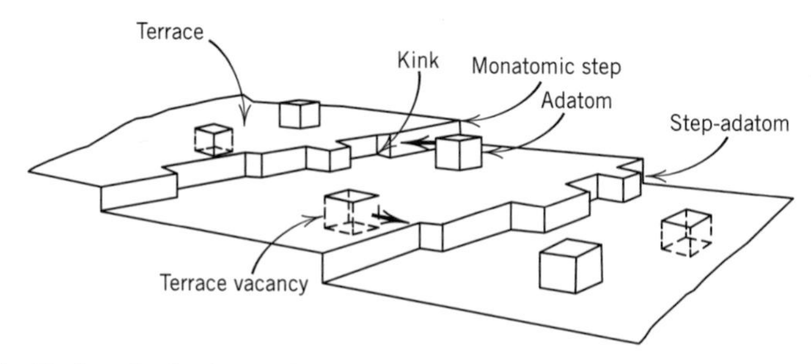

Figure 24.1 Defects in the form of steps, kinks, adatoms, and vacancies on a solid surface. (Reprinted from G. A. Somorjai, *Chemistry in Two Dimensions: Surfaces.* Copyright © 1981 by Cornell University. Used by permission of the publisher, Cornell University Press, Ithaca, NY.)

have different properties because of their different structures. In Section 24.9, we will see that the surface of even a pure atomic solid can be more complicated than the flat surfaces illustrated in Chapter 23, where the positions of surface atoms are simple extensions of the bulk structure. Even simple surfaces may have **defects** of the types shown in Fig. 24.1. These defects have different adsorptive and catalytic properties. As a simplification we will largely ignore defects.

When a molecule approaches a surface, it encounters a net attractive potential that is similar to the potential between two molecules (Section 11.9) and arises for the same reasons. However, a gas molecule near a surface is attracted by many closely spaced surface atoms. The adsorption of a molecule on a solid surface is always an exothermic process. If we represent the gas molecule, the adsorbate, by A and the adsorption site on the surface by S, the process of **adsorption** can be represented as a chemical reaction:

$$\mathrm{A} + \mathrm{S} = \mathrm{AS} \qquad \Delta_{\mathrm{ads}}H < 0 \tag{24.1}$$

In Section 24.7 we will see that this process may involve more than one step, but here we simply want to establish the definition of the enthalpy of adsorption $\Delta_{\mathrm{ads}}H$.

The surface of an atomic solid has about 10^{15} atoms per square centimeter of surface. If we assume that one molecule adsorbs on each atom of the solid surface, then there are 10^{15} sites on which molecules can adsorb. When one molecule is adsorbed on each site, the surface is said to be covered with a monolayer. Kinetic theory makes it possible to calculate the maximum rate at which surface sites can be occupied by gas molecules. In Section 17.6 we saw that the **flux** J_N of molecules of an ideal gas through a hole is given by

$$J_N = \frac{PN_{\mathrm{A}}}{(2\pi MRT)^{1/2}} \tag{24.2}$$

The same equation applies to collisions with a surface, and when SI units are used, J_N is the number of molecules striking the surface per square meter per second. If every molecule striking a clean surface is adsorbed and exactly one gas molecule is adsorbed per surface site, this equation provides the means for calculating the time for a surface to become covered with a monolayer when it is exposed to a gas with molar mass M at a specific temperature and pressure.

Example 24.1 ***Rate of formation of a monolayer***

How many molecules of oxygen strike 1 cm^2 of surface in 1 s when the pressure is 10^{-6} torr and the temperature is 298 K? [1 torr $= \frac{1}{760}$ atm $= (\frac{1}{760}$ atm$)(1.013\,25 \times 10^5$ Pa atm$^{-1}) =$ 133.3 Pa.]

Using equation 24.2,

$$\begin{aligned} J_N &= \frac{(133.3 \times 10^{-6}\ \text{Pa})(6.022 \times 10^{23}\ \text{mol}^{-1})}{[2\pi(32 \times 10^{-3}\ \text{kg mol}^{-1})(8.3145\ \text{J K}^{-1}\ \text{mol}^{-1})(298\ \text{K})]^{1/2}} \\ &= (3.60 \times 10^{18}\ \text{m}^{-2}\ \text{s}^{-1})(0.01\ \text{m cm}^{-1})^2 \\ &= 3.60 \times 10^{14}\ \text{cm}^{-2}\ \text{s}^{-1} \end{aligned}$$

For a surface with 10^{15} sites per cm^2, the exposure of a clean surface to oxygen at 10^{-6} torr for 1 s is sufficient to form 36% of a monolayer, if every molecule sticks. To express the exposure of a surface to a gas that is adsorbed, surface scientists have developed the unit 10^{-6} torr s, which is called the **langmuir.** Thus, exposure of the surface to 1 langmuir of oxygen results in 36% of a monolayer of adsorbed O_2 at 298 K.

By use of calculations of this type it is readily shown that to keep a surface clean even for a few minutes, it is necessary to evacuate the chamber containing the sample surface to a pressure less than 10^{-8} torr. Thus, surfaces that we encounter daily are always covered with adsorbed molecules. To study surfaces, special vacuum chambers are used; the term *ultrahigh vacuum* is used to refer to pressures less than 5×10^{-10} torr. By directing a molecular beam (Section 19.5) on crystallographically distinct surfaces, it is possible to determine the rate of filling surface sites as a function of beam energy and angle.

It is convenient to distinguish between **physical adsorption** and **chemisorption.** The forces causing physical adsorption are of the same type as those that cause the condensation of a gas to form a liquid and are generally referred to as van der Waals forces. The heat evolved in a physisorption process is of the order of magnitude of the heat evolved in the process of condensing the gas, and the amount adsorbed may correspond to several monolayers at a high pressure. The extent of physisorption is smaller at higher temperatures.

Chemisorption involves the formation of chemical bonds. However, it is usually not possible to make a sharp distinction between these two kinds of adsorption, except to say that the enthalpy change in chemisorption is much larger than for physical adsorption, lying in the range 40 to 200 kJ mol^{-1}.

Physical adsorption and chemisorption may often be distinguished by the rates at which these processes occur. Equilibrium in physical adsorption is generally achieved rapidly and is readily reversible. Physical adsorption is reversed by lowering the pressure of the gas or raising the temperature of the surface. Chemisorption, on the other hand, may not occur at an appreciable rate at low temperatures if the chemisorption reaction has an activation energy (Section 18.6). In this case, the rate of chemisorption increases rapidly as the temperature is raised. In chemisorption the bonding may be so tight that the original species may not desorb. For example, heating a graphite surface after adsorbing atomic oxygen results in desorption of carbon monoxide.

In the following section, we consider a very simple model for adsorption presented by Langmuir.

24.2 LANGMUIR ADSORPTION ISOTHERM

The simplest equation for adsorption under equilibrium conditions was derived by Langmuir using kinetic theory. He considered a surface with a specific number of binding sites that are identical and can each adsorb one molecule. Thus, the uptake is limited to a **monolayer.** If a uniform surface has n_0 equivalent sites and n are occupied, the **surface coverage** Θ is defined by $\Theta = n/n_0$. Langmuir also assumed that binding at a site has no influence on the properties of neighboring sites; this means that the enthalpy of adsorption is independent of coverage.

To derive the expression for the Langmuir adsorption isotherm, we will set the rate of adsorption equal to the rate of desorption; in other words, the equilibrium is assumed to be dynamic. The rate of adsorption is taken to be equal to the product of the rate J_N of collisions of molecules of molar mass M with the surface, the fraction of the sites that are not covered $(1 - \Theta)$, the **sticking coefficient** s^*, and the fraction $\exp(-E_{\text{ads}}/RT)$ of the molecules with the **activation energy for adsorption** E_{ads}. As the molecule approaches the surface it usually encounters a potential barrier with a height E_{ads} that has to be overcome before the molecule can be adsorbed. The sticking coefficient s^* is the fraction of the molecules with energy in excess of the activation energy E_{ads} that actually stick. Thus,

$$\text{rate of adsorption} = \frac{PN_{\text{A}}}{(2\pi MRT)^{1/2}}(1 - \Theta)s^* \exp\left(-\frac{E_{\text{ads}}}{RT}\right) \tag{24.3}$$

This equation for the rate of adsorption of molecules on a surface is based on the simple view of the potential for the interaction between a molecule and the surface that is shown in Fig. 24.2; more details are given in Section 24.7.

For an adsorbed molecule to be desorbed, it has to overcome a **potential barrier** E_{des}, as shown in Fig. 24.2. The rate of desorption is taken to be equal to the product of the specific rate constant for desorption k_{d}, the fraction Θ of the sites that are occupied, and the fraction of the molecules with the activation energy E_{des} for desorption. Thus,

$$\text{rate of desorption} = k_{\text{d}}\Theta \exp\left(-\frac{E_{\text{des}}}{RT}\right) \tag{24.4}$$

At equilibrium, the rate of adsorption is equal to the rate of desorption, so the pressure and surface coverage Θ are related by

$$P = \frac{(2\pi MRT)^{1/2}k_{\text{d}}\Theta}{N_{\text{A}}s^*(1 - \Theta)} \exp\left(\frac{\Delta_{\text{ads}}H}{RT}\right) \tag{24.5}$$

where $\Delta_{\text{ads}}H$ is the **enthalpy of adsorption** $(E_{\text{ads}} - E_{\text{des}})$. The enthalpy of adsorption is negative, so $E_{\text{ads}} < E_{\text{des}}$. The Langmuir adsorption isotherm can be written as

$$P = \frac{\Theta}{K(1 - \Theta)} \quad \text{or} \quad \Theta = \frac{KP}{1 + KP} \tag{24.6}$$

where the constant K is given by

$$K = \frac{N_{\text{A}}s^* \exp(-\Delta_{\text{ads}}H/RT)}{k_{\text{d}}(2\pi MRT)^{1/2}} \tag{24.7}$$

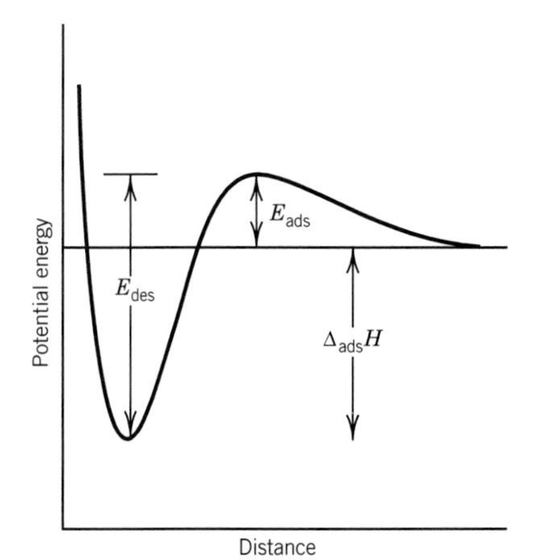

Figure 24.2 Potential energy for the interaction of a molecule with a surface assumed in the derivation of the Langmuir adsorption isotherm.

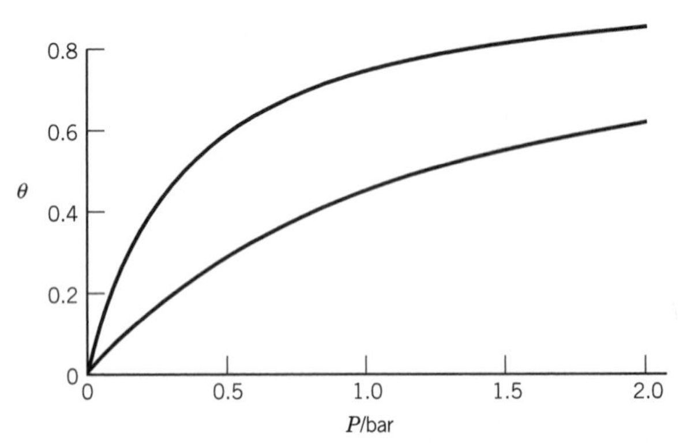

Figure 24.3 Fractional coverage Θ of the surface of a solid at 298 K (upper curve) and 350 K (lower curve). The equilibrium constant for this adsorption is 3 at 298 K for a standard state pressure of 1 bar, and $\Delta H_{\text{ads}} = -20\ \text{kJ mol}^{-1}$. (See Computer Problem 24.B.)

The enthalpy of adsorption introduced in equation 24.1 and appearing in the Langmuir adsorption isotherm (equation 24.5) can be determined from measurements of equilibrium pressures at two or more temperatures and at the same surface coverage Θ. Adsorption isotherms following the Langmuir equation are shown for two temperatures in Fig. 24.3. If we write equation 24.5 for the equilibrium pressures at T_1 and T_2, and ignore the $T^{1/2}$ dependence in comparison with the exponential dependence involving $\Delta_{\text{ads}}H$, we can readily derive

$$\left(\ln \frac{P_1}{P_2}\right)_{\Theta} = \frac{(\Delta_{\text{ads}}H)(T_2 - T_1)}{RT_1T_2} \tag{24.8}$$

Note that this is very nearly the Clausius–Clapeyron equation (Section 6.3), but the enthalpy of adsorption has the opposite sign from the enthalpy of vaporization; $\Delta_{\text{vap}}H$ applies to the process liquid $\rightarrow$ vapor, while $\Delta_{\text{ads}}H$ applies to the process $\text{A} + \text{S} \rightarrow \text{AS}$ (equation 24.1).

Example 24.2 ***Competition of A and B for adsorption sites***

Use the Langmuir method to derive expressions for the fractions Θ_{A} and Θ_{B} of a surface covered by adsorbed molecules A and B, assuming that the molecules compete for the same sites.

The Langmuir equation for pure A can be written

$$r_{\text{A}}\Theta_{\text{A}} = k_{\text{A}}(1 - \Theta_{\text{A}})P_{\text{A}} \tag{a}$$

where the Langmuir constant is written as the ratio of a rate constant k_{A} for adsorption and rate constant r_{A} for desorption: $K_{\text{A}} = k_{\text{A}}/r_{\text{A}}$. Similarly, for pure B,

$$r_{\text{B}}\Theta_{\text{B}} = k_{\text{B}}(1 - \Theta_{\text{B}})P_{\text{B}} \tag{b}$$

When gases A and B are both present, these equations become

$$r_{\text{A}}\Theta_{\text{A}} = k_{\text{A}}(1 - \Theta_{\text{A}} - \Theta_{\text{B}})P_{\text{A}} \tag{c}$$

$$r_{\text{B}}\Theta_{\text{B}} = k_{\text{B}}(1 - \Theta_{\text{A}} - \Theta_{\text{B}})P_{\text{B}} \tag{d}$$

The ratio of these equations is

$$\frac{r_A \Theta_A}{r_B \Theta_B} = \frac{k_A P_A}{k_B P_B} \tag{e}$$

Solving this equation for Θ_B and substituting in equation c yields

$$\Theta_A = \frac{(k_A/r_A)P_A}{1 + (k_A/r_A)P_A + (k_B/r_B)P_B} = \frac{K_A P_A}{1 + K_A P_A + K_B P_B} \tag{f}$$

Similarly, for B,

$$\Theta_B = \frac{(k_A/r_B)P_B}{1 + (k_A/r_A)P_A + (k_B/r_B)P_B} = \frac{K_B P_B}{1 + K_A P_A + K_B P_B} \tag{g}$$

Example 24.3 ***Dissociative adsorption***

If a molecule dissociates on being adsorbed, the process is referred to as dissociative adsorption. Derive the Langmuir adsorption isotherm for dissociative adsorption.

When dissociation occurs on the surface, two sites are required, so the probability of sticking is proportional to the pressure and to the availability of adjacent sites, $k(1-\Theta)^2 P$. The probability of desorption is proportional to the probability that adjacent sites are occupied, $r\Theta^2$, since recombination has to take place on the surface prior to desorption. At equilibrium,

$$r\Theta^2 = k(1-\Theta)^2 P$$

$$\frac{\Theta}{1-\Theta} = \frac{k^{1/2} P^{1/2}}{r^{1/2}}$$

$$\Theta = \frac{KP^{1/2}}{1 + KP^{1/2}}$$

where $K = (k/r)^{1/2}$.

Comment:

The plot of adsorption versus pressure of a gas that does not dissociate on the surface has the same shape as a plot of equilibrium extent of reaction for a reaction $A(g) + B(g) = C(g)$ versus the partial pressure of B or as a plot of binding of oxygen by myoglobin versus the partial pressure of molecular oxygen (Fig. 8.7). Thus the underlying phenomena are very much the same, and standard thermodynamic properties can be calculated in each case.

24.3 USE OF ADSORPTION MEASUREMENTS TO DETERMINE SURFACE AREA

The adsorption of a gas by a solid can be determined by admitting known quantities of a gas into a chamber and measuring the volume and pressure of the gas at equilibrium. If a monolayer is formed, the fractional surface coverage Θ is equal to the ratio of the volume v of the gas adsorbed to the volume v_m required to form

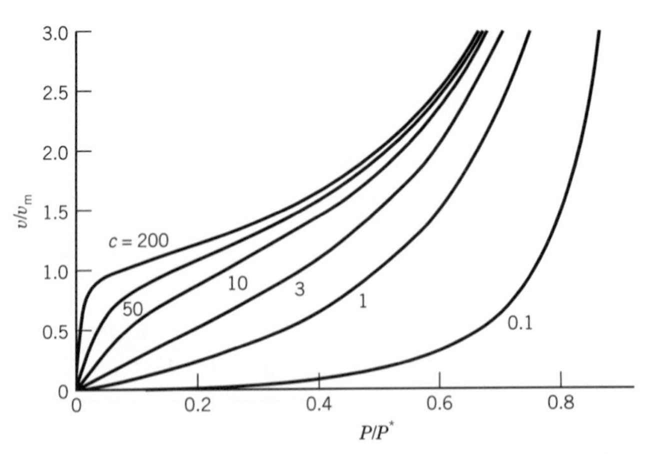

Figure 24.4 Shapes of adsorption isotherms obtained when multilayer adsorption of a gas occurs. These curves were calculated using the BET theory (equation 24.10) with various values of the constant c. (See Computer Problem 24.A.)

a monolayer, that is, $\Theta = v/v_m$, where the volumes are at the same temperature and pressure. Thus, equation 24.6 can be written

$$v = \frac{v_m}{1 + 1/KP} \tag{24.9}$$

If the experimental data follow the Langmuir equation, the parameters v_m and K can be determined from the slope and intercept of the plot of $1/v$ versus $1/P$, which is linear if equation 24.9 is followed. In general, however, there are deviations from the Langmuir equation, especially at higher pressures. If the gas being adsorbed is below its critical point, it is customary to plot the amount adsorbed per gram of adsorbent versus P/P^*, where P^* is the vapor pressure of the bulk liquid adsorbate at the temperature of the experiment. Figure 24.4 shows the shapes of plots of the ratio of the volume v of gas adsorbed (under standard conditions) to the volume v_m required to form a monolayer versus the pressure, specifically P/P^*. These curves have been calculated using the BET theory, which has an adjustable constant c, but, before discussing this theory, we need to reconsider the assumptions of the Langmuir theory.

The derivation of the Langmuir adsorption isotherm involves five implicit assumptions: (1) the gas is ideal, (2) the adsorbed gas is confined to a monolayer, (3) the $\Delta_{ads}H$ of each binding site is the same, (4) there is no lateral interaction between adsorbate molecules, and (5) the adsorbed gas molecules stay at the position where they first collide with the surface. Although the first assumption is good at low pressure, it is not valid if the pressure approaches the critical pressure. The second assumption fails when there is adsorption on top of the monolayer. The third assumption is poor when $\Delta_{ads}H$ is different for different kinds of binding sites. The nonuniversality of the fourth assumption was first shown experimentally when it was found that, in certain cases, the enthalpy of adsorption may increase with the amount adsorbed. This effect is caused by lateral attractions between adsorbed molecules. The fifth assumption is incorrect because there is much evidence that certain adsorbed molecules can be mobile. In spite of these problems, the Langmuir adsorption isotherm is very useful for many systems at low pressures where only monolayer adsorption is involved.

We now return to the increase in physical adsorption beyond that required to form a monolayer, which is observed at high pressures. This is referred to as **multilayer adsorption,** and Brunauer, Emmett, and Teller (BET) developed a theory for it. They assumed that the surface possesses uniform, localized sites and that adsorption on one site does not affect adsorption on neighboring sites, just as in the Langmuir theory. It is further assumed that molecules can be adsorbed in second, third,..., and nth layers, with the surface area available for the nth layer equal to the coverage of the $(n-1)$th layer. The energy of adsorption of the first layer is E_1, and the energy of adsorption in succeeding layers is assumed to be E_L, the energy of liquefaction of the gas. By use of these assumptions it is possible to derive the following equation for the ratio of the volume v of gas adsorbed (under standard conditions) to the volume required to form a monolayer:

$$\frac{v}{v_m} = \frac{cx}{(1-x)[1+(c-1)x]} \qquad x = \frac{P}{P^*} \tag{24.10}$$

where P^* is the vapor pressure of the bulk liquid. This equation can be written in the form

$$\frac{x}{v(1-x)} = \frac{1}{cv_m} + \frac{(c-1)x}{cv_m} \tag{24.11}$$

and so it is possible to determine the parameters v_m and c from a plot of $x/v(1-x)$ versus x. The value of v_m, the volume of gas required to form a monolayer, is of considerable interest because it makes it possible to calculate the surface area of a porous solid. The surface area occupied by a single molecule of adsorbate on the surface can be estimated from the density of the liquefied adsorbate. For example, the area occupied by a nitrogen molecule at $-195\,°C$ is estimated to be 16.2×10^{-20} m^2 on the assumption that the molecules are spherical and that they are close-packed in the liquid. Thus, from the value of v_m obtained from the BET theory, if multilayer adsorption is involved, the surface area of the adsorbent can be calculated. The surface areas of porous solids can be as large as several hundred square meters per gram of solid.

24.4 LOW-ENERGY ELECTRON DIFFRACTION (LEED)

Many advances in surface science have been made using methods that are surface sensitive; that is, they are sensitive only to the outermost atomic layers of the bulk solid. These methods, which are discussed in this section and the next, largely employ **low-energy electrons.** Electrons with energies in the range of 10 to 250 eV are sensitive to the surface because they have a low mean free path in solids. They penetrate the surface about 0.5 to 2 nm, which is about one to four atomic layers. Figure 24.5 shows the mean free path of an electron in a solid as a function of its energy. The mean free path for electrons of a given energy is nearly independent of the atomic weight of the element making up the solid, so this curve is often called the "universal curve." The curve also shows us that in the emission of electrons with energies in the range 10 to 250 eV from solids, the electrons must originate

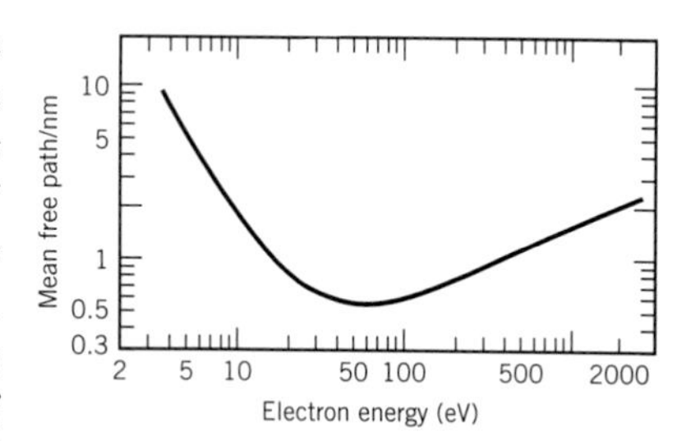

Figure 24.5 Mean free path of an electron in a solid as a function of electron kinetic energy.

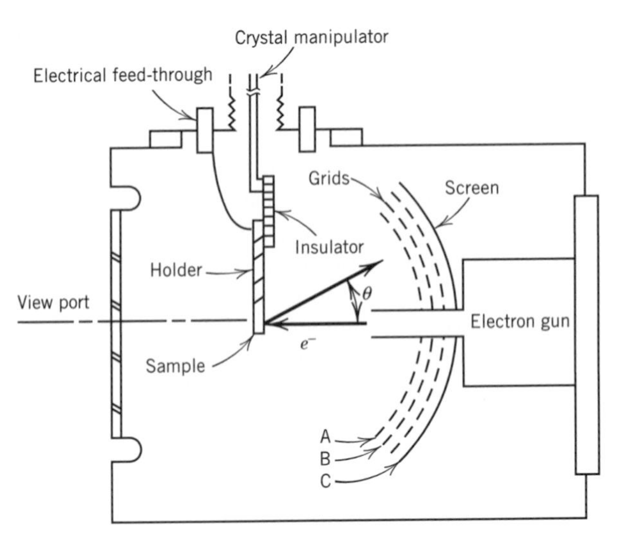

Figure 24.6 Schematic diagram of an LEED apparatus. [From D. G. Castner and G. A. Somorjai, *Chem. Rev.* **79:**233 (1979). Copyright © 1979 American Chemical Society.]

from the top few atomic layers. Therefore, the electron spectroscopies discussed in the next section, which depend on emission from the solid, also provide information on the surface layer, rather than bulk properties.

When a surface is bombarded with electrons, the electrons may be scattered elastically (that is, no energy is lost) or inelastically. Electrons scattered elastically will be diffracted if their de Broglie wavelength is small enough. **Low-energy electron diffraction** (LEED) therefore provides a means for studying the atomic geometry of a surface. In an LEED apparatus electrons are accelerated in an electron gun and strike the surface normally, as shown in Fig. 24.6. The surface backscatters a portion of the electrons. The grids in front of the screen are used to remove the inelastically backscattered electrons, while the elastically backscattered electrons are postaccelerated onto the phosphorescent screen for viewing the diffraction pattern. The presence of sharp diffraction spots indicates that the surface is ordered on an atomic scale.

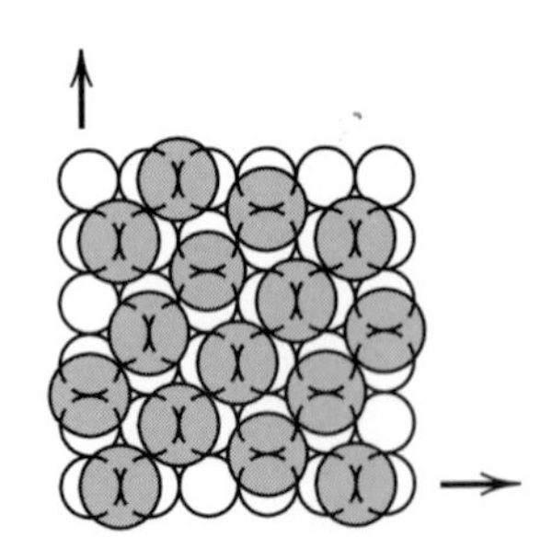

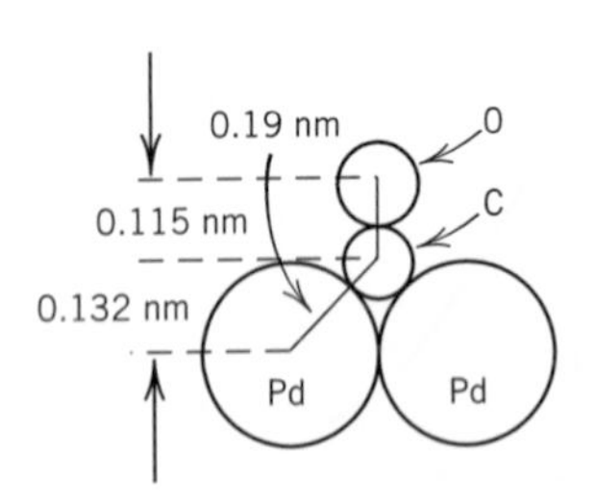

Figure 24.7 Structural model for CO adsorbed on palladium (100). The shaded circles in the top view are CO molecules. [From G. Ertl, *Pure Appl. Chem.* **52:**2051 (1980).]

It is evident from the Bragg equation (Section 23.3) that for there to be diffraction, the de Broglie wavelength of the electrons must be less than $2d$, where d is the distance between atomic planes. The **de Broglie wavelength** λ of electrons accelerated through a potential difference ϕ is given by

$$\lambda/\text{nm} = \left(\frac{1.504\ \text{V}}{\phi}\right)^{1/2} \tag{24.12}$$

As in X-ray diffraction from crystals, the angles at which X-rays are scattered give information about the symmetry and type of surface unit cell, but intensity measurements are required to obtain atomic coordinates. For some crystals the surface structure is quite different from that of the bulk crystal (Section 24.9). LEED can also be used to determine the structure and order of an adsorbed layer. For example, the structural model for CO adsorbed on palladium (100), as determined by LEED, is shown in Fig. 24.7. The CO is bound at an interstitial site with the carbon atom down; this is referred to as a bridge-bonded site.

Example 24.4 ***The minimum accelerating potential for electron diffraction from a crystal***

Derive equation 24.12 and calculate the minimum accelerating potential required to obtain electron diffraction from a crystal with an interplanar spacing of 0.1227 nm. The de Broglie wavelength must therefore be less than 0.2454 nm.

From Example 9.1, the momentum p of an electron with energy E is given by

$$p = (2mE)^{1/2} = \frac{h}{\lambda}$$

or

$$\lambda = \left(\frac{h^2}{2mE}\right)^{1/2}$$

Since the energy of an electron that has been accelerated through a potential difference ϕ is $E = e\phi$, the de Broglie wavelength is given by

$$\lambda = \left(\frac{h^2}{2me\phi}\right)^{1/2}$$

$$= \left(\frac{1.504 \times 10^{-18}\ \mathrm{V\ m^2}}{\phi}\right)^{1/2}$$

which is equation 24.12. For the crystal considered, the minimum accelerating potential is therefore

$$\phi = \frac{1.504 \times 10^{-18}\ \mathrm{V\ m^2}}{(0.2454 \times 10^{-9}\ \mathrm{m})^2} = 24.97\ \mathrm{V}$$

24.5 ELECTRON EMISSION FROM SURFACES

When photons are used to eject electrons, the technique is called **photoelectron spectroscopy** (PES). When these photons are in the ultraviolet region, the abbreviation UPS is used, and when the photons are in the X-ray region, the abbreviation XPS is used. We have already discussed the use of ultraviolet photoelectron spectroscopy (UPS) of gas molecules in Section 14.10; this is the most direct method for determining the ionization potential I of a molecule, and it provides additional information about molecular electronic structure. The kinetic energy of the ejected electron E_e is given by

$$E_e = \tfrac{1}{2}m_e v^2 = h\nu - I - E_v - E_r \qquad (24.13)$$

where I is the **adiabatic ionization energy,** the energy required to remove an electron from the molecule in its ground vibrational and rotational state to produce a molecular ion in its ground vibrational and rotational state. Since the ionization process may leave the molecular ion in an excited vibrational or rotational state, the vibrational energy E_v and rotational energy E_r above the ground state must also be subtracted from the energy of the incident photon.

Photons with energies in the ultraviolet range eject electrons from the valence orbitals of an adsorbed molecule or the valence electron bands of a solid, as shown

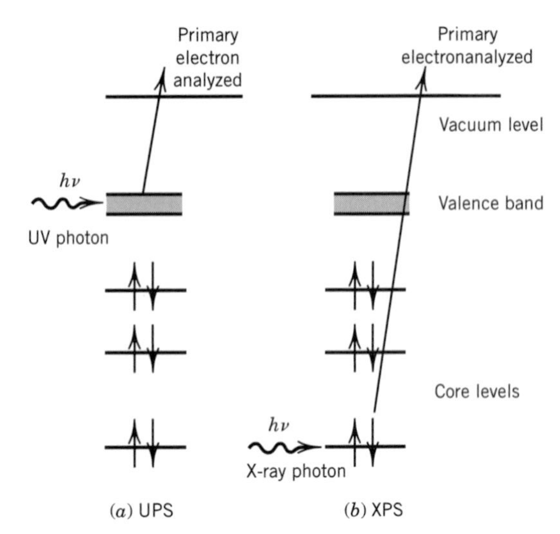

Figure 24.8 (*a*) In ultraviolet photoelectron spectroscopy (UPS), the energies of electrons ejected from the valence band are determined. (*b*) In X-ray photoelectron spectroscopy (XPS), the energies of electrons ejected from core levels are determined.

in Fig. 24.8*a*. Photons at these energies do not provide enough energy to eject electrons from atom cores.

UPS experiments are usually carried out with the He I line (21.2 eV) or the He II line (40.8 eV). Ultraviolet photon spectroscopy of a molecule physisorbed on a solid is a superposition of the ultraviolet photon spectrum of the valence band of the solid and the gas-phase molecular orbitals of the adsorbed molecule. Since the energy spread of the valence band of a solid is about 10 eV, the electrons ejected from the solid have energies spread over a similar range. The ultraviolet photon spectrum of the physisorbed molecule is quite similar to that of the corresponding gas molecule because the interaction with the surface is weak. Chemisorption, however, affects valence orbitals of the molecule and the valence band of the solid, and so the spectrum is complicated and reflects the type of bonding between the molecule and the surface. Important structural and chemical bonding information about chemisorbed species can be obtained from UPS.

In **X-ray photoelectron spectroscopy** (XPS) core electrons are ejected from the metal and from adsorbed molecules, as illustrated in Fig. 24.8*b*. The usual sources of X-rays are Mg $K\alpha$ (1253.6 eV) or Al $K\alpha$ (1486.6 eV). Since the energies of atomic core levels are characteristic of each element, XPS can be used to obtain an elemental analysis of the surface. This type of spectroscopy is referred to as ESCA (electron spectroscopy for chemical analysis) because the relative amounts of elements in a sample can be determined. The ionization potential of an atomic core level depends to some extent on the chemical environment of the atom, so that information can also be obtained about the type of bonding between the adsorbate and the surface.

When a surface is irradiated with soft X-rays, the Auger effect may occur, and the energies of the secondary electrons can be determined. This effect, which can also be produced by bombarding with high-energy electrons, is the basis of another type of surface spectroscopy referred to as **Auger electron spectroscopy** (AES). In AES, an electron is ejected by an X-ray photon, as in XPS, but the emission of the secondary electrons is analyzed, rather than the primary electrons. When a core electron is ejected by absorption of an X-ray photon (or high-energy electron), as shown in Fig. 24.9*a*, an electron from a higher energy level may drop into the core vacancy. The energy liberated in this way can lead to the emission of a second, or Auger, electron, as illustrated in Fig. 24.9*b*. The energies of the Auger electrons are characteristic of the core levels, so information about the atoms present and their bonding is obtained. Note that the energy of the Auger electron is independent of the energy of the exciting radiation. Auger electron spectroscopy is often used to measure the coverage of adsorbed species and to check the cleanliness of a surface. AES can be studied using the same electron optics and collector as used in an LEED experiment.

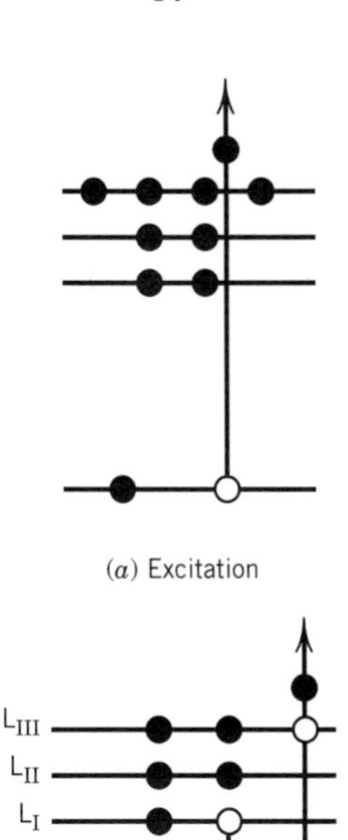

Figure 24.9 Emission of an Auger electron with an energy equal to the difference in energy levels L_I and L_{III}. (*a*) An electron in the K level is ejected by an X-ray photon. (*b*) An electron falls from the L_I level to the K level, and as a consequence, an electron is ejected from the L_{III} level.

Comment:

The development of low-energy electron diffraction, photoelectron spectroscopy, ultraviolet photoelectron spectroscopy, X-ray photoelectron spectroscopy, Auger electron spectroscopy, and related methods have made possible tremendous advances in surface science because their sensitivities are so great. The study of surfaces is important because they are used for catalysis on a very large scale, and of course corrosion starts on surfaces. Surface adsorption is also involved in various kinds of chromatography.

24.6 SCANNING TUNNELING MICROSCOPY (STM) AND ATOMIC FORCE MICROSCOPY (AFM)

The 1986 Nobel Prize in physics was given in part to G. Binnig and H. Rohrer for their development of the scanning tunneling microscope (STM), which makes it possible to "see" a single molecule adsorbed on a surface. A very sharp metal tip is moved over the surface of an electrical conductor at a height of about 500 pm, as shown in Fig. 24.10*a*. The tip is so close to the surface that there is an overlap of the wavefunctions of the atoms of the tip and those of the surface. If a potential difference is applied between the tip and the surface, quantum mechanical tunneling (Section 9.15) permits a current to flow through the gap. The potential of

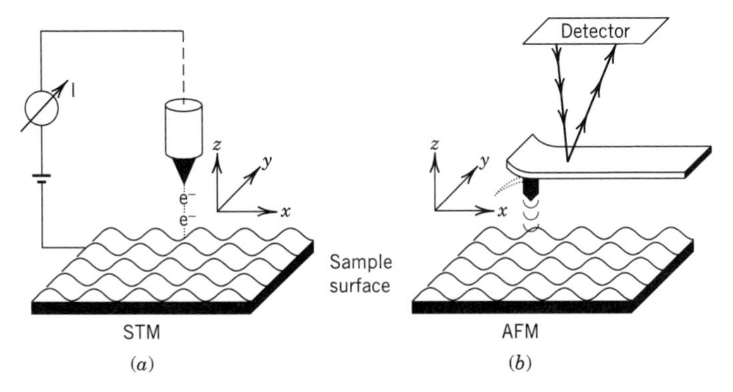

Figure 24.10 Schematic illustration of (*a*) the scanning tunneling and (*b*) the atomic force microscope. A localized probe is used to scan the surface and collect data on the tunneling current in the STM and the deflection of a cantilever-type spring in the AFM. [From J. Frommer, *Angew. Chem. Int. Ed. Engl.* **31:**1298 (1992), with permission.]

the tip with respect to the surface is held constant with a feedback system that regulates piezoelectric tubes to control the vertical motion of the tip so that the tunneling current is kept constant. The image of the topography of the surface that is obtained in this way can reveal single atoms because of the extremely narrow stream of tunneling electrons, which is spatially confined between a few atoms on the tip and a few atoms in the sample.

The atomic force microscope (AFM) measures interactions between the scanned tip and the surface. In contrast with STM, the AFM does not require an electrically conductive surface. The interactions between the tip and the surface may be van der Waals, electrostatic, or magnetic. The AFM image is a map of the forces detected over each point on the surface. Forces in the range 10^{-13} to 10^{-6} N are measured with a probing tip attached to a cantilever-type spring, as illustrated in Fig. 24.10*b*. The AFM image of a freshly cleaved surface of tetracene is shown in Fig. 24.11. Tetracene is made up of four benzene rings fused in a line. The AFM image makes it possible to distinguish between the two translationally nonequivalent molecules that make up the tetracene unit cell, which is shown in the figure. The AFM probe "sees" different surface orientations for the two tetracene molecules; their broad, flat σ systems are not parallel to one another,

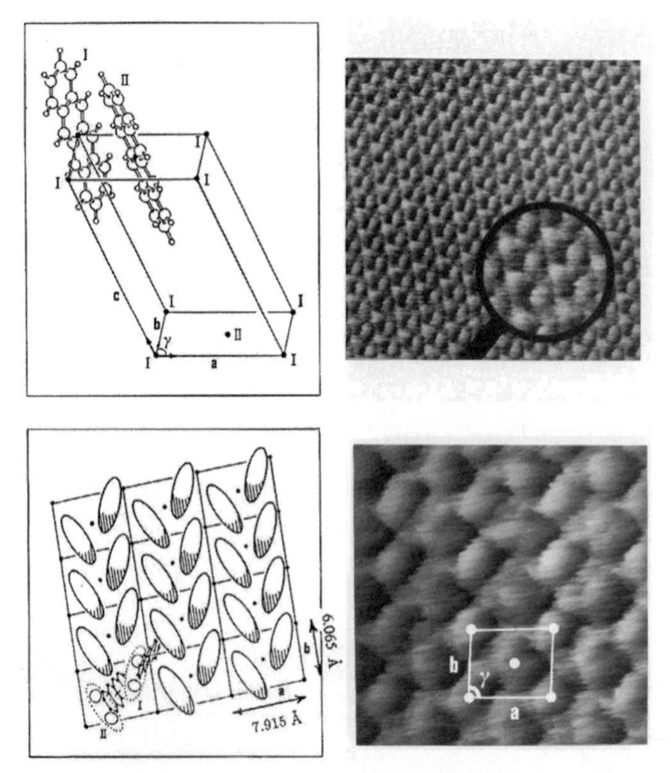

Figure 24.11 AFM image and crystal packing of tetracene. The scanned area is 3×3 nm^2 of the *ab* cleavage plane. The unit cell parameters measured on the surface of the crystal are marked on the image to show that they correspond to those of the bulk. [From J. Frommer, *Angew. Chem. Int. Ed. Engl.* **31:**1298 (1992), with permission.]

and the molecules are tilted in such a way that their short axes present two different angles to the scanned cleavage plane.

24.7 THEORY OF SURFACE REACTIONS

According to the Langmuir theory, a molecule is either bound to the surface or not. However, it is found experimentally that a molecule may initially be bound in a precursor state, which is much like a physisorbed state, and then later makes a transition to a chemisorbed state. Figure 24.12 gives schematic diagrams of the potential energy as a function of distance perpendicular to the surface for three cases. In Fig. 24.12*a*, an AB molecule is physisorbed on the surface but rapidly passes over the low potential barrier to dissociate to form a state in which atoms are chemisorbed; this is called **dissociative chemisorption.** If atoms existed in the gas phase, rather than AB molecules, the potential energy of the system would be given by the dot–dash line. The continuation of the potential energy curve for physisorbed AB molecules is shown as a dashed line. In Fig. 24.12*b*, the potential barrier between physisorbed AB molecules and chemisorbed atoms is much higher, so that only physisorption occurs at low temperatures. At higher temperatures, AB molecules pass over the barrier, spontaneously dissociate, and the atoms are chemisorbed. In this case, the physisorbed state is a **precursor state.** In Fig. 24.12*c*, a very low potential barrier separates the physisorbed state from a state of molecular chemisorption.

Since encounters of gaseous species with a solid surface can be considered to be a chemical reaction, the concepts developed earlier for the dynamics of chemical reactions in the gas phase can be used. Figure 24.13, which shows potentials for two types of surface reaction, is useful for thinking about the dynamics of surface reactions. If the incident particle has a very high kinetic energy, it will probably scatter back into the gas phase, as shown in Fig. 24.13*a*. If the incident particle has less energy, it can lose some energy to the surface and become trapped in an adsorption well, as also shown in Fig. 24.13*a*. If there is a barrier to adsorption, incident particles with low energy cannot be bound, but particles with high energies can be, as shown in Fig. 24.13*b*.

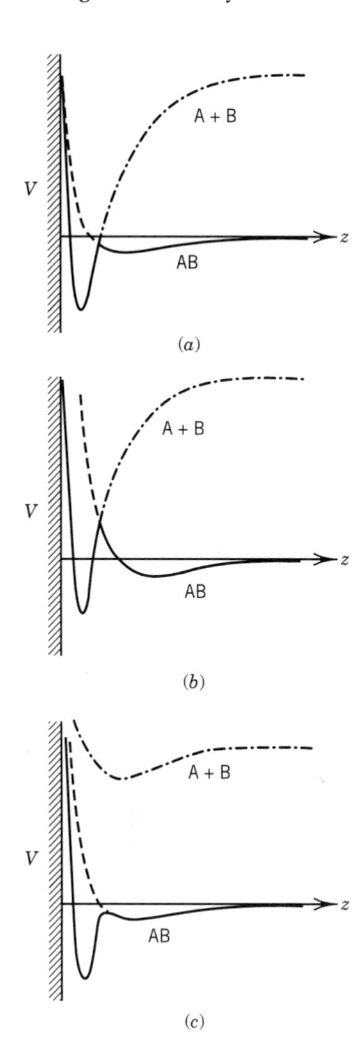

Figure 24.12 Schematic diagrams of an adsorbate–substrate complex for three different ground-state configurations: (*a*) dissociative chemisorption, (*b*) molecular physisorption, and (*c*) molecular chemisorption. [From A. Zangwill, *Physics at Surfaces.* Cambridge, UK: Cambridge University Press, 1988.]

24.8 HETEROGENEOUS CATALYSIS

Reactions that occur in a single phase are referred to as homogeneous reactions, and reactions that occur on a surface are referred to as heterogeneous reactions. The chemistry of a reaction on a surface is quite different from the chemistry in a gas or liquid phase, and many reactions can be carried out at lower temperatures in this way. Solid-state catalysts are used in many large-scale processes in the chemical industry. The production of NH_3 from H_2 and N_2 is catalyzed by iron promoted with Al_2O_3 and K. The oxidation of NH_3 to NO in the production of HNO_3 is catalyzed by Pt-Rh. Enormous quantities of solid catalysts are used in the petroleum industry for "cracking," which is necessary to produce gasoline from petroleum, and for "reforming," which causes the rearrangement of the molecular structures and raises the octane rating of gasoline. To facilitate the regeneration of these catalysts, they may be circulated as fine particles in the gas stream.

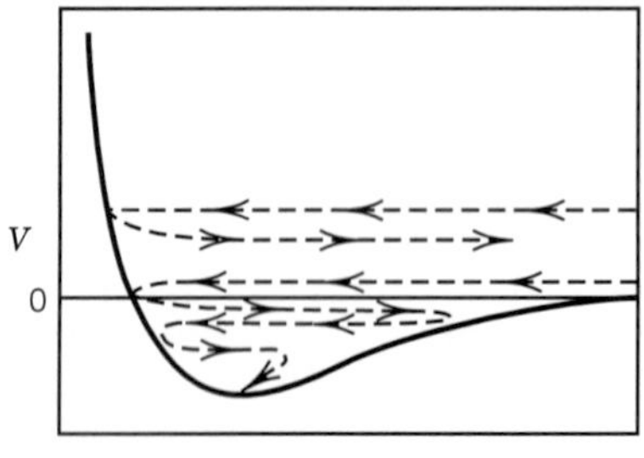

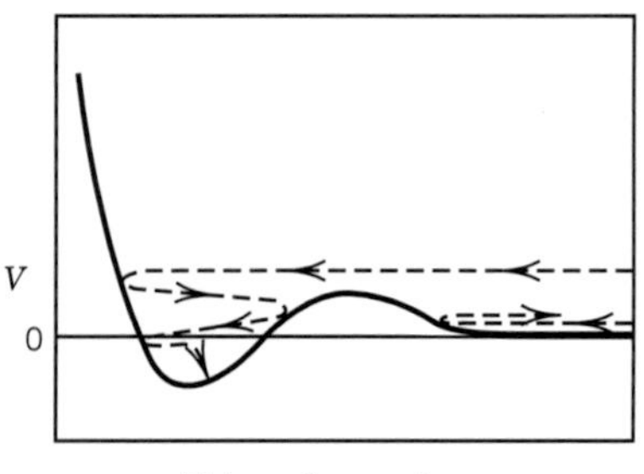

Figure 24.13 Schematic one-dimensional view of the approach of a gas particle to a solid surface: (*a*) simple adsorption well and (*b*) well with a barrier to adsorption. [From J. C. Tully, *Surf. Sci.* **111:**461 (1981).]

The development of catalysts is especially important for slow exothermic reactions. A reaction of this type is $N_2 + 3H_2 = 2NH_3$, which is the basis for the Haber process for the manufacture of ammonia. Although the equilibrium constant may be favorable at room temperature, the rate may be so low that the reaction cannot be used practically. When the temperature is raised, the rate is increased, but at high temperatures the equilibrium constant may be unfavorable, as is readily surmised from Le Châtelier's principle.

The catalyst that is actually used in producing ammonia is a porous structure consisting of small Fe particles (doped with partially reduced adsorbed K_2O) interspersed with Al_2O_3. The purpose of the K_2O is to further enhance the catalytic activity of the Fe particles.

Substances that enhance the activity of catalysts are called **promoters,** and substances that inhibit catalytic activity are called **poisons.** Since only a fraction of the surface of the catalyst may be involved, it is easy to see how relatively small amounts of promoters and poisons may be effective. In the manufacture of sulfuric acid, the presence of a very minute amount of arsenic completely destroys the catalytic activity of the platinum catalyst by forming platinum arsenide at the surface.

The effectiveness of various metals as catalysts for a particular reaction often varies in a regular way with the position of the metal in the periodic table. For example, Fig. 24.14 shows the relative specific activities of rhenium, osmium, iridium, and platinum for the reaction $H_2 + C_2H_6 = 2CH_4$. These metals have atomic numbers 75, 76, 77, and 78. There are several possible explanations as to why catalytic activity might go through a maximum in this way. For example, significantly exothermic adsorption may be required for a significant amount of adsorption. On the other hand, if the metal binds the adsorbate too strongly, it may not be an effective catalyst.

The catalysis of a reaction by a surface involves several steps: (1) diffusion of reactants to the surface; (2) adsorption, usually chemisorption, on the surface; (3) reaction on the surface; (4) desorption of products; and (5) diffusion of products away from the surface. If the third step is the slowest, the adsorption of various reactants and products will be in equilibrium. When this is the situation the Langmuir equation (equation 24.6) is often useful in deriving the rate equation for the surface-catalyzed reaction, as shown in the following paragraph.

First, let us consider a reaction with a single reactant and a single product:

$$A \rightarrow P \tag{24.14}$$

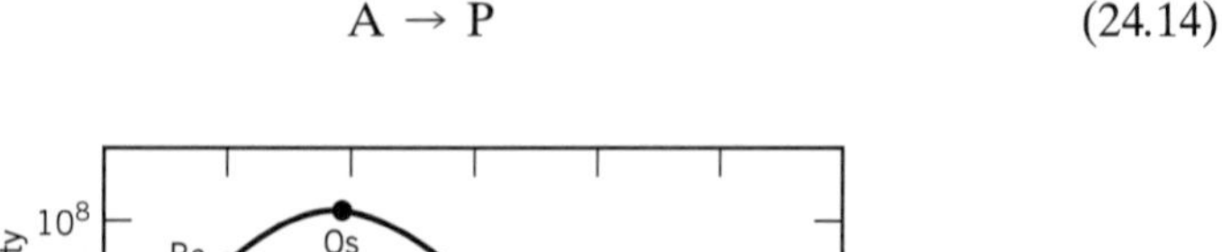

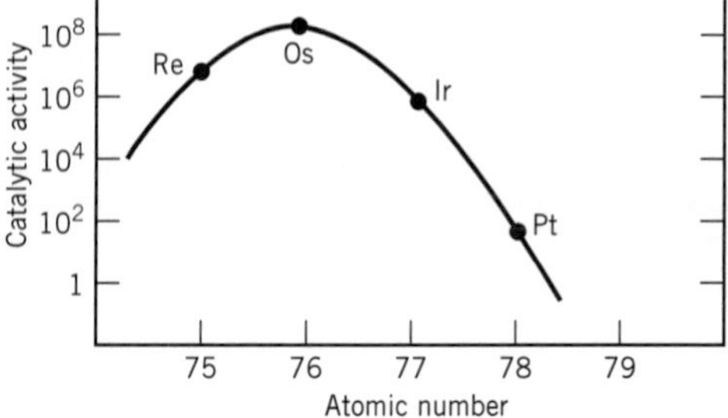

Figure 24.14 Relative specific activities for various metals as catalysts for the reaction $H_2 + C_2H_6 = 2CH_4$. [From J. H. Sinfelt, *Catal. Rev.* **9:**147 (1974).]

If the reaction on the surface is rate determining, the rate of the surface-catalyzed reaction is expected to be proportional to the surface area S_0 of the catalyst and the fraction Θ_A of the surface covered by reactant:

$$-\frac{dn_A}{dt} = kS_0\Theta_A \tag{24.15}$$

where n_A is the number of moles of reactant in the system. If both reactant and product are adsorbed on the surface, the fraction of the surface occupied by A is given by

$$\Theta_A = \frac{K_A P_A}{1 + K_A P_A + K_P P_P} \tag{24.16}$$

according to the Langmuir adsorption isotherm, as shown in Example 24.2. In this case the rate of the heterogeneous reaction is given by

$$-\frac{dn_A}{dt} = \frac{kS_0 K_A P_A}{1 + K_A P_A + K_P P_P} \tag{24.17}$$

According to this rate equation the reaction will slow down as the product P accumulates because it competes with reactant A for the catalytic sites. When initial rates are studied for a reaction following this rate law, it is found that the rate increases linearly with P_A at low pressures and approaches a limiting value of kS_0 as P_A is increased.

Second, let us consider a reaction with two reactants and two products that are adsorbed on the same type of site:

$$A + B \rightarrow P + Q \tag{24.18}$$

The reaction rate is assumed to be proportional to the surface area and to the product of the fractions of the sites occupied by the two reactants:

$$-\frac{dn_A}{dt} = kS_0\Theta_A\Theta_B \tag{24.19}$$

If all of the reactants and products are adsorbed, then

$$\Theta_A = \frac{K_A P_A}{1 + K_A P_A + K_B P_B + K_P P_P + K_Q P_Q} \tag{24.20}$$

$$\Theta_B = \frac{K_B P_B}{1 + K_A P_A + K_B P_B + K_P P_P + K_Q P_Q} \tag{24.21}$$

so that the reaction rate is

$$-\frac{dn_A}{dt} = \frac{kS_0 K_A K_B P_A P_B}{(1 + K_A P_A + K_B P_B + K_P P_P + K_Q P_Q)^2} \tag{24.22}$$

According to this rate equation, if all partial pressures except P_A are held constant and P_A is increased, the rate will pass through a maximum.

Third, let us consider a reaction with two reactants that are adsorbed on different types of sites. Reactant A may be adsorbed on one type of site and reactant B may be adsorbed on another type of site. In the case of a metal oxide catalyst, one reactant might interact with the metal ions and the other reactant might interact with the oxide ions. If the reaction rate is proportional to the surface area and the product of the fraction Θ_{A1} of sites of type 1 occupied by A and the fraction Θ_{B2} of sites of type 2 occupied by B, then

$$-\frac{dn_A}{dt} = kS_0\Theta_{A1}\Theta_{B2}$$

$$= \frac{kS_0K_{A1}K_{B2}P_AP_B}{(1 + K_{A1}P_A)(1 + K_{B2}P_B)} \tag{24.23}$$

where the possible binding of product has been omitted. In contrast with rate equation 24.22, the rate in this case does not pass through a maximum when P_A is increased at constant P_B.

Many additional types of rate equations for heterogeneous reactions may be derived. The possibility that a reactant dissociates on the surface may also have to be included. It is important to have the right rate equation in designing an industrial process.

24.9 SPECIAL TOPIC: SURFACE RECONSTRUCTION

When a crystalline solid is cleaved to expose a fresh surface, some "dangling" bonds are formed, and this may lead to a reconstruction of the surface. For example, when a diamond is cleaved, one hybrid orbital of each carbon atom dangles into the vacuum. Each such orbital is half-occupied. These dangling bonds cause an adjustment of the positions of the atoms in the surface in which some atoms may rise above the surface plane and other atoms sink below the surface plane. Surface atoms may also be displaced horizontally. As the positions of surface atoms change to the geometrical configuration with the lowest Gibbs energy, energy gained by local bond formation is balanced by the work of elastic distortion. Furthermore, this type of surface reconstruction may also occur when atoms or molecules are adsorbed on the surface.

Arrangements of surface atoms can be described by the two-dimensional analogue of equation 23.2, namely,

$$\boldsymbol{T} = h'\boldsymbol{b}_1 + k'\boldsymbol{b}_2 \tag{24.24}$$

where h' and k' are integers. The 14 Bravais lattices of three-dimensional structures are replaced by the five types of two-dimensional nets of equivalent lattice points that are shown in Fig. 24.15. The nomenclature for surface structures involves a comparison of the basis vectors ($\boldsymbol{b}_1$ and $\boldsymbol{b}_2$) of the surface lattice with the corresponding substrate lattice vectors ($\boldsymbol{a}_1$ and $\boldsymbol{a}_2$). The substrate lattice is defined as that plane parallel to the surface below which the three-dimensional (bulk) periodicity is found. The relation between the surface lattice and the substrate lattice is expressed by the ratios of the lengths of the basis vectors $|\boldsymbol{b}_1|/|\boldsymbol{a}_1|$ and $|\boldsymbol{b}_2|/|\boldsymbol{a}_2|$ and the angle of rotation between the two lattices. The nomenclature for surface

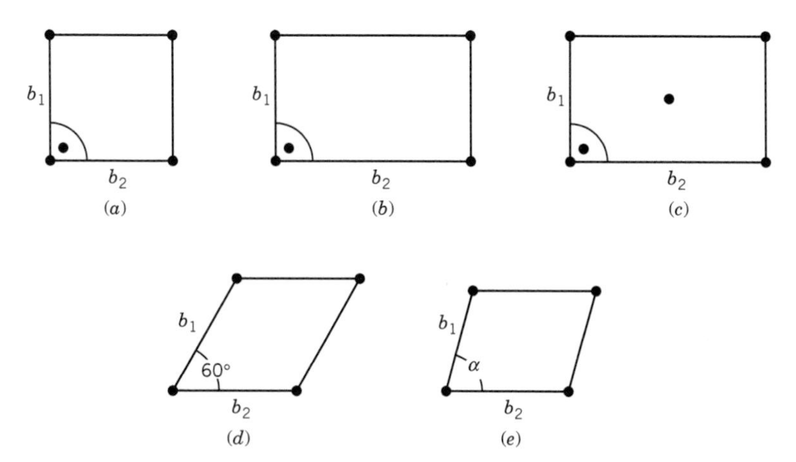

Figure 24.15 The five types of two-dimensional Bravais lattices: (*a*) square, $b_1 = b_2$, $\alpha = 90°$; (*b*) primitive rectangular, $b_1 \neq b_2$, $\alpha = 90°$; (*c*) centered rectangular, $b_1 \neq b_2$, $\alpha = 90°$; (*d*) hexagonal, $b_1 = b_2$, $\alpha = 60°$; (*e*) oblique, $b_1 \neq b_2$, $\alpha \neq 90°$. [From G. Ertl and J. Küppers, *Low Energy Electrons and Surface Chemistry,* 2nd ed. Weinheim: VCH, Verlag Chemie, 1985.]

lattices is illustrated in Fig. 24.16. The atoms in the surface lattice in Fig. 24.16*a* are twice as far apart as in the substrate structure in two dimensions; the angle of rotation is omitted because it is zero. In the overlayer structure in Fig. 24.16*b* the designation $c\,(2\times2)$ indicates that the unit cell of the surface structure is centered. In Fig. 24.16*c* the ratios are $\sqrt{3}$, and the surface lattice is rotated 30° with respect to the substrate.

The surfaces of crystals do not necessarily resemble the surfaces that would correspond to a simple termination of the bulk structure. For example, the $1\bar{1}0$ surface of platinum is shown in Fig. 24.17*b*. This reconstructed surface (labeled 1×2) has a missing row because it is a more stable structure than the simply terminated bulk surface (labeled 1×1). However, when CO is bound by this reconstructed surface, it becomes less stable than the 1×1 surface with CO adsorbed. As CO is added to the surface, the 1×2 to 1×1 transformation takes place as soon as the CO coverage exceeds about 0.2.

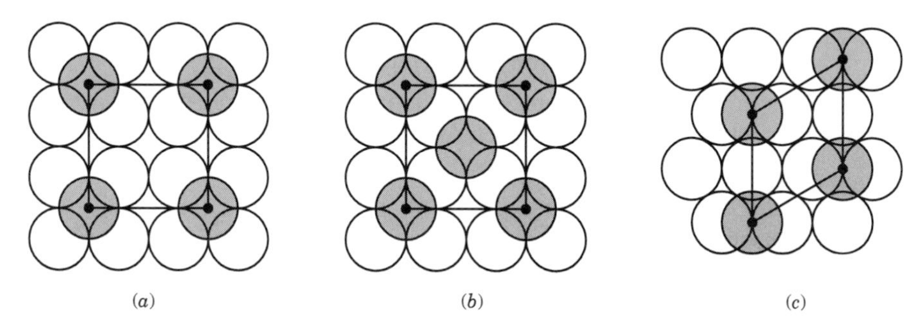

Figure 24.16 Examples for overlayer structures: (*a*) 2×2; (*b*) $c\,(2\times2)$; (*c*) $\sqrt{3}\times\sqrt{3}R\,30°$. [From G. Ertl and J. Küppers, *Low Energy Electrons and Surface Chemistry,* 2nd ed. Weinheim: VCH, Verlag Chemie, 1985.]

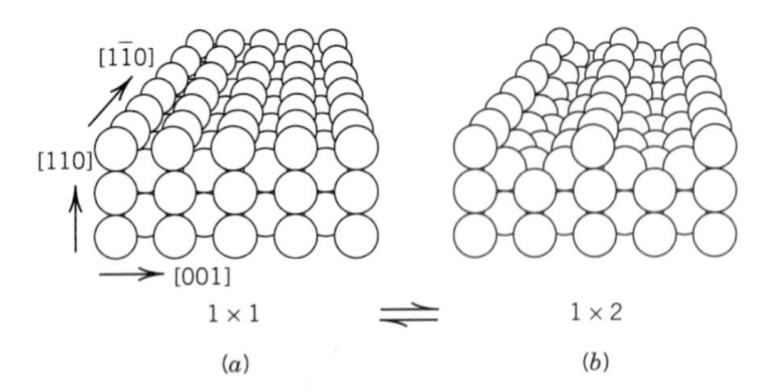

Figure 24.17 (*a*) The $1\bar{1}0$ surface of Pt that would correspond to a simple termination of the bulk structure (1 × 1). (*b*) Actual $1\bar{1}0$ surface of Pt that is energetically more stable (1 × 2). [Reprinted with permission from G. Ertl, *Science* **254**:1750 (1991). Copyright © 1991 American Association for the Advancement of Science.]

Eight Key Ideas in Chapter 24

1. It is usually possible to make a clear distinction between physical adsorption ($\Delta_{ads}H = -10$ to -40 kJ mol^{-1}) and chemisorption ($\Delta_{ads}H = -40$ to -200 kJ mol^{-1}) because chemisorption has a significant activation energy for desorption, in other words, a slower rate.
2. Langmuir derived an adsorption isotherm on the assumptions that (a) the rate of adsorption is proportional to the flux of gas molecules, the fraction of vacant sites, a sticking coefficient, and an exponential dependence on $1/T$ and (b) the rate of desorption is proportional to the fraction of occupied sites and a different exponential dependence on $1/T$.
3. The BET theory takes into account the possibility of multilayer adsorption and makes it possible to determine the volume of gas required to form a monolayer, so that the surface area of the adsorbant can be calculated.
4. Low-energy electron diffraction (LEED) is based on the use of accelerated electrons with de Broglie wavelengths less than twice the interplanar spacing in the crystal to obtain information about the symmetry and type of the surface unit cell.
5. Photoelectron spectroscopy (PES), X-ray photoelectron spectroscopy (XPS), and Auger electron spectroscopy (AES) provide information about the identity and bonding of surface atoms.
6. Scanning tunneling microscopy (STM) and atomic force microscopy (AFM) make it possible to "see" the structure of a surface.
7. The effectiveness of different metals as catalysts of a particular reaction often depends in a regular way on the position of the metal in the periodic table because significant exothermic adsorption may be required for a significant amount of adsorption, but if the metal binds the adsorbate too strongly, it may not be an effective catalyst.
8. When a crystalline solid is cleaved to expose a fresh surface, some "dangling" bonds are formed, and this may lead to a reconstruction of the surface in which surface atoms are displaced. Surface reconstruction may also occur when atoms are adsorbed on the surface.

REFERENCES

A. W. Adamson and A. P. Gast, *Physical Chemistry of Surfaces,* Hoboken, NJ: Wiley, 1997.

G. C. Bond, *Heterogeneous Catalysis: Principles and Applications.* Oxford, UK: Clarendon, 1987.

L. J. Clarke, *Surface Crystallography.* Hoboken, NJ: Wiley-Interscience, 1985.

G. Ertl and J. Küppers, *Low Energy Electrons and Surface Chemistry,* 2nd ed. Weinheim: VCH, Verlag Chemie, 1985.

R. P. H. Gasser, *An Introduction to Chemisorption and Catalysis by Metals.* Oxford, UK: Clarendon, 1985.

G. A. Somorjai, *Chemistry in Two Dimensions: Surfaces.* Ithaca, NY: Cornell University Press, 1981.

A. Zangwill, *Physics at Surfaces.* Cambridge, UK: Cambridge University Press, 1988.

PROBLEMS

Problems marked with an icon may be more conveniently solved on a personal computer with a mathematical program.

24.1 In an ultrahigh vacuum chamber ($P = 5 \times 10^{-10}$ torr), how many molecules strike 1 cm^2 of surface in one second at 298 K if (*a*) the gas is helium and (*b*) the gas is mercury vapor?

24.2 A readily oxidized metal surface with 10^{15} metal atoms per square centimeter is exposed to molecular oxygen at 10^{-5} Pa at 298 K. How long will it take to completely oxidize the surface if the oxide formed is MO?

24.3 In Problem 17.19, we found that at 1 bar and 298 K, 1.074×10^{28} molecules of molecular hydrogen strike a surface per square meter per second. When the 100 plane of metallic copper is exposed to molecular hydrogen under these conditions, what is the rate of collisions with atoms of copper? Copper forms face-centered cubic crystals with the length of the side of the unit cell equal to 361 pm.

24.4 For the adsorption of nitrogen molecules on a certain sample of carbon, the pressure required to half-saturate the surface at 298 K is 2×10^{-5} Pa. If the enthalpy of adsorption is -10 kJ mol^{-1} and the sticking coefficient s^* is unity, what is the rate constant of desorption k_d?

24.5 The pressures of nitrogen required for adsorption of 1.0 cm^3 g^{-1} (25 °C, 1.013 bar) of gas on graphitized carbon black are 24 Pa at 77.5 K and 290 Pa at 90.1 K. Calculate the enthalpy of adsorption at this fraction of surface coverage.

24.6 A mixture of A and B is adsorbed on a solid for which the adsorption isotherm follows the Langmuir equation. If the mole fractions in the gas at equilibrium are y_A and y_B, what is the equation for the adsorption isotherm in terms of total pressure? What is the expression for the mole fraction of A in the adsorbed gas in terms of K_A, K_B, y_A, and y_B?

 **24.7** The following table gives the volume of nitrogen (reduced to 0 °C and 1 bar) adsorbed per gram of active carbon at 0 °C at a series of pressures:

P/Pa	524	1731	3058	4534	7497
v/cm^3 g^{-1}	0.987	3.04	5.08	7.04	10.31

Plot the data according to the Langmuir isotherm, and determine the constants.

24.8 According to Problem 24.7, the Langmuir constant for the adsorption of molecular nitrogen on active carbon at 0 °C is $K = 4.8 \times 10^{-5}$ Pa^{-1}. What pressures of molecular nitrogen are required to cover 10%, 50%, and 90% of the surface at 0 °C?

24.9 According to Problem 17.43, the rate with which oxygen molecules strike a surface at 1 bar and 25 °C is 2.69×10^{23} cm^{-2} s^{-1}. If the oxygen molecules are striking a platinum surface, what are the frequencies of collisions per atom on the (100), (110), and (111) planes? (See Problem 23.32.)

24.10 One gram of activated charcoal has a surface area of 1000 m^2. If complete surface coverage is assumed, as a limiting case, how much ammonia, at 25 °C and 1 bar, could be adsorbed on the surface of 45 g of activated charcoal? The diameter of the NH_3 molecule is 3×10^{-10} m, and it is assumed that the molecules just touch each other in a plane so that four adjacent spheres have their centers at the corners of a square.

24.11 Calculate the surface area of a catalyst that adsorbs 10^3 cm^3 of nitrogen (calculated at 1.013 bar and 0 °C) per gram in order to form a monolayer. The adsorption is measured at -195 °C, and the effective area occupied by a nitrogen molecule on the surface is 16.2×10^{-20} m^2 at this temperature.

24.12 The de Broglie wavelength of an electron that has been accelerated through a potential difference of ϕ is given by

$$\lambda = \left(\frac{h^2}{2me\phi}\right)^{1/2}$$

Derive this equation and verify that it is correct to write equation 24.12.

24.13 In LEED experiments, acceleration voltages of 10 to 200 V are generally used. (*a*) Calculate the energies of electrons accelerated by these voltages in kJ mol^{-1}. (*b*) Calculate the wavelengths of electrons accelerated by these voltages in nm.

24.14 What volume of oxygen, measured at 25°C and 1 bar, is required to form an oxide film on 1 m^2 of a metal with atoms in a square array 0.1 nm apart? Assume that one oxygen atom combines with each metal atom.

24.15 The following table gives data on the adsorption of benzene by graphitized carbon black (P-33) at two surface coverages and two temperatures:

v (cm^3 g^{-1} at 25°C, 1 bar)	0.2	0.4
t/C°	*Pressure* (Pa)	
0	13	27
35.0	80	170

Calculate the enthalpy of adsorption $\Delta_{ads}H$ at each coverage. (S. Ross and J. P. Oliver, *On Physical Adsorption*, p. 238. New York: Wiley-Interscience, 1964.)

24.16 The adsorption of ammonia on charcoal is studied at 30 and 80 °C. It is found that the pressure required to adsorb a certain amount of NH_3 per gram of charcoal is 14.1 kPa at 30 °C and 74.6 kPa at 80 °C. Calculate the enthalpy of adsorption.

24.17 Hydrogen is dissociatively adsorbed on a metal, and the pressure required to obtain half of the saturation coverage of the surface is 10 Pa. (*a*) What pressure will be required to reach $\Theta = 0.75$? (*b*) What pressure would have been required if the adsorption were not dissociative?

24.18 A fresh metal surface with 10^{15} atoms per square centimeter is prepared. This surface is exposed to oxygen at 10^{-2} Pa. If every oxygen molecule that strikes the surface reacts so that there is one oxygen atom per metal atom in the surface, how long will it take for half of the surface to become oxidized at 25 °C?

24.19 Show that the SI units of the diffusion coefficient for surface diffusion are $m^2\ s^{-1}$, the same as for three-dimensional diffusion.

24.20 The volume of nitrogen gas v_m (measured at 1.013 bar and 0 °C) required to form a complete monolayer on a sample silica gel is 129 $cm^3\ g^{-1}$ of gel. Calculate the surface area per gram of the gel if each nitrogen molecule occupies $16.2 \times 10^{-20}\ m^2$.

24.21 The diameter of the hydrogen molecule is about 0.27 nm. If an adsorbent has a surface of 850 $m^2\ cm^{-3}$, what volume of H_2 (measured at 25 °C and 1 bar) could be adsorbed by 100 mL of the adsorbent? It may be assumed that the adsorbed molecules just touch in a plane and are arranged so that four adjacent spheres have their centers at the corners of a square.

Computer Problems

24.A According to the BET theory, plot v/v_m, where v_m is the volume of gas required to form a monolayer, versus P/P^*, where P^* is the saturation vapor pressure of the gas. Show the effect of changing the constant c from 0.10 to 200.

24.B Suppose that at 298.15 K, the adsorption isotherm for a gas on a solid is given by

$$\Theta = KP/(1 + KP)$$

where Θ is the fractional coverage of the surface, $K(298.15\ \mathrm{K}) = 3$ bar, and $\Delta H_{ads} = -20$ kJ mol^{-1}. Plot Θ versus P at 298.15 K and 350 K.

24.C Plot the fractional coverage Θ of a surface for dissociative adsorption with $K = 1$ and for Langmuir adsorption with $K = 1$. Superimpose these plots so that the differences will be clearer.

24.D Plot the de Broglie wavelength λ in nanometers as a function of the accelerating potential difference ϕ in volts from 0 to 100 volts. Check that the result in Example 24.4 is confirmed.

Appendix

A Physical Quantities and Units

The measurement of any physical quantity consists of a comparison with a standard amount of that quantity, which is referred to as a unit. Thus, the statement of a physical measurement consists of two parts: (1) a number that represents the number of times the unit has to be used to give the physical quantity and (2) the unit itself. The SI system of units is founded on the seven base units listed in the following table. (SI stands for Système International d'Unités.) This particular system was defined and given official status by the 11th Conférence Générale des Poids et Mesures in 1960.*

Physical Quantity	*Symbol for Quantity*[a]	*Name of Unit*	*Symbol for SI Unit*[a]	*Definition*
Length	l	meter	m	The meter is the length of the path traveled by light in vacuum during a time interval of 1/299 792 458 of a second.
Mass	m	kilogram	kg	The kilogram is equal to the mass of the international prototype of the kilogram.
Time	t	second	s	The second is the duration of 9 192 631 770 periods of the radiation corresponding to the transition between the two hyperfine levels of the ground state of the cesium-133 atom.
Electric current	I	ampere	A	The ampere is that constant current which, if maintained in two straight parallel conductors of infinite length, of negligible cross section, and placed 1 meter apart in a vacuum, would produce between these conductors a force equal to 2×10^7 newtons per meter of length.

[a] Note that symbols for quantities are always printed in italic type, and symbols for units are always printed in roman type.

*I. Mills, T. Cvitas, K. Homann, N. Kallay, and K. Kuchitsu, *Quantities, Units, and Symbols in Physical Chemistry.* Oxford, UK: Blackwell Scientific, 1993.

Physical Quantity	*Symbol for Quantity*	*Name of Unit*	*Symbol for SI Unit*	*Definition*
Thermodynamic temperature	T	kelvin	K	The kelvin is the fraction 1/273.16 of the thermodynamic temperature of the triple point of water.
Amount of substance	n	mole	mol	The mole is the amount of substance of a system that contains as many elementary entities as there are atoms in 0.012 kilogram of carbon-12. When the mole is used, the elementary entities must be specified and may be atoms, molecules, ions, electrons, other particles, or specified groups of such particles.
Luminous intensity	I_v	candela	cd	The candela is the luminous intensity, in a given direction, of a source that emits monochromatic radiation of frequency 540×10^{12} hertz and has a radiant intensity in that direction of $\frac{1}{683}$ watt per steradian.

In addition, there are two supplementary units: For plane angle, the supplementary SI unit is the radian, with the symbol rad; for solid angle, the supplementary SI unit is the steradian, with the symbol sr. In 1980 the International Committee of Weights and Measures decided to interpret the class of supplemental units as dimensionless derived units. Although the coherent unit for both quantities is the number 1, it is convenient to use the special names radian and steradian instead of the number 1 in many practical cases.

The value of a physical quantity is equal to the product of a numerical value and a unit. Physical quantity = numerical value $\times$ unit.

All quantities may be expressed in SI units or in terms of derived units obtained algebraically by multiplication and division. Some derived units have their own special symbols. For example, the joule, which is the unit of work in the SI, is defined in terms of base units by kg m^2 s^{-2} and is represented by the special symbol J. The principal derived units used in physical chemistry are given in the following table.

Quantity	*Unit*	*Symbol*	*Definition*
Force	newton	N	kg m s^{-2}
Work, energy, heat	joule	J	N m (= kg m^2 s^{-2})
Power, radiant flux	watt	W	J s^{-1}
Pressure	pascal	Pa	N m^{-2}
Electric charge	coulomb	C	A s
Electric potential	volt	V	kg m^2 s^{-3} A^{-1} (= J A^{-1} s^{-1} = J C^{-1})
Electric resistance	ohm	Ω	kg m^2 s^{-3} A^{-2} (= V A^{-1})
Electric capacitance	farad	F	A s V^{-1} (= m^{-2} kg^{-1} s^4 A^2)
Frequency	hertz	Hz	s^{-1} (cycles per second)
Magnetic flux density	tesla	T	kg s^{-2} A^{-1} (= N A^{-1} m^{-1})

It is important to be able to convert from other systems of units to SI units. Some of the conversion factors are as follows:

Physical Quantity	*Name of Unit*	*Symbol*	*Equivalent in SI Units*
Length	angstrom	Å	10^{-10} m (10^{-1} nm)
Energy	electron volt	eV	$1.602\,177\,33 \times 10^{-19}$ J
	wave number	cm^{-1}	$1.986\,447 \times 10^{-23}$ J
	calorie (thermochemical)	cal	4.184 J
	erg		10^{-7} J
Force	dyne		10^{-5} N
Pressure	bar	bar	10^5 N m^{-2}
	atmosphere	atm	101.325 kN m^{-2}
	torr		133.322 N m^{-2}
Electric charge	esu		3.334×10^{-10} C
Dipole moment	debye (10^{-18} esu cm)		3.334×10^{-30} C m
Magnetic flux density	gauss	G	10^{-4} T

In an equation relating physical quantities, the symbols represent numbers and associated units. If units are chosen arbitrarily, additional numerical factors may appear in equations relating different physical quantities. In practice it is more convenient to choose a system of units so that the equations between physical quantities have exactly the same form as the corresponding equations between pure numbers. A system of units that has this property is said to be coherent. The SI system is coherent. This means that if all quantities in a calculation are expressed in SI base units, the result will be expressed in SI base units without including any numerical factors. It is, however, a good habit to check a calculation to see that units cancel to yield the correct units for the final result.

A physical quantity may be converted from one unit to another by multiplying by a conversion factor. To find a conversion factor, it is necessary to express one unit in terms of another. For example, one calorie is equal to 4.184 joules; that is, 1 cal = 4.184 J. Dividing both sides of the equation by 1 cal yields 1 = 4.184 J cal^{-1}. To convert the change in enthalpy of a reaction from calories to joules, simply multiply the change in enthalpy in calories by 4.184 J cal^{-1}; note that this is equivalent to multiplying by 1. A number of useful energy conversion factors are listed inside the back cover, and some frequently used conversion factors are given inside the front cover.

The following examples illustrate the procedure for calculating conversion factors in more complicated cases. The first step is to write down the equation relating the quantities of interest. The conversion factor for converting energies in J mol^{-1} to wave numbers in cm^{-1} is based on the following equation:

$$E = N_A hc\tilde{\nu}$$

The ratio of wave numbers in cm^{-1} to energy in J mol^{-1} is given by

$$\begin{aligned}\frac{\tilde{\nu}}{E} &= (N_A hc)^{-1} \\ &= [(6.022\,136\,7 \times 10^{23}\ \text{mol}^{-1})(6.626\,075\,5 \times 10^{-34}\ \text{J s}) \\ &\quad \times (2.997\,924\,58 \times 10^{8}\ \text{m s}^{-1})]^{-1}\end{aligned}$$

$$
\begin{aligned}
&= 8.359\,346 \text{ mol J}^{-1}\text{ m}^{-1} \\
&= (8.359\,346 \text{ mol J}^{-1}\text{ m}^{-1})(0.01 \text{ m cm}^{-1}) \\
&= 8.359\,346 \times 10^{-2}\ (\text{J mol}^{-1})^{-1}\text{ cm}^{-1}
\end{aligned}
$$

The conversion factor for converting energies in J mol^{-1} to electron volts is based on the following equation:

$$U = N_A Ee$$

The potential in volts required to accelerate an electron to a given energy divided by that energy is

$$
\begin{aligned}
\frac{E}{U} &= (N_A e)^{-1} = 1/F = [(6.022\,136\,7 \times 10^{23} \text{ mol}^{-1})(1.602\,177\,33 \times 10^{-19} \text{ C})]^{-1} \\
&= 1.036\,427 \times 10^{-5} \text{ V } (\text{J mol}^{-1})^{-1}
\end{aligned}
$$

since V = J C^{-1}.

Finally, since kT is an energy and the combination ϵ/kT is frequently found in statistical mechanical calculations, it is sometimes convenient to express energies in terms of temperature by dividing a molecular energy by the Boltzmann constant k. This makes it possible to write $\exp(-\epsilon/kT)$ as $\exp(-\theta/T)$, which is convenient both for making calculations and in thinking about the magnitude of an exponential term in an equation. If an energy E is known in J mol^{-1}, the conversion to kelvins is based on the following equation:

$$E = N_A kT = RT$$

The ratio of the temperature in kelvins to energy in J mol^{-1} is given by

$$
\begin{aligned}
\frac{T}{E} &= (N_A k)^{-1} = R^{-1} \\
&= [(6.022\,136\,7 \times 10^{23} \text{ mol}^{-1})(1.380\,658 \times 10^{-23} \text{ J K}^{-1})]^{-1} \\
&= 0.120\,271\,7 \text{ K } (\text{J mol}^{-1})^{-1}
\end{aligned}
$$

B

Values of Physical Constants

Quantity	*Symbol*	*Value*[a,b]	*Units*
Acceleration due to gravity	g	9.806 65 (exact)	$m\ s^{-2}$
Speed of light in vacuum	c	299 792 458 (exact)	$m\ s^{-1}$
Permeability of vacuum	μ_0	$4\pi \times 10^{-7}$ (exact)	$N\ A^{-2}$
		= 12.566 370 614...	$10^{-7}\ N\ A^{-2}$
Permittivity of vacuum	ϵ_0	$1/\mu_0 c^2$ (exact)	$C^2\ N^{-1}\ m^{-2}$
		= 8.854 187 817...	$10^{-12}\ C^2\ N^{-1}\ m^{-2}$
Planck constant	h	6.626 075 5(40)	10^{-34} J s
$h/2\pi$	$\hbar$	1.054 572 66(63)	10^{-34} J s
Elementary charge	e	1.602 177 33(49)	10^{-19} C
Bohr magneton, $e\hbar/2m_e$	μ_B	9.274 015 4(31)	$10^{-24}\ J\ T^{-1}$
Nuclear magneton, $e\hbar/2m_p$	μ_N	5.050 786 6(17)	$10^{-27}\ J\ T^{-1}$
Rydberg constant, $m_e e^4/8h^3 c\epsilon_0$	R_∞	10 973 731.534(13)	m^{-1}
Bohr radius, $h^2\epsilon_0/\pi m_e e^2$	a_0	0.529 177 249(24)	10^{-10} m
Hartree energy, $e^2/4\pi\epsilon_0 a_0$	E_h	4.359 748 2(26)	10^{-18} J
Electron mass	m_e	9.109 389 7(54)	10^{-31} kg
Proton mass	m_p	1.672 623 1(10)	10^{-27} kg
Neutron mass	m_n	1.674 928 6(10)	10^{-27} kg
Deuteron mass	m_d	3.343 586 0(20)	10^{-27} kg
Avogadro constant	N_A	6.022 136 7(36)	$10^{23}\ mol^{-1}$
Atomic mass constant, $m_u = (1/12)m(^{12}C)$	m_u	1.660 540 2(10)	10^{-27} kg
Faraday constant	F	96 485.309(29)	$C\ mol^{-1}$
Gas constant	R	8.314 510(70)	$J\ K^{-1}\ mol^{-1}$
		0.083 145 1	$L\ bar\ K^{-1}\ mol^{-1}$
		1.987 216	$cal\ K^{-1}\ mol^{-1}$
		0.082 057 8	$L\ atm\ K^{-1}\ mol^{-1}$
Boltzmann constant, R/N_A	k	1.380 658(12)	$10^{-23}\ J\ K^{-1}$

[a] E. R. Cohen and B. N. Taylor, The 1986 CODATA Recommended Values of the Fundamental Physical Constants. *J. Phys. Chem. Ref. Data* **17**:1795 (1988).

[b] Digits in parentheses are the one-standard-deviation uncertainty in the last digits of the given value.

[c] More recent values of physical constants are available on the Web site of the National Institute of Standards and Technology (http://physics.nist.gov/constants).

C

Tables of Physical Chemical Data

Table C.1 List of Tables in the Chapters

Table 1.1	Second and Third Virial Coefficients at 298.15 K
Table 1.2	Critical Constants and Boyle Temperatures
Table 1.3	Van der Waals Constants
Table 2.1	Some Conjugate Pairs of Thermodynamic Variables
Table 2.2	Molar Heat Capacity at Constant Pressure as a Function of Temperature from 300 to 1800 K: $\overline{C}_P = \alpha + \beta T + \gamma T^2 + \delta T^3$
Table 3.1	Entropy of Sulfur Dioxide
Table 4.1	Criteria for Irreversibility and Reversibility for Processes Involving Only Pressure–Volume Work
Table 5.1	Pressures of $CO_2(g)$ in Equilibrium with $CaCO_3(s)$ and $CaO(s)$
Table 6.1	Vapor Pressures of Ice and Water
Table 6.2	Enthalpies and Entropies of Fusion at the Melting Point and Vaporization at the Boiling Point
Table 6.3	Vapor Pressures of Toluene and Benzene
Table 6.4	Henry's Law Constants ($K_i/10^9$ Pa) for Gases at 25 °C
Table 6.5	Activity Coefficients for Acetone–Ether Solutions at 30 °C
Table 6.6	Surface Tensions of Some Liquids
Table 7.1	Mean Ionic Activity Coefficients $\gamma_\pm$ in Water at 25 °C
Table 7.2	Standard Electrode Potentials at 25 °C
Table 8.1	Standard Thermodynamic Quantities for Acid Dissociation at 25 °C
Table 8.2	$pH_a - pH_c$ as a Function of Ionic Strength and Temperature
Table 8.3	pK's of Weak Acids at 298.15 K in Dilute Aqueous Solution at Three Ionic Strengths
Table 8.4	Standard Gibbs Energies of Formation, Standard Enthalpies of Formation, Charge Numbers, and Numbers of Hydrogen Atoms in Species at 298.15 K and Zero Ionic Strength
Table 8.5	Standard Transformed Gibbs Energies of Formation in kJ mol^{-1} of Reactants at 298.15 K, 0.25 M Ionic Strength, and Various pH Values
Table 8.6	Apparent Equilibrium Constants of Six Enzyme-Catalyzed Reactions at 298.15 K, 0.25 M Ionic Strength, and Five pH Values
Table 8.7	Changes in the Binding of Hydrogen Ions in Six Enzyme-Catalyzed Reactions at 298.15 K, 0.25 m Ionic Strength and 5 pH Values
Table 8.8	Standard Transformed Gibbs Energies of Hydrolysis at 25 °C, pH 7, pMg 4, and $I = 0.2$ M

(*continued*)

Table C.1 *(continued)*

(continued)

Table C.1 *(continued)*

Table 20.2	Electric Mobilities at 25 °C in Water at Infinite Dilution
Table 21.1	Intrinsic Viscosities of Macromolecules in Water at 25 °C
Table 21.2	Parameters in $[\eta] = KM^a$ for Polymer–Solvent Systems
Table 21.3	Physical Constants of Proteins at 20 °C in Water
Table 22.1	Relative Permittivities ϵ_r of Gases and Liquids
Table 22.2	Dipole Moments and Polarizabilities of Gaseous Molecules
Table 22.3	Magnetic Susceptibilities and Molar Magnetic Susceptibilities of Paramagnetic Substances at 293 K and 1 bar
Table 22.4	Magnetic Susceptibilities of Diamagnetic Substances at 293 K and 1 bar
Table 23.1	Crystallographic Point Groups
Table 23.2	Crystal Structure Data
Table 23.3	Covalent Radii for Atoms in pm
Table 23.4	Characteristics of Cubic Lattices

Table C.2 Chemical Thermodynamic Properties at 298.15 K and 1 bar[a]

Substance	$\Delta_f H^\circ$ / kJ mol^{-1}	$\Delta_f G^\circ$ / kJ mol^{-1}	$\overline{S}^\circ$ / J K^{-1} mol^{-1}	$\overline{C}_P^\circ$ / J K^{-1} mol^{-1}
O(g)	249.170	231.731	161.055	21.912
O_2(g)	0	0	205.138	29.355
O_3(g)	142.7	163.2	238.93	39.20
H(g)	217.965	203.247	114.713	20.784
H^+(g)	1536.202			
H^+(ao)	0	0	0	0
H_2(g)	0	0	130.684	28.824
OH(g)	38.95	34.23	183.745	29.886
OH^-(ao)	−229.994	−157.244	−10.75	−148.5
H_2O(l)	−285.830	−237.129	69.91	75.291
H_2O(g)	−241.818	−228.572	188.825	33.577
H_2O_2(l)	−187.78	−120.35	109.6	89.1
He(g)	0	0	126.150	20.786
Ne(g)	0	0	146.328	20.786
Ar(g)	0	0	154.843	20.786
Kr(g)	0	0	164.082	20.786
Xe(g)	0	0	169.683	20.786
F(g)	78.99	61.91	158.754	22.744
F^-(ao)	−332.63	−278.79	−13.8	−106.7

(continued)

[a] The values in Table C.2 are from The NBS Tables of Chemical Thermodynamic Properties (1982). The standard state pressure is 1 bar (0.1 MPa). The compounds are in the order of elements used in these tables. For the elements represented in Table C.2, this order is O, H, He, F, Cl, Br, I, S, N, P, C, Pb, Al, Zn, Cd, Hg, Cu, Ag, Fe, Ti, Mg, Ca, Li, Na, K, Rb, and Cs. The standard state for a strong electrolyte in aqueous solution is the ideal solution at unit mean molality (unit activity). The thermodynamic properties of the completely dissociated electrolyte are designated by ai. The thermodynamic properties of undissociated molecules in water are designated by ao. The properties of organic substances with more than two carbon atoms are from D. R. Stull, E. F. Westrum, and G. C. Sinke, *The Chemical Thermodynamics of Organic Compounds* (Hoboken, NJ: Wiley, 1969). The NBS Tables of Chemical Thermodynamic Properties have been published as a supplement to Volume II (1982) of the *Journal of Physical and Chemical Reference Data* and may be ordered from the American Chemical Society, 1155 Sixteenth St., NW, Washington, DC 20036. The conversion to the new standard state pressure is described by R. D. Freeman, *J. Chem. Educ.* **62**:681 (1985).

Table C.2 *(continued)*

Substance	$\Delta_f H°$ kJ mol^{-1}	$\Delta_f G°$ kJ mol^{-1}	$\overline{S}°$ J K^{-1} mol^{-1}	$\overline{C}_P°$ J K^{-1} mol^{-1}
F_2(g)	0	0	202.78	31.30
HF(g)	−271.1	−273.2	173.779	29.133
Cl(g)	121.679	105.680	165.198	21.840
Cl^-(ao)	−167.159	−131.228	56.5	−136.4
Cl_2(g)	0	0	223.066	33.907
ClO_4^-(ao)	−129.33	−8.52	182.0	
HCl(g)	−92.307	−95.299	186.908	29.12
HCl(ai)	−167.159	−131.228	56.5	−136.4
HCl in 100H_2O	−165.925			
HCl in 200H_2O	−166.272			
Br(g)	111.884	82.396	175.022	20.786
Br^-(ao)	−121.55	−103.96	82.4	−141.8
Br_2(l)	0	0	152.231	75.689
Br_2(g)	30.907	3.110	245.463	36.02
HBr(g)	−36.40	−53.45	198.695	29.142
I(g)	106.838	70.250	180.791	20.786
I^-(ao)	−55.19	−51.57	111.3	−142.3
I_2(cr)	0	0	116.135	54.438
I_2(g)	62.438	19.317	260.69	36.90
HI(g)	26.48	1.70	206.594	29.158
S(rhombic)	0	0	31.80	22.64
S(monoclinic)	0.33	0.1	32.6	23.6
S(g)	278.805	238.250	167.821	23.673
S_2(g)	128.37	79.30	228.18	32.47
S^{2-}(ao)	33.1	85.8	−14.6	
SO_2(g)	−296.830	−300.194	248.22	39.87
SO_3(g)	−395.72	−371.06	256.76	50.67
SO_4^{2-}(ao)	−909.27	−744.53	2.01	−293
HS^-(ai)	−17.6	12.08	62.8	
H_2S(g)	−20.63	−33.56	205.79	34.23
H_2SO_4(l)	−813.989	−690.003	156.904	138.91
H_2SO_4(ai)	−909.27	−744.53	20.1	−293
N(g)	472.704	455.563	153.298	20.786
N_2(g)	0	0	191.61	29.125
NO(g)	90.25	86.57	210.761	29.844
NO_2(g)	33.18	51.31	240.06	37.20
NO_3^-(ao)	−205.0	−108.74	146.4	−86.6
N_2O(g)	82.05	104.20	219.85	38.45
N_2O_4(l)	−19.50	97.54	209.2	142.7
N_2O_4(g)	9.16	97.89	304.29	77.28
NH_3(g)	−46.11	−16.45	192.45	35.06
NH_3(ao)	−80.29	−26.50	111.3	
NH_4^+(ao)	−132.51	−79.31	113.4	79.9
HNO_3(l)	−174.10	−80.71	155.60	109.87
HNO_3(ai)	−207.36	−111.25	146.4	−86.6
NH_4OH(ao)	−366.121	−263.65	181.2	
P(s, white)	0	0	41.09	23.840
P(g)	314.64	278.25	163.193	20.786

(continued)

Table C.2 *(continued)*

Substance	$\Delta_f H^\circ$	$\Delta_f G^\circ$	$\overline{S}^\circ$	$\overline{C}_P^\circ$
	kJ mol^{-1}	kJ mol^{-1}	J K^{-1} mol^{-1}	J K^{-1} mol^{-1}
P_2(g)	144.3	103.7	218.129	32.05
P_4(g)	58.91	24.44	279.98	67.15
PCl_3(g)	−287.0	−267.8	311.78	71.84
PCl_5(g)	−374.9	−305.0	364.58	112.8
C(graphite)	0	0	5.74	8.527
C(diamond)	1.895	2.900	2.377	6.113
C(g)	716.682	671.257	158.096	20.838
C_2(g)	0	−0.0330	144.960	29.196
CO(g)	−110.525	−137.168	197.674	29.142
CO_2(g)	−393.509	−394.359	213.74	37.11
CO_2(ao)	−413.80	−385.98	117.6	
CO_3^{2-}(ao)	−677.14	−527.81	−56.9	
CH(g)	595.8			
CH_2(g)	392.0			
CH_3(g)	138.9			
CH_4(g)	−74.81	−50.72	186.264	35.309
C_2H_2(g)	226.73	209.20	200.94	43.93
C_2H_4(g)	52.26	68.15	219.56	43.56
C_2H_6(g)	−84.68	−32.82	229.60	52.63
HCO_3^-(ao)	−691.99	−586.77	91.2	
HCHO(g)	−117	−113	218.77	35.40
HCO_2H(l)	−424.72	−361.35	128.95	99.04
H_2CO_3(ao)	−699.65	−623.08	187.4	
CH_3OH(l)	−238.66	−166.27	126.8	81.6
CH_3OH(g)	−200.66	−161.96	239.81	43.89
$CH_3CO_2^-$(ao)	−486.01	−369.31	86.6	−6.3
C_2H_4O(l, ethylene oxide)	−77.82	−11.76	153.85	87.95
CH_3CHO(l)	−192.30	−128.12	160.2	
CH_3CO_2H(l)	−484.5	−389.9	159.8	124.3
CH_3CO_2H(ao)	−485.76	−396.46	178.7	
C_2H_5OH(l)	−277.69	−174.78	160.7	111.46
C_2H_5OH(g)	−235.10	−168.49	282.70	65.44
$(CH_3)_2O$(g)	−184.05	−112.59	266.38	64.39
C_3H_6(g, propene)	20.42	62.78	267.05	63.89
C_3H_6(g, cyclopropane)	53.30	104.45	237.55	55.94
C_3H_8(g, propane)	−103.89	−23.38	270.02	73.51
C_4H_8(g, 1-butene)	−0.13	71.39	305.71	85.65
C_4H_8(g, 2-butene, *cis*)	−6.99	65.95	300.94	78.91
C_4H_8(g, 2-butene, *trans*)	−11.17	63.06	296.59	87.82
C_4H_{10}(g, butane)	−126.15	−17.03	310.23	97.45
C_4H_{10}(g, isobutane)	−134.52	−20.76	294.75	96.82
C_6H_6(g)	82.93	129.72	269.31	81.67
C_6H_{12}(g, cyclohexane)	−123.14	31.91	298.35	106.27
C_6H_{14}(g, hexane)	−167.19	−0.07	388.51	143.09
C_7H_8(g, toluene)	50.00	122.10	320.77	103.64
C_8H_8(g, styrene)	147.22	213.89	345.21	122.09
C_8H_{10}(g, ethylbenzene)	29.79	130.70	360.56	128.41
C_8H_{18}(g, octane)	−208.45	16.64	466.84	188.87

(continued)

Table C.2 *(continued)*

Substance	$\Delta_f H°$ kJ mol^{-1}	$\Delta_f G°$ kJ mol^{-1}	$\overline{S}°$ J K^{-1} mol^{-1}	$\overline{C}_P°$ J K^{-1} mol^{-1}
Si(s)	0	0	18.83	20.00
SiO_2(s, alpha)	−910.94	−856.64	41.84	44.43
Sn(s, white)	0	0	51.55	26.99
Sn^{2+}(ao)	−8.8	−27.2	−17	
SnO(s)	−285.8	−256.9	56.5	44.31
SnO_2(s)	−580.7	−519.6	52.3	52.59
Pb(s)	0	0	64.81	26.44
Pb^{2+}(ao)	−1.7	−24.43	10.5	
PbO(s, yellow)	−217.32	−187.89	68.70	45.77
PbO_2(s)	−277.4	−217.33	68.6	64.64
Al(s)	0	0	28.33	24.35
Al(g)	326.4	285.7	164.54	21.38
Al_2O_3(s, alpha)	−1675.7	−1582.3	50.92	79.04
$AlCl_3$(s)	−704.2	−628.8	110.67	91.84
Zn(s)	0	0	41.63	25.40
Zn^{2+}(ao)	−153.89	−147.06	−112.1	46
ZnO(s)	−348.28	−318.30	43.64	40.25
Cd(s, gamma)	0	0	51.76	25.98
Cd^{2+}(ao)	−75.90	−77.612	−73.2	
CdO(s)	−258.2	−228.4	54.8	43.43
$CdSO_4 \cdot \frac{8}{3}H_2O$(s)	−1729.4	−1465.141	229.630	213.26
Hg(l)	0	0	76.02	27.983
Hg(g)	61.317	31.820	174.96	20.786
Hg^{2+}(ao)	171.1	164.40	−32.2	
HgO(s, red)	−90.83	−58.539	70.29	44.06
Hg_2Cl_2(s)	−265.22	−210.745	192.5	102
Cu(s)	0	0	33.150	24.435
Cu^{+}(ao)	71.67	49.98	40.6	
Cu^{2+}(ao)	64.77	65.49	−99.6	
Ag(s)	0	0	42.55	25.351
Ag^{+}(ao)	105.579	77.107	72.68	21.8
Ag_2O(s)	−31.05	−11.20	121.3	65.86
AgCl(s)	−127.068	−109.789	96.2	50.79
Fe(s)	0	0	27.28	25.10
Fe^{2+}(ao)	−89.1	−78.90	−137.7	
Fe^{3+}(ao)	−48.5	−4.7	−315.9	
Fe_2O_3(s, hematite)	−824.2	−742.2	87.40	103.85
Fe_3O_4(s, magnetite)	−1118.4	−1015.4	146.4	143.43
Ti(s)	0	0	30.63	25.02
TiO_2(s)	−939.7	−884.5	49.92	55.48
U(s)	0	0	50.21	27.665
UO_2(s)	−1084.9	−1031.7	77.03	63.60
UO_2^{2+}(ao)	−1019.6	−953.5	−97.5	
UO_3(s, gamma)	−1223.8	−1145.9	96.11	81.67
Mg(s)	0	0	32.68	24.89
Mg(g)	147.70	113.10	148.650	20.786
Mg^{2+}(ao)	−466.85	−454.8	−138.1	
MgO(s)	−601.70	−569.43	26.94	37.15

(continued)

Table C.2 *(continued)*

Substance	$\Delta_f H°$ kJ mol^{-1}	$\Delta_f G°$ kJ mol^{-1}	$\overline{S}°$ J K^{-1} mol^{-1}	$\overline{C}_P°$ J K^{-1} mol^{-1}
$MgCl_2$(ao)	−801.15	−717.1	−25.1	
Ca(s)	0	0	41.42	25.31
Ca(g)	178.2	144.3	154.884	20.786
Ca^{2+}(ao)	−542.83	−553.58	−53.1	
CaO(s)	−635.09	−604.03	39.75	42.80
$CaCl_2$(ai)	−877.13	−816.01	59.8	
$CaCO_3$ (calcite)	−1206.92	−1128.79	92.9	81.88
$CaCO_3$ (aragonite)	−1207.13	−1127.75	88.7	81.25
Li(s)	0	0	29.12	24.77
Li^+(ao)	−278.49	−293.31	13.4	68.6
Na(s)	0	0	51.21	28.24
Na^+(ao)	−240.12	−261.905	59.0	46.4
NaOH(s)	−425.609	−379.494	64.455	59.54
NaOH(ai)	−470.114	−419.150	48.1	−102.1
NaOH in $100H_2O$	−469.646			
NaOH in $200H_2O$	−469.608			
NaCl(s)	−411.153	−384.138	72.13	50.50
NaCl(ai)	−407.27	−393.133	115.5	−90.0
NaCl in $100H_2O$	−407.066			
NaCl in $200H_2O$	−406.923			
K(s)	0	0	64.18	29.58
K^+(ao)	−252.38	−283.27	102.5	21.8
KOH(s)	−424.764	−379.08	78.9	64.9
KOH(ai)	−482.37	−440.50	91.6	−126.8
KOH in $100H_2O$	−481.637			
KOH in $200H_2O$	−481.742			
KCl(s)	−436.747	−409.14	82.59	51.30
KCl(ai)	−419.53	−414.49	159.0	−114.6
KCl in $100H_2O$	−419.320			
KCl in $200H_2O$	−419.191			
Rb(s)	0	0	76.78	10.148
Rb^+(ao)	−251.17	−283.98	121.50	
Cs(s)	0	0	85.23	32.17
Cs^+(ao)	−258.28	−292.02	133.05	−10.5

Table C.3 Chemical Thermodynamic Properties at Several Temperatures and 1 bar[a]

T/K	$\overline{C}_P°$ J K^{-1} mol^{-1}	$\overline{S}°$ J K^{-1} mol^{-1}	$\overline{H}_T° - \overline{H}_{298}°$ kJ mol^{-1}	$\Delta_f H°$ kJ mol^{-1}	$\Delta_f G°$ kJ mol^{-1}
			C (graphite)		
0	0.000	0.000	−1.051	0.000	0.000
298	8.517	5.740	0.000	0.000	0.000

(continued)

[a] M. Chase et al., JANAF Thermochemical Tables, 3rd ed., *J. Phys. Chem. Ref. Data* **14,** Supplements 1 and 2 (1985). The values on hydrocarbons, other than CH_4, are from D. R. Stull, E. F. Westrum, and G. C. Sinke, *The Chemical Thermodynamics of Organic Compounds.* Hoboken, NJ: Wiley, 1969.

Table C.3 *(continued)*

T/K	$\overline{C}_P^\circ$ J K^{-1} mol^{-1}	$\overline{S}^\circ$ J K^{-1} mol^{-1}	$\overline{H}_T^\circ - \overline{H}_{298}^\circ$ kJ mol^{-1}	$\Delta_f H^\circ$ kJ mol^{-1}	$\Delta_f G^\circ$ kJ mol^{-1}
500	14.623	11.662	2.365	0.000	0.000
1000	21.610	24.457	11.795	0.000	0.000
2000	24.094	40.771	35.525	0.000	0.000
3000	26.611	51.253	61.427	0.000	0.000
			C(g)		
0	0.000	0.000	−6.536	711.185	711.185
298	20.838	158.100	0.000	716.670	671.244
500	20.804	168.863	4.202	718.507	639.906
1000	20.791	183.278	14.600	719.475	560.654
2000	20.952	197.713	35.433	716.577	402.694
3000	21.621	206.322	56.689	711.932	246.723
			CH_4(g)		
0	0.000	0.000	−10.024	−66.911	−66.911
298	35.639	186.251	0.000	−74.873	−50.768
500	46.342	207.014	8.200	−80.802	−32.741
1000	71.795	247.549	38.179	−89.849	19.492
2000	94.399	305.853	123.592	−92.709	130.802
3000	101.389	345.690	222.076	−91.705	242.332
			CO(g)		
0	0.000	0.000	−8.671	−113.805	−113.805
298	29.142	197.653	0.000	−110.527	−137.163
500	29.794	212.831	5.931	−110.003	−155.414
1000	33.183	234.538	21.690	−111.983	−200.275
2000	36.250	258.714	56.744	−118.896	−286.034
3000	37.217	273.605	93.504	−127.457	−367.816
			CO_2(g)		
0	0.000	0.000	−9.364	−393.151	−393.151
298	37.129	213.795	0.000	−393.522	−394.389
500	44.627	234.901	8.305	−393.666	−394.939
1000	54.308	269.299	33.397	−394.623	−395.886
2000	60.350	309.293	91.439	−396.784	−396.333
3000	62.229	334.169	152.852	−400.111	−395.461
			C_2H_4(g)		
0	0.000	0.000	−10.518	60.986	60.986
298	43.886	219.330	0.000	52.467	68.421
500	63.477	246.215	10.668	46.641	80.933
1000	93.899	300.408	50.665	38.183	119.122
			C_2H_6(g)		
298	52.63	229.60	0.00	−84.68	−32.86
500	78.07	262.91	13.22	−93.89	4.96
1000	122.72	332.28	64.56	−105.77	109.55
			C_4H_{10}(g, *n*-butane)		
298	97.45	310.23	0.00	−126.15	−17.02
500	147.86	372.90	24.94	−140.21	61.10
1000	226.86	502.86	120.96	−155.85	270.31

(continued)

Table C.3 *(continued)*

T/K	$\overline{C}_P^\circ$ J K^{-1} mol^{-1}	$\overline{S}^\circ$ J K^{-1} mol^{-1}	$\overline{H}_T^\circ - \overline{H}_{298}^\circ$ kJ mol^{-1}	$\Delta_f H^\circ$ kJ mol^{-1}	$\Delta_f G^\circ$ kJ mol^{-1}
			C_6H_6(g)		
298	81.67	269.31	0.00	82.93	129.73
500	137.24	325.42	22.43	73.39	164.29
1000	209.87	446.71	112.01	62.01	260.76
			CH_3OH(g)		
298	43.89	239.81	0.00	−201.17	−162.46
500	59.50	266.13	10.42	−207.94	−134.27
1000	89.45	317.59	48.41	−217.28	−56.16
			Cl(g)		
0	0.000	0.000	−6.272	119.621	119.621
298	21.838	165.189	0.000	121.302	105.306
500	22.744	176.752	4.522	122.272	94.203
1000	22.233	192.430	15.815	124.334	65.288
2000	21.341	207.505	37.512	127.058	5.081
3000	21.063	216.096	58.690	128.649	−56.297
			HCl(g)		
0	0.000	0.000	−8.640	−92.127	−92.127
298	29.136	186.901	0.000	−92.312	−95.300
500	29.304	201.989	5.892	−92.913	−97.166
1000	31.628	222.903	21.046	−94.388	−100.799
2000	35.600	246.246	54.953	−95.590	−106.631
3000	37.243	261.033	91.478	−96.547	−111.968
			Cl_2(g)		
0	0.000	0.000	−9.180	0.000	0.000
298	33.949	223.079	0.000	0.000	0.000
500	36.064	241.228	7.104	0.000	0.000
1000	37.438	266.764	25.565	0.000	0.000
2000	38.428	293.033	63.512	0.000	0.000
3000	40.075	308.894	102.686	0.000	0.000
			H(g)		
0	0.000	0.000	−6.197	216.035	216.035
298	20.786	114.716	0.000	217.999	203.278
500	20.786	125.463	4.196	219.254	192.957
1000	20.786	139.871	14.589	222.248	165.485
2000	20.786	154.278	35.375	226.898	106.760
3000	20.786	162.706	56.161	229.790	46.007
			H^+(g)		
0	0.000	0.000	−6.197	1528.085	
298	20.786	108.946	0.000	1536.246	1516.990
500	20.786	119.693	4.196	1541.697	1502.422
1000	20.786	134.101	14.589	1555.084	1457.958
2000	20.786	148.509	35.375	1580.520	1350.840
3000	20.786	156.937	56.161	1604.198	1230.818

(continued)

Table C.3 *(continued)*

	$\overline{C}_P^\circ$	$\overline{S}^\circ$	$\overline{H}_T^\circ - \overline{H}_{298}^\circ$	$\Delta_f H^\circ$	$\Delta_f G^\circ$
T/K	J K^{-1} mol^{-1}	J K^{-1} mol^{-1}	kJ mol^{-1}	kJ mol^{-1}	kJ mol^{-1}
			$H^-(g)$		
0	0.000	0.000	−6.197	143.266	
298	20.786	108.960	0.000	139.032	132.282
500	20.786	119.707	4.196	136.091	128.535
1000	20.786	134.114	14.589	128.692	123.819
2000	20.786	148.522	32.375	112.557	125.012
3000	20.786	156.950	56.161	94.662	135.055
			$HI(g)$		
0	0.000	0.000	−8.656	28.535	28.535
298	29.156	206.589	0.000	26.359	1.560
500	29.736	221.760	5.928	−5.622	−10.088
1000	33.135	243.404	21.641	−6.754	−14.006
2000	36.623	267.680	56.863	−7.589	−21.009
3000	37.918	282.805	94.210	−10.489	−27.114
			$H_2(g)$		
0	0.000	0.000	−8.467	0.000	0.000
298	28.836	130.680	0.000	0.000	0.000
500	29.260	145.737	5.883	0.000	0.000
1000	30.205	166.216	20.680	0.000	0.000
2000	34.280	188.418	52.951	0.000	0.000
3000	37.087	202.891	88.740	0.000	0.000
			$H_2O(g)$		
0	0.000	0.000	−9.904	−238.921	−238.921
298	33.590	188.834	0.000	−241.826	−228.582
500	35.226	206.534	6.925	−243.826	−219.051
1000	41.268	232.738	26.000	−247.857	−192.590
2000	51.180	264.769	72.790	−251.575	−135.528
3000	55.748	286.504	126.549	−253.024	−77.163
			$I(g)$		
0	0.000	0.000	−6.197	107.164	107.164
298	20.786	180.786	0.000	106.762	70.174
500	20.786	191.533	4.196	75.990	50.203
1000	20.795	205.942	14.589	76.937	24.039
2000	21.308	220.461	35.566	77.992	−29.410
3000	22.191	229.274	57.332	77.406	−82.995
			$I_2(g)$		
0	0.000	0.000	−10.116	65.504	65.504
298	36.887	260.685	0.000	62.421	19.325
500	37.464	279.920	7.515	0.000	0.000
1000	38.081	306.087	26.407	0.000	0.000
2000	42.748	332.521	66.250	0.000	0.000
3000	44.897	351.615	110.955	0.000	0.000

(continued)

Table C.3 *(continued)*

T/K	$\overline{C}_P^\circ$ J K^{-1} mol^{-1}	$\overline{S}^\circ$ J K^{-1} mol^{-1}	$\overline{H}_T^\circ - \overline{H}_{298}^\circ$ kJ mol^{-1}	$\Delta_f H^\circ$ kJ mol^{-1}	$\Delta_f G^\circ$ kJ mol^{-1}
			N(g)		
0	0.000	0.000	−6.197	470.820	470.820
298	20.786	153.300	0.000	472.683	455.540
500	20.786	164.047	4.196	473.923	443.584
1000	20.786	178.454	14.589	476.540	412.171
2000	20.790	192.863	35.375	479.990	346.339
3000	20.963	201.311	56.218	482.543	278.946
			NO(g)		
0	0.000	0.000	−9.192	89.775	89.775
298	29.845	210.758	0.000	90.291	86.606
500	30.486	226.263	6.059	90.352	84.079
1000	33.987	248.536	22.229	90.437	77.775
2000	36.647	273.128	57.859	90.494	65.060
3000	37.466	288.165	94.973	89.899	52.439
			NO_2(g)		
0	0.000	0.000	−10.186	35.927	35.927
298	36.974	240.034	0.000	33.095	51.258
500	43.206	260.638	8.099	32.154	63.867
1000	52.166	293.889	32.344	32.005	95.779
2000	56.441	331.788	87.259	33.111	159.106
3000	57.394	354.889	144.267	32.992	222.058
			N_2(g)		
0	0.000	0.000	−8.670	0.000	0.000
298	29.124	191.609	0.000	0.000	0.000
500	29.580	206.739	5.911	0.000	0.000
1000	32.697	228.170	21.463	0.000	0.000
2000	35.971	252.074	56.137	0.000	0.000
3000	37.030	266.891	92.715	0.000	0.000
			N_2O_4(g)		
0	0.000	0.000	−16.398	18.718	18.718
298	77.256	304.376	0.000	9.079	97.787
500	97.204	349.446	17.769	8.769	158.109
1000	119.208	425.106	72.978	15.189	305.410
2000	129.030	511.743	198.518	33.110	588.764
3000	131.200	564.555	328.840	49.178	862.983
			NH_3(g)		
0	0.000	0.000	−10.045	−38.907	−38.907
298	35.652	192.774	0.000	−45.898	−16.367
500	42.048	212.659	7.819	−49.857	4.800
1000	56.491	246.486	32.637	−55.013	61.910
2000	72.833	291.525	98.561	−54.833	179.447
3000	78.902	322.409	174.933	−50.433	295.689

(continued)

Table C.3 (*continued*)

T/K	$\overline{C}_P^\circ$ J K^{-1} mol^{-1}	$\overline{S}^\circ$ J K^{-1} mol^{-1}	$\overline{H}_T^\circ - \overline{H}_{298}^\circ$ kJ mol^{-1}	$\Delta_f H^\circ$ kJ mol^{-1}	$\Delta_f G^\circ$ kJ mol^{-1}
			O(g)		
0	0.000	0.000	−6.725	246.790	246.790
298	21.911	161.058	0.000	249.173	231.736
500	21.257	172.197	4.343	250.474	219.549
1000	20.915	186.790	14.860	252.682	187.681
2000	20.826	201.247	35.713	255.299	121.552
3000	20.937	209.704	56.574	256.741	54.327
			O^-(g)		
0	0.000	0.000	−6.571	105.814	105.814
298	21.692	157.790	0.000	101.846	91.638
500	21.184	168.860	4.318	98.926	85.532
1000	20.899	183.426	14.817	90.723	75.219
2000	20.816	197.878	35.661	72.545	66.619
3000	20.800	206.314	56.467	53.146	67.810
			O_2(g)		
0	0.000	0.000	−8.683	0.000	0.000
298	29.376	205.147	0.000	0.000	0.000
500	31.091	220.693	6.084	0.000	0.000
1000	34.870	243.578	22.703	0.000	0.000
2000	37.741	268.748	59.175	0.000	0.000
3000	39.884	284.466	98.013	0.000	0.000
			e^-(g)		
0	0.000	0.000	−6.197	0.000	0.000
298	20.786	20.979	0.000	0.000	0.000
500	20.786	31.725	4.196	0.000	0.000
1000	20.786	46.133	14.584	0.000	0.000
2000	20.786	60.541	35.375	0.000	0.000
3000	20.786	68.969	56.161	0.000	0.000

Table C.4 Ion Product of Water at a Series of Temperatures

°C	$K_w \times 10^{14}$	°C	$K_w \times 10^{14}$
0	0.114	35	2.09
5	0.185	40	2.92
10	0.292	45	4.02
15	0.450	50	5.47
20	0.681	55	7.20
25	1.006	60	9.61
30	1.47		

Table C.5 pK_1 and pK_2 for Acids in Water at 25 °C[a]

Acid	pK_1	pK_2
Oxalic	1.271	4.266
Phosphoric	2.148	7.198
Glycine (protonated)	2.350	9.780
p-Aminobenzoic	2.45	4.85
Malonic	2.855	5.606
Chloroacetic	2.865	
Tartaric	3.033	4.366
Citric	3.128	4.761
2,6-Dinitrophenol	3.72	
Formic	3.739	
Glycolic	3.831	
Lactic	3.860	
2,4-Dinitrophenol	4.09	
Benzoic	4.201	
Succinic	4.207	5.638
Anilinium ion	4.60	
Acetic	4.756	
Propionic	4.874	
Carbonic	6.352	10.329
Triethanolammonium ion	7.762	
Tris(hydroxymethyl)-methylammonium ion	8.075	
Hydrogen cyanide	9.216	
NH_4^+	9.245	
Phenol	9.998	

[a] These values are for a standard state of $m° = 1$ mol kg^{-1} and apply at zero ionic strength.

Table C.6 Atomic Properties[a]

Z	*Atom*	*Configuration*	*Ground-State Term Symbol*	*First Ionization Energy*, eV	*Orbital Radius*, pm
37	Rb	[Kr]5s	$^2S_{1/2}$	4.176	228.7
38	Sr	[Kr]$5s^2$	$^1S_0$	5.692	183.6
39	Y	[Kr]$5s^24d$	$^2D_{3/2}$	6.6	169.3
40	Zr	[Kr]$5s^24d^2$	$^3F_2$	6.95	159.3
41	Nb	[Kr]$5s4d^4$	$^6D_{1/2}$	6.77	158.9
42	Mo	[Kr]$5s4d^5$	$^7S_3$	7.18	152.0
43	Tc	[Kr]$5s^24d^5$	$^6S_{5/2}$	—	139.1
44	Ru	[Kr]$5s4d^7$	$^5F_5$	7.5	141.0
45	Rh	[Kr]$5s4d^8$	$^4F_{9/2}$	7.7	136.4
46	Pd	[Kr]$4d^{10}$	$^1S_0$	8.33	56.7
47	Ag	[Kr]$5s4d^{10}$	$^2S_{1/2}$	7.574	128.6
48	Cd	[Kr]$5s^24d^{10}$	$^1S_0$	8.991	118.4
49	In	[Kr]$5s^24d^{10}5p$	$^2P_{1/2}$	5.785	138.2
50	Sn	[Kr]$5s^24d^{10}5p^2$	$^3P_0$	7.332	124.0
51	Sb	[Kr]$5s^24d^{10}5p^3$	$^4S_{3/2}$	8.64	119.3
52	Te	[Kr]$5s^24d^{10}5p^4$	$^3P_2$	9.01	111.1
53	I	[Kr]$5s^24d^{10}5p^5$	$^2P_{3/2}$	10.44	104.4
54	Xe	[Kr]$5s^24d^{10}5p^6$	$^1S_0$	12.127	98.6
55	Cs	[Xe]6s	$^2S_{1/2}$	3.893	251.8
56	Ba	[Xe]$6s^2$	$^1S_0$	5.210	206.0
57	La	[Xe]$6s^25d$	$^2D_{3/2}$	5.61	191.5
58	Ce	[Xe]$(6s^24f5d)$	$(^3H_5)$	6.91	197.8
59	Pr	[Xe]$(6s^24f^3)$	$(^4I_{9/2})$	5.76	194.2
60	Nd	[Xe]$6s^24f^4$	$^5I_4$	6.31	191.2
61	Pm	[Xe]$(6s^24f^5)$	$(^6H_{5/2})$	—	188.2
62	Sm	[Xe]$6s^24f^6$	$^7F_0$	5.6	185.4
63	Eu	[Xe]$6s^24f^7$	$^8S_{7/2}$	5.67	182.6
64	Gd	[Xe]$6s^24f^75d$	$^9D_2$	6.16	171.3
65	Tb	[Xe]$(6s^24f^9)$	$(^6H_{15/2})$	6.74	177.5
66	Dy	[Xe]$(6s^24f^{10})$	$(^5I_8)$	6.82	175.0
67	Ho	[Xe]$(6s^24f^{11})$	$(^4I_{15/2})$	—	172.7
68	Er	[Xe]$(6s^24f^{12})$	$(^3H_6)$	—	170.3
69	Tm	[Xe]$6s^24f^{13}$	$^2F_{7/2}$	—	168.1
70	Yb	[Xe]$6s^24f^{14}$	$^1S_0$	6.2	165.8
71	Lu	[Xe]$6s^24f^{14}5d$	$^2D_{3/2}$	5.0	155.3
72	Hf	[Xe]$6s^24f^{14}5d^2$	$^3F_2$	5.5	147.6
73	Ta	[Xe]$6s^24f^{14}5d^3$	$^4F_{3/2}$	7.88	141.3
74	W	[Xe]$6s^24f^{14}5d^4$	$^5D_0$	7.98	136.0
75	Re	[Xe]$6s^24f^{14}5d^5$	$^6S_{5/2}$	7.87	131.0
76	Os	[Xe]$6s^24f^{14}5d^6$	$^5D_4$	8.7	126.6
77	Ir	[Xe]$6s^24f^{14}5d^7$	$^4F_{9/2}$	9.2	122.7
78	Pt	[Xe]$6s4f^{14}5d^9$	$^3D_3$	8.96	122.1
79	Au	[Xe]$6s4f^{14}5d^{10}$	$^2S_{1/2}$	9.223	118.7
80	Hg	[Xe]$6s^24f^{14}5d^{10}$	$^1S_0$	10.434	112.6

(continued)

[a] This table is a continuation of Table 10.3. From J. T. Waber and D. T. Cromer, *J. Chem. Phys.* **42**:4116 (1965).

Table C.6 *(continued)*

Z	*Atom*	*Configuration*	*Ground-State Term Symbol*	*First Ionization Energy*, eV	*Orbital Radius*, pm
81	Tl	$[Xe]6s^2 4f^{14} 5d^{10} 6p$	$^2P_{1/2}$	6.106	131.9
82	Pb	$[Xe]6s^2 4f^{14} 5d^{10} 6p^2$	$^3P_0$	7.415	121.5
83	Bi	$[Xe]6s^2 4f^{14} 5d^{10} 6p^3$	$^4S_{3/2}$	7.287	129.5
84	Po	$[Xe]6s^2 4f^{14} 5d^{10} 6p^4$	$^3P_2$	8.43	121.2
85	At	$[Xe](6s^2 4f^{14} 5d^{10} 6p^5)$	$(^2P_{3/2})$	—	114.6
86	Rn	$[Xe]6s^2 4f^{14} 5d^{10} 6p^6$	$^1S_0$	10.746	109.0
87	Fr	$[Rn](7s)$	$(^2S_{1/2})$	—	244.7
88	Ra	$[Rn]7s^2$	$^1S_0$	5.277	204.2
89	Ac	$[Rn]7s^2 6d$	$^2D_{3/2}$	6.9	189.5
90	Th	$[Rn]7s^2 6d^2$	$^3F_2$	—	178.8
91	Pa	$[Rn](7s^2 5f^2 6d)$	$(^4K_{11/12})$	—	180.4
92	U	$[Rn]7s^2 5f^3 6d$	$^5L_6$	4	177.5
93	Np	$[Rn](7s^2 5f^4 6d)$	$(^6L_{11/12})$	—	174.1
94	Pu	$[Rn](7s^2 5f^6)$	$(^7F_0)$	—	178.4
95	Am	$[Rn](7s^2 5f^7)$	$(^8S_{7/2})$	—	175.7
96	Cm	$[Rn](7s^2 5f^7 6d)$	$(^9D_2)$	—	165.7
97	Bk	$[Rn](7s^2 5f^9)$	$(^6H_{15/2})$	—	162.6
98	Cf	$[Rn](7s^2 5f^{10})$	$(^5I_8)$	—	159.8
99	Es	$[Rn](7s^2 5f^{11})$	$(^4I_{15/2})$	—	157.6
100	Fm	$[Rn](7s^2 5f^{12})$	$(^3H_6)$	—	155.7
101	Md	$[Rn](7s^2 5f^{13})$	$(^2F_{7/2})$	—	152.7
102	No	$[Rn](7s^2 5f^{14})$	$(^1S_0)$	—	158.1
103	Lr	$[Rn](7s^2 5f^{14} 6d)$	$(^2D_{3/2})$	—	—

Table C.7 Ionization Energies E_i of Atoms and Ions in eV

H	13.597	Ne	21.56	K^+	31.71
He	24.587	Ne^+	41.08	Ca	6.11
He^+	54.42	Ne^{2+}	63.45	Ca^+	11.87
Li	5.39	Na	5.14	Sc	6.54
Li^+	75.60	Na^+	47.30	Ti	6.82
Be	9.32	Mg	7.65	V	6.74
B	8.30	Mg^+	15.04	Cr	6.77
C	11.26	Al	5.99	Mn	7.43
C^+	24.38	Al^+	18.83	Fe	7.87
C^{2+}	47.88	Si	8.15	Fe^+	16.18
N	14.53	P	10.49	Fe^{2+}	30.65
N^+	29.60	S	10.36	Co	7.86
N^{2+}	47.44	S^+	23.41	Co^+	17.06
O	13.62	Cl	12.97	Co^{2+}	33.59
O^+	35.12	Ar	15.75	Ni	7.64
O^{2+}	54.9	Ar^+	27.63	Ni^+	18.15
F	17.42	K	4.34	Ni^{2+}	35.20

Table C.8 Bond Lengths and Angles in Polyatomic Molecules

Molecule	*Bond*	*Length*/pm	*Bond*	*Angle*/°
CO_2	C—O	115.98	O—C—O	180
SO_2	S—O	143.21	O—S—O	119.5
SO_3	S—O	143	O—S—O	120
H_2O	O—H	95.8	H—O—H	104.45
H_2S	S—H	134.55	H—S—H	93.3
NH_3	N—H	100.8	H—N—H	107.3
CH_4	C—H	109.3	H—C—H	109.5
C_2H_2	C—H	109.3	H—C—H	180
	C—C	120.3		
C_2H_4	C—H	108.4	H—C—H	115.5
	C—C	133.2		
C_2H_6	C—H	109.3	H—C—H	109.75
	C—C	153.4		

Source: G. W. C. Kaye and T. H. Laby, *Tables of Physical and Chemical Constants*, 1973 ©. Reprinted with permission of Pearson Education, Inc., Upper Saddle River, NJ.

D

Mathematical Relations

D.1 LOGARITHMS AND EXPONENTIALS

The natural logarithm of a number x is the power to which $\mathrm{e} = 2.718281\ldots$ must be raised to yield x. This definition and the properties of natural logarithms are summarized by

$$\begin{aligned} \mathrm{e}^{\ln x} &= x \\ \ln(xy) &= \ln x + \ln y \\ \ln(x/y) &= \ln x - \ln y \\ \ln x^y &= y \ln x \end{aligned}$$

The base 10 logarithm of a number x is the power to which 10 must be raised to yield x.

$$\begin{aligned} 10^{\log x} &= x \\ \ln x &= \ln(10) \log x = 2.303 \log x \end{aligned}$$

Exponential functions have the following properties:

$$a^{m+n} = a^m a^n$$
$$a^m/a^n = a^{m-n}$$
$$(a^m)^n = a^{mn}$$

D.2 SERIES

It is often of interest to see the form of an equation when one of the quantities becomes indefinitely small or indefinitely large. Since most functions we deal with can be expressed by infinite series, the series expression can be used with higher-order terms omitted. The series expression for a function f can be calculated with the Maclaurin series:

$$f(x) = f(0) + \left(\frac{\mathrm{d}f}{\mathrm{d}x}\right)_{x=0} x + \frac{1}{2!}\left(\frac{\mathrm{d}^2 f}{\mathrm{d}x^2}\right)_{x=0} x^2 + \cdots$$

The following infinite series are examples of Maclaurin series:

$$\sin x = x - \frac{x^3}{3!} + \frac{x^5}{5!} + \cdots \qquad [\text{all } x]$$
$$\cos x = 1 - \frac{x^2}{2!} + \frac{x^4}{4!} - \cdots \qquad [\text{all } x]$$
$$\mathrm{e}^x = 1 + x + \frac{x^2}{2!} + \frac{x^3}{3!} + \cdots \qquad [\text{all } x]$$
$$\ln(1+x) = x - \frac{x^2}{2} + \frac{x^3}{3} - \frac{x^4}{4} + \cdots \qquad [x^2 < 1]$$
$$(1+x)^{-1} = 1 - x + x^2 - x^3 + \cdots \qquad [x^2 < 1]$$
$$(1-x)^{-1} = 1 + x + x^2 + x^3 + \cdots \qquad [x^2 < 1]$$
$$(1-x)^{-2} = 1 + 2x + 3x^2 + 4x^3 + \cdots \qquad [x^2 < 1]$$
$$(1+x)^{1/2} = 1 + \frac{x}{2} - \frac{x^2}{8} + \frac{x^3}{16} - \cdots \qquad [x^2 < 1]$$

The series $(1 + x)^n$ is referred to as the binomial series. If n is an integer, the series terminates after $(n + 1)$ terms, but when n is not an integer, the series is infinite:

$$(1+x)^n = 1 + nx + \frac{n(n-1)}{2!}x^2 + \frac{n(n-1)(n-2)}{3!}x^3 + \cdots \qquad [x^2 < 1]$$

A Maclaurin series is an expansion about the point $x = 0$. A Taylor series is an expansion about $x = x_0$. The Taylor series is

$$f(x) = f(x = x_0) + \left(\frac{\mathrm{d}f}{\mathrm{d}x}\right)_{x_0} (x - x_0) + \frac{1}{2!}\left(\frac{\mathrm{d}^2 f}{\mathrm{d}x^2}\right)_{x_0} (x - x_0)^2 + \cdots$$

D.3 CALCULUS

Some basic derivatives are

$$\frac{\mathrm{d}u^n}{\mathrm{d}x} = nu^{n-1}\frac{\mathrm{d}u}{\mathrm{d}x}$$

$$\frac{\mathrm{d}\,\mathrm{e}^u}{\mathrm{d}x} = \mathrm{e}^u\frac{\mathrm{d}u}{\mathrm{d}x}$$

$$\frac{\mathrm{d}\ln x}{\mathrm{d}x} = \frac{1}{x}$$

$$\frac{\mathrm{d}\sin x}{\mathrm{d}x} = \cos x$$

$$\frac{\mathrm{d}\cos x}{\mathrm{d}x} = -\sin x$$

$$\frac{\mathrm{d}z\,[y(x)]}{\mathrm{d}x} = \frac{\mathrm{d}z}{\mathrm{d}y}\frac{\mathrm{d}y}{\mathrm{d}x} \qquad \text{(chain rule)}$$

$$\left(\frac{\partial z}{\partial x}\right)_y\left(\frac{\partial x}{\partial y}\right)_z\left(\frac{\partial y}{\partial z}\right)_x = -1 \qquad \text{(cyclic rule)}$$

$$\frac{\mathrm{d}(u/v)}{\mathrm{d}x} = \frac{v(\mathrm{d}u/\mathrm{d}x) - u(\mathrm{d}v/\mathrm{d}x)}{v^2}$$

Some basic indefinite integrals are

$$\int x^\alpha \mathrm{d}x = \frac{1}{\alpha+1}x^{\alpha+1} \qquad (\alpha \neq -1)$$

$$\int \mathrm{e}^{ax}\,\mathrm{d}x = \frac{1}{a}\,\mathrm{e}^{ax}$$

$$\int \frac{\mathrm{d}x}{x} = \ln|x|$$

$$\int \ln ax\,\mathrm{d}x = x\ln ax - x$$

$$\int x^2\,\mathrm{e}^{bx}\,\mathrm{d}x = \mathrm{e}^{bx}\left(\frac{x^2}{b} - \frac{2x}{b^2} + \frac{2}{b^3}\right)$$

Some basic definite integrals are

$$\int_0^\infty \mathrm{e}^{-ax}\,\mathrm{d}x = \frac{1}{a}$$

$$\int_0^\infty x^n\,\mathrm{e}^{-qx}\,\mathrm{d}x = \frac{n!}{q^{n+1}} \qquad (n > -1, q > 0)$$

$$\int_0^\infty e^{-bx^2}\,dx = \frac{1}{2}\left(\frac{\pi}{b}\right)^{1/2}$$

$$\int_0^\infty x^{2n}e^{-bx^2}\,dx = \frac{1\cdot 3\cdots(2n-1)}{2^{n+1}}\left(\frac{\pi}{b^{2n+1}}\right)^{1/2} \qquad (n = 1, 2, 3, \ldots)$$

$$\int_t^\infty z^n e^{-az}\,dz = \frac{n!}{a^{n+1}}e^{-at}\left(1 + at + \frac{a^2t^2}{2!} + \cdots + \frac{a^n t^n}{n!}\right) \qquad (n = 0, 1, 2, \ldots)$$

$$\int_0^{\pi/2} \sin^2 nx\,dx = \int_0^{\pi/2} \cos^2 nx\,dx = \frac{\pi}{4} \qquad (n = 1, 2, 3, \ldots)$$

$$\int_0^{2\pi} \sin mx \sin nx\,dx = \int_0^{2\pi} \cos mx \cos nx\,dx = 0 \qquad (m \neq n)$$

Also see Table 17.1 for definite integrals.

D.4 SPHERICAL COORDINATES

The choice of a coordinate system is a matter of convenience. When a system has some kind of a natural center, as in the case of an atom, spherical coordinates are convenient, as indicated by Fig. D.4.1. The angle θ is the declination from the north pole, so $0 \leq \theta \leq \pi$. Since there is not a natural zero value for ϕ, the angle around the equator is measured from the x axis, as indicated in the figure, and $0 \leq \phi \leq 2\pi$. Since r is the distance from the origin, $0 \leq r \leq \infty$.

The Cartesian coordinates x, y, and z are related to the spherical coordinates r, θ, and ϕ by

$$x = r \sin\theta \cos\phi \qquad \text{(D.4.1)}$$

$$y = r \sin\theta \sin\phi \qquad \text{(D.4.2)}$$

$$z = r \cos\theta \qquad \text{(D.4.3)}$$

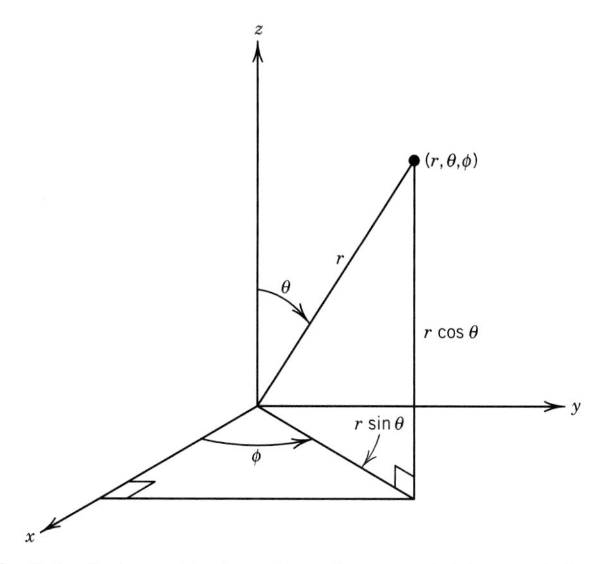

Figure D.4.1 Spherical coordinate system where a point is specified by r, θ, and ϕ.

It is readily shown that the spherical coordinates are related to the Cartesian coordinates by

$$r = (x^2 + y^2 + z^2)^{1/2} \tag{D.4.4}$$

$$\cos\theta = \frac{z}{(x^2 + y^2 + z^2)^{1/2}} \tag{D.4.5}$$

$$\tan\phi = \frac{y}{x} \tag{D.4.6}$$

Figure D.4.2 shows that the differential volume element in spherical coordinates is

$$dV = (r \sin\theta \, d\phi)(r \, d\theta) \, dr = r^2 \sin\theta \, dr \, d\theta \, d\phi \tag{D.4.7}$$

It also shows that in spherical coordinates the differential area is given by

$$dA = r^2 \sin\theta \, d\theta \, d\phi \tag{D.4.8}$$

The volume of a sphere of radius a is given by

$$V = \int_0^a r^2 \, dr \int_0^\pi \sin\theta \, d\theta \int_0^{2\pi} d\phi = \left(\frac{a^3}{3}\right)(2)(2\pi) = \frac{4\pi a^3}{3} \tag{D.4.9}$$

The surface area of a sphere of radius a is given by

$$A = a^2 \int_0^\pi \sin\theta \, d\theta \int_0^{2\pi} d\phi = a^2(2)(2\pi) = 4\pi r^2 \tag{D.4.10}$$

We can integrate a function $f(r, \theta, \phi)$ over the full range of these coordinates by use of

$$F = \int_0^\infty r^2 \, dr \int_0^\pi \sin\theta \, d\theta \int_0^{2\pi} d\phi \, f(r, \theta, \phi) \tag{D.4.11}$$

An example of this type of integral is the orthogonality relation of atomic wavefunctions.

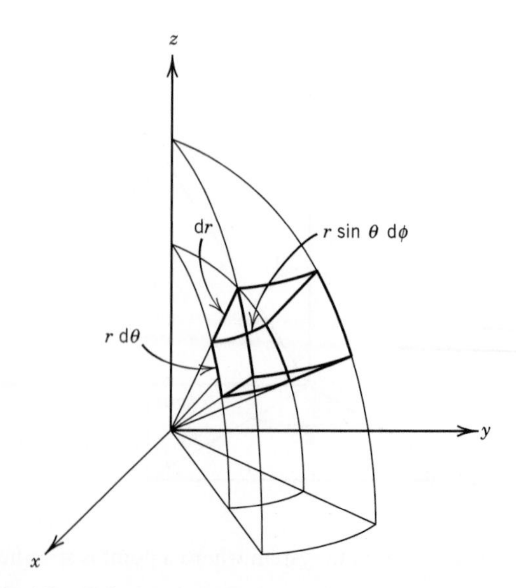

Figure D.4.2 Volume element in a spherical coordinate system.

D.5 LEGENDRE TRANSFORMS

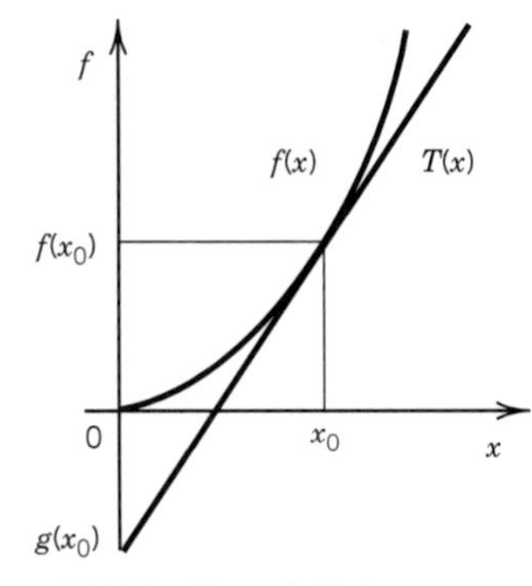

Figure D.5.1 Plot of $f(x)$ versus x.

The variables in a function can be changed by simply substituting an expression for a variable in terms of a new variable. For example, the thermodynamic temperature T in an equation can be replaced using $T = t + 273.15$ to obtain the equation written in terms of the Celsius temperature t. Another way to change variables involves defining a new property that depends on a derivative of the original function, rather than a new variable such as t in this example. This method, which is especially useful in thermodynamics, is referred to as a **Legendre transform.**

Consider a function $f(x)$ that is differentiable for all x; this function is plotted in Fig. D.5.1. The total differential of f is given by

$$\mathrm{d}f = \frac{\mathrm{d}f}{\mathrm{d}x}\mathrm{d}x = p(x)\,\mathrm{d}x \tag{D.5.1}$$

where the function $p(x)$ is the slope $f'(x)$ of $f(x)$ at every value of x. The objective of the Legendre transform is to find a function $g(p)$ of the new variable $p = f'(x)$ that is equivalent to $f(x)$. By equivalent we mean that $f(x)$ and $g(p)$ contain the same information; in short, $g(p)$ can be calculated from $f(x)$, and $f(x)$ can be calculated from $g(p)$. The new function $g(p)$ can be obtained by use of Fig. D.5.1. The value of p at any point along $f(x)$ is the slope $f'(x)$. It is evident from the figure that the equation for the tangent $T(x)$ at any point x_0 along the curve is

$$T(x) = f(x_0) + f'(x_0)(x - x_0) \tag{D.5.2}$$

The intersection of the tangent with the vertical axis is given by

$$g(x_0) = f(x_0) - x_0 f'(x_0) \tag{D.5.3}$$

The value of function g depends on x_0 and, for a general x_0,

$$g = f - xf'(x) = f - xp \tag{D.5.4}$$

The new function g is referred to as the Legendre transform of f, and this equation shows that g is obtained from f by subtracting xp, which is $xf'(x)$. This process can be generalized to functions of two or more variables.

In Chapter 4 we found that the internal energy U is a function of S and V. However, S and V may not be experimentally convenient variables. The following Legendre transform was used to define the Gibbs energy G:

$$G = U + PV - TS = U - V\left(\frac{\partial U}{\partial V}\right)_S - S\left(\frac{\partial U}{\partial S}\right)_V \tag{D.5.5}$$

As shown in Chapter 4, the Gibbs energy is a function of T and P, which are convenient independent variables for work in the laboratory.

D.6 DETERMINANTS

A determinant is a square array of numbers. Its value is defined as a certain sum of products of subsets of the elements. If the determinant has n rows and columns, each term in the sum will have n factors in it. For a determinant of order 2,

$$\begin{vmatrix} a_1 & b_1 \\ a_2 & b_2 \end{vmatrix} = a_1 b_2 - a_2 b_1$$

The value of a large determinant may be obtained by expanding by minors. For a determinant of order 3,

$$\begin{vmatrix} a_1 & b_1 & c_1 \\ a_2 & b_2 & c_2 \\ a_3 & b_3 & c_3 \end{vmatrix} = a_1 \begin{vmatrix} b_2 & c_2 \\ b_3 & c_3 \end{vmatrix} - b_1 \begin{vmatrix} a_2 & c_2 \\ a_3 & c_3 \end{vmatrix} + c_1 \begin{vmatrix} a_2 & b_2 \\ a_3 & b_3 \end{vmatrix}$$

$$= a_1b_2c_3 + a_2b_3c_1 + a_3b_1c_2 - a_3b_2c_1 - a_2b_1c_3 - a_1b_3c_2$$

Determinants of higher order are defined by an analogous row (or column) expansion.

Simultaneous linear equations can be solved using determinants. For example, consider the set

$$\begin{aligned} a_{11}x + a_{12}y + a_{13}z &= c_1 \\ a_{21}x + a_{22}y + a_{23}z &= c_2 \\ a_{31}x + a_{32}y + a_{33}z &= c_3 \end{aligned}$$

The determinant of the coefficients of x, y, and z is

$$D = \begin{vmatrix} a_{11} & a_{12} & a_{13} \\ a_{21} & a_{22} & a_{23} \\ a_{31} & a_{32} & a_{33} \end{vmatrix}$$

It can be shown that

$$x = \frac{1}{D}\begin{vmatrix} c_1 & a_{12} & a_{13} \\ c_2 & a_{22} & a_{23} \\ c_3 & a_{32} & a_{33} \end{vmatrix}$$

$$y = \frac{1}{D}\begin{vmatrix} a_{11} & c_1 & a_{13} \\ a_{21} & c_2 & a_{23} \\ a_{31} & c_3 & a_{33} \end{vmatrix}$$

$$z = \frac{1}{D}\begin{vmatrix} a_{11} & a_{12} & c_1 \\ a_{21} & a_{22} & c_2 \\ a_{31} & a_{32} & c_3 \end{vmatrix}$$

Note that the numerators of these equations are obtained by replacing the column in the denominator that is associated with the unknown quantity with the coefficients on the right-hand side of the simultaneous equations. This way of writing the solution of a set of simultaneous linear equations is referred to as **Cramer's rule.**

If $c_1 = c_2 = c_3 = 0$, the equations are said to be homogeneous. If the equations are homogeneous, there is a trivial solution $x = y = z = 0$. There is a nontrivial solution only if the determinant in the denominator is equal to zero. Section 11.3, on the hydrogen molecule ion, shows that the LCAO method yields two homogeneous equations, and so multiplying out the determinant of coefficients yields a quadratic equation in the energy. The two solutions of the quadratic equation yield the energies of the bonding molecular orbital and the antibonding molecular orbital. Section 11.7 shows that there are four homogeneous equations for 1,3-butadiene, so multiplying out the determinant yields the energies of two bonding and two antibonding Hückel molecular orbitals.

D.7 VECTORS

A vector quantity has direction as well as magnitude. A vector $\boldsymbol{A}$ in a Cartesian coordinate system can be represented by

$$\boldsymbol{A} = A_x\boldsymbol{i} + A_y\boldsymbol{j} + A_z\boldsymbol{k} \tag{D.7.1}$$

where $\boldsymbol{i}, \boldsymbol{j}$, and $\boldsymbol{k}$ are vectors of unit length that point along the x, y, and z axes of the coordinate system. The coordinate systems in this book are right-handed; this means that if you move the fingers of your right hand from $\boldsymbol{i}$ to $\boldsymbol{j}$, your thumb points along $\boldsymbol{k}$. The quantities A_x, A_y, and A_z are referred to as components of A; they can be positive or negative. It follows from the Pythagorean theorem that the length of $\boldsymbol{A}$ is given by

$$A = |\boldsymbol{A}| = (A_x^2 + A_y^2 + A_z^2)^{1/2} \tag{D.7.2}$$

When vectors are added, their components in the three directions add separately. For example, if $\boldsymbol{A} = \boldsymbol{i} + \boldsymbol{j} - \boldsymbol{k}$ and $\boldsymbol{B} = \boldsymbol{i} + 2\boldsymbol{j} + 3\boldsymbol{k}$, the sum of the two vectors is $\boldsymbol{A} + \boldsymbol{B} = 2\boldsymbol{i} + 3\boldsymbol{j} + 2\boldsymbol{k}$. The addition of vectors is illustrated in Fig. 10.16.

There are two ways to form the product of two vectors: scalar product and vector product. The scalar product yields a number (a scalar), and the vector product yields a vector. The **scalar product** of $\boldsymbol{A}$ and $\boldsymbol{B}$ is defined by

$$\boldsymbol{A} \cdot \boldsymbol{B} = |\boldsymbol{A}||\boldsymbol{B}| \cos\theta \tag{D.7.3}$$

where θ is the angle between $\boldsymbol{A}$ and $\boldsymbol{B}$. This is often referred to as the dot product. The scalar product is commutative because $\boldsymbol{A} \cdot \boldsymbol{B} = \boldsymbol{B} \cdot \boldsymbol{A}$. Equation D.7.3 can be used to show that $\boldsymbol{i} \cdot \boldsymbol{i} = \boldsymbol{j} \cdot \boldsymbol{j} = \boldsymbol{k} \cdot \boldsymbol{k} = |1||1| \cos 0° = 1$ and $\boldsymbol{i} \cdot \boldsymbol{j} = \boldsymbol{j} \cdot \boldsymbol{i} = \boldsymbol{i} \cdot \boldsymbol{k} = \boldsymbol{k} \cdot \boldsymbol{i} = \boldsymbol{j} \cdot \boldsymbol{k} = \boldsymbol{k} \cdot \boldsymbol{j} = |1||1| \cos 90° = 0$. When $\boldsymbol{A}$ and $\boldsymbol{B}$ are expressed in terms of components and these equations are used, it can be shown that

$$\boldsymbol{A} \cdot \boldsymbol{B} = A_xB_x + A_yB_y + A_zB_z \tag{D.7.4}$$

There are examples of scalar products in Sections 2.1 and 15.2.

The vector product of $\boldsymbol{A}$ and $\boldsymbol{B}$ is defined by

$$\boldsymbol{A} \times \boldsymbol{B} = |\boldsymbol{A}||\boldsymbol{B}|\, \boldsymbol{c} \sin\theta \tag{D.7.5}$$

where θ is the angle between $\boldsymbol{A}$ and $\boldsymbol{B}$ and $\boldsymbol{c}$ is a unit vector perpendicular to the plane formed by $\boldsymbol{A}$ and $\boldsymbol{B}$. The direction of $\boldsymbol{c}$ is given by the right-hand rule: If the fingers of your right hand move from $\boldsymbol{A}$ to $\boldsymbol{B}$, then $\boldsymbol{c}$ is in the direction of your thumb. This is often referred to as the cross product. The cross product is not commutative because $\boldsymbol{A} \times \boldsymbol{B} = -\boldsymbol{B} \times \boldsymbol{A}$. Equation D.7.5 can be used to show that $\boldsymbol{i} \times \boldsymbol{i} = \boldsymbol{j} \times \boldsymbol{j} = \boldsymbol{k} \times \boldsymbol{k} = |1||1|\, \boldsymbol{c} \sin 0° = 0$, $\boldsymbol{i} \times \boldsymbol{j} = -\boldsymbol{j} \times \boldsymbol{i} = |1||1|\, \boldsymbol{k} \sin 90° = \boldsymbol{k}$, $\boldsymbol{j} \times \boldsymbol{k} = -\boldsymbol{k} \times \boldsymbol{j} = \boldsymbol{i}$, and $\boldsymbol{k} \times \boldsymbol{i} = -\boldsymbol{i} \times \boldsymbol{k} = \boldsymbol{k}$. When $\boldsymbol{A}$ and $\boldsymbol{B}$ are expressed in terms of components and these equations are used, it can be shown that

$$\boldsymbol{A} \times \boldsymbol{B} = (A_yB_z - A_zB_y)\boldsymbol{i} + (A_zB_x - A_xB_z)\boldsymbol{j} + (A_xB_y - A_yB_x)\boldsymbol{k} \tag{D.7.6}$$

This equation can be conveniently expressed as a determinant:

$$\boldsymbol{A} \times \boldsymbol{B} = \begin{vmatrix} \boldsymbol{i} & \boldsymbol{j} & \boldsymbol{k} \\ A_x & A_y & A_z \\ B_x & B_y & B_z \end{vmatrix} \tag{D.7.7}$$

There are examples of cross products in Sections 9.12.

The following operator can be used in different ways:

$$\nabla = \boldsymbol{i}\left(\frac{\partial}{\partial x}\right)_{y,z} + \boldsymbol{j}\left(\frac{\partial}{\partial y}\right)_{x,z} + \boldsymbol{k}\left(\frac{\partial}{\partial z}\right)_{x,y} \tag{D.7.8}$$

1. If a function f is a function of x, y, and z, then ∇f (gradient of f or "grad f") is a vector:

$$\nabla f = \boldsymbol{i}\frac{\partial f}{\partial x} + \boldsymbol{j}\frac{\partial f}{\partial y} + \boldsymbol{k}\frac{\partial f}{\partial z} \tag{D.7.9}$$

2. The scalar product of ∇ with a vector $\boldsymbol{v}$ yields the divergence ("div") of that vector:

$$\nabla \cdot \boldsymbol{v} = \left(\frac{\partial v_x}{\partial x}\right) + \left(\frac{\partial v_y}{\partial y}\right) + \left(\frac{\partial v_z}{\partial z}\right) \tag{D.7.10}$$

3. The vector product of ∇ with a vector $\boldsymbol{v}$ yields the curl of the vector:

$$\nabla \times \boldsymbol{v} = \text{curl } \boldsymbol{v} = \boldsymbol{i}\left(\frac{\partial v_z}{\partial y} - \frac{\partial v_y}{\partial z}\right) + \boldsymbol{j}\left(\frac{\partial v_x}{\partial z} - \frac{\partial v_z}{\partial x}\right) + \boldsymbol{k}\left(\frac{\partial v_y}{\partial x} - \frac{\partial v_x}{\partial y}\right) \tag{D.7.11}$$

The operator ∇^2 (the Laplacian) is given in Cartesian coordinates as

$$\nabla^2 = \frac{\partial^2}{\partial x^2} + \frac{\partial^2}{\partial y^2} + \frac{\partial^2}{\partial z^2} \tag{D.7.12}$$

In spherical coordinates, the Laplacian operator is

$$\nabla^2 = \frac{1}{r^2}\frac{\partial}{\partial r}\left(r^2\frac{\partial}{\partial r}\right) + \frac{1}{r^2 \sin^2\theta}\frac{\partial^2}{\partial \phi^2} + \frac{1}{r^2 \sin\theta}\frac{\partial}{\partial \theta}\left(\sin\theta\frac{\partial}{\partial \theta}\right) \tag{D.7.13}$$

D.8 MATRICES*

A matrix is an array of numbers. If a matrix has m rows and n columns it may be represented by

$$\boldsymbol{A} = \begin{bmatrix} a_{11} & a_{12} & \cdots & a_{1n} \\ a_{21} & a_{22} & & \\ \vdots & & & \\ a_{m1} & a_{m2} & \cdots & a_{mn} \end{bmatrix}$$

The sum of two matrices is defined by

$$\boldsymbol{C} = \boldsymbol{A} + \boldsymbol{B}$$

where $c_{ij} = a_{ij} + b_{ij}$ for every i and j.

The product of a scalar c and a matrix is defined by

$$\boldsymbol{B} = c\boldsymbol{A}$$

where $b_{ij} = ca_{ij}$ for every i and j.

*G. Strang, *Linear Algebra and Its Applications,* New York: Academic, 1980; R. G. Mortimer, *Mathematics for Physical Chemistry,* New York: Macmillan, 1981.

The product of two matrices is similar to the scalar product of two vectors. If $\boldsymbol{C}$ is the product $\boldsymbol{AB}$, then

$$c_{ij} = \sum_{k=1}^{n} a_{ik} b_{kj}$$

where n is the number of columns in $\boldsymbol{A}$. If $\boldsymbol{B}$ is to be multiplied by $\boldsymbol{A}$, it must have as many rows as $\boldsymbol{A}$ has columns. For example,

$$\boldsymbol{AB} = \begin{bmatrix} a_{11} & a_{12} \\ a_{21} & a_{22} \\ a_{31} & a_{32} \end{bmatrix} \begin{bmatrix} b_{11} & b_{12} \\ b_{21} & b_{22} \end{bmatrix} = \begin{bmatrix} a_{11}b_{11} + a_{12}b_{21} & a_{11}b_{12} + a_{12}b_{22} \\ a_{21}b_{11} + a_{22}b_{21} & a_{21}b_{12} + a_{22}b_{22} \\ a_{31}b_{11} + a_{32}b_{21} & a_{31}b_{12} + a_{32}b_{22} \end{bmatrix}$$

Matrix multiplication is not commutative, as illustrated by

$$\boldsymbol{AB} = \begin{bmatrix} 2 & 1 \\ 3 & -2 \end{bmatrix} \begin{bmatrix} 1 & 4 \\ 0 & 2 \end{bmatrix} = \begin{bmatrix} 2 \times 1 + 1 \times 0 & 2 \times 4 + 1 \times 2 \\ 3 \times 1 - 2 \times 0 & 3 \times 4 - 2 \times 2 \end{bmatrix} = \begin{bmatrix} 2 & 10 \\ 3 & 8 \end{bmatrix}$$

$$\boldsymbol{BA} = \begin{bmatrix} 1 & 4 \\ 0 & 2 \end{bmatrix} \begin{bmatrix} 2 & 1 \\ 3 & -2 \end{bmatrix} = \begin{bmatrix} 1 \times 2 + 4 \times 3 & 1 \times 1 - 4 \times 2 \\ 0 \times 2 + 2 \times 3 & 0 \times 1 - 2 \times 2 \end{bmatrix} = \begin{bmatrix} 14 & -7 \\ 6 & -4 \end{bmatrix}$$

Simultaneous linear equations may be solved by use of matrices. For example, the set

$$\begin{aligned} a_{11}x_1 + a_{12}x_2 + a_{13}x_3 &= c_1 \\ a_{21}x_1 + a_{22}x_2 + a_{23}x_3 &= c_2 \\ a_{31}x_1 + a_{32}x_2 + a_{33}x_3 &= c_3 \end{aligned}$$

may be written in matrix notation as

$$\begin{bmatrix} a_{11} & a_{12} & a_{13} \\ a_{21} & a_{22} & a_{23} \\ a_{31} & a_{32} & a_{33} \end{bmatrix} \begin{bmatrix} x_1 \\ x_2 \\ x_3 \end{bmatrix} = \begin{bmatrix} c_1 \\ c_2 \\ c_3 \end{bmatrix}$$

or

$$\boldsymbol{AX} = \boldsymbol{C}$$

The inverse of a matrix $\boldsymbol{A}^{-1}$ has the property

$$\boldsymbol{A}^{-1}\boldsymbol{A} = \boldsymbol{A}\boldsymbol{A}^{-1} = \boldsymbol{E}$$

where $\boldsymbol{E}$ is the identity matrix:

$$\boldsymbol{E} = \begin{bmatrix} 1 & 0 & \cdots & 0 \\ 0 & 1 & \cdots & 0 \\ \vdots & & & \\ 0 & 0 & & 1 \end{bmatrix}$$

If we multiply both sides of $\boldsymbol{AX} = \boldsymbol{C}$ by $\boldsymbol{A}^{-1}$, we obtain

$$\boldsymbol{A}^{-1}\boldsymbol{AX} = \boldsymbol{X} = \boldsymbol{A}^{-1}\boldsymbol{C}$$

Thus, the solution $\boldsymbol{X}$ of the simultaneous equations is obtained by multiplying $\boldsymbol{C}$ by the inverse of $\boldsymbol{A}$. Small matrices may be inverted by hand using Gauss elimination, and large matrices may be inverted with a computer to obtain the solution of the simultaneous linear equations.

If $\boldsymbol{AB} = \mathbf{0}$, $\boldsymbol{B}$ is the null space of $\boldsymbol{A}$. The null space can be calculated by hand for a small $\boldsymbol{A}$ matrix or by use of a computer. Correspondingly, $\boldsymbol{B}^{\mathrm{T}}\boldsymbol{A}^{\mathrm{T}} = \mathbf{0}$, where the superscript T indicates the transpose. The transpose $\boldsymbol{A}^{\mathrm{T}}$ of a matrix $\boldsymbol{A}$ has columns that are taken directly from the rows of $\boldsymbol{A}$; thus it can be constructed without any calculations. Thus, $\boldsymbol{A}^{\mathrm{T}}$ is the null space of $\boldsymbol{B}^{\mathrm{T}}$.

D.9 COMPLEX NUMBERS

A complex number z can be written $z = x + \mathrm{i}y$, where $\mathrm{i} = (-1)^{1/2}$ is the imaginary unit and x and y are real numbers. x is referred to as the real part of z, and y is referred to as the imaginary part of z. It is convenient to write $x = \mathrm{Re}(z)$ and $y = \mathrm{Im}(z)$. Complex numbers arise naturally in solving certain quadratic equations.

Two complex numbers can be summed by adding the real parts and the imaginary parts separately:

$$z_1 + z_2 = (x_1 + x_2) + \mathrm{i}(y_1 + y_2) \tag{D.9.1}$$

They can be subtracted as well:

$$z_1 - z_2 = (x_1 - x_2) + \mathrm{i}(y_1 - y_2) \tag{D.9.2}$$

When complex numbers are multiplied, the two quantities are multiplied as binomials and i^2 is replaced by -1 to obtain

$$z_1 z_2 = (x_1 x_2 - y_1 y_2) + \mathrm{i}(x_2 y_1 + x_1 y_2) \tag{D.9.3}$$

To divide complex numbers, it is convenient to introduce the complex conjugate z^*. The **complex conjugate** z^* is obtained by changing i to $-$i. The complex conjugate of $z = x + \mathrm{i}y$ is $z^* = x - \mathrm{i}y$. The product of a complex number and its complex conjugate is a real number. For example,

$$zz^* = (x + \mathrm{i}y)(x - \mathrm{i}y) = x^2 - \mathrm{i}^2 y^2 = x^2 + y^2 \tag{D.9.4}$$

The square root of this quantity is referred to as the **absolute value** of z and is represented by $|z|$.

$$|z| = (zz^*)^{1/2} = (x^2 + y^2)^{1/2} \tag{D.9.5}$$

The ratio of two complex numbers can be written as a complex number by multiplying numerator and denominator by the complex conjugate of the denominator. For example, to find the expression for $z = (1 + 2\mathrm{i})/(2 + 3\mathrm{i})$, multiply the numerator and denominator by $(2 - 3\mathrm{i})$ to obtain

$$z = \frac{(8 + \mathrm{i})}{(4 + 9)} = \frac{8}{13} + \frac{1}{13}\mathrm{i} \tag{D.9.6}$$

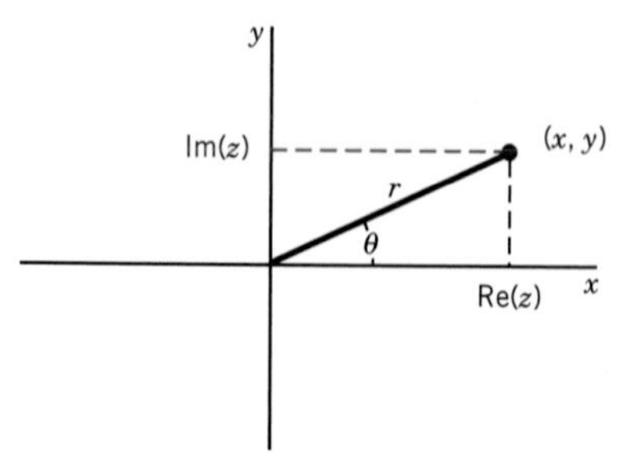

Figure D.9.1 Plot of a complex number.

A complex number can be represented as a point in a plot of $\mathrm{Im}(z)$ versus $\mathrm{Re}(z)$, as shown in Fig. D.9.1. The plane of this figure is referred to as the complex plane. The vector $\boldsymbol{r}$ from the origin to a point in the complex plane makes an angle θ with the x axis; this angle is referred to as the phase angle. The vector is represented by $\boldsymbol{r}$, and its magnitude is represented by r. It is useful to be able to write complex numbers in their polar forms. Figure D.9.1 shows that

$$x = r\cos\theta \quad \text{and} \quad y = r\sin\theta \tag{D.9.7}$$

Therefore,

$$z = r(\cos\theta + \mathrm{i}\sin\theta) \tag{D.9.8}$$

The series expansions of e^x, $\cos x$, and $\sin x$ (see Appendix D.2) can be used to derive **Euler's formula:**

$$e^{\mathrm{i}\theta} = \cos\theta + \mathrm{i}\sin\theta \tag{D.9.9}$$

If we substitute this into equation D.9.8, we obtain

$$z = r e^{\mathrm{i}\theta} \tag{D.9.10}$$

for which the complex conjugate is

$$z^* = r e^{-\mathrm{i}\theta} \tag{D.9.11}$$

Taking the square root of the product zz^* yields the absolute value of z, which is equal to

$$|z| = r \tag{D.9.12}$$

Two relations that are useful in connection with Fourier transforms (Section 15.8) are

$$\cos\frac{2\pi n x}{a} = (e^{\mathrm{i}2\pi n x/a} + e^{-\mathrm{i}2\pi n x/a})/2 \tag{D.9.13}$$

$$\sin\frac{2\pi n x}{a} = (e^{\mathrm{i}2\pi n x/a} - e^{-\mathrm{i}2\pi n x/a})/2\mathrm{i} \tag{D.9.14}$$

D.10 MATHEMATICAL CALCULATIONS WITH PERSONAL COMPUTERS

The use of mathematical applications in personal computers is producing a revolutionary change in solving physical chemical problems. These applications include Mathematica, MathCad, MATLAB, and MAPLE. The existence of these applications has made it possible to include more difficult problems in this edition as Computer Problems. The complete solutions of these problems in Mathematica are provided in the Solutions Manual and on the web at http://wiley.com/college/silbey. These programs not only make it possible to solve the particular problems, but they also make it possible to make similar calculations over different ranges of temperature, pressure, wavelength, etc., and to substitute the properties of other substances without one's being an expert Mathematica programmer. The primary reference on Mathematica is

S. Wolfram, *The Mathematica Book,* 4th ed. New York: Cambridge University Press, 1999.

There are two books on solving physical chemistry problems using Mathematica:

W. H. Cropper, *Mathematica Computer Programs for Physical Chemistry.* New York: Springer, 1998.
J. H. Noggle, *Physical Chemistry Using Mathematica.* New York: HarperCollins, 1996.

Several books have been written to help new users of Mathematica get started. These include

C.-K. Cheung, G. E. Keough, C. Landraitis, and R. H. Gross, *Getting Started with Mathematica.* Hoboken, NJ: Wiley, 1998.

K. R. Coombes, B. R. Hunt, R. L. Lipsman, J. E. Osborn, and G. J. Stuck, *The Mathematica Primer.* New York: Cambridge University Press, 1998.

H. F. W. Höft and M. H. Höft, *Computing with Mathematica.* San Diego: Academic, 1998.

B. F. Torrence and E. A. Torrence, *The Student's Introduction to Mathematica.* New York: Cambridge University Press, 1999.

E

Greek Alphabet

Α, α	Alpha
Β, β	Beta
Γ, γ	Gamma
Δ, δ	Delta
Ε, ϵ	Epsilon
Ζ, ζ	Zeta
Η, η	Eta
Θ, ϑ, θ	Theta
Ι, ι	Iota
Κ, κ	Kappa
Λ, λ	Lambda
Μ, μ	Mu
Ν, ν	Nu
Ξ, ξ	Xi
Ο, o	Omicron
Π, π	Pi
Ρ, ρ	Rho
Σ, σ	Sigma
Τ, τ	Tau
Υ, υ	Upsilon
Φ, φ, ϕ	Phi
Χ, χ	Chi
Ψ, ψ	Psi
Ω, ω	Omega

F

Useful Information on the Web

The National Institute of Standards and Technology (NIST) provides several kinds of useful information on the Web:

Fundamental Physical Constants
http://physics.nist.gov/cuu/Constants/

International System of Units (SI)
http://physics.nist.gov/cuu/Units/index.html

Conversion Factors for Energy Equivalents
http://physics.nist.gov/cuu/Constants/energy.html

NIST Chemistry WebBook
http://webbook.nist.gov/chemistry/

Thermodynamics of Enzyme-Catalyzed Reactions
http://wwwbmed.nist.gov:8080/enzyme.html

Ground Levels and Ionization Energies for the Neutral Atoms
http://physics.nist.gov/PhysRefData/IonEnergy/ionEnergy.html

NIST X-Ray Photoelectron Spectroscopy Database
http://srdata.nist.gov/xps

G

Symbols for Physical Quantities and Their SI Units*

Symbol	Name[a]	Symbol for SI Unit
$\boldsymbol{a}$	acceleration vector (2.1)	$\mathrm{m\ s^{-2}}$
a	activity (4.5)	1
a	hyperfine splitting constant (15.9)	T
a	van der Waals' constant (1.8)	$\mathrm{m^6\ Pa\ mol^{-2}}$
a_0	Bohr radius (10.1)	m
A	absorbance (14.4)	1
A	area (1.3)	$\mathrm{m^2}$
$\boldsymbol{A}$	conservation matrix (5.11)	1
A	Debye–Hückel constant (7.5)	$\mathrm{kg^{1/2}\ mol^{-1/2}}$
A	Helmholtz energy (4.2)	J
$\overline{A}$	partial molar Helmholtz energy (4.4)	$\mathrm{J\ mol^{-1}}$
A	pre-exponential factor in the Arrhenius equation (18.6)	$\mathrm{(mol\ m^{-3})^{1-n}\ s^{-1}}$
A_s	area of a surface (2.5)	$\mathrm{m^2}$
A_{12}	Einstein probability of spontaneous emission (13.2)	$\mathrm{s^{-1}}$
b	van der Waals constant (1.8)	$\mathrm{m^3\ mol^{-1}}$
b	impact parameter (17.8)	m

(*continued*)

[a] The number in parentheses is the section number where the quantity is defined. For thermodynamic quantities a superscript degree sign indicates that the quantity has the value for the standard state. A superscript * is used to designate the molar thermodynamic quantity for a pure substance. Molar thermodynamic quantities and partial molar thermodynamic quantities are indicated by an overbar. The overbar is not used in delta quantities such as $\Delta_{vap}H°$, $\Delta_f H°$, and $\Delta_r G°$. Concentrations are represented by use of square brackets. Quantum mechanical operators are indicated by a circumflex, as in $\hat{A}$ (Section 9.4). Quantities with tildes over them are in wave number units. The SI unit for an operator is the same as for the physical quantity the operator represents. The expectation value of F is represented by $\langle F \rangle$. Vectors and matrices are represented by boldface italic symbols; the corresponding magnitudes are italicized and are not included here. A 1 in the symbol column indicates that the physical quantity does not have SI units.

*Recommendations of the International Union of Pure and Applied Chemistry are given in I. Mills, T. Cvitas, K. Homann, N. Kallay, and K. Kuchitsu, *Quantities, Units, and Symbols in Physical Chemistry.* Oxford, UK: Blackwell Scientific, 1993.

Symbol	*Name*	*Symbol for SI Unit*
b_{max}	distance along line of centers when molecules are in contact (19.1)	m
B	empirical constant in extended Debye–Hückel equation (7.5)	$kg^{1/2}$ $mol^{1/2}$
$\mathbf{B}$	magnetic field strength (magnetic flux density) (10.4)	$T = kg\ s^{-2}\ A^{-1}$
B	magnitude of magnetic field strength (magnetic flux density) (10.4)	$T = kg\ s^{-2}\ A^{-1}$
B	rotational constant (13.4)	m^{-1}
$B(T)$	second virial coefficient (1.5)	$m^3\ mol^{-1}$
B_{21}	Einstein probability of stimulated emission (13.2)	$m\ kg^{-1}$
B_{12}	Einstein probability of stimulated absorption (13.2)	$m\ kg^{-1}$
c	molar concentration (5.1)	$mol\ m^{-3}$
c	speed of light in a vacuum (9.1)	$m\ s^{-1}$
c°	standard concentration (5.7)	$mol\ m^{-3}$
C	capacitance (22.1)	$F = C\ V^{-1} = A\ s\ V^{-1}$
C	number of components (5.9)	1
C_n	rotation element and operation (12.1)	1
C_P	heat capacity at constant pressure (2.7)	$J\ K^{-1}$
$\overline{C}_P$	molar heat capacity at constant pressure (2.8)	$J\ K^{-1}\ mol^{-1}$
$C(T)$	third virial coefficient (1.5)	$m^6\ mol^{-2}$
C_V	heat capacity at constant volume (2.6)	$J\ K^{-1}$
$\overline{C}_V$	molar heat capacity at constant volume (2.6)	$J\ K^{-1}\ mol^{-1}$
$\Delta_r C_P^\circ$	reaction heat capacity at constant pressure (2.12)	$J\ K^{-1}\ mol^{-1}$
d_{12}	collision diameter (17.7)	m
d_{hkl}	interplanar spacing (23.2)	m
$\tilde{D}$	centrifugal distortion constant (13.4)	m^{-1}
D	diffusion coefficient (17.9)	$m^2\ s^{-1}$
D	minimum number of variables to describe the extensive state of a system (1.9)	1
D_e	equilibrium dissociation energy (13.6)	J
D_0	spectroscopic dissociation energy (dissociation energy from ground state) (13.6)	J
e	charge of a proton (10.1)	C
$\mathbf{E}$	electric field strength (7.1)	$V\ m^{-1}$
E	electromotive force (potential difference for a cell) (7.3)	V
E	energy of a particle (9.3)	J
$\tilde{E}$	energy in wave numbers (10.1)	m^{-1}
E	identity element or operator (12.1)	1
E_a	activation energy in the Arrhenius equation (18.6)	$J\ mol^{-1}$
E_{ads}	activation energy for adsorption (24.2)	$J\ mol^{-1}$
E_{des}	activation energy for desorption (24.2)	$J\ mol^{-1}$

(*continued*)

Symbol	*Name*	*Symbol for SI Unit*
E_e	electron kinetic energy (24.5)	J
E_{ea}	electron affinity (10.11)	$J\ mol^{-1}$
E_h	Hartree energy (10.1)	J
E_i	ionization energy (10.2)	J
E°	standard electrode potential (7.7)	V
f	electrostatic factor (20.5)	1
$\boldsymbol{f}$	force vector (2.1)	$kg\ m\ s^{-2}$
f	frictional coefficient (20.1)	$kg\ s^{-1}$
f	fugacity (4.5)	Pa
f	oscillator strength (14.5)	1
f_i	atomic scattering factor (23.6)	1
f_i	fraction of molecules in the *i*th energy state (13.4)	1
$f(v_x)$	probability density for velocity in the *x* direction (17.1)	$s\ m^{-1}$
$f(v_x, v_y, v_z)$	probability density for velocity (17.1)	$s^3\ m^{-3}$
F	Faraday constant (7.1)	$C\ mol^{-1}$
F	number of degrees of freedom (1.9)	1
$F(hkl)$	structure factor (23.6)	1
$\tilde{F}(J)$	rotational term value for quantum number *J* (13.4)	m^{-1}
$F(v)$	speed distribution function (Maxwell) (17.3)	$s\ m^{-1}$
$F(\epsilon)$	energy distribution function (17.3)	J^{-1}
g	acceleration of gravity (1.3)	$m\ s^{-2}$
g_e	electron *g* factor (10.5)	1
g_i	degeneracy (10.2)	1
g_i	orbital in Hartree–Fock method (10.9)	1
g_N	nuclear *g* factor (15.1)	1
g	dimensionless *g* factor (15.9)	1
G	Gibbs energy (4.2)	J
$\overline{G}_i$	molar or partial molar Gibbs energy of species *i* (4.2)	$J\ mol^{-1}$
$\tilde{G}(v)$	vibrational term value for a diatomic molecule (13.6)	m^{-1}
G'	transformed Gibbs energy (7.2)	J
$\Delta G^\ddagger$	Gibbs energy for the formation of the activated complex (19.4)	$J\ mol^{-1}$
$\Delta_f G_i^\circ$	standard Gibbs energy of formation of species *i* (5.4)	$J\ mol^{-1}$
$\Delta_r G_i'^\circ$	standard transformed reaction Gibbs energy (8.5)	$J\ mol^{-1}$
$\Delta_{mix} G$	Gibbs energy of mixing (4.7)	$J\ mol^{-1}$
$\Delta_r G$	reaction Gibbs energy (5.1)	$J\ mol^{-1}$
$\Delta_r G^\circ$	standard reaction Gibbs energy (5.1)	$J\ mol^{-1}$
$\Delta_f G'^\circ$	standard transformed Gibbs energy of formation of species *i* (8.5)	$J\ mol^{-1}$
h	height (1.11)	m
h	Planck's constant (9.1)	J s
$\hbar$	$h/2\pi$ ("*h* bar") (9.2)	J s
h_i	radial factor in orbital (10.9)	1

(continued)

Symbol	*Name*	*Symbol for SI Unit*
H	enthalpy (2.7)	J
H	classical Hamiltonian function (9.4)	J
$\hat{H}$	Hamiltonian operator (9.4)	J
H_{AA}	Coulomb integral (11.3)	J
H_{AB}	resonance integral (11.3)	J
$\overline{H}_i$	molar enthalpy of species i (2.11)	J mol^{-1}
H_{ij}	Hamiltonian matrix elements (11.7)	J
$\overline{H}^\circ_T - \overline{H}^\circ_{298}$	molar enthalpy increment (2.12)	J mol^{-1}
$H_v(x)$	Hermite polynomials (9.10)	1
H'	transformed enthalpy (8.3)	J
$\Delta_{ads}H$	enthalpy of adsorption (24.1)	J mol^{-1}
$\Delta_f H^\circ$	standard enthalpy of formation (2.12)	J mol^{-1}
$\Delta_f H_i'^\circ$	standard transformed enthalpy of formation of reactant i (8.5)	J mol^{-1}
$\Delta_r H$	reaction enthalpy (2.11)	J mol^{-1}
$\Delta_r H'^\circ$	standard transformed enthalpy of reaction (8.5)	J mol^{-1}
$\Delta_{vap}H$	enthalpy of vaporization (6.3)	J mol^{-1}
i	center of symmetry or inversion center (12.1)	1
I	adiabatic ionization potential (24.5)	V
I	intensity of transmitted light (14.4)	J m^{-2} s^{-1}
I	ionic strength (7.5)	mol kg^{-1}
I	moment of inertia (9.11)	kg m^2
I	nuclear spin quantum number (15.1)	1
$\boldsymbol{I}$	total spin angular momentum of a nucleus (15.1)	J s
I_a	intensity of electromagnetic radiation absorbed (19.6)	mol m^{-3} s^{-1}
$\hat{I}_z$	operator for the z component of angular momentum in NMR (15.2)	J s
$I_{xy} = I_{yz}$	product of inertia (13.5)	kg m^2
I_z	moment of inertia about the z axis (13.5)	kg m^2
I_z	z component of nuclear spin (15.1)	J s
J	photodissociation coefficient (19.9)	s^{-1}
J	quantum number of the total angular momentum of an atom (10.12)	1
J	rotational quantum number (9.12)	1
J	spin–spin coupling constant (15.5)	s^{-1}
$\boldsymbol{J}$	total angular momentum vector for an atom (10.5)	J s
J_N	flux of particles (17.6)	m^{-2} s^{-1}
J_{iz}	flux of i in the z direction (17.9)	mol m^{-2} s^{-1}
J_z	z component of the total angular momentum of an atom (10.12)	J s
k	Boltzmann constant (3.6)	J K^{-1}
k	force constant (9.9)	N m^{-1} = kg s^{-2}
k	rate constant (18.2)	(mol m^{-3})$^{1-n}$ s^{-1}
k_{cat}	turnover number (20.10)	s^{-1}
k_d	rate constant for desorption (24.2)	s^{-1}

(*continued*)

Symbol	Name	Symbol for SI Unit
k°	conditional rate constant (20.8)	s^{-1}
k'	apparent rate constant at specified pH (20.10)	varies
$k^\ddagger$	rate constant for the activated complex crossing the barrier (19.4)	s^{-1}
K	constant in Langmuir adsorption isotherm (24.2)	Pa^{-1}
K	equilibrium constant (5.1)	1
K	quantum number of the component of angular momentum along the a axis of a symmetric molecule (13.5)	1
K_a	acid dissociation constant (8.1)	1
K_b	base dissociation constant (8.1)	1
K_c	equilibrium constant expressed in terms of c/c° (5.7)	1
K_f	freezing point constant (6.7)	$K\ kg\ mol^{-1}$
K_i	Henry's law constant of species i (6.5)	Pa
K_I	inhibition constant for an enzymatic reaction (20.10)	$mol\ m^{-3}$
K_M	Michaelis constant (20.10)	$mol\ m^{-3}$
K_P	equilibrium constant expressed in terms of P_i/P°, when needed to distinguish from K_c (5.7)	1
K_w	ion product for water (8.1)	1
K_x, K_y	equilibrium constants in terms of mole fractions (5.6)	1
K'	apparent equilibrium constant at specified pH (8.3)	1
$K^\ddagger$	equilibrium constant for the formation of a transition state (19.4)	1
ℓ	angular momentum quantum number (9.11)	1
ℓ_i	angular momentum vector of electron i in an atom (10.12)	J s
ℓ_x	angular momentum in the x direction (9.4)	J s
ℓ_{zi}	z component of the angular momentum of electron i in an atom (10.12)	J s
L	angular momentum (9.11)	J s
$\mathbf{L}$	angular momentum vector for an atom (9.12)	J s
L	length (2.1)	m
$\mathbf{L}$	vector length (2.1)	m
$\hat{L}_q$	operator for the angular momentum in the q direction (9.4)	J s
$\hat{L}^2$	operator for the square of the angular momentum (9.12)	$J^2\ s^2$
$\mathbf{m}$	magnetic dipole moment vector (22.5)	$A\ m^2$
m	magnetic quantum number (9.12)	1
m	mass (1.3)	kg
m	molality (7.4)	$mol\ kg^{-1}$

(continued)

Symbol	*Name*	*Symbol for SI Unit*
m_e	rest mass of electron (10.1)	kg
m_i	quantum number for z component of the orbital angular momentum of electron i in an atom (10.12)	1
m_I	quantum number for z component of nuclear spin (15.1)	1
m_N	mass of nucleus (10.1)	kg
m_p	mass of proton (15.1)	kg
m_s	spin quantum number for the z component of the spin angular momentum (10.5)	1
m_{si}	quantum number for the z component of the spin of electron i in an atom (10.12)	1
m_u	atomic mass constant (16.3)	kg
m°	standard molality (7.4)	mol kg^{-1}
$m_\pm$	mean ionic molality (7.4)	mol kg^{-1}
$\boldsymbol{M}$	magnetization (magnetic dipole moment per unit volume) (22.5)	A m^{-1}
M	magnetic quantum number for a molecule (11.4)	1
M	molar mass (1.3)	kg mol^{-1}
$\boldsymbol{M}_z$	magnetization in the direction of the field (15.2)	T
M_J	quantum number for z component of total angular momentum for an atom (10.12)	1
M_L	quantum number for z component of the orbital angular momentum of an atom (10.12)	1
$\overline{M}_m$	mass average molar mass (21.1)	kg mol^{-1}
$\overline{M}_n$	number average molar mass (21.1)	kg mol^{-1}
M_r	relative molar mass (21.5)	1
M_S	quantum number for z component of the spin angular momentum of an atom (10.12)	1
n	amount of substance (1.3)	mol
n	order of a reaction (18.2)	1
n	order of reflection in the Bragg equation (23.3)	1
n	principal quantum number (9.6)	1
n	refractive index (21.5)	1
n'	amount of pseudoisomer group (8.3)	mol
$n_c(\mathrm{H})$	amount of hydrogen component (8.3)	mol
N	number of particles (1.1)	1
N	number of species (4.1)	1
N'	number of pseudoisomer groups (8.3)	1
N_A	Avogadro constant (1.1)	mol^{-1}
$\overline{N}_H$	average number of protons bound (8.2)	1
$\overline{N}_{Mg}$	average number of magnesium ions bound (8.2)	1

(continued)

Symbol	*Name*	*Symbol for SI Unit*
$N_{\mathrm{H}}(i)$	number of hydrogen atoms bound by species i (8.5)	1
N_s	number of different species in system (1.1)	1
p	momentum of a particle (9.1)	$\mathrm{kg\ m\ s^{-1}}$
p	number of phases (5.9)	1
$\boldsymbol{p}$	relative momentum of a particle (9.12)	$\mathrm{kg\ m\ s^{-1}}$
pH	$-\log a_{\mathrm{H}^+}$ (7.8)	1
pK	$-\log K$ (8.2)	1
$p_{n\ell}(r)$	radial probability density (10.3)	$\mathrm{m^{-3}}$
$\hat{\boldsymbol{p}}_q$	linear momentum operator (9.4)	$\mathrm{kg\ m\ s^{-1}}$
$\boldsymbol{P}$	dielectric polarization (22.1)	$\mathrm{C\ m^{-2}}$
P	pressure (1.3)	Pa
P	probability (19.1)	1
P°	standard state pressure (5.7)	Pa
P_c	critical pressure (1.7)	Pa
P_i	partial pressure of i (1.4)	Pa
P_i^*	equilibrium vapor pressure of i (6.4)	Pa
P_m	molar polarization (22.3)	$\mathrm{m^3\ mol^{-1}}$
q	heat absorbed by a system (2.1)	J
q	molecular partition function (16.1)	1
$\hat{q}$	coordinate operator (9.4)	m
Q	canonical ensemble partition function (16.10)	1
Q	electric charge (2.5)	C
Q	reaction quotient (5.1)	1
Q_i	charge on the ith ion (11.7)	C
r	radius of curvature (6.3)	m
r	resistivity (20.3)	Ω m
$\boldsymbol{r}$	vector distance (7.1)	m
$\hat{\boldsymbol{r}}$	unit vector in a particular direction (7.1)	m
r_i	mole fraction of i within an isomer group (5.10)	1
$\langle r\rangle_{n\ell}$	expectation value for radius in an atom (10.3)	m
$(\overline{r^2})^{1/2}$	root-mean-square end-to-end distance (21.3)	m
R	electric resistance (20.3)	Ω
R	gas constant (1.2)	$\mathrm{J\ K^{-1}\ mol^{-1}}$
R	number of independent reactions (5.9)	1
$\boldsymbol{R}$	position of a nucleus in a coordinate system (11.1)	m
R	Rydberg constant (10.2)	$\mathrm{m^{-1}}$
R_∞	Rydberg constant for $m_{\mathrm{nuc}} \to \infty$ (10.2)	$\mathrm{m^{-1}}$
R_e	equilibrium internuclear distance in a diatomic molecule (13.6)	m
$R_{n\ell}(r)$	hydrogenlike radial wavefunction (10.3)	$\mathrm{m^{-3/2}}$
$\hat{R}$	transformation operator (12.9)	1

(continued)

Symbol	*Name*	*Symbol for SI Unit*
$\langle R_G^2 \rangle^{1/2}$	radius of gyration (21.3)	m
s	electron spin quantum number (generally referred to as the spin) (10.5)	1
s_i	quantum number of spin angular momentum of electron i in an atom (10.12)	1
$\mathbf{s}_i$	spin angular momentum vector for electron i in an atom (10.12)	J s
s_i	stoichiometric number for elementary step i in a mechanism (18.7)	1
s_{zi}	z component of the spin angular momentum of electron i in an atom (10.12)	J s
s^*	sticking coefficient (24.2)	1
S	entropy (3.2)	J K^{-1}
S	quantum number of spin angular momentum of an atom (10.12)	1
S	reaction cross section (19.1)	m^2
S	sedimentation coefficient (21.5)	s
$\mathbf{S}$	spin angular momentum vector (10.5)	J s
$\hat{S}$	spin angular momentum operator (10.5)	J s
S_{AB}	overlap integral (11.2)	1
$\overline{S}_i$	molar or partial molar entropy of i (3.3)	J K^{-1} mol^{-1}
S_n	improper rotation axis or operator (12.1)	1
$S_{vv'}$	Franck–Condon overlap integral (14.2)	1
S_z	z component of the spin angular momentum (10.5)	J s
$\hat{S}_z$	operator for the z component of the spin angular momentum (10.5)	J s
S'	transformed entropy (8.3)	J K^{-1}
S_0	surface area of catalyst (24.8)	m^2
$\hat{S}^2$	square of operator of spin angular momentum (10.5)	1
$\Delta_{mix}S$	entropy of mixing (3.5)	J K^{-1} mol^{-1}
$\Delta_r S^\circ$	reaction entropy (3.8)	J K^{-1} mol^{-1}
$\Delta_r S'^\circ$	standard transformed reaction entropy (8.5)	J K^{-1} mol^{-1}
$t_{1/2}$	half-life for a reaction (18.2)	s
T	kinetic energy (9.4)	J
T	thermodynamic temperature (1.3)	K
T_B	Boyle temperature (1.5)	K
T_c	critical temperature (1.7)	K
$\hat{T}_x$	operator for the kinetic energy in the x direction (9.4)	J
T_1	spin–lattice relaxation time (15.7)	s
T_2	transverse relaxation time (15.7)	s
T°	reference temperature (3.3)	K
u	electric mobility (20.3)	m^2 V^{-1} s^{-1}
U	internal energy (2.1)	J

(continued)

Symbol	*Name*	*Symbol for SI Unit*
$\overline{U}_i$	molar or partial molar internal energy of i (2.8)	J mol^{-1}
$\overline{U}_t$	translational energy (2.8)	J
v	rate of reaction (18.1)	mol m^{-3} s^{-1}
v	specific volume (21.5)	m^3 kg^{-1}
v	magnitude of velocity of a particle (17.1)	m s^{-1}
$\boldsymbol{v}$	velocity vector (17.1)	m s^{-1}
v	vibrational quantum number (9.10)	1
v_{mp}	most probable speed (17.4)	m s^{-1}
v_s	speed of sound (17.4)	m s^{-1}
$\langle v \rangle$	mean speed (17.4)	m s^{-1}
$\langle v_{12} \rangle$	mean relative speed (17.7)	m s^{-1}
$\langle v^2 \rangle^{1/2}$	root-mean-square speed (17.4)	m s^{-1}
V	potential energy (9.1)	J
V	volume (1.1)	m^3
$\overline{V}$	molar volume (1.1)	m^3 mol^{-1}
$\overline{V}_c$	critical molar volume (1.7)	m^3 mol^{-1}
$\overline{V}_i$	molar volume or partial molar volume of i (1.10)	m^3 mol^{-1}
$\overline{V}_i^*$	molar volume of a pure substance (6.9)	m^3 mol^{-1}
V_S	maximum rate of enzyme-catalyzed reaction (20.10)	mol m^{-3} mol^{-1}
w	work done on a system (2.1)	J
$W(x, y, z)$	probability of coordinates x, y, and z after random walk (21.3)	1
x_i	mole fraction of i in liquid phase (6.4)	1
$\hat{x}$	operator for position (9.4)	m
$\langle x \rangle$	expectation value for x (9.5)	m
$\overline{X}_n$	number average degree of polymerization (21.4)	1
y_i	mole fraction of i in a gas (1.4)	1
Y	fractional saturation of binding (8.7)	1
$Y_\ell^m(\theta, \phi)$	spherical harmonic (9.11)	1
z_i	charge number of ion i (signed) (7.1)	1
z_{11}	collision frequency for like molecules (17.7)	s^{-1}
z_{12}	collision frequency for unlike molecules (17.7)	s^{-1}
Z	atomic number or proton number (10.1)	1
Z	compressibility factor (1.5)	1
Z'	effective nuclear charge (10.7)	1
Z_{11}	collision density for like molecules (17.7)	m^{-3} s^{-1}
Z_{12}	collision density for unlike molecules (17.7)	m^{-3} s^{-1}
α	Coulomb integral (11.7)	J
α	cubic expansion coefficient (4.9)	K^{-1}
α	optical rotation (14.11)	deg
α	molecular polarizability (13.9)	C^2 m^2 J^{-1}

(continued)

Symbol	Name	Symbol for SI Unit
$\boldsymbol{\alpha}$	molecular polarizability matrix (13.9)	$C^2\ m^2\ J^{-1}$
α	spin function for an electron (spin up function) (10.5)	1
α_e	vibration–rotation coupling constant (13.7)	m^{-1}
$[\alpha]$	specific rotation (14.11)	$deg\ m^2\ kg^{-1}$
β	resonance integral (11.7)	J
β	spin function for an electron (spin down function) (10.5)	1
γ	magnetogyric ratio (10.4)	$A\ m^2\ J^{-1}\ s^{-1}$
γ	ratio of C_P to C_V (2.10)	1
γ	surface tension (2.5)	$N\ m^{-1}$
γ_i	activity coefficient of i based on deviations from Raoult's law (6.6)	1
γ_i'	activity coefficient of i based on deviations from Henry's law (6.6)	1
$\gamma_\pm$	mean ionic activity coefficient (7.4)	1
δ	chemical shift in NMR (15.4)	1
δ	path difference (23.6)	m
δ_{ij}	Kronecker delta (9.3)	1
ΔE	uncertainty in energy (9.2)	J
Δt	uncertainty in time (9.2)	s
Δx	extent of wave packet in space (9.2)	m
$\Delta\nu$	range in frequency (9.2)	s^{-1}
ϵ	efficiency (3.9)	1
ϵ	energy of a particle (10.9)	J
ϵ	molar absorption coefficient (14.4)	$m^2\ mol^{-1}$
ϵ	parameter in Lennard-Jones equation (11.9)	J
ϵ_r	relative permittivity (dielectric constant) (7.1)	1
ϵ_r	relative energy of collision (19.1)	$J\ mol^{-1}$
ϵ_0	permittivity of vacuum (7.1)	$C^2\ N^{-1}\ m^{-2}$
ϵ_F	Fermi energy (23.10)	J
ϵ_k	energy along line of centers (19.1)	$J\ mol^{-1}$
ϵ^*	minimum energy along line of centers for reaction (19.1)	$J\ mol^{-1}$
$\Delta\epsilon_0^\ddagger$	energy barrier at absolute zero (19.4)	K
η	number of subsystems in a canonical ensemble (16.10)	1
η	viscosity (17.9)	$kg\ m^{-1}\ s^{-1} = Pa\ s$
η_{sp}	specific viscosity (21.5)	1
$[\eta]$	intrinsic viscosity (21.5)	$m^2\ kg^{-1}$
Θ	surface coverage (24.2)	1
Θ	theta temperature (21.2)	K
Θ_D	Debye temperature for a monatomic crystal (16.12)	K
Θ_e	characteristic electronic temperature (16.6)	K
Θ_E	Einstein temperature for a monatomic crystal (16.12)	K

(continued)

Symbol	*Name*	*Symbol for SI Unit*
Θ_r	characteristic rotational temperature (16.5)	K
Θ_v	characteristic vibrational temperature (16.4)	K
κ	electric conductivity (20.3)	$\Omega^{-1}\ m^{-1}$
κ	isothermal compressibility (1.7)	Pa^{-1}
κ	naperian molar absorption coefficient (14.4)	$m^2\ mol^{-1}$
κ	reciprocal wavelength (9.2)	m
κ	thermal conductivity (17.9)	$J\ m^{-1}\ s^{-1}\ K^{-1}$
λ	mean free path (17.7)	m
λ	quantum number for angular momentum around the internuclear axis (11.3)	1
λ	wavelength (9.1)	m
λ_D	Debye wavelength (16.12)	s^{-1}
Λ	quantum number for electronic angular momentum along molecular axis (11.4)	1
Λ	thermal wavelength (16.3)	m
$\boldsymbol{\mu}$	electric dipole moment (11.8)	C m
$\boldsymbol{\mu}$	magnetic dipole moment (10.4)	$J\ T^{-1} = A\ m^2$
μ	magnitude of dipole moment (11.8)	C m
μ	reduced mass (9.9)	kg
$\hat{\boldsymbol{\mu}}$	quantum mechanical dipole moment operator (13.2)	C m
μ_B	Bohr magneton (10.4)	$J\ T^{-1} = A\ m^2$
μ_i	chemical potential of i (4.1)	$J\ mol^{-1}$
μ_i'	transformed chemical potential of i (8.3)	$J\ mol^{-1}$
μ_{JT}	Joule–Thomson coefficient (2.9)	$K\ Pa^{-1}$
μ_N	nuclear magneton (15.1)	$J\ T^{-1}$
$\boldsymbol{\mu}_{nm}$	transition dipole moment (13.2)	C m
μ_r	relative permeability (22.4)	1
$\boldsymbol{\mu}_s$	magnetic dipole moment vector for spin angular momentum (10.5)	$A\ m^2$
$\boldsymbol{\mu}_\ell$	magnetic dipole moment vector for angular momentum (10.5)	$A\ m^2$
μ_{sz}	z component of the magnetic dipole moment for an electron (15.1)	$A\ m^2$
μ_z	magnetic dipole moment in the z direction (15.1)	$J\ T^{-1} = A\ m^2$
μ_0	permeability of vacuum (22.1)	$N\ A^{-2}$
$\mu_i^*(1)$	reference chemical potential for solute i in a dilute real solution (6.6)	$J\ mol^{-1}$
ν	frequency (9.1)	s^{-1}
$\boldsymbol{\nu}$	stoichiometric number matrix (5.11)	1
ν	sum of stoichiometric numbers for a reaction (5.7)	1
$\tilde{\nu}$	wave number (13.1)	m^{-1}
$\nu_\pm$	number of ions in an electrolyte molecule $(\nu_+ + \nu_-)$ (7.4)	1

(*continued*)

Symbol	Name	Symbol for SI Unit
$\|\nu_e\|$	stoichiometric number of the electron in an electrochemical reaction (7.3)	1
ν_i	stoichiometric number for reactant i (positive for products, negative for reactants) (2.11)	1
ν_i'	apparent stoichiometric number (8.3)	1
$\tilde{\nu}_R$	Raman frequency (13.9)	m^{-1}
$\tilde{\nu}_e x_e$, $\tilde{\nu}_e y_e$	anharmonicity constants (13.6)	m^{-1}
ξ	extent of reaction (2.11)	mol
ξ'	apparent extent of reaction (8.3)	mol
ξ'	dimensionless extent of reaction (5.3)	1
π_i	probability of forming a chain of size i (21.4)	1
Π	osmotic pressure (6.7)	Pa
ρ	number density (17.7)	m^{-3}
ρ	density (1.1)	kg m^3
$\rho(T)$	total radiant energy density (9.16)	J m^{-3}
$\rho_\lambda(\lambda, T)$	energy density as a function of wavelength (9.16)	J m^{-4}
ρ_ν	energy density in a cavity as a function of frequency (9.1)	J m^{-3} s
σ	absorption cross section (14.4)	m^2
$\sigma(\epsilon_r)$	reaction cross section as a function of relative energy of collision (19.1)	m
σ	parameter in Lennard-Jones equation (11.9)	m
σ	shielding constant (15.4)	1
σ	standard deviation (9.5)	varies
σ	symmetry number (16.5)	1
σ	symmetry plane or reflection operator (12.1)	1
σ_g	molecular orbital (11.4)	
σ_x	standard deviation in x (9.5)	m
σ_ν	standard deviation of Doppler broadening (17.4)	s^{-1}
σ_x	variance (9.5)	varies
Σ	magnetic quantum number for spin angular momentum of a molecule (11.4)	1
τ	lifetime (13.2)	s
τ	relaxation time (18.2)	s
τ	turbidity (21.5)	m^{-1}
ϕ	electric potential difference (2.5)	V
ϕ	electric potential (7.2)	V
ϕ	fugacity coefficient (4.5)	1
ϕ	phase difference in scattering (23.6)	deg
ϕ	quantum yield (19.6)	1
ϕ	volume fraction (21.5)	1
$\Delta\phi$	membrane potential (7.10)	V
χ_{el}	electric susceptibility (22.1)	1
χ_{mag}	magnetic susceptibility (22.5)	1

(continued)

Symbol	*Name*	*Symbol for SI Unit*
$\chi_{mag,m}$	molar magnetic susceptibility (22.6)	$m^3\ mol^{-1}$
ψ	wavefunction in three dimensions (9.3)	$m^{-3/2}$
ψ^*	complex conjugate of ψ (9.3)	$m^{-3/2}$
ψ_e	electronic wavefunction (11.1)	$m^{-3/2}$
ψ_n	nuclear motion wavefunction (11.1)	$m^{-3/2}$
$\Psi(\mathbf{r}, t)$	time-dependent wavefunction (9.14)	$m^{-3/2}$
ω	angular frequency ($2\pi\nu$) (9.9)	$rad\ s^{-1}$
Ω	number of microstates in a macrostate (3.6)	1
Ω	quantum number for the total angular momentum along the internuclear axis of a diatomic molecule (14.1)	1

Answers to the First Set of Problems

H

Chapter 1 Thermodynamic State of a Gas

1.1 (1) P, V, T; (2) P, n, T; (3) P, V, n; (4) V, n, T.

1.2 (*a*) 0.696, 0.304; (*b*) 0.522 bar, 0.228 bar; (*c*) 17.00 L.

1.3 21.0 g mol^{-1}; 0.643.

1.4 (*a*) 0.6884, 0.3116; (*b*) 18.49%.

1.5 B'.

1.6 $B = -183$ cm^3 mol^{-1}, $M = 30.07$ g mol^{-1}.

1.7 0.014 L mol^{-1}.

1.8 276 cm^3 mol^{-1}.

1.9 $b = 2\pi d^3 N_A/3$, 21.7 cm^3 mol^{-1}.

1.10 (*a*) 0.603 L mol^{-1}; (*b*) 0.39 L mol^{-1}.

1.12 -0.048 L mol^{-1}, -0.026 L mol^{-1}; Fig. 1.9 yields -0.040 L mol^{-1} and -0.020 L mol^{-1}.

1.13 0.4719 L mol^{-1}.

1.15 $B = 2.38 \times 10^{-5}$ m^3 mol^{-1}, $C = 1.01 \times 10^{-9}$ m^6 mol^{-2}.

1.16 $\overline{V}_c = 0.1914$ L mol^{-1}, $T_c = 310.671$ K, $P_c = 50.609$ bar.

1.17 $\alpha = 1/T$, $\kappa = 1/P$.

1.18 $V = K \exp(\beta T)\exp(-\kappa P)$.

1.19 $V = V_0 \exp[\alpha(T - T_0)]$, $V = V_0[1 + \alpha(T - T_0)]$.

1.20 (*a*) $(\partial P/\partial V)_T = -nRT/(V - nb)^2$, $(\partial P/\partial T)_V = nR/(V - nb)$;
(*b*) $(\partial^2 P/\partial V\, \partial T) = -nR/(V - nb)^2$, $\partial^2 P/(\partial T\, \partial V) = -nR/(V - nb)^2$.

1.22 0.472 bar.

1.22 0.333 bar, $y(O_2) = 0.177$, $y(N_2) = 0.823$.

Chapter 2 First Law of Thermodynamics

2.1 914 m.

2.2 $w = -1.24$ kJ mol^{-1}. The work on the atmosphere is 1.24 kJ mol^{-1}.

2.3 (*a*) 4269 m; (*b*) 6.97 min; (*c*) 1.05 min; (*d*) 1.28 g.

2.4 The second function is exact because $(\partial M/\partial y)_x = (\partial N/\partial x)_y = -1/y^2$.

2.7 (*a*) y, x; (*b*) $1/y, -x/y^2$; (*c*) $1/x, 1/y$; (*d*) $-1/x, -1/y$; (*e*) $y\, e^{xy}, x\, e^{xy}$.

2.8 (*a*) 720 J; (*b*) 735 J; (*c*) −100 J.

2.9 $(\partial/\partial P)(\partial V/\partial T)_P = (\partial/\partial T)(\partial V/\partial P)_T = K\,e^{\alpha T}\,e^{-\alpha T}(-\kappa\alpha)$.

2.10 (*a*) -5.03 kJ mol^{-1}; (*b*) -2.12 kJ mol^{-1}.

2.11 (*a*) $w = nRT \ln[(V_2 - nb)/(V_1 - nb)] + an^2[(1/V_1) - (1/V_2)]$; (*b*) (1) −9697 J; (2) −9575 J.

2.12 (*a*) -3.10 kJ mol^{-1}; (*b*) 40.69 kJ mol^{-1}; (*c*) 37.59 kJ mol^{-1}; (*d*) 40.69 kJ mol^{-1}.

2.13 (*a*) -5.70 kJ mol^{-1}; (*b*) 5.70 kJ mol^{-1}; (*c*) 0; (*d*) 0; (*e*) -2.23 kJ mol^{-1}.

2.14 41.572 kJ mol^{-1}.

2.15 21.468 kJ mol^{-1}.

2.16 (*a*) $w = (P_2V_2 - P_1V_1)/(\gamma - 1)$; (*b*) -1260 J mol^{-1}.

2.17 $\Delta T = 160.4\,°\mathrm{C}$, $P = 1.609$ bar.

2.18 $T = 118.70$ K, -2238 J mol^{-1}.

2.19 (*a*) −46.1 kJ; (*b*) −24.1 kJ.

2.20 0, −899.8, 899.8 J mol^{-1}.

2.21 0, −748, −748, −1247 J mol^{-1}.

2.23 (*a*) −542.2; (*b*) −184.62; (*c*) −103.71; (*d*) −9.48 kJ mol^{-1}.

2.24 (*a*) −13.4, −7.98, −13.6 MJ kg^{-1}; (*b*) −0.744, −0.299, −0.680.

2.25 493.580, 498.346, 513.482 kJ mol^{-1}.

2.26 267.66, −39.84, −214.65, 13.17 kJ mol^{-1}; single reactor.

2.27 −5619 J mol^{-1}.

2.28 45 054 J mol^{-1}.

2.29 −89.849 kJ mol^{-1}.

2.30 493.486 kJ mol^{-1}, 5.115 eV.

2.31 41.4 kJ mol^{-1}.

2.32 −41.6 kJ mol^{-1}.

2.33 −81.21 kJ mol^{-1}.

2.34 −97.12 kJ mol^{-1}.

2.35 (*a*) −25 968 kJ mol^{-1}; (*b*) 2359 kJ mol^{-1}; (*c*) 677.4 kJ mol^{-1}; (*d*) 714.78 kJ mol^{-1}.

Chapter 3 Second and Third Laws of Thermodynamics

3.5 (*a*) 109.04; (*b*) 0 J K^{-1} mol^{-1}.

3.6 55.42 J K^{-1} mol^{-1}.

3.7 (*a*) 6.371; (*b*) 10.618 J K^{-1} mol^{-1}.

3.8 (*a*) 26.4 kJ mol^{-1}; (*b*) −4.96 kJ mol^{-1}; (*c*) 26.4 kJ mol^{-1}; (*d*) 21.4 kJ mol^{-1}; (*e*) 46.99 J K^{-1} mol^{-1}.

3.9 Since this ΔS is always positive, the change is always spontaneous.

3.10 (*a*) 15.56; (*b*) 0 J K^{-1} mol^{-1}.

3.11 19.14 J K^{-1} mol^{-1}, 0.

3.12 (*a*) $\Delta U = 0$ kJ, $\Delta S = 30.03$ J K^{-1}, $w = -9.01$ kJ, $q = -w = 9.01$ kJ; (*b*) $\Delta\overline{U} = 0$ kJ mol^{-1}, $\Delta\overline{S} = 10.01$ J K^{-1} mol^{-1}, $w = -3.00$ kJ mol^{-1}, $q = 3.00$ kJ mol^{-1}; (*c*) $\Delta\overline{U} = 0$ kJ mol^{-1}, $\Delta\overline{S} = 10.01$ J K^{-1} mol^{-1}, $w = 0$ kJ mol^{-1}, $q = 0$ kJ mol^{-1}.

3.13 (*a*) 19.14 J K^{-1} mol^{-1}, the same; (*b*) 0, 19.14 J K^{-1} mol^{-1}.

3.14

	(*a*) *Reversible*	(*b*) *Irreversible*	(*c*) *Isol. Rev.*	(*d*) *Isol. Irrev.*
w/kJ mol^{-1}	−5.71	0	0	0
q/kJ mol^{-1}	5.71	0	0	0
$\Delta\overline{U}$/kJ mol^{-1}	0	0	0	0
$\Delta\overline{H}$/kJ mol^{-1}	0	0	0	0
$\Delta\overline{S}$/J K^{-1} mol^{-1}	19.1	19.1	0	19.1

3.15 19.26 J K^{-1} mol^{-1}.

3.16 (*a*) −39.52; (*b*) 41.92; (*c*) 2.40 J K^{-1}, irreversible because $\Delta S_{syst} = 0$.

3.17 1.12 J K^{-1} mol^{-1}.

3.18 44.95 J K^{-1}.

3.20 12.51 J K^{-1} mol^{-1}.

3.22 14.92 J K^{-1} mol^{-1}.

3.23 −154.4 kJ mol^{-1}.

3.24 150.67 J K^{-1} mol^{-1}.

3.25 (*a*) $\Delta S(H_2) = -\Delta S(\text{surr}) = 45.31$ J K^{-1} mol^{-1}, $\Delta S(H_2 + \text{surr}) = 0$;
(*b*) $\Delta S(H_2) = \Delta S(\text{surr}) = \Delta S(H_2 + \text{surr}) = 0$.

3.26 17 000 ft.

3.27 (*a*) 214 J; (*b*) 307 J, or 93 J more than (*a*).

3.28 (*a*) $\dfrac{\Delta T}{\Delta h} = -\dfrac{Mg}{C_P}$; (*b*) $\Delta T = -16.8\,°\text{C}$.

Chapter 4 Fundamental Equations of Thermodynamics

4.1 (*a*) 24.91; (*b*) 25.55 J K^{-1} mol^{-1}.

4.2 $\overline{C}_P - \overline{C}_V = R/\{1 - (2a/RT)[(\overline{V} - b)^2/\overline{V}^3]\}$.

4.3 $dS = (C_P/T)\,dT - \alpha V\,dP$.

4.4 −10.99 J K^{-1} mol^{-1}.

4.5 $\overline{C}_P - \overline{C}_V = R/(1 - 2a/RT\overline{V})$.

4.6 0.51 J K^{-1} mol^{-1}.

4.7 (*a*) $b - 2a/RT$; (*b*) −25 J bar^{-1} mol^{-1}.

4.8 $(\partial\overline{U}/\partial\overline{V})_{\overline{V}} = (RT^2/\overline{V}^2)(\partial B/\partial T)_{\overline{V}}$.

4.9 $\Delta G° = -213.9$ J mol^{-1}. This is negative, as expected for a spontaneous process at constant T and P.

4.10 (*a*) $\Delta G_2 = \Delta G_1 T_2/T_1 + \Delta H(1 - T_2/T_1)$; (*b*) $\Delta G_2 = \Delta G_1 T_2/T_1 + (\Delta H_1 + T_1\Delta C_P)(1 - T_2/T_1) + T_2\Delta C_P \ln(T_2/T_1)$.

4.11 1.8 kJ mol^{-1}.

4.12 (*a*) −5708 J mol^{-1}; (*b*) −5708 J mol^{-1}.

4.13 166.848 J K^{-1} mol^{-1}, 11.41 kJ mol^{-1}.

4.14 (*a*) −4993 J mol^{-1}; (*b*) 4993 J mol^{-1}; (*c*) 4993 J mol^{-1};
(*d*) 4993 J mol^{-1}; (*e*) 0; (*f*) 0; (*g*) −13.38 J K^{-1} mol^{-1}.

4.15 (*a*) −3193 J mol^{-1}; (*b*) 33 340 J mol^{-1}; (*c*) 33 340 J mol^{-1}; (*d*) 30 147 J mol^{-1};
(*e*) 0; (*f*) 86.8 J K^{-1} mol^{-1}.

4.16 $S = -a + c/T^2$; $H = b + 2c/T$.

4.17 −65.1 J mol^{-1}.

4.18 -63.6 J mol^{-1}.

4.19 $\mu(\text{real}) - \mu(\text{ideal}) = 1.72 \text{ kJ mol}^{-1}$.

4.20 $\overline{G} = \overline{G}^\circ + RT \ln(P/P^\circ) + (b - a/RT)P$

$\overline{S} = \overline{S}^\circ - RT \ln(P/P^\circ) + aP/RT^2$

$\overline{A} = \overline{G}^\circ + RT \ln(P/P^\circ) - RT$

$\overline{U} = \overline{U}^\circ - aP/RT$

$\overline{H} = \overline{H}^\circ + (b - 2a/RT)P$

$\overline{V} = RT/P + b - a/RT$

4.21 $\Delta A = -RT \ln[(V_2 - b)/(V_1 - b)] - a(1/V_2 - 1/V_1), \Delta U = a(1/V_2 - 1/V_1)$.

4.22 $dX = T\,dS - P\,dV - n\,d\mu$, $dY = -S\,dT - P\,dV - n\,d\mu$, $dZ = T\,dS + V\,dP - n\,d\mu$.

4.23 106.9 bar.

4.24 51.7 bar.

4.25 $29.3 \text{ cm}^3 \text{ mol}^{-1}$.

4.26 -1239 J mol^{-1}; $4.159 \text{ J K}^{-1} \text{ mol}^{-1}$.

4.27 3.4 kJ.

Chapter 5 Chemical Equilibrium

5.1 (*a*) 48.9 kJ mol^{-1}; (*b*) $-8.78 \text{ kJ mol}^{-1}$; (*c*) yes.

5.2 6.47×10^{-3}.

5.3 (*a*) 0.503; (*b*) 1.36; (*c*) 0.879.

5.4 1.83.

5.5 1.64×10^{-3}.

5.6 5.73 kJ mol^{-1}.

5.7 (*a*) 9.520 g; (*b*) 1.206, 0.032, 1.440 bar.

5.8 $K = \xi^2(2 - 2\xi)^2/(1 - \xi)(1 - 3\xi)^3(P/P^\circ)$.

5.9 0.0814, 0.0335, 94.6%.

5.10 1.44, 0.545.

5.11 0.0055.

5.12 (*a*) 10.3; (*b*) 22.0 bar.

5.16 0.311.

5.17 231 bar.

5.18 (*a*) 12.7%; (*b*) no effect; (*c*) 54 bar.

5.19 (*a*) 35.6; (*b*) 0.084 bar.

5.20 (*a*) 2.03×10^{-3}; (*b*) increase; (*c*) increase; (*d*) no; (*e*) decrease.

5.21 0.973.

5.22 (*a*) decreases; (*b*) increases; (*c*) no effect; (*d*) decreases; (*e*) increases.

5.23 1023 K.

5.24 75 kJ mol^{-1}.

5.25 0.0145, $K_c = [C_2H_4][H_2]/[C_2H_6]c^\circ$.

5.26 (*a*) $y(N_2) = 0.1266$, $y(H_2) = 0.3798$, $y(NH_3) = 0.493$, $V = 22.27$ L; (*b*) $y(N_2) = 0.990$, $y(H_2) = 0.2970$, $y(NH_3) = 0.302$, $P = 0.698$ bar.

5.28 $P(H_2O) = 0.084$, $P(CO) = 0.458$, $P(H_2) = 0.458$ bar.

5.29 3.47×10^{-5} bar.

5.31 -80.67 J K^{-1} mol^{-1}. Since the ions polarize the neighboring water molecules, the product state is more ordered than the reactant state.

5.32 (*a*) 0.0204, 0.187; (*b*) 154 kJ mol^{-1}.

5.33 425 K.

5.35 3.2 bar.

5.36 1.82×10^{-13} bar.

5.37 $-\Delta_r H/T$ is the increase in the entropy of the reservoir, and $\Delta_r S$ is the increase in the entropy of the reaction. Thus, $-\Delta_r H/T + \Delta_r S = -\Delta_r G$ is the global increase in entropy.

5.38 $\Delta\nu RT > |\Delta U°|$.

5.39 9.2×10^{-5} bar.

5.40 (*a*) 1.4×10^{-5} bar; (*b*) 6.83×10^{-3} bar; (*c*) 0.72 bar.

5.41 $n(CuO) = 0.82$ μmol, $n(Cu_2O) = 0.14$ μmol, $n(O_2) = 0.02$ μmol.

5.42 $\xi = K/[4(P/P°) + K]$.

5.43 0.249.

5.44 (*a*) One, for example, $CO_2 + H_2 = CO + H_2O$;
(*b*) two, for example, $CO_2 + H_2 = CO + H_2O$ and $2CO = CO_2 + C$.

5.45 $3C_2H_2 = C_6H_6$
$5C_2H_2 = C_{10}H_8 + H_2$

$$\begin{bmatrix} 2 & 0 & 6 & 10 \\ 2 & 2 & 6 & 8 \end{bmatrix} \begin{bmatrix} -3 & -5 \\ 0 & 1 \\ 1 & 0 \\ 0 & 1 \end{bmatrix} = \begin{bmatrix} 0 & 0 \\ 0 & 0 \end{bmatrix}$$

5.46 $dG = -S\,dT + V\,dP + \mu_A\,dn'_I + \mu_B\,dn'_B$, where $n'_I = n_A + n_C$ and $n'_B = n_B + n_C$.

5.47 There are four components, so there are $R = N - C = 6 - 4 = 2$ independent reactions. These reactions can be taken to be

$$ClO_3^- + 5Cl^- + 6H^+ = 3H_2O + 3Cl_2$$
$$5ClO_3^- + Cl^- + 6H^+ = 3H_2O + 6ClO_2$$

5.48 (*a*)

$$\mathbf{A} = \begin{array}{c} \begin{matrix} C_2H_4 & C_3H_6 & C_4H_8 \end{matrix} \\ \begin{bmatrix} 2 & 3 & 4 \\ 4 & 6 & 8 \end{bmatrix} \end{array}$$

(*b*) The row-reduced form is [1, 1.5, 2]; (*c*) there is one component because there is one independent row; (*d*) the two independent reactions are

$$R1: \quad 1.5C_2H_4 = C_3H_6$$
$$R2: \quad 2C_2H_4 = C_4H_8$$

5.49 (*a*) 1; (*b*) 2; (*c*) 4; (*d*) 2.

5.50 3; $T, P, n_c(H)/n_c(O)$.

5.51 (*a*) 3; (*b*) 2; (*c*) 4.

Chapter 6 Phase Equilibrium

6.1 The boiling point is reduced 0.4 °C to 68.3 °C.

6.2 90.6 °C.

6.3 92.4 °C.

6.4 (*a*) −38.81; (*b*) −16.6 °C.

6.5 0.258 bar.

6.6 11.548 kJ mol^{-1}, 6.190 × 10^{-3} bar.

6.7 3780 bar.

6.8 (*a*) 44.8 kJ mol^{-1}; (*b*) 97.03 °C; (*c*) 97.36 °C.

6.9 64.0 °C, 152.2 kPa.

6.10 166 Pa.

6.11 (*a*) 31.4 kJ mol^{-1}; (*b*) 13.9 kJ mol^{-1}; (*c*) 17.5 kJ mol^{-1}; (*d*) −88 °C.

6.12 (*a*) 6.85 °C; 5249 Pa; (*b*) 10.48 kJ mol^{-1}.

6.13 3.35 × 10^{-2} cm.

6.14 (*a*) 15.7 cm; (*b*) 7.86 mm.

6.15 (*a*) 200 Pa; (*b*) 20 Pa.

6.16 (*a*) 3.17 × 10^3 Pa; (*b*) 373.5 K.

6.17 3, 4, and 5.

6.18 25.552 mm of mercury or 3406 Pa.

6.20 (*a*) 0.590, 0.410; (*b*) 6980, 2430, 9410 Pa; (*c*) 0.742.

6.21 (*a*) 29.4 kPa; (*b*) 0.581.

6.22 (*a*)

t/°C	88	94	100
x_{benz}	0.633	0.422	0.244
y_{benz}	0.814	0.644	0.439

(*b*) Bubble point, 92 °C; $y_{benz} = 0.72$.

6.23 (*a*) 0.560, 0196; (*b*) 0.884, 0.577.

6.24 Three theoretical plates.

6.25 20.7%.

6.26

$$\Delta_{mix}G = RT(x_1 \ln x_1 + x_2 \ln x_2) + wx_1x_2$$
$$\Delta_{mix}S = -R(x_1 \ln x_1 + x_2 \ln x_2)$$
$$\Delta_{mix}H = wx_1x_2$$
$$\Delta_{mix}V = 0$$

6.27 (*a*) $y_{EtOH} = 0.69$; (*b*) $y_{EtOH} = 1$; (*c*) $y_{EtOH} = 0.69$, $y_{EtOH} = 0.90$.

6.28 −0.00233 °C.

6.30

y_{EtOH}	0	0.2	0.4	0.6	0.8	1.0
γ_{EtOH}	—	2.045	1.316	1.065	0.982	1.000
$\gamma_{(CHCl_3)}$	1	1.111	1.333	1.627	1.854	—

6.32 $\gamma_1 = \exp(wx_2^2/RT)$, $\gamma_2 = \exp(wx_1^2/RT)$.

6.33

$$\frac{w}{RT} = \frac{P - (x_1P_1^* + x_2P_2^*)}{x_1x_2[P_1^* + x_1(P_2^* - P_1^*)]}$$

6.34

x_2	0	0.2	0.4	0.6	0.8	1.0
γ_2	—	0.24	0.13	0.093	0.080	—
γ_1	1.00	1.15	1.52	2.09	2.82	—

6.35 (*a*) 2.3066 kPa; (*b*) −0.372 °C.

6.36 (*a*) 719 Pa; (*b*) 73.4 mm of water.

6.37 255 kg mol^{-1}.

6.38 27.3 bar.

6.39 0.365, 0.516.

6.40 0.297.

6.41 0.513 K.

6.42 0.0569, yes.

6.43 Sb_2Cd_3 is formed.

6.44 In the liquid region $\upsilon = 2$, in the two-phase regions $\upsilon = 1$, and at the eutectic points $\upsilon = 0$.

6.46 461 cm^3 ethanol, 570 cm^3 water; the shrinkage is 31 cm^3.

6.47 0.143 bar.

6.48 $x_{Bi} = 0.700$.

6.50 $0 = x_1\,d\overline{V}_1 + x_2\,d\overline{V}_2$.

6.51 (*a*) $\overline{V} = 18.023 + 53.57x_2 + 1.45x_2^2$; (*b*) $\overline{V}_1 = 18.023 - 1.45x_2^2$, $\overline{V}_2 = 71.60 + 2.90x_2 - 1.45x_2^2$.

Chapter 7 Electrochemical Equilibrium

7.1 14.398 J C^{-1}; 1389.3 kJ mol^{-1}.

7.2 277.8 kJ mol^{-1}.

7.3 5.04 miles.

7.4 (*a*) 6.34×10^{-3}; (*b*) 7.44×10^{-5}.

7.5 0.872.

7.6 (*a*) 0.24; (*b*) 0.80; (*c*) 0.06 mol kg^{-1}.

7.7 (*a*) 0.964; (*b*) 0.880; (*c*) 0.762.

7.8 1.140 V.

7.9 (*a*) 0.710 V; (*b*) 0.804.

7.10 (*a*) Na | NaOH(m) | H_2 | Pt
At 25 °C, $E = E^\circ - 0.0591 \log[m^2 \gamma_\pm^2 P(H_2)^{1/2}]$
(*b*) Pt | H_2 | $H_2SO_4(m)$ | Ag_2SO_4 | Ag
At 25 °C, $E = E^\circ - 0.0296 \log[4m^3 \gamma_\pm^3 P(H_2)^{-1}]$

7.11 (*a*) $Pb(s) + Hg_2SO_4(s) = PbSO_4(s) + 2Hg(l)$; (*b*) −186.16 kJ mol^{-1}, 33.58 J K^{-1} mol^{-1}, −176.15 kJ mol^{-1}.

7.12 (*a*) −131.260 kJ mol^{-1}, −167.127 kJ mol^{-1}, −120.3 kJ mol^{-1};
(*b*) −131.260 kJ mol^{-1}, −167.127 kJ mol^{-1}, 56.6 J K^{-1} mol^{-1}.

7.13 (*a*) 81.902 kJ mol^{-1}, 5.69×10^{-14}; (*b*) 81.673 kJ mol^{-1}, 6.28×10^{-14}.

7.14 0.828 V.

7.15 (*a*) 212.55; (*b*) −262.46; (*c*) −47.40; (*d*) −25.71;
(*e*) −2.9 kJ mol^{-1}.

7.17 −131.258, −157.26, −261.86 kJ mol^{-1}.

7.18 (*a*) 3.1×10^{51}; (*b*) 1.8×10^4; (*c*) 7.8×10^{-77}.

7.19 0.1801 V.

7.20 (*a*) There is an additional independent variable beyond T and P;
(*b*) $F = 2\,(T, P)$; (*c*) $F = 3\,(T, P, E)$.

7.21 (*a*) $K_1 = 4.35 \times 10^{-7}$; $K_2 = 4.69 \times 10^{-11}$;
(*b*) $K_a[H_2CO_3(aq)] = 1.68 \times 10^{-4}$.

7.22 -0.152 V.

7.23 1.33×10^{-5} mol kg^{-1}.

7.24 (*a*) -3.040; (*b*) 2.889; (*c*) 1.458 V.

7.25 261.905 kJ mol^{-1}, 240.12 kJ mol^{-1}, -73.132 J K^{-1} mol^{-1}, -32.572 J K^{-1} mol^{-1}.

7.27 (*a*) 73.08; (*b*) 73.13 J K^{-1} mol^{-1}.

7.28 (*a*) 6.49; (*b*) 3.24×10^{-7}.

7.29 1.30.

7.30 0.60.

7.31 (*a*) 1.086 V; (*b*) 1.032 V.

7.32 (*a*) -817.90 kJ mol^{-1}, 1.0596 V; (*b*) -890.4 kJ mol^{-1}, 356 kJ mol^{-1}.

7.33 3.474 V.

7.34 0.018 V; the β phase is positive.

7.35 -0.036 V.

7.36 (*a*) 96.8×10^3 C mol^{-1}; (*b*) 6.04×10^{23} mol^{-1}.

Chapter 8 Thermodynamics of Biochemical Reactions

8.3 79.885 kJ mol^{-1}, 55.836 kJ mol^{-1}, -80.66 J K^{-1} mol^{-1}, 1.010×10^{-14}.

8.4 For the acetic acid dissociation, $\Delta_r S° = 91$ J K^{-1} mol^{-1}. The increase in order is due to the hydration of the ions that are formed. For aniline, $\Delta_r S° = -18$ J K^{-1} mol^{-1}. This number is smaller because there is no change in the number of ions.

8.5 2.026, 6.831.

8.6 $[His^-] = 6.3 \times 10^{-4}$ mol L^{-1}, $[HisH] = 9.03 \times 10^{-2}$ mol L^{-1}, $[HisH_2{}^+] = 9.03 \times 10^{-3}$ mol L^{-1}, $[HisH_3{}^{2+}] = 6 \times 10^{-8}$ mol L^{-1}.

8.7 4.35.

8.8 $dG = -S\,dT + V\,dP + \mu(H^+)\,dn'(H^+) + \mu(A^-)\,dn'(A)$.

8.9 Equation d can be written

$$G' = n'(P_i)\{r_1\mu_1'^{\circ} + r_2\mu_2'^{\circ} + RT\ln(r_1\ln r_1 + r_2\ln r_2) + RT\ln([P_i]/c°)\}$$

8.10

$$\Delta_f G'^{\circ}(A) = -RT\ln\{\exp[-\Delta_f G°(A^-)/RT] + \exp\{-[\Delta_f G°(HA) - \Delta_f G°(H^+) - RT\ln([H^+]/c°)]/RT\}$$

$$\Delta_f H'^{\circ} = r(A^-)\Delta_f H°(A^-) + r(HA)[\Delta_f H°(HA) - \Delta_f H°(H^+)]$$

8.11 No, 1.1×10^{-2} mol L^{-1}.

8.12 5.77 kJ mol^{-1}.

8.13 135 g.

8.14 -20.3 kJ mol^{-1}.

8.15 11.2.

8.16 -41.0 kJ mol^{-1}.

8.17 (*a*) 3.6 kJ mol^{-1}; (*b*) 5.2 kJ mol^{-1}.

8.18 $Y = ([A]^n/K)/(1 + [A]^n/K)$.

8.19 $K_1 = ([ML] + [LM])/[M]P_L = K_1^* + K_2^*$; $K_2 = [LML]/\{([ML] + [LM])P_L\} = 1/(1/K_3^* + 1/K_4^*)$. Since there are five species and two components (protein and ligand), the number of independent equilibria is 3; $C = N - R$ is $2 = 5 - 3$. Since there are two paths from M to LML, $K_1^*K_3^* = K_2^*K_4^*$.

8.21 If the initial concentration of hemoglobin is 0.0025 M, $[\alpha\beta] = 0.511\times 10^{-5}$ M and $[(\alpha\beta)_2] = 0.002\,47$ M (1.04% dissociated). If the initial concentration of hemoglobin is 0.000 25 M, $[\alpha\beta] = 1.60 \times 10^{-5}$ M and $[(\alpha\beta)_2] = 0.000\,242$ M (3.2% dissociated).

8.22 $2.4, 4 \times 10^7$.

8.23 (*a*) $F = 3, D = 4, (dG)_{T,P,n_{cA},n_{cB}} \le 0$; (*b*) $F = 3, D = 4$, $(dG')_{T,P,n_{cA},\mu_B} \le 0$.

8.26 (*a*) 3919, 0.769; (*b*) 327, 0.969.

Chapter 9 Quantum Theory

9.1 $6.037 \times 10^{17}\ s^{-1}$; X-rays.

9.2 (*a*) 3.61×10^{-19} J, $2.77 \times 10^{20}\ s^{-1}$;
(*b*) 1.99×10^{-23} J, $5.03 \times 10^{25}\ s^{-1}$.

9.3 4.64 eV.

9.4 $4.668 \times 10^7, 4.634 \times 10^7, 4.473 \times 10^7$ m s^{-1}.

9.5 (*a*) 0.0387 nm; (*b*) 2.24 nm.

9.6 0.012 J mol^{-1}.

9.7 0.0259 nm, 3.46 nm.

9.8 0.1452 nm.

9.9 (*a*) 2.21×10^{-33} m; (*b*) 6.626×10^{-19} m; (*c*) 6.626×10^{-10} m;
(*d*) 3.1×10^{-9} m.

9.10 5.27×10^{-25} J, 0.318 J mol^{-1}.

9.11 5.

9.12 (*a*) $-2ax\,e^{-ax^2}$, $4ax^2 e^{-ax^2} - 2a\,e^{-ax^2}$; (*b*) $-b \sin bx$, $-b^2 \cos bx$ (therefore, $\cos bx$ is an eigenfunction of d^2/dx^2 with eigenvalue $-b^2$);
(*c*) $ik\,e^{ikx}$, $-k^2 e^{ikx}$ [therefore, e^{ikx} is an eigenfunction of both d/dx and d^2/dx^2 with eigenvalues ik and $(ik)^2 = -k^2$, respectively].

9.13 $(\hat{x}\hat{p}_x - \hat{p}_x\hat{x}) = -\hbar/i$.

9.14 (*a*) 1/2; (*b*) 0.609.

9.15 (*a*) 580.5, 2322, 5225 kJ mol^{-1}; (*b*) 68.7 nm.

9.16 (*a*) Approximately 27; (*b*) 2.7×10^4; (*c*) 2.7×10^8.

9.18 1, 3, 3.

9.19 (*a*) $0.568\hbar, 1.67\hbar, 2.63\hbar$, compared with $0.500\hbar$; (*b*) 0.03615 nm, 0.3615 nm.

9.20 $(3/2\alpha)^{1/2}$.

9.21 $(3\alpha/2)^{1/2}\hbar$.

9.22 $3\hbar/2$ compared with $\hbar/2$.

9.23 1903.01 N m^{-1}.

9.24 2117.5 cm^{-1}.

9.25 3.364 pm or 2.98% of the equilibrium bond length.

9.26 (*a*) 323.3, 122.5, 575.1 N mol^{-1}; (*b*) 545 cm^{-1}.

9.28 $\frac{3}{2}h\nu_0$.

9.29 $6.93 \times 10^{13}\ s^{-1}, 4.52 \times 10^{13}\ s^{-1}$.

9.30 $\langle x \rangle = 0, \langle x^2 \rangle = \frac{3}{2}(\hbar^2/k\mu)^{1/2}, \Delta x = [\frac{3}{2}(\hbar^2/k\mu)^{1/2}]^{1/2}$.

9.31 (*a*) 1902 N m^{-1}; (*b*) 3.381×10^{-12} m; (*c*) 1.560×10^{-23} kg m s^{-1};
(*d*) 5.273×10^{-35} J s.

9.32 $1.269 \times 10^{12}\ s^{-2}$; 42.3 cm^{-1}.

9.33 $E = \hbar^2/I$; the angular momentum is $[1(1+1)]^{1/2}\hbar = 2^{1/2}\hbar$.

9.34 8.67×10^{-24}, 2.60×10^{-23} J.

9.35 $\mathrm{i}\hbar\hat{L}_z$, $-\mathrm{i}\hbar\hat{L}_y$.

9.36 (*a*) 1.139×10^{-26} kg; (*b*) 1.449×10^{-46} kg m^2; (*c*) 2.58×10^{-3} m.

9.37 (*a*) 4830 K; (*b*) 3623 K.

9.38 (*a*) $hc/4.965k$; (*b*) 5.9×10^{23} mol^{-1}.

Chapter 10 Atomic Structure

10.1 13.598 42 eV.

10.2 (*a*) 145.9; (*b*) 82.05; (*c*) 82.05; (*d*) 36.47 kJ mol^{-1}.

10.3 $E/V = 54.44, 122.49, 217.76, 340.25, 489.96$.

10.4 (*a*) 13.598 48, 13.602 18 eV; (*b*) 656.470, 656.291 nm.

10.6 $1.646\,397 \times 10^{-28}$ kg, $1.983\,35 \times 10^9$ m^{-1}, $E = -(2461/n^2)$ eV, 0.2928 pm.

10.7 $Z = 5$, $n_1 = 2$, and $\lambda^{-1} = 25(0.010\ 973\ 732)(\frac{1}{4} - 1/n_2^2)$; $n_2 = 2 + m$, $m = 1, 2, \ldots$.

10.8 1.8756, 1.2822. 1.0941 μm.

10.9 4.49 cm.

10.10 10 967 758.56 m^{-1}, 10 970 742.75 m^{-1}.

10.13 $\langle r \rangle = 3a_0/2Z$, $r_{\mathrm{mp}} = a_0/Z$.

10.15 1, 0, −1.

10.16 (*a*) 2; (*b*) 8; (*c*) 18 (*note:* degeneracy $= 2n^2$).

10.17 105.8, 176.4 pm.

10.18 For 2s, $r_{\mathrm{node}} = 2a_0/Z$. For 3s, $r_{\mathrm{node}} = 7.098\,a_0/Z$, $1.902\,a_0/Z$.

10.19 (*a*) 317.5, 264.6 pm; (*b*) 105.8, 88.2 pm.

10.20 $\langle r \rangle_{2,0} = (2^2 a_0/Z)(1 + \frac{1}{2}) = 6a_0/Z$; $\langle r \rangle_{2,1} = (2^2 a_0/Z)[1 + \frac{1}{2}(1 - \frac{2}{4})] = 5a_0/Z$.

10.21 408.5 MHz; 0.00508 mm.

10.22 $\langle r \rangle = 5a_0/Z$.

10.24 For 2p, $|\boldsymbol{L}| = 1.491 \times 10^{-34}$ J s, $L_z/\hbar = 1, 0, -1$. For 3d, $|\boldsymbol{L}| = 2.583 \times 10^{-34}$ J s, $L_z/\hbar = -2, -1, 0, +1, +2$.

10.25

	3s	3p	3d
$\lvert\boldsymbol{L}\rvert = \sqrt{\ell(\ell+1)}\hbar$	0	$\sqrt{2}\hbar$	$\sqrt{6}\hbar$
Radial nodes	2	1	0
Angular nodes	0	1	2

10.26 The number of angular nodes is 1; the number of radial nodes is $n - 1 - 1$; the total number of nodes is $n - 1$.

	1s	2s	2p	3p	3d
n	1	2	2	3	3
ℓ	0	0	1	1	2
Angular	0	0	1	1	2
Radial	0	1	0	1	0
Total	0	1	1	2	2

10.27 $N = 2^{-1/2}$.

10.28 1.8569×10^{-23} J, 1.0698×10^{-2} m, microwave.

10.30 $1s^2$, $1s^2$, $1s^2 2s^2 2p^6$, $1s^2 2s^2 2p^6$, $1s^2 2s^2 2p^6$, $1s^2 2s^2 2p^6$.

10.31 2, 2, 6, 2, 6, 10.

Chapter 11 Molecular Electronic Structure

11.1 13.6 eV, 19.4 eV, −300 eV, −51.4 eV.

11.2 941.49 kJ mol^{-1}.

11.3 $[2(1+s)]^{-1/2}$, $[2(1-s)]^{-1/2}$.

11.4 See Fig. 11.3.

11.5 Maximum at $R = 0$, minimum at $R = \infty$, $S = 0.766$.

11.6

$$\psi_1 = \frac{1}{2^{1/2}}\begin{vmatrix} 1s_A(1)\alpha(1) & 1s_A(2)\alpha(2) \\ 1s_B(1)\beta(1) & 1s_B(2)\beta(2) \end{vmatrix} - \frac{1}{2^{1/2}}\begin{vmatrix} 1s_A(1)\beta(1) & 1s_A(2)\beta(2) \\ 1s_B(1)\alpha(1) & 1s_B(2)\alpha(2) \end{vmatrix}$$

$$\psi_2 = \frac{1}{2^{1/2}}\begin{vmatrix} 1s_A(1)\alpha(1) & 1s_A(2)\alpha(2) \\ 1s_B(1)\alpha(1) & 1s_B(2)\alpha(2) \end{vmatrix}$$

ψ_3 is the same as ψ_2 with β replacing α. ψ_4 is obtained from ψ_1 by replacing the minus sign with a plus sign.

11.8 For example,

$$\psi_{sp^2(i)} = \frac{1}{4\sqrt{2\pi}}[(2-\sigma) + \sqrt{2}\sigma\cos\theta]\frac{e^{-\sigma}}{\sqrt{3}}$$

$$\theta = 0^\circ: \quad \psi = \frac{1}{4\sqrt{6\pi}}(2 + 0.4.14\sigma)\,e^{-\sigma}$$

$$\theta = 90^\circ: \quad \psi = \frac{1}{4\sqrt{6\pi}}(2 - \sigma)\,e^{-\sigma}$$

$$\theta = 180^\circ: \quad \psi = \frac{1}{4\sqrt{6\pi}}(2 - 2.414\sigma)\,e^{-\sigma}$$

11.10 $E = (\alpha + \beta)/(1 + S)$ and $E = (\alpha - \beta)/(1 - S)$.

11.11 $\frac{1}{2}$; $\frac{5}{2}$; 3; 2.

11.12 The CH_3 radical is planar with the 3 two-electron bonds due to the overlapping of the sp^2 orbitals of C with the s orbitals of H. The odd electron is in the remaining unhybridized p orbital, which is perpendicular to the plane.

11.13 (*a*) $10\sigma, 2\pi$; (*b*) $14\sigma, 0\pi$; (*c*) $18\sigma, 4\pi$; (*d*) $24\sigma, 6\pi$.

11.14 −153 kJ mol^{-1}.

11.15 ψ_1, no nodes; ψ_2, one node between atoms 2 and 3; ψ_3, two nodes between 1 and 2 and between 3 and 4; ψ_4, three nodes between 1 and 2, 2 and 3, and 3 and 4.

11.16 Butadiene, 0.472β; benzene, 2β; $\beta = -76.5$ kJ mol^{-1}; the extra stabilization of butadiene is −36.1 kJ mol^{-1}.

11.17 0.472β; 2β.

11.19 $r_m = 0.309$ nm.

11.20 5.70 eV, 34.7×10^{-30} C m.

11.21 3.78×10^{-29} C m. The ions polarize each other.

11.22 20.4×10^{-30} C m. The charges are not completely separated in HCl.

11.23 0.17, 0.12, 0.039. These are in accord with the electronegativities.

Chapter 12 Symmetry

12.1 C_{2v}

12.2 C_{3v}

12.3 D_{4h}

12.4 C_2

12.5 D_{3h}

12.6 C_1

12.7 C_{4v}

12.8 D_{3d}

12.9 D_{2h}

12.10 D_{2h}

12.11 D_{2d}

12.12 C_{2v}

12.13 D_{2h}

12.14 C_i

12.15 D_{3h}

12.16

	E	C_2^1	σ_h	i
E	E	C_2^1	σ_h	i
C_2^1	C_2^1	E	i	σ_h
σ_h	σ_h	i	E	C_2^1
i	i	σ_h	C_2^1	E

12.17

S_6^1	S_6^2	S_6^3	S_6^4	S_6^5	S_6^6
	C_3^1	i	C_3^2		E

12.18 1,1-Dichloroethylene, C_{2v}, dipole; *trans*-1,2-dichloroethylene, C_{2h}, no dipole; *cis*-1,2-dichloroethylene, C_{2v}, dipole.

12.19 D_{2d}, no dipole moment.

12.20 H_2S, PCl_3, $C_2H_4Cl_2$, $C(CH_3)ClBrH$, IF_5, thiophene.

12.21 $C(CH_3)ClBrH$.

12.22 1,2-Dichlorobenzene, C_{2v}, dipole; 1,3-dichlorobenzene, C_{2v}, dipole; 1,4-dichlorobenzene, D_{2h}, no dipole.

12.23

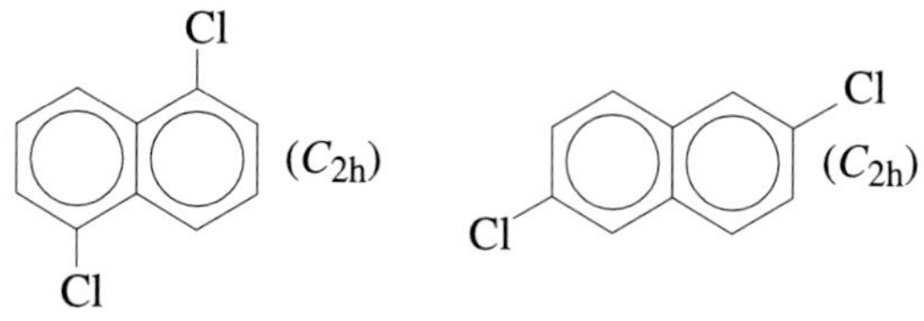

Cl
(C_{2h})
Cl

Cl
(C_{2h})
Cl

12.24 *Cis*-bent, C_{2v}; *trans*-bent, C_{2h}.

Chapter 13 Rotational and Vibrational Spectroscopy

13.1 14.4, 1440, 144 000 K.

13.2 (*a*) 2991, 299.1 nm; (*b*) 3343, 33 430 cm^{-1}; (*c*) 0.415, 4.15 eV.

13.3 (*a*) 207 cm^{-1}; (*b*) 4.83×10^{-3} cm.

13.5 3.162×10^{-27} kg, 5.141×10^{-47} kg m^2.

13.6 1.917×10^{-47} kg m^2.

13.7 The following molecules have permanent dipole moments: HBr, CH_3CH_2OH, H_2O.

13.8 0.091 83 cm.

13.9 112.8 pm, 11.589 cm^{-1}, 15.452 cm^{-1}.

13.10 3.693, 3.596, 3.766, 3.863 cm^{-1}.

13.11 (*a*) 4.37×10^{-47} kg m^2; (*b*) 163 pm.

13.12 At 300 K, 1, 2.710, 3.686, 3.805;
at 1000 K, 1, 2.910, 4.563, 5.830.

13.14 (*a*) 3, 3.79; (*b*) 7, 8.90.

13.15 (*a*) 0.1162; (*b1*) 8.071×10^{-46} kg m^2; (*b2*) same.

13.16 $I = \frac{8}{3}mR^2$.

13.17 6.35 cm^{-1}, 9.95 cm^{-1}, $\lambda_{0\to1} = 1.01 \times 10^{-3}$ m, $\lambda_{1\to2}^{K=0} = 2.51 \times 10^{-4}$ m, $\lambda_{1\to2}^{K=\pm1} = 2.51 \times 10^{-4}$ m.

13.19 (*a*) 2.5 nm; (*b*) 3.5 nm.

13.20

	I_2	Br_2	Cl_2
D_0/eV	1.54	1.97	2.48
k/N m^{-2}	172.2	246.3	323.0

13.21 −879 J mol^{-1}.

13.22 3112.1, 3811.6 cm^{-1}.

13.23 (*a*) 0.0133, 0.0018, 2×10^{-6}; (*b*) 0.0128, 0.0120, 0.0014.

13.25 (*a*) 6.2×10^{-10}; (*b*) 0.0165, 0.034, 0.007.

13.26 2886.30, 9.98, 10.19, 10.30, 0.21 cm^{-1}.

13.27 216.088 kJ mol^{-1}.

13.28 (*a*) 2632.72, 2666.61 cm^{-1};
(*b*) 3.798 34, 3.750 08 μm.

13.29 (*a*) 3; (*b*) 6; (*c*) 7; (*d*) 30.

13.30 (*a*) 3, 0, 0; (*b*) 3, 2, 1; (*c*) 3, 2, 4; (*d*) 3, 3, 6.

13.31 (*a*) $\tilde{\nu}_4$ and $\tilde{\nu}_5$; (*b*) $\tilde{\nu}_3$ and $\tilde{\nu}_5$; (*c*) $\tilde{\nu}_1$, $\tilde{\nu}_2$, and $\tilde{\nu}_4$.

13.32 (*a*) 2143.24 cm^{-1}, 4.665 84 μm;
(*b*) 278.4 cm^{-1}, 35.92 μm.

13.33 (*a*)

Vib. mode	*Changing* μ	*Changing* α	*IR*	*Raman*
ν_1	No	Yes	No	Yes
ν_2	Yes	No*	Yes	No
ν_3	Yes	No*	Yes	No
ν_4	Yes	No*	Yes	No

*The exclusion rule is useful because it is difficult to judge qualitatively whether a vibrational mode involves a change in polarizability.

(*b*)

Vib. mode	*Changing* μ	*Changing* α	*IR*	*Raman*
ν_1	Yes	Yes	Yes	Yes
ν_2	Yes	Yes	Yes	Yes
ν_3	Yes	Yes	Yes	Yes

13.34 $f_0 = 0.5530, f_1 = 0.2472, f_2 = 0.1105, f_3 = 0.0494$.

13.35

$\tilde{\nu}_R/\text{cm}^{-1}$	214	312	454	759
$\lambda/\mu\text{m}$	46.8	32.0	22.0	13.2

13.36 110 pm.

13.37 2.3436, 3.9060, 5.4684, 7.0308 cm^{-1}.

13.38 All of these molecules have a pure Raman spectrum except for CCl_4.

Chapter 14 Electronic Spectroscopy of Molecules

14.1 13.93 eV; 1.012×10^4 m s^{-1}.

14.2 (*a*) 239.5 kJ mol^{-1}; (*b*) 90.53 kJ mol^{-1}, 0.938 eV.

14.3 5080 eV or 490.14 kJ mol^{-1}, $\Delta_f H°(0\text{ K}) = 493.57$ kJ mol^{-1}.

14.4 (*a*) 804.35 nm; (*b*) 433.7 nm; (*c*) 94.32 pm.

14.5 (*a*) 1.2%; (*b*) 0.311 mol L^{-1}; (*c*) 32.8%; (*d*) 10.3 cm.

14.8 (*a*) Yes, 5.81×10^4 L mol^{-1} cm^{-1}; (*b*) 81%.

14.9 2.17×10^{-5}, 3.37×10^{-5} mol L^{-1}.

14.10 1.08, 0.152, and 0.002 2.

14.11

	Integrated Absorption Coeff.	*F*
Strong	1.06×10^7 to 5.30×10^8 L mol^{-1} cm^{-2}	0.046 – 2.29
Weak	1.06×10^3 L mol^{-1} cm^{-2}	4.58×10^{-6}

14.12

	$\lvert\mu_{12}\rvert$/cm	$\lvert R_{12}\rvert$/pm
Strong	9.72×10^{-30}	60.7
Moderate	3.08×10^{-31}	1.92
Weak	9.72×10^{-33}	0.061

14.13 Yes; there is no way to make the predictions agree for all four orbitals; $\beta = -5h^2/8(1.236)ma^2$.

14.14 5.3 cm^{-1}.

14.15 (*a*) 5.3×10^{-4} cm^{-1}, 16 MHz; (*b*) 5.3×10^{-15} cm^{-1}, 1.6×10^{-4} MHz.

14.16 2.517×10^{16}.

14.17 531, 53.1 cm^{-1}.

14.18 (*a*) 10^{-8} J; (*b*) 3.5×10^{10}; (*c*) 4.03×10^{18}.

14.19 0.199%.

14.20 5.14, 4.09, 3.27 km s^{-1}.

14.21 553 eV.

14.22 (*a*) Since F is more electronegative than C and H, it pulls electrons toward itself, thereby decreasing the shielding at the carbon nucleus, and increasing the binding energy of the 1s electrons. (*b*) The binding energy should be greatest for the CF_3 carbon, next largest for the COO carbon, next for the OCH_2 carbon, and smallest for the CH_3 carbon.

14.23 67%.

Chapter 15 Magnetic Resonance Spectroscopy

15.1 1.409, 11.74 T.

15.2 2.394×10^{-5}, 1.995×10^{-2} kJ mol^{-1}; $E/RT = 0.960 \times 10^{-5}$, 0.800×10^{-2}.

15.3 $E/10^{-26}$ J $= -2.240, -0.7465, 0.7456, 2.240$; $\nu = 22.53$ MHz.

15.4 2.675×10^{8} s^{-1} T^{-1}.

15.5 3.4485×10^{-6}.

15.6 0.500 003 43.

15.7 1.69×10^{-2}, 0.665.

15.8 (*a*) 2.67×10^{8} s^{-1} T^{-1}; (*b*) 425.8 MHz.

15.9 (*a*) 17.39 T; (*b*) 91.76 T.

15.10

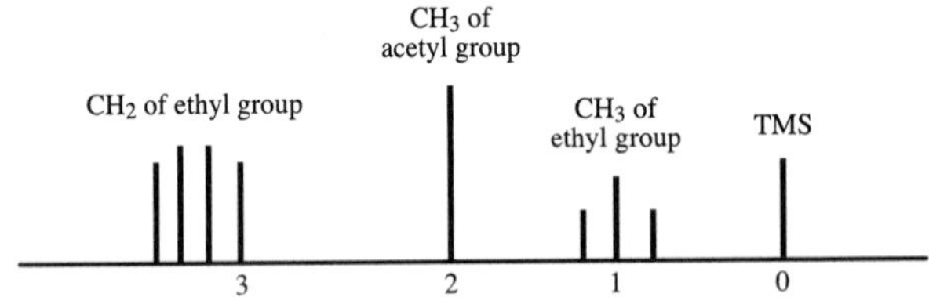

15.11 2.35, 11.7 T.

15.12 (*a*) 14.52; (*b*) 726 Hz.

15.13 7.270×10^{-6}, which is usually given as 7.270.

15.14 (*a*) -2.42×10^{-6}; (*b*) -2.42 μT; (*c*) -4.82 μT.

15.15 (*b*) -145.2 Hz.

15.16

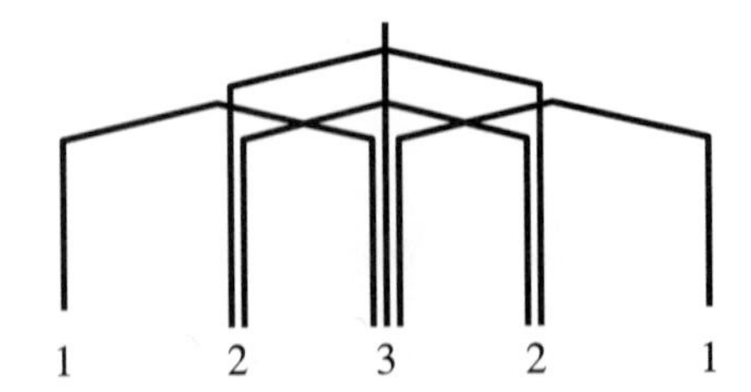

15.17 (*a*) 5.70 Hz; (*b*) 0.318 ppm; (*c*) 576.18, 581.88, 603.84, and 609.54 Hz.

15.19

Transistion	*Relative Frequency/s*$^{-1}$	*Relative Intensity I*
1 → 2	−8.090	0.382
3 → 4	−3.090	1
1 → 3	3.090	1
2 → 4	8.090	0.382

15.22 At room temperature the rate of conversion of cyclohexane from boat to chair forms is so fast that the protons are at the average local magnetic field.

15.23 24.1 s^{-1}.

15.24 2.80 cm^{-1}.

15.25 2.80×10^4 MHz.

Chapter 16 Statistical Mechanics

16.1 (*a*) 0.0080; (*b*) 0.0177.

16.2 (*a*) 1.253×10^{-17}; (*b*) 1.147×10^{-169}; (*c*) 1.100×10^{-4}, 2.596×10^{-40}.

16.7 $\mu = -kT \ln(q/N)$.

16.8 (*a*) 1.23; (*b*) 2.83×10^{-11}.

16.9 1.414×10^{30}.

16.10 8.84×10^{28}, 2.50×10^{29}.

16.11 (*a*) 139.86; (*b*) 82.4 J K^{-1} mol^{-1}.

16.12 146.3 J K^{-1} mol^{-1}.

16.13 $(x + x^2 + x^3)(1 + y)$.

16.14 $x^2 + 2x^3 + x^4$.

16.15 RT, as expected classically.

16.17 1.684.

16.18 12.472 J K^{-1} mol^{-1}, 3.718 kJ mol^{-1}, 47.822 J K^{-1} mol^{-1}, −10.54 kJ mol^{-1}.

16.19 1.43×10^{-6}.

16.20 5.00×10^{29}, 1.42×10^{30}, 2.60×10^{30}, 5.72, 36.8.

16.21 (*a*) 18; (*b*) 9; (*c*) 81; (*d*) 972.

16.22 For CH_4, $\sigma = 12$. For C_2H_4, $\sigma = 4$.

16.23 (*a*) 6; (*b*) 18.

16.24 11.53 J K^{-1} mol^{-1}; −3.44 kJ mol^{-1}.

16.25 $q_e = 8.627$; $f_1 = 0.1159$, $f_2 = 0.3396$, $f_3 = 0.5445$.

16.26 19.92 J K^{-1} mol^{-1}, −17.92 kJ mol^{-1}.

16.27 114.718, 153.301 J K^{-1} mol^{-1}.

16.28

	298 K	1000 K
P_0	0.919	0.625
P_1	0.081	0.303
P_2	0.001	0.072

16.29 4.029, 4.568.

16.30 0.617, 0.294, 0.089.

16.31 0.076, 1.301 kJ mol^{-1}.

16.32 4×10^{-41}.

16.33

T/K	P_1	P_2
0	1	0
$\Delta\epsilon/k$	0.73	0.27
∞	$\frac{1}{2}$	$\frac{1}{2}$

16.34 55.67 J K^{-1} mol^{-1}.

16.35 191.52 J K^{-1} mol^{-1}.

16.36 62.36 J K^{-1} mol^{-1}.

16.37 7.18.

16.38 3.23.

16.40 $\overline{C}^{\circ}_{V}/R = 1.5, 3.5, 6.5, 12; \overline{C}^{\circ}_{P}/R = 2.5, 4.5, 7.5, 13.$

16.41 33.26 J K^{-1} mol^{-1}, 58.20 J K^{-1} mol^{-1}.

Chapter 17 Kinetic Theory of Gases

17.1 (*a*) 8.6; (*b*) 390.

17.2

$v/10^2$ m s^{-1}	1	3	5	7	10
$f(v)/10^{-4}$ s m^{-1}	3.47	18.64	18.42	7.67	0.58

17.3 0.0877.

17.4 (*a*) $(15/4)(kT)^2$; (*b*) $\sqrt{\frac{3}{2}}\,kT$; (*c*) $\sqrt{\frac{2}{3}}$.

17.5 928, 982 m s^{-1}.

17.6 393, 444, 481 m s^{-1}.

17.7 2502 m s^{-1}. Use the construction of Fig. 17.12.

17.8 (*a*) 2.099×10^{-2}; (*b*) 3.721×10^{-5}.

17.9 746 m s^{-1}, compared with a mean speed of 1769 m s^{-1}.

17.10 $\frac{3}{2}RT$, $\frac{3}{2}R$.

17.11 352 m s^{-1} = 787 miles per hour.

17.12 (*a*) 428 m s^{-1}; (*b*) 269 m s^{-1}.

17.13 (*a*) 7.21×10^9 s^{-1}; (*b*) 8.75×10^{28} cm^{-3} s^{-1}; (*c*) 0.354; (*d*) 4.

17.14 (*a*) 1.52×10^{-7}, 0.152 m; (*b*) 3.77×10^{-8}, 0.037 m.

17.15 (*a*) 0.724 cm^{-3}; (*b*) 5.92×10^{-10} s^{-1}; (*c*) 4.83×10^9 miles.

17.16 9.52×10^9 s^{-1}, 1.921×10^8 mol L^{-1} s^{-1}.

17.17 (*a*) 4.63×10^9 s^{-1}; (*b*) 9.258×10^9 s^{-1}; (*c*) 1.244×10^8 mol L^{-1} s^{-1}.

17.18 1.58×10^{-10} s, 7590.

17.19 (*a*) 1.075×10^{28}, 5.866×10^{27} m^{-2} s^{-1}; (*b*) 2.698×10^{27}, 1.473×10^{27} m^{-2} s^{-1}.

17.21 21.3 Pa.

17.22 326 kJ mol^{-1}.

17.23 347 s.

17.24 0.0711 g.

17.25 (*b*) 11 200 m s^{-1}.

17.26 (*a*) 0.217 nm; (*b*) 1.28×10^{-4} m^2 s^{-1}.

17.27 1.26×10^{-5} m^2 s^{-1}.

Chapter 18 Experimental Kinetics and Gas Reactions

18.1 -1.5×10^{-4}, 3.0×10^{-4}, 0.75×10^{-4} mol L^{-1} s^{-1}.

18.2 (*a*) 0.75×10^{-2} mol L^{-1} s^{-1}; (*b*) 1.5×10^{-2} mol L^{-1} s^{-1}; (*c*) 2.25×10^{-2} mol L^{-1} s^{-1}.

18.3 2310 s.

18.4 First order; 6.27×10^{-5} s^{-1}.

18.5 1.32×10^{-4} s^{-1}.

18.7 (*a*) First; (*b*) 0.128 h^{-1}; (*c*) 0.325.

18.8 $k = 0.251$ min^{-1}; $t_{1/2} = 2.76$ min; $\tau = 3.98$ min.

18.9 15.7 s^{-1}.

18.10 (*a*) $t_{1/2} = 1/2k[\text{HI}]_0$; (*b*) 4.01×10^{-2} mol L^{-1} s^{-1}; (*c*) 6.17×10^{-4} bar^{-1} s^{-1}; (*d*) 6.66×10^{-23} cm^3 s^{-1}.

18.11 Second order; 0.675 L mol^{-1} min^{-1}.

18.12 (*a*) Second order; (*b*) 0.59 L mol^{-1} min^{-1}.

18.13 $t_{1/2} = (2^{n-1} - 1)/(n-1)k[\text{A}]_0^{n-1}$.

18.14

t/h	P_A/bar	P_B/bar	P/bar
1	0.50	0.25	0.75
2	0.33	0.33	0.66
∞	0	0.50	0.50

18.15 1.70×10^{-3} s.

18.16 (*a*) 6.25; (*b*) 14.3; (*c*) 0%.

18.17 $a = 0, b = [\text{R}]_0 k, c = [\text{R}]_0 k^2/2$.

18.18 1.000 ± 0.005 mol L^{-1} h^{-1}.

18.19 $d[\text{OI}^-]/dt = (60\ \text{s}^{-1})[\text{I}^-][\text{OCl}]/[\text{OH}^-]$.

18.20 2.50×10^{-3} mol L^{-1}.

18.21

t/d	10	20	40	80
F	3.65×10^{-12}	6.39×10^{-12}	9.98×10^{-12}	13.14×10^{-12}

18.23 $k_1 k_4 = k_2 k_3$.

18.24 $v = k'_A[\text{A}]$, where $k'_A = k_A/(1 + [\text{H}^+]/K_{HA}) + k_{HA}/(1 + K_{HA}/[\text{H}^+])$.

18.25 The time B goes through its maximum is $t = [1/(k_1 - k_2)]\ln(k_1/k_2)$. For a given value of k_2, you would wait the longer time in case (*b*) for B to go through its maximum concentration.

18.26 (*b*) 90.3 kJ mol^{-1}; (*c*) 1.99×10^{12} s^{-1}.

18.27 (*a*) 76 °C; (*b*) −3 °C.

18.28 1420 s.

18.29 2.76×10^{-39} cm^6 s^{-1}.

18.31 (*a*) $d[\text{D}]/dt = k_1 k_3[\text{A}][\text{B}]/(k_2 + k_3)$; (*b*) $E_{app} = E_{a1} + E_{a3} - E_{a2}$.

18.33 $d[\text{D}]/dt = k_1 k_3[\text{A}][\text{B}]/(k_2 + k_3[\text{C}])$.

18.34 (*a*) $k_1 = 5.20 \times 10^{-4}$ s^{-1}; (*b*) $t = 1333$ s.

18.35 (*b*) 3.4×10^3 cm^3 mol^{-1} s^{-1}.

18.37 4.5×10^8 mol L^{-1} s^{-1}.

18.38 $k' = (1.1 \times 10^{13}\ \text{mol L}^{-1}\ \text{s}^{-1})\exp(-104\,000/8.3145T)$.

18.40 $d[CH_3CHO]/dt = -k_2(k_1/2k_4)^{1/2}[CH_3CHO]^{3/2}$.

Chapter 19 Chemical Dynamics and Photochemistry

19.1 2.40×10^{-20} m^2, 87.4 pm.

19.2 (*a*) 8.00×10^{11} mol L^{-1} s^{-1}; (*b*) 8.1×10^{-11} s; (*c*) 6.2×10^{-5} s.

19.4 1.4×10^{11} L mol^{-1} s^{-1}.

19.5 (*a*) 3.14×10^{14} s^{-1}; (*b*) 10 500 cm^{-1}; (*c*) 925 nm; (*d*) 1.31 eV.

19.6 1.7×10^{-3} mol.

19.7 0.167.

19.8 1.3×10^6.

19.10 $v_{CH_3} = 5595$ m s^{-1}, $v_I = 662$ m s^{-1}.

19.11 6.4×10^{-8} s.

19.12 10^{-7} mol L^{-1}.

19.13 (*a*) 10.1; (*b*) a chain reaction is involved.

19.14 1.25×10^{-2} J s^{-1}.

19.15 83 kW L^{-1}.

19.16 $d[CCl_4]/dt = k_2 I_a^{1/2}[Cl_2]^{1/2}/k_3^{1/2} + 2I_a$.

19.17 $I_f = k_f I_a/(k_f + k_q[Q])$, which can be rearranged to $I_a/I_f = 1 + (k_q/k_f)[Q]$ so that k_q/k_f is the slope of the plot of I_a/I_f versus [Q]. The rate constant for fluorescence k_f can be calculated from the half-life of the fluorescence, and so k_q can be calculated from the slope.

19.18

$$A + h\nu \longrightarrow A^* \qquad I_a$$

$$A^* + A \longrightarrow A_2 \qquad k$$

$$A^* \longrightarrow A + h\nu_f \qquad k_f$$

$$\phi = k[A]/(k_f + k[A])$$

The stoichiometric number for step 4 is 2; $s_4 = 2$.

19.20 1.36×10^{17} molecules cm^{-2}.

Chapter 20 Kinetics in the Liquid Phase

20.1 3.46 Pa s.

20.2 3.01×10^{-5} m s^{-1}.

20.3 2.36 kJ mol^{-1}, 1.41×10^{-3} Pa s.

20.5 -0.61 m s^{-1}.

20.6 (*a*) 5.23×10^{-10} m^2 s^{-1}; (*b*) 6.14×10^{-3} m.

20.7 (*a*) 1.75 h; (*b*) 56.5 h.

20.8 2.00×10^{-5} cm s^{-1}, 0.54 cm.

20.10 0.0426 Ω^{-1} m^{-1}.

20.11 (*a*) 2.16×10^{-3} Ω^{-1} m^{-1}; (*b*) 1.09×10^{-3} A.

20.12 9.26×10^4, 9.26×10^3, 9.26×10^2 Ω.

20.15 7.4×10^9 J K^{-1} mol^{-1}.

20.16 0.89 nm.

20.17 4.42×10^{10} mol L^{-1} s^{-1}, 7.65×10^5 s^{-1}.

20.19 $1/\tau = k_1 + k_1' K_A + k_2 + k_2' K_B$.

20.20 7.9×10^5, 1.8×10^3, 25 s^{-1}.

20.21 4.93×10^{-5} s^{-1}, 0.0102 s^{-1}.

20.22 $d[Cl^-]/dt = kK_1[I^-][OCl]/[OH]$.

20.23 1.35×10^{-4} m^{-1}, 5.5×10^{-5} L mol^{-1} min^{-1}.

20.24 0.748, 3.20.

20.25 6.7 mol h^{-1}.

20.26 (*a*) $V_F = 6.5 \times 10^{-7}$ mol L^{-1} s^{-1}, $K_F = 3.9 \times 10^{-6}$ mol L^{-1}, $V_M = 4.0 \times 10^{-7}$ mol L^{-1} s^{-1}, $K_M = 1.03 \times 10^{-5}$ mol L^{-1}, $k_2 = 3.3 \times 10^2$ s^{-1}, $k_{-1} = 2.0 \times 10^2$ s^{-1}

(*b*) $k_1 = 1.4 \times 10^8$ mol L^{-1} s^{-1}, $k_{-2} = 5.1 \times 10^7$ mol L^{-1} s^{-1}.

(*c*) 4.4.

20.27 See equation 20.84.

20.28 15.1, 0.38×10^{-3} mol L^{-1}, 8.6×10^{-6} mol L^{-1}.

20.29 When $v/[E]_0[S]$ is plotted versus $v/[E]_0$, the intercept on the ordinate is k_{cat}/K_M, the slope is $-1/K_M$, and the intercept on the abscissa is k_{cat}.

20.30 234, 4.41×10^{-8} mol L^{-1}, 3.67×10^{-9} mol L^{-1}.

20.31 1.73×10^{-5}, 5.72×10^{-10}, 6.11×10^{-8}; $k_1 = 1.47 \times 10^{-5}$, $k_2 = 10^{-7}$.

20.32 (*a*) $K = 0.00264$, $k_2(25° \text{ C}) = 14.21 \text{ s}^{-1}$, $k_2(0° \text{ C}) = 0.947 \text{ s}^{-1}$; (*b*) $E_{af} = 78.07$ kJ mol^{-1}, $E_{ab} = 73.34$ kJ mol^{-1}.

Chapter 21 Macromolecules

21.1 100 000 g mol^{-1}, 103 000 g mol^{-1}, 1.03.

21.2 $(\overline{r^2})^{1/2} = 31.6$.

21.3 12.7 nm, 17.8 nm.

21.4 $\langle r \rangle = (8/3\pi)^{1/2} N^{1/2} l$.

21.5 $\langle r^2 \rangle = N^{1/2} l$.

21.6 (*a*) 100; (*b*) 3.70×10^{-3}; (*c*) 3.70×10^{-3}.

21.7 (*a*) 0.9943; (*b*) 175; (*c*) 39 900 g mol^{-1}.

21.8 $T_c = \Delta_r H°/\Delta_r S°$. The ceiling temperature for polystyrene is 69 900/104 = 672 K. The ceiling temperature for poly-α-methylstyrene is 35 200/104 = 338 K. The bonding is not as strong as for styrene, as indicated by $\Delta_r H°$, and so depolymerization occurs at a lower temperature. The methyl group apparently prevents as close packing as in polystyrene. The ceiling temperature for tetrafluoroethylene is 154 800/112 = 1382 K, and so Teflon is used on cooking utensils. The bonding is very strong, as indicated by $\Delta_r H°$.

21.10 500 000 g mol^{-1}.

21.11 230 000 g mol^{-1}.

21.12 2.30×10^{-3} g cm^{-3}.

21.13 15 500 g mol^{-1}.

21.14 67 600 g mol^{-1}, 4.

21.15 0.236×10^{-13} s.

21.16 0.9938.

21.17 10.5×10^{-13} s.

Chapter 22 Electric and Magnetic Properties of Molecules

22.2 The electric susceptibility is dimensionless.

22.3 4.45×10^{-40} C^2 m^2 J^{-1}.

22.4 $\alpha' = 2.61 \times 10^{-30}$ m^3 or 17.6 expressed in units of a_0^3. The volume calculated from the collision diameter is 3.72×10^{-26} m^3.

22.5 For two spin $\frac{1}{2}$, $S(S+1) = 2$. The value calculated from the magnetic susceptibility is 2.2, indicating a small contribution from orbital angular momentum.

22.6 0.44.

22.7 2.47.

Chapter 23 Solid-State Chemistry

23.1 $d = ab/(a^2 + b^2)^{1/2}$.

23.2 5.74°, 11.54°, 17.46°.

23.3 $F_{hkl} = f_A + f_B(-1)^{h+k+l}$. If $h + k + l$ is even, the reflection will be strong.

23.4 3.62 g cm^{-3}.

23.5 21.32×10^3 kg m^{-3}.

23.6 (*a*) 316 pm; (*b*) $d_{200} = 158$, $d_{110} = 223$, $d_{222} = 21.2$ pm.

23.7 (*a*) Body-centered; (*b*) face-centered; (*c*) 286.8 pm; (*d*) 57.4 pm; (*e*) 7.84×10^3 kg m^{-3}.

23.8 (*a*) 3.995×10^3 kg m^{-3}; (*b*) 176 pm.

23.9 $6.022\,093 \times 10^{23}$ mol^{-1}.

23.10 39.7 kg mol^{-1}.

23.11 273 pm.

23.12 2.33×10^3 kg m^{-3}.

23.13 12.

23.14 9.95°, 11.50°, 16.38°, 19.31°, 20.21°, 23.51°.

23.15 $F = f[1 + (-1)^{h+k+l}]$.

23.16 0.414.

23.17 282.8 pm.

23.18 4.517×10^3 kg m^{-3}.

23.19 0.0818 eV.

Chapter 24 Surface Dynamics

24.1 (*a*) 5.08×10^{-1} cm^{-2} s^{-1}; (*b*) 7.17×10^{10} cm^{-2} s^{-1}.

24.2 18.5 s.

24.3 7.00×10^8 s^{-1}.

24.4 3.26×10^{19} m^{-2} s^{-1}.

24.5 -11.6 kJ mol^{-1}.

24.6 $v = v_m(y_A K_A + y_B K_B)P/[1 + (y_A K_A + y_B K_B)P]$; $x_A = y_A K_A/(y_A K_A + y_B K_B)$.

24.7 $v_m = 40$ cm^3 g^{-1}, $K = 4.8 \times 10^{-5}$ Pa^{-1}.

24.8 2.3×10^3 Pa, 20.8×10^3 Pa, 188×10^3 Pa.

24.9 2.08×10^8, 2.94×10^8, 1.80×10^8 s^{-1}.

24.10 20.6 L.

24.11 449 m^2.

Index

G

H

I

N

O

RELATIVE ATOMIC MASSES AND ISOTOPIC ABUNDANCES[a,b]

Z	*Symbol*	A	*Relative Atomic Mass,* m_a/u	*Isotopic Abundance,* x/%
1	H	1	1.007 825 037(10)	99.985(1)
		2	2.014 101 787(21)	0.015(1)
		3*	3.016 049 286(32)	
2	He	3	3.016 029 297(33)	0.000 138(3)
		4	4.002 603 25(5)	99.999 862(3)
3	Li	6	6.015 123 2(8)	7.5(2)
		7	7.016 004 5(9)	92.5(2)
4	Be	9	9.012 182 5(4)	100
5	B	10	10.012 938 0(5)	19.9(2)
		11	11.009 305 3(5)	80.1(2)
6	C	12	12 (by definition)	98.90(3)
		13	13.003 354 839(17)	1.10(3)
		14*	14.003 241 993(24)	
7	N	14	14.003 074 008(23)	99.634(9)
		15	15.000 108 978(38)	0.366(9)
8	O	16	15.994 914 64(5)	99.762(15)
		17	16.999 130 6(8)	0.038(3)
		18	17.999 159 39(32)	0.200(12)
9	F	19	18.998 403 25(14)	100
10	Ne	20	19.992 439 1(5)	90.51(9)
		21	20.993 845 3(12)	0.27(2)
		22	21.991 383 7(6)	9.22(9)
11	Na	23	22.989 769 7(9)	100
12	Mg	24	23.985 045 0(8)	78.99(3)
		25	24.985 839 2(12)	10.00(1)
		26	25.982 595 4(10)	11.01(2)
13	Al	27	26.981 541 3(7)	100
14	Si	28	27.976 928 4(7)	92.23(1)
		29	28.976 496 4(9)	4.67(1)
14	Si	30	29.973 771 7(10)	3.10(1)
15	P	31	30.975 363 8(11)	100
16	S	32	31.972 071 8(6)	95.02(9)
		33	32.971 459 1(8)	0.75(1)
		34	33.967 867 74(29)	4.21(8)
		36	35.967 079 0(16)	0.02(1)
17	Cl	35	34.968 852 729(68)	75.77(5)
		37	36.965 902 62(11)	24.23(5)
35	Br	79	78.918 336 1(38)	50.69(5)
		81	80.916 290(6)	49.31(5)
53	I	127	126.904 477(5)	100

[a] IUPAC Commission on Atomic Weights, *Pure Appl. Chem.* **56**:653 (1984).

[b] An asterisk denotes an unstable nuclide. The standard error in parentheses is applicable to the last digits quoted.